DER ZÜCHTER

erscheint in Einzel- oder Mehrfachheften, die später zu einem Band im Umfang von 24 Druckbogen vereinigt werden. Das Einzelheft umfaßt 2 Druckbogen.

Für die Redaktion bestimmte **Sendungen und Zuschriften** sind zu richten an

Herrn Professor Dr. H. Stubbe,
Versuchsgut Gatersleben, Bezirk Dessau (19b).

Es empfiehlt sich, Manuskripte unter „Einschreiben" zu senden.
In die Zeitschrift werden aufgenommen:
a) Originalarbeiten aus dem Gebiet der Pflanzen- und Tierzüchtung.
b) Originalarbeiten aus dem Gebiet der Entstehung der Kulturpflanzen und Haustiere.
c) Genetische und zytologische Originalarbeiten an züchterisch wichtigen Objekten.
d) Physiologische Arbeiten an züchterisch wichtigen Objekten.
e) Arbeiten über neue Methoden, Apparate u. dgl., soweit diese für den Züchter und Züchtungsforscher Bedeutung haben.
f) Zusammenfassende Darstellungen aus für die Züchtung wichtigen Gebieten der Grundlagenforschung.
g) Mitteilungen über Verordnungen, Gesetze und Patente, die die Züchtungsarbeit betreffen.
h) Mitteilungen über die Organsation der Pflanzenzüchtung in und außerhalb Deutschlands sowie Ankündigungen von Züchtertagungen und Kongressen.
i) Gedenktage und Nachrufe.
k) Referatenteil.

Redaktionelle Anweisungen:
1. Die Manuskripte sind in Maschinenschrift druckfertig einzureichen.
2. Abbildungen dürfen nicht aufgeklebt sein, sie sind auf der Rückseite zu numerieren, die Legenden zu den Abbildungen auf einem Sonderblatt der Reihe nach aufzuführen.
3. Autorennamen sind glatt zu unterstreichen, wissenschaftliche Namen von Pflanzen und Tieren gewellt zu unterstreichen.
4. Jeder Arbeit ist ein Literaturverzeichnis anzufügen in der Reihenfolge Name und Vorname (abgekürzt) des Autors, Titel der Arbeit, Erscheinungsort, Band, Seite, Jahreszahl.
5. Der Entstehungsort der Arbeit ist über dem Titel der Arbeit anzugeben.
6. Die Zahl der Abbildungen muß auf das Notwendigste beschränkt werden. Nur wirklich gut reproduzierte Fotos sind einzureichen.
7. Die Autoren erhalten zwei Fahnenabzüge, von denen der eine korrigiert umgehend an den Springer-Verlag, Berlin W 9, Linkstr. 23—24, Schriftleitung: „Der Züchter" zurückzusenden ist. Die zweite Korrektur wird von der Redaktion gelesen. Revisionsfahnen werden den Autoren nur in Ausnahmefällen übersandt.

Sonderabdrucke: Die Verfasser von Originalarbeiten erhalten bei rechtzeitiger Bestellung bis zu 50 Sonderdrucken ihrer Arbeiten kostenlos. Weitere Exemplare können nur gegen Berechnung geliefert werden.

Bezugsbedingungen: „Der Züchter" kann im In- und Auslande durch jede Sortimentsbuchhandlung bezogen werden. Preis des Bandes (12 Einzelhefte) 40.— DM, Einzelheft 4.— DM. Bei Bezug unter Kreuzband kommen die Versandspesen hinzu.

Die Lieferung läuft weiter, wenn nicht unmittelbar nach Lieferung des Schlußheftes eines Bandes Abbestellung erfolgt. Der Bezugspreis ist im voraus zahlbar.

Nachdruck: Der Verlag behält sich das ausschließliche Recht der Vervielfältigung und Verbreitung aller in dieser Zeitschrift zum Abdruck gelangenden Beiträge sowie ihrer Verwendung für fremdsprachliche Ausgaben vor.

Anzeigen werden vom Verlag angenommen. Die Preise wolle man unter Angabe der Größe und des Platzes erfragen.

**Springer-Verlag
Berlin W 9**
Linkstr. 23—24. Fernsprecher 42 81 19.

DER ZÜCHTER

Sonderheft 1948

Genetisch=pflanzenzüchterische Bibliographie 1939—1946 (1947).

Der Plan zur Veröffentlichung einer möglichst umfassenden Liste der während des Krieges erschienenen Literatur auf genetisch-pflanzenzüchterischem Gebiet und der angrenzenden Disciplinen entstand aus dem Wunsch, allen Fachkollegen ein Hilfsmittel zur Orientierung über die einschlägigen wissenschaftlichen Arbeiten aus den Jahren der Isolierung zu geben und damit zur Wiederherstellung internationaler Verbindungen beizutragen. Wenngleich die nachstehende Liste über 13 000 Zitate umfaßt, dürfen wir uns nicht der Hoffnung hingeben, daß sie auch nur annähernd vollständig sei. Eine wesentliche Unterlage für unsere Arbeit waren die mir durch freundliche Vermittlung von Frau Prof. Dr. E. Schiemann zur Verfügung gestellten Plant Breeding Abstracts des Commonwealth Bureau of Plant Breeding and Genetics in Cambridge. Ferner wurden einige andere Literaturzusammenstellungen benutzt, so die Resumptio Genetica, die von Buzatti-Traverso: Genetic Research in Italy and Germany (Heredity, Vol. I), die vom Institute of Genetics, University of Lund: List of Publications 1939—1945 und die von A. Ernst und M. Ernst-Schwarzenbach zusammengestellte Bibliographie Genetica Helvetica (1939—1946).

Alle verfügbaren Zeitschriften wurden durchgesehen, wobei wir mit Bedauern feststellen müssen, daß die russische Literatur nur in geringem Umfang von uns eingesehen werden konnte. Erst in den letzten Monaten ist es erfreulicherweise möglich geworden, russische Zeitschriften zu abonnieren und damit die Arbeiten der russischen Kollegen zu verfolgen. Da die Transskription der russischen Autorennamen in westeuropäischen und amerikanischen Berichten sehr uneinheitlich ist, wurden alle Autorennamen einheitlich, ähnlich dem Damianisystem nach serbokroatischem Schema umgeschrieben. Größere Abweichungen, besonders für Anfangsbuchstaben, wurden in unumgänglichen Fällen im Alphabet berücksichtigt.

Bei Arbeiten allgemein wichtigen genetischen Inhalts wurden auch an zoologischen Objekten durchgeführte Untersuchungen zitiert. Ein Mangel unserer Liste besteht zweifellos darin, daß die Institute, in denen die Verfasser arbeiten, nicht angegeben werden, so daß es vielfach schwer sein wird, mit den Verfassern direkt in Verbindung zu treten. Für viele Arbeiten kann von uns auf Wunsch eine Abschrift des in den Plant Breeding Abstracts veröffentlichten Referates gegeben werden. Aus zeitbedingten Gründen war es nicht möglich, einzelne Arbeiten in verschiedenen Kapiteln zu zitieren. Jedes Zitat konnte nur einmal aufgeführt werden.

Die vorstehende Liste möge eine Brücke sein zwischen den ausführlichen, aber dennoch lückenhaften Literaturnachweisen in den zuständigen Referatenblättern bis zum Kriegsende und den bei uns jetzt erst kürzlich wieder eröffneten gleichen Publikationsorganen, die nicht in der Lage sind, die noch nicht zitierte Literatur der Kriegsjahre und der ersten Nachkriegszeit nachträglich zu verarbeiten.

Dem Springer-Verlag werden alle Fachkollegen Dank wissen für seine Bereitschaft, sich der Mühe des Druckes unterzogen zu haben. Meinen unermüdlichen Mitarbeiterinnen bei dem Abschreiben und der Zusammenstellung der Zitate gilt mein besonderer Dank.

Gatersleben, November 1948.

H. Stubbe.

ISBN 978-3-540-01387-7   ISBN 978-3-642-45812-5 (eBook)
DOI 10.1007/978-3-642-45812-5

Inhaltsverzeichnis.

A. Allgemeiner Teil.

a) Genetik 3
b) Artentstehung und Populationsgenetik 21
c) Allgemeine Pflanzenzüchtung 25
d) Botanik, Angewandte Botanik, Entwicklungsphysiologie 36
e) Cytologie 43
f) Biophysik 57
g) Biochemie 57
h) Pflanzenkrankheiten, Resistenzzüchtung, Bakterien und Pilze 58
i) Mikroskopische Technik 69
k) Statistik und Feldversuchswesen 71
l) Acker- und Pflanzenbau 78
m) Neue Zeitschriften 79
n) Verschiedenes 80

B. Spezieller Teil.

a) Getreide 81
   1. Weizen 84
   2. Roggen 108
   3. Hafer 110
   4. Gerste 114
   5. Mais 120
   6. Hirsen 131
   7. Reis 134
b) Futtergräser 138
c) Futterleguminosen 144
d) Wurzel- und Knollengewächse 151
e) Faserpflanzen 164
f) Zuckerpflanzen 176
g) Stärkepflanzen 187
h) Buchweizen 187
i) Stimulantien 188
k) Gewürzpflanzen 200
l) Ölpflanzen 202
m) Kampferpflanzen 207
n) Gerbstoffpflanzen 208
o) Farbpflanzen 208
p) Heilpflanzen 208
q) Kautschukpflanzen 209
r) Obst 214
   1. Obst, Allgemeines 214
   2. Kernobst 219
   3. Steinobst 223
   4. Schalenobst 227
   5. Beerenobst 229
   6. Agrumen (Citrus) 234
   7. Palmen 236
   8. Sonstiges tropisches und subtropisches Obst 236
   9. Weinbau 240
s) Gemüse 242
   1. Gemüse, Allgemeines 242
   2. Leguminosen 245
   3. Tomaten 253
   4. Kohl 259
   5. Zwiebeln 260
   6. Kürbisgewächse 262
   7. Mohrrüben 265
   8. Salat 265
   9. Sonstige Gemüse 266
t) Forstwirtschaft 269
u) Berichte der Versuchsstationen, Kongresse, Konferenzen usw. 276

A. Allgemeiner Teil.

a) Genetik.

Abraham, E. P., D. Callow u. K. Gilliver: Adaptation of Staphylococcus aureus to growth in the presence of certain antibiotics. Nature, Lond. **158**, 818 bis 821 (1946).

Addison, G.: The chemical formulation of gene structure and gene action. Advances Enzymol. **4**, 1—39 (1944).

Agar, W. E.: A contribution to the theory of the living organism. Melbourne and London, Pp. 207 (1943).

Alihanjan, S. J.: (Die „Paarerscheinung" bei der Entstehung von neuen Genen). C. R. (Doklady) Acad. Sci. URSS. **58**, 1777 (1947). [Russisch.]

Allen, C. E.: The genotypic basis of sex-expression in angiosperms. Bot. Rev. **6**, 227—300 (1940).
— Regeneration, development and genotype. Amer. Nat. **76**, 227—238 (1942).

Altenburg, E.: Genetics. Henry Holt and Co. New York, Pp. XII+452, 148 figs (1945).
— The symbiont theory in explanation of the apparent cytoplasmic inheritance in Paramecium. Amer. Nat. **80**, 661—662 (1946).

D'Amato, E.: Sul corredo cromosomico di Euphorbia pubescens Vahl. N. Giorn. Botan. Ital. n. s. 52 (1945).

Anderson, E.: The hindrance to gene recombination imposed by linkage: an estimate of its total magnitude. Amer. Nat. **73**, 185—188 (1939).
— The technique and use of mass collections in plant taxonomy. Ann. Mo. Bot. Gdn. **28**, 287—292 (1941).
— u. R. P. Ownbey: The genetic coefficients of specific difference. Ann. Mo. Bot. Gdn. **26**, 325—346 (1939).

Andersson, E. u. O. Tedin: The effect upon the mean and variability of the dependent variate of a selection according to the independent. Hereditas, Lund **30**, 249 bis 253 (1944). (Abstr.)

Andrés, J. M.: La consanguinidad y los métodos modernos de mejoramiento. (Consanguinity and modern methods of breeding.) Jorn. Agron, Vet., B. Aires 549 bis 572, 1939 (1940).

Angeli, G.: Xenie, carpocenie o mutazioni somatiche in frutticoltura. (Xenia, carpoxenia or somatic mutations in fruit cultivation.) Ann. Tec. Agr., Roma **12**, 9—36 (1939).

Anonymus: De toepassing van colchicine in het erfolijkheidsonderzoek. (The utilization of colchicine in research on inheritance.) Erfelijkheid in Praktijk, Leiden **4**, 161—163 (1939).
— El Instituto de Genetica. (The Institute of Genetics.) Inst. Genet., Fac. Agron. Vet., Univ. B. Aires **1**, Fasc. 2, 35 (1939).
— The symbolizing of genes and of chromosome aberrations. Chronica Botanica **5**, 479—480 (1939).
— Varietal adaptability Rhod. Agric. J. **36**, 340—341 (1939).
— Prof. E. W. MacBride, F. R. S. Nature, London **146**, 831—832 (1940).
— Symbolisierung der Gene u. der Chromosomabweichungen. Z. Pflanz. **23**, 485—486 (1940).
— The symbolizing of genes and of chromosome aberrations. Genetica **22**, 264—268 (1940).
— The symbolizing of genes and of chromosome aberrations. J. Hered. **31**, 27—28 (1940).
— The symbolizing of genes and of chromosome aberrations. Amer. Nat. **74**, 287 (1940).
— (Work on grafting and vegetative hybridization). Jarovizacija Nr. 5 (32), 141—143 (1940). [Russisch.]
— Genes and chromosomes, structure and organization. Cold Spring Harbor symposia on quantitative biology. George Banta Publishing Company, Menasha, Wisconsin **9** (1941).
— Lucien Daniel. Rev. Bot. Appl. **21**, 149—156 (1941).
— Genetics since 1939. Mon. Sci. News. No. 1, 2—3 (1945).
— Genetics and science in the USSR. Brit. Med. J. Nr. 4528, 616—618 (1947).

Aschenbrenner, A.: Die Berechnung des Inzuchtgrades. Arch. Rassenbiol. **33**, 506—510 (1940).

Asmous, V. C.: Facts, feelings, and freedom of science. Science **103**, 760—761 (1946).
— Boris Aleksandrovich Keller 1874—1945. Science **104**, 339 (1946).

Auerbach, C.: The effect of sex on the spontaneous mutation rate in Drosophila melanogaster. Journ. of Genet. **41**, 255—265 (1941).
— Chemically induced Mosaicism in Drosophila melanogaster. Proc. of the Roy. Soc., Edinburgh, **25**, 211 bis 222 (1946).
— Abnormal segregation after chemical treatment of Drosophila. Genetics **32**, 3—7 (1947).
— u. J. M. Robson: Production of mutations by allyl isothiocyanate. Nature, Lond. **154**, 81 (1944).
—, — Chemical production of mutations. Nature, Lond. **157**, 302 (1946).
— — u. J. G. Carr: The chemical production of mutations. Science **105**, 243—247 (1947).

Avakjan, A. A. u. N. G. Jastreb: (Hybridization by means of grafts). Jarovizacija Nr. 1 (34), 50—77 (1941). [Russisch.]

Avery, A. G. u. A. F. Blakeslee: Mutation rate in Datura seed which had been buried 39 years. Genetics **28**, 69—70 (Abstr.) (1943).

Avery, G. S. (jun.), H. B. Creighton u. B. Shalucha: Growth hormones and heterosis. Amer. J. Bot. **26**, 22s (1939). (Abstr.)

Ayachit, G. R.: Progress of homozygosity due to backcrossing. Curr. Sci. **13**, 79—81 (1944).

Azevedo, J. P. de: A heterosis e o melhoramento. (Heterosis and breeding.) Rev. Agron., Lisboa **27**, 231—234 (1939).

Azevedo, P. de: Ideas sôbre heterosis. (Ideas on heterosis.) Palestras Agronómicas, Lisboa **2**, 127—140, 1939 (1940).

Babcock, E. B. u. J. A. Jenkins: Chromosomes and phylogeny in Crepis. III. The relationships of one hundred and thirteen species. Univ. Calif. Publ. Bot. **18**, 241—291 (1943).
— Stebbins (jr.) u. J. A. Jenkins. Genetic evolutionary processes in Crepis. Amer. Natural. **76**, 337—363 (1942).

Badenhuizen, N. P.: Colchicine-induced Tetraploids obtained from plants of economic value. Nature (Lond.) **1**, 577 (1941).

Baker, H. G.: A note on natural hybridisation, illustrated by reference to Melandrium dioicum (L. emend). Coss. and Germ., and M. album (Mill.) Garcke. Naturalist Oct.-Dec. 129—130 (1945).
— Criteria of hybridity. Nature, Lond. **159**, 221—223 (1947).
Ball, C. R.: More plant study: fewer plant names. J. Arnold Arbor. **27**, 371—385 (1946).
Balli A.: La genetica e la sue applicazioni nel campo vegetale ed animale. Ateneo parm. **14**, 309—321 (1942).
Bamford, R., G. B. Reynard u. J. M. Bellows (jr.): Chromosome number in some tulip hybrids. Bot. Gaz. **101**, 482—490 (1939).
Barber, H. N.: Evolution in the genus Paeonia. Nature, Lond. **2**, 227—228 (1941).
— The pollen-grain division in the Orchidaceae. J. Genet. **43**, 97—103 (1942).
Barthelmess, A.: Mutationsversuche mit einem Laubmoos Physcomitrium piriforme. II. Morphologische und physiologische Analyse der univalenten Protonemen einiger Mutanten. Z. Vererbungsl. **79**, 153—170 (1941).
Bateman, A. J.: Genetical aspects of seed-growing. Nature, Lond. **157**, 752—755 (1946).
Battaglia, E.: Poliploidi da colchicina in Bellis perennis, Bellis annua, Antirrhinum Orontium, Mimosa pudica, Nigella sativa, Helianthus annuus, Ricinus communis, Cucurbita Pepo. Mem. R. Acc. Scient. Lett. Arti., Modena **5**, 115—147 (1941).
— Il poliploidismo da colchicina e la sua trasmissibilità per seme (Ricerche su Bellis annua ed Antirrhinum Orontium). Mem. R. Acc. Scien. Lett. Arti., Modena **5**, 343—350 (1941).
— L'assetto genetico dalla var. alsinoides Neg. (n. var.) di Lychnis coeli rosa L. albiflora. N. Giorn. Botan. Ital. **48**, 389—395 (1941).
— Giganti stabili non poliploidi in Plantago major L. N. Giorn. Bot. Ital. **48**, 396—403 (1941).
Bauch, R.: Experimentelle Mutationsauslösung bei der Hefe durch chemische Stoffe. Wschr. Brauerei 1—7, 9—11 (1942).
— Über Beziehungen zwischen polyploidisierenden carcinogenen und phytohormonalen Substanzen. Auslösung von Gigas-Mutationen der Hefe durch pflanzliche Wuchsstoffe. Naturwiss. **30**, 420—421 (1942).
— Chemogenetische Untersuchungen an der Hefe. Ber. dtsch. bot. Ges. **60**, 42—63 (1943).
Bauer, H.: Röntgenauslösung von Chromosomenmutationen bei Drosophila melanogaster. I. Bruchhäufigkeit, -verteilung und -rekombination nach Speicheldrüsenuntersuchung. Chromosoma **1**, 343—390 (1939).
— Chromosomenforschung. Fortschr. d. Zool. **5**, 279 bis 296, 356—359 (1941).
— Röntgenauslösung von Chromosomenmutationen bei Drosophila melanogaster. 2. Die Häufigkeit des primären Bruchereignisses nach Untersuchungen am Ring-X-Chromosom. Chromosoma (Berlin) **2**, 407 bis 458 (1942).
— Die Entstehung von Chromosomenmutationen durch Röntgenbestrahlung. Eine Stellungnahme zu den Arbeiten von H. Marquardt. Ztschr. f. Bot. **38**, 26—41 (1942).
— Auslösung von Polyploidie durch Kälte bei Drosophila melanogaster. Z. Naturforsch. **1**, 35—38 (1946).
Beadle, G. W.: Physiological aspects of genetics. Annu. Rev. Physiol. **1**, 41—62 (1939).
— The gene. Proc. Amer. Phil. Soc. **90**, 422—431 (1946).
— u. E. L. Tatum: Neurospora. II. Methods of producing and detecting mutations concerned with nutritional requirements. Amer. J. Bot. **32**, 678—686 (1945).
Beadnell, C. M.: Dictionary of scientific terms as used in the various sciences. Watts and Co., London, 2nd Ed. 2s 0d. Pp. X +232 +13. (The Thinker's Library, Nr. 65.) (1942).

Beal, J. M.: Some results of cross-pollination on Lilium regale. Bot. Gaz. **108**, 526—530 (1947).
Beale, G. H.: The genetics of Verbena. 1. J. Genet. **40**, 337—358 (1940).
— Gene relations and synthetic processes. J. Genet. **42**, 197—214 (1941).
— Timiriazev, founder of Soviet genetics. Nature, Lond. **159**, 51—53 (1947).
— u. A. C. Fabergé: Effect of Temperature on the Mutation Rate of an unstable Gene in Portulaca grandiflora. Nature **1**, 356—357 (1941).
Becker, E.: Was wissen wir von den Chromosomen als Träger der Erbanlagen? Ztschr. f. ärztl. Fortbildung **36**, 364—369 mit 10 Textfig. (1939).
Beleites, I.: Untersuchungen zur Mutationsauslösung durch Alphateilchen. Fund. Radiobiologica **5**, 142 bis 152 (1939).
Belgovskij, M. L.: (Zur Frage des Realisationsmechanismus der Mosaikbildung im Zusammenhang mit heterochromatischen Chromosomenteilen.) Žurnal Obščej Biologii (Journ. of Gener. Biology) **5**, 325 (1944). [Russisch.]
— V. S. Kirpičnikov u. A. A. Prokofjeva-Belgovskaja: (Organization of the cell and the chromosome theory of heredity.) Bull. Acad. Sci. U.R.S.S. Sér. Biol. Nr. 5, 662—687 (1940).
Belozerova, N. A. u. S. P. Hačaturov: (The influence of cross-polination on the yield of plants.) Jarovizacija, Nr. 3 (36), 63—66 (1941). [Russisch.]
Bergner, A. D., A. G. Avery u. A. F. Blakeslee: Chromosomal deficiencies in Datura stramonium induced by colchicine treatment. Amer. J. Bot. **27**, 676 bis 683 (1940).
Bernström, P.: Two new hybrids in Lamium. Hereditas **30**, 257—260 (1944).
Bertalanffy, L. von: Bemerkungen zum Modell der biologischen Elementareinheiten. Naturwiss. **32**, 26 (1944).
Bhaduri, P. N.: Application of new technique to cytogenetical reinvestigation of the genus Tradescantia. J. Genet. **44**, 87—127 (1942).
Bianchi, R.: Partenogenesi sperimentale nei bachi da seta. Boll. Soc. Ital. Biol. Sperim, p. 16 (1941).
Billingham, R. E. u. P. B. Medawar: The „cytogenetics" of black and white guinea pig skin. Nature, Lond. **159**, 115—117 (1947).
Blakeslee, A. F. u. S. Satina: New hybrids from incompatible crosses in Datura through culture of excised embryos on malt media. Science **99**, 331—334 (1944).
—, — u. A. G. Avery: Genetic evidence suggestion that egg cells in Datura may sometimes develop from the epidermal layer. Amer. J. Bot. Suppl. **33**, 818 (Abstr.) (1946).
Blanc, R.: Dominigenes of the vestigial series in Drosophila melanogaster. Genetics **31**, 395—420 (1946).
Blaringhem, L.: Sur l'hérédité unilatérale dans les croisements interspecifiques. C. R. Acad. Sci. Paris **209**, 272—276 (1939).
— Seiitiro Ikeno (1866—1943). Rev. Hort. Paris **116**, 44—45 (1944).
— Les espèces jordaniennes et la disjonction des espèces. Bull. Soc. Bot. Fr. **92**, 20—23 (1945).
Boas, R. u. R. Gistl: Über einige Colchicinwirkungen. Protoplasma **33**, 301—310 (1939).
Boerger, A.: Centros de genes en la América del Sur. (Gene centres in South America.) Pensamiento Peruano, Lima, p. 58—63 (1945).
Boivin, A., A. Delaunay, R. Vendrely u. Y. Lehoult: Sur certaines conditions de la transformation du type antigénique et de l'équipement enzymatique d'un colibacille, sous l'effet d'une principe inducteur de nature thymonucléique issu d'un autre colibacille (mutation „dirigée"). Experientia, Basel **2**, 139—140 (1946).

Bonner, D., E. L. Tatum u. G. W. Beadle: The genetic control of biochemical reactions in Neurospora: a mutant strain requiring Isoleucine and Valine. Arch. of Biochem. **3**, 71—91 (1943).

Bonnier, G.: Cangiamenti di frequenza dei geni nell' allevamento di popolazioni. (Change of frequency of genes in the improvement of populations.) Scientia Genetica **1**, 282—289 (1939).

— The genetic effects of breeding in small populations. A demonstration for use in genetic teaching. Hereditas, Lund **33**, 143—151 (1947).

— B. Rasmuson u. M. Rasmuson: „Gene divisibility", as studied by differences in Bar facet numbers in Drosophila melanogaster. Hereditas, Lund **33**, 348—366 (1947).

Boost, Ch.: Genetische Untersuchungen zur Chiasmabildung und Interferenzwirkung bei Drosophila. Z. Vererbungsl. **77**, 386—449 (1939).

Borkovskaja, V. A.: (Graft hybrids and chimaeras.) Jarovizacija Nr. 1 (34), 78—83 (1941). [Russisch.]

Born, H. J. u. K. G. Zimmer: Zur Frage der Steigerung der mutationsauslösenden Wirkung der Röntgenstrahlen durch Einbringung schweratomiger Salze in den Organismus. II. Z. Vererbungsl. **78**, 246—250 (1940).

Bounoure, L.: L'origine des cellules reproductrices et le problème de la lignée germinale. Gauthier-Villars, Paris, Pp. xii+271, 85 figs (1939).

Boyce, S. W.: Maximum rate of selection for dominant quantitative genes. Nature, Lond. **157**, 699 (1946).

Bräm, H.: Untersuchungen zur Phaenanalyse und Entwicklungsgeschichte der Blüten von Primula pulverulenta, P. Cockburnia und ihrer F_1-Bastarde. Arch. d. J. Klaus-Stiftg. **18**, 235—359 (1943).

Braun, W.: Some thoughts on „gene action". Science **104**, 38 (1946).

Brehme, K. S.: Development of the Minute phenotype in Drosophila melanogaster. A Comparative study of the growth of three minute mutants. J. of exper. Zool. **88**, 135—160 (1941).

Breslavec, L. P.: (Die Pflanze und die Röntgenstrahlen). Moskva-Leningrad (Verlag Akad. d. Wiss. UdSSR) (1946). [Russisch.]

Breznev, D. D.: (The capacity of selective fertilization in tomatoes.) Selekcija i Semenovodstvo Nr. 5, 14—15 (1939). [Russisch.]

Bridges, C. B. u. K. S. Brehme (Editor): The mutants of Drosophila melanogaster. Washington, D. C. Nr. 552 VII+257, 128 figs, 3 plates (1944).

Briquet, Junior, R.: Falácias in genética. (Fallacies in genetics.) Ceres, Brasil **5**, 302—312 (1944).

Brito da Cunha, A.: Polymorphism in natural populations of a species of Drosophila. J. Hered. **37**, 253 bis 256 (1946).

Brücher, H.: Die reziprok verschiedenen Art- und Rassenbastarde von Epilobium und ihre Ursachen. II. Das genetisch selbständige Zellplasma als Ursache der reziproken Unterschiede. Z. Vererbungsl. **77**, 455—487 (1939).

— Die Bedeutung des Zellplasmas für die Vererbung. Biologe **8**, 160—168 (1939).

— Spontanes Verschwinden der Entwicklungshemmungen eines Artbastards. Flora, N. F. **34**, 215—228 (1940).

— Einfache Mendelspaltung einer Mutation der Blattfarbe. Biologe **9**, 75—80 (1940).

— Vitalitätssteigerung bei Mutanten in künstlichem Klima. Naturwissenschaften **29**, 422—423 (1941).

Buchmann, W. u. G. Sydow: Weitere Versuche an Drosophila melanogaster über den Einfluß von Schwermetallsalzen auf die Mutationsauslösung durch Röntgenstrahlen. Versuche mit Uranylacetat. Biol. Zbl. **60**, 137—142 (1940).

— u. K. G. Zimmer: Zur Frage der Steigerung der mutationsauslösenden Wirkung der Röntgenstrahlen durch Einbringung schweratomiger Salze in den Organismus. I. Z. Vererbungsl. **78**, 148—154 (1940).

Buchmann, W. u. K. G. Zimmer: Zur Frage der Steigerung der mutationsauslösenden Wirkung der Röntgenstrahlen durch Einbringung schweratomiger Salze in den Organismus. III. Z. Vererbungsl. **79**, 192—198 (1941).

Burgeff, H.: Konstruktive Mutationen bei Marchantia. Vererbungsversuche zur Frage der Evolution der Lebermoose. Naturwiss. **29**, 289—299 (1941).

— Genetische Studien an Marchantia. Einführung einer neuen Pflanzenfamilie in die genetische Wissenschaft. Pp. 296, Gustav Fischer, Jena. (1943).

Burns, W.: Genetics, taxonomy and ecology. Indian J. Genet. Pl. Breed. **4**, 2—7 (1944).

Burr, H. S.: Electrical correlates of pure and hybrid strains of sweet corn. Proc. Nat. Acad. Sci. Wash. **29**, 163—166 (1943).

Burri, R.: Die bakterielle Dissoziation im Rahmen der allgemeinen Vererbungslehre. Ber. Schweiz. Bot. Ges. **53**, A, 277—298 (1943).

Butenandt, A., P. Karlson u. G. Hannes: Über den „Anti-Bar-Stoff", einen genabhängigen, morphogenetischen Wirkstoff bei Drosophila melanogaster. Biol. Zbl. **65**, 41—51 (1946).

Buzzati-Traverso, A.: Genetica di popolazioni in Drosophila. I. Eterozigosi in Drosophila subobscura Collin. Sci. Genetica **2**, 190—223 (1942).

— Genetica di popolazioni in Drosophila. II. I cromosomi di 5 specie del „gruppo obscura" e la incrociabilità di varie razze geografiche. Sci. Genetica **2**, 224 bis 241 (1942).

— Genetica di popolazioni in Drosophila. III. Un maschio mutante raccolto in natura ed alta mutabilità in una popolazione liberamente vivente. Sci. Genetica **2**, 242—251 (1942).

— La determinazione di frequenze geniche in una popolazione specifica. Ricerca Scientifica **13**, 448—461 (1942).

— La meccanica cromosomica determinante la variegazione in white-mottling in Drosophila melanogaster. Rend. Ist. Lombardo Sc. e Lett., Pp. 77 (1943).

— L'estrinsecazione fenotipica dell'effetto di posizione white-mottling in Drosophila melanogaster. Rend. Ist. Lombardo Sc. e Lett., Pp. 77 (1943).

— A bibliography of genetics research published in Italy and Germany, 1939—1945. Heredity **1**, 19—51 (1947).

C...... R.: The question of Dr. Vavilov's death. J. Hered. **37**, 15 (1946).

— Dr. Muller receives Nobel medicine award. J. Hered. **37**, 325—326 (1946).

Calvin, M., M. Kodani u. R. Goldschmidt: Effects of certain chemical treatments on the morphology of salivary gland chromosomes and their interpretation. Proc. Nat. Acad. Sci. USA. **26**, 340—349 (1940).

Câmara, A. de Souza da: Genética e selecção. (Genetics and selection.) Rev. Agron. Lisboa **27**, 410—421 (1939).

— A genética continua a ser atacada. (Genetics continues to be attacked.) Rev. Agron. Lisboa **28**, 330 bis 332 (1940).

Canella, M. F.: Genetica, mutazionismo e neolamarckismo. (Genetik, Mutationismus und Neolamarckismus.) Riv. Biol. **31**, 63—100 (1941).

Carles, J.: Problèmes d'hérédité. Beauchesne et Ses Fils. Paris, Pp. 258, 37 figs. 1944 (1945).

Castle, W. E.: Genes which divide species or produce hybrid vigor. Proc. Nat. Acad. Sci., Wash. **32**, 145 bis 149 (1946).

Catcheside, D. G.: The mechanism of radiation-induced chromosome rearrangements. Proc. 7th Int. Genet. Congr. Edinburgh 23.—30. August, Pp. 86 1939 (1941). (Abstr.)

Catcheside, D. G.: Gene and chromosome theory and cytology at the Seventh International Genetical Congress, Edinburgh 1939. Chronica Botanica 6, 9—11 (1940).
— Polarized segregation in an ascomycete. Ann. Bot. Lond. 8, 119—130 (1944).
— The P-locus position effect in Oenothera. Journ. of Genet. 48, 31—42 (1947).
— A duplication and deficiency in Oenothera. Journ. of Genet. 48, 99—110 (1947).
Catsch, A., A. Kanellis u. Gh. Radu: Über den Einfluß des Alterns bestrahlter Spermien auf die Rate röntgeninduzierter Translokationen bei Drosophila melanogaster. Naturwiss., 31 (1943).
—, —, — u. P. Welt: Über die Auslösung von Chromosomenmutationen bei Drosophila melanogaster mit Röntgenstrahlen verschiedener Wellenlänge. Naturwiss. 32, 228 (1944).
— u. G. Radu: Über die Abhängigkeit der röntgeninduzierten Translokationsrate vom Reifezustand der bestrahlten Gameten bei Drosophila melanogaster. Naturwiss. 31, 368—369 (1943).
—, — Die Abhängigkeit der röntgeninduzierten Translokationsrate bei Drosophila melanogaster von der Intensität der angewandten Bestrahlungsdosis. Naturwiss., H. 35—36 (1943).
—, — u. A. Kanellis: Die Dosisproportionalität der durch Röntgenbestrahlung ausgelösten Translokationen zwischen II. und III. Chromosom bei Drosophila melanogaster. Naturwiss., 31, 368 (1943).
Černjaev, I.: (Vegetative rapprochement.) Sovetskaja Botanika, Nr. 2, 66—69 (1939). [Russisch.]
Chandler, C.: Microsporogenesis in triploid and diploid plants of Hemerocallis fulva. Bull. Torrey bot. Club 67, 649—672 (1940).
Cheesman, E. E. u. K. S. Dodds: Genetical and cytological studies of Musa. 4. Certain triploid clones. J. Genet. 43, 337—357 (1942).
Chiarugi, A.: Sul numero dei cromosomi della Primula Palimuri Petagna. N. Giorn. Botan. Ital. n. s. 47, 519 (1940).
Chopinet, R.: Sur quelques hybrides expérimentaux interspécifiques et intergénériques chez les cruciferes. C. R. Acad. Sci., Paris 215, 545—547 (1942).
Christiansen-Weniger, F.: Sammlung und Erhaltung von wertvollen Genen der Primitiv- und der Wildformen. Forsch.dienst II, 645—648 (1941).
Christoff, M.: Über die Fortpflanzungsverhältnisse bei einigen Arten der Gattung Hieracium nach einer experimentell induzierten Chromosomenvermehrung. Planta 31, 73—90 (1940).
— Die genetische Grundlage der apomiktischen Fortpflanzung bei Hieracium aurantiacum L. Z. Vererbungsl. 80, 103—125 (1942).
— u. G. Papasova: Die genetischen Grundlagen der apomiktischen Fortpflanzung in der Gattung Potentilla. Z. indukt. Abstamm. u. Vererblehre 81, 1—27 (1943).
Ciaccio, G.: Ricerche sulle ghiandole salivari dei Ditteri. I cromosomi giganti studiati con la microfotografia a raggi ultravioletti. Protoplasma 37, 161 (1942).
Cicin, N. V.: (Hybridization-a powerful method in Michurin's plant breeding.) Vestnik Gibridizacii (Hybridization), Nr. 1, 10—19 (1941). [Russisch.]
Cole, E. J.: The relation of genetics to geographical distribution and speciation. 1. Introduction. Amer. Natural. 74, 193—197 (1940).
Colin, E. C.: Elements of genetics. Mendel's laws of heredity with special application to man. Blakiston Co., Philadelphia (1941).
— Elements of genetics. Blakiston Co., Pa., Pp. XIII +402, 90 figs, 18 tables (1946).
Cooper, D. C.: Macrosporogenesis and embryology of Portulaca oleracea. Amer. J. Bot. 27, 326—330 (1940).

Cooper, D. C. u. R. A. Brink: The endosperm as a barrier to interspecific hybridization in flowering plants. Science 95, 75—76 (1942).
Crane, M. B.: Cultivated plants of the past, present and future. Endeavour 2, 111—116 (1943).
Cuénot, L.: La transmission héréditaire dans les croisements entre espèces. Rev. Sci., Paris 79, 317—319 (1941).
Cueti y Rui-Diaz, E.: Die biogenetische organische Evolution. Rev. Acad. Ci. exact., Madrid 35, 415—441, 515—549 (1941).
Cummings, J. M.: Chromosomes of Datura ceratocaula in hybrids obtained by embryo dissection, an advance report. Genetics 32, 84 (1947). (Abstr.)

Daniel, L.: L'hérédité chez les descendants du topinambour greffé (Analogie avec la vigne greffée). Progr. Agric. Vitic. 113, 90—92 (1940).
Danielli, J. F. u. D. G. Catcheside: Phosphatase on chromosomes. Nature, Lond. 156, 294 (1945).
Darlington, C. D.: The evolution of genetic systems. University Press, Cambridge 10s, 6d, Pp. X+149 (1939).
— Cytology. Nature 144, 816—817 (1939).
— The substance of heredity. Endeavour, Lond. 1, 102 bis 105 (1942).
— Heredity, development and infection. Nature, Lond. 154, 164—169 (1944).
— Paracrinkle virus and inheritance. Nature, Lond. 154, 489 (1944).
— The retreat from science in Soviet Russia. Nineteenth Century 142, 157—168 (1947).
— A revolution in Soviet science. Discovery 8, 40—43 (1947).
— u. R. A. Fisher: The polygene concept. Nature, Lond. 150, 154 (1942).
— u. L. La Cour: Nucleic acid starvation of chromosomes in Trillium. J. Genet. 40, 185—213 (1940).
—, — The detection of inert genes. J. Hered. 32, 115 bis 121 (1941).
Davies, R. G.: Genetics in the U.S.S.R. Modern Quart. 2, 336—346 (1947).
Deakin, A.: Genetics and biological theory. Science 103, 570—571 (1946).
Delbrück, M.: Radiation and the hereditary mechanism. Amer. Naturalist 74, 350—362 (1940).
Dellazoppa, J. G.: La uniformidad da las variedades seleccionadas. (The uniformity of selected varieties.) Arch. Fitotécn. Uruguay 3, 336—343 (1940—1941).
Delone, L. N. u. V. I. Didusj: (The efficacy of intravarietal selection in line varieties of selfpollinated crops.) Selekcija i Semenovodstvo (Breeding and Seed Growing) 13, Nr. 7/8, 16—24 (1946). [Russisch.]
Demerec, M.: Production of mutations in Drosophila by treatment with some carcinogens. Science 105, 634 (Abstr.) (1947).
— Mutations in Drosophila induced by a carcinogen. Nature, Lond. 159, 604 (1947).
Dempster, E. R.: „Mock dominance". Science 97, 464—465 (1943).
Dermen, H.: (Durch Colchicin ausgelöste Polyploidie und die Versuchstechnik.) Izvestija Akad. Nauk Arm. SSR. 2, 75—78 (1945). [Russisch.]
— Histogenetic basis of some bud sports and variegations. Genetics 32, 84—85 (1947). (Abstr.)
Dice, L. R.: The effectiveness of adverse selection. Genetics 24, 68—69 (1939). (Abstr.)
Dixon, T. F.: Autosynthetic molecules. Nature, Lond. 155, 596—598 (1945).
Dobzhansky, T.: Studies on the genetic structure of natural populations. Yearb. Carneg. Instn. 1941 bis 1942, Nr. 41, 228—235 (1942).
— O gen como unidade auto — reprodutiva da fisiologia celular. (The gene as the self reproducing unit of cell

physiology.) Rev. Agric. Piracicaba 18, 387—396 (1943).

Dobzhansky, T.: Heterosis. Rev. Agric. Piracicaba. 18, 397—398 (1943).

— Genetics of natural populations. XIII. Recombination and variability in populations of Drosophila pseudoobscura. Genetics 31, 269—290 (1946).

— A directional change in the genetic constitution of a natural population of Drosophila pseudoobscura. Heredity 1, 53—64 (1947).

— u. C. Epling: Contributions to the genetics, taxonomy, and ecology of Drosophila pseudoobscura and its relatives. Washington, D. C., Nr. 554, Pp. III + 183, figs, 4 plates, tables (1944).

— u. A. M. Holz: A re-examination of the problem of manifold effects of genes in Drosophila melanogaster. Genetics 28, 295—303 (1943).

Döring, H.: Über die Dominanzverhältnisse von Allelen verschiedener Mutabilität. Biol. Zbl. 61, 65—68 (1941).

Dotterweich, H.: Über Beeinflussung der Mutabilität von Drosophila melanogaster durch Chemikalien. Zool. Anz., Suppl. 12, 244—253 (1939).

— Die Veränderlichkeit der Mutationsrate von Drosophila melanogaster nach generationenlanger chemischer Beeinflussung. Z. Vererbungsl. 78, 261—272 (1940).

— Mutabilität und Umwelt. Versuche über den Einfluß von Follikelhormon auf die Mutationsauslösung bei Drosophila melanogaster. Biol. generalis (Wien) 15, 109—123 (1941).

— u. L. Schmidtke: Auslösung spezifischer Mutationen bei Drosophila melanogaster nach chemischer Dauerbeeinflussung. I. Versuche mit Follikelhormon. Z. Vererbungsl. 79, 220—231 (1941).

Dubinin, N. P., V. V. Hvostova u. V. V. Mansurova: Chromosomal aberrations, lethal mutations and X-ray dosage. C. R. (Doklady) Acad. Sci. URSS., N. s. 31, 387—389 (1941).

Dunn, L. C. u. S. Gluecksohn-Schoenheimer: Dominance modification and physiological effects of genes. Proc. Nat. Acad. Sci., Wash. 31, 82—84 (1945).

Dvorjankin, F.: (Summing up.) Jarovizacija, Nr. 5 (32), 23—34 (1940). [Russisch.]

Eberhardt, K.: Über den Mechanismus strahleninduzierter Chromosomenmutationen bei Drosophila melanogaster. Chromosoma I, 317—335 (1939).

— Vergleich der chromosomenbruchauslösenden Wirkung von Röntgen- und Neutronenstrahlen bei Drosophila melanogaster. Naturwiss. 31, 23 (1943).

Ellenhorn, J. E.: On the genesis of the chromosome set of Rhoeo discolor Hence. C. R. Acad. Sci. URSS., n. s. 27, 361—364 (1940).

— Über Unterschiede der Chromosomeren bei allelomorphen Trabantenchromosomen von Allium cepa. C. R. (Doklady) Acad. Sci. URSS, 27, 357—360 (1940).

— (Über die verschiedene Färbung der Elternchromosomen bei Bastarden.) C. R. (Doklady) Acad. Sci. URSS. 56, 961—963 (1947). [Russisch.]

Emerson, S.: Growth of incompatible pollen tubes in Oenothera organensis. Bot. Gaz. 101, 890—911 (1940).

— Linkage relationship of two gametophytic characters in Oenothera organensis. Genetics 26, 469—473 (1941).

— The induction of mutations by antibodies. Proc. Nat. Acad. Sci. Wash. 30, 179—183 (1944).

— Genetics as a tool for studying gene structure. Ann. Mo. Bot. Gdn 32, 243—249 (1945).

Ephrussi, B. u. E. Sutton, A reconsideration of the mechanism of position effect. Proc. Nat. Acad. Sci. Wash. 30, 183—197 (1944).

Epling, C.: Actual and potential gene flow in natural populations. Amer. Nat. 81, 104—113 (1947).

— u. H. Lewis: Fertility and natural hybridization in Delphinium and its bearing upon gene exchange and the origin of diploid species. Amer. J. Bot. (Suppl.) 33, 234 (1946). (Abstr.)

Erikson, R. O.: Mass collections: Camossia scilloides. Ann. Mo. Bot. Gdn 28, 293—298 (1941).

Erlandsson, S.: The chromosome numbers of three Artemisia forms. Hereditas 25, 27—30 (1939).

Ernst, A.: Von den Anfängen der Vererbungs- und Mutationsforschung in der Schweiz. Arch. Klaus-Stiftg. 16, 608—620 (1941).

— Vererbung durch labile Gene. Arch. Klaus-Stiftg. Vererb.-Forsch. Zürich 17, Pp. 567, 58 figs, 199 tables (1942).

— Intra- und interspezifische Bestäubungen an dimorphen Primula-Arten der Sektion Candelabra und ihre Aufschlüsse zum Heterostylieproblem. Arch. d. Jul. Klaus-Stiftg. 19 (1944).

Ernst, H.: Cytogenetische Untersuchungen an haploiden Pflanzen von Antirrhinum majus L. I. Die Meiosis. Z. Bot 35, 161—190 (1940).

Ernst-Schwarzenbach, M.: Genetik und Evolution. Vjschr. Naturforsch. Ges. Zürich 85, 35—50 (1940).

Espinasse, P. G.: Genetics in the USSR. Nature, Lond. 148, 739—743.

Etherington, I. M. H.: Non-associative algebra and the symbolism of genetics. Proc. Roy. Soc. Edinb. B 61, 24—42 (1941).

Fabergé, A. C.: The equivalent effect of X-rays of different wave-lenght on the chromosomes of Tradescantia. J. Genet. 40, 379—384 (1940).

— The concept of polygenes. Nature 151, 643 (1943).

Fábián, G. u. G. Matoltsy: Test of a cancerogenic substance in respect to the „non-disjunction' frequency of the X-chromosome in Drosophila. Nature, Lond. 158, 911—912 (1946).

Fano, U.: Note on the theory of radiation-induced lethals in Drosophila. Science 106, 87—88 (1947).

Fassett, N. C.: Mass collections: Rubus odoratus and R. parviflorus. Ann. Mo. Bot. Gdn 28, 299—374 (1941).

Fernandes, A. u. J. B. Neves: Sur l'origine des formes de Narcissus bulbocodium L. à 26 chromosomes. Bol. Soc. Broteriana 15, II. Sér., 43—132 (1941).

Finck, E. v.: Die Allelenserie des Gens ss („spineless") bei Drosophila melanogaster. Biol. Zbl. 62, 379—400 (1942).

Fischer, A.: Biology of tissue cells. Cambridge Pp. VIII + 348. 55 figs. 15 tables (1946).

Fisher, R. A.: The precision of the product formula for the estimation of linkage. Ann. Eugen., Lond. 9, 50—54 (1939).

— Average excess and average effect of a gene substitution. Ann. Eugen. 11, 53—63 (1941).

— Allowance for double reduction in the calculation of genotype frequencies with polysomic inheritance. Ann. Eugen. 12, 169—171 (1944).

— A system of coring linkage data, with special reference to the pied factors in mice. Amer. Nat. 80, 568 bis 578 (1946).

— The theory of linkage in polysomic inheritance. Philos. Trans. 233, Ser. B 55—87 (1947).

Ford, E. B.: Polymorphism. Biol. Rev. 20, 73—88. (1945).

Francini, E.: Ibridazione interspecifica nel genere Paphiopedilum. Cariologia di Paph. Spicerianum e di Paph. Lathamianum (Paph. Spicerianum ♀-Paph. villosum♂). N. Giorn. Botan. Ital. 52, 21—29 (1945).

Frandsen, K. J.: Fremstilling af Planter med forøget Kromosombesaetning ved Hjaelp af Colchicin. (The production of plants with increased chromosome number by means of colchicine.) Naturhistorisk Tidende, København 4, 113—115 (1940).

— Iagttagelser over Indavl og Udavl hos fremmedbefrugtende Planter. (Observations on inbreeding and outbreeding in cross-fertilized plants.) Nord. Jordbr. Forskn. Nos. 5/6, 218—236 (1943).

Freisleben, R. u. A. Lein: Möglichkeiten und praktische Durchführung der Mutationszüchtung. Kühn-Archiv 1943/44 **60**, 211—225 (1944).
— E. A. W. Müller, u. R. v. Sengbusch: Röntgenologische Untersuchungsmethode von Pflanzen und Pflanzenteilen für züchterische Zwecke. Züchter **15**, 3 (1943).
Friedberg, L.: Notions comparées de fluctuation et de variation dans l'amélioration des plantes. Ann. Agron. Paris **12**, 391—420 (1942).
Fries, N.: Über röntgen-induzierte physiologische Mutationen bei Ophiostoma multiannulatum (Hedgc. et Davids). Ark. Bot. **32** A, Nr. 8, 1—9 (1945).
Frimmel, F. v.: Die Bedeutung der Züchtung von Heterosissorten. Z. Pflanzenz. **23**, 638—660 (1941).
Frizzi, G.: Mutazioni e mappa cromosomica in Drosophila subobscura. Boll. Soc. Ital. Biol. Sperim. 16 (1941).
Fröier, K. u. Å. Gustafsson: The influence of X-ray dosage on germination and sprouting ability in barley and wheat. Svensk Bot. Tidskr. **35**, 43—56 (1941).
Frolova, S. L.: Chromosome structure after Removal of thymonucleic Acid by Action of Enzymes. C. R. (Doklady) Acad. Sci. URSS. **30**, 459—462 (1941).
Fuji, S.: (Crossing-over in the sixth chromosome of Drosophila virilis.) Jap. J. Genet. **16**, 29—291 (1940). [Japanisch.]
Fürtauer, R.: Untersuchungen über die Beziehungen zwischen Photoperiode, Lichtintensität sowie Temperatur und der Plasmavererbung bei Epilobium. Mit 9 Abb. Jahrb. f. wiss. Bot. **89**, 412—460 (1940).
Fyfe, J. L.: The action and use of colchicine in the production of polyploid plants. Imperial Bureau of Plant Breeding and Genetics, Cambridge ls. Od. Pp. 10, figs. (1939).
— The Soviet genetics controversy. Modern Quart. **2**, 347—351 (1947).

Garboe, A.: Fra Arvelighedsforskningens Udviklingsgang. (On the origins and development of genetic research.) Nat. Verd. Kbh. **24**, 424—428 (1941).
Garlan, P.: Fundamentals of genetics. Wallerstein Laboratories Communications **9**, 5—27 (1946).
Gates, R. R.: Some recent aspects of genetics. Proc. N. S. Inst. Sci. **20**, 127—140 (1940/41).
Gause, G. F. u. W. W. Alpatov: On the inverse relation between inherent and acquired properties of organisms. Amer. Nat. **79**, 478—480 (1945).
Geigy, R. u. A. N. Aboim: Gonadenentwicklung bei Drosophila nach frühembryonaler Ausschaltung der Geschlechtszellen. Rev. Suisse de Zool. **51**, 410—417 (1944).
Geitler, L.: Die innere Polyploidie pflanzlicher und tierischer Gewebe. Forsch. und Fortschr. **18**, 255—256 (1942).
Gentcheff, G. u. Å. Gustafsson: Parthenogenesis and pseudogamy in Potentilla. Bot. Not. 109—132 (1940).
—, — The balance system of meiosis in Hieracium. Hereditas **26**, 209—249 (1940).
— — The cultivation of plant species from seed to flower and seed in different agar solutions. Hereditas, Lund **26**, 250—256 (1940).
George, W.: Genes and development. Sci. Progr. **35**, 447—469 (1947).
Georlette, R.: Sur la signification des hybrides interspécifiques et intergénériques. Ann. Gembl. **45**, 217 bis 223 (1939).
Gerasimova-Navašina, E. N.: (Über das Verhalten der Spermien im Pollenschlauch bei Crepis.) C. R. (Doklady) Acad. Sci. URSS. **57**, 285 (1947). [Russisch.]
— (Eine mitotische Hypothese der doppelten Befruchtung.) C. R. (Doklady) Acad. Sci. URSS. **57**, 395 (1947). [Russisch.]

Geršenson, S. M.: (Die Natur der sogenannten „genetisch inerten" Chromosomenteilen.) Kijev (Verlag Ukrain. Acad. d. Wiss.) (1940). [Russisch.]
—, R. A. Silberman, O. L. Levočkina, A. M. Paškovskij, P. O. Sitjko, u. N. D. Tarnavskij: (Mutationsauslösung durch Dichlorgiethylsulfid.) C. R. (Doklady) Acad. Sci. URSS. **58**, 1495—1496 (1947). [Russisch.]
Gersh, E. S. u. B. Ephrussi: The mechanism of position effect-Experiments on the phenotypic expression of position effects in relation to changes in pairing of neighboring chromosome regions. Proc. Nat. Acad. Sci. Wash. **32**, 87—94 (1946).
Giles, N.: Spontaneous chromosome aberrations in triploid Tradescantia hybrids. Genetics, **26**, 632—649 (1941).
— The effect of fast neutrons on the chromosomes of Tradescantia. Journ. appl. Physics **12**, 347 (1941).
Glivenko, V. I.: (Studies on mathematical Genetics.) Bull. Acad. Sci. URSS., Sér. biol. 615—635 (1939). [Russisch.]
Goldschmidt, R.: A mutant of Drosophila melanogaster resembling the so-called unstable genes of Drosophila virilis. Proc. Nat. Acad. Sci. Wash. **29**, 203—206 (1943).
— On spontaneous mutation. Proc. Nat. Acad. Sci. Wash. **30**, 297—298 (1944).
— One- or two-dimensional action of mutant loci? Amer. Nat. **79**, 97—103 (1945).
— Position effect and the theory of the corpuscular gene. Experientia, Basel **2**, 197—203, 250—256 (1946).
Golovcov, L. A.: (Fifteen years inbreeding with sugarbeet.) Jarovizacija Nr. 2 (29), 62—67 (1940). [Russisch.]
Goodspeed, T. H. u. F. M. Uber: Radiation and plant cytogenetics. Bot. Rev. **5**, 1—48 (1939).
Gordon, C. u. J. H. Sang: Polygenic inheritance and the Drosophila culture. Nature, Lond. **149**, 610—611 (1942).
Gottschewski, G.: Eine Analyse bestimmter Drosophila pseudoobscura — Rassen- und Artkreuzungen. Z. Vererbungsl. **78**, 338—398 (1940).
— Der heutige Stand der Vererbungswissenschaft. Biologe **12**, 53—64 (1943).
Gourlay, W. B.: The lost scent of Mimulus moschatus. J. R. Hort. Soc. **72**, 285—287 (1947).
Gowen, J. W.: An analysis of the genic or cytoplasmic basis of heterosis. Genetics **30**, 7 (1945). (Abstr.)
— Significance of additive, dominant, complementary gene action to hybrid vigour in Drosophila. Genetics **31**, 217—218 (1946). (Abstr.)
—, J. Stadler u. L. E. Johnson: On the mechanism of heterosis-the chromosomal or cytoplasmic basis for heterosis in Drosophila melanogaster. Amer. Nat. **80**, 506—531 (1946).
Grandjean, F.: Les méthodes pour établir de listes de priorité et la concordance de leur résultats. C. R. Acad. Sci., Paris **214**, 729—733 (1942).
— La signification évolutive des écarts individuels. C. R. Acad. Sci., Paris **215**, 216—220 (1942).
Graner, E. A.: O citoplasma na hereditariedade. (The cytoplasm in inheritance.) Rev. Agric., S. Paulo **16**, 481—486 (1941).
— Thomas Hunt Morgan. Rev. Agric., Piracicaba **21**, 97—98 (1946).
Greenstein, J. P. u. H. W. Chalkley: The influence of nucleic acid on dehydrogenase systems. A contribution to the problem of gene mechanism. Ann. Mo. Bot. Gdn. **32**, 179—185 (1945).
Gregor, J. W.: Ecotypic differentiation. New Phytol. **45**, 254—270 (1946).
Gustafsson, Å.: Mutationsforschung und Züchtung. Züchter **14**, 57—64 (1942).

Gustafsson, Å.: The plastid development in various types of chlorophyll mutations. Hereditas 28, 483—492 (1942).
— The effect of heterozygosity on variability and vigour. Hereditas, Lund 32, 263—286 (1946).
Györffy, B.: Chromosomaszámlálások colchicinnel elöállitott polyploidoknál. (Chromosome numbers in colchicine induced polyploids.) Arb. Ung. Biol. Forsch.-Inst. 12, 326—329 (1940).
— (Die Physiologie der polyploiden Pflanzen). Arb. ung. biol. Forschgsinst. 13, 362—446 (1941). [Ungarisch.]
— Gének és enzymek. (The interrelations of genes and enzymes.) Arb. Ung. Biol. Forschgsinst. 15, 450—461 (1943).

Hačaturov, S. P.: (Differences in the progenies of hybrids.) Jarovizacija Nr. 6, (33), 29—44 (1940). [Russisch.]
Hackbarth, J., P. Michaelis u. G. Scheller: Untersuchungen an dem Antirrhinum-Wildsippen-Sortiment von E. Baur. I. Das Wildsippen-Sortiment und die von E. Baur durchgeführten Kreuzungen. Z. Vererbungsl. 80, 1—102 (1942).
Haddow, A.: Transformation of cells and viruses. Nature, Lond. 154, 194—198 (1944).
Hadorn, E.: Zur Pleiotropie der Genwirkung. Arch. d. Jul. Klaus-Stiftg. Erg.-Bd. zu Bd. 20, 82—95 (1945).
— Thomas Hunt Morgan. Experientia, Basel 2, 75—76 (1946).
— Mutationsversuche mit Chemikalien an Drosophila. I. Wirkung von Colchicin auf transplantierte Larven-Ovarien nach Behandlung in vitro. Revue Suisse de Zoologie 15, 486—494 (1946).
— u. Gloor: Cryptocephal, ein spätwirkender Letalfaktor bei Drosophila melanogaster. Rev. suisse Zool. 50, 256—261 (1943).
— u. H. Niggli: Mutations in Drosophila after chemical treatment of gonads in vitro. Nature 157, 162—164 (1946).
Hagedoorn, A.: Over de oorzaken van mutatie. Autoreferaat genetikadag 3 Juni 1939. (On the causes of mutation. Report of the genetics conference on 3rd June 1939.) Erfelijkheid in Praktijk, Leiden 4, 170—171 (1939).
Haldane, J. B. S.: The theory of the evolution of dominance. J. Genet. 37, 365—374 (1939).
— Lysenko and genetics. Sci. and Soc. 4, 433—437 (1940).
— New paths in genetics. George Allen and Unwin, Ltd., London 7s., 6d., Pp. 206, 17 figs, 10 tables (1941).
— New Paths in Genetics. London, Allen u. Umwin, Pp. 260 (1942).
— Heredity, development and infection. Nature, Lond. 154, 429 (1944).
— Soviet Genetics. Guardian, Nr. 5266, 542 (1946).
Halifman, I. A.: (Darwinian meeting of the Lenin Academy of Agricultural Science.) Priroda (Nature) Nr. 4, 107—113 (1940).
Hall, H. F.: A noty on terminology regarding intersexes. J. Hered. 32, 309—314 (1941).
Hämmerling, J.: Fortpflanzung im Tier- und Pflanzenreich. Sammlung Göschen 1138 (1941).
Hargitt, G. T.: What is germ plasm? Science 1944 100, 343—348 (1944).
Harland, S. C. u. C. D. Darlington: Prof. N. I. Vavilov, For. Mem. R. S. Nature, Lond. 156, 621—622 (1945).
Harnly, M. H.: Flight capacity in relation to phenotypic and genotypic variations in the wings of Drosophila melanogaster. J. of exper. Zool. 88, 263—273 (1941).

Harrington, J. B.: How should varieties of annual self-fertilized crops be perpetuated? J. Amer. Soc. Agron. 31, 472—474 (1939).
Harte, C.: Meiosis und crossing-over. Weitere Beiträge zur Zytogenetik von Oenothera. Z. Bot. 38, 65—137 (1942).
Hartmann, M.: Geschlecht und Geschlechtsbestimmung im Tier- und Pflanzenreich. Sammlung Göschen, 1127, Walter de Gruyter u. Co., Berlin (1939).
Haustein, E.: Über reziproke Verschiedenheiten bei Lobelien-Kreuzungen. Z. Vererbungsl. 79, 245—252 (1941).
Hawkes, J. G.: Some effects on the drug colchicine on cell division. J. Genet. 44, 11—22 (1942).
Haya, T.: A. Critique on the Conception of the Genom. Japan. Journ. Genet. 16, 211—227 (1940).
Hayes, H. K.: Forrest Rhinehart Immer 1899—1946. Science 103, 751 (1946).
— Yield genes, heterosis and combining ability. Amer. Nat. 80, 430—445 (1946).
Hazel, L. N.: The genetic basis for constructing selection indexes. Genetics 28, 476—490 (1943).
Hazina, E. P.: (Alteration of the fruit in the seed progeny of grafts.) Jarovizacija Nr. 1 (34) 84—89 (1941). [Russisch.]
Hecht, A.: Induced tetraploids of a self sterile Oenothera. Genetics 29, 69—74 (1944).
Heilborn, O.: Chromosome studies in Cyperaceae. Hereditas 25, 224—240 (1939).
Heilbronn, A. u. C. Kosswig: Principia genetica. Grundbegriffe und Grundtatsachen der Vererbungswissenschaft. J. Unified Sci. (Erkenntnis) 8, 229—255 (1939).
Henderson, I. F. u. W. D. Henderson: A dictionary of scientific terms. Pronunciation, derivation, and definition of terms in biology, botany, zoology, anatomy, cytology, embryology, physiology. Oliver and Boyd, Edinburgh and London 3rd ed. (revised by F. W. Kenneth). 16s. Od Pp. XII + 383 (1939).
Henderson, J. H. M.: Biochemical genetics. Science 105, 522 (1947).
Henke, K., E. v. Finck u. Ma, Sung-Yun: Über sensible Perioden für die Auslösung von Hitze-Modifikationen bei Drosophila und die Beziehungen zwischen Modifikationen und Mutationen. Z. Vererbungsl. 79, 267—316 (1941).
L'Héritier, P. L. u. F. Hugon de Scoeux: Transmission of the carbon dioxide susceptibility of Drosophila by grafting. Nature, Lond. 157, 728 (1946).
Herskowitz, I. H.: The relationship of X-ray induced recessive lethals to chromosomal breakage. Amer. Nat. 80, 588—592 (1946).
Herzog, G.: Genetische und cytologische Untersuchungen über 15-chromosomige Mutanten von Oenothera biennis und Oenothera Lamarckiana. Flora, N. F. 34, 377—432 (1940).
Hesin, R. V.: (Mütterlicher Effect bei Drosophila melanogaster.) C. R. (Doklady) Acad. Sci., URSS. 58, 667 (1947). [Russisch.]
Heyn, F. A.: Het opwekken van mutaties door straling. (The induction of mutations by radiation.) Vakbl. Biol. 21, 81—88, 101—105 (1941).
Hiorth, G.: Eine Rückkreuzung mit einem tetraploiden Artbastard. Z. Vererbungsl. 78, 155—156 (1940).
— Zur Genetik und Systematik der Gattung Godetia. Z. Vererbungsl. 79, 199—219 (1941).
— Zur Genetik und Systematik der Amoena-Gruppe der Gattung Godetia. Z. Vererbungsl. 80, 289—349 (1942).
— Eine Translokation zwischen einem Godetia Whitneyi und ein G. deflexa-Chromosom. Z. Vererbungsl. 80, 565—569 (1942).

Hiorth, G.: Über das Vorkommen von Hemmungsgenen in Inzuchtlinien von Godetia Whitneyi. Züchter **17/18**, 69—78 (1946).
— Zur Genetik des Artbastardes Godetia amoena × G. Whitneyi. Züchter **17/18**, 109—121 (1947).
Hoffmann, W.: Vererbungsversuche mit Wunderblume, Löwenmäulchen und Mais. Biologe **11**, 313 bis 326 (1942).
Hollaender, A.: The mechanism of radiation effects and the use of radiation for the production of mutations with improved fermentation. Ann. Mo. Bot. Gdn. **32**, 165—178 (1945).
— u. C. P. Swanson: Modification of the X-ray induced mutation rate in fungi by pretreatment with near infrared. Genetics **32**, 90 (1947). (Abstr.)
Holmogorov, A. N.: On a new confimation of Mendel's laws. C. R. (Doklady) Acad. Sci. URSS. **27**, 37—41 (1940).
Horowitz, B.: A note on the interaction between heterosis and photoperiodic response. Aust. J. Sci. **7**, 126—127 (1945).
Horowitz, H. N., D. Bonner, H. K. Mitchell, E. L. Tatum u. G. W. Beadle: Genetics control of biochemical reaktions in Neurospora. The Amer. Naturalist **79**, 304—317 (1945).
Howard, H. W.: The effect of polyploidy an hybridity on seed size in crosses between Brassica chinensis, B. carinata, amphidiploid B. chinensis carinata and autotetraploid B. chinensis. J. Genet. **43**, 105—119 (1942).
Hubricht, L. u. E. Anderson: Vicinism in Tradescantia. Amer. J. Bot. **28**, 957 (1941).
Hudson, J. W.: A device for visualizing the solution of genetics problems. Trans. Ill. Acad. Sci. **34**, 93—94 (1941).
Hudson, P. S. u. R. H. Ricnens: The new genetics in the Soviet Union. Imperial Bureau of Plant Breeding and Genetics, Cambridge, Pp. 88 (1946).
Hull, F. H.: Theoretical regression of F_N on homozygous parents with additive or complementary gene action. Genetics **32**, 90—91 (1947). (Abstr.)
Huskins, C. L.: Polyploidy and mutations. Amer. Naturalist **75**, 329—346 (1941).
Huxley, J.: Evolutionary biology and related subjects. Nature, Lond. **156**, 254—256 (1945).
Hvostova, V. V.: (Neue Angaben über die Zusammenhänge der Gene und des Zytoplasmas in den Erscheinungen der Entwicklung und Vererbung). Uspjehi Sovremennoj Biologii (Advances in Mod. Biol.) **19**, Nr. 3, 403—407 (1945). [Russisch.]

Ibsen, H. L.: Principles of genetics. St. Louis-Chicago-New-York-Cincinnati, Pp. 118, figs., tables., illus. (1942). [Mimeographed.]
Ignatjev, M. V. u. N. I. Šapiro: Ways of stabilization of genotype. I. Selection of stable allelomorphs. C. R. (Doklady) Acad. Sci. URSS. **45**, 206—208 (1944).
— (Die Stabilisationswege des Genotypus. II. Die Auslese der Mutabilitätsmodifikatoren.) C. R. (Doklady) Acad. Sci. URSS. **46**, Nr. 3, 133—137 (1945). [Russisch.]
— (Eine Analyse der Variabilität der Geschwindigkeit des natürlichen Mutationsprozesses bei Drosophila melanogaster.) C. R. (Doklady) Acad. Sci. URSS. **58**, 891 (1947). [Russisch.]
Iltis, H.: Gregor Mendel and his work. Sci. Mon. N. Y. **56**, 414—423 (1943).
— A visit to Gregor Mendel's home. J. Hered. **38**, 163 166 (1947).
Ipatjev, A. N.: Accurate morphological methods for studying varieties. C. R. (Doklady) Acad. Sci. URSS. **42**, 401—403 (1944).
Irwin, M. R.: Antigens, antibodies and genes. Biol. Rev. **21**, 93—100 (1946).
— Immunogenetics. Advances Genet., N. Y. **1**, 133—159 (1947).
— u. R. W. Cumley: Speciation from the Point of View of Genetics. Amer. Natural. **74**, 222—231 (1940).
—, — Immunogenetic studies of species relationships. Amer. Nat. **77**, 211—233 (1943).
Ives, P. T.: The genetic structure of American populations of Drosophila melanogaster. Genetics **30**, 167—196 (1945).

Jaeger, E. C.: A source-book of biological names and terms. Springfield, Illinois, Baltimore, Maryland, Pp. XXVI + 256, 26 figs. (1944).
Jaeger, E.: Contribution á l'étude de la gynodioecie. Bull. Soc. Bot. Fr. **86**, 395—403 (1939).
Jakovlev, P. N.: (Along Darwin's great path.) Sadovodstvo (Horticulture), Nr. 6, 18—24 (1940). [Russisch.]
— (I. V. Michurin.) Sovetskaja Botanika (Soviet Botany), Nr. 1/2, 5—13 (1941). [Russisch.]
Janaki Amal, E. K.: Chromosome diminution in a plant. Nature, Lond. **2**, 839—840 (1940).
Jean, Y.: Production artificielle de mutations par irradiations. (Artificial production of mutations by irradiation.) La Bonne Terre, Canada **22**, 171—186 (1941).
Johnson, N. L.: Parabolic test for linkage. Ann. Math. Statist. **11**, 227—253 (1940).
Jones, D. F.: Somatic segregation. Bot. Rev. **7**, 291 bis 307 (1941).
— Chromosome degeneration in relation to growth and hybrid vigour. Proc. Nat. Acad. Sci. Wash. **28**, 38—44 (1942).
Jones, K. L.: Studies on Ambrosia: III. Pistillate Ambrosia elatior × A. trifida and its bearing on matroclinic sex inheritance. Bot. Gaz. **105**, 226—232 (1943).
Jucci, C.: I fattori ereditari. Vol. I Relaz. 28 Riun. Sips, Pisa (1940).
— Introduzione allo studio della genetica per medici, agrari e naturalisti. (Introduction to the study of genetics for physicians, agriculturalists and naturalists.) Milano, Pp. XXIII + 419, 146 figs., 11 pls. (1944).

Kadam, B. S.: Tables of theoretical frequencies of common Mendelian ratios for n = 1 to 500 with limits of deviations. Poona Agric. Coll. Mag. **35**, 60—71 (1944).
Kamšilov, M. M.: (Heteropolyspermie — eine neue Methode der Mutationsauslösung.) C. R. (Doklady) Acad. Sci. URSS. **57**, 617 (1947). [Russisch.]
Kanellis, A.: Zur Frage der Steigerung der Mutationsrate nach generationenlanger Vorbehandlung mit Follikel Hormon und anschließender Röntgenbestrahlung bei Drosophila. Z. Vererbungsl. **81**, 191—196 (1943).
— u. Gh. Radu: Über die Auslösung von Translokationen durch Röntgenbestrahlung weiblicher Gameten von Drosophila melanogaster. Naturwiss. **31** (1943).
Kaplan, R.: Über die Häufigkeiten phänotypisch abweichender Pflanzen in der F_1-Generation aus verschieden gequollenen und bestrahlten Pollen von Antirrhinum majus. Z. Vererbungsl. **77**, 568—579 (1939).
— Experimentelle und theoretische Untersuchungen über den Mechanismus der Keimungsschädigung durch Röntgenstrahlen beim Pollen von Antirrhinum majus L. Biol. Zbl. **60**, 298—318 (1940).
Kappert, H.: Austauschbesonderheiten im S-Chromosom der immerspaltenden Levkojen (Matthiola incana). Z. Vererbungsl. **78**, 273—293 (1940).
— Vererbungslehre. 98. Sammelband der Schriftenreihe „Soldatenbrief zur Berufsförderung", Pp. 217 (1943).

Kappert, H.: Untersuchungen über Plasmonwirkungen bei Aquilegia. (Gynodioecie — Heterosis — Gestalt des Sporns.) Flora, N. F. **37**, 95 (1943).

Karp, M. L.: Inbreeding and Heterosis. Bull. Acad. Sci. URSS., sér. biol. Pp. 219—249 (1940).

Kaufmann, B. P.: Modification of the frequency of chromosomal rearrangements induced by X-rays in Drosophila. III. Effect of supplementary treatment at the time of chromosome recombination. Genetics **31**, 449—453 (1946).

—, A. Hollaender u. H. Gay: Modification of the frequency of chromosomal rearrangements induced by X-rays in Drosophila. I. Use of near infrared radiation. Genetics **31**, 349—367 (1946).

—, — Modification of the frequency of chromosomal rearrangements induced by X-rays in Drosophila. II. Use of ultra-violet radiation. Genetics **31**, 368 bis 376 (1946).

Kausche, G. A. u. H. Stubbe: Zur Frage der Entstehung röntgenstrahleninduzierter Mutationen beim Tabakmosaikvirusprotein. Naturwiss. **28**, 824 (1940).

Keller, B.: (Genetics on new principles.) Proc. Lenin Acad. Agric. Sci. USSR. Nr. 10, 3—10 (1944). [Russisch.]

Kerkis, J. J. u. A. A. Ljapunov: Segregation in hybrids. C. R. (Doklady) Acad. Sci. URSS. **31**, 47—50 (1941).

Khačaturov, S. P. siehe Hačaturov.

Khazina, S. P. siehe Hazina.

Kholmogoroff, A. N. siehe Holmogorov.

Kihara, H.: (Development of industry and genetics.) A lecture delivered before the second general meeting of the Society. Report of General Meeting on Population Problems, published by the Society for the Study of Population Problems 1939, Pp. 991—997 (1939).

— (What kind of study is the geneticist engaged in?) J. Insur. Med., Japan **39**, Nr. 2 (1940).

— u. K. Matsumoto: Nachkommen mit einem Genomtyp von Aegilops variabilis in der F_4-Generation des Bastards Ae. ovata × variabilis. Japan. Journ. Genet. **16**, 291—294 (1940).

King, E. D.: The effect of low temperature upon the frequency of X-ray induced mutations. Genetics **32**, 161—164 (1947).

Kirsanov, B. A.: (Frequency of breaks of heterochromatin during the translocation process in Drosophila melanogaster. Communication I. Frequency of breaks in the Y-cnromosome.) Izvestija Akademii Nauk SSSR. (Bull. Acad. Sci. URSS. Sér. biol.) Nr. 5, 501—515 (1946). [Russisch.]

Kirsanov, V. A. u. P. R. Martynova: (Zur Frage des Einflusses des 20-Methylholantrens auf die Mutabilität bei Drosophila melanogaster.) C. R. (Doklady) Acad. Sci. URSS., **55**, 765—768 (1947). [Russisch.]

Kisselbach, T. A. u. N. F. Petersen: Mutations for waxy and sugary endosperm in inbred lines of dent corn. J. Amer. Soc. Agron. **36**, 89—91 (1944).

Knapp, E.: Haploide Pflanzen von Antirrhinum majus. Ber. dtsch. bot. Ges. **57**, 371 (1939).

— Die Bezeichnung der „Gene" von Antirrhinum majus, nebst Bemerkungen zur genetischen Nomenklatur und Begriffsbildung. Z. Vererbungsl. **79**, 253 bis 266 (1941).

— u. R. Kaplan: Beeinflussung der Mutationsauslösung und anderer Wirkungen der Röntgenstrahlen bei Antirrhinum majus durch Veränderung des Quellungszustandes der zu bestrahlenden Samen. Z. indukt. Abstamm- u. Vererbungslehre **80**, 501—550 (1942).

Kobel, F.: Vererbungsforschung und Pflanzenzüchtung. Arch. d. Jul. Klaus-Stiftg. **16**, 621—629 (1941).

— Blütenbiologie und Genetik. Mitt. natf. Ges. Bern. Sitz.-Ber. bern. bot. Ges. Pp. 21—24 (1941, 1942).

Kolman, E.: Is it possible to prove or disprove Mendelism by mathematical and statistical methods. C. R. (Doklady) Acad. Sci. URSS. **28**, 834—838 (1940).

Komai, T.: Inversion, a Review. Japan. Journ. Genet. **17**, 65—81 (1941).

Kosambi, D. D.: The estimation of map distances from recombination values. Ann. Eugen **12**, 172—175 (1944).

Kosswig, C.: Über Substitutionsgene und Transfer der Genfunktion. (Substitution genes and transfer of the gene function.) Experientia, Basel **3**, 404—410 (1947).

Kostrjukova, K. J. u. M. V. Černojarov: (In favour of Michurin's cytobiology.) Jarovizacija Nr. 3 (30), 140—151 (1940). [Russisch.]

Kotval, J. P. u. L. H. Gray: Structural changes produced in microspores of Tradescantia by α-radiation. J. Genet. **48**, 135—154 (1947).

Kovalevskaja, L. J.: (Overcoming the unfavourable influence of inbreeding in crosspollinated plants.) Ovoščevodstvo (Vegetable Growing) Nr. 7, 21—25 (1940). [Russisch.]

Koyenuma, N.: Über die Verdoppelung des Zellkerns bzw. Gens und die gesamte Genzahl im Kern bzw. Chromosomen. Ztschr. für Physik **119**, 522—526 (1942).

Krajevoj, S. J.: (Über mögliche Ursachen der Heterosis bei Pflanzen.) C. R. (Doklady) Acad. Sci. URSS. **54**, 825—828 (1946). [Russisch.]

Kružilin, A. S.: (The physiology of variation and inheritance in plants [A working hypothesis]). Socialistic Grain Farming, Saratov, Nr. 1, 142—149 (1941). [Russisch.]

Krythe, J. M. u. S. J. Wellensieck: Five years of colchicine research. Bibliogr. Genet. **14**, 132 (1942).

Kühl, O.: Crossing-over bei Antirrhinum-Speciesbastarden. Bull. Sect. sci. Acad. roum. **19**, 92—99 (1939).

Kuhn, E.: Selbstbestäubungen subdiöcischer Blütenpflanzen, ein neuer Beweis für die genetische Theorie der Geschlechtsbestimmung. Planta **30**, 457—470 (1939).

— Polyploidie und Geschlechtsbestimmung bei zweihäusigen Blütenpflanzen. Naturwiss. **30**, 189—198 (1942.)

Kühn, A.: Grundriß der Vererbungslehre. Verlag von Quelle u. Meyer, Leipzig, Pp. 164 (1939).

— Über eine die Schuppenformbildung und Schuppenpigmentierung beeinflussende Mutation (vd) von Ephestia kühniella Z. Biol. Zbl. **64**, 81—97 (1944).

Lamarque, P.: Les mutations roentgéniennes. Journ. de Radiol. **24**, 193—199 (1941).

Lamm, R.: Notes on an octoploid Solanum punae plant. Hereditas (Lund) **29**, 193—195 (1943).

— Cytogenetic studies in Solanum, sect. tuberarium. Hereditas **31**, 1—128 (1945).

Lamprecht, H.: Lokale Umwandlung von Rezessivität in Dominanz durch die Wirkung eines besonderen Gens bei Phaseolus vulgaris. Z. Vererbungsl. **77**, 186—194 (1939).

— Durch Komplexmutation bedingte Sterilität und ihre Vererbung. Arch. d. Jul. Klaus-Stiftg. Erg.-Bd. zu Bd. 20, 126—141 (1945).

Landauer, W.: Shall we lose or keep our plant and animal stocks? Science **101**, 497—499 (1945).

Lang, A.: Untersuchungen über einige Verwandtschafts- und Abstammungsfragen in der Gattung Stachys L. auf cytogenetischer Grundlage. Bibliotheca Botan. Stuttgart, Verlag Schweizerbartsch (1940).

Langendorff, H. u. K. Sommermeyer: Strahlenwirkungen auf Drosophila-Eier. IV. Die exponentielle Schädigungskurve und der biologische Zeitfaktor bei der Einwirkung von Röntgenstrahlen auf Drosophila-Eier. Strahlenther. **68**, 42—52 (1940).

Langham, D. G.: The effect of light on growth habit of plants. Amer. J. Bot. **28**, 951—956 (1941).

Laptev, I.: (Antipatriotic attacks under the flag of „scientific" criticism.) Pravda, Sept. 2, 2 (1947). [Russisch.]

Lawrence, W. J. C.: Studies on Streptocarpus. II. Complementary sublethal genes. J. Genet. **48**, 16—30 (1947).
— u. J. R. Price: The genetics and chemistry of flower colour variation. Biol. Rev. **15**, 35—58 (1940).
Lea, D. E. u. D. G. Catcheside: The bearing of radiation experiments on the size of the gene. J. Genet. **47**, 41 bis 50 (1945).
Lehmann, E.: Polyploidie und geographische Verbreitung der Arten der Gattung Veronica. Jb. wiss. Bot. **89**, 461—542 (1940).
— Kampf um den Lebensraum und Chromosomenzahl am Beispiel einer Veronica-Gruppe betrachtet. Umschau **45**, 708—710 (1941).
— Zur Genetik der Entwicklung in der Gattung Epilobium. 3. Die Tübinger hirsutum-Biotypen. Jb. wiss. Bot. **89**, 637—686 (1941).
— Zur Genetik der Entwicklung in der Gattung Epilobium. 4. Das „Plasmon" in der Gattung Epilobium: A. Die Grundlagen. Jb. wiss. Bot. **89**, 687—753 (1941).
— Zur Genetik der Entwicklung in der Gattung Epilobium. 4. Das „Plasmon" in der Gattung Epilobium: B. Die Analyse. Jb. wiss. Bot. **90**, 49—98 (1941).
— Kampf um den Lebensraum und Chromosomenzahl am Beispiel einer Veronica-Gruppe betrachtet. Umschau **45**, 708—710 (1941).
Lein, A.: Die genetische Grundlage der Kreuzbarkeit zwischen Weizen und Roggen. Z. Vererbungsl. **81**, 28—61 (1943).
Lerner, I. M. u. L. N. Nazel: Population genetics of a poultry flock under artificial selection. Genetics **32**, 325—339 (1947).
Levan, A.: Studien über die Vererbung der Blütenscheckung bei Petunia. Hereditas **25**, 145—184 (1939).
Lewis, D.: Useful X-ray mutations in plants. Nature, Lond. **158**, 519—520 (1946).
Lewis, J.: A footnote on the Soviet genetics controversy. Modern Quart. **2**, 352—356 (1947).
Lima-de-Paria, A.: The effect of gama radiation upon Mimosa pudica L. Portug. Acta Biol. (A) **1**, 95—109 (1945).
Lindegren, C. C.: A new gene theory and an explanation of the phenomenon of dominance to Mendelian segregation of the cytogene. Proc. Nat. Acad. Sci. Wash. **32**, 68—70 (1946).
Little, C. C. u. K. P. Hummel: A reverse mutation to a „remote" allele in the house mouse. Proc. Nat. Acad. Sci. Wash. **33**, 42—43 (1947).
Loe, L.: Årsberetning fra Norges Landbrukshøgskole for budsjettåret 1. Juli 1940 — 30. Juni 1941. (Annual Report of the Agricultural College of Norway for the year ending June 30, 1941.) Oslo, Pp. 145 (1942).
Loeb, L.: The biological basis of individuality. Charles C. Thomas, Springfield, Illiois, Pp. XIII + 711, tables (1945).
Lorbeer, G.: Struktur und Inhalt der Geschlechtschromosomen. Ber. dtsch. bot. Ges. **59**, 369—418 (1941).
Löve, D.: Some studies on sex-determination in Melandrium rubrum. Svensk Bot. Tidskr. **34**, 234—247 (1940).
Löve, Á. u. D. Löve: Experimental sex reversal in plants. Svensk Bot. Tidskr. **34**, 248—252 (1940).
Ludwig, W.: Notiz zu der unternormalen Streuung in den Moewusschen Chlamydomonas-Versuchen. Z. Vererb.lehre **80**, 612—615 (1942).
— u. R. Freisleben: Übernewere statistische Methoden zur Auswertung von Koppelungsversuchen, vor allem in der Pflanzenzüchtung. Z. Pflanzenz. **24**, 523—538 (1942).
Lüers, H. u. G. Schubert: Untersuchungen zur Frage der selektiven Befruchtung an Drosophila melanogaster und Drosophila funebris. Biol. Zbl. **60**, 69—78 (1940).
— u. A. E. Stubbe: Vergleichende Untersuchungen der Wildtypen verschiedener Drosophila-Arten an Hand von Transplantationen der Augenanlagen. II. Vergleich der Wildtypen von Drosophila melanogaster und Drosophila virilis. Z. Vererbungsl. **79**, 146—152 (1940).
Lüers, H. u. A. E. Stubbe: Vergleichende Untersuchungen der Wildtypen verschiedener Drosophila-Arten an Hand von Transplantationen der Augenanlagen. IV. Vergleich der Wildtypen von Drosophila melanogaster und Drosophila immigrans. Z. Vererbungsl. **79**, 396—400 (1941).
—, — Vergleichende Untersuchungen der Wildtypen verschiedener Drosophila-Arten an Hand von Transplantationen der Augenanlagen. VI. Vergleich der Wildtypen von Drosophila melanogaster und Drosophila repleta. Z. Vererbungsl. **79**, 493—497 (1941).
Lukjanenkó, J.: (The basis of seed selection.) Socialističeskoe Seljskoe Hozjajstvo, Moskva, Nr. 1, 71—85 (1940). [Russisch.]
Lysenko, T. D.: In response to an article by A. N. Kolmogoroff. C. R. (Doklady) Acad. Sci. URSS. **28**, 832 bis 833 (1940).
— (The ways of controlling plant organisms.) Jarovizacija, Nr. 3 (30), 27—44 (1940). [Russisch.]
— (What is Michurin's genetics?) Jarovizacija, Nr. 6 (33), 3—19 (1940). [Russisch.]
— The nature of heredity. Proc. Lenin Acad. Agric. Sci. USSR., Nos 11/12, 3—5 (1942).
— (On inheritance and its changeability.) Socialističeskoe Seljskoe Hozjajstvo (Socialistic Agriculture), Moskva, Nos 1/2 47—69; Nos 3/4, 36—51 (1943). [Russisch.]
— u. T. Dobzhansky (translator): Heredity and its variability. New York, Pp. 65 (1946).

Macarthur, J. W.: Genetics of body size and related characters. I. Selecting small and large races of the laboratory mouse. Amer. Nat. **78**, 142—157 (1944).
McAtee, W. L.: The mechanism of heredity in relation to the theory of natural selection. Ohio J. Sci. **43**, 117—120 (1943).
McIlwain, H.: The magnitude of microbial reactions involving vitamin-like compounds. Nature, Lond. **158**, 898—902 (1946).
Magržikovskaja, K. V. u. V. V. Saharov: Rôle of internal factors in the mutation process. Influence of hybridization on the rate of mutation. C. R. (Doklady) Acad. Sci. URSS **32**, 697—698 (1941).
Maguinness, O. D.: Environment and heredity. Thomas Nelson and Sons, Ltd., London (1940).
Maheshwari, P.: Some recent discoveries in applied biology. Sci. and Cult. **10**, 532—535 (1945).
Mainx, F.: Die Wirkung von Röntgenstrahlen auf die Trennung der attached-X-Chromosomen bei Drosophila melanogaster. Z. Vererbungsl. **78**, 238—245 (1940).
Maksimov, N. A.: Prof. B. A. Keller. Nature, Lond. **157**, 69—70 (1946).
Malécot, G.: Mendélisme et consanguinité. C. R. Acad. Sci., Paris **215**, 313—314 (1942).
Malinovskij, A.: The Rôle of genetic and phenogenetic Phenomena in the Evolution of Species. Bull. Acad. Sci., URSS., Sér. biol., Pp. 575—612 (1939).
Mampell, K.: Genic and nongenic transmission of mutator activity. Genetics **31**, 589—597 (1946).
Mangelsdorf, P. C.: Genetics in Russia. Chronica Botanica **6**, 256—257 (1941).
Manton, I.: Comments on chromosome structure. Nature, Lond. **155**, 470—472, 510 (1945).
Mariani, M.: La scomparsa del grande genetista Italiano Nazareno Strampelli. (The death of the great Italian geneticist, Nazareno Strampelli.) Z. Pflanzenz. **25**, 61—64 (1943).
Marquardt, H.: Zur Analyse röntgeninduzierter Chromosomenveränderungen und Chromosomenmutationen. Eine Erwiderung. Ztschr. f. Bot. **38**, 42—63 (1942).

Marquardt, H.: Über eine röntgeninduzierte Mutation mit abweichendem Erbverhalten bei Oenothera Hookeri (Vorl. Mitt.). Flora (Jena), N. F. **37**, 152—165 (1943).

Marquette, W.: On the question of Russian scientists. Science **104**, 332 (1946).

Marsden-Jones, E. M. u. W. B. Turrill: A quantitative study of petal size and shape in Saxifraga granulata L. J. Genet. **48**, 206—218 (1947).

Marshak, A.: The stage of mitosis at which chromosomes are rendered less sensitive to X-rays by ammonia. Genetics **24**, 103—104 (1939).

Martynova, R. P. u. B. A. Kirsanov: (Über den Einfluß der cancerogenen Stoffe auf den Mutationsprozeß bei Drosophila melanogaster.) C. R. (Doklady) Acad. Sci. URSS **55**, 647—649 (1947). [Russisch.]

Mather, K.: Plant breeding in the light of genetics. Nature **144**, 820 (1939).
— Selection for polygenic characters. Proc. 7th Int. Genet. Congr. Edinburgh 23—30 August, Pp. 211—212, (1939) 1941. (Abstr.)
— Variation and selection of polygenic characters. J. Genet. **41**, 159—193 (1941).
— Genetics and the Russian controversy. Nature, Lond. **149**, 427—430 (1942).
— Polygenes in development. Nature, Lond. **151**, 560 (1943).
— Polygenic inheritance and natural selection. Biol. Rev. **18**, 32—64 (1943).
— Dominicance and heterosis. Amer. Nat. **80**, 91—96 (1946).
— The genetical requirements of bio-assays with higher organisms. Analyst **71**, 407—411 (1946).
— u. P. M. J. Edwards: Specific differences in Petunia. III. Flower colour and genetic isolation. J. Genet. **45**, 243—260 (1943).
— u. P. G. Espinasse: The polygene concept. Nature, Lond. **149**, 731—732 (1942).

Matthey, R.: Les spirales chromosomiques. Arch. Klaus-Stiftg. **16**, 630—644 (1941).

Mehta, B. K.: The role of heterosis in plant breeding and agriculture. Poona Agric. Coll. Mag. **30**, 159—172 (1939).

Meijere, J. C. H. de: Heeft Darwin afgedaan? (Is Darwin finished with?) Vakbl. Biol. **21**, 146—148 (1940).

Melchers, G.: Genetik und Evolution. (Bericht eines Botanikers.) Z. Vererbungsl. **76**, 229—259 (1939).
— Die Ursache für die bessere Anpassungsfähigkeit der Polyploiden. Z. f. Naturforsch. **1**, 160—165 (1946).

Mendel, G.: Versuche über Pflanzen-Hybriden. (Originalhandschrift.) Züchter **13**, 221—268 (1941).

Mendiola, N. B.: Behaviour of some hidden bud variations. Philipp. Agric. **39**, 439—464 (1941).

Miani, G.: Trabanti e nucleoli nel gen Allium. (Trabanten und Nucleolen bei der Gattung Allium.) Ann. di Bot. **22**, 11—27 (1941).

Michaelis, P.: Über den Einfluß des Plasmons auf die Manifestation der Gene. Z. Vererbungsl. **77**, 548—567 (1939).
— Über reziprok verschiedene Sippen-Bastarde bei Epilobium hirsutum. I. Die reziprok verschiedenen Bastarde der Epilobium hirsutum-Sippe Jena. Z. Vererbungsl. **78**, 187—222 (1940).
— Über reziprok verschiedene Sippen-Bastarde bei Epilobium hirsutum. II. Über die Konstanz des Plasmons der Sippe Jena. Z. Vererbungsl. **78**, 223—237 (1940).
— Über reziprok verschiedene Sippen-Bastarde bei Epilobium hirsutum. III. Über die genischen Grundlagen der im Jena-Plasma auftretenden Hemmungsreihe. Z. Vererbungsl. **78**, 295—337 (1940).
— Über reziprok verschiedene Sippen-Bastarde bei Epilobium hirsutum. VI. In welcher Weise sind an der Manifestation der im Jena-Plasma auftretenden Entwicklungstendenz die Gene dieser Sippe beteiligt? Z. Vererbungsl. **80**, 454—499 (1941).

Michaelis, P.: Über reziprok verschiedene Sippen-Bastarde bei Epilobium hirsutum. V. Über die Bedeutung der Genquantität für die Manifestation reziproker Unterschiede Z. Vererbungsl. **80**, 429—453 (1942).
— M. v. Dellingshausen: Über reziprok verschiedene Sippen-Bastarde bei Epilobium hirsutum. IV. Weitere Untersuchungen über die genischen Grundlagen der extrem stark gestörten Bastarde der E. hirsutum-Sippe Jena. Z. Vererbungsl. **80**, 373—428 (1942).
— u. H. Ross: Untersuchungen an reziprok verschiedenen Artbastarden bei Epilobium. 2. Über Abänderungen an reziprok verschiedenen und reziprok gleichen Epilobium-Artbastarden. Flora (Jena), N. F. **37**, 24 bis 56 (1943).

Miège, Em.: L'hérédité de la composition chimique chez hybrides intergénériques. Étude de la descendance d'Aegilops ovata L. var. nigra × Triticum vulgare H. et d'Aegilops ovata × Triticum durum Desf. Sci. Genet. (Torino) **2**, 82—90 (1940).

Milani, R.: Una nuova mutazione del II cromosoma in Drosophila melanogaster. Boll. Soc. It. Biol. Sperim, p. 22 (1946).

Milovidov, P. F.: Über die Chromosomenzahlen bei einigen Leguminosen und anderen Pflanzen. Planta **32**, 38—42 (1941).

Mitchell, J. S.: Dr. D. E. Lea. Nature, Lond. **160**, 81 bis 82 (1947).

Mitin, M.: (On advances in Soviet genetical science.) Selekcija i Semenovodstvo (Breeding and Seed Growing), Nr. 2, 6—18 (1940). [Russisch.]

Moewus, F.: Über Mutationen der Sexual-Gene bei Chlamydomonas. Biol. Zbl. **60**, 597—626 (1940).
— Die Analyse von 42 erblichen Eigenschaften der Chlamydomonas eugametis-Gruppe. I. Teil: Zellform, Membran, Geissel, Chloroplast, Pyrenoid, Augenfleck, Zellteilung. Z. Vererbungsl. **78**, 418—462 (1940).
— Die Analyse von 42 erblichen Eigenschaften der Chlamydomonas eugametos-Gruppe. II. Teil: Zellresistenz, Sexualität, Zygote, Besprechung der Ergebnisse. Z. Vererbungsl. **78**, 463—500 (1940).
— Die Analyse von 42 erblichen Eigenschaften der Chlamydomonas eugametos-Gruppe. III. Teil: Die 10 Koppelungsgruppen. Z. Vererbungsl. **78**, 501—522 (1940).

Möglich, F., R. Rompe u. N. W. Timoféeff-Ressovsky: Bemerkungen zu physikalischen Modellvorstellungen über Energieausbreitungsmechanismen im Treffbereich bei strahlenbiologischen Vorgängen. Naturwiss. **30**, 409—419 (1942).

Mol, W. E. de: Das frühere Blühen der Tulpen, ein neuer Fall einer somatischen Mutation. Züchter **12**, 88—92 (1940).
— Fortgesetzte Untersuchungen betreffs somatischer Tulpenmutationen, welche sich durch frühe Blüte unterscheiden, nebst einer Betrachtung über die Ursachen ihrer Entstehung. Gartenbauwiss. **16**, 70—89 (1941).
— Die früher blühenden somatischen Tulpenmutanten (Untersuchungen 1940—1941) und die Teilungshypothese. Gartenbauwiss. **17**, 106—132 (1942).
— Na het veertiende jaar roentgenbestraling van tulpen en hyacinten ter opwekking van somatische mutaties. (Fourteen years of X-ray irradiation of tulips and hyacinths to induce somatic mutations.) Genetica **23**, 329—352 (1943).
— Die erste gezüchtete triploide Scilla sibirica Andrews. Gartenbauwiss. **17**, 227—239 (1943).
— Na het vijftiende jaar röntgenbestraling van tulpen ter verkrijging van knopmutaties. (After fifteen years of X-ray irradiation of tulips in order to obtain bud mutations.) Landbouwk. Tijdschr. Wageningen **56**, 173—190 (1944).

Moore, A. R.: The individual in simpler forms. University Press, University of Oregon, Eugene 9 figs, 4 tables, 13 plates (1945).
Morgan, L. V.: A variable phenotype associated with the fourth chromosome of Drosophila melanogaster and affected by heterochromatin. Genetics 32, 200 bis 219 (1947).
Morgan, W. T. J.: Transformation of pneumococcal types. Nature, Lond. 153, 763—764 (1944).
Moriya, A.: Preliminary Note on the Chromosome Numbers of Sugarcane Varieties F 108 and some others. Japan. Journ. Genet. 17, 62—64 (1941).
Morozov, V. K.: (Nineteen years work with sunflower by the method of inbreeding.) Jarovizacija, Nr. 2 (29), 33—48 (1940). [Russisch.]
Muhin, N. D.: (Variability in the yielding capacity of the seed progeny from different parts of the ear.) Jarovizacija, Nr. 3 (36), 104—106 (1941). [Russisch.]
Muller: Gene and chromosome theory. Nature 144, 814 bis 816 (1939).
Muller, H. J.: An analysis of the process of structural change in chromosomes of Drosophila. J. Genet. 40, 1—66 (1940).
— Thomas Hunt Morgan 1866—1945. Science 103, 550 bis 551 (1946).
— The gene. Proc. Roy. Soc. 134, 1—37 (1947).
— u. G. Pontecorvo: Recombinants between Drosophila species the F_1 hybrids of which are sterile. Nature, Lond. 2, 199—200 (1940).
Müntzing, A.: Ras och rasblandning. (Race and race mixture.) Tidsspegel, Pp. 144—161 (1942).
— Kritisk översikt över inavelsteorierna. (A critical survey of theories about inbreeding.) Nord. Jordbr-Forskn. Nos. 5/6, 237—249 (1943).
— Nya resultat inom genetiken av intresse för vätförädlingen. (New results in genetics of interest for planting breeding.) Sverig. Utsädesfören. Tidskr. 54, 84—90 (1944).
— On the causes of inbreeding degeneration. Archiv d. Julius Klaus-Stiftg. Erg.-Bd. zu Bd. 20, 153—163 (1945).
Murbeck, S.: Bastarder och artsystematik. (Hybrids and the taxonomy of species.) Bot. Notiser, Pp. 314 bis 332 (1943).
Musijko, A. S.: (Supplementary pollination - a method of increasing the yield of farm crops, and improving their seed types.) Jarovizacija, Nr. 3 (36), 56—62 (1941). [Russisch.]

Navašin, M., H. Gerasimova u. G. M. Belajeva: On the course the process of mutation in the cells of the dormant embryo within the seed. C. R. (Doklady) Acad. Sci. URSS. 26, 948—951 (1940). [Russisch.]
Neel, J. V.: Studies on the mutations affecting the chaetae of Drosophila melanogaster. 1. The interaction of hairy, polychaetoid, and Hairy wing. Genetics 26, 52—68 (1941).
Negodi, G.: Poliploidi da colchicina in Bellis perennis, Bellis annua, Antirrhinum Orontium, Mimosa pudica, Nigella sativa, Helianthus annuus, Ricinus communis, Cucurbita Pepo. Atti Mem. Accad. Sci. Modena, V. s. 5, 115—147 (1941).
Neilson Jones, W. u. F. Freier (translator): Quimeras vegetales e hibridos de injerto. (Plant chimaeras and graft hybrids.) B. Aires, Pp. 157, 21 figs, 2 tables (1946).
Neuhaus, M. E.: On the manifold effect of the cinnabar gene in Drosophila melanogaster. C. R. Acad. Sci. URSS. N. s. 30, 238—241 (1941).
Newcombe, H. B.: The action of X-rays on the cell. 1. The chromosome variable. J. Genet. 43, 145—171 (1942).
Nicholson, C.: Fitogeografia genética. (Phytogeographical genetics.) Rev. Univ. Arequipa 12, 19—56 (1939).
Nilsson, H.: Totale Inventierung der Mikrotypen eines Minimiareals von Taraxacum officinale. Hereditas, Lund. 33, 119—142 (1947).
Noujdin, N. I. siehe Nuždin.
Novák, J.: Genetische Polynome als Grundlage zur Ableitung der Genotypen und Phänotypen und ihrer Häufigkeiten bei allen Arten genetischer Veranlagung. Z. indukt. Abstamm.- u. Vererbungsl. 81, 343—362 (1943).
Nuždin, N. I.: (Der Einfluß der „inerten" Chromosomenabschnitte auf die Manifestierung der Mosaikmerkmale.) C. R. (Doklady) Acad. Sci. URSS. 22, 668 (1939). [Russisch.]
— (Die Gesetzmäßigkeit des Heterochromatineinflusses auf die Mosaikbildung.) Žurnal Obščej Biologii (Journ. of Gener. Biology) 6, 357—388 (1944). [Russisch.]
— (Die erblichen Veränderungen und die Ontogenese.) Žurnal Obščej Biologii (Journ. of Gener. Biology) 6, 381 (1945). [Russisch.]
— (Das Heterochromatin, die Geschwindigkeit der Embryoentwicklung und die Mosaikbildung.) C. R. (Doklady) Acad. Sci. URSS. 54, 61 (1946). [Russisch.]
— (Mosaics and its manifestation.) Izvestija Akademii Nauk SSSR. (Bull. Acad. Sci. URSS. Sér. Biol.), Nr. 5, 519—544 (1946). [Russisch.]
— (The role of hybridization in variation. III. Influence of the Y-chromosome on the variability of the yellow and achaete loci in Drosophila melanogaster.) Žurnal Obščej Biologii (Journ. of General Biology) 8, Nr. 2, 101—124 (1947). [Russisch.]

Oehler, E.: Neuere Methoden zur Erzielung erhöhter Chromosomenzahlen bei Pflanzen (Mehrlingskeimlinge, Temperaturschock, Colchicinbehandlung). (Ref.) Ber. d. Schweiz. Bot. Ges. 51, 457—459 (1940).
Oehlkers, F.: Meiosis und crossing-over. Cytogenetische Untersuchungen an Oenotheren. Z. Vererbungsl. 78, 157—186 (1940).
— Meiosis and crossing-over. Biol. Zbl. 60, 337—348 (1940).
— Genetisch-physiologische Untersuchungen zum Vitalitätsproblem. I. Z. Bot. 35, 271—297 (1940).
— Faktorenanalytische Ergebnisse an Artbastarden. Biol. Zbl. 62, 280—289 (1942).
— Die Auslösung von Chromosomenmutationen in der Meiosis durch Einwirkung von Chemikalien. Z. Vererbungsl. 81, 313—341 (1943).
— Weitere Versuche zur Mutationsauslösung durch Chemikalien. Biol. Zbl. 65, 176—186 (1946).
— u. C. Harte: Über die Aufhebung des Gonen- und Zygotenausfalls bei Oenothera. Flora (Jena), N. F. 37, 106—124 (1943).
Oláh, L.: Über die Ursachen der durch Inzucht entstehenden Depression. Bull. Hung. Coll. Hort. 7, 204 bis 221 (1941).
Olenov, J. M.: (Über das Zusammenwirken der Erbfaktoren einer natürlichen Population.) C. R. (Doklady) Acad. Sci. URSS. 58, 903 (1947). [Russisch.]
— u. J. S. Harmac: (On the origination of obligate heterozygosity.) C. R. (Doklady) Acad. Sci. URSS. 30, 745—747 (1941). [Russisch.]
Olson, P. J.: Scientific independence in Russia. Science 103, 656 (1946).
Ome, K. B. de: Discussion of Paper by J. H. Martin entitled „Heredity and Lymphomatosis". Poultry Sc. 19, 105 (1940).
Ono, T.: (Polyploidy and sex determination in Melandrium. 3. Intersex in M. album.) Botanic. Mag. (Tokyo 54, 348—356 (1940). [Japanisch.]
Östergren, G.: (Elastische Chromosomenabstoßung.) Hereditas, Lund 29, 444—450 (1943).
Ownbey, M. u. W. A. Weber: Natural hybridization in the genus Balsamorhiza. Amer. J. Bot. 30, 179—187 (1943).

Pal, B. P.: Genes: atoms of heredity. Indian Fmg 1, 270—273 (1940).
Palma, S. E.: Estudion de algunas quimeras vegetales. (Study of some plant chimaeras.) An. Inst. Fitotéc. Santa Catalina 4, 75—103 (1942).
Panšin, I., A. N. Panšina u. P. P. Peyrou: Die Dosisabhängigkeit der röntgeninduzierten Chromosomenmutationen mit kleinen Bruchabständen bei Drosophila melanogaster. Naturwiss. 33, 27 (1946).
Pasquini, D.: Microsporogenesi e cariologia di Chilianthus oleaceus Burck (Loganiaceae). Atti Soc. Natur. Matem. Modena 72, 20—21 (1941).
Pätau, K. u. N. W. Timoféeff-Ressovsky: Statistische Prüfung des Unterschiedes der Temperaturkoeffizienten hoher und normaler Mutationsraten nebst einem Beispiel für die Planung von Temperaturversuchen. Z. indukt. Abstamm. u. Vererblehre 81, 62—71 (1943).
—, — Die Genauigkeit der Bestimmung spontaner und strahleninduzierter Mutationsraten nach CLB Kreuzungsmethode bei Drosophila melanogaster. Z. Vererbungsl. 81, 181—190 (1943).
Pedelaborde, J. L.: Método gráfico para análysis rápido de las distribuciones mendelianas. (Graphical method for the rapid analysis of Mendelian distributions.) Rev. Argent. Agron 6, 35—40 (1939).
Pennell, F. W.: Concerning „genotypes". Science 99, 320—321 (1944).
Penrose, L. S.: A further note on the sib-pair linkage method. Ann. Eugen., Camb. 13, 25—29 (1946).
Pereira, C. u. M. P. de Castro: Sobre e equilíbrio genético estavel. (Stable genetic equilibrium.) Arch. Inst. Biol. Def. Agric. Anim. S. Paulo 17, 267—268 (1946).
Peter, J.: Hérédité et selection. J. Forest. suisse 92, 121—126 (1941).
Petrová, J.: Über den Vergleich der α-Strahlenempfindlichkeit von Kern und Plasma. (Vorl. Mitt.) Ber. dtsch. bot. Ges. 60, 148—151 (1942).
Pfankuch, E., G. A. Kausche u. H. Stubbe: Über die Entstehung, die biologische und physikalisch-chemische Charakterisierung von Röntgen- und γ-Strahleninduzierten „Mutationen" des Tabakmosaikvirusproteins. Biochem. Ztsch. 304, 238—258 (1940).
Pierantoni, U.: Nozioni di biologia (compresa la genetica e la biologia delle razze). (Principles of biology [including genetics and biology of races].) Unione Tipografico-Editrice Torinese, Torino (1940).
Pirovano, A.: Progress and directives regarding electrogenetics. Int. Rev. Agric. 36, 173T—189T (1945).
Pirschle, K.: Morphologische und physiologische Dominanz bei 2n- und 4n-Bastarden zwischen der Normalform und zwei monohybriden Mutanten von Impatiens balsamina. Biol. Zbl. 65, 69—80 (1946).
Piza, S. de Toledo (jr.): Bases para una futura compreensão dos cromossômios. (Basis for a future understanding of the chromosomes.) Rev. Agric., S. Paulo 16, 104—117 (1941).
— Cromossômios e citoplasma em hereditariedade. (Chromosomes and cytoplasm in heredity.) Rev. Agric. Piracicaba 17, 51—62 (1942).
— Em tôrno do gen corpuscular. (Regarding the corpuscular gene.) Rev. Agric. São Paulo 19, 26—50 (1944).
— Hereditariedade não-cromosômica e dissociação „Mendeliana" de caracteres veiculados pelo citoplasma. (Non-chromosomic inheritance and Mendelian dissociation of cytoplasmic characters.) Rev. Agric., Piracicaba 21, 32—39 (1946).
Pontecorvo, G.: Microbiology, biochemistry, and the genetics of micro-organisms. Nature, Lond. 157, 95—96 (1946).
— The genetical aspects of bio-assays with micro-organisms. Analyst 71, 411—413 (1946).

Popravko, A. V.: (The methods of selecting perennial herbage plants.) Jarovizacija Nr. 5 (32), 119—121 (1940). [Russisch.]
Potašnikova, B. G.: (The effect of free intra- and inter-varietal fertilization, and its causes.) Jarovizacija, Nr. 2 (35), 98—100 (1941). [Russisch.]
Poulson, D. F.: Genes as physiological agents. Chromosomal control of embryogenesis in Drosophila. Amer. Nat. 79, 340—363 (1945).
Powers, L.: Formulas for determining theoretical effects of certain genetic factors upon inheritance of quantitative characters, with special reference to a study of a Lycopersicon hybrid. J. Agric. Res. 59, 555—577 (1939).
— An expansion of Jones's theory for the explanation of heterosis. Amer. Nat. 78, 275—280 (1944).
Preer, J. R.: Some properties of a genetic cytoplasmic factor in Paramecium. Proc. Nat. Acad. Sci. Wash. 32, 247—253 (1946).
Prezent, I. I.: (Pseudoscientific theories in genetics.) Jarovizacija Nr. 2, 87—117 (1939). [Russisch.]
— (Selective fertilization.) Jarovizacija Nr. 5 (32), 137 bis 139 (1940). [Russisch.]
Prokofjeva-Belgovskaja, A. A.: (Die Heterozyklität der Elternchromosomen.) C. R. (Doklady) Acad. Sci. URSS. 54, 169 (1946). [Russisch.]
— u. M. L. Belgovskij: Change in the crossover properties of a chromosome under the influence of mutation. C. R. (Doklady) Acad. Sci. URSS. 38, 251—253 (1943).
Propach, H.: Cytogenetische Untersuchungen in der Gattung Solanum, Sect. Tuberarium. V. Diploide Artbastarde. Z. Vererbungsl. 78, 115—128 (1940).
— Einige Chromosomenzahlen von Delphinien und ihre Auswertung für die Entstehung der Gartenformen. Gartenbauwiss. 14, 642—650 (1940).
Puhalskij, A.: (The efficiency of some new methods in seed breeding work.) Proc. Lenin Acad. Agric. USSR Nr. 4, 36—42 (1944). [Russisch.]

Rabaud, E.: La notion d'espèce et la génétique. Sciences, Paris 67, 85—91 (1940).
Radhakrishna, Rao, C.: The problem of classification and distance between two populations. Nature, Lond. 159, 30—31 (1947).
Rajewsky, B. N. u. N. W. Timoféeff-Ressovsky: Höhenstrahlung und die Mutationsrate von Drosophila melanogaster. Z. Vererbungsl. 77, 488—500 (1939).
Randolph, L. F.: An evaluation of induced polyploidy as a method of breeding crop plants. Amer. Naturalist 75, 347—365 (1941).
Rapoport, I. A.: (Phänogenetische Analyse der Discretion.) Žurnal Obščej Biologii (Journ. of Gener. Biol.) 2, Nr. 3 (1941). [Russisch.]
— Mutations restituting the telomere. C. R. (Doklady) Acad. Sci. URSS., n. s. 31, 266—269 (1941).
— On roentgenomutations in repetitions and X-ray mutations as conditioned by breaks. C. R. (Doklady) Acad. Sci. URSS., n. s. 31, 270—272 (1941).
— (On the law of expansion and extinction of genetic action.) C. R. (Doklady) Acad. Sci. URSS. 31, 393 bis 396 (1941). [Russisch.]
— (Oxidation and the mechanism of action of mutagenous factors.) J. Gen. Biol. URSS. 4, 65—72 (1943). [Russisch.]
— (Karbonilstoffen und chemischer Mechanismus der Mutationen.) C. R. (Doklady) Acad. Sci. URSS 54, 65 (1946). [Russisch.]
— On the synthesis of gene-products in equimolecular quantities. Amer. Nat. 81, 30—37 (1947).
Reed, S. C.: Interaction between the autosomes of Drosophila melanogaster as measured by viability and rate of development. Canad. J. Res. 19, Sect. D, 75—84 (1941).

Reinders, D. E.: Species crosses in the genus Nigella. Genetica 23, 22—30 (1942).

Reinholz, E.: Auslösung von Röntgenmutationen bei Arabidopsis Thaliana (L.) Heynh. und ihre Bedeutung für die Pflanzenzüchtung und Evolutionstheorie. Office of Military Government for Germany (U. S.) Fiat Rep. Nr. 1006, 70 (1947).

Reinig, W. F.: Die genetisch-chorologischen Grundlagen der gerichteten geographischen Variabilität. Z. Vererbungsl. 76, 260—308 (1939).

Reitz, L. P. u. a.: Nomenclature of genetic factors in wheat. J. Amer. Soc. Agron. 34, 1154 (1942).

Renner, O.: Kurze Mitteilungen über Oenothera. IV. Über die Beziehungen zwischen Heterogamie und Embryosackentwicklung und über diplarrhene Verbindungen. Flora, n. F., 34, 145—158 (1940).

— Zur Kenntnis der 15 chromosomigen Mutanten von Oenothera Lamarckiana. Flora, N. F., 34, 257—310 (1940).

— Über die Entstehung homozygotischer Formen aus komplexheterozygotischen Oenotheren. Flora, N. F. 35, 201—238 (1941).

— Europäische Wildarten von Oenothera. Ber. dtsch. bot. Ges. 60, 448—466 (1942).

— Über das crossing-over bei Oenothera. Flora, N. F. 36, 117—214 (1942).

— Beiträge zur Kenntnis des cruciata-Merkmals der Oenotheren. IV. Gigas-Bastarde. Labilität und Konversibilität der Cr-Gene. Z. Vererbungsl. 80, 590—611 (1942)

— Zur Analyse des Pollenkomplexes percurvans der Oenothera ammophila Focke. Arch. d. Jul. Klaus-Stiftg. Erg.-Bd. zu Bd. 20, 164—184 (1945).

— u. R. Sensenhauer: Versuche über den Erbgang des cruciata-Merkmals der Oenotheren. III. Weitere Belege für somatische Konversion. Z. Vererbungsl. 80, 570—589 (1942).

Resende, F.: (Vielkernige Mikrosporen bei einer Rasse von Antirrhinum majus L. ?) Bol. Soc. Broteriana 15, 5—7 (1941). [Portugiesisch.]

— (Karyokinese und Chromonemata.) Bol. Soc. Broteriana 15, 21—27 (1941). [Portugiesisch.]

Richey, F. D.: Mock-dominance and hybrid vigor. Science 96, 280—281 (1942).

— Bruce's explanation of hybrid vigor. J. Hered 36, 243—244 (1945).

Richharia, R. H.: Soviet Genetics. Guardian, Nr. 5268, 563 (1946).

Rick, C. M.: On the nature of X-ray induced deletions in Tradescantia chromosomes. Genetics 25, 466—482 (1940).

— The genetic nature of X-ray induced changes in pollen. Proc. Nat. Acad. Sci. Wash. 28, 518—525 (1942).

— The X-ray induced mutation rate in pollen in relation to dosage and the nuclear cycle. Genetics 28, 237 bis 252 (1943).

Ridley, H. N.: The scent of Mimulus moschatus. J. R. Hort. Soc. 72, 373 (1947).

Riehl, N., N. W. Timoféeff-Ressovsky u. K. G. Zimmer: Mechanismus der Wirkung ionisierender Strahlen auf biologische Elementareinheiten. Naturwiss. 29, 625—639 (1941).

Robinson, R.: The red and blue colouring matters of plants. Endeavour, Lond. 1, 92—100 (1942).

Roland, G.: Méthodes artificielles permettant de provoquer des mutations et de produire des variétés polyploides. Publ. Inst. Belge Amélior. Better. 7, 188 (1939).

Rose, M.: Les lois de l'hérédité et leur application à l'élevage. Bull. Agric. Algér. Nr. 84, 31—40 (1942).

Ross, H.: Über die Verschiedenheiten des dissimilatorischen Stoffwechsels in reziproken Epilobium-Bastarden und die physiologisch-genetische Ursache der reziproken Unterschiede. I. Die Aktivität der Peroxydase in reziproken Epilobium-Bastarden mit der Sippe Jena. Z. Vererbungsl. 79, 503—529 (1941).

Ross, H.: Über die Natur der Enthemmungen von plasmogengehemmten Epilobium hirsutum ♀ × parviflorum ♂-Bastarden. Naturwiss. 30, 492—493 (1942).

Roulet, E.-L.: Hérédité Mendélienne et analyse combinatoire. Georg et Cie, Genéve Pp. 193, 3 tables (1941).

Rudorf, W.: Die Bedeutung der Polyploidie für die Evolution und die Pflanzenzüchtung. Angew. Bot. 25, 92—114 (1943).

Ruiz Santaella, J.: Métodos empleados en genética vegetal. (Methods employed in plant genetics.) Hojas Direcc. Gen. Agric. Madrid 33, 1—6 (1941).

Russell, E. S.: The directiveness of organic activities. Cambridge University Press., Pp. VIII + 196, 23 figs. (1944).

Rutishauser, A.: Konstante Art- und Rassenbastarde in der Gattung Potentilla. Mitteil. der natf. Ges. Schaffhausen 18, 111—134 (1943).

— Über Kreuzungsversuche mit pseudogamen Potentillen. Arch. d. Jul. Klaus-Stiftg. 21, 469—472 (1946).

Ryžkov, V. L.: (Der labile Zustand der Gene des Genoms und des Cytoplasmas.) Uspjehi Sovremennoj Biologii (Advances in Modern Biology) 2, Nr. 2 (1941). [Russisch.]

Saharov, V. V.: Mutation process in hibernating Drosophila melanogaster. C. R. (Doklady) Acad. Sci. URSS, n. s. 30, 347—349 (1941).

Sansome, F. W. u. J. Philip: Recent advances in plant genetics. J. and A. Churchill., Ltd., Lond. (revised and rewritten by F. W. Sansome) (1939).

Šapiro, N. I.: On the nature of crossing-over induced by X-rays in males of Drosophila melanogaster. C. R. (Doklady) Acad. Sci. URSS. 49, 292—295 (1945).

— (The problem of directed production of mutation in modern genetics.) Uspjehi Sovremennoj Biologii (Advances in Modern Biology) 23, 87—108 (1947). [Russisch.]

—, M. A. Arsenjeva u. S. N. Ardašnikov: (Vergleichende Charakteristik der mutagenen Wirkung der Röntgenstrahlen bei verschiedenen Drosophila-Arten.) C. R. (Doklady) Acad. Sci. URSS. 58, 1785 (1947). [Russisch.]

Savčenko, P. F.: Experimental modification of the chromosomes of a phyletically primitive type. C. R. (Doklady) Acad. Sci. URSS. 27, 1024—1027 (1940).

Savčenko-Beljskij, A.: (Something new in the analysis of breeding material.) Socialističeskoe Zemledelie (Socialist. Agriculture) Nr. 111, 2 (1944). [Russisch.]

Sax, K.: The physiological and genetic effects of X-rays. Amer. J. Bot. 26, 4s (1939). (Abstr.)

— The distribution of X-ray induced chromosomal aberrations. Proc. Nat. Acad. Sci. Wash. 28, 229—233 (1942).

— Diffusion of gene products. Proc. Nat. Acad. Sci. Wash. 28, 303—306 (1942).

— Soviet biology. Science 99, 298—299 (1944).

— Soviet biology. Science 102, 649 (1945).

Ščerbak, S. N.: (Six years of inbreeding of sunflower.) Jarovizacija Nr. 2 (29), 49—61 (1940). [Russisch.]

Schafer, B.: The genetics of Aquilegia vulgaris. J. Genet. 41, 339—347 (1941).

Scheerer, H.: Chromosomenzahlen aus der Schleswig-Holsteinischen Flora. II. Planta 30, 716—725 (1940).

Schiemann, E.: Gedanken zur Gencentrentheorie Vavilovs. Naturwiss. 27, 377—384, 394—401 (1939).

— Antirrhinum majus, mut. filiforme, zugleich ein Beitrag zur Chimärenfrage. Z. Vererbungsl. 79, 50—82 (1940).

Schmalhausen, I. I.: (Kampf ums Dasein und Divergenz der Merkmale.) Žurnal Obščej Biologii (Journ. of Gener. Biol.) 1, Nr. 1 (1940). [Russisch.]

Schmuck, A. siehe Šmuk.
Schopfer, W. H.: Recherches sur l'hérédité l'hermaphroditisme mâle chez Melandrium. Verh. der Schweiz. Naturforsch.-Ges. 120, 159—160 (1940).
Schramm, G. u. L. Rebensburg: Zur vergleichenden Charakterisierung einiger Mutanten des Tabakmosaikvirus. Naturwiss. 30, 48—51 (1942).
Schütze, R.: Die Beeinflussung der Mutationsrate bei Drosophila melanogaster durch kombinierte Behandlung mit Arsen und Röntgenstrahlen. Z. indukt. Abstamm- u. Vererbungsl. 81, 484—499 (1943).
Schwanitz, F.: Über die Pollenkeimung einiger diploider Pflanzen und ihre Autotetraploiden in künstlichen Medien. Züchter 14, 273—282 (1942).
— Genetik und Evolutionsforschung bei Pflanzen. In: G. Heberer: Die Evolution der Organismen, Pp. 430 bis 478 (1943).
Schwartz, V.: Über die Manifestierung der Mutation he in Abhängigkeit vom genotypischen Milieu bei Ephestia kühniella. Biol. Zbl. 61, 478—483 (1941).
Schwemmle. J.: Weitere Untersuchungen an Eu-Oenotheren über die genetische Bedeutung des Plasmas und der Plastiden. Z. Vererbungsl. 79, 321—335 (1941).
— Plastiden und Genmanifestation. Flora, N. F. 37, 61—72 (1943).
Scott-Moncrieff, R.: The genetics and biochemistry of flower colour variation. Ergebn. Enzymforsch. 8, 277—306 (1939).
Sell-Beleites, I. u. A. Catsch: Mutationsauslösung durch ultraviolettes Licht bei Drosophila. I. Dosisversuche mit filtriertem Ultraviolett. Z. indukt. Abstamm.- und Vererbungsl. 80, 551—557 (1942).
Semenza, L.: Azione combinata di alcune mutazioni che agiscono sulla forma dell 'ochio in Drosophila melanogaster. Ricerca Scient. 14, 439 (1943).
Sengbusch, R. v.: Pärchenzüchtung, unter Ausschaltung von Inzuchtschäden. Forschungsdienst 10, 545—549 (1940).
Sengun, A.: Eine neue Mutation „kurzflügelig" (kfl) bei der Mehlmotte Ephestia kühniella Zeller. Biol. Zbl. 60, 23—34 (1940).
Serra, J. A.: An attempt at a synthesis of the physiological and cytological concepts of the gene. (Bol. Soc. Broteriana 19, Sér. 2a, 327—369 (1944).
Shapiro, N. L. siehe Šapiro.
Sherman, M.: The inheritance of self-sterility in various species and varieties of Antirrhinum. Z. Vererbungsl. 77, 3—17 (1939).
Shkvarnikov, P. K. siehe Škvarnikov.
Shlykow, G. siehe Šlykov.
Shull, G. H.: Genetics, the unifying science in biology. Torreya, 43, 126—131 (1943).
Sidky, A. R.: Symbols for backcross and backcross generations. J. Hered. 31, 8 (1940).
Sidorov, B. N.: (Die genotypische Kontrolle der Mosaikbildung bei Drosophila melanogaster.) C. R. (Doklady) Acad. Sci. URSS. 58, 2081—2084 (1947). [Russisch.]
Siherman, H. J. u. Z. F. Samorhina: (Dominant characters and uniformity in the first hybrid generation.) Jarovizacija Nr. 3 (36), 108—110 (1941). [Russisch.]
Sikka, S. M.: Cytogenetics of Brassica Hybrids and Species. Journ. of Genet. 40, 441—509 (1940).
Sinnott, E. W.: Genetics and geometry: mathematicians and biologists in studies of form. Yale Sci. Mag. 18, Nr. 4 (1944).
— u. L. C. Dunn: Principles of genetics. McGraw-Hill Publishing Company, Ltd., London (1939).
Sinskaja, E. N.: Phylogenetic Taxonomy as a Basis for Genetic and Breeding work. Z. Vererbungsl. 78, 399—417 (1940).
Sirks, M. J.: Genotypical predetermination in Datura. Genetica ('s-Gravenhage) 22, 197—214.

Sirks, M. J.: Genenstoffen (Gene substances). Vakbl. Biol. 23, 109—114 (1942).
— Het geslacht. Uitingen en oorzaken. (Sex. Its expression and causes.) Gorinchem Nr. 23, Pp. 176, 75 figs. (1946).
— Lysenko's genetica. (Lysenko's genetics.) Vakbl. Biol. 27, 8—13. (Also Landbouwk. Tijdschr. 59, 17—21.) (1947).
Skolko, A. J.: A cultural and cytological investigation of a two-spored basidiomycete, Aleurodiscus canadensis n. sp. Canad. J. Res. 22, Sect. C, 251—271 (1944).
Škvarnikov, P. A.: (Ein Fall der Chromosomengrundzahländerung bei Crepis capillaris.) C. R. (Doklady) Acad. Sci. URSS. 56, 301—304 (1947). [Russisch.]
Škvarnikov, P. K.: (Mutation in seeds and its significance in seed production and plant breeding.) Bull. Acad. Sci. URSS, Sér. Biol. Pp. 1009—1054 (1939). [Russisch.]
Sleggs, G. F.: The significance of diploidity and crossing-over (theory of differential periodicity). J. Genet. 40, 385—392 (1940).
Šlykov, G.: (In the fetters of false science.) Soviet Subtropics Nr. 6 (58), 54—56 (1939). [Russisch.]
— (Soviet Darwinism at the Federal Agricultural Show.) Sovetskaja Agronomija (Soviet Agronomy) Nr. 8/9, 90—95 (1940). [Russisch.]
Smargonj, E. H.: (Über Teilsterilität bei Arachis hypogaea L.) C. R. (Doklady) Acad. Sci. URSS. 54, 743—745 (1946). [Russisch.]
Smith, H. H.: Recent studies on inheritance of quantitative characters in plants. Bot. Rev. 10, 349—382 (1944).
Smith, L.: Possible practical method for producing hybrid seed of selfpollinated crops trough the use of male-sterility. J. Amer. Soc. Agron. 39, 260—261 (1947).
— A simplified method for establishing the three-point order of genes from F_3 data. J. Amer. Soc. Agron 39, 353—355 (1947).
Šmuk, A.: (Biochemical changes of grafted plants and plant transplantation as a method to pose and solve physiological and biochemical problems.) Doklady Vsesojuz. Akad. Seljsk. Nauk im. V. I. Lenina (Proc. Lenin Acad. Agric. Sci. USSR.): Nos 1/2, 3—13 (1945). [Russisch.]
— u. A. Guseva: Activity of polyploidogenic Compounds as influenced by Hydrogenation. C. R. Acad. Sci. URSS. 30, 642—643 (1941).
Snyder, L. H.: The principles of heredity. D. C. Heath and Company, New York (1940).
Sokolovskaja, A. u. O. Strelkova: Die geographische Verbreitung Polyploider. I. Eine Untersuchung der Flora des Pamirs. Ann. Univ. Leningr. 35, 42—63 (1939).
Solomon, S.: Hybrid vigour in plants and its significance in plant breeding and agriculture. Agric. Liva-Stk. India 9, 139—148 (1939).
Sommermeyer, K.: Bemerkung zur strahlenbiologischen Bestimmung der Genzahlen im X-Chromosom von Drosophila. Z. Vererbungsl. 79, 240—244 (1941).
Sonneborn, T. M.: A new system of determination and inheritance of characters. Rec. Genet. Soc. America Nr. 12, 53—54 (1943). (Abstr.)
— Gene and cytoplasm. I. The determination and inheritance of the killer character in variety 4 of Paramecium aurelia. Proc. Nat. Acad. Sci., Wash. 29, 329—338 (1943).
— Gene and cytoplasm. II. The bearing of the determination and inheritance of characters in Paramecium aurelia on the problems of cytoplasmic inheritance, Pneumococcus transformations, mutations and development. Proc. Nat. Acad. Sci., Wash. 29, 338—343 (1943).

Sonneborn, T. M: Evidence for a bipartite structure of the gene. Genetics 30, 22—23 (1945). (Abstr.)
— Gene action in Paramecium. Ann. Mo. Bot. Gdn. 32, 213—221 (1945).
— Genes as physiological agents. The dependence of the physiological action of a gene on a primer and the relation of a primer to gene. Amer. Nat. 79, 318—339 (1945).
— Inert nuclei inactivity of micronuclear genes in variety 4 of Paramecium aurelia. Genetics 31, 231 (1946). (Abstr.)
— Recent advances in the genetics of Paramecium and Euplotes. Advances Genet., N. Y. 1, 263—358 (1947).
—, R. V. Dippell u. W. Jacobson: Some properties of kappa (killer cytoplasmic factor) and of paramecin (killer substance) in Paramecium aurelia, variety 4. Genetics 32, 106 (1947). (Abstr.)
Sornay, M. A. de: Les gènes. Rev. Agric. Maurice 21, 181—196 (1942).
Spencer, W. P.: Iso-alleles at the bobbed locus in Drosophila hydei populations. Genetics 29, 520—536 (1944).
Spiegelman, S. u. M. D. Kamen: Genes and nucleoproteins in the synthesis of enzymes. Science 104, 581—584 (1946).
Sprague, G. F.: The problem of heterosis. Chronica Botanica 7, 418—419 (1943).
Srinath, K. V.: Mechanism of crossing-over. Natufe, Lond. 158, 840 (1946).
Stadler, L. J.: The experimental alteration of heredity. Growth 3, 321—322 (1939). (Abstr.)
— Genetic studies with ultra-violet radiation. Proc. 7th Int. Genet. Congr. Edinburgh 23.—30. August, Pp. 269—276, 1939 (1941).
— Some observations on gene variability and spontaneous mutation. Spragg Memor. Lectures Plant Breed. Mich. St. Coll. Pp. 3—15 (1939) 1942.
— The effects of x-rays upon dominant mutation in maize. Proc. Nat. Acad. Sci. Wash. 30, 123—128 (1944).
Stebbins, G. L. (jr.): The cytological analysis of species hybrids. II. Bot. Rev. 11, 463—486 (1945).
Stein, E.: Über einige Propfversuche mit erblichen, durch Radiumbestrahlung erzeugten Varianten von Antirrhinum majus, Antirrhinum siculum and Solanum lycopersicum (Tomate König Humbert). Biol. Zbl. 59, 59—78 (1939).
Stern, C.: Genic action as studied by means of the effects of different doses and combinations of alleles. Genetics 28, 441—475 (1943).
— u. E. W. Schaeffer: On wild-type iso-alleles in Drosophila melanogaster. Proc. Nat. Acad. Sci., Wash. 29, 361—367 (1943).
— u. G. Heidenthal: A comparison between the position effects of normal and mutant alleles. Proc. Nat. Acad. Sci. Wash. 32, 26—33 (1946).
Stevens, W. L.: Tables of the recombination fraction estimated from the product ratio. J. Genet. 39, 171—180 (1939).
Stewart, R. N. u. R. Bamford: The chromosomes and nucleoli of Medeola virginiana. Amer. J. Bot. 29, 301—303 (1942).
Stomps, Th. J.: Über die künstliche Herstellung von Oenothera Lamarckiana gigas de Vries. Ber. dtsch. bot. Ges. 60, p. 125—131 (1942).
Stout, A. B. u. C. Chandler: Change from self-incompatibility to self-compatibility accompanying change from diploidy to tetraploidy. Science (N. Y.) 2, 118 (1941).
Straub, J.: Cytogenetik. Fortschr. Bot., Berl. 10, 246—259 (1941).
— Die Nachkommenschaft der tetraploiden Antirrhinum majus Sippe 50. (Vorläufige Mitteilungen.) Ber. dtsch. bot. Ges. 59, 110—113 (1941).

Straub J.: Die Beseitigung der Selbststerilität durch Polyploidisierung. Ber. dtsch. bot. Ges. 59, 296—304 (1941).
— Wege zur Polyploidie. Eine Anleitung zur Herstellung von Pflanzen mit Riesenwuchs. Gebr. Borntrāger, Berlin, Pp. 27 (1941).
— Chromosomenmutationen nach UV.-Bestrahlung. Naturwiss. 29, 13—15 (1941).
— Die Züchtung von Polyploiden mit positivem Selektionswert. Z. Naturforsch. 1, 342—345 (1946).
Strong, L. C.: Genetic analysis of the induction of tumours by methylcholanthrene. XI. Germinal mutations and other sudden biological changes following the subcutaneous injection of methylcholanthrene. Proc. Nat. Acad. Sci. Wash. 31, 290—293 (1945).
— The induction of germinal mutations by chemical means. Amer. Nat. 81, 50—59; also Genetics 32, 108—109 (1947). (Abstr.)
Stubbe, A. E.: Vergleichende Untersuchungen der Wildtypen verschiedener Drosophila-Arten an Hand von Transplantationen der Augenanlagen. VIII. Vergleich der Wildtypen von Drosophila melanogaster und Drosophila buskii. Z. Vererbungsl. 80, 205—209 (1942).
— u. H. Lüers: Vergleichende Untersuchungen der Wildtypen verschiedener Drosophila-Arten an Hand von Transplantationen der Augenanlagen. VI. Vergleich der Wildtypen von Drosophila melanogaster und Drosophila azteca. Z. Vererbungsl. 79, 487—492 (1941).
—, — Vergleichende Untersuchungen der Wildtypen verschiedener Drosophila-Arten an Hand von Transplantationen der Augenanlagen. VII. Vergleich der Wildtypen von Drosophila melanogaster und Drosophila hydei. Z. Vererbungsl. 79, 498—502 (1941).
—, — u. A. Kanellis: Vergleichende Untersuchungen der Wildtypen verschiedener Drosophila- Arten an Hand von Transplantationen der Augenanlagen. III. Vergleich der Wildtypen von Drosophila melanogaster und Drosophila simulans. Z. Vererbungsl. 79, 188—191 (1941).
— u. M. Vogt: Über die Homologie einiger Augenfarbgene bei Drosophila funebris und Drosophila melanogaster an Hand von Augentransplantationen. Z. Vererbungsl. 78, 251—254 (1940).
—, — Vergleichende Untersuchungen der Wildtypen verschiedener Drosophila-Arten an Hand von Transplantationen der Augenanlagen. I. Vergleich der Wildtypen von Drosophila melanogster, Drosophila funebris und Drosophila pseudoobscura. Z. Vererbungsl. 78, 255—260 (1940).
Stubbe, H.: Änderungen des Erbgutes durch physiologische Einflüsse. Umschau. 43, 391—394 (1939).
— Der Einfluß der Ernährung auf die Entstehung erblicher Veränderungen. Angew. Chemie 52, 599 (1939).
— Nährstoffhaushalt und Mutabilität. Forsch. Fortschr. dtsch. Wiss. 15, 189 (1939).
— Kritische Bemerkungen zu Antirrhinum rhinanthoides Lotsy. Biol. Zbl. 60, 590—597 (1940).
— Nutrition and Mutability. Research and Progress 6, Nr. 1, 20—23 (1940).
— L'influenza della nutrizione sull'insorgere delle mutazioni. (The influence of nutrition on the occurrence of mutations.) Scientia Genetica 1, 370—384 (1940).
— Neue Forschungen zur experimentellen Erzeugung von Mutationen. Biol. Zbl. 60, 113—129 (1940).
— Erich v. Tschermak-Seysenegg zum 70. Geburtstage. Naturwiss. 29, 696 (1941).
— Die Gene von Antirrhinum majus IV. (Zur Angleichung der Antirrhinum-Nomenklatur an die Vorschläge der Nomenklatur-Kommission des VII. Internationalen Genetiker-Kongresses, Edinburgh 1939.) Ztschr. für ind. Abst.- u. Vererbungsl. 79, 401—443 (1941).
— Pflanzenzüchtung und Mutationsforschung. Forschungsdienst 15, 333 (1942).

Stubbe, H. u. K. Pirschle: Über einen monogen bedingten Fall von Heterosis bei Antirrhinum majus. Ber. dtsch. bot. Ges. **58**, 546—558 (1940).
— u. F. v. Wettstein: Über die Bedeutung von Klein- und Großmutationen in der Evolution. Biol. Zbl. **61**, 265—297 (1941).
Sturtevant, A. H.: Thomas Hunt Morgan. Amer. Nat. **80**, 22—23 (1946).
— Thomas Hunt Morgan. Genetics **32**, 1—2 (1947).
— u. G. W. Beadle: An introduction to genetics. W. B. Saunders Company, Philadelphia and London (1939).
— u. E. Novitski: The homologies of the chromosome elements in the genus Drosophila. Genetics **26**, 517 bis 541 (1941).
Swanson, C. R.: The distribution of inversions in Tradescantia. Genetics **25**, 438—465 (1940).
Swingle, W. T.: Physiological implications of completely inhibited but not entirely lost characters. Amer. J. Bot. **27**, Nr. 10 Suppl. (1940). (Abstr.)
Synge, R. L. M.: (Proteins and evolutionary theory.) Biohimija (Biochemistry) **10**, 179—188 (1945).

Tatarincev, A. S. u. I. G. Silantjev: (I. V. Michurin's relationship to Mendel's „Laws".) Selekcija i Semenovodstvo (Breeding and Seed Growing), Nr. 1, 11—14 (1940). [Russisch.]
Tatum, E. L.: Are gene mutations responsible for the growth factor requirements of micro-organismus ? J. Bact. **49**, 202—203 (1945).
Tchernaev, I siehe Černjaev.
Teissier, G.: Vitalité et fécondité relatives de diverses combinaisons génétiques comportant un gène léthal, chez la drosophile, C. r. Acad. Sci., Paris **214**, 241—244 (1942).
— Persistance d'un gène léthal dans une population de drosophiles. C. r. Acad. Sci., Paris **214**, 327—330 (1942).
Thoday, J. M. u. D. E. Lea: The effects of ionizing radiations on the chromosomes of Tradescantia bracteata. A comparison between neutrons and X-rays. J. of Genet. **43**, 189—210 (1942).
Thomas, J. A. u. S. Chevais: Production expérimentale de mutations par les trois aminophenylsulfamides isomères, chez la mouche Drosophile. Action sur les cellules males. C. R. Soc. Biol., Paris **137**, 185—187 (1945).
Thompson, W. P.: The causes of hybrid sterility and incompatibility. Trans. Roy Soc. Canada **34**, Nr. 5, 1—13 (1940).
Tihonov, P. M.: (The Control of Dominancy and Segregationin Hybrids.) Jarovizacija Nr. 4, 101—106 (1939). [Russisch.]
Tikhonov, P. M. siehe Tihonov.
Timm, E. W.: Theoretical genetics. Proc. Iowa Acad. Sci. **48**, 51—58 (1941).
Timoféeff-Ressovsky, N. W.: Eine biophysikalische Analyse des Mutationsvorganges. Nova Acta Leopoldina, N. F. **9**, 209—240 (1940).
—: N. K. Koltzoff. Naturwiss. **29**, 121—124 (1942).
— Il meccanismo di mutazione e la natura del gene. (The mechanism of mutation and the nature of the gene.) Scientia Genetica **2**, 126—152 (1942).
— u. K. Zimmer: Mutationsauslösung durch Röntgenbestrahlung unter verschiedener Temperatur bei Drosophila melan. Biol. Zbl. **59**, 358—362 (1939).
— u. K. G. Zimmer: Über Zeitproportionalität und Temperaturabhängigkeit der spontanen Mutationsrate von Drosophila. Z. Vererbungsl. **79**, 530—537 (1941).
—, — Strahlengenetik. Strahlenther. p. 74 (1944).
Tinjakov, G. G.: (Das Ernährungsregime und Variabilität bei Drosophila funebris.) C. R. (Doklady) Acad. Sci. URSS **58**, 307 (1947). [Russisch.]
— (Die Querteilung der Chromosomen.) C. R. (Doklady) Acad. Sci. URSS **58**, 473 (1947). [Russisch.]

Tischler, G.: Über die Siedlungsfähigkeit von Polyploiden. Z. f. Naturforsch. **1**, 157—159 (1946).
Trapani, E.: Azione combinata di due geni non allelomorfi sull'aspetto dell'ala nella Drosophila (Drosophila melanogaster Meig.). R. Ist. Lomb. Sc. e Lett. Rend. Sc. **76**, 343 (1943).
Trofimov, I. E.: (On the increased variation in mutant stocks.) C. R. (Doklady) Acad. Sci. URSS. **33**, 370 bis 372 (1941). [Russisch.]
Turbin, N. V.: (Mixed inheritance and somatic segregation in hybrid plants.) Jarovizacija, Nr. 2 (35), 85 bis 97 (1941). [Russisch.]
— Genetically heterogeneous tissues in plants and vegetative segregation. Agrobiologija (Agrobiology), Nr. 1 136—147 (1946). [Russisch.]
Turrill, W. B.: Genetics in relation to evolution and systematics. Nature **144**, 822—823 (1939).
— Genetics in relation to evolution and systematics at the Seventh International Genetical Congress, Edinburgh 1919. Kew Bull., Nr. 9, 500—504 (1939).
— Taxonomy and phylogeny. Bot. Rev. **8**, 247—270 (Parts 2 and 3 to appear later) (1942).
Tysdal, H. M.: Controlled heterosis as a method of forage crop improvement. Spragg Memor. Lectures Plant Breed. Mich. St. Coll., Pp. 28—38 (1942) 1942.
Tzitzin, N. V. siehe Cicin.

Udoljskaja, N. L.: (Conditions of development of seed in the ear, and the process of segregation in spring wheat hybrids.) Jarovizacija, Nr. 3 (36), 101—104 (1941). [Russisch.]

V., A. S.-L.: La nuova genetica nell' URSS. (The new genetics in the USSR.) Tabacco **51**, 23—30 (1947).
Valadares, A.: Réarrangements somatiques spontanés dans les chromosomes des glandes salivaires de Drosophila melanogaster. Sci. Genetica **2**, 92—100 (1940).
Valdeyron, G.: Où en est le problème de la métaxénie ? Ann. Serv. Bot. Tunis **18**, 43—55 (1941).
Vasiljev, V. N.: (Quelques mots à propos des liens génétiques de Polygonum Marial V. Vassil. de la section Komaroviella V. Vassil.) Botaničeskij Žurnal (Journ. Botanique de l'URSS.) **31**, Nr. 2, 18 (1946). [Russisch.]
Vavilov, N. I.: Genetics in the USSR. Chronica Botanica **5**, 14—15 (1939).
— (Reply to the article by G. N. Shlykow „Formal genetics and consistent Darwinism".) Soviet Subtropics, Nr. 6 (58), 54—56 (1939). [Russisch.]
Vaz, Z.: A contribuição do prof. Dobzhansky ao desenvolvimento da genética. (The contribution of Professor Dobzhansky to the development of genetics.) Bol. Minist. Agric., Rio de J. **33**, Nr. 1, 1—24 (1944).
Vogt, M.: Zur Produktion gonadotropen Hormones durch Ringdrüsen des ersten Larvenstadiums bei Drosophila. Biol. Zbl. **63**, 467—470 (1943).
— Zur labilen Determination der Imaginalscheiben von Drosophila. I. II. III. V. Biol. Zbl. **65**, 223—254, **66**, 81—105, 388—395 (1946—47).
De Vries, L.: German-English science dictionary for students in the agricultural, biological and physical sciences. McGraw-Hill, Publishing Company, Ltd., London (1939).
— French-English science dictionary for students in agricultural, biological, and physical sciences. McGraw-Hill Publishing Company, Ltd., London (1940).

Waddington, C. H.: The physicochemical nature of the chromosome and the gene. Amer. Nat. **73**, 300—314 (1939).
— An introduction to modern genetics. George Allen and Unwin., Ltd., London (1939).

Waddington, C. H.: The raw materials of evolution. Discovery 2, 283 bis 287 (1939).
— Organisers and genes. Cambridge Biological Studies. University Press, Cambridge (1940).
— Canalization of development and the inheritance of acquired characters. Nature, Lond. 150, 563—565 (1942).
— Polygenes and oligogenes. Nature, Lond. 151, 394 (1943).
Wahl, H. A.. Chromosome numbers and meiosis in the genus Carex. Amer. J. Bot. 27, 458—470 (1940).
Wanner, H.: Austausch-Heterozygotie bei Petunia pulverulenta Duthie. Arch. d. J. Klaus-Stift. 17, 453 bis 458 (1942).
Wanscher, J. H.: Om Begreberne Sort og Stamme m. v. samt om Principperne for Navngivningen og Katalogiseringen af Køkkenurter og for Begraensningen af Sorts- og Stammeantallet. (On the concepts, variety and strain etc. and on the principles relating to the naming and cataloguing of vegetables and to the limitation of the number of varieties and strains.) Nord. JordbrForskn., No. 7, 341—356 (1942).
Warmke, H. E. u. A. F. Blakeslee: The establishment of a 4n dioecious race in Melandrium. Amer. J. Bot. 27, 751—762 (1940).
Watanabe, K.: Die Chromosomenzahl der Süß-Kartoffel (Ipomola batatas) und der mit ihr verwandten Pflanzen, welche in Japan wildwachsend oder kultiviert sind. Proc. Crop Sc. Soc., Japan 11, 124—134 (1939).
Weddle, C.: Two colchicine-induced polyploids of the greenhouse Chrysanthemum and their progeny. Proc. Amer. Soc. Horticult. Sci. 38, 658—660 (1941).
Weir, J. A.: A source of genes for evolutionary progress. Genetics. 32, 111—112 (1947). (Abstr.)
Weiss, F. E.: Masters memorial lectures, 1940. Graft hybrids and chimaeras: I, II. J. R. Hort. Soc. 65, 212—217, 237—243 (1940).
Weiss, J.: Biological action of radiations. Nature, Lond. 157, 584 (1946).
Wellensiek, S. J.: De invloed van colchicine op de kerndeeling. (The influence of colchicine on nuclear division.) Hand. Ned. Nat.-en Geneesk. Congr. 27, 2 (1939).
— Indrukken van het zevende internationale genetische congres in Engeland. (Impressions of the Seventh International Genetic Congress in England.) Erfelijkheid in Praktijk, Leiden 4, 177—180 (1939).
— Het kunstmatig verwekken van mutaties met colchicine. (The artifical induction of mutations with colchicine.) Landbouwk. Tijdschr., Wageningen 53, 535—543 (1941).
— Colchicine-mutaties en hun beteekenis voor de plantenveredeling. (Colchicine mutations and their importance in plant breeding.) Jaarb. Algem. Bond Oudleerl. Middelbaar Landbouwonderwijs, Wageningen, Pp. 3—12 (1942).
Westergard, M.: Studies on cytology and sex determination in polyploid forms of Melandrium album. Kobenhavn: Ejnar Munksgaard, Pp. 131, skr. 9 (1940).
Wettstein, F. v.: Über verschiedene Fragen der Vererbungsforschung und Entwicklungsphysiologie. Biologe 8, 3—9 (1939).
— Oenothera — ein klassisches Objekt der Vererbungsforschung. Naturwiss. 31, 177—180 (1943).
Whaley, W. G.: A developmental analysis of heterosis in Lycopersicon. I. The relation of growth rate to heterosis. II. The role of the apical meristem in heterosis. Amer. J. Bot. 26, 609—616, 682—690 (1939).
— Heterosis. Bot. Rev. 10, 461—498 (1944).
White, E. G.: Principles of Genetics. Pp. 352, St. Louis: Mosby (1940).
White, M. J. D.: The chromosomes. Methuen and Co., Ltd., London (1942).

White, O. E.: Genes, species variability and plantbreeding. Amer. Nat. 76, 191—207 (1942).
Wigan, L. G.: Balance and potence in natural populations. J. Genet. 46, 150—160 (1944).
— u. K. Mather: Correlated response to the selection of polygenic characters. Ann. Eugen. 11, 354—364 (1942).
Wilczyński, J.: Über die allgemeine Gleichung der Mendel-Gesetze. Zum Teil ein Beitrag zur Deutung des Dominanzwesens. Biol. Gen. 14, 447—455 (1939).
— Mendel and Darwin. A new link in the old quarell. Acta biotheor., Leiden 7, 81—86 (1943).
Willis, J. C.: Evolution in plants by kaleidoscopic mutation. Proc. Roy. Soc. Ser. B. 131, 161—169 (1942).
Woods, M. W. u. H. G. DuBuy: Cytoplasmic diseases and cancer. Science 102, 591—593 (1945).
Wright, S.: The distribution of self-sterility alleles in populations. Genetics 24, 538—552 (1939).
— On the probability of fixation of reciprocal translocations. Amer. Naturalist 75, 513—522 (1941).
— The physiology of the gene. Physiol. Rev. 21, 487 bis 527 (1941).
— Isolation by distance. Genetics 28, 114—137 (1943).
— Genes as physiological agents. General consideration. Amer. Nat. 79, 289—303 (1945).
— The differential equation of the distribution of gene frequencies. Proc. Nat. Acad. Sci., Wash. 31, 382—389 (1945).
— u. T. Dobzhansky: Genetics of natural populations. XII. Experimental reproduction of some of the changes caused by natural selection in certain populations of Drosophila pseudoobscura. Genetics 31, 125—156 (1946).
Wulff, H. D.: Der Ölgehalt verschiedenchromosomiger Rassen von Kalmus (Acorus Calamus L.). Z. f. Naturforsch. 1, 600—603 (1946).

Yoyohuku, T.: The Direction of Coiling on the Chromosomes in Tradescantia. W. 3 pls., Japan. Journ. Genet. 15, 310—312 (1939).

Zamenhof, S.: Studies on factors influencing mutability. Rec. Genet. Soc. America, No. 12, 57—58 (1943). (Abstr.)
— Studies on induction of mutations by chemicals. II. Experiments with unstable genes. J. Genet. 47, 69—75 (1945).
Zayas y Muñoz, F. de: Qué es la genética y cuál su aplicación práctica a la agricultura. (What is genetics and what its practical application to agriculture.) Rev. Agric. Habana 22, Nr. 7, 103—110 (1939).
Zhukovsky, P. M. siehe Žukovskij.
Zimmer, K. G.: Zur biophysikalischen Analyse des Vorgangs der Tötung von Drosophila-Eiern durch Strahlung. Biol. Zbl. 60, 287—298 (1940).
— u. N. W. Timoféeff-Ressovsky: Über einige physikalische Vorgänge bei der Auslösung von Genmutationen durch Strahlung. Z. Vererbungsl. 80, 353—372 (1942).
—, — Nachtrag zu der Arbeit „Über einige physikalische Vorgänge bei der Auslösung von Gemutationen durch Strahlung". Z. indukt. Abstamm.- u. Vererbungsl. 80, 619 (1942).
Zujtin, A. I.: Influence of the change from the natural complex of developmental conditions to the laboratory one on the mutation rate in Drosophila melanogaster. C. R. (Doklady) Acad. Sci. URSS, n. s. 29, 610—611 (1940). [Russisch.]
Žukovskij, P. M.: (Studies on hybridization and immunity of plants.) Trudy Moskovskoj Seljskohozjajstvennoj Akademii im. K. A. Timirjazeva (Trans. K. A. Timiriaseff Acad. Agric. Moscow) 6, 48 (1944). [Russisch.]

Zündorf, W.: Zytogenetisch-entwicklungsgeschichtliche Untersuchungen in der Veronica-Gruppe Biloba der Sektion Alsinebe Griseb. Z. Vererbungsl. **77**, 195—238 (1939).
— Ein weiterer Beweis für die Bedeutung des Plasmas bei Epilobium-Kreuzungen. Art-Kreuzungen mit Epilobium palustre. Z. Vererbungsl. **77**, 533—547 (1939).

b) Artentstehung und Populationsgenetik.

Allen, C. E.: The evolution and determination of sexual characters in the angiosperm sporophyte. Torreya **43**, 6—15 (1943).
Amadon, D.: Specialization and evolution. Amer. Nat. **77**, 133—141 (1943).
Anderson, E. u. L. Hulbricht: The histological basis of a specific difference in leaf texture. Amer. Nat. **77**, 285—287 (1943).
Anonymus: Parallelbildung und Stammesgeschichte. Biologe **8**, 44—48 (1939).

Babcock, E. B.: New light on evolution from research on the genus Crepis. Amer. Nat. **78**, 385—409 (1944).
— Cytogenetics and speciation in Crepis. Advances Genet. N. Y. **1**, 69—93 (1947).
Ball, G. H.: Parasitism and evolution. Amer. Nat. **77**, 345—364 (1943).
Beadle, G. W. u. E. L. Tatum: A Contribution to the Theory of Evolution by natural Selection. Amer. Natural. **75**, 213—230 (1941).
Berg, R. L.: (Eine genetische Analyse von zwei natürlichen Populationen.) Žurnal Obščej Biologii (Journ. of Gener. Biol.), **2**, Nr. 1 (1941). [Russisch.]
— (Der Zusammenhang zwischen Mutabilität und Isolationsgrad der Populationen von Drosophila melanogaster.) C. R. (Doklady) Acad. Sci. URSS. **36**, Nr. 2 (1942). [Russisch.]
— (Dominanz der schädlichen Mutationen in Populationen von Drosophila melanogaster.) C. R. (Doklady) Acad. Sci. URSS. **36**, Nr. 7 (1942). [Russisch.]
— (Bedeutung der Isolation für die Evolution der Dominanz in natürlichen Populationen von Drosophila melanogaster.) C. R. (Doklady) Acad. Sci. URSS. **36**, Nr. 9 (1942). [Russisch.]
— (Die Korrelation zwischen der Mutabilität und der regulatorischen Fähigkeit des Organismus und ihre evolutionistische Bedeutung.) Izvestija Akad. Nauk SSSR. ser. biol. (Bull. Acad. Sci. URSS. Ser. biol.) **3**, 367—376 (1945). [Russisch.]
Bergdolt, E.: Über Formwandlungen — zugleich eine Kritik von Artbildungstheorien. Biologe **9**, 398—407 (1940).
Bond, T. E. T.: Robert Wight (1796—1872), Dr. Freke and the „Origin of Species". Nature, Lond. **154**, 566—569 (1944).
Burgeff, H.: Konstruktive Mutationen bei Marchantia. Vererbungsversuche zur Frage der Evolution der Lebermoose. Naturwiss. **29**, 289—299 (1941).

Camp, W. H. u. C. L. Gilly: The structure and origin of species. Brittonia **4**, 323—385 (1942).
Cattell, J. (Editor): Biological symposia. Volume I. The Jaques Cattell Press, Lancaster (1940).
— Biological Symposia-Volume IV. Population problems in Protozoa. Experimental control of development and differentiation. Theoretical and practical aspects of polyploidy in crop plants. The species problem. Jaques Cattell Press, Lancaster, Pa. Pp. VI + 293, figs, tables (1941).
Cave, A. J. E. u. R. W. Haines: Meristic variation and reversibility of evolution. Nature, Lond. **154**, 579—580 (1944).

Charles, D. R. u. R. H. Goodwin: An estimate of the minimum number of genes differentiating two species of golden-rod with respect to their morphological characters. Amer. Nat. **77**, 53—69 (1943).
Clark, H. W.: Genes and Genesis. Pacific Press Publ. Assoc., California Pp. 155 (1940).
Clausen, J., D. D. Keck u. W. M. Hiesey. The concept of species based on experiment. Amer. J. Bot. **26**, 103—106 (1939).
—, —, — Experimental studies on the nature of species. I. Effect of varied environments on western North American plants. Carnegie Inst. Wash. Publ., Nr. 520, Washington D. C. (1940).
—, —, — Experimental studies on the nature of species. II. Plant evolution through amphidiploidy and autoploidy, with examples from the Madiinae. Wash., D. C. Publ., Nr. 564, Pp. VII + 174, 86 figs, 14 tables (1945).
—, —, — Heredity of geographically and ecologically isolated races. Amer. Nat. **81**, 114—133 (1947).
Cornman, I.: A basis for ostensible reversal of evolution. Amer. Nat. **77**, 90—93 (1943).

Danne, H. A.: The life energy of species. Engineering Laboratory, New York, Pp. 27, 3 figs, table, charts (1944).
Demerec, M. et al: Cytology, genetics and evolution. University of Pennsylvania Press, Philadelphia (1941).
Dillon Ripley, S.: Suggested terms for the interpretation of speciation phenomena. J. Wash. Acad. Sci. **35**, 337—341 (1945).
Dingler, H.: Die philosophische Begründung der Deszendenztheorie. In G. Heberer: Die Evolution der Organismen, Pp. 2—19 (1943).
Dobzhansky, T. (Editor): Biological Symposia-Volume VI. Temperature and evolution. Isolating mechanisms. Genetic control of embryonic development. Jaques Cattell Press, Lancaster, Pa., Pp. XII + 355, figs, plates, tables (1942).
— Die genetischen Grundlagen der Artbildung. Fischer, Jena, Pp. 252 (1939).
— Genetics and the origin of species. Columbia University Press, New York, 2nd Ed., Pp. XVIII + 446, 24 figs, 24 tables (1941).
— The species concept. Rev. Agric. Piracicaba **18**, 441—442 (1943).
— Genetics of natural populations. IV. IX. XIII. XIV. Genetics **24**, 391—412, **28**, 162—186, **31**, 269—296, **32**, 142—160 (1939—1947).
— u. B. Spassky: Genetics of natural populations. XI. Manifestation of genetic variants in Drosophila pseudoobscura in different environments. Genetics **29**, 270—290 (1944).
— u. S. Wright: Genetics of natural populations. V. X. Genetics **26**, 23—51, **28**, 304—340 (1941—43).
—, — Genetics of natural populations. XV. Rate of diffusion of a mutant gene through a population of Drosophila pseudoobscura. Genetics **32**, 303—324 (1947).
Droogleever Fortuyn, A. B.: Het evolutieprobleem bezien in het licht der erfelijkheidsleer. (The evolution problem regarded in the light of genetics.) Erfelijkheid in Praktijk, Leiden **4**, 155—157 (1939).
Dubinin, N. P.: (Darwinismus und Populationsgenetik.) Uspjehi Sovremennoj Biologii (Advances in Modern Biology) **13**, Nr. 2 (1940). [Russisch.]
— (On the seasonal cycle and the distribution the lethals in populations.) Zoologičeskij Žurnal (Zool. Journ.) **25**, Nr. 5, 495 (1946). [Russisch.]
— On lethal mutations in natural populations. Genetics **31**, 21—38 (1946).
— (Physiological mutations in populations.) Izvestija Akademii Nauk SSSR. (Bull. Acad. Sci. URSS. Ser. biol.) Nr. 4, 487—510 (1947). [Russisch.]

Dubinin, N. P.: (Die Genetik des adoptiven aberrativen Polymorphismus in den Populationen von Drosophila melanogaster.) C. R. (Doklady) Acad. Sci. URSS. 58, 299 (1947). [Russisch.]
— (Genetische Grundlagen des adaptiven Polymorphismus und seine Bedeutung für die Evolution.) C. R. (Doklady) Acad. Sci. URSS. 58, 1497 (1947). [Russisch.]
— u. G. G. Tinjakov: (Der Fruchtbarkeitzyklus der Populationen und die Selektion.) C. R. (Doklady) Acad. Sci. URSS. 51, 309—311 (1946). [Russisch.]
—, — (Die natürliche Auslese in Experimenten mit Populationsinversionen.) C. R. (Doklady) Acad Sci. URSS. 51, 715—718 (1946). [Russisch.]
—, — (Der Saisonzyklus und die Konzentration der Inversionen in den Populationen.) C. R. (Doklady) Acad. Sci. URSS. 52, 77—79 (1946). [Russisch.]
—, — Natural selection and chromosomal variability in populations of Drosophila funebris. J. Hered. 37, 39—44 (1946).
—, — Inversion gradients and natural selection in ecological races of Drosophila funebris. Genetics 31, 537—545 (1946).
—, — Natural selection in experiments with population inversions. J. Genet. 48, 11—15 (1947).
—, — (Migration und natürliche Auslese im Experiment mit natürlichen Populationen.) C. R. (Doklady) Acad. Sci. URSS. 55, 541—544 (1947). [Russisch.]
—, — (Inversionen auf den Grenzen der ökologischen Rassen von Drosophila funebris.) C. R. (Doklady) Acad. Sci. URSS. 55, 643—645 (1947). [Russisch.]
—, — (Stadtökologie und Verbreitung der Inversionen bei Drosophila funebris.) C. R. (Doklady) Acad. Sci. URSS. 56, 865 (1947). [Russisch.]
—, — (Das Klima und die Verbreitung der Inversionen im Artareal von Drosophila funebris.) C. R. (Doklady) Acad. Sci. URSS 56, 965 (1947). [Russisch.]

Elias, M. K.: Structure of grass and its evolution. Amer. J. Bot. Suppl. 29, 7s—8s (1942).

Fabian, G.: Mutation in einer freilebenden Drosophila melanogaster-Population. Magy. Biol. Kutat. Munk. 14, 269—275 (1942).
Flaksberger, K. A.: (Species-a constantly changing category [as exemplified by a study of wheat]). Vestnik Socialističeskogo Rastenievodstva (Soviet Plant Industry Record), Nr. 1, 153—156 (1941). [Russisch.]

Gates, R. R.: Processes of organic evolution. Science 93, 335—339 (1941).
Gause, G. F.: (Das Problem der stabilisierenden Auslese.) Žurnal Obščej Biologii (Journ. of Gener. Biology) 2, Nr. 2 (1941). [Russisch.]
Geršenzon, S. M.: (Eine experimentelle Untersuchung der natürlichen Auslese in einer mutanten Population von Drosophila virilis.) Žurnal Obščej Biologii (Journ. of Gener. Biology) 2, Nr. 3 (1941). [Russisch.]
Goldschmidt, R.: The material basis of evolution. Yale University Press, New Haven and Oxford University Press, London (1940).
— Mimetic polymorphism, a controversial chapter of Darwinism. Quart. Rev. Biol. 20, 147—164, 205—230 (1945).
— „An empirical evolutionary generalization" viewed from the standpoint of phenogenetics. Amer. Nat. 80, 305—317 (1946).
Grassé, P.-P.: Adaptation et évolution. Presse méd. 48, 821—823 (1940).
Gregor, J. W.: The units of experimental taxonomy. Chronica Botanica 7, 193—196 (1942).
— The ecotype. Biol. Rev. 19, 20—30 (1944).
Grimes, C. W.: A story outline of evolution. Bruce Humphries, Inc., Boston, Pp. VIII + 244 (1944).

Haas, O. u. G. G. Simpson: Analysis of some phylogenetic terms, with attempts at redefinition. Proc. Amer. Phil. Soc. 90, 319—349 (1946).
Haase-Bessell, G.: .Evolution. Biologe 10, 233—247 (1941).
— Der Evolutionsgedanke in seiner heutigen Fassung. G. Fischer, Jena, Pp. 76 (1941).
Haldane, J. B. S.: Radioactivity and the origin of life in Milne's cosmology. Nature, Lond. 153, 555 (1944).
Hämmerling, J.: Entwicklung und Regeneration von Acetabularia crenulata. Z. indukt. Abstamm. u. Vererblehre 81, 84—113 (1943).
— Ein- und zweikernige Transplanate zwischen Acetabularia mediterranea und A. crenulata. Z. induct. Abstamm.- u. Vererbungsl. 81, 114—180 (1943).
Heberer, G.: Das Typenproblem in der Stammesgeschichte. In G. Heberer: Die Evolution der Organismen, Pp. 545—588 (1943).
Herrera, A. L.: A new theory of the origin and nature of life. Science 96, Nr. 2479, 14 (1942).
Hiesey, W. M., J. Clausen u. D. D. Keck: Relation between climate and intraspecific variation in plants. Amer. Nat. 76, 5—22 (1942).
Hohlov, S. S.: Historical conditions and evolutionary significance of apomixis in angiosperms. C. R. (Doklady) Acad. Sci. URSS. 52, Nr. 9, 805—807 (1946).
— (The centres of origin or the geographical foci of the formation of species.) Botaničeskij Žurnal SSSR. (Journ. botanique de l'URSS.) 32, Nr. 1, 33—41 (1947). [Russisch.]
Horowitz, N. H.: On the evolution of biochemical syntheses. Proc. Nat. Acad. Sci., Wash. 31, 153—157 (1945).
Hovanitz, W.: A genetic study of wild populations and evolution. Caldasia, Bogota, Nr. 10, 459—464 (1944).
Huxley, J. (Editor): The new systematics. Clarendon Press, Oxford (1940).
— Evolution, the modern synthesis. George Allen and Unwin, Ltd., London (1942).

Iljinskij, A. P.: (Ecotypes, polyploids and natural selection.) Priroda (Nature), Nr. 1, 63—67 (1943). [Russisch.]
— (Some problems and methods of the present day biocoenology.) Žurnal Obščej Biologii (Journ. of General Biology) 6, 355—362 (1945). [Russisch.]

Just, T. K.: The rates of evolutionary processes. Proc. Ind. Acad. Sci. 53, 14—27 (1944).

Kamšilov, M. M.: (Dominanz und Auslese.) C. R. (Doklady) Acad. Sci. URSS. 22, Nr. 6 (1939). [Russisch.]
— (Korrelation und Auslese.) Žurnal Obščej Biologii (Journ. of Gener. Biology) 2, Nr. 1 (1941). [Russisch.]
Kaškarov, D. N.: (Ist die Evolution adoptiv und wie sind die Artmerkmale?) Zoologičeskij Žurnal (Zool. Journ.) 18, Nr. 4 (1939). [Russisch.]
Keller, B. A.: (Evolution of plants on the ground of the particularities of their feeding.) Priroda (Nature), Nr. 1, 26—35 (1945). [Russisch.]
Khohlov, S. S. siehe Hohlov.
Kirpičnikov, V. S.: (Die Bedeutung der Anpassungsmodifikationen für die Evolution.) Žurnal Obščej Biologii (Journ. of Gener. Biology), 1, Nr. 1 (1940). [Russisch.]
— (On hypotheses of fixation of modifications by heredity.) Uspehi Sovremennoj Biologii (Advances in Modern Biology) 18, 314—339 (1944). [Russisch.]
— The adaptive character of intraspecific systematic variation. I. On the adaptive significance of mutations. J. Gen. Biol. 5, 172—192 (1944).
Kleiner, E.: A földrajzi fajtak elhatárolása. (On the delimination of geographical races.) Arb. Ung. Biol. Forsch.-Inst. 13, 447—460 (1941).

Krogh, C. von: Immer wieder: Abstammung oder Schöpfung? Biologe **9**, 414—417 (1940).

Krug, C. A.: Genética e evolução (Genetics and evolution). Rev. Agric., S. Paulo **15**, 271—288 (1940).

Lam, H. J.: Over de theorie der arealen (Chorologie). [On the theory of areas (Chorology).] Vakbl. Biol. **20**, 77—87 (1939).

— Hedendaagsche opvattingen over evolutie en phylogenie. (Present day views of evolution and phylogeny.) Vakbl. Biol. **24**, Nr. 7, 73—81 (1943).

Lamprecht, H.: Die genisch-plasmatische Grundlage der Artbarriere. Agri Hortique Genetica, Landskrona **2**, 75—141 (1944).

— Intra-and inter-specific genes. Agri Hortique Genetica, Landskrona **3**, 45—60 (1945).

Larroque, P.: Principes et méthodes de la sélection rapide des plantes, basée sur la considération des complexes héréditaires. Agron. Trop., Nos. 11/12, 563 bis 581 (1946).

Lewis, D.: Male sterility in natural populations of hermaphrodite plants. The equilibrium between females and hermaphrodites to be expected with different types of inheritance. New Phytologist **40**, 56—63 (1941).

Lucas, C. E.: Excretion, ecology and evolution. Nature, Lond. **153**, 378—379 (1944).

Ludwig, W.: Selektion und Stammesentwicklung. Naturwiss. **28**, 689—705 (1940).

— Über die Rolle des Mutationsdrucks bei der Evolution. Biol. Zbl. **62**, 374 (1942).

— Die Selektionstheorie. In G. Heberer: Die Evolution der Organismen, Pp. 479—520 (1943).

Lukin, E. I.: (Darwinismus und die geographischen Gesetzmäßigkeiten in der Veränderung der Organismen.) Moskva-Leningrad (Verlag Akad. d. Wiss.). [Russisch.]

Lysenko, T. D.: (Natural selection and intraspecific competition.) Selekcija i Semenovodstvo (Breeding and Seed Growing), Nos. 1/2, 4—26 (1946). [Russisch.]

— (Natural selection and intraspecific competition.) Agrobiologija (Agrobiology), Nr. 2, 3—27 (1946). [Russisch.]

— (Soviet Darwinism.) Agrobiologija (Agrobiology), Nr. 1, 7—18 (1946). [Russisch.]

McNair, J. B.: Energy and evolution. Phytologia **2**, 33—50 (1941).

— Advance in phylogenetic position in the cryptogams as indicated by their fats. (Lloydia, Cincinnati **6**, 155 bis 156 (1943).

— Some comparisons of chemical ontogeny with chemical phylogeny in vascular plants. Lloydia, Cincinnati **8**, 145—169 (1945).

— Plant fats in relation to environment and evolution. Bot. Rev. **11**, 1—59 (1945).

Malinovskij, A.: (Die Rolle der genetischen und phänogenetischen Erscheinungen in der Evolution der Art. I. Pleiotropie.) Izvestija Akad. Nauk SSSR, ser. biol. (Bull. Acad. Sci. URSS), Nr. 4 (1939). [Russisch.]

Manevič, E. D.: (Saisonänderungen in der Konzentration der Letalen und Herausspaltung der Homozygoten in der natürlichen Population von Drosophila melanogaster.) C. R. (Doklady) Acad. Sci. URSS. **58**, 899 (1947). [Russisch.]

Marsh, F. L.: Evolution, creation and science. Review and Herald Publishing Association, Washington, D. C. Pp. 304 (1944).

Mather, K.: Outbreeding and separation of the sexes. Nature, Lond. **145**, 484—486 (1940).

Matthew, W. D.: Climate and Evolution. 2. ed. rev. and enl. XI, 223 p. New York Acad. of Sci. (1939).

Mayr, E.: Systematics and the origin of species from the viewpoint of a zoologist. Columbia University Press, New York, Oxford University Press, London, E. C. 4, Pp. 334, 14 tables, 29 figs (1942).

Mayr, E.: Symposium on age of the distribution pattern of the gene arrangements in Drosophila pseudoobscura. Some evidence in favour of a recent date. Lloydia, Cincinnati **8**, 70—83 (1945).

Mečnikov, I. I.: (On Darwinism.) Moskva-Leningrad, Pp. 242 (1943). [Russisch.]

Metz, C. W.: Duplication of chromosome parts as a factor in evolution. Amer. Nat. **81**, 81—103 (1947).

Meyer-Abich, A.: Beiträge zur Theorie der Evolution der Organismen. 1. Das typologische Grundgesetz und seine Folgerungen für Phylogenie und Entwicklungsphysiologie. Acta biotheor., Leiden **7**, 1—80 (1943).

Miller, H.: The genus and species in relation to evolution and to system. Amer. Nat. **73**, 93—95 (1939).

Morrison, G.: Origin of species and varieties. Seed World **50**, Nr. 8, 10—11, 34—35 (1941).

Muller, H. J.: Reversibility in evolution considered from the standpoint of genetics. Biol. Rev. **14**, 261—280 (1939).

Muretov, G. D.: (Die Entstehung der physiologischen Mutationen und ihre Verbreitung in den Populationen.) Žurnal Obščej Biologii (Journ. of Gener. Biology) **2**, Nr. 2 (1941). [Russisch.]

Narodickaja, M. B.: Visible mutations in a natural population of Drosophila fasciata. C. R. (Doklady) Acad. Sci. URSS., n. s. **30**, 169—171 (1941).

Novikov, A. B.: Continuity and discontinuity in evolution. Science **102**, 405—406 (1945).

Olenov, J. M.: (Über den Einfluß der vorangehenden Geschichte der Art auf ihre weitere Entwicklung.) C. R. (Doklady) Acad. Sci. URSS. **31**, Nr. 2, 157—160 (1941). [Russisch.]

— u. I. S. Harmac: (Transformation des normalen Genotypus in natürlichen Populationen von Drosophila melanogaster.) C. R. (Doklady) Acad. Sci. URSS. **24** (1939). [Russisch.]

Paramonov, A. A.: (A course of Darwinism.) Sovetskaja Nauka, Moskva, Pp. 432, 176 figs (1945). [Russisch.]

Poljanskij, V. I.: (Antidarwinism in the native country of Darwin. [The recent work of Prof. Punnett.]) Priroda (Nature), Nr. 11, 87—96 (1939). [Russisch.]

— (The problem of species in botany and the works of Academician V. L. Komarov.) Priroda (Nature), Nr. 5—6, 11—21 (1944). [Russisch.]

Prezent, I. I.: (Lability and stability of properties of plant organisms in connexion with their method of reproduction.) Agrobiologija (Agrobiology), Nr. 1, 63 bis 82 (1946). [Russisch.]

Quintanilha, A.: O problema da delimitação e origem das espécies do ponto de vista da biologia experimental. (The problem of the delimitation and origin of species from the view-point of experimental biology.) Bol. Soc. Brot. **17**, 159—165 (1943).

Rabaud, E.: La notion d'espèce et la génétique. Sciences, Paris **67**, 85—91 (1940).

Rabel, G.: Lamarck: In honour of the 200th anniversary of his birth. The Nineteenth Century and After **137**, 258—264 (1945).

Rahn, O.: Building stones to a chemistry of evolution. Amer. Nat. **73**, 26—43 (1939).

Rapoport, I. A.: (Mehrfache lineare Wiederholungen von Chromosomenstrecken und ihre evolutionistische Bedeutung.) Žurnal Obščej Biologii (Journ. of Gener. Biol.) **1**, Nr. 2 (1940). [Russisch.]

Renner, O. u. M. Voss: Zur Entwicklungsgeschichte randpanaschierter Formen von Prunus, Pelargonium, Veronica, Dracaena. Flora, N.F. **35**, 356—376 (1942).

Rensch, B.: Neuere Probleme der Abstammungslehre. Verlag F. Enke, Stuttgart (1947).
Rothmaler, W.: Artentstehung in historischer Zeit, am Beispiel der Unkräuter des Kulturleins (Linum usitatissimum). Züchter 17/18, 89—92 (1946).
Rozanova, M. A.: (Die Wege und Entstehungen der Rassen und Arten der Pflanzen.) Uspjehi sovremennoj Biologii (Advances in Modern Biology) 12, Nr. 2 (1940). [Russisch.]
— (The species an ecological problem.) Uspjehi sovremennoj biologii (Advances in modern Biology) 23, 69—86 (1947). [Russisch.]
Rytz, W.: Der Weg des „Auch anders" als Theorie für die prinzipielle Richtung bei der Phylogenie (Artentstehung). Arch. d. Jul. Klaus-Stiftg. Erg.-Bd. zu Bd. 20, 268—276 (1945).

Schiemann, E.: Entstehung der Kulturpflanzen. Ergebnisse der Biologie 19, 409—552 (1943).
Schmalhausen, I. I.: (Die Wege und die Gesetzmäßigkeit des Evolutionsprozesses.) Moskva-Leningrad (Verlag Akad. d. Wiss. UdSSR.), Pp. 212 (1939). [Russisch.]
— (Variabilität und Wechsel der Adaptionsformen im Evolutionsprozeß.) Žurnal Obščej Biologii (Journ. of Gener. Biol.) 1, Nr. 4 (1940). [Russisch.]
— (Stabilisierende Auslese und ihre Stellung zwischen den Evolutionsfaktoren. I und II.) Žurnal Obščej Biologii (Journ. of Gener. Biol.) 2, Nr. 3 (1941). [Russisch.]
— (Das Evolutionstempo und die es bestimmenden Faktoren.) Žurnal Obščej Biologii (Journ. of Gener. Biology) 6, Nr. 5 (1943). [Russisch.]
— (Evolutionsfaktoren. — Theorie der stabilisierenden Auslese.) Moskva-Leningrad (Verlag Akad. d. Wiss. UdSSR.), Pp. 396 (1946). [Russisch.]
— (The problems of Darwinism.) Sovetskaja Nauka, Moskva, Pp. 528, 200 figs (1946). [Russisch.]
Schmidt, K. P.: Corollary and commentary for „climate and evolution". Amer. Midl. Nat. 30, 241—253 (1943).
Severcov, A. N.: (Morphologische Gesetzmäßigkeiten der Evolution.) Moskva (Verlag Akad. d. Wiss.) (1939). [Russisch.]
Simpson, G. G.: The Role of the Individual in Evolution. Mit 8 Abb. Journ. Wash. Acad. Sc. 31, 1—20 (1941).
— Tempo and mode in evolution. Columbia University Press, N. Y., Pp. XVIII+237, 36 figs, 19 tables (1944).
— Symposium on age of the distribution pattern of the gene arrangements in Drosophila pseudoobscura. Evidence from fossils and from the application of evolutionary rate distributions. Lloydia, Cincinnati 8, 103 bis 108 (1945).
Sinskaja, E.: (What is a centre of species formation from the viewpoint of Ch. Darwin's theory?) Priroda (Nature), Nr. 11, 51—58 (1939). [Russisch.]
— (The species problem in modern botanical literature.) Uspehi Sovremennoj Biologii (Advances in Modern Biology) 15, 326—359 (1942). [Russisch.]
Sirks, M. J.: The genetic nature of racial and specific differences. Boll. Soc. Ital. Biol. Sper. 15, 184—190 (1940).
Sizov, I. A.: (A great theorist of our time.) Vestnik Socialističeskogo Rastenievodstva (Soviet Plant Industry Record), Nr. 2, 4—11 (1940). [Russiscn.]
Šlykov, G. N.: (On the origin of our cultivated plants.) Priroda (Nature), Nr. 1, 34—41 (1941). [Russisch.]
Smaragdova, N. P.: (Studium der natürlichen Auslese bei Protozoa. VI. Zum Entstehungsmechanismus der geographischen Variabilität.) Zoologičeskij Žurnal (Zool. Journ.) 23, Nr. 1 (1944). [Russisch.]
Solodovnikov, V. B.: (Über den natürlichen Modifikationsprozeß in natürlichen Populationen bei Drosophila melanogaster.) C. R. (Doklady) Acad. S i. URSS. 56, 297—300 (1947). [Russisch.]

Spencer, W. P.: Mutations in wild populations in Drosophila. Advances Genet. N. Y. 1, 359—402 (1947).
Stebbins, G. L.: The genetic approach to problems of rare and endemic species. Madroño 6, 241—258 (1942).
Stebbins, G. L. (jr.): The role of isolation in the differentiation of plant species. Biol. Symp. 6, 217—233 (1942).
— Role of isolation in the differentiation of plant species. Nature, Lond. 155, 150—151 (1945).
— Symposium on the age of the distribution pattern of the gene arrangements in Drosophila pseudoobscura. Evidence for abnormally slow rates of evolution, with particular reference to the higher plants and the genus Drosophila. Lloydia, Cincinnati 8, 84—102 (1945).
Stoletov, V. N.: (Trofim Denisovič Lysenko.) Sovhoznoe Proizvodstvo (State Farming), Nr. 10, 34—48 (1945). [Russisch.]
Stubbe, H.: Mutation und Artentstehung. Umschau 46, 116—118 (1942).

Timoféeff-Ressovsky, N. W.: Sulla questione dell'isolamento biologico entro popolazioni specifiche. (The question of biological isolation within species populations.) Scientia Genetica 1, 317—325 (1940).
— u. E. A. Timoféeff-Ressovsky: Populationsgenetische Versuche an Drosophila. I. Zeitliche und räumliche Verteilung der Individuen einiger Drosophila-Arten über das Gelände. Z. Vererbungsl. 79, 28—34 (1940).
—, — Populationsgenetische Versuche an Drosophila. II. Aktionsbereiche von Drosophila funebris und Drosophila melanogaster. Z. Vererbungsl. 79, 35—43 (1940).
—, — Populationsgenetische Versuche an Drosophila. III. Quantitative Untersuchungen an einigen Drosophila-Populationen. Z. Vererbungsl. 79, 44—49 (1940).
Tinjakov, G. G.: (Eine hochmutable Sippe aus einer Wildpopulation von Drosophila melanogaster.) C. R. (Doklady) Acad. Sci. URSS. 22 (1939). [Russisch.]
Toxopeus, H. J.: Herkomst, vormenrijkdom en veredeling van enkele onzer cultuurgewassen. (Origin, wealth of forms and improvement of some of our crop plants.) Natuurwet. Tijdschr. Ned.-Ind. 101, 19—30 (1941).
Tischler, G.: Polyploidie und Artbildung. Naturwiss. 30, 713—718 (1942).
Travin, I. S.: (The present centres of an intensive formation of species of plants.) Botaničeskij Žurnal (J. Bot. URSS.) 30, 245—250 (1945). [Russisch.]
Tretjakov, D. K.: (The methods of modern phylogenetics.) J. Gen. Biol. USSR. 4, 116—128 (1943). [Russisch.]
Turbin, N. V.: Prof. Boris Keller. Nature, Lond. 155, 702—703 (1945).
Turrill, W. B.: Taxonomy and phylogeny. Bot. Rev. 8, 473—532, 655—707 (1942).

Vasconcellos, J. de Carvalhoe: O conceito de espécie e das suas subdivisões. (The concept of the species and of its subdivisions.) Rev. Agron., Lisboa 27, 214—230 (1939).

Wanscher, J. H.: Enhedsbegreber sideordnede meg og underordnede det almindelige Artsbegreb. (Concepts of units of systematic classification, that are parallel with and subordinate to the general concept of the species. Naturhistorisk Tidende, København 7, 1—4 (1943).
Watson, E. V.: The dynamic approach to plant structure and its relation to modern taxonomic botany. Biol. Rev. 18, 65—77 (1943).
Wellensiek, S. J.: Het streven naar geluk. (The search for happiness.) H. Veeman and Zonen, Wageningen, Pp. 14 (1946).
Wettstein, F. v. u. J. Straub: Experimentelle Untersuchungen zum Artbildungsproblem. III. Weitere

Beobachtungen an polyploiden Bryum-Sippen. Z. Vererbungsl. **80**, 271—280 (1942).
White, M. J. D.: Animal cytology and evolution. Cambridge University Press, London, Pp. VIII+375, 121 figs, 18 tables (1945).
Widmark, E. M. P.: The theory of evolution from a biochemical point of view. Hereditas, Lund **31**, 383 bis 390 (1945).
Willis, J. C.: How plants have found their homes, I, II. J. R. Hort. Soc. **64**, 259—271, 299—313 (1939).
— The course of evolution by differentiation or divergent mutation rather than by selection. University Press, Cambridge (1940).
Wright, S.: On the probability of fixation of reciprocal translocations. Amer. Nat. **75**, 513—522 (1941).
— The material Basis of Evolution. Scient. Monthly, August, Pp. 165—170 (1941).
— Statistical genetics and evolution. Bull. Amer. math. Soc. **48**, 223—246 (1942).
Zimmermann, W.: Über phylogenetische Methoden. Biologe **10**, 47—49 (1941).
Zirkle, C.: The early history of the idea of the inheritance of acquired characters and of pangenesis. Trans. Amer. Philos. Soc. (N. S.) **35**, Pt. II, 91—151 (1946).
Zujtin, A. I. u. M. T. Pavlovec. Mutation in several populations of Drosophila melanogaster under natural conditions. C. R. (Doklady) Acad. Sci. URSS., n. s. **29**, 483—486 (1940).
Žukovskij, P. M.: (Darwinism in a crooked mirror.) Selekcija i Semenovodstvo (Breeding and Seed Growing), Nr. 1, 71—79 (1946). [Russisch.]

c) Allgemeine Pflanzenzüchtung.

Aamodt, O. S.: Breeding crops for production at high fertility levels. J. Amer. Soc. Agron. **35**, 173—174 (1943).
Åberg, E.: Ökade krav på sortkännedom och sortbeskrivning. (Increased need for a knowledge of varieties and for varietal descriptions.) Lantmannen **24**, 503 bis 504 (1940).
— Sortkännedom och sortbeskrivning. (Knowledge of varieties and varietal description.) Lantmannen **24**, 581 (1940).
— Studies of crop production in the United States 1940 to 1943. Publ. Inst. Plant Husb. Roy. Agric. Coll. Sweden, Nr. 1, 87 (1944).
— Forskningstendenser inom Förenta Staternas växtodling. (Research trends in plant breeding in the United States.) Årsb. Jordbruksforskning, Stockholm, Pp. 97 bis 111 (1946).
Akemine, M.: (General reference books and periodicals on genetics and plant breeding.) J. Sapporo Soc. Agric. and For. **32**, 37—61 (1941).
Åkerberg, E.: Aktuella resultat och förädlingsuppgifter för Västernorrlandsfilialens arbetsområde. (Present results and breeding problems for the Västernorrland Branch Stations sphere of work.) Sverig. Utsädesfören. Tidskr. **54**, 91—103 (1944).
Åkerman, Å.: Några synpunkter rörande filialernas betydelse i Utsädesföreningens verksamhet. (Some aspects relating to the importance of the affiliated stations in the work of the Seed Association.) Sverig. Utsädesfören. Tidskr. **49**, 265—270 (1939).
— Växtförädling och folkförsörjning. Föredrag vid Sveriges Utsädesförenings extra möte i Stockholm den 19/3. 1940. (Plant breeding and national supplies — Lecture at the Swedish Seed Association's extraordinary meeting in Stockholm on 19/3. 1940.) Sverig. Utsädesfören. Tidskr. **50**, 57—65 (1940).
— Årsberättelse över Sveriges Utsädesförenings verksamhet under år 1939. (Annual report on the work of the Swedish Seed Association during the year 1939.) Sverig. Utsädesfören. Tidskr. **50**, 153—207 (1940).
Åkerman, Å.: Årsberättelse över Sveriges Utsädesförenings verksamhet under år 1940. (Annual report on the work of the Swedish Seed Association during the year 1940.) Sverig. Utsädesfören. Tidskr. **51**, 125—173 (1941).
— Förädling av åkerbruksväxter för sandjord. (The breeding of agricultural crop plants for sandy soil.) Sverig. Utsädesfören. Tidskr. **51**, 184—188 (1941).
— Årsberättelse över Sveriges Utsädesförenings verksamhet under år 1941. (Annual report on the work of the Swedish Seed Association during the year 1941.) Sverig. Utsädesfören. Tidskr. **52**, 161—211 (1942).
— Årsberättelse över Sveriges Utsädesförenings verksamhet under år 1942. (Annual report of the work of the Swedish Seed Union during the year 1942.) Sverig. Utsädesfören Tidskr. **53**, 117—169 (1943).
— Årsberättelse över Sveriges Utsädesförenings verksamhet under år 1943. (Annual report on the work of the Swedish Seed Association during 1943.) Sverig. Utsädesfören. Tidskr. **54**, 115—172 (1944).
— Det mest aktuella på årets försöksfält i Svalöf. (The most important information concerning this year's experiment plots at Svalöf.) Sverig. Utsädesfören. Tidskr. **54**, 221—227 (1944).
— Några viktigare produktionsmedels utnyttjande efter kriget. (The exploitation of some of the more important natural resources after the war.) Sverig. Utsädesfören. Tidskr. **55**, 14—19 (1945).
— Sveriges utsädesförening inför det nya verksamhetsåret. (The Swedish Seed Association confronts the work of the new year.) Sverig. Utsädesfören. Tidskr. **56**, 2 (1946).
— Arbetsplan för Sveriges Utsädesförening för år 1946. (Plan of work for the Swedish Seed Association for the year 1946.) Sverig. Utsädesfören. Tidskr. **56**, 111—147 (1946).
— Donationer till Sveriges Utsädesförening. (Donations to the Swedish Seed Association.) Sverig. Utsädesfören Tidskr. **56**, 573—574 (1946).
— Sveriges Utsädesföreningens verksamhet och framtida uppgifter. (The work and future tasks of the Swedish Seed Association.) Sverig. Utsädesfören. Tidskr. **56**, 575—581 (1946).
— Brödsädesodlingens fredsplanering och därav föranledda krav på växtförädlingen. (The planning of the growing of bread cereals in peacetime and the consequent requirements in plant breeding.) Årsb. Jordbruksforskning, Stockholm, Pp. 64—68 (1946).
Almeida, J. M. de: Bases fisiológicas do melhoramento das plantas. (Physiological basis of plant breeding.) Rev. Agron., Lisboa **29**, 433—449 (1941).
Andersson, G.: Sveriges Utsädesförenings årsmöte och 60-årsfest den 12—13 juli 1946. (Annual Meeting and 60th Jubilee of the Swedish Seed Association, 12.—13. July, 1946.) Sverig. Utsädesfören Tidskr. **56**, 543 bis 572 (1946).
— Ytterligare donationer till Utsädesföreningens nya laboratoriebyggnad. (Further donations to the new laboratory building of the Swedish Seed Association.) Sverig. Utsädesfören Tidskr. **57**, 71 (1947).
Anonymus: A review of the plant breeding in India with suggestions for the future. (Proceedings of the Second Meeting of the Crops and Soils Wing of the Board of Agriculture and Animal Husbandry in India, held at Lahore from the 6th to 9th December, 1937 with appendices, Delhi **59**—70, 240—251.
— Comunicari si referate în anul 1938. (Institutul de Cercetari Agronomice al României.) (Communications and abstracts for 1938. [Rumanian Institute for Agronomic Research.]). Anal. Inst. Cerc. Agron. Romăn. **10**, 664—682, 1938 (1939).
— La Station de Recherches pour l'Amélioration des Plantes, de Gembloux (Belgique) 1913—1939. Gembloux, Pp. 14 (1939).

Anonymus: Försöksresultat från Finska Mosskulturföreningens försöksstationer år 1938. (Experimental results from the Experimental Stations of the Finnish Association for the Reclamation of Bogland, 1938.) Finska Mossk Fören. Årsb. **43**, 44—93 (1939).
— Some prominent Danish varieties and strains of agricultural and horticultural plants, 1937. Roy. Agric. Soc. Denmark, Copenhagen, Pp. 7 + Suppl. (1939).
— Comunicari si referate în anul 1939. (Institudul de Cercetări Agronomice al României.) (Communications and abstracts for 1939. [Rumanian Institute for Agronomic Research.]). Anal. Inst. Cerc. Agron. Român. **11**, 443—487, 1939 (1940).
— Overheidszorg voor den landbouw en de binnenvisscherij in 1939. (Government provision for agriculture and inland fisheries in 1939.) Landbouw **16**, 557—603 (1940).
— The Institute of Plant Industry, Indore (India). Chronica Botanica **6**, 16—17 (1940).
— La protection juridique des variétés des plantes. 15th Actes Inst. Int. Agric., Rome, Pp. 267—288 (1940).
— Zestiende beschrijvende rassenlijst, opgemaakt ingevolge de beschikking van den Minister van Binnenlandsche Zaken en Landbouw van 21 October 1924, gewijzigd 28 Maart 1933. (16th descriptive list of varieties, drawn up in pursuance of the Order of the Minister of Internal Affairs and Agriculture of the 21st. October, 1924 amended 28th March, 1933.) Inst. Plantenveredeling, Wageningen, p. 245 (1940).
— Crop varieties accessioned since January 3, 1939. Cereal Div., Cent. Exp. Fm. Ottawa, Ont. Pp. 4 (1940). [Mimeographed.]
— Food crop varieties available for distribution. E. Afr. Agric. Res. Sta., Amani, January, p. 8 (1941).
— Zeventiende beschrijvende rassenlijst voor landbouwgewassen. (17th descriptive variety list for agricultural plants.) Inst. Plantenveredeling, Wageningen, Pp. 271 (1941).
— Mobilizing seed power (from „Science in War", a Penguin Special, Harmondsworth/Penguin books, 1940, p. 85 seqq.). Chronica Botanica **6**, 411—413 (1941).
— Weibullsholm 70 år. En minnesskrift. (Seventy years at Weibullsholm. A memorial volume.) W. Weibull. A.-B., Landskrona, Pp. 71 illus. (1941).
— Year book of the Carnegie Institution of Washington 1941—1942. Nr. 41, 309 (1942).
— Testing Lend-Lease seed shipments. USDA going to extreme lengths to see that only quality seed, properly labelled, is sent to Allies. Seed World **52**, Nr. 12, 12—13 (1942).
— Lend-Lease seeds mainly true to variety. Sth. Seedsman **5**, Nr. 11, 30 (1942).
— Achttiende beschrijvende rassenlijst voor landbouwgewassen. (18th descriptive variety list for agricultural plants.) Inst. Platenveredeling, Wageningen, Pp. 264 (1942).
— Agricultural research in New Hampshire. Bull. N. H. Agric. Exp. Sta. Nr. 345, 69 (1942).
— New economic crops. Indian Fmg **3**, 277—280 (1942).
— Field plantings verify variety names. New York AES tests 968 stocks to check on performance as well as on varietal purity. Seed World **·51**, 20—21 (1942).
— Guia dos ensaios de campo. (Guide to the field tests.) Estac. Melhor. Plantas, Elvas, Pp. 20 (1942).
— Haiti: Agricultural Research and development. Chronica Botanica **7**, 179—180 (1942).
— Odlingsvärdet i olika landsdelar hos våra viktigaste sorter och stammar av vårsådda åkerbruksväxter-korta översikter. (Suitability in different parts of the country of our most important varieties and strains of spring sown agricultural crops-brief surveys.) Sverig. Utsädesfören. Tidskr. **52**, 4—44 (1942).
— Horticultural problems on farms of Mississippi. Miss. Fm. Res. **5**, Nr. 9, 4—5 (1942).

Anonymus: Science serving agriculture. Bienn. Rep. Okla. Agric. Exp. Sta., Pp. 92 (1942).
— List of food crop varieties available for distribution. E. Afr. Agric. Res. Sta., Amani, Pp. 12 (1943).
— Conferencias sobre Genética Vegetal Aplicada, pronunciadas por el Profesor Dr. W. Rudorf. (Lectures on applied plant genetics delivered by Dr. W. Rudorf.) Bol. Inst. Invest. Agron., Madr., Nr. 8, 149—237 (1943).
— Seasonal notes by the Department of Field Crops. Pr. Bull. Univ. Alberta **28**, 1—8 (1943).
— (Six years' work of the State Breeding Stations.) Socialističeskoe Seljskoe Hozjajstvo (Socialistic Agriculture) Moskva, Nr. 8, 42—50 (1943). [Russisch.]
— Year book of the Carnegie Institution of Washington 1942—1943. Nr. 42, 208 (1943).
— Bigger harvests. Soviet War News., Nr. 517, 3 (1943).
— Crops and plant breeding. J. R. Agric. Soc. **104**, 8—17 (1943).
— Negentiende beschrijvende rassenlijst voor landbouwgewassen. (19th descriptive variety list for agricultural plants). Inst. Plantenveredeling, Wageningen, Pp. 148 (1943).
— (Principal varieties of farm and herbage crops for the eastern districts.) Sovhoznoe Proizvodstvo (State Farming), Nr. 10/11, 49 (1943). [Russisch.]
— Plant breeding in relation to food production. Curr. Sci. **12**, 239—243 (1943).
— Specialist and research work of the Department of Agriculture, Tanganyika Territory 1943. Pp. 35 (1943). [Mimeographed.]
— Försöksdag på Ultuna. (Research day at Ultuna.) Lantmannen **28**, 689—690 (1944).
— Ergänzung der Sortenliste deutscher Kulturpflanzen. Forschungsdienst **17**, 52 (1944).
— Förädlingsarbetets organisation och kontrollen över utsädesvarorna. Sortäkthetskontrollen utövas av statens Centrala Frökontrollanstalt. (The organization of breeding operations and control over seed for sale. Control of authenticity is exercised by the State Central Seed Control Institute.) Weibulls III. Årsb. **39**, 4—5 (1944).
— Improvement of economic crop plants. Nature, Lond. **153**, 499 (1944).
— Application of genetics to plant and animal breeding. Nature, Lond. **153**, 780—783 (1944).
— Year book of the Carnegie Institution of Washington 1943—1944. Nr. 43, 206 (1944).
— Studiekring voor plantenveredeling. (Study circle for plant breeding.) Landbouw Tijdschr. Wageningen **56**, 211—214 (1944).
— (A short account of the research work done during 1944 at the Institute of Grain Husbandry S.E., USSR. of the Red Labour Banner.) Bjulletenj Instituta Zernovogo Hozjajstvà Jugo-Vostoka SSSR. (Bull. Inst. Grain Husb. S.E., USSR.) Saratov, Nr. 4, 3—18 (1944). [Russisch.]
— Ur „Arbetsplan för Weibullsholms Växtförädlingsanstalt for år 1944." (From the „Programme for Weibullsholm Plant Breeding Institute for 1944"). Weibulls III Årsb. **39**, 35—41 (1944).
— Much delay in search for better plants. N. S. Dak. Hort. **17**, 157 (1944).
— Twintigste beschrijvende rassenlijst voor landbouwgewassen. (20th. descriptive variety list for agricultural plants.) Inst. Plantenveredeling, Wageningen, Pp. 200 (1944).
— Varieties of farm crops for Montana 1945. Circ. Mont. Agric. Exp. Sta., Nr. 182, 35 (1945).
— At the USSR. Lenin Agricultural Academy (Moscow). Agriculture, Moscow, Nr. 12, 1—8 (1945). [Mimeographed.]
— Plant industry. 2. At the Lgov Plantbreeding Station. Agriculture, Moscow, Nr. 5, 5 (1945). [Mimeographed.]

Anonymus: Förädlingsarbetets organisation och kontrollen över utsädesvaror. (The organization of the breeding work and the control of seed for sale.) Weibulls Ill. Årsb. **40**, 4—5 (1945).
— Notes and news. Biology and genetics. Sci. and Cult. **10**, 539 (1945).
— New varieties of agricultural plants. Agriculture, Moscow, Nr. 3, 2—3 (1945).
— Plant breeding. Agriculture, Moscow, Nr. 6, 3—6 (1945). [Mimeographed.]
— Horticulture and vegetable growing. Agriculture, Moscow, Nr. 9, 7—10 (1945). [Mimeographed.]
— The John Innes Horticultural Institution. Nature, Lond. **156**, 586—588 (1945).
— Catalogue des variétés de blés, avoines, orges, maïs, pommes de terre, topinambours, betteraves, fourragerès, soyas, cultivées en France. Minist. Agric., Fr., Pp. 30 (1946).
— Key to varieties of field and vegetable crops under trial, observation and propagation. Nat. Inst. Agric. Bot., Cambridge, Pp. 31 (1946).
— Liste des variétés des espèces agricoles susceptibles d'être soumises au contrôle du S.G.C. Serv. Gen. Contr. Semences Plants Agric. Hort., Pp. 60 (1946).
— Prof. F. R. Immer. Nature, Lond. **157**, 362 (1946).
— Professor H. J. Muller. Mon. Sci. News., Nr. 1, 1—2 (1946).
— Dr. C. D. Darlington, F. R. S. Mon. Sci. News, Nr. 12, 1 (1946).
— Een en twintigste beschrijvende rassenlijst voor landbouwgewassen. (21st descriptive variety list for agricultural plants.) Inst. Plantenveredeling, Wageningen, Pp. 221 (1946).
— Försöksverksamheten på jordbrukets område. (Research activities in the sphere of agriculture.) Lantmannen **31**, 601—602 (1947).
— Harry Weibull. K. LantbrAkad. Tidskr. **86**, 36 (1947).
— Stalin Prize awards for 1946. Soviet News, Nr. 1727, 1 (1947).
— Investigation of horticultural and other problems. Qd Agric. J. **64**, 131—132 (1947).
Arnold, H. C.: Annual report of experiments, season 1941—1942. Agricultural Experiment Station Salisbury. Rhod. Agric. J. **40**, 242—252 (1943).
— Agricultural Experiment Station, Salisbury. Annual report of experiments, season 1942—1943. Rhod. Agric. J. **41**, 208—225 (1944).
— Agricultural Experiment Station, Salisbury. Annual report of experiments, season 1943—1944. Rhod. Agric. **42**, 129—142 (1945).
— Agricultural Experiment Station, Salisbury. Annual report of experiments: season 1944—1945. Rhod. Agric. J. **43**, 344—354 (1946).
Arny, A. C., et al: Varietal standardization and registration. J. Amer. Soc. Agron. **34**, 1154—1155 (1942).
Arzuga, J. G.: Homenaja al Ing. Agr. Enrique Klein en el 25° aniversario de la fundación de su ,,Criadero Argentino de Plantas agrícolas". (Homage to Enrique Klein on the 25th anniversary of the foundation of his ,,Argentine Centre for the Production of Agricultural Plants".) Rev. Argent. Agron. **11**, 328—330 (1944).
Ashton, T.: The use of heterosis in the production of agricultural and horticultural crops. Cambridge, Pp. 30 (1946).
Asmous, V. C.: Freedom of science in Soviet Union. Science **103**, 281—282 (1946).
Avanzi, E.: L'impiego delle razze elette in rapporto al progresso delle coltivazioni erbacee. (The use of selected races in relation to progress in herbaceous crop plants.) Ann. Fac. Agrar. Univ. Pisa **5**, (N. S.) 749—775 (1942).
Avdonin, N.: (Some results of scientific work during a year of war [1941].) Socialističeskoe Seljskoe Hozjajstvo (Socialistic Agriculture) Moscow, Nr. 6/7, 45—52 (1942). [Russisch.]

B.: Öländsk växtförädling. (Plant improvement in Öland.) Lantmannen **25**, 443—444 (1941).
Babcock, E. B.: Recent progress in plant breeding. Sci. Mon., N. Y. **393**—400 (1939).
Backe, H.: Carl Sigismund von Treskow-Friedrichsfelde-Stiftung. Z. Pflanzenz. **24**, 134—135 (1941).
Ball, R. S.: Some South African investigations on fodder crops, field crops and animal husbandry. E. Afr. Agric. J. **5**, 380—382 (1940).
Ballantyne, J. P. S.: Results of experiments 1936—1940. Experimental Station, Kapuskasing, Ontario. Dep. Agric., Dom. Canad., Pp. 43 (1942).
Bayles, B. B.: Improved grains bring higher yields. Agric. Situat. **28**, 13—15 (1944).
Beaumont, J. H.: Report of the Havaii Agricultural Experiment Station, Pp. 94, 1940 (1941).
Beckman, I.: Trabalhos fitotécnicos da Estaçao Experimental de Bagé. (Plant breeding work of the Bagé Experimental Station.) Rev. Fac. Agron. Univ. Montevideo, Nr. 33, 9—48 (1943).
Bell, G. D. H.: Crops and plant breeding. J. R. Agric. Soc. **100**, 5—36 (1939).
— Crops and plant breeding. J. R. Agric. Soc. **101**, 1—33 (1940).
— Crops and plant breeding. J. R. Agric. Soc. **103**, 1—11 (1942).
— Crops and plant breeding. J. R. Agric. Soc. **105**, 1—12 (1944).
— Crops and plant breeding. J. R. Agric. Soc. **106**, 1—12 (1945).
— Crops and plant breeding. J. R. Agric. Soc. **107**, 1—15 (1946).
Benediktov, I. A.: (Reorganization of the work of the scientific research institutes in connexion with agriculture.) Selekcija i Semenovodstvo (Breeding and Seed Growing), Nr. 8/9, 6—8 (1940). [Russisch.]
Berkner, F.: Züchtung und Auslese von Futterpflanzen. Forsch. dienst, Sonderh. **16**, 425 (1942).
Bettoney, S.: Colchicine to aid the plant breeder. Missouri Bot. Gard. Bull. **28**, 119—120 (1940).
Björkman, T. V. E.: Årsberättelse avgiven den 28 januari 1941.) (Annual report 28th January, 1941.) K. LantbrAkad. Tidskr. **80**, 11—29 (1941).
Black, W., C. D. Darlington, P. S. Hudson u. T. J. Jenkin: Plant breeding. British Intelligence Objectives Sub-Committee, London: Final Rep., Nr. 502. Item, Nr. 22. Trip., Nr. 966, Pp. 56 (1946.) [Mimeographed.]
Blizzard, W. L. u. L. E. Hawkins: Science serving agriculture. Part I. Bienn. Rep. Okla. Agric. Exp. Sta. 1942—1944, Pp. 95 (1944).
Blohm: Professor Roemer 60 Jahre. Forscher und Lehrer im Dienste der Landwirtschaft. Forschungsdienst **16**, 209—210 (1943).
Boekholt, K.: Das Standortproblem in der Pflanzenzucht. Forschungsdienst **17**, 449—454 (1944).
Boerger, A.: Aplicación creciente de la química orgánica en la genética vegetal. (The increasing application of organic chemistry in plant genetics.) Arch. Fitotécn. Uruguay **3**, 319—329 (1940—1941).
Boeuf, F.: The problem of wheat production in France and in French North Africa. Int. Rev. Agric. **33**, 221T—240T (1942).
Bondar, G.: Rumos de lavoura no estado de Espirito Santo e culturas tropicais na Bahia. Bol. Inst. Cent. Fom. Econ. Bahia, Nr. 10, 41 (1942).
Bonvicini, M.: Miglioramento genetico delle pianta agrarie. (Genetic improvement of agricultural plants.) Unione Tipografico-Editrice Torinese, Torin. Pp. VIII + 306, 172 figs (1942).
— Problemi organizzativi della genetica agraria. (Problems of organization in plant breeding.) Genetica Agraria, Roma **1**, 112—115 (1946).

Boonstra, A. E. H. R.: Physiologisch onderzoek en plantenveredeling. (Physiological research and plant breeding.) Landbouwk. Tijdschr. Wageningen **54**, 437 bis 456 (1942).

Bormans, P.: La génétique et l'amélioration des plantes. Vie Agric. Rur. **29**, 284—292 (1939).

Boyce, S. W.: Estimation of genes in inheritance of quantitative characters. Nature, Lond. **157**, 657 (1946).

Boyes, J. W.: A new method of plant breeding. Pr. Bull. Dep. Ext. Univ. Alberta **26**, 5 (1941).

Boza Barducci, T.: Memoria anual de 1941 de la Estación Experimental Agrícola de la Molina. (Annual report of the Estación Experiment Agricola de la Molina for 1941.) Lima, Peru, Pp. 112 (1941).

— Plan para centralizar la orientación técnica y el control de los trabajos de mejoramiento vegetal, por selección genética, que realizan las dependencias de la Dirección de Agricultura del Ministerio del Ramo, y para organizar semilleros oficiales u. oficializados, en los principales valles de la costa y sierra del pais, a fin de abastecer con semilla mejorada a los agricultores. (Plan for centralizing the technical orientation and control of the work of plant breeding by genetical selection which is carried out by the dependencies of the Office of the Director of Agriculture of the Ministry and for organizing official or semi-official seed supplies in the main valleys of the coastal and highland region of the country with the object of supplying the cultivators with improved seed.) An. 2a Conv. Agron. Regional, Lima, Peru, Pp. 24 (1944).

— Los progresos alcanzados por el Perú en genética vegetal. (The progress achieved by Peru in plant genetics.) Bol. Estac. Exp. Agric. La Molina, Nr. 27, 34 (1945).

Brandl, M.: Das Werden der Pflanzenzüchtung in der Ostmark. Z. Pflanzenz. **24**, 395—397 (1941).

Briggs, F. N.: The use of the backcross in plant breeding. Proc. 7th Int. Genet. Congr. Edinburgh 23.—30. August, Pp. 81—82, 1939 (1941). (Abstr.)

— Plant breeding and seed control stations in the British Isles. Seed World **48**, Nr. 9, 14 (1940).

Broekema: Over veredeling van landbouwgewassen. (On the improvement of agricultural crops.) Erfelijkheid in Praktijk, Leiden **4**, 182—183 (1939).

Burbank, L.: An architect of nature. Watts and Co., London ls. Od. (The Thinker's Library, Nr. 76) (1939).

— Partner of nature. D. Appleton-Century Company, Incorporated, New York and London (1939).

Burns, W.: The progress of agricultural science in India during the past twentyfive years. Misc. Bull. Imp. Coun. Agric. Res. Delhi, Nr. 26, 49 (1939).

— Some ideas and opportunities for plant geneticists in India. Indian J. Genet. Pl. Breed. **1**, 1—3 (1941).

— u. B. P. Pal: The relationship of agricultural science with taxonomy and cytology. Roy. Bot. Gdn., Calcutta **150**, 23—1—23—6 (1942).

Burr, W. W.: Fifty-fifth Annual Report of the Agricultural Experiment Station of the University of Nebraska College of Agriculture 1942, Pp. 96 (1942).

— Nebraska agriculture. 58th Rep. Neb. Agric. Exp. Sta., Pp. 124 (1944).

Bustarret, J.: Variétés et variations. Ann. Agron., Paris (N. S.) **14**, 336—362 (1944).

Call, L. E.: Tenth Biennial Report of the Director of the Kansas Agricultural Experiment Station for the Biennium July 1, 1938 to June 30, 1940, Pp. 159 (1940).

Câmara, A.: A produtividade no quadro da genética actual. (Yield in the modern genetical picture.) Rev. Agron., Lisboa **29**, 28—55 (1941).

— u. J. Azevedo: Centros produtores de sementes. (Seed production centres.) Rev. Agron., Lisboa **29**, 345—352 (1941).

Camara, A. u. R. Castro: Protecção das zonas de interêsse genético. (Protection of regions of genetic interest. Rev. Agron., Lisboa **29**, 524—529 (1941).

— u. M. Lorena: A influência do meio no „ennobrecimento" das variedades culturais. (Influence of environment on the „nobilization" of crop varieties.) Rev. Agron., Lisboa **29**, 98—99 (1941).

—, — A influência do meio no „ennobrecimento" das variedades culturais. (Influence of enfironment on the „nobilization" of crop varieties.) Rev. Agron., Lisboa **29**, 251—258 (1941).

Castle, W. E.: Size inheritance. Amer. Nat. **75**, 488 bis 498 (1941).

Černogolovin, V.: (At the Far Eastern Institute of Agriculture and Animal Husbandry.) Socialističeskoe Seljskoe Hozjajstvo (Socialist Agriculture) Moscow, Nr. 11/12, 59—60 (1945). [Russisch.]

Charles, D. R. u. H. H. Smith: Distinguishing between two types of gene action in quantitative inheritance. Genetics **24**, 34—48 (1939).

Chavarriaga, Misas, E.: El Departamento de Genética del Ministerio de Agricultura y Cria de los Estados Unidos de Venezuela. (The Genetics Department of the Ministry of Agriculture and Stock of the United States of Venezuela.) Rev. Fac. Nac. Agron., Colombia **5**, 239—243 (1944).

Chew, A. P.: The United States Department of Agriculture. Its structure and functions. Misc. Publ. U. S. Dep. Agric., Nr. 88, 242 (1940).

Cicin, N. V. et al. (Editors): (The regionalization of approved varieties of cereal plants, sunflower, soya bean, lucerne and clover.) Gosudarstvennaja Komissija po Sortoispytaniju Zernovyh Kuljtur pri N.K.Z. SSSR. (State Commission for the Organization of Varietal Tests of Cereal Crops attached to the Peoples' Commissariat of Agriculture of the USSR.) Seljhozgiz, Moskva, Pp. 200 (1939). [Russisch.]

— (The results of the state varietal tests for 1940. Part I. Red clover, lucerne, Sudan grass, Setaria italica, vetches.) Gosudarstvennaja Komissija po Sortoispytaniju Zernovyh Kuljtur pri N.K.Z. SSSR. (State Commission for the Organization of Varietal Tests of Cereal Crops attached to the Peoples Commissariat of Agriculture of the USSR.) Seljhozgiz, Moskva, Pp. 384 (1942). [Russisch.]

—, P. N. Konstantinov, V. E. Pisarev, V. J. Jurjev u. I. V. Jakuškin (Editors): (The results of the state varietal tests for 1939. Part IV. Millet, buckwheat, maize, sorghum and sunflower.) Gosudarstvennaja Komissija po Sortoispytaniju Zernovyh Kuljtur pri N.K.Z. SSSR. (State Commission for the Organization of Varietal Tests of Cereal Crops attached to the Peoples' Commissariat of Agriculture of the USSR.) Seljhozgiz, Moskva, Part IV, Pp. 200 (1941). [Russisch.]

—, P. I. Lisicyn u. V. J. Jurjev: (The results of the state varietal tests for 1939. Part. I: Winter wheat, rye, barley; Part. II: Spring wheat, oats, spring barley, rice; Part. III: Peas, lentils, Phaseolus, soya bean, Lathyrus sativus, Cicer arietinum.) Gosudarstvennaja Komissija po Sortoispytaniju Zernovyh Kuljtur pri N.K.Z. SSSR. (State Commission for the Organization of Varietal Tests of Cereal Crops attached to the Peoples' Commissariat of Agriculture of the USSR.). Moscow, Pt. I. Pp. 260; Pt. II. Pp. 400; Pt. III. Pp. 134 (1941). [Russisch.]

— u. P. E. Marinič (Editors): (Varieties of field crops.) Moskva, Pp. 344, 158 figs. (1944). [Russisch.]

Clark, R. T.: Value of agricultural research in wartime. 48th and 49th Rep. Mont. Agric. Exp. Sta., Pp. 63 (1940—1942).

Coïc, Y.: Hérédité des caractères biochimiques chez les végétaux. Ann. Agron., Paris **12**, 361—390 (1942).

Conner, A. B.: 52nd Annual Report of the Texas Agricultural Experiment Station. Pp. 304 (1939).
— Fifty-fourth annual report of the Texas Agricultural Experiment Station, 1941. Pp. 202 (1941).
— Agricultural research serves farmers, ranchers and industry. 55th and 56th Rep. Tex. Agric. Exp. Sta. 1942—1943. Pp. 45 (1943).
Coolhaas, C.: Een overzicht van de werkzaamheden van het Proefstation M. O. J. in 1938. (A survey of the activities of the Central and East Java Experiment Station in 1938.) Bergcultures 13, 556—566 (1939).
— Een overzicht van de werkzaamheden van het Proefstation Midden- en Oost-Java in 1939. (A survey of the activities of the Central and East Java Experiment Station in 1939.) Bergcultures 14, 454—461 (1940).
— Veredeling van Nederlandsch Indische cultuurgewassen. (Improvement of Netherlands Indies cultivated crops.) Bergcultures 14, 514—524 (1940).
— Een overzicht van de werkzaamheden van het Proefstation Midden- en Oost-Java in 1940. (A survey of the activities of the Central and East Java Experimental Station in 1940.) Bergcultures 15, 676—685 (1941).
Cooper, H. P.: Fifty-fith Annual Report of the South Carolina Experiment Station of Clemson Agricultural College for the year ended June 30, 1942, Pp. 186 (1943).
Cooper, T. P.: Fifty-sixth Annual Report of the Agricultural Experiment Station of the University of Kentucky, 1943. Pp. 64 (1943).
Corbett, R. B.: Science at work for the farmer. 54th. Ann. Rep. Md. Agric. Exp. Sta., Pp. 70 (1940—1941).
— Research - a war effort. 55th. Rep. Md. Agric. Exp. Sta. 1941—1942, Pp. 360 (1942).
— Science serves in war. 56th. Ann. Rep. Md. Agric. Exp. Sta. 1942—1943, Pp. 47 (1944).
Corns, J. B.: Results of 1940 field trials in central and nothern Illinois. Trans. Ill. Hort. Soc. 74, 301—310 (1940).
Coster, C.: Algemeene beschouwingen over selectie van overjarige gewassen. (General considerations on selection of perennial crops.) Bergcultures 14, 1328—1335 (1940).
— Het werk van het Proefstation West-Java in 1940. (The work of the West Java Experiment Station in 1940.) Bergcultures 15, 1124—1133 (1941).
— The work of the West-Java Research Institute, Buitenzorg. 1938—1941. Emp. J. Exp. Agric. 10, 22—30 (1942).
Crane, M. B.: Seed and food in war-time. J. R. Hort. Soc. 65, 321—326 (1940).
Crawford, M. D. C.: Art of the ancients. A panorama of cotton and other textiles from earliest days. J. N. Y. Bot. Gdn. 43, 285—293 (1942).
Crépin, C.: Rapport sommaire sur les travaux poursuivis en 1941 par les stations d'amélioration des plantes. Ann. Agron., Paris 12, 633—669 (1942).
Cross, W. E.: Memoria annual del año 1939. (Annual report for 1939.) Rev. Industr. Agric. Tucumán 30, 107 (1940).
— Memoria anual den ano 1940. (Annual report for 1940.) Rev. Industr. Agric. Tucumán 31, 5—106 (1941).
— Memoria anual del año 1944. (Annual report for 1941.) Rev. Industr. Agric. Tucumán 32, 5—114 (1942).
— Notas sobre el progreso de la Agricultura y las industrias agropecuarias de Tucumán durante los ultimos sesenta años. (Notes on the progress of agriculture and the rural industries in Tucumán during the last 60 years.) Bol. Estac. Agric. Tucumán, 36, 75 (1942).
— Memoria anual del año 1943. (Annual report for the year 1943.) Rev. Industr. Agric. Tucumán 34, 111—190 (1944).
— Memoria anual del año 1941. (Annual Report for the year 1944.) Rev. Industr. Agric. Tucumán 36, 5—85 (1946).

Dahlberg, G.: Nature, nurture and probability. Nature, Lond. 156, 539 (1945).
Darlington, C. D.: Production genetics in Sweden. Nature, Lond. 151, 183—185 (1943).
Das, C. M.: What's doing in All-India-United Provinces. Indian Fmg. 3, 94—96 (1942).
Davidson, W. A.: Questions and answers concerning the Federal Seed Act. Seed World 47, Nr. 7, 12—13 (1940).
Deb, N.: What's doing in All-India-Bengal. Indian Fmg. 3, 91—93 (1942).
Dellazoppa, J. G. u. M. O. Bentancur: Empleo de fitohormonas en la multiplicación vegetativa por estacas, su aplicación en genética vegetal. (Use of plant hormones in vegetative propagation by cuttings and its application in plant genetics.) Arch. Fitotécn. Uruguay 3, 262—275 (1940—1941).
Dewan, P. M.: Report of the Minister of Agriculture, Province of Ontario, for the year ending March 31st, 1943, Pp. 106 (1943).
Dias, B. P.: Método de seleção das „raças ecológicas". (Method of selection of ecological race.) Rev. Agron., Brasil. 5, 390—392 (1941).
Dobzhansky, T.: What is heredity? Science 100, 406 (1944).
Domingo, W. R. u. A. C. Schuffelen: Landbouwscheikundig onderzoek en plantenveredeling. (Research in agricultural chemistry in relation to plant breeding.) Landbouwk. Tijdschr. Wageningen 54, 457—474 (1942).
Dorasami, L. S.: Brief sketch of work done in the botanical section. Mysore Agric. J. 23, 61—62 (1944 bis 1945).
Dorman, C.: Highlights of the work of the Mississippi Experiment Station. 54th Ann. Rep. Miss. Agric. Exp. Sta., Pp. 59 (1941).
— Highlights of the work of the Mississippi Experiment Station. 56th Rep. Dir. Miss. Agric. Exp. Sta., Pp. 54 (1943).
Dorst, J. C.: Plantenveredeling. (Plant breeding.) Landbouwk. Tijdschr. Wageningen 54, 500—506 (1942).
Druce, G.: Some Czechoslovak contributions to genetics (1866—1938). Nature, Lond. 151, 495—496 (1943).
Dubinin, N. P.: Work of Soviet biologists: theoretical genetics. Science 105, 109—112 (1947).

Eastman, M. G.: Agricultural research in New Hampshire. Rep. Dir. N. H. Agric. Exp. Sta. 1939, Bull., Nr. 319, 46 (1940).
Ejhfeljd, I. G.: (The establishment of food-production in the far north.) Proc. Lenin Acad. Agric. Sci. USSR., Nr. 1, 3—8 (1944). [Russisch.]
Eikeland, H. J.: Planteforedlingsarbeidet i jordbruket. (The work of plant breeding in agriculture.) Tidsskr. Norske Landbr. 46, 71—82 (1939).
Engledow, F. L.: The place of plant physiology and of plant-breeding in the advancement of British agriculture Emp. J. Exp. Agric. 7, 145—149 (1939).
— Science and the land. Chem. Ind., Lond. 61, 239—245. (1942).

F....., O. H.: Genetics and plant breeding. Chronica Botanica 6, 157—158 (1940).
Franck, W. J.: Bescherming van kweekersrechten. Opbrengstvermeerdering door het gebruik van gekeurd zaaizaad. (The protection of growers' rights. Increasing the yield by the use of inspected seed.) Landbouwk. Tijdschr., Wageningen 54, 125—139 (1942).
Frandsen, H. N.: Planteforaedlingsarbejdets Opgaver og Muligheder. (The tasks and possibilities of plant breeding.) Tidsskr. Landøkon, Nr. 10, 553—574 (1942).
— N. J. F.'s Sektion for Avlsbiologi. (N. J. F's Section for the study of the biology of breeding.) Nord. Jordbr. Forskn., Nr. 1, 52—54 (1944).

Frandsen, K. J.: Iagttagelser over polyploide Former af nogle Kulturplanter. (Beta, Brassica, Sinapis, Trifolium og Medicago.) (Observations on polyploid forms of some cultivated plants [Beta, Brassica, Sinapis, Trifolium and Medicago].) Tidsskr. Planteavl **49**, 445 bis 496 (1945).

Frankel, O. H.: The theory of plant breeding for yield. Heredity **1**, 109—120 (1947).

Fraser, J. G. C., W. Kalbfleisch u. J. M. Armstrong: New miniature thresher. Sci. Agric. **23**, 183—186 (1942).

Frenkel, A. I u. N. N. Seleznev: (Summary of the work of the State Breeding Stations for 1938—1939.) Selekcija i Semenovodstvo (Breeding and Seed Growing), Nr. 7, 3—9 (1940). [Russisch.]

Frimmel, F. v.: Hofrat Prof. Dr. h. c. Erich v. Tschermak-Seysenegg zum 70. Geburtstag. Züchter **13**, 217 bis 219 (1941).

Fuchs, W. H.: Physiologische Methoden in der Pflanzenzüchtung. Kühn-Archiv 1943—1944 **60**, 288—314 (1944).

Funchess, M. J.: Agronomic problems of the South. J. Amer. Soc. Agron. **32**, 96—106 (1940).

G..., R., C..., Fr.: Den skånska växtförädlingen arbetar målmedvetet. Ett. besök på Weibullsholm. (Scanian plant breeding follows a plan. A visit to Weibullsholm.) Lantmannen **24**, 621—622 (1940).

—, — Utsädesföreningen inför nya uppgifter-Ett besök på Svalöf. (The Seed Association confronting new tasks. A visit to Svalöf.) Lantmannen **24**, 674—676 (1940).

Garber, R. J.: Plant breeding in relation to human nutrition. Science **101**, 288—293 (1945).

Gardiner, J. G.: Report of the Minister of Agriculture for the Dominion of Canada for the year ended March 31, 1945 Pp. 212 (1945).

Gates, R. R.: Plant genetics and human welfare. Lancet **236**, 1472—1475 (1939).

George, L. V.: New crops while you wait. Sth. Seedsman **10**, Nr. 6, Pp. 15, 54, 58 (1947).

Gescher, N.: L'amélioration des plantes par l'hétérosis. Bull. mens renseign. techn. **32**, 297—321 (1941).

Gescher, N. V.: Protection of new products discovered by plant breeders. Int. Rev. Agric. **37**, 1 T—10 T (1946).

Gilbert, B. E.: Fifty-fifth Annual Report of the Rhode Island State College Agricultural Experiment Station, Kingston, R. I. 1943. Contr. **615**, 60 (1943).

Gilbert, W. W.: Jesse Baker Norton, 1877—1938. J. Hered. **31**, 273—276 (1940).

Giovannelli, B.: Nazareno Strampelli (29 Maggio 1866 — 23 Gennaio 1942). (Nazareno Strampelli, 29. May 1866 — 23. January 1942.) Genetica Agraria, Roma **1**, 5—8 (1946).

Gizbertówna, W.: Udział hodowli w aklimatyzacji roślin. (The role of breeding in plant acclimatization.) Życie Rolnicze **4**, Nr. 4, 14—16 (1939).

Gökgöl, M.: Yurdumuzda tohum islâhciliğinin esas gayeleri ve bunun inkisaf yollari. (The principal aims of seed improvement and its ways of development in our country.) Ziraat Dergīsī **4**, 10—13 (1943).

— Türkiye tohum islâhciliiğnin amaçlari ve bunlara ulaştiracak yollar. (Aims of seed improvement in Turkey and ways of achieving these aims.) Cankaya Matbaasi, Ankara (Genel Sayi: 629), Nr. 7, 40 (1946).

Goodale, H. D.: Progress report on possibilities in progeny-test breeding. Science **94**, 442—443 (1941).

Goulden, C. H.: Problems in plant selection. Proc. 7th Int. Genet. Congr. Edinburgh 23.—30. August, Pp. 132—133, 1939 (1941). (Abstr.)

Govaert: L'amélioration des plantes au Congo Belge. Rapp. Comm. Sect., Résumés, VIII Congr. Int. Agric. Trop. Subtrop. Rome (48), 63—65 (1939).

Grew, E. S.: Herencia y crianza de plantas. (Heredity and breeding of plants.) Rev. Inst. Defensa Café, Costa Rica **16**, 421—424 (1946).

Griesinger, R.: Die Bedeutung der Ergebnisse der Polyploidieforschung für die Pflanzenzüchtung. Ber. dtsch. bot. Ges. **60**, 36—41 (1942).

— Die Bedeutung der Ergebnisse der Polyploidieforschung für die Pflanzenzüchtung. Ber. dtsch. bot. Ges. **60**, 36—41 (1943).

Grishko, N.: The biological section of the Ukrainian Academy of Sciences. Nature, Lond. **152**, 404—405 (1943).

Groenewolt, J. K.: Ter inleiding. (Introduction.) Zesde Jaarboekje Nationaal Comité voor Brouwgerst, Wageningen, Pp. 5—9 (1941).

Grossman, J. J.: (F. F. Hallett's methods of breeding.) Socialistic Grain Farming, Saratow, Nr. 4, 164—167 (1940). [Russisch.]

Györffy, B.: A polyploid növények élettana. (The physiological and chemical conditions in polyploid plants.) Arb. Ung. Biol. Forsch.-Inst. **13**, 362—446 (1941).

H......, A.: The aims of plant breeding in German agriculture. Int. Rev. Agric. **31**, T, 119—123 (1940).

Haan, D. de: Klimaat en plantenveredeling. (Climate and plant breeding.) Landbouwk. Tijdschr. Wageningen **54**, 483—499 (1942).

Haan, H. de: Raseigenschappen en rassenkeuze. Varietal characters and the choice of varieties.) Jaarb. Algem. Bond Oud-leerl. Inricht. Middelbaar Landbouwonderwijs, Wageningen, Pp. 3—10 (1944).

— De ontwikkeling en de beteekenis van de rassenlijst voor landbouwgewassen (1924—1944). (Development and importance of the list of varieties of agricultural crops [1924—1944].) Landbouwk. Tijdschr. **56**, 78—86 (1944).

Haddon, C. B.: Biennial report of the Northeast Louisiana Experiment Station 1941—1942, Pp. 40 (1942).

Hagedoorn: Het tusschenschakelen van extra wintergeneraties bij de plantenveredeling. (The insertion of extra winter generations in plant improvement.) Erfelijkheid in Praktijk, Leiden **5**, 199—200 (1940).

Haldane, J. B. S.: The interaction of nature and nurture. Ann. Eugen. **13**, 197—205 (1946).

Hall, D.: How the plant breeder goes to work: I, II. J. R. Hort. Soc. **65**, 283—288, 327—333 (1940).

Hansen, N. E.: New hardy fruits for the Northwest. Bull. S. Dak. Agric. Exp. Sta., Nr. 339, 31 (1940).

— Fifty years work as agricultural explorer and plant breeder. 42nd Ann. Rep. S. Dak. State Hort. Soc. for the year ending June 30, 1945, 119—136 (1945).

Harrington, J. B.: Observations on plant breeding and seed distribution in the United States. Rep. 11th. Annu. Mtg. Univ. Sask. Canad. Seed Gr. Ass., Pp. 20—29 (1940).

Hastings, W. R.: 1944 All-Americas show trade activity in wartime. Seed World **54**, 8—9, 14, 16 (1943).

Hawkins, R. S. (Editor): Fifty-second Annual Report of the Arizona Agricultural Experiment Station for the year ending June 30, 1941, Pp. 99 (1942).

Hayes, H. K.: Some illustrations of methods in plant breeding. Proc. Amer. Soc. Sug. Beet. Technol., Pp. 1 bis 17 (1940). [Mimeographed.]

— u. F. R. Immer: Methods of plant breeding. McGraw-Hill Publishing Company, Ltd. London (1942).

Hazel, L. N. u. J. L. Lush: The efficiency of three methods of selection. J. Hered. **33**, 393—399 (1942).

Heinicke, A.: Sixty-second Annual Report of the New York State Agricultural Experiment Station for the year ended June 30, 1943, 75 (1944).

Heinisch, O.: Die Pflanzenzüchtung im Protektorat Böhmen-Mähren. Mitt. Landw. **55**, 11—12, 30—31 (1940).

— Neue Grundlagen für die Pflanzenzüchtung und das Sortenwesen im Protektorat Böhmen und Mähren. Z. Pflanzenz. **24**, 398—408 (1941).

Heinrich, W.: Die Pflanzenzüchtung des Donaulandes in ihrer Bedeutung für den Südostraum. Wien. Landw. Ztg. **92**, 127—128 (1942).

Hellbo, E.: Allmänheten och sortkännedomen. (The public and knowledge of varieties.) Lantmannen **24**, 562—563.

Henney, H. J.: Director's Annual Report of the Colorado Agricultural Experiment Station for the fifty-fifth fiscal year 1941—1942. (1942).

— Fifty-fifth Annual Report of the Colorado Agricultural Experiment Station 1941—1942, Pp. 58 (1942).

— Fifty-sixth Annual Report of the Colorado Agricultural Experiment Station 1942—1943. 1943, Pp. 39 (1943).

Henrath, H.: In memoriam Prof. IrC. Broekema. (In memory of Prof. C. Broekema.) Landbouw **16**, 381 bis 384 (1940).

Hildebrand, A.: Die Pflanzenzüchtung im Warthegau. Mitt. Landw. **55**, 469—470 (1940).

Hill, A. F.: Recent investigations in tropical and subtropical agriculture. Chronica Botanica **6**, 441—443 (1941).

Hill, A.G.: The improvement of native food crops. A précis of the more important work done in East Africa during 1944. E. Afr. Agric. Res. Inst., Amani Nr. DF/5/2, Pp. 15 (1945).

— Selection and improvement of food plants in relation to better nutrition. E. Afr. Agric. J. **12**, 125—127 (1946).

Hill, J. A.: Fifty-first Annual Report of the University of Wyoming, Agricultural Experiment Station 1940 bis 1941, Pp. 47 (1941).

Hind, H. L.: Brewing science and practice. Vol. II. Brewing processes. Chapman and Hall, Ltd., London (1940).

Hino, I.: (The variation in resistance to the toxic action of potassium chlorate and its diagnostic application in distinguishing the diseased plants from the healthy.) Agric. & Hort. Japan **14**, 1483—1486 (1939).

Holthorp, H. E.: Tricotyledony. Nature, Lond. **153**, 13—14 (1944).

Honecker, L.: Aufgaben der Pflanzenzüchtung in Kriegs- und Nachkriegszeit. Prakt. Blätter Pflanzenbau u. Pflanzenschutz **19**, 142—164. (Aus Z. Pflkrankh. 1943, **53**, 142—143.) (1941—1942).

Hove, W. Ten: De vermeerdering en de verspreiding van hoogwaardig zaaizaad door de zaadhoeven. (The multiplication and distribution by the seed farms of high grade seed for planting.) Landbouw **16**, 604—614 (1940).

Howard, W. L.: Luther Burbank's plant contributions. Bull. Calif. Agric. Exp. Sta., Nr. 691, 110 (1945).

— Luther Burbank, a victim of hero worship. Chronica Botanica 1945—1946, London **9**, 299—520 (1946).

Hudson, P. S.: Plant breeding and genetics today. Advance. Sci. **3**, 252—267 (1945).

— How new crops are found. Countryman **34**, 43—46 (1946).

Huitema, W. K: Het landbouwkundig onderzoek bij overjarige cultures, in het bijzonder bij eenige nieuwe handelsgewassen. (Agricultural research in regard to perennial crops, in particular some new commercial plants.) Bergcultures **15**, 818—825 (1941).

Hutchinson, J. B.: The application of genetics to plant breeding. I. The genetic interpretation of plant breeding problems. J. Genet. **40**, 271—282 (1940).

Ing, E. G.: Research work at John Innes Horticultural Institution. Worcs. Agric. Quart. Chron. **14**, Nr. 1, 83—88 (1945).

Ipatjev, A. N.: (Underlying regularities in the composition and structure of variety populations.) Trudy Omsk. Seljskohozjajstvennogo Inst. imeni S. M. Kirova (Trans. Kirov Inst. Agric. Omsk, USSR.) **4** (17), 109 bis 126 (1939). [Russisch.]

Ipatjev, A. N.: Selection after elemental characters in plant breeding. C. R. (Doklady) Acad. Sci. URSS. N. s. **31**, 171—172 (1941).

Isbell, C. L.: The value of native material in breeding horticultural crops for Alabama. Proc. Amer. Soc. Hort. Sci. **38**, 599—604 (1941.)

Ivanov, S. S.: A seedling method for testing aphid resistance and its application to breeding and inheritance studies in cucurbits and other plants. J. Hered. **36**, 357 bis 361 (1945).

J....., J.: Utsädesföreningens årsmöte på Svalöf. (Annual Meeting of the Seed Association at Svalöf.) Lantmannen **24**, 653—655 (1940).

Jack, H. W.: Progress notes on the General Experimental Station, Sigatoka. Agic. J., Fiji. **10**, 76—78 (1939).

Jardine, J. T. et al: Report on the agricultural experiment stations, 1944. U. S. Dep. Agric., Wash. 1944, Pp. 130 (1945).

— Report on the agricultural experiment stations, 1945. U. S. Dep. Agric., Wash. 1945, Pp. 172 (1946).

—, F. D. Fromme u. H. L. Knight: Report on the Agricultural Experiment Stations, 1940. U. S. Dep. Agric., Washington, D. C., Pp. 272 (1941).

Jenkin, T. J.: The Welsh Plant Breeding Station. Farming **1**, 141—148 (1946).

Johnson, I. B.: Agricultural research in South Dakota. 56th Ann. Rep. S. Dak. Agric. Exp. Sta., Pp. 58 (1943).

Jørgensen, C. A.: Nye fremskridt i arvelighedslaeren-Nyi veje for planteforaedlingen. (New progress in genetics — New ways of plant improvement.) Tidsskr. Landøkon. Nr. 10, 673—689 (1939).

Kadam, B. S.: Deterioration of varieties of crops and the task of the plant breeder. Indian J. Genet. Pl. Breed. **2**, 159—172 (1942).

Kaftanov, S.: Laureates of science. Soviet News Nr. 1728, 4 (1947).

Kalkus, J. W.: Report of agricultural research and other activities of the Western Washington Experiment Station for the fiscal year ended March 31, 1941, 70 (1941).

— Report of agricultural research and other activities of the Western Washington Experiment Station for the fiscal year ended March 31, 1942, 70 (1942).

Kamat, M. N.: Progress of plant pathological research in Bombay. Poona Agric. Coll. Mag. **33**, 97—100 (1941).

Kappert, H.: Die Vererbungswissenschaft in der gärtnerischen Pflanzenzüchtung unter besonderer Berücksichtigung der Blumenzüchtung. Mit 10 Abb. Forschungsdienst **10**, 533—545 (1940.)

— Die Bedeutung der Polyploidie in der Cyclamenzüchtung. Züchter **13**, 106—114 (1941).

Kartman, L.: Soviet genetics and the „autonomy of science". Sci. Mon., N. Y. July, Pp. 67—70 (1945).

Kennedy, T. L.: Report of the Minister of Agriculture, Province of Ontario, for the year ending March 31st, 1944. Sess. Pap., Nr. 21, 109 (1944).

— Report of the Minister of Agriculture, Province of Ontario, for the year ending March 31st, 1945. Sess. Pap., Nr. 21, 111 (1945).

Kerefov u. Kupermann: (The Kabardino-Balkar State Breeding Station.) Selekcija i Semenovodstvo, Nr. 4, 20—22 (1939). [Russisch.]

Khan, M. A.: What's doing in All-India. The Punjab. Indian Fmg. **1**, 86—90 (1940).

— What's doing in All-India. The Punjab. Indian Fmg. **11**, 590—594 (1941).

Khan, M. S.: Variety improvement work in N.-W. F. province. Quart. Notes Agric. Dep. N.-W. F. Prov. **3**, 1—3 (1941).

Kifer, R. S., B. H. Hurt u. A. A. Thornbrough: The influence of technical progress on agricultural production. Yearb. U. S. Dep. Agric., Pp. 509—532 (1940).

King, G. H.: Twenty-fourth Annual Report. Bull. Ga Coastal Plain Exp. Sta., Nr. 40, 112 (1944).
King, J. R.: Method for covering emasculated flowers in plant breeding. Bot. Gaz. 102, 217—220 (1940).
Knapp, E.: Deutsche Beiträge zur Vererbungsforschung im letzten Jahrzehnt. Forsch. Fortschr., Berlin 19, 202—211 (1943).
Kobel, F.: Vererbungsforschung und Pflanzenzüchtung. Verh. d. Schweiz. Naturforsch.-Ges. 121, 241—242 (1941).
Koblet, R.: Wandlungen und Probleme des Pflanzenbaus seit dem 18. Jahrhundert. Landw. Jb. Schweiz 55, 614—622 (1941).
Kormilicyn, A. M.: (The Institute of the Arid Subtropics at the Pan-Union Exhibition.) Soviet Subtropics, Nr. 9 (72), 11—23 (1940).
Kreutz, H.: Ludwig Kiessling. Züchter 14, 96—98 (1942).
Krickl, M.: Zur Frage der Züchtung einer winterharten Arundo donax L. Züchter 17/18, 67—69 (1946).
Krylow, A.: (Experience gained in the course of work at the Kamenno-Stepnaja Station.) Socialističeskoe Seljskoe Hozjajstvo (Socialistic Agriculture) Moscow, Nr. 1, 35—43 (1944). [Russisch.]
Kuckuck, H.: Pflanzenzüchtung. Sammlung Göschen, 1134, Walter de Gruyter und Co., Berlin (1939).
Kugler, W. F.: Amparo a la propiedad en la creación de nuevas variedades de plantas — Antecedentes en el país y en el extranjero. (Ownership protection for breeding new varieties of plants. Local and foreign antecedents.) „Granos" Semilla Selecta, B. Aires 3, Nr. 11, 3—20 (1939).
Kusatz, H.: Saatgutzüchtung und Sortenbereinigung in der Ostmark. Wien, Landw. Ztg. 91, 72—74 (1941).

L....., H. M.: Limitations in plant breeding. Int. Šug. J. 43, 77—79 (1941).
L....., W.: Ferdinand von Lochow zum Gedenken. Von Landroggen zum Petkuser Roggen. Forschungsdienst 16, 113—114 (1943).
Ladd, C. E.: 55th. Annual Report of the Cornell University Agricultural Experiment Station 1942. Pp. 192 (1942).
— Fifty-sixth Annual Report of Cornell University Agricultural Experiment Station 1943. Pp. 190 (1943).
Lamprecht, H.: Arbetsplan för Weibullsholms Växtförädlingsanstalt för år 1940. (Programme of work for Weibullsholm Plant Breeding Institute for the year 1940.) Weibulls Ill. Årsb. 35, 5—26 (1940).
— Växtförädlingsarbetet på Weibullsholm. Dess omfattning och metoder. (Plant breeding at Weibullsholm. Its extent and methods.) Weibulls Ill. Årsb. 37, 9—12 (1942).
— Ett framgångsrikt växtförädlingsarbete. (Promising plant breeding operations.) Weibulls Ill. Årsb. 38, 24 bis 26 (1943).
Langham, D. G.: Three useful gadgets for plant breeders. J. Hered. 38, 29—32 (1947).
Larter, L. N. H.: Report of the Senior Botanist. Jamaica 27. July, Pp. 9 (1944).
Lasser, E.: Das Sortenproblem im alpinen Pflanzenbau. Kühn-Archiv 1943—1944, 60, 358—368 (1944).
Latčenko, V. N.: (The Šatilovo Breeding Station-participant in the Soviet Agricultural Exhibition.) Selekcija i Semenovodstvo (Breeding and Seed Growing), Nr. 7, 32—35 (1940). [Russisch.]
Lawrence, W. J. C.: Practical plant breeding. George Allen and Unwin, Ltd., London (1939).
Levčenko, L. P.: (The question of the influence of geographical co-ordinates on yield and regional distribution of varieties.) Selekcija i Semenovodstvo (Breeding and Seed Growing), Nos. 11/12, 7—13 (1940). [Russisch.]

Lewin, C. J.: Annual Report of the Department of Agriculture, Northern Rhodesia for the year 1944. Pp. 12 (1945).
Libowitzky, J.: Beachtenswertes bei der Züchtung von Cyclamen persicum. Gartenbauwiss. 16, 4—11 (1941).
Lin, Cheng Yao: (Report of the Agricultural Experiment Station of Fukien University.) Fukien Agric. J. 2, 239—242 (1940).
Lindsey, A. H.: Annual report of the Massachusetts Agricultural Experiment Station 1943—1944. Bull. Mass. Agric. Exp. Sta., Nr. 417, 78 (1944).
Lj....., E.: Sveriges Utsädesförenings årsmöte. (Swedish Seed Associations annual meeting.) Sverig. Utsädesfören. Tidskr. 49, 251—264 (1939).
— Sveriges Utsädesförenings årsmöte. (Annual meeting of the Swedish Seed Association.) Sverig. Utsädesfören Tidskr. 54, 228—235 (1944).
— Utsädesföreningens extra möte under Lantsbruksveckan 1944. (The Seed Associations special meeting during the agricultural week in 1944.) Sverig. Utsädesfören. Tidskr. 54, 55—62 (1944).
Ljung, E. W.: Per Artur Olsson. (Per Artur Olsson.) Sverig. Utsädesfören. Tidskr. 55, 160—163 (1945).
Ludwig, W. u. R. Freisleben: Über neuere statistische Methoden zur Auswertung von Koppelungsversuchen, vor allem in der Pflanzenzüchtung. Z. Pflanzenzücht. 24, 523—528 (1942).
Lukjanenko, P. P.: (Selection according to specific gravity as a means of increasing the yielding capacity of seed.) Selekcija i Semenovodstvo (Breeding and Seed Growing), Nr. 3, 17—20 (1940). [Russisch.]

McCall, M. A.: Crop improvement, a weapon of war and an instrument of peace. J. Amer. Soc. Agron 36, 717—725 (1944).
McDowell, C. H.: Research serves agriculture. 57th Rep. Tex. Agric. Exp. Sta., Pp. 49 (1944).
McKee, C.: Serving Montana agriculture through research. 46th and 47th Rep. Mont. Agric. Exp. Sta., Pp. 67 (1938—1940).
Mackie, J. R.: Annual report of the Agricultural Department, Nigeria 1942, Pp. 17 (1943). [Mimeographed.]
MacLean, C. A.: Report of the Agricultural Department, Bihar, for the period from the 1st April 1938 to the 31st March 1939, Pp. 42 (1941).
McTaggart, A.: Plant introduction. I. A review, with notes on outstanding species. Pamphl. Coun. Sci. Industr. Res. Aust., Nr. 114, 5—14 (1942).
Mahnorylo, V. F.: (The Stalinsk Breeding Station at the Soviet Agricultural Exhibition.) Selekcija i Semenovodstvo (Breeding and Seed Growing), Nr. 7, 21—22 (1939). [Russisch.]
Malandin, G. A. (Editor): (Symposium of research work of the Čeljabinsk Regional Agricultural Research Station.) Ogiz-Čeljabgiz, Čeljabinsk 1, 520 (1939). [Russisch.]
Marotta, F. P.: La Facultad de Agronomía y Veterinaria en la Universidad. (The Faculty of Agronomy and Veterinary Science in the University.) Buenos Aires 1936—1940. Pp. 664 (1943).
Maslennikov: (30 years research and practical work of the Novyi Uren' Agricultural Experiment Station.) Socialistic Grain Farming, Saratov, Nr. 2, 62—68 (1941). [Russisch.]
Mather, K.: The balance of polygenic combinations. J. Genet. 43, 309—336 (1942).
— u. D. de Winton: Adaptation and counter adaptation of the breeding system in Primula. Ann. of Bot., N. s. 5, 297—311 (1941).
Maya Das, C.: What's doing in All-India-United Provinces. Indian Fmg. 3, 602—603 (1942).

Meier, K.: Berichte der Eidgenössischen Versuchsanstalt für Obst-, Wein- und Gartenbau in Wädenswil für die Jahre 1938—1939. Landw. Jb., Schweiz **56**, 97—168 (1942).

— Bericht der Eidgenössischen Versuchsanstalt für Obst-, Wein- und Gartenbau in Wädenswil für die Jahre 1941 und 1942. Landw. Jb., Schweiz **58**, 417 bis 495 (1944).

Méneret, G.: Amélioration et expérimentation de plantes de grande culture au Centre de Recherches Agronomiques de Colmar en 1937—1938. Sélectionneur **8**, 23—40 (1939).

Meyer, H. A.: Die neuen Versuchs- und Forschungsanstalten. Mitt. Landw. **56**, 705—706, 725—726, 745 bis 746 (1941).

Meyer, K.: Unsere Forschungsarbeit im Kriege. Hauptbericht anläßlich der Kriegstagung 1941 des Forschungsdienstes. Forschungsdienst **11**, 253—286 (1941).

Miles, L. G.: Plant-breeding and the production of better seed. Qd Agric. J. **54**, 258—267 (1940).

Miller, M. F., S. B. Shirky u. H. J. L'Hote: Work of the Agricultural Experiment Station during the year ending June 30, 1939. Bull. Mo. Agric. Exp. Sta., Nr. 444, 106 (1942).

—, —, — Investigations for the benefit of the Missouri farmer. Work of the Agricultural Experiment Station during the year ending June 30, 1944. Bull. Mo. Agric. Exp. Sta., Nr. 491, 71 (1945).

Moore, R. P., J. A. Rigney, G. K. Middleton u. L. S. Bennett: Official variety tests-1943. Corn-soybeans-cotton-wheat-oats-barley. Bull. N. C. Agric. Exp. Sta., Nr. 343, 50 (1944).

Mosolov, V.: New achievements of Soviet agricultural science. Soviet News, Nr. 1693, 4 (1947).

Müntzing, A.: Den teoretiska genetiken och växtförädlingen. (Theoretical genetics and plants breeding.) Sverig. Utsädesfören. Tidskr. **56**, 582—587 (1946).

Näf, A.: Die Entwicklung des Saatzuchtwesens in der Schweiz und seine Bedeutung für den inländischen Getreidebau. Ber. Schweiz. Bot. Ges. **53** A, 44—61 (1943).

Neatby, K. B.: Accelerating and guiding plant evolution. C. S. T. A. Rev., Nr. 36, 5—11 (1943).

Nebel, B. R.: Symposium on „Theoretical and practical aspects of polyploidy in crop plants". Introduction. Amer. Nat. **75**, 289—290 (1941).

Newsom, I. E.: Fifty-fourth Annual Report of the Colorado Agricultural Experiment Station 1940—1941, Pp. 61 (1941).

Nicolaisen, W.: Die Pflanzenzucht in der Erzeugungsschlacht. Forschungsdienst **17**, 499—512 (1944).

Nilsson-Ehle, H.: Årsberättelse över Sveriges Utsädesförenings verksamhet under år 1938. (Annual report on the work of the Swedish Seed Association during the year 1938.) Sverig. Utsädesfören. Tidskr. **49**, 183—240 (1939).

Nilsson-Leissner, G.: Organisationen av jordbruksförsöksverksamheten och växtförädlingen i Sverige. (The organisation of agricultural activities and plant breeding in Sweden.) Tidsskr. Landøkon., Nr. 6, 441 bis 455 (1939).

— Foderproduktionen i Sverige. VI. Resultat av växtförädlingen. (Fodder production in Sweden. VI. Plant breeding results.) Nord. JordbrForskn. **23**, 178—180 (1941).

Nizenjkov, N. P.: (Die Anwendung der Elektromethode in der Züchtung.) Selekcija i Semenovodstvo, Nr. 9/10, 34—42 (1946). [Russisch.]

Nolla, J. A. B.: Annual Report of the Agricultural Experiment Station, University of Puerto Rico, for the fiscal year 1939—1940, Pp. 66 (1940).

O'Connell, T.: Production of seed grain in Eire. B. G. A. Rec., Nr. 5, 11—17 (1941).

Odnokonj, J. M.: (The Amur State Breeding Station.) Selekcija i Semenovodstvo, Nr. 4, 22—24 (1939). [Russisch.]

Oehler, E.: Art- und Gattungsbastarde bei Kulturpflanzen und ihre Bedeutung für die Pflanzenzüchtung. (Ref.) Ber. d. Schweiz. Bot. Ges. **51**, 455—456 (1940).

Oljhovoj, G. D.: (The work of the Gory State Breeding Station.) Selekcija i Semenovodstvo (Breeding and Seed Growing), Nr. 4, 6—7 (1940). [Russisch.]

Orton, C. R.: Farm science looks ahead. Bienn. Rep. W. Va Agric. Exp. Sta. 1942—1944, Nr. 317, 56 (1944).

Osipov, V. S.: (The production of élite seed.) Vestnik Ovoščevodstvo i Kartofelj (Vegetable and Potato Journal), Nr. 2, 3—11 (1941). [Russisch.]

Pacheco Herrarte, M.: La hibridación de plantas. (Plant hybridization.) Rev. Agric. Guatemala **17**, 237 bis 241 (1940).

Paguirigan, D. B.: The utilization of first filial generation hybrids in crop improvement in the Philippines. Proc. 6th Pacific Sci. Congr. **4**, 709—716 (1940).

Pahomova, V. P.: (The Kharkov Agricultural Breeding Station, a participator in the Soviet Agricultural Exhibition.) Selekcija i Semenovodstvo (Breeding and Seed Growing), Nr. 7, 16—19 (1939). [Russisch.]

Pal, B. P. u. S. Ramanujam: Recent Advances in plant breeding with special reference to the work of the Imperial Agricultural Research Institute. Roy. Bot. Gdn., Calcutta **150**, Nr. 24, 1—10 (1942).

—, — Plant breeding and genetics at the Imperial Agricultural Research Institute, New Delhi. Indian J. Genet. Pl. Breed. **4**, 43—53 (1944).

Panos, D. A.: Fünf Jahre Pflanzenzüchtung in Griechenland. Züchter **11**, 341—346 (1939).

Panse, V. G.: A statistical study of quantitative inheritance. Ann. Eugen., Camb. **10**, 76—105 (1940).

— Application of genetics to plant breeding. 2. The inheritance of quantitative characters and plant breeding. J. Genet. **40**, 283—302 (1940).

— Methods in plant breeding. Indian J. Genet. Pl. Breed. **2**, 151—158 (1942).

Parrott, P. J.: Sixty-first annual report for the fiscal year ended 30. June 1942. N. Y. St. Agric. Exp. Sta. Pp. 96 (1942).

Pedersen, A.: Krydsningsfaren i Frøavlen. (The danger of crossing in seed growing.) Nord. Jordbr. Forskn., Nos. 5/6, 182—196 (1943).

Pehrson, J.: Försöksfeltet vid lantbrukshögskolans institution för växttodlingslära 1942—1943. (Experiment plots at the Agricultural College Institute for the study of Plant Cultivation 1942—1943.) Lantmannen **27**, 942—944, 960—962 (1943).

Petrov, I. P.: (Device for sowing breeding material.) Selekcija i Semenovodstvo (Breeding and Seed Growing), Nr. 4, 42 (1940). [Russisch.]

— (Turkmenistan Experiment Station of the All-Union Institute of Plant Industry.) Vestnik Socialističeskogo Rastenievodstva (Soviet Plant Industry Record), Nr. 5, 188—191 (1940). [Russisch.]

Philp, J.: On wheat breeding and genetics. Proc. 7th Int. Genet. Congr. Edinburgh 23.—30. August, Pp. 237. 1939 (1941). (Abstr.)

Pincus, J. W.: The genetic front in the USSR. J. Hered. **31**, 165—168 (1940).

Pinto da Silva, A. R.: Um aspecto da etnobotanica Portuguesa: os nomes vernáculos das plantas. (An aspect of Portoguese etnobotany: vernacular plant names.) Rev. Agron. Lisboa **30**, 376—379 (1942).

Pires, D. R. V.: Métodos de melhoramento em Svalöf e Cambridge. (Methods of plant breeding at Svalöf and Cambridge.) Palestras Agron. Lisboa **1**, 12 (1938) 1939.

— O melhoramento de plantase a fisiologia. (Plant breeding and physiology.) Rev. Agron., Lisboa **29**, 111 (1941).

Pires, D. R. V.: O melhoramento de plantase a fisiologia. (Plant breeding and physiology.) Rev. Agron., Lisboa 30, 64—69 (1942).
— Em busca de novas raças de plantas. (In search of new races of plants.) Rev. Agron. Lisboa 30, 153—175 (1942).
Pirovano, A.: Interventi elettrici nella genetica orticola a scopo utilitario. (Electrical intervention in horticultural genetics with a practical object.) Atti III Riunione Soc. Ital. Genet. Eugen. Bologna, p. 7 (1939).
Piza, S. de Toledo (jr.): Breves considerações em tôrno dos vobábulos ,,gen" e ,,cromossômio". (Brief considerations of the words ,,gene" and ,,chromosome".) Rev. Agric. S. Paulo 18, 109—112 (1943).
Popenoe, W.: Plant resources of Guatemala. Chronica Botanica 7, 16—19 (1942).
Povolockaja, E. E.: (The Kuban Experiment Station of the Institute of Plant Industry.) Selekcija i Semenovodstvo, Nr. 4, 24—25 (1939). [Russisch.]
Prijamopoljskij, P. K.: (The Voroshilovsk Breeding Station, participator in the Soviet Agricultural Exhibition.) Selekcija i Semenovodstvo (Breeding and Seed Growing), Nr. 7, 32—33 (1939). [Russisch.]
Puhaljskij, A. V.: (Forty years of scientific activity of the Station.) Selekcija i Semenovodstvo, Nr. 5, 30—32 (1939). [Russisch.]
— (The production of élites at the Šatilovo State Breeding Station in 1939.) Selekcija i Semenovodstvo (Breeding and Seed Growing), Nr. 1, 14—18 (1940). [Russisch.]

R....., G. N.: Plant genetics as applied to the agricultural industry in India. Sci. and Cult. 7, 373—376 (1942).
Ramanujam, S.: Self-sterility and plant breeding. Indian Fmg. 3, 193—196 (1942).
— Genetical research as applied to plant breeding in post-war India. Curr. Sci. 13, 63—65 (1944).
Ramella, R.: La Estación Experimental de Rafaela. Su origen, desenvolvimiento y aportes al progreso agrícola argentino. (The Rafaela Experimental Station. Its origin, development and contributions towards Argentine agricultural progress.) ,,Granos" Semilla Selecta, B. Aires Nos 1, 2 and 3, 14—19 (1941).
Ramiah, K.: Plant breeding and genetical work in India. 28th Indian Sci. Congr., Benares, Sect. XI, Agric., p. 30 (1941).
Randolph, L. F.: Symposium on ,,Theoretical and practical aspects of polyploidy in crop plants". An evaluation of induced polyploidy as a method of breeding crop plants (with discussion by G. M. Darrow). Amer. Nat. 75, 347—365 (1941).
Ranga Rau, D. S.: Importance of physiological studies in modern plant breeding. Poona Agric. Coll. Mag. 34, 123—132 (1942).
Reed, H. J.: Science solves farm problems and aids agricultural production. 55th. Rep. Director Ind. Agric. Exp. Sta., Pp. 109 (1942).
— u. W. V. Lambert: Research solves farm problems. Report of the Director, Purdue University Agricultural Experiment Station for the year ending June 30, 1941, Pp. 114 (1941).
Regel, C.: Beiträge zur Kenntnis von mitteleuropäischen Nutzpflanzen, III. Angew. Bot. 24, 278—302 (1942).
Reinhardt, Fr.: Der Aufbau des Saatgutwesens im General-Gouvernement. Kühn-Archiv 1943/44, 60, 204—210 (1944).
Reiterer, M.: Zur heutigen Saatgutversorgung. Kärnter Bauer 96, 233—234 (1946).
Richharia, R. H.: Plant breeding technique in recent years. (Lectures delivered at the Agricultural Research Institute, Nagpur.) Printed at the Bangalore Press, Mysore Road, Bangalore City (1939).
— Plant breeding and genetics in India. Patna, Pp. 403, figs, tables, photos. (1945).

Riehm, E.: Biologische Reichsanstalt für Land- und Forstwirtschaft in Berlin-Dahlem. Wissenschaftl. Jahresbericht 1938. Landw. Jb. 90, 201—344 (1940).
— Biologische Reichsanstalt für Land- und Forstwirtschaft in Berlin-Dahlem. Wissenschaftlicher Jahresbericht 1939. Mitt. Biol. Reichsanst., Berlin 1941, Nr. 63, 108 (1941).
— Biologische Reichsanstalt für Land- und Forstwirtschaft in Berlin-Dahlem. Wissenschaftlicher Jahresbericht 1940. Mitt. Biol. Reichsanstalt Landw. 65, 1—110 (1941).
Rivoire, P.: Comment lutter contre la dégénérescence des végétaux. Rev. Hort. Agric. Afrique Nord 46, 7—11 (1942).
Rjazanov, J.: (Brief summary of results and the problems of the South-eastern Research Institute for grain culture.) Socialistic Grain Farming, Saratov, Nr. 1, 7—29 (1940). [Russisch.]
Rjazanov, K. D.: (Demonstration at the Soviet Agricultural Exhibition of the achievements of the Tadjik Breeding Station.) Selekcija i Semenovodstvo (Breeding and Seed Growing), Nr. 7, 19—20 (1939). [Russisch.]
Robb, W.: Notes on plant breeding in Sweden. Scot. Agric. 26, 151—157 (1947).
Roberts, J.: A head thresher for plant breeding studies. Agric. Engng. St. Joseph, Mich. 22, 14, 32 (1941).
Robertson, D. W.: Plant breeding in agronomy section outlined: several crops dicussed. Colo. Fm. Bull. 2, Nr. 3, 6—8 (1940).
Robinson, D. H.: Plant breeding remains an art. Fmr's Wkly, London 10 (26), 47 (1939).
Rodrigo, E.: Administration Report of the Acting Director of Agriculture for 1939. Rep. Dep. Agric., Ceylon, Pp. 34. D. (1941).
— Administration report of the acting Direktor of Agriculture for 1941. Part IV — Education, science, and art. (D). Ceylon, Pp. D 15 (1942).
— Administration Report of the Acting Director of Agriculture. Ceylon for 1942. Pp. D. 16 (1943).
Roemer, Th.: Entwicklungslinien der Züchtungsmethoden. Erweiterter Vortrag bei der Tagung des Reichsverbandes deutscher Pflanzenzuchtbetriebe in Wien, 26. Juni 1939. Kühn-Arch. 54, 267—294 (1939).
— Aufgaben und Leistungen der deutschen Pflanzenzucht im Kriege. Mitt. Landw. 59, 463—464 (1944).
— Professor Dr. Rudolf Freisleben. Z. Pflanzenz. 26, 163—164 (1944).
— u. W. Rudorf (Herausgeber): Handbuch der Pflanzenzüchtung. (Nicht abgeschlossen.) Paul Parey, Berlin (1939—1942).
Rössger, W.: Beitrag zur Technik der Weizenkreuzung. Forschungsdienst 14, 330 (1942).
Rudorf, W.: Warum Züchtungsforschung an den Kulturpflanzen? Umschau 43, 743—747 (1939).
— Verleihung des Carl Sigismund von Treskow-Friedrichsfelde-Preis an Herrn Hofrat Prof. Dr. h. c. Erich Tschermak von Seysenegg. Z. Pflanzenz. 24, 413 bis 414 (1941).
— Kreuzung innerhalb der Art. Handb. d. Pflanzenz. 1, 451—502 (1941).
— A importância da hibridação intra- e inter-genérica no melhoramento das plantas cultivadas. (The importance of intrageneric and intergeneric hybridization in the improvement of cultivated plants.) Agron. Lusitana 6, 333—347 (1944).

S....., B. G. L.: Botany and human welfare. Curr. Sci. 15, 17 (1946).
S—s, N.: Professor H. Nilsson-Ehle 70 år. (Professor H. Nilsson-Ehle — 70th birthday.) Svensk Papp. Tidn. 46, 45 (1943).
Sabnis, T. S.: Progress of botany with special reference to economic plants. Proc. 31st Indian Sci. Congr., Delhi, Pt. II, Pp. 60—73 (1944).

Saloheimo, L.: Försöksresultat från Finska Mosskulturföreningens Karelska försöksstation för år 1942. (Results from the Karelian Experimental Station of the Finnish Society for the Cultivation of Bogland, 1942.) Finska Moss Fören. Årsb. **47**, 48—69 (1943).
Salter, R. M.: Integrating soil and crop research. J. Amer. Soc. Agron. **33**, 237—245 (1941).
Sarup, S.: Breeding of disease resistant crops. Allahabad Fmr. **14**, 98—108 (1940).
Schaub, I. O. u. L. D. Baver: Research and farming. 65th Rep. N. C. Agric. Exp. Sta., Pp. 92 (1942).
—, — Research and farming. 67th Rep. N. C. Agric. Exp. Sta. **3**, Progr. Rep., Nr. 4, 111 (1944).
—, R. M. Salter u. L. D. Baver: Agricultural research in North Carolina 1939—1940. Rep. N. C. Agric. Exp. Sta., Pp. 74 (1940).
Scheibe, A.: Ludwig Kiessling †. Leben und Wirken eines Pflanzenbauers und Pflanzenzüchters im Dienste der deutschen Landwirtschaft. Z. Pflanzenz. **24**, 592 bis 598 (1942).
Schelotto, B.: Informe anual correspondiente al año agricola 1940—1941. (Annual Report for the year 1940 bis 1941.) Bol. Chacra Exp. „La Prevision" **3**, 59—111 (1941).
Schiemann, E.: Die genetische Abteilung des Botanischen Gartens in Dahlem. Notizbl. Bot. Gart. Mus. Berlin-Dahlem **15**, 145—163 (1940).
Schlösser, L. A.: Landsorte — Hochzucht — Landeszucht. Forschungsdienst **17**, 343—349 (1944).
Schuster, G. L.: Annual Report of the Director for the fiscal year ending June 30, 1944. Bull. Del. Agric. Exp. Sta., Nr. 251, 46 (1944).
Schwanitz: Artkreuzung im Dienste der Pflanzenzüchtung. Wien. landw. Ztg. **89**, 81—92 (1939).
Schwanitz, F.: Polyploidie und Pflanzenzüchtung. Naturwiss. **28**, 353—361 (1940).
— Untersuchungen über den Ertrag getriebener diploider und tetraploider Gartenkresse (Lepidium sativum). Züchter **13**, 155—160 (1941).
Schwarze, P.: Die Verwendung der refractometrischen Fettbestimmung zu Serienuntersuchungen an Zuchtmaterial. Züchter **12**, 164—167 (1940).
Scott Watson, J. A.: Science and the farmer. Lecture to the conference of Scottish agricultural students. Edinburgh, January 31, 1947. J. Minist. Agric. **54**, 1—9 (1947).
Seleznev, N. N. u. Z. F. Tomašević: Exhibit at the 1940 Agricultural Exhibition of the achievements of the State Breeding Stations and of the most distinguished Soviet breeders.) Selekcija i Semenovodstvo (Breeding and Seed Growing), Nr. 5, 3—5 (1940). [Russisch.]
Sen, S. N.: Plant breeding in India. Sci. and Cult. **10**, 110—111 (1944).
Seneviratne, L. J. de S.: Administration report of the Acting Director of Agriculture. Ceylon for 1943, Pp. D. 18 (1945).
Sengbusch, R. von: Theorie und Praxis der Pflanzenzüchtung. Societäts-Verlag Frankfurt a. M., Pp. 127 (1939).
Singh, M. P.: Crop improvement by selection. Allahabad Fmr. **14**, 117—127 (1940).
Skirm, G. W.: Embryo culturing as an aid to plant breeding. J. Hered. **33**, 211—215 (1942).
Skorohodov, P. I.: (Über die Rationalisierung der Kreuzungsmethodik und -technik der Körnerpflanzen.) C. R. (Doklady) Acad. Sci. URSS. **58**, 123—125 (1947). [Russisch.]
Skvorcov, S. N.: (Demonstration at the Soviet Agricultural Exhibition of achievements in the work of breeding and seed production.) Selekcija i Semenovodstvo (Breeding and Seed Growing), Nr. 7, 6—10 (1939). [Russisch.]

Smirnov, V.: (Immediate problems of breeding.) Soviet Subtropics, Nr. 10, 56—58 (1939). [Russisch.]
Smith, H. C.: Annual report of the Department of Agriculture. Tasmania, for 1945—1946. Nr. 19, 21 (1946).
Soliterman, P. V.: (Zonal conference of workers in breeding stations, variety control centres and the seed variety store in the south.) Selekcija i Semenovodstvo (Breeding and Seed Growing), Nos. 11/12, 1—7 (1940). [Russisch.]
Starr, S. H.: Twenty-first Annual Report 1940—1941. Bull. Ga. Coastal Plain Exp. Sta., Nr. 32, 139 (1941).
Stephens, S. G.: The application of genetics to plant breeding. Trop. Agriculture Trin. **21**, 126—129 (1944).
Stevens, N. E.: How plant breeding programs complicate plant disease problems. Science **95**, 313—316 (1942).
Stockdale, F.: The application of economic botany in the tropics. Bull. Imp. Inst., Lond. **37**, 546—555 (1939).
— The application of economic botany in the tropics. Trop. Agriculturist **94**, 250—256 (1940).
Stojković, L.: (Cereal improvement and the principles of organization of agricultural research in Czechoslovakia.) Arhiv Minist. Poljopr. **6**, 153—160 (1939). [Serbisch.]
Swaine, J. M.: Presidential address. Scientific research, the key to progress in agriculture. Trans. Roy. Soc. Can. **33**, 18 (3. Ser., Sect. V) (1939).
Symons, T. B.: Twenty-sixth Annual Report of the University of Maryland Extension Service for the year 1940. Pp. 87 (1940).

Taggart, W. G.: Agricultural research in Louisiana 1941 bis 1942. Rep. La Agric. Exp. Sta., Pp. 145 (1942).
Telschow, E.: Tätigkeitsbericht der Kaiser Wilhelm-Gesellschaft zur Förderung der Wissenschaften für das Geschäftsjahr 1940—1941. Naturwiss. **29**, 425—433 (1941).
Thadani, K. I.: Annual report of the Department of Agriculture, Sind for the year 1940—1941 (up to 30th June, 1941), Karachi, Pp. 196 (1943).
Tjallama, H. T.: Lijst van de in 1945 en 1946 door de rijkslandbouwconsulenten en andere instanties genomen veldproeven. (List of the field trials conducted by the Government legal advisors and other authorities.) Meded. Landbvoorlicht Dienst, Wageningen, Nr. 44, 100 (1946).
Törnqvist, G. I.: Några resultat av verksamheten vid Sveriges Utsädesförenings Övre-Norrlandsfilial. (Some results of the work of the Upper Norrland Branch Station of the Swedish Seed Association.) Sverig. Utsädesfören. Tidskr. **55**, 397—404 (1945).
Torpe, N. V.: Kort redogörelse för Värmlandsfilialens verksamhet och arbetsuppgifter. Föredrag vid Sveriges Utsädesförenings årsmöte i Karlstad den 27 juli 1943. (Short report on the work and future tasks of the Värmland Branch Station-Report at the meeting of the Swedish Seed Association in Karlstadt. 27 July 1943.) Sverig. Utsädesfören. Tidskr. **53**, 405—411 (1943).
Torssell, R.: Arbetsresultat och framtidsperspektiv vid Ultunafilialens verksamhet. Föredrag vid Sveriges Utsädesförenings årsmöte i Uppsala den 18 juli 1939. (Results and future prospects of the work of the Ultuna Branch-Lecture at the annual meeting of the Swedish Seed Association in Uppsala, 18th July, 1939.) Sverig. Utsädesfören. Tidskr. **49**, 275—306 (1939).
— De senaste resultaten för praktiken från vår växtförädling. (The latest results for the practical man from our plant breeding.) Årsb. Jordbruksforskning, Stockholm, Pp. 39—51 (1945).
Tschermak-Seysenegg, E. von: Theodor Roemer zum 60. Geburtstag. Z. Pflanzenz. **25**, 187—189 (1943).
Tukey, H. B.: Plant breeding by incubator methods. Sci. Mon., N. Y. **58**, 321—322 (1944).

Ullrich, H.: Möglichkeiten der Weberkardenzüchtung. Züchter 11, 44—47 (1939).

Valle, O.: Sort-och standardiseringsfrågan i Finland. (The variety and standardization problems in Finland.) Årsb. Jordbruksforskning, Stockholm, Pp. 145—150 (1945).

Varney, H. R.: Fifty-seventh Annual Report of the Vermont Agricultural Experiment Station-July 1, 1943—June 30, 1944. Bull. Vt Agric. Exp. Sta., Nr. 520, 34 (1944).

Vasconcellos, J. de Carvalho e: Acêrca da carta fitogeográfica. (Regarding phytogeographical maps.) Sér. Estud. Inform. Téc. Minist. Econ., Serv. Edit. Repart. Estud., Inform. Prop., Lisboa, Nr. 20, 19 (1942).

Vavilov, N. I. u. I. L. Nikitin (Editors): (Molotov's Nikita Botanic Garden during 125 years' activity [1812 bis 1937]). Vsesojuznaja Akademija Seljskohozjajstvennyh Nauk im. V. I. Lenina Moscow, Pp. 192 (1939). [Russisch.]

Venkatraman, T. S.: Plant breeding — a solution for food scarcity. Indian Fmg. 7, 299—300 (1946).

Vesikivi, A.: Försöksresultat från Finska Mosskulturföreningens försöksstationer år 1941. (Results from the experimental stations of the Finnish Society for the Cultivation of Bogland, 1941.) Finska MossFören. Årsb. 46, 45—76 (1942).

Viljoen, P. R.: A review of the Union's Agricultural Industry. Annual Report of the Secretary for Agriculture and Forestry for the year ended 31st August, 1939. Fmg S. Afr., Pp. 467—594 (1939).

Vilmorin, R. de: La génétique et son rôle en agriculture. C. R. Acad. Agric. Fr. 32, 640—645 (1946).

Wahlen, F. T.: Bericht über die Tätigkeit der Eidg. landwirtschaftlichen Versuchsanstalt Zürich-Oerlikon für die Jahre 1934—1938. Landw. Jb. Schweiz 54, 271 bis 357 (1940).

Walker, R. H.: Research aids Utah agriculture. Biennial Report 1938—1940. Bull. Utah. Agric. Exp. Sta., Nr. 294, 119 (1940).

Wallace, H. A.: Report of the Secretary of Agriculture Washington, December 1, 1939. U. S. Dep. Agric., Pp. 169 (1939).

— Report of the Secretary of Agriculture Washington, September 4, 1940. U. S. Dep. Agric., Pp. 184 (1940).

Wanner, H.: Strukturelle Hibridität bei Lilium umbellatum. Vjschr. naturforsch. Ges. Zürich 86, 299—306 (1941).

Waters, H. B.: Agriculture in the Gold Coast. Emp. J. Exp. Agric. 12, 83—102 (1944).

Watkins, D. W.: Farming for victory. Rep. Dir. Co-oper. Ext. Wrk. S. Carolina, U.S.D.A., Pp. 156 (1941).

Webster, G. T.: Nebraska outstate crops and soils tests. Variety tests for 1944. Bull. Neb. Agric. Exp. Sta., Nr. 372, 38 (1945).

Weger, N.: Die Temperaturverhältnisse in Isolierkästen. Gartenbauwirtsch. 14, 604—613 (1940).

Weibull, W.: Till vårt lands jordbrukare och trädgårdsodlare. (To the agriculturists and horticulturists of our country.) Weibulls III. Årsb. 39, 2—3 (1944).

Weiss, M. G.: Nursery planter. J. Amer. Soc. Agron. 33, 472—474 (1941).

Wellensiek, S. J.: De selecție der tropische gewassen. (The selection of the tropical crops.) Landbouw Tijdschr., Wageningen 53, 240—253 (1941).

— Vegetatieve vermeerdering en plantenveredeling, speciaal bij rogge. (Vegetative multiplication and plant breeding, especially in rye.) Landbouwk. Tijdschr. Wageningen 54, 422—436 (1942).

— Grondslagen der algemeene plantenveredeling. (The principles of general plant breeding.) H. D. Tjeenk Willink and Zoon N. V. Haarlem, Pp. XVI+492 78 figs, tables (1943).

Wettstein, F. v.: Erich von Tschermak-Seysenegg zum 70. Geburtstag. Z. Pflanzenz. 24, 301—303 (1941).

Wettstein, W. v.: Möglichkeiten der Züchtung neuer Ökotypen nach Kreuzung. Züchter 14, 282—285 (1942).

Wilson, H. K.: Agronomic advances in the agriculture of the corn belt and the Great Plains regions. Science 99, 499—505 (1944).

Winter, F. L., W. H. Pierce u. G. W. Scott: Research and the vegetable seed industry. West Cann. Pack. 31, Nr. 12, 11—12, 46 (1939).

Witte, H.: Sveriges Fröodlareförbunds Riksfröutställningar under åren 1921—1945. (National seed shows of the Swedish Seed Growers Union during the years 1921—1945.) Sverig. Utsädesfören. Tidskr. 55, 201 bis 230 (1945).

Wraae-Jensen, H.: Gennemforelse af lokale Forsog. (The carrying out of local trials.) Tidsskr. Landokon. 10, 497—511 (1944).

X.: Utsädesbolaget i Svalöf år. Vetenskap och praktik i fruktbärande samverkan. (The Seed Company in Svalöf-Jubilee-Science and practice in fruitful collaboration.) Lantmannen 25, 417—418 (1941).

Yarnell, S. H.: Influence of the environment on the expression of hereditary factors in relation to plant breeding. Proc. Amer. Soc. Hort Sci. 41, 398—411 1942, also Science 96, 505—508 (1942).

Žebrak, A. R.: Soviet biology. Science 102, 357—358 (1945).

Zembrovskij, I. M., N. E. Krutikov u. P. K. Ažigoev: (A programme of investigation into agricultural problems to be undertaken at the Kazah Agricultural Institute.) Bjull. Kazah. Naučno-issled. Inst. Zemled. im. Akad. V. R. Viljjamsa (Bull. V. R. Williams Kazah. Inst. Agric.), Nr. 5/6, 39 (1940). [Russisch.]

Zhigach, K.: Achievements of Soviet technique. Soviet News, Nr. 1734, 1 (1947).

Zwoboda, A.: Eine spontane, fertile Artkreuzung. Z. Pflanzenz. 24, 339—340 (1941).

d) Botanik, Angewandte Botanik, Entwicklungsphysiologie.

Åberg, E.: Problems in the classification of cultivated plants. Chronica Botanica 7, 375—378 (1943).

Abrams, L.: Illustrated flora of the Pacific States Washington, Oregon and California. Vol. II. Polygonaceae to Krameriaceae (Buckwheats to Kramerias). Stanford University Press, California and Oxford University Press, London Pp. VIII+635, illus. 1300—2962 (1944).

Adams, G. u. S. L. Smith: Experiment station research on the vitamin content and the preservation of foods. Misc. Publ. U. S. Dep. Agric., Nr. 536, 88 (1944).

Addicott, F. T.: Pollen germination and pollen tube growth, as influenced by pure growth substances. Plant Physiol. 18, 270—279 (1943).

Adler, R.: Das Wesen der Kurz- und Langtagpflanzen. Forschungsdienst 9, 332—367 (1940).

Adriance, G. W. u. F. R. Brison: Propagation of horticultural plants. McGraw-Hill Publishing Company, Ltd., London (1939).

Agresti, O. R.: David Lubin. A study in practical idealism. Univ. Calif. Press (1941).

Åkerman, Å.: Professor H. Nilsson-Ehle och hans insats inom växtförädlingen. (Prof. H. Nilsson-Ehle and his contribution to plant improvement.) Svenska BryggFören. Månadsbl., Nr. 4, 5 (1939).

Alberts, H. W.: The forage resources of Latin America. Peru. Imperial Bureau of Pastures and Forage Crops, Aberytstwyth, Bull. 37, 2s. 6d. Pp. 24 (1947).

Anderson, E.: Mass collections. Chronica Botanica 7, 378—380 (1943).
— u. D. Schregardus: A method of recording and analyzing variations of internode pattern. Ann. Mo. Bot. Gdn. 31, 241—247 (1944).
André, É u. M. Kogane-Charles: Sur quelques caractères chimiques des graines de colza utilisables en vue de sélectioner les meilleures variétés. C. R. Acad. Sci. Paris 215, 587—588 (1942).
Anonymus: Bibliography on cold resistance in plants. Imperial Bureau of Plant Breeding and Genetics, Cambridge, p. 22 (1939).
— The cacti of Arizona. Univ. Ariz. Bull. 11, 1—134, 52 pls. (1940).
— Hormona vegetal de efectos sorprendentes. (Plant hormone with surprising effects.) Ciencia, Mexico, D. F. 2, 363 (1941).
— Nyttevekstboka. (A book of useful plants.) Nyttevekstforeningen Øystese Trykkeri, Pp. 356 figs. photos (1942).
— Cold Spring Harbor Symposia on quantitative biology. Volume X. The relation of hormones to development. Biol. Lab. Cold Spring Harbor L. I., N. Y., Pp. XI +167. illus (1942).
— Differences in the systematics of plants and animals and their dependence on differences in structure, function and behaviour in the two groups. Pros. Linn. Soc., Lond. 153, 272—287 (1942).
— Imperata cylindrica. Taxonomy, distribution, economic significance and control. Imperial Agricultural Bureaux Joint Publication, Great Britain, Nr. 7, 63 (1944).
— Plant industry. 3. Botanists return from expeditions. Agriculture, Moskow, Nr. 11, 3—4 (1945). [Mimeographed.]
— Botanists resume foreign contacts. Soviet News, Nr. 1434, 4 (1946).
Arber, A.: Goethe's botany. Chronica Botanica 10, 67 bis 124. Published by Chronica Botanica Co., Waltham, Mass.; Wm. Dawson and Son, Ltd., London (1946).

Babcock, E. B.: Alice Eastwood Semi-Centennial Publications. Nr. 11. Endemism in Crepis. Proc. Calif. Acad. Sci. 25, 269—289 (1944).
Batalla de Rodriguez, M. A.: Estudio de las plantas cultivadas en la región de Izúcar de Matamoros, Pue. (Study of the cultivated plants in the region of Izúcar de Matamoros, Puebla.) An. Inst. Biol. Univ., Méx. 13, 463—489 (1942).
Bateman, A. J.: Specific differences in Petunia. II. Pollen growth. J. Genet. 45, 236—242 (1943).
— Contamination of seed crops. I. Insect pollination. J. Genet. 48, 257—275 (1947).
Beck, W. A. u. R. A. Joly: Studies in pollen tube culture. Stud. Institutum Divi Thomae, Cincinnati 3, 81—101 (1941).
Beetle, A. A.: Spezific decapitalization. Chronica Botanica 7, 380—381 (1943).
Bertsch, K. u. F.: Geschichte unserer Kulturpflanzen. Wissenschaftliche Verlagsgesellschaft m. b. H., Stuttgart, Pp. 268 (1947).
Bhaduri, P. N.: Improved smear methods for rapid double staining. J. Roy. Micros. Soc. 60, Nr. 1—2, 1—7 (1940).
Biswas, K.: The Royal Botanic Garden, Calcutta. Sci. and Cult. 6, 26—33 (1940).
Blake, S. F. u. A. C. Atwood: Geographical guide to floras of the world. Part I. Misc. Publ. U. S. Dep. Agric., Nr. 401, 336 (1942).
Blakeslee, A. F.: Micro-grafting, a method of securing seedlings from excised embryos which fail to develop roots. Amer. J. Bot. 31, Nr. 8, p. 1s. (1944). (Abstr.)
— Removing some of the barriers to crossability in plants Proc. Amer. Phil. Soc. 89, 561—574 (1945).
Bloch, R.: Irreversible differentiation in certain plant cell lineages. Science 106, 320—322 (1947).
Boerger, A.: Recursos vegetales del Uruguay. (Plant resources of Uruguay.) Chronica Botanica 7, 27—29 (1942).
Bojarkin, A. N.: (Eine neue Methode zur quantitativen Bestimmung der Aktivität der Wuchsstoffe.) C. R. (Doklady) Acad. Sci. URSS. 57, 197 (1947). [Russisch.]
Brink, R. A. u. D. C. Cooper: The endosperm in seed development. Bot. Rev. 13, 423—477, 479—541 (1947).
Bouillenne, R.: Phytobiologie. (Plant biology.) Paris, Pp. 787, 287 figs (1946).
— u. J. de Roubaix: Action des hormones sur la croissance de la betterave sucrière. Bull. Soc. Roy. Sci. Liège 11, 656—675 (1942).
Bounoure, L.: Sexe et intersexualité dans la biologie moderne. Rev. gén. Sc. 50, 178—186 (1939).
Bower, F. O.: Botany of the living plant. MacMillan and Co., Ltd., London (1939).
Brand, D. D.: The origin and early distribution of New World cultivated plants. Agric. Hist. 13, 109—117 (1939).
Breslavec, L.: Induced double flowers in stocks. C. R. (Doklady) Acad. Sci. URSS. 40, 206—207 (1943).
Burkart, A.: Las Leguminosas argentinas silvestres y cultivades. (The wild and cultivated Argentine Leguminosae.) Buenos Aires, Pp. XIX + 590. 123 figs. illus. tables (1943).
Burkland, E. R.: Plants America gave the world. Agric. Amer. 1, Nr. 8, 1—6, 16 (1941).

Cain, S. A.: Criteria for the indication of center of origin in plant geographical studies. Torreya 43, 132—154 (1943).
— Foundations of plant geography. N. Y., Pp. XIV + 556 32 tables. 63 figs (1944).
Camp, W. H.: The herbarium in modern systematics. Amer. Nat. 77, 322—344 (1943).
—, H. W. Rickett u. C. A. Weatherby: International Rules of Botanical Nomenclature. Brittonia 6, 1—120 (1947).
Cannon, W. B. u. R. M. Field: International relations in science: a review of their aims and methods in the past and in the future. Chronica Botanica 9, 253—298 (1945).
Carter, G. F.: Plant geography and culture history in the American Southwest. Publications in Anthropology, Viking Fund, New York, Nr. 5, 140, 27 figs, 6 tables (1945).
Cartter, J. L.: Equipment for maintaining controlled temperature and low humidity in a seed storage room. J. Amer. Soc. Agron. 34, 1017—1027 (1942).
Charetschko-Sawizkaja, H.: Selektive Befruchtung. Z. Pflanzenz. 26, 187—198 (1944).
Chevalier, A.: Les idées de Lamarck sur les plantes cultivées et les sources de ses informations sur leur origine et leurs variations. Rev. Bot. Appl. 26, 245 bis 255 (1946).
Chodat, F.: La colchicine, clef de la fécondité. Rev. hort. suisse 13, 169—173 (1940).
Cicin, N.: Moscow's new botanical gardens. Soviet News, Nr. 1697, 4 (1947).
Cockerell, T. D. A.: Problems of nomenclature. Nature, Lond. 155, 548 (1945).
Cooper, W. C. u. V. T. Stoutemyer: Suggestions for the use of growth substances in the vegetative propagation of tropical plants. Trop. Agriculture, Trin. 22, 21—31 (1945).
Copeland, H. F.: Biological terminology. Science 100, 265—266 (1944).
Core, E. L., E. E. Berkley u. H. A. Davis: West Virginia grasses. Bull. W. Va Agric. Exp. St., Nr. 313, 96, 40 pls (1944).

Coutinho, L. A.: Formação de células polinucleadas pela acção da traumatina. (Formation of polynucleate cells by the action of traumatin.) Rev. Agron. Lisboa **29**, 102 (1941).
Crane, M. B.: The classification of horticultural plants, varieties, synonyms and strains. J. R. Hort. Soc. **71**, 56—61 (1946).
Croizat, L.: What is the trinomial Typicus? II. Bull. Torrey Bot. Cl. **70**, 406—417 (1943).
— The homonym question. Madroño **7**, 159—160 (1944).
— History and nomenclature of the higher units of classification. Bull. Torrey Bot. Cl. **72**, 52—75 (1945).

Dade, H. A.: Colour terminology in biology. Mycological Papers, Nr. 6. Kew, Surrey, Pp. 21. 2 charts (1943).
Dahl, A. O.: A preliminary investigation of the acenaphthene response seen in certain seedlings. Amer. J. Bot. **26** (1939) 5 s (1939). (Abstr.)
Dahlgren, B. E. u. P. C. Standley: Edible and poisonous plants of the Caribbean region. Washington, D. C., Pp. IV+102. 72 illus. tables (1944).
Dickson, B. T.: Standardized plant names. A list of standard common names for the more important Australian grasses, other pasture plants, and weeds. Bull. Coun. Sci. Industr. Res. Aust., Nr. 156, 99 (1942).
Dobzhansky, Th.: The Race Concept in Biology. Scient. Monthly 1941, Febr., Pp. 161—165 (1941).
Doraşami, L. S. u. K. Gopala Iyengar: Vegetables from wild plants. Mysore Agric. J. **21**, 32—35 (1942 bis 1943).
Džaparidze, L. I.: (Aristotle's concept on the existence of sex in plants.) Sovetskaja Botanika (Soviet Botany), Nr. 3, 194—199 (1946). [Russisch.]

East, E. M.: The distribution of self-sterility in the flowering plants. Proc. Amer. Phil. Soc. **82**, 449—518 (1940).
Engard, C. J.: Organogenesis in Rubus. Honolulu, Res. Publ., Nr. 21, XVI+234. 448 figs. 3 tables (1944).
Engel, Chr.: The tocopherol (vitamin E) content of milling products from wheat, rye and barley and the influence of bleaching. Z. Vitaminforsch. (Bern) **12**, 220—222 (1942).
Epling, C.: Taxonomy and genonomy. Science **98**, 515 bis 516 (1944).
Ernst, H.: Die photoperiodische Reaktion bei autotetraploidem Antirrhinum majus L. Ber. dtsch. bot. Ges. **59**, 351—355 (1941).
Eyster, W. H.: The induction of fertility in genetically self-sterile plants. Science **94**, 144—145 (1941).

Fagerlind, F.: Die Samenbildung und die Zytologie bei agamospermischen und sexuellen Arten von Elatostema und einigen nahestehenden Gattungen nebst Beleuchtung einiger damit zusammenhängender Probleme. K. Svenska Vetensk Akad. Handl. **21**, Nr. 4, 130 (1944).
— Der tetrasporische Angiospermen-Embryosack und dessen Bedeutung für das Verständnis der Entwicklungsmechanik und Phylogenie des Embryosacks. Ark. Bot. **31** A, 1—71 (1944).
— Is my terminology of the apomictic phenomena of 1940 incorrect and inappropriate? Hereditas, Lund **30**, 590—596 (1944).
— Makrosporogenese und Embryosackbildung bei agamospermischen Taraxacum-Biotypen. Svensk. Bot. Tidskr. **41**, 365—390 (1947).
Fassett, N. C.: The leguminous plants of Wisconsin. University of Wisconsin Press, Madison, Pp. XII+157. 59 figs. 24 plates. maps (1939).
— A manual of aquatic plants. McGraw-Hill Publishing Company, Ltd., London (1940).

Feilden, G. St. C.: Vegetative propagation of tropical and sub-tropical plantation crops. Techn. Commun. Bur. Hort. Plant. Crops, East Malling, Kent (1940).
Fernald, M. L. u. A. C. Kinsey: Edible wild plants of eastern North America. N. Y., Pp. XIV+452. 129 figs. 25 plates (1943).
Fernandes, A.: Sur la position systématique et l'origine de Narcissus Broussonetii Lag. Bol. Soc. Broteriana **14**, 53—66 (1940).
Fish, S.: The Plant Research Laboratories, Burnley. Wartime activities. J. Dep. Agric. Vict. **43**, 386—388 (1945).
Fisher, R. A. u. V. C. Martin: Spontaneous occurrence in Lythrum salicaria of plants duplex for the shortstyle gene. Nature, Lond. **160**, 541 (1947).
Frenguelli, J.: Rasgos principales de fitogeografía Argentina. (Principale characteristics of the phytogeography of Argentina.) Publ. Didácticas Divulg. Cient. del Museo de La Plata, República Argentina, Nr. 2, 119. plates. figs (1940).
Fröier, K.: Näringsrika vilda växter. (Nutritious wild plants.) Stockholm, Pp. 32. 32 figs (1942).

Gardner, V. R. u. W. D. Baten: Studies in the nature of the clonal variety. II. Selection within a periclinal chimera. Techn. Bull. Mich. Agric. Exp. Sta., Nr. 179, 48 (1942).
Gavaudan, P., N. Gavaudan u. J.-F. Durand: Nouvelles considérations sur l'activité modificatrice de la caryocinnese et de la cytodiérnese exercée sur les végétaux par quelques hydrocarbures cycliques et leurs dérivés. C. R. Acad. Sci., Paris **210**, 114—116 (1940).
Gentry, H. S.: Rio Mayo plants. A study of the flora and vegetation of the valley of the Rio Mayo, Sonora. Carnegie Institution of Washington. Washington, D. C. Publ. **527**, Pp. VII+328. 29 pls. (1942).
Gericke, W. F.: The complete guide to soilless gardening. Putnam and Co., Ltd., London (1940).
Gibbs, R. D.: Comparative chemistry as an aid to the solution of problems in systematic botany. Trans. Roy. Soc. Can. **39**, Sect. 5, 71—103. 1945 (1946).
Gimesi, N.: A Cucurbita pepo pollenexinéjének keletkezése és fejlodése. (Origin and ontogeny of the pollen exine of C. pepo.) Bull. Hung. Coll. Hort. **7**, 139—145 (1941).
Glinyany, N. P.: Variations in the vegetation period. C. R. (Doklady) Acad. Sci. URSS. **25**, 78—82 (1939)
Golubinskij, I. N.: (Gegenseitige Beeinflussung der Pollenkörner verschiedener Arten bei gemeinschaftlicher Keimung auf künstlichen Substraten.) C. R. (Doklady) Acad. Sci. URSS. **54**, 73 (1946). [Russisch.]
— (Influence of pollen mixtures and density of pollen on its germination.) Agrobiologija (Agrobiology), Nr. 3, 59—70 (1946). [Russisch.]
— (Zur Kenntnis der Keimungsphysiologie der Pollen. III. Über den Einfluß der Narben auf die Keimung des Pollenkornes.) C. R. (Doklady) Acad. Sci. URSS. **55**, 773—776 (1947). [Russisch.]
Goodwin, A. J. H.: The origins of certain African foodplants. S. Afr. J. Sci. **36**, 445—463 (1939).
Graves, G.: Trees, shrubs, and vines for the northeastern United States. London, Pp. XI+267, 68 figs (1945).
Gray, W. D.: The existence of physiological strains in Physarum polycephalum. Amer. J. Bot. **32**, 157—160 (1945).
Greenway, P. J.: Origins of some East African food plants. Part. 1. E. Afr. Agric. J. **10**, 34—39 (1944).
Gregory, W. C.: Phylogenetic and cytological studies in the Ranunculaceae Juss. Trans. Amer. Phil. Soc. **31**, 443—521 (1941).
Greulach, V. A.: „Photoperiodism" versus „photoperiodicity". Science **101**, 353—354 (1945).

Gustafson, F. G.: Probable causes for the difference in facility of producing parthenocarpic fruits in different plants. Proc. Amer. Soc. Hort. Sci. **38**, 479—481 (1941).
— Parthenocarpy: natural and artificial. Bot. Rev. **8**, 599—654 (1942).
Gustafsson, Å.: The X-ray resistance of dormant seeds in some agricultural plants. Hereditas, Lund **30**, 165 bis 178 (1944).
— The terminology of the apomictic phenomena. Hereditas, Lund **30**, 145—151 (1944).
— Apomixis in higher plants. Part. I. The mechanism of apomixis. Acta Univ. Lund **57**, 66 (1946).
— Apomixis in higher plants. Part. II. The causal aspect of apomixis. Acta. Univ. Lund. **43**, 71—178 (1947).
Gutiev, G. T.: (Reorganize the chain of subtropical scientific institutes.) Soviet Subtropics, Nr. 11/12 (75 bis 76), 12—14 (1940). [Russisch.]
Guyot, A. L.: Critères biologiques et systématique végétale. Bull. Soc. Bot. Fr. **92**, 143—146 (1945).
— Genèse de la flore terrestre. Paris, Pp. 136 (1946).
Györffy, B.: A colchicin hatásmechanizmusa. (A review on the mechanism of colchicine action.) Arb. Ung. Biol. Forsch. Inst. **12**, 330—351 (1940).

H...., F. S.: Plant Science Research in Iran. Chronica Botanica **6**, 353 (1941).
Hatch, M. H.: The Species Concept. The logical Basis of the Species Concept. Amer. Natural. **75**, 193—212 (1941).
Heim, R.: La réproduction chez les plantes. Armand Colin, Paris, Collection, Nr. 220 (1939).
Heyrovsky, J. u. H. Hasselbach: Die polarographische Carotinbestimmung. Z. Pflanzenz. **25**, 443—450 (1943).
Hlebnikova, N. A.: Growth and development of white poppy on varying daylength. C. R. (Doklady) Acad. Sci. URSS. **32**, 503—504 (1941).
— u. K. G. Moskovec: On the biology of Chenopodium ambrosioides L. C. R. (Doklady) Acad. Sci. URSS. **32**, 161—162 (1941).
Hoedt, T. G. E.: Over proeven en het nemen van proeven in onze bergcultures. (On experiments and the carrying out of experiments in regard to our plantation crops.) Bergcultures **14**, 2—14 (1940).
Hunter, H.: The National Institute of Agricultural Botany. J. Inst. Brew. **51**, 189—194 (1945).
Hurst, E.: The poison plants of New South Wales. Poison Plants Committee of New South Wales, Sydney, Pp. XIV+498. 23 tables. figs. (1942).
Hurwitz, S.: New field crops for Palestine. 1. Forage, pasturage, and green manure. Bull. Agric. Res. Sta. Rehovoth, Nr. 26, 116 (1940).
Hutton, E. M.: A new method for tomato and cucumber seed extraction. J. Coun. Sci. Industr. Res. Aust. **16**, 97—103 (1943).
Hylander, C. J. u. O. B. Stanley: Plants and man. The Blakiston Company, Philadelphia (1941).

Ivanov, N. N. u. E. V. Dodonova: (Determination of albumen content in seeds of leguminous and oil plants with the help of colorimetric method.) Proc. Lenin Acad. Agric. Sci. URSS., Nr. 20, 23—26 (1940).
Ivanovskaja, E. V.: (Die Kultur der Bastardembryone der Gräser auf künstlichem Substrat.) C. R. (Doklady) Acad. Sci. URSS. **54**, 449—452 (1946). [Russisch.]

Jack, H. A.: Biological field stations of the world. Chronica Botanica **9**, 73 (1945).
Jeffrey, E. C.: Hormones in relation to reproduction and mutation. Amer. J. Bot. Suppl. **33**, 822 (1946). (Abstr.)
Jensen, C.: Über die Möglichkeit, mit Hilfe von Lichtbehandlung die Keimfähigkeit von Samen zu verlängern. Z. Bot. **37**, 487—499 (1942).
Jensen, I. u. H. Bogh: (Über die Bedingungen, die auf die Gefahr der Kreuzung bei windbestäubten Kulturpflanzen Einfluß haben.) Tidsskr. Planteavl **46**, 238 bis 266 (1941). [Dänisch.]
Joaquim Barroso, L.: Chaves para a determinação de gêneros Brasileiros e exóticos das dicotiledoneas mais cultivadas no Brasil. (Keys for the determination of the Brazilian and exotic genera of dicotyledons most cultivated in Brazil.) Bol. Soc. Brasil. Agron. **5**, 173 bis 182 (1942).
Johansen, D. A.: A critical survey of the present status of plant embryology. Bot. Rev. **11**, 87—107 (1945).
Jones, G. N.: American species of Amelanchier. University of Illinois Press, Urbana, Illinois, Pp. 126. 23 pls. 14 maps. (1946).
— u. T. Just (Editor): Flora of Illinois. Containing keys for identification of the flowering plants and ferns. American Midland Naturalist Monogr., Nr. 2. University Press, Notre Dame, Ind., Pp. 317, illus. (1945).
Jones, S. G.: Introduction to floral mechanism. Blackie and Son, Ltd., London and Glasgow (1939).

Katarjan, T. G.: (Sub-tropical science and its tasks.) Soviet Subtropics, Nr. 4 (56), 44—49 (1939). [Russisch.]
Kawerau, E.: Ascorbic acid. Part. 3: The ascorbic acid content of fruits and vegetables grown in Eire. Sci. Proc. R. Dublin Soc. **23**, 181—196 (1944).
Kazarjan, B. O.: (Über die Bedeutung der Lichtqualität für die photoperiodische Reaktion der Pflanzen.) C.R. (Doklady) Acad. Sci. URSS. **54**, 77 (1946). [Russisch.]
Kearney, T. H. u. R. H. Peebles: Flowering plants and ferns of Arizona. Misc. Publ. U. S. Dep. Agric., Nr. 423, 1069. 29 pls. (1942).
Kelly, W. C. u. O. Smith: Specific gravity determination as an aid in research. Proc. Amer. Soc. Hort. Sci. **44**, 329—333 (1944).
Kelsey. H. u. W. A. Dayton (Editors): Standardized plant names. Second edition. A revised and enlarged listing of approved scientific common names of plants and plant products in American commerce or use. J. Horace McFarland Company, Harrisburg (1942).
Khlebnikova, N. A. siehe Hlebnikova.
King, J. R. u. R. M. Brooks: The terminology of pollination. Science **105**, 379—380 (1947).
Kirchner, O. von, E. Loew u. C. Schröter: Lebensgeschichte der Blütenpflanzen Mitteleuropas. Strasburg, 4 volumes (incomplete).
Kjaer, A.: Spiringen af nedgravet og tørt opbevaret Frø. I. 1934—1939. (Germination of buried and dry stored seed I. 1934—1939.) Tidsskr. Planteavl. **45**, 486—507 (1941).
Klešnin, A.: Role of spectra of visible light in photoperiodic and formative processes at various developmental phases. C. R. (Doklady) Acad. Sci. URSS. **52**, Nr. 9, 813—816 (1946).
Koller, P. C.: The effects of radiation on pollen grain development, differentiation, and germination. Proc. Roy. Soc. Edinb. **61**, Abt. B, 398—428 (1943).
Komarov, V.: (Die Lehre von der Art.) Moskva-Leningrad (Verlag Akad. Wiss. UdSSR.), Pp. 212 (1940). [Russisch.]
— u. N. A. Busch (Editors): (Flora URSS.) Vol. VIII—IX. Moskva-Leningard **8/9** (1939). [Russisch.]
—, B. K. Šiškin u. S. V. Jusepčuk (Editors): (Flora URSS.) Vol. XII. Moskva-Leningrad **12**, Pp. 918 (1941). [Russisch.]
Konstantinov, N. N.: (Die Tageslänge als Bestimmungsfaktor für die Daseinsmöglichkeit einer Art.) C. R. (Doklady) Acad. Sci. URSS. **47**, Nr. 9 688 bis 690 (1945). [Russisch.]
Kosarev. M. G.: (Žitnjak [Agropyron spp.].) Moscow, Pp. 168, 119 tables. 18 figs. (1941). [Russisch.]

Kovalevskij, G. V.: (The theoretical foundations of the geography of cultivated plants.) Priroda (Nature), Nr. 1, 35—44 (1946). [Russisch.]

Kreier, G. K.: (The theory of phasic development in plant systematics.) Sovetskaja Botanika (Soviet Botany), Nr. 1/2, 39—50 (1941). [Russisch.]

Krenke, N. P.: (The theory of the cycle of senescence and rejuvenation of plants and its practical application. Moscow, Pp. 135 (1940). [Russisch.]

Kudrjašov, S. N.: (Plant resources of Uzbekistan.) J. Bot. URSS. **27**, 109—116 (1942). [Russisch.]

Kuhn, E.: Physiologie und Vererbung der Selbststerilität bei Blütenpflanzen. Naturwiss. **28**, 1—9 (1940).

Kuijper, J.: Onze kultuurgewassen. Hun geschiedenis en hun betekenis voor den mens. (Our cultivated plants. Their history and their importance for man.) Gorinchem, Pp. 220. 17 figs. 51 tables (1945).

Kundu, B. C.: A textbook of botany for junior students. Dasgupta and Co., Calcutta, Pp. VIII + 412. 219 figs. (1944).

La Cour, L. F. u. R. Drew: Partition-chromatography and living cells. Nature, Lond. **159**, 307—308 (1947).

— u. A. C. Faberge: The use of cellophane in pollen tube technic. Stain Technol. **18**, 196 (1943).

Laing, R. M. u. E. W. Blackwell: Plants of New Zealand. Whitcombe and Tombs, Ltd., London, Australia, New Zealand (1940).

Lammerts, W. E.: Embryo culture an effective technique for shortening the breeding cycle of deciduous trees and increasing germination of hybrid seed. Amer. J. Bot. **9**, 166—171 (1942).

Lang, A.: Übertragung der Hemmwirkung der Blätter auf die Blütenbildung bei Hyoscyamus niger in Kurztagbedingungen durch Pfropfung. Naturwiss. **30**, 590 bis 591 (1942).

— u. F. v. Wettstein: Entwicklungsphysiologie. Fortschr. Bot. **10**, 278—307 (1941).

La Rue, C. D.: Regeneration of endosperm of gymnosperms and angiosperms. Amer. J. Bot. **31**, Nr. 8, p. 4s. (1944). (Abstr.)

Lawrence, W. J. C. u. J. Newell: Seed and potting composts, with special reference to soil sterilization. George Allen and Unwin, Ltd., London (1939).

Lebedev, S. I.: (Über die Veränderung des Karotingehaltes der Pflanze.) C. R. (Doklady) Acad. Sci. URSS. **58**, 85—88 (1947). [Russisch.]

Lee, Tsung-Lê u. Hwang, Tsung-Chen: Growth stimulation by manganese sulfate, indole-3-acetic acid and colchicine in pollen germination and pollen tube growth. Acta Brevia Sinénsia, Nr. 8, 21—22 (1944). [Mimeographed.]

Le Gros Clark, W. E. u. P. B. Medawar ((Editors): Essays on growth and form presented to D'Arcy Wentworth Thompson. Clarendon Press, Oxford, Pp. VIII + 408, figs. tables. pls. (1945).

Lelesz, E.: The problem of increase in vitamin content of agricultural products in view of improving the diet of the people. Int. Rev. Agric. **33**, 265T—285T (1942).

Lemos Pereira, A. de: Sôbre o citoplasma e a membrana da célula vegetal. II. Microsporos e outras células, plastos e outras formações celulares perante a reacção de Feulgen. (The cytoplasm and the membrane of the plant cell. II. Microspores and other cells, plastids and other cellular formations and the Feulgen reaction.) Bol. Soc. Brot. **17**, 167—181 (1943).

Levine, M.: Colchicine and X-rays in the treatment of plant and animal overgrowths. Bot. Rev. **11**, 145—180 (1945).

Levitt, J.: Frost killing and hardiness of plants. A critical review. Burgess Publishing Company, Minneapolis, Minn. (1941).

Lewis, D.: The physiology of incompatibility in plants I. The effect of temperature. Proc. Roy. Soc. **131**,, Ser. B, Nr. 862, 13—26 (1942).

— The physiology of incompatibility in plants. II. Linum grandiflorum. Ann. Bot. Lond. **7**, 115—122 (1943).

— Physiology of incompatibility in plants. III. Autopolyploids. J. Genet. **45**, 171—185 (1943).

— Incompatibility in plants: Its genetical and physiological synthesis. Nature, Lond. **153**, 575—578 (1944).

— Splitting the gene. Discovery **8**, 168—173 (1947).

— Competition and dominance of incompatibility alleles in diploid pollen. Heredity **1**, 85—108 (1947).

— u. L. F. La Cour: Collection of pollen and artificial wind pollination. Nature, Lond. **153**, 167—168 (1944).

Loomis, W. E. u. C. A. Shull: Experiments in plant physiology. A laboratory textbook. McGraw-Hill Publishing Company, Ltd., London (1939).

Lowson, J. M.: Textbook of botany. Revised by W. O. Howarth and L. G. G. Warne. University Tutorial Press, Ltd., London, Pp. VIII + 584. 413 figs. 8 pls. (1945).

Luyet, B. J. u. P. M. Gehenio: The mechanism of injury and death by low temperature. A review. Biodynamica **3**, Nr. 60, 33—99 (1940).

McLean, R. C. u. W. R. Cook, Ivimey: Plant science formulae. A reference book for plant science laboratories (including bacteriology). MacMillan and Co., London (1941).

Maheshwari, P.: The role of growth hormones in the production of seedless fruits. Sci. and Cult. **6**, 85—89 (1940).

Maksimov, N. A. u. A. F. Klešnin: (Luminiszenzlampen als Radiationsquelle für die Lichtkultur der Pflanzen.) C. R. (Doklady) Acad. Sci. URSS. **57**, 201 (1947). [Russisch.]

—, R. H. Tureckaja u. M. F. Muhina: (Die Prüfung der physiologischen Aktivität einiger Wuchsstoffe.) C. R. (Doklady) Acad. Sci. URSS. **55**, 659—662 (1947). [Russisch.]

Manciot, A.: Les plantes sauvages utiles. Paris, Pp. 111, illus. (ohne Jahr).

Mangenot, G.: Colchicine et phytohormones. Sciences (Paris) **69**, 25—43 (1942).

Martin, A. C.: Instability in scientific namens of plants. Amer. Midl. Nat. **34**, 799—800 (1945).

— The slighted role of seeds in plant phylogeny. Amer. J. Bot. Suppl. **33**, 842 (1946). (Abstr.)

Martínez, M.: Las Pinaceas Mexicanas. (The Pinaceae of Mexico.) An. Inst. Biol. Univ. Méx. **16**, 345, 300 figs. (1945).

Mason, E. W.: New species and old. Trans. Brit. Mycol. Soc. **25**, 433—434 (1942).

Mason, H. L.: The edaphic factor in narrow endemism. Madroño **7**, 209—226, 241—257 (1946).

Mather, K.: Mating discrimination in plants. Endeavour **2**, Nr. 5, 17—21 (1943).

— Specific differences in Petunia. I. Incompatibility. J. Genet. **45**, 215—235 (1943).

— Genetical control of incompatibility in angiosperms and fungi. Nature, Lond. **153**, 392—394 (1944).

Medsger, O. P.: Edible wild plants. The MacMillan Company, London and New York (1939).

Melchers, G.: Die Auslösung des generativen Entwicklungsabschnittes der höheren Pflanzen. Züchter **14**, 177—182 (1942).

— u. A. Lang: Auslösung von Blütenbildung bei der Langtagpflanze Hyoscyamus niger in Kurztagbedingungen durch Infiltration der Blätter mit Zuckerlösungen. Naturwiss. **30**, 589—590 (1942).

Merrill, E. D.: Some economic aspects of taxonomy. Torreya **43**, 50—64 (1943).

— Plant life of the Pacific World. MacMillan Co., New York, Pp. XV + 295. 256 figs. (1945).

Merrill, E. D.: Merrilleana. A selection from the general writings. Chronica Botanica 10, 127—394 (1946).

Meyer, B. S. u. D. B. Anderson: Plant physiology. A textbook for colleges and universities. Chapman and Hall, Ltd., London (1940).

Monteiro Filho, H.: A nova sistemática. (The new systematics.) Rodriguésia, Rio de J. 6, Nr. 15, 45—52 (1942).

Müntzing, A. u. G. Müntzing: Some new results concerning apomixis, sexuality and polymorphism in potentilla. Bot. Notiser, Pp. 237—278 (1941).

Murneek, A. E.: Sexual reproduction and the carotinoid pigments in plants. Amer. Naturalist 75, 614—620 (1941).

Murphy, J. B.: The influence of magnetic fields on seed germination. Amer. J. Bot. Suppl. 29, 15 s (1942). (Abstr.)

Neergaard, P.: Danish species of Alternaria and Stemphylium. Taxonomy, parasitism, economical significance. Communication from the Phytopathological Laboratory of J. E. Ohlsens Enke, Copenhagen. Humphrey Milford, Oxford University Press, London, Pp. 560. 158 figs. 48+67 tables. (1945).

Nelson, A.: Principles of agricultural botany. Thomas Nelson and Sons, Ltd. London, 35 s. Pp. XVII+556. 182 figs. 128 half-tone pls. 17 coloured pls. (1946).

Newman, I. V.: The living plant. Wellington, N. Z., Pp. 128 (1946).

Novikov, A. B.: The concept of integrative levels and biology. Science 101, 209—215 (1945).

Nozzolini, V.: Cenni sulla teoria degli stadi di sviluppo delle piante agrarie. Jarovizzazione della patata. (Notes on the theory of phasic development of agricultural plants. Vernalization of potatoes.) Genetica Agraria, Roma 1, 60—77 (1946).

O'hanlon, M. E.: Fundamentals of plant science. N. Y., Pp. XI+488. 268 figs. (1941).

Olmsted, C. E.: Growth and development in range grasses. III. Photoperiodic responses in the genus Bouteloua. Bot. Gaz. 105, 165—181 (1943).

Östergren, G.: Note on Elymus arenarius. Bot. Not., p. 99 (1942).

Overbeek, J. van, M. E. Conklin u. A. F. Blakeslee: Factors in coconut milk essential for growth and development of very young Datura embryos. Science 94, 350—351 (1941).

—, —, — Chemical stimulation of ovule development and its possible relation to parthenogenesis. Amer. J. Bot. 28, 647—656 (1941).

Paauw, F. van der: Agronomie en botanie. (Agronomy and botany.) Vakbl. Biol. 24, 107—116 (1943).

Pal, B. P.: Degeneration of improved crops in India. Indian Fmg. 3, 261—264 (1942).

Parker, M. W. u. H. A. Borthwick: Day length and crop yields. Misc. Publ. U. S. Dep. Agric., Nr. 507, 22 (1942).

Parodi, L. R. u. A. I. Pastore: Géneros de plantas cultivadas representados en la flora indigena de la República Argentina. (Genera of cultivated plants represented in the indigenous flora of the Argentine Republic.) Physis, B. Aires 18, 255—268 (1939).

Pirschle, K.: Weitere Untersuchungen über die Auswirkung eines Gen-abhängigen Wirkstoffs bei Petunia in einem Pfropfversuch auf älteren Unterlagen. Z. Abstamm.lehre 76, 512—534 (1939).

— Über den Chlorophyllgehalt autopolyploider Pflanzen. Naturwiss. 29, 45—46 (1941).

— Stickstoff- und Eiweißanalysen an autopolyploiden Pflanzen. Planta 32, 517—534 (1941/42).

Pirschle, K.: Stickstoff- und Aschenanalysen an Wasserkulturen mit polyploiden Pflanzen. Naturwiss. 30, 646—647 (1942).

— Wasserkulturversuche mit polyploiden Pflanzen. I. Stellaria media. Biol. Zbl. 62, 253—279 (1942).

— Wasserkulturversuche mit polyploiden Pflanzen. II. Stellaria media, Einfluß von Spurenelementen. Biol. Zbl. 62, 455—482 (1942).

Poddubnaja-Arnoldi, V.: (Present position as regards the problem of asexual reproduction in angiosperms.) J. Bot. URSS. 25, 75—91 (1940). [Russisch.]

Popenoe, W.: Plant resources of Honduras. Chronica Botanica 7, 217—219 (1942).

Potapenko, A.: (Über die Entstehung der photoperiodischen Reaktion und ihre Anpassungsbedeutung.) C. R. (Doklady) Acad. Sci. URSS. 57, 959—961 (1947). [Russisch.]

Prát, S.: Rostlina pod drobnohledem. (The plant under the microscope.) Praha, Pp. 205, 82 figs. (1945).

Psarev, G. M.: (Zur Frage der Rolle des Heteroauxins bei der Formbildung und den Wachstumsprozessen der Pflanzen.) C. R. (Doklady) Acad. Sci. URSS. 56, 877 bis 880 (1947). [Russisch.]

Rabinowitsch, E. I.: Photosynthesis and related processes. Interscience Publishers, Inc., New York 1, XIV+599. 63 figs. tables. (1945).

Rabotnov, T. A.: (Un essay sur la définition de l'âge des plants herbacées.) Botaničeskij Žurnal (Journ. Botanique de l'URSS.) 31, Nr. 5, 28 (1946). [Russisch.]

Rafinesque, C. S.: A life of travels. Chronica Botanica 8, 293—360. Waltham, Mass. (1944).

Ranzi, F.: Two processes for preserving small animals, herbarium material, phytopathological specimens, etc. Int. Rev. Agric. 33, 86M—89M (1942).

Rapoport, I. A.: (Substances which destroy the symmetry of the organism.) Doklady Acad. Nauk., SSSR. 27, 370—373 (1940). [Russisch.]

Regel, C.: Beiträge zur Kenntnis von mitteleuropäischen Nutzpflanzen. Angew. Bot. 24, 465—484 (1942).

Rehder, A.: On the concept of type. Torreya 44, 6—7 (1944).

Rhind, D.: The grasses of Burma. Baptist Mission Press, Calcutta, Pp. 99 (1945).

Rivera Morales, I. u. F. Miranda: Nombres vulgares de plantas en el S. O. del estado de Puebla. (Vernacular names of plants in the south-west of the State of Puebla.) An. Inst. Biol. Univ. Méx. 13, 493—498 (1942).

Rodgers, A. D.: John Merle Coulter-missionary in science. Princeton University Press, N. J., Pp. VIII +321. photos. (1944).

— American Botany 1873—1892. Decades of transition. Oxford University Press, London, Pp. 340. illus. (1944).

Roodenburg, J. W. M.: Daglengte, bloemvorming en bloei. (Length of day, flower formation and flowering.) Vakbl. Biol. 27, 65—77 (1947).

Rose, P. G.: The Botanic Garden, Brussels. Gdnrs' Chron. 116, 228—229 (1944).

Rozanova, M. A.: (Experimentelle Grundlagen der Pflanzensystematik.) Verl. Akad. Wiss. UdSSR. Moskva-Leningrad, Pp. 255, 52 Abb. (1946). [Russisch.]

— (Problems of intra-species systematics. An analysis of the external and internal structure of some complex species of higher plants.) Žurnal Obščej Biologii (Journal of General Biology) 7, Nr. 4, 285—295 (1946). [Russisch.]

Russell, P. G.: Economic plants of interest to the Americas. Office For. Agric. Relations, U. S. Dep. Agric., Pp. 29 (ohne Jahr).

— Names of crop plants used in the Americas. Div. Pl. Explor Intro. Bur. Pl. Ind., Soils, Agric. Eng., U. S. Dep. Agric., Pp. 29 (1943).

Saint-Hilaire, A. de: Esquisse de mes voyages au Brésil et Paraguay. Chronica Botanica 10, 1—61 (1946).
Salisbury, E. J.: The reproductive capacity of plants. Studies in quantitative biology. G. Bell and Sons, Ltd., London (1942).
Samygin, G. A.: (Über die Abhängigkeit der photoperiodischen Reaktion von der Blattzahl der Pflanze.) C. R. (Doklady) Acad. Sci. URSS. 58, 147—150 (1947). [Russisch.]
Sanders, M.: Abnormalities in the development of interspecific hybrid embryos in Datura. Amer. J. Bot. 31, Nr. 8, 5 s (1944). (Abstr.)
Sanz, C.: Pollen germination and pollen-tube growth of various species of Solanaceae and other families in pistils of Datura Stramonium. Amer. J. Bot. 31, Nr. 8, 5 s (1944). (Abstr.)
Šardakov, V. S.: Reaction for peroxidase as an index of viability of plant pollen. C. R. (Doklady) Acad. Sci. URSS. 26, 267—270 (1940).
Savič, V. P.: (Komarov Botanical Institute of the Academy of Sciences of the USSR. during 1941.) Priroda (Nature), Nr. 3/4, 94—96 (1942). [Russisch.]
Ščeglova, O. A.: (Influence of environment on the development of some monoecious plants. I. Influence of the light factor.) Sovetskaja Botanika (Soviet Botany), Nr. 1/2, 90—94 (1941). [Russisch.]
Ščepotjev, F. L.: (Das Wachstum der Evonimus verrucosa im Zusammenhang mit längerem Einfluß des Kurztages.) C. R. (Doklady) Acad. Sci. URSS. 58, 151—153 (1947). [Russisch.]
— (Neue Angaben über die positive Wirkung des Kurztages auf das Wachstum des Xylems der Pflanzen.) C. R. (Doklady) Acad. Sci. URSS. 58, 2101 (1947). [Russisch.]
Schlösser, L. A.: Physiologische Untersuchungen an polyploiden Pflanzenreihen. Forsch.dienst 10, 28—40 (1940).
Schopfer, W. H.: Plants and Vitamins. Waltham, Mass. Pp. XIV+293. 17 tables. 17 figs. 3 plates. (1943).
Schwarze, P.: Ein neuer Weg der Vorbehandlung des Materials für die refraktometrische Fettbestimmung in Zuchtmaterial. (Fettbestimmung in Zuchtmaterial, 2. Mitteilung.) Züchter 13, 184—191 (1941).
— Über die Methodik der Auslese eiweißreicher Zuchtstämme und die Variabilität der Eiweißqualität in Zuchtmaterial. Z. Pflanzenz. 26, 1—55 (1944).
Schwemmle, J.: Keimversuche mit alten Samen. Z. Bot. 36, 225—261 (1940).
Serpuhova, V. I.: (A study of intensive culture phytocoenoses.) Botaničeskij Žurnal SSSR. (Journ. botanique de l'URSS.) 32, Nr. 2, 90—98 (1947). [Russisch.]
Simonet, M. u. G. Igolen: Action de quelques dérivés de la quinoléine sur la mitose de l'orge. C. R. Soc. Biol., Paris 138, 234—235 (1944).
Smart, J. (Editor): Bibliography of key works for the identification of the British fauna and flora. Linnean Society, Burlington House, London, Pp. VIII+105 (1942).
Smith, A. C.: Vegetational zones of northern South America. Amer. J. Bot. Suppl. 29, 17 s (1942). (Abstr.)
Smith, F. H.: Megagametophyte of Clintonia. Bot. Gaz. 105, 263—267 (1943).
Smith, H. M.: Categories of species names in zoology. Science 102, 185—189 (1945).
Sprague, M. L.: A discussion on the differences in observance between zoological and botanical nomenclature. Proc. Linn. Soc., London 1943—1944 156, 126—146 (1944).
Standley, P. C. u. J. A. Steyermark: Flora of Guatemala. Fieldiana (Botany) 24, Pt. V, Pp. V+502 (1946).
Stebbins, G. L. (jr.): Apomixis in angiosperms. Bot. Rev. 7, 507—542 (1941).

Stout, A. B.: The nomenclature of cultivated plants. Amer. J. Bot. 27, 339—347 (1940).
— Classes and types of intraspecific incompatibilities. Amer. Nat. 79, 481—508 (1945).
— Inactivation of incompatibilities in tetraploid progenies of Petunia axillaris. Torreya 44, 45—51 (1945).
— Types of intra-specific incompatibilities. Bull. Torrey Bot. Cl. 73, 93—95 (1946).
Stuckey, I. H.: Some effects of photoperiod on leaf growth. Amer. J. Bot. 29, 92—97 (1942).
Suhorukov, K. u. N. Bolšakova: (Freies und gebundenes Hormon der Zellteilung in den Pflanzen.) C. R. (Doklady) Acad. Sci. URSS. 53, 475 (1946). [Russisch.]
— u. O. Semovskih: (Über die Wirkungen der Auxine auf die Pflanzenzellen.) C. R. (Doklady) Acad. Sci. URSS. 54, 85 (1946). [Russisch.]
Svenson, H. K.: Modern taxonomy and its relation to geography. Torreya 43, 44—49 (1943).
— On the descriptive method of Linnaeus. Contr. Brooklyn Bot. Gdn., Nr. 103, 273—388 (1945).
— The Linnaean concept of species. Amer. J. Bot. Suppl. 33, 843 (1946). (Abstr.)

Täckholm, V., G. Täckholm u. M. Drar: Flora of Egypt. Vol. I. Pteridophyta, Gymnospermae and Angiospermae, part Monocotyledones: Typhaceae-Gramineae. Bull. Fac. Sci. Fouad Univ., Cairo, Nr. 17, 574 (1941).
Tang, P. S. u. Shih-Wei Loo: Tests on after-effects of auxin seed treatment. Amer. J. Bot. 27, 385—386 (1940).
Thimann, K. V.: III. Symposium on experimental control of development and differentiation. The hormone control of plant development. Biol. Symp. 4, 213—219 (1941).
Thomas, M.: Plant physiology. J. & A. Churchill, Ltd., London (1940).
Thommen, E.: Taschenatlas der Schweizer Flora. Verlag Birkhäuser, Basel, Pp. XIV+294. 3001 figs. (1945).
Thompson, W. P.: The causes of hybrid sterility and incompatibility. Trans. Roy. Soc. Can. 1940, 5 (Biol. Sci.), 34 (1940).
Thorndike, L. u. F. S. Benjamin (jr.): The herbal of Rufinus. University of Chicago Press, Chicago, Illinois, Pp. XLIII+476 (1946).
Tincker, M. A. H.: Recent work on germination. Proc. Linn. Soc., Lond. 154, 167—184 (1943).
Tippo, O.: A modern classification of the plant kingdom. Chronica Botanica 7, 203—206 (1942).
Transeau, E. N., H. C. Sampson u. L. H. Tiffany: Textbook of botany. Harper and Brothers, London (1940).
Tschermak-Seysenegg, E. v.: Fruchtbarkeit ohne Befruchtung. (Spontane und künstlich bewirkte Parthenogenese.) Anz. Akad. Wiss. Wien math.-nat. Klasse 9, 49—56 (1946).
Tumanov, I. I.: (Physiological bases of cold resistance in crop plants.) Seljhozgiz, Moskva-Leningrad (1940). [Russisch.]
Turbin, N. V. u. V. E. Kozlov: (Spektral-photometrische Studien über Pflanzenbastarde und die biologische Spezifität der bevorzugten Absorption des UV-Lichtes bei Pflanzengeweben.) C. R. (Doklady) Acad. Sci. URSS. 58, 1171 (1947). [Russisch.]
Tureckaja, R. H.: (Eine Methode zur Bestimmung der Aktivität von Wuchsstoffen für die Wurzelbildung.) C. R. (Doklady) Acad. Sci. URSS. 57, 295 (1947). [Russisch.]
Turesson, G.: Variation in the apomictic microspecies of Alchemilla vulgaris L. Bot. Notiser, Pp. 413—427 (1943).
Turner, W. I. u. V. M. Henry: Growing plants in nutrient solutions or scientifically controlled growth. John Wiley and Sons, Inc. New York (1939).

Turrill, W. B.: The centenary of the Royal Botanic Gardens, Kew, April 1, 1941. Chronica Botanica 6, 414—417 (1941).
— The ecotype concept. A consideration with appreciation and criticism, especially of recent trends. New Phytol. 45, 34—43 (1946).
Tzitzin, N. siehe Cicin.

Vansell, G. H.: Factors affecting the usefulness of boneybees in pollination. Circ. U. S. Dep. Agric., Nr. 650, 32 (1942).
Vasconcellos, J. de Carvalhoe: Estudo das formas cultivadas. (Study of cultivated forms.) Rev. Agron. Lisboa 29, 462—467 (1941).
Vavilov, N. I.: Entering a new epoch. Chronica Botanica 6, 433—437 (1941).
Verdoorn, F. (Editor): Plants and plant science in Latin America. Waltham, Mass., the Chronica Botanica Co.; London, W. 1, Wm. Dawson and Sons, Ltd., Pp. XXXVII+381. 38 plates. 45 illus. (1945).

Warmke, H. E. u. A. F. Blakeslee: The establishment of a 4n dioecious race in Melandrium. Amer. J. Bot. 27, 751—762 (1940).
Weatherby, C. A.: Changes in botanical names. Amer. Midl. Nat. 35, 795—796 (1946).
Weatherwax, P.: Plant biology. Philadelphia and London, Pp. VI+455. 182 figs. 1 table. (1943).
Webber, J. M.: Polyembryony. Bot. Rev. 6, 575—598 (1940).
Weevers, T.: The relation between taxonomy and chemistry of plants. Blumea, Leiden 5, 412—422 (1943).
Went, F. W.: Thermoperiodicity in plants. Amer. J. Bot. (Suppl.) 33, 224 (1946). (Abstr.)
White, O. E.: Fasciation — its characteristics, distribution, origin, hereditability, relation to environment, and plant-world analogy to cancer. Genetics 24, 89 (1939). (Abstr.)
— The biology and fasciation and its relation to abnormal growth. J. Hered. 36, 11—22 (1945).
White, P. R.: A handbook of plant tissue culture. Lancaster, Pa. Pp. XIII+277 (1943).
Williams, C. B.: Area and number of species. Nature, Lond. 152, 264—267 (1943).
Williams, R. O. u. R. O. Williams (jr.): Revised third edition of the useful and ornamental plants in Trinidad and Tobago. A. L. Rhodes, M. B. E., Government Printer, Trinidad and Tobabo (1941).
Wilson, C. M. (Editor): New crops for the new world. MacMillan Co., N. Y., Pp. VIII+295. 32 pls. tables. (1945).
Winkler, H.: Die Notwendigkeit eines Zusammenfassungsverfahrens zur Überwindung der fortschreitenden Literaturzersplitterung und Unübersehbarkeit unserer Forschungsergebnisse. Ber. dtsch. bot. Ges. 57, (128) bis (138) (1939).
Wittwer, S. H. u. A. E. Murneek: Relation of sexual reproduction to development of horticultural plants. II. Physiological influence of fertilization (gametic union). Proc. Amer. Soc. Hort. Sci. 40, 205—208 (1942).
Wong, Cheong-Yin: Induced parthenocarpy of watermelon, cucumber and pepper by the use of growth promoting substances. Proc. Amer. Soc. Hort. Sci. 36, 632—636. 1939 (1938).
Woods, M. W. u. H. B. DuBuy: Further studies on plastid variegations. Amer. J. Bot. (Suppl.) 33, 225 (1946). (Abstr.)
Wulff, E. V.: An introduction to historical plant geography. Chronica Botanica Co. Waltham, Mass., Wm. Dawson and Sons, Ltd., Pp. XV+223. 35 figs. (1943).
Wyman, D.: The arboretums and botanical gardens of North America. Chronika Botanica Co., Waltham, Mass. 10: Pp. 395—498. Plates 27—43 (1947).

Yeager, A. F. u. W. P. Haubrich: A comparison of the effect of colchicine applications on plants and seeds. Proc. Amer. Soc. Hort. Sci. 45, 251—254 (1944).

Zafren, S. J.: On the theory of obtaining hay rich in provitamin A. C. R. (Doklady) Acad. Sci. URSS. 52, Nr. 8, 717—720 (1946).
Ždanova, L. P.: (Die photoperiodische Reaktion der Kurz- und Langtagpflanzen im Zusammenhang mit dem Pflanzenalter.) C. R. (Doklady) Acad. Sci. URSS. 58, 484 (1947). [Russisch.]
Zimmermann, P. W.: Present status of „plant hormones". Industr. Engng. Chem. 35, 596—601 (1943).
— u. A. E. Hitchcock: Formative effects induced with β-naphthoxyacetic acid. Contr. Boyce Thompson Inst. 12, 1—14 (1941).

e) Cytologie.

Amerio, G. u. G. Dalla Nora: Studi di citologia sperimentale: L'effetto di agenti chimici e fisici diversi sopra la struttura della cellula vegetale. (Experimentell-cytologische Untersuchungen: Die Wirkung verschiedenartiger chemischer und physikalischer Faktoren auf die Struktur der Pflanzenzelle.) Ist. Lombardo, Rend., III., s. 75, 109—118 (1942).
Amin, K. C.: Application of colchicine to cotton. Indian Fmg. 4, 257—258 (1943).
Anonymus: Use of a chemical in plantbreeding. Science (Suppl.) 90, Nr. 2328, 8; also Poona Agric. Coll. Mag. 1939, 31, 145—146. (1939).
— Changing heredity of plants. Indian Fmg. 1 33 (1940).
— Sulfa drugs potent in plants. Seed World 50, Nr. 7, 36—37 (1941).
— Radiation and living cells. Nature, Lond. 157, 738 bis 739 (1946).
Áskell u. D. Löve: Notiser-Till nordiska cytologer och botaniser. (Notes-To Scandinavian cytologists and botanists!) Svensk Bot. Tidskr. 35, 320 (1941).
Astaurov, B. L.: (Ein experimenteller Beweis des Fehlens der direkten schädigenden Wirkung der X-Strahlen auf das Cytoplasma der lebenden Zelle.) C. R. (Doklady) Acad. Sci. URSS. 58, 887 (1947). [Russisch.]
Atabekova, A. J.: (On the mechanism of formation of somatic duplications.) Bull. Acad. Sci. URSS. Sér. Biol., Pp. 510—515 (1939). [Russisch.]
Atchison, E.: A preliminary phylogenetic study of the Myrtaceae. Amer. J. Bot. (Suppl.) 33, 215 (1946). (Abstr.)
Atwood, S. S. u. H. D. Hill: The regularity of meiosis in microsporocytes of Trifolium repens. Amer. Journ. of Bot. 27, 730—735 (1940).
Auerbach, C.: Abnormal segregation after chemical treatment of Drosophila. Genetics. 32, 3—7 (1947).

Babcock, E. B. u. J. A. Jenkins: Chromosomes and phylogeny in Crepis. III. The relationships of one hundred and thirteen species. Univ. Calif. Publ. Bot. 18, 241—292 (1943).
Baker, J. R.: Cytological technique. The principles and practice of methods used to determine the structure of the metazoan cell. London, Pp. VII+211. illus. (1945).
— u. F. K. Sanders: Establishment of cytochemical techniques. Nature, Lond. 158, 129 (1946).
Balzli, J.: Mutation et gigantisme. Rev. Int. Soya 3, 161—162 (1943).
Barber, H. N.: The experimental control of chromosome pairing in Fritillaria. J. Genet. 43, 359—374 (1942).
— u. H. G. Callan: Distribution of nucleic acid in the cell. Nature, Lond. 153, 109 (1944).
Barros, N.: Contribution á l'étude caryologique du genre Leucojum L. Bol. Soc. Broteriana 13, 545—572 (1939).

Barthel, C.: Årsberättelse avgiven den 28 januari 1944. (Annual report made on 28 January 1944.) K. Lantbr-Akad. Tidskr. **83**, 11—23 ((1944).

Bates, G. H.: Colchicine-induced polyploidy in nature. Nature, Lond. **143**, 643 (1939).

— Polyploidy induced by colchicine and its economic possibilities. Nature, Lond. **144**, 315—316 (1939).

Battaglia, E.: Cariologia dei generi Aster, Agathaea, Boltonia ed alcuni dati su i generi Felicia, Erigeron (Asteraceae sez. Astereae subsez. Asterinae). Boll. Soc. adriatica Scient. Nat., Trieste **39**, 7—55 (1941).

— Sulla terminalogia dei processi meiotici. N. Giorn. Botan. Ital. n. s. **52** (1945).

Bauch, R.: Experimentell erzeugte Polyploidreihen bei der Hefe. Naturwiss. **29**, 687—688 (1941).

— Experimentelle Auslösung von Gigas-Mutationen bei der Hefe durch carcinogene Kohlenwasserstoffe. Naturwiss. **30**, 263—264 (1942).

— Über Beziehungen zwischen polyploidisierenden, carcinogenen und phytohormonalen Substanzen. Auslösung von Gigas-Mutationen der Hefe durch pflanzliche Wuchsstoffe. Naturwiss. **30**, 420—421 (1942).

Bauer, H.: Die Chromosomenmutationen. Z. Vererbungsl. **76**, 309—322 (1939).

— Der Aufbau der Chromosomen und seine Abänderung. Naturwiss. **75**, 300 (1942).

Beadle, G. W.: Biochemical genetics. Chem. Rev. **37**, 15—96 (1945).

Beal, J. M.: Induced chromosomal changes and their significance in growth and development. Amer. Nat. **76**, 239—252 (1942).

Beams, H. W. u. R. L. King: The effect of centrifugation on plant cells. Bot. Rev. **5**, 132—154 (1939).

Beatty, A. V.: Mitotic periodicity in leaves. Genetics **27**, 131 (1942). (Abstr.)

— The division of the generative nucleus in Eschscholtzia. Amer. J. Bot. **30**, 378—382 (1943).

Berger, C. A.: SAT-chromosomes. Science **92**, 380—381 (1940).

— Cytological effects of combined treatments with colchicine and naphthalene-acetic acid. Amer. J. Bot. Suppl. **33**, 817—818 (1946). (Abstr.)

— u. E. R. Witkus: A possible source of new evidence on the difference between mitosis and meiosis. Genetics **28**, 70 (1943). (Abstr.)

—, E. R. Witkus u. B. J. Sullivan: The cytological effects of benzene vapor. Bull. Torrey Bot. Cl. **71**, 620 bis 623 (1944).

Bergner, A. D.: Chromosomal interchange among six species of Datura in nature. Amer. J. Bot. **30**, 431—440 (1943).

Bernström, P.: Polyploidy induced by colchicine in Lamium. Bot. Not., Pp. 407—408 (1941).

Bertalanffy, L. von: Die organismische Auffassung und ihre Auswirkungen. Biologe **10**, 247—258, 337—345 (1941).

Bhaduri, P. N.: Cytological analysis of structural hybridity in Rhoeo discolor Hance. J. Genet. **44**, 73—85 (1942).

— u. P. C. Bose: Cyto-genetical investigations in some common cucurbits, with special reference to fragmentation of chromosomes as a physical basis of speciation. J. Genet. **48**, 237—256 (1947).

Biebl, R.: Wirkung der UV-Strahlung auf Allium-Zellen. Protoplasma (Berlin) **36**, 491—513 (1942).

Blakeslee, A. F.: The induction of polyploids and their genetic significance. Proc. 7th Int. Genet. Congr. Edinburgh 23—30 August, Pp. 65—72. 1939 (1941).

— Induced evolution in plants through chromosome changes. Proc. 8th Amer. Sci. Congr., Wash. **3**, 173 (1940).

— Effect of induced polyploidy in plants. Amer. Nat. **75**, 117—135 (1941).

Blakeslee, A. F.: III. Symposium on experimental control of development and differentiation. Effect of induced polyploidy in plants. Biol. Symp. **4**, 183—201 (1941).

—, H. E. Warmke u. A. G. Avery: Characteristics of induced polyploids in different species of angiosperms. Genetics **24**, 66 (1939). (Abstr.)

Bogyó, T.: A növény-és allatnemesítés ujabb sejttani alapjai. (Modern cytological basis of plant and animal breeding.) Mezögazdas. Kutatás **13**, 42—47 (1940).

— A poliploida szerepe a fajok kialakulásában és elterjedésében különös figyelemmel a növénynemesítésre. (The role of polyploidy in the origin and propagation of species with special regard to plant-breeding.) Dolgozatok a M. Kir. József Nádorx Müszaki és Gazdaságtudományi Egyetem Mezögazdasági Növénytani Intézetéböl, Budapest, Nr. 7, 70 (1941).

Boivin, A. u. E. Vendrely: Sur le rôle possible des deux acides nucléiques dans la cellule vivante. Experientia, Basel **3**, 32—34. (1947).

Bolle, L. u. J. Straub: Die Paarungskräfte im Hetero- und Euchromatin von tetraploider Impatiens balsamina. (Vorläufige Mitteilung.) Planta **32**, 489—492 (1941/42).

Boost, C. u. W. Ludwig: Über die Häufigkeit mehrfacher Chiasmen und ihre Beziehung zu einer gerichteten Chiasmabildung. Chromosoma **1**, 300—309 (1939).

Bourne, G. (Editor): Cytology and cell physiology. Clarendon Press, Oxford, 20s., Pp. XII+296. Illus. (1942).

Bowden, W. M.: Diploidy, polyploidy and winter hardiness relationships in the flowering plants. Amer. J. Bot. **27**, 357—371 (1940).

— The chromosome complement and its relationship to cold resistance in the higher plants. Chronica Botanica **6**, 123—125 (1940).

— Chromosome studies in tropical, subtropical, and temperate zone plants with reference to polyploidy and winter hardiness. Genetics **26**, 140—141 (1941). (Abstr.)

— A list of chromosome numbers in higher plants. I. Acanthaceae to Myrtaceae. Amer. J. Bot. **32**, 81—92 (1945).

Boyes, J. W.: A new method of plant breeding. Pr. Bull. Univ. Alberta **26**, 5 (1941).

Brachet, J.: La spécificité de la réaction de Feulgen pour la détection de l'acide thymonucléique. Experientia, Basel **2**, 142—143 (1946).

Breslavec, L. P.: (Methods of detecting polyploid plants at different stages of development.) Bull. Acad. Sci. URSS., Sér. Biol., Nr. 5, 706—716 (1940). [Russisch.]

— (The rise of the cell theory.) J. Gen. Biol. **5**, 96—122 (1944).

Brumberg, E. M. u. L. T. Larionov: Ultra-violet absorption in living and dead cells. Nature, Lond. **158**, 663—664 (1946).

Brumfield, R. T.: Cell-lineage studies in root meristems by means of chromosome rearrangements induced by X-rays. Amer. J. Bot. **30**, 101—110 (1943).

— Effect of colchicine pretreatment on the frequency of chromosomal aberrations induced by X-radiation. Proc. Nat. Acad. Sci., Wash. **29**, 190—193 (1943).

Buchholz, J. T.: Chromosome structure under the electron microscope. Science **105**, 607—610 (1947).

Buck, J. B.: On the origin of chromosome rearrangements. Genetics **24**, 66 (1939). (Abstr.)

— Micromanipulation of salivary gland chromosomes. J. Hered. **33**, 3—10 (1942).

Burton, G. W.: A cytological study of some species in the genus Paspalum. J. agricult. Res. **60**, 193—197 (1940).

— A cytological study of some species in the tribe Paniceae. Amer. J. Bot. **29**, 355—360 (1942).

Callan, H. G.: Distribution of nucleic acid in the cell. Nature, Lond. **152**, 503 (1943).

Câmara, A.: A centrifugação fonte de variações cromosómicas. (Centrifuging, a source of chromosomal variations.) Agron. Lusitana **2**, 181—202 (1940).

— Cromonemata ramificados induzidos pelos raios-X. (Ramification of the chromonemata induced by X-rays.) Boll. Soc. Ital. Biol. Sper. **15**, 61—70 (1940).

— Ruturas e rearranjos cromosómicas induzidos pelos raios-X. (Chromosome breakages and rearrangements induced by X-rays.) Scientia Genetica **1**, 339—353 (1940).

— u. S. Vasconcelos: Não-disjunção provocanda artificialmente com anidrido carbónico. (Non-disjunction induced artificially with carbon dioxide.) Brotéria **14**, 188—195 (1945).

Carey, M. A. u. E. S. McDonough: On the production of polyploidy in Allium with paradichlorbenzene. J. Hered. **34**, 238—240 (1943).

Carr, J. G.: Mechanism of the Feulgen reaction. Nature, Lond. **156**, 143—144 (1945).

Caspersson, T.: Die Eiweißverteilung in den Strukturen des Zellkerns. Chromosoma **1**, 562—604 (1940).

— Nukleinsäureketten und Genvermehrung. Chromosoma **1**, 605—619 (1940).

— Über Eiweißstoffe im Chromosomgerüst. Naturwiss. **28**, 514—515 (1940).

— Bemerkungen zur Arbeit von W. J. Schmidt: Einiges über optische Anisotropie und Feinbau von Chromatin und Chromosomen. Chromosoma **2**, 247—250 (1941).

— Studie über den Eiweißumsatz der Zelle. Naturwiss. **29**, 33—43 (1941).

— „Chromosomin" and nucleic acids. Nature, Lond. **153**, 499—500 (1944).

—, H. Landstrom-Hydén u. L. Aquilonius: Cytoplasmanukleotide in eiweißproduzierenden Drüsenzellen. Chromosoma **2**, 111—131 (1941).

— u. B. Thorell: Der endozelluläre Eiweiß- und Nukleinsäurestoffwechsel in embryonalem Gewebe. Chromosoma **2**, 132—154 (1941).

Castro, D. de: A colchicina na indução de poliplóides. (Colchicine in the induction of polyploids.) Palestras Agronómicas, Lisboa **2**, 141—155 (1939) 1940.

— Alguns efeitos da colchicina. (Certain effects of colchicine.) Agron. Lusitana **2**, 91—103 (1940).

— Nota àcêrca da acção da colquicina sôbre o centrómero (Note on the action of colchicine on the centromere.) Agron. Lusitana **4**, 61—70; also Rev. Agron. Lisboa **30**, 460 (1942).

— u. F. Carvalho Fontes: Primeiro contacto citológico com a flora halófila dos salgados de Sacavém. (First cytological contact with the halophilic flora of the saltings of Sacavém.) Brotéria **15**, 38—46 (1946).

Castronovo, A.: Estudio cariológico de doce especies de Leguminosas argentinas. (Caryological study of twelve species of Argentine Leguminosae.) Darwiniana, B. Aires **7**, 38—57 (1945).

Catcheside, D. G.: The centromere. Chronica Botanica **5**, 160—161 (1939).

— Effects of ionizing radiations on chromosomes. Biol. Rev. **20**, 14—28 (1945).

— The P-locus position effect in Oenothera. J. Genet. **48**, 31—42 (1947).

— A duplication and a deficiency in Oenothera. J. Genet, **48**, 99—110 (1947).

— u. D. E. Lea: The effect of ionization distribution on chromosome breakage by X-rays. J. Genet. **45**, 186 bis 196 (1943).

—, — u. J. M. Thoday: Types of chromosome structural change induced by the irradiation of Tradescantia microspores. J. Genet. **47**, 113—136 (1946).

—, —, — The production of chromosome structural changes in Tradescantia microspores in relation to dosage, intensity and temperature. J. Genet. **47**, 137 bis 149 (1946).

Chopinet, R.: Autopolyploidie expérimentale. Rev. Hort., Paris **116**, 53—54 (1944).

Choudhuri, H. C.: Chemical nature of chromosomes. Nature, Lond. **152**, 475 (1943).

Chu, J.: Reaction of nucleic acid to acetocarmine. Nature Lond. **157**, 513—514 (1946).

Ciferri, R.: L'impiego della colchicina in biologia. (The use of colchicine in biology.) Saggiatore **2**, 261 (1942).

Claude, A.: The constitution of protoplasm. Science **97**, 451—456 (1943).

— u. J. S. Potter: Isolation of chromatin threads from the resting nucleus of leukemic cells. J. Exp. Med. **77**, 345—354 (1943).

Cleveland, L. R.: The origin and evolution of meiosis. Science **105**, 287—289 (1947).

Conger, A. D.: Duration of prophase as an influence of chiasma frequency. Genetics. **32**, 83 (1947). (Abstr.)

Conn, H. J.: Progress in the standardization of stains. Stain Tech. **17**, 145—146 (1942).

Constantinescu, D. Gr.: Über eine Untersuchungsmethode der pflanzlichen Nucleolusstruktur. Bull. Sect. sci. Acad. roum. **23**, 282—291 (1941).

— Sur l'évolution du chondriome du sac embryonnaire de Digitalis purpurea L. C. R. Acad. Sci., Paris **216**, 206—207 (1943).

Cooper, D. C.: Haploid-diploid twin embryos in Lilium and Nicotiana. Amer. J. Bot. **30**, 408—413 (1943).

Cooper, K. W.: Invalidation of the cytological evidence for reciprocal chiasmata in the sex chromosome bivalent of male Drosophila. Proc. Nat. Acad. Sci., Wash. **30**, 50—54 (1944).

Cornman, I.: Susceptibility of Colchicum and Chlamydomonas to colchicine. Bot. Gaz. **104**, 50—62 (1942).

— A summary of evidence in favour of the traction fiber in mitosis. Amer. Nat. **78**, 410—422 (1944).

— Alteration of mitosis by coumarin and parasorbic acid. Amer. J. Bot. (Suppl.) **33**, 217 (1946). (Abstr.)

— The responses of onion and lily mitosis to coumarin and parasorbic acid. J. Exp. Biol. **23**, 292—297 (1947).

Cortvriendt, S. F., J. van Holder u. G. Verkissen: Proeven over het gebruik van groeistoffen bij het stekken van allerlei gewassen. (Experiments on the use of growth substances in raising cuttings of various plants.) Meded. Landbouwhogeschool Gent. **12**, 13—62 (1947).

Courtine, J.: La nature de la cytomyxie. C. R. Acad. Sci., Paris **209**, 234—236 (1939).

Crescini, F.: Genesi sperimentale di nuove forme coltivate e possibilità dei mezzi attuali. (Experimental production of new cultivated forms and the possibilities of the present methods.) Ital. Agric. **78**, 13—20 (1941).

Cunha, A. G. da: Le chondriome végétal et son évolution Brotéria **13**, 49—72 (1944).

Dangeard, P.: Sur les changements de structure réversibles dans le noyau et le cytoplasme des cellules de Bryonia dioica. C. R. Soc. Biol., Paris **135**, 766—768 (1941).

— Sur les changements de structure réversibles des noyaux et du cytoplasme dans les poils aériens de la courge. C. R. Acad. Sc. Paris **212**, 713—716 (1941).

— Sur quelques modifications pathologiques des plastes et sur la mise en évidence d'une membrane plastidaire. C. R. Acad. Sci., Paris **213**, 884—887 (1941).

— Sur la membrane des noyaux, dans les cellules épidermiques des écailles bulbaires d'Allium cepa. C. R. Soc. Biol., Paris **137**, 233—235 (1943).

Danielli, J. F.: Cytology, biophysics and biochemistry. Nature, Lond. **157**, 321—322 (1946).

— Establishment of cytochemical techniques. Nature, Lond. **157**, 755—757 (1946).

Danielli, J. F.: Establishment of cytochemical techniques. Nature, Lond. **158**, 129—130 (1946).

Darlington, C. D.: Misdivision and the genetics of the centromere. J. Genet. **37**, 341—364 (1939).
— Mind and Matter. Nature, Lond. **146**, 808 (1940).
— The origin of iso-chromosomes. J. Genet. **39**, 351 bis 361 (1940).
— The prime variables of meiosis. Biol. Rev. **15**, 307 bis 322 (1940).
— Polyploidy, crossing-over, and heterochromatin in Paris. Ann. of Bot. N. s. **5**, 203—216 (1941).
— Chromosome chemistry and gene action. Nature, Lond. **149**, 66—69 (1942).
— Vitamin C and chromosome number in Rosa. Nature, Lond. **150**, 404 (1942).
— u. L. F. La Cour: The handling of chromosomes. George Allen & Unwin, Ltd., London, Pp. 165 (1942).
—, *— Chromosome breakage and the nucleic acid cycle. J. Genet. **46**, 180—267 (1945).
—, — Nucleic acid and the beginning of meiosis. Nature, Lond. **157**, 875—876 (1946).
— u. E. K. Janaki Ammal: Chromosome atlas of cultivated plants. George Allen and Unwin, Ltd., London, 12s., 6d., Pp. 397. 3 tables (1945).
—, — Adaptive isochromosomes in Nicandra. Ann. Bot., London **9**, 267—281 (1945).
— u. K. Mather: Chromosome balance and interaction in Hyacinthus. J. Genet. **46**, 52—61 (1944).

Decoux, L. u. L. Ernould: Colchicine et polyploidie. (Colchicine and polyploidy.) Publ. Inst. Belge Amélior. Better. **11**, 363—426 (1943).

Dellingshausen, M. von: Zellphysiologische Untersuchungen an Epilobien mit genetisch verschiedenen Plasmen. Planta **34**, 17—33 (1944).

Demerec, M.: Chromosome structure as viewed by a geneticist. Amer. Nat. **73**, 331—338 (1939).
— et al: Cytology, Genetics and Evolution. V, 168 p. Philadelphia: Univ. of Pennsylvania Press. (1941).

Dempster, W. T.: Paraffin compression due to the rotary microtome. Stain. Tech. **18**, 13—24 (1943).

Dermen, H.: Colchicine polyploidy and technique. Bot. Rev. **6**, 599—635 (1940).
— Hormonal polyploidy in plants. Amer. J. Bot. **27**, Nr. 10, Suppl. p. 14s (1940). (Abstr.)
—, H. H. Smith u. S. L. Emsweller: The use of colchicine in plant breeding. Bur. Pl. Ind. Soils, Agric. Res. Admin., U. S. Dep. Agric., Pp. 6 (1943). [Mimeographed.]

Devaux, H.: La theorie electrique de la mitose. C. R. Soc. Biol., Paris **137**, 237—238 (1943).

Dixon, H. H.: Evidence for a mitotic hormone: observations on the mitoses of the embryo-sac of Fritillaria imperiatis. Sci. Proc. R. Dublin Soc. **24**, 119—124 (1946).

Dobson, W. J.: The effects of phosphorus starvation on the nucleic acids of Tradescantia virginiana. Amer. J. Bot. Suppl. **33**, 834 (1946). (Abstr.)

Dobzhansky, T.: A directional change in the genetic constitution of a natural population of Drosophila pseudoobscura. Heredity **1**, 53—64 (1947).

Dodson, E. O.: Some evidence for the specificity of the Feulgen reaction. Stain Techn. **21**, 103—105 (1946).

Doutreligne, J.: Les divers „types" de structure nucléaire et de mitose somatique chez les phanérogames. (The various types of nuclear structure and mitosis in phanerogams.) Cellule **48**, 191—212 (1939).

Dubinin, N. P., V. V. Hvostova u. V. V. Mansurova: (Chromosomal aberrations, lethal mutations and X-ray dosage.) C. R. (Doklady) Acad. Sci. URSS. **31**, 387 bis 389 (1941). [Russisch.]
— u. G. G. Tinjakov: Structural variability of chromosomes in urban and rural populations. C. R. (Doklady) Acad. Sci. URSS. **51**, 155—157 (1946).

Dubos, R. J.: The bacterial cell in its relation to problems of virulence, immunity and chemotherapy. Cambridge, Mass., Pp. XIX + 460. 32 figs. 44 tables. 8 pls. (1946).

Dufrenoy, J.: The fistinction between ribose- and desoxyribose-nucleoproteins and its cytological implications. Biodynamica **4**, 131—152 (1943).

Duncan, R. E.: Production of variable aneuploid numbers of chromosomes within the root tips of Paphiopedilum Wardii. Amer. J. Bot. **32**, 506—509 (1945).

Dustin, P.: Some new aspects of mitotic poisoning. Nature, Lond. **159**, 794—797 (1947).

Ehrenberg, L.: The shape of the spindle at metaphase is conditioned by the shape of its molecules. Hereditas, Lund **31**, 240 (1945). (Abstr.)
— Morphology and chemistry of the metaphase spindle. Hereditas, Lund **32**, 15—36 (1946).
— Influence of temperature on the nucleolus and its coacervate nature. Hereditas, Lund **32**, 407—418 (1946).
— u. G. Östergren: Experimental studies on nuclear and cell division. Bot. Not., Pp. 203—206 (1942).

Eigsti, O. J.: A preliminary study of the effects of certain organic substances upon cell division. Amer. J. Bot. **27**, Nr. 10 Suppl. 14s—15s (1940). (Abstr.)
— Methods for growing pollen tubes for physiological and cytological studies. Proc. Okla. Acad. Sci. **20**, 45—47 (1940).
— A comparative study of mitosis in diploid and tetraploid species. Amer. J. Bot. Suppl. **29**, 7s (1942).
— The pollen tube method for making comparisons of differences in mitotic rates between diploids and tetraploids. Genetics **32**, 85 (1947). (Abstr.)
— u. L. Schnell: A comparison of colchicine treatment with a glycerine base and water base. Genetics **28**, 73 (1943). (Abstr.)
— u. B. Tenney: Range of concentrations and number of applications of colchicine effective for the induction of polyploidy in Vinca rosea. Genetics **28**, 73—74 (1943). (Abstr.)

Ellenhorn, J. E.: (A critical revision of mitosis in connexion with the theory of the spiral chromosome structure.) J. Bot. URSS. **25**, 189—238 (1940).
— (The genesis of the chromosome complex of Rhoeo discolor Hence.) Doklady Acad. Nauk. SSSR. **27**, 361—364 (1940). [Russisch.]

Elvars, I.: On an application of the electron microscope to plant cytology. Acta Hort. Berg. **13**, 149—245 (1943).

Emberger, L.: Alexandre Guilliermond (1876—1945). Rev. Gén. Bot. **53**, 337—361 (1946).

Emsweller, S. L. u. P. Brierley: Effects of high temperature on metaphase pairing in Lilium Longiflorum. Bot. Gaz. **105**, 49—57 (1943).

Fabergé, A. C.: An experiment on chromosome fragmentation in Tradescantia by X-rays. J. Genet. **39**, 229—248 (1940).
— Homologous chromosome pairing: the physical problem. J. Genet. **43**, 121—144 (1942).
— Snail stomach cytase. A new reagent for plant cytology. Stain Technol. **20**, 1—4 (1945).

Fagerlind, F.: Verschiedene Arten von Polyploidie. Chronica Botanica **6**, 25—52 (1941).
— Konstanz und Syndeseverhältnisse der Polyploide. Chronica Botanica **6**, 320—321 (1941).
— Die Eigenschaftsregressionen der Polyploide. Chronica Botanica **6**, 349—350 (1941).
— Der Zusammenhang zwischen Perennität, Apomixis und Polyploidie. Hereditas, Lund **30**, 179—200 (1944).

Fano, V. u. L. D. Marinelli: Note on the time-intensity factor in radiobiology. Proc. Nat. Acad. Sci., Wash. **29**, 59—66 (1943).

Favarger, C.: Etude cariologique sur une espèce tétraploide du genre Silene. Bull. soc. botan. suisse **53**, 210 bis 218 (1943).
Favorskij, M. V.: New polyploidy-inducing chemicals. C. R. (Doklady) Acad. Sci. URSS. **25**, 71—74 (1939).
Feindel, W. H.: Somatic chromosomes in Vicia Faba L. Proc. N. S. Inst. Sci. **20**, 155 (1940/1941).
Fernandes, A.: Sur la caryo-sistématique du groupe Jonquilla du genre Narcissus L. Bol. Soc. Broteriana **13**, 487—544 (1939).
— Sur le comportement des chromosomes surnuméraires hétérochromatiques pendant la méiose. I. Chromosomes longs hétérobrachiaux. Bol. Soc. Broteriana **20**, 93—154 (1946).
— Sur le devenir des micronoyaux formés à la microsporogenèse. Bol. Soc. Broteriana **20**, 181—200 (1946).
— u. J. B. Neves: Sur l'origine des formes de Narcissus bulbocodium L. à 26 chromosomes. Bol. Soc. Brot **15**, 2nd. Ser., 43—132 (1941).
— u. J. A. Serra: Euchromatine et hétérochromatine dans leurs rapports avec le noyau et le nucléole. Bol. Soc. Broteriana **19**, Sér. 2a, 67—117 (1944).
Flovik, K.: Cytologi og planteforedling. (Cytology and plant breeding.) Meld. Stat. Forsøks. Holt, H30—H41. 1939 (1940).
— Cytologi og planteforedling. (Cytology and plant breeding.) Meld. Stat. Forsøksstasjoner Plantekult. 1939 (1941).
— Chromosome numbers and polyploidy within the flora of Spitzbergen. Hereditas, Lund **26**, 430—440 (1940).
Frey-Wyssling, A.: The submicroscopic Structure of the Cytoplasm. Journ. R. Microsc. Soc., Lond. **60**, 128—138 (1940).
— Über Genbau und Gengröße. 4 Jahresb. d. S. S. G. Arch. d. Jul. Klaus-Stiftg. **19**, 451—456 (1944).
Friedrich-Freksa, H.: Bei der Chromosomenkonjugation wirksame Kräfte und ihre Bedeutung für die identische Verdoppelung von Nucleoproteinen. Naturwiss. **28**, 376—379 (1940).
Fröier, K., O. Gelin u. Å. Gustafsson: The cytological response of polyploidy to x-ray dosage. Särtryck ur Botaniska Notiser. Lund, Pp. 200—216 (1941).
—, Å. Gustafsson u. O. Tedin: The relation of mitotic disturbances to X-ray dosage and polyploidy. Hereditas, Lund **28**, 165—170 (1942).
Frolova, S. L.: Study of fine chromosome structure under enzyme treatment. J. Hered. **53**, 235—246 (1944).
Fukushima, E.: Acenaphthene as a polyploidizing agent. Proc. Imp. Acad. Japan **15**, 98—100 (1939).
Gajewski, W.: Cytogenetic investigations on Anemone L. I. Anemone Janczewskii a new amphidiploid species of hybrid origin. Acta Soc. Bot. Polon. **17**, 129—194 (1946).
Garcia, J. G.: (Beitrag zum caryologisch-systematischen Studium der Gattung Lavandula L.) Bol. Soc. Broteriana, II. s. **16**, 183—193 (1942). [Portugiesisch.]
Gates, R. R.: Nucleoli, satellites and sex chromosomes. Nature, Lond. **144**, 794—795 (1939).
— Nucleoli and related nuclear structures. Bot. Rev. **8**, 337—409 (1942).
Gaulden, M. E.: An experimental study of somatic pairing. Genetics **30**, 5 (1945). (Abstr.)
Gäumann, E.: Über die Geschwindigkeit der Kernwanderung bei Pilzen. Ber. d. dtsch. Bot. Ges. **59**, 283 bis 287 (1941).
Gavaudan, P.: Action sur la caryocinèse, la cytodiérèse et la croissance végétales des hydrocarbures cycliques a deux noyaux benzéniques sans atomes de carbone communs et des dérivés nitrés et méthylés du benzène, du naphtalène et de l'acénaphtène. C. R. Soc. Biol., Paris **136**, 383—384 (1942).

Gavaudan, P.: Etude quantitative de l'action mitoinhibitrice des substances aromatiques: definition et terminologie des effets cytologiques utilisés comme test. C. R. Soc. Biol., Paris **137**, 281—283 (1943).
— Etude quantitative de la mito-inhibition dans la cellule végétale: description d'une technique applicable a divers essais pharmacodynamiques. C. R. Soc. Biol., Paris **137**, 342—343 (1943).
— u. N. Gavaudan: Action de l'apiol sur la caryocinèse et la cytodiérèse chez quelques phanérogames. C. R. Acad. Sci., Paris **209**, 805—807 (1939).
—, — Mise en évidence sur les méristèmes radiculaires de Triticum vulgare de l'existence d'une propriété mitoinhibitrice commune aux divers apiols. C. R. Soc. Biol., Paris **131**, 998—1000 (1939).
—, — Action sur la caryocinèse et la cytediérèse des végétaux, des isomères de l'apiol de Persil. C. R. Acad. Sci., Paris **210**, 576—578 (1940).
—, — Action du benzène et de ses homologues sur la caryocinèse et la cytodiérèse végétales. C. R. Soc. Biol., Paris **137**, 50 (1943).
—, — Analogie entre les effets exercés sur la caryocinèse et la cytodiérèse végétales par le chloroforme et les substances aromatiques. C. R. Soc. Biol., Paris **137**, 509—510 (1943).
Geerts, S. J.: Colchicine in het botanisch onderzoek. (Colchicine in botanical research.) Vakbl. Biol. **25**, 1—8 (1944).
Geitler, L.: Kernwachstum und Kernbau bei zwei Blütenpflanzen. Chromosoma **1**, 474—485 (1940).
— Temperaturbedingte Ausbildung von Spezialsegmenten an Chromosomenenden. Chromosoma **1**, 554—561 (1940).
— Neue Ergebnisse und Probleme auf dem Gebiet des Chromosomenbaues. Naturwiss. **28**, 649—656 (1940).
— Die Polyploidie der Dauergewebe höherer Pflanzen. Ber. dtsch. bot. Ges. **58**, 131—142 (1940).
— Neue Untersuchungen über Bau und Wachstum des Zellkerns in Geweben. Naturwiss. **28**, 241—248 (1940).
— Über die Struktur des generativen Kerns im zweikernigen Angiospermenpollen. Planta **32**, 187—195 (1941).
— Das Wachstum des Zellkerns in tierischen und pflanzlichen Geweben. Erg. Biol. **18**, 1—54 (1941).
— Über die Struktur des generativen Kerns im zweikernigen Angiospermenpollen. Planta **32**, 187—195 (1941).
— Morphologie und Entwicklungsgeschichte der Zelle. Fortschr. Bot. **10**, 1—17 (1941).
— Schnellmethoden der Kern- und Chromosomenuntersuchung. 2. umgearb. Auflage. Mit 8 Abb., IV, Pp. 28. Berlin: Borntraeger (1942).
— Kern und Chromosomenbau bei Protisten im Vergleich mit denen höherer Pflanzen und Tiere. Ergebnisse und Probleme. Naturwiss. **30**, 151—156, 162—166 (1942).
— Die Mechanik der Mitose. Naturwiss. **31**, 501—504 (1943).
— u. L. Tschermak-Woess: Cytologie der Wildbestände von Allium carinatum und oleraceum bei Lunz (Niederdonau, nördliche Kalkalpen). Naturwiss., **33**, 27 (1946).
Gerch, M.: Untersuchungen über die Bedeutung der Nucleolen im Zellkern. Z. Zellforsch. **30**, 83—528 (1940).
Ghimpu, V.: Sur les recherches caryologiques des plantes. C. R. Acad. Sc. Roum. **5**, 88—95 (1941).
Giles, N. H. (jr.): Comparative studies of the cytogenetical effects of neutrons and X-rays. Genetics **28**, 398 bis 418 (1943).
— The origin of iso-chromosomes at meiosis. Genetics **28**, 512—524 (1943).

Giles, N. H. (jr.) u. R. F. Humphreys: Comparative effects of X-rays and neutrons in inducing chromosomal rearrangements and mutations, primarily in Tradescantia; measurement of neutron dosages and intensity of neutron radiation. Yearb. Amer. Philos. Soc. 1942, Pp. 152—153 (1943).

Glotov, V.: Effect of colchicine from Colchicum umbrosum Stev. on the camphor basil. C. R. (Doklady) Acad. Sci. URSS. 24, 502—504 (1939).

Goncalves da Cunha, A.: (Noch einiges über die Herkunft der Plastiden.) Bol. Soc. Broteriana II, 16, 161 bis 164 (1942).

Goodspeed, T. H. u. M. V. Bradley: Amphidiploidy. Bot. Rev. 8, 271—316 (1942).

Grafl, I.: Kernwachstum durch Chromosomenvermehrung als regelmäßiger Vorgang bei der pflanzlichen Gewebedifferenzierung. Chromosoma 1, 265—275 (1939).

— Über das Wachstum der Antipodenkerne von Caltha palustris. Chromosoma 2, 1—11 (1941).

Graner, E. A.: Estrutura dos cromosômios. (Structure of the chromosomes.) Rev. Agric. Piracicaba 18, 419 bis 429 (1943).

Gregory, W. C.: Meiosis and the cultivation of excised anthers in nutrient solution. Amer. J. Bot. 26, 3s—4s (1939). (Abstr.)

Guinochet, M.: Sur quelques modifications des réactions physico-chimiques de la cellule végétale, provoquées par les substances mitoinhibitrices. C. R. Acad. Sci., Paris 210, 579—580 (1940).

Gulick, A.: The chemistry of the chromosomes. Bot. Rev. 7, 433—457 (1941).

Gulland, J. M., G. R. Barker u. D. O. Jordan: Structure of nucleic acid in the dividing cell. Nature, Lond. 153, 20 (1944).

—, —, — Terminology of nucleic acids. Nature, Lond. 153, 194 (1944).

Gustafson, F. G.: Growth hormone studies of some diploid and autotetraploid plants. J. Hered. 35, 269 bis 272 (1944).

Gustafsson, Å.: A general theory for the inter-relations of meiosis and mitosis. (Preliminary note.) Hereditas, Lund 25, 31—32 (1939).

— The interrelations of meiosis and mitosis. I. The mechanism of agamospermy. Hereditas 25, 289—322 (1939).

— Meiosis und Mitosis. Eine Erklärung der meiotischen Erscheinungen bei Hieracium. Chromosoma 2, 367 bis 387 (1942).

— The x-ray resistance of dormant seeds in some agricultural plants. Hereditas 30, 165—178 (1944).

— The plant species in relation to polyploidy and apomixis. Hereditas, Lund 32, 444—448 (1946).

— u. A. Nygren: The temperature effect on pollen formation and meiosis in Hieracium robustum. Hereditas, Lund 32, 1—14 (1946).

Györffy, B.: Die Colchicimethode zur Erzeugung polyploider Pflanzen. Züchter 12, 139—149 (1940).

— Untersuchungen über den osmotischen Wert polyploider Pflanzen. Planta 32, 15—37 (1941/42).

Häfliger, E.: Zytologisch-embryologische Untersuchungen pseudogamer Ranunkeln der Auricomus-Gruppe. Ber. schweiz. bot. Ges. 53, 317—382 (1943).

Haga, T.: (A statistical analysis of the chiasma formation.) Jap. J. Genet. 15, 308—310 (1939).

— A critique on the conception of the genom. Jap. J. Genet. 16, 211—227 (1940).

Hagerup, O.: Nordiske Kromosom-Tal. I. (Northern chromosome numbers. I.) Bot. Tidsskr. 45, 385—395 (1941).

— Cytoökologische Bicornes-Studien. Planta 32, 6—14 (1941).

Håkansson, A.: Die Meiosis von verschiedenen Mutanten von Godetia Whitneyi. Lunds universitets årsskrift, N. F. Avd. 2, bd. 36, nr. 5 (37 pp.) (1940).

— Eine tertiäre Trisome von Godetia Whitneyi. Bot. Not., Pp. 395—398 (1940).

— Zytologische Beobachtungen an Kreuzungen zwischen Godetia deflexa und G. Whitney. Bot. Not. Pp. 183 bis 198 (1941).

— Zytologische Studien an Rassen und Rassenbastarden von Godetia Whitneyi und verwandten Arten. Lunds universitets årsskrift. N. F. Avd. 2, bd. 38, nr. 5 (69 pp.) (1942).

— Die Meiosis einiger Godetia-Bastarde. Bot. Not., Pp. 271—283 (1943).

— Meiosis in a nullisomic and in an asyndetic Godetia Whitneyi. Hereditas, Lund 29, 179—190 (1943).

— Meiosis in a hybrid with one set of large and one set of small chromosomes. Hereditas 29, 461—474 (1943).

— Studies on a peculiar chromosome configuration in Godetia Whitneyi. Hereditas 30, 597—612 (1944).

— Überzählige Chromosomen in einer Rasse von Godetia nutans Hiorth. Bot. Not., Pp. 1—19 (1945).

— Zytologische Studien an monosomischen Typen von Godetia Whitneyi. Hereditas 31, 129—162 (1945).

— Zur Zytologie von Godetia-Arten und -Bastarden. Hereditas 27, 319—336 (1941).

Haldane, J. B. S. u. D. E. Lea: A mathematical theory of chromosomal rearrangements. J. Genet. 48, 1—10 (1947).

Hancock, B. L.: Cytological and ecological notes on some species of Galium L. em. Scop. New Phytol. 41, 70—78 (1942).

Haney, W. J.: Nucleolar numbers and attachments in Lilium. Abstr. Diss. Univ. Md. 41, 15—16 (1944). (Abstr.)

Harlova, G. V.: (Beobachtungen an dem Kern, während der Introkynese.) C. R. (Doklady) Acad. Sci. URSS. 53, 573 (1946). [Russisch.]

Harrison, J. W. H., K. B. Blackburn u. E. Bolton: Vitamin C and chromosome number in Rosa. Nature, Lond. 150, 574 (1942).

Hartmair, V.: Cytologische Feststellungen an Primula malacoides. Züchter 12, 32—34 (1940).

Havas, L. J.: A colchicine chronology. J. Hered. 31, 115—117 (1940).

— A colchicin biológiai és gazdasági jelentösége. (The biological and economic significance of colchicine.) Mag. Technika 1, 1—6 (1946).

Hawkes, J. G.: Some effects of the drug colchicine on cell division. J. Genet. 44, 11—22 (1941).

Heuts, M. J.: Influence of humidity on the survival of different chromosomal types in Drosophila pseudoobscura. Proc. Nat. Acad. Sci., Wash. 33, 210—213 (1947).

Hinton, T.: A study of chromosome ends in salivary gland nuclei of Drosophila. Biol. Bull. Wood's Hole 88, 144—165 (1945).

Hintzsche, E. v.: Statistische Probleme aus der Kerngrößenforschung. Experientia, Basel 1, 103—110 (1945).

Hoffman, J. C.: Wright's hypothesis: its relation to volume growth of tissue cells and mitotic index. Science 106, 343—344 (1947).

Hollande, A.-Ch. u. G. Hollande: L'origine du noyau de la cellule. Bull. Soc. Bot. Fr. 92, 57—59 (1945).

Holter, H.: Establishment of cytochemical techniques. Nature, Lond. 158, 917 (1946).

Huskins, C. L.: II. Symposium on theoretical and practical aspects of polyploidy in crop plants. Polyploidy and mutations. Biol. Symp. 4, 133—148 (1941).

— Structural differentiation of the nucleus. Reprinted from: The structure of protoplasm. Monogr. Soc. Pl. Physiol. Iowa St. Coll. Press. Ames, Iowa, Pp. 109—126 (1942).

Iljinskij, A. P.: (Biochemical differences between species within polyploid series.) Priroda (Natur), Nr. 4, 70 (1943).

Janaki-Ammal, E. K.: The breakdown of meiosis in a male-sterile Saccharum. Ann. of Bot., N. s. **5**, Nr. 17, 83—87 (1941).

Jaretzky, R. u. G. Schenk: Versuche mit Acenaphten und Colchicin an Gramineen- und Leguminosenkeimlingen. J. f. Bot. **89**, 13—19 (1940).

Jeener, R.: Sur les liens de la phosphatase alcaline avec les nucléoprotéides du noyau cellulaire et des granules cytoplasmiques. Experientia, Basel **2**, 458—459 (1946).

Jeffrey, E. C.: Chiasmatypy or the doctrine of delayed action fertilization. Science **102**, 653—656 (1945).
— The nucleus in relation to heredity and sex. Science **106**, 305—308 (1947).
— u. E. J. Haertl: The chromosomal crepusculum. Amer. Nat. **73**, 560—564 (1939).
—, — The present status of synapsis and chiasmatypy. Amer. J. Bot. Suppl. **28**, 4s (1941).

Jensen, H. W.: On the questionable existence of sex chromosomes in the angiosperms. Amer. Nat. **74**, 67 bis 88 (1940).

Joffe, E. J.: Polymorphism of karyotype in Helianthus annuus, L. C. R. (Doklady) Acad. Sci. URSS., N. s. **30**, 76—78 (1941).

Johnson, L. P. V.: A rapid squash technique for stem and root tips. Canad. J. Res. **23**, Sect. C., 127—130 (1945).

Kagawa, F.: (The effect of abnormal environment on the pollen formation in certain species and genus hybrids.) Proc. Crop. Sci. Soc., Japan **12**, 5—15 (1940).

Kaplan, R.: Zur Frage der physikochemischen Struktur des Chromosoms. Naturwiss. **28**, 79—80 (1940).

Karasawa, K.: Karyological Studies in Crocus. 2. Japan. Journ. of Bot. **11**, 129—140 (1940).

Kartašova, N. N.: On the treatment of vegetable cells with colchicine. C. R. (Doklady) Acad. Sci. URSS. **46**, 372—374 (1945).

Kaul, K. N., B. Mukerji u. R. N. Chopra: An indigenous mounting medium for microscopic work. Curr. Sci. **10**, 486—488 (1941).

Kelman, M.: The forces influencing chromosome pairing in Drosophila melanogaster. Amer. Nat. **79**, 567—570 (1945).

Kerkis, J.: On the nature of the so-called inert regions of the chromosomes in relation to the problem of the chromosome structure. C. R. (Doklady) Acad. Sci. URSS. **22**, 606—608 (1939).

Khan, R.: Artificial induction of polyploidy with special reference to colchicine. Sci. and Cult. **7**, 480—485 (1942).

Kihara, H.: (Ecological significance of polyploids.) Study of Ecology **5**, Nr. 2, 147—151 (1939).
— (A survey of studies of polyploidy.) Bot. and. Zool. **7**, 123—128 (1939).

Kisch, R.: Morphologie und Cytologie haploider Pflanzen von Epilobium hirsutum. Z. Bot. **36**, 513—537 (1941).

Knaysi, G.: Elements of bacterial cytology. Ithaca, N. Y., Pp. XII+209. 91 figs. 19 tables. X plates (1944).

Koller, P. C.: X-rays and cells. Nature, Lond. **155**, 778—780 (1945).

Kostoff, D.: (Duplication of the chromosome number [polyploidy] as a method of producing new forms of plants.) Selekcija i Semenovodstvo (Breeding and Seed Growing), Nr. 8, 29—32 (1939). [Russisch.]
— Induction of polyploidy by pulp and disintegrating tissues from Colchicum. Nature, Lond. **143**, 287—288 (1939).
— Effect of the fungicide „Granosan" on atypical growth and chromosome doubling in plants. Nature, Lond. **144**, 334 (1939).

Kostoff, D.: Polyploids are more variable than their original diploids. Nature, Lond. **144**, 868—869 (1939).
— Evolutionary significance of chromosome length and chromosome number in plants. Biodynamica, Nr. 51, 1—14 (1939).
— Evolutionary significance of chromosome size and chromosome number in plants. Curr. Sci. **8**, 306—310 (1939).
— Further studies upon the chromosome structure and behaviour. Cellule **48**, 179—186 (1939).
— Heritable variations conditioned by euploid chromosome alterations. Chronica Botanica **5**, 17—19 (1939).
— (New plants obtained by chromosome doubling [polyploidy].) Zemledelie, Sofia **43**, 18—23 (1939). [Bulgarisch.]
— Production de plantes á caractères nouveaux par le doublement du nombre des chromosomes (polyploïdie). (The production of plants with new characters by doubling the number of chromosomes [polyploidy].) Rev. Bot. Appl. **19**, 81—88 (1939).
— (Polyploids obtained by treating germinating seeds.) Priroda (Nature), Nr. 2, 103—104 (1939). [Russisch.]
— Atypical growth, abnormal mitosis and polyploidy induced by ethyl-mercury-chloride. Phytopath. Z. **13**, 91—96 (1940).
— Fertility and chromosome length. Correlations between chromosome length and viability of gametes in autopolyploid plants. J. Hered. **31**, 33—34 (1940).
— The frequency of the cell division in polyploid plants. Curr. Sci. **9**, 277—278 (1940).
— Polyploidie und landwirtschaftliche Produktion. Z. Pflanzenz. **25**, 284—304 (1943).
— (Chromosome changes obtained by treatment with neutrons.) Med. Fakultät, Sofia **23**, 51—61 (1943/44). [Bulgarisch.]
— u. A. Chevalier: Production de plantes á caractères nouveaux par le doublement du nombre des chromosomes (polyploïdie). Rev. Bot. Appl. **19**, 81—88 (1939).

Koževnikov, B. F.: (Die experimentelle Analyse der bevorzugten Trennung der nichthomologen und homologen Chromosomen im Zusammenhang mit der Frage der Natur der trennungsauslösenden Kräfte.) C. R. (Doklady) Acad. Sci. URSS. **58**, 663 (1947). [Russisch.]
— (Bevorzugte Trennung der Chromosomen und die Hypothese der elektrischen Ladungen.) C. R. (Doklady) Acad. Sci. URSS. **58**, 895 (1947). [Russisch.]

Krishnaswamy, N.: Untersuchungen zur Cytologie und Systematik der Gramineen. Beih. z. Bot. Zbl. **60**, 1—56 (1940).

Küster, E.: Über Plasmapfropfungen. Gustav Fischer, Jena (1939).

Kuwada, Y.: Chromosome structure. A critical review. Cytologia, Tokyo **10**, 213—256 (1939).

L....., H. M.: A new agent for stimulating polyploidy. Int. Sug. J. **41**, 454 (1939).

Lamm, R.: Varying cytological behaviour in reciprocal Solanum crosses. Hereditas **27**, 202—208 (1941).
— Cytogenetic studies in Solanum, sect. Tuberarium. Hereditas **31**, 128 (1944).

Lea, D. E.: Actions of radiations on living cells. Cambridge, Pp. XII+402, 61 figs, 83 tables, 4 pls. (1946).

Lefèvre, J.: Similitudes des actions cytologiques exercées par le phényluréthane et la colchicine sur des plantules végétales. C. R. Acad. Sci., Paris **208**, 301 bis 304 (1939).
— Note de biologie végétale. II. — Cytogénétique. Ann. Inst. Nat. Agron **31**, 29—46 (1939)
— Actions similaires sur les mitoses végétales de l'anéthol et des substances du groupe de la colchicine. C. R. Soc. Biol., Paris **133**, 616—618 (1940).

Lehotzky, P. v.: Die Wirkung des elektrischen Stromes auf den Zellkern. Arch. exp. Zellforsch. **25**, 74—78 (1943).

Lemos Pereira, A. de: (Über die Karyologie von Narcissus odorus L. und Narcissus gracilis Sab.) Bol. Soc. Broteriana **14**, 67—96 (1940). [Portugiesisch.]

Lettré, H. u. M. Albrecht: Narcotin, ein Mitosegift. Naturwiss. **30**, 184—185 (1942).

Levan, A.: The effect of colchicine on meiosis in Allium. Hereditas, Lund **25**, 9—26 (1939).

— Cytological phenomena connected with the root swelling caused by growth substances. Hereditas, Lund **25**, 87—96 (1939).

— En ny tillämpning av kemien inom växtförädlingen. (A new application of chemistry in plant breeding.) Sverig. Utsädesfören. Tidskr. **50**, 66—76 (1940).

— The effect of acenaphthene and colchicine on mitosis of Allium and Colchicum. Hereditas **26**, 262—276 (1940).

— Note on the somatic chromosomes of some Colchicum species. Hereditas **26**, 317—320 (1940).

— The cytology of the species hybrid Allium cepa × fistulosum and its polyploid derivatives. Hereditas **27**, 253 bis 272 (1941).

— A gene for the remaining in tetrads of ripe pollen in Petunia. Hereditas **28**, 429—435 (1942).

— The macroscopic colchicine effect - a hormonic action ? Hereditas, Lund **28**, 244 (1942). (Abstr.)

— The pigment content of polyploid plants. Hereditas, Lund **29**, 255—268 (1943).

— On the chromosomes of a new Mahonia-Berberis hybrid. Hereditas, Lund **30**, 401—404 (1944).

— Cytological reactions induced by inorganic salt solutions. Nature, Lond. **156**, 751—752 (1945).

— Polyploidiförädlingens nuvarande läge. Föredrag vid 34: de Svenska Lantbruksveckan den 20 mars 1945. (The present state of plant breeding by induction of polyploidy. Lecture delivered at the 34th Swedish Agricultural Week at Stockholm on March the 20th 1945.) Sverig. Utsädesfören. Tidskr. **55**, 109—143 (1945).

— Aktuelle Probleme der Polyploidiezüchtung. Archiv d. Julius Klaus-Stiftung 20: Erg.-Bd.: Festgabe ... A. Ernst, 142—152 (1945).

— The thresholds of colchicine action in barley, rye, diploid oat, and in their artificial tetraploids. Hereditas, Lund **32**, 294—295 (1946). (Abstr.)

— Heterochromaty in chromosomes during their contraction phase. Hereditas, Lund **32**, 449—468 (1946).

— u. T. Levring: Some experiments on c-mitotic reactions within Chlorophyceae and Phaeophyceae. Hereditas, Lund **28**, 400—408 (1942).

— u. A. Löve: Different chromosome numbers within the collective species Carex polygama. Hereditas **28**, 495 bis 496 (1942).

— u. G. Östergren: The mechanism of c-mitotic action: observations on the naphthalene series. Hereditas, Lund **29**, 381—443 (1943).

— u. E. Steinegger: The resistance of Colchicum and Bulbocodium to the c-mitotic action of colchicine. Hereditas, Lund **33**, 552—566 (1947).

Levickij, G. A.: (Cytological foundations of evolution.) Priroda (Nature), Nr. 5, 33—44 (1939). [Russisch.]

— Die Karyotypen einiger verwandter Artenpaare. J. Bot. URSS. **25**, 292—296 (1940).

Levine, M.: The effects of X-rays on ,,colchicine tumors'' on the root tips of the common onion. Amer. J. Bot. Suppl. **29**, 13s (1942). (Abstr.)

— The effect of colchicine and acenaphthene in combination with X-rays on plant tissue-I. Introduction. Bull. Torrey Bot. Cl. **72**, 563—574 (1945).

— The effect of colchicine and acenaphthene in combination with X-rays on plant tissue-III. Bull. Torrey Bot. Cl. **73**, 167—183 (1946).

— u. S. Gelber: The metaphase stage in colchicinized onion root-tips. Bull Torrey Bot. Cl. **70**, 175—181 (1943).

Lewis, D.: The evolution of sex in flowering plants. Biol. Rev. **17**, 46—67 (1942).

— The incompatibility sieve for producing polyploids. J. Genet. **45**, 261—264 (1943).

Little, T. M.: Tetraploidy in Antirrhinum majus induced by sanguinarine hydrochloride. Science **96**, 188—189 (1942).

— Gene segregation in autotetraploids. Bot. Rev. **11**, 60—85 (1945).

Lorbeer, G.: Struktur und Inhalt der Geschlechtschromosomen. Ber. dtsch. bot. Ges. **59**, 369—418 (1941).

Lorz, A. P.: Heterocyclicity or polysomaty ? J. Hered. **37**, 297, 306 (1946).

Löve, Á.: Cyto-genetic studies in Rumex. Bot. Not. 1940 Pp. 157—169; Polyploidy in Rumex acetosella L. Nature, Pp. 145, 351 (1940).

— Physiological differences within a natural polyploid series. Hereditas, Lund **28**, 504—506 (1942). (Abstr.)

— Rumex lunaria L., a gynodioecious tetraploid species. Nature **151**, 559—560 (1943).

— A Y-linked inheritance of asynapsis in Rumex acetosa. Nature **152**, 358—359 (1943).

— The dioecious forms of Rumex subgenus Acetosa in Scandinavia. Bot. Not., Pp. 237—254 (1944).

— Cytogenetic studies on Rumex subgenus Acetosella. Hereditas, Lund **30**, 1—136 (1944).

— u. D. Löve: Chromosome numbers of Scandinavian plant species. Bot. Notiser, Pp. 19—59 (1942).

—,— The significance of differences in the distribution of diploids and polyploids. Hereditas, Lund **29**, 145 bis 161 (1943).

—,— Cyto-taxonomical studies on boreal plants. II. Some notes on the chromosome numbers of Juncaceae. Ark. Bot. **31** B., 1—6 (1944).

—,— Cyto-taxonomical studies on boreal plants. III. Some new chromosome numbers of Scandinavian plants. Ark. Bot. **31** A., 1—22 (1944).

—,— Experiments on the effects of animal sex hormones on dioecious plants. Ark. Bot. **32** A., Nr. 13, 1—60 (1945).

McCallum, G. A.: The four strand structure of meiotic chromosomes of Aloe and Lilium. Amer. J. Bot. **26**, 672 (1939). (Abstr.)

Mckay, J. W., P. C. Burrell u. L. D. Goodhue: Applying colchicine to plants by the aerosol method. Science **101**, 154—156 (1945).

Maheschwani, P.: Origin of haploid-diploid twin embryos in angiosperms. Nature, Lond. **156**, 173—174 (1945).

Malheiros, N. u. D. de Castro: Chromosome number and behaviour in Luzula purpurea Link. Nature, Lond. **160**, 156 (1947).

Mangenot, G. u. S. Carpentier: Le syndrome mitoclasique. C. R. Soc. Biol., Paris **138**, 105—107 (1944).

Manton, I.: Chromosome length at the early meiotic prophases in Osmunda. Ann. Bot., Lond. **9**, 155—178 (1945).

— New evidence on the telophase split in Todea barbara. Amer. J. Bot. **32**, 342—348 (1945).

— u. J. Smiles: Observations on the spiral structure of somatic chromosomes in Osmunda with the aid of ultraviolet light. Ann. Bot., Lond. **7**, 195—212 (1943).

Marinelli, L. D., B. R. Nebel, N. Giles u. D. R. Charles: Chromosomal effects of low X-ray doses on five-day Tradescantia microspores. Amer. J. Bot. **29**, 866—874 (1942).

Marquardt, H.: Die Röntgenpathologie der Mitose. III. Weitere Untersuchungen des Sekundäreffekts der Röntgenstrahlen auf die haploide Mitose von Bellevalia romana. Z. Bot. **36**, 273—386 (1941).

Marquardt, H.: Über Bau, Häufigkeit und Auswirkungen der spontanen Translokationen. Flora, N. F. **35**, 239—302 (1941).
— Untersuchungen über den Formwechsel der Chromosomen im generativen Kern des Pollens und Pollenschlauches von Allium und Lilium. Planta **31**, 670 bis 725 (1941).
— Die Bestimmung der Dosisabhängigkeit röntgeninduzierter Chromosomenveränderungen bei Bellevalia romana. Z. Bot. **37**, 241—317 (1942).
— Die Verteilung röntgeninduzierter Veränderungen auf den Chromosomen von Bellevalia romana. Ber. d. Dtsch. Bot. Ges. **60**, 98—124 (1942).
— u. A. Ernst: Cytogenetische Röntgenversuche am reifenden Pollen von Antirrhinum majus. I. Z. Bot. **35**, 191—223 (1940).
Marshak, A.: The nature of chromosome division and the duration of the nuclear cycle. Proc. Nat. Acad. Sci., Wash. **25**, 502—510 (1939).
— A comparison of the sensitivity of mitotic and meiotic chromosomes of Vicia faba and its bearing on theories of crossing-over. Proc. Nat. Acad. Sci., Wash. **25**, 510 bis 516 (1939).
— Effects of fast neutrons on chromosomes in mitosis. Proc. Soc. Exp. Biol. **41**, 176—180 (1939).
— u. M. Bradley: X-ray inhibition of mitosis in relation to chromosome number. Proc. Nat. Acad. Sci., Wash. **30**, 231—237 (1944).
Martin, J. N.: The chromosome and the origin of species. Proc. Ia Acad. Sci. **46**, 49—58 (1939) 1940.
Martinoli: Nuova stazione di Chrysanthemum flosculosum L, e studio cariologico della specie. N. Giorn. Botan. Ital. n. s. **49**, 472—474 (1942).
Mascré, M. u. G. Deysson: Propriétés mitoclasiques de l'ésérine. Bull. Soc. Bot. Fr. **91**, 206—209 (1944).
—, — Action mitoclasique du camphre. Bull. Soc. Bot. Fr. **92**, 103—104 (1945).
Mather, K.: The determination of position in crossing-over. III. The evidence of metaphase chiasmata. J. Genet. **39**, 205—223 (1940).
— The genetical activity of heterochromatin. Proc. Roy. Soc. **132**, Ser. B., 308—332 (1945).
Matthey, R.: Les spirales chromosomiques. Verh. d. Schweiz. Naturforsch. Ges. **121**, 242 (1941).
Maude, P. F.: The Merton catalogue. A list of the chromosome numerals of species of British flowering plants. New Phytol. **38**, 1—31 (1939).
Mazia, D.: Enzymes and nucleic acid synthesis in the cell. Proc. Mo. Acad. Sci. **7**, 91 (1942).
Melville, R. u. M. Pyke: Vitamin C and chromosome number in Rosa. Nature **150**, 574 (1942).
Menezes, O. B. de: Poliploidia indução e colquicina. (Polyploidy, induction and colchicine.) Bol. Minist. Agric. Rio de Janeiro **33**, 1—16 (1944).
Mensinkai, S. W.: The conception of the satellite and the nucleolus, and the behaviour of these bodies in cell division. Ann. Bot., London **3**, 763—794 (1939).
Meyer, J. R.: Colchicine-Feulgen leaf smears. Stain Techn. **18**, 53—56 (1943).
Mezzeti-Bambacioni, V.: Osservazioni cariologiche in alcune Lauracee. C. 13 fig. Scientia Genetica **1**, 326 bis 332 (1940).
Michaelis, P.: Experimentelle Untersuchungen über die geographische Verbreitung von Plasmon-Unterschieden und der auf diese Unterschiede empfindlichen Gene, sowie deren theoretische Bedeutung für das Kern-Plasma-Problem. Biol. Zbl. **62**, 170—186 (1942).
Milla, L.: Variabilità della grandezza nucleare in Artemia salina Leach partenogenetica ed anfigonica. R. Ist. Lombardo Sc. e Lett. Rend. Sc. **74**, 287 (1941).
Miller, O. H. u. L. Fischer: Comparative analyses of normal and tetraploid Datura stramonium and Datura tatula. J. Amer. Pharm. Ass. Sci. Ed. **35** (1), 23—27 (1946).
Milovidov, P. F.: Zur Anuclealität des pflanzlichen Nucleolus. Planta **31**, 60—72 (1940).
Minder, W. u. A. Liechti: Über den gegenwärtigen Stand des Hauptproblems der Strahlenbiologie. (Modellversuche zum Primäreffekt der biologischen Strahlenwirkung.) Experientia, Basel **1**, 298—307 (1945).
Mirsky, A. E.: Chromosomes and nucleoproteins. Adv. Enzymol. **3**, 1—34 (1943).
—, A. W. Pollister u. H. Ris: The chemical composition of chromosomes. Genetics **31**, 224—225 (1946). (Abstr).
Mitra, J.: A contribution to the embryology of some Compositae. J. Indian. Bot. Soc. **26**, 105—123 (1947).
Modilevskij, J. S.: (Haploidy in angeospermous plants.) Uspehi Sovremmenoj Biologii (Advances in Modern Biology) **15**, 129—153 (1942). [Russisch.]
Monné, L.: Struktur- und Funktionszusammenhang des Zytoplasmas. Experientia, Basel **2**, 153—159 (1946).
Morgan, T. H. u. J. Schultz: Investigations on the constitution of the germinal material in relation to heredity. Yearb. Carneg. Instn. 1941—1942, Nr. 41, 242—245 (1942).
Mudra, A.: Incercări in vederea obtinerii de forme poliploide la grâu prin tratament cu colchicină. (Experiments on the production of polyploid forms of wheat by colchicine treatment.) Bul. Fac. Agron. Cluj. **8**, 264—268 (1939).
Muller, H. J.: A brief comment on mother-daughter chromosomes. J. Hered. **37**, 246 (1946).
Müntzing, A.: Chromosomenaberrationen bei Pflanzen und ihre genetische Wirkung. Z. Vererbungsl. **76**, 323 bis 351 (1939).
— Imcompatibility and fertility in experimental and natural polyploids. Proc. 7th Int. Genet. Congr. Edinburgh 23—30 August, p. 224 1939 (1941). (Abstr.)
— Polyploidi och växtförädling. (Polyploidy and plant breeding.) Sverig. Utsädesfören-Tidskr. **51**, 307—316 (1941).
— New material and cross combinations in Galeopsis after colchicine-induced chromosome doubling. Hereditas **27**, 193—201 (1941).
— Experimentella kromosomtalsförändringar och deras betydelse för växtförädlingen. (Experimental chromosome number alteration and their importance for plant breeding.) K. Lantbr. Acad. Tidskr. **81**, 97—114 (1942).
— Fertility improvement by recombination in autotetraploid Galeopsis pubescens. Hereditas **29**, 201—204 (1943).
— Hybrid vigour in crosses between pure lines of Galeopsis tetrahit. Hereditas **31**, 391—398 (1945).
— u. A. Levan: Berättelse över verksamheten vid Sveriges Utsädesförenings kromosomavdelning under tiden 1 oktober 1935—30 september 1940. (Report on the activity of the chromosome division of the Swedish Seed Association during the period October 1st 1935-September 30th 1940.) Sverig. Utsädesfören. Tidskr. **51** 83—93 (1941).
— u. G. Müntzing: Recent results in Potentilla. Hereditas **28**, 232—235 (1942).
—, — Spontaneous changes in chromosome number in apomictic Potentilla collina. Hereditas **29**, 451—460 (1943).
—, — A pentaploid F_1 hybrid between two diploid Potentilla species. Hereditas **30**, 631—638 (1944).
—, — The mode of reproduction of hybrids between sexual and apomictic Potentilla argentea. Bot. Not., Pp. 49—71 (1945).
— u. E. Runquist: Note on some colchicine-induced polyploids. Hereditas, Lund. **25**, 491—495 (1939).
Murneek, A. E. u. S. H. Wittwer: Synapsis and syngamy as stimulating processes of plant development. Science **98**, 384—385 (1943).

Navašin, M. S.: (Chromosomenanordnung in der Metaphase und Kerndynamik.) C. R. (Doklady) Acad. Sci. URSS. **57**, 613—616 (1947). [Russisch.]
— u. H. Gerasimova: Production of polyploid plants from leaves treated with colchicine. C. R. (Doklady) Acad. Sci. URSS. **24**, 948—950 (1939). [Russisch.]
—, — Production of polyploids by administering colchicine solution via roots. C. R. (Doklady) Acad. Sci. URSS. **26**, 681—683 (1940).
Nebel, B. R.: Chromosome structure. Bot. Rev. **5**, 563—626 (1939).
— On coiling in chromosomes. Amer. Nat. **73**, 289—299 (1939).
Newcomer, E. H.: The duality of mitochondria in plants. Amer. J. Bot. (Suppl.) **33**, 221 (1946). (Abstr.)
— Concerning the duality of the mitochondria and the validity of the osmiophilic platelets in plants. Amer. J. Bot. **33**, 684—697 (1946).
Nihous, M.: Effects cytophysiologiques d'un traitement colchicique, mortel, chez l'embryon du pois cultivé en germination. C. R. Soc. Biol., Paris **138**, 534—535 (1944).
Nilsson, H.: Eine diploide Form aus der tetraploiden Oenothera gigantea. Hereditas **25**, 1—8 (1939).
Noggle, G. R.: The physiology of polyploidy in plants. I. Review of the literature. Lloydia, Cincinnati **9**, 153 bis 173 (1946).
Noguti, Y.: (On crop-plant breeding.) Bot. and Zool. **7**, 287—294 (1939).
Noujdin, N. j. siehe Nuždin.
Nuždin, N. j.: Role of structural homo- and heterozygosity in mosaic formation. Nature, Lond. **155**, 514 (1945).
Nybom, N. u. B. Knutsson: Investigations on c-mitosis in Allium Cepa. Hereditas, Lund **33**, 220—234 (1947).

Oehlkers, F.: Die Auslösung von Chromosomenmutationen in der Meiosis durch Einwirkung von Chemikalien. Z. indukt. Abstamm.- u. Vererbungsl. **81**, 313—341 (1943).
Oláh, L.: A kromoszóma-kutatás szerepe a növényrendszertanban. (The role of chromosome research in plant systematics.) Bot. Közl. **36**, 144—152 (1939).
O'Mara, J. G.: Observations on the immediate effects of colchicine. J. Hered. **30**, 35—37 (1939).
Ono, T.: (Induced polyploidy in some dioecious plants.) Jap. J. Genet. **15**, 319—321 (1939).
Östergren, G.: Note on the acetocarmine method. Hereditas **28**, 239—240 (1942).
— Chromosome numbers in Anthoxanthum. Hereditas **28**, 242—243 (1942).
— Elastic chromosome repulsions. Hereditas, Lund **29**, 444—450 (1943).
— An efficient chemical for the induction of sticky chromosomes. Hereditas, Lund **30**, 213—216 (1944).
— Colchicine mitosis, chromosome contraction, narcosis and protein chain folding. Hereditas, Lund **30**, 429 bis 467 (1944).
— Equilibrium of trivalents and the mechanism of chromosome movements. Hereditas **31**, 498 (1945).

Paddock, E. F.: A theoretical advantage of using autotetraploids in crop plant breeding. Amer. J. Bot. (Suppl.) **33**, 222 (1946). (Abstr.)
Painter, T. S.: Cell growth and nucleic acid in the pollen of Rhoeo discolor. Bot. Gaz. **105**, 58—68 (1943).
Palmgren, O.: Chromosome numbers in angiospermous plants. Bot. Notiser, Pp. 348—352 (1943).
Panšin, I. B.: Relative frequency of union of chromosome fragments. C. R. (Doklady) Acad. Sci. URSS. **33**, 320—322 (1941).
Pantulu, J. V.: Further chromosome numbers in the Caesalpiniaceae. Curr. Sci. **12**, 274 (1943).

Parthasarathy, N.: An Indian source for colchicine. Curr. Sci. **10**, 446 (1941).
Pätau, K.: Cytologischer Nachweis einer positiven Interferenz über das Centromer. (Der Paarungskoeffizient, I.) Chromosoma **2**, 36—63 (1941).
Pathak, G. N.: Studies in the cytology of Crocus. Ann. of Bot. N. s. **4**, 227—256 (1940).
Pavulans, J.: Über die Nuclealreaktion in Embryosäcken und Pollenkörnern einiger Angiospermen. Protoplasma **34**, 22—29 (1940).
Pease, D. C.: Hydrostatic pressure effects upon the spindle figure and chromosome movement. II. Experiments on the meiotic divisions of Tradescantia pollen mother cells. Biol. Bull. Wood's Hole **91**, 145—169 (1946).
Pěrk, P.: Zur Kenntnis der Mitose. XI. Über den Einfluß von Radium- und Röntgenstrahlen auf die Zellteilung in Gewebekulturen. Z. Zellforsch. usw., A. **32**, 1—32 (1942).
Peters, J. J.: Cytological effects of sulfanilamide on Allium cepa. Bot. Gaz. **107**, 390—392 (1946).
Pfeiffer, H. H.: Experimentelle Cytologie. Chronica Botanica Co., Waltham, Mass. (1940).
Phillips, H. M.: The formation of nucleoli in Erythronium. Amer. J. Bot. (Suppl.) **33**, 222 (1946). (Abstr.)
Pirschle, K.: Weitere Beobachtungen über den Einfluß von langwelliger und mittelwelliger UV-Strahlung auf höhere Pflanzen, besonders polyploide und hochalpine Formen (Stellaria, Epilobium, Arenaria, Silene). Biol. Zbl. **61**, 452—473 (1941).
— Quantitative Untersuchungen über Wachstum und „Ertrag" autopolyploider Pflanzen. Z. indukt. Abstamm.- u. Vererbungsl. **80**, 126—156 (1942).
— Weitere Untersuchungen über Wachstum und „Ertrag" von Autopolyploiden (2n, 3n, 4n) und ihren Bastarden. Z. indukt. Abstamm.- u. Vererbungsl. **80**, 247—270 (1942).
Piza, S. u. de Toledo jr.: Dorso-ventralidade dos cromosômios. (Dorsoventrality of chromosomes.) Rev. Agric. S. Paulo **17**, 154—168 (1942).
— Fatos velhos e novos em favor da teoria do cromossômio-unidade. (New and old facts in favour of the theory of the chromosome-unit.) Rev. Agric. Piraciciba **18**, 191—207 (1943).
— The uselessness of the spindle fibers for moving the chromosomes. Amer. Nat. **77**, 442—462 (1943).
Pohlendt, G.: Cytologische Untersuchungen an Mutanten von Antirrhinum majus L. I. Deletionen im uni-Chromosom. Z. Vererbungsl. **80**, 281—288 (1942).
— Cytologische Untersuchungen an Mutanten von Antirrhinum majus L. II. Mosaikpflanzen mit reziproken Translokationen. Chromosoma **2**, 387—406 (1942).
Pollister, A. W. u. A. E. Mirsky: Terminology of nucleic acids. Nature, Lond. **152**, 692—693 (1943).
— Distribution of nucleic acids. Nature, Lond. **153**, 711. (1944).
Pontecorvo, G.: Structure of heterochromatin. Nature, Lond. **153**, 365—367 (1944).
Prakken, R.: A new trisomic matthiola type. Hereditas **28**, 297—305 (1942).
— u. A. Levan: Notes on the colchicine-meiosis of Allium cernuum. Hereditas, Lund **32**, 123—126 (1946).
Prokofjeva-Belgovskaja, A. A.: (Cytological study of the „simple breaks" in the inert region of the scute[8] chromosome of Drosophila melanogaster.) Izvestija Akademii Nauk SSSR. (Bull. Acad. Sci. URSS. Sér. biol.), Nr. 3, 349—359 (1939). [Russisch.]
— (Die Verteilung der Brüche im X-Chromosom bei Drosophila melanogaster.) C. R. (Doklady) Acad. Sci. URSS. **23**, Nr. 3, 369—371 (1939). [Russisch.]
— (Die „inerten" Abschnitte der inneren Teile des X-Chromosoms bei Drosophila melanogaster.) Izvestija Akad. Nauk SSSR., ser. biol. (Bull. Acad. Sci. URSS., Sér. biol.) Nr. 3, 362—370 (1939). [Russisch.]

Prokofjeva-Belgovskaja, A. A.: Heterocyclicity of the system „maternal chromosome-daughter chromosome". C. R. (Doklady) Acad. Sci. URSS. **49**, 601—604 (1945).
— (Die Heterochromatisierung als Veränderung des Chromosomenzyklus.) Žurnal Obščej Biologii (Journ. of Gener. Biology) **6**, 93—124 (1945). [Russisch.]
— „Mother" and „daughter" chromosomes. Significance of heterocyclic systems in paired nuclei. J. Hered. **37**, 239—246 (1946).
— (Die Heterozyklität des Systems des Zellkerns.) C. R. (Doklady) Acad. Sci. URSS. **53**, 745—748 (1946). [Russisch.]
— Heterochromatization as a change of chromosome cycle. J. Genet. **48**, 80—98 (1947).
— (Heterocyclycity of the system of cell nucleus.) Žurnal Obščej Biologii (Journ. of General Biology) **8**, Nr. 4 (1947). [Russisch.]
Propach, H.: Die Centromeren in der Pollenkornmitose von Tradescantia gigantea Rose. Chromosoma **1**, 521 bis 525 (1940).

Randolph, L. F.: II. Symposium on theoretical and practical aspects of polyploidy in crop plants. An evaluation of induced polyploidy as a method of breeding crop plants. Biol. Symp. **4**, 151—167 (1941).
—, E. C. Abbe u. J. Einset: Comparison of shoot apex and leaf development and structure in diploid and tetraploid maize. J. Agric. Res. **69**, 47—76 (1944).
Rapoport, I. A.: (Chemical reaction with the protein aminogroup in the structure of genes.) Žurnal Obščej Biologii (Journ. of General Biology) **8**, Nr. 5, 359 bis 378 (1947). [Russisch.]
— (Acetylierung des Eiweißes der Gene und Mutationen.) C. R. (Doklady) Acad. Sci. URSS. **58**, 119 (1947). [Russisch.]
Rashevsky, N.: Mathematical biophysics of cell division. Bull. Math. Biophys. **5**, 99—102 (1943).
Resende, F.: Die Nukleolen bei Antirrhinum majus L. Ber. dtsch. bot. Ges. **58**, 460—470 (1940).
— Cariocinese e cromonemata (Nota preleminar). Bol. Soc. Brot. **15**, 2nd Ser., 21—27 (1941).
— Movimento, aglutinação, pontes e distensão dos cromosomas na mitose. (Movement, agglutination, bridges and distension of chromosomes at mitosis.) Bol. Soc. Brot. **15**, 2nd. Ser., 163—196 (1941).
— Hétérochromatine. (Heterochromatin.) Portugaliae Acta Biologica, Lisboa **1**, Ser. A., 139—173 (1945).
— Behaviour of the „nucleolar olistherozone". Nature, Lond. **157**, 266 (1946).
— Sur la constitution histo-chimique probable de la olisthérozone nucléolaire. Portugaliae Acta Biologica, Lisboa **1**, Sér. A., 265—270 (1946).
—, A. de Lemos-Pereira u. A. Cabral: Sur la structure des chromosomes dans les mitoses des méristèmes radiculaires. Portugaliae Acta Biologica, Lisboa **1**, Sér. A. 9—46 (1944).
—, J. Salord u. L. Leite-Rio: Structure of chromosomes as observed in root-tips. IV. Olistherochromatin and chromosomic contraction. Portugaliae Acta Biologica, Lisboa **1**, Sér. A. 412—415 (1946).
Resende-Pinto, M. C. de: Une nouvelle méthode de coloration des nucléoles — le tannin-fer III. Portugaliae Acta Biologica, Lisboa **1**, Sér. A. 309—310 (1946).
Ris, H.: The structure of meiotic chromosomes in the grasshopper and its bearing on the nature of „chromomeres" and „lamp-brush chromosomes". Biol. Bull. Wood's Hole **89**, 242—257 (1945).
— u. H. Crouse: Structure of the salivary gland chromosomes of Diptera. Proc. Nat. Acad. Sci., Wash. **31**, 321—327 (1945).
Rondoni, P.: Cancro denaturazione delle proteine. (Cancer and the denaturation of the proteins.) Experientia, Basel **2**, 127—132 (1946).

Rosen, G. von: A rapid method for sorting polyploid material. Hereditas, Lund **32**, 129—130 (1946). (Abstr.)
— Chromosome determination in root-tips and leaves by the rapid orcein method. Hereditas, Lund. **32**, 551 bis 554 (1946). (Abstr.)
— A rapid nigrosine method for chromosome counts applicable to growing plant tissues. Nature, Lond. **160**, 121—122 (1947).
Rosenberg, O.: The influence of low temperatures on the development of the embryosac mother cell in Lilium longiflorum Thumb. Hereditas, Lund **32**, 65—92 (1946).
Ruch, F.: Zur Schraubenstruktur des Metaphasechromosoms der 1. meiotischen Teilung bei Tradescantia virginica. Vjschr. Naturf.-Ges. Zürich **90**, *214—215 (1945).
Rudorf, W.: Die Bedeutung der Polyploidie für die Evolution und die Pflanzenzüchtung. Angew. Bot. **25**, 92—114 (1943).
Ruttle, M. L.: Colchicine and the production of the new varieties of plants. Rev. Agric. P. Rico **31**, 623 bis 631 (1939).
— u. B. R. Nebel: Cytogenetic results with colchicine. Biol. Zbl. **59**, 79—87 (1939).
Ryan, F. J.: Crossing-over and second division segregation in fungi. Bull. Torrey Bot. Cl. **70**, 605—611 (1943).

Saez, F. A.: Alteraciones experimentales inducidas por la acción de la gravedad en las células somáticas de „Lathyrus Odoratus" (Leguminosae). (Experimental changes induced by the action of gravity on the somatic cells of L. odoratus.) An. Soc. Cient. Argent. **4**, 139 bis 150 (1941).
Sahni, B.: Permanent labels for microscope slides. Curr. Sci. **10**, 485—486 (1941).
Sakai, K.: (Diurnal periodicity of somatic mitosis in the root-tips of several crop-plants.) Jap. J. Genet. **17**, 35—40 (1941).
Salazar, A. L.: Sur la question chondriome et plastidome. Anat. Anz. **94**, 284—287 (1943).
Sampath, S., S. S. Rajan u. S. P. Singh: Chemical structure in relation to action on plant nucleus. Curr. Sci. **15**, 137 (1946).
—, B. N. Singh u. R. K. Bansal: Effect of colchicine on plant cells. Curr. Sci. **8**, 121—122 (1939).
Sancho Peñasco, F.: Las mutaciones provocadas. (Induced mutations.) Bol. Inst. Nac. Invest. Agron. Madr., Nr. 5, 153—170 (1941).
Sansome, E. R.: Pairing behaviour and chiasma formation in plants with a structurally heterozygous pair in Pisum. Genetica **21**, 420—433 (1939).
Sax, K.: The time factor in X-ray production of chromosome aberrations. Proc. Nat. Acad. Sci., Wash. **25**, 225—233 (1939).
— An analysis of X-ray induced chromosomal aberrations in Tradescantia. Genetics **25**, 41—68 (1940).
— The mechanisms of X-ray effects on cells. J. Gen. Physiol. **25**, 533—537 (1942).
— The effect of centrifuging upon the production of X-ray induced chromosomal aberrations. Proc. Nat. Acad. Sci., Wash. **29**, 18—21 (1943).
— Temperature effects on Y-ray induced chromosome aberrations. Genetics **32**, 75—78 (1947).
— u. R. T. Brumfield: The relation between X-ray dosage and the frequency of chromosomal aberrations. Amer. J. Bot. **30**, 564—570 (1943).
— u. E. V. Enzmann: The effect of temperature on X-ray induced chromosome aberrations. Proc. Nat. Acad. Sci., Wash. **25**, 397—405 (1939).
— u. K. Mather: An X-ray analysis of progressive chromosome splitting. J. Genet. **37**, 483—490 (1939).
— u. C. P. Swanson: Differential sensitivity of cells to X-rays. Amer. J. Bot. **28**, 52—59 (1941).

Schmidt, W. J.: Einiges über optische Anisotropie und. Feinbau von Chromatin und Chromosomen. Chromosoma 2, 86—110 (1941).

Schnarf, K.: Vergleichende Cytologie des Geschlechtsapparates der Kormophyten. Berlin-Zehlendorf: Bornträger, Pp. 249 (1941).

Schrader, F.: The structure of the kinetochore at meiosis. Chromosoma 1, 230—237 (1939).

— The present status of mitosis. Amer. Nat. 74, 25—33 (1940).

— Mitosis. The movements of chromosomes in cell division. N. Y., Pp. X+110. 15 figs. 1 table (1944).

Schröderheim, J.: Untersuchungen über den Ascorbinsäuregehalt in Hagebutten. Acta Univ. Lund 37, Nr. 9, 57; K. Fysior. Sällsk. Handl. 52, Nr. 9 (1941).

Schultz, J.: The function of heterochromatin. Proc. 7th Int. Genet. Congr., Edinburgh 23—30 August Pp. 257—262. 1939 (1941).

— u. F. S. Jose: Differentiation of chromosomal proteins by staining techniques. Genetics 30, 20—21 (1945). (Abstr.)

Schwanitz, F.: Polyploidie und Phylogenie. Biologe 8, 323—335 (1939).

— Polyploidie und Pflanzenzüchtung. Naturwiss. 28, 353—361 (1940).

Schwarz, P. A.: Anatomical and cytological changes in seedlings from chemically treated cereal grain. C. R. (Doklady) Acad. Sci. URSS. 28, 354—356 (1940).

Schwemmle, J.: Plastidenmutationen bei Eu-Oenotheren. Z. Vererbungsl. 79, 171—187 (1941).

— Plastiden und Genmanifestation. Flora, N. F. 37, 61 (1943).

Scott, F. M.: Cytology and microchemistry of nuclei in developing seed of Echinocystis macrocarpa. Bot. Gaz. 105, 329—338 (1944).

Seifriz, W.: Protoplasmic streaming. Bot. Rev. 9, 49 bis 123 (1943).

Semmens, C. S.: Improved cytological methods with crystal violet. Stain Tech. 17, 147—148 (1942).

Sengbusch, R. v.: Poylploide Kulturpflanzen (Roggen, Hafer, Stoppelrüben, Kohlrüben und Radieschen). Züchter 13, 132—134 (1941).

Sentein, P.: L'action des toxiques sur la cellule en division. Montpellier, Pp. 252, 4 pls. (1941).

Serra, J. A.: Relations entre la chimie et la morphologie nucléaire. Bol. Soc. Brot. 16, 83—135 (1942).

— Sur la composition protéique des chromosomes et la réaction nucléale de Feulgen. Bol. Soc. Brot. 17, 203 bis 211 (1943).

— u. A. Queiroz Lopes: Données pour une cytophysiologie du nucléole. I. L'activité nucléolaire pendant la croissance de l'oocyte chez des Helicidae. Portugaliae Acta Biologica, Lisboa 1, Sér. A, 51—94 (1945).

Seshachar, B. R. u. K. V. Srinath: The nucleolus. Curr. Sci. 15, 9—11 (1945).

—, — The ciliate macronucleus. Curr. Sci. 16, 83—84 (1947).

Sheffield, F. M. L.: The nucleolus. Nature, Lond. 153, 687—688 (1944).

Sherman, M.: Karyotype evolution: a cytogenetic study of seven species and six interspecific hybrids of Crepis. Univ. Calif. Publ. Bot. 18, 369—408 (1946).

Shimamura, T.: (A cytological experiment with colchicine in plants.) Jap. J. Genet. 15, 158—159 (1939).

— Cytological studies on polyploidy induced by colchicine. Cytologia, Tokyo 9, 486—494 (1939).

— Studies on the effect on the centrifugal force upon nuclear division. (Fuji, K.: On the mechanism of nuclear division and chromosome arrangement. 6.) Cytologia (Tokyo) 11, 186—216 (1940).

Shmuck, A. siehe Šmuk.

Sidorov, B. N.: (Study on the nature of changes caused by structural chromosome mutations.) C. R. (Doklady) Acad., Sci. URSS. 31, 390—392 (1941). [Russisch.]

Simonet, M.: Anomalies de la caryocinè végétale des types colchicinique et paradichlorobenzénique, produites par un dérivé nitré des carbures cycliques: le M. nitroxylène-1-3-5. C. R. Soc. Biol., Paris 133, 561—563 (1940).

— u. F. Armenzoni: Anomalies de la caryocinèse dues à l'action des dérivés iodés des carbures cycliques. C. R. Acad. Sci., Paris 209, 354—356 (1939).

— u. R. Chopinet: Apparition de mutations géantes et polyploidies chez le colza, la pervenche et le lin à grande fleur, après application de colchicine. C. R. Acad. Sci., Paris 209, 238—240 (1939).

— u. G. Igolen: Obtention, sous l'influence de vapeurs d'essence de petit grain mandarinier, d'effets comparables à ceux exercés par la colchicine sur les caryocinèses végétales. C. R. Acad. Sci., Paris 210, 510—511 (1940).

Sinke, N.: Experimental studies in cell-nuclei. Mem. Coll. Sci. Kyoto 15, Ser. B, 1—126 (1940).

Sinnot, E. W. u. R. Bloch: Cytoplasmic behavior during division of vacuolate plant cells. Proc. nat. Acad. Sci. U.S.A. 26, 223—227 (1940).

Sipkov, T. P.: (Changes in the number of chromosomes of plant organisms.) Jarovizacija, Nr. 3 (36), 93—98 (1941). [Russisch.]

Skovstedt, A.: Cytological studies in twin plants. C. R. Lab. Carlsberg, Copenhagen 22, Sér. Physiol., 427 bis 446 (1939).

— Chromosome numbers in the Malvaceae. II. C. R. Lab. Carlsberg, Sér. Physiol. 23, 195—242 (1941).

Smith, B. W. u. M. T. Smith: Sex chromosomes in Rumex hastatulus Baldwith XY_1Y_2 pairing. Genetics 32, 104—105 (1947). (Abstr.)

Smith, S. G.: The reproduction of the nucleus. Sci. Agric. 24, 491—509 (1944).

Šmuk, A.: (Concerning the effect of chemical substances on the change in hereditary properties of plants.) Priroda (Nature), Nr. 3, 74—78 (1939). [Russisch.]

— u. A. Guseva: Active concentrations of acenaphthene inducing alterations in the processes of cell division in plants. C. R. (Doklady) Acad. Sci. URSS. 22, 441—443 (1939).

—, — Chemical-structure of substances inducing polyploidy in plants. C. R. (Doklady) Acad. Sci. URSS. 24, 441—446 (1939).

—, — Polyploidogenic action on plants of naphthol ethers and naphthoic acid esters. C. R. (Doklady) Acad. Sci. URSS. 26, 460—463 (1940).

—, — Haploid derivatives of aromatic hydrocarbons and their polyploidogenic activity. C. R. (Doklady) Acad. Sci. URSS. 26, 674—677 (1940).

—, — Methoxyl derivatives of benzene and naphthalene studied with regard to their polyploidogenic action on plants. C. R. (Doklady) Acad. Sci. URSS. 30, 639—641 (1941).

— u. D. Kostoff: Brome-acenaphthene and bromenaphthaline as agents inducing chromosome doubling in rye and wheat. C. R. (Doklady) Acad. Sci. URSS., 23, 263—266 (1939).

Sokolovskaja, A. P. u. O. S. Strelkova: Polyploidy and karyological races under conditions in the Arctic. C. R. (Doklady) Acad. Sci. URSS. 32, 144—147 (1941).

Sparrow, A. H.: X-ray sensitivity of meiotic and microspore chromosomes of Trillium. Rec. Genet. Soc. Amer. Nr. 12, 54—55 (1943). (Abstr.)

— Changes in sensitivity of chromosomes to X-ray breakage during microsporogenesis. Genetics 32, 106—107 (1947). (Abstr.)

Sparrow, A. H.: u. M. R. Hammond: Desoxyribonucleic-acid-containing bodies in the cytoplasm of pollen mother cells at early meiotic prophase. Genetics 32, 107 (1947). (Abstr.)
—, M. L. Ruttle u. B. R. Nebel: Comparative cytology of sterile intra- and fertile inter-varietal tetraploids of Antirrhinum majus L. Amer. J. Bot. 29, 711—715 (1942).
Spencer, W. R.: Crossingover in spaghetti chromosomes. J. Hered. 35, 8—10 (1944).
Stacey, M., R. E. Deriaz, E. G. Teece u. L. F. Wiggins: Chemistry of the Feulgen and Dische nucleal reactions. Nature, Lond. 157, 740—741 (1946).
Stapp, C.: Der Pflanzenkrebs und sein Erreger Pseudomonas tumefaciens. XI. Mitteilung. Zytologische Untersuchungen des bakteriellen Erregers. Zbl. Bact. 105, 1—14 (1942).
— Der Pflanzenkrebs und sein Erreger Pseudomonas tumefaciens. XIII. Mitteilung. Über die Bedeutung des Colchicins als polyploidisierendes Mittel für den Erreger und als angebliches Bekämpfungsmittel gegen den Wurzelkropf. Zbl. Bact. 106, 338—350 (1944).
Stebbins, G. L. jr.: The effect of anaerobic conditions on mitosis in seedlings of Hordeum. Amer. J. Bot. 26, 674 (1939). (Abstr.)
— The significance of polyploidy in plant evolution. Amer. Nat. 74, 54—66 (1940).
— Polyploid complexes in relation to ecology and the history of floras. Amer. Nat. 76, 36—45 (1942).
— Types of polyploids: their classification and significance. Advances Genet., N. Y. 1, 403—429 (1947).
Stedman, E. u. E. Stedman: Chromosomin, a protein constituent of chromosomes. Nature, Lond. 152, 267 bis 269 (1943).
—, — Distribution of nucleic acid in the cell. Nature, Lond. 152, 503—504 (1943).
—, — Probable function of histone as a regulator of mitosis. Nature, Lond. 152, 556—557 (1943).
—, — „Chromosomin" and nucleic acids. Nature, Lond. 153, 500—502 (1944).
Steedman, H. F.: Ester wax: a new embedding medium. Nature, Lond. 156, 121—122 (1945).
Stein, E.: Cytologische Untersuchungen an Antirrhinum majus mut. cancroidea. Endomitosen-Entwicklung. Chromosoma 2, 308—333 (1942).
— Über einige durch Radiumbestrahlung erzeugte Periklinalchimären von Petunia und Antirrhinum siculum mit Veränderungen der Zellstruktur. Biol. Zbl. 62, 483—508 (1942).
Steinegger, E. u. A. Levan: Constitution and c-mitotic activity of iso-colchicine. Hereditas, Lund 33, 385 bis 396 (1947).
—, — The cytological effect of chloroform and colchicine on Allium. Hereditas, Lund 33, 515—525 (1947).
Steinitz, L. M.: The effect of lack of oxygen on mitosis in barley. Amer. J. Bot. 30, 622—626 (1943).
— The effect of lack of oxygen on meiosis in Tradescantia. Amer. J. Bot. 31, 428—443 (1944).
Stern, C., R. H. MacKnight u. M. Kodani: The phenotypes of homozygotes and hemizygotes of position alleles and of heterozygotes between alleles in normal and translocated positions. Genetics 31, 598—619 (1946).
Stern, H.: The formation of polynucleated pollen mother cells in Trillium erectum. J. Hered. 37, 47—50 (1946).
Steyaert, R. L.: Another superior pith for free-hand sections. Science 103, 695 (1946).
Stomps, T. J.: Over de kernen der bacteriën. (On the nuclei of the bacteria.) Vakbl. Biol. 21, 270—274 (1940).
— Over de cytologie der maligne tumoren. (On the cytology of malignant tumours.) Vakbl. Biol. 25, 9—11 (1944).
Stout, A. B. u. C. Chandler: Hereditary transmission of induced tetraploidy and compatibility in fertilization. Science 96, 257—258 (1942).

Straub, J.: Die Erzeugung von Blütenpflanzen mit verminderter Chromosomenzahl (Hypodiploide). Ber. dtsch. bot. Ges. 57, (155)—(174) (1939).
— Polyploidieauslösung durch Temperaturwirkungen. (Entwicklungsphysiologische Untersuchungen an der reproduktiven Phase von Gasteria.) Z. Bot. 34, 385 bis 481 (1939).
— Chromosomenuntersuchungen an polyploiden Blütenpflanzen. I. Die Chromatinmasse bei künstlich ausgelösten Autopolyploiden. Ber. dtsch. bot. Ges. 57, 531—544 (1939).
— Die Auslösung von polyploidem Pisum sativum. Ber. dtsch. bot. Ges. 58, 430—436 (1940).
— Quantitative und qualitative Verschiedenheiten innerhalb von polyploiden Pflanzenreihen. Biol. Zbl. 60, 659—669 (1940).
— Ergebnisse und Probleme der Polyploidieforschung. Forschungsdienst 12, 318—324 (1941).
— Untersuchungen über die zytologische Grundlage der Komplexheterozygotie. Chromosoma 2, 64—76 (1941).
— Die Cytologie der haploiden Epilobien und die Phylogenie der Gattung. Biol. Zbl. 61, 573—588 (1941).
— Chromosomenstruktur. Naturwiss. 31, 97—108, 396 (1943).
Strub, W.: Untersuchungen zur Phänanalyse und Zytologie des Artbastardes Primula (Auricola L. × Viscosa All.) Arch. d. Jul. Klaus-Stiftg. 15, 105—183 (1940).
Stuart, N. W. u. S. L. Emsweller: Use of enzymes to improve cytological techniques. Science 98, 569—570 (1943).
Subramaniam, M. K.: Is the macronucleus of Ciliates endopolyploid? Curr. Sci. 16, 228—229 (1947).
Sugiura, T.: Studies on the chromosome numbers of higher plants. IV. Cytologia 10, 324—333 (1940).
— Studies on the chromosome number of higher plants. V. Cytologia 10, 363—370 (1940).
Sullivan, B. J. u. H. I. Wechsler: The cytological effects of podophyllin. Science 105, 433 (1947).
Sutton, E.: Trisomics in Pisum sativum derived from an interchange heterozygote. J. Genet. 38, 459—476 (1939).
Suzuki, E.: (Cytological observations on some sugar cane varieties. [A prelim. note].) Jap. J. Genet. 16, 276—278 (1940). [Japanisch.]
Svešnikova, I. N.: New method for comparative cytological study of species. C. R. (Doklady) Acad. Sci. URSS 30, 761—763 (1941).
Swanson, C. P.: Differences in meiotic coiling between Trillium and Tradescantia. Pap. Mich. Acad. Sci. 28, 133—142 (1942).
— Some considerations on the phenomenon of chiasma terminalization. Amer. Nat. 76, 593—610 (1942).
— Differential sensitivity of prophase pollen tube chromosomes to X-rays and ultraviolet radiation. J. Gen. Physiol. 26, 485—494 (1943).
— Secondary association of fragment chromosomes in generative nucleus of Tradescantia and its bearing on their origin. Bot. Gaz. 105, 108—112 (1943).
— The behavior of meiotic prophase chromosomes as revealed through the use of high temperatures. Amer. J. Bot. 30, 422—428 (1943).
— X-ray and ultraviolet studies on pollen tube chromosomes. I. The effect of ultraviolet (2537 A) on X-ray-induced chromosomal aberrations. Genetics 29, 61—68 (1944).
— A consideration of the structure of the prophase chromosomes in the pollen tubes of Tradescantia. Genetics 32, 109 (1947). (Abstr.)
— X-ray and ultraviolet studies on pollen tube chromosomes. II. The quadripartite structure of the prophase chromosomes of Tradescantia. Proc. Nat. Acad. Sci., Wash. 33, 229—232 (1947).

Swanson, C. P. u. A. Hollaender: The frequency of X-ray induced chromatid breaks in Tradescantia as modified by near infrared radiation. Proc. Nat. Acad. Sci., Wash. **32**, 295—302; also Biol. Wood's Hole **91**, 242 (1946). (Abstr.)
— u. R. Nelson: Spindle abnormalities in Mentha. Bot. Gaz. **104**, 273—280 (1942).

Takenaka, Y.: (Sex characters and descent of angiospermae.) Jap. J. Genet. **15**, 45 (1939).
Tang, P. S. u. W. S. Loo: Polyploidy in soybean, pea, wheat and rice, induced by colchicine treatment. Science **91**, 222 (1940).
Taylor, H.: A physiological study of diploid and related tetraploid plants. Proc. Okla. Acad. Sci. **22**, 137—138 (1942).
— Cyto-taxonomy and phylogeny of the Oleaceae. Brittonia **5**, 337—367 (1945).
Thoday, J. M. u. J. Read: Effect of oxygen on the frequency of chromosome aberrations produced by X-rays. Nature, Lond. **160**, 608 (1947).
Thomas, J.-A.: Action du p-aminophénylsulfamide sur la chromatine de l'oeuf d'Oursin en segmentation. C. R. Acad. Sci., Paris **214**, 90—93 (1942).
— Essais sur le mécanisme de l'action mitoclasique des sulfamides et des sulfones, chez l'oeuf d'oursin en segmentation. Evolution de la sensibilité des larves au cours du développement. C. R. Soc. Biol., Paris **137**, 38—39 (1943).
— Evolution nucléaire de l'oeuf d'Oursin en segmentation, après blocage de la mitose par la 4-oxy-4' aminodiphénylsulfone (98 G). C. R. Soc. Biol., Paris **137**, 12—13 (1943).
Thomas, P. T.: Plant improvement by use of drugs. Market Gr., Lond., Pp. 2.
— The use of drugs in plant improvement. Gdnrs' Chron. **112**, 238—239 (1942).
— Experimental imitation of tumour conditions. Nature Lond. **156**, 738—740 (1945).
— u. R. Drew: Chemical control of mitosis. Nature Lond. **152**, 564—565 (1943).
Thompson, R. C.: A technique for treating small seedlings with colchicine. Plant Physiol. **18**, 128—130 (1943).
Togby, H. A.: Cytological methods for Crepis species. Stain Techn. **17**, 171—175 (1942).
— A cytological study of Crepis fuliginosa, C. neglecta, and their F_1 hybrid, and its bearing on the mechanism of phylogenetic reduction in chromosome number. J. Genet. **45**, 67—111 (1943).
Traub, H. P.: Effect of sulfanilamide and other sulfa compounds on nuclear conditions in plants. J. Hered. **32**, 157—159 (1941).
Tschermak, E.: Durch Colchicinbehandlung ausgelöste Polyploidie bei der Grünalge Oedogonium. Naturwiss. **30**, 683—684 (1942).

Ullrich, H. u. P. van Veen: Dichroitische Effekte in pflanzlichem Plasma. Naturwiss. **30**, 1 (1942).

Vaarama, A.: Experimental studies on the influence of DDT insecticide upon plant mitosis. Hereditas, Lund **33**, 191—219 (1947).
Valadares, M. u. I. Regalheiro: „Difference in phase" (1) in the euchromatic cycle of chromosomes of the same karyokinetic phase. Portugaliae Acta Biologica, Lisboa **1**, Sér. A, 312—315 (1946).
Valencia, R. M.: Is there transmission of the effect of heterochromatin on variegation in Drosophila? Genetics. **32**, 109—110 (1947). (Abstr.)
Vanderlyn, L.: On the concepts of mitosis. Science **104**, 514—515 (1946).
Vilkomerson, H.: Chromosomes of Astragalus. Bull. Torrey Bot. Cl. **70**, 430—435 (1943).

Villars, R.: Contribution à l'étude cytologique de l'action des rayons X sur les cellules végétales. C. R. Soc. Biol., Paris **131**, 947—950 (1939).
— Étude cytologique de l'action du thymol sur les cellules végétales. C. R. Soc. Biol., Paris **133**, 206—208 (1940).
— Étude cytologique de l'action des rayons X sur les racines colchicinées. C. R. Soc. Biol., Paris **133**, 424 bis 426 (1940).

Wada, B.: Lebensbeobachtungen über die Einwirkung des Colchicins auf die Mitose, insbesondere über die Frage der Spindelfigur. Cytologia **11**, 93—116 (1940).
— (Effects of colchicine on dividing cells. Atractosomfiguren in somatic cells.) Jap. J. Genet. **17**, 32—34 (1941).
Wanner, H.: Cytologische Analyse der Artbastarde Primula (pulverulenta Duthie × cockburniana Hemsl.) und ihrer Eltern. Arch. J. Klaus-Stiftg. **16**, 495—557 (1941).
Warmke, H. E.: A study of spontaneous breakage of the Y-chromosome in Melandrium. Amer. J. Bot. (Suppl.) **33**, 224 (1946). (Abstr.)
— Sex determination and sex balance in Melandrium. Amer. J. Bot. **33**, 648—660 (1946).
Weichsel, G.: Polyploidie, veranlaßt durch chemische Mittel. Insbesondere Colchicinwirkung bei Leguminosen. Züchter **12**, 25—32 (1940).
Wellensiek, S. J.: The newest fad, colchicine, and its origin. Chronica Botanica **5**, 15—17 (1939).
— De bouwsteenen der organismen. (The building stones of the organisms.) Uitgeversmaatschappij W. de Haan N. V., Utrecht, Pp. 64 (1943).
Werner, G.: Zytologische Untersuchungen über die Wirkung des Colchicins bei zwei verschieden reagierenden Pflanzen: Lein und Erbse. Biol. Zbl. **60**, 86 bis 103 (1940).
Westergaard, M.: Studies on cytology and sex determination in polyploid forms of Melandrium album. Ejnar Munksgaard, Kopenhagen, Pp. 131 (1940).
— Aberrant Y chromosomes and sex expression in Melandrium album. Hereditas, Lund **32**, 419—443 (1946).
Wettstein, F. von: Experimentelle Untersuchungen zum Artbildungsproblem. II. Zur Frage der Polyploidie als Artbildungsfactor. Ber. dtsch. bot. Ges. **58**, 374—388 (1940).
— Warum hat der diploide Zustand bei den Organismen den größeren Selektionswert? Naturwiss. **31**, 574—577 (1943).
White, M. J. D.: Amount of heterochromatin as a specific character. Nature, Lond. **152**, 536—537 (1943).
Whitehouse, H. L. K.: A brief account of the fundamentals of genetics. Sch. Sci. Rev., Nr. 95, 77—90 (1943).
Wichterman, R.: Further evidence of polyploidy in the conjugation of green and colorless Paramecium bursaria. Biol. Bull. Wood's Hole **91**, 234 (1946). (Abstr.)
Wicks, L. F., C. Carruthers u. M. G. Ritchey: The Piccolyte resins as microscopic mounting media. Stain Tech. **21**, 121—126 (1946).
Wiebalck, U.: Untersuchungen zur Physiologie der Meiosis. XI. Reifeteilung und Kohlehydratspiegel der Pflanze. Z. Bot. **36**, 161—212 (1940).
Wilson, G. B. u. E. R. Boothroyd: Temperature-induced differential contraction in the somatic chromosomes of Trillum erectum L. Canad. J. Res. **22**, Sect. C. 105—119.
Witkus, E. R.: Endomitosis in plants. Biol. Bull. Woods Hole **89**, 191—192 (1945). (Abstr.)
— Endomitotic tapetal cell divisions in Spinacia. Amer. J. Bot. **32**, 326—330 (1945).

Witkus, E. R.: Naturally occurring polyploid mitosis in the normal development of Allium cepa. Amer. J. Bot. (Suppl.) **33**, 224—225 (1946). (Abstr.)
— Additional evidence on the role of polyploidy in plant development. Amer. J. Bot. Suppl. **33**, 828 (1946). (Abstr.)
— u. C. A. Berger: Veratrine, a new polyploidy inducing agent. J. Hered. **35**, 131—133 (1944).
—, — Polyploid mitosis in the normal development of Mimosa pudica. Bull. Torrey. Bot. Cl. **74**, 279—282 (1947).
Wulff, H. D.: Chromosomenstudien an der schleswig-holsteinischen Angiospermen-Flora. Ber. dtsch. bot Ges. **57**, 424—431 (1939).

Yamaha, G.: (How to observe chromonemata.) Bot. and Zool. **7**, 111 (1939).
— u. R. Ueda: Über den Einfluß der Ultraschallwellen auf die Wurzelspitzenzellen von Vicia Faba L. — Ein Orientierungsversuch. Cytologia, Tokyo **9**, 524—532 (1939).

Zeiger, K.: Neuere Anschauungen über den Feinbau des Protoplasmas. Klin. Wschr. **1**, 201—205 (1943).

f) Biophysik.

Bless, A. A.: The biological effects of X-rays as a function of intensity. Proc. Nat. Acad. Sci., Wash. **30**, 118—121 (1944).
Born, H. J., G. Melchers, K. Pätau u. K. G. Zimmer: Inaktivierungsversuche an Tabak-Mosaik-Virus mit ionisierenden Strahlen. Im Druck (1944).
—, N. W. Timoféeff-Ressovsky u. K. G. Zimmer: Biologische Anwendungen des Zählrohres. Naturwiss. **30**, 600—603 (1942).
Bünning, V. E.: Quantenmechanik und Biologie. Naturwiss. **31**, 194—197 (1943).

Gurvič, A. u. L. Gurvič: (Twenty years of mitogenetic radiation.) Uspehi Sovremennoj Biologii (Advances in Modern Biology) **16**, 305—334 (1943). [Russisch.]

Jordan, P.: Die Physik und das Geheimnis des organischen Lebens. Friedr. Vieweg & Sohn, Braunschweig, Pp. 183 (1941).

Koyenuma, N.: Trefferwahrscheinlichkeit und Variabilität. Bemerkungen zur Arbeit von H. v. Schelling in Naturwiss. 30, 306. Naturwiss. **30**, 732—733 (1942).

Latarjet, R.: Action du froid sur la réparation des radio-lésions chez une levure et chez une bactérie. C. R. Acad. Sci., Paris **217**, 186—188 (1943).
Lazarev, P. P.: (Modern problems of biophysics.) The Academy of Sciences of the USSR., Moscow-Leningrad Pp. 152. 63 figs. 4 tables. illus. (1945). [Russisch.]

Milani, R.: Ricerche sulla espressività e la penetranza dei diversi gradi di questa nel ceppo s. o. in funzione della temperatura. Boll. soc. It. Biol. Sperim. p. 22 (1946).

Rashevsky, N.: Some Remarks on the Movement of Chromosomes during Cell Division. Bull. math. Biophysics **3**, 1—3 (1941).

Schelling, H. von: Trefferwahrscheinlichkeit und Variabilität. Ein Versuch zur Deutung der Wirksamkeit von Antigenen. Naturwiss. **30**, 306—312 (1942).
Schubert, G.: Über die genetische und allgemeinbiologische Strahlenwirkung als Grundlage unserer Anschauungen über die Erbschädigungen. Mit 4 Abb. Röntgenpraxis **13**, 1—14 (1941).

Schubert, G.: Kernphysik und Medizin. Göttingen: Verlag Muster-Schmidt, Pp. 344 (1947).
Smith, F.: Orienterende undersøkelser over innflydelsen av infra-rød stråling på plantene. I. (Preliminary researches on the effects of infra-red radiation on plants. I.) Meld. Norg. LandbrHøgsk. **24**, 69—144 (1944).
Sommermeyer, K.: Bemerkungen zur Theorie des strahlenbiologischen Sättigungseffektes. Naturwiss. **30**, 104—105 (1942).
— Zur Auswertung des strahlenbiologischen Sättigungseffektes. Naturwiss. **31**, 172—173 (1943).
— Zur Auswertung des strahlenbiologischen Sättigungseffektes. Strahlenther. **31**, 173 (1943).

Withrow, R. B.: Radiant energy nomenclature. Plant Physiol. **18**, 476—487 (1943).

Zalkind, S. J.: (The present state of the question concerning the inhibitors of mitogenetic radiation). Uspehi Sovremennoj Biologii (Advances in Modern Biology) **16**, 415—425 (1943). [Russisch.]
Zimmer, K. G.: Zur Berücksichtigung der „biologischen Variabilität" bei der Treffertheorie der biologischen Strahlenwirkung. Biol. Zbl. **61**, 208—220 (1941).
— Zur treffertheoretischen Analyse der Antigenwirkung. Naturwiss. **30**, 452—453 (1942).
— Ergebnisse und Grenzen der treffertheoretischen Deutung von strahlenbiologischen Dosis-Effekt-Kurven. Biol. Zbl. **62**, 72 (1943).
— Über den Mechanismus der Wirkung ionisierender Strahlen auf Lösungen. I. Phys. Ztschr. **45**, 265 (1944).
— u. J. Bauman: Über den Mechanismus der Wirkung ionisierender Strahlen auf Lösungen. II. Phys. Ztschr. **45** (1944).
— u. E. C. Crohn: Über den Mechanismus der Wirkung ionisierender Strahlen auf Lösungen. III. Im Druck. (1944.)

g) Biochemie.

Bennett, H.: The chemical formulary. Volume VII. Chemical Publishing Co., Inc., Brooklyn, N. Y., Pp. XXXII+474 (1945).

Jacobs, M. B. (Editor): The chemistry and technology of food and food products. Interscience Publishers, Inc., New York, Pp. XV+952. 79 figs. 218 tables (1944).
Jordan, P.: Über die Spezifizität von Antikörpern, Fermenten, Viren, Genen. Naturwiss. **29**, 89—100 (1941).

Knjaginičev, M. I.: (The biochemical study of initial and hybrid material.) Selekcija i Semenovodstvo (Breeding and Seed Growing), Nr. 12, 8—12 (1939). [Russisch.]
— (Biochemical variation and its significance in breeding food crops.) Soviet Plant Industry Record, Nr. 1, 89 bis 103 (1940). [Russisch.]

Malisoff, W. M.: Dictionary of biochemistry and related subjects. Philosophical Library, Inc., New York, Pp. 579 (1943).
Martin, H.: The scientific principles of plant protection with special reference to chemical control. Edward Arnold and Co., London (1940).
Matlin, D. R.: Chemical gardening. Latest developments in soilless culture of plants. Chemical Publishing Co. Inc. Brooklyn, N. Y., Pp. VI+159 tables. plates (1942).

Nicol, H.: Plant growth-substances. Their chemistry and applications, with special reference to synthetics. Leonard Hill, Ltd., London (1940).

Pigman, W. W. u. M. L. Wolfrom (Editors): Advances in carbohydrate chemistry. Academic Press Inc., New York 1, Pp. XII + 374 figs. tables. (1945).
—, — Advances in carbohydrate chemistry. Academic Press Inc. N. Y., 2, 36s. 6d. Pp. XIV + 323. tables. (H. K. Lewis & Co., Ltd., London.) (1946).
Prokošev, S. M.: (Methods of determining the titre of the solution of the stain used for estimating ascorbic acid [vitamin C].) Vestnik Socialističeskogo Rastenievodstva (Soviet Plant Industry Record), Nr. 3, 147 bis 148 (1940). [Russisch.]

Young, C. B. F. u. K. N. Coons: Surface active agents, theoretical aspects and applications. Chemical Publishing Co., Inc., Brooklyn, N. Y., Pp. X + 381. figs. tables (1945).

h) Pflanzenkrankheiten, Resistenzzüchtung, Bakterien u. Pilze.

Ainsworth, G. C. u. G. R. Bisby: A dictionary of the fungi. Imperial Mycological Institute, Kew, Surrey, Pp. VIII + 359. 138 figs (1943).
—, — A dictionary of the fungi. Imperial Mycological Institute, Kew, Surrey, 2nd Ed. 20s., Pp. VIII + 431. 139 figs (1945).
Åkerberg, E.: Om angrepp av knäppare. (On attack by wireworm.) Lantmannen 24, 603—604 (1940).
Åkerman, A.: Tre års erfarenheter rörande våra åkerbruksväxter vinterhärdighet. (Three years' experiments on the winterhardiness of our agricultural crops.) Sverig. Utsädesfören. Tidskr. 52, 291—305 (1942).
Altenburg, E.: The „viroid" theory in relation to plasmagenes, viruses, cancer and plastids. Amer. Nat. 80, 559—567 (1946).
— Tumour formation in relation to the origin of viruses. Amer. Nat. 81, 72—76 (1947).
Andersson, G.: Kölden och växtlivet. (Cold and plant life.) Sverig. Utsädesfören. Tidskr. 56, 69—74 (1946).
Andrews, W. B. u. C. F. Briscoe: The reponse of vetch and soybeans to strains of nodule bacteria. J. Amer. Soc. Agron. 35, 271—278 (1943).
Anonymus: (Investigation on Derris cultivation from the chemical point of view.) Monogr. Gov. Agric. Res. Inst. Taiwan, Nippon (Japan), Nr. 73, 40 (1939).
— Recent advances in economic entomology. J. Aust. Inst. Agric. Sci. 8, 80—81 (1942).
— Making new diseases. Mon. Sci. News, Nr. 16, 2—3 (1942).
— What the scientists are doing. Pyrethrum in Mysore. Indian Fmg. 3, 441—442 (1942).
— What the scientists are doing. Testing technique. Indian Fmg. 3, 597 (1942).
— The maize stalk-borer. Fmg. S. Afr. 17, 763—766 (1942).
— Research work and workers. Seed World 53, Nr. 2, 40 (1943).
— List of common British plant diseases. Cambridge University Press, Pp. 61 (1944).
— Virus names used in the Review of Applied Mycology. Imperial Mycological Institute, Kew, Surrey, Pp. 44 (Mimeographed) (1944).
— Gråfläcksjuka. En av Manganbrist orsakad växtsjukdom. (Grey speck. A disease caused by manganese deficiency.) Flygbl. Växtskyddsanst., Stockh., Nr. 75, 4 (1945).
— Heredity and the nucleus in yeasts. J. Inc. Brew. Guild 31, 1—5 (1945).
— Heterogenesis and the origin of viruses. Nature, Lond. 158, 406—407 (1946).
— Genetics for the brewer. Wallerstein Laboratories Communications 9, 1—3 (1946).
— Common names of British insect and other pests. Part I. Mendip Press, Ltd., Bath 2s., Pp. 30 (1947).

Arbuthnot, K. D.: Strains of the European corn borer in the United States. Techn. Bull. U. S. Dep. Agric., Nr. 869 20 (1944).
Ark, P. A.: Studies on bacterial canker of tomato. Phytopathology 34, 394—400 (1944).
Armstrong, G. M., B. S. Hawkins u. C. C. Bennett: Cross inoculations with isolates of Fusaria from cotton, tobacco, and certain other plants subject to wilt. Phytopathology 32, 685—698 (1942).

Babu Naidu, M. u. V. M. Bakshi: The cytology of yeast. Curr. Sci. 15, 231 (1946).
Bagchee, K.: Contributions to our knowledge of the morphology, cytology and biology of Indian coniferous rusts. Part 1. Cronartium himalayense Bagchee and Peridermium orientale Cooke on Pinus longifolia Roxb. Indian For. Rec. Bot. 1, 247—266 (1941).
Baker, G. E.: Heterokaryosis in Penicillium notatum. J. Bact. 47, 581 (1944).
— Nuclear behavior in relation to culture method for Penicillium notatum Westling. Science 99, 436 (1944).
Baldacci, E.: La resistenza delle piante alle malattie. (The disease resistance of plants.) Genova-Roma-Napoli-Città di Castello, Pp. 261. 14 figs (1942).
Barrons, K. C.: Studies of the nature of root knot resistance. J. Agric. Res. 58, 263—271 (1939).
Bawden, F. C.: Problems in breeding for disease resistance. Chronica Botanica 6, 247 (1941). (Abstr.)
— Virus diseases of plants. J. R. Soc. Arts. 94, 136—168 (1946).
Beadle, G. W.: Investigation of genes controlling known synthesis in the ascomycete Neurospora. Yearb. Amer. Philos. Soc. 1942, Pp. 143—145 (1943).
— u. V. L. Coonradt: Heterocaryosis in Neurospora crassa. Genetics 29, 291—308 (1944).
—, H. K. Mitchell u. J. F. Nye: Kynurenine as an intermediate in the formation of nicotinic acid from tryptophane by Neurospora. Proc. Nat. Acad. Sci., Wash. 33, 155—158 (1947).
Bergman, B.: Zytologische Studien über die Befruchtung und Askokarpbildung bei Sphaerotheca Castagnei Lev. Svensk. Bot. Tidskr. 35, 194—210 (1941).
Berkeley, G. H.: Root-rots of certain non-cereal crops. Bot. Rev. 10, 67—123 (1944).
Bessey, E. A.: Some problems in fungus phylogeny. Mycologia 34, 355—379 (1942).
Bever, W. M.: A nompathogenic buff-colored barley smut. Phytopathology 32, 637—639 (1942).
— Hybridization and genetics in Ustilago Hordei and U. nigra. J. Agric. Res. 71, 41—59 (1945).
Bisby, G. R.: Nomenclature of fungi. Mycologia 36, 279—285 (1944).
— An introduction to the taxonomy and nomenclature of fungi. Imperial Mycological Institute, Kew, 5s, Pp. VII + 117 (1945).
Bjaanes, M.: Resistensforedling. (Breeding for resistance.) Tidsskr. Norske Landbr. 46, 512—522 (1939).
Björling, K.: Undersökningar rörande klöverrötan. II. Studier av utvecklingshistoria och variation hos Sclerotinia trifoliorum. (Investigations on clover rot. II. Studies on the developmental history and variation of Sclerotinia trifoliorum.) Medd. Växtskyddsanst., Stockh., Nr. 37, 154 (1942).
Black, L. M.: Different vector specificities for varieties of a plant virus. Phytopatology 33, 17 (1943). (Abstr.)
— Genetic variation in the clover leafhopper's ability to transmit potato yellow-dwarf virus. Genetics 28, 200 bis 209 (1943).
— Some viruses transmitted by Agallian leafhoppers. Proc. Amer. Phil. Soc. 88, 132—144 (1944).
Blagoveščenskij, A. V.: (The biochemical factors of natural selection in plants.) Žurnal Obščej Biologii (Journ. of General Biology), Nr. 4, 217—234 (1945). [Russisch.]

Blanchard, R. A. et al: Developments and methods in plant breeding studies as they relate to insect resistant crops. North Cent. States Ent. Proc., Nr. 19, 33—38 (1940).

Boivin, A., A. Delauney, R. Vendrely u. Y. Lehoult: L'acide thymonucléique polymérisé, principe paraissant susceptible de déterminer la spécificité sérologique et l'équipement enzymatique des bactéries. Signification pour la biochimie de l'hérédité. Experientia, Basel 1, 334—335 (1945).

Bonner, D.: Further studies of mutant strains of Neurospora requiring isoleucine and valine. J. Biol. Chem. 166, 545—554 (1946).

— Production of biochemical mutations in Penicillum. Amer. J. Bot. 33, 788—791 (1946).

— Studies on the biosynthesis of penicillin. Arch. Biochem. 13, 1—9 (1947).

— u. G. W. Beadle: Mutant strains of Neurospora requiring nicotinamide or related compounds for growth. Arch. Biochem. 11, 319—328 (1946).

—, E. L. Tatum u. G. W. Beadle: The genetic control of biochemical reactions in Neurospora: a mutant strain requiring isoleucine and valine. Arch. Biochem. 3, 71—91 (1943).

Born, H. J., A. Lang, G. Schramm u. K. G. Zimmer: Versuche zur Markierung von Tabakmosaikvirus mit Radiophosphor. Naturwiss. 29, 222—223 (1941).

Boughey, A. S.: A preliminary list of plant diseases in the Anglo-Egyptian Sudan. Imperial Mycological Institute, Kew, 3s. Mycol. Pap., Nr. 14, 16 (1946).

Bowden, W. M.: Polyploidy and winter hardiness relationships in the flowering plants with reference to karyogeographical problems. Amer. J. Bot. 26, 5s (1939). (Abstr.)

— Chromosome number and winter hardiness relationships in the higher plants. Proc. Va. Acad. Sci. 2, 173. 1940—1941 (1941).

Braun, H.: Biologische Spezialisierung bei Synchytrium endobioticum (Schildb.) Pèrc. (Vorläufige Mitteilung.) Z. Pfl. Krankh. 52, 481—486 (1942).

Braun, W.: Some recent results of studies on bacterial variation. Genetics 32, 80 (1947). (Abstr.)

Bredemann, G.: Über die Züchtung heuschreckenresistenter Pflanzen. Z. Pflanzenkrkh. 51, 337—342 (1941).

Brett, C. H.: Insecticidal properties of the indigobush (Amorpha fructicosa). J. Agric. Res. 73, 81—96 (1946).

Briggs, F. N.: Breeding disease-resistant crops. Science 96, 60 (1942).

Brooks, F. T.: Disease-resistant plants. Endeavour, Lond. 1, 114—117 (1942).

Brown, N. A.: Tumors on elm and maple trees. Phytopathology 31, 541—548 (1941).

Brunson, A. M.: Cooperation of agronomists and entomologists in plant breeding studies on resistance and tolerance to insects. North. Cent. States Ent. Proc., Nr. 19, 28—29 (1940).

Bucher, P.: Symbiose der Tiere mit pflanzlichen Mikroorganismen. Sammlung Göschen, 1128. Walter de Gruyter and Co., Berlin (1939).

Buller, A. H. R.: The diploid cell and the diploidization process in plants and animals, with special reference to the higher fungi. Bot. Rev. 7, 335—387, 389—431 (1941).

Burkholder, P. R. u. N. H. Giles jr.: Production of biochemical mutants in Bacillus subtilis by means of ultra-violet radiation. Amer. J. Bot. Suppl. 33, 829 (1946). (Abstr.)

—, — Induced biochemical mutations in Bacillus subtilis. Amer. J. Bot. 34, 345—348 (1947).

Burkholder, W. H. u. C. C. Li: Variations in Phytomonas vesicatoria. Phytopathology 31, 753—755 (1941).

Burns, W.: Breeding for disease-resistance in agricultural plants. Sci. & Cult. 2, 619—621.

Čajlahjan, M. H.: (Entwicklung der verschiedenen Sommerwurzarten im Zusammenhang mit Wachstum und Entwicklung der Wirtspflanzen.) C. R. (Doklady) Acad. Sci. URSS. 55, 881—884 (1947). [Russisch.]

— (Anfälligkeit gegenüber den Blütenparasiten und Geschlecht der Pflanzen.) C. R. (Doklady) Acad. Sci. URSS. 56, 321—324 (1947). [Russisch.]

Castellani, E.: Razze di piante resistenti e prevenzione delle malattie. (Resistant races of plants and the prevention of diseases.) Genova Pp. 96, 14 figs, tables (1943).

Cercós, A. P. u. E. A. Favret: „Ustilago maydis", una nueva fuente de radiación mitogenética. (U. maydis, a new source of mitogenetic radiation.) Rev. Argent. Agron. 13, 128—137 (1946).

Česnokov, P. G.: (Resistance of species and varieties of crop plants to insect pests.) Vestnik Socialističeskogo Rastenievodstva (Soviet Plant Industry Record), Nr. 3, 131—144 (1940). [Russisch.]

Chakrabarti, S.: What's doing in All-India-Assam. Indian Fmg. 3, 97—98 (1942).

Chapman, G. H.: The concept of a „strain" in bacteriology. Science 101, 429—430 (1945).

Cherewick, W. J.: Studies on the biology of Erysiphe graminis DC. Canad. J. Res. 22, Sect. C., 52—86 (1944).

Chester, K. S.: The nature and prevention of plant diseases. Philadelphia, Pa., Pp. XII + 584. 12 tables, 207 figs (1942).

Chiarugi, A.: L'eredità in patologia vegetale. (Heredity in plant pathology.) Relaz. IV Congr. Int. Patol. Comp. Roma 1, 155—210 (1939).

Chilton, S. J. P.: Variations in sporulation of different isolates of Collotrichum destructivum. Mycologia 35, 13—20 (1943).

— A heritable abnormality in the germination of chlamydospores of Ustilago zeae. Phytopathology 33, 749—765 (1943).

—, G. B. Lucas u. C. W. Edgerton: Genetics of Glomerella. III. Grosses with a conidial strain. Amer. J. Bot. 32, 549—554 (1945).

— u. H. E. Wheeler: Studies on the nature of „segregation" in certain plus strains of Glomerella. Phytopathology 37, 4 (1947). (Abstr.)

Christensen, J. J.: Genetic variation in Gibberella zeae in relation to adaptation. Phytopathology 36, 396 (1946). (Abstr.)

— u. H. A. Rodenhiser: Physiologic specialization and genetics of the smut fungi. Bot. Rev. 6, 389—425 (1940).

Clausen, R. T.: Yam bean, warm-climate plant is a possible new insecticide. Fm Res. 10, Nr. 3, 14 (1944).

Clayton, E. E. u. J. A. Stevenson: Peronospora tabacina Adam, the organism causing blue mold (downy mildew) disease of tobacco. Phytopathology 33, 101 bis 113 (1943).

Corbett, C. E.: Plantas ictiotóxicas. Farmacologia da rotenona. (Fish poison plants. Pharmacology of rotenone.) Monogr. Fac. Med. Univ. S. Paulo, Nr. 1, 157 (1940).

Coutinho, L. de Azevedo: Os virus como agentes modificadores dos cromosomas. (Viruses as agents modifying the chromosomes.) Rev. Agron., Lisboa 28, 83—100 (1940).

Cowan, J. R.: The value of double cross hybrids involving inbreds of similar and diverse genetic origin. Sci. Agric. 23, 287—296 (1943).

Craigie, J. H.: Heterothallism in the rust fungi and its significance. Trans. Roy. Soc. Canada, Abt. V, Pp. 19 bis 40 (1942).

Croland, R.: Action des rayons X sur la fréquence d'une mutation bactérienne. C. R. Acad. Sci., Paris 216, 616—618 (1943).

Crowell, I. H. u. E. Lavalee: Check list of diseases of economic plants in Canada. Dom. Dep. Agric., Sci. Serv., Canada, Pp. 68 (1942). [Mimeographed.]
Cunningham, I. J.: The chemistry, pharmacology, and toxicology of ergot (Claviceps purpurea). N. Z. J. Sci. Tech. 23, 138A—145A (1941).
Cutter, V. M. jr.: Observations on the chromosomal morphology of Neurospora tetrasperma with the acetocarmine smear technique. Amer. J. Bot. (Suppl.) 33, 217 (1946). (Abstr.)
— The chromosomes of Neurospora tetrasperma. Mycologia 38, 693—698 (1946).

Dahms, R. G. u. F. A. Fenton: Plant breeding and selecting for insect resistance. J. Econ. Ent. 32, 131 bis 134 (1939).
Défago, G.: Seconde contribution à la connaissance des Valsées von Höhnel. Phytopath. Z. 14, 103—147 (1942).
Delbrück, M.: A statistical problem. J. Tenn. Acad. Sci. 19, 177—178 (1944).
— Spontaneous mutations in bacteria. Ann. Mo. Bot. Gdn. 32, 223—233 (1945).
— Bacterial viruses or bacteriophages. Biol. Rev. 21, 30—40 (1946).
Demerec, M.: Genetic aspects of changes in Staphylococcus aureus producing strains resistant to various concentrations of penicillin. Ann. Mo. Bot. Gdn. 32, 131—138 (1945).
— Production of Staphylococcus strains resistant to various concentrations of penicillin. Proc. Nat. Acad. Sci., Wash. 31, 16—24 (1945).
— Induced mutations and possible mechanisms of the transmission of heredity in Escherichia coli. Proc. Nat. Acad. Sci., Wash. 32, 36—46 (1946).
— u. U. Fano: Bacteriophage — resistant mutants in Escherichia coli. Genetics 30, 119—136 (1945).
Deve, P., G. Pontecorvo u. C. Higginbottom: X-ray induced mutations in dried bacteria. Nature, Lond. 160 503—504 (1947).
Dey, N. C.: A preliminary note on the antibacterial substances from Aspergillus flavus. Curr. Sci. 14, 265 bis 267 (1945).
Dickson, J. G.: Outline of diseases of cereals and forage crop plants of the northern part of the United States. Burgess Publishing Company, Minneapolis, Minn (1939).
Dillon Weston, W. A. R. u. K. M. Smith: Crop diseases and their control. J. R. Agric. Soc. 105, 125 bis 132 (1944).
Dimock, A. W.: Studies on ascospore variants of Hypomyces Ipomoeae. Mycologia 31, 709—727 (1939).
Ditlevsen, E.: A case of simple segregation in Saccharomyces italicus. C. R. Lab. Carlsberg 24, Sér. Physiol., 31—37 (1944).
Dodge, B. O.: Some problems in the genetics of the fungi. Science 90, 379—385 (1939).
— Genetic interpretations of cultural variations in the fungi. Proc. 8th. Amer. Sci. Congr., Wash. 3, 199—206 (1940).
— Second-division segregation and crossingover in the fungi. Bull. Torrey. bot. Cl. 67, 467—476 (1940).
— Conjugate nuclear division in the fungi. Mycologia 34, 302—307 (1942).
— Heterocaryotic vigor in Neurospora. Bull. Torrey Bot. Cl. 69, 75—91 (1942).
— A study of the inheritance of the factors for heterocaryotic vigor in Neurospora tetrasperma. Yearb. Amer. Philos. Soc. 1942, Pp. 148—150 (1943).
— Further remarks on mycogenetic terminology. Mycologia 37, 629—635 (1945).
— Further remarks on mycogenetic terminology. Mycologia 37, 784—791 (1945).
— Self-sterility in „bisexual" heterocaryosis of Neurospora. Bull. Torrey Bot. Cl. 73, 410—416 (1946).

Dodge, B. O., M. B. Schmitt u. A. Appel: Inheritance of factors involved in one type of heterocaryotic vigor. Proc. Amer. Phil. Soc. 89, 575—589 (1945).
— u. F. J. Seaver: Species of Ascobolus for genetic study. Mycologia 38, 639—651 (1946).
Drain, B. D., W. A. Simanton u. A. C. Miller: Studies on clonal of pyrethrum. Proc. Amer. Soc. Hort. Sci. 44 521—524 (1944).
DuBoy, H. G. u. M. W. Woods: Evidence for the evolution of phytopathogenic viruses from mitochondria and their derivatives. II. Chemical evidence. Phytopathology 33, 766—777 (1943).
Dulaney, E. L.: Penicillin production by the Aspergillus nidulans group. Mycologia 39, 582—586 (1947).
Dunin, M. S.: (The immunogenesis of plants, and its practical utilization.) Moskovskaja ordena Lenina Seljskohozjajstvennaja Akademija imeni K. A. Timirjazeva, Naučnaja Konferencija 6—13 Dekabrja 1944 g. II (Timirjazeva Agric. Acad. Moscow Rep. Sci. Conf. 6—13 December) 1944, Nr. 2, 56—58 (1945). [Russisch.]

Edgerton, C. W., S. J. P. Chilton u. G. B. Lucas: Genetics of Glomerella. II. Fertilization between strains. Amer. J. Bot. 32, 115—118 (1945).
Eicke, R. u. E. Köhler: Beobachtungen an den Eiweißkristallen der Kartoffelsorte „Juli". Protoplasma 38, 64—70 (1943).
Eigsti, O. J.: Colchicine — a bacterial habitat. Amer. J. Bot. (Suppl.) 33, 218 (1946). (Abstr.)
Elliott, C.: Recent developments in the classification of bacterial plant pathogens. Bot. Rev. 9, 655—666 (1943).
Euler, H. von, L. Ahlström u. B. Högberg: Veränderungen der Hefezellen durch Röntgenstrahlen und durch chemische Substanzen. Hoppe-Seyler's Ztschr. Physiol. Chem. 277, Nr. 1/2, 1—25 (1942).
Exner, B. u. S. J. P. Chilton: Variation in single basidiospore cultures of Rhizoctonia solani. Phytopathology 33, 3 (1943). (Abstr.)
—, — Cultural differences among single basidiospore isolates of Rhizoctonia solani. Phytopathology 33, 171 bis 174 (1943).

Fernandez Valiela, M. V.: Introducción a la fitopatologia. (Introduction to phytopathology.) B. Aires, Pp. XVI + 625. 144 figs. tables (1942).
Filinger, G. A. u. A. B. Cardwell: A rapid method of determining when a plant is killed by extremes of temperatures. Proc. Amer. Soc. Hort. Sci. 39, 85—86 (1941).
Fischer, G. W.: Some evident synonymous relationships in certain graminicolous smut fungi. Mycologia 35, 610—619 (1943).
— u. C. S. Holton: Inheritance of sorus characters in hybrids between Ustilago Avenae and U-perennans Mycologia 33, 555—567 (1941).
—, — Studies of the susceptibility of forage grasses to cereal smut fungi. IV. Crossinoculation experiments with Urocystis tritici, U. occulta, and U. agropyri. Phytopathology 33, 910—921 (1943).
Flor, H. H.: New physiologic races of flax rust. J. Agric. Res. 60, 575—591 (1940).
— Pathogenicity of aeciospores obtained by selfing and crossing known physiologic races of Melampsora lini. Phytopathology 31, 852—854 (1941).
— Inheritance of pathogenicity in Melampsora lini. Phytopathology 32, 653—669 (1942).
Florey, H. W.: Penicillin. J. R. Soc. Arts. 92, 652—658 (1944).
Frandsen, K. J.: Studier over Sclerotinia trifoliorum Eriksson. (Studies on S. trifoliorum Eriksson.) Danske Forlag, København, Pp. 220 (1946).

Frandsen, N. O.: Septoria-Arten des Getreides und anderer Gräser in Dänemark. Medd. Plantepat. Afd., Kgl.,Veterinaerog Landbohojskole, Kobenhabn, Nr. 26, 92 (1943).
Fransen, J. J.: Pyrethrum. Tijdschr. Ned. Heidemaatsch. **52**, 126—151 (1940).
Fries, N.: Über die Sexualität einiger Hydnaceen. Bot. Notiser, Pp. 285—300 (1941).
— X-ray induced mutations in the physiology of Ophiostoma. Nature, Lond. **155**, 757—758 (1945).
— X-ray induced parathiotrophy in Ophiostoma. Svensk Bot. Tidskr. **40**, 127—140 (1946).
— Experiments with different methods of isolating physiological mutations of filamentous fungi. Nature, Lond. **159**, 199 (1947).
— Mutant strains of Ophiostoma multiannulatum requiring components of different nucleiotides. Ark. Bot. **33**, Nr. 3, 1—7 (1947).
— u. L. Jonasson: Über die Interfertilität verschiedener Stämme von Polyporus abietinus (Dicks) Fr. Svensk. Bot. Tidskr. **35**, 177—193 (1941).
— u. U. Trolle: Combination experiments with mutant strains of Ophiostoma multiannulatum. Hereditas, Lund **33**, 377—384 (1947).

Gameleja, N. F.: (The question of variability of microbes.) Agrobiologija (Agrobiology), Nr. 3, 115—118 (1946). [Russisch.]
Garcia-Rada, G., G. J. Vallega, W. Q. Loegering u. E. C. Stakman: An unusually virulent race of wheat stem rust, Nr. 189. Phytopathology **32**, 720 bis 726 (1942).
Garrett, S. D.: Losses to world agriculture through root disease of crops. Chem. Ind., London **58**, 953—958 (1939).
— Losses to world agriculture through root disease of crops. Trop. Agriculture, Trin. **17**, 49—52 (1940).
Gattani, M. L.: Differences in diploid lines of Ustilago zeae. Phytopathology **36**, 398 (1946). (Abstr.)
Gäumann, E.: Immunität und Immunitätsreaktionen bei Pflanzen. Vjschr. Naturf. Ges., Zürich **89**, 221 (1944).
— Pflanzliche Infektionslehre. Basel, Pp. 611, 311 figs. 90 tables (1945).
Georgi, C. D. V. u. G. L. Teik: Preliminary results of analysis of clonal types of Derris under field conditions. Malay, Agric. J. **27**, 302—331 (1939).
—, — Further experiments with selected plants of Derris elliptica, Changi, Nr. 3. Malay. Agric. J. **28**, 44—68 (1940).
Gešele, E. E.: (The principles of phytopatnological assay in selection.) Ogiz. Selhozgiz, Moskva, Pp. 120 (1941). [Russisch.]
Ghose, T. P.: A note on Derris and other rotenone bearing vegetable insecticides, their occurrence and possibilities of cultivation in India. For. Leafl. Dehra Dun, Nr. 20, 9 (1942).
Giddings, N. J.: Additional strains of the sugar-beet curly top virus. J. Agric. Res. **69**, 149—157 (1944).
Giles, N. H. (jr.): Induced biochemical mutants in Absidia glauca. Amer. J. Bot. (Suppl.) **33**, 218—219 (1946). (Abstr.)
Gill, T. u. E. C. Dowling: The Forestry Directory. Washington, D. C., Pp. VII+411. tables, maps (1943).
Ginsburg, J. M., J. B. Schmitt u. T. S. Reid: A rotenone-bearing variety of Tephrosia virginiana in New Jersey. J. Econ. Ent. **35**, 276—280 (1942).
Góis, A. u. E. M. Correia: Influencia da raça da levedura (S. ellipsoideus) como elemento a considerar na correcção da concentração hidrogeniónica dos môstos. (Influence of the race of yeast [S. ellipsoideus] as a factor to be considered in correcting the hydrogen ion concentration of musts.) Rev. Agron. Lisboa **30**, 445 (1942).

Goldin, M. I.: (Die Klassifizierung des „Strik" Virus der Tomaten.) Mikrobiologija (Microbiology) **6**, Nr. 4, 320—322 (1947). [Russisch.]
Gonzales Toriño, H.: Selección de levaduras vínicas especializadas en la procucción de altas granduaciones alcohólicas. (Selection of pure wine yeasts specialized in the production of high degrees of alcohol.) Rev. Fac. Agron., Univ. Montevideo, Nr. 17, 57—145 (1939).
Gorlenko, M. V.: (Twenty Five Years in the Study of Cereal Diseases in USRR. [1917—1941].) Botaničeskij Žurnal (Journ. Botanique de l'URSS.) **31**, Nr. 1, 3 (1946). [Russisch.]
— (Toxine bei Schimmelpilzen.) C. R. (Doklady) Acad. Sci. URSS. **54**, 453—455 (1946). [Russisch.]
— u. I. V. Voronkevič: The causative agent of the slimy bacteriosis of cabbage. C. R. (Doklady) Acad. Sci. URSS. **52**, Nr. 9, 802—812 (1946). [Russisch.]
Goodspeed, T. H.: Experimental induction of heritable and other alterations in the fungus Neurospora tetrasperma. Proc. 8th. Amer. Sci. Congr., Wash. **3**, 223 bis 229 (1940).
Gough, H. C.: A review of the literature on soil insecticides. Imperial Institute of Entomology, London, 10s, Pp. II+161 (1945).
Gračeva, N. P.: (Controlled variation in enteric bacteria.) Agrobiologija (Agrobiology), Nr. 3, 136—141 (1946). [Russisch.]
Greener, B. M.: (Breeding pyrethrum for insecticides.) Proc. Lenin Acad. Agric. Sci. USSR., Nr. 6, 13—16 (1941). [Russisch.]
Greis, H.: Mutations- und Isolationsversuche zur Beeinflussung des Geschlechts von Sordaria fimicola (Rob.). Z. Bot. **37**, 1—116 (1941).
Groves, J. W.: Variations in Botrytis cinerea. Proc. Canad. Phytopath. Soc., Nr. 14, 13 (1946). (Abstr.)
Gunther, F. A. u. F. M. Turrell: The location and state of rotenone in the root of Derris elliptica. J. Agric. Res. **71**, 61—79 (1945).

Haldane, J. B. S. u. H. L. K. Whitehouse: Symmetrical and asymmetrical post-reduction in ascomycetes. Nature, Lond. **154**, 704 (1944).
Hansberry, R., R. T. Clausen u. L. B. Norton: Variations in the chemical composition and insecticidal properties of the yam bean (Pachyrrhizus). J. Agric. Res. **74**, 55—64 (1947).
— u. C. Lee: The yam bean, Pachyrrhizus erosus Urban, as a possible insecticide. J. Econ. Ent. **36**, 351—352 (1943).
Hansen, H. N.: Heterocaryosis and variability. Phytopathology **32**, 639—640 (1942).
— u. W. C. Synder: The dual phenomenon and sex in Hypomyces solani F. Cucurbitae. Amer. J. Bot. **30**, 419—422 (1943).
—, — Inheritance of sex in fungi. Proc. Nat. Acad. Sci., Wash. **32**, 272—273 (1946).
Hansen, H. P.: Studier over Kartoffelviroser i Danmark II. Fortsatte Sortsundersøgelser. (Studies on potato viruses in Denmark II. Variety investigations continued.) Tidsskr. Planteavl **46**, 355—362 (1941).
— Om Nomenklatur for Planteviru samt nogle Synonymer for Kartoffelvira og Kartoffelviroser. (On nomenclature for plant viruses and some synonyms for viruses and virus diseases of the potato.) Tidsskr. Planteavl **46**, 363—372 (1941).
— Orienterende Undersøgelser over nogle Tobaks- og Tomatviroser i Danmark. (Preliminary investigations of some viruses of tobacco and tomatoes in Denmark.) Nord. JordbrForskn. Nos. 5/6, 264—272 (1943).
Hansford, C. G.: Uganda plant diseases. E. Afr. Agric J. **10**, 147—151 (1945).
Hardison, J. R.: Pathogenicity of races of Erysiphe graminis on grasses in the tribe Hordeae. Phytopathology **33**, 5 (1943). (Abstr.)

Harrington, C. D.: The occurrence of physiological races of the pea aphid. J. Econ. Ent. **36**, 118—119 (1943).

Harrison, A. L.: A method for resting resistance of tomatoes to Fusarium wilt. Phytopathology **30**, 86—87 (1940). (Abstr.)

Hassebrauk, K.: Mit Hilfe neuer Testsorten durchgeführte Untersuchungen über die physiologische Spezialisierung von Puccinia triticina Erikss. Arb. biol, Reichsanst., Land- u. Forstw., Berlin **23**, 37—51 (1940).

Henrici, A. T.: The yeasts. Genetics, cytology, variation, classification and identification. Bact. Rev. **5**. 97—179 (1941).

Hermann, F. J.: The Amazonian varieties of Lonchocarpus nicou, a rotenoneyielding plant. J. Wash. Acad. Sci. **37**, 111—113 (1947).

Hershey, A. D.: Mutation of bacteriophage with respect to type of plaque. Genetics. **31**, 620—640 (1946).

Hildebrand, E. M.: Indexing cherry yellows on peach. Phytopathology **32**, 712—719 (1942).

— Strains of yellows virus in Montmorency cherry. Phytopathology **33**, 6 (1943). (Abstr.)

— The genetic designation of „strain" in bacteriology. Science **102**, 101—102 (1945).

Hirschhorn, E.: A cytologic study of several smut fungi. Mycologia **37**, 217—235 (1945).

— u. Hirschhorn, J.: Formas fisiólogicas en „Ustilago zeae" de diversas localidades de la Argentina. Su clasificación geográfica. (Physiological forms in U. zeae from different localities of Argentina: their geographical classification.) Physis, B. Aires **18**, 181—222 (1939).

Hollaender, A., K. B. Raper u. R. D. Coghill: The production and characterization of ultraviolet-induced mutations in Aspergillus terreus. I. Production of the mutations. Amer. J. Bot. **32**, 160—165 (1945).

—, C. P. Swanson u. I. Posner: The sun as a source of mutation producing radiation. Amer. J. Bot. Suppl. **33**, 830 (1946). (Abstr.)

— u. E. M. Zimmer: The effect of ultraviolet radiation and X-rays on mutation production in Penicillium notatum. Genetics **30**, 8 (1945). (Abstr.)

Hollande, A.-Ch. u. G. Hollande: Remarques au sujet de la structure cytologique de quelques microbes. Bull. Soc. Bot. Fr. **90**, 109—110 (1943).

Holman, H. J.: A survey of insecticide materials of vegetable origin. The Imperial Institute, London (1940).

Holton, C. S.: Preliminary investigations on dwarf bunt of wheat. Phytopathology **31**, 74—82 (1941).

— Further studies on the oat smuts, with special reference to hybridization, cytology, and sexuality. J. Agric. Res. **62**, 229—240 (1941).

— Extent of pathogenicity of hybrids of Tilletia tritici and T. levis. J. Agric. Res. **65**, 555—563 (1942).

— Transgressive inheritance of pathogenicity factors in hybrids between the two races of Tilletia tritici. Phytopathology **32**, 9 (1942). (Abstr.)

— Inheritance of chlamydospore and sorus characters in species and race hybrids of Tilletia caries and T. foetida. Phytopathology **34**, 586—592 (1944).

— u. G. W. Fischer: Hybridization between Ustilago avenae and U. perennans. J. Agric. Res. **62**, 121—128 (1941).

— u. H. A. Rodenhiser: New physiologic races of Tilletia tritici and T. levis. Phytopathology **32**, 117 bis 129 (1942).

Honecker, L.: Die wichtigsten Blattfleckenkrankheiten der Gerste, ihre Bedeutung und die Möglichkeiten ihrer Bekämpfung. Prakt. Bl. Pflanzenbau **21**, 1—19 (1943).

Hopkins, R. H.: Some recent advances in the biology of malting and brewing. The artificial hybridisation of yeasts. J. Inst. Brew. **46**, 68—74 (1940).

— Some prospects in yeast research. J. Inst. Brew. **51**, 138—140 (1945).

Horowitz, N. H., D. Bonner, H. K. Mitchell, E. L. Tatum u. G. W. Beadle: Genes as physiological agents. Genic control of biochemical reactions in Neurospora. Amer. Nat. **79**, 304—317 (1945).

—, M. B. Houlahan, M. G. Hungate u. B. Wright: Mustard gas mutations in Neurospora. Science **104**, 233—234 (1946).

Houlahan, M. B. u. H. K. Mitchell: A suppressor in Neurospora and its use as evidence for allelism. Proc. Nat. Acad. Sci., Wash. **33**, 223—229 (1947).

Huber, G. A. u. K. Baur: Brown rot on stone fruits in Western Washington. Phytopathology **31**, 718—731 (1941).

Humphrey, H. B., C. O. Johnston, R. M. Caldwell u. L. E. Compton: Revised register of physiologic races of leaf rust of wheat (Puccinia triticina). U. S. Dep. Agric., Bur. Pl. Ind., Div. Cereal Crops and Diseases, Washington, D. C. January, Pp. 18 (Mimeographed) (1939).

Husfeld, B.: Zur Züchtung krankheitswiderstandsfähiger Kulturpflanzen. Angew. Bot. **25**, 115—125 (1943).

Hvostova, V. V. u. S. J. Goldat: Colchicine induced tetraploids in chrysanthemum. C. R. (Doklady) Acad. Sci. URSS. **31**, 623—624 (1941).

Ingold, C. T.: Genetics of the microfungi. Nature, Lond. **157**, 614—616 (1946).

Izrailskij, V. P.: (Acquired immunity in plants.) Uspehi Sovremennoj Biologii (Advances in Modern Biology) **15**, 162—189 (1942). [Russisch.]

— (Growth promoting substances and their rôle in bacterial tumor development.) Uspjehi Sovremennoj Biologii (Advances in Modern Biology) **23**, 109—126 (1947). [Russisch.]

Jackson, H. S.: Life cycles and phylogeny in the higher fungi. Trans. Roy. Soc. Can. **38**, Sect. V, Ser. III, 32 (1944).

Jones, M. A.: Application of a modified red-color test for rotenone and related compounds to Derris and Lonchocarpus. J. Ass. Off. Agric. Chem., Wash. **28**, 352—359 (1945).

—, W. A. Gersdorff u. E. R. McGovran: A toxicological comparison of Derris and Lonchocarpus. J. Econ. Ent. **39**, 281—283 (1946).

—, D. G. White u. C. Pagán: Evaluation of some clones of Derris elliptica. Trop. Agriculture, Trin. **23**, 89—93 (1946).

Johnson, M. J., J. J. Stefaniak, F. B. Gailey u. B. H. Olson: Penicillin production by a superior strain of mold. Science **103**, 504—505 (1946).

Johnson, T. u. M. Newton: Crossing and selfing studies with physiologic races of oat stem rust. Canad. J. Res. **18**, Sect. C., 54—67 (1940).

—, — Mendelian inheritance of certain pathogenic characters of Puccinia graminis tritici. Canad. J. Res. **18** Sect. C., 599—611 (1940).

—, — The inheritance of a mutant character in Puccinia graminis tritici. Canad. J. Res. **21**, Sect. C., 205—210 (1943).

—, — Specialization, hybridization, and mutation in the cereal rusts. Bot. Rev. **12**, 337—392 (1946).

Johnston, C. L. u. F. J. Greaney: Studies on the pathogenicity of Fusarium species associated with root rot of wheat. Phytopathology **32**, 670—684 (1942).

Johnston, C. O., H. B. Humphrey, R. M. Caldwell u. L. E. Compton: Third revision of the international register of physiologic reces of leaf rust of wheat [Puccinia rubigo-vera tritici (Triticina)]. U. S. Dep. Agric., Bur. Pl. Industr., Washington, D. C., Pp. 20 (1942). [Mimeographed.]

Juganova, O. N.: (Der Einfluß der Temperatur auf die Länge der Inkubationszeit bei der Entwicklung von Venturia inaequalis Winter bei Äpfeln.) Mikrobiologija (Microbiology) **16**, Nr. 4, 315—319 (1947). [Russisch.]

Jukes, T. H. u. A. C. Dornbush: Growth stimulation of Neurospora cholineless mutant by dimethylaminoethanol. Proc. Soc. Exp. Biol., N. Y. **58**, 142—143 (1945).

Kallós-Deffner, L.: Zur serologischen Differenzierung von Hefearten. Ark. Bot. **30** B, 1—2 (1942).

Kausche, G. A. u. H. Stubbe: Über die Entstehung einer mit Röntgenstrahlen induzierten „Mutation" des Tabakmosaikvirus. Naturwiss. **27**, 501—502 (1939).

—, — Zur Frage der Entstehung röntgenstrahleninduzierter Mutationen beim Tabakmosaikvirusprotein Naturwiss. **28**, 824 (1940).

Keitt, G. W., M. H. Langford u. J. R. Shay: Inheritance of pathogenicity and certain mutant characters in Venturia inaequalis. Phytopathology **33**, 19 (1943). (Abstr.)

—, —, — Venturia inaequalis (Cke.) Wint. II. Genetic studies on pathogenicity and certain mutant characters. Amer. J. Bot. **30**, 491—500 (1943).

—, C. C. Leben u. J. R. Shay: Inheritance of pathogenicity and sex reaction in Venturia inaequalis. Phytopathology **36**, 403 (1946). (Abstr.)

Kernkamp, M. F.: The relative effect of environmental and genetic factors on growth types of Ustilago zeae. Phytopathology **32**, 554—567 (1942).

— u. W. J. Martin: The pathogenicity of paired haploid lines of Ustilago zeae versus the pathogenicity of numerous mixed haploids. Phytopathology **31**, 1051—1053 (1941).

Kessler, W. u. W. Ruhland: Über die inneren Ursachen der Kälteresistenz der Pflanzen. Forschungsdienst, Sonderh. 16, 345 (1942).

Khvostova, V. V. siehe Hvostova.

Kirjalova, E. N. u. T. A. Čistovič: (Breeding pure yeast cultures for fruit and berry wine making.) Proc. Lenin. Acad. Sci. USSR., Nr. 18, 30—32 (1940). [Russisch.]

Köhler, E.: Über die Variabilität und Mutabilität pflanzenpathogener Virusarten, dargestellt am Kartoffel-X-Virus und am Tabakmosaikvirus (Sammelbericht). Biol. Zbl. **61**, 298—328 (1941).

— Über die Resistenzeigenschaften von Nicotiana glutinosa gegenüber dem Tabakmosaikvirus. Z. Pflanzenkrankh. **51**, 449—462 (1941).

— Über vergebliche Versuche, beim Tabakmosaikvirus „Mutationen" in Rohsäften zu erzielen. Z. Pflanzenkrankh. **52**, 392—397 (1942).

— u. I. Hauschild: Betrachtungen und Versuche zum Problem der „erworbenen Immunität" gegen Virusinfektionen bei Pflanzen. Züchter **17/18**, 97—105 (1947).

Koppel, C. van de: Lonchocarpus-wortel (cubé of timbo), een waardevol insecticide uit het Amazonegebied, vergeleken met den Derris-wortel van Zuidoost-Azië. (Lonchocarpus root (cubé or timbo), a useful insecticide from the Amazon region, compared with the Derris root of S. E. Asia.) Landbouwk. Tijdschr. **55**, 63—73 (1943).

Kostoff, D.: Studies on atypical growth in plants from a cytogenetic point of view. J. Genet. **39**, 469—484 (1940).

Kotila, J. E.: A new sugar beet leaf blight caused by a strain of Corticium solani. Phytopathology **33**, 6—7 (1943). (Abstr.)

— Rhizoctonia foliage blight of sugar beets. J. Agric. Res. **74**, 289—314 (1947).

Krejer, G. K.: (Pyrethrum as a plant insecticide and its cultivation under the conditions in the USSR.) Vestnik Socialističeskogo Rastenievodstva (Soviet Plant Industry Record), Nr. 3, 95—106 (1940). [Russisch.]

Kreutzer, W. A., E. W. Bodine u. L. W. Durrell: A sexual phenomenon exhibited by certain isolates of Phytophthora capsici. Phytopathology **30**, 951—957 (1940).

Kumar, L. S. S. u. A. Abraham: Cytological studies in Indian parasitic plants. I. The cytology of Striga. Proc. Indian Acad. Sci. **14**, Sect. B., 509—516 (1941).

—, — Cytological studies in Indian parasitic plants. II. The cytology of Loranthus. Proc. Indian Acad. Sci. **15**, Sect. B., 253—255 (1942).

Lampen, J. O. u. M. J. Jones: Studies on the sulfur metabolism of Escherichia coli. II. Interrelations of norleucine and methionine in the nutrition of Escherichia coli and of a methionine-requiring mutant of Escherichia coli. Arch. Biochem. **13**, 47—53 (1947).

—, — u. A. B. Perkins: Studies on the sulfur metabolism of Escherichia coli. I. The growth characteristics and metabolism of a mutant strain requiring methionine. Arch. Biochem. **13**, 33—45 (1947).

—, R. R. Roepke u. M. J. Jones: Studies on the sulfur metabolism of Escherichia coli. III. Mutant strains of Escherichia coli unable to utilize sulfate for their complete sulfur requirements. Arch. Biochem. **13**, 55—66 (1947).

Lange-De La Camp, M.: Blüteninfektionen mit Myzel von Ustilago tritici. Z. Pflanzenkrankh. **50**, 142—150 (1940).

Langford, M. H. u. G. W. Keitt: Heterothallism and variability in Venturia pirina. Phytopathology **32**, 357 bis 369 (1942).

Large, E. C.: The advance of the fungi. Jonathan Cape, London (1940).

Larter, L. N. H.: Breeding plants resistant to disease. J. Jamaica Agric. Soc. **46**, 25—27 (1942).

— Seed corn. J. Jamaica Agric. Soc. **47**, 26—28 (1943).

Lea, D. E. u. M. H. Salaman: Experiments on the inactivation of bacteriophage by radiations, and their bearing on the nature of bacteriophage. Proc. Roy. Soc. **133**, Ser. B, 434—444 (1946).

Lederberg, J. u. E. L. Tatum: Gene recombination in Escherichia coli. Nature, Lond. **158**, 558 (1946).

Lennette, E. H.: Recent advances in viruses. A brief survey of recent work on viruses and virus diseases. Science **98**, 415—423 (1943).

Leonian, L. H. u. L. V. Greene: Induced autotrophism in yeast. J. Bact. **45**, 329—339 (1943).

Levan, A.: On the ubiquity of the camphor reaction of yeast. Hereditas, Lund **30**, 255—256 (1944). (Abstr.)

— Mitotic disturbances induced in yeast by chemicals, and their significance for the interpretation of the normal chromosome conditions of yeast. Nature, Lond. **158**, 626 (1946).

— Studies on the camphor reaction of yeast. Hereditas, Lund. **33**, 457—514 (1947).

Lilly, V. G. u. H. L. Barnett: The inheritance of partial thiamin deficiency in Lenzites trabea (Pers.) Fries. Amer. J. Bot. Suppl. **33**, 831 (1946). (Abstr.)

Lindegren, C. C.: Genetics of yeast. J. Bact. **44**, 623 (1942).

— The use of the fungi in modern genetical analysis. Iowa St. Coll. J. Sci. **16**, 271—290 (1942).

— Nuclear apparatus and sexual mechanism in a micrococcus. Jowa St. Coll. J. Sci. **16**, 307—318 (1942).

— An analysis of the mechanism of budding in yeasts and some observations on the structure of the yeast cell. Mycologia **37**, 767—780 (1945).

— Mendelian and cytoplasmic inheritance in yeasts. Ann. Mo. Bot. Gdn. **32**, 107—123 (1945).

— Heterokaryosis and heterosis in the fungi, their biological and industrial significance. Genetics **30**, 13 (1945). (Abstr.)

Lindegren, C. C.: Cytoplasmic effect and chromosome maps in Saccharomyces cerevisiae. Genetics 31, 223 (1946). (Abstr.)
— Function of volutin (metaphosphate) in mitosis. Nature, Lond. 159, 63—64 (1947).
— u. H. N. Andrews: Cytoplasmic hybrids in Penicillium notatum. Bull. Torrey Bot. Cl. 72, 361—366 (1945).
— u. E. Hamilton: Autolysis and sporulation in the yeast colony. Bot. Gaz. 105, 316—321 (1944).
— u. G. Lindegren: A new method for hybridizing yeast. Proc. Nat. Acad. Sci., Wash. 29, 306—308 (1943).
—, — Legitimate and illegitimate mating in Saccharomyces cerevisiae. Genetics 28, 81 (1943). (Abstr.)
—, — Selecting, inbreeding, recombining, and hybridizing commercial yeasts. J. Bact. 46, 405—419 (1943).
—, — Environmental and genetical variations in yield and colony size of commercial yeasts. Ann. Mo. Bot. Gdn. 30, 71—82 (1943).
—, — Segregation, mutation, and copulation in Saccharomyces cerevisiae. Ann. Mo. Bot. Gdn. 30, 453—463 (1943).
—, — Instability of the mating type alleles in Saccharomyces. Ann. Mo. Bot. Gdn. 31, 203—211 (1944).
—, — Sporulation in Saccharomyces cerevisiae. Bot. Gaz. 105, 304—316 (1944).
—, — Vitamin-synthesizing deficiencies in yeasts supplied by hybridization. Science 102, 33—34 (1945).
— u. C. Raut: The effect of the medium on apparent vitamin-synthesizing deficiencies of microorganisms. Ann. Mo. Bot. Gdn. 34, 75—84 (1947).
—, S. Spiegelman u. G. Lindegren: Mendelian inheritance of adaptive enzymes in yeast. Proc. Nat. Acad. Sci., Wash. 30, 346—352 (1944).
—, —, — Survey of growth and gas production of genetic variants of Saccharomyces cerevisiae on different sugars. Arch. Biochem. 6, 185—198 (1945).
Ling Lih: Host index of the parasitic fungi of Szechuan. Nanking J. 11, Nr. 3, 1—26 (1941).
— u. Juhwa Y. Yang: Stem blight of cotton caused by Alternaria macrospora. Phytopathology 31, 664—671 (1941).
Little, V. A.: Rotenone content, an inherited character in the roots of devil's shoestring, Tephrosia virginiana. J. Econ. Ent. 35, 54—57 (1942).
Lucas, G. B.: Genetics of Glomerella. IV. Nuclear phenomena in the ascus. Amer. J. Bot. 33, 802—806 (1946).
—, S. J. P. Chilton u. C. W. Edgerton: Genetics of Glomerella. I. Studies on the behavior of certain strains. Amer. J. Bot. 31, 233—239 (1944).
Luria, S. E.: Mutations of bacterial viruses affecting the host-range, and their relation to bacterial mutations. J. Bact. 47, 416—417 (1944).
— Genetics of bacterium-bacterial virus relationship. Ann. Mo. Bot. Gdn. 32, 235—242 (1945).
— Mutations of bacterial viruses and of their bacterial hosts. Genetics 30, 13—14 (1945). (Abstr.)
— Non-independent mutations in bacteria. Genetics 32, 95 (1947). (Abstr.)
— u. M. Delbrück: Mutations of bacteria from virus sensitivity to virus resistance. Genetics 28, 491—511 (1943).
Lutman, B. F.: Actinomyces in potato tubers. Phytopathology 31, 702—717 (1941).
— Actinomycetes in various parts of the potato and other plants. Bull. Vt. Agric. Exp. Sta., Nr. 522, 72 (1945).

Malloch, W. S.: The inheritance of induced mutations in Neurospora tetrasperma. Mycologia 34, 325—347 (1942).
Marcilla Arrazola, J. u. E. Feduchy Mariño: Contribución al estudio de una levadura perteneciente al género Saccharomycodes, capaz de fermentar mostos de uva fuertemente sulfitados (mostos azufrados), sin previa desulfitación. (Contribution to the study of a yeast belonging to the genus Saccharomycodes, capable of fermenting sulphurous grape must without previous removal of SO_2.) Bol. Inst. Invest. Agron. Madr., Nr. 9, 1—40 (1943).
Marcilla Arrazola, J. u. E. Feduchy Mariño: Contribución al estudio de una levadura perteneciente al género Saccharomycodes, capaz de fermentar mostos de uva fuertemente sulfitados (mostos azufrados), sin previa desulfitación. (Contribution to the study of a yeast belonging to the genus Saccharomycodes, capable of fermenting sulphuros grape must without previous removal of SO_2.) Bol. Inst. Invest. Agron. Madr., Nr. 10, 293—296 (1944).
Mardjanian, G. M.: (To the question of toxic characters of different Pyrethrum species.) Proc. Lenin Acad. Agric. Sci. USSR., Nr. 10, 26—29 (1941). [Russisch.]
Marshak, A. u. W. N. Takahashi: The effect of pH on inactivation of tobacco mosaic virus by X-rays. Proc. Nat. Acad. Sci., Wash. 28, 211—216 (1942).
Martin, W. J.: A study of the genetics of Sorosporium syntherismae and Sphacelotheca panici miliacei. Phytopathology 33, 569—585 (1943).
Martyn, E. B.: Diseases of plants in Jamaica. Bull. Dep. Sci. Agric., Jamaica, Nr. 32 (NS.), 34 (1942).
Mastenbroek, C.: Enkele veldwaarnemingen over virusziekten van lupine en een onderzoek over haar mozaikziekte. (Some field observations on virus diseases of the lupin, and an investigation on its mosaic diseases.) Tijdschr. PlZiekt. 48, 97—118 (1942).
— Gele roest (Puccinia glumarum) op kropaar (Dactylis glomerata). (Yellow rust [P. glumarum] on cocksfoot [D. glomerata].) Tijdschr. PlZiekt. 52, 66—67 (1946).
— u. A. J. P. Oort: Het voorkomen van moederkoren (Claviceps) op granen en grassen en de specialisatie van de moederkorenschimmel. (The occurrence of ergot [Claviceps] on cereals and grasses, and the physiological specialization of the ergot fungus.) Tijdschr. PlZiekt. 47, 165—185 (1941).
Mather, K.: Heterothally as an out-breeding mechanism in fungi. Nature, Lond. 149, 54—56 (1942).
Mehta, K. C.: Further studies on cereal rusts in India. Sci. Monogr. Imp. Coun. Agric. Res., Nr. 14, VII + 224 (1940).
Melchers, G.: Über einige Mutationen des Tabakmosaikvirus und eine „Parallelmutation" des Tomatenmosaikvirus. Naturwiss. 30, 48 (1942).
Melhus, I. E. u. G. C. Kent: Elements of plant pathology. Macmillan Company, New York (1939).
Mestre Artigas, C. u. A. Mestres Jané: Fermentaciones comparativas con diferentes levaduras. (Comparative fermentations with different yeasts.) Bol. Inst. Nac. Invest. Agron.,Madr., Nr. 14, 1—28 (1946).
Mille, L.: Los Barbascos. (The Barbascos.) Flora, Ecuador 2, Nos. 5/6, 127—138 (1944).
Miller, J. J.: The theories concerning variability in fungi. Proc. Canad. Phytopath. Soc., Nr. 14, 13. (1946). (Abstr.)
Mitchell, H. K. u. M. B. Houlahan: Neurospora. IV. A temperature-sensitive riboflavinless mutant. Amer. J. Bot. 33, 31—35 (1946).
Mittmann-Maier, G.: Untersuchungen über die Anfälligkeit von Apfel- und Birnensorten gegenüber der Moniliafruchtfäule. Gartenbauwiss. 15, 334—361 (1940).
Moore, R. H.: Derris grows in America. Agric. Amer. 5, 10—12, 16, 18 (1945).
— Some effects of altitude and water supply on the composition of Derris elliptica. Bot. Gaz. 107, 467 bis 474 (1946).
Moulton, F. R. (Editor): The genetics of pathogenic organisms. Publ. Amer. Assoc. Advanc. Sci., Nr. 12, 90 (1940).

Moyer, A. J. u. R. D. Coghill: The laboratory-scale production of itaconic acid by Aspergillus terreus. Arch. Biochem., N. Y. **7**, 167—183 (1945).

Mujica, R. F.: Inmunización mediante la formación genética de variedades resistentes a las enfermedades de las plantas. (Immunization by means of the genetical production of varieties resistant to plant diseases.) Bol. Sanid Veg., Chile **3**, 15—30 (1943).

Murphy, D. M.: A great Northern bean resistant to curlytop and common bean-mosaic viruses. Phytopathology **30**, 779—784 (1940).

Myers, W. G. u. H. J. Hanson: New strains of Penicillium notatum induced by bombardment with neutrons. Science **101**, 357—358 (1945).

Nagel, L.: A cytological study of yeasts (Saccharomyces cerevisiae). Ann. Mo. Bot. Gdn. **33**, 249—288 (1946).

Naidu, M. u. V. M. Bakshi: The chromosome number in Torula utilis. Curr. Sci. **15**, 164 (1946).

Naumov, N. A. u. A. K. Zubarev (Editors): (The rusts of cereal crops.) Lenin Acad. Agric. Sci., published by Seljhozgiz, Moscow, Pp. 286 (1939). [Russisch.]

Nègre, E.: Jules Ventre (1880—1939). Ann. Éc. Agric. Montpellier **26**, 61—94 (1941).

Neill, J. C.: Ergot. N. Z. J. Sci. Tech. **23**, 130A—137A (1941).

Newton, M. u. T. Johnson: A mutation for pathogenicity in Puccinia graminis tritici. Canad. J. Res. **17**, 297—299, Sect. C (1939).

Nickerson, W. J.: Some trends in research on yeasts. Chronica Botanica **7**, 409—412 (1943).

— u. K. V. Thimann: The chemical control of conjugation in Zygosaccharomyces. II. Amer. J. Bot. **30**, 94—101 (1943).

Nizenjkov, N. P.: (A new method for determination of cold resistance in winter plants.) Proc. Lenin Acad. Sci. USSR., Nr. 1, 21—22 (1940). [Russisch.]

— (Application of an electrical method in breeding.) Selekcija i Semenovodstvo (Breeding and Seed Growing), Nr. 9/10, 34—42 (1946). [Russisch.]

Nobles, M. K.: Secondary spores in Corticium effuscatum. Canad. J. Res. **20**, Sect. C, 347—357 (1942).

— A contribution toward a clarification of the Trametes serialis complex. Canad. J. Res. **21**, Sect. C, 211 bis 234 (1943).

Norris, D. O.: Strains of spotted wilt virus and the identy of tomato tip-blight virus with spotted wilt. J. Coun. Sci. Industr. Res. Aust. **16**, 91—92 (1943).

Nosov, D. I.: (Interspecific hybridization in Pyrethrum.) Vestnik Gibridizacii (Hybridization), Nr. 2, 89—94 (1941). [Russisch.]

Nyberg, C.: Über sogenannte S- und R-Formen bei den Hefen. Zbl. Bakt. **103**, 277—280 (1941).

— Über sogenannte S- und R-Formen bei den Hefen. II. Zbl. Bakt. **105**, 241—248 (1942).

— Generationsväxling med diploid och haploid fas hos jästsvampar. (Alteration of generations with diploid and haploid phase in yeast fungi.) Nord. Medicin **13**, 26—28 (1942).

Oakberg, E. F. u. S. E. Luria: Mutations to sulfonamide resistance in Staphylococcus aureus. Genetics **32** 249—261 (1947).

Olive, L. S.: Nuclear phenomena involved at meiosis in Coleosporium Helianthi. J. Elisha Mitchell Sci. Soc. **58**, 43—51 (1942).

Oliveira, B. d': Studies on Puccinia anomala Rost. I. Physiologie races on cultivated barleys. Ann. Appl. Biol. **26**, 56—82 (1939).

— Aspectos actuais do problema das ferrugens. (Present aspects of the rust problem.) Palestras Agronómicas, Lisboa **2**, 5—77 (1939) 1940.

Oliveira, B. d': Aspectos do problema da sexualidade nas uredíneas. (Aspects of the problem of sexuality in the Uredineae.) Rev. Agron., Lisboa **29**, 107—108 (1941).

— A fase aecidica e o heterotalismo na Puccinia anomala Rostr. (The aecidial phase and heterothalism in P. anomala Rostr.) Rev. Agron. Lisboa **30**, 460—461 (1942).

— Da necessidade de intensificação do estudo das raças fisiológicas dos ferrugens dos cereais na Peninsula. (The necessity for intensifying the studiy of the physiological races of cereal rusts in Spain and Portugal.) Rev. Agron. Lisboa **30**, 467—468 (1942).

— u. M. C. F. de Sousa: Raças fisiológicas da Puccinia graminis tritici em Portugal. (Physiological races of P. graminis tritici in Portugal.) Agron. Lusitana **2**, 243—252 (1940).

Olson, P. J., T. J. Harrison, J. E. Blakeman u. R. Whiteman: Corn in Manitoba. Publ. Manitoba Dep. Agric. Immigr., Nr. 178, 24 (1942).

Oort, A. J. P.: Ziekteresistentie en plantenveredeling. (Disease resistance and plant breeding.) Landbouwk. Tijdschr. Wageningen **54**, 474—482 (1942).

— Het voorkomen van stuifbrand in tarwerassen, die tot nog toe onvatbaar waren. (The occurrence of loose smut in wheat races hitherto not susceptible.) Tijdschr. PlZiekt. **51**, 89 (1945).

Padwick, G. W.: Recent advances in control of fungous diseases of plants. Indian Fmg. **3**, 478—481 (1942).

Painter, R. H.: The economic value and biologic significance of insect resistance in plants. J. Econ. Ent. **34**, 358—367 (1941).

— Insect resistance of plants in relation to insect physiology and habits. J. Amer. Soc. Agron. **35**, 725—732 (1943).

Pal, B. P. u. B. B. Mundkur: Studies in Indian cereal smuts. I. Cereal smuts and their control by the development of resistant varieties. Proc. Indian Acad. Sci. **9** Sect. B, 267—270 (1939).

Peshkoff, M. A.: Fine structure and mechanism of division of the „nuclei" of the bacterium Caryophanon latum. Nature, Lond. **157**, 137—138 (1946).

Petri, L.: Alcune questioni di fitopatologia generale. (Certain questions of general phytopathology.) Ann. Fac. Agrar. Univ. Pisa **3** (N. S.), 229—261 (1940).

Pirschle, K.: Resistenzversuche mit polyploiden Pflanzen. Naturwiss. **29**, 338—339 (1941).

Plattner, P. A.: Penicillin. Experientia, Basel **1**, 167 bis 179 (1945).

Pontecorvo, G.: Genetic techniques in the development of microbiological assays. Biochem. J. **41**, Nr. 1, XII—XIII (1947). (Abstr.)

— u. A. R. Gemmell: Genetic proof of heterokaryosis in Penicillium notatum. Nature, Lond. **154**, 514—515 (1944).

—, — Colonies of Penicillium notatum and other moulds as models for the study of population genetics. Nature, Lond. **154**, 532—534 (1944).

Pratt, R. u. J. Dufrenoy: Physiological comparison of two strains of Penicillium. Science **102**, 428—429 (1945).

Prostoserdov, N. N.: (Soviet sherry.) Priroda (Nature) Nr. 3, 29—37 (1946). [Russisch.]

Punyasingha, T.: The relation of varieties of the soybean to various strains of the Rhizobia. Thai Sci. Bull. **3** Nr. 1, 11—27 (1941).

Quanjer, H. H.: Specialisatie van plantenparasieten en anatomisch-physiologisch onderzoek ter verklaring daarvan. (The specialization of plant parasites and anatomical and physiological investigations to explain it.) Landbouwk. Tijdschr. **55**, 532—533 (1943).

Quanjer, H. M.: Phytopathologische terminologie, met speciale bespreking van den begrippen biotrophie, premuniteit en antistoffen. (Phytopathological terminology, with a special discussion on the concepts biotrophy, premunity, and anti-substances.) Tijdschr. PlZiekt. **48**, 1—16 (1942); also Z. Pflanzenkrankh. **53** (1943) 200.

Quintanilha, A.: Doze anos de citologia e genética dos fungos. (Twelve years of cytology and genetics of the fungi.) Agron. Lusitana **3**, 241—306 (1941).

— O problema da delimitação e origem das espécies, do ponto de vista da biologia experimental. (The problem of the delimitation and origin of species from the point of view of experimental biology.) Rev. Agron. Lisboa **30**, 473 (1942).

— La conduite sexuelle de quelques espèces d'agaricacées. Bol. Soc. Broteriana **19**, Sér. 2a, 27—65 (1944).

— u. S. Balle: Étude génétique des phénomènes de nanisme chez les Hyménomycètes. Bol. Soc. Broteriana **14**, 17—46 (1940).

Ramachandra Rao, T. N., S. S. Soundar Rajan u. M. Sreenivasaya: Influence of carcinogens on yeast. Curr. Sci. **15**, 283—284 (1946).

Raper, K. B. u. D. F. Alexander: Penicillin. V. Mycological aspects of penicillin production. J. Elisha Mitchell Sci. Soc. **61**, 74—113 (1945).

Razdorskaja, L. A.: (The introduction of insecticidal Tephrosiae in the USSR. A. preliminary communication.) Sovetskaja Botanika (Soviet Botany), Nr. 5/6, 68—78 (1941). [Russisch.]

— (A rapid method of determining the germination of seed of Pyrethrum cinerarifolium [Trev.].) Priroda (Nature) **3**, 66—69 (1943). [Russisch.]

Reboul, J.: Nouvelles expériences sur l'action des rayons sur les levures. C. R. Acad. Sci., Paris **215**, 261 bis 263 (1942).

Reed, G. B.: Problems in the variation of pathogenic bacteria. Publ. Amer. Ass. Advanc. Sci., Nr. 12, 28 bis 33 (1940).

Reed, G. M.: Physiologic races of oat smuts. Amer. J. Bot. **27**, 135—143; also Contr., Nr. 90, Brooklyn Bot. Gdn. (1940).

— Phytopathology 1867—1942. Torreya **43**, 155—169 (1943); also Contr. Brooklyn Bot. Gdn. **1943**, Nr. 99.

— Physiologic specialization of the parasitic fungi. II. Bot. Rev. **12**, 141—164 (1946).

Regnery, D. C.: A leucineless mutant strain of Neurospora crassa. J. Biol. Chem. **154**, 151—160 (1944).

Renaud, J.: Sur la stabilité des races de levures de vin. C. R. Soc. Biol., Paris **131**, 681—682 (1939).

— Les races de levures obtenues par irradiation sont-elles semblables aux races naturelles? C. R. Soc. Biol., Paris **137**, 131—132 (1943).

— Les levures des vins du Val de Loire. Rev. Gén. Bot., Nr. 629, 193—211, 241—274 (1946).

— u. A. Lacassagne: Anomalies fixées de la division, chez une race de levure obtenue par irradiation. C. R. Soc. Biol., Paris **137**, 51—52 (1943).

Rhoades, H. E.: The adaptive enzymes of certain strains of yeast. Wallerstein Laboratories Communications **4**, 86 (1941). (Abstr.)

Richards, O. W.: The Actinomyces of potato scab demonstrated by fluorescence microscopy. Stain Tech. **18**, 91—94 (1943).

Riehm, E.: Über die Zunahme der Pflanzenkrankheiten und Schädlinge. Pflkrankh. **53**, 3—12 (1943).

Ripper, W. E.: Biological control as a supplement of chemical control of insect pests. Nature, Lond. **153**, 448—452 (1944).

Rischkov, V. L. siehe Ryžkov:

Rizet, G.: De l'hérédité du caractère absence de pigment dans le mycélium d'un Ascomycète du genre Podospora. C. R. Acad. Sci., Paris **209**, 771—774 (1939).

Rizet, G.: La valeur génétique des périthèces nés sur des souches polycaryotiques chez le Podospora anserina. Bull. Soc. Bot. Fr. **88**, 517—520 (1941).

Rodriquez Sardiña, J.: La obtencion de plantas resistentes a enfermedades. (The production of plants resistant to diseases.) Bol. Pat. Veg. Entom. Agric., Madrid **11**, 43—96. 1942 (1943).

Roemer, T.: Zehnjährige Versuche über Selektionswirkung in der Symbiose von Pilz und Kulturpflanze. Kühn-Arch. **48**, 169—178 (1939).

— Resistenzzüchtung. Forschungsdienst, Sonderh. 16, 321 (1942).

Rosen, H. R.: Breeding a disease-resistante red climbing Rose. Science **1**, 260—261 (1941).

Rudorf, W.: Methoden zur Prüfung und Züchtung von Kulturpflanzen auf Frostresistenz. Z. ges. Kälteindustr. **48**, 121—127 (1941).

— Resistenzzüchtung, ihre Grundlagen und Methoden. Z. Pflanzenz. **25**, 190—208 (1943).

Ruiz, O. M.: Contribución al conocimiento de las levaduras del aguamiel y del pulque. III. Torulopsis hydromelitis n. sp. (Contribution to the knowledge of the yeasts of hydromel and pulque. III. T. hydromelitis n. sp.) An. Inst. Biol. Univ. Méx. **11**, 539—554 (1940).

— Contribución al conocimiento de las levaduras del aguamiel y del pulque. IV. (Contribution to the knowledge of the yeasts of hydromel and pulque. IV.) An. Inst. Biol. Univ. Méx. **12**, 49—68 (1941).

— Contribución al conocimiento de las levaduras del aguamiel y del pulque. (Contribution to the knowledge of the yeasts of hydromel and pulque.) An. Inst. Biol. Univ. Méx. **13**, 1—21 (1942).

Ryan, F. J. u. J. Lederberg: Reverse-mutation and adaptation in leucineless Neurospora. Proc. Nat. Acad. Sci., Wash. **32**, 165—173 (1946).

—, L. K. Schneider u. R. Ballentine: Mutations involving the requirements of uracil in Clostridium. Proc. Nat. Acad. Sci., Wash. **32**, 261—271 (1946).

Ryžkov, V. L.: The nature of ultra-viruses and their biological activity. Phytopathology **33**, 950—955 (1943).

— (The theoretical basis of virus diseases of plants. General theory of viruses.) Akademija Nauk Sojuza SSR. Institut Mikrobiologii (Acad. of Sciences USSR. Inst. of Microbiology), Pp. 223 (1944). [Russisch.]

— (The problem of a natural system of viruses.) Žurnal Obščej Biologii (Journ. of General Biology) **8**, Nr. 3, 169—181 (1947). [Russisch.]

Saburova, P. V.: (Physiological characteristics of wheat affection by Puccinia triticina under the action of different soil humidity.) Botaničeskij Žurnal (Journ. Botanique) **31**, Nr. 4, 35—47 (1946). [Russisch.]

Šalyt, M. S.: (Wild insecticidal pyrethrums in the USSR.) Sovetskaja Botanika (Soviet Botany), Nr. 3, 97—100 (1941). [Russisch.]

Sansome, E. R.: Heterokaryosis and the mating-type factors in Neurospora. Nature, Lond. **156**, 47 (1945).

— Recent genetical experiments with yeasts. Nature, Lond. **157**, 52—53 (1946).

— Maintenance of heterozygosity in a homothallic species of the Neurospora tetrasperma type. Nature, Lond. **157**, 484—485 (1946).

— Induction of „Gigas" forms of Penicillium notatum by treatment with camphor vapour. Nature, Lond. **157**, 843—844 (1946).

Sansome, E. V.: Heterokaryosis, mating-type factors, and sexual reproduction in Neurospora. Bull. Torrey Bot. Cl. **73**, 397—409 (1946).

Sansome, F. W.: Breeding disease-resistant plants. Nature, Lond. **145**, 690—693 (1940).

Šapovalov, M. u. J. W. Lesley: Wilt resistance of the riverside variety of tomato to both Fusarium and Verticillium wilts. Phytopathology **30**, 760—768 (1940).

Šatova-Zeljaeva, E. V.: (On isolation of high molecular weight protein with the properties of Solanum virus Orton.) Izvestija Akademii Nauk SSSR. (Bull, Acad. Sci. URSS., Sér. biol.), Nr. 5, 497—500 (1946). [Russisch.]

Savulescu, T.: Das Vorkommen und die Verbreitung der in Rumänien den Weizenstinkbrand hervorbringenden Tilletia-Arten. Phytopath. Z. **14**, 148—187 (1942).

— u. C. Sandu-Ville: Încercări pentru stabilirea raselor fiziologice la cele două specii de Tilletia ce produc malura grăului în România. (Attempts to determine the physiological races of the two species of Tilletia which cause wheat smut in Rumania.) Anal. Inst. Cerc. Agron. Român. **10**, 518—631. 1938 (1939).

Schiemann, E.: Über die Züchtung resistenter Rassen der Kulturpflanzen. Chronica Botanica **5**, 161—164 (1939).

Schmidt, H., K. Diwald u. O. Stocker: Plasmatische Untersuchungen an dürreempfindlichen und dürreresistenten Sorten landwirtschaftlicher Kulturpflanzen. Planta **31**, 559—596 (1940).

Schneider-Orelli, O.: Vergleichende Untersuchungen an nord- und südschweizerischem Reblausmaterial. Mitt. schweiz. ent. Ges. **27**, 584—610 (1939).

Schramm, G.: Über die Struktur des Tabakmosaikvirus. Forsch. Fortschr. **19**, 225—226 (1943).

— u. L. Rebensburg: Zur vergleichenden Charakterisierung einiger Mutanten des Tabakmosaikvirus. Naturwiss. **30**, 48—50 (1942).

Schultz, H.: Untersuchungen über die Fußkrankheit der Ackerbohne. Zbl. Bakt. **106**, 38—50 (1943).

Schwalb, H.: Abriß über den derzeitigen Stand der Virusforschung. (Sammelreferat.) Züchter **14**, 167 bis 175 (1942).

Seemann, J.: Über die Temperaturverhältnisse in einem bewetterten Tiefkühlgewächshaus. Gartenbauwiss. **17**, 186—192 (1942).

Seiffert, G.: Virus diseases in man, animal and plant. N. Y., Pp. IX+332. 7 figs. 7 tables (1944).

Shay, J. R.: Genetic studies of certain mutant characters in Venturia inaequalis. Phytopathology **33**, 11 (1943). (Abstr.)

Sievers, A. F. u. E. C. Higbee: Plants for insecticides and rodenticides. Foreign Agric. Rep. U. S. Dep. Agric., Nr. 8, 20 (1943).

—, M. S. Lowman u. G. A. Russell: Factors affecting the rotenone content of devil's shoestring. J. Econ. Ent. **36**, 593—598 (1943).

—, —, — u. W. N. Sullivan: Changes in the insecticidal value of the roots of cultivated Devil's Shoestring, Tephrosia virginiana, at four seasonal growth periods. Amer. J. Bot. **27**, 284—288 (1940).

Simmonds, J. H. u. R. S. Mitchell: Black end and anthracnose of tne banana with special reference to Gloeosporium musarum Cke. and Mass. Bull. Coun. Sci. Industr. Res. Aust., Nr. 131, 63 (1940).

Simmonds, S., E. L. Tatum u. J. S. Fruton: The utilization of phenylalanine and tyrosine derivatives by mutant strains of Escherichia coli. J. Biol. Chem. **169**, 91—101 (1947).

Singh, H. B.: Effect of frost on some economic plants of Delhi. Indian J. Agric. Sci. **13**, 279—282 (1943).

Skovsted, A.: Successive mutations in Nadsonia Richteri Kostka. C. R. Lab. Carlsberg **23**, Sér. Physiol., 409—453 (1943).

Smith, G.: An introduction to industrial mycology. Edward Arnold and Co., London, Pp. XIV+271. 143 figs. tables (1946).

Smith, K. M.: The virus. Life's enemy. University Press, Cambridge (1940).

— Virus diseases of farm and garden crops. Littlebury and Co., Ltd., Worcester, 10s, 6d, Pp. 111. 16 pls. 14 figs (1945?).

Smith, T. E. u. K. J. Shaw: Pathogenicity studies with Fusaria isolated from tobacco, sweet potato and cotton. Phytopathology **33**, 469—483 (1943).

Smith, W. E.: Observations indicating a sexual mode of reproduction in a common Bacterium. (Bacteroides funduliformis.) J. Bact. **47**, 417 (1944).

Sneep, J.: De biochemie van het parasitisme. (The biochemistry of parasitism.) Tijdschr. PlZiekt. **52**, 125—137 (1946).

Snelling, R. O.: Resistance of plants to insect attack. Bot. Rev. **7**, 543—586 (1941).

— The place and methods of breeding for insect resistance in cultivated plants. J. Econ. Ent. **34**, 335—340 (1941).

Spencer, E. L. u. W. C. Price: Accuracy of the local-lesion method for measuring virus activity. I. Tobacco-mosaic virus. Amer. J. Bot. **30**, 280—290 (1943).

Spiegelman, S.: The physiology and genetic significance of enzymatic adaptation. Ann. Mo. Bot. Gdn. **32**, 139—163 (1945).

— u. C. C. Lindegren: A comparison of the kinetics of enzymatic adaptation in genetically homogeneous and heterogeous populations of yeast. Ann. Mo. Bot. Gdn. **31**, 219—233 (1944).

—, — u. L. Hedgecock: Mechanism of enzymatic adaptation in genetically controlled yeast populations. Proc. Nat. Acad. Sci., Wash. **30**, 13—23 (1944).

—, — u. G. Lindegren: Maintenance and increase of a genetical character by a substratecytoplasmic interaction in the absence of the specific gene. Proc. Nat. Acad. Sci., Wash. **31**, 95—102 (1945).

Sprague, R.: A revised check list of the parasitic fungi on cereals and other grasses in Oregon. Plant Dis. Reporter Suppl., Nr. 134, 36 (1942). [Mimeographed.]

Srb., A. M. u. N. H. Horowitz: The ornithine cycle in Neurospora and its genetic control. J. Biol. Chem. **154**, 129—139 (1944).

Srinath, K. V.: The cytology of the yeast. Curr. Sci. **15**, 25 (1946).

— The chromosomes of Saccharomyces cerevisiae. Curr. Sci. **15**, 50—51 (1946).

Srinivasan, B.: The cytology and genetics of yeasts. Curr. Sci. **10**, 90—91 (1941).

Stahmann, M. A. u. J. F. Stauffer: Induction of mutants in Penicillium notatum by methyl-bis (β-chloroethyl) amine. Science **106**, 35—36 (1947).

Stakman, E. C.: The need for research on the genetics of pathogenic organisms. Publ. Amer. Ass. Advanc. Sci., Nr. 12 9—17 (1940).

— The nature and importance of physiologic specialization in phytopathogenic fungi. Science **105**, 627—632 (1947).

—, M. F. Kernkamp, T. H. King u. W. J. Martin: Genetic factors of mutability and mutant characters in Ustilago zeae. Amer. J. Bot. **30**, 37—48 (1943).

—, —, W. J. Martin u. T. H. King: The inhertance of a white mutant character in Ustilago zeae. Phytopathology **33**, 943—949 (1943).

— W. Q. Loegering, R. C. Cassell u. L. Hines: Population trends of physiologic races of Puccinia graminis tritici in the United States for the period 1930 to 1941. Phytopathology **33**, 884—898 (1943).

Steinberg, R. A.: Variants in fungi: formation, reversion and prevention. Science **100**, 10 (1944).

— u. C. Thom: Reversions in morphology of nitrite-induced „mutants" of Aspergilli grown on amino acids. J. Agric. Res. **64**, 645—652 (1942).

Stepanov, V. N.: (Resistance of agricultural crops to frost during various developmental phases.) Moskovskaja ordena Lenina Seljskohozjajstvennaja Akademija imeni K. A. Timirjazeva. Doklady Vypusk III. Naučnaja Konferencija 4—11 ijunja 1945 g. (Timirjazev Agric. Acad. Moscow Proc., Nr. III. Sci. Conf. 4—11 June 1945.) Pp. 28—32 (1946). [Russisch.]

Stephen, W. A.: Studies on the thermal resistance of honey yeasts. Sci. Agric. 22, 705—720 (1942).

Stevenson, J. A.: A preliminary list of authors of plant parasites with recommended abbreviations. Plant Dis. Reporter 28, 366—395 (1944).

— u. A. G. Johnson: The nomenclature of the cereal smut fungi. Plant Dis. Reporter 28, 663—670 (1944).

Stone, W. S., O. Wyss u. F. Haas: The production of mutations in Staphylococcus aureus by irradiation of the substrate. Proc. Nat. Acad. Sci., Wash. 33, 59—66 (1947).

Straib, W.: Weitere Beiträge zur Kenntnis der Spezialisierung der Getreideroste und des Leinrostes. Arb. biol. Reichsanst. Land.- u. Forstw., Berlin 23, 233—263 (1941).

— Untersuchungen zur Biologie und Bekämpfung des Bohnenrostes Uromyces phaseoli (Pers.) Wint. Gartenbauwiss. 17, 397—445 (1943).

Studhalter, R. A. u. W. S. Glock: Apparatus for the production of artificial frost injury in the branches of living trees. Science 96, 165 (1942).

Subramaniam, M. K.: Induction of polyploidy in Saccharomyces cerevisiae. Curr. Sci. 14, 234 (1945).

— Endopolyploidy in yeasts. Curr. Sci. 16, 157—158 (1947).

— u. B. Ranganathan: Mitosis during budding in Saccharomyces cerevisiae. Curr. Sci. 14, 78—79 (1945).

—, — The chromosome number of Saccharomyces cerevisiae. Curr. Sci. 14, 131—132 (1945).

—, — A new mutant of Saccharomyces cerevisiae. Nature, Lond. 157, 49—50 (1946).

—, — Peculiar cytological behaviour of a distillery yeast. Nature, Lond. 157, 50—51 (1946).

—, — Staining the chromosomes of yeast by the Feulgen technique. Nature, Lond. 157, 657 (1946).

Suhov, K. S. u. A. M. Vovk: (Eine neue Viruskrankheit der Tomatenblattkräuselung und ihr Übertrager — Zikade Agallia venosa Fall.) C. R. (Doklady) Acad. Sci. URSS. 56, 433—435 (1947). [Russisch.]

—, — (Abhängigkeit der Reproduktion des Tabakmosaikvirus von der synthetischen Aktivität der Proteasen der Wirtspflanze.) C. R. (Doklady) Acad. Sci. URSS. 57, 625—628 (1947). [Russisch.]

Sullivan, J. T. u. S. J. P. Chilton: The effect of leaf rust on the carotene content of white clover. Phytopathology 31, 554—557 (1941).

Swanson, C. P. u. A. Hollaender: Modification of the ultraviolet mutation rate by pretreatment with near infrared. Amer. J. Bot. Suppl. 33, 832 (1946). (Abstr.)

Takizawa, R.: Study on hybridization of yeast. Jap. J. Genet. 15, 351—352 (1939).

Tapke, V. F.: Occurrence, identification, and species validity of the barley loose smuts, Ustilago nuda, U. nigra and U. medians. Phytopathology 33, 194—209 (1943).

— Physiologic races of Ustilago nigra. Phytopathology 33, 324—327 (1943).

Tatum, E. L.: X-ray induced mutant strains of Escherichia coli. Proc. Nat. Acad. Sci., Wash. 31, 215—219 (1945).

— u. G. W. Beadle: Genetic control of biochemical reactions in Neurospora: an „aminobenzoicless" mutant. Proc. Nat. Acad. Sci., Wash. 28, 234—243 (1942).

—, — Biochemical genetics of Neurospora. Ann. Mo. Bot. Gdn. 32, 125—129 (1945).

— u. T. T. Bell: Neurospora. III. Biosynthesis of thiamin. Amer. J. Bot. 33, 15—20 (1946).

Tempe, J. de: Alkaloidvorming door Claviceps purpurea (Fr.) Tul. in saprophytische cultuur. (Alkaloid formation by C. purpurea (Fr.) Tul in saprophytic culture.) Thesis, Univ. Amsterdam, Pp. 84. (1945).

Tervet, I. W.: Problems in the determination of physiologic races of Ustilago avenae and U. levis. Phytopathology 30, 900—913 (1940).

Thatcher, F. S.: A stem-end rot of potato tubers caused by Rhizoctonia solani. Phytopathology 32, 727—730 (1942).

Thaysen, A. C. u. M. Morris: Preparation of a giant strain of Torulopsis utilis. Nature, Lond. 152, 526—528 (1943).

Thom, C.: Mycology presents penicillin. Mycologia 37, 460—475 (1945).

Thomas, H. R.: A nonchromogenic sporulating variant of Alternaria solani. Phytopathology 33, 729—731 (1943).

Thorne, R. S. W.: Description of a top-fermentation strain of Saccharomyces cerevisiae Hansen, catalogued as number 6479 in the national collection of type cultures. J. Inst. Brew. 50, 222—224 (1944).

— Inheritance in yeast. J. Inst. Brew. 53, 25—36 (1947).

Thorpe, H. C.: Pyrethrum breeding: a progress report. E. Afr. Agric. J. 5, 364—368 (1940).

— Pyrethrum breeding. E. Afr. Agric. J. 5, 479—480 (1940).

Timm, E. W. u. E. W. Lindstrom: Experimental proof of mutation in virulence of the bacterial wilt pathogen of maize. Genetics 28, 94 (1943). (Abstr.)

Topley, W. W. C.: The Croonian Lecture. The biology of epidemics. Proc. Roy. Soc. 130, Sér. B, 337—359 (1942).

Tulasne, R. u. R. Vendrely: Demonstration of bacterial nuclei with ribonuclease. Nature, Lond. 160, 225—226 (1947).

Tyler, J.: Plants reported resistant or tolerant to root knot nematode infestation. Misc. Publ. U. S. Dep. Agric., Nr. 406, 91 (1941).

Ullrich, H.: Über Strukturänderungen beim Gefrieren von Gelen. Kolloidztschr. 96, 348—353 (1941).

— Beziehungen zwischen Struktur und Frostresistenz bei Pflanzen. Forsch.dienst, Sonderh. 16, 280—283 (1942).

— u. J. Seemann: Über die Verwendung von handelsüblichen Kühlschränken für Frostresistenzversuche an Pflanzen, insbesondere Frostresistenzprüfungen. Z. Pflanzenz. 25, 1—7 (1943).

Utter, L. G.: Studies on experimentally produced physiologic races of the oat smuts. Phytopathology 33, 14 (1943). (Abstr.)

V....., M.: Derris et Lonchocarpus. (Derris and Lonchocarpus.) Bull. Agric. Congo Belge 34, 259—260 (1943).

Valleau, W. D. u. E. M. Johnson: An outbreak of plantago virus in burley tobacco. Phytopathology 33, 210—219 (1943).

Vallega, J.: Especialización fisiológica de Puccinia graminis tritici en la Argentina, Chile y Uruguay. (Physiological specialization of P. graminis tritici in Argentina, Chile and Uruguay.) Rev. Argent. Agron. 7, 196 bis 220 (1940).

Vaughn, R. H.: The acetic acid bacteria. Wallerstein Laboratories Communications 5, 5—26 (1942).

Veitch, R. et al.: Farm crops and pastures. Qd Agric. Past. Handb. 1, 448 (1941).

Vincent, J. M.: Variation in the nitrogen-fixing property of Rhizobium trifolii. Nature, Lond. 153, 496 bis 497 (1944).

Virtanen, A. I.: Symbiotic nitrogen fixation. Nature, Lond. 155, 747—748 (1945).

Vloten, H. van: Is verrijking van de mycoflora mogelijk? (Naar aanleiding van de populierenroest.) (Is enrichment of the mycoflora possible? [In relation to poplar rust].) Tijdschr. PlZiekt. 50, 49—62 (1944).

Vogt: Staatliches Weinbauinstitut in Freiburg i. Br. Versuchs- und Forschungsanstalt für Weinbau und Weinbehandlung. Jahresbericht 1939, Pp. 51 (1940).
— Staatliches Weinbauinstitut in Freiburg i. Br. Versuchs- und Forschungsanstalt für Weinbau und Weinbehandlung. Jahresbericht 1940, Pp. 40 (1941).
Vučković, V.: (Comparative investigation of the yeasts Dingač, Pijavičina, M. O. and Sherry.) Arhiv Minist. Poljopr. 7, Nr. 18, 111—122 (1940). [Serbisch.]

Waksman, S. A., H. C. Reilly u. A. Schatz: Strain specificity and production of antibiotic substances. V. Strain resistance of bacteria to antibiotic substances, especially to streptomycin. Proc. Nat. Acad. Sci., Wash. 31, 157—164 (1945).
— u. A. Schatz: Strain specificity and production of antibiotic substances. VI. Strain variation and production of streptothricin by Actinomyces Lavendulae. Proc. Nat. Acad. Sci., Wash. 31, 208—214 (1945).
Wallerstein, J. S. u. A. L. Schade: Some considerations on the nature of yeast. I. The structure and functions of the cell. Wallerstein Laboratories Communications 3, 91—106 (1940).
—, — Some considerations on the nature of yeast. II. Their origin and relation to other organisms. Wallerstein Laboratories Communications 3, 182—198 (1940).
Watson, I. A.: Inheritance of resistance to stem rust in crosses with Kenya varieties of Triticum vulgare Vill. Phytopathology 31, 558—560 (1941).
Weber, A.: Hvem er interesserede i at købe Farvefotografier af Plantesygdomme? (Who is interested in buying coloured photographs of plant diseases?) Nord. JordbrForskn., Nr. 5/6, 273—274 (1943).
Weindling, R. u. G. M. Armstrong: A water-culture infection method used in the study of Fusarium wilt of cotton. Phytopathology 29, 23 (1939). (Abstr.)
Wellman, F. L.: A technique for studying host resistance and pathogenicity in tomato Fusarium wilt. Phytopathology 29, 945—956 (1939).
— Increase of pathogenicity in tomato wilt Fusarium. Phytopathology 33, 175—193 (1943).
— A technique to compare virulence of isolates of Alternaria solani on tomato leaflets. Phytopathology 33, 698—706 (1943).
Whelton, R. u. H. J. Phaff: A nonrespiratory variant of Saccharomyces cerevisiae. Science 105, 44—45 (1947).
Whitaker, Th. W. u. D. E. Pryor: The inheritance of resistance to powdery mildew (Erysiphe Cichoracearum) in lettuce. Phytopathology 31, 534—540 (1941).
White, D. G.: A comparison of the number of protoxylem strands with the rotenone content in derris roots. Proc. Amer. Soc. Hort. Sci. 46, 370—374 (1945).
White, N. H.: The genetics of Ophiobolus graminis Sacc. 1. Heritable variations for culture colour and pathogenicity. J. Counc. Sci. Industr. Res. Aust. 15, 118 bis 124 (1942).
— u. G. A. McIntyre: The pathogenicity of single spore isolates of Ophiobolus graminis under field conditions. J. Coun. Sci. Industr. Res. Aust. 16, 93—94 (1943).
White, O. E.: Temperature reaction, mutation and geographical distribution in plant groups. Proc. 8th Amer. Sci. Congr., Wash. 3, 287—294 (1940).
Whitehouse, H. K. L.: Crossing-over in Neurospora. New Phytol. 41, 23—62 (1942).
— u. J. B. S. Haldane: Symmetrical and asymmetrical reduction in ascomycetes. J. Genet. 47, 208—212 (1946).
Wilkins, W. H.: Investigation into the production of bacteriostatic substances by fungi. Trans. Brit. Mycol. Soc. 28, 110—114 (1945).
Williams, C. B.: Genetics in relation to diseases of animals and plants. Nature, Lond. 155, 674—675 (1945).

Wilson, G. F.: The stem and bulb eelworm, Anguillulina dipsaci (Kühn, 1858): the importance of collating evidence on the behaviour of biologic strains. Ann. Appl. Biol. 30, 364—370 (1943).
Wingard, S. A.: The nature of disease resistance in plants. 1. Bot. Rev. 7, 59—109 (1941).
Winge, Ö.: Sur l'hétérothallisme du Saccharomycodes Ludwigii. Scientia Genetica 2, 167—170 (1941).
— Croisement inter-spécifique chez les champignons. Scientia Genetica 2, 171—189 (1941).
— On segregation and mutation in yeast. C. R. Lab. Carlsberg 24, Sér. Physiol., 79—96 (1944).
— u. O. Laustsen: On a cytoplasmatic effect of inbreeding in homozygous yeast. C. R. Lab. Carlsberg 23, Sér. Physiol., 17—39 (1940).
Witkin, E. M.: A case of inherited resistance to radiation in bacteria. Genetics 31, 236 (1946). (Abstr.)
— Inherited differences in sensitivity to radiation in Escherichia coli. Proc. Nat. Acad. Sci., Wash. 32 59—68 (1946).
— Genetics of resistance to radiation in Escherichia coli. Genetics 32, 221—248 (1947).
Woods, M. W. u. H. G. DuBuy: Evidence for the evolution of phytopathogenic viruses from mitochondria and their derivatives. I. Cytological and genetic evidence. Phytopathology 33, 637—655 (1943).
Worsley, R. R.: The flowering of Derris elliptica. E. Afr. Agric. J. 10, 6 (1944).
Wotčal, A. E.: (Physiological bases of the resistance of plants to hot, dry winds. Artificial drywind apparatus and methods of work.) Soviet Plant Industry Record, Nr. 1, 63—70 (1940). [Russisch.]

Yamamoto, Y.: (Genetical investigations on Saccharomycetes. I. Segregations in Saccharomyces Sake Yabe.) Bot. Mag. Tokyo 53, 449—459 (1939).
— (Varietal hybrids in Japanese Saké yeast.) Jap. J. Genet. 15, 353—355 (1939).
— (On some new yeast-types produced by hybridization.) Jap. J. Genet. 16, 302—304 (1940).
Yarwood, C. E.: Sporulation injury associated with downymildew infections. Phytopathology 31, 741—748 (1941).
Yust, H. R., H. D. Nelson u. R. L. Busbey: Comparative susceptibility of two strains of California red scale to HCN, with special reference to the inheritance of resistance. J. Econ. Ent. 36, 744—749 (1943).

Zagalo, A. C.: Variação do Coryneum Longistipatatum Berl. et Bres. (Variation in C. longistipatatum Berl. et Bres.) Rev. Agron., Lisboa 30, 461 (1942).
Zak, G. A.: (On the fundamentals of phytopathological characteristics of varieties and on the significance of artificial inoculation in selection.) Proc. Lenin Acad. Agric. Sci. USSR., Nr. 6, 12—14 (1940). [Russisch.]
Zamenhof, S.: „Mutations" and cell divisions in bacteria. Genetics 30, 28 (1945). (Abstr.)
— Unstable strains of the colon bacillus. Two new mutants of B. coli-mutabile. J. Hered. 37, 273—275 (1946).
Zirkle, R. E.: Significant differences in X-ray sensitivity among various species and strains of yeast. J. Bact. 43, 119 (1942).

i) Mikroskopische Technik.

Adelhelm, E.: Handling extremely thin paraffin sections. Stain Techn. 18, 144 (1943).

Brown, A. F. u. W. M. Jones: A methyl methacrylate-silica replica technique for electron microscopy. Nature Lond. 159, 635—636 (1947).
Burch, C. R. u. J. P. P. Stock: Phase-contrast microscopy. J. Sci. Instrum. 19, 71—75 (1942).

Carvajal, F.: A superior pith for free-hand sections. Science **103**, 112 (1946).
Conn, H. J.: Biological stains. A handbook on the nature and uses of the dyes employed in the biological laboratory. Biotechn. Publications, Geneva, N. Y. 4th ed., Pp. 308, 5 figs. Tables (1940).
— Progress in the standardization of stains. No further certification of gentian violet. Stain Techn. **6**, 141—142 (1941).
Conn, J. E.: Chlorazol black as a stain for root-tip chromosomes. Stain Techn. **18**, 189—192 (1943).
Cooper, K. W.: A useful accessory to the Zeiss mechanical stage for oil immersion microscopy. Stain Techn. **18**, 177—178 (1943).
Cutter, V. M.: Smear methods for the study of chromosomes in ascomycetes. Stain Techn. **21**, 129—131 (1946).

Darrow, M. A.: Anilin blue as a counterstain in cytology. Stain Techn. **19**, 65—66 (1944).

Elvers, I.: A simple method of making freehand sections. Svensk Bot. Tidskr. **39**, 192—196 (1945).
Emig, W. H.: Stain technique. Science Press. Printing Co., Lancaster, Pa, Pp. 75 (1941).
Emsweller, S. C. u. N. W. Stuart: Improving smear technics by the use of enzymes. Stain Techn. **19**, 109—114 (1944).

Hancock, B. L.: A schedule for chromosome counts in some plants with small chromosomes. Stain Techn. **17**, 79—83 (1942).

Johansen, D. A.: Plant microtechnique. McGraw-Hill Publishing Company, Ltd. London (1940).

Kerns, K. R.: Pasternack's paraffin method modified for plant tissue. Stain Techn. **6**, 155—156 (1941).
Kraft, M. M.: Etude critique des colorations en histologie végétale. Mém., Nr. 48, Soc. Vaud. Sci. Nat. **7**, 91—165. [Contained in Trav. Inst. Bot., Lausanne 1940—1944, **3**.]

La Cour, L.: Acetic-orcein: a new stain-fixative for chromosomes. Stain Techn. **6**, 169—174 (1941).
La Cour, L. F.: Improvements in plant cytological technique. II. Bot. Rev. **13**, 216—240 (1947).
Langlet, O.: A handy field method of fixing root-tips. Svensk Bot. Tidskr. **40**, 425—426 (1946).
Law, A. G.: Root-tip smears following fixation with boiling water. Stain Techn. **18**, 117—120 (1943).
Linfoot, E. H.: Phase difference microscopy. Nature, Lond. **155**, 76 (1945).
Lowe, J.: Root tip smears for maize. Stain Techn. **21**, 127—128 (1946).

McCartney, J. E.: A new immersion („Polyric"). J. Path. Bact. **56**, 265—266 (1944).
McKay, H. H.: The use of enzymes in the preparation of root-tip smears. Stain Techn. **21**, 111—114 (1946).
Martin, L. C.: Phase-contrast methods in microscopy. Nature, Lond. **159**, 827—830 (1947).
Metcalf, R. L. u. R. L. Patton: Fluorescence microscopy applied to entomology and allied fields. Stain Techn. **19**, 11—27 (1944).
Meyer, J. R.: Prefixing with paradichlorobenzene to facilitate chromosome study. Stain Techn. **20**, 121 bis 125 (1945).
Michel, K.: Die Darstellung von Chromosomen mittels des Phasenkontrastverfahrens. Naturwiss. **29**, 61—62 (1941).
Morce, R.: Control of the ferric ion concentration in iron-aceto-carmine staining. Stain Techn. **18**, 103 bis 108 (1944).

Munzos, F. J. u. H. A. Charipper: The microscope and its use. Chemical Publishing Co., Inc., Brooklyn, N. Y., Pp. XII + 334. 122 figs. diagrams (1943).

Pavlovskij, E. N.: (The use of oil from Juniperus turcomanica B. Fedtsch. in microscopy.) Priroda (Nature), Nr. 4, 90 (1943). [Russisch.]
Proskurjakov, N. J. u. A. N. Koževnikova: (A direkt method of determining starch in plant material.) Biohimija **5**, 624—629 (1940). [Russisch.]

Rafalko, J. S.: A modified Feulgen technic for small and diffuse chromatin elements. Stain. Techn. **21**, 91—93 (1946).
Rezende-Pinto, M. C. de: A new cytological technic, tannin-iron III, for nucleoli and plastids. Stain Techn. **22**, 3—4 (1947).
Richards, O. W.: The effective use und proper care of the microscope. Spencer Lens Company, Buffalo, N. Y., Pp. 61, 47 figs (1941).
— The effective use and proper care of the microtome. Spencer Lens Company Buffalo, N. Y., Pp. 88. 40 figs (1942).
— Phase difference microscopy for living unstained protoplasm. Amer. J. Bot. **31**, Nr. 8, 12s (1944). (Abstr.)
—, J. H. Small u. P. W. Collyer: Microscopy with plastic substitutes for cover glasses. Stain Techn. **19**, 59—62 (1944).
Romaniak, T. H.: The use of unsaturated polyester resins for embedding biological material. Science **104**, 601—602 (1946).

Salazar, A. L.: A new technic: tannin-iron II. Stain Techn. **21**, 149—151 (1946).
Sampath, S.: A modified form of acetocarmine. Curr. Sci. **9**, 229—230 (1940).
Šapiro, S.: Warm safranine for plant tissues. Science **105**, 50 (1947).
Sass, J. E.: Elements of botanical microtechnique. McGraw-Hill Publishing Company, Ltd., London (1940).
Semmens, C. S.: Nucleolar stain and nucleal reaction. Bot. Gaz. **104**, 645—649 (1943).
— u. P. N. Bhaduri: Staining the nucleolus. Stain Techn. **16**, 119—120 (1941).
Serra, J. A.: Improvements in the histochemical arginine reaction and the interpretation of the reaction. Portugaliae Acta Biologica, Lisboa **1**, Ser. A, 1—8 (1944).
— Histochemical tests for proteins and amino acids; the characterization of basic proteins. Stain Techn. **21**, 5—18 (1946).
— u. A. Queiroz Lopes: Une méthode pour la démonstration histochimique du phosphore des acides nucléiques. Portugaliae Acta Biologica, Lisboa **1**, Ser. A, 111—122 (1945).
Smith, L.: The acetocarmine smear technic. Stain Techn. **22**, 17—31 (1947).
Speese, B. M.: Chromosome counts for temporary study. Science **102**, 256 (1945).
Steyaert, R. L.: A technique for obtaining quickly permanent mounts of nonembedded botanical material. Science **105**, 47—48 (1947).
Stowel, R. E.: The use of tertiary butyl alcohol microtechnique. Science **96**, 165—166 (1942).
— The specificity of the Feulgen reaction for thymonucleic acid. Stain Techn. **21**, 137—148 (1946).
Svešnikova, I. N.: Analysis of nucleus development and of changes of thymonucleic acid in ontogenesis. C. R. (Doklady) Acad. Sci. URSS. **32**, 216—218 (1941).
Swanson, C. P.: The use of acenaphthene in pollen tube technic. Stain Techn. **15**, 49—52 (1940).

Thomas, P. T.: The aceto-carmine method for ruit material. Stain Techn. **15**, 167—172 (1940).

Warmke, H. E.: Precooling combined with chrom-osmoacetic fixation in studies of somatic chromosomes in plants. Stain Techn. **21**, 87—89 (1946).
Wilson, G. B.: The Venetian turpentine mounting medium. Stain Techn. **20**, 133—135 (1945).
Witlin, B.: Darkfield illuminators in microscopy. Science **102**, 41—42 (1945).
Wittlake, E. B.: Permanent prestaining in botanical microtechnic. Ohio J. Sci. **44**, 36—38 (1944).

k) Statistik und Feldversuchswesen.

Anderson, J. A.: The role of statistics in technical papers. Trans. Amer. Ass. Cereal Chemists **3**, 69—73 (1945).
Anderson, R. L.: Missing-plot techniques. Biometrics Bull. **2**, 41—47 (1946).
Anonymus: Handledning i försöksteknik. (The technique of field experiments.) Medd. Lantbrukshögskolan Jordbruksförsöksanstalten, Norrtälje, Nr. 1, 207 (1939).
Anós, A.: Estadao actual de la teoría de la comprobación de hipótesis estadísticas. (Present state of the theory of testing statistical hypotheses.) Bol. Inst. Invest. Agron. Madr., Nr. 10, 93—135 (1944).
— Un método gráfico para hallar el número necesario de repeticiones en las experiencias comparativas. (A graphical method for determining the number of replications necessary in comparative experiments.) Bol. Inst. Nac. Invest. Agron. Madr., Nr. 13, 1—16 (1945).
-Ansari, M. A. A. u. G. K. Sant: A study of soil heterogeneity in relation to size and shape of plots in wheat field at Raya (Muttra district). Indian J. Agric. Sci. **13**, 652—656 (1943).
Arcaneaux, G., I. E. Stokes, B. A. Belcher u. R. T. Gibbens (jr.): A study of the relation between size of field sample and experiment errors of juice analysis and sugar-yield determination in connection with sugarcane variety tests. Proc. 6th Congr. Int. Soc. Sug. Cane Tech., La, Pp. 744—759 (1938) 1939.
Arkin, H. u. R. R. Colton: An outline of statistical methods as applied to economics, business, education, social and physical sciences etc. Barnes, and Noble, Inc., New York (1939).
Artin, E.: On the theory of complex functions. Notre Dame Mathematical Lectures. Indiana, Nr. 4, 55—70 (1944).

Baker, G. A.: Distribution of the ratio of sample range to sample standard deviation for normal and combinations of normal distributions. Ann. Math. Statist. **17**, 366—369 (1946).
Baldwin, E. M.: Table of percentage points of the t-distribution. Biometrika **33**, 362 (1946).
Bancroft, T. A.: On biases in estimation due to the use of preliminary tests of significance. Ann. Math. Statist. **15**, 190—204 (1944).
Bär, A. L. S.: Interpretatie van proefveld-resultaten. (Interpretation of the results of field experiments.) Landbouwk. Tijdschr. Wageningen **51**, 229—246 (1939).
Barlett, M. S.: The statistical significance of canonical correlations. Biometrika **32**, 29—37 (1940/41).
Barnard, G. A.: Significance tests for 2×2 tables. Biometrika **34**, 123—138 (1947).
— The meaning of a significance level. Biometrika **34**, 179—182 (1947).
— 2×2. A note on E. S. Pearson's paper. Biometrika **34**, 168—169 (1947).
Baten, W. D.: Formulas for finding estimates for twe and three missing plots in randomized layouts. Tech. Bull. Mich. Agric. Exp. Sta., Nr. 165, 16 (1939).
— How to determine which of two variables is better for predicting a third variable. J. Amer. Soc. Agron. **33**, 695—699 (1941).
Baten, W. D.: Advantages of singling out degrees of freedom in analyses of variance. Ann. Math. Statist. **13**, 113 (1942). (Abstr.)
—, J. I. Northam u. A. F. Yeager: Grouping of strains or varieties by use of a latin square. J. Amer. Soc. Agron. **33**, 616—622 (1941).
Beall, G.: The technique of randomization in field work. Canad. Ent. **72**, 45—51 (1940).
Berkson, J.: A note on the X^2-test, the Poisson and the Dinomial. J. Amer. Statist. Ass. **35**, 362—367 (1940).
— Experience with tests of significance: a reply to Professor R. A. Fisher. J. Amer. Statist. Ass. **38**, 242 bis 246 (1943).
— Approximation of chi-square by „probits" and by „logits". J. Amer. Statist. Ass. **41**, 70—74 (1946).
Bhattacharyya, A.: A note on Ramamurti's problem of maximal sets. Sankhyā: Indian J. Statist. **6**, 189 bis 192 (1942).
Bhattacharyya, B. C.: On an aspect of Pearsonian system of curves and a few analogies. Sankhyā: Indian J. Statist. **6**, 415—418 (1944).
Bhattacharyya, K. N.: A note on two-fold triple systems. Sankhyā: Indian J. Statist. **6**, 313—314 (1943).
Biehler, W. u. H. Wollschitt: Beitrag zur statistischen Beurteilung biologischer Wirkungen. Naunyn-Schmiedebergs Arch. **198**, 278—291 (1941).
Bigot, A.: Grafische voorstelling van proefveldresultaten. (Graphic representation of experiment plot results.) Arch. Suikerind. Nederland. Ned.-Ind. **2**, 44—48 (1941).
— Factorieele vakkenproeven in de Javasuikerindustrie. (Factorial plot experiments in the Java sugar industry.) Arch. Suikerind. Nederland. Ned.-Ind. **2**, 361—364 (1941).
Bishop, D. J. u. U. S. Nair: A note on certain methods of testing for the homogeneity of a set of estimated variances. Suppl. J. R. Statist. Soc. **6**, 89—99 (1939).
Bondorff, K. A.: Arbejdsfeljlens Størrelse ded lokale Forsøg. (The magnitude of experimental error in field experiments.) Nord. Jordbr. Forskn. **24**, 151—152 (1942).
Bonnier, G.: The X^2 linkage test. Hereditas, Lund **28**, 230—232 (1942). (Abstr.)
Borden, R. J.: Studies in experimental technique. Selection of layout: blocks versus Latin squares. Hawaii. Plant. Rec. **43**, 7—10 (1939).
— Studies in experimental technique: plot arrangment. Proc. 6th Congr. Int. Soc. Sug. Cane Tech., La, Pp. 733 bis 744 (1938) 1939.
— Yield variations with special reference to border effects in field tests. Hawaii. Plant. Rec. **47**, 195—203 (1943).
Börlin, W.: Gruppenweise Reserverechnung bei Verwendung von Selektions- und Dekremententafeln. Mittschw. Versich. math., H. 38, 53—104, mit 3 Tabellen (1939).
Bose, R. C. u. K. Kishen: On the problem of confounding in the general symmetrical factorial design. Sankhyā: Indian J. Statist. **5**, 21—36 (1940).
Boyce, S. W.: Statistical studies on New Zealand wheat trials. I. The efficiency of lattice design. N. Z. Sci. Tech. **27**, Sect. A, 270—275 (1945).
— Statistical studies on New Zealand wheat trials. II. The analysis of lattice trials with incomplete data. N. Z. J. Sci. Tech. **27**, Sect. A, 276—280 (1945).
Brandt, A. E.: The relation between the design of an experiment and the analysis of variance. J. Amer. Statist. Ass. **36**, 283—292 (1941).
Brieger, F. G.: Sôbre O „X^2-test". (The X^2 test.) J. Agron., S. Paulo **3**, 103—110 (1940).
— Coeficiente de variação e indice de variança. (Variation coefficient and index of variance.) Bragantia, São Paulo **2**, 313—331 (1942).

Brinton, W. C.: Graphic presentation. Brinton Associates, N. Y., Pp. 512 (1939).

Brouwer, E.: Over correlaties van zeer hoogen graad en haar biologische beteekenis. (On high correlations and their biological significance.) Landbouwk. Tijdschr. **56**, 504—507 (1945).

Buros, O. K. (Editor): The second yearbook of research and statistical methodology books and reviews. Gryphon Press, New Jersey, Pp. XX + 383 (1941).

Burton, G. W.: Estimating individual forage plant yields. J. Amer. Soc. Agron. **36**, 709—712 (1944).

Buzzati-Traverso, A.: La determinazione di frequenze geniche in una popolazione specifica. (The determination of gene frequencies in a species population.) Ricerca Scientifica **13**, 448—458 (1942).

Calvet, R. P. u. M. M. de Zulueta: Métodos estadísticos para la comparación de gran número de variedades. (Statistical methods for comparing a large number of varieties.) Bol. Inst. Nac. Invest. Agron., Madr., Nr. 14, 29—62 (1946).

Camp, B. H.: Further comments on Berkson's problem. J. Amer. Statist. Ass. **35**, 368—376 (1940).

— Some recent advances in mathematical statistics, 1. Ann. Math. Statist. **13**, 62—73 (1942).

Cavalli, L. L. u. C. Magni: Methods of analysing the virulence of bacteria and viruses for genetical purposes. Heredity **1**, 127—132 (1947).

Chambers, E. G.: Statistical calculation for beginners. University Press, Cambridge (1940).

— Statistical techniques in applied psychology. Biometrika **33**, 269—273 (1946).

Chan-Choong, P. A.: Testing large numbers of rice varieties by the quasi-factorial method. Agric. J. Brit. Guiana **10**, 78—88 (1939).

Christidis, B. G.: Further studies on competition in yieldtrials with cotton. Emp. J. Exp. Agric. **7**, 111 bis 120 (1939).

Clark, A. u. W. H. Leonard: The analysis of variance with special reference to data expressed as percentages. J. Amer. Soc. Agron. **31**, 55—66 (1939).

Cochran, W. G.: The use of the analysis of variance in enumeration by sampling. J. Amer. Statist. Ass. **34**, 492—510 (1939).

— Note on an approximate formula for significance levels of z. Ann. Math. Statist. **11**, 93—95 (1940). (Abstr.)

— The analysis of variance when experimental errors follow the Poisson or the binomial law. Ann. Math. Statist. **11**, 335—347 (1940).

— Lattice designs for wheat variety trials. J. Amer. Soc. Agron. **33**, 351—360 (1941).

— The distribution of the largest of a set of estimated variances as a fraction of their total. Ann. Eugen **11**, 47—52 (1941).

— The X^2 correction for continuity. J. Iowa St. Coll. Sci. **16**, 421—436 (1942).

— Some additional lattice square designs. Res. Bull. Ia Agric. Exp. Sta., Nr. 318, 731—748 (1943).

— Some developments in statistics. Chronica Botanica **7**, 383—386 (1943).

— The comparison of different scales of measurement for experimental results. Ann. Math. Statist. **14**, 205—216 (1943).

— Analysis of variance for percentages based on unequal numbers. J. Amer. Statist. Ass. **38**, 287—301 (1943).

— Some consequences when the assumptions for the analysis of variance are not satisfield. Biometrics **3**, 22—38 (1947).

Comstock, R. E.: Overestimation of mean squares by the method of expected numbers. J. Amer. Statist. Ass. **38**, 335—340 (1943).

Coolhaas, C.: Proeftuinen, proefvelden en proefveldendienst in de organisatie der Proefstations. (Experimental plantations, test plots and the test plot service in the organization of the Experiment Stations.) Bergcultures **13**, 720—725 (1939).

Copeland, A. H.: The teaching of the calculus of probability. Notre Dame Mathematical Lectures, Indiana, Nr. 4, 31—43 (1944).

Cornish, E. A.: The analysis of quasi-factorial designs with incomplete data. J. Aust. Inst. Agric. Sci. **6**, 31—39 (1940).

— The estimation of missing values in incomplete randomized block experiments. Ann. Eugen., Camb. **10**, 112—118 (1940).

— The estimation of missing values in quasifactorial designs. Ann. Eugen., Camb. **10**, 137—143 (1940).

— The analysis of covariance in quasi-factorial designs. Ann. Eugen. **10**, 269—279 (1940).

— The analysis of quasi-factorial designs with incomplete data. 2. Lattice squares. J. Aust. Inst. Agric. Sci. **7**, 19—26 (1941).

— The recovery of inter-block information in quasi-factorial designs with incomplete data. 1-Square, triple and cubic lattices. Bull. Coun. Sci. Industr. Res., Nr. 158, 22 (1943).

— The recovery of inter-block information in quasi-factorial designs with incomplete data. 2. Lattice squares. Bull. Coun. Sci. Industr. Res., Nr. 175, 19 (1944).

Covas, G. u. E. M. Sivori: Eficiencia en ensayos comparativos de rendimientos planeados según el método de „blocks incompletos simétricos". (Efficiency of comparative yield tests based on the incomplete symmetrical block method.) Rev. Argent. Agron. **6**, 126 bis 130 (1939).

Cowden, D. J.: Correlation concepts and the Doolittle method. J. Amer. Statist. Ass. **38**, 327—334 (1943).

Cox, G. M.: Enumeration and construction of balanced incomplete block configurations. Ann. Math. Statist. **11**, 72—85 (1940).

Craig, C. C.: Recent advances in mathematical statistics. II. Ann. Math. Statist. **13**, 74—85 (1942).

Crist, J. W.: Correlation from ranks, for horticultural research. Proc. Amer. Soc. Hort. Sci. **38**, 593—595 (1941).

— The coefficient of contingency for horticultural research. Proc. Amer. Soc. Hort. Sci. **42**, 484—486 (1943).

Croxton, F. E. u. D. J. Cowden: Applied general statistics. Prentice-Hall, Inc., New York (1939).

Crump, S. L.: The estimation of variance components in analysis of variance. Biometrics Bull. **2**, 7—11 (1946).

Curtiss, J. H.: On transformations used in the analysis of variance. Ann. Math. Statist. **14**, 107—122 (1943).

Dahlberg, G.: Statistical methods for medical and biological students. George Allen and Unwin, Ltd., London (1940).

Davis, J. F.: A comparison between yields calculated from the grain-straw ratio and those calculated from small cut-out areas. J. Amer. Soc. Agron. **31**, 832—840 (1939).

—, R. L. Cook u. W. D. Baten: A method of statistical analysis of a factorial experiment involving influence of fertilizer analysis and placement of fertilizer on stand yield of cannery peas. J. Amer. Soc. Agron. **34**, 521 bis 532 (1942).

Dawson, C. D. R.: An example of the quasi-factorial design applied to a corn breeding experiment. Ann. Eugen., Camb. **9**, 157—173 (1939).

Day, B. B. u. L. Austin: A three-dimensional lattice design for studies in forest genetics. J. Agric. Res. **59**, 101—119 (1939).

De Fina, A. L.: Método gráfico e indice numérico para conocer la seguridad de eficiencia de las variedades agricolas. (A graphical and numerical index for determining the degree of efficiency of agricultural varieties.) Rev. Fac. Agron. La Plata **25**, 21—54 (1943).

De Lury, D. B.: The analysis of Latin squares when some observations are missing. J. Amer. Statist. Ass. **41**, 370—389 (1946).

Deming, W. E.: Some thoughts on statistical inference. J. Wash. Acad. Sci. **31**, 85—92 (1941).

Didusj, V. I.: (Über die Ertragssteigerung bei Eliten.) Selekcija i Semenovodstvo, Nr. 9/10, 3—5 (1946). [Russisch.]

Dijkveld, Stol, J. J.: Het uitschakelen van systematische fouten bij proefvelden in vierkantsvorm. (The elimination of systematic errors in rectangular experimental plots.) Landbouwk. Tijdschr., Wageningen **54**, 185—202 (1942).

— Oogstformuleering. (Formulae for yield.) Landbouwk. Tijdschr. Wageningen **54**, 726—738, 798—817 (1942).

Dixon, W. J.: Further contributions to the problem of serial correlation. Ann. Math. Statist. **15**, 119—144 (1944).

— u. A. M. Mood: The statistical sign test. J. Amer. Statist. Ass. **41**, 557—566 (1946).

Dodd, E. L.: Certain tests for randomness applied to data grouped into small sets. Ann. Math. Statist. **13**, 113 (1942). (Abstr.)

Dorph-Petersen, K.: Fejlberegning paa Forsøg med systematisk Parcelfordeling. (Estimation of standard error in experiments with systematic plot distribution.) Nord. JordbrForskn. **24**, 140—150 (1942).

Down, E. E.: Plot technic studies with small grains. J. Amer. Soc. Agron. **34**, 472—481 (1942).

Dudok van Heel, J. P.: Iets over proefvelden. (Some comments on experiment plots.) Van Zaad tot Suiker, Nr. 9, 136—138 (1939).

Dwyer, P. S.: The calculation of correlation coefficients from ungrouped data with modern calculating machines. J. Amer. Statist. Ass. **35**, 671—673 (1940).

Dyson, E. J.: A note on kurtosis. J. R. Statist. Soc. **106**, 360—361 (1943).

Eckert, W. J.: Punched card methods in scientific computation. The Thomas J. Watson Astronomical Computing Bureau, Columbia University, Pp. VII + 136. 23 figs (1940).

Eisenhart, C.: The assumptions underlying the analysis of variance. Biometrics **3**, 1—21 (1947).

Eleving, G.: The asymptotical distribution of range in samples from a normal population. Biometrika **34**, 111—119 (1947).

Emmert, E. M.: Partial elimination of experimental error from data by the use of significance tests. Proc. Amer. Soc. Hort. Sci. **37**, 272—278 (1939) 1940.

Epstein, B. u. C. W. Churchman: On the statistics of sensitivity data. Ann. Math. Statist. **15**, 90—96 (1944).

Fan, F. R. u. W. F. Koo: (Application of analysis of covariance to quasi-factorial experiment.) Kwangsi Agric. **2**, 14—20 (1941).

Finney, D. J.: The joint distribution of various ratios based on a common error mean square. Ann. Eugen. **11**, 136—140 (1941).

— Standard errors of yields adjusted for regression on an independent measurement. Biometrics Bull. **2**, 53—55 (1946).

— A note on „missing-plot techniques". Biometrics Bull. **2**, 94 (1946).

— Latin squares of the sixth order. Experientia, Basel **2**, 404—405 (1946).

— Orthogonal partitions of the 5 × 5 Latin squares. Ann. Eugen., Camb. **13**, 1—3 (1946).

— Orthogonal partitions of the 6 × 6 Latin squares. Ann. Eugen. **13**, 184—196 (1946).

— The construction of confounding arrangements. Emp. J. Exp. Agric. **15**, 107—112 (1947).

Fisher, R. A.: The comparison of samples with possibly unequal variances. Ann. Eugen., Camb. **9**, 174—180 (1939).

— An examination of the different possible solutions of a problem in incomplete blocks. Ann. Eugen., Camb. **10**, 52—75 (1940).

— The asymptotic approach to Behrens's integral, with further tables for the d-test of significance. Ann. Eugen. **11**, 141—172 (1941).

— The interpretation of experimental four-fold tables. Science **94**, 210—211 (1941).

— New cyclic solutions to problems in incomplete blocks. Ann. Eugen. **11**, 290—299 (1942).

—, A. S. Corbet u. C. B. Williams: The relation between the number of species and the number of individuals in a random sample of an animal population. J. Anim. Ecol. **12**, 42—58 (1943).

Ford, L. R.: Alignment charts. Notre Dame Mathematical Lectures, Indiana, Nr. 4, 1—29 (1944).

Frankena, H. J. u. M. P. Both: Enkele opmerkingen over proefveldtechniek. (Some observations on the technique of field experiments.) Versl. Rijkslandbproefst. 's Grav., Nr. 45 (16) A, 427—437 (1939).

Geary, R. C.: The frequency distribution of $\sqrt{b_1}$ for samples of all sizes drawn at random from a normal population. Biometrika. **34**, 68—97 (1947).

— u. J. P. G. Worlledge: On the computation of universal moments of tests of statistical normality derived from samples drawn at random from a normal universe. Application to the calculation of the seventh moment of b_2. Biometrika **34**, 98—110 (1947).

Geppert, M. D.: Über die Alterskorrektur von Merkmalshäufigkeiten in der Erbstatistik. Arch. math. Wirtsch. u. Sozialforschg. **6**, 80—102 (1940).

Ghosh, B.: Measures of heterogenity in agricultural and similar fields, and their inter-relations. Sci. and Cult. **11** 382—383 (1946).

Gobeil, R.: Importance des statistiques dans les travaux forestiers. La Forêt, Québec **2**, Nr. 2, 31—48 (1940).

Goldberg, H. u. H. Levine: Approximate formulas for the percentage points and normalization of t and X^2. Ann. Math. Statist. **17**, 216—225 (1946).

Goulden, C. H.: Fundamentals of experimentation. Spragg Memor. Lectures Plant Breed., Mich. St. Coll., Pp. 16—27 (1941) 1942.

— A uniform method of analysis for square lattice experiments. Sci. Agric. **25**, 115—136 (1944).

Gries, G. A., J. G. Horsfall u. N. Turner: Polymodal response curves in biological research. Phytopathology **35**, 654—655 (1945). (Abstr.)

Grubbs, F. E.: On the distribution of the radial standard deviation. Ann. Math. Statist. **15**, 75—81 (1944).

Gsell, R.: Über Messungen an Anacampis Pyramidalis (L.) Rich. und anderen europäischen Orchideen. Ber. d. Schweiz. Bot. Ges. **51**, 257—309 (1940).

Gumbel, E. J.: On the reliability of the classical X^2 test. Ann. Math. Statist. **14**, 253—263 (1943).

— Ranges and midranges. Ann. Math. Statist. **15**, 414 bis 422 (1944).

Harrington, J. B.: The number of replicated small plat tests required in regional variety trials. J. Amer. Soc. Agron. **31**, 287—299 (1939).

Harshbarger, B.: On the analysis of a certain six-by-six four-group lattice design using the recovery of inter-block information. Ann. Math. Statist. **16**, 387 bis 390 (1945).

Hartley, H. O.: Recent advances in mathematical statistics. J. Roy. Statist. Soc. **103**, 534—560 (1940).

Hellbo, E.: Fältkontrollen och dess betydelse. (Field control work and its importance.) Årsb. Jordbruksforskning, Stockholm, 166—172 (1946).

Hendricks, W. A., J. P. Quinn u. A. B. Godfrey: Interpretation of Mendelian class frequencies. J. Agric. Res. **58**, 755—760 (1939).

Herchenroder, M. V. M.: A review of modern practical methods of analysis in statistics. Rev. Agric. Maurice **22**, 51—68 (1943).

Hintzsche, E.: Statistische Probleme aus der Kerngrößenforschung. Experientia, **1**, 103—110 (1945).

— Über Normalkurven der Kerngrößenverteilung. Mitt. d. Naturf. Ges., Bern. N. F. **4** (1946).

Hoel, P. G.: Testing the homogeneity of Poisson frequencies. Ann. Math. Statist. **16**, 362—368 (1945).

Hollander, W. F.: Notes on graphic biometric comparisons of samples. Amer. Nat. **80**, 494—496 (1946).

Holme, R. V.: Corn microplots and their interpretation. An interim discussion. J. A. S. T. Quart. **5**, Nr. 2, 3—8 (1941).

Holmes, R. L. u. A. M. O-Neal: A study of the number of stalks of cane required for accuracy in sampling experimental plots. Proc. 6th Congr. Int. Soc. Sug. Cane Tech., La, Pp. 760—764 (1938) 1939.

Hoogland, J. J. u. A. L. S. Bär: Gedetailleerde variatieanalyse van proefveldresultaten. (Detailed analysis of variance of the results of plot trials.) Arch. Suikerind. Nederland. Ned.-Ind. **2**, 402—407 (1941).

Hsu, P. L.: On generalized Analysis of Variance. I. Biometrika **31**, 221—237 (1940).

— Some simple facts about the separation of degrees of freedom in factorial experiments. Sankhyā: Indian J. Statist. **6**, 253—254 (1943).

Hsu, T. H.: Samples from two bivariate normal populations. Ann. Math. Statist. **12**, 279—292 (1941).

Hubbs, C. L. u. A. Perlmutter: Biometric comparison of several samples, with particular reference to racial investigations. Amer. Nat. **76**, 582—592 (1942).

Hudson, H. G.: Population studies with wheat. I. Sampling. J. Agric. Sci. **29**, 76—110 (1939).

— Population studies with wheat. III. Seed rates in nursery trials and field plots. J. Agric. Sci. **31**, 138 bis 144 (1941).

Hussain, Q. M.: Symmetrical incomplete block designs with $\lambda = 2$, k = 8 or 9. Bull. Calcutta Math. Soc. **37**, 115—123 (1945).

Immer, F. R.: Distribution of yields of single plants of varieties and F_2 crosses of barley. J. Amer. Soc. Agron. **34**, 844—850 (1942).

— Some uses of statistical methods in plant breeding. Biometrics Bull. **1**, 13—15, 28 (1945).

Irwin, J. O. u. M. G. Kendall: Sampling moments of moments for a finite population. Ann. Eugen. **12**, 138 bis 142 (1944).

Jacop, W. C.: Análise estatística de experimentos fatoriais. (Statistical analysis of factorial experiments.) Bol. Minist. Agric. Rio de J. **33**, 41—56 (1944).

Jeffreys, H.: Random and systematic arrangements. Biometrika **31**, 1—8 (1939).

Jones, A. E.: A useful method for the routine estimation of dispersion from large samples. Biometrika **33**, 274 bis 282 (1946).

Johnson, I. J. u. H. C. Murphy: Lattice and lattice square designs with oat uniformity data and in varietytrials. J. Amer. Soc. Agron. **35**, 291—305 (1943).

Justesen, S. H.: Toepassing van de strooiingsanalyse bij vakkenproeven. (Application of the analysis of variance in field experiments.) Landbouw **15**, 346 bis 363 (1939).

Kac, M.: A remark on independence of linear and quadratic forms involving independent Gaussian variables. Ann. Math. Statist. **16**, 400—401 (1945).

Kadam, B. S.: How to conduct yield trials improved strains of crops in districts. Poona Agric. Coll. Mag. **34**, 117—123 (1942).

Kalamkar, R. J. u. Dhannalal: Variability of plant density and the estimation of yield of cotton by sampling. Proc. 26th Indian Sci. Congr., Lahore **3**, 198 (1939). (Abstr.)

—, — Sampling studies in a cotton varietal trial. Sankhyā. Indian J. Statist **4**, 567—576 (1940).

Kempthorne, O.: Comments on the note „On a theorem concerning sampling". J. R. Statist. Soc. **107**, 58 (1944).

— The analysis of a series of experiments by the use of punched cards. Suppl. J. R. Statist. Soc. **8**, 118 bis 127 (1946).

— Recent developments in the design of field experiments. IV. Lattice squares with split-plots. J. Agric. Sci. **37**, 156—162 (1947).

Kendall, M. G.: A theory of randomness. Biometrika **32**, 1—16 (1941).

Kerr, H. W.: Notes on plot technique. Proc. 6th Congr. Int. Soc. Sug. Cane Tech., La, Pp. 764—778 (1938) 1939.

Kishen, K.: On a general method of constructing the appropriate compounds in terms of the twelve 3×3 Latin Squares for any of the eight components $N_1K_1P_1$, $N_1K_1P_2$, $N_1K_2P_1$ etc., of the second order interaction in a 3×3 factorial arrangement. Proc. 26th Indian Sci. Congr., Lahore, Pt. III, Pp. 199 (1939). (Abstr.)

— On a simplified method of expressing the components of the second order interaction in a 3^3 factorial design. Sankhyā. Indian J. Statist. **4**, 577—580 (1940).

Kosambi, D. D.: A bivariate extension of Fisher's z-test. Curr. Sci. **10**, 191—192 (1941).

— A test of significance for multiple observations. Curr. Sci. **11**, 271—274 (1942).

Krishna, Iyer, P. V.: Symmetrical incomplete randomized blocks. Proc. 27th Indian Sci. Congr., Madras, Pt. III, Sect. Agric. **58**, 231 (1940). (Abstr.)

— The analysis of simple non-symmetrical experiments. Indian J. Agric. Sci. **10**, 686—690 (1940).

— The analysis of incomplete split plot designs. Sci. and Cult. **6**, 487 (1941).

— Standard error of the difference between two estimates for incomplete block experiments. Curr. Sci. **10**, 165 (1941).

— Studies with wheat uniformity trial data. I. Size and shape of experimental plots and the relative efficiency of different lay-outs. Indian J. Agric. Sci. **12**, 240—262 (1942).

— Studies with wheat uniformity trial data. II. Balanced versus randomized arrangement. Indian J. Agric. Sci. **12**, 263—273 (1942).

— Studies with wheat uniformity trial data. III. Distributions of variances and ratio of variances. Indian J. Agric. Sci. **12**, 274—280 (1942).

— The distribution of the mean of samples from a rectangular population. Curr. Sci. **14**, 18—19 (1945).

Krishna Raoa, C. R.: Familial correlations or the multivariate generalisations of the intraclass correlation. Curr. Sci. **14**, 66—67 (1945).

Krishnaswami Ayyangar, A. A.: Statistical formulae. Curr. Sci. **12**, 145 (1943).

— Interaction formulae in analysis of variance. Curr. Sci. **14**, 35 (1945).

Kulešov, N. N.: (Über die Feld- und Labormethoden zur Erforschung der Aussaat-Aufgang-Periode.) C. R. (Doklady) Acad. Sci. URSS. **54**, 77 (1946). [Russisch.]

LeClerg, E. L.: Relation of field plot design to seed-source tests of Irish potatoes. Amer. Potato J. **19**, 75—79 (1942).

— u. M. T. Henderson: Relative efficiency of the two-dimensional quasi-factorial design as compared with a randomized-block arrangement when concerned with yields of Irish potatoes. Amer. Potato J. **17**, 279—282 (1940).

Leggat, C. W.: Statistical Aspects of Seed Analysis. Bot. Rev. **5**, 505—529 (1939).

Levene, H. u. J. Wolfowitz: The covariance matrix of runs up and down. Ann. Math. Statist. **15**, 58—69 (1944).

Li, J. C. R.: Design and statistical analysis of some confounded factorial experiments. Res. Bull. Ia Agric. Exp. Sta., Nr. 333, 453—492 (1944).

Livermore, J. R.: Report of the committee on standardization of field plot technique. Amer. Potato J. **17**, 114—123 (1940).

Löfvenmark, H.: Ritning av försöksplaner med skrivmaskin. (Delineation of plans of experiments with the typewriter.) Lantmannen **25**, 796—798 (1941).

Lord, E.: The use of range in place of standard deviation in the t-test. Biometrika **34**, 41—67 (1947).

Lynch, P. B.: Sampling methods for the estimation of grain yields in cereal trials. N. Z. J. Sci. Tech. **22**, 151A—157A (1940).

— New harvesting technique with cereal trials. Sampling methods necessitated by use of header-harvester. N. Z. J. Agric. **64**, 173—175 (1942).

MacDonald, D., W. L. Fielding u. D. F. Ruston: Experimental methods with cotton. I. The design of plots for variety trials. J. Agric. Sci. **29**, 35—47 (1939).

Madow, W. G. u. L. H. Madow: On the theory of systematic sampling. I. Ann. Math. Statist. **15**, 1—24 (1944).

Mahalanobis, P. C. (Editor): Proceedings of the Second Session of the Indian Statistical Conference held in Lahore, January 1939. Statist. Publ. Soc. Calcutta, Pp. 168 (1940).

— Statistical definition of standard yield of crops. Sankhyā: Indian J. Statist. **6**, 97—98 (1942).

— Use of small-size plots in sample surveys for crop yields. Nature, Lond. **158**, 798—799 (1946).

Mahoney, C. H. u. W. D. Baten: The use of the analysis of covariance and its limitation in the adjustment of yields based upon stand irregularities. J. Agric. Res. **58**, 317—328 (1939).

Mather, K.: Statistical analysis in biology. Methuan & Co., Ltd., London. 16s 0d., Pp. 247, tables, diagrams (1943).

— Statistical analysis in biology. London, Pp. 267, 61 tables, 10 figs (1946).

Mathisen, H. C.: A method of testing the hypothesis that two samples are from the same population. Ann. Math. Statist. **14**, 188—194 (1943).

Mendonça, P. de Varennese: Das distribuições estatísticas mais usadas em provas de significação X^2 de Pearson, t de Student, z de Fisher. (Statistical distributions most used in tests of significance — Pearson's X^2 method, the t test of Student and Fisher's z test.) Rev. Agron, Lisboa **28**, 32—51 (1940).

Menger, K.: On the relation between calculus of probability. Notre Dame Mathematical Lectures, Indiana, Nr. 4, 44—53 (1944).

Mises, R. v.: On the problem of testing hypotheses. Ann. Math. Statist. **14**, 238—252 (1943).

Molina, E. C.: Some fundamental curves for the solution of sampling problems. Ann. Math. Statist. **17**, 325 bis 335 (1946).

Mood, A. M.: Stratified sampling. Ann. Math. Statist. **13**, 113 (1942). (Abstr.)

Nair, K. R.: Some balanced confounded arrangements for the 5^n type of experiment. Proc. 26th Indian Sci. Congr. Lahore, Pt. III, Pp. 200 (1939). (Abstr.)

— The application of co-variance technique to field experiments with mixed-up yields. Sci. and Cult. **4**, 474 (1939).

— The application of the technique of analysis of covariance to field experiments with several missing or mixed-up plots. Sankhyā: Indian J. Statist. **4**, 581 bis 588 (1940).

Nair, K. R.: Balanced confounded arrangements for the 5^n type of experiment. Sankhyā: Indian J. Statist. **5**, 57—70 (1940).

— Efficiency of the adjustment for concomitant characters in biological experiments. Sankhyā: Indian J. Statist. **6**, 167—174 (1942).

— Certain inequality relationships among the combinatorial parameters of incomplete block designs. Sankhyā: Indian J. Statist. **6**, 255—259 (1943).

— Statistical notes for agricultural workers, Nr. 27. Calculation of standard errors and tests of significance of different types of treatment comparisons in split-plot and strip arrangements of field experiments. Indian J. Agric. Sci. **14**, 315—319 (1944).

— The recovery of inter-block information in incomplete block designs. Sankhyā: Indian J. Statist. **6**, 383—390 (1944).

—, S. C. Chakravarti u. P. C. Mahalanobis: A 10×10 quasi-factorial experiment at Chinsurah with 100 strains of rice. Proc. 26th Indian Sci. Congr., Lahore **3**, 199 (1939). (Abstr.)

— u. P. C. Mahalanobis: A simplified method of analysis of quasi factorial experiments in square lattice with a preliminary note on joint analysis of yield of paddy and straw. Indian J. Agric. Sci. **10**, 663—685 (1940).

— u. C. R. Rao: A general class of quasi-factorial designs leading to confounded designs for factorial experiments. Sci. and Cult. **7**, 457—458 (1942).

— u. M. P. Shrivastava: On a simple method of curve fitting. Sankhyā: Indian J. Statist. **6**, 121—132 (1942).

Newman, D.: The distribution of range in samples from a normal population, expressed in terms of an independent estimate of standard deviation. Biometrika **31**, 20—30 (1939).

Neyman, J.: Conception of equivalence in the limit of tests and its application to certain λ and X^2-tests. Ann. Math. Statist. **11**, 477—478 (1940).

Nissen, Ø.: Feilberegning på forsøk med systematisk parsellfordeling. (The calculation of error in experiments with a systematic distribution of plots.) Nord. JordbrForskn., Nr. 7, 357—358 (1942).

O'Neil, J. B. u. H. S. Gutteridge: A note on the calculation of standard errors for treatment means after adjustment for regression. Sci. Agric. **21**, 358—359 (1941).

Panse, V. G.: Studies in the technique of field experiments. V. Size and shape of blocks and arrangement of plots in cotton trials. Indian J. Agric. Sci. **11**, 850 bis 865 (1941).

— Plot-size in yield surveys. Nature, Lond. **159**, 820 (1947).

— u. G. R. Ayachit: Ten per cent probability of z and the variance ratio. Indian J. Agric. Sci. **14**, 244—247 (1944).

Papadakis, J. S.: Comparaison de différentes méthodes d'expérimentation phytotechnique. Rev. Argent. Agron. **7**, 297—363 (1940).

Pätau, K.: Eine statistische Bemerkung zu Moewus' Arbeit „Die Analyse von 42 erblichen Eigenschaften der Chlamydomonas eugametos-Gruppe. III. Teil". Z. Vererbungsl. **79**, 317—320 (1941).

— Eine neue X^2-Tafel. Z. Vererbungsl. **80**, 558—564 (1942).

Paterson, D. D.: Statistical technique in agricultural research. A simple exposition of practice and procedure in biometry. McGraw-Hill Publishing Company, Ltd., London 18s. 0d., Pp. IX + 263 (1939).

Paul, W. R. C. u. M. Fernando: Field-plot technique with chillies (Capsicum annuum L). Trop. Agriculturist **93**, 270—275 (1939).

Pearce, S. C.: Sampling methods for the measurement of fruit crops. J. R. Statist. Soc. **107**, 117—126 (1944).
Pearson, E. S.: The choise of statistical tests illustrated on the interpretation of data classed in a 2×2 table. Biometrika **34**, 139—163 (1947).
— u. J. Wishart (Editors): „Student's" collected papers. Biometrika Office, University College, London, Pp. XIV +224, tables, figs, graphs, diagrams (1942).
Penrose, L. S.: Some notes on discrimination. Ann. Eugen., Camb. **13**, 228—237 (1947).
Pérez Calvet, R.: Un estudio sobre las experiencias de uniformidad y su empleo en la elección de parcela de repetición. (A study on uniformity experiments and their use in the choice of replication plots.) Bol. Inst. Invest. Agron. Madr., Nr. 12, 329—348 (1945).
—, M. M. Zulueta u. A. Anós: Experimentación agrícola. Fundamentos estadísticos y métodos operatorios. (Agricultural experimentation. Statistical principles and methods of operation.) Instituto Nacional de Investigaciones Agronómicas, Madrid, Pp. VII+272. 19 figs. tables (1943).
Peters, C. C. u. W. R. van Voorhis: Statistical procedures and their mathematical bases. McGraw-Hill Publishing Company, Ltd., London (1940).
Phipps, I. F., A. T. Pugsley, S. R. Hockley u. E. A. Cornish: The analysis of cubic lattice designs in varietal trials. Bull. Coun. Sci. Industr. Res. Aust., Nr. 176, 40 (1944).
Plackett, R. L.: Limits of the ratio of mean range to standard deviation. Biometrika **34**, 120—122 (1947).
Platt, J. R.: A mechanical determination of correlation coefficients and standard deviations. J. Amer. Statist. Ass. **38**, 311—318 (1943).
Plummer, H. C.: Probability and frequency. Macmillan and Co., Ltd., London (1940).

Ramamurti, B. u. B. Sitaraman: On maximal sets of confounded interactions in a $(2^n, 2^k)$ confounded design. Sankhyā: Indian J. Statist. **6**, 183—188 (1942).
Rao, C. R.: Quasi-latin squares in experimental arrangements. Curr. Sci. **12**, 322—323 (1943).
Rider, P. R.: An introduction to modern statistical methods. John Wiley and Sons, Inc., New York 13s. 6d, Pp. IX+220 (Chapman and Hall, London) (1939).
Robbins, H.: On distribution-free tolerance limits in random sampling. Ann. Math. Statist. **15**, 214—216 (1944).
Roessler, E. B.: Valid estimates of variance in the analysis of pooled data. Proc. Amer. Soc. Hort. Sci. **42**, 481—483 (1943).
— u. L. D. Leach: Analysis of combined data for identical replicated experiments. Proc. Amer. Soc. Hort. Sci. **44**, 323—328 (1944).

Salmon, S. C.: The use of modern statistical methods in field experiments. J. Amer. Soc. Agron. **32**, 308—320 (1940).
Sant, G. K.: Some features in the analysis of covariance with split-plot designs. Proc. 27th Indian Sci. Congr., Madras, Pt. III, Sect. Agric. **57**, 231 (1940). (Abstr.)
Satterthwaite, F. E.: A generalized analysis of variance. Ann. Math. Statist. **13**, 34—41 (1942).
Saunders, A. R.: Statistical methods with special reference to field experiments. Sci. Bull. Dep. Agric. S. Afr. 2nd. ed., Nr. 200, 112 (1939).
— Efficiency of design in field experiment at Potchefstroom, South Africa. Emp. J. Exp. Agric. **12**, 157 bis 162 (1944).
Savur, S. R.: A note on the arrangement of incomplete blocks, when $k = 3$ and $\lambda = 1$. Ann. Eugen., London **9**, 45—49 (1939).
Sawkins, D. T.: Simple regression and correlation. J. Roy. Soc. N. S. W. **77**, 85—95 (1944).

Schad, C., G. Meneret u. R. Mayer: Application des méthodes Fisher et Student à la comparaison des variétés en petites parcelles. Ann. Phytogénét., Paris **7**, 91—107 (1941).
Scheffé, H.: A note on the Behrens-Fisher test. Ann. Math. Statist. **15**, 430—434 (1944).
— u. J. W. Tukey: A formula for sample sizes for population tolerance limits. Ann. Math. Statist. **15**, 217 (1944).
Schelling, H. von: Über die exakte Behandlung des Zusammenhanges zwischen biologischen Merkmalsreihen. Arb. Staatl. Inst. exper. Ther. Frankf. **39**, 35—71 (1940).
— Die Bedeutung der statistischen Methodik für die Biologie. Erg. Hyg. **24**, 87—149 (1941).
— Fehlerrechnung bei biologischen Messungen. Klin. Wschr. **2**, 741—743 (1941).
— Mutungsgrenzen für die Differenz zweier unabhängig voneinander beobachteter Häufigkeiten. Arb. Staatl. Inst. exp. Ther. Frankfurt **41**, 47—86 (1941).
— Statistische Schätzungen auf kombinatorischer Grundlage. Z. angew. Math. Mech. **21**, 52—58 (1941).
— Eine Formel für die Teilsummen gewisser hypergeometrischer Reihen und deren Bedeutung für die Wahrscheinlichkeitstheorie. Naturwiss. **30**, 757 (1942).
Sekar, C. C.: Distribution of Fisher's g_1, for samples of three from a continuous rectangular distribution. Curr. Sci. **13**, 10—11 (1944).
Sheppard, W. F.: British Association for the Advancement of Science Mathematical Tables: Volume VII. The probability integral. University Press, Cambridge (1939).
Siao, F. u. C. C. Cheng: (A study on the efficiency of quasifactorial design.) Kwangsi Agric. **3**, 371—384 (1942).
Sillitto, G. P.: The distribution of Kendall's coefficient of rank correlation in rankings containing ties. Biometrika **34**, 36—40 (1947).
Simon, H. A.: Symmetric tests of the hypothesis that the mean of one normal population exceeds that of another. Ann. Math. Statist. **14**, 149—154 (1943).
Simpson, G. G.: Note on graphic biometric comparison of samples. Amer. Nat. **79**, 95—96 (1945).
— u. A. Roe: A standard frequency distribution method. Amer. Mus. Novitates, Nr. 1190, 1—19 (1942).
Snedecor, G. W.: Statistical methods applied to experiments in agriculture and biology. Collegiate Press. Inc. Ames, Iowa, Pp. XVI+485. tables (1946).
— u. E. S. Haber: Statistical methods for an incomplete experiment on a perennial crop. Biometrics Bull. **2**, 61—67 (1946).
Sommermeyer, K.: Zur statistischen Analyse der Wirkung harter Strahlen auf biologische Objekte. II. III. Strahlenther. **69**, 715; **70**, 184 (1941).
Stevens, W. L.: The completely orthogonalized Latin square. Ann. Eugen., London **9**, 82—93 (1939).
Accuracy of mutation rates. J. Genet. **43**, 301—307 (1942).
Stoffels, A.: De berekening van middelbare fouten bij niet-homogene proefvelden. (The calculation of the mean errors in non-homogeneous field plots.) Landbouwk. Tijdschr., Wageningen **52**, 165—174 (1940).
Subbaiya, M.: Sampling in sugar cane experimental work. Proc. 26th Indian Sci. Congr., Lahore **3**, 197 bis 198 (1939). (Abstr.)
Subramonia Iyer, S.: A supplementary note on the analysis of 3^3 and 3^4 designs (with three factor interactions confounded) in field experiments in agriculture. Indian J. Agric. Sci. **10**, 691—692 (1940).
Sukhatme, P. V.: Moments and product moments of momentstatistics for samples of finite and infinite populations. Sankhyā: Indian J. Statist. **6**, 363—382 (1944).
— Bias in the use of small-size plots in sample surveys for yield. Curr. Sci. **15**, 119—120 (1946).

Sukhatme, P. V.: The problem of plot size in large-scale yield surveys. J. Amer. Statist. 42, 297—310 (1947).

Syzrancev, P. I.: (The mathematical treatment of a small number of experimental data.) Socialistic Grain Farming, Saratov, Nr. 2, 163—182 (1941). [Russisch.]

Tedin, O.: Small samples of a Poisson series. Hereditas, Lund 31, 238—240 (Abstr.) (1945).

Tharp, W. H., C. H. Wadleigh u. H. D. Barker: Some problems in handling and interpreting plant disease data in complex factorial designs. Phytopathology 31, 26—48 (1941).

Thompson, C. M.: Tables of percentage points of the incomplete beta-function. Biometrika 32, 151—181 (1941).

— Table of percentage points of the x^2 distribution. Biometrika 32, 187—191 (1941).

— u. M. Merrington: Tables for testing the homogeneity of a set of estimated variances. Biometrica 33, 296—304 (1946).

Tintner, G.: The variate difference method. Principia Press, Inc., Bloomington, Indiana (1940).

Tippett, L. H. C.: The methods of statistics. London, 3rd Ed., Pp. 284. tables. 13 figs (1941).

Torrie, J. H. u. J. G. Dickson: The use of statistical methods in quality evaluation of barley and malta data. Cereal Chem. 20, 579—594 (1943).

—, H. L. Shands u. B. D. Leith: Efficiency studies of types of design with small grain yield trials. J. Amer. Soc. Agron. 35, 645—661 (1943).

Tsai, H. u. C. Y. Chow: Studies on field plot technique in wheat. Chinese J. Agric. 1, 117—118 (1943).

Upholt, W. M.: Observations on wartime biometrics. Biometrics Bull. 1, 47—52 (1945).

Vaidyanathan, M. u. S. Subramonia Iyer: A note on the analysis of 3^3 and 3^4 designs (with three factor interactions confounded) in field experiments in agriculture. Indian J. Agric. Sci. 10, 233—236 (1940).

Vajda, S.: Technique of the analysis of variance. Nature, Lond. 160, 27 (1947).

Van Uven, M. J.: Mathematical treatment of the results of agricultural and other experiments. P. Noordhoff N, V. Groningen and Batavia 2nd Ed., Pp. VI + 310 (1946).

Visser, W. C.: Over de bruikbaarheid van de grafisch-statistische bewerkingstechniek. (On the applicability of the graphical and statistical method of treatment.) Landbouwk. Tijdschr., Wageningen 54, 403—416 (1942).

Wald, A.: On the analysis of variance in case of multiple classifications with unequal class frequencies. Ann. Math. Statist. 12, 346—349 (1941).

— On the principles of statistical inference. Notre Dame Mathematical Lectures I. Edwards Bros., Inc., Ann. Arbor Michigan, Pp. 49 (1942). [Mimeographed.]

— On the efficient design of statistical investigation. Ann. Math. Statist. 14, 131—140 (1943).

— On a statistical problem arising in the classification of an individual into one of two groups. Ann. Math. Statist. 15, 145—162 (1944).

— On cumulative sums of random variables. Ann. Math. Statist. 15, 283—296 (1944).

Wallis, W. A.: The correlation ratio for ranked data. J. Amer. Statist. Ass. 34, 533—538 (1939).

— Compounding probabilities from independent significance tests. Ann. Math. Statist. 13, 111—112 (1942). (Abstr.)

Walsh, J. E.: On the power function of the sign test for slippage of means. Ann. Math. Statist. 17, 358 bis 362 (1946).

— Concerning the effect of intraclass correlation on certain significance tests. Ann. Math. Statist. 18, 88—96 (1947).

Wanner, H.: Streuungszerlegung, ein Hilfsmittel der modernen Statistik zur Beurteilung von Sortenanbauversuchen. Landw. Jb. Schweiz 55 (1941).

Waugh, F. V.: The computation of partial correlation coefficients. J. Amer. Statist. Ass. 41, 543—546 (1946).

Weiss, M. G. u. G. M. Cox: Balanced incomplete block and lattice square designs for testing yield differences among large numbers of soybean varieties. Res. Bull. Ia Agric. Exp. Sta., Nr. 257, 291—316 (1939).

Welch, B. L.: The generalization of „Student's" problem when several different population variances are involved. Biometrika 34, 28—35 (1947).

Welker, E. L. u. F. L. Wynd: Influence of unknown factors on the validity of mathematical correlations of biological data. Plant Physiol. 18, 498—507 (1943).

Wellensiek, S. J.: Een studiekring voor proeftechniek. (A study circle for the technique of field experiments.) Landbouwk. Tijdschr., Wageningen 53, 743 (1941).

Wilks, S. S.: Mathematical statistics. Princeton University Press, N. J., Pp. XI + 284 (1943).

— Sample criteria for testing equality of means, equality of variances and equality of covariances in a normal multivariate distribution. Ann. Math. Statist. 17, 257 bis 281 (1946).

Willcox, O. W.: How to make a standard yield diagram. J. Amer. Soc. Agron. 39, 74—77 (1947).

Williams, C. B.: Yule's „characteristic" and the „index of diversity". Nature, Lond. 157, 482 (1946).

Wilm, H. G.: Notes on analysis of experiments replicated in time. Biometrics Bull. 1, 16—20 (1945).

Wilson, E. B.: The controlled experiment and the fourfold table. Science 93, 557—560 (1941).

Wishart, J.: Field trials: their lay-out and statistical analysis. Imperial Bureau of Plant Breeding and Genetics, Cambridge, 2s. 6d, Pp. 36 (1940).

— The cumulants of the z and of the logarithmic X^2 and t distribution. Biometrika 34, 170—178 (1947).

Wolfowitz, J.: Asymptotic distribution of runs up and down. Ann. Math. Statist. 15, 163—172 (1944).

Yates, F.: The recovery of inter-block information in variety trials arranged in three-dimensional lattices. Ann. Eugen., Camb. 9, 136—156 (1939).

— The comparative advantages of systematic and randomized arrangements in the design of agricultural and biological experiments. Biometrika 30, 440—466 (1939).

— Lattice squares. J. Agric. Sci. 30, 672—687 (1940).

— Modern experimental design and its function in plant selection. Emp. J. Exp. Agric. 8, 223—230 (1940).

— Methods and purposes of agricultural surveys. J. R. Soc. Arts 91, 367—379 (1943).

— A review of recent statistical developments in sampling and sampling surveys. J. R. Statist. Soc. 109, 12 (1946).

— u. R. W. Hale: The analysis of Latin squares when two or more rows, columns, or treatments are missing. Suppl. J. R. Statist. Soc. 6, 67—79 (1939).

Youden, W. J.: Experimental designs to increase accuracy of greenhouse studies. Contr. Boyce Thompson Inst. 11, 219—228 (1940).

Young, L. C.: On randomness in ordered sequences. Ann. Math. Statist. 12, 293—300 (1941).

Zimmer, K. G.: Statistische Ultramikrometrie mit Röntgen-Alpha- und Neutronenstrahlung. Phys. Ztschr. 44, 233 (1943).

Zuber, M. S.: Relative efficiency of incomplete block designs using corn uniformity trial data. J. Amer. Soc. Agron. 34, 30—47 (1942).

Zubin, J.: Nomographs for determining the significance of the differences between the frequencies of events in two contrasted series or groups. J. Amer. Statist. Ass. 34, 539—544 (1939).

1) Acker- und Pflanzenbau.

Andersen, J. C. u. A. Poulsen: Avl af Markfrø. (Raising seed of field crops). Kgl. Danske Landhusholdningsselskab, København, Pp. 226, 73 figs. 9 tables. (1946).

Anonymus: A handbook of Philippine agriculture, issued in commemoration of the thirtieth anniversary. Published by the College of Agriculture, University of the Philippines (1939).

— Bibliographie d'agriculture tropicale, 1938. Int. Inst. Agric., Rome (1939).

— Bibliothèques agricoles dans le monde et bibliothèques spécialisées dans les sujets se rapportant à l'agriculture. Int. Inst. Agric., Rome (1939).

— Fifty years of progress on Dominion Experimental Farms, 1886—1936. J. O. Patenaude, Ottawa (1939).

— Agriculture in the twentieth century. Essays on research, practice, and organization presented to Sir Daniel Hall. Clarendon Press, Oxford (1939).

— Agriculture and animal husbandry in India 1937 bis 1938. Imp. Coun. Agric. Res. (1940).

— Bibliography of soil science, fertilizers and general agronomy. 1937—1941. Imp. Bur. Soil Science, Harpenden (1941).

— The Queensland Agricultural and Pastoral Handbook. 1941: Vol. I, Pp. 448; 1940: Vol. II, Pp. 386; 1938: Vol. III, Pp. 254; 1939: Vol. IV, Pp. 199.

— Climate and man. Yearb. U. S. Dep. Agric., Washington, D. C. (1941).

— Farming handbook. Jarrold & Sons, Ltd., Norwich, 5 s. Pp. 220. 52 illus. (1942).

— Agronomy handbook for South Carolina. Bull. Clemson Agric. Coll. U. S. D. A. Nr. 104, 137 (1942).

— L'expérimentation au champ dans les stations de recherches agronomiques Françaises. Ann. Agron., Paris, 13, 295—331 (1943).

— Agricultural research in Great Britain. H. M. Stationery Office, London (1943).

— Co-operative systems in European agriculture. Advance. Sci. Lond., 2, 356—359 (1943).

— Speeding up plant improvement. Mon. Sci. News., Nr. 39, 2—3 (1944).

— Sveriges Lantbruksförbund. Årsbok 1944. (Swedish Agricultural Union. Yearbook 1944.) Stockholm, Pp. 272, 27 tables. Illus. (1944).

— An agricultural expedition to Iran. Agriculture, Moscow, Nr. 6, 10—11 (1945). [Mimeographed.]

— Sveriges Lantbruksförbund. Årsbok 1945, 1946 bis 1947 (Swedish Agricultural Union, Yearbook 1945, 1946—47). Lantbruksförbundets Tidskriftsaktiebolag, Stockholm 1945: Pp. 256, 20 tables. illus.; 1946 bis 1947: Pp. 225, 22 tables. illus. (1945).

— Bibliography of Soil Science, Fertilizers and General Agronomy 1940—1944. Imperial Bureau of Soil Science, Harpenden, Pp. 567 (1946).

— Nytt utifrån (News from abroad). Fruktodlaren, Nr. 1, 15—16 (1946).

— Plant husbandry. Agricultural Chron., Moscow, Nr. 11, 3—4 (1946). [Mimeographed.]

— Guide to Boghall Experimental Farm. Edinb. and E. Scot. Coll. Agric., Edinburgh, Pp. 30 (1946).

— Developing village India. Imperial Council of Agricultural Research, New Delhi, Pp. XVI + 291, 118 pls. (1946).

— Annuaire National de l'Agriculture 1947. Horizons de France, Paris, Pp. XVI + 711, tables (1947).

— Bulletins Agronomiques Minist. France d'Outre Mer. (1947).

— The use and misuse of shrubs and trees as fodder. Imperial Agricultural Bureaux: Joint Publication, Nr. 10, 9 s., XXXIV + 231, 70 figs. 28 tables. (1947).

Avdonin, N.: (Apropos of G. T. Gutiev's article: „Reorganize the chain of subtropical scientific institutes".) Soviet Subtropics, Nr. 11/12, (75/76), 14—15 (1940). [Russisch.]

— (Achievements of science- in the service of socialistic agriculture.) Socialističeskoe Seljskoe Hozjajstvo (Socialistic Agriculture) Moscow: Nos, 4/5, 54—58 (1945). [Russisch.]

Beck, F. V.: The field seed industry in the United States. University of Wisconsin Press, Madison, Wisconsin, Pp. XXII + 230, 49 figs. tables (1944).

Boerger, A.: Investigaciones Agronómicas (Agronomic Research). Buenos Aires, Vol. I, Pp. 758; Vol. II, Pp. 1043; Vol. III, Pp. 443, 112 tables, 71 figs. (1943).

— Agronomía. Consejos metodológicos. (Agronomy. Counsels on method.) Casa A. Barreiro y Ramos, S.A. Montevideo, Pp. XIX + 538, 37 figs., 4 tables (1946).

Bottazzi, G. B.: Il centro di conservazione di ceppi agrari. (The centre for the maintenance of agricultural strains.) Genetica Agraria, Roma, 1, 254—275 (1947).

Caldwell, J. S., C. W. Culpepper u. M. C. Hutchins: Further comparative studies of varieties of certain fruits and vegetables for dehydration. Proc. Amer. Soc. Hort. Sci. 46, 375—387 (1945).

Castetter, E. F. u. W. H. Bell: Pima and Papago Indian agriculture. Inter-Americana studies I. University of New Mexico Press, Albuquerque, N.Mex., Pp. XV + 245, 5 figs. (1942).

Coolhaas, C.: Zijn er mogelijkheden voor het planten van andere gewassen dan de thans in Oost-Java gecultiveerde? (Are there possibilities of planting other crops than those now cultivated in East Java?) Bergcultures, 15, 1252—1259 (1941).

Cross, W. E.: Notas sobre el progreso de la agricultura y las industrias agropecuarias de Tucumán durante los últimos sesenta años. (Notes on the progress of agricultural and rural industries in Tucumán during the last 60 years.) Bol. Estac. Exp. Agric. Tucumán, Nr. 36, 75 (1942).

Dahlberg, G. u. I. Johansson: Svenskt Lantbrukslexikon. (Swedish Agricultural Lexicon.) A.—B. Svensk Litteratur, Stockholm, Teil I, Pp. 572; Teil II, Pp. 573—1084 (1941).

Dix, W.: Acker- und Pflanzenbaufragen in der Türkei. Landw. Jb. 93, 1—47 (1943).

Ede, R.: The principles of agriculture. London, Pp. XIII + 272, 23 figs. tables (1945).

Eggebrecht, H.: Vereinfachung der Saatgutuntersuchung. Forschungsdienst 16, 299—301 (1943).

Fergusson, D. et. al.: Agricultural Improvement Council for England and Wales. First Report. Rep. Minist. Agric. Fish., London, Pp. 18 (1943).

Grainger, J.: Garden science. University of London Press. Ltd. (1940).

H....., F.: The p'ant breeder and the soil. Trop. Agriculture, Trin. 21, 181—183 (1944).

Hall, A. D.: Reconstruction and the land. An approach to farming in the national interest. Macmillan & Co., Ltd., London (1941).

Handy, E. S. C.: The Hawaiian Planter-Volume I, his plants, methods and areas of cultivation. Bernice P. Bishop Museum, Honolulu, Hawaii, Bull. 161 (1940).

Harkness, D. A. E.: War and British agriculture. P. S King and Son. Ltd., London (1941).

Haskell, G.: Spatial isolation of seed crops. Nature, Lond. **152**, 591—592 (1943).

Hutchison, C. B. (Editor): California agriculture. University of California Press, California, Pp. VIII + 444 figs. tables. pls. (1946).

Jakuškin, I. V.: (Pflanzenbau.) Selhozgiz. Moskva, Pp. 680, 125 Abb. (1947). [Russisch.]

Kanitkar, N. V.: Dry farming in India. New Delhi, Nr. 15, 352, tables. figs. (1944).

Kellar, H. A.: Living agricultural museums. Agric. Hist. **19**, 186—190 (1945).

Klages, K. H. W.: Ecological crop geography. MacMillan Co., New York, Pp. XVIII + 615, 66 tables, 108 figs. (1942).

Lek, H. A. A. van der, u. E. Krijthe: Bevordering van de wortelvorming van stekken door middel van groeistoffen. (Promotion of root formation of cuttings by means of growth substances.) Meded. Landb-Hoogesch., Wageningen **44**, Nr. 7, 91 (1940).

— u. J. Krijthe: Over groeistoffen en hare toepassing in den tuinbouw, in het bijzonder bij het stekken. (On growth substances and their use in horticulture, in particular in raising cuttings.) Meded. LandbvoorlichtDienst, Wageningen, Nr. 25, 119 (1943).

Leonard, W. H. u. A. G. Clark: Field plot technique. Burgess Publishing Company, Minneapolis, Minn. (1939).

Lott, R. V.: The terminology of fruit maturation and ripening. Proc. Amer. Soc. Hort. Sci. **46**, 166—172 (1945).

Lysenko, T. D.: (About some of the main aims of agricultural science.) Proc. Lenin Acad. Agric. Sci. USSR., Nos 5/6, 3—12 (1942). [Russisch.]

— (Some fundamental problems of agricultural science.) Socialističeskoe Seljskoe Hozjajstvo (Socialist Agriculture), Moscow, Nr. 5, 18—26 (1942). [Russisch.]

MacEvan, J. W. G. u. A. H. Ewen: General agriculture. Thomas Nelson and Sons, Limited, Toronto and London (1939).

Massingham, H. J. (Editor): England and the farmer. B. T. Batsford, Ltd., London (1941).

Morrison, B. Y.: The technique of plant exchange. Agric. Amer. **3**, Nr. 1, 3—6 (1943).

Musijko, A. C.: (The application of supplementary artificial pollination to agricultural crops.) Pan Soviet Lenin Acad. Agric. Sci., Moscow, Pp. 17 (1941). [Russisch.]

Nickerson, D.: Color measurement and its application to the grading of agricultural products. Misc. Publ. U. S. Dep. Agric., Nr. 580, 62 (1946).

Osvald, H.: Spånads- och oljeväxter. (Fibre and oil crops.) Nordisk Rotogravyrs Handböcker för Jordbrukare, Stockholm, Pp. 259. illus. pls. tables. (1944).

Porter, R. H.: Testing the quality of seeds for farm and garden. Res. Bull. Ia Agric. Exp. Sta., Nr. 334, 495—586 (1944).

Rather, H. C.: Field crops. McGraw-Hill Book Company, London and New York, 26 s., Pp. IX + 454 (1942).

Schilletter, J. C. u. H. W. Richey: Textbook of general horticulture. McGraw-Hill Publishing Company Ltd., London (1940).

Skovgaard, K. u. A. Pedersen: Survey of Danish agriculture with a supplement on Danish horticulture. National Danish F. A. O. Committee, Denmark, Pp. 169, 42 tables. photos. (1946).

Stockdale, F. A.: Report by Sir Frank Stockdale, K. C. M. G., C. B. E. Agricultural Adviser to the Secretary of State for the Colonies on a visit to Malaya, Java, Sumatra and Ceylon 1938. Colon. Adv. Coun. Agric. Anim. Health C. A. C. **454**, 108 (1939).

Tallarico, G.: L'Italia, centro mediterraneo di diffusione delle sementi elette. (Italy, the Mediterranean centre for distribution of selected seed.) Ital. Agric. **79**, 455—464 (1942).

Torsell, R.: Biodlingens betydelse i lanthushållningen. (The importance of apiculture in agriculture and horticulture in Sweden.) K. Lantbr. Akad. Tidskr. **83**, 218—239 (1944).

Tothill, J. D. (Editor): Agriculture in Uganda. Oxford University Press. (1940).

Wards, H. C. A.: International Institute of Agriculture. N. Z. J. Agric. **74**, 59—62 (1947).

Whyte, R. O.: Crop production and environment. Faber and Faber Ltd., London, 25 s., Pp. 372, 53 figs. 32 pls. 36 tables. (1946).

Wood, R. C.: A note-book of tropical agriculture. Trinidad, Pp. 147, tables, illus. (1945).

Zimmerman, F. L. u. P. R. Read: Numerical list of the current publications of the United States Department of Agriculture. Compiled by comparison with the originals. Superintendent of Documents, Washington, D. C., Pp. V + 929. (Misc. Publ. Nr. 450.) (1941).

m) Neue Zeitschriften.

Acta Agriculturae Suecana, Stockholm.

Advances in Genetics. Is edited by M. Demerec and is published by the Academic Press Inc., 125 East 23rd St. New York 10, N. Y.

Agri Hortique Genetica, Landskrona.

Agrobiologija (Agrobiology). The well-known Soviet journal Jarovizacija (Vernalization) has apparently now been replaced by a new publication bearing the name Agrobiologija (Agrobiology) (1946). [Russisch.]

Bragantia. Secretaria da Agricultura, Industria e Comércio do Estado de Sao Paulo, Instituto Agronomico, Caixa Postal, 28, Campinas, Estado de Sao Paulo, Brasil.

Canadian Grain Journal. Journal Publishing Company, 548 Grain Exchange Building, Winnipeg.

Chromosoma. Julius Springer, Berlin.

Corn. Corn Industries Research Foundation, 5 East 45th Street, New York 17, N. Y. (1945).

Economic Botany. New York Botanical Garden, New York, N. Y. (1947).

Endeavour. C. H. Waddington on the Epigenotype, J. G. Crowther on Science in the USSR. and Al. L. Bacharach on the Manufacture and Use of Vitamins. (Price 5 s.) (1942).

Experientia, Basel. New journal providing information on current scientific research in many fields. Basel (1946).

Genetica Agraria, Roma. New journal, instituted as a companion to Scientia Genetica. Roma (1946).

Heredity. The journal is to be published three times a year. Annual subscription, 40 s. Publishers: Oliver and Boyd, Ltd., 98 Great Russell Street, London, W.C

Imperial Forestry Bureau. Imperial Forestry Bureau 39, Museum Road Oxford, England.

Indian Farming. Imperial Council of Agricultural Research, India.

Indian Cotton Growing Review. Communications concerning the journal should be addressed to the Secretary, Indian Central Cotton Committee, Nicol Road, Ballard Estate, Bombay.

Journal of the British Grassland Society. Aberystwyth, Wales (1947).

Journal of the Institute of Corn and Agricultural Merchants Limited. The journal is obtainable from the Secretary, The Institute of Corn and Agriculture Merchants Limited 5, Copthall Chambers, London, E.C.2; price per copy 5s.

Pacific Science. University of Hawaii, Honolulu 10, Hawaii.

Palestine Journal of Botany (Jerusalem Series). Dep. of Botany, Hebrew University, Jerusalem, Palestine.

Palestras Agronómicas. (Published by Estação Agronómica Nacional, Lisbon) (1939).

Portugaliae Acta Biologica, Lisboa (A). A new journal, published under the auspices of the Institute of Botany of the Faculty of Sciences, Lisbon.

Publications of the Malt Research Institute, Madison, Wisconsin. A new journal.

Revista de Investigaciones Agrícolas (published by the Ministry of Agriculture of Argentina) (1947).

Scientia Genetica. Published by Rosenberg and Sellier, Via Andrea Doria, Torino, Italy.

The Chinese Journal of Scientific Agriculture. Published by the Ministry of Agriculture and Forestry at Chunking.

The Indian Journal of Genetics and Plant Breeding. The Indian Society of Genetics and Plant Breeding, Imperial Agricultural Research Institute, New Delhi (1941).

The Queensland Journal of Agricultural Science. New Journal published by the Queensland Department of Agriculture and Stock, Brisbane (1944).

n) Verschiedenes.

Akenhead, D.: The information service of the Imperial Bureau of Horticulture and Plantation Crops. Proc. Brit. Soc. Int. Bibl. 1, 17—24 (1939) 1940.

Anonymus: Science and UNESCO. London, Pp. 64 (ohne Jahr).

— (Scientific records of the Moscow State University, Jubilee Series (1755—1940), 54, Biology.) Moscow, Pp. 371 (1940). [Russisch.]

— Proceedings of the Eighth American Scientific Congress held in Washington, May 10—18, 1940. Organization, activities, resolutions, delegations. Proc. 8th. Amer. Sci. Congr. Washington (1940).

— Latin American Scientific Societies and Institutions. Pan American Sanitary Bureau, Washington, D.C., Nr. 141, 146 (1942). [Mimeographed.]

— Catalogue of Lewis's Medical, Scientific and Technical Lending Library, Part I and Part II. London (1944), Pp. 928 (1943).

— (General Assembly of the Academy of Sciences of the USSR. held on 25th to 30th September of 1943.) Academy of Sciences, USSR., Moskva-Leningrad, Pp. 246 (1944). [Russisch.]

— The place of science in industry. Advance Sci. 3, 106—156 (1945).

— Emulsion technology, theoretical and applied. Chemical Publishing Co., Inc., N.Y., 2nd Edition, Pp. XIII+360. figs. tables. (1946).

Anonymus: Supplement to the Catalogue of Lewis's Medical Scientific and Technical Lending Library, Supplement 1944—1946. H. K. Lewis and Co., Ltd., London, Price: to subscribers 2s. 6d., to non-subscribers 5s, Pp. IV+176 (1947).

— The world of learning. London, Pp. 520 (1947).

Arroyo, R.: Studies on rum. Res. Bull. P. R. Agric. Exp. Sta., Pp. IX+272. 38 figs. 47 tables. (1945).

Ashby, E.: Scientist in Russia. Penguin Books, Harmondsworth, Middlesex and New York, 1s. 0d. Pp. 252, 4 figs. (1947).

Bajkov, A. A.: (The Academy of Sciences of the USSR. during the last 25 years.) Jubilee Session of the Academy of Sciences of the USSR. Moscow-Leningrad, Pp. 28—53 (1943). [Russisch.]

Bates, R. S.: Scientific societies in the United States. John Wiley and Sons, Inc., N.Y., Pp. VII+246. tables. (1945).

Browne, C. A.: Thomas Jefferson and the scientific trends of his time. Waltham, Mass., Pp. 63. illus. maps. (1943).

Cusset, F.: English-French and French-English technical dictionary. Brooklyn, N.Y., Pp. 590 (1946).

Fersman, A.: Science in the USSR. Advanc. Sci. 3, 62—77 (1944).

Hartmann, M.: Allgemeine Biologie. III. Aufl. Fischer-Jena, Pp. 869 (1947).

Hill, A. V.: A report to the government of India on scientific research in India. The Royal Society, London, Pp. 55 (1945).

Hiss, P. H: A selective guide to the English literature on the Netherlands West Indies with a supplement on British Guiana. New York City, Booklet, Nr. 9, XIII+129 (1943).

Honig, P. u. F. Verdoorn (Editors): Science and scientists in the Netherlands Indies. Board for the Netherlands Indies, Surinam and Curaçao, New York City 102, Special. Suppl. Pp. XXII+491. 134 figs. (1945).

Klaauw, C. J. van der (Editor): Bibliographia Biotheoretica 1930—1934. Vol. II: 14 guilders, Pp. 310, 1935—1939; Vol. III: 21 guilders, Pp. 371, 1944; Vol. IV: Pt 1,5 guilders, Pp. 82. Published by E. J. Brill, Leiden (1930—1944).

Lawrence, A. S. C.: The scientific photographer. Cambridge University Press, London (1941).

Magnussen, J., O. Madsen u. H. Vinterberg: The English-Danish and Danish-English Dictionaries. London and Copenhagen, Pp. 362 (1943).

Mayer, A. W.: Chemical-technical dictionary. (German-English-French-Russian.) Chemical Publ. Co., Inc., Brooklyn, N.Y., Pp. 872 (1942).

Needham, J. u. J. S. Davies (Editors): Science in Soviet Russia. Watts and Co., London (1942).

Ogden, C. K.: Basic for science. London, Pp. 314. 10 figs. (Psyche miniatures-General Series Nr. 95) (1942).

Orbelli, L. A.: (Development of biological sciences in the USSR. during the last 25 years.) Jubilee Session of the Academy of Sciences of the USSR. Moscow-Leningrad, Pp. 216—227 (1943). [Russisch.]

Savory, T. H.: Latin and Greek for biologists. London, Pp. 42 (1946).

Snyder, E. E.: Biology in the making. McGraw-Hill Book Company, Inc., New York and London (1940).

Tory, H. M. (Editor): A history of science in Canada. The Ryerson Press, Toronto (1939).

Tweney, C. F. u. L. E. C. Hughes (Editors): Chambers's technical dictionary. Comprising terms used in pure and applied science: medicine: the chief manufacturing industries: engineering: construction: the mechanic trades. With definitions by recognised authorities. W. and R. Chambers, Ltd. London and Edinburgh (1940).

B. Spezieller Teil.

a) Getreide.

Åkerman, Å.: Några erfarenheter rörande höstsädens övervintring samt om vinterhärdigheten hos olika höstvetesorter. (Some findings regarding the overwintering of autumn cereals and the winter-hardiness of different autumn wheat varieties.) K. LantbrAkad. Tidskr. 84, 192—215 (1945).

—, J. Jakobsson u. J. E. Lindberg: Undersökningar av kvaliteten hos 1941 års brödsädesskörd. (Investigations of the quality of the bread cereal harvest of 1941.) Sverig. Utsädesfören. Tidskr. 52, 306—342 (1942).

—, —, — Undersökningar av kvaliteten hos 1942 års brödsädesskörd. (Investigations of the quality of the bread cereals harvest of 1942.) Sverig. Utsädesfören. Tidskr. 53, 239—274 (1943).

Anderson, J. A. u. T. R. Aitken: Eighteenth Annual Report of the Board of Grain Commissioners Grain Research Laboratory, Winnipeg, Manitoba 1944, Pp. 69 (1945).

Anonymus: Bibliography of baking quality tests. (Supplement). Imperial Bureau of Plant Breeding and Genetics, Cambridge 1s. 6d. Pp. 32 (1939).

— Cereal laboratory methods with reference tables. Amer. Ass. Cereal Chem., Lincoln, Nebraska (1941).

— The Scottish Plant Breeding Station, Craigs House, Corstorphine, Edinburgh. Trans. Highl. Agric. Soc. 53, 103—106 (1941).

— En arbetets fest på Weibullsholm. (A research celebration at Weibullsholm.) Lantmannen 25, 601—602 (1941).

— Manitoba field crops: their culture and hazards. Winnipeg, Pp. 47 (1942).

— Saskatchewan cereal variety recommendations for 1943. Saskatch. Cereal Var. Comm., Sask. December 14—15, Pp. 7 (1942).

— Qualitätszüchtung bei Weizen und Roggen. Forschungsdienst 15, 106 (1943).

— Probleme der Brotnahrung im Kriege. Forschungsdienst 16, 145—146 (1943).

— Zulassung neuer Sorten deutscher Kulturpflanzen. Mitt. Landw. 58, 893 (1943).

— Improved strains of important cereal, pulse and oilseed crops evolved by the Department of Agriculture, Bombay Province. Leafl. Dep., Agric., Bombay Province, Nr. 1, 10 (1944).

— Descripción de las principales variedades de trigo, avena, cebada, centeno, lino y maíz, cultivadas en la República Argentina. (Description of the principal varieties of wheat, oats, barley, rye, flax and maize, grown in the Argentine Republic.) Publ. Minist. Agric. Republica Argentina, Nr. 20, 262 (1945).

— Plant industry. 1. New high-harvest crops. Agriculture, Moscow, Nr. 4, 3—5 (1945).

— Field husbandry. Agriculture, Moscow, Nr. 10, 7—9 (1945). [Mimeographed.]

— Pedigree seed. Cereal varieties available for distribution. J. Agric. W. Aust. 22, 232—236 (1945).

— Varieties of wheat, oats and barley recommended for 1945 sowing. Agric. Gaz. N.S.W. 56, 3—7 (1945).

— Russia makes magic wheat, perennial, resists smut and rust. Sth. Seedsman 8, 32 (1945).

Anonymus: Mallee Research Station experimental work and results. J. Dep. Agric. Vict. 44, 409—416, 434—440 (1946).

— Varieties of cereals for spring sowing. Fmrs' Leafl. Nat. Inst. Agric. Bot., Cambridge (Revised Ed.), Nr. 2, 4 (1946).

Armstrong, S. F.: Cereal varieties for autumn sowing. J. Minist. Agric., London 52, 295—297 (1945).

— Cereal varieties for spring sowing. J. Minist. Agric. 52, 481—482 (1946).

Atkins, I. M. u. R. G. Dahms: Reaction of small-grain varieties to green bug attack. Tech. Bull. U. S. Dep. Agric., Nr. 901, 30 (1945).

— u. P. C. Mangelsdorf: The isolation of isogenic lines as a means of measuring the effects of awns and other characters in small grains. J. Amer. Soc. Agron 34, 667—668 (1942).

Atkinson, R. E.: Mosaic of wheat in the Carolinas. Plant Dis. Reporter 29, 86 (1945). [Mimeographed.]

Ausemus, E. R.: Breeding for disease resistance in wheat, oats, barley and flax. Bot. Rev. 9, 207—260 (1943).

Bayles, B. B.: Technical cooperation in small grain improvement. J. Amer. Soc. Agron., Nr. 39, 207—213 (1947).

Bell, G. D. H.: Cereal breeding and research at the Cambridge University plant Breeding Station. Proc. 7th Int. Genet. Congr. Edinburgh, 23—30 August, Pp. 62 (1939) 1941. (Abstr.)

Bertelli, J. C.: Control de las enfermedades de los cereales y del lino. (Control of cereal and flax diseases.) Rev. Asoc. Ingenieros Agron., Montevideo 18, Nr. 73, 9—32 (1946).

Boeuf: L'organisation de la production et du commerce des grains en Argentine. C. R. Acad. Agric. Fr. 26, 374—383 (1940).

Boewe, G. H.: Disease of small grain crops in Illinois. Circ. Ill. Nat. Hist. Surv. Nr. 35, 130 (1939).

Bonnett, O. T. u. W. M. Bever: Head-hill method of planting head selections of small grains. J. Amer. Soc. Agron. 39, 442—445 (1947).

Brandl, M.: Zur Sortenbeschreibung unserer Getreidearten. Forschungsdienst 10, 606—607 (1940).

Breakwell, E. J. u. E. M. Hutton: Cereal breeding and variety trials, at Roseworthy College, 1937—1938. J. Dep. Agric., S. Aust. 42, 632—645 (1939).

— u. D. H. S. Mellor: Plant breeding and some experimental work at Roseworthy College, 1939—1940. J. Dep. Agric. S. Aust. 44, 252—261 (1940).

Brooks, F. T.: Biologic specialization in the cereal rusts. Ann. Appl. Biol. 31, 362—366 (1944).

Calvo, M. Q. u. M. F. Segura: Ensayo para la obtención de quenopodio en Costa Rica. (Attempt to produce chenopodium in Costa Rica.) Rev. Méd. (San José) 5, 185—187 (1942).

Cárdenas, M.: Descripción preliminar de las variedades de Chenopodium quinoa de Bolivia. (Preliminary description of the Bolivian varieties of Chenopodium quinoa.) Rev. Agric. Bolivia 2, Nr. 2, 13—26 (1944).

Cevallos Tovar, W.: La Quinua-el centeno. (Quinoa-rye). Publ. Univ. Autón. Cochabamba, Cuad. Agric., Nr. 6/7, 32 (1945).

Chang, S. C.: Lenght of dormancy in cereal crops and its relationship to afterharvest sprouting. J. Amer. Soc. Agron. **35**, 482—490 (1943).

Cherewick, W. J.: A method of establishing rust epidemics in experimental plots. Sci. Agric. **26**, 548—551 (1946).

Chester, K. S.: Methods of appraising intensity and detructiveness of cereal rusts with particular reference to Russian work on wheat leaf rust. Plant Dis. Reporter Suppl., Nr. 146, 99—121 (1944). [Mimeographed.]

Christie, G. I.: Report of the President for the year ending March 31st, 1942. 76th Rep. Ont. Agric. Coll. Exp. Fm 1941, Pp. 69 (1942).

Ciferri, R.: Osservazioni ecologico-agrarie e sistematiche su piante coltivate in Etiopia. (Ecologo-agrarian and systematic observations on plants cultivated in Etiopia.) Atti Ist. Bot. e Lab. Crittogamico, Univ. Pavia **2**, Ser. 5, 121—232 (1944).

—, u. I. Baldrati: I cereali dell'Africa Italiana. II. Il ,,Teff'' (Eragrostis Teff) cereale da panificazione dell'Africa Orientale Italiana montana. (The cereals of Italian Africa. II. Teff [E. Teff.], a bread cereal from the mountains of Italian East Africa.) Bibl. Agrar. Colon., Firenze, Pp. 106 (1939).

—, — La cerealicoltura in Africa orientale. VI.—Il ,,Teff'' (Eragrostis Teff). (Cereal cultivation in East Africa. VI.—Teff [E. Teff].) Ital. Agric. **77**, 170—176 (1940).

Cochran, W. C.: The estimation of the yields of cereal experiments by sampling for the ratio of grain to total produce. J. Agric. Sci. **30**, 262—275 (1940).

Crawford, D. C., P. J. Hamersma u. B. W. Marloth: The chemical composition of some South African cereals and their milling products. Sci. Bull. Nr. 20 Dep. Agric. For S. Afr. Chem. Ser., Nr. 171, 100 (1942).

Dillon Weston, W. A. R. u. R. E. Taylor: Observations on ergot in cereal crops. J. Agric. Sci. **32**, 457—464 (1942).

Dorsey, E.: Chromosome doubling in the cereals. J. Hered. **30**, 393—395 (1939).

Drahorad, F.: Gebirgspflanzenzüchtung mit besonderer Berücksichtigung des Getreidebaues. Z. Pflanzenz. **24**, 352—362 (1941).

— Züchtung alpiner Getreidearten. Forschungsdienst, Sonderh. 16, 364 (1942).

— u. L. Dimitz: Zur Verbreitung der Getreidesorten in der Ostmark, unter Berücksichtigung der geographisch-ökologischen Verhältnisse. Züchter **12**, 9—16 (1940).

Eichmeyer: Norwegische Getreidesorten. Mitt. Landw. **58**, 338 (1943).

Ellerton, S.: What is vernalisation? Fmr's Wkly, Lond. **24**, 33 (1946).

Engelke, H.: Weizenertrag und -güte als Züchtungs- und Düngungsfrage. Züchter **13**, 49—59 (1941).

Ericsson, G. u. M. Genchel: Härkomstförsök med korn och havre utförda vid Statens Försöksgård Offer 1935—1942. (Provenance experiments with barley and oats, carried out at the Offer State Experiment Farm, 1935—42.) Medd. Lantbr. Jordbruksförsöksanstalten, Nr. 12, 48 (1945).

Flaksberger, K. A. (Editor): (Key to the cereals proper-wheat, rye, barley, oats.) All-Union Lenin Acad. Agric. Sci. Inst. Pl. Ind., Moscow and Leningrad (1939). [Russisch.]

—, F. H. Bahteev, A. I. Mordvinkina, E. S. Jakuševskij, N. P. Ivanov u. G. M. Popova: (Material for formulating directions for provinces and republics in regard to approval of grain.) Vestnik Socialističeskogo Rastenievodstva (Soviet Plant Industry Record), Nr. 2, 159—167 (1940). [Russisch.]

Forlani, R.: Sulle spighe ramificate o ,,del miracolo''. (Branched or ,,miracle'' ears.) Genetica Agraria, Roma **1**, 78—94 (1946).

— Allogamia in alcuni cereali. (Cross-fertilization in certain cereals.) Genetica Agraria, Roma **1**, 130—146 (1947).

Freisleben, R.: Die Gersten und Weizen der deutschen Hindukusch-Expedition 1935. Angew. Bot. **22**, 105 bis 132 (1940).

Fröier, K.: Stråsädessorter i våra skandinaviska grannländer. (Varieties of cereals in the countries of our Scandinavian neighbours.) Årsb. Jordbruksforskning, Stockholm, Pp. 151—163 (1945).

— Växtförädling av stråsäd i Tyksland och England under andra världskriget. (Breeding of straw cereals in Germany and England during the Second World War.) Årsb. Jordbruksforskning, Stockholm, Pp. 112 bis 121 (1946).

— u. Å. Gustafsson: The influence of seed size and hulls on X-ray susceptibility in cereals. Hereditas, Lund **30**, 583—589 (1944).

Fuchs, W. H.: Die Grundlagen der Züchtung auswuchsfester Getreidesorten. Forschungsdienst, Sonderh. 16, 339 (1942).

Gallagher, P. H. u. T. Walsh: The susceptibility of cereal varieties to manganese deficiency. J. Agric. Sci. **33**, 197—203 (1943).

Garrett, S. D.: The take-all disease of cereals. Tech. Commun. Bur. Soil Sci., Harpenden, Nr. 41, 40 (1942).

Geoffroy, R.: Le blé, la farine, le pain. Dunod, Paris (1939).

Glick, D.: The choline content of pure varieties of wheat, oats, barley, flax, soybeans, and milled fractions of wheat. Cereal Chem. **22**, 95—101 (1945).

Gorlenko, M. V.: (Twenty-five year of studying diseases of bread cereals in the USSR [1917—1942]). Botaničeskij Žurnal (J. Bot. URSS) **31**, 3—17 (1946). [Russisch.]

Graiff, G. L.: Contributo alla cerealicoltura libica. (Contribution to cereal culture in Libya.) Agricoltura Libica. **10**, 1—24 (1941).

— Contributo allo studio della ,,Primaverilizzazione'' dei cereali nei paesi caldoaridi con speciale riguardo al grano. (A contribution to the study of the ,,vernalization'' of cereals in countries with a hot, dry climate, with special reference to the grain.) Ann. Cent. Sper. Agrar. Zootec. Libia 1941 **4**, 9—34 (1942).

Groenewolt, J. K.: Opbrengst en kwaliteit. (Yield and quality.) Landbouwk. Tijdschr., Wageningen **51**, 843 bis 862 (1939).

Guyot, A. L., M. Massenot u. A. Saccas: Etudes expérimentales sur les rouilles des graminées et des céréales en 1944. Ann. Ec. Agric. Grignon 1945—1946 **5**, Sér. 3, 33—81 (1946).

—, —, — Considérations morphologiques et biologiques sur l'espèce Puccinia graminis Pers. sensu lato. Ann. Ec. Agric. Grignon 1945—1946 **5**, Sér. 3, 82—146 (1946).

—, —, — Etudes expérimentales sur les rouilles des graminées et des céréales em 1945. Ann. Ec. Agric. Grignon 1945—1946 **5**, Sér. 3, 213—266 (1946).

Haan, H. de: Rassenstatistiek. Overzicht van de verschuivingen in het rassensortiment. (The distribution of varieties. Changes in the varietal composition.) Landbouwk, Tijdschr., Wageningen **51**, 809—843 (1939).

Harrington, J. B.: The effect of having rows different distances apart in rod row plot tests of wheat, oats and barley. Sci. Agric. **21**, 589—606 (1941).
— Cereal variety zone co-ordination in the Prairie Provinces. Sci. Agric. **25**, 279—284 (1945).
— u. J. Whitehouse: Cereal variety results at Saskatoon, 1937—1943. Circ. Coll. Agric. Univ. Saskatchewan, Nr. 539, 9 (1943). [Mimeographed.]
Haugum, O.: Våre kornsorter. En beskrivelse av de alminneligst dyrkede sorter. (Our cereals. A description of the most commonly grown varieties.) C. Dahls Bok- and Kunsttrykkeri A/S, Oslo, Pp. 29, 3 tables. (1940).
Hunter, H.: Winter hardiness in cereals. J. R. Agric. Soc. **101**, Nr. 2, 28—36 (1941).
Hunziker, A. T.: Las especies alimenticias de Amaranthus y Chenopodium cultivadas por los Indios de América. (The edible species of Amaranthus and Chenopodium cultivated by the American Indians.) Rev. Argent. Agron. B. Aires **10**, 297—354 (1943).
Huskins, C. L.: Symposium on ,,Theoretical and practical aspects of polyploidy in crop plants". Polyploidy and mutations (with discussion by H. E. Warmke). Amer. Nat. **75**, 329—346 (1941).

Johansson, E.: Studier och försök rörande de på gräs och sädesslag levande tripsarnas biologi och skadegörelse. II. Tripsarnas frekvens och spridning i jämförelse med andra sugande insekters samt deras fröskadegörande betydelse. (Studies and investigations on the biology and damage to the thrips living on grass and cereals. II. Frequency and distribution of the thrips in comparison with other suctorial insects and the significance of their injury to seed.) Med. Växtskyddanst., Stockholm, Nr. 46, 59 (1946).
Johnson, I. J.: Recent advances in applied plant genetics. Proc. Iowa Acad. Sci. **48**, 59—63 (1941).
Johnson, L. P. V., G. A. Young u. J. B. Marschall: A note on the production of vitamin C by sprouting seeds. Sci. Agric. **25**, 499—503 (1945).
Johnson, T., B. Peturson u. W. J. Cherewick: Physiologic races of cereal rust in Canada in 1945. 25th Rep. Canad. Pl. Dis. Surv. 1945, Pp. 18—19 (1946). [Mimeographed.]

Kagawa, F.: Alteration of characters in crop plants induced by X-ray irradiation. Jap. J. Bot. **10**, 35—41 (1939).
Kar, B. K.: Vernalization of crops cultivated in India. Nature, Lond. **157**, 811—812 (1946).
Kent-Jones, D. W.: Modern cereal chemistry. Northern Publishing Co., Ltd., Liverpool (1939).
— u. A. J. Amos: Modern cereal chemistry. Northern Publishing Co., Ltd. 4th Ed. 50s. Pp. vii + 651. 78 figs. 195 tables. (1947).
Kihara, H.: Anwendung der Genomanalyse für die Systematik von Triticum und Aegilops. Jap. J. Genet. **16** 309—320 (1940).
— Verwandtschaft der Aegilops-Arten im Lichte der Genomanalyse. Ein Überblick. Züchter **12**, 49—62 (1940).
Koblet, R.: Ergebnisse und Ziele der getreidebaulichen Versuchsarbeit. Schweiz. landw. Mh., Nr. 3, 57—87 (1944).
Kondo, N.: (Chromosome doubling in Secale, Haynaldia and Aegilops by colchicine treatment). Jap. J. Genet. **17**, 46—54 (1941).
Krotov, A. S.: (Overcoming sterility of a hybrid by prolonging its life). Agrobiologija (Agrobiology), Nr. 2, 123—125 (1946). [Russisch.]
Kuznecov, A. V.: (Local varieties of winter crops in the eastern regions of Kazakhstan SSR.). Selekcija i Semenovodstvo (Breeding and Seed Growing), Nr. 8/9, 19—20 (1940). [Russisch.]

Lakon, G.: Topographischer Nachweis der Keimfähigkeit der Getreidefrüchte durch Tetrazoliumsalze. Ber. dtsch. bot. Ges. **60**, 299 (1942).
Lamas, P. J. A.: Como se juzga el valor agrícola de una variedad para grano. (How the agricultural value of a variety od a grain crop is judged.) ,,Granos" Semilla Selecta, B. Aires **6**, Nr. 10/12, 34—35 (1942).
Larmour, W. T.: Never ending search continues for new and better varieties of cereal grains. Dominion Cereals Division has many projects underway. Canad. Grain J. **1**, 13 (1946).
Lenglen: La fécondation artificielle des céréales: le procédé Hooibrenck. (Artificial fertilization of cereals. Hooibrenck's method.) C. R. Acad. Agric. Fr. **29**, 326—333 (1943).
Lockwood, J. F.: Flour milling. London 1945, Pp. 511, 224 figs. tables. (1946).

McBeth, C. W.: Tests on the susceptibility and resistance of several southern grasses to the root-knot nematode Heterodera marioni. Proc. Helminthol. Soc. Wash. **12**, 41—44 (1945).
McFadden, E. S.: New developments in small grain breeding for grain and forage. Pap. Tex. Agric. Work Assoc., Pp. 126—128 (1942).
— u. C. H. McDowell: New rust-resistant small grains for south and central Texas. Progr. Rep. Tex. Agric. Exp. Sta., Nr. 850, 3 (1943). [Mimeographed.]
Mayr, E.: Beiträge zur Sortenfrage im Bergbauernbetrieb der Alpengaue. Züchter **14**, 249—252 (1942).
Meurman, H.: Kevätviljakokeiden tulokset maatalouskoelaitoksen Puutarhaosastolla vv. 1927—1938. (The results of spring grain trials at the Horticultural Department of the Agricultural Research stations, years 1927—1938.) Valt. Maatalousk. Tiedon, Nr. 165, 30 (1939.)
Miller, W. B.: Recommended cereal varieties. J. Dep. Agric. Vict. **45**, 63—66 (1947).
Moore, W. C.: Cereal diseases. Nature, Lond. **154**, 139—141 (1944).
Moormann, B.: Untersuchungen über Keimruhe bei Hafer und Gerste. Kühn-Arch. **56**, 41 (1942).
Mühle, E.: Die Rostpilze der wichtigsten zur Samengewinnung angebauten Futtergräser. Phytopath. Z. **14**, 83—101 (1942).

Naumov, N. A., E. E. Gešele u. A. A. Šitikova-Rusakova: (The rusts of cereals in the USSR. A monographical study). Seljhozgiz, Moskva-Leningrad (1939). [Russisch.]
Newman, L. H.: The best varieties of grain. Fmrs' Bull. 96, Dep. Agric. Canada Publ. **701**, 13 (1940).

Oort, A. J. P.: De vatbaarheid voor stuifbrand van in Nederland verbouwde of beproefde rassen van tarwe en gerst. (The susceptibility to loose smut of varieties of wheat and barley cultivated or tested in the Netherlands.) Meded. LandbHoogesch., Wageningen **44**, 54 (1940).

Packard, C. M.: Insect enemies of our cereal crops. Rep. Smithson. Instn. 1942, Pp. 323—338 (1943).
Pelshenke, P.: Ziele, Methoden und Erfolge der Qualitätszüchtung bei Weizen und Roggen. Z. Pflanzenz. **25**, 343—361 (1943).
— Zwölf Jahre Schrotgärmethode. Kühn-Archiv **60**, 265—287 (1943/1944).
Pisarev, V. E.: (Plant breeding for high yields). Trudy Zonaljnogo Inst. Zernovogo Hozjajstva Rajonov Nečernozemnoj Polosy (Trans. Zonal Inst. Grain Husbandry Non-Black-Soil-Districts), Nr. 10, 3—34 (1941). [Russisch.]
— u. N. M. Vinogradova: Trigeneric hybrids Elymus x wheat x rye. C. R. (Doklady) Acad. Sci. USSR. **49**, 218—219 (1945).

Pizarro, C. M.: Indice bibliográfico de las Gramineas Chilenas. (Bibliographical index to the Chilean Gramineae.) Bol. Téc. Minist. Agric. Dep. Genet. Fitotec., Santiago, Nr. 2, 88 (1941).
Plotnikov, I. G.: (Vegetative hybridization of cereals and its importance in breeding work). Socialistic Grain Farming, Saratov, Nr. 6, 69—88 (1940). [Russisch.]
Pohjakallio, O.: Frågan om växternas motståndskraft mot torrperioder i Finland. (The question of the resistance of plants to drought periods in Finland.) Nord. JordbrForskn. Nr. 5/6, 206—226 (1945).
Puhaljskij, A. V.: (Improvement of self-pollinated varieties). Jarovizacija, Nr. 5 (32), 48—56 (1940). [Russisch.]

Roemer, Th.: Mehr und besseres Brotgetreide. Pflanzenbauliche und pflanzenzüchterische Fragen unserer Zeit. Forschungsdienst 16, 115—123 (1943).
Rusakov, L. F.: (Breeding cereals for disease resistance). Selekcija i Semenovodstvo (Breeding and Seed Growing), Nr. 1/2, 48—61 (1946). [Russisch.]

Schulerud, A.: Das Roggenmehl. Moritz Schäfer, Leipzig (1939).
Simmonds, P. M.: A review of the investigations conducted in western Canada on root rots of cereals. Sci. Agric. 19, 565—582 (1939).
Smith, D. C.: Intergeneric hybridization of cereals and other grasses. J. Agric. Res. 64, 33—47 (1942).
Smith, L.: Relation of polyploidy to heat and X-ray effects in the cereals. J. Hered. 34, 131—134 (1943).
— A comparison of the effects of heat and X-rays on dormant seeds of cereals, with special reference to polyploidy. J. Agric. Res. 73, 137—158 (1946).
Spangenberg, J.: El trigo „Klein 157" en el Uruguay. (The wheat Klein 157 in Uruguay.) Rev. Fac. Agron. Univ. Montevideo, Nr. 32, 73—102 (1943).
Sundelin, G.: Sortvalet för stråsädesodlingen. (The work of selecting varieties of cereals for cereal cultivation.) Årsb. Jordbruksforskning, Stockholm, Pp. 124—144 (1945).
— u. S. Eliasson: Den lokala sortförsöksverksamheten. I. Sammanställningar av resultaten av sortförsöken med höstsäd under åren 1929—1938. (Local variety trials. I. Survey of the results of autumn sown grain variety trials in 1929—1938.) Medd. Lantbrukshögskolan Jordbruks, Nr. 3, 131 (1940).
—, — Den lokala sortförsöksverksamheten. II. Sammanställningar av resultaten av sortförsöken med vårstråsäd under åren 1929—1939. (Local variety trias. II. Survey of the results of spring sown grain variety trials in the years 1929—1939.) Medd. Lantbrukshögskolan Jordbruks, Nr. 5, 216 (1941).
Suneson, C. A.: Frost injury to cereals in the heading stage. J. Amer. Soc. Agron 33, 829—834 (1941).

Thomas, B.: Glasigkeit und innere Kornausbildung. D. Mühlenlab. Leipzig 11, 57—64 (1941).
Thomas, I., A. J. Millington u. H. G. Cariss: The utilisation of cereals in Western Australia. J. Agric. W. Aust. 21, 206—226 (1944).
Tschermak von Seysenegg, E.: Weitere Beobachtungen über die hybridogene Pseudoparthenogenesis. Proc. 7th Int. Genet. Congr. Edinburgh, 23—30 August, Pp. 299—301, 1939 (1941). (Abstr.)

Vallega, J.: Aspectos de la lucha contra las royas. (Aspects of the struggle against the rusts.) Rev. Fac. Agron. Univ. Montevideo, Nr. 34, 9—35 (1943).
Vesikivi, A.: Försöksresultat från Finska Mosskulturföreningens försöksstationer år 1940. (Experimental results from the experimental stations of the Finnish Association for the cultivation of Bogland, 1940.) Finska MossFören. Årsb. 45, 47—87 (1941).

Volkart, A.: Getreidezucht. Schweiz. landw. Mh., Nr. 5, 144—147 (1939).
Wahlen, E. T.: Wandlungen und Ziele des schweizerischen Pflanzenbaues. Ber. Schweiz. Bot. Ges. 53 A, 21—43 (1943).
Wahlen, F. T.: Bericht über die Tätigkeit der Eidg. Landwirtschaftlichen Versuchsanstalt Zürich-Oerlikon für die Jahre 1938—1942. Landw. Jb. Schweiz 58, 317—416 (1944).
Wallin, J. R.: Parasitism of Xanthomonas translucens. (J. J. and R.) Dowson on grasses and cereals. Iowa St. Coll. J. Sci. 20, 171—193 (1946).
Warren, C. H.: Corn country. B. T. Batsford, Ltd. London (1940).
Waterhouse, W. L.: Some aspects of plant pathology. Rep. Aust. Ass. Adv. Sci. 24, 234—259 (1939).
Whitcomb, W. O.: The grain inspection laboratory. Twenty-five years-service to Montana. Bull. Mont. Agric. Exp. Sta., Nr. 396, 19 (1941).
Woodward, R. W.: How new varieties of small grains are produced. Fm Home Sci., Utah 6, 6—7, 16 (1945).

Young, L. W.: Obituary-Edward Franklin Gaines. Science 100, 241—242 (1944).

Ziegenbein, G.: Arbeitsweise und Erfolge in der Hafer- und Gerstenzüchtung an der Pflanzenzuchtstation der Universität Halle. Kühn-Archiv 60, 440—454 (1943 bis 1944).

1. Weizen.

Adams, W. E.: Inheritance of resistance to leaf rust in common wheat. J. Amer. Soc. Agron. 31, 35—40 (1939).
Afanasjeva, A. S.: (Comparative investigation of microsporogenesis in wheat [normal and X-rayed]). Bull. Acad. Sci. URSS., Sér. Biol., Pp. 224—243 (1941). [Russisch.]
Aitken, T. R. u. M. H. Fisher: Mixing tolerances of varieties of hard red spring wheat. Cereal Chem. 22, 392—406 (1945).
Ajroldi, P.: Nuovo contributo allo studio biologico delle „Tilletia". (A new contribution to the biological study of Tilletia.) Riv. Patol. Veg. 30, 149—157 (1940).
Åkerberg, E.: Resultat från 25 års försök med höstvete vid Sveriges Utsädesförenings Västernorrlandsfilial. (The results of 25 years of experiments with autumn wheat at the Västernorrland Branch Station of the Svedis Seed Association.) Sverig. Utsädesfören Tidskr. 53, 32—47 (1943).
Åkerman, Å.: Svalöfs Skandiahöstvete II. (01090 b$_2$). (Svalöf's winter wheat Skandia II. [01090 b$_2$].) Sverig. Utsädesfören. Tidskr. 49, 157—160 (1939).
— Spring-wheat breeding in Sweden. Proc. 7th Int. Genet. Congr. Edinburgh 23—30 August, Pp. 46—47, 1939 (1941). (Abstr.)
— Försök till stegrande av vårvetets avkastning. I. Korsningar med Brunt Schlanstedtervårvete jämte beskrivning av Svalöfs Progressvårvete. (Experiments to increase the yield from spring wheat. I. Crosses with Brunt Schlanstedter [Brown Schlanstedter] spring wheat with a description of Svalöfs Progress spring wheat.) Sverig. Utsädesfören. Tidskr. 53, 51—66 (1943).
— Svalöfs Skandiavete III. Nytt Skandiavete med bättre vinterhärdighet än hos Skandia II. (Svalöfs Skandia wheat III. A new Skandia wheat with better winterhardiness than Skandia II.) Sverig. Utsädesfören. Tidskr. 54, 63—65 (1944).
— u. K. Fröier: Vårveteodling och vårvetesorter. (Spring wheat growing and spring wheat varieties.) Lantmannen 25, 308—310 (1941).

Åkerman, Å. u. K. Fröier: Varveteödlingen i Sverige-dess utveckling och framtidsutsikter. (Spring wheat cultivation in Sweden — its development and future prospects.) Sverig. Utsädesfören Tidskr. **53**, 4—12 (1943).

—, J. E. Lindberg u. S. Augustin: Undersökningar av kvaliteten hos 1945 års brödsädesskörd. (Investigations on the quality of the 1945 bread cereal harvest.) Sverig. Utsädesfören. Tidskr. **56**, 495—529 (1946).

—, — u. J. Jakobsson: Undersökningar av kvaliteten hos 1943 års brödsädesskörd. (Investigations on the quality of the 1943 bread cereal harvest.) Sverig. Utsädesfören. Tidskr. **54**, 185—220 (1944).

—, —, — Undersökningar av kvaliteten hos 1944 års brödsädesskörd. (Investigations on the quality of the 1944 cereal harvest.) Sverig. Utsädesfören. Tidskr. **55**, 343—375 (1945).

Alberts, H. W.: Wheat in Peru. Agric. Amer. **3**, 233 bis 235 (1943).

Aleev, N. P.: (Die Transplantation der Weizenembryonen an die vegetativen Organe der Pflanze und ihr Einfluß auf die Entwicklung der Weizen). C. R. (Doklady) Acad. Sci. URSS. **58**, 295 (1947). [Russisch.]

Aleksandrov, V. G. u. O. G. Aleksandrova: (The early stages of development of the wheat endosperm and embryo). J. Bot. URSS. **24**, 383—396 (1939). [Russisch.]

—, — (The question as to wether the polar nuclei of the embryo sacs have their own protoplasm.) Sovetskaja Botanika (Soviet Botany), Nr. 1, 19—33 (1944).

Allen, G.: Grain growers visit Brandon Experimental Farm; see new Redman wheat development. Canadian Grain J. **2**, Nr. 1, 15 (1946).

— Redman, wheat outyields Regent. Canad. Grain. J. **2**, 15 (1946).

Almeida, J. M. de: Hereditariedade do carácter aristado dos trigos. (Inheritance of awning in wheats.) Agron. Lusitana **1**, 327—351 (1939).

— Hereditariedade do comprimento das aristas da terceira flôr, no Triticum vulgare. (Inheritance of the length of the awns on the third floret in T. vulgare.) Agron. Lusitana **1**, 394—400 (1939).

— Acêrca da hereditariedade do carácter ,,dente apical das glumas'' no trigo. (The inheritance of the character ,,apical glume tooth'' in wheat.) Agron. Lusitana **2**, 233—242 (1940).

Améen, G.: Bakningsdugligheten och diastatiska tillståndet hos vetemjöl. (Backfähigkeit und diastatische Kraft von Weizenmehl.) Agri Hort. Gen. **4**, 68—73 (1946).

Anderson, J. A.: Enzymes and their role in wheat technology. Interscience Publishers, Inc., N. Y., Pp. IX + 371. tables. figs. (1946).

— u. T. R. Aitken: The quality of western Canadian wheat, 1942 crop. Bd. Grain Comm. Grain Res. Lab., Winnipeg, Manitoba, Pp. 22 (1941).

—, — Nineteenth Annual Report of the Board of Grain Commissioners, Grain Research Laboratory, Winnipeg, Manitoba 1945, Pp. 63 (1946).

—, — Twentieth Annual Report of the Grain Research Laboratory, Board of Grain Commissioners, Winnipeg, Manitoba, 1946, Pp. 79 (1947).

— u. R. L. Cunningham: Micro tests of alimentary pastes. III. The differential response of varieties to processing methods. Canad. J. Res. **21**, Sect. C, 265—275 (1943).

— u. W. J. Eva: Protein survey of western Canadian wheat 1942 crop. Bd. Grain Comm. Grain Res. Lab., Winnipeg, Manitoba, Pp. 32 (1942).

—, — Starch content of western Canadian wheat. II. Its Estimation from protein content, and some estimated data. Canad. J. Res. **21**, Sect. C, 323—331 (1943).

Andersson, G.: Gas change and frost hardening studies in winter cereals. Håkan Ohlsson Bocktryckeri, Lund, Pp. 163 (1944).

D'Andre, E.: La calidad industrial de los trigos. (The industrial quality of wheats.) ,,Granos'' Semilla Selecta, B. Aires **10**, Nr. 1/3, 46—49 (1946).

— Aptitudes y características segun zonas de veinticinco variedades de trigo que se incluyeron en el ensayo comparativo de rendimiento de 1943 (cosecha 1943—1944). Capabilities and characteristics according to geographical zones of twenty-five varieties of wheat which were included in the comparative yield trial of 1943 (1943—1944 crop). ,,Granos'' Semilla Selecta, B. Aires **10**, Nr. 4/6, 5—32 (1946).

D'André, H.: Calidad industrial de los trigos. Valor medio y variabilidad de los factores que la determinan. Datos de la experimentación oficial de variedades, realizada en distintas regiones, durante el quinquenio 1935—1940. (Industrial quality of wheats. Average value and variability of the factors that determine it. Data on the official variety tests carried out in different regions from 1935 to 1940.) ,,Granos'' Semilla Selecta, B. Aires **4**, Nr. 5/6, 3—34 (1940).

— La calidad industrial de los trigos provenientes del ensayo ,,standard'' conducido por la red oficial de ensayos territoriales, en la region cerealera Argentina, cosecha 1939—1940. (The industrial quality of the wheats from the standard test carried out by the official variety testing service in the cereal region of Argentina. Harvest 1939—40). ,,Granos'' Semilla Selecta, B. Aires **4**, Nr. 9/10, 3—25 (1940).

Anodin, P.: (The breeding of spring wheat immune to smut). Sovhoznoe Proizvodstvo. (State Farming.) Nr. 12, 21—23 (1944). [Russisch.]

Anonymus: Steadfast winter wheat. Nat. Inst. Agric. Bot., Cambridge, Pp. 4.

— A new wheat rust in Kenya. Rhod. Agric. J. **36**, 3—4 (1939).

— Report of the sixth hard spring wheat conference. Minneapolis, Minnesota, February 2 and 3 (1939).

— Trigo variedad ,,Kanhard selección la Prevision''. (The wheat variety Kanhard selección la Prevision.) Bol. Chacra Exp. ,,La Prevision'' **2**, 240 (1939).

— Trigo variedad ,,Candeal selección la Prevision''. (The wheat variety Candeal seleccion la Prevision.) Bol. Chacra Exp. ,,La Prevision'' **2**, 241 (1939).

— The new Michels hybrid grass. Seed World **47**, Nr. 11, 11 (1940).

— Improved wheat variety: selection from Braemar Velvet. Certified seed available for coming season. Tasm. J. Agric. **11**, 41 (1940).

— New varieties of wheat, barley, and field pea. J. Dep. Agric. S. Aust. **43**, 534—535 (1940).

— Variedades de trigos genéticos del Ministerio de Agricultura. (Genetical wheat varieties of the Ministry of Agriculture.) Minist. Agric. Dep. Genet. Fitotec., Santiago, Pp. 16 (1940—1941).

— Dictionary of spring wheat varieties. Northw. Crop Improvement Ass. Minn., Pp. 92 (1941).

— A new winter hardy wheat. J. Dep. Agric. Eire **38**, 175—176 (1941).

— I frumenti ,,Italo Balbo'', ,,Comandante Baudi'', ,,Comandante Novaro''. (The wheats Italo Balbo, Comandante Baudi and Comandante Novaro.) Ist. Naz. Genet. Cerealicoltura, Roma **19**, 7 (1941).

— The wheat contest-1941. Circ. Clemson Agric. Coll. U.S.D.A., Nr. 199, 16 (1941).

— Wheat varieties in New Zealand. Bull. Canterbury Agric. Coll. Nr. 141, 4 (1941).

— China: National Agricultural Research Bureau. Chronica Botanica **7**, 83—84 (1942).

— I frumenti ,,Bruno'', ,,Eia'', ,,Alalà''. (The wheat varieties Bruno, Eia and Alalà.) Ist. Naz. Genet. Cerealicoltura, Roma **20**, 9 (1942).

Anonymus: New rust-resistant wheat. Pastoral Rev. **52**, 45 (1942).
— Varieties of cereals for autumn sowing. Fmrs'Leafl. Nat. Inst. Agric. Bot., Nr. 1, 4 (1942).
— Wheat. N. Z. Off. Yearb., Pp. 335—336 (1942).
— Wheat varieties in the Wimmera. J. Dep. Agric. Vict. **40**, 222—224 (1942).
— Distribución de las variedadas de trigo aconsejadas para la siembra del año 1943 por el Tribunal de Fiscalización de Semillas. (Distribution of the wheat varieties recommended by the Seed Certification Tribunal for sowing in 1943.) Noticioso, B. Aires **8**, 41—44 (1943).
— A valuable new wheat. Corn Tr. News **168**, 233 (1943).
— Agronomic aspects of hard spring wheat breeding. 6. Objectives in breeding programs and outlook for next three years. Rep. Mill. Baking Mtg., 7th Hard Spring Wheat Conf., Minneapolis, Minn., February 28 and 29 — March 1, 1944, Pp. 47—48 (1944). [Mimeographed.]
— Agronomic aspects of hard spring wheat breeding. 8. Newer technics, equipment, and experimental designs. Rep. Mill. Baking Mtg., 7th Hard Spring Wheat Conf., Minneapolis, Minn., February 28 and 29 — March 1, 1944, Pp. 50—51 (1944). [Mimeographed.]
— Varieties of winter wheat. Fmrs' Leafl. Nat. Inst. Agric. Bot., Cambridge, Nr. 8, 4 (1944).
— O trigo nas estações experimentais do suldo país. (Wheat in the experimental stations in the south of our country.) Bol. Minist. Agric., Rio de J. **33**, Nr. 5, 145 (1944).
— A new wheat for South Australia. J. Dep. Agric. S. Aust. **47**, 302 (1944).
— A giant-grained hybrid wheat. Science Suppl. **99**, 10—12 (1944).
— Amphidiploids of wheat and Agropyron. J. Hered. **35**, 128 (1944).
— New smut resistant wheat released by station. Fm. Home Sci. Utah **5**, 5 (1944).
— Giant grained hybrid wheat. Sth. Seedsman **7**, 53 (1944).
— Wheat quality and new varieties. Wheat Advisory Committee Meeting. J. Dep. Agric. Vict. **42**, 350—355 (1944).
— Agricultural Research Institutions. Agriculture, Moscow, Nr. 10, 14—15 (1945). [Mimeographed.]
— Field experiments, 1944. J. Dep. Agric. Eire **42**, 288—289 (1945).
— Yields of winter wheat 1943—1944. Circ. Univ. Ill. Agric. Exp. Sta., Nr. 588, 4 (1945).
— William Farrer centenary. J. Dep. Agric. Vict. **43**, 202 (1945).
— 100 new varieties of wheat tested on prairies for sawfly resistance. Canad. Grain J. **1**, Nr. 4, 13 (1945).
— Personalities in plant breeding. J. Dep. Agric. W. Aust. **22**, 302—303 (1945).
— Como multiplicar a produção do trigo no País. (How to increase the production of wheat in our country.) Bol. Minist. Agric., Rio de J. **34**, 108—109 (1945).
— Westar-promising star in Texas winter wheat show. Sth. Seedsman **8**, 48 (1945).
— Baking quality of the 1944—45 crop. N. Z. Wheat Rev. 1945—1946: Publ., Nr. 42, 6—7 (1946).
— Work in progress. N. Z. Wheat Rev. 1945—1946, Publ., Nr. 42, 8—10 (1946).
— Four new recommended wheats which combine stem rust resistance with yield and quality. Agric. Gaz. N. S. W. **57**, 3—6 (1946).
— Varieties of wheat recommended for 1947 sowing. Agric. Gaz. N. S. W. **57**, 626—627 (1946).
— „Rescue" new wheat variety forward step in fighting sawfly ravage in west. Canad. Grain. J. **1**, 8 (1946).

Anonymus: Sawfly resistant wheat on way. Canad. Grain. J. **2**, 16 (1946).
— Fife-Tuscan resistant to loose smut. N. Z. Wheat Rev. 1945—1946, Publ., Nr. 42, 17 (1946).
— New wheat varieties for the Wimmera and Mallee. J. Dep. Agric. Vict. **44**, 148—149 (1946).
— Cereal Crops. Commonw. Agric. **17**, 3—9 (1946).
— Nuove razze di grano. (New races of wheat.) Terra e Sole, Roma, Nr. 22, 2 (1946).
— Work in progress at the Wheat Research Institute. N. Z. Wheat Rev., 1946, Pp. 49—53 (1947).
— The work of other institutions. N. Z. Wheat Rev. 1946, Pp. 53—55 (1947).
— Recommend new wheat for Peace River area. Canad. Grain J. **2**, Nr. 8, 19 (1947).
— New strain of Apex wheat developed at Saskatchewan „U". Canad. Grain. J. **2**, Nr. 7, 12 (1947).
— New wheat-rye hybrid. Soviet News, Nr. 1636, 2 (1947).
— Prospect for creation of perennial wheat not bright says Saskatchewan Professor. Canad. Grain. J. **2**, Nr. 6, 8 (1947).
Araratjan, A. G.: Über die Meiosis zweier Getreidearten. C. R. (Doklady) Acad. Sci. URSS. **28**, 645—648 (1940).
Armstrong, J. M.: Investigations in Triticum-Agropyron hybridization. Emp. J. Exp. Agric. **13**, 41—53 (1945).
— u. H. A. McLennan: Amphidiploidy in Triticum-Agropyron hybrids. Sci. Agric. **24**, 285—298 (1944).
— u. T. M. Stevenson: The effects of continuous line selection in Triticum-Agropyron hybrids. Emp. J. Exp. Agric. **15**, 51—64 (1947).
Artemova, A. u. V. Jakovleva: (Wheat-Agropyron hybrids in Kazahstan.) Sovhoznoe Proizvodstvo (State Farming), Nr. 8/9, 25—26 (1944). [Russisch.]
Arzuaga, J. G.: Perspectivas de producción de nuevas variedades de trigo en el Instituto Fitotécnico de Santa Catalina. (Prospects of producing new varieties of wheat in the Santa Catalina Institute of Applied Botany.) An. Inst. Fitotéc. Santa Catalina **1**, 9—15 (1939) 1940.
Atkins, I. M.: Reaction of some varieties and strains of winter wheat to artificial inoculation of loose smut. J. Amer. Soc. Agron. **35**, 197—204 (1943).
— The present status of wheat improvement work in north Texas. Report of the Fifth Hard Red Winter Wheat Improvement Conference, Manhattan, Kansas, February 12, 13 and 14, 1945. Div. Cereal Crops Dis., Plant Ind. Sta., Beltsville, Md., p. 3 (1945). [Mimeographed.]
— The use of nursery plots vs. field plots for varietal evaluation. Report of the Fifth Hard Red Winter Wheat Improvement Conference, Manhattan, Kansas, February 12, 13 and 14, 1945. Div. Cereal Crops Dis., Plant Ind. Sta., Beltsville, Md., Pp. 15—16 (1945). [Mimeographed.]
— Observations on resistance to green bugs in cereal crops. Report of the Fifth Hard Red Winter Wheat Improvement Conference, Manhattan, Kansas, February 12, 13 and 14, 1945. Div. Cereal Crops Dis., Plant Ind. Sta., Beltsville, Md., p. 23 (1945). [Mimeographed.]
— Resistance of winter wheat varieties to artificial inoculation with loose smuts. Report of the Fifth Hard Red Winter Wheat Improvement Conference, Manhattan, Kansas, February 12, 13 and 14, 1945. Div. Cereal Crops Dis., Plant Ind. Sta., Beltsville, Md., Pp. 30—31 (1945). [Mimeographed.]
—, E. D. Hansing u. W. M. Bever: Reaction of varieties and strains of winter wheat to loose smut. J. Amer. Soc. Agron. **39**, 363—377 (1947).
— u. E. S. McFadden: Rust resistant Austin. Sth. Seedsman **9**, 22, 38 (1946).

Atkins, I. M.: u. J. R. Quinby: „Comanche", disease-resistant wheat, for planters of the plains. Sth. Seedsman 6, 17, 44 (1943).
—, — u. P. B. Dunkle: New, early maturing Wichita wheat. Sth. Seedsman 7, 24 (1944).
Aufhammer, G.: Erfolge der Kleberweizenerzeugung. Die Ergebnisse der Qualitätsuntersuchungen von 1935 bis 1942. Forschungsdienst 17, 397—407 (1944).
Ausemus, E. R.: Wheat meal fermentation tests with hard red spring wheats. Rep. Mill. Baking Mtg, 7th Hard Spring Wheat Conf., Minneapolis, Minn., February 28 and 29 — March 1, 1944, p. 22 (1944). [Mimeographed.]
— Stem and leaf rust in relation to wheat breeding. 7. Breeding for stem rust resistance in spring wheat. Rep. Mill. Baking Mtg, 7th Hard Spring Wheat Conf., Minneapolis, Minn., February 28 and 29 — March 1, 1944, Pp. 28—30 (1944). [Mimeographed.]
— Bunt, rootrots, scab and other diseases in relation to wheat breeding. 3. Scab. Scab studies with spring wheat. Rep. Mill. Baking Mtg, 7th Hard Spring Wheat Conf., Minneapolis, Minn., February 28 and 29 — March 1, 1944, Pp. 37 (1944). [Mimeographed.]
— Breeding winter hardy wheats for the northern great plains area. Report of the Fifth Hard Red Winter Wheat Improvement Conference, Manhattan, Kansas, February 12, 13 and 14, 1945. Div. Cereal Crops Dis., Plant Int. Sta: Beltsville, Md., Pp. 10—11 (1945). [Mimeographed.]
— Breeding rust resistant winter wheats for the northern great plains area. Report of the Fifth Hard Red Winter Wheat Improvement Conference, Manhattan, Kansas, February 12, 13 and 14, 1945. Div. Cereal Crops Dis., Plant Ind. Sta., Beltsville, Md., Pp. 26—28 (1945). [Mimeographed.]
— et al.: A summary of genetic studies in hexaploid and tetraploid wheats. J. Amer. Soc. Agron. 38, 1082 bis 1099 (1946).
— u. R. H. Bamberg: Breeding hard red winter wheats for the Northern Great Plains area. J. Amer. Soc. Agron. 39, 198—206 (1947).
— u. E. C. Stakman: Newthatch. What it is and what it took to make it. Minn. Fm. Home. Sci. 2, Nr. 1.
—, —, E. W. Hanson, W. F. Geddes u. P. P. Merritt: Newthatch wheat. Tech. Bull. Minn. Agric. Exp. Sta., Nr. 166, 20 (1944).

Babcock, E. B.: The chronology of Hope wheat. J. Hered. 31, 132—133 (1940).
Bailey, R. S.: A method for the formulization of farinograph curves. Cereal. Chem. 17, 701—766 (1940).
Baker, G. A. u. F. N. Briggs: Wheat bunt field trials. J. Amer. Soc. Agron. 37, 127—133 (1945).
Balašev, I. S. u. I. P. Petrov: (Growing spring wheat and Triticum-Agropyron hybrids in greenhouses as a method of accelerating breeding.) Theses and scientific papers read at the 4th District Conference of Workers of Universities and Research Institutions, Omsk, Nr. 1, Agron. Sect., 47—48 (1941). [Russisch.]
Baldacci, E. u. R. Ciferri: Studi sulla „stretta" dei cereali. (Studies om premature ripening in cereals.) Atti Ist. Bot. e Lab. Crittogamico, Univ. Pavia 1, Ser. 5, 217—276 (1944).
Balley, C. H.: The constituents of wheat and wheat products. N. Y., Pp. 332. 17 figs. 115 tables. (1944).
Bamberg, R. H.: The dwarf bunt problem. Rep. Mill. Baking Mtg, 7th Hard Spring Wheat Conf., Minneapolis, Minn., February 28 and 29 — March 1, 1944, Pp. 33—34 (1944). [Mimeographed.]
— Problems in breeding winter wheat for winterhardiness in Montana. Report of the Fifth Hard Red Winter Wheat Improvement Conference, Manhattan, Kansas, February 12, 13 and 14, 1945. Div. Cereal Crops Dis., Plant Ind. Sta., Beltsville, Md., Pp. 11 bis 12 (1945). [Mimeographed.]
Bamberg, R. H.: Breeding for resistance to bunt. Report of the Fifth Hard Red Winter Wheat Improvement Conference, Manhattan, Kansas, February 12, 13 and 14, 1945. Div. Cereal Crops Dis., Plant Ind. Sta., Beltsville, Md., Pp. 28—29 (1945). [Mimeographed.]
Bartel, A. T.: Some physiological characteristics of four varieties of spring wheat presumably differing in drought resistance. J. Agric. Res. 74, 97—112 (1947).
Bartlett, H.: New wheats for the North-west. Agric. Gaz. N. S. W. 52, 184—188 (1941).
Bassarskaja, M. A., N. I. Ermolaeva u. L. E. Hodjkov: (Description of wheat plants obtained from seed of intravarietal crosses.) Jarovizacija, Nr. 1 (28), 23—26 (1940). [Russisch.]
Bates, J. C.: Varietal differences in anatomy of cross-section of wheat grain. Bot. Gaz. 104, 490—493 (1943).
Bayfield, E. G.: The effect of wheat variety upon maltose values. J. Amer. Soc. Agron. 38, 624—629 (1946).
Bayles, B. B.: General aspects of the soft wheat improvement program. Rep. 8th East. Wheat Conf. Cincinnati, Ohio, November 10, 1943, Pp. 16 (1944). [Mimeographed.]
— Hard winter wheat improvement in the future. Report of the Fifth Hard Red Winter Wheat Improvement Conference, Manhattan, Kansas, February 12, 13 and 14, 1945. Div. Cereal Crops Dis., Plant Ind. Sta. Beltsville, Md., Pp. 52—57 (1945). [Mimeographed.]
— u. C. A. Suneson: Effect of awns on kernel weight, test weight, and yield of wheat. J. Amer. Soc. Agron. 32, 382—388 (1940).
— u. J. W. Taylor: Wheat improvement in the eastern United States. Cereal Chem. 16, 208—223 (1939).
—, — Report of the uniform soft winter wheat nurseries 1938—39. U. S. Dep. Agric., Bur. Pl. Ind., Div. Cereal Crops and Dis., Washington, D. C., Pp. 11 (1940). [Mimeographed.]
—, — Report of the uniform soft winter wheat nurseries 1938—39. Div. Cereal Crops and Diseases, Bur. Pl. Ind. U.S. Dep. Agric. March 7, Pp. 11 (1940). [Mimeographed.]
—, — Report of the uniform eastern soft winter wheat nurseries 1939—40. U. S. Dep. Agric., Bur. Pl. Ind., Div. Cereal Crops and Dis., Washington, D. C., Pp. 11 (1941). [Mimeographed.]
Becker, H.: Ergebnisse und Erfahrungen bei der Resistenzzüchtung gelbrostwiderstandsfähiger Weizen. Z. Pflanzenz. 24, 539—568 (1942).
— Arbeitsweise und Erfolge in der Weizenzüchtung an der Pflanzenzuchtstation der Universität Halle. Kühn-Archiv 60, 369—401 (1943/1944).
Becker, J., W. H. Fuchs u. B. Japha: Grundlagen und Erfahrungen der Züchtung winterfester Weizen. Züchter 17/18, 235—240 (1947).
Beckmann, I.: El problema del trigo en el Brasil. (The wheat problem in Brazil.) Rev. Fac. Agron. Univ., Montevideo, Nr. 19, 21—36 (1940).
— Os trabalhos de melhoramento em trigo na Estação Experimental de Bagé. (Work on the improvement of wheat at the Experimental Station of Bagé.) Bol. Agron. Brasil 9, 45—48 (1945).
Belmonte-Freixa, J.: Algunos métodos para determinar rapidamente el valor industrial de los trigos. (Certain methods for the determination of the industrial value of wheats.) Arch. Fitotécn. Uruguay 3, 276—290 (1940—1941).
Berg, S. O.: Arbetet vid Weibullsholm för höjandet av höstvetets bakningskvalitet. (Work at Weibullsholm to improve the baking quality of winter wheat.) K. LantbrAkad. Tidskr. 79, 97—123 (1940).

Berg, S. O.: Weibulls Äringvete III. Ny höstvetesort för södra och östra Götalands slättbygder. (Weibull's Äring III wheat. A new variety of winter wheat for southern and eastern Götaland plains.) Weibulls Ill. Årsb. **35**, 31—40 (1940).
— Stråsädesodlingens äggviteproblem. (The protein problem in cereal cultivation.) Weibulls Ill. Årsb. **36**, 14—32 (1941).
— Über die Beziehungen zwischen Körnerertrag, Rohproteingehalt und Rohproteinertrag verschiedener Weizensorten sowie ihre züchterische Bedeutung. Z. Pflanzenz. **23**, 542—561 (1941).
— Weibulls Eroicavete. Ny höstvetesort för södra och östra Götalands slättbygder. (Weilbulls Eroica wheat. A new winter wheat for the plains of south and east Götaland.) Weibulls Ill. Årsb. **37**, 28—31 (1942).
— Weibulls Original Atlevårvete. Ny elit med tidigare mognad utlämnas våren 1943. (Weibull's original Atle spring wheat. A new earlier maturing élite released in spring 1943.) Weibulls Ill. Årsb. **38**, 10—11 (1943).
— Weibulls Original Eroicavete. Ny, mycket högavkastande höstvetesort för södra och östra Götalands slättbygder. (Weibull's original Eroica wheat. A new, very high yielding autumn wheat for southern and eastern plains of Götaland.) Weibulls Ill. Årsb. **38**, 12—23 (1943).
— Eroica, ny mycket högavkastande höstvetesort för södra och östra Götalands slättbygder. (Eroica, a new very high yielding autumn wheat variety for the lowlands of southern and eastern Gotäland.) Agri Hortique Genetica, Landskrona **1**, 15—28 (1943).
— Ny elit av Atlevårvete med tidigare mognad. (An earlier ripening, new élite of Atle spring wheat.) Agri Hortique Genetica, Landskrona **1**, 29—30 (1943).
— Weibulls Ergovete. Resultat från prövning i officiella försök tillsammans med andra sorter 1930—1942. (Weibull's Ergo wheat. Result from testing in official trials together with other varieties, 1930—1942.) Agri Hortique Genetica, Landskrona **2**, 43—62 (1944).
— Nya resultat av veteförädlingen vid Weibullsholm. (New results of wheat breeding at Weibullsholm.) Weibulls Ill. Årsb. **39**, 22—23 (1944).
— Weibulls Bronsvete. Ny vårvetesort för Götalands slättbygder. (Weibulls Brons wheat. A new spring wheat variety for the plains of Götaland.) Agri Hortique Genetica, Landskrona **3**, 1—13 (1945).
— Är vetemjölets örighetsgrad huvudsakligen en sortegenskap? (Ist die Griffigkeit des Weizenmehls hauptsächlich eine Sorteneigenschaft?) Agri Hort. Gen. **4**, 1—14 (1946).
Bertsch, F.: Der Dinkel. Landw. Jb. **92**, 241—252 (1942).
Bever, W. M.: Physiologic specialization in Ustilago tritici and the effect of vernalization on the incidence of loose smut in artificially inoculated winter wheat. Rep. 8th East. Wheat Conf. Cincinnati, Ohio, November 10, 1943, Pp. 1—4 (1944). [Mimeographed.]
Birdsall, J. E. u. K. W. Neatby: Researches on drought resistance in spring wheat. III. Size and frequency of stomata in varieties of Triticum vulgare and other Triticum species. Canad. J. Res. **22**, Sect. C, 38—51 (1944).
Blodgett, E. C. u. H. K. Schultz: Stem distortion of wheat. J. Amer. Soc. Agron. **38**, 717—722 (1946).
Blunt, D. L.: Agricultural Department Annual Report, Colony and Protectorate of Kenya 1943, Pp. 6 (1944).
Boerger, A.: Weizenzüchtung in Uruguay. Züchter **11**, 265—277 (1939).
— Angewandte Genetik als entscheidender Faktor für das Vordringen des Weizenbaues im subtropischen Osten Südamerikas. Proc. 7th Int. Genet. Congr. Edinburgh, 23—30 August, Pp. 72—73, 1939 (1941). (Abstr.)

Boerger, A.: Las perspectivas del cultivo triguero en el Brasil. (Wheat growing prospects in Brazil.) Arch. Fitotécn. Uruguay **3**, 239—261 (1940—1941).
Boeuf, F.: The problem of wheat production in France and in French North Africa. Int. Rev. Agric. **33**, 221T—240T (1942).
— La valeur industrielle des blés durs de Tunisie. C. R. Acad. Agric. Fr. **32**, 401—404 (1946).
Bonnett, O. T., C. M. Woodworth, G. H. Dungan u. B. Koehler: Prairie: a new soft winter wheat for Illinois. Bull. Ill. Agric. Exp. Sta., Nr. 513, 595—600 (1945).
Bottazzi, G.: Indagini su la coltivazione dei frumenti di razze elette nelle 9 provincie Lombarde. (Studies on the cultivation of élite wheat races in the 9 provinces of Lombardy.) Ist. Naz. Genet. Cerealicoltura, Roma, Pp. 16 (1941).
Bottazzi, G. B.: Espercienze di jarovizzazione del frumento. (Experiments on vernalizing wheat.) Genetica Agraria, Roma **1**, 147—166 (1947).
Boyes, J. W.: Division of cereal crops. Pr. Bull. Univ. Alberta **27**, 2—3 (1942).
Breakwell, E. J., E. M. Hutton u. D. H. S. Mellor: Cereal breeding and variety trials at Roseworthy College, 1938—39. J. Dep. Agric. S. Aust. **43**, 524—533 (1940).
Bremer, H. u. Özkan: Türkiye'de sürme hastaliğinin mevkii ve ehemmiyeti. (Position and importance of bunt in Turkey.) Ziraat Dergisi **4**, 124—130 (1943).
—, — Türkiye'de hububatin pas hastaliklari hakkinda tetkik ve mütalâalar. (Investigations and observation of rust diseases on cereals in Turkey.) Ziraat Dergisi **5**, 5—12 (1944).
Briggs, F. N.: Linkage between the Martin and Turkey factors for resistance to bunt, Tilletia tritici, in wheat. J. Amer. Soc. Agron. **32**, 539—541 (1940).
Britten, E. J. u. W. P. Thompson: The artificial synthesis of a 42-chromosome wheat. Science **93**, 479 (1941).
Brojakovskij, N. V.: (Susceptibility of winter wheat varieties to brown rust and of oat varieties to crown rust.) Naučnye Zapiski Saharnoj Promyšlennosti (Sci. Trans. Sug. Ind.), Nr. 1/2, 87—99 (1940). [Russisch.]
Builin, D. P.: (New varieties of winter wheat.) Socialistic Grain Farming, Saratov, Nr. 3, 57—59 (1941). [Russisch.]
— u. E. S. Builina: (Breeding winter wheat in the central Volga basin.) Selekcija i Semenovodstvo (Breeding and Seed Growing), Nr. 7, 27—30 (1939). [Russisch.]
Bujanov, J. M.: (The accuracy of technological investigations in breeding wheat.) Selekcija i Semenovodstvo (Breeding and Seed Growing), Nr. 8, 18—19 (1939). [Russisch.]
Bumbesti, A. I. D.: Ereditatea densitatii spicului la grâu. (The inheritance of ear density in wheat.) Viata Agric. **30**, 145—148 (1939).
Bustarret, J.: Orientation à donner au perfectionnement de nos blés. Comm. Sta. Nat. Encouragement Agric., Paris, Pp. 15 (1943).

Caffera, R.: El trigo Kanred en los Estados Unidos de Norte América y en la Republica Argentina. (Kanred wheat in the USA. and in the Argentine Republic.) „Granos" Semilla Selecta, B. Aires, **7**, Nr. 10, 11 u. 12, 3—16 (1943).
— Cuatro nuevas variedades de trigo de inscrición definitiva: Benvenuto Inca, Buck Claromecó, Klein 157 y Klein Alberti. (Four new varieties of wheat definitively registered: Benvenuto Inca, Buck Claromecó, Klein 157 and Klein Alberti.) „Granos" Semilla Selecta, B. Aires **9**, Nr. 7/9, 35—37 (1945).
Caffrey, M.: Wheat growing. J. Dep. Agric. Eire **41**, 38—42 (1944).

Caffrey, M.: The identification of varieties of bread wheat (Triticum vulgare). J. Dep. Agric. Eire **42**, 6—19 (1945).

Caldwell, R. M.: Disease resistance in the soft winter wheats. Rep. 6th East. Wheat Conf. Wooster, Ohio, June 16, Pp. 4—9 (1939). [Mimeographed.]

—, W. B. Cartwright u. L. E. Compton: Status of breeding for hessian fly resistance in the soft red winter wheats. Rep. 8th East. Wheat Conf. Cincinnati, Ohio, November 10, 1943, Pp. 4—5 (1944). [Mimeographed.]

—, —, —: Inheritance of hessian fly resistance derived from W 38 and durum P. I. 94587. J. Amer. Soc. Agron. **38**, 398—409 (1946).

— u. L. E. Compton: Complementary lethal genes in wheat. J. Hered. **34**, 67—70 (1943).

—, — Correlated inheritance of resistance to eigth races of wheat leaf rust, Puccinia rubigovera tritici, powdery mildew, Erysiphe graminis tritici, and glume color. Phytopathology **37**, 4 (1947). (Abstr.)

—, — Inheritance of resistance to loose smut of wheat, Ustilago tritici, in the varietal cross Trumbull x Wabash. Phytopathology **37**, 4 (1947). (Abstr.)

Câmara, A. de Sousa da: Accao do calor sobre os fenotipos do „Triticum monococcum L." (Action of heat on the phenotypes of T. monococcum L.) Rev. Agron. Lisboa **28**, 229—241 (1940).

— O problema da fragmentação cromosónica, operada pelos raios X, estudado no Triticum monococcum. Agron. Lusitana **3**, 341—359 (1941).

— Roturas dos cromosomas provocadas pelos raios X. (Chromosome ruptures caused by X-rays.) Rev. Agron., Lisboa **29**, 96—97 (1941).

— Estudo comparativo de cariotipos no género Triticum. (Comparative study of the caryotypes in the genus Triticum.) Agron. Lusitana **5**, 95—117 (1943).

— Transplantação de embriões. (Transplantation of embryos.) Agron. Lusitana **5**, 375—386 (1943).

— Cromosomas dos trigos hexaplóides. (Chromosomes of the hexaploid wheats.) Agron. Lusitana **6**, 221—251 (1944).

— Un estudo citológico dos trigos durum Portugueses. (A cytological study of the Portuguese Triticum durum forms.) Bol Soc. Broteriana **19**, Sér. 2a, 273—287 (1944).

—, R. Castro u. J. M. Almeida: Problemas de selecção dos trigos. (Problems of breeding wheat.) Rev. Agron. Lisboa **30**, 470—471 (1942).

— u. L. A. Coutinho: Citologia dos trigos tetraploides. (Cytology of the tetraploid wheats.) Agron. Lusitana **1**, 268—314 (1939).

Čanov, G. V. u. V. V. Kobeljkov: (Promising varieties of winter wheat). Selekcija i Semenovodstvo (Breeding and Seed Growing, Nr. 8/9, 22 (1940). [Russisch.]

Cardosa, C. P.: Genetica do trigo e sua influencia no melhoramento da lavoura triticea do paiz. (Genetics of wheat and its influence on the wheat production of the country.) Rev. Soc. Rur. Brasil. **19**, Nr. 222, 12—20 (1939).

Cartwright, W. B.: Breeding for resistance to Hessian fly. Rep. 6th East. Wheat Conf. Wooster, Ohio, June 16, Pp. 2—3 (1939). [Mimeographed.]

—, R. M. Caldwell u. L. E. Compton: Relation of temperature to the expression of resistance in wheats to hessian fly. J. Amer. Soc. Agron. **38**, 259—263 (1946).

— u. D. W. La Hue: Testing wheats in the greenhouse for Hessian fly resistance. J. Econ Ent. **37**, 385—387 (1944).

— u. R. G. Shands: Wheat varieties resistant to the hessian fly and their reactions to stem and leaf rusts. Tech. Bull. U. S. Dep. Agric., Nr. 877, 6 (1944).

—, — Resistance to the Hessian fly in crosses of some common spring wheats. J. Amer. Soc. Agron. **38**, 845—847 (1946).

Castiglioni, J. M.: Influencia de la época y densidad de siembra sobre algunos caracteres morfológicos en ocho variedades de trigo. Análisis biométrico. (Influence of the date and density of sowing on some morphological characters in eight varieties of wheat. Biometric analysis.) „Granos" Semilla Selecta, B. Aires **5**, Nr. 7/9, 3—46 (1941).

Česnokov, P. G.: (Spring wheats resistant to Hessian fly [Phytophaga destructor Say.].) Selekcija i Semenovodstvo (Breeding and Seed Growing), Nr. 8, 12—13 (1939). [Russisch.]

— Geographic variation of wheats in their resistance to Hessian fly (Phytophaga destructor Say). C. R. (Doklady) Acad. Sci. URSS. **23**, 287—291 (1939).

— (Methods of characterization selected plants for resistance to Oscinella frit L. and Phytophaga destructor Say.) Proc. Lenin Acad. Sci. USSR., Nr. 5, 23—30 (1940). [Russisch.]

Chabrolin, C. u. J. Miège: Essais comparatifs régionaux de variétés de blé. Ann. Serv. Bot. Tunis. **18**, 145—210 (1941).

Chang, S. C.: Morphological causes for varietal differences in shattering of wheat. J. Amer. Soc. Agron. **35**, 435—441 (1943).

Chao, Jen Jung: (Report of a preliminary study on the grading of 33 samples of Chinese wheat.) Fukien Agric. J. **2**, 276—279 (1940).

Chesnokov, P. G. siehe Česnokov.

Chester, K. S.: Breeding wheat for disease resistance in Oklahoma. Report of the Fifth Hard Red Winter Wheat Improvement Conference, Manhattan, Kansas, February 12, 13 and 14, 1945. Div. Cereal Crops Dis., Plant Ind. Sta., Beltsville, Md., Pp. 25 (1945). [Mimeographed.]

— The genes of plenty. Sci. Mon. N. Y., Pp. 483—485 (1945).

Chevrette, J. E.: Inheritance of earliness and other characters in spring wheat. Cornell Univ. Abstr. Thes., Pp. 322—323. 1941 (1942).

Chin, K.-C.: Disjonctions singulières des hybrides interspécifiques de blés, engrains et froments (Monococcum x Tr. vulgare). C. R. Acad. Sci. Paris **209**, 240—242 (1939).

— Modification des grains dans l'hybride de blé monococcum x vulgare. C. R. Acad. Sci. Paris **215**, 305 bis 306 (1942).

Chin, T.C.: (Cytology of wheat and its application.) Chinese J. Agric. Sci. **1**, 66—92 (1943).

— The cyto-genetical studies of the Indian dwarf wheat hybrids. Acta Brevia Sinensis, Nr. 6, 10—11 (1944). (Abstr.) [Mimeographed.]

— The inheritance of some quantitative characters in the interspecific crosses of wheat. Chinese J. Sci. Agric. **1**, 204—217 (1944).

— Wheat-rye hybrids. J. Hered. **37**, 195—196 (1946).

— u. C. S. Chwang: The cytology of „blue" wheat hybrids. Indian J. Agric. Sci. **12**, 661—678 (1942).

—, — Cytogenetic studies of hybrids with „Makha" wheat. Bull. Torrey Bot. Cl. **71**, 356—366 (1944).

—, — The cyto-genetical studies of the „makka" wheat hybrids. Acta Brevia Sinensia, Nr. 6, 9—10 (1944). (Abstr.) [Mimeographed.]

Chinoy, J. J.: Correlation between yield of wheat and temperature during ripening of grain. Nature, Lond. **159**, 442—444 (1947).

Christian, C. S. u. S. G. Gray: Interplant competition in mixed wheat populations and its relation to single plant selection. J. Coun. Sci. Industr. Res. Aust. **14**, 59—68 (1941).

Christiansen-Weniger, F.: Ablauf und Modifizierbarkeit des Entwicklungsrhythmus bei Weizen. Z. Pflanzenz. **25**, 305—342 (1943).

Christiansen-Weniger F. u. E. Emre: Anadoluda Triticum durum' da meydana gelen „dönme" danelerin sebepleri ve buna karşi tedbirler. (Causes of the occurrence of „Dönme" grains in T. durum in Anatolia and countermeasures.) T. C. Yüksek Ziraat Enstitüsü Çalişmalarinda, Ankara, Nr. 102, 18 + 16 (1940).

—, — Ursachen für das Auftreten von „Dönme"-Körnern bei Triticum durum in Anatolien und Gegenmaßnahmen. Züchter 12, 81—88 (1940).

Chrony, E.: Orígen e historia del trigo Guatraché M. A. (Origin and history of the wheat Guatraché M. A.) „Granos" Semilla Selecta, B. Aires 4, Nr. 7/8, 21 bis 30 (1940).

Chu, Vong-May: The prevalence of the wheat nematode in China and its control. Phytopathology 35, 288—295 (1945).

Cicin, N. V.: Wheat and couch grass hybrids. Sci. and Cult 6, 18—20 (1940).

— (Extending winter crops towards the eastern regions of the USSR. and producing new varieties of winter crops.) Socialističeskoe Seljskoe Hozjajstvo (Socialistic Agriculture), Moscow, Nr. 10, 48—61 (1940). [Russisch.]

— (Distant hybridization-the chief method of breeding.) Selekcija i Semenovodstvo (Breeding and Seed Growing, Nr. 10, 4—7 (1940). [Russisch.]

— (Transformation of the nature of cultivated plants.) Sovhoznoe Proizvodstvo (State Farming), Nr. 1/2, 39—41 (1943). [Russisch.]

Ciferri, R.: Frumenti e granicoltura indigena in Etiopia. (Native wheats and wheat cultivation in Abyssinia.) Atti R. Accad. Georgofili, Firenze, Pp. 12 (1939).

— Frumenti e granicoltura indigena in Etiopia. (Native wheats and wheat cultivation in Abyssinia.) Agricoltura Colon 33, 337—349 (1939).

— La cerealicoltura in Africa Orientale. V. Complessità del problema frumentario e prospettive di produzione. (Cereal cultivation in East Africa. V. The complexity of the wheat problem and the prospects of production.) Ital. Agric. 77, 31—42 (1940).

— Vecchie razze locali di frumento coltivate in Italia nell' anno XVIII. (Ancient local races of wheat cultivated in Italy in the year XVIII.) Ital. Agric. 78, 641—646 (1941).

— Relazione sull' attività del Laboratorio Crittogamico, dell' Osservatorio Fitopatologico e del Centro Studi sugli Anticrittogamici durante gli anni 1944 e 1945. (Report on the activities of the Cryptogamic Laboratory, the Phytopathological Observatory and the Centre of Studies on Anticryptogamic Substances for 1944 and 1945.) Atti Ist. Bot. e Lab. Crittogamico, Univ. Pavia 5, Ser. 5, 277—321 (1946).

— u. M. Caravini: La cerealicoltura in Africa Orientale. VII.-I frumenti oasicoli della Libia in rapporto a quelli etiopici. (Cereal cultivation in East Africa. VII. The wheats of the Libyan oases in relation to those of Ethiopia, Ital. Agric. 77, 409—415 (1940).

— u. G. R. Gigliolo: I frumenti di Rodi. (The wheats of Rhodes.) Relaz. Monogr. Agr.-Colon., Firenze, Nr. 57, 20 (1939).

—, — I cereali dell'Africa Italiana. I. I frumenti dell' Africa Orientale Italiana studiati su materiali originali. (The cereals of Italian Africa. I. The wheats of Italian East Africa studied on original material.) Bibl. Agrar. Colon., Firenze, Pp. 298 (1939).

—, — La cerealicoltura in A.O.I. I. — I frumenti duri. (Cereal culture in Italian East Africa. I. The hard wheats.) Ital. Agric. 76, 247—257 (1939).

—, — La cerealicoltura in Afrika Orientale. II. I frumenti piramidali, turgidi, polacchi e dicocchi. (Cereal cultivation in East Africa. II. The pyramidale, turgidum, polonicum and dicoccum wheats.) Ital. Agric. 76, 387—397 (1939).

Ciferri, R. u. G. R. Gigliolo: La cerealicoltura in Africa Orientale. (III. Triticum vulgare and T. compactum.) Ital. Agric. 76, 765—774 (1939).

—, — La cerealicoltura in Africa Orientale. IV. Caratteristiche dei gruppi minori di frumenti etiopici e loro „formulazione". (Cereal cultivation in East Africa. IV. Characteristics of the smaller groups of Ethiopian wheats and a „formula" for them.) Ital. Agric. 76, 837—844 (1939).

—, — I frumenti dell'isola di Rodi. (The wheats of the island of Rhodes.) Ital. Agric. 77, 767—770 (1940).

Clark, J. A.: Registration of improved wheat varieties, XIII. J. Amer. Soc. Agron 32, 72—75 (1940).

— Results of spring wheat varieties grown in cooperative plot and nursery experiments in the spring wheat region in 1940 with averages for 1929 to 1940. U.S. Dep. Agric., Bur. Pl. Ind., Div. Cereal Crops and Dis., Washington, D. C., Pp. 1—41 (1941). [Mimeographed.]

— Registration of improved wheat varieties, XIV. J. Amer. Soc. Agron 33, 254—256 (1941).

— Registration of improved wheat varieties, XV. J. Amer. Soc. Agron 35, 245—248 (1943).

— Registration of improved wheat varieties, XVI. J. Amer. Soc. Agron. 36, 447—452 (1944).

— Agronomic aspects of hard spring wheat breeding. 2. Results of uniform varieties in plot experiments, 1929 to date. Rep. Mill. Baking Mtg, 7th Hard Spring Wheat Conf., Minneapolis, Minn., February 28 and 29 — March 1, 1944, Pp. 43—45 (1944). [Mimeographed.]

— Agronomic aspects of hard spring wheat breeding. 4. Results from uniform regional nurseries, 1932 to date. Rep. Mill. Baking Mtg, 7th Hard Spring Wheat Conf., Minneapolis, Minn., February 28 and 29 — March 1, 1944, Pp. 46 (1944). [Mimeographed.]

— u. B. B. Bayles: Classification of wheat varieties grown in the United States in 1939. Tech. Bull. U. S. Dep. Agric., Nr. 795, 146 (1942).

— u. K. S. Quisenberry: Distribution of the varieties and classes of wheat in the United States in 1939. Circ. U. S. Dep. Agric., Nr. 634, 75 (1942).

Claassen, C. E., O. A. Vogel u. E. F. Gaines: The inheritance of reaction of Turkey-Florence-1 x Oro-1 to race 8 of Tilletia levis. J. Amer. Soc. Agron: 34, 687—694 (1942).

Cook, L. J. u. A. J. Farquhar: Baking quality of wheat. Influence of environment on varieties. J. Dep. Agric. S. Aust. 49, 191—196, 291—295, 399—412 (1945).

Cortázar Sagarminaga, R.: Estudio de la herencia de la resistencia al Puccinia graminis tritici y otros caracteres en los crūzamientos de las variedades de trigo Sinvalocho x Premier y Sinvalocho x Regent. (Study of the inheritance of resistance to P. g. tritici and other characters in crosses of the wheat varieties Sinvalocho x Premier and Sinvalocho x Regent.) Agric. Tec. Chile 4, 88—97 (1944).

Craigie, J. H.: Increase in production and value of the wheat crop in Manitoba and eastern Sakatchewan as a result of the introduction of rust resistant wheat varieties. Sci. Agric. 25, 51—64 (1944).

Crépin, C.: Le blé au cours de la campagne 1941—42. Acad. Agric. Fr., Pp. 6 (1943).

—, J. Bustarret u. R. Chevalier: Création pour la France de blés résistant à la Carie. Proc. 7th Int. Genet. Congr. Edinburgh, 23—30 August, Pp. 91—92 1939 (1941). (Abstr.)

Cugnac, A. de u. H. Belval: Identification de la lévosine (lévoholoside) dans l'extrait glucidique soluble des grains d'un Aegilops. Bull. Soc. Bot. Fr. 87, 97—101 (1940).

Cutler, G. H.: Soft winter wheat breeding project, Department of Agronomy, Purdue University, June 1941. Report of the Seventh Eastern Wheat Conference, Lafayette, Indiana, June 19—20, Pp. 3—7 (1941).

— Fairfield wheat. Circ. Ind. Agric. Exp. Sta., Nr. 276, 8.

Dadswell, I. W. u. J. F. Gardner: The relation of alpha-amylase and susceptible starch to diastatic activity. Cereal Chem. **24**, 79—99 (1947).

De Grado, A.: Ensayos comparativos de variedades de trigos, selección por líneas puras y estudios relativos a diferentes formas de siembra. (Comparative tests of wheat varieties; pure line selection and relative studies under different sowing conditions.) Bol. Inst. Invest. Agron. Madr., Nr. 10, 273—292 (1944).

Dellazoppa, J. G.: Aspectos fundamentales de los trabajos fitogenéticos realizados en ,,La Estanzuela". (Fundamental aspects of the plant breeding work accomplished at La Estanzuela.) Rev. Fac. Agron. Univ. Montevideo, Nr. 34, 37—50 (1943).

Demjanovič, N. I.: (Results of the work of the L'gov breeding station on the production of resistant varieties of winter wheat.) Naučnye Zapisky Saharnoj Promyšlennosti (Sci. Trans. Sug. Ind.), Nr. 2/3, 92—100 (1939). [Russisch.]

Derevickij, N. F.: (The technique of crossing wheat by means of wind pollination.) Selekcija i Semenovodstvo (Breeding and Seed Growing), Nr. 5, 15 (1940). [Russisch.]

— (Terms for sowing and approval of winter wheat varieties.) Proc. Lenin Acad. Agric. Sci. USSR., Nr. 6, 11—12 (1941). [Russisch.]

Dines, F. T.: The determination of wheat varieties by kernel characteristics and some of its uses. 23rd Rep. Int. Crop. Impr. Ass., Chicago, Pp. 45—54 (1941).

Dionigi, A.: Gara genetica di precocità col frumento. (Genetischer Wettstreit der Frühreife beim Weizen.) Riv. Biol. **33**, 257—258 (1942).

— Eredita del carattere ,,letargo delle cariossidi" in incroci reciproci di frumento. Meccanismo di letargo dei semi. (Inheritance of the character ,,delayed germination of the caryopsis" in reciprocal crosses of wheat. Mechanism of delayed germination of the seed.) Genetica Agraria, Roma **1**, 32—37 (1946).

Dolgušin, D. A.: (The seed production work of the breeding and genetics institute.) Selekcija i Semenovodstvo, Nr. 5, 16—17 (1939). [Russisch.]

— (Seed production of grain crops.) Jarovizacija, Nr. 1 (34), 6—26 (1941). [Russisch.]

— (,,Odessa 13".) Jarovizacija, Nr. 3 (36), 32—37 (1941). [Russisch.]

Draghetti, A.: Natura genetico-morfologica-ambientale della resistenza del frumento alle puccinie sp. (The genetical, morphological and environmental nature of resistance of wheat to Puccinia sp.) Genetica Agraria, Roma **1**, 103—111 (1946).

Dumon, A. G.: Het genetisch gedrag van den F_1 — soortbastaard Triticum vulgare lutescens x Aegilops ventricosa. (The genetic behaviour of the F_1 interspecific hybrid T. vulgare lutescens x Ae. ventricosa.) Agricultura, Louvain **43**, 1—6 (1942).

— u. J. van Wonterghem: Onderzoek in verband met de wintervastheid van tarwerassen in 1938—1939. Een nieuw wintervast bastaardras: Astra. (Research on the winter-hardiness of wheat varieties in 1938—39. A new winterhardy hybrid variety, Astra.) Agricultura, Louvain **42**, 26—47 (1939).

Dusseau, A.. Blé dur issu du croisement de deux tendres italiens. Scientia Genetica **2**, 79—81 (1940).

Eifrig, H.: Die Sichtbarmachung des mehligen Anteils bei Hartweizen. Forschungsdienst **11**, 187—189 (1941).

El Khishen, A. A. u. F. N. Briggs: Inheritance of resistance to bunt (Tilletia caries) in hybrids with Turkey wheat selections C. I. 10015 and 10016. J. Agric. Res. **71**, 403—413 (1945).

Elgueta, G. M.: Influencias que determinan la composición mineral del trigo. (Influences which determine the mineral composition of wheat.) Agric. Tec. Chile **4**, 7—16 (1944).

Elgueta G. M., S. R. Cortazar, A. E. Letelier u. C. A. Coronel: Informe preliminar sobre investigaciones culturales en trigo 1940—1941. (Preliminary report on cultural investigations with wheat 1940—1941.) Bol. Téc. Minist. Agric. Dep. Genet. Fitotec., Santiago, Nr. 1, 99 (1941).

Eliasson, S. u. G. Jacobson: Sortförsök med höstsäd. Sammanställningar av resultaten av de av jordbruksförsöksanstalten ledda försöken med höstvete och höstråg under åren 1939 (1929)—1943. (Variety trials with autumn sown cereals. A collocation of the results of the variety trials of winter wheat and winter rye conducted by the Agricultural Experiment Station in 1939(1929)—1943.) Medd. Lantbr. Jordbruksförsöksanstalten, Nr. 13, 199 (1945).

Ellerton, S.: The origin and geographical distribution of Triticum sphaerococcum Perc. and its cytogenetical behaviour in crosses with T. vulgare Vill. J. Genet. **38**, 307—324 (1939).

— Genetics and gene distribution in Triticum vulgare Vill. and other Triticum species. Abstr. Diss. Univ. Camb. (1938—1939), Pp. 22—23 (1940).

— Reaction of wheat varieties grown in Britain to Erysiphe. Nature, Lond. **153**, 776—777 (1944).

— A wheat yield trial in Maldon. Agric. Bull., Essex **2**, 84—86 (1944).

— The Maldon wheat variety trial, second year's results. Agric. Bull. Essex **3**, 95—96 (1945).

— In search of bigger wheat crops. Fmr's Wkly, Lond. **24**, 31 (1946).

— Harvests in peril. Fmr's Wkly, Lond. **25**, 32—33 (1946).

— Applied natural selection. Fmr's Wkly, Lond. **24**, 37 (1946).

Engelke, H.: Können Klebergehalt und Kleberertrag durch Züchtung und Düngung erhöht werden? Landw. Jb. **89**, 190—201 (1939).

Erdtman, G.: Sädesslagens pollenmorfologi. (Pollen morphology of the cereals.) Svensk Bot. Tidskr. **38**, 73—80 (1944).

Eritzian, A. A.: (On the study of the form building processes in inter-species crossings of wheat.) Trav. Inst. Bot. Tbilissi **7**, 135—180 (1940).

Eršov, V. I.: (On the technique of determining drought resistance in winter wheat.) Selekcija i Semenovodstvo, Nr. 1, 22—24 (1939). [Russisch.]

Fahmi, T.: (A technical method of selecting wheat restsing black rust [Puccinia graminis].) Egypt. Agric. Rev. **19**, 184—192 (1941).

Fang, C. T.: Physiologic specialization of Puccinia glumarum Erikss. and Henn. in China. Phytopathology **34**, 1020—1024 (1944).

Fauvel, J. H.: Une mutation inattendue du blé mahon. Rev. Hort. Agric. Afrique Nord **46**, 119—121 (1942).

Favret, E. A.: Presencia de la raza 15 de ,,Puccinia rubigo-vera tritici" en la Argentina. (The presence of race 15 of P. rubigo-veratritici in Argentina.) Rev. Invest. Agríc., B. Aires **1**, 63—64 (1947).

Feekes, W.: Verslagen van de Technische Tarwe Commissie. XVII. De tarwe en haar milieu. (Report of the Technical Commission on Wheat. XVII. Wheat and its environment.) Drukkerij Gebroeders Hoitsema N.V., Groningen, Pp. 523—888 (1941).

— u. W. H. van Dobben: Verslagen van de Technische Tarwe Commissie. XIV. Bakonderzoek bij de rassenproeven 1933—38. (Report of the Technical Commission on Wheat. XIV. Baking studies in variety trials 1933—38.) Drukkerij Gebroeders Hoitsema N.V., Groningen, Pp. 339—429 (1940).

—, — Verslagen van de Technische Tarwe Commissie. XV. Vochtgehalte en bewaarbaarheid van tarwe. (Report of the Technical Commission on Wheat. XV. Moisture content and storage property of wheat.)

Drukkerij Gebroeders Hoitsema N.V., Groningen, Pp. 435—463 (1941).

Feekes W. u. W. H. van Dobben: Verslagen van de Technische Tarwe Commissie. XVI. De oriënteerende methoden ter bepaling van den bakaard van tarwe. (Report of the Technical Commission on Wheat. XVI. The exploratory methods for determining the baking quality of wheat.) Drukkerij Gebroeders Hoitsema N.V., Groningen, Pp. 469—517 (1941).

Fellows, H.: Resistance to Septoria speckled leaf blotch. Report of the Fifth Hard Red Winter Wheat Improvement Conference, Manhattan, Kansas, February 12, 13 and 14, 1945. Div. Cereal Crops Dis., Plant Ind. Sta., Beltsville, Md., Pp. 32 (1945). [Mimeographed]

Fifield, C. C. et al: Chemical, milling and baking experiments with hard red spring wheats, 1940 crop. U. S. Dep., Agric., Bur. Pl. Ind., Washington D. C., Pp. 28 (1941). [Mimeographed]

— Objectives in quality testing of varieties by the U. S. Department of Agriculture, Beltsville, Maryland. Rep. Mill. Baking Mtg, 7th Hard Spring Wheat Conf., Minneapolis, Minn., February 28 and 29 — March 1, 1944, Pp. 9—13 (1944). [Mimeographed]

— A summary of the collaborative study of test baking procedures by the Minnesota, North Dakota and U.S.D.A. Laboratories. Rep. Mill. Baking Mtg, 7th Hard Spring Wheat Conf., Minneapolis, Minn., February 28 and 29 — March 1, 1944, Pp. 13—16 (1944). [Mimeographed]

— Quality characteristics of wheat varieties grown in the western United States. Tech. Bull. U. S. Dep. Agric., Nr. 887, 35 (1945).

Filipčenko, S. A.: (The biological characteristics of seeds resulting from intravarietal crossing.) Jarovizacija, Nr. 2 (35), 65—67 (1941). [Russisch.]

— u. N. A. Šelomova: (The identity of different forms of shattered inheritance.) Agrobiologija (Agrobiology), Nr. 1, 83—88 (1946). [Russisch.]

Finney, K. F.: Methods of estimating and the effect of variety and protein level on the baking absorption of flour. Cereal Chem. 22, 149—158 (1945).

— Water remaining hydrated against centrifugal force as an index of the protein quality of hard wheats. Report of the Fifth Hard Red Winter Wheat Improvement Conference, Manhattan, Kansas, February 12, 13 and 14, 1945. Div. Cereal Crops Dis., Plant Ind. Sta., Beltsville, Md. Pp. 38—39 (1945). [Mimeographed.]

— u. M. A. Barmore: Yeast variability in wheat variety test baking. Cereal Chem. 20, 194—200 (1943).

—, — Varietal responses to certain baking ingredients essential in evaluating the protein quality of hard winter wheats. Cereal Chem. 22, 225—243 (1945).

— u. W. T. Yamazaki: Water retention capacity as an index of the loaf volume potentialities and protein quality of hard red winter wheats. Cereal Chem. 23, 416—427 (1946).

Fischer, G. J., V. Gheorghianov u. D. G. Larriera: Ensayos con ocho trigos Uruguayos y catorce trigos Argentinos realizados en los departamentos de Soriano, Rio Negro, Durazno, Maldonado, Colonia y Canelones en el año 1938. (Tests with 8 Uruguay wheats and 14 Argentine wheats carried out in the departments of Soriano, Rio Negro, Durazno, Maldonado, Colonia and Canelones in 1938.) Arch. Fitotécn. Uruguay 3, 111 bis 137 (1939).

Fittschen, H. H.: Weitere Beiträge zur Züchtung steinbrandresistenter Weizensorten. Phytopath. Z. 12, 169—218 (1939).

Flaksberger, K. A.: (On the question of the nomenclature of wheat varieties.) Selekcija i Semenovodstvo (Breeding and Seed Growing), Nr. 2, 18—19 (1940). [Russisch.]

Flaksberger, K. A.: (Origin of the wheat Batetskaja.) Soviet Plant Industry Record, Nr. 1, 105 (1940). [Russisch.]

— (On wheats in the western Ukraine and in western White Russia.) Vestnik Socialističeskogo Rastenievodstva (Soviet Plant Industry Record), Nr. 2, 40 bis 50 (1940). [Russisch.]

Fondard u. Cabasson: Fécondation croisée, rayons X, ondes courtes et mutations chez les blés de Provence. C. R. Acad. Agric. Fr. 25, 503—509 (1939).

Forlani, R.: Il rinvigorimento delle sementi. La fecondazione incrociata intrarazza del grano. (Rejuvenation of seeds. Intra-racial cross-fertilization in wheat.) Ital. Agric. 78, 127—133 (1941).

— Sterilità in alcune graminacee e infezioni da Claviceps sp. (Sterility in certain graminaceous plants and infection by Claviceps sp.) Genetica Agraria, Roma 1, 218—224 (1947).

— Ibridi Triticum x Aegilops. (Triticum x Aegilops hybrids). Genetica Agraria, Roma 1, 237—253 (1947).

Forster, H. C. u. R. D. Croll: Varietal resistance to foot-rots in wheat. J. Aust. Inst. Agric. Sci. 7, 121 bis 123 (1941).

Frankel, O. H.: Analytical yield investigations on New Zealand wheat. IV. Blending varieties of wheat. J. Agric. Sci. 29, 249—261 (1939).

— Tainui, a new spring wheat variety. N. Z. J. Sci. Tech. 20, 319A—323A (1939).

— Some reflexions on breeding wheat for baking quality. Proc. 7th Int. Genet. Congr. Edinburgh, 23—30 August, Pp. 122—123, 1939 (1941).

— A critical survey of breeding wheat for baking quality. J. Agric. Sci. 30, 98—112 (1940).

— „Fife-Tuscan" wheat. A new variety for „Tuscan land". N. Z. J. Sci. Tech. 22, 303A—308A (1941).

Fröier, K.: Keimung und Triebkraft bei Hafer und Weizen nach verschiedenen Röntgendosen. Hereditas, Lund 27, 360—370 (1941).

—, O. Gelin u. Å. Gustafsson: The cytological response of polyploidy to X-ray dosage. Bot. Notiser, Pp. 199—216 (1941).

Fuggles-Couchman, N. R. u. G. B. Wallace: Cultivation and diseases of wheat. Pamphl. Dep. Agric. Tanganyika, Nr. 38, 19 (1945).

Fuller, P.: An examination of the 1944 Southern Rhodesian wheat crop. Rhod. Agric. J. 42, 459—466 (1945).

Gadea Loubriel, M.: Los trigos convenientes para la cuenca del Ebro. (The wheats suitable for the Ebro basin.) Agricultura, Madrid 11, 132—136 (1942).

— Estudios previos en torno al problema del tizón o caries del trigo (género Tilletia). (Preliminary studies on the problem of bunt or caries of wheat [genus Tilletia].) Bol. Inst. Nac. Invest. Agron. Madr., Nr. 13, 241—267 (1945).

Geddes, W. F.: Objectives in breeding for improved quality in hard wheat. J. Amer. Soc. Agron 33, 490 bis 503 (1941).

— Milling and baking research in relation to wheat breeding. 1. Objectives in the quality testing of wheat samples. Rep. Mill. Baking Mtg, 7th Hard Spring Wheat Conf., Minneapolis, Minn., February 28 and 29 — March 1, 1944, Pp. 5—7 (1944). [Mimeographed.]

Geslin, H.: Etude des lois de croissance d'une plante en fonction des facteurs du climat. (Température et radiation solaire.) Contribution à l'étude du climat du blé. Thèses Fac. Sci. Univ. Paris, Sér. A, Nr. 2087, 116 (1944).

Giacanelli, E.: Ricerche sul contenuto in catalasi di varie razze di frumenti. (Studies on the catalase content of various wheat races.) Genetica Agraria, Roma 1, 225—236 (1947).

Gilmer, W. E., H. A. Frieson u. J. B. Harrington: The resistance of wheat varieties to seed bleaching. Sci. Agric. **26**, 437—447 (1946).

Giordano, H. J.: El trigo Sinvalocho M. A. (Sin Rival x 38 M. A.-No. 32 Rafaela.) (The wheat Sinvalocho M. A. [Sin Rival x 38 M. A.-No. 32 Rafaela.]) „Granos" Semilla Selecta, B. Aires **3**, 3—16 (1939).

Giovannelli, B.: Osservazioni sui frumenti perenni russi Triticum orientale x Agropyrum glaucum e T. lutescens 062 x A. gl. e loro derivati da incroci. (Observations on Russian perennial wheats T. orientale x A. glaucum and T. lutescens 062 x A. glaucum and their derivatives from crossing.) Genetica Agraria, Roma **1**, 125—129 (1947).

Giovannini, J. M.: Estudio de diversas variedades de trigo adquiridas por el Servicio Oficial de Distribución de Semillas, con y sin inspección de cultivos en pié, en ralación a su identidad. (Study of various wheat varieties acquired by the Official Seed Distribution Service, with and without an inspection of the standing crops, in relation to their identity.) Rev. Fac. Agron., Montevideo, Nr. 39, 19—40 (1945).

Glynne, M. D. u. F. J. Moore: Eyespot and lodging in wheat. J. Minist. Agric. **53**, 305—308 (1946).

Gökğöl, M.: Zur Frage des Ursprungsgebietes der Weizen. Proc. 7th Int. Genet. Congr. Edinburgh, 23 bis 30 August, Pp. 130—131, 1939 (1941). (Abstr.)

— Über die Genzentrentheorie und den Ursprung der Weizen. Z. Pflanzenz. **23**, 562—578 (1941).

Gorlač, A. A.: (A new method of selecting wheat for resistance to brown rust.) Selekcija i Semenovodstvo (Breeding and Seed Growing), Nr. 6, 14 (1939). [Russisch.]

Gorter, G. J. M. A.: Wheat stunt — a new cereal disease. Fmg S. Afr. **22**, 29—32 (1947).

Graiff, G. L.: Contributo alla cerealicoltura libica. (A contribution to the cultivation of cereals in Libya.) Ann. Cent. Sper. Agrar. Zootec. Libia 1940 **3**, 123 bis 155 (1941).

Granate, L.: Studiul grâului din experienţe cu soiurişi lucrări culturale prin metoda farinografului Brabender. (The study of wheat by the Brabender farinograph method in variety tests and cultural tests.) Viaţa Agric. **32**, 200—204 (1941).

Granhall, I.: Genetical and physiological studies in interspecific wheat crosses. Hereditas, Lund **29**, 269 bis 380 (1943).

— On heterosis effects in Triticum vulgare. Hereditas, Lund **32**, 287—293 (1946).

Greaney, F. J. u. H. A. H. Wallace: Varietal susceptibility to kernel smudge in wheat. Phytopathology **33**, 4—5 (1943). (Abstr.)

—, — Varietal susceptibility to kernel smudge in wheat. Sci. Agric. **24**, 126—134 (1943).

Grebennikov, P. E.: (The multiple mass selection as an agrotechnical method of increasing the yield of winter wheat.) Ann. White Russian Agric. Inst., Gorki **10** (32), 1—12 (1939). [Russisch.]

Greer, E. N. u. J. B. Hutchinson: Dormancy in British-grown wheat. Nature, Lond. **155**, 381—382 (1945).

Grekov, P. I.: Verstärkte (verdoppelte) Ernährung des Embryo der Sommerweizen mit Endosperm und ihr Einfluß auf die erste Generation. C. R. (Doklady) Acad. Sci. URSS. **27**, 834—837 (1940).

Gruzl, F.: A búza, illetöleg a liszt minöségének és minösitésének alapelvei. (Principles of grading and valuation of wheat and flour.) Rep. Hung. Agric. Exp. Sta. **42**, 1—10 (1939).

Guerzi, E.: I cromosomi somatici di cinque razze di frumento. (The somatic chromosomes of five races of wheat.) Ital. Agric. **76**, 639—657 (1939).

Gugnin, J. E.: (A new variety of winter wheat Lutescens 116.) Selekcija i Semenovodstvo (Breeding and Seed Growing), Nr. 8/9, 25—26 (1940). [Russisch.]

Gull, P. W.: Wheat and barley show possibilities if, when needed. Miss. Fm. Res. **5**, Nr. 8, 3 (1942).

Hackwell, A. B. u. B. F. McKeon: Wimmera wheat tests. Trials at Longerenong. J. Dep. Agric. Vict. **39**, 1—8 (1941).

Hagborg, W. A. F. u. R. F. Peterson: A plot inoculation method for determining the resistance of wheat varieties to bacterial black chaff. Phytopathology **37**, 8 (1947). (Abstr.)

Hamilton, J. W.: Pawnee „pays off". Sth. Seedsman, **8**, Nr. 10; 16, 55 (1945).

Hanisch, H.: Klimabedingtheit der Weizenzüchtung in Südmähren. Z. Pflanzenz. **24**, 341—351 (1941).

Hansing, E. D.: Studies on bunt and loose smut in Kansas. Report of the Fifth Hard Red Winter Wheat Improvement Conference, Manhattan, Kansas, February 12, 13 and 14, 1945. Div. Cereal Crops Dis., Plant Ind. Sta., Beltsville, Md,, Pp. 31 (1945). [Mimeographed.]

Hanson, E. W.: Varietal reaction to diseases, St. Paul, Minn. Rep. Mill. Baking Mtg. 7th Hard Spring Wheat Conf., Minneapolis, Minn., February 28 and 29 — March 1, 1944, Pp. 40—41 (1944). [Mimeographed.]

Harrar, J. G., W. Q. Loegering u. E. C. Stakman: Relation of physiologic races of Puccinia graminis tritici to wheat improvement in Southern Mexico. Phytopathology **34**, 1002 (1944). (Abstr.)

Harrington, J. B.: Yielding capacity of wheat crosses as indicated by bulk hybrid tests. Canad. J. Res. **18**, Sct. C., 578—584 (1940).

— Intra-varietal crossing in wheat. J. Amer. Soc. Agron. **36**, 990—991 (1944). (Abstr.)

— u. P. F. Knowles: Dormancy in wheat and barley varieties in relation to breeding. Sci. Agric. **20**, 355 bis 364 (1940).

—, — The breeding significance of after-harvest sprouting in wheat. Sci. Agric. **20**, 402—413 (1940).

Harris, R. H.: New equipment for wheat quality testing. Bi-m. Bull. N. Dak. Agric. Exp. Sta., **4**, Nr. 1, 4—6 (1941).

— Baking quality of flours of five hard red spring wheats. Bi-m. Bull. N. Dak. Agric. Exp. Sta. **7**, Nr. 2, 15 bis 18 (1944).

— Milling and baking research in relation to wheat breeding. 2. Studies and methods used in attaining objectives. Rep. Mill. Baking Mtg, 7th Hard Spring Wheat Conf., Minneapolis, Minn., February 28 and 29 — March 1, 1944, Pp. 7—8 (1944). [Mimeographed.]

—, R. V. Olson u. J. Johnson (jr.): Viscosity changes in sodium salicylate dispersion of hard red spring wheat gluten in relation to variety and environment. Cereal Chem. **19**, 748—763 (1942).

— u. L. D. Sibbitt: Comparative baking quality of wheat starches. Bi-m. Bull. N. Dak. Agric. Exp. Sta. **4**, Nr. 2, 1—6 (1941).

—, — A method of „fingerprinting" North Dakota wheat flour in respect to their baking strength. Bi-m. Bull. N. Dak. Agric. Exp. Sta. **4**, Nr. 5, 2—5 (1942).

—, — u. G. M. Scott: Comparative effects of variety and environment on some properties of North Dakota hard red spring wheat flours. Cereal Chem. **22**, 75—81 (1945).

— u. L. R. Waldron: The relation of test weight and protein content to the milling and baking quality of hard red spring wheat hybrids. J. Agric. Res. **72**, 131 bis 136 (1946).

—, — u. L. D. Sibbitt: Comparative yields and quality data for five hard red spring wheat varieties as affected by growth location. Bi-m. Bull. N. Dak. Agric. Exp. Sta. **7**, Nr. 5, 8—11 (1945).

Hart, H.: Stem and leaf rust in relation to wheat breeding. 4. Artificial field epidemics with stem rust. Rep. Mill. Baking Mtg, 7th Hard Spring Wheat Conf. Minneapolis, Minn., February 28 and 29 — March 1, 1944, Pp. 27 (1944). [Mimeographed.]
— Stem rust on new wheat varieties and hybrids. Phytopathology 34, 884—899 (1944).
— u. J. L. Allison: A browning reaction to stem rust in wheat. Phytopathology 33, 484—496 (1943).
Harvey, P. H.: Hereditary variation in plant nutrition. Genetics 24, 74 (1939). (Abstr.)
Haussmann, G.: Il grano perenne; la sua origine e e'utilizzazione. (Perennial wheat; its origin and utilization.) Bologna-Edizioni Agricole. Pp. 82 (Undated). 14 figs. tables.
Helemskij, A. V.: (The quality as seed material and the winter hardiness of wheat regenerated by intravarietal crossing.) Selekcija i Semenovodstvo (Breeding and Seed Growing), Nr. 8/9, 38—39 (1940). [Russisch.]
Hetherington, E. V. u. G. S. Smith: Constancy of rank of durum wheats in macaroni color. Cereal Chem. 20, 345—351 (1943).
Heyne, E. G.: Effect of awns on yield and other characters in hard red winter wheat. Report of the Fifth Hard Red Winter Wheat Improvement Conference, Manhattan, Kansas, February 12, 13 and 14, 1945. Div. Cereal Crops Dis., Plant Ind. Sta., Beltsville, Md. 1—15, Pp. 7—8 (1945). [Mimeographed.]
— Use of new experimental designs. Report of the Fifth Hard Red Winter Wheat Improvement Conference, Manhattan, Kansas, February 12, 13 and 14, 1945. Div. Cereal Crops Dis., Plant Ind. Sta., Beltsville, Pp. 17 (1945). [Mimeographed.]
— u. L. P. Reitz: Characteristics and origin of Blackhull wheats. J. Amer. Soc. Agron. 36, 768—778 (1944).
—, G. A. Wiebe u. R. H. Painter: Complementary genes in wheat. J. Hered. 34, 243—245 (1943).
Hilgendorf, F. W.: Wheat in New Zealand. Whitecombe and Tombs, Limited, London (1939).
Holodnyj, T.: The bird of perennial wheat. Soviet News Nr. 1562, 2 (1946).
Holton, C. S. u. F. D. Heald: Bunt or stinking smut of wheat. (A world problem.) Burgess Publishing Company, Minneapolis Minn. (1941). [Mimeographed.]
— u. A. G. Johnson: Physiologic races in Urocystis tritici. Phytopathology 33, 169—171 (1943).
— u. C. A. Suneson: Varietal reaction to bunt in the western wheat region of the United States. J. Amer. Soc. Agron. 34, 63—71 (1942).
—, — Wheat varietal reaction to dwarf bunt in the western wheat region of the United States. J. Amer. Soc. Agron. 35, 579—583 (1943).
Holzwarth, F.: Hankóczy Jenö 1939. Rep. Hung. Agric. Exp. Sta. 42, 298—301 (1939).
Hörberg, Y.: „Bakningsdugligheten och diastatiska tillståndet hos vetemjöl". (Backfähigkeit und diastatische Kraft von Weizenmehl.) Agri Hort. Gen. 4, 74—78 (1946).
Hore, H. L., H. J. Sims u. C. G. Webb: Mallee Research Station. Experiments and results. J. Dep. Agric. Vict. 38, 523—532, 552—559 (1940).
Horovitz, N.: Algunas características de variedades de trigo. Observaciones hechas en los años 1937 y 1938, en la colección de especies y variedades de la Estación Experimental de Pergamino. (Certain characteristics of wheat varieties. Observations made in the years 1937 and 1938 on the collection of species and varieties of the Pergamino Experimental Station.) Publ. Minist. Agric. Nac., Estac. Exp. Pergamino Nr. 3, 22; also „Granos" Semilla Selecta, B. Aires 3, Nr. 6, 3—22 (1939).
— Descripción de variedades agrícolas de trigo por sus caracteres morfológicos. (Description of agricultural varieties of wheat by their morphological characters.) „Granos" Semilla Selecta, B. Aires 3, 5—130 (1939).
Horovitz, N.: Los principales trigos argentinos de invierno y prima- vera. (The principal Argentine winter and spring wheats.) „Granos" Semilla Selecta, B. Aires 8, Nos. 10—12, 37—38 (1944).
— Descripción de algunas variedades de trigo cultivadas en la Argentina. (Description of some wheat varieties cultivated in Argentina.) Rev. Argent. Agron. 11, 195—204 (1944).
Horovitz, S.: Nuevo tipo de híbrido constante de trigo × centeno („Triticum vulgare" × „Secale cereale".) (New type of constant wheat × rye hybrid. [T. vulgare × S. cereale].) Physis, B. Aires 18, 285—290 (1939).
Hoyle, S. S.: Wheat growing in Nyasaland. Nyasald Agric. Quart. J. 5, Nr. 1, 14—19 (1945).
Humphrey, H. B. u. J. A. Clark: Summary of uniform spring wheat rust nursery for 1940. U. S. Dep. Agric. Bur. Pl. Ind., Div. Cereal Crops and Dis., Washington, D. C.; Pp. 42—43 (1941). [Mimeographed.]
Huskins, C. L.: Fatuoid, speltoid and related mutations of oats and wheat. Bot. Rev. 12, 457—514 (1946).
— u. S. G. Smith: Compactoid and speltoid mutations in Triticum vulgare. Collecting Net 15, 171 (1940). (Abstr.)

Ibrahim Shah, M.: Better variety of wheat for the unirrigated lands in the Province. Quart. Notes Agric. Dep. N. W. F. Prov. 3, Nr. 3, 5—7 (1941).
Ikeda, T.: (Breeding wheat varieties for bread making.) Bot. and Zool. 7, 219—226 (1939).
Illarionov, V. F.: (Grafting wheat on rye.) Vestnik Gibridizacii (Hybridization), Nr. 2, 101—102 (1941). [Russisch.]
Immer, F. R.: Testing and recommending varieties of field crops in Minnesota. Chronica Botanica 7, 10—12 (1942).
Isenbeck, K.: Beobachtungen, Erfahrungen und Gedanken zur Dürreresistenz des Weizens als Züchtungsproblem. Chronica Botanica 5, 165—166 (1939).
Ivanov, P. K.: (The root system and drought resistance of different varieties of wheat.) Socialističeskoe Zernovoe Hozjajstvo Saratov, Nr. 6, 129—141 (1939). [Russisch.]

Jackson, S. H. u. A. G. O. Whiteside: Investigations on the thiamin content of Canadian wheat and flour. Sci. Agric. 22, 366—377 (1942).
Jacometti, G.: Autarchia anche nella produzione di grani di forza. (Autarchy also in the production of strong wheats.) Ital. Agric. 78, 513—518 (1941).
Jakovlev, M. S. u. R. A. Ergesjan: (Structural peculiarities of the wheat grain as a result of intravarietal crossing.) Jarovizacija, Nr. 2 (35), 56—61 (1941). [Russisch.]
Japha, B.: Studien über Frühjahrsschäden an Winterweizen. Kühn-Archiv 60, 429—439 (1943/1944).
Jasny, N.: The wheats of classical antiquity. Baltimore, Pp. 176, 2 pls. tables (1944).
Johansson, A.: Höstvetesorternas vinterhärdighet. (The winter hardiness of varieties of autumn wheat.) Lantmannen 31, 509 (1947).
Johnson, J. A.: Summary of baking quality of certain new varieties. Report of the Fifth Hard Red Winter Wheat Improvement Conference, Manhattan, Kansas, February 12, 13 and 14, 1945. Div. Cereal Crops Dis., Plant Ind. Sta., Beltsville, Md., Pp. 41 (1945). [Mimeographed.]
—, C. O. Swanson u. E. G. Bayfield: The correlation of mixograms with baking results. Cereal Chem. 20, 625—644 (1943).
Johnson, L. P. V. u. H. W. Holtz: Colchicine treatment techniques for sprouted seeds and seedlings. Canad. J. Res. 24, Sect. C., 303—304 (1946).

Johnson, L. P. V., H. A. McLennan u. J. M. Armstrong: Fertility and morphological characters in Triticum-Agropyron hybrids. Genetics 24, 91—92 (1939). (Abstr.)

Johnson, T.: Recent changes in the physiologic races of certain of the cereal rusts. Proc. Canad. Phytopath. Soc., Nr. 14, 13 (1946). (Abstr.)

— u. W. A. F. Hagborg: Melanism in wheat induced by high temperature and humidity. Canad. J. Res. 22, Sect. C., 7—10 (1944).

— u. M. Newton: The influence of light and certain other environmental factors on the matureplant resistance of Hope wheat to stem rust. Canad. J. Res. 18, Sect. C., 357—371 (1940).

—, —: The effect of high temperature on the stem rust resistance of wheat varieties. Canad. J. Res. 19, Sect. C., 438—445 (1941).

—, —: The occurrence of new strains of Puccinia triticina in Canada and their bearing on varietal reaction. Sci. Agric. 26, 468—478 (1946).

Johnston, C. O.: Progress in breeding for rust resistance in wheat. Report of the Fifth Hard Red Winter Wheat Improvement Conference, Manhattan, Kansas, February 12, 13 and 14, 1945, Div. Cereal Crops Dis., Plant Ind. Sta., Beltsville, Md., Pp. 24—25 (1945). [Mimeographed.]

Jones, E. T.: A discussion of Hessian fly resistance in certain wheat varieties. Trans. Kans. Acad. Sci. 43, 243—265 (1940).

— Insect resistance in wheat. J. Amer. Soc. Agron. 35, 695—703 (1943).

— Sources of resistance to the hessian fly. Report of the Fifth Hard Red Winter Wheat Improvement Conference, Manhattan, Kansas, February 12, 13 and 14, 1945. Div. Cereal Crops Dis., Plant Ind. Sta., Beltsville, Md., Pp. 21 (1945). [Mimeographed.]

Kagawa, F.: The effect of abnormal temperature on the course of pollen formation in a genus hybrid Triticum compactum × Secale cereale. Jap. J. Bot. 10, 55—68 (1939).

— (High temperature treatments in Triticum and the character and chromosomes of the next generation.) Proc. Crop. Sci. Soc. Japan 12, 90—93 (1940).

Kahidze, N. T.: Chromomere structure of mitotic chromosomes in wheats. C. R. (Doklady) Acad. Sci. URSS. 26, 468—470 (1940).

Kakizaki, Y. u. S. Suzuki: (Earliness as influenced by seasonal growth habit in wheat hybrids.) Jap. J. Genet. 16, 59—63 (1940).

Kale, G. T.: Breeding rust resistant wheat. Int. Rev. Agric. 30, 325—330 (1939).

Kalinin, P. K.: (Breeding wheat at Khibiny.) Selekcija i Semenovodstvo, Nr. 2/3, 26—28 (1939). [Russisch.]

Kamçioglu, H. I.: Türkiye ve Ecnebî buğdaylarinin teknolojik evsafi ve bilhassa ekmeklik kabiliyetlerinin tetkik ve mukayeseleri. (The study and comparison of the technical properties and especially the baking qualities of foreign and Turkish wheats.) T. C. Ziraat Vekâleti Yüksek Ziraat Enstit. Çalişmal., Nr. 107, 163 (1941).

Kar, B. K.: Vernalization of Indian crops. II. Photostage in wheat (Triticum vulgare) and oat (Avena sp). Proc. 31st. Indian Sci. Congr., Delhi, Pt III, Pp. 87 bis 88 (1944). (Abstr.)

Kargopolova, N. N.: (Intra-varietal crossing and the increase in spring wheats of resistance to smut.) Jarovizacija, Nr. 3 (36), 67—69 (1941). [Russisch.]

Kasparjan, A. S.: A new amphidiploid-einkorn × Persian wheat (Triticum monococcum Hornemanni Clem. × Triticum persicum fuliginosum Zhuk.). C. R. (Doklady) Acad. Sci. URSS. 26, 166—169 (1940).

Kattermann, G.: Über heterogenomatische amphidiploide Weizenroggenbastarde. Z. Pflanzenz. 23, 179 bis 209 (1939).

Kenway, C. B. u. H. B. Peto: Researches on drought resistance in spring wheat. I. A machine for measuring the resistance of plants to artificial drought. Canad. J. Res. 17, Sect. C, 294—296 (1939).

Khan, M. A.: What's doing in All-India. The Punjab. Indian Fmg 3, 152—155 (1942).

Khelemskii, A. V. siehe Helemskij.

Kholodny, T. siehe Holodnyj.

Kihara, H.: (A new classification of the genus Aegilops on the basis of genom analysis.) Jap. J. Genet. 15, 336—337 (1939).

— (Genetics of interspecific hybrids.) Kagaku (Science) 9, 454—460 (1939).

— (Polyploidy of wheat.) Bot. and Zool. 7, 211—218 (1939).

— (Formations of haploids by means of delayed pollination in Triticum monococcum.) Bot. Mag. Tokyo 54, 178—185 (1940).

— (Haploids produced by delayed-pollination in einkorn wheat.) Agric. and Hort. 15, Nr. 5, 194 (1940).

— Verwandtschaft der Aegilops-Arten im Lichte der Genomanalyse. Ein Überblick. Züchter 12, 49—62 (1940).

— Anwendung der Genomanalyse für die Systematik von Triticum und Aegilops. Japan. Journ. Genet. 16, 309—320 (1940).

— u. K. Matsumoto: (Morphology of species hybrids in the genus Aegilops.) Jap. J. Genet. 15, 334—336 (1939).

—, — Nachkommen mit einem Genomtyp von Aegylops variabilis in der F_4 Generation des Bastardes Ae. ovata × variabilis. Jap. J. Genet. 16, 291—294 (1940).

— u. S. Matsumura: Weitere Untersuchungen über die pentaploiden Triticum-Bastarde. XII. Schlußmitteilung. Jap. J. Bot. 11, 27—39 (1940).

—, — Genomanalyse bei Triticum und Aegilops. VIII. Rückkreuzung des Bastards Ae. caudata × Ae. cylindrica zu den Eltern und seine Nachkommen. Cytologia, Tokyo 11, 493—506 (1941).

Kirilenko, M. K.: (Work on hybridizing wheat.) Selekcija i Semenovodstvo (Breeding and Seed Growing, Nr. 5, 28 (1940). [Russisch.]

Kiseleva, A. K.: (A new variety of winter wheat „Velutinum 081".) Selekcija i Semenovodstvo (Breeding and Seed Growing), Nr. 9, 15 (1939). [Russisch.]

Klein, E.: Tres nuevas variedades culturales de trigo. (Three new cultivated varieties of wheat.) Rev. Argent. Agron. 8, 154—160 (1941).

Kneen, E. u. M. J. Blish: Carbohydrate metabolism and winter hardiness of wheat. J. Agric. Res. 62, 1—26 (1941).

— u. H. L. Hads: Effects of variety and environment on the amylases of germinated wheat and barley. Cereal Chem. 22, 407—418 (1945).

Knjaginičev, M. I.: (On the importance of variety in raising the protein content of the grain of wheats in the regions of the USSR.) Vestnik Socialističeskogo Rastenievodstva (Soviet Plant Industry Record), Nr. 3, 117—130 (1940). [Russisch.]

— (The accumulation and the physicochemical properties of the protein in the grain of different species and varieties of wheats.) Vestnik Socialističeskogo Rastenievodstva (Soviet Plant Industry Record), Nr. 5, 118—132 (1940). [Russisch.]

—, I. F. Mutulj u. J. K. Palilova: Wheat species characterized according to actibity and quality of amylase in their grain. C. R. (Doklady) Acad. Sci. URSS. 27, 1020—1023 (1940).

—, —, — (The activity of carbohydrase in the ripening grain of different wheat varieties.) Biohimija 5, 288 bis 300 (1940). [Russisch.]

Knjaginičev, M. I. u. J. K. Palilova: (The action and quality of catalase in wheats, barleys and plants of genera related to that of wheat.) Biohimija **5**, 55—64 (1940). [Russisch.]

Knowles, P. F.: A second factor for awn barbing in durum wheat. Canad. J. Res. **21**, Abt. C, 198—204 (1943).

— u. J. B. Harrington: Breeding smooth-awned durum and vulgare wheats. Sci. Agric. **23**, 697—707 (1943).

Koblet, R.: Über die Eiweißbildung im Getreide. Schweiz. landw. Mh. **19**, 122—134 (1941).

Kolesnikov, P.: Branches of the Academy of Sciences of the USSR. II. Science **98**, 231—233 (1943).

Kondo, M. u. Y. Kasahara: (Variety-distinction of wheat and barley by means of phenol coloration.) Proc. Crop. Sci. Soc. Japan **11**, 230—252 (1939).

Kondratenko, F.: (Breeding winter crops at the Barnaul State Breeding Station.) Selekcija i Semenovodstvo (Breeding and Seed Growing), Nr. 4, 9—10 (1940). [Russisch.]

Kononenko, M. V.: (The effect of the dates of seeding on the nature of winter wheat in successive generations.) Proc. Lenin Acad. Agric. Sci. USSR., Nr. 1/2, 41—44 (1942). [Russisch.]

Konovalov, I. N.: (The physiological characteristics of the influence of vernalization on the growth of plants.) Sovetskaja Botanika (Soviet Botany), Nr. 3, 21—36 (1944). [Russisch.]

Kostjučenko, I. A.: (The question of organization and technique of intravarietal crossing in spring wheat.) Selekcija i Semenovodstvo (Breeding and Seed Growing), Nr. 5, 16—18 (1940). [Russisch.]

— (First results of work of the Siberian breeding stations with winter crops.) Jarovizacija, Nr. 5 (32), 62—68 (1940). [Russisch.]

Kostoff, D.: Das Verhalten der Chromosomen in Weizenbastarden. IV. Die natürlichen Gattungsbastarde Triticum Timopheevi x Agropyrum repens. Züchter **13**, 269—272 (1941).

— Haploide Triticum vulgare und die Variabilität ihrer diploiden Nachkommenschaften. Züchter **15**, 121—125 (1943).

Kovaleva, P. G.: (Notes on work on the crossing of Agropyron species with wheat.) Vestnik Gibridizacii (Hybridization), Nr. 2, 99—100 (1941). [Russisch.]

Krasovskij, I.: (Effect of conditions at time of setting and ripening of seeds on the drought resistance of wheat.) Bull. Acad. Sci. URSS. Sér. Biol., Nr. 4, 495—503 (1940). [Russisch.]

Kravčenko, J. S.: (On varietal names.) Selekcija i Semenovodstvo (Breeding and Seed Growing), Nr. 10, 27—29 (1940). [Russisch.]

Krishna Iyer, P. V. u. S. Azizuddin Ahmad: Efficiency of some sampling methods for wheat crop. Curr. Sci. **12**, 258—259 (1943).

Krishnaswamy, N.: Cytological studies in a haploid plant of Triticum vulgare. Hereditas, Lund **25**, 77—86 (1939).

Krotov, A. S.: (Selection as a method of improving self pollinated plants.) Selekcija i Semenovodstvo (Breeding and Seed Growing), Nr. 5, 6—9 (1940). [Russisch.]

— (On improving self-pollinated varieties.) Soviet Plant Industry Record, Nr. 1, 22—26 (1940). [Russisch.]

Kühl, O. u. N. Sârbu: Rezultatele experientelor cu şase soiuri de grau de toamnă pe anii 1933—36. (Results of experiments with six varieties of winter wheat in the years 1933—1936.) Anal. Inst. Cerc. Agron. Român. **10**, 214—232. 1938 (1939).

Kupcov, A. I.: An attempt of synthesizing winter wheats for the sub-tayga zone of west Siberia. C. R. (Doklady) Acad. Sci. URSS. **43**, 166—169 (1944).

Kuperman, F. M.: (Variability and structure of winter wheat yield at high altitude zones of Kabarda-Balkarian Autonomous Republic). Proc. Lenin Acad. Agric. Sci. USSR., Nr. 10, 3—7 (1941). [Russisch.]

Kuprijanov, S. I. u. P. E. Sudnov: (Nature of the development of secondary shoots in A. I. Deržavin's wheat [perennial wheat-rye hybrid].) Selekcija i Semenovodstvo (Breeding and Seed Growing). Nr. 7, 33—37 (1939). [Russisch.]

Kupzow, A. I. siehe Kupcov.

Kuznecov, A. V.: (Let us create suitable varieties of winter wheat for Siberia.) Jarovizacija, Nr. 5 (32), 76—78 (1940). [Russisch.]

Lamprecht, H.: Förädlingsarbetet med höstvete pä Weibullsholm. (Breeding work with autumn wheat at Weibullsholm.) Agri Hortique Genetica, Landskrona **1**, 3—14 (1943).

Lange de la Camp, M.: Die Weizen der Deutschen Hindukusch-Expedition 1935. Landw. Jb. **88**, 14—133 (1939).

Lapčenko, G. D.: (Hybrids between Agropyrum and wheat.) Vestnik Gibridizacii (Hybridization), Nr. 1, 20—33 (1941). [Russisch.]

Lathouwers, V.: Manuel de l'amélioration des plantes cultivées, Tome II. L'amélioration du froment. (Cytologie, morphologie, physiologie, génétique, phylogénie.) Librairie Agricole de la Maison Rustique, Paris, Pp. XVI + 579. tables. 66 figs. (1942).

Laude, H. H.: Long-time wheat variety yield comparisons. J. Amer. Soc. Agron. **36**, 152—161 (1944).

— Problems of varietal purity. Report of the Fifth Hard Red Winter Wheat Improvement Conference, Manhattan, Kansas, February 12, 13 and 14, 1945. Div. Cereal Crops Dis., Plant Ind. Sta., Beltsville, Md. Pp. 5—6 (1945). [Mimeographed.]

— Field vs. nursery plots for varietal evaluation. Report of the Fifth Hard Red Winter Wheat Improvement Conference, Manhattan, Kansas, February 12, 13 and 14, 1945. Div. Cereal Crops Dis., Plant Ind. Sta., Beltsville, Md. Pp. 16—17 (1945). [Mimeographed.]

— u. A. F. Swanson: Natural selection in varietal mixtures of winter wheat. J. Amer. Soc. Agron. **34**, 270 bis 274 (1942).

Lavrušin, M.: (Zoning of varieties of winter wheat.) Socialističeskoe Seljskoe Hozjajstvo (Socialistic Agriculture), Moskva, Nr. 7/8, 88—90 (1941). [Russisch.]

Leckie, W.G.: Annual Report of the Department of Agriculture, Basutoland, for the year ending 30th September 1944, Pp. 16 (1944).

Lehmann, H.: Herkunft, Verbreitung und Züchtung des Weizens. einer unserer wichtigsten und ältesten Kulturpflanzen. Mühle **76**, 113—115 (1939).

Lein, A.: Die genetische Grundlage der Kreuzbarkeit zwischen Weizen und Roggen. Z. indukt. Abstamm. u. Vererblehre **81**, 28—59 (1943).

— Die Wirksamkeit von Kreuzbarkeitsgenen des Weizens in Kreuzungen von Roggen ♀ mit Weizen ♂. Züchter **15**, 1—3 (1943).

— Über Rückkreuzungsversuche eines amphidiploiden Weizen × Roggen-Bastards mit Roggen. Kühn-Archiv **60**, 226—237 (1943/1944).

Leonard, E. H.: Progress in wheat improvement in the Pacific Northwest. Bull. Ass. Operative Millers, p. 1320 (1943).

Letelier, A. E.: La densidad de siembra en relacion con los ensayos multi-varietales de cereales. (The density of sowing in relation to trials of cereals involving many varieties.) Bol. Téc. Minist. Agric. Chile, Nr. 5, 18 (1943).

Levine, M. N.: Stem and leaf rust in relation to wheat breeding. 1. Relation of leaf rust to spring wheat improvement. Rep. Mill. Baking Mtg, 7th Hard Spring Wheat Conf., Minneapolis, Minn., February 28 and 29 — March. 1, 1944, Pp. 24—26 (1944). [Mimeographed.]

Levitskij, G. A., M. A. Sizova u. V. A. Poddubnaja-Arnoldi: Comparative morphology of the chromosomes in wheat. C. R. (Doklady) Acad. Sci. URSS. **25**, 142—145 (1939).

Li, C. H. u. H. W. Li: Cytological studies of a haploid wheat plant. Chinese J. Sci. Agric. **1**, 183—189 (1944).

Li, H. W., W. K. Pao u. C. H. Li: Desynapsis in the common wheat. Amer. J. Bot. **32**, 92—101 (1945).

Litovčenko, A. G.: (Über die kritische Ernährungsperiode bei Winterweizen.) C. R. (Doklady) Acad. Sci. URSS. **55**, 65 (1947). [Russisch.]

Livingstone, J. E. u. J. C. Swinbank: Two types of late spring frost injury to winter wheat. J. Amer. Soc. Agron. **39**, 536—544 (1947).

Li-Ying Shen: (Breeding of the National Research 28 wheat.) Special Publication of the National Agricultural Research Bureau, Szechuan, Nr. 21 (1939).

Ljaščenko, I. F.: (Mutationen der Winterweizen.) C. R. (Doklady) Acad. Sci. URSS. **58**, 1501 (1947). [Russisch.]

Loegering, W. Q.: Stem and leaf rust in relation to wheat breeding. 3. Greenhouse inoculations with stem rust. Rep. Mill. Baking Mtg, 7th Hard Spring Wheat Conf., Minneapolis, Minn., February 28 and 29 — March 1, 1944, Pp. 27 (1944). [Mimeographed.]

Lomejko, S.: (The toutes by which wheat (T. vulgare Vill.) has penetrated into Europe from its centre of origin-a contribution to the study of the origin of domestic wheats.) Arhiv Minist. Poljopr. **6**, Nr. 14, 97—126 (1939).

Love, H. H. u. W. T. Craig: Better wheat for New York. Bull. Cornell Agric. Exp. Sta., Nr. 828, 27 (1946).

Love, R. M.: Cytogenetics of vulgare-like derivatives of pentaploid wheat crosses. Genetics **24**, 92 (1939). (Abstr.)

— The role of cytology in wheat improvement. Proc. 7th Int. Genet. Congr. Edinburgh 23—30 August, p. 197, 1939 (1941). (Abstr.)

— A cytologically deficient speltoid of hybrid origin. Genetics **25**, 126 (1940). (Abstr.)

— Chromosome number and behaviour in a plant breeder's sample of pentaploid wheat hybrid derivatives. Canad. J. Res. **18**, Sect. C, 415—434 (1940).

— Chromosome behaviour in F_1 wheat hybrids. I. Pentaploids. Canad. J. Res. **19**, 351—369 (1941).

— A cytogenetic study of offtypes in a winter wheat, Dawson's Golden Chaff, including a white chaff mutant. Canad. J. Res. **21**, Sect. C, 257—264 (1943).

— u. C. A. Suneson: Cytogenetics of certain Triticum-Agropyron hybrids and their fertile derivatives. Amer. J. Bot. **32**, 451—456 (1945).

Lukjanenko, P. P.: (Production of a new variety of winter wheat in Kuban „Novo-Ukrainka Nr. 83".) Selekcija i Semenovodstvo, Nr. 5, 7—10 (1939). [Russisch.]

— (Methods of breeding of winter wheats resistant to brown rust.) Jarovizacija, Nr. 3 (36), 38—47 (1941). [Russisch.]

— (The influence of the maternal form on the winter hardiness of interspecies hybrids of wheat.) Jarovizacija, Nr. 3 (36), 99—101 (1941). [Russisch.]

Lukjanjuk, V. I.: (New promising variety of spring wheat.) Selekcija i Semenovodstvo (Breeding and Seed Growing), Nr. 7, 30—31 (1939). [Russisch.]

Luneva, M. F.: (New varieties of winter wheat of the Čakinsk State Breeding Station.) Selekcija i Semenovodstvo (Breeding and Seed Growing), Nr. 8/9, 20—21 (1940). [Russisch.]

McCalla, A. G.: Varietal differences in the relation between protein content of wheat and loaf volume of bread. Canad. J. Res. **18**, Sect. C, 111—121 (1940).

— u. W. G. Corns: Effects of variety and environment on the starch content of wheat and barley. Canad. J. Res. **21**, Sect. C, 307—321 (1943).

McCluggage, M. E.: Micro milling and baking of small samples of wheat. Cereal Chem. **20**, 185—193 (1943).

— Factors influencing the pearling test for kernel hardness in wheat. Cereal Chem. **20**, 686—700 (1943).

McEwen, J., G. W. Walker, W. L. Waterhouse u. G. L. Sutton: Commonwealth Farrer Memorial „Lambrigg", Canberra, A. C. T. Unveiling ceremony and dedication, 16th January 1939. Published under the authority of the Farrer Memorial Trust, Sydney, N. S. W., Pp. 23.

McFadden, E. S.: Brown necrosis, a discoloration associated with rust infection in certain rust-resistant wheats. J. Agric. Res. **58**, 805—819 (1939).

— What is being done with species hybrids in Texas-some new synthetic hexaploid wheats. Report of the Fifth Hard Red Winter Wheat Improvement Conference, Manhattan, Kansas, February 12, 13 and 14, 1945. Div. Cereal Crops Dis., Plant Ind. Sta. Beltsville, Md. Pp. 2 (1945). [Mimeographed.]

— u. E. R. Sears: The artificial synthesis of Triticum spelta. Genetics **30**, 14 (1945). (Abstr.)

—, — The origin of Triticum spelta and its free-threshing hexaploid relatives. J. Hered. **37**, 81—89, 107 bis 116 (1946).

Macindoe, S. L.: William Farrer's contribution to our knowledge of inheritance. J. Aust. Inst. Agric. Sci. **5**, 208—212; also Agric. Gaz. N. S. W. **51**, 63—64 (1940). (Abstr.)

— Science and a better way of life for farmers. J. Aust. Inst. Agric. Sci. **9**, 115—121 (1943).

— New stem rust resistant wheats to replace Eureka. Agric. Gaz. N. S. W. **56**, 530—531 (1945).

— Centenary of the birth of William Farrer, 1845—1906. J. Aust. Inst. Agric. Sci. **11**, 72—73 (1945).

McKeon, B. F.: Wheat varieties in the Wimmera. J. Dep. Agric. Vict. **41**, 243—245 (1943).

McLoughlin, D. E.: Wheat production in Southern Rhodesia. Rhod. Agric. J. **36**, 260—274 (1939).

— u. T. K. Sansom: Wheat production in Southern Rhodesia. Rhod. Agric. J. **42**, 222—242 (1945).

Makarovskij, A. F.: (Variation of morphological characters in wheat varieties.) Socialistic Grain Farming, Saratov, Nr. 1, 162—170 (1941). [Russisch.]

Maljcev, T. S.: (Paper read at the Narkomzem board meeting.) Jarovizacija, Nr. 2 (35), 148 (1941). [Russisch.]

Mallik, A. K., V. Satakopan u. S. Gopal Rao: A study on the estimation of the yield of wheat by sampling. Indian J. Agric. Sci. **15**, 219—225 (1945).

Mann, H. H.: Wheat in the Middle East. Emp. J. Exp. Agric. **14**, 31—42 (1946).

Manunta, C.: Sul colore delle farine. Studio comparato del contenuto in pigmenti delle farine di varie razze di Triticum coltivate in varie località. (Flour colour. A comparative study of the pigment content of the flours from different Triticum races cultivated in different localities.) Genetica Agraria, Roma **1**, 38—59 (1946).

— Analisi cromatografica e spettroscopica dei pigmenti carotenoidi delle farine e studio comparato del contenuto in pigmenti delle cariossidi di varie razze di Triticum coltivate in varie località. (Chromotographic and spectroscopic analysis of the carotenoid pigments of flours and a comparative study of the pigment content of the grains of various Triticum races cultivated in various localities.) Genetica Agraria, Roma **1**, 167 bis 186 (1947).

Marani, M. u. G. Goia: Ricerche su alcune provenienze di frumento „Damiano Chiesa". (Studies on certain provenances of the wheat Damiano Chiesa.) Ital. Agric. **78**, 780—784 (1941).

Marinič, P.: (Early maturing spring wheat varieties for the east.) Socialističeskoe Seljskoe Hozjajstvo (Socialistic Agriculture), Moskva, Nr. 4, 54—56 (1942). [Russisch.]

Marušev, A. I.: (Industrial qualities of varieties of wheat, winter rye and millet bred at the S. E. Institute of Grain Culture). Socialistic Grain Farming, Saratov, Nr. 2, 51—61 (1940). [Russisch.]

— (The quality of the grain from wheat varieties of the Institute compared with that of varieties from other plant breeding stations in the USSR.) Bjulletenj Instituta Zernovogo Hozjajstva Jugo-Vostoka SSSR. (Bull. Inst. Grain Husb. S. E. USSR.), Saratov, Nr. 3, 29—34 (1945). [Russisch.]

Matsumura, S.: 20jährige zytogenetische Untersuchung des pentaploiden Weizenbastards zwischen Emmer- und Dinkelreihen. Züchter 11, 289—301 (1939).

— Weitere Untersuchungen über die pentaploiden Triticum-Bastarde. IX. Aequations- und Zertationskreuzungen des Bastards T. durum × T. vulgare. Jap. J. Bot. 9, 353—371 (1939).

— (The missing chromosome in B type speltoid wheats.) Jap. J. Genet. 15, 323—324 (1939).

— Weitere Untersuchungen über die pentaploiden Triticum-Bastarde X. Kreuzungsversuche mit gemischtem Pollen. Jap. J. Bot. 10, 477—487 (1940).

— Weitere Untersuchungen über die pentaploiden Triticum-Bastarde XI. Jap. J. Bot. 11, 17—25 (1940).

— (Induced haploidy and autotetraploidy in Aegilops ovata L.) Bot. Mag., Tokyo 54, 404—413 (1940).

—, K. Matsumoto, K. Yamashita u. Y. Nakamura: Häufigkeit der verschiedenchromosomigen F_2-Pflanzen bei den pentaploiden Weizenbastarden. M. 1 Textfig. u. 1 Taf. Bot. a. Zool. 7, 1719—1725 (1939).

Mayr, E.: Sortenfragen und Sortengebiete in der Ostmark. Angew. Bot. 22, 86—97 (1940).

Medvedeva, G. B.: (The problem of dominance in wheat.) Proc. Lenin Acad. Agric. Sci. USSR., Nr. 8/9, 29—36 (1944). [Russisch.]

— A contribution to the study of the inheritance of awnedness in wheat. C. R. (Doklady) Acad. Sci. URSS. 47, 507—509 (1945).

— (Directed variation in wheat hybrids.) Agrobiologija (Agrobiology), Nr. 1, 89—94 (1946). [Russisch.]

Mehta, K. C. u. B. P. Pal: Rust-resistant wheats for India. Nature, London 146, 98 (1940).

Mejster, N. G.: (Breeding and raising seed of winter wheat.) Naučnyj Otčet Inst. Zernovogo Hozjajstva Jugo-Vostoka SSSR za 1941—42 gg. (Sci. Rep. Inst. Grain Husbandry South-Eastern USSR for 1941—42), Pp. 167—177 (1944). [Russisch.]

— (Varieties of winter wheat from Saratov.) Bjulletenj Instituta Zernovogo Hozjajstva Jugo-Vostoka SSSR. (Bull. Inst. Grain Husb. S. E. USSR.), Saratov, Nr. 2, 6—11 (1945). [Russisch.]

Meljničenko, J. S. u. M. J. Tregubenko: (Effectiveness of intravarietal crossing in winter wheat.) Selekcija i Semenovodstvo (Breeding and Seed Growing), Nr. 10, 9—13 (1940). [Russisch.]

Meredith, W. O. S., W. J. Eva u. J. A. Anderson: Effect of variety and environment on some qualities of malted wheat flour. Cereal Chem. 21, 233—240 (1944).

Merritt, P. P.: The Minnesota Station milling and baking methods and the philosophy for using them. Rep. Mill. Baking Mtg, 7th Hard Spring Wheat Conf., Minneapolis, Minn., February 28 and 29 — March 1, 1944, Pp. 8—9 (1944). [Mimeographed.]

— u. W. F. Geddes: The complementary baking properties of Minnesota spring and winter wheat varieties. Cereal Chem. 20, 98—103 (1943).

Miczyński, K.: The inheritance of some characters in the intervarietal crosses of Aegilops. Proc. 7th Int. Genet. Congr. Edinburgh, 23—30 August, Pp. 219 (1939) (1941). (Abstr.)

Middleton, G. K., W. H. Chapman, J. W. Hendricks u. D. W. Colvard: Wheat varieties for North Carolina. Bull. N. C. Agric. Exp. Sta., Nr. 328, 11 (1940).

Miège, E.: L'hérédité de la composition chimique chez les hybrides intergénériques. Étude de la descendance d'Aegilops ovata L. var. nigra x Triticum vulgare et d'Aegilops ovata x Triticum durum Desf. Proc. 7 th Int. Genet. Congr. Edinburgh 23—30 August, Pp. 219 bis 220, 1939 (1941). (Abstr.)

— L'hérédité de la composition chimique chez les hybrides intergénériques. Scientia Genetica 2, 82—91 (1940).

— u. J. Courtine: Méthodes de diagnose spécifique dans le genre „Triticum". Application à l'identification de quelques blés intermédiaires. 4th Congr. Fed. Soc. Savantes Afr. N., Pp. 37.

Milan, A.: Sensibilità per la Ustilago tritici (Pers.) Jens. di alcuni ibridi normali di Frumento. (The susceptibility to U. tritici (Pers.) Jens. of some normal hybrids of wheat). Riv. Patol. Veg. 29, 71—84 (1939).

Miller, E. G., H. G. Gauch u. G. A. Gries: A study of the morphological nature and physiological functions of the awns of winter wheat. Tech. Bull. Kans. Agric. Exp. Sta., Nr. 57, 82 (1944).

Miller, W. B., H. L. Hore u. H. J. Sims: Mallee Research Station, Walpeup. Experiments and results. J. Dep. Agric. Vict. 37, 511—519, 535—546 (1939).

Mirzojan, A. I.: (New early varieties of spring wheat for the forest and sub-forest regions of Siberia.) Selekcija i Semenovodstvo (Breeding and Seed Growing), Nr. 7, 25—27 (1939). [Russisch.]

— (New promising varieties of spring wheat in the northern region of the USSR. excluding the Black Soil Zone.) Selekcija i Semenovodstvo (Breeding and Seed Growing), Nr. 9, 10—12 (1939). [Russisch.]

Moran, T.: Report on the quality-for bread-making purposes- of wheat harvested in 1939 at the headquarters and sub-stations of the National Institute of Agricultural Botany. J. Nat. Inst. Agric. Bot. 5, 9—27 (1944).

— u. C. R. Jones: Nutrients in British-grown and imported wheat. Nature, Lond. 157, 643—646 (1946).

Mori, H.: (On the method of testing the quality of wheat flour.) Proc. Crop. Sci. Soc. Japan 11, 112—123 (1939).

Morris, H. E. u. A. M. Schlehuber: Studies on control of bunt of wheat. Bull. Mont. Agric. Exp. Sta., Nr. 393, 18 (1941).

Morris, V. H.: Tests useful in evaluating soft wheat varieties. Rep. 8th East. Wheat Conf. Cincinatti, Ohio November 10, 1943, Pp. 6—10 (1944). [Mimeographed.]

Mudra, A.: Contributiuni la genetica grâului. III. Forma spícului şi lungimea paiului. (Contribution to the genetics of wheat. III. Ear shape and straw length.) Bul. Fac. Agron. Cluj 7, 176—192 (1938) 1939.

— Cercetări asupra raportului dintre insuşirile elitelor şi insuşirile primelor descendenţe la grâul de toamnă. (Investigations on the correlations between the characters of élite plants and of their progeny in winter wheat.) Bul. Fac. Agron. Cluj 8, 1—8 (1939).

Mundkur, B. B. u. B. P. Pal: Studies in Indian cereal smuts. II. Varietal resistance of Indian and other wheats to loose smut. Indian J. Agric. Sci. 11, 675 bis 686 (1941).

Müntzing, A.: Studies on the properties and the ways of production of rye-wheat amphidiploids. Hereditas, Lund 25, 387—430 (1939).

Muravjev, P. A.: (The grain quality of certain annual Triticum Agropyron hybrids of N. V. Tzitzin and of A. I. Deržavin's perennial wheat.) Selekcija i Semenovodstvo (Breeding and Seed Growing), Nr. 8, 22—24 (1939). [Russisch.]

Muravjev, P. A.: (An experiment on improving the grain quality of a winter wheat élite.) Selekcija i Semenovodstvo (Breeding and Seed Growing), Nr. 11 bis 12, 40—42 (1940). [Russisch.]

Nalivkin, A. A.: (Zur Frage der Agrotechnik und Düngung der Sommerweizensorten.) Trudy Vologodskogo Seljskohoz. Instituta, Nr. 1, 139—145 (1940). [Russisch.]

Nanda, K. K. u. J. J. Chinoy: Effect of photoperiodic treatment on pollen fertility. Curr. Sci. 14, 241 (1945).

Navalihina, N. K.: Restitution of fertility in a wheat-rye hybrid through colchicine treatment. C. R. (Doklady) Acad. Sci. URSS. 27, 587—589 (1940).

Nazarenko, S. I.: (Drought resistance of plants as influenced by the conditions under which the seeds were raised.) Sovetskaja Botanika (Soviet Botany), Nr. 1 bis 2, 72—79 (1941). [Russisch.]

Neatby, K. W.: New varieties of spring wheat resistant to stem rust in the Canadian west, and their genetical background. Emp. J. Exp. Agric. 10, 245—252 (1942).

Nemlienko, N. E.: (Results of work with Triticum-Agropyron hybrids.) Selekcija i Semenovodstvo, Nr. 4, 16—19 (1939). [Russisch.]

Nevano, G.: Due nuove razze elette di frumento: ,,Anna Nevano" ed ,,Ariano Irpino". (Two élite races of wheat, Anna Nevano and Ariano Irpino.) Ital. Agric. 79, 42—45 (1942).

Newman, L. H.: New wheat creations and their significance to Canada. Canad. Georgr. J. 18, 208—216 (1939).

Newton, M. u. W. J. Cherewick: Erysiphe graminis in Canada. Canad. J. Res. 25, Sect. C, 73—93 (1947).

— u. T. Johnson: Adult plant resistance in wheat to physiologic races of Puccinia triticina Erikss. Canad. J. Res. 21, Sect. C, 10—17 (1943).

—, — Physiologic races of Puccinia graminis tritici in Canada, 1919 to 1944. Canad. J. Res. 24, Sect. C, 26—38 (1946).

—, — u. B. Peturson: Seedling reactions of wheat varieties to stem rust and leaf rust and of oat varieties to stem rust and crown rust. Canad. J. Res. 18, Sect. C, 489—506 (1940).

Nieves, R.: Herencia del color del grano en el trigo. (Inheritance of grain colour in wheat.) Jorn. Agron. Vet., B. Aires, Pp. 129—154, 1939 (1940).

Nikolaenko, E. I.: A short-day wheat from China. C. R. (Doklady) Acad. Sci. URSS. 30, 353—355 (1941).

Newton, M. u. W. J. Cherewick: Erysiphe graminis in Canada. Canad. J. Res. 25, Sect. C, Pp. 73—93 (1947).

Noble, W. B., W. B. Cartwright u. C. A. Suneson: Inheritance of resistance to hessian fly. J. Econ. Ent. 33, 580—581 (1940).

— u. C. A. Suneson: Differentiation of the two genetic factors for resistance to the Hessian fly in Dawson wheat. J. Agric. Res. 67, 27—32 (1943).

Nover-Schlichting, I.: Untersuchungen über den Weizenmehltau, Erysiphe graminis tritici, im Rahmen der Resistenzzüchtung. Z. Pflanzenz. 24, 71—103 (1941).

Novickij, A.: (A promising variety of spring wheat.) Socialističeskoe Seljskoe Hozjajstvo (Socialistic Agriculture) Moscow, Nr. 2, 88 (1941). [Russisch.]

Novickij, S. P.: (Resistance of Triticum-Agropyron hybrids to the main cereal diseases.) Theses and scientific papers read at the 4th District Conference of Workers of Universities and Research Institutions, Omsk, Nr. 1: Agron. Sect., 58—60 (1941). [Russisch.]

Novogrudskij, D. M.: Relation of diurnal variation in water content of leaves and of tillering nodes of wheat to variation in their water holding capacity. C. R. (Doklady) Acad. Sci. URSS. 52, Nr. 9, 817—820 (1946).

Novogrudskij, D. M.: On the moisture content, moisture-holding capacity and hydrophily of dry matter of the leaf series in wheat. C. R. (Doklady) Acad. Sci. URSS. 53, Nr. 8, 721—723 (1946).

Oláh, L.: A búzafajták rozsdaellenállóképességének átöröklése. (The inheritance of rust resistance of wheat varieties.) Rep. Hung. Agric. Exp. Sta. 42, 203—234 (1939).

— Hogyan lehetne a bánkúti búzák fagyállóságát megjavítani. (On the possibility of improving the winterhardiness of Bánkut wheat.) Rep. Hung. Agric. Exp. Sta. 42, 235—240 (1939).

Oliva, A.: La resistenza al freddo dei frumenti montani nel collaudo invernale 1941—42. (The cold resistance of the mountain wheats in the winter test of 1941—42.) Ital. Agric. 79, 290—293 (1942).

O'Mara, J. G.: Cytogenetic studies on Triticale. I. A method for determining the effects of individual Secale chromosomes on Triticum. Genetics 25, 401—408 (1940).

— The substitution of a specific Secale cereale chromosome for a specific Triticum vulgare chromosome. Genetics 32, 99—100 (1947). (Abstr.)

Oort, A. J. P.: Onderzoekingen over stuifbrand. II. Overgevoeligheid van tarwe voor stuifbrand (Ustilago tritici). (Investigations on loose smut. II. Hypersensitiveness of wheat to loose smut [U. tritici].) Tijdschr. PlZiekt. 50, 73—106 (1944).

Oparin, A. J.: (Biochemical indices of the baking quality of grain and flour.) Bull. Acad. Sci. URSS., Sér. Biol., Pp. 43—70 (1939). [Russisch.]

Oseledec, P. I.: (An interspecific hybrid consisting of three species [preliminary communication]). Selekcija i Semenovodstvo (Breeding and Seed Growing), Nr. 6, 8—10 (1940). [Russisch.]

Östergren, G.: A hybrid between Triticum turgidum and Agropyron junceum. Hereditas, Lund 26, 395 bis 398 (1940).

Packard, C. M.: Breeding wheat and alfalfa for resistance to insect attack. J. Econ. Ent. 34, 347—352 (1941).

Painter, R. H.: Breeding for resistance to hessian fly in hard red winter wheat. Report of the Fifth Hard Red Winter Wheat Improvement Conference, Manhattan, Kansas, February 12, 13 and 14, 1945. Div. Cereal Crops Dis., Plant Ind. Sta., Beltsville, Md., Pp. 21—22 (1945). [Mimeographed.]

— u. E. T. Jones: The Hessian fly resistance of Pawnee wheat. J. Kans. Ent. Soc. 18, 130—149 (1945).

—, —, C. O. Johnston u. J. H. Parker: Transference of Hessian fly resistance and other characteristics of Marquillo spring wheat to winter wheat. Tech. Bull. Kans. Agric. Exp. Sta., Nr. 49, 55 (1940).

Paiva, O.: Notas sôbre fisiologia e seleção de trigo. (Notes on physiology and selection of wheat.) Rev. Agron., Brazil 6, 535—536 (1942).

Pal, B. P.: The Pusa wheats: the wheat-breeding work of the Imperial Agricultural Research Institute. Emp. J. Exp. Agric. 12, 61—73 (1944).

— et al.: The description of crop-plant characters and their ranges of variation. III. The variability of Indian wheats. Indian J. Agric. Sci. 11, 477—492 (1941).

— u. B. B. Mundkur: Studies in Indian cereal smuts. III. Varietal resistance of Indian and other wheats to flag smut. Indian J. Agric. Sci. 11, 687—694 (1941).

Palienko, V. I.: (New hybrid varieties of winter wheat from the Rostov Breeding Station.) Selekcija i Semenovodstvo (Breeding and Seed Growing), Nr. 6, 33 (1939). [Russisch.]

Palmova, E.: (Kind of inheritance of quantitative characters in hybridization of hard wheat.) Proc. Lenin Acad. Sci. USSR., Nr. 1, 10—13 (1940). [Russisch.]

Pan, C. L.: A genetic study of mature plant resistance in spring wheat to black stem rust, Puccinia graminis tritici and reaction to black chaff, Bacterium translucens, var. undulosum. J. Amer. Soc. Agron. 32, 107—115 (1940).

Pančenko, N. P.: (Varieties of winter wheat for the Leningrad region.) Soviet Plant Industry Record, Nr.1, 104 (1940). [Russisch.]

— (Controlled conversion of varieties of winter wheat in breeding operations and in nature.) Vestnik Socialističeskogo Rastenievodstva (Soviet Plant Industry Record), Nr. 3, 49—58 (1940). [Russisch.]

Panse, E.: Die züchterischen Grundlagen zur Steigerung der Winterfestigkeit bei Weizen mit besonderer Berücksichtigung der Hallenser Züchtungsarbeiten. Kühn Archiv 60, 402—428 (1943/1944).

Pao, W. K. u. H. W. Li: (On the inheritance of pentaploid wheat hybrids. A critique.) Chinese J. Agric. Sci. 1, 23—32 (1943).

—, — On the inheritance of pentaploid wheat hybrids. A critique. Acta Brevia Sinensia, Nr. 7, 21—22 (1944). (Abstr.) [Mimeographed.]

—, — Maternal inheritance of variegation in common wheat. Chinese J. Sci. Agric. 1, 166—171 (1944).

—, — Maternal inheritance of variegation in common wheat. J. Amer. Soc. Agron. 38, 90—94 (1946).

—, C. H. Li, C. W. Chen u. H. W. Li: Inheritance of dwarfness in common wheat. J. Amer. Soc. Agron. 36, 417—428 (1944).

—, —, T. W. Ching u. H. W. Li: (Studies on the inheritance of dwarfness in common wheat.) Chinese J. Sci. Agric. 1, 1—12 (1943).

—, —, —, — Studies on the inheritance of dwarfness in common wheat. Acta Brevia Sinensia, Nr. 7, 20—21 (1944). (Abstr.) [Mimeographed.]

Papadakis, J. S.: The relation of the number of tillers per unit area to the yield of wheat and its bearing on fertilizing and breeding this plant — the space factor. Soil. Sci. 50, 369—388 (1940).

Paremud, L. H.: (New forms of wheat-rye hybrids.) Selekcija i Semenovodstvo (Breeding and Seed Growing), Nr. 4, 4—5 (1940). [Russisch.]

Parodi, L. R.: Una especie de trigo que debe cambiar de nombre: Triticum paradoxum, nov. nom. (A wheat species which ought to change its name: T. paradoxum, nov. nom.) Rev. Argent. Agron. 7, 49—50 (1940).

Pathak, G. N.: Studies in the cytology of Crocus and cereals with special reference to satellites and nucleoli. Proc. 7th Int. Genet. Congr. Edinburgh 23—30 August, Pp. 232—233, 1939 (1941). (Abstr.)

— Studies in the cytology of cereals. J. Genet. 39, 437—467 (1940).

— A preliminary study of the cytology of interspecific hybrids in Triticum and an intergeneric hybrid, T. vulgare x Aegilops caudata. Indian J. Genet. Pl. Breed. 2, 37—42 (1942).

Pavlov, K.: Wheat No. 11-Agronomic and Botanical description. Rev. Inst. Rech. Agron. Bulg. 9, Nr. 4, 45—78 (1939).

Pence, R. O.: Summary of the milling of 1944 U.S.D.A. samples. Report of the Fifth Hard Red Winter Wheat Improvement Conference, Manhattan, Kansas, February 12, 13 and 14, 1945. Div. Cereal Crops Dis., Plant Ind. Sta., Beltsville, Md., Pp. 40 (1945). [Mimeographed.]

Perak, J. T.: Triticum durum tetraploide obtenido por colchicina. (Tetraploid T. durum obtained by colchicine.) An. Inst. Fitotec. Santa Catalina 2, 7—8 (1940).

Pereira, A.: Proteínas do trigo. I-contribuição para o estudo da variação da proteína e gluten segundo a variedade e a região. (Proteins of wheat. I-Contribution to the study of the variation of protein and gluten according to variety and locality.) Agron. Lusitana 6, 367—396 (1944).

Peterson, R. F.: Inheritance of resistance of H-44 and Hope wheats to stem rust. J. Hered. 31, 272 (1940).

—, T. Johnson u. M. Newton: Varieties of Triticum vulgare practically immune in all stages of growth to stem rust. Science 91, 313 (1940).

—, A. J. Lejeune u. H. C. Laidlaw: Identification of grain samples of hard red spring wheat varieties grown in western Canada. Sci. Agric. 25, 711—717 (1945).

— u. R. M. Love: A study of the transference of immunity to stem rust from Triticum durum var. Iumillo to T. vulgare by hybridization. Sci. Agric. 20, 608 bis 623 (1940).

— u. W. O. S. Meredith: Agronomic and quality characteristics of Carleton durum wheat grown in the durum wheat area of Western Canada. Sci. Agric. 25, 107—113 (1944).

Peto, F. H.: Chromosome doubling induced by temperature shocks in hybrid zygotes of Triticum vulgare pollinated with Agropyron glaucum. Genetics 24, 93 (1939). (Abstr.)

— Fertility and meiotic behaviour in F_1 and F_2 generations of Triticum-Agropyron hybrids. Genetics 24, 93 (1939). (Abstr.)

— Cytology of Triticum-Agropyron glaucum backcrosses. Proc. 7th Int. Genet. Congr. Edinburgh 23—30 August, Pp. 235—236, 1939 (1941). (Abstr.)

— u. J. W. Boyes: Hybridization of Triticum and Agropyron. VI. Induced fertility in vernal emmer x A. glaucum. Canad. J. Res. 18, Sect. C, 230—239 (1940).

— u. G. A. Young: Hybridization of Triticum and Agropyron. VII. New fertile amphidiploids. Canad. J. Res. 20, Sect. C, 123—129 (1942).

Petrov, G. G. u. A. P. Sokolovskij: (The problem of drought resistance. I. On the transpiration of xeromorphic wheats.) Trudy Omsk. Seljskohozjajstvennogo Inst. imeni S. M. Kirova (Trans. Kirov Inst. Agric. Omsk, USSR.) 4 (17), 3—22 (1939). [Russisch.]

Peturson, B.: Epidemiology of rust in western Canada as influenced by the introduction of stem-rust-resistant varieties. Phytopathology 37, 18 (1947). (Abstr.)

—, M. Newton u. A. G. O. Whiteside: The effect of leaf rust on the yield and quality of wheat. Canad. J. Res. 23, Sect. C, 105—114 (1945).

—, —, — The effect of leaf rust (Puccinia triticina) on the yield, grade, and quality of wheat. Phytopathology 37, 18 (1947). (Abstr.)

Philp, J.: A new wheat variety „Mabrook". Egypt. Agric. Rev. 19, Part 4, 280—284 (1941).

— u. A. G. Selim: Rust-resistant wheats for Egypt. Nature, London 147, 209 (1941).

Phipps, I. F., S. R. Hockley u. A. T. Pugsley: Warigo-a disease-resistant wheat. J. Aust. Inst. Agric. Sci. 9, 17—20 (1943).

Piacco, R.: Un ibrido fra frumento e segale. (A hybrid of wheat and rye.) Risicoltura 29, 151 (1939).

Pirovano, A.: Stimoli mutative sul grano. (Stimuli inducing mutation in wheat.) Ricerca Scientifica 10, 693—703 (1939).

Pisarev, V. E. u. E. S. Malinovskaja: (Breeding spring wheat for resistance to Fusarium.) Selekcija i Semenovodstvo (Breeding and Seed Growing), Nr. 8, 13—18 (1939). [Russisch.]

—, — (The breeding of spring wheats resistant to Fusarium.) Trudy Zonaljnogo Inst. Zernovogo Hozjajstva Rajonov Nečernozemnoj Polosy (Trans. Zonal Inst. Grain Husbandry Non-Black-Soil Districts), Nr. 10, 35—38 (1941). [Russisch.]

— u. N. M. Vinogradova: Hybrids between wheat and Elymus. C. R. (Doklady) Acad. Sci. URSS. 45, 129—132 (1944).

Platt, A. W.: The influence of some environmental factors on the expression of the solid stem character in certain wheat varieties. Sci. Agric. 22, 139—151 (1941).

Platt, A. W.: Breeding wheats for sawfly resistance. Canad. Geogr. J. 23, 138—141 (1946).
— u. J. G. Darroch: The seedling resistance of wheat varieties to artificial drought in relation to grain yield. Sci. Agric. 22, 521—527 (1942).
—, — u. H. J. Kemp: The inheritance of solid stem and certain other characters in crosses between varieties of Triticum vulgare. Sci. Agric. 22, 216—224 (1941).
— u. C. W. Farstad: The reaction of wheat varieties to wheat stem sawfly attack. Sci. Agric. 26, 231—247 (1946).
— u. R. Larson: An attemp to transfer solid stem from Triticum durum to T. vulgare by hybridization. Sci. Agric. 24, 214—220 (1944).
Plotnikov, N. J.: (Promising varieties of spring wheat for the Transural, Western Siberia and Kazakhstan.) Selekcija i Semenovodstvo (Breeding and Seed Growing), Nr. 7, 23—25 (1939). [Russisch.]
— (Introduction of new varieties of spring wheat in the east. Socialističeskoe Seljskoe Hozjajstvo (Socialistic Agriculture) Moskva, Nr. 12, 16—19 (1942).
Poddubnaja-Arnoljdi, V.: On hybridization between Triticum and Elymus. C. R. (Doklady) Acad. Sci. URSS. 24, 378—381 (1939).
— A contribution to the chromosome morphology of hard wheat. C. R. (Doklady) Acad. Sci. URSS. 48, 278—280 (1945).
Poehlman, J. M. u. F. Bowman: The production of quality in Missouri soft wheat. Bull. Mo. Agric. Exp. Sta., Nr. 487, 15 (1945).
Pohilj, I. F.: (A new variety of winter wheat [„Lutescens 09"].) Selekcija i Semenovodstvo (Breeding and Seed Growing), Nr. 9, 16 (1939). [Russisch.]
— (The winter wheat Efremovka.) Selekcija i Semenovodstvo (Breeding and Seed Growing), Nr. 8/9, 22—23 (1940). [Russisch.]
— (The new variety of winter wheat Stakhanovka.) Selekcija i Semenovodstvo (Breeding and Seed Growing), Nr. 10, 13—14 (1940). [Russisch.]
Ponomarenko, F. A.: (An experiment on breeding winter wheat and rye in the East.) Jarovizacija, Nr. 5 (32), 69—75 (1940). [Russisch.]
— (The role of selective fertilization in the winter hardiness of the Siberian wheat Skala.) Vestnik Socialističeskogo Rastenievodstva (Soviet Plant Industry Record), Nr. 1, 10—19 (1941). [Russisch.]
Popescu, S.: Rezultatele culturilor comparative cu soiuri de grâu, orz, ovăz, în Moldova întregită in anii 1933—1937. (Results of comparative tests of varieties of wheat, barley and oats in Moldavia in the years 1933—1937.) Anal. Inst. Cerc. Agron. Român. 10, 141—191, 1938 (1939).
Popova, G. I.: Cytologische Untersuchung eines neuen Bastardes Triticum Timopheevi x Agropyrum elongatum. Cytologia, Tokyo 9, 495—498 (1939).
— (Fertility in wheat x Agropyrum hybrids.) Vestnik Gibridizacii (Hybridization), Nr. 2, 15—20 (1941). [Russisch.]
Povolotskij, P. K.: (A new variety of winter wheat.) Selekcija i Semenovodstvo, Nr. 2/3, 25—26 (1939). [Russisch.]
Pozdnjakov, M. M.: (Variation in the ear colour of wheat under the highland conditions of Balkaria.) Jarovizacija, Nr. 6 (33), 113 (1940). [Russisch.]
Pridham, J. T.: A successful cross between Triticum vulgare and Triticum Timopheevi. J. Aust. Inst. Agric. Sci. 5, 160—161 (1939).
Przyborowski, J. u. Z. Nawrocki: Doswiadczenia z odmianami pszenicy jarej przeprowadzone w Polsce w latach 1936—1938. (Experiments with spring sown varieties of wheat carried out in Poland during 1936 bis 1938.) Prace Nauk. Rol. Suppl. to Przegl. Doswiad. Roln., Nr. 8, 32 (1939).

Psarev, G. M. u. H. A. Veselovskaja: (Über den Einfluß einiger synthetischen Stoffe auf die Entwicklung der Embryowurzeln des Winterweizens.) C. R. (Doklady) Acad. Sci. URSS. 56, 973 (1947). [Russisch.]
Pugsley, A. T.: Varietal resistance of wheat to loose smut. J. Aust. Inst. Agric. Sci. 9, 86—88 (1943).
— Breeding for bunt resistance in Australian wheats. J. Aust. Inst. Agric. Sci. 11, 28—34 (1945).
— u. S. R. Hockley: Wheat variety and crossbred trials at Pallamana, 1944—46. J. Dep. Agric. S. Aust. 50, 289—291 (1947).
— u. I. F. Phipps: The combination of stem rust resistant genes from the wheats Eureka and Warigo. J. Aust. Inst. Agric. Sci. 9, 130—132 (1943).
Punič, A. E.: (Results of 3 years' tests of annual spring forms of Triticum-Agropyron hybrids.) Theses and scientific papers read at the 4th District Conference of Workers of Universities and Research Institutions, Omsk, Nr. 1, Agron. Sect., 57—58 (1941). [Russisch.]
Pustovojt, E. S.: (Work on the improvement of the winter wheat Buivolinka.) Selekcija i Semenovodstvo (Breeding and Seed Growing), Nr. 7, 40—41 (1939).
Putnam, H. O.: Report of milling and baking results by the Northwest Crop Improvement Association. Wheat Variety Committee and Plant Breeders Meeting. Rep. Mill. Baking Mtg, 7th Hard Spring Wheat Conf., Minneapolis, Minn., February 28 and 29 — March 1, 1944, Pp. 1—3 (1944). [Mimeographed.]
Putnam, L. G.: A study of the inheritance of solid stems in some tetraploid wheats. Sci. Agric. 22, 594—607 (1942).
Quisenberry, K. S.: New wheats of the future. Grain Feed J. 88, 346—347 (1943).
— Pawnee, the new winter wheat for Nebraska. Rep. Neb. Bd. Agric. Pp. 127—138 (1944).
—, H. A. Rodenhiser u. C. O. Johnston: Bunt reaction of hard red winter wheats in 1938—42. J. Amer. Soc. Agron. 37, 514—522 (1945).
—, O. J. Webster u. T. A. Kiesselbach: Varieties of winter wheat in Nebraska. Bull. Neb. Agric. Exp. Sta. Nr. 326, 28 (1940).
—, —, — Varieties of winter wheat for Nebraska. Bull. Neb. Agric. Exp. Sta. Nr. 367, 20 (1944).
Ranjan, S.: A preliminary note on the X-ray mutants of Pusa (52) wheat. Proc. Indian Acad. Sci. 12, Sect. B., 62—66 (1940).
— The agricultural problem of India. J. Indian Bot. Soc. 26, 1—8 (1947).
Raw, A. R.: Intergeneric hybridization. A preliminary note of investigations on the use of colchicine in inducing fertility. J. Dep. Agric. Vict. 37, 50—52 (1939).
— Wheat breeding. Progress and development. J. Dep. Agric. Vict. 38, 1—2, 25 (1940).
Redfearn, C. u. N. R. Fuggles-Couchman: Large-scale wheat production at Oldeani, Tanganyika Territory, in 1943 and 1944. E. Afr. Agric. J. 11, 122—129 (1945).
Reed, E. W.: Need for additional research from the standpoint of the miller and baker. Report of the Fifth Hard Red Winter Wheat Improvement Conference, Manhattan, Kansas, February 12, 13 and 14, 1945. Div. Cereal Crops Dis., Plant Ind. Sta., Beltsville, Md., Pp. 46—51 (1945). [Mimeographed.]
Reichert, I.: On the disease resistance of wild emmer. Palest. J. Bot. Ser. 4, 179—183 (1944).
Reitz, L. P.: Breeding hard red winter wheat in Kansas. Report of the Fifth Hard Red Winter Wheat Improvement Conference, Manhattan, Kansas. February 12, 13 and 14, 1945. Div. Cereal Crops Dis., Plant Ind. Sta., Beltsville, Md., Pp. 6—7 (1945). [Mimeographed.]
— u. E. G. Heyne: Wheat planting and wheat improvement in Kansas. 33rd Bienn. Rep. Kans. Bd. Agric. 1941—1942, 38, 168—207 (1942).

Reitz, L. P., E. T. Jones, C. O. Johnston u. R. H. Painter: Agronomic tests of new resistant varieties and hybrids of hard red winter wheat in the presence of stem rust and Hessian fly. J. Amer. Soc. Agron. **35**, 216—229 (1943).

—, C. O. Johnston u. K. L. Anderson: New combinations of genes in wheat × wheatgrass hybrids. Trans. Kans. Acad. Sci. **48**, 151—159 (1945—46).

— u. H. H. Laude: Comanche and Pawnee: new varieties of hard red winter wheat for Kansas. Bull. Kans. Agric. Exp. Sta., Nr. 319, 16 (1943).

Revilla, V. A.: Razas fisiológicas de la roya negra del trigo (Puccinia gráminis tritici) encontradas en el Perú. (Physiological races of the black rust of wheat (P. g. tritici) encountered in Peru). Bol. Estac. Exp. Agric. „La Molina" Nr. 26, 16 (1945).

Risso Patron, R.: Descripción de 35 variedades de trigo del país, con observaciones sobre la constancia de algunos caracteres morfológicos. (Description of 35 local varieties of wheat, with observations on constancy of certain morphological characters). Rev. Fac. Agron. La Plata **24**, 57—234 (1939).

Roberts, T. C.: Milling and baking research in relation to wheat breeding. 3. Trends in commercial milling and baking practices and their probable effects upon wheat quality requirements. Rep. Mill. Baking Mtg., 7th Hard Spring Wheat Conf., Minneapolis, Minn., February 28 and 29—March 1, 1944, Pp. 17—19 (1944.) [Mimeographed.]

— Wheat and flour quality requirements. Report of the Fifth Hard Red Winter Wheat Improvement Conference, Manhattan, Kansas, February 12, 13 and 14, 1945. Div. Cereal Crops Dis., Plant Ind. Sta., Beltsville, Md., Pp. 33—36 (1945). [Mimeographed.]

Roberts, W.: Work on Punjab wheat since 1906. Indian Fmg. **2**, 10—11 (1941).

Robertson, D. W.: Winter wheat breeding program at Fort Collins. Report of the Fifth Hard Red Winter Wheat Improvement Conference, Manhattan, Kansas, February 12, 13 and 14, 1945. Div. Cereal Crops Dis., Plant Ind. Sta., Beltsville, Md., Pp. 12 (1945.) [Mimeographed.]

Rodenhiser, H. A. u. J. A. Clark: Summary of uniform spring wheat bunt nursery, 1940. U. S. Dep. Agric., Bur. Pl. Ind. Div. Cereal Crops and Dis., Washington, D. S. Pp. 44—45 (1941). [Mimeographed.]

— u. C. S. Holton: Variability in reaction of wheat differential varieties to physiologic races of Tilletia levis and T. tritici. Phytopathology **32**, 158—165 (1942).

—,— Distribution of races of Tilletia foetida and T. caries in relation to the wheat improvement program in the United States. Phytopathology **35**, 488—489 (1945). (Abstr.)

—, — Distribution of races of Tilletia caries and Tilletia foetida and their relative virulence on certain varieties and selections of wheat. Phytopathology **35**, 955—969 (1945).

— u. R. W. Smith: Bunt, rootrots, scab and other diseases in relation to wheat breeding. 1. Bunt. Varietal tests and physiologic race surveys. Bunt resistance studies at Dickinson, No. Dak. Rep. Mill. Baking Mtg., 7th Hard Spring Wheat Conf., Minneapolis, Minn., February 28 and 29—March 1, 1944, Pp. 31—32 (1944). [Mimeographed.]

— u. J. W. Taylor: The effect of photoperiodism on the development of bunt in two spring wheats. Phytopathology **33**, 240—244 (1943).

Rodriguez, V. J.: Observations on leaf and stripe rust of wheat in Mexico. Phytopathology **36**, 410 (1946). (Abstr.)

Roemer, Th.: Ausgangsmaterial für die Resistenzzüchtung bei Getreide. Ergebnisse 20jähriger Arbeit der Pflanzenzuchtstation Halle a. S. Z. Pflanzenz. **24**, 304 bis 332 (1941).

Roemer, Th.: Erzeugung von Qualitätsweizen. Forschungsdienst, Sonderh. 16, 368 (1942).

Romanovici, A.: Originele varietații de grâu „Mentana". (The origins of the wheat variety „Mentana"). Viața Agric. **30**, 55—57 (1939).

Rössger, W.: Beitrag zur Technik der Weizenkreuzung. Forsch.dienst **14**, 330—334 (1942).

Rusakov, I. F.: (The loss of rust resistance by wheat varieties in North Caucasus.) Agrobiologija (Agrobiology) Nr. 2, 52—62 (1946). [Russisch.]

Ryžej, I. P.: (Work on breeding winter wheat at the Kirgiz Breeding Station.) Selekcija i Semenovodstvo (Breeding and Seed Growing) Nr. 7, 31—32 (1940). [Russisch].

S...., B. C.: A synthesis of a 42-chromosome wheat. Nature, Lond. **152**, 575—576 (1943).

Safronov, A.: (Crimean wheat.) Sovhoznoe Proizvodstvo (State Farming) Nos. 4/5, 21—22 (1945). [Russisch.]

Sakalo, V. D.: (On the causes of different response of spring wheat varieties to fertilizers.) Proc. Lenin Acad. Agric. Sci. USSR. Nr. 2, 10—14 (1941). [Russisch.]

Saks, A. I.: (On several direct methods of evaluating wheat varieties as regards their drought resistance.) Bull. Acad. Sci. URSS. Sér. Biol. Nr. 1, 163—175 (1941). [Russisch.]

Salmon, S. C.: Agronomic aspects of hard spring wheat breeding. 9. Where do we go from here. Rep. Mill. Baking Mtg. 7th. Hard Spring Wheat Conf., Minneapolis, Minn., February 28 and 29—March 1, 1944, Pp. 51—53 (1944). [Mimeographed.]

— Techniques and progress in breeding wheat. Report of the Fifth Hard Red Winter Wheat Improvement Conference, Manhattan, Kansas, February 12, 13 and 14, 1945. Div. Cereals Crops Dis., Plant Ind. Sta., Beltsville, Md. Pp. 14—15 (1945). [Mimeographed.]

Saltykovskij, M. I. u. E. S. Sapryguina: On selecting pairs in crossing and breeding winter hardy wheat. C. R. (Doklady.) Acad. Sci. URSS. **25**, 766—769 (1939).

—, — A method for determination of cold-resistance of winter cereals. C. R. (Doklady) Acad. Sci. URSS. **28**, 544—457 (1940).

—, — Winterness in the second generation of wheat hybrids. C. R. (Doklady) Acad. Sci. URSS. **31**, 628 bis 631 (1941).

—, — Synthesis of winter-hardy wheats. C. R. (Doklady) Acad. Sci. URSS. **39**, 280—283 (1943).

—, — A method for increasing winterness in wheats. C. R. (Doklady) Acad. Sci. URSS. **45**, 349—352 (1944).

Sande Bakhuyzen, H. L. van de: De ontwikkelingsgeschiedenis der tarweplant. (The ontogeny of the wheat plant.) Landbouwk. Tijdschr. **55**, 533—548 (1943).

Sandstedt, R. M. u. K. Fortman: Effect of environment during the growth and development of wheat on the baking properties of its flour. Cereal Chem. **21**, 172—188 (1944).

Sansom, T. K.: Wheat. Brief characteristics of varieties tested at the Plant Breeding Station, Salisbury, and available for distribution. Rhod. Agric. J. **39**, 11—14 (1942).

— A description of the more common Rhodesian wheat varieties. Rhod. Agric. J. **39**, 477—483 (1942).

— Wheat varieties tested at the Plant Breeding Station, Salisbury, and available for distribution. Rhod. Agric. J. **40**, 6 (1943).

— Wheat. Varieties tested at the Plant Breeding Station. Salisbury, and available for distribution. Rhod. Agric. J. **42**, 19—20 (1945).

— Wheat. Varieties tested at the Plant Breeding Station, Salisbury, and available for distribution. Rhod. Agric. **42**, 486—487 (1945).

Sapryguina, E. S.: (The use of natural cold in breeding winter cereals.) Selekcija i Semenovodstvo (Breeding and Seed Growing) Nr. 8/9, 13—16 (1940). [Russisch.]
— Cold-resistance of wheat F_2 hybrids with reference to phasic characteristics of their parents. C. R. (Doklady) Acad. Sci. URSS. 30, 840—843 (1941).
Sârbu, N. u. O. Kühl: Studiu comparativ al soiurilor de grâu de toamnă „Sandu-Aldea" obtinute prin hibridare. (Comparative study of lines of the autumn wheat „Sandu-Aldea" obtained by hybridization). Anal. Inst. Cerc. Agron. Român. 10, 254—281, 1938 (1939).
Savickij, M. S.: (New and promising varieties of winter and spring wheats at the Soviet Agricultural Exhibition.) Selekcija i Semenovodstvo (Breeding and Seed Growing) Nr. 8, 5—9. [Russisch.]
Ščerbakov, A. P.: (Hastening the germination of seeds.) Sovetskaja Botanika (Soviet Botany) Nos 4/5, 60—70 (1944). [Russisch.]
Ščerbina, D. R.: (Effect of intra-varietal crossing and conditions of rearing of the spring wheat TZA/32.) Jarovizacija Nr. 2 (35), 62—64 (1940). [Russisch.]
Schad, C., R. Mayer, u. G. Méneret: Contribution à l'étude de la qualité du blé et de son amélioration dans la région du Centre. Ann. Agron., Paris, 11, 223 bis 269 (1941).
Scharnagel, Th.: Fortschritte der Weizenzüchtung unter Beachtung der Auswinterung. Mitt. Landw. 58, 23—24 (1943).
Schiemann, E.: Zur Demonstration eines Weizen-„stammbaums". Biologe 8, 148 (1939).
— Die Getreidefunde der neolithischen Siedlung Trebus, Krs. Lebus/Mark. Ber. dtsch. bot. Ges. 58, 446—459 (1940).
— Weizenstammbäume. Bot. Jb. 71, Nr. 1, 1—31 (1940).
Schneidermann, J. A.: (Intercrossing capacity of wheat with Agropyron and the fertility of their hybrids in relation to the conditions of development.) Socialistic Grain Farming, Saratow Nr. 3, 13—33 (1940). [Russisch.]
Schribaux, M.: Le froid et les blés en terre. C. R. Acad. Agric. Fr. 25, 29—44 (1939).
Sears, E. R.: Amphidiploids in the Triticinae induced by colchicine. J. Hered. 30, 38—43 (1939).
— Cytogenetic studies with polyploid species of wheat. I. Chromosomal aberrations in the progeny of a haploid of Triticum vulgare. Genetics 24, 509—523 (1939).
— Monosomes, trisomes and segmental interchanges from a haploid of Triticum vulgare. Genetics 24, 84 (1939). (Abstr.)
— Monofactorially conditioned inviability of an intergeneric hybrid in the Triticinae. Genetics 25, 134 (1940). (Abstr.)
— Amphidiploids in the seven-chromosome Triticinae. Res. Bull. Mo. Agric. Exp. Sta. Nr. 336, 46 (1941).
— Chromosome pairing and fertility in hybrids and amphidiploids in the Triticinae. Res. Bull. Mo. Agric. Exp. Sta. Nr. 337, 20 (1941).
— Nullisomics in Triticum vulgare. Genetics 26, 167 bis 168 (1941). (Abstr.)
— Inviability of intergeneric hybrids involving Triticum monococcum and T. aegilopoides. Genetics 29, 113 bis 127 (1944).
— Cytogenetic studies with polyploid species of wheat. II. Additional chromosomal aberrations in Triticum vulgare. Genetics 29, 232—246 (1944).
— The amphidiploids Aegilops cylindrica × Triticum durum and A. ventricose × T. durum and their hybrids with T. aestivum. J. Agric. Res. 68, 135—144 (1944).
— Isochromosomes and telocentrics in Triticum vulgare. Genetics 31, 229—230 (1946). (Abstr.)
— The sphaerococcum gene in wheat. Genetics 32, 102 bis 103 (1947). (Abstr.)

Séguéla, J. M.: Technique de la sélection du blé en Tunisie. Ann. Serv. Bot. Agron. Tunis 18, 71—143 (1941).
Šehurdin, A. P.: (New varieties of awnless Triticum durum.) Selekcija i Semenovodstvo Nr. 4, 14—15 (1939). [Russisch.]
— (New Saratow varieties of spring wheat.) Sovhoźnoe Proizvodstvo (State Farming) Nr. 1/2, 26—28 (1943). [Russisch.]
— u. V. N. Mamontova: (Breeding and seed production of spring wheat.) Naučnyj Otčet Inst. Zernovogo Hozjajstva Jugo-Vostoka SSSR za 1941—42 gg. (Sci. Rep. Inst. Grain Husbandry South-Eastern USSR. for 1941—42), Pp. 116—137 (1944). [Russisch.]
Semenjuk, W.: Chromosomal stability in certain rust resistant derivatives from a T. vulgare × T. Timopheevi cross. Sci. Agric. 27, 7—20 (1947).
Sen, B. u. S. C. Chakravarti: Vernalisation response of Indian wheats. Curr. Sci. 14, 124—125 (1945).
—, —, B. P. Pal u. G. S. Murty: Vernalisation response of cultivated Indian wheat. Curr. Sci. 15, 351 bis 352 (1946).
Sergeev, V. Z. u. A. A. Starkov: (A new promising variety of spring soft wheat, Pionerka [D-SP-21-44]). Socialističeskoe Zernovoe Hozjajstvo (Socialistic Grain Farming) Saratov, Nr. 4, 118—123 (1939). [Russisch.]
Šestakov, V. E.: (The winter wheat variety Hostianum 122/76). Selekcija i Semenovodstvo (Breeding and Seed Growing) Nr. 8/9, 24—25 (1940). [Russisch.]
— (Results of breeding winter wheat at the Petrovskii Experiment Station). Bjulletenj Instituta Zernovogo Hozjajstva Jugo-Vostoka SSSR. (Bull. Inst. Grain Husb. S. E. USSR.) Saratov, Nr. 2, 12—15 (1945). [Russisch.]
Ševčuk, T. N.: (A new promising variety of winter wheat [Ul'janovka]). Selekcija i Semenovodstvo (Breeding and Seed Growing) Nr. 1, 29—30 (1940). [Russisch.]
Shands, R. G.: Scab resistance in winter wheat. Rep. 6th East. Wheat Conf. Wooster, Ohio June 16, Pp. 9—10 (1939). [Mimeographed.]
— Disease resistance of Triticum Timopheevi transferred to common winter wheat. J. Amer. Soc. Agron. 33, 709—712 (1941).
— Scab resistance of spring wheats at Madison, Wis. Rep. Mill. Baking Mtg, 7th Hard Spring Wheat Conf., Minneapolis, Minn., February 28 and 29—March 1, 1944, Pp. 39 (1944). [Mimeographed.]
Sharbakoff, C. D.: Wheats for Tennessee growers. Circ. Tenn. Agric. Exp. Sta. Nr. 86, 7 (1943).
Sharman, B. C.: Branched heads in wheat and wheat hybrids. Nature, Lond. 153, 497—498 (1944)
— 'Coloured anthers' — A new monofactorial character in wheat, T. vulgare, Host. Nature, Lond. 154, 675 (1944).
— Agropyron-like segregates from a cross between Triticum vulgare Host. and T. durum Desf. J. Hered. 37, 55 (1946).
Shellenberger, J. A.: Problems in evaluating quality. Report of the Fifth Hard Red Winter Wheat Improvement Conference, Manhattan, Kansas, February 12, 13 and 14, 1945. Div. Cereal Crops Dis., Plant Ind. Sta., Beltsville, Md. p. 37 (1945). [Mimeographed.]
Shen, T. H.: Adaptability of wheat varieties in relation to the various regions and breeding centres in China. Proc. 7th Int. Genet. Congr. Edinburgh 23—30 August, Pp. 262—263, 1939 (1941). (Abstr.)
Sibbitt, L. D. u. R. H. Harris: Comparisons between some properties of mixograms from flour and unsifted whole wheat meal. Cereal Chem. 22, 531—538 (1945).
Sibilia, C.: Alcune razze fisiologiche di „Puccinia graminis tritici" Erikss. et Henn. nell'Africa Orientale Italiana. (Some physiological forms of P. graminis Tritici Erikss. et Henn. in Italian East Africa.) Bol. Patol. Veg. Roma 20, 115—118 (1940).

Sibilia, C.: Notizie sulla specializzazione fisiologica di Puccinia triticina Erikss. in Libia. (Information on the physiological specialization of P. triticina Erikss. in Libya.) Agricultura Colon. **34**, 100—101 (1940).
— Sulla resistenza alle ruggini di alcuni grani di montagna. (On the resistance of some highland wheats to the rusts.) Ital. Agric. **21**, 638—641 (1942).
— Due razze fisiologiche di Puccinia triticina Erikss. et Henn. del campo sperimentale di Filipiomboli. (Two races of P. triticina Erikss. et Henn. from the experimental field at Filipiomboli.) Ann. Ente Consorziale Interprov. Toscano Sementi, Firenze 1939—1942 **3**, 5 (1942).
— Determinazione di alcune razze fisiologiche italiane di „Puccinia triticina" Erikss. e di „Puccinia graminis tritici" Erikss. et Henn. (Determination of certain Italian physiological races in P. triticina Erikss. and P. g. tritici Erikss. et Henn.). Bol. Staz. Pat. Veg. Roma 1942. **22**, 193—196 (1943).
Sim, J. T. R., P. D. Henning u. C. B. Henning: Kernel characteristics of the bread-wheat varieties grown in South Africa. Bull. Dep. Agric. S. Afr. Nr. 253, 64 (1945).
Šimanskij, N. K.: (Controlled conversion of the nature of the spring wheat Erythrospermum 1160 into winter wheat.) Jarovizacija Nr. 4 (31) 25—29 (1940). [Russisch.]
Simmonds, P. M. u. B. J. Sallans: Testing wheat seedlings for resistance to Helminthosporium sativum. Sci. Agric. **26**, 25—33 (1946).
Sims, H. J.: Cereal variety trials in the Mallee, 1944. J. Dep. Agric. Vict. **43**, 199—202 (1945).
— u. C. G. Webb: Cereal trials in the Mallee. J. Dep. Agric. Vict. **42**, 101—107 (1944).
—,— u. G. Blachburn: Mallee wheat experiments. Variety trials at Walpeup. J. Dep. Agric. Vict. **41**, 169—172 (1943).
Šinarev, I. T.: (Natural hybrids and their progeny.) Jarovizacija Nr. 1 (29), 100—101 (1940). [Russisch.]
Singh, H.: Some useful wild plants of the Delhi Province. Indian J. Agric. Sci **15**, 297—308 (1945).
Singh, R. u. C. H. Bailey: A biochemical and technological study of Punjab wheat varieties. Cereal Chem. **17**, 169—203 (1940).
Širkevič, E. I.: (Breeding winter wheat at the Kamalinsk State Breeding Station.) Selekcija i Semenovodstvo (Breeding and Seed Growing) Nr. 4, 8—9 (1940). [Russisch.]
Sisakjan, N. u. A. Kobjakova: (The prevailing direction of enzyme action as an indication of drought resistance in cultivated plants. IV. The effect of wilting upon esterification and the hydrolysis of phosphoric esters in plants.) Biohimija **5**, 225—233 (1940). [Russisch.]
Sizova, M. A.: Structural chromosome alterations in Triticum durum. C. R. (Doklady) Acad. Sci. URSS **25**, 75—77 (1939).
Skosyreva, A. N.: (Question of the salt resistance of perennial wheat.) Vestnik Akademii Nauk USSR. (Record of the Academy of Sciences USSR.) Nr. 6, 80 bis 87 (1944). [Russisch.]
Skripčinskij, V. V.: (Dynamik der Sortenreinheit der Winterweizen.) Selekcija i Semenovodstvo Nr. 9/10, 43—45 (1946). [Russisch.]
Šmelev, I. H.: (Frost resistance of the main groups of varieties of winter wheat under conditions at Puškin.) Vestnik Socialističeskogo Rastenievodstva (Soviet Plant Industry Record) Nr. 2, 78—86 (1940). [Russisch.]
Smith, D. C.: Intergeneric hybridization. Chronica Botanica **7**, 417—418 (1943).
— Intergeneric hybridization of Triticum and other grasses, principally Agropyron. J. Hered. **34**, 219—224 (1943).

Smith, G. S.: Durum wheat breeding. Bi-m.Bull. N. Dak. Agric. Exp. Sta. **1**, Nr. 4, 10—12 (1939).
— Two new durum wheat varieties. Bi-m. Bull. N. Dak. Agric. Exp. Sta. **5**, Nr. 4, 2—3 (1943).
— Agronomic aspects of hard spring wheat breeding. 7. The durum wheat breeding program, 1944. Rep. Mill. Baking Mtg. 7th Hard Spring Wheat Conf. Minneapolis, Minn., February 28 and 29—March 1, 1944. Pp. 49—50 (1944). [Mimeographed.]
— M. A. Carleton, pioneer durum wheat scientist. Bi-m. Bull. N. Dak. Agric. Exp. Sta. **7**, Nr. 5, 3—7 (1945).
Smith, L.: Reciprocal translocations in Triticum monococcum. Genetics **24**, 86 (1939).
— Mutants and linkage studies in Triticum monococcum and T. aegilopoides. Res. Bull. Mo. Agric. Exp. Sta. Nr. 298, 26 (1939).
— Hereditary susceptibility to X-ray injury in Triticum monococcum. Amer. J. Bot. **29**, 189—192 (1942).
— Haploidy in einkorn. J. Agric. Res. **73**, 291—301 (1946).
— Chromosomal fragments in diploid wheat and their usefulness in genetic studies. Genetics **32**, 105 (1947). (Abstr.)
Smith, W.: Origin of Michels grass and its uses. Seed World **48**, Nr. 2, 9 (1940).
Smith, W. P. C. u. A. J. Millington: Stem rust of wheat and its control by breeding resistant varieties. J. Dep. Agric. W. Aust. **21**, 1—16 (1944).
Šmuk, A., V. Pisarev u. N. Vinogradova: (Changes in the characteristics of wheat germinated on rye endosperm). Proc. Lenin Acad. Agric. Sci. USSR. Nr. 7, 9—11 (1944). [Russisch.]
Šnajderman, J. A.: (Breeding wheat-Agropyron hybrids.) Naučnyj Otčet Inst. Zernovogo Hozjajstva Jugo-Vostoka SSSR. za 1941—42 gg. (Sci. Rep. Inst. Grain Husbandry South-Eastern USSR. for 1941—42) Pp. 178—189 (1944). [Russisch].
Sokoljskij, D. P.: (The improvement of the local varieties of winter wheat in the Ivanovo region and the production of élites.) Selekcija i Semenovodstvo (Breeding and Seed Growing) Nr. 10/11, 17—22 (1939). [Russisch.]
Sologub, S. D.: (Promising varieties of spring wheat.) Selekcija i Semenovodstvo Nr. 5, 33—34 (1939).
Sorokin, K. A., (Distribution of wheat pollen by the wind). Proc. Lenin Acad. Agric. Sci. URSS. Nr. 16, 8—15 (1939). [Russisch.]
— (The influence of methods of emasculation and pollination of spring wheat on the set of grain and the 1,000 grain weight). Selekcija i Semenovodstvo (Breeding and Seed Growing) Nr. 6, 13 (1940). [Russisch.]
— (Variability in characters of glume of wheat.) Proc. Lenin Acad. Agric. Sci. USSR. Nr. 10, 8—12 (1941). [Russisch.]
Soutter, R. E.: Choosing wheat varieties for grain. Qd. Agric. J. **60**, 197—209 (1945).
— Production of seed wheat. Qd. Agric. J. **61**, 75 (1945).
Spasojević, V.: Beziehungen zwischen der Zahl der Chromosomen (n) und der Größe der Pollenkörner beim Genus Triticum. Züchter **14**, 215—217 (1942).
Stakman, E. C.: Stem and leaf rust in relation to wheat breeding. 2. Races of stem rust. Rep. Mill. Baking Mtg, 7th Hard Spring Wheat Conf., Minneapolis, Minn., February 28 and 29—March 1, 1944, Pp. 26—27 (1944). (Mimeographed)
— Plant pathologist's merry-go-round. J. Hered. **37**, 259—265 (1946).
—, M. N. Levine u. W. Q. Loegering: Identification of physiologic races of Puccinia graminis tritici. Agric. Res. Admin., Bur. Entomol. Plant Quar., U. S. Dep. Agric. Nr. E/617, 27 (1944). [Mimeographed.]
— u. W. Q. Loegering: Recent changes in prevalence of physiologic races of Puccinia graminis tritici in the United States. Phytopathology **32**, 17 (1942). (Abstr.)

Stanford, E. H.: A new factor for resistance to bunt, Tilletia tritici, linked with the Martin and Turkey factors. J. Amer. Soc. Agron. **33**, 559—568 (1941).

Starkov, A. A.: (The best varieties of spring wheat in the Rostov region). Socialistic Grain Farming, Saratov Nr. 3, 44—49 (1940). [Russisch.]

Starling, T. M., S. A. Wingard u. M. H. McVicker: Vahart wheat a new variety for Virginia. Bull. Va Agric. Exp. Sta. Nr. 386, 4 (1946).

Stefanovskij, I. A.: (Resistance of spring wheat to hot, dry winds.) Vestnik Socialističeskogo Rastenievodstva (Soviet Plant Industry Record) Nr. 2, 87—97 (1940). [Russisch.]

— u. E. M. Večeslova: (Prospects of breeding spring wheat in the arid regions of the Southeast.) Selekcija i Semenovodstvo Nr. 5, 10—12 (1939). [Russisch.]

Stoa, T. E.: Agronomic aspects of hard spring wheat breeding. 1. Introductory remarks. Rep. Mill. Baking Mtg, 7th Hard Spring Wheat Conf., Minneapolis, Minn., February 28 and 29—March 1, 1944, Pp. 42 (1944). [Mimeographed.]

— Cadet- a new beardless wheat. Bi-m. Bull. N. Dak. Agric. Exp. Sta. **8**, 21—22 (1946).

— u. R. H. Harris: Which rust resistant wheats are most desirable? Bi-m. Bull. N. Dak. Agric. Exp. Sta. **1**, Nr. 3, 31—35 (1939).

Straib, W.: Die Faktorenbeziehungen im Verhalten des Weizens gegen verschiedene Gelbrostrassen. Z. Vererbungsl. **77**, 18—62 (1939).

— Die Feststellung der Rostresistenz beim Getreide und Lein. Forschungsdienst **13**, 24—29 (1942).

— u. A. Noll: Untersuchungen über den Einfluß der Hitze auf den Rostparasitismus. Zbl. Bakt. **106**, 257 bis 277 (1944).

Strampelli, N.: Frumento „Tiriamo diritto" („T. D."). (The wheat „Tiriamo diritto" [„T. D."].) Ann. Tec. Agr. Roma, **12**, 1—7 (1939).

Sudnov, P.: (The after-harvest growing of perennial wheat and perennial rye.) Vestnik Gibridizacii (Hybridization) Nr. 1, 113—114 (1941). [Russisch.]

Šuljga, M. S.: (The question of breeding winter and spring wheats resistant to fungous diseases and pests.) Selekcija i Semenovodstvo (Breeding and Seed Growing) Nr. 6, 11—14 (1939). [Russisch.]

Sullivan, B.: Milling and baking research in relation to wheat breeding. 4. Quality characteristics of an ideal hard red spring wheat. Rep. Mill. Baking Mtg, 7th Hard Spring Wheat Conf., Minneapolis, Minn., February 28 and 29—March 1, 1944, Pp. 19—21 (1944). [Mimeographed.]

Šulyndin, A. F.: (On the question of the selection of mother plants in hybridization). Selekcija i Semenovodstvo (Breeding and Seed Growing) Nr. 9, 7—10 (1939). [Russisch.]

— (On choosing the mother plant in hybridization.) Soviet Plant Industry Record Nr. 1, 50—57 (1940). [Russisch.]

— (Further notes on differences in first generation hybrids.) Jarovizacija Nr. 4 (31), 104—105 (1940). [Russisch.]

— („Secondary segregation" in hybrids of hard wheats.) Jarovizacija Nr. 3 (36), 118—119 (1941). [Russisch.]

— (Some examples of dominance and absorption in wheat hybrids.) Jarovizacija Nr. 3 (36), 119—120 (1941). [Russisch.]

Suneson, C. A. u. F. N. Briggs: Wheat production in California. Bull. Calif. Agric. Exp. Sta. Nr. 659, 18 (1941).

— u. W. K. Pope: Progress with Triticum × Agropyron crosses in California. J. Amer. Soc. Agron. **38**, 956—963 (1946).

—, O. C. Riddle u. F. N. Briggs: Yields of varieties of wheat derived by backcrossing. J. Amer. Soc. Agron. **33**, 835—840 (1941).

Suneson, C. A. u. G. A. Wiebe: Survival of barley and wheat varieties in mixtures. J. Amer. Soc. Agron. **34**, 1052—1056 (1942).

Svetozarova, V. V.: Second genom of Triticum Timopheevi Zhuk. C. R. (Doklady) Acad. Sci. URSS. **23**, 473—477 (1939).

Svinarev, V. I.: (An investigation of the lenght of the vegetative period in wheat. On the dominance in F_1). Bull. Acad. Sci. URSS., Sér. Biol. Nr. 3, 173—177 (1942). [Russisch.]

— (The segregation of F_2 and F_3 wheat hybrids with respect to the lenght of the vegetative period.) Bull. Acad. Sci. URSS., Sér. Biol. Nr. 3, 178—180 (1942). [Russisch.]

Swanson, A. F.: Present status of winter wheat work at the Fort Hays Station. Report of the Fifth Hard Red Winter Wheat Improvement Conference, Manhattan, Kansas, February 12, 13 and 14, 1945. Div. Cereal Crops Dis., Plant Ind. Sta., Beltsville, Md. Pp. 8 (1945). [Mimeographed.]

Swanson, C. O.: Effects of moisture on the physical and other properties of wheat. IV. Exposure of five varieties to light rains during harvest. Cereal Chem. **20**, 703 bis 714 (1943).

Swen, C. I.: Über Variation und Erblichkeit einiger Weizenkreuzungen in bezug auf Proteingehalt. Kühn's Archiv **54**, 369—430 (1940).

— (Variation and inheritance of protein content, gluten quality, and delayed germination in some wheat crosses.) Kwangsi Agric. **3**, 1—37 (1942).

T...., M. J.: Smut diseases of wheat. Curr. Sci. **11**, 78 (1942).

Tang Siang-Yu u. Li Chia-Wan: Studies on kernel characteristics of different varieties of wheat with reference to bread quality. Nanking J. **11**, Nr. 3 (1942). (Abstr.)

Tavčar, A.: Glasigkeit des Kornes und ihr Verhalten zu anderen Kornmerkmalen beim Weizen (Triticum vulgare L.) Polj. Znanstvena Smotra, Zagreb Nr. 4, 5 (1943).

Taylor, J. W., B. B. Bayles u. C. C. Fifield: A simple measure of kernel hardness in wheat. J. Amer. Soc. Agron. **31**, 775—784 (1939).

Tennberg, F. u. J. Jokihaara: Paikallisten kevätvehnän laatukokeiden tulokset vuosilta 1933—1937. (The results of the local spring wheat tests carried out in the years 1933—1937.) Valt. Maatalousk. Julk. Nr. 105, 68 (1939).

Thatcher, F. S.: Cellular studies in relation to rust resistance. Phytopathology, **33**, 14 (1943). (Abstr.)

Thomas, I.: New cereal varieties in Australia. J. Agric. W. Aust. **23**, 272—273 (1946).

— u. A. J. Millington: Wheat and oat variety trials 1941. J. Dep. Agric. W. Aust. **19**, 66—69 (1942).

—, — Wheat and oat variety trials 1942. J. Dep. Agric. W. Aust. **20**, 91—96 (1943).

—, — Recent developments in Australian wheat-breeding. J. Dep. Agric. W. Aust. **22**, 277—293 (1945).

Thompson, E. G.: Spring wheat. J. Minist. Agric. **53**, 423—425 (1947).

Thompson, W. P., E. J. Britten u. J. C. Harding: The artificial synthesis of a 42-Chromosome species resembling common wheat. Canad. J. Res. **21**, Abt. C, 134—144 (1943).

— u. I. Hutcheson: Chromosome behaviour and fertility in diploid wheat with translocation complexes of four and six chromosomes. Canad. J. Res. **20**, Sect. C, 267—281 (1942).

Throckmortion, R. I.: Need for additional research from the agronomic point of view. Report of the Fifth Hard Red Winter Wheat Improvement Conference, Manhattan, Kansas, February 12, 13 and 14, 1945. Div. Cereal Crops. Dis., Plant Ind. Sta., Beltsville, Md. Pp. 42—46 (1945). [Mimeographed.]

Tihonov, P. M.: (Genetical heterogeneity of the different stems of a hybrid.) Jarovizacija Nr. 3 (36), 106 bis 108 (1941). [Russisch.]

Tikhonov, P. M.: siehe Tihonov.

Timm, E. u. K. Trognitz: Die Kleberweizenuntersuchungen 1942/43 der Reichsanstalt für Getreideverarbeitung. Z. ges. Getreidew. 30, 102—105 (1943).

Timofeeva-Tyulina, M.: (Training of plants — a potent means of increasing the hardiness of winter wheat.) Soviet Plant Industry Record Nr. 1, 37—39 (1940). [Russisch.]

Tingey, D. C.: A new wheat variety resistant to smut, rust and mildew developed at station. Fm & Home Sci. 2, 12 (1941).

Titov, P. I.: (Inheritance of awns in wheat.) Socialistic Grain Farming, Saratov Nr. 1, 122—141 (1940). [Russisch.]

— (Inheritance of grain colour in wheat.) Socialistic Grain Farming, Saratov Nr. 3, 34—43 (1940). [Russisch].

Tomula, E. S.: Bericht über die Untersuchungen bezüglich der Weizenernte 1940. Acta Agr. Fenn. Helsinski Nr. 51, 68 (1943).

Tsai, H.: (Studies on Szechuan wheat.) Tech. Bull. Szechuan Agric. Exp. Sta. Nr. 2, 170 (1942).

Tschermak-Seysenegg, E. von: Über muttergleiche Scheinbastarde (hybridogene Parthenogenesis). Züchter 11, 337—341 (1939).

Tsitsin, N. V. siehe Cicin.

Tumanjan, M. G.: Über die Rolle des Ustilago bei der Formbildung der Weizen und seine Bedeutung für die Selektion. C. R. (Doklady) Akad. Sci. URSS. 30, 172—174 (1941).

— (Obtaining soft wheats from hard by artificial means.) Jarovizacija, Nr. 2 (35), 13—18 (1941). [Russisch.]

Tyner, L. E. u. W. C. Broadfoot: Field tests of the differential reaction of wheat varieties to root rot. Sci. Agric. 24, 153—163 (1943).

Uchikawa, I.: (Cytogenetical studies on dwarf compactoid wheat with 42 chromosomes.) Jap. J. Genet. 15, 315—317 (1939).

Umar, S. M.: A case of maternally-inherited variegation in wheat. Indian J. Genet. Pl. Breed. 3, 61—66 (1943).

Uppal, B. N. u. V. P. Gokhale: A new race of Puccinia graminis tritici and two biotypes of race 42. Curr. Sci. 16, 61 (1947).

Vallega, J.: Especialización fisiológica de Puccinia graminis Tritici, en Brasil. (Physiological specialization of P. graminis Tritici in Brazil.) An. Inst. Fitotec. Santa Catalina 3, 29—36 (1940).

— Razas fisiológicas de „Puccinia triticina" procedentes de Ipanema, San Pablo, Brasil. (Physiological races of P. triticina coming from Ipanema, San Pablo, Brazil.) Rev. Argent. Agron. 8, 57—59 (1941).

— Razas fisiológicas de Puccinia triticina y P. graminis tritici comunes en Chile. (Physiological races of P. triticina and P. graminis tritici common in Chile.) Bol. Téc. Minist. Agric. Dep. Genet. Fitotec., Santiago, Nr. 3, 32 (1942).

— Razas fisiológicas de Puccinia rubigo-vera tritici, comunes en Argentina. (Physiological races of P. r.-v. tritici, common in Argentina.) An. Inst. Fitotec. Santa Catalina 4, 40—57 (1942).

— Reacción de algunos trigos con respecto a las razas fisiológicas de Puccinia rubigo-vera tritici, comunes en Argentina. (Reaction of certain wheats to the physiological races of P. rubigo-vera tritici common in Argentina.) Rev. Fac. Agron. B. Aires 11, 91—115 (1944).

— u. H. Cenoz: Reacción de algunos trigos a las razas fisiológicas de Erysiphe graminis Tritici comunes en Argentina. (Reaction of certain wheat varieties to the physiological races of E. graminis Tritici common in Argentina.) An. Inst. Fitotec. Santa Catalina 1941, 3, 45—58 (1943).

Vannuccini, G.: Nazareno Strampelli e le sue creazioni. (N. Strampelli and his creations.) Ital. Agric. 79, 121—126 (1942).

Vasiljev, B. I.: Wheat-rye hybrids. II. Genetical analysis of crossability of rye with various species of wheat. C. R. (Doklady) Acad. Sci. URSS. 27, 598 bis 600 (1940).

Venault: La culture du blé au Tchad. Bull. Agric. Congo Belge 32, 118—125 (1941).

Vik, K.: Sortsforsok, såtidsforsok og forgrodeforsok med hosthvete og hostrug. (Variety trials, sowing time trials and comparison of different places in crop rotation for winter wheat and winter rye.) 48. Årsmeld. Norges LandbrHøisk. Åkervekstforsok, Pp. 17 bis 86 (1939).

Vinogradova, E. I.: (Variability of varietal characters in wheats.) Selekcija i Semenovodstvo (Breeding and Seed Growing), Nr. 2, 37—38 (1940). [Russisch.]

— (A case of uniformity of F_2 hybrids from natural crosses.) Jarovizacija, Nr. 5 (32), 124—125 (1940). [Russisch.]

Vogel, O. A.: Studies of the relationships of some features of wheat glumes to resistance to shattering and of the use of glume strength as a tool in selecting for high resistance to shattering. Res. Stud. St. Coll. Wash. 7, 199—200 (1939). (Abstr.)

— Relation of glume strength and other characters to shattering in wheat. J. Amer. Soc. Agron. 33, 583 bis 589 (1941).

— u. O. E. Barbee: Comparative performance of wheat varieties in eastern Washington. Bull. Wash. Agric. Exp. Sta., Nr. 450, 28 (1944).

—, C. E. Classen u. E. F. Gaines: The inheritance of reaction of Turkey-Florence- 1 x Oro- 1 to race 11 of Tilletia tritici. J. Amer. Soc. Agron. 36, 473—479 (1944).

—, S. P. Swenson u. C. S. Holton: Orfed wheat. Bull. Wash. Agric. Exp. Sta., Nr. 451, 10 (1944).

Volosky Yadlin, D.: Identificación de razas fisiológicas del Puccinia graminis tritici y P. triticina algunos estudios efectuados en Chile. (Identification of physiological races of P. g. tritici and P. triticina, some investigations made in Chile.) Agric. Tec. Chile 5, 70—78 (1945).

Waal, D. de: Het onderscheiden van de nederlandsche tarwerassen aan het zaad door de kleurreactie met phenol. (Distinguishing the Dutch varieties of wheat by the colour reaction of the seed with phenol.) Landbouwk. Tijdschr., Wageningen 52, 753—760 (1940).

Waddell, W. H.: A study of the relation between the seedling and mature-plant reaction to Puccinia graminis tritici in durum wheat crosses involving Jumillo. Canad. J. Res. 18, Sect. C, 258—272 (1940).

Wagner, S.: Qualitätsprüfungen an Winterweizen. Landw. Jb. Schweiz 55, 739—772 (1941).

— Der heutige Stand der Getreidezüchtung in der Schweiz und Vorschläge für deren Ausbau. Ber. Schweiz. Bot. Ges. 53, 299—312 (1943).

— Sortenprüfungen bei Winterweizen. Land. Jb. Schweiz 45, 590—604 (1944).

Waldron, L. R.: A new wheat variety for western N. D. Bi-m. Bull. N. Dak. Agric. Exp. Sta. 4, Nr. 4, 10 (1942).

— Comparison of large and small kerneled wheat as to yield, grown alone and intermixed in the row. Bi-m. Bull. N. Dak. Agric. Exp. Sta. 5, 27—32 (1943).

— Study of two hard red spring wheat varieties grown comparably but differing in kernel weight. Bi-m. Bull. N. Dak. Agric. Exp. Sta. 6, 25—30 (1943).

— How wheats behave in composition with one another. Bi-m. Bull. N. Dak. Agric. Exp. Sta. 6, 7—14 (1944).

Waldron, L. R: The new Mida wheat variety. Bi-m. Bull. N. Dak. Agric. Exp. Sta. **6**, Nr. 4, 9 (1944).
— Perennial wheats. Bi-m. Bull. N. Dak. Agric. Exp. Sta. **7**, 26—27 (1944).
— u. R. H. Harris: New varieties of rust resistant wheat. Bi-m. Bull. N. Dak. Agric. Exp. Sta. **1**, Nr. 3, 35—37, also Northw. Miller **198**, Nr. 7, 4 (1939).
—, —, T. E. Stoa u. L. D. Sibbitt: Protein and quality in hard red spring wheat with respect to temperature and rainfall. Bull. N. Dak. Agric. Exp. Sta. Nr. 311, 20 (1942).
—, —, —, — Mida wheat. A new hard red spring variety. Circ. N. Dak. Agric. Exp. Sta., Nr. 68, 16 (1944).
Walster, H. L.: North Dakota wheat breeders in action. Bi-m. Bull. N. Dak. Agric. Exp. Sta. **4**, Nr. 3, 1—6 (1942).
Wang, T. H.: (Inoculation experiments with loose smut of wheat.) New Agric. J. Fukien **2**, 396—403 (1942).
Wang, Y.-C.: Rust reactions of Chinese wheat varieties and certain Canadian hybrid strains. Canad. J. Res. **20**, Sect. C, 108—115 (1942).
Waterhouse, W. L.: A note on crossing wheat and rye. Aust. J. Sci. **2**, 63 (1939).
— u. I. A. Watson: Australian rust studies. VI. Comparative studies of biotypes of race 34 of Puccinia graminis tritici. Proc. Linn. Soc. N.S.W. **66**, 269—275 (1941).
Watkins, A. E.: The inheritance of glume shape in Triticum. J. Genet. **39**, 249—264 (1940).
— u. S. Ellerton: Variation and genetics of the awn in Triticum. J. Genet. **40**, 243—270 (1940).
Watson, D. J.: Comparative physiological studies on the growth of field crops. I. Variation in net assimilation rate and leaf area between species and varieties, and within and between years. Ann. Bot., Lond. **11**, 40—76 (1947).
Watson, I. A.: Inheritance of resistance to stem rust in crosses with Kenya varieties of Triticum vulgare Vill. Phytopathology **31**, 558—560 (1941).
— u. W. L. Watherhouse: A third factor for resistance to Puccinia graminis tritici. Nature, Lond. **155**, 205 (1945).
Webster, O. J.: Winter wheat improvement in Nebraska. Report of the Fifth Hard Red Winter Wheat Improvement Conference, Manhattan, Kansas, February 12, 13 and 14, 1945. Div. Cereal Crops Dis., Plant Ind. Sta., Beltsville, Md, Pp. 9—10 (1945). [Mimeographed.]
Weibel, R. O. u. K. S. Quisenberry: Field versus controlled freezing as a measure of cold resistance of winter wheat varieties. J. Amer. Soc. Agron. **33**, 336—343 (1941).
Weibull, W.: Vad säga de officiella försöksresultaten om Weibulls höstvetesorter och deras odlingssäkerhet under ogynnsamma betingelser? (What do the official experimental results say about Weibull's winter wheat varieties and their reliability under unfavourable conditions?) Lantmannen **25**, 668 (1941).
Wells, D. G. u. S. P. Swenson: Inheritance and interaction of genes governing reaction to stem rust, leaf rust, and powdery mildew in a spring wheat cross. J. Amer. Soc. Agron. **36**, 991—992 (1944). (Abstr.)
Wells, J. G. jr., D. L. Clanahan u. B. R. Churchill: Regent wheat for the Upper Peninsula. Quart. Bull. Mich. Agric. Coll. **27**, 298—300 (1945).
Wenholz, H.: New wild species of wheat: value to plant breeders. Agric. Gaz. N.S.W. **50**, 543 (1939).
— Farrer oration 1946. J. Aust. Inst. Agric. Sci. **12**, 88—92 (1946).
—, J. T. Pridham, C. K. Vears u. W. M. Curteis: Wheat varieties in Australia. Agric. Gaz. N.S.W. 1938: **49**, 583—586, 649—652; 1939: **50**, 13—17, 71—74, 86, 131—135, 181—184 (1938—1939).

Wenholz, H., J. T. Pridham, C. P. Vears u. W. M. Curteis: Wheat varieties in Australia. Agric Gaz. N.S.W. **50**, 308—311, 361—365, 417—420 (1939).
—, —, —, — Wheat varieties in Australia. Agric. Gaz. N.S.W. **50**, 539—543; **51**, 11—14, 30 (1939).
—, —, —, — Wheat varieties in Australia. Agric. Gaz. N.S.W. **51**, 65—68, 133—137 (1940).
—, —, —, — Wheat varieties in Australia. Agric. Gaz. N.S.W. **51**, 195—198, 242—244, 312—314, 347 (1940).
—, —, —, — Wheat varieties in Australia. Agric. Gaz. N.S.W. **51**, 371—373, 397, 485—488 (1940).
—, —, —, — Wheat varieties in Australia. Purple straw. Important influence on Australian wheat-growing. Agric. Gaz. N.S.W. **51**, 605—610 (1940).
—, —, —, — Wheat varieties in Australia. Agric. Gaz. N.S.W. **52**, 205—209, 260—264, 305—308, 355 bis 358 (1941).
Whiteside, A. G. O. u. S. H. Jackson: The thiamin content of Canadian hard red spring wheat varieties. Cereal Chem. **20**, 542—551 (1943).
Whitney, D. E., H. Herren u. B. D. Westerman: The thiamine and riboflavin content of the grain and flour of certain varieties of Kansas-grown wheat. Cereal Chem. **22**, 90—95 (1945).
Wichmann, W.: Experimentelle Untersuchungen zur Frage der Dürreresistenz bei Sommerweizen und Hafer. Kühn's Arch. **56**, 125—162 (1942).
Widenin, K. F.: (Annual wheat × Agropyron hybrids.) Vestnik Gibridizacii (Hybridization), Nr. 1, 111—112 (1941). [Russisch.]
Wiklund, K.: Vårveteodlingen i södra Norrland. (The growing of spring wheat in southern Norrland.) Sverig. Utsädesfören. Tidskr. **57**, 62—69 (1947).
Williamson, J.: Las hibridaciones naturales en el trigo. (Natural crosses in wheat.) „Granos" Semilla Selecta, B. Aires, Nos. 1, 2 and 3, 20—22 (1941).
Wingard, S. A. u. F. D. Fromme: Susceptibility of wheat varieties and selections to loose smut. Tech. Bull. Va Polytechn. Inst., Agric. Exp. Sta., Nr. 70, 26 (1941).
Woodforde, A. H.: The improvement of Tasmanian wheat varieties. Tasm. J. Agric. **10**, 13—18 (1939).
Woodward, R. W. u. D. C. Tingey: Cache, a beardless, smut-resistant winter wheat. Bull. Utah Agric. Exp. Sta., Nr. 312, 10 (1944).
Woolford, B. C.: Deep south gets new, anti-rust recruits. Sth. Seedsman **6**, 9, 53, 57 (1943).
Worzella, W. W.: The technic of producing a new soft wheat. Cereal Chem. **16**, 188—194 (1939).
— A preliminary summary of genetic studies in hexaploid wheats and proposed symbols for the characters. Report of the Seventh Eastern Wheat Conference, Lafayette, Indiana, June 19—20, Pp. 13—17 (1941).
— Some objectives in breeding for yield and other agronomic characters in wheat. J. Amer. Soc. Agron. **33**, 174—180 (1941).
— Inheritance and interrelationship of components of quality, cold resistance, and morphological characters in wheat hybrids. J. Agric. Res. **65**, 510—522 (1942).
— Response of wheat varieties to different levels of soil productivity: I. Grain yield and total weight. J. Amer. Soc. Agron. **35**, 114—124 (1943).
— The effect of level of soil fertility on wheat quality. Rep. 8th East. Wheat Conf. Cincinnati, Ohio, November 10, 1943, Pp. 5—6 (1944). [Mimeographed.]
— u. G. H. Cutler: Inheritance of quality and cold resistance in wheat and their interrelationship. Rep. 6th East. Wheat Conf. Wooster, Ohio, June 16, Pp. 3 bis 4 (1939). [Mimeographed.]
—, — Factors affecting cold resistance in winter wheat. J. Amer. Soc. Agron. **33**, 221—230 (1941).
Wurfel, D. A.: Wheat varieties. J. Dep. Agric. S. Aust. **50**, 99—102 .

Yu, T. F., H. R. Wang u. C. T. Fang: Varietal resistance and susceptibility of wheat to flag smut (Urocystis tritici Köern.). IV. Further studies on physiologic specialization in Urocystis tritici Köern. Phytopathology 35, 332—338 (1945).
—, —, — u. S. Y. Yin: Preliminary studies on physiologic specialization in Tilletia tritici and T. levis in China. Phytopathology 35, 879—884 (1945).
—, —, —, — Studies on physiologic specialization in Tilletia tritici and T. levis in China. Chinese J. Sci. Agric. 1, 281—287 (1944).

Zabluda, G. V.: (A method for the comparative study of drought resistance in wheat.) Sovetskaja Botanika (Soviet Botany), Nr. 5/6, 154—166 (1940). [Russisch.]
Zaharjevskij, A. A.: (Oversoming sterility in hybrids T. durum x T. Timopheevi.) Jarovizacija, Nr. 3 (30), 90—105 (1940). [Russisch.]
— (Control of sterility and the breeding of hybrids of Tr. Durum x Tr. Timopheevi.) Proc. Lenin Acad. Agric. Sci. USSR., Nr. 2, 5—9 (1941). [Russisch.]
Zažurilo, V. K. u. G. M. Sitnikova: Natural ways of transmission of the winter wheat mosaic virus. C. R. (Doklady) Acad. Sci. URSS. 29, 429—432 (1940).
Žebrak, A. R.: Amphidiploids of hard wheat and einkorn produced through colchicine treatment. C. R. (Doklady) Acad. Sci. URSS. 25, 53—55 (1939).
— Production of amphidiploid of Tr. durum x Tr. Timopheevi. C. R. (Doklady) Acad. Sci. URSS. 25, 56—59 (1939).
— Experimental production of Triticum polonicum x Tr. durum amphidiploid through colchicine treatment. C. R. (Doklady) Acad. Sci. URSS. 29, 400—403 (1940).
— On the fertility of the amphidiploid hybrid of hard wheat with einkorn. C. R. (Doklady) Acad. Sci. URSS. 29, 480—482 (1940).
— Production of a T. Timopheevi x T. durum v. hordeiforme 010 amphidiploid by colchicine treatment. C. R. (Doklady) Acad. Sci. URSS. 29, 604—607 (1940).
— (Experimental production of amphidiploids in sterile wheat hybrids.) Trudy Moskovskoj ordena Lenina Seljskohozjajstvennoj Akademii imeni K. A. Timirjazeva 4, Nr. 1, 161—173 (1940). [Russisch.]
— (Colchicine induced amphidiploids obtained from the cross Triticum Gibridizacii Timopheevi x T. vulgare.) Vestnik Gibridizacii (Hybridization), Nr. 1, 92—98 (1941). [Russisch.]
— Comparative fertility of amphihaploid and amphidiploid hybrids T. Timopheevi x T. durum v. hordeiforme 010. C. R. (Doklady) Acad. Sci. URSS. 30, 54—56 (1941).
— (Production of T. persicum x T. Timopheevi amphidiploids.) C. R. (Doklady) Acad. Sci. URSS. 31, 485—487 (1941). [Russisch.]
— Colchicine induced amphidiploids of Triticum turgidum x Triticum Timopheevi. C. R. (Doklady) Acad. Sci. URSS. 31, 617—619 (1941).
— (Synthesis of new wheat species.) Socialističeskoe Seljskoe Hozjajstvo (Socialistic Agriculture), Moscow, Nr. 8, 51—57 (1943). [Russisch.]
— Production of amphidiploids of Triticum orientale x Triticum Timopheevi by Colchicine treatment. C. R. (Doklady) Acad. Sci. URSS. 42, 352—354 (1944).
— Production of amphidiploids of Triticum polonicum x Triticum Timopheevi. C. R. (Doklady) Acad. Sci. URSS. 43, 120—121 (1944).
— (The synthesis of new species of wheats.) Trudy Moskovskoj Seljskohozjajstvennoj Akademii im. K. A. Timirjazeva. (Trans. K. A. Timiriaseff Acad. Agric. Moscow) 6, 5—54 (1944). [Russisch.]
— Synthesis of new species of wheats. Nature, Lond. 153, 549—551 (1944).

Žebrak, A. R.: New amphidiploid species of wheat and their significance for selection and evolution. Amer. Nat. 80, 271—279 (1946).
Zhebrak, A. R. siehe Žebrak.
Žigalov, S. A.: (Investigation of the winter-hardiness of hybrids from crosses of winter wheat with couch.) Theses and scientific papers read at the 4th District Conference of Workers of Universities and Research Institutions, Omsk, Nr. 1, Agron. Sect., 53—56 (1941). [Russisch.]
Žogolev, A. M.: (A new variety of winter wheat in the fields of the collective and state farms.) Socialistic Grain Farming, Saratov, Nr. 2, 100 (1941). [Russisch.]
Žukovskij, P. M.: (Notes.) Sovhoznoe Proizvodstvo (State Farming), Nr. 3/4, 47 (1943). [Russisch.]

2. Roggen.

Åkerberg, E. u. K. Wiklund: Erfarenheter fran förädling och försök med höstråg vid Sveriges Utsädesförenings Västernorrlandsfilial under 10-årsperioden 1935—1944. (Results obtained from breeding and trials with winter rye at the Västernorrland Branch Station of the Swedish Seed Association during the ten year period 1935—1944.) Sverig. Utsädesfören. Tidskr. 55, 431—443 (1945).
Anonymus: Contributions to the study of vernalization. Mon. Sci. News Nos 9/10, 8 (1945).
Aust, S.: Erhöhte Saatgutgewinnung bei Roggen durch vegetative Vermehrung. Züchter 13, 84—87 (1941).
— u. Ossent, H. P.: Qualitätszüchtung beim Roggen. Züchter 13, 78—84 (1941).

Borisenko, S. F.: (Family analysis in the work of breeding and seed production with rye.) Jarovizacija, Nr. 5 (32) 57—61 (1940). [Russisch.]
Breslavec, L. P.: Polyploidsin rye induced by X-rays. C. R. (Doklady) Acad. Sci. URSS. Nr. 22, 354—357 (1939).
— Polyploid forms of spring rye. C. R. (Doklady) Acad. Sci. URSS. 29, 328—331.

Camara, A.: Efeitos do calor sôbre a microsporogénese do Secale cereale. (Effects of heat on microsporogenesis in S. cereale). Scientia Genetica, 1, 86—102 (1939).
Chin, T. C.: Cytology of the autotetraploid rye. Bot. Gaz. 104, 627—632 (1943).
Cicin, N. V.: (Prospects of breeding winter hardy forms of winter crops.) Vestnik Akademii Nauk SSSR. (Record of the Academy of Sciences USSR.) Nr. 3, 116 bis 122 (1944). [Russisch.]

Fetisov, A. I.: Reduction division in the 16-chromosome rye. C. R. (Doklady) Acad. Sci. URSS 25, 146 bis 147 (1939).
Frimmel, F.: Beitrag zur Xenienfrage bei Roggen. Züchter 11, 301—307 (1939).

Gluščenko, I. E.: (Results of twice repeated cross-pollination of rye in the plots of the collection.) Jarovizacija Nr. 1 (34), 27—33 (1941) [Russisch.]
— (Intervarietal hybridization of rye.) Doklady Vsesojuz. Akad. Seljsk. Nauk im. V. I. Lenina (Proc. Lenin Acad. Agric. Sci. USSR.) Nr. 4/5, 11—23 (1945). [Russisch.]

Hačaturov, S. P.: (The inheritance of the capacity for selective fertilization in gametes on fertilization.) Jarovizacija Nr. 4 (31), 21—24 (1940). [Russisch.]
— (The possibility of planned improvement of the breed in rye seeds.) Jarovizacija Nr. 5 (32), 35—47 (1940). [Russisch.]
Håkansson, A.: Some observations on the seed development in two strains of Triticale. Acta Agriculturae Suecana, Stockholm, 1, 377—384 (1946).

Jakimovič, E. D.: (Zur Frage der mehrjährigen Ustilago hypodites (Schl.) am Roggen-Queckebastard.) C. R. (Doklady) Acad. Sci. URSS. **57**, 731—32 (1947). [Russisch.]

Jeliseev, N.: White rye. Soviet News 1680, P. 4 (1947).

Jermoljev, E.: Die Brüchigkeit des Roggens. (The brittleness of rye.) Z. Pflanzenz. **24**, 59—70 (1941). [Russisch.]

Kahidze, N. T.: Meiosis in inbred rye. C. R. (Doklady) Acad. Sci. URSS. **25**, 68—70 (1939).

Kattermann, G.: Ein neuer Krayotyp bei Roggen. Chromosoma **1**, 284—299 (1939).

Khačaturov, S. P. siehe Hačaturov.

Kolokolov, K. V.: (The local variety of rye „Manyč"). Selekcija i Semenovodstvo (Breeding and Seed Growing) Nr. 8/9, 26 (1940.) [Russisch.]

Kondo, N.: (Chromosome doubling in Secale, Haynaldia and Aegilops by colchicine treatment.) Jap. J. Genet. **17**, 46—53 (1941). [Japanisch.]

Kondratenko, F.: (Analysis of population of winter-rye according to length of vernalization stage.) Soviet Plant Industry Record Nr. 1, 27—34 (1940). [Russisch.]

Kostjučenko. I.: (Winter rye varieties for the east of the USSR.) Socialističeskoe Seljskoe Hozjajstvo (Socialistic Agriculture) Moskva Nr. 1, 47—50 (1942). [Russisch.]

Kostoff, D.: The frequency of polyembryony and chlorophyll deficiency in rye. Curr. Sci. **8**, 356—358 (1939).

— Frequency of polyembryony and chlorophyll deficiency in rye. C. R. (Doklady) Acad. Sci. URSS. **24**, 479—482 (1939).

— A case of vivipary in rye. Curr. Sci. **9**, 279—280 (1940).

Kostjučenko, N. A.: (Results of two years' work with winter rye in Kazahstan and southern Siberia.) Vestnik Socialističeskogo Rastenievodstva. (Soviet Plant Industry Record) Nr. 1, 3—9 (1941). [Russisch.]

Kotthoff, P.: Die Resistenz von Roggensorten gegen Anguillulina (Ditylenchus) dipsaci (Kühn). Angew. Bot. **24**, 79—99 (1942).

Krasnjuk, A. A.: (Rye-Agropyrum hybrids.) Vestnik Gibridizacii (Hybridization) Nr. 2, 3—14 (1941). [Russisch.]

— (Twelve years inbreeding rye.) Socialistic Grain Farming, Saratov, Nr. 2, 26—46 (1941). [Russisch.]

— (Breeding and seed raising of winter rye.) Naučnyj Otčet Inst. Zernovogo Hozjajstva Jugo-Vostoka SSSR. za 1941—42 gg. (Sci. Rep. Inst. Grain Husbandry South-Eastern USSR. for 1941—42), Pp. 153—166 (1944). [Russisch.]

— (A new and promising variety of winter rye, Volžanka.) Bjulletenj Instituta Hozjajstva Jugo-Vostoka SSSR. (Bull. Inst. Grain Husb. S. E. USSSR.) Saratow Nr. 2, 3—5 (1945). [Russisch.]

— (Supplementary artificial pollination in rye.) Agrobiologija (Agrobiology) Nr. 1, 133—135 (1946). [Russisch.]

— (Breeding rye by the method of composite populations.) Agrobiologija (Agrobiology) Nr. 2, 44—51.(1946). [Russisch.]

— (Vorläufige Ergebnisse der Roggenzüchtung.) Selekcija i Semenovodstvo, Nr. 9/10, 20—22 (1946). [Russisch.]

Lamm, R.: Chromosome behaviour in a triploid rye plant Hereditas, Lund, **30**, 137—144 (1944).

Landes, M.: The causes of self-sterility in rye. Amer. J. Bot. **26**, 567—571. (1939).

Levan, A.: Studies on the meiotic mechanism of haploid rye. Hereditas, Lund, **28**, 177—211 (1942).

Ling, L.: The histology of infection of susceptible and resistant selfed lines of rye by the rye smut fungus, Urocystis occulta. Phytopathology **30**, 926—935 (1940).

Ljung, E. W.: Sveriges Utsädesförenings rågförädlingsarbeten. Föredrag vid Sveriges Utsädesföreningsårsmöte den 20. Juli 1944. (The rye breeding work of the Swedish Seed Association. Lecture at the annual meeting of the Swedish Seed Association, 20 July, 1944). Sverig. Utsädesfören. Tidskr. **54**, 295—315. (1944).

Lundqvist, A.: On self-sterility and inbreeding effect in tetraploid rye. Hereditas, Lund, **33**, 570—571 (1947). (Abstr.).

Mokrov, S. V.: (Stability of characters in crosspollinated plants.) Jarovizacija Nr. 1 (34), 34—36 (1941). [Russisch.]

Müntzing, A.: Aneuploidy and seed shrivelling in tetraploid rye. Hereditas, Lund, **29**, 65—75 (1943).

— Genetical effects of duplicated fragment chromosomes in rye. (Hereditas, Lund, **29**, 91—112 (1943).

— Några försöksresultat med rågvete och tetraploidt korn. (Some experimental results with rye-wheat and tetraploid barley.) Nord. Jordbr. Forskn. Nos. 5/6, 250—262 (1943).

— Double crosses of inbred rye. Bot. Notiser Pp. 333 bis 345 (1943).

— Rågvetet erbjuder många problem av både teoretisk och praktisk natur. (Rye-wheat presents many problems of both a theoretical and practical nature.) Lantmannen **28**, 1023—1026 (1944).

— Cytological studies of extra fragment chromosomes in rye. 1. Isofragments produced in misdivision. Hereditas, Lund **30**, 231—248 (1944).

— Cytological studies of extra fragment chromosomes in rye. II. Transmission and multiplication of standard fragments and iso-fragments. Hereditas, Lund **31**, 457 bis 477 (1945).

— Cytological studies of extra fragment chromosomes in rye. III. The mechanism of non-disjunction at the pollen mitosis. Hereditas, Lund **32**, 97—119 (1946).

— Sterility in rye populations. Hereditas, Lund **32**, 521 bis 549 (1946).

— u. Prakken, R.: Chromosomal aberrations in rye populations. Hereditas **27**, 273—308 (1941).

Neel, L. R.: Balbo rye-Tennessee's discovery. Sth. Seedsman **7**, 17, 32, 33 (1944).

Nilsson, F.: Rågförsök och rågförädling vid Sveriges Utsädesförenings Västernorrlandsfilial. (Rye experimentation and rye improvement at the Västernorrland Branch of the Swedish Seed Association.) Sverig. Utsädesfören. Tidskr. **50**, 4—30 (1940).

Nordenskiöld, H.: Studies of a haploid rye plant. Hereditas, Lund **25**, 204—210 (1939).

O'mara, J. G.: A photoperiodism accompanying autotetraploidy. Amer. Nat. **76**, 386—393 (1942).

— Meiosis in autotetraploid Secale cereale. Bot. Gaz. **104**, 563—575 (1943).

Östergren, G. u. R. Prakken: Behaviour on the spindle of the actively mobile chromosome ends of rye. Hereditas, Lund **32**, 473—494 (1946).

Pehl, P.: Ermittlung früh schnittreifer Futterroggen für den Zwischenfruchtbau. Züchter **13**, 59—63 (1941).

Pelshenke, P.: Studien über die Backfähigkeit von Roggensorten. Z. Pflanzenz. **24**, 1—58(1941).

Pogosjan, S. A.: (Overcoming the depression in the progeny of inbred plants.) Agrobiologija (Agrobiology) Nr. 1, 123—129 (1946). [Russisch.]

Popoff, A.: Über die Auswuchsneigung des Roggens. (The tendency towards germination in rye.) Z. Pflanzenz. **23**, 535—541 (1941).

Prakken, R.: Studies of asynapsis in rye. Hereditas, Lund **29**, 475—495 (1943).

— u. A. Müntzing: A meiotic peculiarity in rye, simulating a terminal centromere. Hereditas, Lund, **28**, 441 bis 482 (1942).

Prjanišnikova, Z. D.: (The question of utilizing the products of inbreeding in rye breeding.) Seleckcija i Semenovodstvo (Breeding and Seed Growing) Nr. 6, 15—16 (1939). [Russisch.]

Purvis, O. N.: Studies in vernalisation of cereals. V. The inheritance of the spring and winter habit in hybrids of Petkus rye. Ann. Bot., London 3, 719—729 (1939).

Rezničuk, S. P.: (Perennial rye [preliminary communication]). Socialističeskoe Zernovoe Hozjajstvo (Socialistic Grain Farming) Saratow Nr. 4, 87—90 (1939). [Russisch.]

Rosenstiel, K. von u. L. Mittelstenscheid: Über die Erzeugung amphidiploider Roggen-Weizen-Bastarde (Secalotrica). Züchter 15, 173—183. (1943).

Rudnickij, N. V. u. K. A. Gluhih: (Inter-varietal cross-pollination in rye.) Jarovizacija Nr. 2 (35), 35 bis 42 (1941). [Russisch.]

Saltykovskij, M. I. u. E. S. Sápryguina: (Winter-hardy winter-rye.) Selekcija i Semenovodstvo (Breeding and Seed Growing) Nr. 6, 10—11 (1939) [Russisch].

Ščerbina, D. R.: (The influence of artificial supplementary pollination on the quality of the seed of winter rye.) Selekcija i Semenovodstvo (Breeding and Seed Growing) Nr. 5, 13—14 (1940). [Russisch.]

— (On the artificial pollination of winter-rye.) Soviet Plant Industry Record Nr. 1, 35—36 (1940). [Russisch.]

Seinhorst, J. W.: Een laboratoriummethode voor de bepaling van de vatbaarheid van rogge voor aantasting door het stengelaaltje (Ditylenchus dipsaci (Kühn) Filipjev.). (A laboratory method for determining the degree of susceptibility of rye to infestation by the stem eelworm [D. dipsaci (Kühn) Filipjev]). Tijdschr. PlZiekt. 51, 39—52 (1945).

Sengbusch, R. v.: Polyploider Roggen. Züchter 12, 185—189 (1940).

Shcherbina, D. R. siehe Ščerbina.

Sims, H. J.: Ryecorn - a cereal for winter grazing and drift control in the Mallee. J. Dep. Agric. Vict. 42, 151—154, 162 (1944).

Smagin, G. D.: (An experiment on breeding double-eared rye.) Jarovizacija Nr. 6 (33), 105—107 (1940). [Russisch.]

Šmargon, E. N.: Chromomere structure of the chromosome set of rye. C. R. (Doklady) Acad. Sci. URSS. 23, 267—269 (1939).

Thompson, W. P. u. D. Johnston: The cause of incompatibility between barley and rye. Canad. J. Res. 23, Sect. C., 1—15 (1945).

Tosun, O.: Bugday × çavdar melezleri. (Wheat × rye crosses.) T. C. Yüksek Ziraat Enstitüsü Çalişmalarindan, Ankara, Nr. 139, 122 (1943).

Tschermak-Seysenegg, E. v.: Ein neuer fruchtbarer Getreideartbastard: Agropyrum intermedium × Secale cereale = Agrosecale. Anz. Akad. d. Wissensch. Wien, math.-naturw. Kl. 78, 1—3 (1941).

Vasconcellos, J. de Carvalho e: Considerações acêrca do melhoramento do centeio. (Considerations regarding the improvement of rye.) Minist. Agric., Direc. Ger. Serv. Agric., Belem Nr. 12, Sér. Estud. Infor. Téc. Pp. 26 (1940).

Wagner, S.: Die Übertragung nicht mendelnder Buntblättrigkeit durch den Pollen bei Roggen. Arch. Klaus-Stift. Vererb.-Forsch. 17, 449—453 (1942).

Wellensiek, S. J.: Nieuwe methoden voor roggeselectie, met behulp van vegetative vermeerdering. (New methods of rye-selection, with the aid of vegetative propagation.) Zaaizaad en Pootgoed, 2, Nr. 7, 9—11; Nr. 8, 8—11 (1940).

Wellensiek, S. J.: Ras, zaaitijd en bemesting bij rogge. (Race, sowing time and manuring in rye.) Meded. Landbvoorlicht Dienst, Wageningen Pp. 145—153 (1944).

Whiteside, A. G. O.: Report of uniform winter rye variety trials. 1941 crop. Rep. Cereal Div., Dom. Exp. Fm., Ottawa Nr. 54, Pp. 9 (1942). [Mimeographed].

Yeliseyev, N. siehe Jeliseev.

Žukov, I.: (Breeding winter varieties in the eastern regions of the USSR.). Jarovizacija Nr. 2 (35), 149—150 (1941). [Russisch.]

Zwoboda, A.: Eine spontane, fertile Artkreuzung. Z. Pflanzenz. 24, 339—340 (1941).

3. Hafer.

Å....., Å.: Ett par försök med Bøtøhavre. (A few experiments with the Bøtø oat.) Sverig. Utsädesfören. Tidskr. 55, 147 (1945).

Adair, E. O.: Arkansas oat hybrid shows great promise. Extensive field trials prove De Soto has marked resistance to rust and smut in addition to being high-yielding strain. Sth. Seedsman 5, Nr. 8; 7, 31 (1942).

Åkerman, Å.: Havreskördarna böra ökas. (The oat harvests must be increased.) Lantmannen 28, 1096 bis 1097 (1944).

— Nya erfarenheter rörande Samehavrens odlingsvärde. (New findings as regards the cultivation value of the Same oat.) Sverig. Utsädesfören. Tidskr. 55, 101—108 (1945).

— Nya iakttagelser rörande olika havresorters motståndskraft mot gråfläcksjuka. (New observations on the resistance of different varieties of oats to grey spot.) Sverig. Utsädesfören. Tidskr. 56, 159—172 (1946).

— Svalöfs Siriushavre II. (The Svalöf Sirius II oat.) Sverig. Utsädesfören. Tidskr. 57, 42—45 (1947).

— u. K. Fröier: Studien über eine spontane chlorina-Mutation in Avena sativa. Heredita, Lund. 27, 371 bis 401 (1941).

—, — Die Faktorenverteilung der lutescens-Sippe von Avena sativa auf die Elternsorten. Hereditas, Lund 28, 171—176 (1942).

Anonymus: Oat variety trials on the Dominion Experimental Farms 1939. Rep. Cereal Division, Cent. Exp. Fm, Ottawa, Ont., Nr. 39, 32 (1939).

— Picton winter oat. Nat. Inst. Agric. Bot., Cambridge, Pp. 4 (1940).

— Marion, a new oat variety, has advanteges. Seed World 49, Nr. 1, 9 (1941).

— Forsøg med Havresorter 1939—1942. (Trials with oat varieties 1939—42.) Tidsksr. Planteavl. 47, 720 bis 724 (1943).

— Frit fly. Adv. Leafl. Minist. Agric. Fish., Lancs., Nr. 110, 4 (1943).

— Oat varieties. Mon. Rep. Minist. Agric. N. Ire 17, 342—343 (1943).

— Sandford oat. Nat. Inst. Agric. Bot. Cambridge, January, Pp. 3 (1944).

— Varieties of winter oats and winter barley. Fmrs' Leafl. Nat. Inst. Agric. Bot. Cambridge, Nr. 9, 4 (1945).

— Cereal varieties of barley and oats. Agricultura Chron., Moscow, Nr. 6, 4—5 (1946). [Mimeographed.]

— Oat variety trials. J. Dep. Agric. Eire 43, 88—89 (1946).

— New oats variety Beacon particularly adaptable to fertile soils of Ontario. Canad. Grain J. 2, Nr. 7, 14 (1947).

— Garry oats shows up well in 1946 co-operative tests. Canad. Grain J. 2, Nr. 8, 14 (1947).

— Results of oat variety trials. Mon. Rep. Minist. Agric. N. Ire 22, 9—10 (1947).

Åsander, F.: Havreförsök vid Sveriges Utsädesförenings Jämtlandsfilial 1918—1945. (Oat trials at the Jämtland Station of the Swedish Seed Association 1918 bis 1945.) Sverig. Utsädesfören. Tidskr. 56, 148—158 (1946).

Ashby, E. u. V. May: Physiological studies in drought resistance. I. Technique. Proc. Linn. Soc. N.S.W. 66, 107—112 (1941).

Atkins, R. E.: Factors affecting milling quality in oats. J. Amer. Soc. Agron. 35, 532—539 (1943).

Atkinson, R. E.: A new mosaic chlorosis of oats in the Carolinas. Plant Dis. Reporter 29, 86—89 (1945). [Mimeographed.]

Bose, R. D. u. B. B. Mundkur: Studies in Indian cereal smuts. IV. Varietal resistance of Indian and other oats to smuts. Indian J. Agric. Sci. 11, 695—702 (1941).

Brandwein, P. F.: Infection studies on the covered smut of oats. Bull. Torrey Bot. Cl. 67, 673—691 (1940).

Caffrey, M.: Glasnevin Ardri oats. J. Dep. Agric. Eire 37, 354—361 (1940).

Caldwell, R. M., R. R. Mulvey u. L. E. Compton: Benton and Clinton new disease resistant spring oats. Bull. Ind. Agric. Exp. Sta., Nr. 514, 10 (1946).

Carqué, E.: Origen y propagación de la avena loca. (Origin and propagation of Avena fatua.) Agricultura, Madrid 12, 69—70 (1943).

Carson, G. P.: The inheritance of certain quantitative characters in oats. Abstr. Diss. Univ. Camb. (1938 bis 1939), Pp. 17—18 (1940).

Česnokov, P. G.: World's assortment of oat varieties in relation to the Swedish fly. C. R. (Doklady) Acad. Sci. URSS. 27, 718—721 (1940).

Cochran, G. W., C. O. Johnston, E. G. Heyne u. E. D. Hansing: Inheritance of reaction to smut, stem rust, and crown rust in four oat crosses. J. Agric. Res. 70, 43—61 (1945).

Coffman, F. A.: Heat resistance in oat varieties. J. Amer. Soc. Agron. 31, 811—817 (1939).

— Results from the Cooperative Coordinated Oat Breeding Nurseries for 1939, and the Uniform Winterhardiness Nurseries for 1939—1940. U. S. Dep. Agric., Bur. Pl. Ind. Div. Cer. Crops and Dis., Washington, D. C., Pp. 127 (1940). [Mimeographed.]

— Results from the cooperative coordinated oat breeding nurseries for 1940, and the uniform winter-hardiness nurseries for 1940—1941. U. S. Dep. Agric. Bur. Pl. Ind., Div. Cereal Crops and Diseases, Pp. 115 (1941). [Mimeographed.]

— The comparative winter hardiness of oat varieties. Circ. U. S. Dep. Agric., Pp. 35 (1941).

— Survival of oats grown in winterhardiness nurseries, 1937 to 1941. J. Amer. Soc. Agron. 34, 651—658 (1942).

— Origin of cultivated oats. J. Amer. Soc. Agron. 38, 983—1002 (1946).

—, E. G. Heyne, C. O. Johnston, H. Stevens u. H. C. Murphy: Improvement and distribution of spring-sown red oats. J. Amer. Soc. Agron. 37, 479 bis 498 (1945).

—, H. B. Humphrey u. H. C. Murphy: New red oats for fall seeding resistant to rusts and smuts. J. Amer. Soc. Agron. 33, 872—882 (1941).

— u. T. R. Stanton: Dormancy in fatuoid and normal oat kernels. J. Amer. Soc. Agron. 32, 459—466 (1940).

Cunliffe, N. u. D. J. Hodges: Studies on Oscinella frit Linn. Notes on the resistance of cereals to infestation. Ann. Appl. Biol. 33, 339—360 (1946).

Derick, R. A, u. D. G. Hamilton: Shattering in oats. Sci. Agric. 25, 426—431 (1945).

Down, E. E. u. J. W. Thayer: Huron, a new oat variety for Michigan. Quart. Bull. Mich. Agric. Exp. Sta. 22, 209—212 (1940).

—, — jr.: Eaton oats. Quart. Bull. Mich. Agric. Exp. Sta. 29, 190—193 (1947).

Dungan, G. H., O. T. Bonnett u. W. L. Burlison: Spring oat varieties for Illinois. Bull. Ill. Agric. Exp. Sta., Nr. 481, 443—471 (1942).

Ellison, W.: The cytology of certain diploid and tetraploid Avena hybrids. Proc. 7th Int. Genet. Congr. Edinburgh 23—30 August, Pp. 109—110 1939, (1941). (Abstr.)

— The cytology of certain diploid, triploid and tetraploid Avena hybrids. Genetica 22, 409—418 (1940).

Emme, E. K.: (Hybrids of naked oats. Hybrids of the 42-chromosome naked oats.) Bull. Acad. Sci. URSS., Sér. Biol., Pp. 516—530 (1939). [Russisch.]

— u. A. I. Mordvinkina: (Hybrids of the 14-chromosome naked oats.) Bull. Acad. Sci. URSS., Sér. Biol., Pp. 530—540 (1939). [Russisch.]

Ericsson, G.: Svalöfs Vidarhavre. Ny, mycket tidig vithavresort för Norrland. (Svalöf's Vidar oata new very variety of white oat for Norrland.) Sverig. Utsädesfören. Tidskr. 49, 397—400 (1939).

Fetisov, A. I.: Chromosome doubling by colchicine and crossability of tetraploids in Avena brevis (Roth). C. R. (Doklady) Acad. Sci. URSS. 27, 705—709 (1940).

Friedberg, R.: Les avoines cultivées en France. Centre Nat. Rech. Agron., Paris, Imprimerie Nationale, Pp. 172 (1942).

Fröier, K.: Nyare förädlingsframsteg av betydelse för havreodlingen i Norrland. (Recent progress in breeding of importance for oat cultivation in Norrland.) Lantmannen 28, 271—273 (1944).

— Genetical studies on the chlorophyll apparatus in oats and wheat. Hereditas, Lund 32, 297—406 (1946).

— u. Å. Åkerman: Die Faktorenverteilung der Lutescens-Sippe von Avena sativa auf die Elternsorten. Hereditas, Pp. 171—176 (1942).

Gelin, O. u. S. Undenäs: Weibulls Bambuhavre 1934 bis 1939. (Weibull's Bambu oats 1934—1939.) Weibull's Ill. Årsb., Nr. 35, 30 (1940).

—, — Weibulls Original Triohavre. Ny förädling för Svea- och Götaland. (Weibull's original Trio oat. A new strain for Svealand and Götaland.) Weibulls. Ill. Årsb. 38, 8—10 (1943).

Goffart, H.: Untersuchungen am Hafernematoden (Heterodera schachtii, Schm.) unter besonderer Berücksichtigung der schleswig-holsteinischen Verhältnisse. Arb. biol. Reichsanst. Land- u. Forstw., Berl. 23, 141—161 (1941).

Hagborg, W. A. F.: Investigations on grey speck of oats in Manitoba. Phytopathology 37, 8 (1947). (Abstr.)

Hamilton, D. G.: Certain oat culm characters and their relationship to lodging. Sci. Agric. 21, 646—676 (1941).

Hansing, E. D., E. G. Heyne u. L. E. Melchers: Studies on smut-resistant oats for Kansas. J. Amer. Soc. Agron. 37, 499—508 (1945).

—, — u. T. R. Stanton: A new race of Ustilago avenae. Phytopathology 36, 400 (1946). (Abstr.)

—, —, — Reactions of oat varieties and selections to four races of loose smut. Phytopathology 36, 433—435 (1946).

Harrington, J. B.: Oat varieties and their production. Bull. Saskatch. Agric. Ext. Dep., Nr. 104, 8 (1942).

— The differential response of spring-sown varieties of oats and barley to date of seeding and its breeding significance. J. Amer. Soc. Agron. 38, 1073—1081 (1946).

Hayes, H. K., M. B. Moore u. E. C. Stakman: Studies of inheritance in crosses between Bond, Avena byzantina, and varieties of A. sativa. Tech. Bull. Univ. Minn. Agric. Exp. Sta., Nr. 137, 38 (1939).
— Breeding for resistance to crown rust, stem rust, smut, and desirable agronomic characters in crosses between Bond, Avena byzantina, and cultivated varieties of Avena sativa. J. Amer. Soc. Agron. 33, 164—173 (1941).
Hirschhorn, E.: Adiciones y correcciones a las espécies del género „Ustilago" en la Argentina. (Additions and corrections to the species of the genus „Ustilago" in Argentina.) An. Soc. Cient. Argent. 133, 217—218 (1942).
Holman, W. I. M. u. W. Godden: The aneurin (vitamin B_1) content of oats. I. The influence of variety and locality. II. Possible losses in milling. J. Agric. Sci. 37, 51—57 (1947).
Howard, H. W.: Meiotic irregularities in hexaploid oats. I. Univalent frequencies in spring x winter variety hybrids of Avena sativa. J. Agric. Sci. 37, 139—144 (1947).
Huskins, C. L., G. F. Sander u. Love: Chromosome mutations in Avena. Collecting Net 15, 170—171 (1940). (Abstr.)

Jones, E. T.: A comparison of the segregation of wild versus cultivated base in the grain of diploid, tetraploid and hexaploid species of oats. Proc. 7th Int. Genet. Congr. Edinburgh 23—30 August, Pp. 170 bis 171 1939 (1941). (Abstr.)
— A comparison of the segregation of wild versus normal or cultivated base in the grain of diploid, tetraploid and hexaploid species of oats. Genetica 22, 419—434 (1940).
— Aberystwyth-bred varieties of oats and pure line selections of Hen Gymro wheat. Leafl. Ser. Welsh Pl. Breed. Sta., Nr. 5, 10 (1945).
Johnson, L. P. V. u. H. A. McLennan: An attempt to hybridize annual and perennial Avena species. Canad. J. Res. 17, Sect. C, 35—37 (1939).
Judkins, W. P.: The influence of kernel size, age, location in panicle, and variety of oat, on the variability of the Avena test. Amer. J. Bot. (Suppl.) 33, 181—184 (1946).

Keese, H. v.: Kann die Methode des Gefäßversuches zur Prüfung der Leistungsfähigkeit neuer Getreidezüchtungen herangezogen werden? Z. Pflanzenz. 26, 199—213 (1944).
Kingsolver, C. H.: Reaction of varieties and selections of oats to Pseudomonas coronafaciens. Phytopathology 32, 12 (1942). (Abstr.)
— Pathogenicity on Avena and growth response of Pseudomonas coronafaciens (Elliott) Stapp. Iowa St. Coll. J. Sci. 19, 29—31 (1944).
Kopeljkievskij, G. V.: (New promising lines of oats of the Šatilovo Breeding Station.) Selekcija i Semenovodstvo, Nr. 1, 9—12 (1939). [Russisch.]
Kornilov, A.: (Two crops a year.) Sovhoznoe Proizvodstvo (State Farming), Nr. 7, 34 (1944). [Russisch.]
Kugler, W. F. u. S. Foucault: Descripción de variedades agrícolas de avena por sus caracteres morfológicos. (Description of agricultural varieties of oats by their morphological characters.) „Granos" Semilla Selecta, B. Aires 3, 169—182 (1939).

Lacroix, L.: L'avoine à Gembloux, de 1932 á 1939. Bull. Inst. Agron. Gembloux 10, 53—83 (1941).
Larmour, W. T.: „No. 601" new promising oat variety has many outstanding qualities. Canad. Grain. J. 1, 10 (1946).

Lewicki, S.: Badania had wartośćia ziarna owsa Plonu 1936 r. (The quality of oats cultivated in Poland [crop of 1936 year].) Prace Laborat. Badania Jakościowy Wartości Zbóż, Pulawy 23, 48 (1939).

MacKey, J.: Vithavresorter för fastmarksjordar i södra och mellersta Sverige. (White oat varieties for mineral soils of south and central Sweden.) Sverig. Utsädesfören. Tidskr. 57, 46—61 (1947).
Mándy, G. von: Beiträge zur Morphologie einiger gezüchteter Hafersorten Ungarns. Züchter 14, 65—73 (1942).
Meehan, F. u. H. C. Murphy: A new Helmintosporium blights of oats. Science 104, 413—414 (1946).
Middleton, G. K.: Oats-with-an-overcoat. Sth. Seedsman 9, Nr. 9; 12, 36 (1946).
Miles, H. W.: Frit fly on oats in the west of England. Rep. Agric. Hort. Res. Sta. Long Ashton, Pp. 119 bis 123 (1943).
Moore, M. B., H. K. Hayes u. E. C. Stakman: The field reactions of certain oats hybrids and varieties to stem rust. Phytopathology 35, 549—551 (1945).
Moormann, B.: Untersuchungen über Keimruhe bei Hafer und Gerste. Kühn-Arch. 56, 41—79 (1942).
Mordvinkina, A. I.: (Eco-geographical classification of cultivated and weed oats.) Proc. Lenin Acad. Agric. Sci. USSR., Nr. 9, 3—10 (1939). [Russisch.]
Mulvey, R. R.: A new era in oat production. Circ. Ind. Agric. Exp. Sta., Nr. 296, 4 (1944).
Murphy, H. C. u. L. C. Burnett: More oats, fewer acres! Fm. Sci. Reporter 4, Nr. 1, 6—7 (1943).
—, — Clinton oats arrive. Farm Sci. Reporter 6, Nr. 4, 3—7 (1945).
—, —, C. H. Kingsolver, T. R. Stanton u. F. A. Coffman: Relation of crown-rust infection to yield, test weight, and lodging of oats. Phytopathology 30, 808—819 (1940).
— u. F. Meehan: Reaction of oat varieties to a new species of Helminthosporium. Phytopathology 36, 407 (1946). (Abstr.)
—, T. R. Stanton u. F. A. Coffman: Breeding for disease resistance in oats. J. Amer. Soc. Agron. 34, 72—89 (1942).

Newton, M. u. T. Johnson: Physiologic specialization of oat stem rust in Canada. Canad. J. Res. 22, Sect. C, 201—216 (1944).
Niederhauser, J. S.: Oat disease survey in northern Pennsylvania. Plant Dis. Reporter 28, 1072—1077 (1944). [Mimeographed.]
Nishiyama, I.: Cytogenetical studies in Avena. II. On the progenies of pentaploid Avena hybrids. III. Experimentally produced eu- and hyperhexaploid aberrants in oats. Cytologia, Tokyo 10, 88—100, 101—104 (1939).
— (On the evolution of some grain characters in Avena.) Jap. J. Genet. 15, 321—323 (1939).
— (Disturbed segregation ratios caused by certation.) Jap. J. Genet. 16, 300—301 (1940).

O'Kelly, J. F.: New oat varieties found of value in hill station tests. Miss. Fm. Res. 5, Nr. 8, 2 (1942).

Patel, N. M.: Independence in inheritance of the loose smut reaction and lemma colouration in an oat cross. Bot. Ser. Bull. Inst. Agric. India, Nr. 1, 27 (1943).
— Inheritance of loose smut reaction in crosses with Victoria and Smut Resistant (Cornell)-6 under field conditions of growth and infection. Abstr. Thes. Cornell Univ. 1942, Pp. 378—379 (1943).
— Dimeric transmission of disease resistance to Ustilago Avenae (Pers.) Jens. Race 21 (Reed) in Avena hybrid of interspecific derivation. Proc. 31st Indian Sci. Congr., Delhi Pt. III, Pp. 157—158 (1944). (Abstr.)

Peturson, B.: Adult plant resistance of some oat varieties to physiologic races of crown rust. Canad. J. Res. 22, Sect. C, 287—289 (1944).
Poehlman, J. M.: Growing good crops of oats in Missouri. Bull. Mo. Agric. Exp. Sta., Nr. 439, 12 (1942).
Pohjakallio, O.: Über die Brandresistenzzüchtung von Hafer. (On the breeding of smut resistant oats.) Maataloust. Aikakausk. 11, 97—107 (1939).
— Beiträge über die Untersuchungen über die Brandresistenzzüchtung des Hafers. Maataloust. Aikakausk. 11, 183—197 (1939).
Prince, F. S., L. J. Higgins u. P. T. Blood: Small grain and corn variety tests. Sta. Circ. N. H. Agric. Exp. Sta., Nr. 67, 11 (1944).

Rasmussen, K. J.: Undersøgelser over skridningen hos havre 1927—39. (Investigations on shooting in oats 1927—39.) K. VetHøjsk. Aarsskr., Pp. 1—37 (1941).
Reed, G. M.: Inheritance of smut resistance in some oat hybrids. Amer. J. Bot. 6, 451—457 (1941).
— Reports on research for 1940. Plant pathology. Brooklyn Bot. Gdn. Rec. 30, 81—87 (1941).
— Inheritance of smut resistance in hybrids of Navarro oats. Amer. J. Bot. 29, 308—314 (1942).
— Reports on research for 1941: Plant pathology. Brooklyn Bot. Gdn Rec. 31, 90—94 (1942).
— Reports on research for 1942. Plant pathology. Brooklyn Bot. Gdn. Rec. 32, 75—78 (1943).
— Reports on research for 1943. Plant pathology. Brooklyn Bot. Gdn. Rec. 33, 9—11 (1944).
— u. T. Stanton: Susceptibility of Lee x Victoria oat selections to loose smut. Phytopathology 32, 100—102 (1942). (Abstr.)
Robinson, J. F.: A summary of oat variety trials at the College Farm. Aber, 1933—44. Welsh J. Agric. 18, 59—71 (1945).
Rosen, H. R.: The 1944 ,,Arkansas Traveler". Sth. Seedsman 7, 11, 54 (1944).
— Breeding oats to combine resistance to race 45 and other races of crown rust common in Arkansas. Phytopathology 35, 143—144 (1945).
— Incidence of crown rust of oats from October to March, 1944—1945, in Arkansas. Plant Dis. Reporter 29, 286—287 (1945). [Mimeographed.]
— Helminthosporium blight of oats in Arkansas. Phytopathology 37, 19 (1947). (Abstr.)
—, L. M. Weetman u. C. K. McClelland: Hybridizing oats to combine growth for winter pasture, hardiness, and resistance to rusts and smuts. J. Amer. Soc. Agron. 32, 12—14 (1940).

Saloheimo, L.: Jämförelse av havresorter på Finska Mosskulturföreningens Karelska försöksstation under tioårsperioden 1933—1942. (A comparison of oat varieties of the Karelian Experimental Station of the Finnish Society for the Cultivation of Bogland, during the ten year period 1933—1942.) Finska MossFören. Årsb. 47, 103—128 (1943).
Sander, H. G. F.: Chromosome aberrations as the cause of fatuoid, steriolid and subfatuoid mutations in oats. Genetics 24, 94 (1939). (Abstr.)
Schlehuber, A. M., J. J. Sturm u. R. H. Bamberg: Oat variety tests in Montana. Bull. Mont. Agric. Exp. Sta., Nr. 399, 20 (1942).
Šepeleva, E. M.: Karyosystematic study on cultivated and wild oats. C. R. (Doklady) Acad. Sci. URSS. 25, 228—231 (1939).
Shands, H. L. u. B. D. Leith: Vicland Oats. Bull. Wisc. Agric. Exp. Sta., Nr. 462, 15 (1944).
Shepeleva, E. M. siehe Šepeleva.
Siang, Yin Ko, J. H. Torrie u. J. G. Dickson: Inheritance of reaction to crown rust and stem rust and other characters in crosses between Bond, Avena byzantina, and varieties of A. sativa. Phytopathology 36, 226 bis 235 (1946).
Sims, H. J.: Mallee oat variety trials. Results at Walpeug Research Station. J. Dep. Agric. Vict. 41, 1—5 (1943).
Stakman, E. C. u. W. Q. Loegering: The Potential importance of race 8 of Puccinia graminis avenae in the United States. Phytopathology 34, 421—425 (1944).
Staniszkis, W.: Doświadczenia z odmianami owsa przedrowadzone w Polsce w latach 1923—1935. (Experiments with oat varieties carried out in Poland in the years 1923 to 1935. Prace Nauk. Pol. (Pap. Agric. Res.) Suppl. to Przegl. Doświad. Roln. (Rev. Agric. Res.), Nr. 5, 65 (1939).
Stanton, T. R.: Registration of varieties and strains of oats, IX. J. Amer. Soc. Agron 32, 76—82 (1940).
— Registration of varieties and strains of oats, X. J. Amer. Soc. Agron 33, 246—251 (1941).
— Registration of varieties and strains of oats, XI. J. Amer. Soc. Agron 34, 275—279 (1942).
— That ,,someday" is today for oat breeders. Sth. Seedsman 6, Nr. 5; 9, 45, 48 (1943).
— Registration of varieties and strains of oats, XII. J. Amer. Soc. Agron 35, 242—244 (1943).
— Registration of varieties and strains of oats, XIII. J. Amer. Soc. Agron 36, 445—446 (1944).
— The present status of breeding for disease resistance in oats. Phytopathology 36, 688 (1946). (Abstr.)
— u. F. A. Coffman: Grow disease-resistant oats. Fmrs'Bull. U. S. Dep. Agric., Nr. 1941, 13 (1943).
— u. H. C. Murphy: Field studies of smut resistance in oats. J. Amer. Soc. Agron 34, 248—258 (1942).
Stapledon, R. G.: The Welsh Plant Breeding Station. Country Life 89, 236—237 (1941).
Stefanovskij, J. A. u. E. M. Večeslova: (Drought resistance of different agroecological groups of oats.) Proc. Lenin Acad. Agric. Sci. USSR., Nr. 11, 3—8 (1939). [Russisch.]
Stephens, S. G.: Yield characters of selected oat varieties in relation to cereal breeding technique. J. Agric. Sci. 32, 217—254 (1942).
Stoa, T. E.: Oats varieties and rust. Bi-m. Bull. N. Dak. Agric. Exp. Sta. 7, 8—10 (1945).
— A new oat disease. Bi-m. Bull. N. Dak. Agric. Exp. Sta. 9, Nr. 2, 46—47 (1946).
— u. C. M. Swallers: Varieties of oats for North Dakota. Bi-m. Bull. N. Dak. Agric. Exp. Sta. 4, Nr. 3, 24—30 (1942).
—, — New varieties of oats for North Dakota. Bi-m. Bull. N. Dak. Agric. Exp. Sta. 5, Nr. 3, 17—22 (1943).
Stuart, A.: Oats. N. Z. J. Agric. 72, 461—465 (1946).
Suhov, K. S. u. A. M. Vovk: (Damage caused by ,,Zakuklivaniye", a mosaik disease of oats, and mode of its distribution in nature.) Bull. Acad. Sci. URSS. Sér. Biol., Pp. 121—144 (1939). [Russisch.]
Sundelin, G.: Försöksverksamhetens bedrag till lösandet av vårens växtodlingsproblem. (The contribution made by research to the solution of the problem of spring crops.) Lantmannen 31, 243—245 (1947).
Swenson, S. P.: Miomark oats. Circ. S. Dak. Agric. Exp. Sta., Nr. 32, 4 (1941).

Taborda de Morais, A.: Estudos nas aveias II. As aveias portuguesas da secção Euavena Griseb. (Studies in oats II. The Portuguese oats of the section Euavena Griseb.) Bol. Soc. Broteriana 13, 573—709 (1938—39).
Taysi, M. V.: Türkiye yabani yulaflari, formlarin toplanma ve teshisi, kültür yulafi islahina yarayislarinin tetkiki. (Turkish wild oats. Collection and determination of the forms and testing their suitability as initial breeding material.) T. C. Yüksek Ziraat Enstitüsü Çalişmalarindan, Ankara, Nr. 115, 56 (1941).

Tervet, I. W.: The relative susceptibility of different lots of oat varieties to smut. Phytopathology **31**, 672—673 (1941).
— The relation of seed quality to the development of smut in oats. Phytopathology **34**, 106—115 (1943).
— u. H. Hart: Variation in reaction of Anthony oats to stem rust, Puccinia graminis avenae. Phytopathology **32**, 1087—1090 (1942).
Thomas, I. u. A. J. Millington: Oat variety trials in the wheatbelt in 1944. J. Agric. W. Aust. **21**, 365—371 (1944).
Tingey, D. C., R. W. Woodward u. T. R. Stanton: Uton. A new high-yielding white oat resistant to loose and covered smuts. Bull. Utah Agric. Exp. Sta., Nr. 296, 15 (1941).
Tohtuev, A.V.: (Reaction to ,,Zakuklivanie" of different varieties of barley and oats.) Theses and scientific papers read at the 4th District Conference of Workers of Universities and Research Institutions, Omsk, Nr. 1, Agron. Sect., 69—71 (1941).
Törnqvist, G. I.: Svalöfs Samehavre (01341). Ny havresort för nordligaste Sverige. (Svalöf's Same oats (01341). A new variety of oats for the most northerly parts of Sweden.) Sverig. Utsädesfören. Tidskr. **51**, 383—385 (1941).
— Svalöfs Samehavre (01341). Ny havresort för nordligaste Sverige. (Svalöf's ,,Same" oat 01341: a new oat variety for northernmost Sweden.) Sverig. Utsädesfören. Tidskr. **53**, 48—50 (1943).
Torrie, J. H.: Correlated inheritance in oats of reaction to smuts, crown rust, stem rust, and other characters. J. Agric. Res. **59**, 783—804 (1939).
Toulson, G. A.: Winter oat varieties. Seale-Hayne Annu. **1**, 17—24 (1938—1939).
Troickaja, E.: (The black oat Džety-su.) Selecija i Semenovodstvo (Breeding and Seed Growing), Nr. 2, 25—26 (1940). [Russisch.]

Vallega, J.: Especialización fisiológica de Puccinia coronata avenae, en Argentina. (Physiological specialization of P. coronata Avenae in Argentina.) An. Inst. Fitotec. Santa Catalina **2**, 53—82 (1940).
— Razas fisiológicas de Puccinia graminis avenae halladas en Argentina. (Physiological races of P. graminis Avenae found in Argentina. Rev. Fac. Agron. B. Aires **3**, 517—529 (1943).
Vidme, T.: Om dei morfologiske avlingskomponentane hjå nokre havresortar. (On the morphological components of yield in some varieties of oats.) Meld. Norg. LandbrHøgsk. **20**, 203—228 (1940).
Voss, J.: Zur Prüfung der Resistenz von Hafersorten gegen Flugbrand (Ustilago avenae [Persoon] Jensen). Z. Pflanzenz. **23**, 20—46 (1939).

Waller, E.: Svalöfs Solhavre II. (Vg. 01534 a). Ny medeltidig, mycket strastyv vithavresort för södra och mellersta Sverige. (Svalöfs Solhavre II (Vg. 01534a) — A new medium early, very stiff strawed white oat variety for southern and central Sweden.) Sverig. Utsädesfören. Tidskr. **52**, 370—381 (1942).
Warner, J. D.: New oat varieties for the southeast. Florida's Quincy Nos 1 and 2, highly resistant to rust and smut, also show promise in Georgia and Alabama. Sth. Seedsman **5**, Nr. 12, 9, 29 (1942).
Weetman, L. M.: Genetic studies in oats of resistance to two physiologic races of crown rust. Phytopathology **32**, 19 (1942). (Abstr.)
Welch, A.: Pythium root necrosis of oats. Iowa St. Coll. J. Sci. **19**, 361—399 (1945).
Welsh, J. N.: Vanguard oats-origin, description and performance. Publ. Dep. Agric. Can., Nr. 651, 14 1939).
— History, description, distribution and performance of Ajax and Exeter oats. Sci. Agric. **25**, 96—106 (1944).

Werth, E.: Der Hafer, eine urnordische Getreideart. (Zur Geographie und Geschichte der Kulturpflanzen und Haustiere XXV.) Z. Pflanzenz. **26**, 92—102 (1944).
Wexelsen, H.: Kombinerte foredlings- og kvelstoffgjødslingsforsøk i havre. (Combined breeding and nitrogen manuring experiments with oats.) Tidsskr. Norske Landbr. **50**, 25—36 (1943).
Williams, M. F.: Out of the Bayou Country ... Camellia oats. Sth. Seedsman, **6**, Nr. 6; 11, 38 (1943).

Zacks, M. B.: Oats and climate in southern Ontario. Canad. J. Res. **23**, Sect. C, 45—75 (1945).

4. Gerste

Åberg, E.: The taxonomy and phylogeny of Hordeum L. Sect. Cerealia Ands. with special reference to Thibetan barleys. A.-B. Lundequistska Bokhandeln, Uppsala, Pp. 156. 20 pls. maps. (Symb. Bot. Upsaliens. IV, 2) (1940).
— Effect of vernalizazion on the development of stripe in barley. Phytopathology **35**, 367—368 (1945).
— u. G. A. Wiebe: Irregular barley, Hordeum irregulare, sp. nov. J. Wash. Acad. Sci. **35**, 161—164 (1945).
—, — Classification of barley varieties grown in the United States and Canada in 1945. Tech. Bull. U.S. Dep. Agric., Nr. 907, 190 (1946).
—, — u. A. D. Dickson: Ash content of barley awns and kernels as influenced by location, season, and variety. J. Amer. Soc. Agron. **37**, 583—586 (1945).
Almeida, J. M. de: O desenvolvimento da espiga nas primeiras idades, como processo de distinção de formas de inverno e formas de primavera na cevada. (The development of the ear in the early stages, as a means of distinguishing winter and spring forms in barley.) Rev. Agron. Lisboa **30**, 462 (1942).
Anderson, J. A.: Factors related to the protein content of malting barley. Wallerstein Laboratories Communications **7**, 179—191 (1944).
— Canadian research on malting barley. Wallerstein Laboratories Communications **8**, 5—22 (1945).
— Canadian research on malting barley. Part II. The relations between certain barley and malt properties. Emp. J. Exp. Agric. **13**, 1—10 (1945).
—, C. A. Ayre u. W. O. S. Meredith: Varietal differences in barleys and malts. V. Wort nitrogen and malt extract and their correlations with barley nitrogen fractions. Canad. J. Res. **17**, Sect. 3, 25—34 (1939).
— u. W. J. Eva: Protein survey of western Canadian barley 1942 crop. Bd. Grain Comm. Grain Res. Lab., Winnipeg, Manitoba, Pp. 16 (1942).
—, W. O. S. Meredith u. H. R. Sallans: Malting quality of Canadian barleys. IV. A summary of information of special interest to plant breeders. Sci. Agric. **23**, 297—314 (1943).
—, H. R. Sallans u. W. O. S. Meredith: Varietal differences in barley and malts. XII. Summary of correlations between 18 major barley, malt and malting properties. Canad. J. Res. **19**, Sect. C, 278—291 (1941).
Andersson, G.: Vergleichende Untersuchungen der Assimilationsintensität diploider und tetraploider Gerste. Svensk. Bot. Tidskr. **37**, 175—199 (1943).
Anonymus: Brouwgerstrassen. (Malting barley varieties.) Vijfde Jaarboekje Nationaal Comité voor Brouwgerst, Wageningen, Pp. 30—42 (1940).
— De voornaamste eischen welke aan brouwgerst gesteld worden. (The principal requirements demanded in malting barley.) Vijfde Jaarboekje Nationaal Comité voor Brouwgerst, Wageningen, Pp. 106—107 (1940).
— The history of malting barley in Ireland. B. G. A. Rec., Nr. 4, 7—8 (1940).

Anonymus: Methoden ter beoordeling van gerst, mout en bier, vastgesteld door het National Comité voor Brouwgerst. (Methods for the appraisal of barley, malt and beer, set up by the National Committee for Brewing Barley.) Meded. Nat. Comité v. Brouwgerst, Wageningen, Nr. 3, 21 (1941).
— Industrial evaluation of two barley varieties grown in 1939. Malt. Res. Inst. Madison, Nr. 1, 64 (1941).
— Industrial evaluation of two barley varieties grown in 1940. Malt Res. Inst. Madison, Nr. 2, 67 (1942).
— Industrial evaluation of two barley varieties grown in 1941. Malt Res. Inst. Madison, Nr. 3, 59 (1942).
— Barley varieties resistant to stripe, Helminthosporium gramineum Rabh. J. Amer. Soc. Agron. 35, 736—737 (1943).
— Camton barley. Nat. Inst. Agric. Bot. Cambridge, January, Pp. 4 (1943).
— Pioneer winter barley. Leafl. Nat. Inst. Agric. Bot., Pp. 2 (1943).
— Smooth bearded barley available for farmers. Sth. Seedsman 6, 57 (1943).
— Prefect six-row winter barley. Leafl. Nat. Inst. Agric. Bot. Cambridge, Pp. 2 (1944).
— Industrial evaluation of two barley varieties grown in 1942. Malt Res. Inst. Madison, Nr. 4, 48 (1944).
— U. S. breeds better barleys for special needs; aphid-resistant strain due for Texas, Oklahoma. Sth. Seedsman 7, 32 (1944).
— Large scale barley variety experiments. J. Dep. Agric. Éire 42, 99—105 (1945).
— Long search for dual purpose barley now appears on road to success. Canad. Grain J. 1, 13 (1946).
— The reconstruction of the research organization. Inst. Brew. (1946?), Pp. 32 (1946).
— Industrial evaluation of Bay barley a new variety released in Michigan. Publ. Malt Res. Inst. Madison, Wisconsin, Nr. 5, 22, 1946 (1947).
— Brewing and malting industries announce formation of Barley Improvement Institute. Canad. Grain J. 2, Nr. 11, 18 (1947).
— New feed barley at Brandon Experimental Farm shows promise: christened Vantage. Canad. Grain J. 2, Nr. 6, 8 (1947).
Ansari, M. A. A.: Improved barley. Indian Fmg. 4, 569—570 (1943).
Arnason, T. J., J. B. Harrington u. H. A. Friesen: Inheritance of variegation in barley. Canad. J. Res. 24, Sect. C, 145—157 (1946).
Arny, D. C.: Inheritance of resistance to barley stripe. Phytopathology 35, 781—804 (1945).
Atkins, I. M. u. P. B. Dunkle: Barley production in Texas. Bull. Tex. Agric. Exp. Sta., Nr. 605, 35 (1941).

Bahteev, F. H. u. E. M. Darevskaja: An intergeneric hybrid between barley and Elymus. C. R. (Doklady) Acad. Sci. URSS. 47, 300 (1945).
— (Data on the intravarietal crossing of barley.) Izvestija Akademii Nauk SSSR. Serija biologičeskaja. (Bull. Acad. Sci. URSS., Sér. Biol.), Nr. 4, 471—484 (1945). [Russisch.]
— (Stellt Hordeum agriocrithon eine Art dar?) C. R. (Doklady) Acad. Sci. URSS. 57, 195 (1947). [Russisch.]
— u. I. A. Palamarčuk: (Cytologische Untersuchungen der ersten Generation von Hordeum-Elymusbastard.) C. R. (Doklady) Acad. Sci. URSS. 56, 751 (1947). [Russisch.]
Baker, J. L. u. T. J. Ward: Malting and brewing trials with three types of English barleys. J. Inst. Brew. 52, 17—25 (1946).
Barbee, O. E.: Barley production in Washington. Bull. Wash. St. Agric. Exp. Sta., Nr. 382, 48 (1939).
Becker: Die Gerstensorte. Dtsch. landw. Pr. 67, 77 (1940).

Bell, G. D. H.: A study on the date of ear emergence in barley. J. Agric. Sci. 29, 175—228 (1939).
— The breeding of two-row winter-hardy barley. J. Agric. Sci. 34, 223—238 (1944).
— u. G. P. Carson: The inheritance of rachilla length in barley. J. Agric. Sci. 31, 246—279 (1941).
Bergal, P. u. L. Friedberg: Essai d'identification des orges cultivées en France. Ann. Épiphyt. Phytogénét. 6, 157—306 (1940).
Bishop, L. R.: Second memorandum on barley germination. J. Inst. Brew. 51, 215—224 (1945).
— Third memorandum on barley germination. J. Inst. Brew. 52, 273—282 (1946).
Blinova, N. P.: (On the technique of crossing spring barley.) Selekcija i Semenovodstvo, Nr. 2/3, 36—39 (1939).
Briggs, F. N.: Production and improvement of malting barley in California. Amer. Brewer, 72, 37, also Brewers Digest 14, 30—31 (1939).
— The history and improvement of malting barley in California. J. Inst. Brew. 46, 31—33 (1940).
— Linkage relations of factors for resistance to mildew in barley. Genetics 30, 115—118 (1945).
— u. B. A. Madson: The production and improvement of malting barley in California. Wallerstein Laboratories Communications, Nr. 7, 63 (1939). (Abstr.)
— u. E. H. Stanford: Linkage relations of the Goldfoil factor for resistance to mildew in barley. J. Agric. Res. 66, 1—5 (1943).
Brink, R. A. u. D. C. Cooper: Embryo viability and development of the seed following interspecific hybridization. Rec. Genet. Soc. America, Nr. 12, 43—45 (1943). (Abstr.)
—, — The antipodals in relation to abnormal endosperm behavior in Hordeum jubatum x Secale cereale hybrid seeds. Genetics 29, 391—406 (1944).
—, — u. L. E. Ausherman: A hybrid between Hordeum jubatum and Secale cereale. J. Hered. 35, 67—75 (1944).
Brown, B. M.: Report on the fermentation industries for 1944. Soc. Chem. Ind. & Inst. Brew., Pp. 19 (1944).
— u. J. Wilmot: Malting and brewing trials with three types of English barleys. J. Inst. Brew. 53, 10—14 (1947).
Buchli, M.: Anbauversuche mit Wintergerstensorten. Landw. Jb. Schweiz 45, 701—714 (1944).
Burkart, A.: Noticia sobre dos variedades de cebada desnuda, cultivadas en pequeña escala en la Argentina. (Note on two naked barley varieties cultivated on a small scale in Argentina.) Rev. Argent. Agron 12, 216—221 (1945).
Burnham, C. R.: A gene for „long" chromosomes in barley. Genetics 31, 212—213 (1946). (Abstr.)

Caffera, R.: Tipos y variedades de cebada. (Types and varieties of barley.) „Granos" Semilla Selecta. B. Aires 9, Nr. 4/6, 37—43 (1945).
Caffrey, M.: Barley. J. Dep. Agric. Eire 40, 31—44 (1943).
Chen, Shao-Lin, Shu-Min Shen u. P. S. Tang: Studies on colchicine-induced autotetraploid barley. I and II. Cytological and morphological observations. Amer. J. Bot. 32, 103—106 (1945).
Chin, T.-C.: The cytology and genetics of Hordeum. Abstr. Diss. Univ. Camb., Pp. 13—14 (1939/40) 1941.
— The cytology of some wild species of Hordeum. Ann. Bot., London 5, 535—545 (1941).
Christoff, M. A.: Untersuchungen über die Kältefestigkeit der Wintergerste. Z. Pflanzenz. 23, 47—90 (1939).
Comrie, A. A. D.: Report on the Fermentation Industries for 1939. Prepared for the Society of Chemical Industry and the Institute of Brewing, Pp. 30.

Cooper, D. C. u. R. A. Brink: Collapse of the seed following the mating of Hordeum jubatum x Secale cereale. Genetics **29**, 370—390 (1944).

Copertini, S.: Esame chimico-technologico di orzi della Libia. (Chemical and technological study of the barleys of Libia.) Agricoltura Colon. **33**, 626—632 (1939).

Das, C. M.: What's doing in all-India. United Provinces. Indian Fmg. **3**, 498—500 (1942).

Dasananda, S.: Quantitative and qualitative inheritance in barley. Abstr. Thes. Cornell Univ., Pp. 330 bis 332, 1943, (1944).

Dickson, J. G.: Scab of wheat and barley and its control. Fmrs' Bull. U. S. Dep. Agric., Nr. 1599, 22 (1942).
— Helminthosporium foot rot of barley. Phytopathology **36**, 397 (1946). (Abstr.)

Ekdahl, I.: Comparative studies in the physiology of diploid and tetraploid barley. Ark. Bot. **31 A**: Häfte 2, Nr. 5, 45 (1944).

Farstad, C. W. u. A. W. Platt: The reaction of barley varieties to wheat stem sawfly attack. Sci. Agric. **26**, 216—224 (1946).

Freisleben, R.: Die Gersten und Weizen der Deutschen Hindukusch-Expedition, 1935. Angew. Bot. **22**, 105 bis 132 (1940).
— Die phylogenetische Bedeutung asiatischer Gersten. Züchter **12**, 257—272 (1940).
— Anschauungen über die Abstammung der Kulturgersten auf Grund des Materials der deutschen Hindukusch-Expedition 1935. Forsch. u. Fortschr., Pp. 140 bis 142 (1941).
— Untersuchungen an tetraploiden Kulturgersten. Forschungsdienst, Sonderh. **16**, 361 (1942).
— Ein neuer Fund von Hordeum agriocrithon Åberg. Züchter **15**, 25—29 (1943).
— u. A. Lein: Über die Auffindung einer mehltauresistenten Mutante nach Röntgenbestrahlung einer anfälligen reinen Linie von Sommergerste. Naturwissenschaften **30**, 608 (1942).
—, — Vorarbeiten zur züchterischen Auswertung röntgeninduzierter Mutationen. I. Die in der Behandlungsgeneration (X) sichtbare Wirkung der Bestrahlung ruhender Gerstenkörner. Z. Pflanzenz. **25**, 235—254 (1943).
—, — Vorarbeiten zur züchterischen Auswertung röntgeninduzierter Mutationen. II. Mutationen des Chlorophyllapparates als Testmutationen für die mutationsauslösende Wirkung der Bestrahlung bei Gerste. Z. Pflanzenz. **25**, 255—83 (1943).
—, u. I. Metzger: Genetische Studien zur Gerstenzüchtung. I. Vererbung und Koppelung der Mehltauresistenz und der Spindelgliedzahl. Z. Pflanzenz. **24**, 507—522 (1942).

Friesen, H. A.: Awn-barbing in barley. Canad. J. Res. **24**, Sect. C, 292—297 (1946).

Gelin, O. E. V.: The cytological effect of different seed-treatments in X-ray barley. Hereditas, Lund **27**, 209 bis 219 (1941).

Glick, D.: The enzymes of barley. Some recent contributions to our knowledge of their nature and properties. Wallerstein Laboratories Communications **4**, 20—24 (1941).

Gombert, J.: De veredeling van gerst in het bijzonder met het oog op de industrueele verwerking. (The breeding of barley especially with a view to industrial manufacture.) Achtste Jaarboekje Nationaal Comité voor Brouwgerst, Wageningen, Pp. 26—31 (1943).

Göpp, K. u. W. Sauer: Die deutschen Braugerstensorten. Zusammenfassende Bearbeitung mehrjähriger Untersuchungsergebnisse. Wschr. Brau. **56**, 209—216, 219—223 (1939).

Gramolin, B. A.: (Changes in the reproductive organs of barley as a result of hybridization and conditions of rearing.) Jarovizacija, Nr. 2 (35), 19—30 (1941). [Russisch.]

Granhall, I.: Svalöfs Ymerkorn (Sv 40/13 b_1). (Svalöfs Ymer barley [Sv 40/13 b_1].) Sverig. Utsädesfören Tidskr. **54**, 237—239 (1944).

Gregory, F. G. u. Purvis, O. N.: Abnormal flower development in barley involving sex reversal. Nature, Lond. **160**, 221—222 (1947).

Greis, H.: Vergleichende physiologische Untersuchungen an diploiden und tetraploiden Gersten. Züchter **12**, 62—73 (1940).

Grieben, H. u. Cipolla, G.: Resultado de los ensayos comparativos de rendimiento „standard" entre variedades de cebada cervecera. Años agricolas 1939/40 a 1941/42. (Results of the comparative „standard" yield trials of malting barley varieties. Agricultural years 1939/40 to 1941/42.) „Granos" Semilla Selecta, B. Aires **6**, 19—31 (1942).

Gudkov, A. N.: (Determination of the colour of barley seeds by chemical methods.) Selekcija i Semenovodstvo (Breeding and Seed Growing) Nr. 6, 32 (1940). [Russisch.]

Gustafsson, Å.: The mutation system of the chlorophyll apparatus. Acta Univ. Lund. **36**, Nr. 11, 40, K. Fysiogr. Sällsk. Handl. **51**, Nr. 11 (1940).
— Mutation experiments in barley. Hereditas, Lund **27**, 225—242 (1941).
— Preliminary yield experiments with ten induced mutations in barley. Hereditas, Lund, **27**, 337—359 (1941).
— Mutationsforskning och växtförädling. (Plant breeding and mutation.) Sverig. Utsädesfören. Tidskr. **51**, 344—362 (1941).
— Mutationsforschung und Züchtung. Züchter **14**, 57 bis 64 (1942).
— The plastid development in various types of chlorophyll mutations. Hereditas, Lund **28**, 483—492 (1942).
— Drastic morphological mutations in barley. Hereditas, Lund **32**, 120—122 (1946).
— Mutations in agricultural plants. Hereditas, Lund **33**, 1—100 (1947).
— The advantageous effect of deleterious mutations. Hereditas, Lund **33**, 573—575 (1947). (Abstr.)
— u. Åberg, E.: Two extreme X-ray mutations of morphological interest. Hereditas, Lund **26**, 257—261 (1940).

Harlan, H. V., M. L. Martini u. H. Stevens: A study of methods in barley breeding. Tech. Bull. U. S. Dep. Agric. Nr. 720, 26 (1940).
—,—,— The effect of temperature on seed set in barley crosses. J. Amer. Soc. Agron. **35**, 316—320 (1943).

Harrison, T. J.: Barley and linseed flax committee pays tribute to Montcalm founder. Canad. Grain J. **2**, Nr. 7, 17 (1947).

Hartong, B. D.: Teelt en oderzoek van browgerst in het buitenland. (Growing and investigation of malting barley abroad.) Zevenbe Jaarboekje Nationaal Comité voor Brouwgerst, Wageningen, Pp. 74—87 (1942).

Hayes, H. K.: Barley varieties registered, V. J. Amer. Soc. Agron. **32**, 84 (1940).
— Barley varieties registered, VI. J. Amer. Soc. Agron. **33**, 252—254 (1941).
— Barley varieties registered, VII. J. Amer. Soc. Agron. **34**, 281—282 (1942).
— Barley varieties registered, VIII. J. Amer. Soc. Agron. **35**, 240 (1943).
— Barley varieties registered, IX. J. Amer. Soc. Agron. **36**, 444 (1944).
— Barley varieties registered, H. J. Amer. Soc. Agron **37**, 645 (1945).

Hertzman, N.: Weibulls Balderkorn. Ny förädling med högsta avkastningsförmåga och strästyvhet. (Weibull's Balder barley. A new improved variety with the highest yield capacity and straw stiffness.) Weibulls Ill. Årsb. **37**, 25—27 (1942).

Hlynka, K.: „Montclam", new Canadian malting barley is result of 22 years of research work. Canad. Grain J. **1**, Nr. 6, 10—11 (1946).

Hodjkov, L. E.: (The appearance of „winter" and „semiwinter" plants in the progeny of a cross between spring cereals.) Jarovizacija Nr. 1 (34), 47—49 (1941). [Russisch.]

Hoffmann, W.: Die Vererbung der Winter-Sommer-Form und der Winterfestigkeit der Gerste. Z. Pflanzenz. **26**, 56—91 (1944).

Honecker, L.: Erbanalytische Untersuchungen über das Verhalten der Gerste gegenüber verschiedenen physiologischen Rassen des Mehltaues. Z. Pflanzenz. **24**, 429—506 (1942).

— Resistenzzüchtung gegen Mehltau und Rost bei Gerste. Erfahrungen und Ergebnisse vierzigjähriger Züchtungsarbeit. Z. Pflanzenz. **25**, 209—234 (1943).

Hunter, H.: Various aspects of malting barley breeding. J. Inc. Brew. Guild **25**, 177—192 (1939).

— Various aspects of malting barley breeding. J. Inst. Brew. **45**, 286—298 (1939).

— Dr. E. S. Beaven. Nature, Lond. **148**, 776—777 (1941).

— Malting barley and how to grow it. Red Cross Agric. Fund Leafl. Pp. 8 (1942).

— The improvement of malting barleys, in retrospect and prospect. J. Inst. Brew. **49**, 296—302 (1943).

— Appendix. J. Inst. Brew. **52**, 25—26 (1946).

Huyskes, J. A.: Brouwgerstrassen. (Varieties of malting barley.) Zesde Jaarboekje Nationaal Comité voor Brouwgerst, Wageningen, Pp. 96—109 (1941).

— Brouwgerstrassen. (Malting barley varieties.) Zevende Jaarboekje Nationaal Comité voor Brouwgerst, Wageningen, Pp. 95—112 (1942).

— Brouwgerstrassen. (Malting barley varieties.) Achtste Jaarboekje Nationaal Comité voor Brouwgerst, Wageningen, Pp. 62—74 (1943).

Immer, F. R.: Relation between yielding ability and homozygosis in barley crosses. J. Amer. Soc. Agron. **33**, 200—206 (1941).

— u. J. J. Christensen: Studies on susceptibility of varieties and strains of barley to Fusarium and Helminthosporium kernel blight when tested under muslin tents or in nurseries. J. Amer. Soc. Agron. **35**, 515—522 (1943).

—. — u. W. Q. Loegering: Reaction of strains and varieties of barley to many physiologic races of stem rust. Phytopathology **33**, 253—254 (1943).

— u. M. T. Henderson: Linkage studies in barley. Genetics **28**, 419—440 (1943).

Jensen, N. F.: Powdery mildew of barley. Studies of yield losses and the inheritance of disease resistance. Abstr. Thes. Univ. Cornell, Pp. 333—334, 1943 (1944).

Johnson, I. J. u. E. Åberg: The inheritance of brittle rachis in barley. J. Amer. Soc. Agron. **35**, 101—106 (1943).

Johnston, W. H.: License Vantage barley; may replace Plush. Canad. Grain J. Nr. 8, 10 (1947).

Judin, A. F.: (Continuation of the work of converting hulled barleys into neked barley.) Jarovizacija, Nr. 3 (30), 207—209 (1940). [Russisch.]

Karnauhov, I. P.: (New methods of identifying varieties of barley by the grain.) Vestnik Socialističeskogo Rastenievodstva (Soviet Plant Industry Record) Nr. 5, 109—117 (1940). [Russisch.]

Karpečenko, G. D.: On the transverse division of chromosomes as a result of colchicine treatment. C. R. (Doklady) Acad. Sci. URSS. **2**, 404—406 (1940).

Karpečenko, G. D.: Tetraploid six-rowed barleys obtained by colchicine treatment. C. R. (Doklady) Acad. Sci. URSS. **27**, 47—50 (1940).

Kartamyšev, V.: (Many rowed barley.) Jarovizacija Nr. 6 (33), 109—110 (1940). [Russisch.]

Kattermann, G.: Sterilitätsstudien bei Hordeum distichum. Z. Vererbungsl. **77**, 63—103 (1939).

Keegan, R.: Barley in Ontario. Bull. Ont. Dep. Agric. Nr. 442, 11 (1944).

Khod'kov, L. E. siehe Hodjkov.

Koudelka, V.: Udio pljevica u sastavu našeg proljetnog i ozimog pivarskog ječma. (The proportion of husk in the composition of our summer and winter malting barleys.) Archiv Minist. Poljopr. **6**, 47—54 (1939).

Krajevoj, S. J.: On natural selection in populations. C. R. (Doklady) Acad. Sci. URSS. **24**, 716—719 (1939).

— Experimental production of awned varieties of barley from awnless varieties. C. R. (Doklady) Acad. Sci. URSS. **29**, 419—421 (1940).

— (Viable plants from hybridization of defective mutants of barley.) Vestnik Gibridizacii (Hybridization) Nr. 2, 103—105 (1941). [Russisch.]

— Chimeras in barley. C. R. (Doklady) Acad. Sci. URSS. **30**, 448—450 (1941).

— Selective fertilization in barley. C. R. (Doklady) Acad. Sci. URSS. **36**, 105—107 (1942).

Kravčenko, J. M. u. L. E. Hodjkov: (Cases of branching of the barley ear and stalk.) J. Bot. Acad. Sci. Ukraine, **1**, 275—277 (1940). [Russisch.]

Kugler, W. F. u. S. Foucault: Descripción de variedades agricolas de cebada forrajera y cervecera por sus caracteres morfológicos. (Description of agricultural varieties of feeding and malting barley by their morphological characters.) „Granos" Semilla Selecta, B. Aires, **3**, 150—168 (1939).

Kuyk, P. G.: Aan welke eischen moet goede pelgerst voldoen? (To what requirements must good pearl barley conform?) Achtste Jaarbcekje Nationaal Comité voor Brouwgerst, Wageningen, Pp. 40—46 (1943).

Lamprecht, H. u. N. Hertzman: Weibulls Original Balderkorn. Ny förädling. (Weibull's original Balder barley. A new production.) Weibulls Ill. Årsb. **38**, 4—7 (1943).

Lavallard, M. L.: New cold-hardy barley released by Arkansas. Sth. Seedsman **10**, Nr. 8, 21, 44 (1947).

Lejeune, A. J.: Correlated inheritance of stem rust reaction, nitrogen content of grain and karnel weight in a barley cross. Sci. Agric. **26**, 198—211 (1946).

— A note on the reaction of certain barley varieties to races 15 B of stem rust (Puccinia graminis Tritici Erikss. and Henn.). Sci. Agric. **27**, 183—185 (1947).

Leonard, W. H.: Inheritance of fertility in the lateral spikelets of barley. Genetics **27**, 299—316 (1942).

— Inheritance of reduced lateral spikelet appendages in the Nudihaxtoni variety of barley. J. Amer. Soc. Agron. **34**, 211—221 (1942).

— Barley culture in Japan. J. Amer. Soc. Agron. **39**, 643—658 (1947).

Litzenberger, S. C.: Compana and Glacier barley. Bull. Mont. Agric. Exp. Sta. Nr. 422, 18 (1944).

Livingston, J. E.: The inheritance of resistance to Ustilago nuda. Phytopathology **32**, 451—466. (1942).

Lomako, A. Z.: (A. F. Judin's naked barleys.) Selekcija i Semenovodstvo (Breeding and Seed Growing) Nr. 8, 32—34 (1939). [Russisch.]

Love, H. H. u. W. T. Craig: Wong, a winter barley for New York. Bull. Cornell. Agric. Exp. Sta. Nr. 796, 15 (1943).

Ma, R. H.: (The yield and quality trials of smooth-awned barley varieties.) J. Agric. Ass. China, Nr. 177, 63—75 (1944).

McCalla, A. G.: Titan: a new feed barley. Pr. Bull., Alberta 29, 1—2 (1943).

Marschall, F.: Der Braugerstenbau in der Schweiz. Landw. Jb. Schweiz 55, 783—802 (1941).

Martini, M. L. u. H. V. Harlan: Barley freaks. J. Hered. 33, 339—343 (1942).

Mayr, E.: Die ostmärkischen Gerstenzuchtsorten. Züchter 12, 16—19 (1940).

Meredith, W. O. S.: Prediction of malt extract of hybrid barleys. Sci. Agric. 23, 355—361 (1943).
— Malting quality of Canadian barleys: V. Summary of seven years tests on Montcalm, a new smooth awned variety. Sci. Agric. 26, 560—565 (1946).
—, P. J. Olson u. H. Rowland: Cultural studies with barley. III. The effects of cultural practices on malting quality. Sci. Agric. 23, 135—153 (1942).
—, H. Rowland u. H. R. Sallans: Malting quality of Canadian barleys. III. Twenty-eight varieties, 1938, 1939 and 1940 trials. Sci. Agric. 22, 584—593 (1942).
— u. H. R. Sallans: Varietal differences in barleys and malts. XIII. Wort attenuation, viscosity, and turbidity, and their inter-relations. Canad. J. Res. 21, Sect. C., 351—362 (1943).
—, — Varietal differences in barleys and malts. XIV. Intervarietal relations between wort properties and barley and malt properties. Canad. J. Res. 23, Sect. F, 132 bis 142 (1945).
—, — u. H. Rowland: Prediction of malt diastatic power of hybrid barleys. Sci. Agric. 22, 761—771 (1942).

Middleton, G. K. u. W. H. Chapman: Winter survival and yield of rough and smooth awned segregates in fall-sown barley. Proc. 41st. Annu. Conv. Ass. Sth. Agric. Wkrs, Birmingham, Ala, February 7—9, p. 86 (1940). (Abstr.)
—, — An association of smooth-awnedness and spring growth habit in barley strains. J. Amer. Soc. Agron. 33, 361—366 (1941).
—, — Resistance to floral-infecting loose smut (Ustilago nuda) in fall-sown barley varieties at Statesville, North Carolina. Phytopathology 31, 351—353 (1941).
—, —, R. W. McMillen, J. W. Hendricks u. D. W. Colvard: Winter barley in North Carolina. Bull. N. C. Agric. Exp. Sta. Nr. 336, 19 (1942).
— u. R. W. McMillen: Winter survival of rough-and smooth-awned barleys. J. Amer. Soc. Agron. 36, 626 bis 627 (1944).

Mukerji, B. K. u. R. R. Agarwal: Preliminary study on the influence of variety, manures and irrigation on the composition and quality of barley. Indian J. Agric. Sci. 14, 109—116 (1944).

Müntzing, A.: Differential response to X-ray treatment of diploid and tetraploid barley. Acta Univ. Lund 11, (6), 10 (1941).
— Frequency of induced chlorophyll mutations in diploid and tetraploid barley. Hereditas, Lund 28, 217—221 (1942).

Myler, J. L.: Awn inheritance in barley. J. Agric. Res. 65, 405—412 (1942).
— u. E. H. Stanford: Colour inheritance in barley. J. Amer. Soc. Agron. 34, 427—436 (1942).

Neatby, K. W.: Seasonal notes by the Department of Field Crops. I. Varietal differences in wheat and barley with respect to hail damage. Pr. Bull. Dep. Ext. Univ. Alberta 25, 1—2 (1940).

Newton, M., B. Peturson u. W. O. S. Meredith: The effect of leaf rust of barley on the yield and quality of barley varieties. Canad. J. Res. 23, Sect. C., 212 bis 218 (1945).

Nilsson-Ehle, H.: Kornförädlingen vid Svalöf och dess resultat. (Barley breeding at Svalöf and its results.) Sverig. Utsädesfören. Tidskr. 52, 365—369 (1942).

Nilsson-Ehle, H.: Kornförädlingen på Svalöf 1926 bis 1944 och dess resultat. (Barley breeding at Svalöf 1926—1944 and its results.) Årsb. Jordbruksforskning, Stockholm, Pp. 52—70 (1945).

Olov, E. V. G.: The cytological effect of different seedtreatments in X-rayed barley. Hereditas 27, 209—219 (1941).

Parker, J. H.: Hybrid or crossbred barley. Brewers Digest 21, Nr. 12, 51—57 (1946).
— More and better barley. Wallerstein Laboratories Communications 9, 59 (1946). (Abstr.)

Perak, J. T.: Número de cromosomas de algunas espécies de Hordeum espontáneas en Argentina. (Chromosome number in certain wild Hordeum species in Argentina.) An. Inst. Fitotec. Santa Catalina 1941, 3, 7—11 (1943).

Percival, J.: The origin of barley. Suppl. Book of Dunns Farm Seeds p. 2 (1941).

Pisarev, V. E., N. M. Vinogradova u. B. A. Poddubnaja-Arnoldi: A haploid barley plant produced by remote hybridization. C. R. (Doklady) Acad. Sci. URSS. 49, 372 (1945).

Poehlmann, J. M.: Sources of resistance to loose smut, Ustilago nuda, in winter barleys. J. Amer. Soc. Agron. 39, 430—437 (1947).

Pope, M. N.: Viability of pollen and ovules of barley after cold storage. J. Agric. Res. 59, 453—463 (1939).
— Artificially induced vivipary in barley. J. Amer. Soc. Agron. 33, 850—851 (1941).
— Cleavage polyembryony in barley. J. Hered. 34, 153 bis 154 (1943).
— The temperature factor in fertilization and growth of the barley ovule. J. Agric. Res. 66, 389—402 (1943).
— Some notes on technique in barley breeding. J. Hered. 35, 99—111 (1944).
— Ash content of barley plant parts when grown on two different soils. J. Amer. Soc. Agron. 37, 582—583 (1945).
— Seminal root number in cultivated barley. J. Amer. Soc. Agron. 37, 771—778 (1945).
— u. E. Brown: Induced vivipary in three varieties of barley possessing extreme dormancy. J. Amer. Soc. Agron. 35, 161—163 (1943).

Pugsley, A. T. u. A. Vines: Breeding Australian barleys resistent to covered smut. J. Aust. Inst. Agric. Sci. 12, 44—47 (1946).

Quincke, F. L.: Interspecific and intergeneric crosses with Hordeum. Canad. J. Res. 18, Sect. C. p. 372—373 (1940).

Raw, A. R.: Research barley. A new variety. J. Dep. Agric. Vict. 40, 521—523 (1942).
— Some aspects of barley breeding. J. Aust. Inst. Agric. Sci. 12, 142—144 (1946).

Riddle, O. C. u. C. A. Suneson: Crossing studies with male-sterile barley. J. Amer. Soc. Agron. 36, 62—65 (1944).

Rigoni, V. A.: Apricación del fenol y otros reactivos en la identificación de variedades agrícolas de cebada, avena y centeno. Manifestaciones de fluorescencia en cebada y avena. (Application of phenol and other reagents in the identification of agricultural varieties of barley, oats and rye. Manifestations of fluorescence in barley and oats.) „Granos" Semilla Selecta, B. Aires 7, Nr. 1/3, 3—22 (1943).

Robertson, D. W.: Genetics of barley. J. Amer. Soc. Agron. 31, 273—283 (1939).
— Studies of barley genetics in Colorado. Proc. 7th Int. Genet. Congr. Edinburgh 23—30 August, p. 252, 1939 (1941). (Abstr.)
— u. O. H. Coleman: The addition of two factor pairs for chlorophyll—deficient seedlings to the linkage groups of barley. J. Genet. 39, 401—410 (1940).

Robertson, D. W., F. R. Immer, G. A. Wiebe u. H. Stevens: The location of two genes for mature plant characters in barley in linkage group No. 1. J. Amer. Soc. Agron. 36, 66—72 (1944).
—, A. M. Lute u. H. Kroeger: Germination of 20-year-old wheat, oats, barley, corn, rye, sorghum, and soybeans. J. Amer. Soc. Agron. 35, 786—795 (1943).
— u. G. A. Wiebe: Genetic factors in barley. Published privately. Pp. 25. [Mimeographed.]
—,— u. F. R. Immer: A summary of linkage studies in barley. J. Amer. Soc. Agron. 33, 47—64 (1941).
—,— u. R. G. Shands: A summary of linkage studies in barley: Supplement I, 1940—1946. J. Amer. Soc. Agron. 39, 464—473 (1947).
Rosendahl, G.: Cytologische Untersuchungen an tetraploiden Gersten. Kühn-Archiv 60, 238—252 (1944).

Sadovnikov, G. T.: (A local naked barley.) Selekcija i Semenovodstvo (Breeding and Seed Growing) Nr. 9, 24 (1939). [Russisch].
Sallans, H. R. u. J. A. Anderson: Observations on the study of varietal differences in the malting quality of barley. Part. IV. Canad. J. Res. 17, Sect. C, 57—71 (1939).
—, W. O. S. Meredith u. J. A. Anderson: Varietal differences in barleys and malts. 11. Simultaneonus relations between malt extract and two or more barley properties. Canad. J. Res. 19, Sect. C, 234—250 (1941).
Saloheimo, L.: Jämförelse av kornsorter på Finska Mosskulturföreningens Karelska försöksstation under åren 1930—1940. (A comparison of barley varieties at the Karelian Experiment Station of the Finnish Society for the Cultivation of Bogland during 1930—1940). Finska Moss Fören. Årsb. 46, 104—123 (1942).
Salygin, I. N.: Genetic analysis of albinism mutations induced in barley by X-radiation. C. R. (Doklady) Acad. Sci. URSS. 25, 60—63 (1939).
Savinskaja, V. F.: (A case of natural hybridization in barley.) Vestnik Gibridizacii (Hybridization) Nr. 2, 106—107 (1941). [Russisch.]
Schiemann, E.: Neue Probleme der Gerstenphylogenie. Züchter 11, 145—147 (1939).
Scott, R. C.: The origin of Prior Barley. J. Dep. Agric. S. Aust. 44, 397—398 (1941).
Shalygin, I. N. siehe Šalygin.
Shands, H. L. u. D. C. Arny: Stripe reaction of spring barley varieties. Phytopathology 34, 572—585 (1944).
— u. C. W. Schaller: Response of spring barley varieties to floral loose smut inoculation. Phytopathology 36, 534—548 (1946).
Shands, R. G.: Chevron, a barley variety resistant to stem rust and other diseases. Phytopathology 29, 209 bis 211 (1939).
— An apparent linkage of resistance to loose smut and stem rust in barley. J. Amer. Soc. Agron. 38, 690—692 (1946).
Šibaev, P. N.: (The gluten of barley.) Zonaljnyj Naučno-Issledovateljskij Institut Zernovogo Hozjajstva Nečernozemnoj Polosy SSSR. (Zonal Res. Inst. of Grain Husbandry of the non-Black-soil Belt of the USSR.) Moskva, Pp. 10 (1945). [Russisch.]
Simonet, M. u. A. Fardy: Tétraploïdie chez l'orge nue à deux rangs (Hordeum distichum L. var. nudum L. 2 n = 14) provoquée par l'action de la colchicine sur le coléoptile. Rev. Sci. 81, 453—454 (1943).
Smith, L.: An inversion, a reciprocal translocation, trisomics, and tetraploids in barley. J. Agric. Res. 63, 741—750 (1941).
— Cytogenetics of a factor for multiploid sporocytes in barley. Amer. J. Bot. 29, 451—456 (1942).
Stamberg, O. E. u. J. W. McBain: The protein fractions of some barleys and malts. Wallerstein Laboratories Communications 5, 35—39 (1942).

Stanford, E. H. u. F. N. Briggs: Two additional factors for resistance to mildew in barley. J. Agric. Res. 61, 231—236 (1940).
Stoa, T. E.: A new variety of barley-Tregal. Bi-m. Bull. N. Dak. Agric. Exp. Sta. 5, Nr. 3, 25—26 (1943).
Stubbe, H. u. G. Bandlow: Mutationsversuche an Kulturpflanzen. I. Röntgenbestrahlungen von Winter- und Sommergersten. Züchter 17/18, 365—374 (1947).
Šuljga, P. M.: (Raising the percentage of fertilization in barley.) Selekcija i Semenovodstvo (Breeding and Seed Growing) Nr. 5, 24—25 (1940). [Russisch.]
Suneson, C. A.: A male sterile character in barley. A new tool for the plant breeder. J. Hered. 31, 213—214 (1940).
— The use of male-sterile in barley improvement. J. Amer. Soc. Agron. 37, 72—73 (1945).
— Effect of barley stripe, Helminthosporium gramineum Rab., om yield. J. Amer. Soc. Agron. 38, 954—955 (1946).
— u. B. R. Houston: Male-sterile barley for study of floral infection. Phytopathology 32, 431—432 (1942).
— u. O. C. Riddle: Hybrid vigor in barley. J. Amer. Soc. Agron. 36, 57—61 (1944).
Suryanarayana Murty, G.: Segregation and correlated inheritance of rust-resistance and epidermal characters in a barley cross. Indian J. Genet. Pl. Breed. 2, 73—75 (1942).
Swanson, A. F. u. H. H. Laude: Barley production in Kansas. Bull. Kans. Agric. Exp. Sta. Nr. 318, 38 (1943).
Swenson, S. P.: Genetic and cytologic studies of a brachytic mutation in barley. J. Agric. Res. 60, 687—713 (1940).
— u. D. G. Wells: The linkage relations of four genes in chromosome I of barley. J. Amer. Soc. Agron. 36, 426 bis 435 (1944).

Tapke, V. F.: Occurrence and distribution of physiologic races of Ustilago hordei in the United States. Phytopathology 35, 489 (1945). (Abstr.)
— New physiologic races of Ustilago Hordei. Phytopathology 35, 970—976 (1945).
— u. W. M. Bever: Effective methods of inoculating seed barley with covered smut (Ustilago hordei). Phytopathology 32, 1015—1021 (1942).
Tavčar, A.: Otpadanje osja kod ječma Hordeum sativum L., nasljedjivanje i biljnogojska važnost. (Deciduous awn in H. sativum L., its inheritance and importance in breeding.) Poljopr. Naučna Smotra, Zagreb 1, 51—65 (1939).
— Beitrag zur Vererbung der Spindelbrüchigkeit bei einigen Nacktgersten. Z. Pflanzenz. 24, 333—338 (1941).
— The inheritance of the firmness of the spikelets on naked barley (Hordeum sat. nudum). Poljodjelska Znanstvena Smotra, Zagreb Nr. 8, 41—56. (1944).
Tedin, O. u. E. Andersson: Urvalsstudier hos korn. (Selection studies in barley.) Sverig. Utsädesfören Tidskr. 53, 98—110 (1943).
Thayer, J. W. u. E. E. Down: Bay, a new barley variety. Quart. Bull. Mich. Agric. Exp. Sta. 28, 270—271 1946).
Thomson, W. P.: The frequency of fertilization and the nature of embryo and endosperm development in intergeneric crosses in cereals. Proc. 7th Int. Genet. Congr. Edinburgh 23—30 August, p. 281, 1939 (1941). (Abstr.)
Thren, R.: Zur Frage der physiologischen Spezialisierung des Gerstenflugbrandes Ustilago nuda (Jensen) Kellerm. et Sw. und der Entstehung neuer Gerstenbrand-Rassen. Phytopath. Z. 13, 539—571 (1941).
Thunaeus, H.: Nya maltkornssorter genom röntgenbestrålning. (New varieties of malting barley by means of X-irradiation.) Svensk BryggTidskr. 61, 73—83 (1946).

Tidd, J. S.: Inheritance of resistance to Erysiphe graminis hordei in a cross between Featherstone and Nepal barley. Phytopathology 30, 24—25 (1940). (Abstr.)

Tohtuev, A. V.: (Breeding barley for yield and earliness by the hybridization method.) Selekcija i Semenovodstvo (Breeding and Seed Growing) Nos. 11/12, 13—16 (1940). [Russisch.]

— Inheritance of the length of growing period in barley. C. R. (Doklady) Acad. Sci. URSS. 27, 147—150 (1940).

Tokhtuev, A. V. siehe Tohtuev.

Tometorp, G.: Cytological studies on haploid Hordeum distichum. Hereditas, Lund, 25 241—254 (1939).

Ullrich, H. u. M. Canel: Über das photoperiodische Wirkungsspektrum bei Isaria-Sommergerste. Naturwissenschaften 27, 367 (1939).

Vik, K.: Ulike reaksjon for sommervarme og nedbør hos toradsbygg og seksradsbygg. (Different reaction of two-rowed and six-rowed barleys to warm summer and rainfall.) Meld. Norg. LandbrHøgsk. 21, 127—180; also 50. Arsmeld. Norges LandbrHøgsk. Åkervekstforsøk 1942 Pp. 1—54 (1941).

Vinogradova, N. M. u. V. E. Pisarev: (Hybridization of cultivated barley with wild barleys.) Vestnik Akademii Nauk SSSR. (Record of the Academy of Sciences USSR.) Nos 4/5, 65—66 (1944). [Russisch.]

Waal, V. A. V. D.: De zaaizaadvoorziening bij zomergerst. (The supple of spring barley seed for sowing.) Zesde Jaarboekje Nationaal Comité voor Brouwgerst, Wageningen, 72—81 (1941).

Wei, T. C., W. Y. Hsu u. J. T. Chen: (Physiological specialization of covered smut of barley.) J. Agric. Ass. China Suppl. Nr. 50, 26—27 (1945).

Weibel, R. O.: Winter barley in West Virginia. Bull. Va. Agric. Exp. Sta. Nr. 314, 27 (1944).

Wellensiek, S. J.: Iets over erfelijkheid en afstamming van gerst. (On the genetics and phylogeny of barley.) Fijfde Jaarboekje Nationaal Comité voor Brouwgerst, Wageningen, Pp. 83—94 (1940).

Wiebe, G. A.: Some problems in breeding barley for industrial uses. Brew. Dig. N. Y. 15 (11), 43—45 (1940).

—, P. R. Cowan u. L. Reinbach-Welch: Yields of barley varieties in the United States and Canada 1937 bis 1941. Tech. Bull. U. S. Dep. Agric. Nr. 881, 83 (1944).

Woodward, R. W.: Inheritance of a melaninlike pigment in the glumes and caryopses of barley. J. Agric. Res. 63, 21—28 (1941).

— Linkage relationships between the allelomorphic series B. B^{mb}, B^g and $A_t a_t$ factors in barley. J. Amer. Soc. Agron. 34, 659—661, (1942).

— The I^h, I, i allels in Hordeum deficiens genotypes of barley. J. Amer. Soc. Agron. 39, 474—482 (1947).

— u. D. C. Tingey: Velvon, a new amooth-awned barley. Bull. Agric. Exp. Sta. Utah Nr. 293 11. (1940).

—, — Inoculation experiments with covered smut of barley. J. Amer. Soc. Agron. 33, 632—642 (1941).

Yu, T. F.: (Breeding hulled barley for resistance to covered smut [Ustilago Hordei (Pers.) K. and S.] in Kiangsu Province.) Nanking J. Bd. 9, 281—292 (1940).

— u. C. T. Fang: A preliminary report on further studies of physiologic specialization in Ustilago hordei. Phytopathology 35, 517—520 (1945).

5. Mais.

Abbe, E. C. u. B. O. Phinney: The effect of the gene d_1 on the developmental pattern and cellular constitution of the stem in maize. Amer. J. Bot. 27, Nr. 10, Suppl. Pp. 1s. (1940). (Abstr.)

—, — The action of the gene $dwarf_1$ in the ontogeny of the stem in maize. Genetics 27, 129 (1942). (Abstr.)

Abbe, E. C. u. B. O. Phinney: Interaction of genes for size and form in maize. Genetics 30, 1 (1945). (Abstr.)

—, L. F. Randolph u. J. Einset: Developmental pattern of the leaf blade in 2n and 4n Zea mays as related to the structure of the growing point. Amer. J. Bot. 26, 4s (1939). (Abstr.)

Albright, J. L.: Sweet corn hybrids and inbreds-results of canning and variety tests in Wisconsin. Canner 88, Nr. 12, 88—89 (1939).

Alfaro, A. C.: Un nuevo carácter adherente en maíz. (A new adherent character in maize.) An. Inst. Fitotéc. Santa Catalina 4, 28—39 (1942).

Anderson, D. C.: 1942 report of the yield trials with corn hybrids and varieties in Missouri. Manual Mo. Agric. Ext. Serv., Nr. 30, 15 (1943).

—, L. L. McHoney u. R. F. Powell: 1941 report of the yield trials with corn hybrids and varieties in Missouri. Manual Agric. Ext. Serv. Univ. Mo., Nr. 24, 23 (1942).

—, E.: A cytological, taxonomic, and genetic monograph of the genus Tripsacum with reference to its allies Zea and Euchlaena. Yearb. Amer. Philos. Soc. 1942, Pp. 142—143 (1943).

— A variety of maize from the Rio Loa. Ann. Mo. Bot. Gdn. 30, 469—474 (1943).

— Homologies of the ear and tassel in Zea Mays. Ann. Mo. Bot. Gdn. 31, 325—344 (1944).

— Two collections of prehistoric corn tassels from southern Utah. Ann. Mo. Bot. Gdn. 31, 345—354 (1944).

— The sources of effective germ-plasm in hybrid maize. Ann. Mo. Bot. Gdn. 31, 355—361 (1944).

— Maize in Mexico — a preliminary survey. Ann. Mo. Bot. Gdn. 33, 147—247 (1946).

— u. F. D. Blanchard: Prehistoric maize from Cañon del Muerto. Amer. J. Bot. 29, 832—835 (1942).

— u. H. C. Cutler: Races of Zea Mays: I. Their recognition and classification. Ann. Mo. Bot. Gdn. 29, 69—88 (1942).

— u. R. O. Erickson: Antithetical dominance in North American maize. Proc. Nat. Acad. Sci. Wash. 27, 436—440 (1941).

—, E. G.: Translocations in maize involving chromosome 8. Genetics 24, 385—390 (1939).

— Translocations in maize involving the short arm of chromosome I. Genetics 26, 452—459 (1941).

— u. R. A. Brink: Translocations in maize involving chromosome 3. Genetics 25, 299—309 (1940).

— u. L. F. Randolph: Location of the centromeres on the linkage maps of maize. Genetics 30, 518—526 (1945).

—, G.: Om majsodling i Sverige. (Maize cultivation in Sweden.) Sverig. Utsädesfören. Tidskr. 52, 151—160 (1942).

Andrés, J. M.: Análisis genético del color de endosperma en algunos maices comerciales Argentinos. (Genetical analysis of endosperm colour in certain Argentine commercial maize varieties.) Inst. Genét., Fac. Agron. Vet., Univ. B. Aires 1, Fasc. 3, 25 (1939).

— u. P. C. Bascialli: Híbridos comerciales de maíz. Primeros resultados de los ensayos comparativos de rendimiento — Año 1938—1939. (Commercial maize hybrids. First results of comparative yield tests, 1938 bis 1939.) Inst. Genét., Fac. Agron. Vet., Univ. B. Aires 1, Nr. 5, 20 (1940).

—, — Caracteres hereditarios aislados en maíces cultivados en la Argentina. (Hereditary characters isolated in maizes cultivated in Argentina) Inst. Genét., Fac. Agron. Vet., Univ. B. Aires 2, Nr. 1, 26 (1941).

—, — u. A. Lotti: Híbridos comerciales de maíz. Resultados de las experiencias realizadas en el año agrícola 1940—1941. (Commercial maize hybrids. Results of experiments carried out in the agricultural year 1940—1941.) J. Agron. Vet. 1941, Pp. 15 (1941).

Andrés, J. M. u. F. Saura: Maíces argentinos tetraploides obtenidos por tratamiento con calor. (Tetraploid Argentine maizes obtained by heat treatment.) Rev. Fac. Agron. B. Aires 11, 17—30 (1944).

Andrew, R. H., R. A. Brink u. N. P. Neal: Some effects of the waxy and sugary genes on endosperm development in maize. J. Agric. Res. 69, 355—371 (1944).

Anonymus: Recent linkage studies in maize. Genetics 24, 59—63 (1939).
— Hybrid seed corn — binder twine. Rep. Kans. St. Bd. Agric. 59, Nr. 238, 24 (1940).
— Hybrid corn. Science 93, 10 (Suppl.) (1941).
— Nation-wide survey shows third of U. S. corn acreage is hybrid. Seed World 50, Nr. 7, 24—26 (1941).
— Sweet corn for canning. Publ. Nat. Canners Assoc., Pp. 60 (1941).
— Corn hybrids in Tennessee. Seed World 51, Nr. 5, 24 (1942).
— Science for the farmer. Farm crops. 55th. Rep. Pa. Agric. Exp. Sta., Bull., Nr. 429, 10—11 (1942).
— Hybrid seed increase aids war program. Sth. Seedsman 5, Nr. 12, 17 (1942).
— Sweet corn. Adv. Leafl. Minist. Agric. Fish. Lancs., Nr. 297, 4 (1943).
— Conferência sôbre a genética do milho. (Lecture on maize genetics.) Rev. Agron. São Paulo 7, 499—501 (1943).
— Louisiana hybrids make good out-of-state showing. Sth. Seedsman 6, Nr. 2, 39 (1943).
— Nueva variedad de maíz breve mejorada por una selección y cultivo cuidadoso. (New improved variety of early maize by selection and careful cultivation.) Rev. Agric. Guatemala 20, Nr. 10, 11 u. 12, 64—65 (1943).
— Mais in uw moestuin. (Maize in your kitchen garden.) Rijksuitgeverij dienst van de Nederlandsche Staatscourant's-Gravenhage, Pp. 59, 29 figs. (1944).
— Corn for tropics being developed. Agric. Amer. 4, 76 (1944).
— O milho hibrido nos E. U. A. (Hybrid maize in the USA.) Bol. Agron. Brasil. 9, 32—34 (1945).
— Corn varieties for general planting in South Georgia. Mimeo Pap. Ga Coastal Plain Exp. Sta., Nr. 36, 1 (1945). [Mimeographed.]
— Piracicaba sweet corn developed in Brazil. Agric. Amer. 6, 169 (1946).
— Corn hybrids recommended for production in Ontario 1946. Suppl. Pamphl. Dom. Can. Dep. Agric., Nr. 22, 2 (1946).

Avery, G. S. jr., J. Berger u. B. Shalucha: Auxin content of maize kernels during ontogeny, from plants of varying heterotic vigor. Amer. J. Bot. 29, 765 bis 772 (1942). (Abstr.)

Baker, W. A.: The corn borer program for 1943. Rep. 6th Corn Improv. Conf. N. Cent. Reg., St. Louis, Mo., November 12, 1942, Pp. 29—31 (1942).

Bascialli, P. C.: E. maíz Colorado Manfredi M. A. Su origen, historia y valor agrícola. (The maize Colorado Manfredi M. A. Its origin, history and agricultural value.) „Granos" Semilla Selecta, B. Aires 3, Nr. 9, 3—8 (1939).
— La producción de semillas de híbridos comerciales de maíz. (The production of seeds from commercial maize hybrids.) „Granos" Semilla Selecta, B. Aires 4, Nr. 9—10, 26—28 (1940).
— Características agrícolas de las variedades comerciales de maíz. (Agricultural characteristics of the commercial varieties of maize.) „Granos" Semilla Selecta, B. Aires 10, 39—43 (1946).

Beadle, G. W.: Teosinte and the origin of maize. J. Hered. 30, 245—247 (1939).

Beard, D. F.: Relative values of unrelated single crosses and an open-pollinated variety as testers of inbred lines of corn. Abstr. Doct. Diss. Ohio Univ., Nr. 33, 9—18 (1940).
—, F. C.: The germination capacity of maize pollen having abberrant nuclei. Bull. Torrey Bot. Cl. 70, 449—456 (1943).

Beattie, J. H.: Growing sweet corn for the cannery. Fmr's Bull. U. S. Dep. Agric., Nr. 1634, 18 (1945).

Berkner, F. W.: Beiträge zur Kenntnis der Maispflanze. (Anregungen für die Auslese bei der Maiszüchtung.) Z. Pflanzenz. 23, 210—238 (1939).

Bernstein, L.: Hybrid vigor in corn and the mobilization of endosperm reserves. Amer. J. Bot. 30, 801 bis 809 (1943).

Bigger, J. H.: Breeding corn for resistance to insect attack. J. Econ. Ent. 34, 341—347 (1941).
— Insect resistance in corn. J. Amer. Soc. Agron. 35, 689—694 (1943).
—, R. O. Snelling u. R. A. Blanchard: Resistance of corn strains to the southern corn rootworm, Diabrotica duodecimpunctata F. J. Econ. Ent. 34, 605—613 (1941).

Blanchard, R. A., J. H. Bigger u. R. O. Snelling: Resistance of corn strains to the corn ear worm. J. Amer. Soc. Agron. 33, 344—350 (1941).

Boceta Durán, V.: Nuevo dispositivo para realicar fecundaciones artificiales en el maíz. (New device for effecting artificial maize pollinations.) Bol. Inst. Invest. Agron. Madr. Nr. 9, 277—287 (1943).

Boerger, A., M. Canel u. D. Burdenski: Investigaciones con maiz efectuadas en La Estanzuela desde 1934 bis 1935 a 1938/39. (Investigations on maize carried out at La Estanzuela between 1934 and 1939.) Arch. Fitotécn. Uruguay 3, 186—229 (1939).

Borgeson, C.: Methods of detasseling and yield of hybrid seed corn. J. Amer. Soc. Agron. 35, 919—922 (1943).

Borlaug, N. E.: Diseases of teosinte in Mexico. Phytopathology 36, 395 (1946) (Abstr.).

Bradshaw, I. R.: The steady march of southern corn. Southern inbreds crossed with best Corn Belt varieties make high yields in the south. Sth. Seedsman 4, Nr. 2, 8, 24 (1941).

Brieger, F. G.: Problemas de melhoramentos do milho. (Problems of maize improvement.) Rev. Soc. Rur. Brasil. 19, Nr. 222, 38—42 (1939).
— Origem do milho. (Origin of maize.) Rev. Agric. Piracicaba 18, 409—418 (1943).
— Estudos sôbre a inflorescência de milho com referência especial aos problemeas filogenéticos. (Studies on the inflorescence of maize with special reference to phylogenetic problems.) Bragantia, São Paulo 5, 659—716 (1945).

Brimhall, B., G. F. Sprague u. J. E. Sass: A new waxy allel in corn and its effect on the properties of the endosperm strach. J. Amer. Soc. Agron. 37, 937 bis 944 (1945).

Brown, W. L.: Those new sweet corn varieties. Science and not accident has brought us the improved types of to-day. Seed World 53, Nr. 2, 8—9 (1943).

Brunson, A. M. u. G. M. Smith: Hybrid popcorn. J. Amer. Soc. Agron. 37, 176—183 (1945).

Bryan, A. A. u. J. E. Sass: Heritable characters in maize 51 — „knotted leaf". J. Hered. 32, 343—346 (1941).

Bryan, W. W. u. A. J. Schindler: Hybrid maize for Queensland. Qd. Agric. J. 54, 353—360 (1940).

Burkholder, P. R., I. McVeigh u. D. Moyer: Niacin in maize. Yale J. Biol. Med. 16, 559—563 (1944).

Burnham, C. R.: Cytogenetic studies of an interchange between chromosomes 1 and 7 in maize. Genetics 26, 143 (1941). (Abstr.)

Burnham, C. R.: Cytogenetic studies of a case of pollen abortion in maize. Genetics **26**, 460—468 (1941).
— Chromosome disjunction in maize interchanges. Genetics **30**, 2 (1945). (Abstr.)
— An „Oenothera" or multiple translocation method of establishing homogygous lines. J. Amer. Soc. Agron. **38**, 702—707 (1946).
— u. J. L. Cartledge: Linkage relations between smut resistance and semisterility in maize. J. Amer. Soc. Agron. **31**, 924—933 (1939).

Caldwell, J. S. u. C. W. Culpepper: Suitability for dehydration of 34 varieties and strains of sweet corn. Part. III. Canner **96**, 15—16, 28 (1943).
—, — Suitability for dehydration of 34 varieties and strains of sweet corn. Part. IV. Canner **96**, 20—21, 28 (1943).
Cameron, J. W.: Chromosomes of a maize relative, Polytoca macrophylla Benth. Amer. J. Bot. **30**, 776—778 (1943).
— A study of the genic control of carbohydrates in maize endosperm. Genetics **32**, 80—81 (1947). (Abstr.)
Capinpin, J. M.: A lethal-linked kernel variation of Lagkit corn. Philipp. Agric. **27**, 866—874 (1939).
— u. A. Nakornthap: The value of first generation hybrid seed of some regional strains of Lagkitan corn. Philipp. Agric. **28**, 271—285 (1939).
— u. A. O. Rollan: Hybrid vigor in the first generation crosses between strains of Cebu corn. Philipp. Agric. **28**, 491—503 (1939).
Cárdenas, M.: Consideraciones morfológicas y filogenéticas sobre algunas razas de maíz, procedentes del Perú y Bolivia. (Morphological and phytogenetical consederations on certain races of maize from Peru and Bolivia). Rev. Univ., Cuzco **33**, 178—185 (1944).
Carter, G. F. u. E. Anderson: A preliminary survey of maize in the southwestern United States. Ann. Mo. Bot. Gdn. **32**, 297—322 (1945).
Castro, C.: Informe rendido al Vicepresidente Wallace sobre el mejoramiento del maiz. (Report presented to Vice-president Wallace regarding maize breeding.) Rev. Fac. Nac. Agron., Colombia **5**, 15—21. (1943).
Castro, D. Duarte de: Qual é a origem do milho? (What is the origin of maize?) Rev. Agron., Lisboa **27**, 235—236 (1939).
Celestre, M. R.: Ricerche citologiche e genetiche su un granturco nano elettrogenito. (Cytogenetic and genetic researches on dwarf maize produced electrogenetically.) Genetica Agraria, Roma **1**, 204—217 (1947).
Cheu, S. P.: (Correlation of corn borer damage with growth condition of corn and its significance on corn breeding work.) Kwangsi Agric. **2**, 126—133 (1941).
Clapp, A. L., E. G. Heyne, C. D. Davis u. W. O. Scott: Kansas corn tests, 1944. Bull. Kans. Agric. Exp. Sta. Nr. 325, 35 (1945).
Clark, F. J.: A gene for abnormal meiotic spindle formation in maize. Genetics **24**, 68 (1939). (Abstr.)
— Cytogenetic studies of divergent meiotic spindle formation in Zea mays. Amer. J. Bot. **27**, 547—559 (1940).
— Preliminary investigations in Zea mays of the germination capacity of pollen with aberrant nuclei. Genetics **27**, 137 (1942). (Abstr.)
— u. F. C. Copeland: Chromosome aberrations in the endosperm of maize. Amer. J. Bot. **27**, 247—250 (1940).
Clark, D. G., H. Hecht, O. F. Curtis u. J. I. Shafer jr. Stomatal behaviour in inbred and hybrid maize. Amer. J. Bot. **28**, 537—541 (1941).
Copeland, F. C.: Growth rates in inbred and hybrid corn embryos. Collecting Net **15**, 169 (1940). (Abstr).
Copper, R. R.: Hybrid corn stabilizes corn yields. Trans. Ill. Acad. Sci. **32**, 43—45 (1939).
— et. al: Eighth annual Illinois corn performance tests 1941. Bull. Ill. Agric. Exp. Sta. Nr. 482, 475—527 (1942).

Coulter, F. C.: The story of garden vegetables. VII: Sweet corn a distinctively American crop. Seed World **50**, 12—13 (1941).
Crim, R. F., et al.: Maturity ratings of corn hybrids registered for sale in Minnesota in 1944. Bull. Minn. Agric. Exp. Sta. Nr. 383, Pp. 19 (1945).
Cunningham, J. C.: Maize bibliography for the years 1917 to 1936, inclusive. Contra. Ia Corn Res. Inst., Ia Agric. Exp. Sta. **2**, 364 (1941).
Curtis, J. J. u. F. R. Earle: Analyses of double-cross hybrid corn varieties produced on farms. Cereal Chem **23**, 88—96 (1946).
Cutler, H. C.: Medicine men and the preservation of a relict gene in maize. J. Hered. **35**, 291—294 (1944).

Darlington, C. D. u. M. B. Upcott: The activity of inert chromosomes in Zea Mays. J. Genet. **41**, 275—295 (1941).
Dawson, C. D. R.: Sweet corn in England. J. R. Hort. Soc. **70**, 111—119 (1945).
Del Valle, G.: Tipos cubanos de maíz (maíz dulce). (Cuban maize types [sweet corn].) Rev. Minist. Agric., Habana **28**, 4—10 (1945).
Delwiche, E. J. u. A. M. Strommen: Corn for Northern Wisconsin. Spec. Circ. Ext. Serv. Coll. Agric. Univ. Wis. April p. 4 (1941). [Mimeographed.]
Dias, B.: O melhoramento do milho. (Maize breeding.) Rev. Agron., São Paulo **7**, 424—427 (1943).
Dicke, F. F. u. M. T. Jenkins: Susceptibility of certain strains of field corn in hybrid combinations to damage by corn earworms. Tech. Bull. U. S. Dep. Agric. Nr. 898, 36 (1945).
Dodds, K. S. u. N. W. Simmonds: A cytological basis of sterility in Tripsacum laxum. Ann. Bot. **10**, 109 bis 116 (1946).
Doty, D. M., M. S. Bergdoll u. S. R. Miles: The chemical composition of commercial hybrid and open-pollinated varieties of dent corn and its relation to soil, season, and degree of maturity (a preliminary report.) Cereal Chem. **20**, 113—120 (1943).
—, —, H. A. Nash u. A. M. Brunson: Amino acids in corn grain from several single cross hybrids. Cereal Chem. **23**, 199—209 (1946).
—, G. M. Smith, J. R. Reach u. J. T. Sullivan: The effects of storage on the chemical composition of some inbred and hybrid strains of sweet corn. Bull. Ind. Agric. Exp. Sta. Nr. 503, 31 (1945).
Drummond, G. A.: Algumas explicações sobre os milhos híbridos. (Some explanations on hybrid maize.) Ceres, Brasil **7**, Nr. 37, 34—41 (1946).
Dungan, G. H., J. H. Bigger, A. D. Lang, B. Koehler u. O. Bollin: Illinois hybrid corn tests 1944. Bull. Ill. Agric. Exp. Sta. Nr. 509, 455—484 (1945).
—, C. M. Woodworth, A. L. Lang, J. H. Bigger u. R. O. Snelling: Developments in hybrid corn production. Illinois Fmrs'Inst., Springfield, Pp. 51 (1939).

Eckhardt, R. C.: 1942 Tennessee corn performance tests. Tennessee Sta. Pp. 5 (1943).
— u. A. A. Bryan: Effect of the method of combining the four inbred lines of a double cross of maize upon the yield and variability of the resulting hybrid. J. Amer. Soc. Agron. **32**, 34—53 (1940).
—, — Effect of the method of combining two eazly and two late inbred lines of corn upon the yield and variability of the resulting double crosses. J. Amer. Soc. Agron. **32**, 645—656 (1940).
—, W. A. Douglas u. A. L. Hamner: Tests of corn hybrids and varieties in Mississippi 1945. Bull. Miss. Agric. Exp. Sta. Nr. 427, 10 (1946).
Edwards, E. T.: The American hybrid maize programme. J. Aust. Inst. Agric. Sci. **6**, 146—153 (1940).

Einset, J.: Characteristics of parthenogenetic diploids derived from tetraploid maize. Genetics 26, 150 (1941). (Abstr.)
— Chromosome lenght in relation to transmission frequency of maize trisomes. Genetics 28, 349—364 (1943).
— A cytological and genetic study of primary trisomic types in Zea mays. Abstr. Thes. Cornell Univ. 1942, 361—362 (1943).
Eldredge, J. C. u. P. J. Lyerly: Popcorn in Iowa. Bull. Ia Agric. Exp. Sta. Nr. P54, 753—778. (1943).
Elliott, C.: Bacterial wilt of dent corn. Phytopathology 31, 7—8 (1941). (Abstr.)
— Bacterial wilt of dent corn inbreds. Phytopathology 32, 262—265 (1942).
— Relative susceptibility to Pythium root rot of twelve dent corn inbreds. J. Agric. Res. 64, 711—723 (1942).
— A Pythium stalk rot of corn. J. Agric. Res. 66, 21—39 (1943).
— Helminthosporium turcicum leaf blight of inbred lines and crosses of dent corn in 1942. Div. Cereal Crops Dis., Bur. Pl. Industr. U. S. D. A. Pp. 8 (1943). [Mimeographed]
— Helminthosporium turcicum leaf blight of field corn inbreds and hybrids. Phytopathology 33, 18 (1943). (Abstr.)
— u. M. T. Jenkins: Helminthosporium turcicum leaf blight of corn. Phytopathology 35, 485 (1945). (Abstr.)
—, — Helminthosporium turcicum leaf blight of corn. Pbytopathologiy 36, 660—666 (1946).
Ellis, G. H., L. F. Randolph u. G. Matrone: A comparison of the chemical composition of diploid and tetraploid corn. J. Agric. Res. 72, 123—130 (1946).
Emerson, R. A.: A zygotic lethal in chromosome I of maize and its linkage with neighboring genes. Genetics 24, 368—384 (1939).
Enzie, W. D.: The relation of spacing to yield and to plant and ear development of some yellow sweet corn hybrids in New York. Bull. N. Y. St. Agric. Exp. Sta. Nr. 700, 19 (1942).
— Some new sweet corn varieties for New York growers. Frm. Res. N. Y. 8 (1), 7, 9. (1942). (Abstr.)
Erwin, A. T.: Anent the origin of sweet corn. Zea Mays, L. var. rugosa, Bonaf. Ia. St. Coll. J. Sci. 16, 481—485 (1942).
— Sweet corn not an important Indian food plant in the pre-Columbian period. J. Amer. Soc. Agron. 39, 117 bis 121 (1947).
Etchecopar, J. A. u. M. C. Illia: Descripción de las principales variedades de maiz cultivadas en la Argentina. (Description of the principal varieties of maize cultivated in Argentina.) Rev. Argent. Agron. 11, 175 bis 194 (1944).
Eyster, H. C.: Theoretical aspects of hybrid corn genetics and hybrid vigor. 31, 215 (1946). (Abstr.)

Fauvel, J. H.: Etude sur l'origine du maïs. Rev. Hort. Agric. Afrique Nord 45, 77—83 (1941).
Federer, W. T. u. G. F. Sprague: A comparison of variance components in corn yield trials: I. Error, tester X line, and line components in top-cross experiments. J. Amer. Soc. Agron. 39, 453—463 (1947).
Fennell, J. L.: Milho resistente ao gusano da raiz. (Maize resistant to rootworm.) Bol. Agron. Brazil 10, 156—157 (1946).
Fischer, H. E.: Causes of sterility in autotetraploid maize. Genetics 26, 151 (1941). (Abstr.)
Fleischmann, R.: 33 Jahre Maiszüchtung. Z. Pflanzenz. 24, 363—373 (1941).
Flint, W. P., G. H. Dungan, J. H. Bigger u. A. L. Young: Corn-borer control. A three-point program. Circ. Univ. Ill. Nr. 521, 12 (1942).
Fogel, S.: Gene action and histological specificity of pigmentation patterns of certain R. alleles. Genetics 31, 215—216 (1946). (Abstr.)
Fogel, S.: Allelic differentiation and correlations in gene action. Genetics 32, 86 (1947). (Abstr.)
Fraser, A. C.: Some materials for genetic instruction. J. Hered. 30, 375—378 (1939).

Gaessler, W. G., R. M. Hixon u. E. S. Haber: The quantity of pericarp in several hybrid and inbred strains of sweet corn. Iowa St. Coll. J. Sci. 14, 379 bis 383 (1940).
Giles, N. H. (jun.), P. R. Burkholder, I. McVeigh u. K. S. Wilson: Comparative studies on the B-vitamin content of trisomic and disomic maize. Genetics 31, 216—217 (1946). (Abstr.)
Giles, W. F.: Sweet or sugar corn. J. R. Hort. Soc. 70, 54—56 (1945).
Gilly, C. L. u. I. E. Melhus: Distribution and variability in teosinte. Amer. J. Bot (Suppl.) 33, 235 (1946). (Abstr.)
Gini, E.: Estudios sobre esterilidad en maíces regionales de la Argentina. (Studies on sterility in the regional maize forms of Argentina.) Ann. Inst. Fitotéc. Santa Catalina 1, 135—158 1939, (1940).
Gomes, R.: O milho e os „indios coroados" do Paraná. (Maize and the coloured Indians of the Paraná.) Chacaras e Quintais, São Paulo 66, 588—589 (1942).
Graner, E. A.: Variações do valor de „linkage". (Variations in the linkage value.) Rev. Agric. S. Paulo 15, 168—175 (1940).
Graner, E. A.: Observações sôbre o estudo da genética nos estados unidos da América do Norte. (Observations on the study of genetics in the USA.) Bol. Soc. Brasil. Agron. 6, 293—308 (1943).
— Genética da côr amarela-laranja nas sementes do milho. (Genetics of yellow-orange colour in maize seeds.) Rev. Agric. Piracicaba 18, 443—445 (1943).
— The yellow-orange endosperm of maize. Amer. Nat. 79, 187—192 (1945).
— Testes para a localização de fatores genéticos no milho. Testes de ligação. (Tests for the localization of genetical factors in maize. Linkage tests.) Rev. Agric., Piracicaba 21, 8—20 (1946).
Green, J. M.: Comparative rates of pollen tube establishment in diploid and tetraploid maize. J. Hered. 37, 117—121. (1946).
Grieben, H. u. H. A. Devoto: Resultado de los ensayos comparativos de rendimiento „standard" entre variedades de maíz, réalizados durante el año agricola 1940/41. (Result of the comparative „standard" yield trials of maize varieties effected during the agricultural year 1940—1941.) „Granos" Semilla Selecta, B. Aires 6, 6—27 (1942).
Gudkov, I. N.: (Drought and the flowering process in maize.) Selekcija i Semenovodstvo Nr. 2/3, 21—24. (1939). [Russisch.]

Haagen-Smit, A. J., R. Siu u. G. Wilson: A method for the culturing of excised, immature corn embryos in vitro. Science 101, 234. (1945).
Haber, E. S.: Sweet corn hybrids. Bull. Ia Agric. Exp. Sta. Nr. P 15 (N. S.), 436—468 (1940).
— Dent, flint, flour and waxy maize for improvement of sweet corn inbreds. Proc. Amer. Soc. Hort. Sci. 46, 293—294 (1945).
— New sweet corn hybrids for canners. Fm. Sci. Reporter, Iowa 6, 24 (1945).
— u. W. G. Gaessler: Sugar content of sweet corn pollen and kernels of inbred and hybrid strains susceptible to tassel infestation by aphis. Proc. Amer. Soc. Hort. Sci. 40, 429—431 (1942).
Hadžinov, M. I.: A dominant mutable gene for purple colour in the R series of multiple allelomorphs in maize. C. R. (Doklady) Acad. Sci. URSS. 23, 366—369 (1939).

Hadžinov, M. I.: Selective fertilization in maize when pollinated with a pollen mixture. C. R. (Doklady) Acad. Sci. URSS. **29**, 410—414 (1940).

Harland, S. C.: A new method of maize improvement. Trop. Agriculture, Trin. **23**, 114 (1946).

— A new method of maize improvement. Trop. Agriculture, Trin. **23**, 130 (1946).

Harper, R. E.: USDA-34 — a tropical sweet corn. Agric. Amer. **6**, 74—75 (1946).

Harvey, P. H.: Hereditary variation in plant nutrition. Genetics **24**, 437—461 (1939).

— The occurrence of deficient kernels in southern corn varieties. J. Elisha Mitchell Sci. Soc. **56**, 223. (1940).

— Preliminary results of the utilization of Corn Belt inbreds in the N. C. corn breeding program. Proc. 42nd Annu. Conv. Ass. Sth. Agric. Wkrs, Atlanta, Ga., February 5—7, Pp. 96—97 (1941).

— u. G. K. Middleton: The performance of corn hybrids in North Carolina. Agron. Inform. Circ. N. C. Agric. Exp. Sta. Nr. 124, 12 (1940). [Mimeographed.]

Hayes, H. K. u. I. J. Johnson: The breeding of improved self lines of corn. J. Amer. Soc. Agron. **31**, 710 bis 724 (1939).

—, R. P. Murphy u. E. H. Rinke: A comparison of the actual yield of double crosses of maize with their predicted yield from single crosses. J. Amer. Soc. Agron. **35**, 60—65 (1943).

—, —, — u. C. Borgeson: Minhybrid corn varieties for Minnesota. Bull. Minn. Agric. Exp. Sta. Nr. 354, 40 (1941).

—, — u. Tsiang Y. S.: The development of a synthetic variety of corn from inbred lines. J. Amer. Soc. Agron. **36**, 998—1000 (1944). (Abstr.)

—, —, — The relationship between predicted performance of double crosses of corn in one year with predicted and actual performance of double crosses in later years. J. Amer. Soc. Agron. **38**, 60—67 (1946).

—, —, — Experimental study convergent improvement and backcrossing in corn. Tech. Bull. Minn. Agric. Exp. Sta. Nr. 172, 40 (1946).

Helsel, P. E.: Development of sweet corn hybrids. West. Cann. Pack. **33** (12), 13—16 (1941).

Henderson, P.: Some new hybrid sweet corn. How Cream O'Gold and Sunny-Gold were bred. Market Gr. J. **66**, 29, 31—32 (1940).

Heyne, E. G. u. A. M. Brunson: Genetic studies of heat and drought tolerance in maize. J. Amer. Soc. Agron. **32**, 803—814 (1940).

—, A. L. Clapp, C. R. Porter, W. O. Scott u. C. D. Davis: Kansas corn tests, 1945. Bull. Kans. Agric. Exp. Sta. Nr. 329. 40 (1946).

— u. H. H. Laude: Resistance of corn seedlings to high temperatures in laboratory tests. J. Amer. Soc. Agron. **32**, 116—126 (1940).

Ho, W.-C. u. J. M. Koepper: Host response of maize seedlings to Pythium graminicolum. Phytopathology, **32**, 9 (1942). (Abstr.)

Hoegemeyer, L. C.: An association of root injury by white grubs, Phyllophaga spp., and lodging of crossbred strains of corn. J. Amer. Soc. Agron. **33**, 1100 bis 1107 (1941).

Horovitz, S.: Nuevo gen del cuarto cromosoma de maiz (luteomaculata). (A new gene of the fourth chromosome in maize (luteomaculata)). An. Inst. Fitotec. Santa Catalina 1941 **3**, 13—19 (1943).

— u. A. H. Marchioni: Herencia de la resistencia a la langosta en el maíz ,,Amargo". (Inheritance of locust resistance in ,,Amargo" maize.) An. Inst. Fitotec. Santa Catalina **2**, 27—52 (1940).

—, — Herencia de la resistencia a la langosta en el maiz ,,Amargo". (Mode of inheritance of the resistance to the locust in the maize Amargo. ,,Granos" Semilla Selecta, B. Aires **6**, 37—38 (1942).

Horovitz, S., A. H. Marchioni u. H. G. Fisher: Mejoramiento del maíz dulce para la industria del envasado. Aumento del contenido de azúcar. (Improvement of sweet maize for the liquor trade. Increasing the content of sugar.) ,,Granos" Semilla Selecta, B. Aires **6**, Nos. 10—12, 23 (1942).

—, —, — El factor su_x y el aumento del contenido de azúcar, en el maíz para choclo. (The factor su_x and the increase in sugar content in sweet corn.) An. Inst. Fitotéc. Santa Catalina 1941, **3**, 37—44 (1943).

Hoskins, J. D.: Fifty-third Annual Report of the Agricultural Experiment Station of the University of Tennessee, for 1940 (1941). Pp. 110 (1941).

Huber, L. L.: Thin stands of corn produce bigger ears but lower yields than thicker plantings. Suppl. Nr. 2, Bull. Pa. Agric. Exp. Sta. Nr. 464, 10.

— Later planting reduces corn borer demage. Bi-m. Bull. Ohio Agric. Exp. Sta. **27**, 72—73 (1942).

— u. G. H. Stringfield: Strain susceptibility to the European corn-borer and the corn-leaf aphid in maize. Science **92**, 172 (1940).

—, — Aphid infestation of strains of corn as an index of their susceptibility to corn borer attack. J. Agric. Res. **64**, 283—291 (1942).

Huelsen, W. A.: Sweet-corn inbreds and crosses released by the Illinois Station. Bull. Ill. Agric. Exp. Sta. Nr. 466, 279—355 (1940).

— Sweet corn hybrids for canning and market released by the Illinois Station. Circ. Ill. Agric. Exp. Sta. Nr. 504, 20 (1940).

— Sweet corn hybrids, 1940 - the outlook for the canner. Canner **92** (12), 54 (1941).

Hull, F. H.: Recurrent selection for specific combining ability in corn. J. Amer. Soc. Agron. **36**, 989—990 (1944).

— Recurrent selection for specific combining ability in corn. J. Amer. Soc. Agron. **37**, 134—145 (1945).

— Regression analyses of corn yield data. Genetics **31**, 219 (1946). (Abstr.)

— Overdominance and corn breeding where hybrid seed is not feasible. J. Amer. Soc. Agrion. **38**, 1100—1103 (1946).

— Cryptic homozygous lines. J. Amer. Soc. Agron. **39**, 438—439 (1947).

—, J. D. Warner u. W. A. Carver: Corn varieties and hybrids and corn improvement. Bull. Fla. Agric. Exp. Sta. Nr. 355, 50 (1941).

Hume, A. N.: A possible new method for the control of pollination of corn. J. Amer. Soc. Agron. **33**, 265—266 (1941).

Humlum, J.: Zur Geographie des Maisbaus. Ursprung, Verbreitung, heutige Ausdehnung des Maisbaus und seine Bedeutung für den Welthandel. Anforderungen des Maises an das Klima, mit besonderem Hinblick auf Rumänien. Einar Harcks Forlag, København Pp. 317, 57 tables. 140 plates (1942).

Hunziker, A. T.: Granos hallados en el yacimiento arqueológico de Pampa Grande (Salta, Argentina). (Grains discovered in the archaeological site of Pampa Grande, Salta, Argentina). Rev. Argent. Agron. **10**, 146—154 (1943).

Hurt, E. F.: Maize for table and poultry feeding, a dual purpose hybrid. J. R. Hort. Soc. **71**, 138—141. (1946).

Hutchcroft, C. D., J. L. Robinson u. F. Reiss: The 1946 Iowa corn yield test. Bull. Ia Agric. Exp. Sta. Nr. P 84, 752—804 (1947).

Ionescu, M. V., O. G. Popescu, H. Slușanschi u. L. Gaal: Contribuțiuni la studiul chimic al porumburilor Românești (recoltele 1936 și 1937). (Contribution to the chemical study of the Rumanian maize varieties of the harvest of 1936 and 1937.) Anal. Inst. Cerc. Agron. Român. **10**, 282—357 (1938, 1939).

Ionescu, M. V., O. G. Popescu, H. Sluşanschi u. L. Gaal: Contribuţiuni la studiul chimic alporumburilor Românești recolta 1938. (Contribution to the chemical study of the Rumanian maize varieties of the harvest of 1938.) Anal. Inst. Cerc. Agron. Român. 11, 266—313, 1939 (1940).
— u. Sluşanschi, H.: Contribuţiuni la studiul determinării amidonului in cereale. Nota IIa: Relaţii intre conţinutul. in amidon şi materile celulozice. (Contributions to the study of the starch determination in cereals. Note II. Relationship between starch content and cellulose materials.) Anal. Inst. Cerc. Agron. Român. 10, 358—376 (1938, 1939).

Janetzki, C.: Probleme der Maiszüchtung. Forsch.-dienst 11, 648—660 (1941).
Jenkins, M. T.: New developments that may affect the corn industries. The importance of corn hybrids to the corn industry. Contr. Ia Corn Res. Inst. 1, 208—212 (1939).
— The segregation of genes affecting yield of grain in maize. J. Amer. Soc. Agron. 32, 55—63. (1940).
— Report of the Second Southern Corn Improvement Conference held at Knoxville, Tennessee August 26 and 27, 1940. Washington D. C. 49 (1940). [Mimeographed.]
— Reports of the Fourth and Fifth Corn Improvement Conferences of the North Central Region. Washington. D. C. February 18, Pp. 91 (1942). [Mimeographed.]
— (Compiler): Results of the cooperative uniform comparisons of corn hybrids, 1941.- Div. Cereal Crops. Dis., Bur. Pl. Ind. U. S. Dep. Agric. Pp. 81 (1942). [Mimeographed.]
— Report of the Third Southern Corn Improvement Conference held at Memphis, Tennessee, February 5, 1942. Div. Cereal Crops Dis., Bur. Pl. Ind. Sta., Beltsville, Maryland, Pp. 22 (1942). [Mimeographed.]
— The corn breeding program in relation to the war effort. Rep. 6th Corn Improv. Conf. N. Cent. Reg., St. Louis, Mo., November 12, 16—19 (1942). (Mimeographed.]
— Breeding corn for war. „No other cereal crop will produce as many food units per acre". Seed World 52, Nr. 12, 10—11 (1943).
— Report of the Seventh Corn Improvement Conference of the North Central Region September 1—2, 1943 Lafayette, Indiana. Bur. Pl. Ind. Sta., Beltsville, Pp. 13 (1943). [Mimeographed.]
— Results of the cooperative uniform comparisons of corn hybrids 1942. Bur. Pl. Industr. Sta., Beltsville, Maryland, U. S. D. A. Pp. 97 (1943). [Mimeographed.]
— Report of the First Northeastern Corn Improvement Conference Connecticut Agricultural Experiment Station, New Haven, Connecticut, February 2—3, 1945. Div. Cereal Crops Dis., Bur. Pl. Ind. Soils, Agric. Eng., Pl. Ind. Sta., Beltsville Md. Pp. 21 (1945). [Mimeographed.]
— Report of the Fourth Southern Corn Improvement Conference Birmingham, Alabama, January 24—25, 1945. Div. Cereal Crops Dis., Bur. Pl. Ind., Soils, Agric. Eng., Pl. Ind. Sta., Beltsville Md., Pp. 43 (1945). [Mimeographed.]
— Report of the Eight Corn Improvement Conference of the North Central region. U. S. Dep. Agric. Pl. Ind. Beltsville, Md. 1946. Pp. 26, (1947). [Mimeographed.]
— u. A. L. Robert: Helmithosporium turcicum leaf blight ratings on corn at Plant Industry Station, Beltsville, Md. 1946. U. S. Dep.-Agric., Agric. Res. Admin. Bur. Pl. Industr. Soils, Agric. Engin. Beltsville. Md. Pp. 19 (1946). [Mimeographed.]

Johann, H. u. A. D. Dickson: A soluble substance in cornstalks that retards growth of Diploidia zeae in culture. J. Agric. Res. 71, 89—110 (1945).
Johnson, I. J. u. Hayes, H. K.: The value in hybrid combinations of inbred lines of corn selected from single crosses by the pedigree method of breeding. J. Amer. Soc. Agron. 32, 479—485 (1940).
— u. E. S. Miller: Immediate effect of cross pollination on the carotenoid pigments in the endosperm of maize. Cereal Chem. 16, 88—92 (1939).
Jones, D. F.: Growth changes associated with chromosome breakage and reattachment. Genetics 24, 77 (1939). (Abstr.)
— Variable effect of the C locus in maize following translocation. Genetics 24, 100 (1939). (Abstr.)
— Continued inbreeding in maize. Genetics 24, 462 bis 473 (1939).
— Sex intergrades in dioecious maize. Amer. J. Bot. 26, 412—415 (1939).
— Segmental exchange in somatic cells of maize. Proc. 7th Int. Genet. Congr. Edinburgh 23—30 August, p.170, 1939 (1941). (Abstr.)
— Growth changes resulting from chromosome rearrangement. Science 91, 423 (1940). (Abstr.)
— Natural and induced changes in chromosome structure in maize endosperm. Proc. Nat. Acad. Sci. Wash. 27, 431—435 (1941).
— Growth changes associated with chromosome aberrations. Genetics 28, 78 (1943). (Abstr.)
— Growth changes in maize endosperm associated with the relocation of chromosome parts. Genetics 29, 420 bis 427 (1944).
— Heterosis resulting from degenerative changes. Genetics 30, 527—542 (1945).
— The importance of degenerative changes in living organism. Science 102, 209 (1945).
— Effect of temperature on the growth and sterility of maize. Science 105, 390—391 (1947).
Jugenheimer, R. W.: The Kansas hybrid corn program. Rep. Kans. St. Bd. Agric. 1940, 59, Nr. 236, 36—47 (1940).

Kadam, B. S.: Chromosome studies in relation to fertility and vigor in inbred and openpollinated strains of autotetraploid maize. Abstr. Thes. Cornell Univ. Pr. Ithaca, N. Y., 338—341 (1941).
Karp, M. L.: Improvement of varieties of Indian corn through inbreeding combined with cross-pollination. C. R. (Doklady) Acad. Sci. URSS. 22, 613—614 (1939).
— (The practical utilization of inbreeding in the USA.). Selekcija i Semenovodstvo Nr. 2—3, 28—31 (1939). [Russisch.]
Kelly, I. u. E. Anderson: Sweet corn in Jalisco. Ann. Mo. Bot. Gdn. 30, 405—412 (1943).
Kemmerer, A. R., G. S. Fraps u. P. C. Mangelsdorf: The relation between the vitamin-A-active carotenoids in corn and the number of genes for yellow color. Cereal Chem. 19, 525—528 (1942).
Kemp, W. B. u. R. G. Rothgeb: Selection and genetic responses in a segregating maize population. Bull. Md. Agric. Exp. Sta. Nr. A 26, Pp. 33 (1943).
Kempton, J. H. u. J. W. McLane: Hybrid vigor and weight of germs in the Seeds of maize. J. Agric. Res. 64, 65—80 (1942).
— u. L. R. Maxwell: Effect of temperature during irradiation on the X-ray sensitivity of maize seed. J. Agric. Res. 62, 603—618 (1941).
Kerle, W. D.: Hybrid maize will substantially increase yields. Agric. Gaz. N. S. W. 52, 403—404 (1941).
— Maize variety recommendations for 1946 sowing. Agric. Gaz. N. S. W. 57, 411—413, 443 (1946).

Kibizov, V. P.: (Breeding maize by the method of crossing inbred lines.) Selekcija i Semenovodstvo (Breeding and Seed Growing) Nr. 6, 17—20 (1939). [Russisch.]

Kiesselbach, T. A.: The detasseling hazard of hybrid seed corn production. J. Amer. Soc. Agron. 37, 806 bis 811 (1945).

Kinman, M. L. u. Sprague, G. F.: Relation between number of parental lines and theoretical performance of synthetic varieties of corn. J. Amer. Soc. Agron. 37, 341—351 (1945).

Kiričenko, E. G.: (Selecting maize for low and high weights of core in the cob.) Jarovizacija Nr. 3 (36), 70—73 (1941.) [Russisch.]

— u. A. S. Musijko: (Improvement of the quality of the breed in maize seeds.) Jarovizacija Nr. 1 (34) 37 bis 39 (1941). [Russisch.]

Koch, L. W. u. H. F. Murwin: The hybrid corn industry in Ontario: pathological and other problems. Emp. J. Exp. Agric. 13, 100—111 (1945).

Koehler, B., O. Bolin u. R. R. Copper: Corn inbreds ranked according to their resistance to damage from Diplodia stalk rot as determined in single crosses in 1939. Dep. Agron., Ill. Agric. Exp Sta. January, Pp. 4 (1940). [Mimeographed.]

Kozubenko, V. E.: (Breeding maize for high yield and drought resistance.) Selekcija i Semenovodstvo (Breeding and Seed Growing) Nr. 8, 27—29 (1939). [Russisch.]

Kožuhov, I. V.: (An experiment in construction of a system of evolution of Zea mays.) Proc. Lenin Acad. Agric. Sci. USSR. Nr. 20, 14—17 (1939). [Russisch.]

Kravčenko, S. K.: (Questions of intervarietal crosspollination in maize.) Selekcija i Semenovodstvo (Breeding and Seed Growing) Nr. 5, 11—13 (1940). [Russisch.]

Krickl, M.: Zur Züchtung mehrkolbiger Maissorten und der Einfluß des Temperaturfaktors auf die Reife der Einzelpflanze. Züchter 15, 125—134 (1943).

Krug, C. A., G. P. Viégas u. L. Paoliéri: Híbridos comerciais de milho. (Commercial maize hybrids.) Bragantia, S. Paulo 3, 367—552 (1943).

Kulešov, N.: (Maize in the fields of Siberia.) Socialističeskoe Seljskoe Hozjajstvo (Socialistic Agriculture) Moskva, Nr. 1, 56—62 (1944). [Russisch.]

Kvakan, P.: O nekim gospodarskim svojstvima križanaca kukuruza I generacije. (Some industrial qualities of first generation maize hybrids.) Arhiv Minist. Poljopr. 6, 27—46 (1939).

— Prirod križanaca kukuruza I. generacije u Maksimiru godine 1939. (The 1939 Maksimir hybrid corn yield test.) Poljopr. Naučna Smotra, Zagreb, 2, 120—130 (1940). [Kroatisch.]

Langham, D. G.: The inheritance of intergeneric differences in Zea-Euchlaena hybrids. Genetics 24, 78 (1939). (Abstr.)

— The inheritance of intergeneric differences in Zea-Euchlaena hybrids. Genetics 25, 88—107 (1940).

— Venezuela-1, una selección de maíz recomendable. (Venezuela-1, a maize selection to be recommended.) Circ. Minist. Agric. Cría, Dep. Genét. Inst. Exp. Agric. Zootec., El Valle, D. F. Nr. 2, 8 (1942).

— Maíz dulce, Venezuela-2-una nueva clase de maíz. (Sweet corn Venezuela-2, a new type of maize.) Circ. Minist. Agric. Cría, dep. Genét. Inst. Exp. Agric. Zootec., El Valle, D. F., Nr. 3, 4 (1942).

— u. O. Gorbea: Maiz blanco Venezuela-3, una selección de alto rendimiento. (The white maize Venezuela-3, a high yielding selection.) Circ. Minist. Agric. Cría, Dep. Genét. Inst. Exp. Agric. Zootec., El Valle, D. F., Nr. 5, 3 (1944).

Laughnan, J. R.: Chemical studies concerned with the action of the gene A_1 in maize. Genetics 31, 222 (1946). (Abstr.)

Lebedev, G. A.: Failure of cytokinesis during microsporogenesis in Zea mays following heat treatment. Cytologia, Tokyo 10, 434—442 (1940).

— El maiz en la defensa. (Maize in defence.) Agric. Exp. P. R. 1, 4—5 (1941).

Lesik, I. L.: Capacity for selective fertilization in maize.) Jarovizacija, Nr. 4 (31), 105—106 (1940). [Russisch.]

Lincoln, R. E.: Host-parasite interactions with bacterial wilt of maize. Science Bd. 89, 159—160 (1939).

— Bacterial wilt resistance and genetic host-parasite interactions in maize. J. Agric. Res. 60, 217—239 (1940).

— u. E. W. Lindström: Micro-evolution of host-parasite interactions in bacterial wilt of maize. Genetics 24, 78 (1939). (Abstr.)

Lindstrom, E. W.: Analysis of modern maize breeding principles and methods. Proc. 7th Int. Genet. Congr. Edinburgh 23—30 August, 191—196, 1939 (1941).

— Inheritance of seed lingevity in maize inbreds and hybrids. Genetics 27, 154 (1942). (Abstr.)

— Experimental data on the problem of dominance in quantitative character inheritance in maize and tomatoes. Genetics 28, 81—82 (1943). (Abstr.)

Longley, A. E.: Knob positions on corn chromosomes. J. Agric. Res. 59, 475—490 (1939).

— Chromosome morphology in maize and its relatives. Bot. Rev. 7, 263—289 (1941).

— Knob positions on teosinte chromosomes. J. Agric. Res. 62, 401—413 (1941).

— Abnormal segregation during megasporogenesis in maize. Genetics 30, 100—113 (1945).

Lonnquist, J. H. u. R. W. Jugenheimer: Factors affecting the success of pollination in corn. J. Amer. Soc. Agron. 35, 923—933 (1943).

Lowe, J. u. O. E. Nelson (jr.): Miniature seed-a study in the development of a defective caryopsis in maize. Genetics 31, 525—533 (1946).

Ludbrook, W. V.: A yield trial of top crosses between a Victorian strain of Funk's Yellow Dent maize and twelve American inbreds. J. Aust. Inst. Agric. Sci. 12, 139—141 (1946).

Lyerly, P. J.: Some genetic and morphologic characters affecting the popping expansion of popcorn. J. Amer. Soc. Agron. 34, 986—999. Lowa St. Coll. J. Sci. 17, 98—99 (1942).

McClintock, B.: The behavior in successive nuclear divisions of a chromosome broken at meiosis. Proc. Nat. Acad. Sci. Wash. 25, 405—416 (1939).

— The stability of broken ends of chromosomes in Zea mays. Genetics 26, 234—282 (1941).

— The association of mutants with homozygous deficiencies in Zea mays. Genetics 26, 542—571 (1941).

— Spontaneous alterations in chromosome size and form in Zea Mays. Cold Spring Harbor Symposia on Quantitative Biology 9, 72 (1941).

— The fusion of broken ends of Chromosomes following nuclear fusion. Proc. Nat. Acad. Sci. Wash. 28, 458 bis 463 (1942).

— The relation of homozygous deficiencies to mutations and allelic series in maize. Genetics 29, 478—502 (1944).

McKeon, C. J.: Maize seed selection. Qd. Agric. J. 60, 261—268 (1945).

McRostie, G. P. u. J. D. MacLachlan: Hybrid corn studies I. Sci. Agric. 22, 307—313 (1942).

MacVickar, M. H. u. P. M. Phillippe: More stalks, more corn per acre. Sth. Plant. 107, Nr. 3, 20—21 (1946).

— u. G. M. Shear: Variations in response of different varieties and hybrids of field corn to planting rate. J. Amer. Soc. Agron. 38, 933—935 (1946).

Maliani, C.: Studio comparativo sui granturchi coltivati nelle Venezie nel 1942 — Communicazione preliminare. (Comparative studies of the maize varieties cultivated in the Venice provinces in 1942. Preliminary communication.) Genetica Agraria, Rome 1, 95—102 (1946).

Mangelsdorf, P. C.: Origin of maize. Nature, London **146**, 338 (1940).
— The origin of maize. Proc. 8th. Amer. Sci. Congr., Wash. **3**, 267—274 (1940).
— The origin and nature of the ear of maize. Bot. Mus. Leafl. Harv. **12**, 33—75. plates (1945).
— The genetic nature of teosinte. Genetics **32**, 95—96 (1947). (Abstr.)
— Treating genetic data with punched cards. Genetics **32**, 96 (1947). (Abstr.)
— The origin and evolution of maize. Advances Genet., N. Y. **1**, 161—207 (1947).
— u. J. W. Cameron: Western Guatemala a secondary center of origin of cultivated maize varieties. Bot. Mus. Leafl. Harv. **10**, 217—252 (1942).
— u. R. G. Reeves: The origin of Indian corn and its relatives. Bull. Tex. Agric. Exp. Sta. Nr. 574, 315 (1939).
—, — L'origine du maïs. Rev. Bot. Appl. **20**, 628—633 (1940).

Marchioni, A. H.: Micropruebas de resistencia a la langosta en el maíz ,,Amargo". (Microtests of resistance to locust of Amargo maize.) An. Inst. Fitotéc. Santa Catalina **1**, 159—166 (1939) 1940.

Marino, A. E.: Una variación ,,tardía" en maíz. (A ,,late" variation in maize.) Rev. Argent. Agron. **6**, 237—240 (1939).
— Herencia del color de aleurona en el maíz ,,piamontés". (Inheritance of aleurone colour in the maize ,,Piamontés".) Physis, B. Aires **18**, 47—67 (1939).
— Relación entre el número de hojas y la precocidad cuando varían las épocas de siembra en el maíz. (Relation between the number of leaves and earliness when the time of sowing is varied in maize.) Rev. Argen. Agron. **10**, 239—243 (1943).
— u. Luna, J. T.: Análisis estadístico del número de hileras en espigas de maíces comerciales. (Statistical analysis of number of rows per cob in commercial maize varieties.) Rev. Argent. Agron. **8**, 131—137 (1941).
—, — Planeos de ensayos en blocks incompletos (lattice y lattice balanceado). Análisis de resultados en ensayos con maíces. (Designs for trials laid down in incomplete blocks (lattice and balanced lattice). Analysis of results of trials with maize.) Rev. Argent. Agron. **8**, 281—316 (1941).
—, — Ensayos de variedades comerciales de maíces y épocas de siembra. (Resultados de 3 años.) (Trials of commercial varieties of maize and times of sowing. [Three years' results.]) Rev. Argent. Agron. **9**, 96—109 (1942).
—, — La precocidad de las variedades comerciales de maíces en Santa Fe. (The earliness of the commercial varieties of maize in Santa Fe.) Rev. Argent. Agron. **10**, 155—168 (1943).
—, — Dos híbridos comerciales de maíz. (Two commercial maize hybrids.) Rev. Argent. Agron. **14**, 50—60 (1947).

Mather, K.: Chiasma frequencies in trisomic maize. Genetics **24**, 104 (1939). (Abstr.)
— Competition for chiasmata in diploid and trisomic maize. Chromosoma **1**, 119—129 (1939).
— u. E. C. Barton-Wright: Nicotinic acid in sugary and starchy maize. Nature, Lond. **157**, 109—110 (1946).

Mayer, L. S. u. A. B. Strand: A hybrid sweet corn for Tennessee. Circ. Tenn. Agric. Exp. Sta. Nr. 75, 4 (1941).

Mazoti, L. B.: Estudio genético sobre maices amilaceos de Argentina. (Genetical study on starchy Argentine maize.) An. Inst. Fitotéc. Santa Catalina **2**, 17—26 (1940).
— Contribución a la genética del maiz. Rev. Argent. Agron. **12**, 174—202j (1945).

Melhus, I. E., G. Semenuk u. J. R. Wallin: The influence of climate on distribution of corns in Guatemala. Amer. J. Bot. Suppl. **33**, 825 (1946). (Abstr.)
—, —, —, G. M. Watkins u. G. J. Goodman: Comparative development of some United States, Mexican and Central American corns at different latitudes and altitudes. Amer. J. Bot. (Suppl.) **33**, 220—221 (1946). (Abstr.)

Mendes, C. T.: A selecção na cultura do milho. (Selection in maize growing.) Bol. Agric., S. Paulo **40**, 125 bis 148 (1939).

Menezes, O. B. de: Tempo de germinação do grão de polen e mitose de um milho brasileiro. (Rate of germination of pollen grains and mitosis in a Brazilian maize.) Bol. Soc. Brasil. Agron. Rio de J. **7**, 27—32 (1944).

Michels, C. A.: Hybrid corn and its apparent advantages over open-pollinated varieties. Mimeo Leafl. Idaho Agric. Exp. Sta. Nr. 32, 6 (1939).

Millang, A. u. G. F. Sprague: The use of punched card equipment in predicting the performance of corn double crosses. J. Amer. Soc. Agron. **32**, 815—816 (1940).

Miller, E. S. u. I. J. Johnson: Inheritance of chlorophyll in F_1 crosses made reciprocally between selfed lines of corn. Proc. Soc. Exp. Biol., N. Y. **44**, 26—28 (1940).

Moll, A. C.: Evergreen type hybrids lead in Ohio sweet corn trials. Canner **102**, Nr. 3, 12 (1945).

Molotkovskij, G.: (The question of transforming the nature of maize.) Sovetskaja Botanika (Soviet Botany) Nr. 5/6, 325—331 (1940). [Russisch.]

Montelaro, J.: Corn fit for the gods. Sth. Seedsman **9**, 14 (1946).

Morgan, D. T. (jr.): The formation of chromocenters in interkinetic nuclei of maize by knobs and B. chromosomes. J. Hered. **34**, 195—198 (1943).

Morrison, G.: The story of sweet corn and some of the leading varieties - their uses and adaptabilities. Nat. Seedsman **8** (2), 22—23 (1941).
— Grand old man of hybrid corn. Seed World **55**, 16, 18, 20 (1944).
— Hybrid corn. Science in practice. Econ. Bot. **1**, 5—19 (1947).

Mostert, J. F. T.: Hybrid maize prospects in South Africa. Fmr's Wkly, Bloemfontein **71**, 2400—2401 (1946).

Mudra, A.: Însușirea ,,forma bobului" in ameliorarea porumbului. (The character ,,grain shape" in maize breeding.) Bul. Fac. Agron. Cluj **7**, 166—168 (1938) 1939.
— Experiențe cu privire la păstrarea polenului de porumb. (Experiments on pollen storage in maize.) Bul. Fac. Agron. Cluj. **7**, 172—175 (1938) 1939.
— Über die Reichweite des Pollenstaubes beim Mais in geschlossenem Verband. Züchter **15**, 29—31 (1943).

Murdoch, H. A.: Hybrid vigor in maize embryos. J. Hered. **31**, 361—363 (1940).

Murphy, R. P.: Convergent improvement with four inbred lines of corn. J. Amer. Soc. Agron. **34**, 138—150 (1942).

Neal, N. P.: Wisconsin corn hybrids 1941. Spec. Circ. Wis. Coll. Agric. Ext. Serv. March, Pp. 12 (1941).
— Informe sobre cultivo de maiz en el Uruguay. (Report on maize cultivation in Uruguay.) Admin. Nac. Combust. Alcohol y Portland, Repub. Orient. Uruguay, Montevideo, Pp. 56 (1944).
— Hybrid corn and its contribution to American agriculture. Annu. Mtg. Ont. Crop Improv. Ass., Toronto, February 5, 6 and 7th. Pp. 9 (1945). [Mimeographed.]

Nevens, W. B. u. G. H. Dungan: Yields of corn hybrids harvested for silage: and methods to determine best time to harvest. Bull. Ill. Agric. Exp. Sta. Nr. 494, 387—412 (1942).

Olson, P. J.: Exchange of certain alternative stable characters in crosses between dent and flint corn. Tech. Bull. N. Dak. Agric. Exp. Sta. Nr. 291, 38 (1939).

O'Mara, J. G.: Cytological observations on Zea-Euchlaena hybrids. Genetics **24**, 82—83 (1939). (Abstr.)

— A cytogenetic study of Zea and Euchlaena. Res. Bull. Mo. Agric. Exp. Sta. Nr. 341, 16 (1942).

Ospitaletche, A. E.: La conservación de la pureza varietal en el maíz. (The preservation of varietal purity in maize.) „Granos" Semilla Selecta, B. Aires **6**, Nr. 10 bis 12, 30—33 (1942).

Paddick, M. E.: Vegetative development of inbred and hybrid maize. Res. Bull. Ia Agric. Exp. Sta. Nr. 331, 376—399 (1944).

— u. H. B. Sprague: Maize seed characters in relation to hybrid vigor. J. Amer. Soc. Agron. **31**, 743—750 (1939).

Painter, R. H. u. A. M. Brunson: Differential injury within varieties, inbred lines, and hybrids of field corn caused by the corn earworm, Heliothis armigera (Hbn.). J. Agric. Res. **61**, 81—100 (1940).

Parenti, E.: Varietà di granoturco nane precoci, Bianco dentato precoce friulano. (Early dwarf varieties of maize-Friuli Early White Dent.) Ital. Agric. **21**, 651 bis 658 (1942).

Parodi, L. R.: Una nueva especie de Maidea sudamericana del género Tripsacum. (A new species of South American Maydeae of the genus Tripsacum.) Rev. Argent Agron. **9**, 249 (1942).

Patch, L. H.: Survival, weight, and location of European corn borers feeding on resistant and susceptible field corn. A. Agric. Res. **66**, 7—19 (1943).

— u. R. T. Everley: Resistance of dent corn inbred lines to survival of first-generation European corn borer larvae. Tech. Bull. U. S. Dep. Agric. Nr. 893, 10 (1945).

—, J. R. Holbert u. R. T. Everly: Strains of field corn resistant to the survival of the European corn borer. Tech. Bull. U. S. Agric. Nr. 823, 22 (1942).

—, G. W. Still, B. A. App u. C. A. Crooks: Comparative injury by the European corn borer to open-pollinated and hybrid field corn. J. Agric. Res. **63**, 355 bis 368 (1941).

—, —, M. Schlosberg u. G. T. Bottger: Factors determining the reduction in yield of field corn by the European corn borer. J. Agric. Res. **65**, 472—482 (1942).

Pavlov, B. J.: (Further notes on deterioration from cross breeding in maize.) Jarovizacija, Nr. 5 (32), 127 bis 128 (1940). [Russisch.]

Pepper, B. B. u. C. S. Garrison: The performance of hybrid field corns under European corn borer conditions in New Jersey. J. Econ. Ent. **34**, 281—284 (1941).

Perry, H. S.: The Ga gene as a means of reducing contamination of sweet corn. J. Hered. **36**, 131—134 (1945).

Philp, J.: A comparative test of the yield of F_1 hybrids between inbred lines of maize. Bull. Minist. Agric. Egypt. Nr. 202, 6 (1939).

Phinney, B. O.: Gene action in the development of the leaf in Zea mays. Amer. J. Bot. **31**, Nr. 8, 4s. (1944) (Abstr.).

— Cell lenght in the parenchyma of the midrib of normal and dwarf-1 maize at various stages of development. Amer. J. Bot. (Suppl.) **33**, 222—223 (1946). (Abstr.)

Pickett, B. S.: Sho' nuf Dixie sweet corn. Sth. Seedsman **7**, 15, 40 (1944).

Pinnell, E. L.: The variability of certain quantitative characters of a double cross hybrid in corn as related to the method of combining the four inbreds. J. Amer. Soc. Agron. **35**, 508—514 (1943).

Poole, C. F.: Corn earworm resistance and plant characters. Proc. Amer. Soc. Hort. Sci. **38**, 605—609 (1940).

Porter, J. W., F. M. Strong, R. A. Brink u. N. P. Neal: Carotene content of the corn plant. J. Agric. Res. **72**, 169—187 (1946).

Pound, F. J.: A few facts about maize for the Trinidad grower. Proc. Agric. Soc. Trin. Tob. **40**, 241—262 (1940).

Randolph, L. F.: Genetic characteristics of the B chromosomes in maize. Genetics **26**, 608—631 (1941).

— The influence of heterozygosis in fertility and vigor in autotetraploid maize. Genetics, **27**, 163 (1942). (Abstr.)

— u. H. E. Fischer: The occurrence of parthenogenetic diploids in tetraploid maize. Proc. Nat. Acad. Sci. Wash. **25**, 161—164 (1939).

— u. D. B. Hand: Relation between carotenoid content and number of genes per cell in diploid and tetraploid corn. J. Agric. Res. **60**, 51—64 (1940).

Rattray, A.: Hybrid-maize. Rhod. Agric. J. **38**, 535 bis 537 (1941).

Reeves, R. G.: Chromosome knobs in relation to the origin of maize. Genetics **29**, 141—147 (1944).

— u. P. C. Mangelsdorf: A proposed taxonomic change in the tribe Maydeae (family Gramineae). Amer. J. Bot. **29**, 815—817 (1942).

Rhoades, M. M.: On the high mutation rate of the a_1 allele in maize induced by the Dt gene. Proc. 7th Inz. Genet. Congr. Edinburgh 23—30 August, 247—248, 1939 (1941). (Abstr.)

— Studies of a telocentric chromosomen in maize with reference to the stability of its centromere. Genetics **25**, 483—520 (1940).

— Different rates of crossing over in male and female gametes of maize. J. Amer. Soc. Agron. **33**, 603—615 (1941).

— Preferential segregation in maize. Genetics **27**, 395 bis 407 (1942).

— On the anaphase movement of chromosomes. Proc. Nat. Acad. Sci. Wash. **28**, 433—436 (1942).

— Genic induction of an inherited cytoplasmic difference. Proc. Nat. Acad. Sci. Wash. **29**, 327—329 (1943).

— On the genetic control of mutability in maize. Proc. Nat. Acad. Sci. Wash. **31**, 91—95 (1945).

— Crossover chromosomes in unreduced gametes of asynaptic maize. Genetics **32**, 101 (1947). (Abstr.)

— u. V. H. Rhoades: Genetics studies with factors in the tenth chromosome in maize. Genetics **24**, 302—314 (1939).

Richey, F. D.: Maize hybrids susceptible to earworm. Heritable differences in susceptibility of corn hybrids to early attack. J. Hered. **35**, 327—328 (1944).

— Isolating better foundation inbreds for use in corn hybrids. Genetics **30**, 455—471 (1945).

— Hybrid vigor and corn breeding. J. Amer. Soc. Agron. **38**, 833—841 (1946).

— Multiple convergence as a means of augmenting the vigor and yield inbred lines of corn. J. Amer. Soc. Agron. **38**, 936—940 (1946).

— Corn cob fur. J. Hered. **37**, 251—252 (1946).

— Corn breeding: gamete selection, the oenothera method, and related miscellany. J. Amer. Soc. Agron. **39**, 403—411 (1947).

Riess, F. u. J. L. Robinson: The 1944 Iowa corn yield test. Bull. Ia Agric. Exp. Sta. Nr. P71, 372—416 (1945).

Rife, D. C.: A mutation in corn pericarp. Ohio J. Sci. **44**, 143—144 (1944).

Robbins, W. J.: Growth substances in a hybrid corn and its parents. Bull. Torrey Bot. Cl. **67**, 565—574 (1940).

— Factor Z in hybrid maize. Bull. Torrey Bot. Cl. **68**, 222—228 (1941).

Roberts, E. u. I. R. Hoener: Causes of preferences exhibited by animals for certain inbred lines of corn. J. Amer. Soc. Agron. **33**, 448—453 (1941).

Roberts, L. M.: The effects of translocations on growth in Zea mays. Genetics 27, 166 (1942). (Abstr.)
— The effect of translocation on growth in Zea mays. Genetics 27, 584—603 (1942).
Robinson, J. L. u. F. Reiss: The 1943 Iowa corn yield test. Bull. Ia. Agric. Exp. Sta. Nr. P58, 852—903 (1944).
—, — The 1945 Iowa corn yield test. Bull. Ia Agric. Exp. Sta. Bull. Nr. P79, 600—642 (1946).
Rogers, J. S.: Speaking of yields ... and Texas hybrid corn. Hybrids outstrip open-pollinated varieties in tests at four Texas substations and in Arkansas. Sth. Seedman 5, Nr. 11; 9, 32, 36 (1942).
Roman, H.: Translocations involving „B" chromosomes in maize. Genetics 27, 167 (1942). (Abstr.)
Rusconi, C.: El maíz en las tumbas indígenas de Mendoza. (Maize in the native tombs of Mendoza.) Darwiniana, B. Aires 7, 117—119 (1945).

Saboe, L. C. u. H. K. Hayes: Genetic studies of reactions to smut and of firing in maize by means of chromosomal translocations. J. Amer. Soc. Agron. 33, 463 bis 470 (1941).
Salamov, A. B.: (Inbreeding maize.) Selekcija i Semenovodstvo (Breeding and Seed Growing) Nr. 10/11, 22—27 (1939). [Russisch.]
— (On the spatial isolation of maize.) Selekcija i Semenovodstvo (Breeding and Seed Growing) Nr. 3, 25—27 (1940). [Russisch.]
Sanguineti, M. E.: Estudio del caracter „siamensis" en maíz (Zea mays L.). [Study of the character „siamensis" in maize (Z. mays L.)]. An. Inst. Fitotéc. Santa Catalina 1, 17—134 (1939) 1940.
Sansom, T. K.: Breeding Diplodia resistant varieties of maize. Rhod. Agric. J. 37, 442—444 (1940).
Sass, J. E.: Schedules for sectioning maize kernels in paraffin. Stain Tech. 20, 93—98 (1945).
— Development of endosperm and antipodal tissue in „Argentine waxy" maize. Amer. J. Bot. 33, 223 (1946). (Abstr.)
— u. A. A. Bryan: Histology of a genetic malformation in corn. Proc. Ia. Acad. Sci. 46, 180 (1939) 1940. (Abstr.)
— u. J. M. Green: Cytohistology of the reaction of maize seedlings to colchicine. Bot. Gaz. 106, 483—488 (1945).
— u. G. F. Sprague: A „twin-embryo" abnormality in maize. Proc. Iowa Acad. Sci. 47, 155—156 (1940).
Saunders, A. R.: Should we also take up hybrid maize? A full review of breeding methods and considerations which show that there would be serious snags in applying the American practice in South Africa. The Department's policy. Fmr's Wkly, Bloemfontein 59, 1492 bis 1493, 1507 (1940).
Schelotto, B. u. A. H. Marchioni: Algunos datos sobre el maíz en Venezuela. Distribución de las variedades botánicas. (Some data on maize in Venezuela. Distribution of the botanical varieties.) Rev. Argent. Agron. 8, 49—56 (1941).
Schindler, A. J.: Insect transmission of Wallaby ear disease of maize. J. Aust. Inst. Agric. Sci. 8, 35—37 (1942).
Schlosberg, M.: Sweet corn resistance to the European corn borer. Canner, 92, (12), 54—56 (1941).
Semeniuk, G., J. R. Wallin u. I. E. Melhus: Rootnecrosis resistance in maize. Phytopathology 37, 20 (1947). (Abstr.)
Shafer, J. (jr.): Water loss from excised leaves. Amer. J. Bot. 29, 89—91 (1942).
— The relation of embryo axis weights to heterosis. Amer. J. Bot. 31, 503—506 (1944).
— u. R. G. Wiggans: Correlation of total dry matter with grain yield in maize. J. Amer. Soc. Agron. 33, 927—932 (1941).

Shank, D. B.: Top-root ratios of inbred and hybrid maize. Genetics 25, 134 (1940). (Abstr.)
— Top-root ratios of inbred and hybrid maize. J. Amer. Soc. Agron. 35, 976—987 (1943).
— Effects of phosphorus, nitrogen, and soil moisture on top-root ratios of inbred and hybrid maize. J. Agric. Res. 70, 365—377 (1945).
Sharman, B. C.: A twin seedling in Zea mays L. Twinning in the Gramineae. New. Phytol. 41, 125—129 (1942).
— A six-stamened flower in Zea mays L. New Phytol. 41, 130—133 (1942).
— Short nights: an unappreciated hindrance to maize cultivation in England. J. R. Hort. Soc. 72, 195—202 (1947).
Shoemaker, J. S. u. C. Walkof: Sweet corn in Alberta. Bull. Univ. Alberta Nr. 38, 75 (1941).
Shull, G. H.: Hybrid seed corn. Science 103, 547—550 (1946).
Singleton, W. R.: Hybrid vigour and its utilization in sweet corn breeding. Amer. Nat. 75, 48—60, also Proc. 7th. Int. Genet. Congr. Edinburgh 23—30 August, 264 bis 265, 1939 (1941). (Abstr.)
— Influence of female stock on the functioning of small pollen male gametes. Proc. Nat. Acad. Sci. Wash. 26, 102—104 (1940).
— New corn hybrids. Canning Age, 22, 559—560 (1941).
— Breeding behavior of C30 a diminutive P39 mutant whose hybrids show increased vigor. Genetics 28, 89 (1943). (Abstr.)
— Hybrid vigor in the intra-inbred cross P39 × C30 in maize. Rec. Genet. Soc. America Nr. 12, 52—53 (1943). (Abstr.)
— „Long husk" sterility in maize. J. Hered. 37, 29—30 (1946).
— Inheritance of indeterminate growth in maize. J. Hered. 37, 61—64 (1946).
— Mutations in maize inbreds. Genetics 32, 104 (1947). (Abstr.)
— u. F. J. Clark: Cytological effects of treating maize pollen with ultra-violet light. Genetics 25, 136 (1940). (Abstr.)
— u. P. C. Mangelsdorf: Gametic lethals on the fourth chromosome of maize. Genetics 25, 366—390 (1940).
Skrabal, R.: Standraum und Sorte bei Maisanbau in Trockengebieten. Mitt. Landw. 59, 417—418 (1944).
Smith, G. M.: Sweet corn hybrids, 1940. Canner 92, (12), 54 (1941).
— Improved Golden Cross Bantam and Purgold sweet corn. Bull. Ind. Agric. Exp. Sta. Nr. 513. 4 (1946).
— u. A. M. Brunson: Hybrid popcorn in Indiana. Bull. Ind. Agric. Exp. Sta. Nr. 510, 18 (1946).
Smith, L. M.: Mrs. Peppard hybridizes popcorn. Seed World 49, Nr. 2, 9 (1941).
Snelling, R. O., R. A. Blanchard u. J. H. Bigger: Resistance of corn strains to the leaf aphid, Aphis maidis Fitch. J. Amer. Soc. Agron. 32, 371—381 (1940).
Sokolov, B. P.: (Maize hybrids of the Dnepropetrovsk State Breeding Station.) Selekcija i Semenovodstvo (Breeding and Seed Growing) Nr. 3, 27—30 (1940). [Russisch.]
— (Maize hybrids, their production and utilization in agriculture.) Selekcija i Semenovodstvo (Breeding and Seed Growing) Nos. 1/2, 30—45 (1946). [Russisch.]
Spencer, J. T.: A comparative study of the seasonal root development of some inbred lines and hybrids of maize. J. Agric. Res. 61, 521—538 (1940).
Sprague, G. F.: An estimation of the number of topcrossed plants required for adequate representation of a corn variety. J. Amer. Soc. Agron. 31, 11—16 (1939).
— Corn hybrids for Missouri. Circ. Mo. Agric. Exp. Sta. Nr. 201, 27 (1939).
— Heritable characters in maize. 50-Vestigial glume. J. Hered. 30, 143—145 (1939).

Sprague, G. F.: The location of dominant favorable genes in maize by means of an inversion. Genetics 26, 170 (1941).
— Building new corn hybrids. Fm. Sci. Rep. Iowa 2, 7—9 (1941).
— Production of hybrid corn Bull. Ia. Agric. Exp. Sta. Nr. P 48, 556—582 (1942).
— Early testing of inbred lines of corn. J. Amer. Soc. Agron. 38, 108—117 (1946).
— The experimental basis for hybrid maize. Biol. Rev. 21, 101—120 (1946).
—, B. Brimhall u. R. M. Hixon: Some effects of the waxy gene in corn on properties of the endosperm starch. J. Amer. Soc. Agron. 35, 817—822 (1943).
— u. R. M. Hixon: Waxy corn — a new crop. Fm. Sci. Reporter 3, Nr. 4, 10—11 (1942).
—, — A new starch from corn. War experiments develop waxy corns to replace tapioca starch. Seed World 53, Nr. 2, 20—21 (1943).
— u. M. T. Jenkins: A comparison of synthetic varieties, multiple crosses, and double crosses in corn. J. Amer. Soc. Agron. 35, 135—147 (1943).
— u. J. E. Sass: Heritable characters in maize: „accessory blade". Proc. Iowa Acad. Sci. 49, 256 (1942).
— u. L. A. Tatum: General vs. specific combining ability in single crosses of corn. J. Amer. Soc. Agron. 34, 923—932 (1942).
Stadler, L. J.: Gene action in anthocyanin synthesis in maize. Amer. J. Bot. Suppl. 29, 17s—18s (1942). (Abstr.)
— Gamete selection in corn breeding. J. Amer. Soc. Agron. 36, 988—989 (1944). (Abstr.)
— Spontaneous mutation at the R locus in maize. I. The aleuronecolor and plant-color effects. Genetics 31, 377 bis 394. (1946)
— u. S. Fogel: Gene variability in maize. I. Some alleles of R (R^r series). Genetics 28, 90—91 (1943). (Abstr.)
—, — Gene variability in maize. II. The action of certain R. alleles. Genetics 30, 23—24 (1945). (Abstr.)
— u. H. Roman: The genetic nature of X-ray and ultra-violet induced mutations affecting the gene A in maize. Genetics 28, 91 (1943). (Abstr.)
— u. F. M. Uber: Genetic effects of ultraviolet radiation in maize. IV. Comparison of monochromatic radiations. Genetics 27, 84—118 (1942).
Standen, J. H.: Variability of Nigrospora on maize. J. Sci. Iowa St. Coll. 17, 263—275 (1943).
Steele, L. u. I. R. Bradshaw: New hybrids foil old foes. Insects, diseases and drouth are being whipped to develop better hybrids for south. Sth. Seedsman 6, Nr. 3; 10, 39 (1943).
Stevenson, G. C.: Notes pratiques sur l'amélioration du maïs par la sélection. Rev. Agric. Maurice 23, 21—25 (1944).
Stoneberg, H.: Louisiana-adepted hybrid corns. Sth. Seedsman 8, Nr. 3, 18 (1945).
Stringfield, G. H.: Hybrid corn in Ohio-historical notes. Bi-m. Bull. Ohio Agric. Exp. Sta. 26, Nr. 208, 3—5 (1941).
— u. D. H. Bowman: Breeding corn hybrids for smut resistance. J. Amer. Soc. Agron. 34, 486—494 (1942).
—, R. D. Lewis u. H. L. Pfaff: The Ohio cooperative corn performance tests. Spec. Circ. Ohio Agric. Exp. Sta. Nr. 59, 27 (1940).

Tapke, V. F.: New methods of artificial seed inoculation for testing the resistance of spring and winter barleys against stripe, Helminthosporium gramineum. Phytopathology 37, 21 (1947). (Abstr.)
Tatum, L. A.: The effect of genetic constitution and processing methods on the ability of maize seed to germinate in cold soil. J. Sci. Iowa St. Coll. 17, 138 bis 140 (1942).

Tavčar, A.: Inheritance of 2-, 3-, 4- and 6-articulate leaf whorls in Zea Mays L. Proc. 7th Int. Genet. Congr. Edinburgh 23—30 August, 280—281, 1939 (1941) (Abstr.).
— Svojstva prve sjemenske generacije od sitnozrnih i krupnozrnih genotipova kod kukuruze-Zea Mays L.. (The immediate effect of crossing of small and large seeded genotypes on the character of kernels of Zea Mays L.). Poljopr. Naučna Smotra, Zagreb 2, 77—96 (1940).
— Le maïs huit rangs dans la création d'hybrides productifs. Inst. Amélioration Plantes Univ. Roy. Zagreb (Yougoslavie) Pp. 217—224.
Teixeira, E. F.: Milho hibrido (Hybrid maize). Bol. Agron. Brazil. 10, 170—173 (1946).
Thompson, R. W.: The corn borer situation in Ontario in 1941 with notes on hybrid and broom corn infestation. 72nd Rep. Ent. Soc. Ont. 12—15 (1943).

Uber, F. M.: Ultra-violet spectrophotometry of Zea mays pollen with the quartz microscope. Amer. J. Bot. 26, 799—807 (1939).
Ullstrup, A. J.: Inheritance of susceptibility to infection by Helminthosporium maydis race 1 in maize. J. Agric. Res. 63, 331—334 (1941).
— The corn disease situation in 1942. Rep. 6th Corn Improv. Conf. N. Cent. Reg. St. Louis, Mo., November 12, 1942, 20—21 (1942). [Mimeographed.]
— Disease of dent corn in the United States. Circ. U. S. Dep. Agric. Nr. 674, 34 (1943).
— u. A. M. Brunson: Linkage relationships of a gene determining susceptibility to a disease in corn. Phytopathology 36, 412 (1946). (Abstr.)
—, — Linkage relationships of a gene in corn determining susceptibility to a Helminthosporium leaf spot. J. Amer. Soc. Agron. 39, 606—699 (1947).

Valle, C. G.: Estudios genéticos sobre el maíz. 2. La prueba y utilización de las líneas homogéneas. (Genetical studies on maize. 2. The testing and utilization of pure lines.) Mem. Soc. Cubana Hist. Nat. 16, 29—36 (1942).
Valle, C. I. del: El maiz dulce. (Sweet corn). Bol. Estac. Exp. Agron., Santiago, Nr. 62, 18 (1945).
Vallega, J.: Observaciones preliminares sobre especialización fisiológica de Puccinia sorghi, en Argentina. (Preliminary observations on physiological specialization of P. sorghi in Argentina.) An. Inst. Fitotéc. Santa Catalina, 4, 14—16 (1942).
Van Lanen, J. M., F. W. Tanner (jr.) u. S. E. Pfeiffer: Composition of hybrid corn tassels. Cereal Chem. 23, 428—432 (1946).
Venta, J. de la: La mejora genética del maíz. (Genetic improvement of maize.) Agricultura, Madrid 10, 7—10 (1941).

Wagenaar, G. A. W.: Massaselectie bij maïs. (Mass selection in maize.) Landbouw 17, 307—313 (1941).
Walker, C.: Maize growing for grain. N. Z. J. Agric. 72, 281—299 (1946).
Walker, E. A. u. J. W. Magruder: Maryland field corn leaf blight disease survey—1942. Plant Dis. Rep. 27, 126—135 (1943).
Walter, E. V. u. A. M. Brunson: Differential susceptibility of corn hybrids to Aphis maidis. J. Econ. Ent. 33, 623—628 (1940).
—, — Selection for aphid resistance within inbred lines of maize. J. Amer. Soc. Agron. 38, 974—977 (1946).
Wang, F. H.: Embryological development of inbred and reciprocal hybrid Zea mays L. Amer. J. Bot. (Suppl.) 33, 224 (1946). (Abstr.)
— Embryological development of inbred and hybrid Zea mays L. Amer. J. Bot. 34, 113—125 (1947).

Weatherwax, P.: Corn for morphological and genetic work. Genetics **31**, 235 (1946) (Abstr.).
— Primitive characteristics in a Peruvian variety of maize. Amer. J. Bot. Suppl. **33**, 827 (1946) (Abstr.).
Weaver, H. L.: A developmental study of maize with particular reference to hybrid vigor. Amer. J. Bot. **33**, 615—624 (1946).
Welch, J. E.: Linkage in autotetraploid maize. Abstr. Thes. Cornell Univ. 1942, 400—403 (1943).
Whaley, W. G. u. A. L. Long: The behavior of excised roots of heterotic hybrids and their inbred parents in culture. Bull. Torrey Bot. Cl. **71**, 267—275 (1944).
Wiidakas, W.: Hybrid corn for North Dakota. Bi-m. Bull. N. Dak. Agric. Exp. Sta. **1**, Nr. 4, 3—6 (1939).
— Early North Dakota corn hybrids. Bi-m. Bull. N. Dak. Agric. Exp. Sta. **4**, Nr. 4, 13—15 (1942).
— Planning the fight against the European corn borer in the north central states. Bi-m. Bull. N. Dak. Agric. Exp. Sta. **7**, 29—31 (1945).
— u. L. A. Jensen: 1945 hybrid corn field trials. Agron. Mimeo Circ. N. Dak. Agric. Exp. Sta. Nr. 77, 20 (1946). [Mimeographed.]
— u. W. J. Leary: 1942 hybrid corn field trials. Agron. Mimeo. Circ. N. Dak. Agric. Exp. Sta. Nr. 74, 19 (1943).
Woodworth, C. M. u. O. Bolin: Brief description of ear and kernel characters of various single crosses. Dep. Agron. Agric. Exp. Sta., Ext. Serv. Agric. Home Econ., Univ. Ill., Coll. Agric., Urbana, Ill. January 4 (1939). [Mimeographed.]
—, — Brief description of inbred lines of corn. Dep. Agron. Agric. Exp. Sta., U. S. Dep. Agric., Univ. Ill., Coll. Agric., Urbana, Ill. Sept. p. 8 (1940). [Mimeographed.]
Wouters, W.: Contribution à la biologie florale du maïs. Sa pollinisation libre et sa pollinisation controlée en Afrique centrale. Publ. Inst. Agron. Congo Belge, Sér. Sci. Nr. 23, 51 (1941).
Wu, Shao-Kwei: The relationship between the origin of selfed lines of corn and their value in hybrid combination. J. Amer. Soc. Agron. **31**, 131—140 (1939).

Young, H. C. (jr.): The toothpick method of inoculating corn for ear and stalk rots. Phytopathology **33**, 16 (1943). (Abstr.)

Zapparoli, T. V.: Il granoturco „Marano". (The maize „Marano" Ital. Agric. **76**, 155—159 (1939).
— Il granoturco „Scagliolo 23 A". (The maize „Scagliolo 23 A"). Ital. Agric. **76**, 239—245 (1939).
— Derivati del granoturco Nostrano dell'Isola. Nostrano dell'Isola, Finardi (S. M.) Isola Basso (S. M.) — Letizia (S. M.). [Derivatives of the maize variety Nostrano dell'Isola: Nostrano dell'Isola, Finardi (S. M.) — Isola Basso (S. M.) — Letizia (S. M.).] Ital. Agric. **76**, 317—326 (1939).
— Due altri granoturchi selezionati, Rostrato Cajo Duilio (S. M.), Giallo tondo S. Pancrazio B. (S. M.). [Two further selected maize varieties, Rostrato Cajo Duilio (S. M.) and Giallo tondo S. Pancrazio B. (S. M.)]. Ital. Agric. **76**, 609—614 (1939).
— Il granuturco Nano precoce del „Teso", o Semetta „Teso". (The early dwarf maize „Teso" or „Semetta Teso".) Ital. Agric. **77**, 335—338 (1940).
— Il granoturco cinquantino „Davini". (The early maize Davini.) Ital. Agric. **77**, 455—459 (1940).
— Il granoturco Precoce di Gagnolo. (The maize Bagnolo Early.) Ital. Agric. **78**, 579—583 (1941).
— u. G. Battaglini: Varietà di granoturco nane precoci selezionate. Il granoturco „Trentinella" di Pescara. (Varieties of early dwarf maize obtained by selection. The Trentinella maize from the province of Pescara. Opus Staz. Sper. Maiscolt. Bergamo Nr. 34, 577—579 (1943).

Zapparoli, T. V. u. E. Parenti: Lo studio sistematico dei granoturchi da foraggio. (Systematic study of forage maize. Opus. Staz. Sper. Maiscolt. Bergamo Nr. 37, 183—192 (1943).
Zuber, M. S. u. J. L. Robinson: The 1938 Iowa corn yield test. Bull. Ia. Agric. Exp. Sta. Nr. 379, 77 (1939).

6. Hirsen.

Althaus, W. G.: „60-day" milos! Sth. Seedsman **10**, Nr. 4, 13 (1947).
Anonymus: Tests show „New Grain" is not superior sorghum. Seed World **48**, Nr. 11, 17—19 (1940).
— „Bonita", grain sorghum hybrid, makes bow. Sth. Seedsman **6**, 20 (1943).
— Starch from grain sorghum. 34th Bienn. Rep. Kans. St. Bd. Agric. 1943/44 Pp. 134—135 (1944).
— Midland, new grain sorghum for upper South likes pre-harvest frost for best threshing. Sth. Seedsman **10**, Nr. 3, 46 (1947).
— Texas drylands are tailor-made to hold bulk of nation's grain sorghum acreage. Sth. Seedsman **10**, Nr. 4; 15, 58—62 (1947).
— 10 years of better sorghums. Sth. Seedsman **10**, Nr. 4; 28, 32—33, 36 (1947).
Arenkova, D. N.: Polyploid races in millet (Panicum miliaceum L.). C. R. (Doklady) Acad. Sci. URSS. **29**, 332—335 (1940).
Aseeva, L. M.: (On the method and technique of hybridization of millet.) Selekcija i Semenovodstvo (Breeding and Seed Growing) Nr. 3, 33 (1940). [Russisch.]
Ayyangar, G. N. R. u. A. K. K. Nambiar: Lethal green seedlings in sorghum. Current Sc. **8**, 417—418 (1939).

Barham, H. N., J. A. Wagoner, C. L. Campbell u. E. H. Harclerode: The chemical composition of some sorghum grains and the properties of their starches. Tech. Bull. Kans. Agric. Exp. Sta. Nr. 61, 47 (1946).
Bennett, H. W. u. P. G. Hogg: The F_2 and F_3 generations of a sorghum × Johnson grass hybrid. Proc. 43rd Conv. Assoc. Sth. Agric. Wrks. Memphis, Tenn, Pp. 84—85 (1942).
Berkner: Die Hirsearten und ihre Kultur. Tsch. landw. Pr. **67**, 29 (1940).
Brown, M. S.: Haploid plants in sorghum. J. Hered. **34**, 163—166 (1943).
Burley, R. u. H. Graham: New sorghums are valuable grain crop. Successful Fmg. **45**, Nr. 2, 53—54 (1947).

Chin, T. C.: The cytology of polyploid sorghum. Amer. J. Bot. **33**, 611—614 (1946).
Ciferri, R.: La cerealicoltura in Africa Orientale. III-Generalità botanico-agrarie sui sorghi. (Cereal cultivation in East Africa. III-Botanicoagrarian discussions concerning the sorghums.) Ital. Agric. **78**, 79—86 (1941).
— La cerealicoltura in A. O. I. IX-Le durre. (Cereal cultivation in Italian East Africa. IX-Sorghums.) Ital. Agric. **79**, 294—304 (1942).
Cirk, G. P.: (Extending the range of millet towards the north.) Vestnik Socialističeskogo Rastenievodstva (Soviet Plant Industry Record), Nr. 2, 108—111 (1940). [Russisch.]
Clydesdale, C. S.: Field crops. Grain sorghums. Qd. Agric. J. **57**, 133—141 (1943).
Cooper, D. C. u. R. A. Brink: The antipodals as a factor in seed failure following an interspecific mating in the Gramineae. Rec. Genet. Soc. America Nr. 12, 45—47 (1943). (Abstr.)
Couch, J. F., R. R. Briese u. J. H. Martin: Hydrocyanic acid content of sorghum varieties. J. Wash. Acad. Sci. **29**, 146—161 (1939).

Dahms, R. G.: Insect resistance in sorghum and cotton. J. Amer. Soc. Agron. **35**, 704—715 (1943).
— u. J. H. Martin: Resistance of F_1 sorghum hybrids to the chinch bug. J. Amer. Soc. Agron. **32**, 141—147 (1940).
Darlington, C. D. u. P. T. Thomas: Morbid mitosis and the activity of inert chromosomes in Sorghum. Proc. Roy. Soc. **130**, Ser. B. 127—150 (1941).
Doughty, L. R.: A note on Striga investigations at Amani. E. Afr. Agric. **8**, 33—38 (1942).

Fennell, J. L.: Grain for tropical America. Agric. Amer. **6**, 202—204 (1946).
Fiorentini, G.: Contributo alla conoscenza delle caratteristiche chimico-technologiche delle cariossidi di dura dell'Africa Orientale Italiana. (Contribution to the study of the chemical and technological characteristics of the grains of durra from Italian East Africa.) Agricoltura Colon **36**, 134—139 (1942).
Firsova, M. A.: (Characteristics of some recommended varieties of millet as regards extraction and quality of grits.) Selekcija i Semenovodstvo (Breeding and Seed Growing) **13**, Nr. 6, 58—61 (1946).

Garber, E. D.: A cytological study of the genus Sorghum: subsections Para-sorghum and Eu-sorghum. Amer. Nat. **78**, 89—93 (1944).
Gorst, G. F.: (The method of millet breeding and seed production must be changed.) Selekcija i Semenovodstvo (Breeding and Seed Growing) Nr. 3, 37—38 (1940). [Russisch.]

Hamilton, J. W.: „Old faithful".... sorghum triple-threater. Sth. Seedsman **9**, 12, 40 (1946).
Heller, V. G. u. Seiglinger. J. B.: Chemical composition of Oklahoma grain sorghums. Bull. Okla. Agric. Exp. Sta. Nr. B-274, 5 (1944).
Heuser, W. u. H. Schleip: Die Korn- und Strohleistung der Hirse, insbesondere der Stämme der Deutschen Hindukusch-Expedition und ihre Verwendungsmöglichkeit im praktischen Anbau. Forschungsdienst **9**, 176—183 (1940).
Heyne, E. G., L. E. Melchers u. A. E. Lowe: Reaction of F_1 sorghum plants to milo disease in the greenhouse and field. J. Amer. Soc. Agron. **36**, 628—630 (1944).
Hoffmaster, D. E. u. E. C. Tullis: Susceptibility of sorghum varieties to Macrophomina dry rot (Charcoal rot). Plant Dis. Reporter **28**, 1175—1184 (1944). (Mimeographed.]

Jakuševskij, E. S.: (New promising varieties of millet.) Soviet Plant Industry Record Nr. 1, 106 (1940).
— (Varieties of millet raised by the Kuban Experimental Station of the Institute of Plant Industry.) Vestnik Socialističeskogo Rastenievodstva (Soviet Plant Industry Record) Nr. 2, 105—107 (1940). [Russisch.]
— (Prospects of raising sorghum in arid regions of the USSR.). Vestnik Socialističeskogo Rastenievodstva (Soviet Plant Industry Record) Nr. 5, 178—180 (1940). [Russisch.]
— (Japanese millet as a new fodder and industrial crop.) Vestnik Socialističeskogo Rastenievodstva (Soviet Plant Industry Record) Nr. 5, 188 (1940). [Russisch.]
Janaki Ammal, E. K.: Supernumerary chromosomes in para-sorghum. Curr. Sci. **8**, 210—211 (1939).
— Chromosome diminution in a plant. Nature, London **146**, 839—840 (1940).
Jones, D. L.: Texans worth knowing: Plainsman, Caprock. Sth. Seedsman **7**, 13, 44 (1944).

Kadam, B. S., S. M. Patel u. R. K. Kulkarni: Consequences of inbreeding in bajri. J. Hered. **31**, 201—207 (1940).

Karper, R. E.: Registration of improved sorghum varieties, II. J. Amer. Soc. Agron. **33**, 257—258 (1941).
— New wartime crop may prove bonanza. Waxy endosperm sorghum producing special type of starch for industrial uses may also meet synthetic rubber needs. Sth. Seedsman **6**, Nr. 1; 9, 36 (1943).
— Registration of sorghum varieties, IV. J. Amer. Soc. Agron. **36**, 453 (1944).
— Combine sorghums. Sth. Seedsman **8**, 14, 24, 30—31 (1945).
— u. J. R. Quinby: Successful sorghums. Sth. Seedsman **9**, 11, 34, 46 (1946).
—, — The history and evolution of milo in the United States. J. Amer. Soc. Agron. **38**, 441—453 (1946).
Kendrick, J. B. u. F. N. Briggs: Pythium root rot of milo and the development of resistant varieties. Bull. Calif. Agric. Exp. Sta. Nr. 629, 18 (1939).
Kida, T.: (A classification of Italian millet in Formosa.) J. Taihoku Soc. Agric. For. **5**, 150—163 (1941).
Koehler, B.: Disease treatening broom corn production in Illinois. Plant Dis. Rep. **27**, 70—73 (1943).
Krishnaswamy, N. u. G. N. Rangaswami Ayyangar: Note on Sorghum Stapfii C. E. C. Fischer. Curr. Sci. **9**, 461—462 (1940).
—, — Chromosomal alterations induced by X-rays Bajri (Pennisetum typhoides Stapf & Hubbard). J. Indian Bot. Soc. **20**, 111—117 (1941).
—, — An autotriploid in the pearl millet (Pennisetum typhoides S. & H.) Proc. Indian Acad. Sci. **13**, Sect B, 9—23 (1941).
—, — Certain abnormalities in millets induced by X-rays. Proc. Indian Acad. Sci. **16**, Sect. B. 1—9 (1942).
—, — Anatomical studies in the leaves of the millets. J. Indian Bot. Soc. **21**, 249—262 (1942).
Kublan, A.: Aussichten und Möglichkeiten des Hirseanbaus. Ergebnis der züchterischen Bearbeitung. Forschungsdienst **16**, 276—283 (1943).
Kumar, L. S. S.: Flowering plants which attack economic crops. 1. Striga. Indian Fmg. **1**, 593—595 (1940).
— Striga research in Bombay. Indian Fmg. **1**, 609—610 (1940).
— Non-heritable polyembryony in Andropogon Sorghum Curr. Sci. **11**, 241 (1942).
— u. W. V. Joshi: Experiments on the effect of X-ray on Pennisetum typhiodeum Rich., Nicotiana tabacum Linn. and Brassica juncea Hk. f. and T. Indian J. Agric. Sci. **9**, 675—684 (1939).

Laubscher, F. X.: A genetic study of sorghum relationships. Sci. Bull. Dep. Agric. S. Afr. Nr. 242, 22 (1945).
Leukel, R. W., J. H. Martin u. C. L. Lefebvre: Sorghum diseases and their control. Fmrs' Bull. U. S. Dep. Agric. Nr. 1959, 46 (1944).
Li, C. H. u. H. W. Li: Supernumerary chromosomes in pearl millet (Pennisetum typhoideum Rich.) Chinese J. Sci. Agric. **1**, 139—141 (1943).
—, W. K. Pao u. H. W. Li: Interspecific crosses in Setaria. II. Cytological studies of interspecific hybrids involving: 1, S. faberii and S. italica, and 2, a three way cross, F_2 of S. italica × S. viridis and S. faberii. J. Hered. **33**, 351—355 (1942).
Li, H. W., C. H. Li u. W. K. Pao: (Cytological and genetical studies of the interspecific cross Setaria italica × S. viridis.) Chinese J. Sci. Agric. **1**, 229—248 (1944).
—, —, — Cytological and genetical studies of the interspecific cross of the cultivated foxtail millet, Setaria italica (L.) Beauv., and the green foxtail millet, S. viridis L. J. Amer. Soc. Agron. **37**, 32—54 (1945).
—, J. C. Meng u. C. H. Li: Genetic studies with foxtail millet, Setaria italica (L.) Beauv. J. Amer. Soc. Agron. **32**, 426—438 (1940).
Lindenbein, W.: Ursprungszentrum, Kulturstromverbreitung und Einzelwanderung bei kultivierten Andropogon-Arten. Engl. Bot. Jahr. **71**, 337—364 (1940).

Mac Vicar, R. M. u. H. R. Parnell: The inheritance of plant colour and the extent of natural crossing in foxtail millet. Sci. Agric. **22**, 80—84 (1941).

Mann, H. H.: Millets in the Middle East. Emp. J. Exp. Agric. **14**, 208—216 (1946).

Martin, J. H.: Breeding sorghums for social objectives. J. Hered. **36**, 99—106 (1945).

Maruševˇ, A. I.: (Methods and investigations of the technological qualities of millet and sorghum.) Socialistic Grain Farming, Saratov, Nr. 5, 74—80 (1940). [Russisch.]

Melchers, L. E.: The reaction of a group a Sorghums to the covered and loose kernel smuts. Amer. J. Bot. **27**, 789—791 (1940).

— On the cause of the milo disease. Phytopathology **32**, 640—641 (1942).

— u. E. D. Hansing: The effect of sorghum kernel smuts on the development of the host. J. Agric. Res. **66**, 145—165 (1943).

— u. A. E. Lowe: The reaction of sorghum varieties and hybrids to milo disease. Plant Dis. Reporter Suppl. 126, 165—175 (1940). [Mimeographed.]

Mihajlovskij, V.: (A new cultivated plant-perennial sorghum.) Selekcija i Semenovodstvo (Breeding and Seed Growing) Nr. 7, 39—40 (1939).

Miles, L. G.: Field crops. Breeding grain sorghums for Queensland. Qd. Agric. J. **57**, 261—265 (1943).

Miller, C. C.: Miller's Dwarfs: custom-built broom corn. Sth. Seedsman **9**, 12, 47 (1946).

Mogileva, A. M.: (The choice of original material from which to breed early varieties of millet.) Trudy Zonaljnogo Inst. Zernovogo Hozjajstva Rajonov Nečernozemnoj Polosy (Trans. Zonal Inst. Grain Husbandry Non-Black-Soil Districts) Nr. 10, 59—78 (1941). [Russisch.]

Narasimhan, M. J.: Early maturing ragi. Mysore Agric. Cal. p. 20—21 (1940).

Olive, L. S., C. L. Lefebvre u. H. S. Shervin: The fungus that causes sooty stripe of sorghum spp. Phytopathology **36**, 190—200 (1946).

Parodi, L. R.: Una nueva especie de ,,Sorghum" cultivada en la Argentina. (A new Sorghum species cultivated in Argentina.) Rev. Argent. Agron. B. Aires **10**, 361—372 (1943).

— Las especies de Sorghum cultivadas en la Argentina. (The species of Sorghum cultivated in Argentina.) Rev. Argent. Agron. **13**, 1—35 (1946).

Patel, Z. H.: Occurrence of xenia in pearl millet (Pennisetum typhoideum) Stapf and Hubbard. Curr. Sci. **8**, 363—364 (1939).

— A sugary mutant in pearl millet (Pennisetum typhoideum.) Proc. 28th Indian Sci. Congr. Benares Part. III: Sect. Agric.: abst. 40, P. 258 (1941).

Plotnikov, N. J.: (Varietal sowings of millet.) Selekcija i Semenovodstvo (Breeding and Seed Growing) **13**, Nr. 6, 62—65 (1946). [Russisch.]

Popov, G. I.: (Constructive selection.) Selekcija i Semenovodstvo (Breeding and Seed Growing) Nr. 5, 9—11 (1940). [Russisch.]

— (The importance of diversity in millet.) Agrobiologija (Agrobiology) Nr. 2, 28—43 (1946). [Russisch.]

Price, C. A.: The merits of Bonita. Sth. Seedsman **7**, Nr. 11; 16, 19 (1944).

Pruthi, H. S. u. M. Singh: Stored grain pests and their control. III. Natural resistance of different grains to insect attack. Misc. Bull. Imp. Coun. Agric. Res. New Delhi Nr. 57, 13 (1943).

Quinby, J. R.: A solution of one farm labor problem. Combine varieties step up harvesting of grain sorghums; will cause expansion of 1943 acreage and aid war effort. Sth. Seedsman **6**, Nr. 1; 11, 50 (1943).

Quinby, J. R. u. J. C. Gaines: The effect of parasites and predators on corn eaworms in Sumac Sorgo heads. J. Econ. Ent. **35**, 264 (1942).

— u. R. E. Karper: Inheritance of mature plant characters in sorghum induced by radiation. J. Hered. **33**, 323—327 (1942).

—, — Sweet Sudan is a comer ... and then some, Sth. Seedsman **6**, Nr. 4, 11, 39 (1943).

—, — The inheritance of three genes that influence time of floral initiation and maturity date in milo. J. Amer. Soc. Agron. **37**, 916—936 (1945).

—, — Heterosis in sorghum resulting from the heterozygous condition of a single gene that affects duration of growth. Amer. J. Bot. **33**, 716—721 (1946).

Ramaswami, S., P. S. Sarma u. M. Sreenivasaya: Studies in insect nutrition-assay of ,,quality" in crops. Curr. Sci. **11**, 53—54 (1942).

Rangaswamy Ayyangar G. N. et al.: The description of crop plant characters and their ranges of variation. IV. The variability of Indian sorghum (jowar). Indian J. Agric. Sci. **12**, 528—263 (1942).

— u. N. Krishnaswami: Studies on the histology and colouration of the pericarp of the sorghum grain. Proc. Indian Acad. Sci. **14**, Sect. B., 114—136 (1941).

— u. A. Kunhi Koran Nambiar: Genic differences governing the distribution of stigmatic feathers in sorghum. Curr. Sci. **8**, 214—216 (1939).

—, — Lethal green seedlings in sorghum. Curr. Sci. **8**, 417—418 (1939).

—, — The inheritance of the manifestation of purple colour at the pulvinar regions in the panicles of sorghum. Curr. Sci. **10**, 80—82 (1941).

— The inheritance of depth of green colour in the leaves of sorghum. Curr. Sci. **10**, 492—493 (1941).

—, V. Panduranga Rao u. A. Kunhi Koran Nambiar: The occurrence and inheritance of purple blotched grains in sorghum. Curr. Sci. **8**, 213—214 (1939).

— u. B. W. X. Ponnaiya: Hairiness of the midrib edges in sorghum. Curr. Sci. **8**, 115—116 (1939).

—, — The occurrence and inheritance of panicletip sterility in sorghum. Curr. Sci. **8**, 116—117 (1939).

—, — Cleistogamy and its inheritance in sorghum. Curr. Sci. **8**, 418—419 (1939).

—, — The occurrence and inheritance of a bloomless sorghum. Curr. Sci. **10**, 408—409 (1941).

—, — Two new genes conditioning the tint of the colour on the glumes of sorghum. Curr. Sci. **10**, 410—411 (1941).

—, — Wavy peduncle — the basic cause of goosenecking in sorghum. Curr. Sci. **10**, 528—530 (1941).

—, — Sorghums with felty glumes. Curr. Sci. **10**, 533 bis 534 (1941).

— u. M. A. Sankara Ayyar: The inheritance and linkage affinities of the yellow coloured midrib in sorghum. Curr. Sci. **9**, 542—543 (1940).

—, — u. A. Kunhikoran Nambiar: Recurrent pseudo-mutations in sorghum. Curr. Sci. **8**, 171—172 (1939).

—, —, — The inheritance of purple pigment at the base of anthers in sorghum. Curr. Sci. **10**, 491—492 (1941).

—, —, T. R. Narayanan u. A. Kunhi Koran Nambiar: The grain sorghums of the Durra group. Proc. Indian Acad. Sci. **15**, Sect. B., 133—147 (1942).

— u. U. L. Srinivasa Rao: Studies in the Barnyard millet — Echinochloa colona var. frumentacea, C. E. C. Fischer. Madras Agric. J. **29**, 3—12 (1941).

— u. T. Venkataramana Reddy: The occurrence and inheritance of hairiness of leaf tip in sorghum. Madras Agric. J. **27**, 210—214 (1939).

—, — The inheritance of a new type of purple pigmentation manifesting on the lumes at anthesis. Curr. Sci. **9**, 228—229 (1940).

Rangaswamy Ayyangar G. N. u. T. Venkataramana Reddy: The inheritance of hairy styles (and barbed collumns of awns) in sorghum. Curr. Sci. 9, 282—283 (1940).
—, — Sorghum-awns of inconstant length and their inheritance. Curr. Sci. 9, 283—284 (1940).
—, — Sorghum, spikelet — awn relationships and inheritance. Madras Agric. J. 28, 306—308 (1940).
—, — Seedling-adult colour relationships and inheritance in sorghum. Indian J. Agric. Sci. 12, 341—363 (1942).
Romanov, V. A.: (New varieties of millet.) Bjulletenj Instituta Zernovogo Hozjajstva Jugo-Vostoka SSSR. (Bull. Inst. Grain Husb. S. E. USSR.) Saratov Nr. 2, 3—5 (1944). [Russisch.]
Romanov, V. L.: (Seed raising, breeding and cultivation of millet.) Naučnyj Otčet Inst. Zernovogo Hozjajstva Jugo-Vostoka SSSR. za 1941—42 gg. (Sci. Rep. Inst. Grain Husbandry South-Eastern USSR for 1941—42), 209 bis 214 (1944). [Russisch.]

Sankaram, A.: A plea for more millets. Indian Fmg. 7, 72—76 (1946).
Scheibe, A.: Die Hirsen im Hindukusch. Ein Beitrag zur Kenntnis von Kulturpflanzen in geographischer Rückzugsposition. (Ergebnisse der Deutschen Hindukusch-Expedition VI.) Z. Pflanzenz. 25, 392—436(1943).
Sieglinger, J. B.: A sorghum seed color chimera. J. Hered. 31, 363—364 (1940).
— Progress in sorghum improvement. Bull. Okla. Agric. Exp. Sta. Nr. B-295, 61—64 (1946).
Skarien, K.: Sorghum history: saga of the giants and dwarfs. Sth. Seedsman 10, Nr. 4 20, 64 (1947).
Stansel, R. H.: Humidity holds no terror for this kafir. Sth. Seedsman 5, Nr. 10; 12, 36 (1942).
Stephens, J. C.: Linkage of green-striped-2 in sorghum. J. Amer. Soc. Agron. 36, 469—470 (1944).
— A second factor for subcoat in sorghum seed. J. Amer. Soc. Agron. 38, 340—342 (1946).
— An allele for recessive red glume color in sorghum. J. Amer. Soc. Agron. 39, 784—790 (1947).
— u. J. R. Quinby: The D Rs P linkage group in sorghum. J. Agric. Res. 59, 725—730 (1939).
—, — The Ms_2 A V_{10} linkage group in sorghum. J. Agric. Res. 70, 209—218 (1945).
Swanson, A. F. u. H. H. Laude: Sorghums for Kansas. Bull. Kans. Agric. Exp. Sta. Nr. 304, 63 (1942).

Vaheeduddin, S.: The pathogenicity and genetics of some sorghum smuts. Tech. Bull. Minn. Agric. Exp. Sta. Nr. 154, 46 (1942).
Viégas, G. P.: A seleção do sorgo vassoura. (Selection of broom corn.) Bragantia, São Paulo 1, 177—232 (1941).
Viguier, P.: Les sorghos à grain et leur culture au Soudan français. Rev. Bot. Appl. 25, 163—222 (1945).
Vijayaraghavan, C. u. V. Panduranga Rao: The main cause of crop failure in the Black Soil Tract of the Bellary District. Madras Agric. J. 27, 271—274 (1939).

Wang, C. D.: Physiologic specialization and the control of millet smut. Phytopathology 34, 1050—1055 (1944).
Werth, E.: Zur Herkunft der Kultur der tropischen Hirsen in Afrika. Zugleich eine Entgegnung auf die Ausführungen von W. Lindenbein. (Zur Geschichte und Geographie der Kulturpflanzen und Haustiere XXI.) Bot. Jb. 73, 106—112 (1943).

Yu, T. F.: Inheritance of kernel smut resistance in millet crosses. Science Rec., Chungking 1, Nr. 1 u. 2, 248 bis 250 (1942).
— Reaction of improved millet varieties to infection with downy mildew (Sclerospora) graminicola Schroet). Chinese J. Sci. Agric. 1, 199—203 (1944).

Zakladnyj, G. A.: (Results of work with millet.) Selekcija i Semenovodstvo Nr. 5, 36—37 (1939). [Russisch.]

7. Reis.

Adair, C. R.: Inheritance in rice of reaction to Helminthosporium oryzae and Cercospora oryzae. Tech. Bull. U. S. Dep. Agric. Nr. 772, 19 (1941).
— u. J. W. Jones: Effect of environment on the characteristics of plants surviving in bulk hybrid populations of rice. J. Amer. Soc. Agron. 38, 708—716 (1946).
Akemine, M.: (Analysis of variance as applied to the intervarietal difference of sterility in paddy rice.) J. Sapporo Soc. Agric. For. 31, Nr. 149, 14 (1939).
Anonymus: Una nuova razza di riso: il ,,Settantuno". (A new rice variety: Settantuno.) Ital. Agric. 77, 262 (1940).
— Blast-resistant rice. Indian Fmg. 2, 33 (1941).
— Rice. Dept. Agric. Burma, Markets Sect. Surv. Nr. 9, 112 (1941).
— I risi perennati (Oryza perennis e O. latifolia) della Republica Dominicana. (The perennial rices (O. perennis and O. latifolia) of the Dominican Republic.) Atti Ist. Bot. e Lab. Crittogamico, Univ. Pavia 7, Ser. 5, 7—17 (1946).
Ayyangar, C. R. S., N. Parthasarathy u. K. Ramaswami: Inter-specific hybridization in Oryza. Proc. 28th Indina Sci. Congr. Benares Part III: Sect. Agric. abst. 37, 257 (1941).

Backer, C. A.: The wild species of Oryza in the Malay archipelago. Blumea, Leiden, Suppl. III, 45—55 (1946).
Baldacci, E.: Ricerche ed esperienze sulle malattie del riso. IV. Alcune manifestazioni di sterilità della pannocchia. Researches and experiments on rice diseases. IV. Some manifestations of panicle sterility.) Atti Ist. Bot. e Lab. Crittogamico, Univ. Pavia 5, Ser. 5, 249 bis 274 (1945).
Beachell, H. M.: Effect of photoperiod on rice varieties grown in the field. J. Agric. Res. 66, 325—340 (1943).
— u. J. W. Jones: Tetraploids induced in rice by temperature and colchicine treatments. J. Amer. Soc. Agron. 37, 165—175 (1945).
Berwick, E. J. H.: Pure strains of padi in Krian. Malay. Agric. J. 28, 429—435 (1940).
Birkinshaw, F.: A review of field experiments on padi in Malaya. Malay Agric. J. 28, 507—516 (1940).
— A review of field experiments on padi in Malaya. Malay. Agric. J. 29, 3—43 (1941).
Bobone, M.: Contribuição para o estudo e classificação botánica das variedades de ,,Oryza sativa" L. (Contribution to the study and botanical classification of the varieties of O. sativa L.). Rev. Agron. Lisboa 27, 272 bis 285 (1939).
Both, M. P.: Rijstselectie en bodemvruchtbaarheid. (Rice selection and soil fertility.) Landbouw 17, 735 bis 747 (1941).
Bragadin, E. A.: Características agrícolas de las variedades de arroz cultivadas en la República Argentina. (Agricultural characteristics of the varieties of rice cultivated in the Argentine Republic.) ,,Granos" Semilla Selecta, B. Aires 6, 28—29 (1942).
Bravo, A. F.: El cultivo del arroz en el noroeste de la provincia de Corrientes. (The cultivation of rice in the north-east of the province of Corrientes.) ,,Granos" Semilla Selecta, B. Aires 10, Nr. 10, 11 u. 12, 26—32 (1946).
Brown, W. V.: Cytological studies in the Oryzeae and Zizaneae. J. Elisha Mitchell Sci. Soc. 57, 201—202 (1941). (Abstr.)
Burgos, J. J.: La distribución de los pigmentos antociánicos en el arroz y su comportamiento hereditario. (The distribution of the anthocyanin pigments in rice and their hereditary behaviour.) Rev. Fac. Agron. 25, 175—214 (1943).

Cabal Martinez, E.: Estudio comparativo de los caracteres biométricos, agronómicos e industriales de noventa variedades de arroz. (Comparative study of the biometric, agronomic and industrial characters of ninety varieties of rice.) Rev. Fac. Nac. Agron., Colombia 5, 263—427 (1944).

Cañete, L.: Instrucciones para las labores de cultivos en los arrozales. (Instructions for work of cultivation in the rice fields.) Rev. Minist. Agric. Cuba 28, 41—47, 50—52 (1945).

Chakrabarti, S.: What's doing in All-India. Andrew Sail: prolific winter paddy. Indian Fmg. 4, 318 (1943).

Chadam, G. V.: An anatomical study of the shedding and non-shedding characters in the genus Oryza. J. Indian Bot. Soc. 21, 339—349 (1942).

Chambliss, C. E.: The botany and history of Zizania aquatica L. („wild rice"). Rep. Smithson. Instn. 1940, 369—382 (1941).

Chao, Lien-Fang u. Hsu, Kuan-Jen: A study of the time of heading in Oryza sativa L. Indian J. Genet. Pl. Breed. 4, 69—74 (1944).

Chiappelli, R.: La varietà di riso „Stirpe 136" (The rice variety „Strain 136"). Risicoltura 29, 346—350 (1939).

Chilton, S. J. P. u. E. C. Tullis: A new race of Cercospora oryzae on rice. Phytopathology 36, 950—952 (1946).

Chow, J. K. u. W. C. Li: (A study of the methods of artificial hybridization of rice.) Kwangsi Agric. 1, 3—8 (1940).

Ciferra, R.: Relazione sull' attività del R. Laboratorio Crittogamico e del R. Osservatorio di Fitopatologia di Pavia durante l'anno 1942. (Report on the activities of the Royal Cryptogamic Laboratory and the Royal Phytopathological Observatory at Pavia for the year 1942.) Atti Ist. Bot. e Lab. Crittogamico, Univ. Pavia 1, Ser. 5, 1—83 (1943).

Dash, J. S.: Report of the Department of Agriculture' British Guiana for the year 1943. Pp. 12 (1944).

Dass, C. M.: Notes on padi varieties grown in Fiji. Agric. J. Fiji 13, 94—95 (1942).

Dave, B. B.: The wild rice problem in the Central Provinces and its solution. Indian J. Agric. Sci. 13, 46—53 (1943).

— Improved rice strains in the Central Provinces. Indian J. Agric. Sci. Bd. 13, 479—488 (1943).

Dillewijn, C. van: Een proef omtrent colchicinebehandeling van rijst. (An experiment on colchicine treatment of rice.) Landbouwk. Tijdschr., Wageningen 53, 544—545 (1941).

Dixit, P. D.: Some improved paddy varieties for dalua cultivation in Orissa. Proc. 27 th Indian Sci. Congr., Madras Pt. III, Sect. Agric. Abst. 24, 217 (1940).

— Some interesting features in a cross between a purple and a green coloured variety of paddy. Proc. 28th Indian Sci. Congr. Benares Part III: Sect. Agric.: abst. 39, 257 (1941).

— A dwarf mutation in winter paddy. Proc. 29 th Indian Sci. Congr. Baroda Part. III; Sect. Agric. Abstr. 41, 216 (1942).

Donald, D. A.: Further notes on rice varieties. Agric. J. Fiji 15, 43—45 (1944).

Douglas, W. A. u. J. W. Ingram: Rice-field insects. Circ. U. S. Dep. Agric., Pp. 32 (1942).

Fotidar, M. R.: Natural cross-pollination in paddy. Indian Fmg. 6, 15—16 (1945).

Fuke, Y.: (Improvements of lowland varieties.) Bot. and Zool. 7, 202—210 (1939).

Ganguli, P. M. u. J. L. Sen: Variability in Boro paddy. Proc. 28th Indian Sci. Congr. Benares Part III: Sect. Agric. abst. 35, 256 (1941).

Germek, E.: A hibridação no arroz. (Hybridization in rice.) J. Agron., S. Paulo 2, 329—342 (1939).

Germek, E.: Processo eficiente para produção de mudas para progénies de arroz. (An efficient method for producing seedlings of rice progenies.) Bragantia, São Paulo 2, 515—520 (1942).

— Métodos de cultura e variedades de arroz. (Methods of cultivation and varieties of rice.) Rev. Agric. Piracicaba 21, 264—270 (1946).

Grant, J. W.: Rice breeding in Burma. Indian Fmg. 1, 606—608 (1940).

Guščin, G. G.: Ensayos de classification botánica de los arroces cultivados (Oryza sativa L.). (An attempted botanical classification of the cultivated forms of rice, O. sativa L.) „Granos" Semilla Selecta, B. Aires 6, 23—53 (1942).

Hamid, S. u. J. A. Baker: Notes on an attempt to estimate threshing qualities in padi. Malay. Agric. J. 27, 269—270 (1939).

Hedayetullah, S. u. A. K. Chakravorty: A comparative study of the mechanical system in five species of the genus Oryza. J. Dep. Sci., Calcutta 1, 21—28 (1941).

— u. B. N. Ghosh: The effect of colchicine on rice. Curr. Sci. 15, 74—75 (1946).

—, S. Sen u. K. R. Nair: Statistical notes for agricultural workers. Nr. 26. Influence of dates of planting and spacings on some winter varieties of rice. Indian J. Agric. Sci. 14, 248—259 (1944).

Houssaye, D. A. De La: Independent assortment, interaction of factors and linkage studies in the F_2 of a rice cross. Proc. La Acad. Sci. 6, 52—59 (1942).

— A chlorophyll deficiency in rice. Proc. La Acad. Sci. 7, 27—34 (1943).

Hsu, K. J.: On sterility resulting from crossing different types of rice. Indian J. Genet. Pl. Breed. 5, 51—57 (1945).

— u. H.-J. Lu: A genetical study of botanical characters in Oryza sativa. Indian J. Genet. Pl. Breed. 3, 108 bis 114 (1943).

— T. S.: (An experiment on intervarietal competition in rice.) Kwangsi Agric. 1, 14—20 (1940).

Ishikawa, J.: (Studies on partially sterile rice plant. Proc. Crop. Sci. Soc. Japan 12, 25—30 (1940).

Jagoe, R. B.: Padi selection and varietal trials 1938 bis 1939. Malay. Agric. J. 27, 468—512 (1939).

Jenkins, J. M.: Biennial report of the Rice Experiment Station Crowley, Louisiana 1937—1938. La St. Univ., Agric. Mech. Coll., Agric. Exp. Sta. Pp. 32.

Jodon, N. E.: Advances toward new and improved rice varieties. Bienn. Rep. Rice Exp. Sta., Crowley, La, 1939—1940, 20—24 (1940).

— Inheritance and linkage relationships of a chlorophyll mutation in rice. J. Amer. Soc. Agron. 32, 342—346 (1940).

— The inheritance of flower fragrance and other characters in rice. J. Amer. Soc. Agron. 36, 844—848 (1944).

— u. H. M. Beachell; Rice dwarf mutations and their inheritance. J. Hered. 34, 155—160 (1943).

— u. S. J. P. Chilton: Some characters inherited independently of reaction to physiologic races of Cercospora oryzae in rice. J. Amer. Soc. Agron. 38, 864—872 (1946).

—, T. C. Ryker u. S. J. P. Chiltou: Inheritance of reaction to phystologic races of Cercospora oryzae in rice. J. Amer. Soc. Agron. 36, 497—507 (1944).

Jones, J. W. u. A. E. Longley: Sterility and aberrant chromosome numbers in Caloro and other varieties of rice J. Agric. Res. Bd. 62, 381—399 (1941).

— u. M. N. Pope: Adventitious roots on panicles of rice. J. Hered. 33, 55—58 (1942).

Juliano, J. B.: Progress in rice research in the Philippines. Philipp. J. Agric. 12, 125—196 (1941).

Kadam, B. S.: Genic analysis of rice. II. Chlorophyll deficiencies. Indian J. Genet. Pl. Breed. 1, 13—27 (1941).
— Rice in Bombay. Indian Fmg. 6, 51—54 (1945).
—, M. V. Gadkari u. G. G. Patil: A long-glumed mutation in rice. Curr. Sci. 10, 331—333 (1941).
— u. K. Ramiah: Symbolization of genes in rice. Indian J. Genet. Pl. Breed. 3, 7—27 (1943).
—, — Bibliography on genetics of rice. Indian J. Genet. Pl. Breed. 3, 125—132 (1943).
Kagawa, F.: Studies on the inheritance of a type of large-grained, partially sterile rice plant. Jap. J. Bot. 10, 1—33 (1939).
Kawamura, E.: (Reaction of certain species of the genus Oryza to the infection of Piricularia Oryzae.) Bult. Sci. Fak. Terk. Kjusu Univ. 9, 157—166 (1940).
Kik, M. C. u. F. B. van Landingham: Riboflavin in products of commercial rice milling and thiamin and riboflavin in rice varieties. Cereal Chem. 20, 563—569 (1943).
Kiričenko, K. S., A. P. Džulaj u. I. S. Kosenko: (The cultivation of rice in the province of Krasnodar.) Kraevoe Knigoizdateljstvo (Provincial Publishing House) Krasnodar, Pp. 60 (1939). [Russisch.]
Kondo, M. u. Y. Kasahara: (On the discrimination of rice varieties by means of colouring reaction of hulled kernels to phenol-fuchsine.). Proc. Crop. Sci. Soc. Japan 12, 122—128 (1940).
—, — (Variety identification of hulled rice kernels by means of the potassium iodide test.). Proc. Crop. Sci. Soc. Japan 12, 333—338 (1941).
—, — (Variety identification of rice kernels by means of alcali-test.) Proc. Crop Sci. Soc. Japan 12, 325—332 (1941).
Kosaka, H. u. D. Yasukawa: (Studies in the response to unfavourable natural conditions of certain varieties of rice plant. 2. Comparison of growth of the plant under the sunlight of various intensities.) Proc. Crop Sci. Soc. Japan 10, 394—408 (1939).
Kuang, H. H., Y. H. Chang u. D. S. Tu: Studies on the variation of polyhusks in cultivated rice. Acta Brevia Sinensia Nr. 9, 15 (1945). (Abstr.) [Mimeographed.]
—, —, — Studies on the variation of polyhusks in cultivated rice (Oryza sativa L.). J. Genet. 47, 260—270 (1946).
—, T. M. Feng u. D. S. Tu: Studies on the percentage of seed setting in varietal crosses in rice. Acta Brevia Sinensia Nr. 9, 17 (1945). (Abstr.) [Mimeographed.]
—, D. S. Tu u. Y. H. Chang: Genetical studies on the polyhusks in cultivated rice (Oryza sativa L.) Chinese J. Sci. Agric. 1, 125 (1943).
—, —, — Genetical studies on the polyhusks in cultivated rice (Oryza sativa L.). Natural Sci. Soc. China Nr. 5, 15—16 (1943). [Mimeographed.]
—, —, — Tung Pu Loa, a natural hybrid in rice. Chinese J. Sci. Agric. 1, 182 (1944).
—, —, — Tung Pu Loa, a natural hybrid of rice. Acta Brevia Sinensia Nr. 9, 16 (1945). (Abstr.) [Mimeographed.]
—, —, — Linkage studies of awn in cultivated rice (Oryza sativa L.). J. Genet. 47, 249—259 (1946).
Kuilman, L. W.: Wortelstudies aan tropische landbouw-gewassen. 1. Wortelontwikkeling en vruchtbaarheid. (Root studies of tropical agricultural crops. I. Root development and fertility.) Landbouw. 17, 673—690 (1941).
— Wortelstudies aan tropische landbouw-gewassen. 1. Wortelontwikkeling en vruchtbaarheid. (Root studies on tropical agricultural crops. 1. Root development and fertility.) Meded. Alg. Proefst. Landb., Buitenz. Nr. 48, 18 (1941).

Lin, Chen Yao: (Study on the blooming habit of early, late and medium varieties of rice.) Fukien Agric. J. 3, 1—24 (1941).
—, C. Y. u. H. K. Chen: (Effect of photoperiodic treatment on the growth of rice.) Fukien Agric. J. 4, 93 bis 123 (1942).
—, Cheng Yao u. Shih Kai Chen: (Length of day and the growth of rice.) Fukien Agric. J. 3, 162—167, 1942: 4, 93—123 (1941/1942).
Liu, T. T.: The nature of hybrid sterility between „Oryza japonica" and „Oryza indica". Acta Brevia Sinensi, Nr. 8, 17 (1944). [Mimeographed.]
Lo, T.-Y. u. C.-H. Wu: The buffer action of the Chinese ricemeal extracts. Chinese J. Sci. Agric. 1, 195—198 (1944).

Majima, I.: (Observations on some characters of tetraploid rice plants.) Jap. J. Genet. 16, 190—191 (1940).
Martin, A. L.: Rice investigations. Rice J., Texas 55, 161—167 (1940).
Meulen, J. G. J. van der: Over de natuurlijke kruisbestuiving bij rijst en resultaten van een onderzoek daarover op Java. (Natural crossing in rice and results of experiments on this subject made in Java.) Landbouw 15, 649—732, Also Meded. Alg. Proefst. Landb. Buitzenzorg Nr. 38, 84 (1939).
— Massa-selectie bij sawahrijst. (Mass selection in sawah rice.) Landbouw 15, 733—735 (1939).
— De rijstselective in Nederlandsch Indië. (Rice selection in the Netherlands East Indies.) Landbouw 17, 943—1030 (1941).
Midusima, U. u. T. Yamada: (Some considerations on the relation between F_1 sterility and mesocotyl length among Japanese and foreign rice varieties, Oryza sativa L.). Jap. J. Genet. 15, 14—18 (1939).
Mitra, A. K. u. P. S. Gupta: Rice in the United Provinces. Indian Fmg. 3, 482—485 (1942).
—, — Production of more rice per acre in the United Provinces. Indian Fmg. 6, p. 398—402 (1945).
Moh, P. C.: (Deep water paddy of Kwangtung and Kwangsi.) Nung Pao 9, 390—391 (1944).
Morinaga, T.: Cytogenetical studies on Oryza sativa L. IV. The cytogenetics of F_1 hybrid of O. sativa L. and O. minuta Presl. Jap. J. Bot. 11, 1—16 (1940).
—, K. Nakajima u. T. Yumen: (The size and form of rice caryopses, and their mode of inheritance. [Preliminary note].) Jap. J. Genet. 15, 225—235 (1939).
— u. T. Tajiri: (The inheritance of polycaryoptic rice, with special reference to the germination structure of the lemma.) Jap. J. Genet. 17, 57—62 (1941).
Musset, R. u. L. Musset: Le riz dans de monde. Rev. Bot. Appl. 22, 151—180, 263—306; 24, 71—83, 221 bis 282 (1942).

Nagamatu, T.: (Cyto-ecological studies on rice, Oryza sativa L.). Jap. J. Genet. 16, 278—281 (1940).
Nakayama, K.: (On the heritable dwarf characters in rice, Oryza sativa L.). Jap. J. Genet. 15, 170—178 (1939).
— (The relation of chlorophyll content to the anatomical structure of leaves in three dwarf types of rice plant.) Jap. J. Genet. 15, 272—280 (1939).
— (Comparative studies on the panicle development in normal and dwarf types of rice plant.) Jap. J. Genet. 16, 139—148 (1940).
Nandi, H. K.: Rice in Assam. Indian Fmg. 5, 505—508 (1944).
— u. P. M. Ganguli: Inheritance of the flowering character in rice. Proc. 27th Indian Sci. Congr., Madras Pt. III, Sect. Agric. Abst. 33, 221 (1940).
—, — Inheritance of earliness in Surma Valley rices. Indian J. Agric. Sci. 11, 9—20 (1941).

Nandi, H. K. u. P. M. Ganguli: Inheritance of height of plants in Sail paddy. Proc. 28th Indian Sci. Congr. Benares Part III: Sect. Agric. :abst. 34, Pp. 256 (1941).

Nelson, M. u. C. R. Adair: Rice variety experiments in Arkansas. Bull. Ark. Agric. Exp. Sta. Nr. 403, 28 (1940).

Okura, E.: (Diploid F_1 hybrids produced from the cross between tetraploid and diploid race on Oryza sativa L. [Preliminary note].) Jap. J. Genet. 16, 228—233 (1940).

Opsomer, J.-E.: L'amélioration de la culture du riz au Congo Belge. (Improvement of rice growing in the Belgian Congo.) Riz. et Rizic. 13, 117—147 (1939).

— L'amélioration du riz à Yangambi (Congo Belge). The breeding of rice at Yangambi [Belgian Congo].) Agric. Élev. Congo Belge, 13, 51—54 (1939).

— Contribution à l'étude de l'hétérosis chez le riz. Publ. Inst. Nat. Agron. Congo Belge Nr. 24, 30 (1942).

Pal, B. P. u. S. Ramanujam: A new type of variegation in rice. Indian J. Agric. Sci. 11, 170—176 (1941).

—, — A new type of variegation in rice. Proc. 28th Indian Sci. Congr. Benares, Part. III: Sect. Agric. abst. 38, 257 (1941).

Pan, C. L. u. P. Kung: (Cultural factors affecting the yield of rice.) New Agric. J. Fukien 2, 320—329 (1942).

Parija, P., P. D. Dixit u. G. V. Challam: Some anatomical peculiarities in the stem of some flood resistant paddy selections. Proc. 27th Indian Sci. Congr., Madras Pt. III, Sect. Agric. Abst. 30, 220 (1940).

Patankar, Y. K.: Improvement of Ratnagiri rices by breeding. Part. 1. Bull. Dep. Agric. Bombay Nr. 181, 61, 1938 (1939).

Paul, W. R. C.: Paddy cultivation. Department of Civil Defence, Colombo, Pp. 90, 45 figs. tables (1945).

Piacco, R.: Varietà di riso ottenute per ibridazione artificiale alla Stazione di Risicoltura. (Varieties of rice obtained by artificial hybridization at the Stazione di Risicoltura.). Risicoltura 29, 11—20 (1939).

— Comportamento genetico di una mutazione dell' „Oryza sativa". (The genetical behaviour of a mutation of O. sativa.) Risicoltura 29, 134—137 (1939).

— Fenotipi ottenuti dall'incrocio Nano X Nanetto. (Phenotypes obtained from the cross Nano X Nanetto.) Risicoltura 29, 164—169 (1939).

— L'allettamento nel riso. (Lodging in rice.) Ital. Agric. 77, 565—572 (1940).

Poggendorff, W. H.: Rice culture in Australia. J. Aust. Inst. Agric. Sci. 5, 193—199 (1939).

Portères, R.: Sur le ségrégation géographique des gènes de l'Oryza glaberrima Steudel dans l'Ouest africain et sur les centres de culture de cette espèce. C. R. Acad. Sci. Paris 221, 152—153 (1945).

— A propos du polyphylétisme d'Oryza sativa L. et sur un groupe de variétés de riz cultivés à épillets glabres de cette espèce. Rev. Bot. Appl. 26, 54—57 (1946).

— Systématique intraspécifique chez Oryza glaberrima Steud. Rev. Bot. Appl. 26, 57—59 (1946).

— Les riz à encoches. (The notched rices.) Agron. Trop. Nr. 1/2, 69—72 (1946).

— Les riz flottants de l'espèce O. sativa L. et leurs possibilités d'exploitation en Afrique. Agron. Trop. Nos. 9 bis 10, 467—503 (1946).

Ramiah, K. u. K. Ramaswami: Floating habit in rice. Indian J. Agric. Sci. 11, 1—8 (1941).

—, — Hybrid vigour in rice (Oryza sativa L.). Indian J. Genet. Pl. Breed. 1, 4—12 (1941).

Rjabov, I. E.: (Rice breeding according to the theory of phasic development of plants.) Jarovizacija Nr. 3 (36), 48—55 (1941). [Russisch.]

Roddan, G. M.: A report of rice cultivation in the Gambia (with suggestions for development). Develop. Welfare Gambia: Chap. VII Agric., Sess. Pap. Nr. 2, 1—18 (1943).

Ru, S. K. u. T. L. Chu: (Effect of row distance, hill distance and number of plants per hill on the yield and some other characters of rice.) J. Agric. Ass. China, Nr. 177, 9—42 (1944).

Ryker, T. C.: Physiological specialization in Cercospora oryzae. Phytopathology 30, 21 (1940). (Abstr.)

— Linkage in rice of two resistant factors to Cercospora oryzae. Phytopathology 31, 19—20 (1941). (Abstr.)

— Physiologic specialization in Cercospora oryzae. Phytopathology 33, 70—74 (1943).

— New pathogenic races of Cercospora oryzae affecting rice. Phytopathology 37, 19—20 (1947). (Abstr.)

— u. S. J. P. Chilton: Inheritance and linkage of factors for resistance to two physiologic races of Cercospora oryzae in rice. J. Amer. Soc. Agron. 34, 836—840 (1942).

— u. N. E. Jodon: Inheritance of resistance to Cercospora oryzae in rice. Phytopathology 30, 1041—1047 (1940).

Sakai, K.: (Chromosome studies in Oryza sativa L. II. An unexpected asynapsis of the pollen mother-cell chromosomes.) Jap. J. Genet. 16, 193—202 (1940).

Saran, A. B.: On the viability of paddy seeds Oryza sativa. Curr. Sci. 14, 271 (1945).

— Studies on the effect of „short" and „long day" treatment on the growth period and the flowering dates of different paddy varieties. J. Indian Bot. Soc. 24, 153 bis 161 (1945).

Sen Gupta, J. C. u. N. K. Sen: Effect of vernalization and photoperiod on late sown Aman paddy-II. Sci. and Cult. 11, 273—274 (1945).

Sharma, S. V.: Paddy cultivation in Mysore. Bull. Dep. Agric. Mysore Nr. 20, 42 (1939).

Singh, R. D. u. S. Khan: 246 Palman-Suffaid — a fine, high-yielding variety of rice. Seas. Notes Punjab Agric. Deop. 18, Nr. 2, 15—18 (1939).

Soyza, D. J. de: Hill paddy cultivation in Ceylon. Trop. Agriculturist 100, 211—218 (1944).

Sreenivasan, A.: Nutritive value of the protein and mineral constituents of rice varieties. Cereal Chem. 19, 36—47 (1942).

Sukhatme, P. V.: Random sampling for estimating rice yield in Madras Province. Indian J. Agric. Sci. 15, 308 bis 318 (1945).

Swarup Paracer, C. u J. Chand Luthra: Further studies on the stem-rot disease of rice caused by Sclerotium oryzae Catt. in the Punjab. Indian J. Agric. Sci. 14, 44—48 (1944).

Terao, H., Y. Kondo, Y. Doi u. S. Izumi: (Physiological studies of the rice plant with special reference to the crop failure caused by the occurrence of unseasonable low temperature. (IV) Varietal differences of rice in regard to the anthesis and fertilization under low temperature.) Proc. Crop. Sci. Soc. Japan, 12, 203 bis 208 (1940).

— u. U. Midusima: Some considerations on the classification of Oryza sativa L. into two subspecies, so-called „Japonica" and „Indica". Jap. J. Bot. 10, 213 bis 258 (1939).

—, — (Some considerations on the classification of Oryza sativa L. into two subspecies, so-called Japonica and Indica.) Jap. J. Genet. 15, 10—13 (1939).

—, Y. Otani u. Y. Doi: (Physiological studies of the rice plant with special reference to the crop failure caused by the occurrence of unseasonable low temperature. (V) Anthesis and fertilization as affected by the low temperature treatment on heading.) Proc. Crop. Sci. Soc. Japan 12, 209—215 (1940).

—, —, — u: Z. Tyo: (Physiological studies of the rice plant with special reference to the crop failure caused by the occurence of unseasonable low temperature. III. On the impotency of pollen and pistills under low temperature.) Proc. Crop. Sci. Soc. Japan 12, 196—202 (1940).

Terao, H., Y. Ōtani, Y. Doi u. T. Huziwara: (Physiological studies of the rice plant with special reference to the crop failure caused by the occurence of unseasonable low temperature. (VI) Fertilization in the spikelets subjected to the low temperature treatment after flowering.) Proc. Crop. Sci. Soc. Japan 12, 216—227 (1940).

Toxopeus, H. J.: Aanteekeningen over de rijstcultuur in Fransch-Indochina. (Notes on rice cultivation in French Indo-China.) Landbouw 15, 291—303 (1939).

Vaidya, V. G.: Improvement in the yield of paddy. Indian Fmg. 5, 449—453 (1944).

Vasconcellos, J. de Carvalho E.: Formas cultivadas de arroz existentes em Portugal (seu estudo botânico e classificação. (The cultivated forms of rice existing in Portugal [Their botanical study and classification].) Com. Regulad. Comér. Arroz. Minist. Comér. Indústr., Lisboa, Pp. 107 (1939).

Venkata Subramanian, M. K.: Artificial germination of rice pollen. Curr. Sci. 12, 208—209 (1943).

Venkitasubba Iyer, C. S.: What's doing in All-India. Cochin. Indian Fmg. 6, 184—186 (1945).

Viégas, G. P., E. B. Germek u. H. S. Miranda: Contribuição para a melhoria da rizicultura no estado de São Paulo. (Contribution towards the improvement of rice cultivation in the state of São Paulo.) Bragantia, São Paulo 5, 187—196 (1945).

Viguier, P.: Note sur le problème de l'intensification de l'intensification de la riziculture dans le bassin du Niger. Agron. Trop. Nr. 7—8, 375—387 (1946).

Villalba, O. P.: El cultivo del arroz en Cuba. (The cultivation of rice in Cuba.) Rev. Minst. Agric. Cuba 29, 64—71 (1946).

Wickizer, V. D. u. M. K. Bennett: The rice economy of Monsoon Asia. Food Research Institute, Stanford University, California (1941).

Wickramasekera, G. V.: A multiple seeded variety of rice (Oryza sativa L.). Trop. Agriculturist 92, 279—280 (1939).

Williams, V. R., W. C. Knox u. E. A. Fieger: A study of some of the vitamin B-complex factors in rice and its milled products. Cereal Chem. 20, 560—563 (1943).

Wulff, A.: De rijstcultuur in Malaka. (Rice cultivation in Malaya.) Landbouw 15, 101—135 (1939).

Yamaguti, Y.: (The genetics of rice.) Jap. J. Genet. 15, 1—9 (1939).

Yasui, K.: Diploid-bud formation in a haploid Oryza with some remarks on the behaviour of nucleolus in mitosis. Cytologia, Tokyo, 11, 515—525 (1941).

Yoshi, H.: (Studies on the nature of rice blast resistance. I. The effect of silicic acid to the resistance. II. The effect of combined use of silicic acid and nitrogenous manure to the toughness of the leaf blade of rice and its resistance to rice blast. III. Relation between rice blast resistance and some physical and chemical properties of the different portions of the leaf blade of rice.) Bult. Sci. Fak. Terk. Kyušu Univ. 9, 277—291, 292 bis 296, 297—307 (1941).

b) Futtergräser.

Aamodt, O. S.: Report of the First Southern Pasture and Forage Crop Improvement Conference July 23 and 24, 1940. Tifton, Georgia, Pp. 20. [Mimeographed.]

Åkerberg, E.: Apomictic and sexual seed formation in Poa pratensis. Hereditas, Lund 25, 359—370 (1939).

— Cytogenetic studies in Poa pratensis and its hybrid with Poa alpina. Hereditas, Lund 28, 1—126 (1942).

— Further studies of the embryo and endosperm development in Poa pratensis. Hereditas, Lund 29, 199—201 (1943).

Albrecht, H. E.: Strain differences in tolerance to 2,4-D. in creeping bent grasses. J. Amer. Soc. Agron. 39, 163—165 (1947).

Andersen, S.: Poa supina Schrad. paavist i Danmark (Bornholm). (P. supina Schrad. discovered in Denmark [Bornholm].) Bot. Tidsskr. 46, 269—271 (1944).

Andrés, J. M.: Número de cromosomas en las especies del género Hordeum espontáneas en los alrededores de Buenos Aires. (Chromosome numbers in the wild species of the genus Hordeum in the vicinity of Buenos Aires.) Rev. Fac. Agron. B. Aires 9, 100—108 (1941).

— Número de cromosomas en las especies del género Hordeum espontáneas en los alrededores de Buenos Aires. (Chromosome numbers in the wild species of the genus Hordeum in the vicinity of Buenos Aires.) Inst. Genét., Fac. Agron. Vet. Univ. B. Aires 2, 29—37 (1941).

Anonymus: Cocksfoot seed production. Agric. Bull. Canterburry Chamb. Comm., Nr. 142, 4 (1941).

— A new Bermuda. Ext. Fm-News 28, 3 (1943).

— Odlingsvärdet i olika landsdelar av våra viktigaste vallväxtstammar.-Korta översikter. (Value in different parts of the country of our most important strains of herbage plants-Short surveys.) Sverig. Utsädesfören. Tidskr. 53, 13—31 (1943).

— Advances in grassland husbandry and fodder production. Aberystwyth, Nr. 32, 108, tables. figs. (1944).

— Plant industry. 2. Winter wheat planted in the spring. Agriculture, Moscow, Nr. 11, 2—3 (1945). [Mimeographed.]

— Research and the farmer. VI. Seeds. Mon. Rep. Minist. Agric. N. Ire 20, 162—164 (1945).

— Short-rotation ryegrass. Its breeding and characteristics. N. Z. J. Agric. 71, 465—470 (1945).

— Improving pastures and grasslands for the Northeastern States at the U. S. Regional Pasture Research Laboratory. Misc. Publ. U. S. Dep. Agric., Nr. 590, 29 (1946).

— Advances in grassland husbandry and fodder production. Second Symposium. Imperial Bureau of Pastures and Field Crops, Aberystwyth, Bull. 38: 6 s. Od., Pp. 83, 6 figs. tables. maps. (1947).

— Phalaris tuberosa. Commonw. Agric. 17, 62—66 (1947).

Atwood, S. S.: Cytogenetics and breeding of forage crops. Advances Genet., N. Y. 1, 1—67 (1947).

— u. H. A. MacDonald: Selecting plants of bromegrass for ability to grow at controlled high temperature. J. Amer. Soc. Agron 38, 824—832 (1946).

Bailey, R. Y.: The narrow-leaf Bahia grasses. Soil Conservation, Washington D. C. 11, 227—230 (1946).

Banfield, W. G. u. I. H. Stuckey: Extent of intercrossing among Agrostis species. Amer. J. Bot. Suppl. 29, 2s (1942). (Abstr.)

Belval, H. u. A. de Cugnac: Recherches phylétiques sur le genre Bromus. XI. Sur la valeur du genre Ceratochloa P. B. d'après la nature du glucide lévogyre du Brome de Schrader. Bull. Soc. Bot. Fr. 89, 4—5 (1942).

Benoist, R.: Une nouvelle espèce du genre Andropogon. Bull. Soc. Bot. Fr. 87, 340 (1940).

Berkner, F.: Züchtung und Auslese von Futterpflanzen. Forschungsdienst, Sonderh. 16, 425 (1942).

Blake, S. T.: Monographic studies in the Australian Andropogoneae, Part 1, including revisions of the genera Bothriochloa, Capillipedium, Chrysopogon, Vetiveria and Spathia. Univ. Qd Pap., Dep. Biol. 2, Nr. 3, 62 (1944).

— Two new grasses from New Guinea. Blumea, Leiden Suppl. III, Pp. 56—62 (1946).

Blanchard, R. A.: Insect resistance in forage plants. J. Amer. Soc. Agron. 35, 716—724 (1943).

Bor, N. L.: A new species of Dactyloctenium from India. Blumea, Leiden, Suppl. III, Pp. 44 (1946).

Boyle, W. S.: Cytological evidence for the taxonomic position of Schizachne purpurascens. Madroño **7**, 129—130 (1944).
— A cyto-taxonomic study of the North American species of Melica. Madroño **8**, 1—26 (1945).
Brittingham, W. H.: An artificial hybrid between two species of bluegrass, Canada Bluegrass (Poa compressa L.) and Kentucky Bluegrass (P. pratensis L.). J. Hered. **32**, 57—63 (1941).
— An artificially produced hybrid between Poa compressa L. and P. pratensis L. Genetics **26**, 141—142 (1941). (Abstr.)
— The nature and extent of variation in Kentucky bluegrass as criteria for apomictic seed formation. Genetics **27**, 134—135 (1942). (Abstr.)
— Type of seed formation as indicated by the nature and extent of variation in Kentucky bluegrass, and its practical implications. J. Agric. Res. **67**, 225—264 (1943).
Brouwer, W.: Züchtung und Auslese von Luzerne, Rotklee, Gelbklee, Hornklee, Steinklee, Esparsette und Serradella. Forschungsdienst Sonderh. **16**, 422 (1942).
Brown, B. A. u. R. I. Munsell: An evaluation of Kentucky bluegrass. J. Amer. Soc. Agron. **37**, 259—267 (1945).
Brown, W. L.: Chromosome complements of five species of Poa with an analysis of variation in Poa pratensis. Amer. J. Bot. **26**, 717—723 (1939).
— The cytogenetics of Poa pratensis. Ann. Mo. Bot. Gdn. **28**, 493—515 (1941).
Brown, W. V.: A cytological study in the Paniceae. Amer. J. Bot. Supp. **33**, 818 (1946). (Abstr.)
Burton, G. W.: A cytological study of some species in the genus Paspalum. J. Agric. Res. **60**, 193—197 (1940).
— A cytological study of some species in the tribe Paniceae. Amer. J. Bot. **29**, 355—359 (1942).
— Tift Sudan, a utopia grass for Southeast. Sth. Seedsman **5**, Nr. 1; 7, 31, 35 (1942).
— Interspecific hybrids in the genus Paspalum. J. Hered. **34**, 15—23 (1943).
— Hybrids between Napier grass and cattail millet. J. Hered. **35**, 227—232 (1944).
— Dallis grass seed sources. J. Amer. Soc. Agron. **37**, 438—468 (1945).
— Much maligned Bermuda..... now a profit item. Sth. Seedsman **8**, 16, 34 (1945).
— Bahia grass types. J. Amer. Soc. Agron. **38**, 273 bis 281 (1946).
— What about Bahia? Sth. Seedsman **9**, 15, 59, 62 (1946).
Calvert, E. L. u. A. E. Muskett: Blind-seed disease of rye-grass (Phialea temulenta Prill. and Delacr.). Ann. Appl. Biol. **32**, 329—343 (1945).
Camus, A.: Sur quelques graminées. Bull. Soc. Bot. Fr. **87**, 82—84 (1940).
— Un nouvel hybride du genre Bromus. Bull. Soc. Bot. Fr. **1944, 91**, 79—80 (1944).
— Graminées nouvelles de Madagascar. Bull. Soc. Bot. Fr. **92**, 50—53 (1945).
Cariss, H. G.: The utilisation of cereals for green fodder. J. Agric. W. Aust. **22**, 154—162 (1945).
Chamberlain, D. W. u. J. L. Allison: A brown leaf spot on Bromus inermis Leyss. caused by Pyrenophora bromi (Died). Drechsler. Phytopathology **34**, 997 bis 998 (1944). (Abstr.)
Chandrasekaran, S. N. u. D. D. Sundararaj: A very promising drought resistant fodder grass for south India. Madras Agric. J. **34**, 21—24 (1946).
Chase, A.: Popuan grasses collected by L. J. Brass, III. J. Arnold Arbor **24**, 77—89 (1943).
Cheng Chung-Fu: Self-fertility studies in three species of commercial grasses. J. Amer. Soc. Agron. **38**, 873 bis 881 (1946).

Chiao, C. Y.: Kiangsu grasses. Contr. Herbarium, Coll. Agric. For., Univ. Nanking, Pp. 115 (1943).
Chippindall, L.: Contributions to the grass flora of Africa. Blumea, Leiden, Suppl. III, Pp. 25—41 (1946).
— The common names of grasses in South Africa. Bull. Dep. Agric. S. Afr. 1946, Nr. 265, 91 (1947).
Church, G. L.: Cytotaxonomic studies in the Gramineae Spartina, Andropogon and Panicum. Amer. J. Bot. **27**, 263—271 (1940).
— A cytological and morphological approach to the species problem in Glyceria. Amer. J. Bot. Suppl. **29**, 5s. (1942) (Abstr.).
Claridge, J. H.: „HI ryegrass"-short rotation ryegrass renamed. N. Z. J. Agric. **67**, 286 (1943).
Clark, J. W.: The effect of some environmental influences in bulk hybridization of grass. J. Amer. Soc. Agron. **36**, 132—140 (1944).
Clark, M.: Lo, a new buffalo. Sth. Seedsman **8**, Nr. 10; 15, 50 (1945).
Clarke, S. E., J. A. Campbell u. W. Shevkenek: The identification of certain native and naturalized grasses by their vegetative characters. Tech. Bull. Dep. Agric. Dom. Canad., Nr. 50 (Publ. 762), 129 (1944).
Cook, C. W.: A study of the roots of Bromus inermis in relation to drought resistance. Ecology **24**, 169—182 (1943).
Corkill, L. u. R. E. Rose: Observations on susceptibility of perennial rye-grass to blind seed disease. N. Z. J. Sci. Technol. **27**, Sect. A, 14—18 (1945).
Cornelius, D. R.: The effect of source of little bluestem grass seed on growth, adaptation, and use in revegetation seedings. J. Agric. Res. **74**, 133—143 (1947).
Coukos, C. J.: Seed dormancy and germination in some native grasses. J. Amer. Soc. Agron. **36**, 337 bis 345 (1944).
Covas, G.: Número de cromosomas de algunas Gramíneas argentinas. (Chromosome numbers of some Argentine Gramineae.) Rev. Argent. Agron. **12**, 315 bis 317 (1945).
— u. M. Bocklet: Número de cromosomas de algunas Gramineae-Stipinae de la flora argentina. (Chromosome numbers of some Graminae-Stipinae of the Argentine flora.) Rev. Argent. Agron. **12**, 261—265 (1945).
Crombie, B.: The breeding of grasses and clovers. J. Dep. Agric., Éire **40**, 286—290 (1943).
Cugnac, A. de: Sur un cas de dominance inconstante chez un hybride intergénérique de Graminées. C. R. Soc. Biol. Paris **130**, 877—879 (1939).
— Un cas remarquable de pléiotropie dans une hybridation de Graminées. C. R. Soc. Biol. Paris **132**, 404 bis 406 (1939).
— Reconstititution experimentale d'une graminée éteinte par un croisement interspécifique. C. R. Acad. Sci. Paris **209**, 61—63 (1939).
— Conséquences génétiques et phylétiques du croisement de deux Graminées. C. R. Acad. Sci., Paris **209**, 696—698 (1939).
— Recherches phylétiques sur le genre Bromus. VIII. Une double reconstitution expérimentale, faisant revivre une variété éteinte, et ses conséquences systématiques. Bull. Soc. Bot. Fr. **86**, 409—419 (1939).
— Recherches phylétiques sur le genre Bromus. XIII. L'appendice de l'ovaire et du grain chez des Bromes et quelques genres voisins. A propos d'un dessin de l'„Histoire des Plantes" de Baillon. Bull. Soc. Bot. Fr. **92**, 216—222 (1945).
— u. H. Belval: Hybridation entre deux genres de Graminées caractérisés par des glucides différents: fructoholoside et fructoglucoholoside. Nature du glucide formé chez l'hybride. Bull. Soc. Chim. Biol., Paris **21**, 689—694 (1939).

Cugnac u. H. Belval: Nature du glucide d'un hybride obtenu entre deux genres de Graminées différant par leurs réserves glucidiques. C. R. Acad. Sci. Paris **208**, 377—379 (1939).

—, — Recherches phylétiques sur le genre Bromus. IX. Glucides et affinités des Bromes et des Fétuques. Bull. Soc. Bot. Fr. **88**, 402—410 (1941).

—, — Sur les affinités des genres Agrostis et Apera d'après leur constitution glucidique. Bull. Soc. Bot. Fr. **91**, 1—3 (1944).

— u. A. Camus: Un x Festulolium expérimental nouveau: Festulolium Colini-Lolium temulentum x Festuca pratensis. Bull. Soc. Bot. Fr. **91**, 16—19 (1944).

—, — Recherches phylétiques sur le genre Bromus. XII. Un hybride interspécifique nouveau: x Bromus Laagei hyb. nov. = Bromus tectorum x squarrosus. Bull. Soc. Bot. Fr. **91**, 172—174 (1944).

— u. M. Simonet: Recherches phylétiques sur le genre Bromus. X. Quelques nombres de chromosomes et leur signification phylétique et phylogénétique. Bull. Soc. Bot. Fr. **88**, 513—517 (1941).

—, — Les nombres de chromosomes de quelques espèces du genre Bromus (Graminées). C. R. Soc. Biol. Paris **135**, Nr. 9/10, 728—731 (1941).

Dahms, R. G.: Rice stinkbug as a pest of sorghums. J. Econ. Ent. **35**, 945—946 (1942).

Dickason, F. G.: The circumference-length ratio. A new diagnostic character for identifying bamboos. J. Indian Bot. Soc. **21**, 351—353 (1942).

Dillmann, A. C.: The beginning of crested wheatgrass in North America. J. Amer. Soc. Agron. **38**, 237 bis 250 (1946).

Dore, W. G. u. A. E. Roland: The Grasses of Nova Scotia. Proc. N. S. Inst. Sci. 1941—1942, **20**, 177 bis 288 (1942).

Edwards, D. C.: The Nzoia type of Rhodes grass (Chloris gayana) for temporary leys. E. Afr. Agric. J. **9**, 62—68 (1943).

— A new fodder crop: Perennial Kavirondo sorghum. J. Jamaica Agric. Soc. **48**, 286—291 (1944).

Engelbert, V.: Reproduction in some Poa species. Canad. J. Res. **18**, Sect. C, 518—521 (1940).

— The development of twin embryo sacs, embryos, and endosperm in Poa Arctica R. Br. Canad. J. Res. **19**, Sect. C, 135—144 (1941).

Evans, M. u. C. P. Wilsie: Flowering of bromegrass, Bromus inermis, in the greenhouse as influenced by length of day, temperature, and level of fertility. J. Amer. Soc. Agron. **38**, 923—932 (1946).

Faria, A. Lima de: Disturbance in microspore cytology of Anthoxanthum. Hereditas, Lund **33**, 539—551 (1947).

Fitzpatrick, J. M.: A cytological and ecological study of some British species of Glyceria. New Phytol. **45**, 137—144 (1946).

Franzke, C. J.: Rancher, a low hydrocyanic acid forage sorghum. Circ. S. Dak. Agric. Exp. Sta., Nr. 57, 8 (1945).

— Ree wheatgrass. Its culture and use. Circ. S. Dak. Agric. Exp. Sta., Nr. 58, 8 (1945).

—, L. F. Puhr u. A. N. Hume: A study of sorghum with reference to the content of HCN. Tech. Bull. S. Dak. St. Coll., Nr. 1, 51 (1939).

French, M. H.: The compositions of different types of Star grass. E. Afr. Agric. J. **11**, 100—103 (1945).

Fults, J. L.: Somatic chromosome complements in Bouteloua. Amer. J. Bot. **29**, 45—55 (1942).

Garber, E.: The rate of root elongation in diploid and tetraploid Sudan grass and rye. Amer. Nat. **77**, 190 bis 192 (1943).

Garber, R. J. u. S. S. Atwood: Natural crossing in Sudan grass. J. Amer. Soc. Agron. **37**, 365—369 (1945).

— u. S. J. P. Chilton: The occurrence and inheritance of certain leaf „spots" in sudan grass. J. Amer. Soc. Agron. **34**, 597—606 (1942).

Goossens, A. P. u. M. C. Papendorf: A revision of the genus Agrostis Linn. in South Africa. S. Afr. J. Sci. **41**, 172—185 (1945).

Gorman, L. W. u. J. P. Lambert: The Grasslands Division substation at Lincoln. N. Z. J. Sci. Tech. **25**, Sect. A, 73—80 (1943).

Gould, F. W.: Notes on the genus Elymus. Madroño, **8**, 42—47 (1945).

Gregor, J. W.: Phleum alpinum L. emend. Gaud. and P. commutatum Gaud. Trans. Bot. Soc. Edinb. **34**, 368—370 (1946).

Grøntved, J.: Agropyrum junceum (L.) Beauv. x Elymus arenarius L. (A. junceum [L.] Beauv. x. E. arenarius L.) Bot. Tidsskr. **46**, 407—411 (1946).

Grontved, P.: Erfaringer og Forsog med Renholdelse af Froafgroder. (Experience and experiments on the maintenance of purity in crops grown for seed.) Nord. JordbrForskn., Nr. 5/6, 197—202 (1943).

Gustafsson, Å.: The relationship of Calamagrostis neglecta and lapponica in Siberia, as interpreted by Nilsson-Ehle. Hereditas, Lund **32**, 550—551 (1946). (Abstr.).

Hagerup, O.: Studies on the significance of polyploidy. III. Deschampsia and Aira. Hereditas, Lund **25**, 185—192 (1939).

Håkansson, A.: Die Entwicklung des Embryosacks und die Befruchtung bei Poa alpina. Hereditas, Lund **29**, 25—61 (1943).

— Ergänzende Beiträge zur Embryologie von Poa alpina. Bot. Notiser, Pp. 299—311 (1944).

Hardison, J. R.: Specialization of pathogenicity in Erysiphe graminis on wild and cultivated grasses. Phytopathology **34**, 1—20 (1944).

— A leaf spot of tall fescue caused by a new species of Cercospora. Mycologia **37**, 492—494 (1945).

— Observations on grass diseases in Kentucky, September, 1942 to September, 1944 and a preliminary check list. Plant Dis. Reporter **29**, 76—85 (1945). [Mimeographed.]

— Specialization of pathogenicity in Erysiphe graminis on Poa and its relation to bluegrass improvement. Phytopathology **35**, 62—71 (1945).

— Specialization in Erysiphe graminis for pathogenicity on wild and cultivated grasses outside the tribe Hordeae. Phytopathology **35**, 394—405 (1945).

Harlan, J. R.: The breeding structure of Bromus carinatus Hook. & Arn. Amer. J. Bot. Suppl. **28**, 3s (1941). (Abstr.)

— Cleistogamy and chasmogamy in Bromus carinatus Hook and Arn. Amer. J. Bot. **32**, 66—72 (1945).

— Natural breeding structure in the Bromus carinatus complex as determined by population analyses. Amer. J. Bot. **32**, 142—148 (1945).

— Grasses of the plains. Sth. Seedsman **8**, Nr. 12, 28—29 (1945).

Hartung, M. E.: Chromosome numbers in Poa, Agropyron, and Elymus. Amer. J. Bot. **33**, 516—531 (1946).

Hussainy, S. A.: Vegetative sports in the bamboo (Bambusa arundinacea Willd.) Curr. Sci. **9**, 132 (1940).

Hayes, H. K. u. A. R. Schmid: Selection in self-pollinated lines of Bromus inermis Leyss., Festuca elatior L. and Dactylis glomerata L. J. Amer. Soc. Agron. **35**, 934—943 (1943).

Henrard, J. T.: Notes on the nomenclature of some grasses, II. Blumea, Leiden **4**, 496—538 (1941).

— On a new species of Paspalum from the island of Bonaire. Blumea, Leiden **5**, 324—327 (1943).

Hill, H. D. u. W. M. Myers: Isolation of diploid and tetraploid clones from mixoploid plants of ryegrass (Lolium perenne L.). Produced by treatment of germinating seeds with colchicine. J. Hered. **35**, 359—361 (1944),

—, — A schedule including cold treatment to facilitate somatic chromosome counts in certain forage grasses. Stain Tech. **20**: 89—92 (1945).

Hills, K. L.: Material for testing isolation efficiency in perennial rye grass. J. Aust. Inst. Agric. Sci. **8**, 170 (1942).

Hogg, P. G.: Breeding sorghums for low cyanide content. Proc. 43rd Conv. Assoc. Sth. Agric. Wrks, Memphis, Tenn, Pp. 36—37 (1942).

— u. H. L. Ahlgren: Enviromental, breeding, and inheritance studies of hydrocyanic acid in Sorghum vulgare var. Sudanense. J. Agric. Res. **67**, 195—210 (1943).

Hollowell, E. A.: Registration of varieties and strains of grasses, I. J. Amer. Soc. Agron. **37**, 653—654 (1945).

Hubbard, C. E.: Henrardia, a new genus of the Gramineae. Blumea, Leiden, Suppl. III, Pp. 10—21 (1946).

Hurcombe, R.: Chromosome studies in Cynodon S. Afr. J. Sci. **42**, 144—146 (1946).

Jansen, P.: Eragrostis Henrardii, nov. spec. Blumea, Leiden, Suppl. **3**, 42—43 (1946).

Jenkin, T. J.: Evolution in wild populations. Proc. 7th Int. Genet. Congr. Edinburgh 23—30 August, Pp. 167—168 1939 (1941).

— Aberystwyth strains of grasses and clovers. J. Minist. Agric. **50**, 343—349 (1943).

— The breeding of herbage plants in relation to grassland husbandry. J. Brit. Grassland Soc. **1**, 126—133 (1946).

— u. P. T. Thomas: Interspecific and intergeneric hybrids in herbage grasses III. Lolium loliaceum and Lolium rigidum. J. Genet. **37**, 255—286 (1939).

Jenkins, T. A.: K—R is a better grass. Sth. Seedsman **10**, Nr. 6, 18 (1947).

Jones, M. D. u. L. C. Newell: Pollination cycles and pollen dispersal in relation to grass improvement. Res. Bull. Neb. Agric. Exp. Sta., Nr. 148, 42 (1946).

Johnson, B. L.: Cyto-taxonomic studies in Oryzopsis. Bot. Gaz. **107**, 1—32 (1945).

— Natural hybrids between Oryzopsis hymenoides and several species of Stipa. Amer. J. Bot. **32**, 599—608 (1945).

— u. G. A. Rogler: A cyto-taxonomic study of an intergeneric hybrid between Oryzopsis hymenoides and Stipa viridula. Amer. J. Bot. **30**, 49—56 (1943).

Johnsson, H.: Cytological studies in the genus Alopecurus. Acta Univ. Lund **37**, Nr. 3, K. Fysiogr. Sällsk. Handl. **52**, Nr. 3, 3—43.

— Meiotic aberrations and sterility in Alopecurus myosuroides Huds. Hereditas, Lund **30**, 469—566 (1944).

Julander, O.: Drought resistance in range and pasture grasses. Plant Physiol. **20**, 573—599 (1945).

Julén, G.: Om möjligheterna att genom självbefruktning av timotej framställa förbättrade strammar. (On the possibilities that by the selfpollination of timothy improved strains may be produced.) Sverig. Utsädesfören. Tidskr. **52**, 258—282 (1942).

— Nyare försök med hundäxingstammar vid Sveriges Utsädesförening. (Recent experiments with cocksfoot strains by the Swedish Seed Association.) Sverig. Utsädesfören. Tidskr. **53**, 67—81 (1943).

— Redogörelse för de senaste årens försök med timotej. (Report on the experiments of the last few years with timothy.) Sverig. Utsädesfören. Tidskr. **53**, 336—340 (1943).

— Some problems arising in the breeding of herbage plants in Sweden. J. Brit. Graßland Soc. **2**, 57—62 (1947).

Karper, R. E.: Registration of improved sorghum varieties, III. J. Amer. Soc. Agron **34**, 280 (1942).

— u. J. R. Quinby: Breeding new varieties of Sudan grass. Progr. Rep. Tex. Agric. Exp. Sta., Nr. 707, 3 (1940). [Mimeographed.]

—, — Five new hegaris. Sth. Seedsman **9**, 16, 48 (1946).

Keller, W.: Seed production on grass culms detached prior to pollination. J. Amer. Soc. Agron. **35**, 617 bis 624 (1943).

— The bulk hybridization of smooth bromegrass. J. Hered. **35**, 49—56 (1944).

— An evaluation of kraft and parchment paper bags for the control of pollination in grasses. J. Amer. Soc. Agron. **37**, 902—909 (1945).

— Designs and technic for the adaptation of controlled competition to forage plant breeding. J. Amer. Soc. Agron **38**, 580—588 (1946).

Keng, Yi-Liu u. Pai-Chieh Kong: New bamboos from Szechuan Province, China. J. Wash. Acad. Sci. **36**, 76—86 (1946).

Kiellander, C. L.: Studies on apospory in Poa pratensis L. Svensk Bot. Tidskr. **35**, 321—332 (1941).

— A subhaploid Poa pratensis L. with 18 chromosomes and its progeny. Svensk Bot. Tidskr. **36**, 200—220 (1942).

— En ny Poa-hybrid, Poa palustris L. × Poa compressa L. (A new Poa hybrid, P. palustris L. × P. compressa L.). Bot. Notiser, Pp. 363—372 (1944).

Klages, K. H.: Idaho Amber sorgo. Circ. Univ. Idaho Agric. Exp. Sta. Nr. 97, 8 (1944).

Klinkowski, M.: Die Bedeutung der Iberischen Halbinsel für die deutsche Pflanzenzüchtung. (The importance of the Iberian peninsula for German plant-breeding.) Ernähr. Pfl. **35**, 9—13 (1939).

Kloos, A. W. (jr.), An adventitious new Deschampsia species. Blumea, Leiden Suppl. **3**, 22—24 (1946).

Knobloch, I. W.: Morphological variation and cytology of Bromus inermis. Bull. Torrey Bot. Cl. **70**, p. 467 bis 472 (1943).

Knowles, P. F.: Improving an annual brome grass, Bromus mollis L., for range purposes. J. Amer. Soc. Agron. **35**, 584—594 (1943).

— Interspecific hybridizations of Bromus. Genetics **29**, p. 128—140 (1944).

—, R. P. u. W. H. Horner: Methods of selfing and crossing crested wheat grass, Agropyron cristatum (L.) Beauv. Sci. Agric. **23**, 598—611 (1943).

Kramer, H. H. Morphologic and agronomic variation in Poa pratensis L. in relation to chromosome numbers. J. Amer. Soc. Agron. **39**, 181—191 (1947).

Kreitlow, K. W. u. W. M. Myers: Reactions to crown rust in Festuca elatior and F. elatior var. arundinacea. Phytopathology **36**, 404 (1946). (Abstr.)

—, —, Resistance to crown rust in Festuca elatior and F. elatior var. arundinacea. Phytopathology **37**, 59—63 (1947).

Krishnaswamy, N.: Untersuchungen zur Cytologie und Systematik der Gramineen. Beih. Bot. Zbl. **60**, 1—56 (1940).

Law, A. G. u. K. L. Anderson: The effect of selection and inbreeding on the growth of big bluestem (Andropogon furcatus, Muhl.). J. Amer. Soc. Agron. **32**, 931 bis 943 (1940).

Leitzke, B.: Über Einzelpflanzenuntersuchungen im Rahmen der Gräserzüchtung. Z. Pflanzenz. **25**, 38—60 (1943).

Levan, A.: Syncyte formation in the pollen mothercells of haploid Phleum pratense. Hereditas, Lund, **27**, 243 bis 252 (1941).

Lindberg, H.: Iter cyprium. Contributio ad cognitionem florae insulae Cypri. (Cyprian journey. A contribution to the knowledge of the flora of the Island of Cyprus.) Acta Soc. Sci. Fenn., Helsingfors, **2**, Nr. 7, 37 (1946).

Litardière, M. R. de: Recherches sur les Poa annua subsp. exilis et subsp. typica. Relations taxonomiques, chorologiques et caryologiques. Rev. Cytol. Cytophysiol. Vég. **3**, 134—141 (1939).
— Sur les caractères chromosomiques et la systématique des Poa du groupe du Poa annua L. Rev. Cytol. et Cytophysiol. végét. **4**, 81—85 (1939).
Love, R. M.: A haploid Stipa cernua. Rec. Genet. Soc. America Nr. 12, 51 (1943). (Abstr.)
— Interspecific hybridization in Stipa L. I. Natural hybrids. Amer. Nat. **80**, 189—192 (1946).
— Interspecific and intergeneric hybridization in forage crop improvement. J. Amer. Soc. Agron. **39**, 41—46 (1947).
Lund, A.: Andersogelser over Karotininholdet i Foderplanter. (Investigations on the carotene content in fodder plants.) Nord. Jordbr. Forskn. Nr. 7, 321—340 (1942).

McAlister, D. F.: Determination of soil drought resistance in grass seedlings. J. Amer. Soc. Agron. **36**, 324 bis 336 (1944).
McClure, F. A.: New bamboos, and some new records, from French Indo-China. J. Arnold Arbor. **23**, 93—102 (1942).
— Bamboo in Ecuador's highlands. Agric. Amer. **6**, 164 bis 167 (1946).
Matthews, A. C.: Observations on methods of increasing the germination of Panicum anceps Michx. and Paspalum notatum Flügge. J. Amer. Soc. Agron. **39**, 439 bis 442 (1947).
Miège, J.: Contribution à l'étude des Phalaridées. Bull. Soc. Hist. Nat. Afr. N. **30**, 223—245 (1939).
Moffett, A. A.: Note on the cytology of Rhodes grass. Rhod. Agric. J. **41**, 11—13 (1944).
Molostov, A. S., A. V. Popravko u. E. K. Slipčenko, (Plantbreeding and seed production.) Naučnyj Otčët Moršanskoj Gosuďar. Selekc. Stancii po Travam za 1941—1942 (Sci. Rep. Moršansk State Plant Breed. Sta. Herbage Crops 1941—1942) Ogiz, Seljhozgiz, Moskva 9—42 (1945).
Müntzing, A.: Further studies on apomixis and sexuality in Poa. Hereditas, Lund **26**, 115—190 (1940).
— Characteristics of two haploid twins in Dactylis glomerata. Hereditas, Lund, **29**, 134—140 (1943).
— Different chromosome numbers in root tips and pollen mother cells in a sexual strain of Poa alpina. Hereditas Lund **32**, 127—129 (1946). (Abstr.)
— u. R. Prakken: The mode of chromosome pairing in Phleum twins with 63 chromosomes and its cytogenetic consequences. Hereditas, Lund **26**, 463—501 (1940).
Murphy, R. P.: Methode of breeding crested wheatgrass, Agropyron cristatum (L.) Beauv. J. Amer. Soc. Agron. **34**, 553—565 (1942).
Myers, W. M.: Colchicine induced tetraploidy in perennial ryegrass Lolium perenne L. J. Hered. **30**, 499 bis 504 (1939).
— Tetrasomic inheritance in Dactylis glomerata. Genetics **25**, 126 (1940). (Abstr.)
— Genetical consequences of the chromosomal behavior in orchard grass, Dactylis glomerata L. J. Amer. Soc. Agron. **33**, 893—900 (1941).
— Meiotic behavior of Phleum pratense, Phleum subulatum, and their F₁ hybrid. J. Agric. Res. **63**, 649—659 (1941).
— Variations in chromosomal behaviour during meiosis among plants of Lolium perenne L. Cytologia **11**, 388 bis 406 (1941).
— Heritable variations in seed set under bag among plants of orchard grass, Dactylis glomerata L. J. Amer. Soc. Agron. **34**, 1042—1051 (1942).
— Analysis of variance and covariance of chromosomal association and behaviour during meiosis in clones of Dactylis glomerata. Bot. Gaz. **104**, 541—552 (1943).

Myers, W. M.: Second generation progeny tsets of the method of reproduction in Kentucky bluegrass, Poa pratensis L. J. Amer. Soc. Agron. **35**, 413—419 (1943).
— The effects of inbreeding upon meiotic irregularity in plants of Dactylis glomerata. Genetics **28**, 83—84 (1943). (Abstr.)
— Cytological and genetic analysis of chromosomal association and behavior during meiosis in hexaploid timothy (Phleum pratense). J. Agric. Res. **68**, 21—33 (1944).
— Cytological studies of a triploid perennial ryegrass and its progeny. J. Hered. **35**, 17—23 (1944).
— The randomness of chromosome distribution at anaphase I in autotriploid Lolium perenne L. Bull. Torrey Bot. Cl. **71**, 144—151 (1944).
— Meiosis in autotetraploid Lolium perenne in relation to chromosomal behavior in autopolyploids. Bot. Gaz. **106**, 304—316 (1945).
— Effects of cytoplasm and gene dosage on expression of male sterility in Dactylis glomerata. Genetics **31**, 225—226 (1946). (Abstr.)
— u. S. J. P. Chilton: Correlated studies of winterhardiness and rust reaction of parents and inbreed progenies of orchard grass and timothy. J. Amer. Soc. Agron. **33**, 215—220 (1941).
— u. H. D. Hill: Studies of chromosomal association and behaviour and occurrence of aneuploidy in autotetraploid grass species, orchard grass, tall oat grass, and crested wheat grass. Bot. Gaz. **102**, 236—255 (1940).
—, — The association and behavior of chromosomes in autotetraploid grasses. Genetics **25**, 129 (1940). (Abstr.)
—, — Variations in meiotic behavior among plants of the autotetraploid Dactylis glomerata. Genetics **26**, 162 (1941). (Abstr.)
—, — Variations in chromosomal association and behaviour during meiosis among plants from open-pollinated populations of Dactylis glomerata. Bot. Gaz. **104**, 171—177 (1942).
—, — Increased meiotic irregularity accompanying in Dactylis glomerata L. Genetics **28**, 383—397 (1943).
—, — Distribution and nature of polyploidy in Festuca elatior L. Bull. Torrey Bot. Cl. **74**, 99—111 (1947).

Nannfeldt, J. A.: On the polymorphy of Poa arctica R. Br., with special reference to its Scandinavian forms. Symb. Bot. Upsaliens. **4**, Nr. 4, 85 (1940).
Newell, L. C. u. F. D. Keim: Field performance of bromegrass strains from different regional seed sources. J. Amer. Soc. Agron. **35**, 420—434 (1943).
— u. H. M. Tysdal: Numbering and note-taking systems for use in the improvement of forage crops. J. Amer. Soc. Agron. **37**, 736—749 (1945).
Nielsen, E. L.: Grass studies. III. Additional somatic chromosome complements. Amer. J. Bot. **26**, 366—371 (1939).
— Analysis of variation in Panicum virgatum. J. Agric. Res. **69**, 327—353 (1944).
— Cytology and breeding behavior of selected plants of Poa pratensis. Bot. Gaz. **106**, 357—382 (1945).
— Breeding behavior and chromosome numbers in progenies from twin and triplet plants of Poa pratensis. Bot. Gaz. **108**, 26—40 (1946).
— The origin of multiple marcogametophytes in Poa pratensis. Bot. Gaz. **108**, 41—50 (1946).
Nilsson, F.: The hybrid Festuca arundinacea × F. pratensis and some of its derivatives. Bot. Notiser Nr. 1, p. 33—50 (1940).
Nilsson-Leissner, G.: A case increased vitality in sibpollinated later generations of selffertilised Dactylis glomerata strains. Hereditas (Lund) **28**, 222—224 (1942).
— On the possibilities of breeding new, improved strains of timothy by means of selfing. Hereditas, Lund, **28**, 500—502 (1942). (Abstr.)

Nilsson-Leissner, G.: Några synpunkter på förädlingen av betesväxter). (Some views on the improvement of pasture plants. Sverig. Utsädesfören. Tidskr. 55, 164—173 (1945).
— Professor Hernfrid Witte. Nord. JordbrForskn. Nr. 1 bis 2, 45—46 (1945).
Nissen, Ö.: A possible explanation of increased vigour in sib-pollinated later generations after self-fertilisation. Hereditas, Lund 32, 296 (1946). (Abstr.)
Nordenskiöld, H.: Cytological studies in triploid Phleum. Bot. Notiser Nr. 1, 12—32 (1941).
— Cyto-genetic studies in the genus Phleum. Acta Agriculturae Suecana, Stockholm, 1, 1—137 (1945).
Nygren, A.: The genesis of some Scandinavian species of Calamagrostis. Hereditas, Lund 32, 131—262 (1946).

Olah, L.: Interspecific hybrids in the genus Phleum. Proc. 7th Int. Genet. Congr. Edinburgh 23—30 August, 228, 1939 (1941).
Olmsted, C. E.: Growth and development in range grasses. IV. Photoperiodic responses in twelve geographic strains of side-oats grama. Bot. Gaz. 106, 46—74. (1944).
— Growth and development in range grasses. V. Photoperiodic responses of clonal divisions of three latitudinal strains of side-oats grama. Bot. Gaz. 106, 382—401 (1945).
Östergren, G.: Cytology of Agropyron junceum, A. repens and their spontaneous hybrids. Hereditas, Lund, 26, 305—316 (1940).
— On the morphology of Agropyron junceum (L.) PB., A. repens (L.) PB. and their spontaneous hybrids. Bot. Notiser Nr. 2, 133—143 (1940).
— Chromosome numbers in Anthoxanthum. Hereditas, Lund 28, 242—243 (1942). (Abstr.)
— Note on Elymus arenarius. Bot. Notiser Pp. 99 (1942).
— Heterochromatic B-chromosomes in Anthoxanthum. Hereditas, Lund 33, 261—296 (1947).

Parodi, L. R.: Estudio crítico de las gramíneas Austral-Americanas del género Agropyron. (Critical study of the South American grasses of the genus Agropyron.) Rev. Mus. La Plata 3, (N. S.) Sect. Bot. 1—63 (1940).
— Estudio preliminar sobre el género ,,Chusquea" en la Argentina. (Preliminary study on the genus Chusquea in Argentina.) Rev. Argent. Agron. 8, 331—345 (1941).
— Gramíneas austroamericanas nuevas o críticas. II. (New or critical South American Gramineae II.) Notas Mus. La Plata 8, 75—100 (1943).
— Los bambúes cultivados en la Argentina. (The bamboos cultivated in Argentina.) Rev. Argent. Agron. 10, 89—110 (1943).
— The Andean species of the genus Stipa allied to Stipa obtusa. Blumea, Leiden, Suppl. 3, 63—70 (1946).
Parris, G. K. u. J. C. Ripperton: Reactions of Napier grass, Merker grass, and their crosses to Helminthosporium eyespot. Phytopathology 31, 855 (1941).
Parthasarathy, N.: Cytogenetical studies in Oryzeae and Phalarideae. III. Cytological studies in Phalarideae. Ann. Bot., Lond. 3, 43—76 (1939).
— Chromosome numbers in Bambuseae. Curr. Sci. 15, 233—234 (1946).
Pavlenko, E. K.: (Vegetative hybridization of forage grasses.) Vestnik Gibridizacii (Hybridization) Nr. 2, 102—103 (1941). [Russisch.]
Porter, C. L.: Studies in Wyoming grasses III. Tribe Chlorideae, the grama grasses. Univ. Wyo. Publ. Sci. 8, 53—58 (1941).
— Studies in Wyoming grasses. IV. Tribe Hordeae, the barley grasses. Univ. Wyo. Publ. Sci. 9, 17—31 (1942).
— Studies in Wyoming grasses. V. Tribe Festuceae, the fescue tribe. Univ. Wyo. Publ. Sci. 11, 13—36 (1944).
— u. R. Lang: Studies in Wyoming grasses II. Tribe Agrostideae - the red top tribe. Univ. Wyo. Publ. Sci 8, 33—52 (1941).

Ragonese, A. E. u. P. R. Marcó: Observaciones sobre la biología floral de la cebadilla criolla. (Observations on the floral biology of wild barley.) Rev. Argent. Agron. 8, 196—199 (1941).
—, — Influencia del fotoperíodo sobre le formación de flores cleistógamas y chasmógamas en cebadilla criolla. (Influence of photoperiod on the formation of cleistogamic and chasmogamic flowers in wild barley). Rev. Agron. 10, 179—185 (1943).
Rangaswami Ayyangar, G. N. u. B. W. X. Ponnaiya: Studies in Sorghum sudanense, Stapf - The Sudan grass. Proc. Indian Acad. Sci. 10, Sect. B, 237—254 (1939).
—, — The occurrence and inheritance of shoots from the axils of panicle branches in Sorghum sudanense. Curr. Sci. 10, 409—410 (1941).
—, — Studies in Sorghum halepense (Linn.) Pers. — the Johnson Grass. Proc. Indian Acad. Sci. 13, Sect. B, 157 bis 162 (1941).
—, — Studies in Para-sorghum Snowden - the group with bearded nodes. Proc. Indian Acad. Sci. 14, Sect. B., 17—24 (1941).
Reeder, J. R.: Additional evidence of affinities between Eragrostis and certain Chlorideae. Amer. J. Bot. Suppl. 33, 843 (1946). (Abstr.)
Richharia, R. H. u. J. P. Kotwal: Chromosome number in bamboo (Dendrocalamus strictus). Indian J. Agric. Sci. 10, 1033 (1940).
Robertson, J. H. u. L. Weaver: A new tetraploid wheatgrass from Nevada. Bull. Torrey Bot. Cl. 69, 434 bis 437 (1942).
Rogler, G. A.: Response of geographical strains of grasses to low temperatures. J. Amer. Soc. Agron. 35, 547 bis 559 (1943).
— Two new grasses — Mandan wildrye and green stipagrass. Bi-m. Bull. N. Dak. Agric. Exp. Sta. 8, 11—12 (1946).
Rudorf, W.: Über eine Zwerg-compactum-Mutation bei Festuca pratensis L. Ber. dtsch. bot. Ges. 60, 132—147 (1942).

Saez, F. A. u. O. Núñez: La citología de Sorghum almum Parodi nueva especie con 40 cromosomas. I. Los cromosomas somáticos. (The cytology of Sorghum almum Parodi, a new species with 40 chromosomes. I. The somatic chromosomes.) Notas Mus. La Plata 8, 333 bis 348 (1943).
Salomon, E. S.: Sorghum sudanense (Piper) Stapf tetraploide obtenido por colchicina. (Tetraploid S. sudanense [Piper] Stapf obtained by colchicine.) An. Inst. Fitotec. Santa Catalina 2, 13—16 (1940).
Šalygin, I. N.: Production of tetraploids in Lolium by treating germinating seeds with colchicine. C. R. (Doklady) Acad. Sci. URSS. 30, 527—529 (1941).
Santos, J. V.: The Philippine, Chinese, and Indo-Chinese species of the grass genus Garnotia Brongniart. J. Arnold Arbor. 25, 85—96 (1944).
Saura, F.: Cariologia de algunas especies del género Paspalum. (Karyology of certain species of the genus Paspalum.) Inst. Genét., Fac. Agron. Vet., Univ. B. Aires 2, 41—48 (1941).
— Cariología de Gramíneas: géneros Paspalum, Stipa, Poa, Andropogon y Phalaris. (Karyology of the Gramineae: genera Paspalum, Stipa, Poa, Andropogon and Phalaris.) Rev. Fac. Agron. B. Aires 10, 344—353 (1943).
Schmieder, M. v. u. L. Niggl: Im Dienste der Futterpflanzenzüchtung. Forschungsdienst 17, 617—618 (1944).
Schultz, H. K.: A study of methods of breeding orchard grass, Dactylis glomerata L. J. Amer. Soc. Agron. 33, 546—558 (1941).
Scott, L. B.: Winter-fighting Wilmington narrow-leaf Bahia. Sth. Seedsman 8, Nr. 12; 14, 30 (1945).

Scott, L. B.: The South's all-level grass. Sth. Seedsman **9**, Nr. 5; 15, 55 (1946).
Severson, H.: When south goes for livestock grass seed gets the call. Seed World **54**, 8—9, 46 (1943).
Shalygin, I. N. siehe Šalygin.
Smirnova, M. I. u. G. A. Katanskaja: A biochemical method of large scale estimation of prussic acid in sorghum for breeding purposes. C. R. (Doklady) Acad. Sci. URSS. **24**, 592—595 (1939).
Smith, D. C.: Pollination and seed formation in grasses. J. Agric. Res. **68**, 79—95 (1944).
— u. E. L. Nielsen: Morphological variation in Poa pratensis L. as related to subsequent breeding behavior. J. Amer. Soc. Agron. **37**, 1033—1040 (1945).
—, — Comparative breeding behavior of progenies from enclosed and open-pollinated panicles of Poa pratensis L. J. Amer. Soc. Agron. **38**, 804—809 (1946).
—, —, u. H. L. Ahlgren: Variation in ecotypes of Poa pratensis. Bot. Gaz. **108**, 143—166 (1946).
Sodiro, R. A. u. L. Mille: Sertula florae ecuadorensis. Series 3. (A garland of the Ecuadorian flora. 3rd. series.) Bol. Inst. Bot., Ecuador **4**, 7—81 (1945).
Soulier, E. J.: A composite perennial Elymus-wheat-Agropyrum hybrid. C. R. (Doklady) Acad. Sci. URSS. **47**, 578—579 (1945).
Stahl, C.: Beretning fra Statsfrøkontrollen for det 70. Arbejdsaar fra 1. Juli 1940 til 30 Juni 1941. (Report from the State Seed Control Station for the 70th Year od Work from 1 July 1940 to 30 June 1941.) Tidsskr. Planteavl **46**, 569—641 (1942).
Stebbins, G. L. (jr.): Comparative growth rates of diploid and autotetraploid Stipa lepida. Amer. J. Bot. Suppl. **28**, 6s (1941). (Abstr.)
— Artificial synthesis of old and new polyploid species in Bromus. Genetics **32**, 107—108 (1947). (Abstr.)
— u. R. M. Love: A cytological study of California forage grasses. Amer. J. Bot. **28**, p. 371—382 (1941).
— u. H. A. Tobgy: The cytogenetics of hybrids in Bromus. I. Hybrids within the section Ceratochloa. Amer. J. Bot. **31**, 1—11 (1944).
—, — u. J. R. Harlan: Alice Eastwood Semi-Centennial Publications. Nr. 13. The cytogenetics of hybrids in Bromus II. Bromus carinatus and Bromus arizonicus. Proc. Calif. Acad. **25**, 307—321 (1944).
—, J. I. Valencia u. R. M. Valencia: Artificial and natural hybrids in the Gramineae, tribe Hordeae I. Elymus, Sitanion and Agropyron. Amer. J. Bot. **33**, 338 bis 351 (1946).
—, —, —, Artificial and natural hybrids in the Gramineae, tribe Hordeae. II. Agropyron, Elymus and Hordeum. Amer. J. Bot. **33**, 579—586 (1946).
Stewart, W. A., T. J. Jenkin u. J. A. Lindsay: Green pastures. IV. Grass breeding and seed production. J. Minist. Agric. **51**, 407—409 (1944).
Stoimenov, A.: (Content and physiological basis of the production of dhurrin in certain sorghum plants from the standpoint of their utilization as green forage-preliminary communication.) Rev. Inst. Rech. Agron. Bulg. **9**, Nr. 1, 17—32 (1939). [Russisch.]
Stomps, T. J.: Het Ammophila baltica-probleem. (The A. baltica problem.) Vakbl. Biol. **24**, 81—82 (1943).
Stuckey, I. H. u. W. G. Banfield: The morphological variations and the occurence of aneuploids in some species of Agrostis in Rhode Island. Amer. J. Bot. **33**, 185—190 (1946).
Sullivan, J. T.: Further comparisons of plants with different chromosome numbers in respect to chemical composition. J. Amer. Soc. Agron. **36**, 537—543 (1944).
— u. W. M. Myers: Chemical composition of diploid and tetraploid Lolium perenne L. J. Amer. Soc. Agron. **31**, 869—871 (1939).

Susarov, A.: (Some species and varieties of perennial herbage plants for the dry south-eastern parts of the USSR.) Sovhoznoe Proizvodstvo (State Farming) Nr. 4, 28—29 (1944). [Russisch.]

Thomas, H. L. u. H. K. Hayes: A selection experiment with Kentucky bleugrass. J. Amer. Soc. Agron. **39**, 192—197 (1947).
Tiemann, A.: Zwanzig Jahre im Dienst des Futtersamenbaues und der Zucht von Futterpflanzen. Kühn-Archiv **60**, 337—346 (1943/1944).
Tihovskaja, Z. P. u. N. V. Pervuhina: (Elymus arenarius in the far north - can it be brought under cultivation?) Priroda (Nature) Nr. 2, 75—78 (1946). [Russisch.]
Tinney, F. W.: Cytology of parthenogenesis in Poa pratensis. J. Agricl. Res. **60**, 351—360 (1940).
— u. O. S. Aamodt: The progeny test as a measure of the types of seed-development in Poa pratensis L. J. Hered. **31**, 457—464 (1940).
Tryon, R. M. (jr.) u. L. E. Booher: Remarks on the genus Elymus in Minnesota. Amer. J. Bot. Suppl. **33**, 843 (1946). (Abstr.)
Tsiang, Y. S.: Variation and inheritance of certain characters of brome grass, Bromus inermis Leyss. J. Amer. Spoc. Agron. **36**, 508—522 (1944).

Vinogradova, N. M. u. M. A. Novoderežkina: (The breeding of fodder grasses.) Trudy Zonaljnogo Inst. Zernovogo Hozjajstva Rajonov Nečernozemnoj Polosy (Trans. Zonal Inst. Grain Husbandry Non-Black-Soil Districts) Nr. 10, 113—144 (1941). [Russisch.]
De Vries, D. M. u. A. A. Kruijne: Onderscheiding der grassen van ons hooi- en weiland volgens kenmerken der niet bloeiende spruiten. (How to distinguish the grasses of our hay and pasture lands by characteristics of the non-flowering shoots.) Meded. Landbvoorlicht Dienst, Wageningen Nr. 43, 55 (1946).

Webster, C. B.: Side-oats Grama-O. K. Seed World **54**, 20 (1943).
Wenger, L. E.: Buffalo grass. Bull. Kans. Agric. Exp. Sta. Nr. 321, Pp. 78 (1943).
Wernham, C. C. u. St. J. P. Chilton: Typhula snowmold of pasture grasses. Phytopathology **33**, 1157 bis 1165 (1943).
Wexelsen, H.: Chlorophyll-deficient seedlings in timothy (Phleum Pratense L.) J. Hered. **32**, 227—231 (1941).
White, W. J.: Intergeneric crosses between Triticum and Agropyron. Sci. Agric. **21**, 198—232 (1940).

Zazoev, M.: (The breeding of Sudan grass.) Sovhoznol Proizvodstvo (State Farming) Nr. 3, 10—15 (1944). [Russisch.]

c) Futterleguminosen.

Ahlgren, G. H.: Breeding alfalfa to fit the soil. M. J. Agric. **26**, Nr. 2, 3—4 (1944).
Åkerberg, E. u. G. Julén: Vårt svenska rödklövermaterial i belysning av utförda stamförsök. (Our Swedish red clover as judged from results of strain trials.) K. LantbrAkad. Tidskr. **85**, 541—593 (1946).
— u. N. Schwanbom: Weibulls Robusta-vitklöver. En ny vitklöverstam av Strynö-typ. (Weibull's white clover Robusta. A new strain of white clover of the Strynö type.) Weibulls Ill. Årsb. **34**, 21—24 (1939).
Albrecht, H. R.: Earliness of maturity as a factor influencing seed production in vetch. J. Amer. Soc. Agron. **34**, 662—667 (1942).
— Vetches .... tailored for and by the south. Sth. Seedsman **6**, 10, 55 (1943).

Albrecht, H. R. u. T. R. Chamberlain: Instability of resistance to aphids in some strains of alfalfa. J. Econ. Ent. **34**, 551—554 (1941).

Andrews, A. L.: Fighting clover, „Manganese" extends its front. Sth. Seedsman **7**, 13, 40, 41 (1944).

Andrews, W. D.: Pasture investigations-results of 1942 trials. J. Dep. Agric. Vict. **41**, 277—285 (1943).

Anonymus: Red clover. (Trifolium pratense). Tech. Leafl. Res. Adv. Dep., Miln and Co., Chester, Nr. 4, 8.

— (The material for compiling manuals, by regions and republics, relating to certification of cereals, legumes and oil bearing crops.) Vestnik Socialističeskogo Rastenievodstva (Soviet Plant Industry Record), Nr. 3, 149—154 (1940). [Russisch.]

— Hobbyist raises four-leaf clovers. Sci. Observ. **3**, Nr. 2, 3 (1941).

— Results of station research have practical applications to war emergency agricultural production problems. Fm. Home Sci., Utah **3**, Nr. 4, 6—8, 10 (1942).

— Lamb fattening: trials with sweet blue lupins. Bull. Canterbury Agric. Coll., Nr. 154, 4 (1942).

— British clovers for Britain. Welsh Plant Geneticist's conclusions. Fmrs Wkly, Lond. **23**, Nr. 19, 18 (1945).

— Husbandry. 3. Valuable properties of wild clover. Agriculture, Moscow, Nr. 9, 6—7 (1945). [Mimeographed.]

— Simpson's sweet blue strain extolled by „Monticello News". Sth. Seedsman **8**, 11, 52 (1945).

— Lucerne. Fmrs' Leafl. Nat. Inst. Agric. Bot., Cambridge (Revised Ed.), Nr. 4, 4 (1946).

Armstrong, J. M. u. D. R. Gibson: Inheritance of certain characters in the hybrid of Medicago media and Medicago glutinosa. Sci. Agric. **22**, 1—10 (1941).

Arutiunova, A. G.: Chromosome morphology in certain species of clover. C. R. (Doklady) Acad. Sci. URSS. **27**, 825—827 (1940).

Atwood, S. S.: Cytogenetics of incompatibility in Trifolium repens. Genetics **25**, 109 (1940). (Abstr.)

— Genetics of cross-incompatibility among self-incompatible plants of Trifolium repens. J. Amer. Soc. Agron. **32**, 955—968 (1940).

— Controlled self- and cross-pollination of Trifolium repens. J. Amer. Soc. Agron. **33**, 538—545 (1941).

— Cytogenetic basis of self-compatibility in Trifolium repens. Genetics **26**, 137 (1941). (Abstr.)

— Cytological basis for incompatibility in Trifolium repens. Amer. J. Bot. **28**, 551—557 (1941).

— Genetics of self-compatibility in Trifolium repens. J. Amer. Soc. Agron. **34**, 353—364 (1942).

— Genetics of pseudo-self-compatibility and its relation to cross incompatibility in Trifolium repens. J. Agric. Res. **64**, 699—709 (1942).

— The multiple oppositional alleles causing cross-incompatibility in Trifolium repens. Genetics **27**, 129 bis 130 (1942). (Abstr.)

— Oppositional alleles causing cross-incompatibility in Trifolium repens Genetics **27**, 333—338 (1942).

— „Natural crossing" of white clover by bees. J. Amer. Soc. Agron. **35**, 862—870 (1943).

— The value of self-compatibility in breeding white clover. J. Amer. Soc. Agron. **36**, 990 (1944). (Abstr.)

— Colchicine-induced polyploids in white clover. J. Amer. Soc. Agron. **36**, 173—174 (1944).

— Oppositional alleles in natural populations of Trifolium repens. Genetics **29**, 428—435 (1944).

— The behavior of the self-compatibility factor and its relation to breeding methods in Trifolium repens. J. Amer. Soc. Agron. **37**, 991—1004 (1945).

— u. H. D. Hill: The regularity of meiosis in microsporocytes of Trifolium repens. Amer. J. Bot. **27**, 730—735 (1940).

— u. J. T. Sullivan: Inheritance of a cyanogenetic glucoside and its hydrolyzing enzyme in Trifolium repens. J. Hered. **34**, 311—320 (1943).

Atwood, S. S. u. J. T. Sullivan: Inheritance of a cyanogenetic glucoside and its hydrolyzing enzyme in white clover. Genetics **28**, 69 (1943). (Abstr.)

Baur, G.: Kreuzbefruchtung (Pärchenzüchtung) bei Rotklee in ihrer Bedeutung für die praktische Züchtung. Z. Pflanzenz. **23**, 611—637 (1941).

Baženova, M.: (A sweet lupin variety.) Doklady Vsesojuz. Akad. Seljsk. Nauk im. V. I. Lenina (Proc. Lenin Acad. Agric. Sci. USSR.), Nr. 3, 34—36 (1945). [Russisch.]

Becker, R. u. Z. Wehrmacht: Untersuchungen über die Lebensdauer verschiedener Weißkleeformen. Pflanzenbau **20**, 27—40 (1943).

Berkner, F.: Kritische Beiträge zur Kenntnis, Züchtung und Nutzung der Vicia villosa. Forsch. dienst. **10**, 418—442 (1940).

Bird, J. N.: Red clover improvement in Quebec. Agriculture **3**, 8 (Macdonald Coll. J. Ser. Nr. 216) (1946).

Bjälfve, G.: Baljväxternas rotknölar hos olika sorter. (Root nodules of legumes of different varieties.) Lantmannen **31**, 302—304 (1947).

Black, C. A. u. K. Lawton: Iowa soils need nitrogen. Fm Sci. Reporter, Iowa **6**, 14—16 (1945).

Black, M.A.u. J.H.Claridge: Sweet blue lupins. N. Z. J. Agric. **65**, 69—70 (1942).

Bobrov, E. G.: (The origin of cultivated red clover) Sovetskaja Botanika (Soviet Botany), Nr. 1, 27—39 (1940). [Russisch.]

Brink, R. A. u. D. C. Cooper: Somatoplastic sterility in Medicago sativa. Science **90**, 545—546 (1939).

—, — Double fertilization and development of the seed in Angiosperms. Bot. Gaz. **102**, 1—25 (1940).

—, — The significance of double fertilization in flowering plants. Genetics **25**, 113 (1940). (Abstr.)

Č. N.: (New properties of lupin.) Socialističeskoe Seljskoe Hozjajstvo (Socialistic Agriculture) Moskva, Nr. 5, 95 (1941).

Carlson, J. W.: Pollination, lygus-infestation, genotype, and size of plants as affecting seed setting and seed production in alfalfa. J. Amer. Soc. Agron. **38**, 502 bis 514 (1946).

Carsner, E.: The story of blight-resistant beet seed. U. and I. Cultiv., Utah **1**, Nr. 3, 5—6 (1941).

Castro, D. de: Algumas contagens de cromosomas no género Ulex L. (sensu lato). (Some chromosome counts in the genus Ulex L. [sensu lato].) Agron. Lusitana **3**, 103—113 (1941).

Cooper, D. C. u. R. A. Brink: Partial self-incompatibility and the collapse of fertile ovules as factors affecting seed formation in alfalfa. J. Agric. Res. **60**, 453 bis 472 (1940).

Cooper, J. F.: The lupines: for bitter.... or for sweet ? Sth. Seedsman **7**, 26—27 (1944).

— Sweet lupines. Sth. Seedsman **8**, 11, 46 (1945).

Corkill, L.: Cyanogenesis in white clover (Trifolium repens L.). I. Cyanogenesis in single plants. N. Z. J. Sci. Tech. **22**, 65B—67B (1940).

— Cyanogenesis in white clover (Trifolium repens L.). V. The inheritance of cyanogenesis. N. Z. J. Sci. Technol. **23**, 178B—193B (1942).

Coutinho, A.: Tipos cariológicos nas Vicias. (Karyological types in Vicia.) Palestras Agronómicas, Lisboa **2**, 81—95 (1939).

— Raças cariológicas na V. sativa L. (Karyological races in V. sativa L.). Agron. Lusitana **2**, 379—403 (1940).

— O „mosaico das leguminosas" agente modificador de genes ? (Mosaic of leguminous plants as an agent modifying genes ? Rev. Agron. Lisboa **30**, 459—460 (1942).

— O „mosaico das leguminosas" agente perturbador da hereditariedade ? (Mosaic of leguminous plants as an agent disturbing the hereditary mechanism ?) Agron. Lusitana **4**, 273—292 (1942).

Coutinho, A.: Novos subsídios para a cariologia do género Vicia L. (New contributions to the caryology of the genus Vicia L.) Bol. Soc. Broteriana **19**, 449 bis 455 (1945).
— u. M. da Cunha e: Subsídios para o estudo cariológico do género Trigonella L. (Notes for a karyological study of the genus Trigonella L.) Agron. Lusitana **4**, 73—86, also Rev. Agron. Lisboa **30**, 458—459 (1942).
— u. A. Santos: Novas, contribuïções para a cariologia do género Trigonella L. (New contributions towards the caryology of the genus Trigonella L.) Agron. Lusitana **5**, 349—361 (1943).
Crandall, B. H. u. H. D. Tate: The bee, Calliopsis andreniformis, as a faktor in alfalfa seed setting. J. Amer. Soc. Agron. **39**, 161—163 (1947).

Dawson, C. D. R.: Tetrasomic inheritance in Lotus corniculatus L. J. Genet. **42**, 49—72 (1941).
Deloffre, G.: Recherches cytophysiologiques sur Lupinus angustifolius. A. Taffin-Lefort, Paris-Lille, Pp.137, 82 tables. 4 graphs. (1939).
Drewes, H.: Something new has been added to beets. Sth. Seedsman **5**, Nr. 10; 10, 31, 35 (1942).
Dunkle, P. B.: Tight-land clover! Sth. Seedsman **9**, Nr. 9; 14, 40, 48 (1946).
Dworok, K.: Wesen und Möglichkeiten der Luzerne-Zuchtmethoden in Esterháza. Pflanzenbau **18**, 1—8 (1941).

Eardley, C. M.: Tree legumes for fodder. J. Agric. S. Aust. **48**, 342—345 (1945).
Eastwood, A.: New species and varieties of Lupinus. Leafl. West Bot. **3**, 169—175 (1942).
Evans, M. u. I. J. Johnson: The comparative rates of growth of tetraploid and diploid sweetclover, Melilotus alba Desr. J. Amer. Soc. Agron. **37**, 867—875 (1945).

Farley, H. M. u. A. H. Hutchinson: Seed development in Medicago (alfalfa) hybrids. 1. The normal ovule. Canad. J. Res. **19**, Sect. C, 421—437 (1941).
Filippov, A. S.: (Interspecific and intraspecific vegetative hybridization of potatoes.) Vestnik Ovoščevodstvo i Kartofelj (Vegetable and Potato Journal), Nr. 2, 16—39 (1940). [Russisch.]
Fischer, A.: North America as a gene centre of various genera and species of Leguminosae. Res. & Progr., Berlin **5**, 297—299 (1939).
Fotolds, M.: Seed color studies in biennial white sweet clover, Melilotus alba. Journ. Amer. Soc. Agron. **31**, 678—686 (1939).
Fowlds, M.: Seed color studies in biennial white sweet clover, Melilotus alba. J. Amer. Soc. Agron. **31**, 678 bis 686 (1939).
Friend, W. H.: Goin' places . . . . with Emerald. Sth. Seedsman **7**, Nr. 9, 12 (1944).
Fryer, J. R.: The alfalfa seed setting problem. Pr. Bull. Alberta **30**, 10—12. (1945).
Fyfe, J. L.: Polyploidy in sainfoin. Nature, Lond. **158**, 418 (1946).

Garver, S., J. M. Slatensek u. T. A. Kiesselbach: Sweetclover in Nebraska. Bull. Neb. Agric. Exp. Sta., Nr. 352, 47 (1943).
Gescher, N. von: The sweet lupin, a new plant-breeding conquest. Int. Rev. Agric. **34**, 327T—346T (1943).
Gettys, R. E. u. I. J. Johnson: The nature and inheritance of sterility in sweet clover, Melilotus officinalis Lam. J. Amer. Soc. Agron. **36**, 228—237 (1944).
Grandfield, C. O.: Buffalo alfalfa. Circ. Kans. Agric. Exp. Sta., Nr. 226, 7 (1945).
— u. R. I. Throckmorton: Alfalfa in Kansas. Bull. Kans. Agric. Exp. Sta., Nr. 328, 64 (1945).

Griesinger, R. u. M. Klinkowski: Geographie und Cytologie des europäischen Formenkreises der Gattung Ornithopus. Züchter **11**, 147—161 (1939).
Grizzard, A. L. u. T. B. Hutcheson: Experiments with Lespedeza. Bull. Va Agric. Exp., Sta., Nr. 328, 20 (1940).

Hackbarth, J.: Fragen der Vererbung, insbesondere der Koppelung bei Lupinus luteus. Züchter **13**, 34—36 (1941).
— u. H.-J. Troll: Über die Zahl der Gene bei Alkaloidfreiheit bei Lupinus luteus. Züchter **13**, 63—65 (1941).
— Ein neuer Zuchtstamm von gelben Süßlupinen mit schneller Jugendentwicklung. Züchter **13**, 65—68 (1941).
— Züchtung und Anbau der weißen Lupine. Mitt. Landw. **56**, 774—776 (1941).
— u. B. Husfeld: Die Süßlupine. Züchtung, Anbau und Verwertung einer neuen Kulturpflanze. Paul Parey, Berlin (1939).
Ham, W. E. u. H. M. Tysdal: The carotene content of alfalfa strains and hybrids with different degrees of resistance to leafhopper injury. J. Amer. Soc. Agron. **38**, 68—74 (1946).
Hansen, J.: (Versuch mit Stämmen von Weißklee.) Tidsskr. Planteavl **46**, 642—660 (1942). [Dänisch.]
Hanson, C. H.: Cleistogamy and the development of the embryo sac in Lespedeza stipulacea. J. Agric. Res. **67**, 265—272 (1943).
Hardenburg, E. V. u. F. J. Stevenson: Mohawk: a new baking potato. Amer. Potato J. **20**, 79—86 (1943).
Hartwig, E. E.: A new mutant leaf character in sweet clover. J. Hered. **32**, 171—172 (1941).
— Inheritance of growth habit, cotyledon colour, and cup-leaf in Melilotus alba. J. Amer. Soc. Agron. **34**, 160—166 (1942).
— Effects of self-pollination in sweet clover. J. Amer. Soc. Agron. **34**, 376—387 (1942).
Henderson, M. T. u. E. L. LeClerg: Studies of some factors affecting fruit setting in Solanum tuberosum in the field in Louisiana. J. Agric. Res. **66**, 67—76 (1943).
Henson, L.: Breeding for adaptation in red clover. Proc. 41st Annu. Conv. Ass. Sth Agric. Wkrs, Birmingham, Ala., February 7—9, Pp. 84 (1940). (Abstr.)
Hieke, K.: Pflanzenphysiologische Untersuchungen über die Alkaloide: II. Zur Alkaloidführung der Pfropfpartner bei heteroplastischen Solanaceenpfropfungen. Planta **33**, 185—205 (1942).
Hills, K. L.: The reaction of varieties of Trifolium subterraneum to attack by Uromyces trifolii as a heritable charakter. J. Coun. Sci. Industr. Res. Aust. **17**, 74—78 (1944).
— Dormancy and hardseededness in T. subterraneum. 2. The progress of after-harvest ripening. J. Coun. Sci. Industr. Res. Aust. **17**, 186—190 (1944).
— Dormancy and hardseededness in T. subterraneum. 3. The effect upon dormancy of germination at three different constant temperatures. J. Counc. Sci. Industr. Res. Aust. **17**, 191—196 (1944).
— Dormancy and hardseededness in T. subterraneum. 4. Variation between varieties. J. Coun. Sci. Industr. Res. Aust. **17**, 242—250 (1944).
Hollowell, E. A.: Registration of varieties and strains of sweet clover, I. J. Amer. Soc. Agron. **35**, 825—829 (1943).
— Registration of varieties and strains of red clover, I. J. Amer. Soc. Agron. **35**, 830—833 (1943).
— Registration of varieties and strains of alfalfa, I. J. Amer. Soc. Agron. **37**, 649—652 (1945).
— „Dixie" does it . . . . . Sth. Seedsman **9**, 20 (1946).
— Crimson clover. Leafl. U. S. Dep. Agric., Nr. 160, 8 (1947).

Horner, W. H. u. W. J. White: Investigations concerning the coumarin content of sweet clover. III. The inheritance of the low coumarin character. Sci. Agric. 22, 85—92 (1941).

Horošajlov, N. G.: (On cross-pollination of sainfoin.) Vestnik Socialističeskogo Rastenievodstva (Soviet Plant Industry Record). Nr. 2, 112—115 (1940). [Russisch.]

— (On some classifications of fodder plants.) Vestnik Socialističeskogo Rastenievodstva (Soviet Plant Industry Record), Nr. 2, 116—123 (1940). [Russisch.]

— (The root system of sainfoin.) Vestnik Socialističeskogo Rastenievodstva (Soviet Plant Industry Record), Nr. 3, 87—94 (1940). [Russisch.]

— (The breeding of esparcet [sainfoin].) Vestnik Socialističeskogo Rastenievodstva (Soviet Plant Industry Record), Nr. 1, 99—111 (1941). [Russisch.]

Jakuševa, E. I.: (The winter hardiness of lucerne and clover and ways of increasing it.) Vestnik Socialističeskogo Rastenievodstva (Soviet Plant Industry Record), Nr. 5, 54—65 (1940). [Russisch.]

Jannaccone, A.: Selezione della veccia. (Vetch selection.) Servizio delle Sementi, Pp. 14—40 (1939).

Jones, F. R. u. W. K. Smith: Segregation of resistance to bacterial wilt in crosses involving Grimm alfalfa. J. Amer. Soc. Agron. 39, 423—425 (1947).

Johnson, I. J.: Cross fertility relationships of „Golden Annual" sweet clover with common species of Melilotus. J. Amer. Soc. Agron. 34, 259—262 (1942).

— u. J. E. Sass: Tetraploidy in Melilotus alba induced by colchicine. Proc. Iowa Acad. Sci. 49, 254 (1942).

—, — Self- and cross-fertility relationships and cytology of autotetraploid sweet clover, Melilotus alba. J. Amer. Soc. Agron. 36, 214—227 (1944).

Jones, F. R.: Winter injury and longevity in unselected clones from four wiltresistant varieties of alfalfa. J. Amer. Soc. Agron. 37, 828—838 (1945).

Josefsson, A.: Några synpunkter på förädlingsarbetet med lupin. (New points of view on breeding operations with lupins.) Sverig. Utsädesfören. Tidskr. 54, 104—114 (1944).

— Blå sotlupin till mogen skörd. (Sweet blue lupins that ripen seed.) Lantmannen 28, 1005, 1009 (1944).

Julén, G.: Investigations on diploid, triploid and tetraploid lucerne. Hereditas, Lund 30, 567—582 (1944).

— Några rödklöverstammars avkastning och kvalitet i försök vid Svalöf 1933—1942. (The yield and quality of some red clover strains in experiments at Svalöf 1933—1942.) Sverig. Utsädesfören Tidskr. 54, 332 bis 338 (1944).

Kirsanova, V. A.: (The control of vitamin C and carotene in lucerne varieties of the Uzbek S.S.R. Biohimija (Biochemistry) 9, 113—118 (1944). [Russisch.]

Kitch, K.: Seed-keeping clover. Sth. Seedsman 9, Nr. 10; 22, 48 (1946).

Klawitter, G. u. R. v. Sengbusch: Die Züchtung von vollkommen alkaloidfreien Süßlupinen, die sich zur Herstellung von menschlichen Nahrungsmitteln eignen. Züchter 15, 10—12 (1943).

Klinkowski, M.: Zur Züchtung mehltauresistenter Lupinen. Wiss. Jber. Biol. Reichsanst. Land- u. Forstwirtsch. 1940. Mitt. Biol. Reichsanst. 1941, Heft 65, 29 (1941/1942).

— Zur Kalkresistenz der Lupine. Mitt. Biol. Reichsanst., Heft 65, 29 (1941/1942).

— Zur Frage der Fremdbefruchtung der Serradella. Züchter 14, 240—243 (1942).

— u. R. Griesinger: Versuche zur Erzeugung polyploider Rassen bei der Gattung Ornithopus. Züchter 11, 313—317 (1939).

— u. J. Hackbarth: Zur Kenntnis der züchterischen Bedeutung iberischer Wildformen von Lupiteus L. und L. angustifolius L. Z. Pflanzenz. 23, 579—610 (1941).

Koepper, J. M.: Relative resistance of alfalfa species and varieties to rust caused by Uromyces striatus. Phytopathology 32, 1048—1057 (1942).

Koljcov, N. K.: On the methods of artificially inducing polyploids by treatment with colchicine. C. R. (Doklady) Acad. Sci. URSS. 23, 482—485 (1939).

Koperžinskij, V.: New data on the structure of the lucerne flower.) Proc. Lenin Acad. Agric. Sci. USSR., Nr. 7, 19—21 (1944).

Krantz, F. A., A. G. Tolaas, H. O. Werner, H. W. Goss u. J. H. Jensen: The Kasota potato. Amer. Potato J. 20, 25—27 (1943).

Ledingham, G. F.: Cytological and developmental studies of hybrids between Medicago sativa and a diploid form of M. falcata. Genetics 25, 1—15 (1940).

Lepper, R. jr. u. T. E. Odland: Inheritance of flower color in alfalfa. J. Amer. Soc. Agron. 31, 209—216 (1939).

Levan, A.: Framställning av tetraploid rödklöver. (Production of tetraploid red clover.) Sverig. Utsädesfören. Tidskr. 50, 115—124 (1940).

— Plant breeding by induction of polyploidy and some results in clover. Hereditas, Lund 28, 245—246 (1942). (Abstr.)

Lisicyn, P. u. V. Fedorčuk: (The relationship between the percentage of leaf and the contact of crude protein in red clover.) Doklady Vsesojuz. Akad. Seljsk. Nauk im. V. I. Lenina (Proc. Lenin Acad. Agric. Sci. USSR.), Nr. 3, 3—11 (1945). [Russisch.]

Lowig, E.: Gedanken zur Züchtung und Samengewinnung von Wiesenrotklee. Züchter 13, 128—132 (1941).

Lubenec, P. A.: (Improvement of local varieties of lucerne and organization of their seed production by methods of ground-control and field inspection.) Proc. Lenin Acad. Agric. Sci. URSS., Nr. 16, 19—27 (1939). [Russisch.]

McClure, F. A.: New bamboos from venezuela and Colombia. J. Wash. Acad. Sci. 32, 167—183 (1942).

MacDonald, H. A.: Birdsfoot trefoil (Lotus corniculatus). Abstr. Thes. Cornell Univ., Pp. 403—405, 1943 (1944).

— Birdsfoot trefoil (Lotus corniculatus L.). Its characteristics and potentialities as a forage legume. Mem. Cornell Agric. Exp. Sta., Nr. 261, 182 (1946).

McKee, R.: Lespedeza culture and utilization. Fmrs' Bull. U. S. Dep. Agric. 1940 (Revised 1946), Nr. 1852, 14 (1946).

—, H. L. Hyland u. G. E. Ritchey: Preliminary information on sweet lupines in the United States. J. Amer. Soc. Agron. 38, 168—172 (1946).

— u. J. L. Stephens: Kudzu as a farm crop. Fmrs' Bull. U. S. Dep. Agric., Nr. 1923, 13 (1943).

Malheiros, N.: Elementos para e estudo citológico do género „Lupinus". (Cytological study of the genus „Lupinus".) Rev. Agron., Lisboa 29, 105—106 (1941).

— Elementos para o estudo citologico do género Lupinus. (Cytological study of the genus Lupinus.) Agron. Lusitana 4, 231—236 (1942).

Manke, K. F. u. W. H. Friend: Texas „progress report" on Emerald sweetclover. Sth. Seedsman 7, Nr. 9; 12, 52 (1944).

Melville, J., I. E. Coop, B. W. Doak u. I. Reifer: Cyanogenesis in white clover (Trifolium repens L.). IV. Methods of determination and general considerations. N. Z. J. Sci. Tech. 22, 144B—154B (1940).

— u. B. W. Doak: Cyanogenesis in white clover (Trifolium repens L.). II. Isolation of the glucosidal constituents. N. Z. J. Sci. Tech. 22, 67B—71B (1940).

Mettès, M.: Curieux cas de gémellité chez Lupin blanc. Bull. Soc. Bot. Fr. 87, 65 (1940).

Meyer, K.: Photometrische Alkaloidbestimmung zur Untersuchung von Zuchtmaterial der weißen und blauen Lupine. Landw. Jb. **91**, 418—440 (1941).

Middelburg, H. A.: Enkele beschouwingen en mededeelingen over selectie bij Crotalaria juncea. (Some views and information about selection in C. juncea.) Bergcultures **15**, 208—211 (1941).

Miège, E.: La production des „lupins doux". Essais entrepirs au Maroc. (The production of „sweet lupins". Experiments undertaken in Morocco.) Rev. Bot. Appl. **20**, 16—24 (1940).

Milovidov, P. F.: Über die Chromosomenzahlen bei einigen Leguminosen und anderen Pflanzen. Planta **32**, 38—42 (1941/42).

Minderhoud, A.: Over de bestuiving van de roode klaver. (On the pollination of red clover.) Landbouwk. Tijdschr., Wageningen **53**, 755—794 (1941).

Mirošničenko, I.: (Lathyrus sativus — a high-yielding leguminous crop.) Sovhoznoe Proizvodstvo (State Farming), Nr. 4, 29—30 (1944). [Russisch.]

Mogileva, A. M.: (Breeding spring vetch.) Trudy Zonaljnogo Inst. Zernovogo Hozjajstva Rajonov Nečernozemnoj Polosy (Trans, Zonal Inst. Grain Husbandry Non-Black-Soil Districts), Nr. 10, 101—111 (1941). [Russisch.]

Mostovoj, K.: (Virescence of the flowers as a cause of plant sterility.) Ann. Tschech. Akad. Landw. 1940, **15**, 142—146 (1940).

Munro, J. A. u. H. S. Telford: Recent progress in wireworm control. Bi-m. Bull. N. Dak. Agric. Exp. Sta. **5**, Nr. 2, 7—11 (1942).

Musser, H. B. u. J. K. Thornton: Local, domestic and foreign red clover seed. Bull. Pa Agric. Exp. Sta., Nr. 458, 16 (1943).

Nicolaisen, W., B. Leitzke u. I. Witzig: Untersuchungen im Rahmen der Züchtung der Kleearten auf Widerstandsfähigkeit gegen den Kleekrebs (Sclerotinia trifoliorum Erikss.). Phytopath. Z. **12**, 585 bis 645 (1939).

Nijdam, F. E.: The colour of the flowers of Trifolium pratense L. Genetica **21**, 16—28 (1939).

— The heredity of a brownish-black spot on the seedcoat of Trifolium pratense L. caused by melanine. Genetica **22**, 123—130 (1940).

— Massaselectie en stamselectie in roode klaver. (Mass selection and individual selection of red clover.) Landbouwk. Tijdschr., Wageningen **53**, 744—754 (1941).

Nilsson, F. u. E. Andersson: Polyploidi hos släktet Medicago. (Polyploidy in the genus Medicago.) Sverig. Utsädesfören. Tidskr. **51**, 363—382 (1941).

—, — Polyploidy in the genus Medicago. Heredita, Lund **29**, 197—198 (1943).

Nilsson-Leissner, G.: Rödklöverförsök pa klöverrötsjord vid Svalöf. (Red clover experiments on soil infected by clover rot at Svalöf.) Sverig. Utsädesfören. Tidskr. **52**, 352—363 (1942).

Nutman, P. S.: Genetical factors concerned in the symbiosis of clover and nodule bacteria. Nature, Lond. **157**, 463—465 (1946).

Odland, T. E. u. R. Lepper jr.: A crinkled leaf mutation in alfalfa. J. Amer. Soc. Agron. **31**, 128—130 (1939).

Paljčevskij, A.: (The Trans-Caucasian sainfoin.) Sovhoznoe Proizvodstvo (State Farming), Nos. 11/12, 12—13 (1945). [Russisch.]

Panse, E.: Möglichkeiten der Steigerung der Eiweißleistung bei Luzerne durch Züchtung auf hohen Eiweißgehalt. Z. Pflanzenz. **24**, 229—274 (1941).

Parsons, F. L.: Some cold storage of Kansas potatoes. Bull. Kans. Agric. Exp. Sta., Nr. 310, 18 (1942).

Peacock, N. D., A. Meyer u. A. B. Strand: A strain of Nancy Hall sweetpotato selected for color of flesh. Circ. Tenn. Agric. Exp. Sta., Nr. 80, 4 (1942).

Pierce, W. P.: Cytology of the genus Lespedeza. Amer. J. Bot. **26**, 736—744 (1939).

Pohjakallio, O.: Kasvinjalostusbiologisia tutkimuksia apilamädästä. (Investigations on clover rot from the biological aspect of plant breeding.) Suom. Tiedeakat. Esitelm. Pöytäk. 1939 (1940), 115—128 (1940).

Ragonese, A. E. u. P. R. Marcó: Resistencia al nemátode del tallo de diversas lineas y procedencias de alfalfas. (Resistance to stem nematode of various lines and provenances of lucerne.) Rev. Argent. Agron. B. Aires **10**, 370—384 (1943).

Reddick, D.: Development of blight-immune varieties. Amer. Potato J. **20**, 118—126 (1943).

Ricker, P. L.: New and noteworthy Asiatic species of Lespedeza. Amer. J. Bot. **33**, 256—258 (1946).

Rinke, E. H.: Inheritance of coumarin in sweet clover. J. Amer. Soc. Agron. **37**, 635—642 (1945).

— u. I. J. Johnson: Self-fertility in red clover in Minnesota. J. Amer. Soc. Agron. **33**, 512—521 (1941).

Ritchey, G. E.: Hairy indigo, a legume for Florida. Pr. Bull. Fla Agric. Exp. Sta., Nr. 624, 4 (1946).

—, R. McKee, R. B. Becker, W. M. Neal u. P. D. Dix Arnold: Crotalaria for forage. Bull. Fla Agric. Exp. Sta., Nr. 361, 72 (1941).

Rudorf, W.: Eine unifoliata-Mutante bei Medicago media Perss. Züchter **13**, 272—274 (1941).

— u. O. Schröck: Über das Auftreten stark abgeänderter Formen bei Steinklee. Züchter **13**, 1—4 (1941).

Saloheimo, L.: Om klöverodling på torvjord samt resultat fråm försöken med olika klöverstammar på Finska Mosskulturföreningens Karelska försöksstation under åren 1928—1938. (On clover growing on peat soil with results from experiments with different clover strains at the Finnish Experimental Station in Karelia for the Reclamation of Bogland during 1928—38.) Finska MosskFören. Årsb. **43**, 94—107 (1939).

Ščenkova, M. S.: (Zur Biologie des Trifolium pratense L. in den Bedingungen des Nordens.) C. R. (Doklady) Acad. Sci. URSS. **55**, 875—876 (1947). [Russisch.]

Schander, H.: Gedanken über die Unterschiede und Übereinstimmungen der Chloroseerscheinungen von Lupinen und Holzgewächsen. Gartenbauwiss. **17**, 304—309 (1943).

Schelhorn, M. v.: Über eigene und fremde Versuche zur Art- und Gattungsbastardierung bei Vicia, Lens, Pisum und Lathyrus. Forsch.dienst. **9**, 70—78 (1940).

— Zur Cytogenetik von Medicago falcata L. Züchter **17/18**, 92—93 (1946).

— Über eine erbliche Umbildung der Blütenstände bei Vicia villosa. Züchter **17/18**, 153—155 (1947).

Schiel, E. u. A. E. Ragonese: Infección de la alfalfa con „Rhizobium meliloti" en la provincia de Santa Fe. (Inoculation of lucerne with Rh. meliloti in the province of Santa Fe.) Rev. Argent. Agron. **9**, 114 bis 169 (1942).

Schröck, O.: Die Züchtung alkaloidarmen Geissklees, Galega officinalis. Züchter **13**, 115—117 (1941).

— Untersuchungen über die Möglichkeit der Verwendung der Korrelationen in der Züchtung der Luzerne auf Eiweißreichtum. Züchter **14**, 234—240 (1942).

— Genetische Beobachtungen an Luzerne (Medicago media). Z. Pflanzenz. **25**, 81—91 (1943).

— Beobachtungen an einem Bastard zwischen Luzerne (Medicago media) und Gelbklee (Medicago lupulina) und seiner Nachkommenschaft. Züchter **15**, 4—10 (1943).

— Untersuchungen an diploiden und tetraploiden Klonen von Luzerne, Gelbklee und Steinklee. Z. Pflanzenz. **26**, 214—222 (1944).

Schwanbom, N.: Weibulls Original Resistenta rödklöver i jämförelse med andra rödklöverstammar. (Weibull's Original Resistenta red clover in comparison with other red clover strains.) Weibulls Ill. Årsb. **39**, 19—22 (1944).

Sears, O. H. u. W. L. Burlison: Lespedeza: its place in Illinois agriculture. Circ. Univ. Ill. Coll. Agric. Ext. Serv., Nr. 561, 19 (1943).

Sengbusch, R. v. u. H. Kress: Über das Auftreten zweier rezessiver Mutationen bei Lupinus albus in bestimmter Reihenfolge. Biol. Zbl. **59**, 222—224 (1939).

— Die Auffindung einer neuen weißsamigen Mutante im Süßlupine-Stamm 8 (Stamm W 8/37, Lupinus lutens). Züchter **12**, 19—20 (1940).

— Die Züchtung von Süßlupinen mit nichtplatzenden Hülsen. Die Kombination der Eigenschaften „Alkaloidfrei" und „Nichtplatzen der Hülsen" und die Bedeutung der doppelt und dreifach recessiven alkaloidfreien Formen für die Süßlupinenzüchtung. Züchter **12**, 149—152 (1940).

— Süßlupinen und Öllupinen. Die Entstehungsgeschichte einiger neuer Kulturpflanzen. Landw. Jahrb. **91**, 719—880 (1942).

Ščerbakov, C. D.: Breeding disease-resistant red clover. Proc. 41st Annu. Conv. Ass. Sth. Agric. Wkrs, Birmingham, Ala., February 7—9, Pp. 195 (1940). (Abstr.)

Sexsmith, J. J. u. J. R. Fryer: Studies relating to fertility in alfalfa (Medicago sativa L.). I. Pollen viability as affected by seasonal age of the plant. Sci. Agric. **24**, 95—100 (1943).

—, — Studies relating to fertility in alfalfa (Medicago sativa L.). II. Temperature effect on pollen tube growth. Sci. Agric. **24**, 145—151 (1943).

Sherbakov, C. D. siehe Ščerbakov.

Sigrianskaja, N. D.: (Das Welken der Luzerne und ihr Wurzelfaulen). C. R. (Doklady) Acad. Sci. URSS. **55**, 651—654 (1947). [Russisch.]

Singleton, H. P., C. E. Nelson u. C. O. Stanberry: Alfalfa varieties under irrigation. Bull. Wash. St. Agric. Exp. Sta., Nr. 464, 32 (1945).

Sinskaya, E. N.: Phylogenetic taxonomy as a basis for genetic and breeding work. Illustrated by Medicago. (Preliminary communication.) Z. indukt. Abstamm.- u. VererbLehre **78**, 399—417 (1940).

— (Interspecific hybridization in alfalfa.) Vestnik Gibridizacii (Hybridization), Nr. 1, 85—91 (1941). [Russisch.]

— On the diploid species of yellow alfalfa. C. R. (Doklady) Acad. Sci. URSS. **48**, 281—282 (1945).

Skrepinskij, A. I.: (Experiments with sainfoin varieties.) Bjulletenj Instituta Zernovogo Hozjajstva Jugo-Vostoka SSSR. (Bull. Inst. Grain Husb. S.E. USSR.) Saratov, Nr. 1, 28—30 (1944). [Russisch.]

Slatensek, J. M.: The occurrence methods of determination, and inheritance of coumarin in sweetclover, Melilotus. Dep. Agron. Univ. Neb., Pp. 7 (1945). (Abstr.)

Šmatok, E. G.: (Lupins with a low alkaloid concent as a fodder for dairy cattle.) Kormodobyvanie (Fodder Production), Nr. 1, 62—67 (1941). [Russisch.]

Smirnova, M. I. u. B. S. Moškov: Changes of the alkaloid content in lupin grafts. Soviet Plant Industry Record, Nr. 2, 68—77 (1940).

Smith, W. K.: Propagation of chlorophyll-deficient sweetclover hybrids as grafts. J. Hered. **34**, 135—140 (1943).

Šmygun, V. N.: (The method of crossing spring vetch.) Selekcija i Semenovodstvo (Breeding and Seed Growing), Nr. 5, 22—24 (1940). [Russisch.]

— (An investigation into methods of hybridizing spring vetch.) Trudy Zonaljnogo Inst. Zernovogo Hozjajstva Rajonov Nečernozemnoj Polosy (Trans. Zonal Inst. Grain Husbandry Non-Black-Soil Districts), Nr. 10, 93—99 (1941). [Russisch.]

Squibb, R. L.: Desmodiums—„Alfalfas of the tropics". Agric. Amer. **5**, 151—153 (1945).

Stahmann, M. A., C. F. Huebner u. K. P. Link: Studies on the hemorrhagic sweet clover disease. V. Identification and synthesis of the hemorrhagic agent. J. Biol. Chem. **138**, 513—527 (1941).

Stevenson, T. M. u. J. L. Bolton: An evaluation of the self-tripping character in breeding for improved seed-yield in alfalfa. Emp. J. Exp. Agric. **15**, 82—88 (1947).

— u. W. J. White: Investigations concerning the coumarin content of sweet clover. I. The breeding of a low-coumarin line of sweet clover-Melilotus alba. Sci. Agric. **21**, 18—28 (1940).

Stitt, R. E.: Variation in tannin content of clonal and open-pollinated lines of perennial lespedeza. J. Amer. Soc. Agron. **35**, 944—954 (1943).

— Natural crossing and segregation in sericea lespedeza, Lespedeza cuneata (Dumont) G. Don. J. Amer. Soc. Agron. **38**, 1—5 (1946).

Sturkie, D. G.: A clover „worth growing". Sth. Seedsman **7**, Nr. 6; 13, 41 (1944).

Svešnikova, I. N.: Cytogenetical analysis of heterosis in hybrids of Vicia. J. Hered. **31**, 349—360 (1940).

— A new technique for the comparative cytological analysis of species. C. R. (Doklady) Acad. Sci. URSS. **30**, 756—758 (1941).

Swenson, S. P.: Inheritance of seed color in biennial white sweet clover, Melilotus alba. J. Amer. Soc. Agron. **34**, 452—459 (1942).

Sysina, N. A.: (Embryologische Untersuchung von Lathyrus articulatus L. und L. clymenum L.) C. R. (Doklady) Acad. Sci. URSS. **58**, 2097—2100 (1947). [Russisch.]

Telford, H. S.: Wireworm injury and potato varieties. Bi-m. Bull. N. Dak. Agric. Exp. Sta. **4**, Nr. 5, 7—8 (1942).

Tkačenko, P. I.: (Colonal selection in potatoes.) Vestnik Ovoščevodstvo i Kartofelj (Vegetable and Potato Journal), Nr. 2, 63—66 (1940). [Russisch.]

Toevs, J. L.: New regional alfalfa varieties show promise. U. and I. Cultivator **4**, p. 10, 26 (1944).

Tome, G. A. u. I. J. Johnson: Self- and cross-fertility relationships in Lotus corniculatus L. and Lotus tenuis Wald. et Kit. J. Amer. Soc. Agron. **37**, 1011 bis 1023 (1945).

Torssell, R.: Odling, förädling och fröodling av blåluzern. (Cultivation, improvement and seed production of blue lucerne.) Sverig. Utsädesfören. Tidskr. **53**, 275—284 (1943).

Trankowskij, D. A.: (The chromosomes of certain species of vetch. [Lathyrus].) Učenye Zapiski Moskovskogo Gosudarstvennogo Universiteta. Trudy Instituta Botaniki (Scientific Proceedings of the Moscow State University. Transactions of the Institute of Botany) **36**, 102—111 (1940). [Russisch.]

Travin, I. S. u. V. D. Ščerbačeva: (Red clover.) Moskva, Pp. 392 (1941). [Russisch.]

Troll, H. J.: Vegetationsbeobachtungen an Lupinen in verschiedenen geographischen Breiten. Züchter **12**, 129—139 (1940).

— Saatzeitversuche mit Zucht- und Landsorten sowie Wildformen von Lupinus luteus und Lup. angustifolius. Pflanzenbau **16**, 403—430 (1940).

— Kornertragsqualität verbessernde, schnelltrocknende, kahlhülsige, gelbe Lupinen. Züchter **13**, 283—289 (1941).

— Anthocyanmutanten bei Lupinus angustifolius L. Züchter **15**, 73—78 (1943).

Trotter, A. u. Cristinzio: Prove di allevamento e di selezione della patata a mezzo di semi. I—II. (Experiments in improving and selection of potato by means of seeds. I—II.) Ricerche Osserv. e Divulg. Fitopat.

Camp. ed Mezzog. R. Osserv. Reg. Fitopat. Portici **8**, 1—64 (1940).
Tysdal, H. M.: Discussion on yield of self-fertilised lines of alfalfa. Rep. 8th Alfalfa Improvement Conf. July 8 and 9, Fort Collins, Colorado, Pp. 38 (1940). [Mimeographed.]
— 1942 Report of the Uniform Advanced Alfalfa Nurseries. U. S. Dep. Agric. Res. Admin. Bur. Pl. Ind., Soils, Agric. Engng, Div. For. Crops and Diseases, Washington, Pp. 17 (1943). [Mimeographed.]
— Report of the Uniform Alfalfa Nurseries. U. S. Dep. Agric., Agric. Res. Admin. Bur. Pl. Industr. Soils Agric. Engin. Beltsville, Maryland 1944, Pp. 26 (1945). [Mimeographed.]
— Influence of tripping, soil moisture, plant spacing, and lodging on alfalfa seed production. J. Amer. Soc. Agron. **38**, 515—535 (1946).
—, T. A. Kiesselbach u. H. L. Westover: Alfalfa breeding. Res. Bull. Neb. Agric. Exp. Sta., Nr. 124, 46 (1942).
— u. J. R. Garl: A new method for alfalfa emasculation. J. Amer. Soc. Agron. **32**, 405—407 (1940). (Abstr.)
— u. T. A. Kiesselbach: Hybrid alfalfa. J. Amer. Soc. Agron. **36**, 649—667 (1944).

Ufer, M.: Ein züchterisch brauchbares Verfahren zur Auslese cumarinarmer Formen beim Steinklee. Züchter **11**, 317—321 (1939).

Vansell, G. H. u. F. E. Todd: Alfalfa tripping by insects. J. Amer. Soc. Agron. **38**, 470—488 (1946).
Vasilčenko, J. T.: (Generis Medicago species novae [Falcago Rchb. emend. Boiss.].) Botaničeskij Zurnal (Journ. Botanique de l'URSS), **31**, Nr. 3, 23 (1946). [Russisch.]
Voronjuk, B. A.: (Botanical varieties of spring vetch and their productiveness.) Trudy Zonaljnogo Inst. Zernovogo Hozjajstva Rajonov Nečernozemnoj Polosy (Trans. Zonal Inst. Grain Husbandry Non-Black-Soil. Districts), Nr. 10, 79—83 (1941).
Voronjik, B. A.: (The relationship between agriculturally desirable characters in spring vetch and the date of seed production.) Trudy Zonaljnogo Inst. Zernovogo Hozjajstva Rajonov Nečernozemnoj Polosy (Trans. Zonal Inst. Grain Husbandry Non-Black-Soil Districts), Nr. 10, 85—91 (1941). [Russisch.]

Weihing, R. M. u. D. W. Robertson: The comparative performance of alfalfa varieties in nursery and field plots in irrigated soil infested with Phytomonas insidiosa. J. Amer. Soc. Agron. **35**, 125—136 (1943).
Weimer, J. L.: A new species of Colletotrichum on vetch. Phytopathology **35**, 977—990 (1945).
— Lespedeza anthracnose. Phytopathology **36**, 524 bis 533 (1946).
— u. J. M. Elrod: Powdery mildew of annual lespedezas. Plant Dis. Reporter **30**, 13—16 (1946). [Mimeographed.]
Weiss, F.: Viruses described primarily on leguminous vegetable and forage crops. Plant Dis. Reporter Suppl. 32—80 (1945). [Mimeographed.]
Werth, E.: Der Spörgel, eine urgermanische Futterpflanze. (Zur Geographie und Geschichte der Kulturpflanzen und Haustiere XXIV.) Angew. Bot. **25**, 349—354 (1943).
Westover, H. L.: Alfalfa varieties in the United States. Fmrs' Bull. U. S. Dep. Agric., Nr. 1731, 17 (1934) (revised 1945).
— Report of the Ninth Alfalfa Improvement Conderence November 12, 1942, St. Louis, Missouri, Pp. 19 (Mimeographed).
Wexelsen, H.: Studies on fertility, inbreeding and heterosis in red clover (Trifolium pratense L.) Norske Vid. Akad. Oslo Mat.-Naturv. Klasse, Nr. 1, 141 (1945).
Wexelsen, H.: Undersøkelser over blomstring, frosetting og frøsetting og frøavl i luserne. (Investigations on flowering, seed setting and breeding in lucerne.) Tidsskr. Norske Landbr. **53**, 125—161 (1946).
— u. S. Skaare: Rødkløverens blomstring of bestøvning. (Flowering and pollination of red clover.) Tidsskr. Norske Landbr. **47**, 53—67, 109—122 (1940).
White, W. J.: An improved method of rooting alfalfa cuttings. Sci. Agric. **26**, 194—197 (1946).
Willard, C. J.: Kinds and varieties of sweetclover. Bi-m. Bull. Ohio Agric. Exp. Sta. **30**, 22—26 (1945).
Williams, R. D.: Genetics of chlorophyll deficiencies in red clover (Trifolium pratense L.). I. Albinos. J. Genet. **37**, 441 (1939).
— Genetics of chlorophyll deficiencies in red clover (Trifolium pratense. L.). II. Yellow seedling factors. J. Genet. **37**, 459 (1939).
— Genetics of cyanogenesis in white clover (Trifolium repens). J. Genet. **38**, 357—365 (1939).
— Incompatibility alleles in Trifolium pratense L.; their frequency and linkage relationships. Proc. 7th Int. Genet. Congr. Edinburgh 23—30 August, Pp. 316 1939 (1941). (Abstr.)
Williams, W.: Varieties and strains of red and white clover-British and foreign. Welsh Plant Breed. Sta. Aberystwyth, Ser. H, Nr. 16, 26 (1945).
— Genetics of red clover (Trifolium pratense L.) compatibility. II. (a) Homozygous self-sterile S_xS_x genotypes obtained as a result of pseudo-fertility; (b) self-fertility. J. Genet. **48**, 51—68 (1947).
— Genetics of red clover (Trifolium pratense L.) compatibility. III. The frequency of incompatibility S alleles in two nonpedigree populations of red clover. J. Genet. **48**, 69—79 (1947).
Wilson, J. K.: Variation in seed as shown by symbiosis. Mem. Cornell Agric. Exp. Sta., Nr. 272, 21 (1946).
Winkler, H.: Gul sötlupin—en ny, rikgivande foderbaljväxt för lättare jordar. (Yellow sweet lupin—a new productive fodder legume for lighter soils.) Svenska Vall-MosskFören. Kvart. **5**, 393—412 (1943).
Wipf, L.: Chromosome numbers in root nodules and root tips of certain Leguminosae. Bot. Gaz. **101**, 51—67 (1939).
— u. D. C. Cooper: Somatic doubling of chromosomes and nodular infection in certain Leguminosae. Amer. J. Bot. **26**, 3 s (1939). (Abstr.)
—, — Somatic doubling of chromosomes and nodular infection in certain Leguminosae. Amer. J. Bot. **27**, 821—824 (1940).
Wuttke, H.: Einfache Alkaloiduntersuchungsmethoden von gelben und blauen Lupinen. Züchter **14**, 83—86 (1942).
— Gegen Fusarium oxysporum resistente Stämme der gelben Lupine. Züchter **15**, 31—33 (1943).

Young, J. O: Cytological investigations in Desmodium and Lespedeza. Bot. Gaz. **101**, 839—850 (1940).

Zahnley, J. W.: Madrid sweet clover, a new crop. 32nd Bienn. Rep. Kansas Bd. Agric. **37**, 224—227 (1940).
Zimmermann, K.: Histologische und entwicklungsgeschichtliche Untersuchungen an Hülsen von Lupinus luteus, insbesondere der Neuzüchtung 3535 A mit nichtplatzenden Hülsen. Züchter **14**, 129—132 (1942).
— Weitere Untersuchungen zu den meteorologischen Grundlagen für die Auslese u. Prüfung von Lupinen mit nichtplatzenden Hülsen (Lupinus luteus und Lupinus angustifolius.) Züchter **14**, 165—167 (1942).
— Einige Versuche zum Problem der Hartschaligkeit bei Lupinus angustifolius. Züchter **14**, 182—185 (1942).

d) Wurzel- und Knollengewächse.

Abegg, F. A.: List of characters and gene symbols reported for the species Beta vulgaris L. Proc. Amer. Soc. Sug. Beet Technol. Pp. 109—113 (1940). [Mimeographed.]

Adams, J. B.: Aphid resistance in potatoes. Amer. Potato J. **23**, 1—22 (1946).

Agerberg, L. S.: Sortförsök med potatis i Norrbotten 1914—1938. (Variety trials with potatoes in Norbotten 1914—1938.) Sverig. Utsädesfören. Tidskr. **49**, 307 bis 325, 377—396 (1939).

Akeley, R. V. u. F. J. Stevenson: Yield, specific gravity, and starch content of tubers in a potato breeding program. Amer. Potato J. **20**, 203—217 (1943).

—, — The inheritance of dry-matter content in potatoes Amer. Potato J. **21**, 83—89 (1944).

Akimoto, S.: (Some experiments on hastening flowering in the sweet potato [Ipomoea Batatas]). Agric. Hort. Bd. **11**, 993—998 (1939).

Allen, T. C. u. G. H. Rieman: Occurrence of hopperburn resistance and susceptibility in the potato. Amer. Potato J. **16**, 139—142 (1939).

Anderson, W. S. et al.: Regional studies of time of planting and hill spacing of sweet potatoes. Circ. U. S. Dep. Agric. Nr. 725, 20 (1945).

Anonymus: Growing ,,roots" for seed. ,,Growmore" Leafl. Nr. 85, 4.

— Potatoes from Denmark. Potato Export Cttee Minist. Agric. in collaboration with Agric. Coun. Denmark Pp. 45 (ohne Jahr).

— Potato-breeding experiments in Russia. J. Dep. Agric. Eire, **36**, 305—306 (1939).

— Schurftziekte bij aardappelen. (Scab in potatoes.) Dep. Econ. Zaken Direct. Landb. 'S-Gravenhage, Pp. 118 (1939).

— Electric lights mature hard-to-get potato seed. Market Gr. J. **66**, 199 (1940).

— New potato needed in Maine. Market Gr. J. **67**, 495 (1940).

— New Sequoia potato is widely acclaimed. Ext. Farm-News **25**, Nr. 12, 4. (1940).

— New blight-resistant potato varieties. Int. Rev. Agric. **32**, 201 M. (1941).

— Beet scab in New York fields. Plant. Dis. Rep. **26**, 438 (1942).

— Kräftimmuna potatissorter. (Wart immune varieties of potatoes.) Flygbl. Växtskyddsanst. Stockh. Nr. 62, 4 (1942).

— Where did the potato come from? Mon. Sci. News Nr. 7, 2—3 (1942).

— Yellow sweet potato high in vitamin ,,A". Sth. Seedsman **5**, Nr. 12, 36 (1942).

— Two potato crops in one year. Mon. Sci. News Nr. 8, 3 (1942).

— Items. A fine new potato named Mohawk. Science Suppl. **98**, 10 (1943).

— Les maladies de la pomme de terre. (The diseases of the potato.) Minist. Agric. Ravitaillement, Paris, Pp. 67 (1943).

— Plant recording. Potato ,,Dominion", raised by James Flemming, ,,Sunny Mount Ohakune Junction. J. Roy. N. Z. Inst. Hort. **13**, 10 (1943).

— Lord Derby gold medal trials, 1939—1942 (inclusive). J. Nat. Inst. Agric. Bot. **5**, 70—80 (1944).

— Dry rot of potatoes. Adv. Leafl. Minist. Agric. Fish., Lancs. Nr. 218, 4 (1944).

— Sequoia potato shows leafhopper resistance. Sth. Seedsman **7**, 49 (1944).

— Plant industry. 4. Potato yields increased. Agriculture, Moskau, Nr. 4, 9—10 (1945).

— Åtgärder för bekämpande av bladmögel och brunröta hos potatis. (Measures for combating blight and brown rot in potatoes.) Flygbl. Växtskyddsanst., Stockh. Nr. 70, 8 (1945).

Anonymus: New potatoes raised in Northern Ireland. Gdnrs' Chron. **117**, 14 (1945).

— Potatoes. Bull. Me Agric. Exp. Sta. Nr. 438, 505—602 (1945).

— Strains of swedes. Fmrs' Leafl. Nat. Inst. Agric. Bot., Cambridge Nr. 10, Pp. 4 (1946).

— Trials of potatoes for immunity from wart disease. J. Minist. Agric., Lond. **52**, 475—476 (1946).

— Vers une pomme de terre résistante aux pucerons. La Pomme de Terre Française **9**, Nr. 85, 3—6 (1946).

— La pomme de terre en Tchécoslovaquie. La Pomme de Terre Française **9**, Nr. 86, 14—16 (1946).

— Varieties of potatoes. Fmrs' Leafl. Nat. Inst. Agric. Bot., Cambridge (Revised Ed.) Nr. 3, 4 (1946).

— What the scientists are doing. Potatoes. Indian Fmg. **7**, 84—86 (1946).

— Potato-growing trials. Leafl. Min. Agric. N. Ire Nr. 7, 11 (1946).

— Potato variety trials. Leafl. Min. Agric. N. Ire. Nr. 7, 12 (1947).

— Original varieties of potatoes. Sativa, Havlížkův Brod, Pp. 40 (1947).

— Premières observations sur les variétés étrangères cultivées en 1946. La Pomme de Terre Française **10**, Nr. 90, 10—13 (1947).

— Riktsortlista över önskvärda matpotatissorter uppgjord av Sveriges Potatisodlares Riksförbund år 1947. (A guide list of varieties of Swedish Potato Growers, 1947.) Lantmannen **31**, 157—158 (1947).

Armstrong, J. M.: Production and value of polyploid field roots. Sci. Agric. **22**, 787—798 (1942).

Arnason, T. J.: Sterility in potatoes. Canad. J. Res. **19**, Sect. C. 145—155 (1941).

— Female sterility in potatoes. Canad. J. Res. **21**, Abt. C, 41—56 (1943).

Arnaud, G.: La ,,gale noire" ou ,,galle verruqueuse" de la pomme de terre. Ann. Épiphyt. **8**, 89—98 (1942).

Arnautov, V. V.: (Transmutation of the potato plant by controlled nutrition.) Ovoščevodstvo (Vegetable Growing) Nr. 12, 24—26 (1939). [Russisch.]

— (Changing the nature of the potato plant by controlled nutrition.) Vestnik Ovoščevodstvo i Kartofelj (Vegetable and Potato Journal) Nr. 1, 6—9 (1940). [Russisch.]

Babb, M. F., J. E. Kraus u. G. H. Starr: Tolerance to psyllid yellows of potato varieties as reflected in yields. Amer. Potato J. **21**, 321—341 (1944).

Baensch, H.-U.: Methodische und vergleichende Untersuchungen über die Fleischfestigkeit von Betarüben. Z. Pflanzenz. **25**, 8—37 (1943).

Bald, J. G.: A report on agricultural features of the Australian potato industry. Pamphl. Coun. Sci. Industr. Res. Aust. Nr. 106, 72 (1941).

— Transmission of potato virus diseases. 4. Ground work studies on the growth of normal potato foliage. J. Coun. Sci. Industr. Res. Aust. **17**, 94—111 (1944).

— Development of differences in yield between FX and virus X-infected Up-to-Date potatoes. J. Coun. Sci. Industr. Res. Aust. **17**, 263—273 (1944).

— A plan of growth, maturity and yield of the potato plant. Emp. J. Exp. Agric. **14**, 43—48 (1946).

— Potato virus diseases in Australia. Farming, Lond. **1**, 182—185 (1947).

— u. G. A. H. Helson: Estimation of damage to potato foliage by potato moth, Gnorimoschema operculella (Zell). J. Coun. Sci. Industr. Res. Aust. **17**, 30—48 (1944).

—, D. O. Norris u. G. A. H. Helson: Transmission of potato virus diseases 5. Aphid populations, resistance, and tolerance of potato varieties to leaf roll. Bull. Coun. Sci. Industr. Res. Aust. Nr. 196, 32 (1946). [Mimeographed.]

— u. C. E. W. Oldaker: Reactions of Tasmanian Bismark and Brownell potatoes to the commoner virus

diseases. J. Coun. Sci. Industr. Res. Aust. **18**, 209 bis 218 (1945).
Bald, J. G. u. A. T. Pugsley: The main virus diseases of the potato in Victoria. Pamphl. Coun. Sci. Industr. Res. Aust. Nr. 110, 40 (1941).
— u. N. H. White: Potato virus X: The average severity of strain mixtures in three varieties of potato. J. Coun. Sci. Industr. Res. Aust. **15**, 300—306 (1942).
Balls, E. K.: Expedition to the Andes, 1938—1939. J. R. Hort. Soc. **65**, 289—295 (1940).
— Potatoes and other plants in the Andes. Gdnrs'Chron. **107**, 9—10, 128—129, **108**: 52—53, 100—101 (1940).
Barabanov, P. N. u. I. M. Piunowskij: (New sorts of potatoes in the BSSR.) Ann. White Russian Agric. Inst., Gorki **10** (32), 59—66 (1939). [Russisch.]
Barham, H. N., G. Kramer u. G. N. Reed: Influence of various factors on the starch content of Kansas-grown potatoes and sweet potatoes. J. Agric. Res. **67**, 395—406 (1943).
—, J. A. Wagoner, B. M. Williams u. G. N. Reed: A comparison of the viscosity and certain microscopical properties of some Kansas starches. J. Agric. Res. **68**, 331—345 (1944).
Barnell, H. R. u. D. W. P. Greenham: The effect of variety and source on the quality of dehydrated potatoes. Chem. Ind. Lond. Nr. 23, 315—318 (1947).
Bates, G. H.: Propagation of potato seed tubers from stems. Nature, Lond. **152**, 135 (1943).
Bawden, F. C.: Potato virus diseases. Nature, Lond. **150**, 476—477 (1942).
— u. B. Kassanis: Varietal differences in susceptibility to potato virus Y. Ann. Appl. Biol. **33**, 46—50 (1945).
Bazavluk, V. J.: (The process of fusion and chimaera formation in potatos grafts.) Bull. Acad. Sci. URSS. Ser. Biol. Nr. 2, 181—197 (1940). [Russisch.]
Becker, C. L.: Inheritance studies in the interspecific cross Solanum demissum Lindl. × S. tuberosum L. J. Agric. Res. **59**, 23—39 (1939).
Berggren, A. T. u. K. Vik: Forsøk med ulike rotvekstslag på klumprotsmittet jord. (Trials with different root crops on club root infected soil.) Meld. Norg. Landbr-Høgsk. **23**, 146—168 (1943).
Berkner: Eine „Studie" zu einem Rübensortenversuch. Pflanzenbau **16**, 430—434, 458—466 (1940).
Berljand, S. S.: (The inheritance by the potato of aerial tuber formation as a result of grafting.) Jarovizacija, Nr. 2 (35), 107—109 (1941). [Russisch.]
Björling, K.: Från Växtskyddsanstaltens Potatiskräftlaboratorium i Svalöv. (From the Plant Protection Institute Potato Wart Disease Laboratory in Svalöf.) Växtskyddsnotiser Nr. 6, 93—97 (1939).
— Fältförsök med potatiskräfta år 1944. (Field trials on potato wart disease 1944.) Växtskyddsnotiser Nr. 6, 81—84 (1944).
Black, W.: Science has answer to potato blight. Immune varieties — in due course. Fmg News **93**, Nr. 28, 5, 14 (1941).
— Inheritance of resistance to two strains of blight (Phytophthora infestans de Bary) in potatoes. Trans. Roy. Soc. Edinb. 1942—1943 **61**, 137—147 (1943).
— Inheritance of resistance to blight (Phytophthora infestans) in potatoes: unbalanced segregations. Proc. Roy. Soc. Edinburgh **62**, Sect. B: 171—181 (1945).
— u. G. Cockerham: Some modern aspects of potato production. Trans. Highl. Agric. Soc. Scot. Pp. 17 (1943).
— u. C. M. Driver: Potato breeding. British Intelligence Objectives Sub-committee, London, Final Rep. Nr. 1248, Item Nr. 22, Trip Nr. 2606, Pp. 31 (1946). [Mimeographed.]
Blodgett, F. M. u. F. J. Stevenson,: The new scab-resistant potatoes, Ontario, Seneca and Cayuga. Amer. Potato J. **23**, 315—329 (1946).

Boczkowska u. P. Grison: Vitesse de destruction des pousses de pommes de terre de différentes variétés par les insectes printaniers du doryphore. C. R. Acad. Agric. Fr. **29**, 333—334 (1943).
Bonde, R., F. J. Stevenson u. C. F. Clark: Resistant of certain potato varieties and seeding progenies to late blight in the tubers. Phytopathology **30**, 733—748 (1940).
—, —, — u. R. V. Arkeley: Resistance of certain potato varieties and seedling progenies to ring rot. Phytopathology **32**, 813—819 (1942).
Boonstra, A. E. H. R.: Rasverschillen bij bieten VI. Het verloop van den groei bij 7 bietenrassen, waaronder voederbieten zoowel als suikerbieten. (Race differences in beets. VI. The course of growth in 7 races of beets, both fodder and sugar beets.) Meded. Inst. Suikerbiet., Bergen-o.-Z. Nr. 2, 13—95 (1942).
Boswell, V. R. et al.: Place and season effects on yields and starch content of 38 kinds of sweet-potatoes. Circ. U. S. Dep. Agric. Nr. 714, 15 (1944).
Boza Barducci, T.: Plan genético para la producción de papas-semilla. (Genetical plan for the production of seed potatoes.) Bol. Estac. Exp. Agric. La Molina Nr. 24, 18 (1944).
Bryun, H. L. G. de: Methode voor het vastellen van de vatbaarheidsgraad van aardappelknollen voor de aardappelziekte. (Method for the determination of the susceptibility of potato tubers to late blight.) Tijdschr. PlZiekt. **49**, 77—99 (1943).
— Aardappelschurft en vruchtopvolging. (Potato scab and crop rotation.) Tijdschr. PlZiekt. **49**, 100—108 (1943).
Bukasov, S. M.: El origen de las especies de papa. (The origin of potato species.) Rev. Argent. Agron. **6**, 230 bis 236 (1939).
— The origin of potato species. Physics, B. Aires **18**, 41 bis 46 (1939).
— Interspecific hybridization in the potato. Physis, B. Aires **18**, 269—284 (1939).
— (The successes and failures of interspecific hybridization of the potato.) Vestnik Socialističeskogo Rastenievodstva (Soviet Plant Industry Record) Nr. 3, 39—48 (1940). [Russisch.]
— (New wild potato species from Argentine and Uruguay. Soviet Plant Industry Record Nr. 4, 3—12 (1940). [Russisch.]
— (Classification of potato species on dry matter content.) Soviet Plant Industry Record, Nr. 4, 144—145 (1940). [Russisch.]
— Una nueva especie de „Solanum" del subgénero „Tuberarium" de la República Argentina. (A new species of Solanum of the sub-genus Tuberarium from the Argentine Republic.) Rev. Argent. Agron **7**, 363 (1940).
— The geography of the endemic potatoes of South America. Rev. Argent. Agron. **8**, 83—104 (1941).
— (The origin of species of potatoes.) Vestnik Socialističeskogo Rastenievodstva (Soviet Plant Industry Record) Nr. 1, 157—164 (1941). [Russisch.]
Burger, I. J.: Studies on the processing of vegetables. (a) Some problems of the industrial utilization of vegetables. Sci. Bull. Dep. Agric. S. Afr. Nr. 253, 1—2 (1946).
Burnham, J. C. u. D. J. MacLeod: Varietal susceptibility of potatoes to aphid injury. Canad. Ent. **74**, 36 (1942).
Burton, W. G.: The characteristics of certain varieties of potato with special reference to their suitability for drying. Ann. Appl. Biol. **31**, 89—96 (1944).
— Anther and petal colour of potato varieties. Nature, Lond. **155**, 733 (1945).
Bushnell, J.: Experiments with early potatoes on sandy loam in southern Ohio. Bi-m. Bull. Ohio Agric. Exp. Sta. **27**, 63—70. (1942).

Bushnell, J., J. P. Sleesman u. F. J. Stevenson: Erie, a late potato, adapted to Ohio. Amer. Potato J. **22**, 29—32 (1945).

Cadman, C. H.: Nature of tetraploidy in cultivated European potatoes. Nature, Lond. **152**, 103—104 (1943).
— Autotetraploid inheritance in the potato: some new evidence. J. Genet. **44**, 33—52 (1942).

Caldwell, J. S., B. C. Brunstetter, C. W. Culpepper u. B. D. Ezell: Causes and control of discoloration in dehydration of white potatoes. Canner **100**, Nr. 13, 35—39, 112, 114, 118, 120, 122; Nr. 14, 15—16, 18, 30—32, 34; Nr. 15, 14, 16, 24, 26—27 (1945).
—, C. W. Culpepper u. P. M. Lombard: Suitability for dehydration in white potatoes as determined by the factors of variety and place of production. I. Amer. Potato J. **21**, 211—216 (1944).
—, — u. F. J. Stevenson: Suitability for dehydration in white potatoes as determined by the factors of variety, place of production, and stage of maturity. II. Amer. Potato J. **21**, 217—229 (1944).
—, P. M. Lombard u. C. W. Culpepper: Variety and place of production as factors in determining suitability for dehydration in white potatoes. Canner **97**, Nr. 3, 30, 32, 34—35, 42, 44; Nr. 4, 14—17, 24; Nr. 5, 15—16, 18—19, 28 (1943).

Cárdenas, M.: Enumeración de las papas silvestres de Bolivia. Descripción de dos especies nuevas de Cochabamba. (List of the wild potatoes of Bolivia. Description of two new species from Cochabamba.) Rev. Agric. Bolivia **2**, Nr. 2, 27—36 (1944).
— Notas sobre taxonomía de plantas económicas de Bolivia. Algunas especies bolivianas de Solanum (Sección Tuberarium) poco conocidas en los herbarios. (Notes on the taxonomy of economic plants of Bolivia. Some Bolivian species of Solanum [Section Tuberarium] little known in herbaria.) Rev. Agric., Bolivia **2**, Nr. 3, 78—84 (1945).
— u. J. G. Hawkes: New and little-known wild potato species from Bolivia and Peru. J. Linn. Soc. (Bot.) **53**, 91—108 (1946).

Carson, G. P. u. H. W. Howard: Self-incompatibility in certain diploid potato species. Nature, Lond. **150**, 290 (1942).
—, — Inheritance of the „bolter" condition in the potato. Nature, Lond. **154**, 829 (1944).
—, — Note on the inheritance of the King Edward type of colour in potatoes. J. Genet. **46**, 358—360 (1945).

Casas, A. B.: Las mejores variedades de boniato en Cuba. (The best varieties of sweet potato in Cuba.) Rev. Minist. Agric. Cuba **27**, Nr. 28, 41—43 (1944).

Cates, J. S.: New blight-immune potatoes. Co. Gent., p. 17, 67 (1945).

Chevalier, A.: Les espèces de Solanum cultivées venues du Nouveau-Monde. Les origines et les résultats de leur culture. Rev. Bot. Appl. **19**, 825—835 (1939).
— Révision de quelques Oxalis utiles ou nuisibles. Répartition géographique et naturalisation de ces espèces. Rev. Bot. Appl. **20**, 657—694 (1940).
— Un légume tropical à répandre: la petite pomme de terre d'Afrique (Coleus rotundifolius). Rev. Bot. Appl. **26**, 296—300 (1946).

Chitwood, B. G. u. E. M. Buhrer: The life history of the golden nematode of potatoes, Heterodera rostochiensis Wollenweber, under Long Island, New York, conditions. Phytopathology **36**, 180—189 (1946).

Chopinet, R.: Sur quelques hybrides expérimentaux interspécifiques et intergénériques chez les crucifères. C. R. Acad. Sci. **215**, 545—547 (1942).

Choudhuri, H. C.: Cytological studies in the genus Solanum. I. Wild and native cultivated „diploid" potatoes. Trans. Roy. Soc. Edinb. **61**, 113—135 (1943).

Coudhuri, H. C.: Cytological and genetical studies in the genus Solanum. II. Wild and cultivated „diploid" potatoes. Trans. Roy. Soc. Edinb. 1942—1944 **61**, 199—219 (1944).

Clark, C. F.: Potato breeding investigations in 1938: Review of literature. Amer. Potato J. **16**, 212—220 (1939).
— The Calrose potato: a new variety possessing resistance to late blight. Amer. Potato J. **23**, 343—347 (1946).
— u. P. M. Lombard: Descriptions of and key to American potato varieties. Circ. U. S. Dep. Agric., Nr. 741, 50 (1946).
—, — u. E. F. Whiteman: Coking quality of the potato as measured by specific gravity. Amer. Potato J. **17**, 38—45 (1940).
—, F. J. Stevenson u. L. A. Schaal: The inheritance of scab resistance in certain crosses and selfed lines of potatoes. Phytopathology **28**, 878 (1938).

Clarke, A. E., W. C. Edmundson u. P. M. Lombard: Seed-setting in potatoes as affected by spraying with a-naphthaleneacetamide and by light. Amer. Potato J. **18**, 273—279 (1941).
— u. P. M. Lombard: Flower bud formation in the potato plant as influenced by variety, size of seed piece, and light. Amer. Potato J. **19**, 97—105 (1942).

Clinch, P. E. M.: Observations on a severe strain of potato virus X. Sci. Proc. R. Dublin Soc. **23**, 273 bis 299 (1944).
—, J. B. Loughnane u. R. McKay: Notes on the leaf roll and mosaic diseases of potatoes in relation to seed potato production. J. Dep. Agric. Éire **41**, 263 bis 276 (1944).

Cochran, H. L.: The carotene content of sweet potatoes. Proc. Amer. Soc. Hort. Sci. **41**, 259—264 (1942).

Cockerham, G.: Potato breeding for virus resistance. Ann. Appl. Biol. **30**, 105—108 (1943).
— The reactions of potato varieties to viruses X, A, B and C. Ann. Appl. Biol. **30**, 338—344 (1943).

Costa, A. S.: Notas sôbre o melhoramento da batata nos Estados Unidos. (Notes on the improvement of the potato in the United States.) Bragantia, S. Paulo **3**, 347—366 (1943).

Coulter, F. C.: The story of garden vegetables. XIV. The potato: „A potato by analogy and Irish by adoption". Seed World **52**, Nr. 12, 38—39 (1942).
— Story of garden vegetables. XV. Sweet potatoes: one of the first American foods to reach the Old World. Seed World **53**, Nr. 2, 38—39 (1943).

Crépin, C. u. J. Bustarret: Quelques problèmes de l'amélioration de la pomme de terre. Acad. Agric. Fr., Pp. 10 1941 (1942).

Crescini, F.: Il „Rafanobrassica" (Raphanus sat. x Brassica oler.). (Raphanobrassica [R. sativus x B. oleracea].) Ital. Agric. **79**, 253—258 (1942).

Curtis, L. C.: The effect of storage on the betanin and sucrose content of garden beets (Beta vulgaris) and its importance in a breeding program with this crop. Proc. Amer. Soc. Hort. Sci. **41**, 370—374 (1942).

Darpoux, H.: Le mildiou de la pomme de terre — son traitement. La Pomme de Terre Française **10**, Nr. 90, 5—8 (1947).

Davey, V. McM.: Hybridization in Brassicae and the occasional contamination of seed stocks. Ann. Appl. Biol. **26**, 634—636 (1939).

Davidson, W. D.: Famous potato raisers. J. R. Agric. Soc. **100**, Pt. II, Pp. 1—22 (1939).

Delaney, D.: The potato crop. J. Dep. Ageic. Eire **41**, 290—294 (1944).

Díaz de Mendivil, J. M.: La mejora del cultivo de la patata en España. (The improvement of potato cultivation in Spain.) Agricultura, Madrid **13**, 55—68 (1944).

Dorst, J. C.: Schurftaantasting bij nakomelingen van verschillende aardappelkruisingen. (Scab infection in the progenies of different potato crosses.) Tijdschr. PlZiekt. **45**, 157—161 (1939).

Dove, W. F., E. F. Murphy u. R. V. Akeley: Varietal differences and inheritance of vitamins C and A in potatoes. Genetics **28**, 72—73 (1943). (Abstr.)

Driver, C. M.: Recent advances in potato breeding. N. Z. J. Sci. Tech. **23**, 180A—184A (1941).

— u. J. G. Hawkes: Photoperiodism in the potato. Imperial Bureau of Plant Breeding and Genetics, Cambridge, Pp. 36, 3 figs. 3 tabl. (1943).

Dufrénoy, J.: La production aux Etats-Unis d'hybrides de pommes de terre résistants aux maladies à virus. Rev. Hort. Paris **28**, 7—8 (1942).

Dykstra, T. P.: Potato virus diseases: review of literature 1941. Amer. Potato J. **19**, 267—279 (1942).

— Results of experiments in control of bacterial ring rot of potatoes in 1940. Amer. Potato J. **18**, 27—54 (1941).

Edmond, J. B. u. J. A. Martin jr.: The flowering and fruiting of the sweet potato under greenhouse conditions. Proc. Amer. Soc. Hort. Sci. **47**, 391—399 (1946).

Edmundson, W. C.: Response of several varieties of potatoes to different photoperiods. Amer. Potato J. **18**, 100—112 (1941).

— Comparison of Katahdin potato pollen produced in the field and in the greenhouse. Amer. Potato J. **19**, 12—15 (1942).

—, B. J. Landis u. L. A. Schaal: Potato production in the Western states. Fmrs' Bull. U. S. Dep. Agric., Nr. 1843, 58 (1945).

— u. L. A. Schaal: Potato breeding for Fusarium resistance. Amer. Potato J. **17**, 92—95 (1940).

—, — u. A. M. Binkley: Pawnee potato is new variety ‚tailored' for northern Colorado growing conditions. Colo. Fm. Bull. **5**, Nr. 2, 13—15 (1943).

—, —, — The Pawnee potato. Circ. U. S. Dep. Agric., Nr. 665, 6 (1943).

Eig, A.: A revesion of the Chenopodiaceae of Palestine and neighbouring countries. Palest. J. Bot. **3**, 119 bis 137 (1945).

Eikeland, H. J.: Er nepesortane ulikt sterke mot „vattersott"? (Do varieties of turnips differ in resistance to „dropsy".) Nord. Landbruk, Nr. 18 (1939).

Eliasson, S.: Den lokala sortförsöksverksamheten. IV. Sammanställningar av resultaten av sortförsöken med potatis under åren 1931 (1915)—1941. (The work of local variety trials. IV. Comparisons of the results of the variety trials with potatoes during 1931(1915) bis 1941.) Medd. Lantbrukshögskolan Jordbruksförsöksanstalten, Nr. 10, 258 (1944).

Ellenby, C.: Standardization of root excretions for immunity trials on the potato root eelworm. Nature Lond. **154**, 363—364 (1944).

— Susceptibility of South American tuberforming species of Solanum to the potato-root eelworm Heterodera rostochiensis Wollenweber. Emp. J. Exp. Agric. **13**, 158—168 (1945).

— The influence of potato variety on the cyst of the potato-root eelworm, Heterodera rostochiensis Wollenweber. Ann. Appl. Biol. **33**, 433—446 (1946).

Ellis, N. K.: A comparison of stands and yields from seed pieces cut from the apical and stem ends of Irish Cobbler, Russet Rural, and Chippewa potatoes. Proc. Amer. Soc. Hort. Sci. **40**, 516—518 (1942).

Erwin, A. T.: The Federal potato breeding program. Trans. I. a Hort. Soc. **77**, 336—340 (1942).

Esbo, H.: Potatisodlingen och dess viktigaste problem i olika länder. Erfarenheter från studieresor, i Danmark, Tyskland, Holland och England. (Potato cultivation and its most important problems in different countries. Impressions from visits to Denmark, Germany, Holland and England.) K. Lantbr. Acad. Tidskr. **80**, 481—505 (1941).

Esbo, H.: En del vanliga potatissorters reaktion mot vissa vira. (The reaction of some of the common potato varieties to certain viruses.) K. LantbrAkad. Tidskr. **84**, 299—313 (1945).

Ext, W. u. H. Goffart: 10 Jahre Kampf gegen den Kartoffelnematoden in der Provinz Schleswig-Holstein. Angew. Bot. **24**, 1—16 (1942).

Ezell, B. D. u. M. S. Wilcox: The ratio of carotene to carotenoid pigments in sweet-potato varieties. Science **103**, 193—194 (1946).

Federova, N. J.: Isolierung von Phytophthoraresistenten Formen der Kartoffel durch Beurteilung der Nachkommenschaft. C. R. (Doklady) Acad. Sci. URSS. **32**, 282—284 (1941).

— Inoculation tests of potato for resistance to Phytophthora. C. R. (Doklady) Acad. Sci. URSS. **33**, 73—75 (1941).

Feistritzer, W.: Möglichkeiten einer systematischen Resistenzzüchtung gegen die Abbaukrankheiten der Kartoffel. Kühn-Archiv **60**, 347—357 (1943/1944).

Fernandez Valiela, M. V.: Principales virus que afectan a la papa cultivada (con especial referencia a Gran Bretaña). (Principal viruses which affect the cultivated potato [with special reference to Great Britain].) Cent. Estud. Agron., Fed. Univ. B. Aires, Pp. 112 (1946).

Filippov, A. S.: (Mixed inheritance and vegetative segregation in the potato.) Vestnik Ovoščevodstvo i Kartofelj (Vegetable and Potato Journal), Nr. 5, 106 bis 112 (1940). [Russisch.]

— („Vegetative rapprochement" as a means of overcoming cross-incomplatibility in potatoes.) Vestnik Ovoščevodstvo i Kartofelj (Vegetable and Potato Journal), Nr. 1, 36—42 (1941). [Russisch.]

— (Training the species S. semidemissum and its utilization for breeding.) Vestnik Ovoščevodstvo i Kartofelj (Vegetable and Potato Journal), Nr. 2, 60—66 (1941). [Russisch.]

Folsom, D.: Potato varieties: the newly named, the commercial, and some that are useful in breeding. Amer. Potato J. **22**, 229—242 (1945).

— u. F. J. Stevenson: Potato resistance to leaf roll. Phytopathology **34**, 999—1000 (1944). (Abstr.)

—, — Resistance of potato seedling varieties to the natural spread of leaf roll. Amer. Potato J. **23**, 247 bis 264 (1946).

Forbes, A. P. S.: Origin and early history of the potato. Nyasald Agric. Quart. J. **3**, 28—30 (1943).

Foulon, L. A.: El problema economic de la papa. (The economic problem of the potato.) Inst. Econ. Legislac. Rur., Fac. Agron. Vet., Univ. B. Aires **2**, 361 (1939).

Frandsen, H. N.: Kan vi haeve Foderroernes Udbytte 10—20 pCt. ved Anvendelse af Triploider? (Can we increase the yield of fodder beets by 10—20% by the use of triploids?) Tidsskr. Frøavl, Nr. 417, 172—174 (1947).

Frandsen, K. J.: Colchicininduzierte Polyploidie bei Beta vulgaris L. (Polyploidy in B. vulgaris L. induced by colchicine.) Züchter **11**, 17—19 (1939).

— Beiträge zur Cyto-Genetik der Brassica napus L., der Brassica campestris L. und deren Bastarden, sowie der amphidiploiden Brassica napocampestris. K. VetHøjsk. Aarsskr., Pp. 59—90 (1941).

— The experimental formation of Brassica juncea Czern. et Coss. D. Bot. Arkiv. **11**, Nr. 4, 1—17 (1943).

Gardner, M. E., R. Schmidt u. F. J. Stevenson: The Sequoia potato: a recently-introduced insect-resistant variety. Amer. Potato J. **22**, 97—103 (1945).

Gates, R. R.: Polyploidy in the potato. Nature, Lond. **152**, 416—417 (1943).

Gemmell, A. P.: The resistance of potato varieties to Heterodera schachtii, Schmidt, the potato-root eelworm. Ann. Appl. Biol **30**, 67—70 (1943).

Glover, J.: Environment and the growth of the potato (Solanum tuberosum) in tropical East Africa. Emp. J. Exp. Agric. **15**, 9—26 (1947).

Gluščenko, I. E.: (Genotypical variation of the potato resulting from conditions of growth.) Jarovizacija, Nr. 6 (33), 98—101 (1940). [Russisch.]

— (On the problem of genetic heterogeneity in plant tissues.) Proc. Lenin Acad. Agric. Sci. USSR., Nr. 5 bis 6, 35 (1944). [Russisch.]

— (Genetical differences in quality in tissues of potato.) Agrobiologija (Agrobiology), Nr. 1, 18—50 (1946). [Russisch.]

Goffart, H.: Resistenzprüfung von Kartoffelsorten gegenüber Heterodera schachtii Schmidt. Züchter **11**, 123—130 (1939).

Guern, A. P.: (On self-pollinated strains of potatoes.) Proc. Lenin Acad. Agric. Sci. USSR., Nr. 7, 29—36 (1940). [Russisch.]

H....., A.: Les remèdes à la dégénérescence physiologique de la pomme de terre en Union soviétique. Rev. Bot. Appl. **25**, 237—238 (1945).

Haddock, J. L. u. P. T. Blood: Variations in cooking quality of potatoes as influenced by varieties. Amer. Potato J. **16**, 126—133 (1939).

Hagberth, M. O.: Potatisodlareförbundets lista över matpotatissorter. (Potato Growers' Union's list of table varieties of potato.) Lantmannen **31**, 226 (1947).

Hansen, H. R. u. M. Nissen: Beskrivelse af Kartoffelsorter, dyrket i Danmark. (Description of varieties of potatoes grown in Denmark.) Tidsskr. Planteavl **49**, 559—663 (1945).

Harečko-Savickaja, H.: Selbststerilität und Selbstfertilität bei Beta vulgaris L. Z. Pflanzenz. **26**, 103 bis 118 (1944).

Harter, L. L.: Sweetpotato diseases. Fmrs' Bull. U. S. Dep. Agric. Nr. 1059, 26 (1944).

Hartman, R. E. u. R. V. Akeley: Potato wart in Amerika. Amer. Potato J. **21**, 283—288 (1944).

Haudricourt, A.: Les bases de la sélection de la pomme de terre. D'après S. M. Bukasov. Rev. Bot. Appl. **20**, 97—116, 179—189 (1940).

Hawkes, J. G.: Cytogenetic studies on South American potatoes. Abstr. Diss. Univ. Camb. Pp. 18—19, 1940 bis 1941 (1942).

— Potato collecting expeditions in Mexico and South America. Imperial Bureau of Plant Breeding and Genetics, Cambridge (1941).

— Potato Collecting Expeditions in Mexico and South America. II. Systematic classification of the collections. Imperial Bureau of Plant Breeding and Genetics, Cambridge. Pp. 141, tables 3 maps. 93 figs. 2 plates (1944).

— The indigenous American potatoes and their value in plant-breeding. Part I. Resistance to disease. Part II. Physiological properties, chemical composition, and breeding capabilities. Emp. J. Exp. Agric. **13**, 11—40 (1945).

— The story of the potato. Discovery Pp. 38—46 (1945).

— Potato „bolters": an explanation based on photoperiodism. Nature, Lond. **157**, 375—376 (1946).

— u. C. M. Driver: Origin of the first European potatoes and their reaction to length of day. Nature, Lond. **157**, 591 (1946).

—, — Origin of the first European potatoes and their reaction to length of day. Nature, Lond. **158**, 168—169 (1946).

—, — Origin of the first European potatoes and their reaction to length of day. Nature, Lond. **158**, 713 (1946).

Hawkes, J. G. u. H. W. Howard: Salaman's culture of blight resistant „Aya papa". Nature, Lond. **148**, 25 (1941).

Hellbo, E. u. H. Esbo: Våra potatissorter. Systematisk behandling. (Our potato varieties. A systematic study. Statens Centrala Frökontrollanstalt, Lantbruksförbundets Tidskriftsaktiebolag, Stockholm Nr. 1, 128, 3 plates. 15 figs. (1942).

Hendrickx, F. L.: Le Mildeou de la pomme de terre. Centre Afrique **10**, 3 (1943).

Hern, A. P.: (The question of re-examining the schema of potato breeding.) Vestnik Ovoščevodstvo i Kartofelj (Vegetable and Potato Journal) Nr. 2, 57—59 (1941). [Russisch.]

Hertzman, N.: Weibulls Slättbo Barres II, stam 16. Ny foderbetstam med mycket högt odlingsvärde. (Weibull's Slättbo Barres II, strain 16. A new fodder beet strain with very high cultivation value.) Weibulls Ill. Årsb. **36**, 5—7 (1941).

— Landets förråd av kålrotsfrö är otillräckligt. I hur stor utsträckning kan man odla foderbetor i stället för kålrötter? (The country's supply of rutabaga seed is inadequate. How far can fodder beet be grown instead of swedes?) Weibulls Ill. Årsb. **38**, 27—29 (1943).

— Weibulls-Original Slättbo Barres II, stam 16. Landets mest odlingsvärda foderbeta. (Weibull's Original Slättbo Barres II, strain 16. The fodder beet most worth growing in the country.) Weibulls Ill. Årsb. **39**, 8—11 (1944).

— Weibulls Original Drottning Kålrot, stam 38. En idealiskt vacker kålrot med hög avkastning. (Weibulls Original Drottning swede, strain 38. An ideal, fine swede with high yield. Weibulls Ill. Årsb. **39**, 14—16 (1944).

— Weibullsholms Växtförädlingsanstalts Original rotfruktsstammar. (Original strains of root crops from the Weibullsholm Plant Breeding Institute.) Weibulls Ill. Årsb. **40**, 30—37 (1945).

Hirst, F. u. W. B. Adam: Potatoes for canning. Effect of soil and variety. Rep. Fruit Veg. Preserv. Res. Sta. Campden Pp. 11—15 (1943).

Hodge, W. H.: Algunos tubérculos olvidados. (Some overlooked tuberiferous plants.) Rev. Fac. Nac. Agron., Colombia **6**, Nr. 22, 1—17 (1946).

Hollar, V. E. u. E. S. Haber: Factors related to stem-end shrink of the sweet potato. Proc. Amer. Soc. Hort. Sci. **46**, 359—369 (1945).

Holmberg, C.: Potatiskräftans bekämpande och potatissortfrågan. (Combating potato wart and the question of potato variety.) Lantmannen **25**, 195—198 (1941).

Howard, H. W.: The nomenclature of Brassica species. Curr. Sci. **9**, 494—495 (1940).

— Self-incompatibility in polyploid forms of Brassica and Raphanus. Nature, Lond. **149**, 302 (1942).

Hudson, P. S.: Work on the South American potato collection up to 31st December, 1944. Fifteenth Ann. Rep. Exec. Coun. Imp. Agric. Bur. 1943—1944, p. 22—23 (1945).

— Work on the South American potato collection up to 31st December, 1945. 16th Rep. Imp. Agric. Bureaux Executive Coun., London 1944—1945, p. 26—27 (1946).

Hutton, E. M.: Breaking dormancy of the potato. J. Coun. Sci. Industr. Res. Aust. **15**, 262—267 (1942).

— The relationship between necrosis and resistance to virus Y in the potato. 2. Some genetical aspects. J. Coun Sci. Industr. Res. Aust. **18**, 219—224 (1945).

— The relationship between necrosis and resistance to virus Y in the potato. 3. Interrelation with virus C. J. Coun. Sci. Industr. Res. Aust. **19**, 273—282 (1946).

— u. J. G. Bald: The relationship between necrosis and resistance to virus Y in the potato. J. Coun. Sci. Industr. Res. Aust. **18**, 48—52 (1945).

Ivanov, V. I.: Formation of polyploid forms in Solanum sect. Tuberarium. C. R. (Doklady) Acad. Sci. URSS. **24**, 486—488 (1939).
— A cytological survey of the reciprocal hybrids of the potato (Solanum Antipoviczii Buk. × S. tuberosum) × S. tuberosum L. C. R. (Doklady) Acad. Sci. URSS. **27**, 51—54 (1940).

Ivanovskaja, E. V.: Cytological study of Solanum Millanii Buk. et Lechn. C. R. (Doklady) Acad. Sci. URSS. **24**, 389—391 (1939).
— A haploid plant of Solanum tuberosum L. C. R. (Doklady) Acad. Sci. URSS. **24**, 517—520 (1939).
— Cytological analysis of hybrids between diploid and tetraploid species of potatoes.) Bull. Acad. Sci. URSS. Sér. Biol. Nr. 1, 21—33 (1941).

Janse, L. C.: Droge stof en refractie van voerbieten en suikerbieten in 1941. (Dry matter and refractive index of fodder beets and sugar beets in 1941.) Landbouwk. Tijdschr. Wageningen, **54**, 796—797 (1942).

Jaščuk, A. P.: (Reaction of potato varieties to manuring.) Ovoščevodstvo (Vegetable Growing) Nr. 1 39 bis 42 (1940). [Russisch.]

Jehle, R. A. u. F. J. Stevenson: The Potomac potato. Amer. Potato J. **22**, 261—266 (1945).

Jenkins, J. M. (jr.): The performance of certain potato varieties in South Carolina in 1942. Amer. Potato J. **19**, 213—216 (1942).

Jensen, J. H. u. R. W. Goss: Infection of first-year potato seedlings with Fusarium solani var. eumartii. Amer. Potato J. **19**, 239—242 (1941).

Jermoljev, E.: Serologie bei Kartoffelzüchtung. Z. Pflanzenz. **24**, 104—107 (1941).

Johnstone, F. E. (jr.): Chromosome doubling in potatoes induced by colchicine treatment. Amer. Potato J. **16**, 288—304 (1939).
— Experimentally induced chromosome doubling in Solanum tuberosum L. and related tuber-bearing species. Cornell Univ. Abstr. Thes. p. 331—334, 1940 (1941).

Jones, L. K., C. L. Vincent u. E. F. Burk: The resistance of progeny of Katahdin potatoes to viroses. J. Agric. Res. **60**, 631—644 (1940).

Jordán de Urries, F.: Obtención de nuevas variedades de patata. (Production of new varieties of potato.) Agricultura, Madrid **13**, 72—76 (1944).

Jørstad, I.: Potetkreftens utbredelse i Norge og fortegnelse over potetsorter prøvd mot kreft. (The extent of potato wart disease in Norway and a list of varieties of potatoes tested for resistance.) Meld. Skadeinsekter Plantesykdommer, Oslo 1938. Pp. C 56 (1940).

Josefsson, A.: C-vitaminundersökningar på potatis. (Vitamin-C investigations on potatoes.) Sverig. Utsädesfören Tidskr. **54**, 37—54 (1944).
— Torkresistens och rotmassa hos några potatissorter. (Drought resistance and root mass in some varieties of potatoes.) Sverig. Utsädesfören. Tidskr. **54**, 325—331 (1944).
— Rostfläcksjuka hos potatis. (Spraing in potatoes.) Sverig. Utsädesfören. Tidskr. **55**, 34—42 (1945).

Kagawa, F.: Chimeras in sweet potatoes. Jap. J. Bot. **10**, 43—54 (1939).
— (Chimeras in sweet potato plant.) Proc. Crop. Sci. Soc. Japan **10**, 377—383 (1939).

Kameraz, A. J.: (Wild species as initial material in potato breeding.) Soviet Plant Industry Record Nr. 4, 13—30 (1940). [Russisch.]
— (The use of the forms of S. andigenum Juz. et Buk. in potato breeding.) Vestnik Socialističeskogo Rastenievodstva (Soviet Plant Industry Record) Nr. 5, 165 bis 177 (1940). [Russisch.]
— (Wild Chilean species S. leptostigma Juz. and S. Molinae Juz. in potato breeding.) Vestnik Socialističeskogo Rastenievodstva (Soviet Plant Industry Record) Nr. 1, 180—187 (1941). [Russisch.]

Kangas, J. T.: Marygold: two-crop potato. Sth. Seedsman **8**, Nr. 12; 13, 32 (1945).

Kapenga, C.: De mogelijkheid van de kruising Solanum tuberosum ♀ × S. chacoense ♂.) (The possibility of the cross S. tuberosum ♀ × S. chacoense ♂.) Genetica **23**, 537—538 (1943).

Karikka, K. J., L. T. Dudgeon u. H. M. Hauck: Influence of variety, location, fertilizer, and storage on the ascorbic acid content of potatoes grown in New York State. J. Agric. Res. **68**, 49—63 (1944).

Kasparova, S. A. u. I. V. Glazunov: Variation of biochemical processes in resting potato tubers grown in the Arctic. C. R. (Doklady) Acad. Sci. URSS. **31**, 679—682 (1941).

Katin-Jarcev, L. V.: (Chief results and trends of work on improved potato varieties in Western Siberia.) Theses and scientific papers read at the 4th District Conference of Workers of Universities and Research Institutions, Omsk, Nr. 1, Agron. Sect. 60—64 (1941). [Russisch.]

Khan, Shame-Ul-Islam: Pollen sterility in different varieties of „Solanum tuberosum". Acta Brevia Sinensia Nr. 8, 17 (1944). [Mimeographed.]

Killinger, C. B. u. W. E. Stokes: Chufas in Florida. Bull. Fla. Agric. Exp. Sta. Nr. 419, 16 (1946).

Kimbrough, W. D.: Comparative yield and composition of the seedling L 4—5 and Triumph sweet potatoes. Proc. Amer. Soc. Hort. Sci. **40**, 444 (1942). (Abstr.)

Kirste, A.: Die Abhängigkeit des Stärkeprozents der Kartoffel von den Wachstumsbedingungen und Anbaumaßnahmen. Landw. Jb. **93**, 289—305 (1943).

Knudsen, V. E.: Forsøg med Kartoffelsorter 1937—1941. (Experiments with varieties of potatoes 1937—1941.) Tidsskr. Planteavl. **47**, 143—170 (1942).

Köhler, E.: Untersuchungen über Y-Virus-Resistenz bei Kartoffeln. Züchter **12**, 273—275 (1940).
— Die Überempfindlichkeitsreaktion bei Solanum nodiflorum Jacq. gegenüber Stämmen des Tabakmosaik- und des Kartoffel-X-Virus. Z. PflKrankh. **52**, 450 bis 454 (1942).
— Solanum demissum Lindh. als mögliche Testpflanze des A-Virus der Kartoffel. Nachr.Bl. dtsch. Pfl.sch.-dienst **22**, 77 (1942).
— u. O. Bode: Untersuchungen über die Variabilität der Stärkekornform in der Gattung Solanum, sectio Tuberarium. Züchter **15**, 135—143 (1943).
— u. J. Paukšens: Solanum demissum Lindl. als Testpflanze verschiedener Mosaikviren. Züchter **16**, 8—11 (1944).

Kolotova, S. S. u. A. P. Volodin: (Intensifying the drought-resistance of potatoes before sowing them.) Izvestija Biol. Naučno-issled. Inst. pri Molotovskom Gosud. Univ. imeni M. Gorjkogo (Bull. Inst. Rech. Biol. Molotov) **12**, 19—34 (1941). [Russisch.]

Kopetz, L. M.: Zeitstufenversuche mit der Kohlrabisorte „Roggli". Ein Beitrag zur Frage der Auslösung von Schossern. Züchter **14**, 136—137 (1942).

Kovalenko, G. M.: (Summer plantings of potatoes.) Soviet Plant Industry Record Nr. 4, 47—50 (1940). [Russisch.]

Koverga, A.: (The vernalization of forage turnip for purposes of seed production.) Doklady Vsesojuz. Akad. Seljsk. Nauk im. V. I. Lenina (Proc. Lenin. Acad. Agric. Sci. USSR.) Nr. 11/12, 42—48 (1944). [Russisch.]

Krantz, F. A.: Twenty-five years in the history of the potato. Amer. Potato **16**, 25—31 (1939).
— Potato breeding methods. III. A suggested procedure for potato breeding. Tech. Bull. Minn. Agric. Exp. Sta. Nr. 173, 24 (1946).
—, C. L. Becker u. Z. M. Fineman: Incidence and inheritance of pollen sterility in the potato. J. Agric. Res. **58**, 593—601 (1939.)
— u. C. J. Eide: Inheritance of reaction to common scab in the potato. J. Agric. Res. **63**, 219—231 (1941).

Krantz, F. A. u. A. G. Tolaas: The Red Warba potato. Amer. Potato J. **16**, 185—190 (1939).

Krasočkin, V. T.: (Breeding of beets resistant to bolting.) Vestnik Socialističeskogo Rastenievodstva (Soviet Plant Industry Record). Nr. 3, 77—86 (1940). [Russisch.]

Kreutz, H. u. M. v. Schelhorn: Brassica-Pormen als Grünfutterpflanzen. Forschungsdienst **9**, 440—452 (1940).

Krickl, M.: Zur Züchtung von frostwiderstandsfähigen und nicht holzig werdenden frühen Kohlrabi. Gartenbauwiss. **18**, 185—203 (1944).

Kristensen, R. K.: Kartoflernes Vaegtfylde og deres Indhold af Torstorf og Stivelse. (The specific gravity of potatoes and their content of dry matter and starch.) Tidsskr. Planteavl **46**, 661—672 (1942).

Kružilin, L. S. u. S. V. Ananjeva: (New methods of potato breeding and methods of cultivating.) Naučnyj Otčet Inst. Zernovogo Hozjajstva Jugo-Vostoka SSSR za 1941—42 gg. (Sci. Rep. Inst. Grain Husbandry South Eastern USSR for 1941—42), 226—242 (1944). [Russisch.]

L......, A.: Frühkartoffelzucht im Donauland. Obst. u. Gem. **6**, 227 (1942).

L......, G. N.: P. A. Olsson †. Nord. Jordbr. Forskn. Nr. 1/2, 44 (1945).

Lamm, R.: Varying cytological behaviour in reciprocal Solanum crosses. Proc. 7th Int. Genet. Congr. Edinburgh 23—30 August, p. 779, 1939 (1941). (Abstr.)

— Varying cytological behaviour in reciprocal Solanum crosses. Hereditas, Lund, **27**, 202—208 (1941).

— Notes on an octoploid Solanum punae plant. Hereditas, Lund **29**, 193 (1943).

— Cytogenetic studies in Solanum, sect. Tuberarium. Hereditas, Lund **31**, 1—128 (1945).

Lampitt, L. H., L. C. Baker u. T. L. Parkinson: Vitamin-C content of potatoes. II. The effect of variety, soil, and storage. J. Soc. Chem. Ind., Lond. **64**, 22 bis 26 (1945).

Lamprecht, H.: Rotfruktsförädlingen och dess mål. (Root crop breeding and its objects.) Årsb. Jordbruksforskning, Stockholm Pp. 70—80 (1945).

— u. N. Hertzman: Nya förädlingsresultat med rotfrukter vid Weibullsholms Väytförädlingsanstalt. (New breeding results with root crops at Weibullsholm Plant Breeding Institute.) Weibulls Ill. Årsb. **34**, 5—11 (1939).

—, — Weibullsholms Växtförädlingsanstalts Original-Rotfruktsstammar. (The original strains of root crops at Weibullsholm Plant Breeding Institute.) Weibulls Ill. Årsb. **34**, 12—20 (1939).

—, — Immuna II, ny mot klumprotsjuka mycket motståndskraftig stam av rova. (Immuna II, a new strain of turnip, highly resistant to club-root.) Agri Hortique Genetica, Landskrona **1**, 31—33 (1943).

—, — Weibulls Original Immuna rova II, stam 26. En ny mot klumprotsjuka mycket motståndskraftig stam. (Weibull's Original Immuna II turnip, strain 26. A new strain, very resistant to club-root.) Weibulls Ill. Årsb. **38**, 30—31 (1943).

—, — Weibulls Original Regia, stam 24, ny fodersockerbeta. (Weibull's Original Regia, strain 24, a new sugar mangle.) Weibulls Ill. Årsb. **39**, 12—13 (1944).

—, — Weibulls Original Tellus Bortfelder, stam 21. En ny stam, som i odlingsvärde är överlägsen alla i marknaden befintliga rovstammar. (Weibull's Original Tellus Bortfelder, strain 21. A new strain, superior in cultivation value to all existing turnip strains on the market.) Weibull's Ill. Årsb. **39**, 17—18 (1944).

Larson, R. H.: A foliar mottle and necrosis in Chippewa potatoes associated with infection by a strain of the potato X virus. Phytopathology **33**, 1216—1217 (1943).

Larson, R. H.: Resistance in potato varieties to yellow dwarf. J. Agric. Res. **71**, 441—451 (1945).

— u. A. R. Albert: Physiological internal necrosis of potato tubers in Wisconsin. J. Agric. Res. **71**, 487 bis 505 (1945).

Leach, J. G., P. Decker u. H. Becker: Pathogenic races of Actinomyces scabies in relation to scab resistance. Phytopathology **29**, 204—209 (1939).

Lechnovicz, B. C.: Nuevas variedas peruanas del Solanum andigenum. (New Peruvian varieties of S. andigenum.) Bol. Mus. Hist. Nat. „Javier Prado" Peru **5**, 427—430 (1941).

LeClerg, E. L.: Non virus leafroll of Irish potatoes. Amer. Potato J. **21**, 5—13 (1944).

— Genetic leaf roll of Irish potato seedlings. Phytopathology **35**, 877—878 (1945).

— Breeding for resistance to early blight in the Irish potato. Phytopathology **36**, 1011—1015 (1946).

— Just around the corner: South' new seed potatoes. Sth. Seedsman **9**, 11, 50 (1946).

— Association of specific gravity with dry-matter content and weight of Irish potato tubers. Amer. Potato J. 1947, **24**, 6—9 (1947).

— Comparative dry-matter content of varieties of Irish potatoes grown in Louisiana. Amer. Potato J. **24**, 73 (1947).

Lehmann, H.: Untersuchungen über die Genetik und Physiologie der Resistenz der Kartoffel gegen Phytophthora infestans de Bary. Die genetische Analyse der Resistenz von Solanum demissum sp. (vorl. Mitteilung). Züchter **13**, 33—34 (1941).

Lehnovič, V. S.: (New forms of the potato of the Andes from Central Peru.) Vestnik Socialističeskogo Rastenievodstva (Soviet Plant Industry Record) Nr. 5, 186 bis 187 (1940). [Russisch.]

Leiper, R. T.: The potato eelworm problem of to-day. J. R. Agric. Soc. **100**, Pt. III, 63—73 (1940).

Lep, P. M. (Editor): (Beet growing. [A symposium].) Seljhozgiz, Moskva, Pp. 436, 60 figs. (1941). [Russisch.]

Lepigre, A.-L.: Essais d'égermage des pommes de terre par les vapeurs d'oxyde d'éthylène. Ann. Inst. Agric. Serv. Rech. Expériment. Agric. Algérie **2**, 35—61 (1945).

Levan, A.: Abnorm utformning av pollen inom ett pollenfack av Beta. (Abnormal pollen formation in a pollen sac of Beta.) Bot. Notiser 463—465 (1939).

Lindsay, M. A.: Sectional notes. California. Amer. Potato J. **22**, 289 (1945).

Livermore, J. R.: Correlation of seedling performance in the greenhouse and subsequent yield in the field. Amer. Potato J. **16**, 41—43 (1939).

— Naming selections from established varieties. Amer. Potato J. **21**, 72—74 (1944).

— u. F. E. Johnstone (jr.): The effect of chromosome doubling on the crossability of Solanum chacoense, S. Jamesii and S. bulbocastanum with S. tuberosum. Amer. Potato J. **17**, 170—173 (1940).

Locke, S. B.: Field resistance to leafroll infection in potato varieties. Phytopathology **37**, 14 (1947). (Abstr.)

Lombard, P. M. et al.: Report of potato variety nomenclature committee. Amer. Potato J. **19**, 68—71 (1942).

Lunden, A. P.: Mål og metoder ved foredlingsarbeidet for sykdomsresistens hos poteten (Solanum tuberosum). (Objects and methods in breeding for disease resistance in the potato [S. tuberosum].) 48: Årsmeld. Norges LandbrHøisk, Åkervekstdorsøk p. 1—16 (1939).

Lunden, A. P.: Forsøk med tidligpotetsorter 1933 bis 1940. (Trials with early varieties of potatoes 1933 bis 1940.) Meld. Norg. LandbrHøgsk. **22**, 57—82; also 50. Årsmeld. Norges Landbr. Høgsk. Åkervekstforsøk 1942, p. 55—79 (1942).

— Synonymer i potetavlen. (Synonyms in potato growing.) Meld. Norg. Landbr. Høgsk. **25**, 101—139 (1954).

McIntosh, T.: The importance of the variety in potato production. Gdnrs'Chron. **107**, 116—117, 132; also Advanc. Sci. 1940 **1**, 313—314 (1940).
— Variations in potato varieties. Scot. J. Agric. **25**, 125 bis 132 (1945).
McKay, R. u. P. E. M. Clinch: Leaf roll infection in the potato varieties Skerry Champion, Shamrock and Matador. J. Dep. Agric. Éire **41**, 200—208 (1944).
M'Master Davey, V.: Classification in the swede. Scot. J. Agric. **26**, 39—43 (1946).
Macmillan, H. G.: Note on the origin of the potato. Gartenbauwiss. **14**, 308—325 (1939—40).
— The aid of exploration in potato improvement. Amer. Potato J. **19**, 255—266 (1942).
Madsen, S. B.: Om Bekaempelse af Kartoflens virussygdomme saerlig ved Foraedling og Fremavl. (On the control of virus diseases of the potato especially in breeding and propagation.) Medd. VetHojsk. Landbr. Plant Nr. 23, 85 (1944).
Magruder, R. et al: Descriptions of types of principal American varieties of red garden beets. Misc. Publ. U. S. Dep. Agric. Nr. 374, 60 (1940).
Maksimovič, M. M.: (New methods of breeding work with potato.) Vestnik Ovoščevodstvo i Kartofelj (Vegetable and Potato Journal) Nr. 1, 10—14 (1940). [Russisch.]
Marani, M., G. Goia u. L. Rossi: Prova d'orientamento di varietà di petate. (Preliminary test of potato varieties.) Riv. Frutticolt. **7**, 98—99 (1943).
Matthews, R. E. F.: Status of potato virus B. Nature, Lond. **159**, 713—714 (1947).
Mattson, H.: Breeding potatoes of North Dakota. Bi-m. Bull. N. Dak. Agric. Exp. Sta. **1**, Nr. 3, 37—39 (1939).
Mendiola, N. B.: A search for hidden and traumatic bud variations in sweet potato. Philipp. Agric. **27**, 726—754 (1939).
Metcalf, H. N. u. E. V. Hardenburg: Potato culture and storage investigations reported during 1941 and 1942. Amer. Potato J. **21**, 91—115 (1944).
Meunissier, A.: Les problèmes de sélection chez la pomme de terre. Rev. Hort. Paris **27**, 177—180, 202—203 (1940/1941).
— Les problèmes de sélection chez la pomme de terre. Rev. Bot. Appl. **21**, 139—141 (1941).
Meyer, G.: Zellphysiologische und anatomische Untersuchungen über die Reaktion der Kartoffelknolle auf den Angriff der Phytophthora infestans bei Sorten verschiedener Resistenz. Arb. biol. Reichsanst. Land- u. Forstw. Berl. **23**, 97—132 (1940).
Miller, J. C.: Further studies and technic used in sweet potato breeding in Louisiana. J. Hered. **30**, 485—492 (1939).
— Further studies and technic in sweet potato breeding in Louisiana. Proc. Amer. Soc. Hort. Sci. **36**, 665 bis 667 (1938) 1939.
— Technic and methods of breeding Irish potatoes in the south. Proc. 41st Annu. Conv. Ass. Sth. Agric. Wkrs, Birmingham, Ala., February 7—9, Pp. 142 bis 143 (1940). (Abstr.)
— Breeding sweet potatoes for table stock and for starch. Proc. 42nd Annu. Conv. Ass. Sth. Agric. Wkrs, Atlanta, Ga., February 5—7, Pp. 158—159 (1941).
— u. H. M. Covington: Some of the factors affecting the carotene content of sweet potatoes. Proc. Amer. Soc. Hort. Sci. **40**, 519—522 (1942).
—, F. McGoldrich u. E. L. LeClerg: Relation of some growth characters to stoloniferous condition in seedling Irish potatoes. Amer. Potato. J. **17**, 140—147 (1940).
Mohammad, A. u. S. M. Sikka: Cytological investigations in raya (Brassica juncea Coss), toria (Brassica napus L. var. dichotoma, Prain) and F_1 hybrid between them. Proc. 28th Indian Sci. Congr. Benares, Part III: Sect. Agric. abst. 43, Pp. 259 (1941).

Moldenhawer, K.: Genealogia polskich odmian ziemniaków wpisanych do rejestru odmian oryginalnych. (The genealogy of Polish potato varieties entered in the register of original varieties.) Przegl. Doświad. Roln. (Rev. Agric. Res.) **2**, 1—5 (1939).
Molodožnikov, M. M.: (Sweet potato. A valuable industrial, fodder and technological crop.) Vsesojuznaja Selekcionnaja Stancija Vlažno-Subtropičeskih Kuljtur (State Breeding Station for Crops of the Humid Sub-tropics) Suhumi, Pp. 46 (1943). [Russisch.]
Molotkovskij, G. H.: (Dekapitierung der Keimsprosse der jarovisierten Kartoffelknollen.) C. R. (Doklady) Acad. Sci. URSS. **53**, 861—863 (1946). [Russisch.]
— (Zur Frage der Veränderung der Eigenschaften der Kartoffeln durch Transplantation.) C. R. (Doklady) Acad. Sci. URSS. **58**, 1821—1823 (1947). [Russisch.]
Montaldo, Bustos, A.: Determinación de la calidad culinaria de las papas mediante su peso específico. (Determination of culinary quality of potatoes by specific weight.) Agric. Tec. Chile **4**, 78—87 (1944).
— Obervaciones sobre el período de reposo de la papa. (Observations on the dormancy period of the potato.) Agric. Tec. Chile **6**, 93—108 (1946).
Moreau, R. E.: The climatic background to the problem of potato varieties for East Africa. Part. I. E. Afr. Agric. J. **9**, 127—135 (1944).
— The climatic background to the problem of potato varieties for East Africa. Part II. E. Afr. Agric. J. **9**, 203—212 (1944).
— The yield and maturity period of potatoes (Solanum tuberosum) at low latitudes. Emp. J. Exp. Agric. **12**, 13—30 (1944).
Morinaga, T.: (The sterile mutants found in a certain strain of the rape.) Jap. J. Genet. **16**, 72—74 (1940).
Morton, C. V.: Taxonomic studies of tropical American plants. Some South American species of Solanum. Contr. U. S. Nat. Herb. **29**, 41—72 (1944).
Moščenko, S. I.: (Potato growing in the central Pamirs.) Vestnik Socialističeskogo Rastenievodstva (Soviet Plant Industry Record), Nr. 1, 133—137 (1941). [Russisch.]
Mujica, R. F.: Patogenicidad de algunas cepas del Verticillium albo-atrum Rei. y. Berth. (Pathogenicity of certain strains of V. albo-atrum Rei. et Berth.) Bol. Sanid Veg., Chile **1**, 7—20 (1941).
— Susceptibilidad de variedades de papas a la sarna polvorienta causada por la Spongospora subterranea (Wallr.) John. (Susceptibility of potato varieties to powdery scab caused by Spongospora subterranea [Wallr.] John.) Bol. Sanid. Veg., Santiago **2**, 17—19 (1942).
Müller, K. O.: Über die Abbauresistenz der Kartoffel und die Züchtung abbaufester Kartoffelsorten. Z. Pflanzenz. **23**, 1—19 (1939).
— Physiologisch-genetische Untersuchungen zur Analyse der Phytophthora-Resistenz der Kartoffel. Proc. 7th Int. Genet. Congr. Edinburgh 23—30 August, Pp. 222—223, 1939 (1941). (Abstr.)
— Die Erfolge der Züchtung phytophthoraresistenter Kartoffelsorten. Nachrbl. Ges. dtsch. Pflschdienst. **21**, 17 (1941).
— u. H. Börger: Experimentelle Untersuchungen über die Phytophthora-Resistenz der Kartoffel — zugleich ein Beitrag zum Problem der ,,erworbenen Resistenz" im Pflanzenreich. Arb. biol. Reichsanst. Land- u. Fortsw. Berl. **23**, 189—231 (1941).
— u. R. Griesinger: Der Einfluß der Temperatur auf die Reaktion von anfälligen und resistenten Kartoffelsorten gegenüber P. infestans. Angew. Bot. **24**, 130 bis 149 (1942).
—, G. Meyer u. M. Klinkowski: Physiologisch-genetische Untersuchungen über die Resistenz der Kartoffel gegenüber Phytophthora infestans. Naturwiss. **27**, 765—768 (1939).

Müller, K. O. u. H. Orth: Über einen Spätpflanzversuch mit Kartoffeln. Ernähr. Pfl. **37**, 37—40 (1941).
—, E. Pfeil u. F. Piekenbrock: Über die Beziehungen zwischen dem spezifischen Gewicht und dem Stärkegehalt der Knollen bei Kartoffelsorten und -zuchtstämmen verschiedener Genealogie. Angew. Bot. **25**, 178—195 (1943).
— u. K. Sellke: Beiträge zur Frage der Züchtung kartoffelkäferwiderstandsfähiger Kartoffelsorten. Z. Pflanzenz. **24**, 186—228 (1941).
—, — Über die Aussichten der Züchtung von „käferfesten" Kartoffelsorten. Mitt. Biol. Reichsanst. H. d. **64**, 10—23 (1941).
Munerati, O.: Die Vererbung der Weißblättrigkeit bei Beta vulgaris L. Züchter **14**, 214—215 (1942).
— Über die Möglichkeit bei Beta vulgaris L. Rassen mit zahlreichen Anomalien der Keimlinge getrennt zu züchten. Züchter **14**, 253 (1942).
Murphy, E.: Storage conditions which affect the vitamin C content of Maine-grown potatoes. Amer. Potato J. **23**, 197—218 (1946).
—, W. F. Dove u. R. V. Akeley: Observations on genetic, physiological, and environmental factors affecting the vitamin C content of Mainegrown potatoes. Amer. Potato J. **22**, 62—83 (1945).

Nandi, H. K.: Potato in Assam. Indian Fmg. **5**, 551 bis 554 (1944).
Nash, L. B.: Relation of variety and environmental condition to partial composition and cooking quality. Amer. Potato J. **18**, 91—99 (1941).
Neuweiler, E.: Kartoffelanbauversuche der Vereinigung schweizerischer Versuchs- und Vermittlungsstellen für Saatkartoffeln. Landw. Jb. Schweiz **45**, 865—889 (1944).
Nicolaisen, W.: Züchtung von Raps. Z. Pflanzenz. **25**, 362—379 (1943).
Nilsson, F.: Potatisförädling för Norrland. (Potatobreeding for Norrland.) Sverig. Utsädesfören. Tidskr. **49**, 137—147 (1939).
Noll, A.: Untersuchungen über die Biologie und Bekämpfung des Kartoffelschorfes (Actinomyces.) Landw. Jb. **89**, 41—113.
Novikov, F. A.: (Influence of external conditions on the seed qualities of potato tubers.) Vestnik Ovoščevodstvo i Kartofelj (Vegetable and Potato Journal), Nr. 1, 24—33 (1940). [Russisch.]
Nozzolini, V.: Il problema delle patate da seme. (The problem of seed potatoes.) Genetica Agraria, Roma **1**, 276—284 (1947).

Olsson, P. A.: Stocklöpningen i rotfruktskulturer under vegetationsperioden 1938. (Bolting in root crops during the vegetation period 1938.) Sverig. Utsädesfören. Tidskr. **49**, 326—335 (1939).
— Klumprotsjuka (Plasmodiophora brassicae Wor.) på rovor och kålrötter samt åtgärder mot densamma speciellt ur växtförädlingssynpunkt. (Club root disease [P. brassicae Wor.] on turnips and swedes and also protective measures to be taken against it especially from the point of view of plant breeding.) Sverig. Utsädesfören. Tidskr. **49**, 4—76 (1939).
— Klumprotsjuka (Plasmodiophora brassicae Wor.) på rovor och kålrötter samt åtgärder mot densamma, speciellt ur vaxtförädlingssynpunkt. II. Fortsatta undersökningar samt försök med resistansförädling. (Club root disease [Plasmodiophora brassicae Wor.] in turnips and swedes and the means of control, especially by plant breeding. II. Further investigations, and experiments in breeding resistant strains.) Sverig. Utsädesfören Tidskr. **50**, 287—360 (1940).
— Svensk rotfruktsodling. En sammanställning av resultat från Sveriges Utsädesförenings försök till ledning vid val av rotfruktsslag och-sort. (Swedish root crop cultivation - A survey of results from the Swedish Seed Association's experiments as a guide in the choice of the kind and variety of root crop.) Sverig. Utsädesfören. Tidskr. **52**, 45—104 (1942).
Olsson, P. A.: Om fodderrotfrukternas vinterhärdighet. (On the winter-hardiness of fodder root crops.) Sverig. Utsädesfören. Tidskr. **52**, 343—351 (1942).
Oortwijn Botjes, J. G.: Deinvloed van bladrolziekte op de opbrengst van verschillende aardappelrassen. (The effect of leaf roll on the yield of different potato varieties.) Tijdschr. PlZiekt. **47**, 25—31 (1941).
Opitz, K.: Weitere Versuche über den durch Viruskrankheiten herbeigeführten Abbau der Kartoffel. Pflanzenbau **16**, 323—342 (1940).
Ostanin, S. N.: (The biochemical characteristics of starch of new species and varieties of potatoes.) Proc. Lenin Acad. Agric. Sci. USSR., Nr. 2, 26—30 (1940). [Russisch.]
Osvald, H.: Potatis. (The potato.) Kooperativa Förbundets Bokförlag, Stockholm, Pp. 128, photos, illus. (ohne Jahr.)

Paddock, E. F.: The backcross transmission of an inversion in an interspecific hybrid of Solanum. Genetics **28**, 85 (1943). (Abstr.)
Pal, B. P.: A controlled-illumination chamber for maintaining short day species of potatoes. Indian J. Genet. Pl. Breed. **2**, 183 (1942).
— u. Pushkarnath: The Simla Potato Breeding Station. Indian Fmg. **1**, 25—28 (1940).
—, — Genetic nature of self- and cross-incompatibility in potatoes. Nature, Lond. **149**, 246—247 (1942).
—, — Self- and cross-incompatibility in some diploid species of Solanum. Curr. Sci. **13**, 235—236 (1944).
Pančenko, J. I.: (Upbringing and selection in breeding forms of root crops.) Jarovizacija, Nr. 2 (29), 105 bis 108 (1940). [Russisch.]
— (Breeding fodder root crops.) Trudy Zonaljnogo Inst. Zernovogo Hozjajstva Rajonov Nečernozemnoj Polosy (Trans. Zonal Inst. Grain Husbandry Non-Black-Soil Districts), Nr. 10, 161—179 (1941). [Russisch.]
Parris, G. K.: Does the Sequoia variety of potato possess resistance to leaf roll virus and to frost? Plant Dis. Reporter **29**, 126—127 (1945). [Mimeographed.]
Peeling, B. A.: Breeding new sweet potatoes sexually. Agrarian, Clemson, S. C. **1**, Nr. 2, 13 (1939).
Perlova, R. L.: Production of an autohexaploid Solanum Vallis Mexici Juz. by means of its cultivation at the Pamir. C. R. (Doklady) Acad. Sci. URSS. **25**, 419—422 (1939).
— (Behaviour of South American potatoes in the Pamirs.) Soviet Plant Industry Record, Nr. 4, 33—46 (1940). [Russisch.]
— Production of tetraploid plants in triploid potato species, group Andigena by cultivating in the Pamirs. C. R. (Doklady) Acad. Sci. URSS. **27**, 55—58 (1940).
— Spontaneous occurrence of diploid plants in the offspring of the triploid Solanum Maglia Schlechtd. grown in the Pamirs. C. R. (Doklady) Acad. Sci. URSS. **27**, 710—713 (1940).
— Seedling progeny of diploid species of potato. C. R. (Doklady) Acad. Sci. URSS. **29**, 336—339 (1940).
— Production of the original species of the Chile autotriploid Solanum maglia Schlechtd. at Pamir. C. R. (Doklady) Acad. Sci. URSS. **48**, 56—58 (1945).
— (The self-fertility of South-American potatoes in the Western Pamir.) Izvestija Akademii Nauk. SSSR. Serija biologičeskaja. (Bull. Acad. Sci. URSS., Sér. Biol.), Nr. 4, 460—470 (1945). [Russisch.]

Perlova, R. L.: (The morphology of the berries as a taxonomic character of tuber-bearing species of Solanum L. [section Tuberarium Bitt.].) Botaničeskij Žurnal (J. Bot. URSS.) **31**, Nr. 2, 19—32 (1946). [Russisch.]

Petersson, G. u. L. Fredriksson: Potatisutsädets klyvning. (Cutting of seed potatoes.) Lantmannen **31**, 147—149, 157 (1947).

Pettersson, M. L. R.: Xeromorphy and its bearing on disease resistance in plants. Abstr. Diss. Univ. Camb., Pp. 19—20, 1939/40 (1941).

Piekarski, A.: Dobór odmian rakoodpornych nach Slasku. (Selection of wart-resistant varieties for Silesia.) Roczn. Ochrony Roślin **6**, Nr. 2, 19—27 (1939).

Pierce, W. D.: A few remarks on the possible origin of the sweet potato. Bull. S. Calif. Acad. Sci. **39**, 229 bis 230 (1941).

Piettre, L.: Anomalies des inflorescences et des fleurs développées prématurément chez Solanum tuberosum. Ann. Sci. Nat. (11e Sér. Botanique) 1939—1940, **1**, 359—373 (1939—1940).

Plank, J. E. van der: Origin of the first European potatoes and their reaction to length of day. Nature, Lond. **157**, 503—505 (1946).

— Origin of the first European potatoes and their reaction to length of day. Nature, Land. **158**, 168 (1946).

— Origin of the first European potatoes and their reaction to length of day. Nature, Lond. **158**, 712—713 (1946).

— Some climatic factors determining high yields of potatoes. Part. I. Temperature and length of growing-season. Emp. J. Exp. Agric. **14**, 217—223 (1946).

— Some climatic factors determining high yields of potatoes. Part II. The potato at low latitudes and high altitudes. Emp. J. Exp. Agric. **15**, 1—8 (1947).

Podlipenko, F.: (How to transplant potatoes.) Sovhoznoe Proizvodstvo (State Farming), Nos. 4/5, 23 bis 25 (1945). [Russisch.]

Pollard, A., M. E. Kieser, A. Crang u. T. Wallace: Factors affecting quality in potatoes. Rep. Agric. Hort. Res. Sta., Long Ashton, Bristol, Pp. 184—199 (1944).

Priebus, K.: Die Keimlingsfarbe bei Herbstrüben- und Kohlsorten als Mittel zur Sortenunterscheidung. Gartenbauwiss. **18**, 27—31 (1944).

Prokošev, S. M.: (Biochemistry of the potato.) Biohimija Kuljturnyh Rastenij **4**, 5—58 (1939). [Russisch.]

— u. A. J. Kameraz: (Inheritance of chemical composition of tubers in interspecific potato crosses.) Soviet Plant Industry Record, Nr. 4, 51—60 (1940). [Russisch.]

— u. N. L. Mattison: (Biochemical characteristics of new species of potatoes.) Soviet Plant Industry Record, Nr. 4, 61—74 (1940). [Russisch.]

Propach, H.: Cytogenetische Untersuchungen in der Gattung Solanum, Sect. Tuberarium. V. Diploide Artbastarde. Z. indukt. Abstamm- u. Vererbungsl. **78**, 115—127 (1940).

Pushkarnath: Studies on sterility in potatoes. I. The genetics of self- and crossincompatibilities. Indian J. Genet. Pl. Breed. **2**, 11—36 (1942).

— Studies on sterility in potatoes. II. Abnormalities in flowering. Indian J. Genet. Pl. Breed. **3**, 121—124 (1943).

— Potato sprouts as a source of „seed". Curr. Sci. **14**, 236—237 (1945).

Ramanujam, S.: An apetalous mutation in turnip (Brassica campestris L.). Nature, Lond. **145**, 552 bis 553 (1940).

Ramanujam, S. u. D. Srinivasachar: Cytogenetic investigations in the genus Brassica and the artificial synthesis of B. juncea. Indian J. Genet. Pl. Breed. **3**, 73—88 (1943).

Ranninger, R.: Die Züchtung der Krautrübe für Futterzwecke und als Ölfrucht. Züchter **15**, 39—43 (1943).

Ratera, E. L.: Contribución al estudio del polen de papas. Observaciones en algunas variedades de papas cultivadas y en especies salvajes afines. (Contribution to the study of potato pollen. Observations on certain varieties of cultivated potatoes and on related wild species.) Inst. Genét. Fax. Agron. Vet., Univ. B. Aires **1**, Nr. 4, 19 (1940).

— Determinación del número de cromosomas de los Solanum aculeados de los alrededores de Buenos Aires. (Determination of the chromosome number of the aculeate Solanums from the environs of Buenos Aires.) Inst. Genét. Fac. Agron. Vet., Univ. B. Aires **1**, Nr. 6, 7 (1940).

— Cariología de algunas variedades cultivadas de papas Sudamericanas. (Karyology of certain cultivated S. American potatoes.) Rev. Fac. Agron. B. Aires **9**, 254—261 (1942).

— Los „Solanum" aculeados de la Capital Federal y sus alrededores. (The aculeate Solanum species from the Federal Capital and its environs.) Agronomía, B. Aires **30**, 360—368 (1942).

— „Solanum chacoense" Bitter espontaneo en los alrededores de Buenos Aires. (Solanum chacoense Bitter, growing wild in the environs of Buenos Aires.) Agronomía, B. Aires **31**, 56—60 (1942).

— Observationes sobre la biologia floral de Solanum chacoense Bitter. (Observations on the floral biology of Solanum chacoense Bitter.) Rev. Fac. Agron. B. Aires **10**, 87—93 (1943).

— Número de cromosomas de algunas Solanáceas Argentinas. (Chromosome numbers in some Argentine Solanaceae.) Rev. Fac. Agron. B. Aires **10**, 318—325 (1943).

— Ensayo de variedades cultivadas de Solanum andigenum Juz. et Buk., en la Facultad de Agronomía y Veterinaria de Buenos Aires. (Test of cultivated varieties of S. andigenum Juz. et Buk. in the Agriculture and Veterinary Faculty at Buenos Aires.) Rev. Fac. Agron. B. Aires **11**, 63—77 (1944).

Rawlins, W. A., Some varietal differences in wireworm injury to potatoes. Amer. Potato J. **20**, 156 bis 158 (1943).

Rayner, R. W.: Notes on the effect of day length on potato yields. E. Afr. Agric. J. **11**, 25—28 (1945).

Razumov, V. I.: (Vegetative hybrids.) Soviet Plant Industry Record, Nr. 1, 104 (1940). [Russisch.]

— (Features in the development of plants from tubers obtained as a result of grafting.) Vestnik Socialističeskogo Rastenievodstva (Soviet Plant Industry Record), Nr. 3, 21—30 (1940). [Russisch.]

Reddick, D.: Scab immunity. Amer. Potato J. **16**, 71—76 (1939).

— Problems in breeding for disease resistance. Chronica Botanica **6**, 73—77 (1940).

— u. W. R. Mills: Blight immune versus blight resistant potatoes. Amer. Potato J. **16**, 220—224 (1939).

— u. L. C. Peterson: Empire - a blight resistant variety. Amer. Potato J. **22**, 357—362 (1945).

Reestman, A. J., M. van Eekelen, H. Fontein u. T. F. Hendriks: Het ascorbinezuurgehalte van de nederlandsche aardappelrassen. (The ascorbic acid content of the Dutch varieties of potatoes.) Landbouwk. Tijdschr. **55**, 574—598 (1943).

Reimers, F. E.: (The effect of producing turnip seed within a single season [by means of vernalization] upon the biological characters of the progeny.) Vestnik Ovoščevodstvo i Kartofelj (Vegetable and Potato Journal), Nr. 5, 47—50 (1940). [Russisch.]

Rhodes, W. E. u. A. F. Davies: The selection and preprocessing of potatoes for canning with special reference to control of texture by calcium chloride. Chem. Ind. Lond., Nr. 21, 162—163 (1945).

Richard, J. G.: The maintenance of seed stock of the Porto Rico sweet potato. Proc. Amer. Soc. Hort. Sci. **36**, 691 (1938) 1939.

Riche, F. J. H. le: Studies on the processing of vegetables. (b) Ascorbic acid (vitamin C) content of some pea, turnip and potato varieties. Sci. Bull. Dep. Agric. S. Afr., Nr. 253, 3—8 (1946).

Riedl, W. A., F. J. Stevenson u. R. Bonde: The Teton potato: a new variety resistant to ring rot. Amer. Potato J. **23**, 379—389 (1946).

Rieman, G. H. u. J. S. McFarlane: The resistance of the Sebago variety to yellow dwarf. Amer. Potato J. **20**, 277—283 (1943).

—, W. E. Tottingham u. J. S. McFarlane: Potato varieties in relation to blackening after cooking. J. Agric. Res. **69**, 21—31 (1944).

Roivainen, H.: Perunarutto ja sen torjuminen. (Potato blight and its prevention.) Valt. Maatalousk. Tiedon, Nr. 155, 9 (1939).

Ross, A. F.: Susceptibility of Grean Mountain and Irish Cobbler commercial strains to stem-end browning. Amer. Potato J. **23**, 219—234 (1946).

Rožalin, L. V.: (Degeneration of the potato and the insufficiency of the theory of infection by mosaic virus diseases.) Vestnik Ovoščevodstvo i Kartofelj (Vegetable and Potato Journal), Nr. 1, 15—23 (1940). [Russisch.]

Rubin, B. A. u. V. E. Sokolova: (Temperaturkurven der Stärkesynthese bei den Kartoffeln im Zusammenhang mit der Pflanzenentwicklung.) C. R. (Doklady) Acad. Sci. URSS. **54**, 335 (1946). [Russisch.]

Rybin, V. A.: Tetraploid Solanum Ryninii Juz. et Buk. produced by colchicine treatment. C. R. (Doklady) Acad. Sci. URSS. **27**, 151—154 (1940).

Salaman, R. N.: Breeding for immunity to blight and other diseases in the potato. Proc. 7th Int. Genet. Congr. Edinburgh 23—30 August, Pp. 253—254, 1939 (1941). (Abstr.)

— Potato expedition. 10th Rep. Imp. Agric. Bur. Lond. (1938—1939), Pp. 52—54 (1940).

— The biology of the potato. With special reference to its use as a prime war-time food. Chem. Ind., London **59**, 735—737 (1940).

— The Ormskirk Potato Research Station. Nature, London **146**, 634—637 (1940).

— Recent research in potato breeding. Emp. J. Exp. Agric. **11**, 125—139 (1943).

— Report of the Potato Synonym Committee of the National Institute of Agricultural Botany on the potatoes sent for immunity trials to the Potato Testing Station, Ormskirk, Lancashire, 1939. J. Nat. Inst. Agric. Bot. **5**, 60—63 (1944).

— Report of the Potato Synonym Committee of the National Institute of Agricultural Botany on the trials of new seedling potatoes, carried out at the Midland Agricultural College, Sutton Bonington, in 1941. J. Nat. Inst. Agric. Bot., **5**, 64—67 (1944).

— Report of the Potato Synonym Committee of the National Institute of Agricultural Botany on the trials of new seedling potatoes, carried out at the Midland Agricultural College, Sutton Bonington, in 1942. J. Nat. Inst. Agric. Bot. **5**, 68—69 (1944).

— The early European potato: its character and place of origin. J. Linn. Soc. (Bot.) **53**, 1—27 (1946).

Saljkova, A. K.: (The biochemical investigation of the sweet potato under the conditions of Apsheron.) Soviet Subtropics, Nr. 9 (72), 51—53 (1940). [Russisch.]

Samuel, G.: Some factors affecting the yield of the potato crop. J. R. Soc. Arts **92**, 562—573 (1944).

Satyanarayana Rao, N.: Chromosome studies in the genus Ipomaea. Curr. Sci. **16**, 156 (1947).

Schaal, L. A.: Variation and physiologic specialization in the common scab fungus (Actinomyces scabies.) J. Agric. Res. **69**, 169—186 (1943).

Schaper, P.: Arbeiten und Probleme zur züchterischen Bekämpfung des Kartoffelkäfers. I. Ergebnisse der Prüfung deutscher Kultursorten auf das Verhalten gegen den Kartoffelkäfer im freien Befall. II. Untersuchungen über das Verhalten verschiedener Solanum-Arten gegen den Kartoffelkäfer. III. Resistenz und Anfälligkeit verschiedener Hybriden der Gattung Solanum gegen den Befall und Fraß des Kartoffelkäfers. Z. Pflanzenz. **23**, 239—322 (1939).

Schermerhorn, L. G.: Jersey Orange sweet potato. Hort. News N. J. **28**, 1920 (1947).

Schlumberger, O.: Die Zuverlässigkeit der Kartoffelkrebs-Prüfungen. Forschungsdienst **16**, 215—220 (1943).

— Kartoffelsortenprüfung auf Schorfwiderstandsfähigkeit 1939. Mitt. Landw. **55**, 9—11 (1940).

Schultz, E. S., C. F. Clark u. F. J. Stefenson: Resistance of potato to viruses A and X, components of mild mosaic. Phytopathology **30**, 944—951 (1940).

Schwanitz, F.: Beiträge zur Züchtung und Genetik selbstfertiler Rüben (Beta vulgaris L.). I. Erste Ergebnisse von Kreuzungen zwischen selbststeriler Beta vulgaris L. und selbstfertiler Beta maritima L. Züchter **12**, 167—178 (1940).

— Über die Pollenkeimung einiger diploider Pflanzen und ihrer Autotetraploiden in künstlichen Medien. Züchter **14**, 273—282 (1942).

Schwartz, M.: Bericht über die von Dr. Sellke im Jahre 1938 in Ahun (Creuse) durchgeführten Arbeiten. Prüfung der Resistenzeigenschaften von Kartoffelhybriden. C. R. 4th Conf. Com. Int. Ét. Comm. Lutte Contre Doryphore, Wageningen, Pp. 49—52 (1939).

Schwartz, M.: Die Kartoffelkäfer-Forschungsstation der Biologischen Reichsanstalt in Kruft. Mitt. Biol. Reichsanst. Landw. **64**, 5—10 (1941).

Seinhorst, J. W. u. M. J. Dunlop: De aantasting van enige Solanumsoorten en enige kruisingen tussen Solanum demissum en S. tuberosum door het stengelaaltje Ditylenchus dipsaci (Kuhn) Filipjev. (The infestation of some species of Solanum and some crosses between S. demissum and S. tuberosum by stem eelworm, D. dipsaci (Kuhn) Filipjev.) Tijdschr. PlZiekt. **51**, 73—81 (1945).

Sellke, K.: Über im Sommer 1938 im Kartoffelkäfer-Feldlaboratorium Ahun (Frankreich) durchgeführte Versuche zur Prüfung von Hybriden auf Kartoffelkäfer-Widerstandsfähigkeit. Arb. biol. Reichsanst. Land- u. Forstw., Berlin **23**, 1—20 (1940).

Sengbusch, R. v.: Neue Stärkewaagen zur Schnellbestimmung des Stärkegehalts von Kartoffeln für züchterische Zwecke. Forschungsdienst **13**, 19—24 (1942).

Sixta, J.: Odrůdy bramborů keřkovského šlechtěni. (Potato varieties bred at Keřkov.) Česk Zeměd., Prag **23**, 1—3 (1941).

Sleesman, J. P.: Resistance in wild potatoes to attack by the potato leafhopper and the potato flea beetle. Amer. Potato J. **17**, 9—12 (1940).

— u. F. J. Stevenson: Breeding a potato resistant to the potato leafhopper. Amer. Potato J. **18**, 280—298 (1941).

Smith, O., L. B. Nash u. A. L. Dittman: Potato quality VI. Relation of temperature and other factors to blackening of boiled potatoes. Amer. Potato J. **19**, 229—254 (1942).

Snell, K.: Die Lichtkeime der im Jahre 1942 zugelassenen Kartoffelsorten. Angew. Bot. **24**, 249—258 (1942).

Solodovnikov, F. S.: (Significance of age and origin of grafted potato plants in vegetative hybridization.) Vestnik Ovoščevodstvo i Kartofelj (Vegetable and Potato Journal) Nr. 3, 20—27 (1940). [Russisch.]

Solodovnikov, F. S.: (The effect of the time of planting on the berry formaton in potato.) Proc. Lenin Acad. Agric. Sci. USSR. Nr. 5/6, 24—25 (1944). [Russisch.]

Sørensen, H.: Stammeforsøg med Rødbeder 1941 bis 1943. (Strain trials with beetroot 1941—43.) Tidsskr. Planteavl. 49, 252—263 (1944).

Sosa-Bourdouil, C.: Sur les baies de quelques solanées et leur teneur en acide ascorbique (Vitamine C). Bull. Soc. Bot. Fr. 87, 322—324 (1940).

Soukup, V.: La clasificación de las papas del departamento de Puno. (The classification of the potatoes from the department of Puno. Bol. Minist. Fom., Lima Nr. 18, 72 (1939).

Steinbauer, C. E. et al.: Cooperative tests of sweetpotato varieties, introductions, and seedlings for starch production and market purposes. Circ. U. S. Dep. Agric. Nr. 653, 42 (1942).

Stelzner, G.: Colchicininduzierte Polyploidie bei Solanum tuberosum L. Züchter 13, 121—128 (1941).

— Zur Frage der Virusübertragung durch Samen, insbesondere des X-, Y- und Blattrollvirus der Kartoffel. Züchter 14, 225—234 (1942).

— Die Bestimmung des Stärkegehaltes von Kartoffeln. Forschungsdienst 16, 189—191 (1943).

— Wege zur züchterischen Nutzung des Solanum chacoense Bitt. in Hinblick auf die Züchtung käferresistenter Kartoffelsorten. Züchter 15, 33—38 (1943).

— Über die Fertilitätsverhältnisse bei Bastardierungen zwischen der frostfesten Wildkartoffel (Solanum acaule Bitt. und der Kulturkartoffel Solanum tuberosum L. Züchter 15, 143—144 (1943).

— u. C. Prohl: Der Einfluß später zusätzlicher Stickstoffdüngung auf Eiweißgehalt und Eiweißertrag der Kartoffel. Pflanzenbau 20, 129—136 (1944).

— u. H. Schwalb: Reaktion einer Reihe von Solanaceen auf Infektion mit A-, Y- und X-Virus der Kartoffel unter Berücksichtigung ihrer Brauchbarkeit als Testpflanze. Phytopath. Z. 14, 497—511 (1942).

—, — Die Virusanfälligkeit von Solanum demissum-Herkünften. Züchter 15, 187—190 (1943).

— u. M. Torka: Tageslänge, Temperatur und andere Umweltsfaktoren in ihrem Einfluß auf die Knollenbildung der Kartoffel. Züchter 12, 233—237 (1940).

Stevenson, F. J.: Potato disease control by breeding resistant varieties. Trans. Iowa Hort. Soc. 74, 221—235 (1939).

— Potato varieties better suited to a specific set of growing conditions. Proc. 24th Annu. Mtg. Ohio Veg. and Potato Gr. Ass. Pp. 14—18 (1939).

— Variability in cooking quality in potatoes. Proc. 24th Annu. Mtg. Ohio Veg. and Potato Gr. Ass. p. 42—46 (1939).

— Potato varieties recently distributed to growers in the United States. Amer. Potato J. 17, 217—235 (1940).

— Genetics, cytogenetics and breeding in the potato: review of literature. Amer. Potato J. 17, 299—314 (1940).

— Potato breeding, genetics, and cytology: review of literature, 1940. Amer. Potato J. 18, 317—329 (1941).

— Potato varieties recently introduced to growers. Amer. Potato J. 19, 41—42 (1942).

— Potato breeding, genetics, and cytology: review of recent literature. Amer. Potato J. 20, 267—279 (1943).

— Potato breeding, wither bound? Amer. Potato J. 21, 192—199 (1944).

— Potato breeding, genetics, and cytology: review of recent literature. Amer. Potato J. 22, 36—52 (1945).

— Results and outlook in potato breeding. Proc. 30th Annu. Mtg. Ohio Veg. and Potato Gr. Ass. p. 148—158 (1945).

— Breeding potatoes resistant to disease. Nat. Hort. Mag. 25, 18—28 (1946).

Stevenson, F. J.: Spud-in-the-bud Sth. Seedsman, 9, Nr. 9; 15, 34, 39, 42, 46 (1946).

—, D. Folsom u. T. P. Dykstra: Virus leaf roll resistance in the potato. Amer. Potato J. 20, 1—10 (1943).

—, L. A. Schaal, C. F. Clark u. R. V. Akeley: Potato-scab gardens in the United States. Phytopathology 32, 965—971 (1942).

—, E. S. Schultz u. C. F. Clark: Inheritance of immunity from virus X (latent mosaic) in the potato. Phytopathology 29, 362—365 (1939).

—, —, R. V. Ekeley u. L. C. Cash: Breeding for resistance to late blight in the potato. Amer. Potato J. 22, 203—223 (1945).

Stuart, W.: Parental identity in varietal selection names. Amer. Potato J. 21, 139—140 (1944).

Sugawara, R.: Inducing the flowering and the fruiting of sweet potato by water culture. Jap. J. Bot. 10, 335 bis 342 (1940).

Sun, V. G.: A note on „Tsontsai". Chinese J. Sci. Agric. 1, 143—146 (1943).

— (A study of the size of pollen grains in the genus Brassica.) Abstr. J. Agric. Ass. China Nr. 175, 20—34 (1943).

— (Hybrid vigor in Brassica.) Abstr. J. Agric. Ass. China Nr. 175, 35—58 (1943).

— The evaluation of some taxonomical characters of the cultivated Brassicas and a suggested key for the identification of varieties and species. Acta Brevia Sinensia, Nr. 7, 6 (1944). [Mimeographed.]

— A study on the size of pollen grains in the genus Brassica. J. Agric. Ass. China 175, 20—34 (1944).

— Hybrid vigor in Brassica. J. Agric. Ass. China 175, 35—58 (1944).

— The evaluation of taxonomic characters of cultivated Brassica with a key to species and varieties — I. The characters. Bull. Torrey Bot. Cl. 73, 244—281 (1946).

— The evaluation of some taxonomic characters of cultivated Brassica with a key to species and varieties — II. The key. Bull. Torrey Bot. Cl. 73, 370—377 (1946).

— u. L. C. Sze: A note on „Tsontsai". Acta Brevia Sinensia, Nr. 7, 23—24 (1944). [Mimeographed.]

—, — A note on „Tsontsai". Proc. Amer. Soc. Hort. Sci. 46, 295—298 (1945).

Sutton, E. P. F.: Farm roots and their development. J. Bath W. S. Co. Soc. 6th Ser. 13, 87—93 (1938—39).

Tamargo, M. A.: A study of seedlings and varieties of the Irish potato in Cuba. Amer. Potato J. 17, 323—327 (1940).

— Investigation program for the production of seed potato in tropical countries. Proc. 16th. Annu. Conf. Asoc. Tecn. Azucareros, Cuba Pp. 83—87 (1942).

Tan, C. C. u. L. C. Sze: On the pollen sterility of „Solanum tuberosum". Acta Brevia Sinensia Nr. 8, 16 (1944). [Mimeographed.]

Tedin, O.: Fortsatta C-vitaminundersökningar på potatis. (Further vitamin C researches on potato.) Sverig. Utsädesfören Tidskr 51, 238—255 (1941).

— Mål och vägar vid Sveriges Utsädesförenings potatisförädling för södra Sverige. (Aims and means in the Swedish Seed Association's potato breeding for southern Sweden.) Sverige. Utsädesfören. Tidskr. 52, 105 bis 115 (1942).

— Nytillväxt hos potatis. (New growth in potatoes.) Sverig. Utsädesfören. Tidskr. 54, 316—324 (1944).

Telford, H. S.: Wireworm injuri and potato varieties. Bi-m. Bull. N. Dare. Agric. Exp. Sta 4, Nr. 5, 7—8 (1942.)

Thénard, J.: La lutte contre le Doryphore à l'aide de variétés de pommes de terre résistantes. Rev. Hort. Paris 28, 135—136 (1942).

Thomas, P. T.: ‚Bolters' in potatoes. Nature, Lond. 155, 242 (1945).

Thompson, E. G.: Mangold trials, 1936—1938. J. Nat. Inst. Agric. Bot. **5**, 28—44 (1944).

Thunberg, T.: Der Citratgehalt der Kartoffel. K. Fysiogr. Sällsk. Lund Förh. **15**, Nr. 8, 1—5 (1945).

Timofeev, N. N.: (Inheritance of characters in vegetables in connexion with the origin of seeds from different branches.) Moskovskaja ordena Lenina Seljskohozjajstvennaja Akademija imeni K. A. Timirjazeva. Doklady Vypusk III. Naučnaja Konferencija 4—11 Ijunja 1945 g. (Timirjazev Agric. Acad. Moscow Proc. No. III. Sci Conf. 4—11 June 1945) Pp. 64—68 (1946). [Russisch].

Tkačenko, P. I.: (Colonal Selection in potatoes.) Vestnik Ovoščevodstvo i Kartofelj (Vegetable and Potato Journal) Nr. 2, 63—66 (1940). [Russisch.]

— (New interspecific hybrids with a wild potato species.) Vestnik Ovoščevodstvo i Kartofelj (Vegetable and Potato Journal) Nr. 2, 93—94 (1941). [Russisch.]

— (A high yielding and early clone of Wohltmann.) Vestnik Ovoščevodstvo i Kartofelj (Vegetable and Potato Journal) Nr. 2, 95 (1941). [Russisch.]

Torka, M.: Die Widerstandsfähigkeit eines Solanum chacoense-Klons gegenüber dem Kartoffelkäfer (Leptinotarsa decemlineata Say). Züchter **15**, 145—148 (1943).

Tottingham, W. E., R. Nagy, A. F. Ross, J. W. Marek, u. C. O. Clagett: A primary cause of darkening in boiled potatoes as revealed by greenhouse cultures. J. Agric. Res. **67**, 177—193 (1943).

—, —, —, —, — Blackening indices of potatoes grown under various conditions of field culture. J. Agric. Res. **74**, 145—164 (1947).

Trouvelot, B.: Les phénomènes de résistance naturelle des plantes aux attaques des insectes et essai de leur utilisation pour la lutte contre le doryphore. Verh. VII Int. Kongr. Ent. **4**, 2726—2730 (1939).

— Les recherches faites en 1938 sur les plantes résistantes aux attaques du Doryphore. C. R. 4th Conf. Com. Int. Et. Comm. Lutte Contre Doryphore, Wageningen, Pp. 46—49 (1939).

— u. Müller-Böhme: Etude sur la valeur alimentaire, pour les larves du doryphore, d'hybrides S. demissum × S. tuberosum. Verh. VII Int. Kongr. Ent. **4**, 2731 bis 2741 (1939).

Tucker, J.: Potatoes in Mexico. Amer. Potato J. **20**, 218 bis 223 (1943).

Turlapova, A.: (Development of an early industrial variety of potato by grafting.) Proc. Lenin Acad. Agric. Sci. USSR. Nr. 5/6, 28—30 (1944). [Russisch.]

Tušnjakova, M.: (Grafting of alkaloid plants.) Proc. Lenin Acad. Agric. Sci. USSR. Nr. 10, 24—31 (1944). [Russisch.]

Tussing, E. B.: Potato demonstration results in 1938. Proc. 24th Annu. Mtg. Ohio Veg. and Potato Gr. Ass. Pp. 5—9 (1939).

Van Duyne, F. O., J. T. Chase u. J. I. Simpson: Effect of various home practices on ascorbic acid content of potatoes. Food Res., Illinois **10**, 72—83 (1945).

Vargas, C. C.: Nuevas especies de papas silvestres del Perú. (New species of wild potatoes from Peru.) Rev. Argent. Agron. **10**, 396—398 (1943).

Veselovskij, I. A.: Biochemical and anatomical properties of starch of different varieties of potatoes and their importance for industrial purposes. Amer. Potato J. **17**, 330—339 (1940).

— (Growing potatoes from seeds.) Socialističeskoe Seljskoe Hozjajstvo (Socialistic Agriculture) Moscow, Nr. 7, 58 (1944). [Russisch.]

Vinogradova, N. B.: (Methods of breeding potatoes for blight resistance.) Vestnik Ovoščevodstvo i Kartofelj (Vegetable and Potato Journal) Nr. 3, 39—71 (1940). [Russisch.]

Voss, J.: Zur Schoßauslösung und Prüfung der Schoßneigung von Rübensorten (Beta vulg. L. und Brassica Napus L. var. Napobrassica (L.) Reichenb.). Züchter **12**, 34—44, 73—77 (1940).

Walker, J. C., J. P. Jolivette u. W. W. Harke: Varietal susceptibility in garden beets to boron deficiency. Soil Sci. **59**, 461—464 (1945).

—, — u. J. G. Mclean: Boron deficiency in garden and sugar beet. J. Agric. Res. **66**, 97—123 (1943).

Wallace, J. O.: N. Z. resistant swede. N. Z. J. Agric. **67**, 341, 343 (1943).

— Turnip and swede crop. N. Z. J. Agric. **73**, 215—223 (1946).

Wang, C. M.: (Physiological specialization in Peronospora parasitica and reaction of hosts.) Chinese J. Sci. Agric. **1**, 249—257 (1944).

Watanabe, K.: (Chromosomes of Japanese sweet potatoes and their allied plants.) Proc. Crop. Sci. Soc. Japan **11**, 124—134 (1939).

Weibull, W.: En maning till vårt lands rotfruktsodlare. (A warning to our root crop growers.) Weibulls Ill. Årsb. **39**, 6—7 (1944).

Weiss, F.: Viruses, virus diseases, and similar maladies of potatoes, Solanum tuberosum L. Plant Dis. Reporter Suppl. **155**, 82—140 (1945). [Mimeographed.]

Werner, H. O.: Performance of clonal strains of Triumph potatoes. I. Triumph strains on dry land in western Nebraska. Amer. Potato J. **17**, 66—80 (1940).

— Relation of length of photoperiod and intensity of supplemental light to the production of flowers and berries in the greenhouse by several varieties of potatoes. J. Agric. Res. **64**, 257—274 (1942).

— Relative response of several varieties of potatoes to progressively changing temperatures and photoperiods controlled to simulate „northern" and „southern" conditions. Amer. Potato J. **19**, 30—40 (1942).

Whalley, M. E.: Abstracts on potato research. National Research Council of Canada, Ottawa Nr. 1013, 122 (1944). [Mimeographed.]

Wheeler, E. J., F. J. Stevenson u. H. C. Moore: The Menominee potato: a new variety resistant to common scab and late blight. Amer. Potato J. **21**, 305—311 (1944).

Whitehead, T., T. P. McIntosh u. W. M. Findlay: The potato in health and disease. Oliver and Boyd, Edinburgh, 2nd ed. revised, 25s. Pp. xv + 400. 19 tables. 31 figs. (1945).

Willigen, A. H. A. de: De chemische samenstelling van den aardappel. (The chemical composition of the potato.) Landbouwk. Tijdschr. Wageningen **54**, 693—725 (1942).

— u. A. J. Reestman: Rassenproeven met fabrieksaardappelen in 1943. (Variety tests of industrial potatoes in 1943.) Versl. Landbouwk. Onderzoek. Nr. 50 (11) B, 619—653 (1945).

Wolf, M. J. u. B. M. Duggar: Estimation and physiological role of solanine in the potato. J. Agric. Res. **73**, 1—32 (1946).

Wollenweber, H. W.: Über die Lebensdauer von Kartoffelsamen. Angew. Botanik **24**, 259—260 (1942).

Wright, R. E.: Three steps to improve sweet potatoes. Fm. & Ranch **63**, Nr. 3, 17 (1944).

Wright, R. C., J. S. Caldwell, T. M. Whiteman u. C. W. Culpepper: The effect of previous storage temperatures on the quality of dehydrated potatoes. Amer. Potato J. **22**, 311—323 (1945).

Yarwood, C. E.: Increased yield and disease resistance of giant hill potatoes. Amer. Potato J. **23**, 352—369 (1946).

Younkin, S. G.: Purple-top wilt of potatoes caused by the aster yellows virus. Amer. Potato J. **20**, 177—183 (1943).

Zaikovskaja, N. E.: Reduction division in interspecific hybrids in the genus Beta L. II. Cytological evidence on the question on experimental synthesis of the species Beta trigyna W. et K. (2n = 54). C. R. (Doklady) Acad. Sci. URSS. 23, 944—948 (1939).

Zosimovič, V. P.: Eco-geographical characteristic of the wild species of beet (Beta L.). C. R. (Doklady) Acad. Sci. URSS. 24, 69—72 (1939).

— Evolution of cultivated beet B. vulgaris L. C. R. (Doklady) Acad. Sci. URSS. 24, 73—76 (1939).

Zuckerman, H. K.: Selection of parent seed potatoes. Amer. Potato J. 19, 61—67 (1942).

Zvereva, P. A.: (New methods of breeding frost resistant potatoes.) Doklady Vsesojuz. Akad. Seljsk. Nauk im. V. I. Lenina (Proc. Lenin Acad. Agric. Sci. USSR.) Nr. 4/5, 31—33 (1945). [Russisch.]

— (Overcoming uncrossability in potatoes by means of vegetative rapprochement.) Agrobiologija (Agrobiology), Nr. 2, 126—128 (1946). [Russisch.]

e) Faserpflanzen.

A....., N.: What the scientists are doing. Combing of Indian and Kampala cottons. Indian Fmg 6, 369 bis 370 (1945).

Abaeva, S. S.: Influence of high temperatures on viability of pollen in cotton (in relation to premature fruit dropping.) C. R. (Doklady) Acad. Sci. URSS. 32, 443—445 (1941).

Abraham, P.: Cytological studies in Gossypium. I. Chromosome behaviour in the interspecific hybrid G. arboreum x G. Stocksii. Indian J. Agric. Sci. 10, 285—298 (1940).

— Morphology of the somatic chromosomes of three Asiatic cottons. Indian J. Agric. Sci. 10, 299—302 (1940).

Admad, N. u. A. N. Gulati: The effect of storage under certain specified conditions on the quality of Indian cottons. Technol. Bull. Indian Cott. Cttee, Nr. 31, 21 (1942).

Afify, A.: Chromosome pairing and chiasma formation in Aloe. Bull. Fac. Sci., Fouad I Univ. Cairo, Nr. 25, 97—110 (1945).

Afzal, M.: Research on cotton in the Punjab. Indian Fmg 7, 276—280 (1946).

— Research on cotton in the Punjab. II. American cotton. Indian Fmg 7, 341—345 (1946).

— Problems in cotton improvement in the Punjab. Indian Cott. Gr. Rev. 1, 50—59 (1947).

— u. M. Abdul Ghani: Cotton jassid in the Punjab. Indian Fmg 7, 407—410 (1946).

—, D. Nath Nanda u. M. Abbas: Studies on the cotton jassid (Empoasca devastans Distant) in the Punjab. IV. A note on the statistical study of jassid population. Indian J. Agric. Sci. 13, 634—638 (1943).

— u. S. M. Nawaz: The staple characteristics of the Punjab-American cotton crop. Indian Cott. Gr. Rev. 1, 22—25 (1947).

—, S. Rajaraman u. M. Abbas: Studies on the cotton jassid (Empoasca devastans Distant) in the Punjab. III. Effect of jassid infestation on the development and fibre properties of the cotton plant. Indian J. Agric. Sci. 13, 192—203 (1943).

Ahmad, N.: Annual Report of the Director, Technological Laboratory for the year ending 31st May, 1940. Indian Cent. Cott. Comm., Bombay, p. 40 (1940).

— Technological Reports on trade varieties of Indian cottons, 1940. Technol. Bull. Indian Cott. Comm. Series A, Nr. 51, 107 (1940).

— Technological Reports on standard Indian cottons, 1940. Technol. Bull. Indian Cott. Comm. Series A, Nr. 52, 110 (1940).

— Technological Reports on trade varieties of Indian cottons, 1941. Technol. Bull. Indian Cott. Comm. Series A, Nr. 53, 103 (1941).

Ahmad, N.: Technological Reports on standard Indian cottons, 1941. Technol. Bull. Indian Cott. Comm. Series A, Nr. 54, 115 (1941).

— Technological Reports on trade varieties of Indian cottons, 1942. Technol. Bull. Indian Cott. Comm. Series A, Nr. 55, 106 (1942).

— Technological Reports on standard Indian cottons, 1942. Technol. Bull. Indian Cott. Comm. Series A, Nr. 56, 106 (1942).

— Technological research on cotton in India being an account of the work done at the Indian Central Cotton Committee Technological Laboratory, 1924—1941. Indian Centr. Cott. Comm. Technol. Lab., Bombay, p. 182 (1942).

— Technological Reports on trade varieties of Indian cottons, 1943. Technol. Bull. Indian Cott. Comm. Series A, Nr. 57, 84 (1943).

— Technological Reports on standard Indian cottons 1943. Technol. Bull. Indian Cott. Comm. Series A, Nr. 58, 103 (1943).

Technological reports on trade varieties of Indian cottons, 1944. Technol. Bull. Indian Cott. Comm. Ser. A, Nr. 61, 83 (1944).

— Technological reports on standard Indian cottons, 1944. Technol. Bull. Indian Cott. Comm. Ser. A, Nr. 62, 103 (1944).

— Annual report of the Director of the Technological Laboratory, Indian Central Cotton Committee, for the year ending 31st May 1945, Pp. 47 (1945).

— Technological reports on standard Indian cottons, 1945. Technol. Bull. Indian Cott. Comm. Ser. A, Nr. 63, 107 (1945).

Åkerman, Å.: Anförande vid Lanlaboratoriets invigning den 16/12 1940. (Speech at the inaugural ceremony of the Flax Laboratory on 16/12 1940.) Sverig. Utsädesfören. Tidskr. 50, 361—364 (1940).

— Det nya linlaboratoriet. En historisk överblick över dess tillkomst. (The new flax laboratory. A historical survey of its origin.) Sverig. Utsädesfören. Tidskr. 55, 86—91 (1945).

Allan, H. H. u. L. M. Cranwell: Vivipary in phormium. Rec. Auck. Inst. Mus. 2, 269—279 (1942).

Almeida, J. L. Ferreira de: Sobre a cariologia de Salix salviifolia Brot. e S. babylonica L. (The caryology of S. salviifolia Brot. and S. babylonica L.) Bol. Soc. Broteriana 20, 201—242 (1946).

Amin, K. C.: A preliminary note of interspecific hybridization and use of colchicine in cotton. Curr. Sci. 9, 74—75 (1940).

— Interspecific hybridization between Asiatic and New World cottons. Indian J. Agric. Sci. 10, 404—413 (1940).

Anonymus: Les résultats de la sélection cotonnière. Agric. Élev. Congo Belge 13, 139—140 (1939).

— Halo-length measurement. Emp. Cott. Gr. Rev. 16, 272—275 (1939).

— New pure strains of flax. Leafl. Min. Agric. N. Ire, Nr. 74, 9 (1939).

— (Achievements and prospects of Soviet linen flax breeding.) Len i Konoplja (Flax and Hemp), Nr. 10, 22—26 (1940). [Russisch.]

— Observations sur les essais d'Urena lobata à la station expérimentale de l'Inéac à M'Vuazi. Agric. Élev. Congo Belge 14, 19—20 (1940).

— Indian jute research. Indian Fmg 1, 81—82 (1940).

— Cotton breeding at Surat. Indian Fmg 1, 234—235 (1940).

— Difference between Capsularis and Olitorius. Bull. Indian Cent. Jute Cttee 4, 368—369 (1941).

— Flax variety trials on Dominiom experimental farms and stations 1941. Mimeogr. Rep. Cereal Div. Centr. Exp. Farm, Ontario, Nr. 53, 11 (1941).

— Indian jute grown in Brazil. Statesman, Calcutta 107, Nr. 21172, 3 (1941).

Anonymus: Two new Cambodia strains. Indian Fmg 2, 83 (1941).
— New Swedish plant. Text. Merc., May 15 (1942).
— „Tifton Station 21". A new cotton for south Georgia. Mimeogr. Paper Ga Coastal Plain Exp. Sta., Nr. 15, 2 (1942).
— Whate the scientists are doing. Breeding wilt-resistance. Indian Fmg. 3, 442—443 (1942).
— Monthly notes on progress of schemes. Agricultural research. Bull. Indian Cent. Jute Comm. 5, 433 (1943).
— Trabalhos na Estação Experimental de Sete Lagoas. (Work at the Experimental Station of Sete Lagoas.) Bol. Minist. Agric., Rio de J. 32, Nr. 11, 117—118 (1943).
— A melhor fibra de algodão produzida no Brasil. (The best cotton fibre produced in Brazil.) Bol. Minist. Agric. Rio de J. 32, 144—145 (1943).
— Items-Black cotton. Science, Suppl. 97, Nr. 2509, 9 (1943).
— New strain of Sea Island cotton shows promise. Sth. Seedsman 6, Nr. 4, 41 (1943).
— O linho em Portugal. (Flax in Portugal.) Direcção Geral dos Servicos Agricolas, Minist. Econ., p. 292 (1943).
— What the scientists are doing. New cotton for Broach. Indian Fmg 4, 37 (1943).
— What the scientists are doing: Dholleras cotton improvement. Indian Fmg 4, 468—469 (1943).
— (The areas that have been planted with different types of cotton for the year 1943.) Egypt. Agric. Rev. 21, 221—227 (1943).
— Progress of technical schemes. Bull. Indian Cent. Jute Ctte 6, 138—139 (1943).
— Progress of technical schemes. Bull. Indian Cent. Jute Ctte 6, 226—227 (1943).
— Progress of technical schemes. Bull. Indian Cent. Jute Comm. 7, 130—131 (1944).
— Novo híbrido de algodão nas culturas paraibanas. (A new cotton hybrid for cultivation in Paraíba.) Bol. Minist. Agric., Rio de J. 33, Nr. 9, 127—128 (1944).
— Nova linhagem de algodão em São Paulo. (A new strain of cotton in São Paulo.) Bol. Minist. Agric., Rio de J. 33, 157—158 (1944).
— Progress of technical schemes. Bull. Indian Cent. Jute Comm. 7, 172—173 (1944).
— Progress of technical schemes. Agricultural research. Bull. Indian Cent. Jute Cttee 8, 401—402 (1945).
— Progress of technical schemes. Agricultural research. Botanical. Bull. Indian. Cent. Jute Cttee 9, 177 (1946).
— Progress of technical schemes. Agricultural research. Botanical. Bull. Indian. Cent. Jute Cttee 9, 206 (1946).
— Progress of technical schemes. Bull. Indian Cent. Jute Cttee 9, 249—250 (1946).
— Progress of technical schemes. Agricultural research. Bull. Indian Cent. Jute Cttee 12, 640—641 (1946).
— Coloured cottons grow in Uzbekistan. Soviet War News, Nr. 855, 3 (1944).
— (New varieties of cotton.) Socialističeskoe Zemledelje (Socialist Agriculture), Nr. 122, 2 (1944). [Russisch.]
— Uses of the common stinging nettle (Urtica dioica). Mon. Sci. News, Nr. 30, 2 (1944).
— What the scientists are doing. Improvement of Cocanadas cotton. Indian Fmg. 5, 80—81 (1944).
— Cotton varieties for South Georgia farms. Mimeo. Pap. Ga Coastal Plain Exp. Sta., Nr. 34, 1 (1945). [Mimeographed.]
— Plant industry. 1. New achievements in selection. Agriculture, Moscow, Nr. 11, 1—2 (1945). [Mimeographed.]
— What the scientists are doing. A new cotton for Mathio tract. Indian Fmg. 6, 273 (1945).
Anonymus: What the scientists are doing. Red leaf in American cotton (G. hirsutum). Indian Fmg. 6, 469—470 (1945).
— What the scientists are doing. Jassids in cotton. Indian Fmg. 7, 83—84 (1946).
— What the scientists are doing. X-ray treatment of cotton seed. Indian Fmg. 7, 360—361 (1946).
— Progress of technical schemes. Bull. Indian Cent. Jute Cttee 9, 462—463 (1947).
Ansari, M. A. A.: Indigenous and exotic cottons of Iran. Indian J. Agric. Sci. 10, 522—533 (1940).
Anson, R. R.: Local cotton history in the Sudan, 1942 bis 1946. Emp. Cott. Gr. Rev. 23, 77—82 (1946).
Arndt, C. H.: Viability and infection of light and heavy cotton seeds. Phytopathology 35, 747—753 (1945).
Arny, A. C.: Registration of improved flax varieties, II. J. Amer. Soc. Agron. 36, 454—457 (1944).
— Registration of improved flax varieties, III. J. Amer. Soc. Agron. 37, 646—648 (1945).
Arutjunova, L. G.: (Germination of cotton pollen in intravarietal crosses.) Jarovizacija, Nr. 1 (28), 18—22 (1940). [Russisch.]

Balasubrahmanyan, R. u. T. V. Rangaswami: Contabescence of anthers in the American cottons grown at the Agricultural Research Station, Palur (Madras Province). Indian Cott. Gr. Rev. 1, 28—29 (1947).
Ball, C. R.: Far western novelties in Salix. Madroño 6 (7), 227—239 (1942).
— Salix floridana Chapman, a valid species. J. Arnold Arbor. 24, 103—106 (1943).
Ballard, W. W.: Empire ... crown prince of cotton. Sth. Seedsman 9, 15, 32, 44 (1946).
Baltazar, E. P. u. M. C. Chakrabandhu: A study of some of the important characters of cotton varieties grown from selfed and unselfed seeds. Philipp. Agric. 29, 150—172 (1940).
Barker, H. E. u. E. E. Berkley: Fiber and spinning properties of cotton, with special reference to varietal and environmental effects. Tech. Bull. U. S. Dep. Agric., Nr. 931, 36 (1946).
Bartels, K.: Untersuchungen über die Vererbung quantitativer Eigenschaften: die Stengellänge und Blütezeit des Leins. Z. indukt. Abstamm. u. Vererb-Lehre 78, 14—58 (1940).
Baylis, G. T. S.: Stem-break and browning (Polyspora lini) of flax in New Zealand. N. Z. J. Sci. Tech. 23, 1A—8A (1941).
Beasley, J. O.: The origin of American tetraploid Gossypium species. Amer. Nat. 74, 285—286 (1940).
— The production of polyploids in Gossypium. J. Hered. 31, 39—48 (1940).
— Hybridization of American 26-chromosome and Asiatic 13-chromosome species of Gossypium. J. Agric. Res. 60, 175—181 (1940).
— Hybridization, cytology, and polyploidy of Gossypium. Chronica Botanica 6, 394—395 (1941).
— Meiotic chromosome behaviour in species, species hybrids, haploids, and induced polyploids of Gossypium. Genetics 27, 25—54 (1942).
— Wild species may aid in breeding better cottons. Cott. & Cott. Oil Pr. 43, 5—6 (1942).
— u. M. S. Brown: Asynaptic Gossypium plants and their polyploids. J. Agric. Res. 65, 421—427 (1942).
—, — Asynaptic plants of Gossypium and their polyploids. Genetics 27, 131 (1942). (Abstr.)
—, — The production of plants having an extra pair of chromosomes from species hybrids of cotton. Rec. Genet. Soc. America, Nr. 12, 43 (1943). (Abstr.)
Belovickaja, N. A. u. E. I. Grečuhin: (On the monoecious hemp.) Bull. Acad. Sci. URSS., Sér. Biol. Pp. 311—334 (1939). [Russisch.]

Bereznjakovskaja, A. V.: (Production of early maturing Sea Island cotton.) Jarovizacija Nr. 1 (34), 40—46 (1941). [Russisch.]

Black, C. A.: Effect of commercial fertilizers on the sex expression of hemp. Bot. Gaz. 107, 114—120 (1945).

Bolhuis, G. G.: Bloeiwaarnemingen bij Hibiscus Sabdariffa L. en Hibiscus cannabinus L. (Some observations in flowering with H. Sabdariffa L. and H. cannabinus L.) Landbouw 16, 404—412 (1940).

Borges, M. de L. V.: Uma doença do linho nova para Portugal. (A flax disease new to Portugal.) Brotéria 15, 129—136 (1946).

Borlaug, N. E.: Variation and variability of Fusarium lini. Tech. Bull. Minn. Agric. Exp. Sta. Nr. 168, 40 (1945).

Borodič, Z. N.: (On the problem of biology of flowering of flax.) Proc. Lenin Acad. Agric. Sci. USSR. Nr. 7, 19—26 (1940). [Russisch.]

Boza Barducci, T.: Vistas del nuevo pabellón de genética de algodon inaugurado en Noviembre 1939. (Views of the new cotton genetics pavilion opened in November 1939.) (Minist. Fom. Direcc. Agric. Canad. Estac. Exp. Agric. La Moline, Peru Pp. 9 (1939).

— u. R. M. Madoo: Relationship of Gossypium Raimondii Ulbr. Nature, Lond. 145, 553 (1940).

—, — Investigaciones acerca del parentesco de la especie Peruana de algodonero Gossypium Raimondii, Ulbrich. (Investigations on the relationship of the Peruvian cotton species G. Raimondii Ulbrich.) Bol. Minist. Fom., Lima Nr. 22, 29 (1941).

—, G. García Ráda u. J. Wille: Control of internal boll rot of the cotton plant, caused by insect punctures (Dysdercus sp.), through selection of resistant strains. Nature, Lond. 156, 235—236 (1945).

Braga, O. de S. u. R. E. Kalckmann: Competição de variedades, de densidades e adubações de linha. (Competition of varieties, spacing and manuring in flax.) Bol. Soc. Brasil. Agron. Rio de J. 7, 7—20 (1944).

Bredemann, G.: Züchtung auf Fasergehalt bei Hanf (Cannabis sativa L.). Züchter 14, 201—213 (1942).

— Die Bestimmung des Fasergehaltes bei Massenuntersuchungen von Hanf, Flachs, Fasernesseln und anderen Bastfaserpflanzen. Faserforschung 16, 14 (1942).

Breslavec, L. P.: (Morphological changes in hemp induced by X-radiation. I. Macroscopic observations.) Izvestija Akademii Nauk SSSR. (Bull. Acad. Sci. URSS. Sér. biol.) Nr. 3, 335—348 (1939). [Russisch.]

Brixhe, A.: Rapport sur une mission d'étude effectuée aux Etats-Unis du 5 août au 18 octobre 1939. Bull. Agric. Congo Belge 32, 89—117; 33, 9—54 (1941, 1942).

Brown, C. H.: Cotton-breeding technique as evolved at Giza. Bull. Minist. Agric. Egypt. Nr. 218, 13 (1939).

— Selection and hybridization. Emp. Cott. Gr. Rev. 16, 111—114 (1939).

Brown, H. B.: Registration of improved cotton varieties, II. J. Amer. Soc. Agron. 32, 83 (1940).

— Results from inbreeding Upland cotton for a ten-year period. J. Amer. Soc. Agron. 34, 1084—1089 (1942).

— Registration of improved cotton varieties, III. J. Amer. Soc. Agron. 35, 241 (1943).

— u. C. B. Haddon: Influence of varietal differences on the grade of cotton. J. Amer. Soc. Agron. 35, 249 bis 255 (1943).

Camara, A.: The effect of X-radiation on the chromosomes of Aloë arborescens. Proc. 7th Int. Genet. Congr. Edinburgh 23—30 August, p. 83, 1939 (1941). (Abstr.)

Camargo, F. C.: Vida e utilidade das Bromeliáceas. Biology and utility of the Bromeliaceae.) Bol. Téc. Inst. Agron. Norte, Brasil Nr. 1, 31 (1943).

Čanev, K.: (The determination of the quantity of fibre in breeding of hemp for fibre.) Rev. Inst. Rech. Agron. Bulg. 9, Nr. 2, 3—18 (1939). [Bulgarisch.]

Capinpin, J. M. u. I. Khambanonda: Studies on the cytology and crosscompatibility of some cotton varieties. Philipp. Agric. 28, 163—186 (1939).

Cass Smith, W. P. u. H. L. Harvey: Flax rust in Western Australia. J. Dep. Agric. W. Aust. 23, 42—45 (1946).

Castro, D. R. de: Posição actual de sistemática dos linhos portugueses. (Present position of the systematics of Portuguese flaxes.) Bol. Soc. Broteriana 19, Sér. 2a, 223—232 (1944).

Ceitlin, G.: (Southern ripening hemp.) Len i konoplja (Flax and Hemp) Nr. 5, 54—55 (1940). [Russisch.]

Chakrabarti, S.: What's doing in All-India. Better cotton for Assam. Indian Fmg. 4, 317 (1943).

Chandratane, M. F.: Cotton in Ceylon. Trop. Agriculturist 101, 164—169 (1945).

Charton, R. O. u. G. Roberty: Le cotonnier dans la colonie du Tchad. Rev. Bot. Appl. 26, 300—304 (1946).

Chen, Y. S.: (Hopei cotton improvement and the future of the Stoneville Nr. 4.) Agric. Sci., Peking, p. 77 (1939).

Cheo, M. T.: (Experiments on the resistance of Chicken Foot [Gossypium arboreum var. neglectum and G. arboreum var. roseum] to the cotton leaf-roller [Sylepta derogata].) New Agric. J. Fukien 3, 62—69 (1943).

Chevalier, A.: Les espèces, variétés et hybrides de cotonniers, spécialement les cotonniers d'origine asiatique cultivés en Afrique tropicale. C. R. Acad. Sci. Paris 208, 22—25 (1939).

— Les espèces, variétés et hybrides de cotonniers d'Amerique actuellement cultivées en Afrique tropicale et leur amelioration. C. R. Acad. Sci. Paris 208, 241—245 (1939).

— L'origine, la culture et les usages de cinq Hibiscus de la section Abelmoschus. Rev. Bot. Appl. 20, 319 bis 328, 402—419 (1940).

— Histoire de deux plantes cultivées d'importance primordiale. Le lin et le chanvre. Rev. Bot. Appl. 24, 51—71 (1944).

Chi Pao Yu: The inheritance and linkage relations of yellow seedling, a lethal gene in Asiatic cotton. J. Genet. 39, 61—68 (1939).

— The inheritance and linkage relations of curly leaf and virescent bud, two mutants in Asiatic cotton. J. Genet. 39, 69—77 (1939).

Choudhury, J. K.: Growth and yield of jute plants in relation to watering. Sci. and Cult. 11, 445 (1946).

Cinda, K.: (The Egyptian cotton variety 213). Sovetskij Hlopok (Soviet Cotton) Nr. 3, 35—37 (1940). [Russisch.]

— (Intra-varietal crossing in cotton.) Jarovizacija Nr. 2 (35), 51—55 (1941). [Russisch.]

Čiževskij, I. u. I. Tihonov: (10 years' work of the State Flax Research Institute.) Len i Konoplja (Flax and Hemp) Nr. 10, 6—9 (1940). [Russisch.]

Conrad, C. M. u. J. W. Neely: Heritable relation of wax content and green pigmentation of lint in Upland cotton. J. Agric. Res. 66, 307—312 (1943).

Crane, J. C.: Roselle (Hibiscus sabdariffa L.) as a fiber crop. Div. Latin Amer. Agric., U. S. Dep. Agric. Pp. 47 (1943). [Mimeographed.]

— Kenaf (Hibiscus cannabinus L.) as a fiber crop. Off. For. Agric. Relations, U. S. Dep. Agric. Div. Latin Amer. Agric., Pp. 39 (1943). [Mimeographed.]

— u. J. B. Acuña: Varieties of kenaf (Hibiscus cannabinus) a bast fiber plant, in Cuba. Bot. Gaz. 106, 349 bis 355 (1945).

—, — The comparative evaluation of fourteen types of ramie under Cuban conditions. J. Amer. Soc. Agron. 38, 152—167 (1946).

— u. R. E. Alonso: Fiber content in relation to length and age of Sansevieria Thunb. leaves. J. Amer. Soc. Agron. 37, 953—961 (1945).

Das, G. M.: Studies on the jute stem-weevil Apion corchori Marshall. 1. Bionomics and life history. Indian J. Agric. Sci. **14**, 295—303 (1944).

Demkin, A.: (The main forms and varieties of hemp.) Len i Konoplja (Flax and Hemp) Nr. 10, 26—29 (1940). [Russisch.]

Deshpande, R. B.: A note on the occurrence of a chlorophyll deficiency in linseed (Linum usitatissimum L.). Curr. Sci. **8**, 168—169 (1939).

— Inheritance of leaf-lobe in Hibiscus cannabinus L. Indian J. Genet. Pl. Breed. **2**, 181—182 (1942).

Dewey, L. H.: Fibras vegetales. (Plant fibres.) Publ. Agric. Un. Panamer. Wash. Nr. 137/140, 101 (1941).

Dharma Rajulu, K.: Introduction of Cambodia cotton in the Ceded districts of Madras Province. Indian Cott. Gr. Rev. **1**, 84—86 (1947).

Dillman, A. C.: Cold tolerance in flax. J. Amer. Soc. Agron. **33**, 787—799 (1941).

— u. T. E. Stoa: Flaxseed production on the North Central States. Fmrs.' Bull. U. S. Dep. Agric. Nr. 1747, 19 (1942).

Dona' dalle Rose, A.: La colchicina come stimolante mutativo su lino (L. usitatissimum). [Colchicine as an agent stimulating mutation in flax (L. usitatissimum).] Ital. Agric. **76**, 695—702 (1939).

Doop, J. E. A. den: Nicht-blühende Sisalpflanzen. Faserforschung **14**, 9—27 (1939).

Dorasami, L. S.: Improved varieties of cotton grown in the Mysore State. Mysore Agric. J. **21**, 55—59 (1942 bis 1943).

— Review of work done on cotton in Mysore. Indian Cott. Gr. Rev. **1**, 39 (1947).

Dorman, C.: A year of research in Mississippi farm problems. Cotton investigation: breeding, genetics, varieties. Miss. Fm. Res. **5**, Nr. 10, 3—4 (1942).

Dunlap, A. A.: Light, drought, and heat as factors in cotton boll-shedding. Phytopathology **34**, 999 (1944). (Abstr.)

— Fruiting and shedding of cotton in relation to light and other limiting factors. Bull. Tex. Agric. Exp. Sta. Nr. 677, 104 (1945).

Edwards, H. T.: The introduction of abacá (manila hemp) into the western hemisphere. Rep. Smithson Instn. 1945 Publ. 3817, 327—349 (1946).

ElKilani, M. A.: (New varieties of cotton.) Special Suppl. Egypt. Agric. Mag. **1941**, 19—24 (1941).

Elkin, S.: (Use wild hemp as breeding material.) Len i Konoplja (Flax and Hemp) Nr. 9, 28 (1940). [Russisch.]

Elladi, K. V.: (Inheritance of length of vegetative period in flax hybrids.) Bull. Acad. Sci. URSS., Sér. Biol. p. 371—388 (1939). [Russisch.]

Engledow, F. L.: Mr. M. A. Bailey. Nature, Lond. **144**, 890—891 (1939).

Fahmi, T.: (A technical method of selection in cotton for immunity against wilt.) Egypt. Agric. Rev. **19**, 6—17 (1941).

Fang, W. P.: A new species of Salix from Szechwan. J. W. China Border Res. Soc. **15**, Ser. B. 178—180 (1945).

Floderus, B.: Some new Salix species and hybrids. Bot. Notiser Nr. 2, 227—230 (1940).

— Salix rotundifolia Trautv. and Salix nummularia Anderss. Svensk Bot. Tidskr. **35**, 351—352 (1941).

— Salix helvetica Villars and its subspecies and hybrids. Svensk. Bot. Tidskr. **37**, 73—80 (1943).

— Salix Starkeana Willdenow. Svensk Bot. Tidskr. **37**, 81—82 (1943).

Flor, H. H.: Inheritance of rust reaction in a cross between the flax varieties Buda and J. W. S. J. Agric. Res. **63**, 369—388 (1941).

— Flax rust. Bi-m. Bull. N. Dak. Agric. Exp. Sta. **3**, Nr. 6, 7—9 (1941).

Flor, H. H.: Pasmo disease of flax. Bi-m. Bull. N. Dak. Agric. Exp. Sta. **6**, 31—33 (1943).

— Genetics of pathogenicity in Melampsora lini. J. Agric. Res. **73**, 335—357 (1946).

— Inheritance of reaction to rust in flax. J. Agric. Res. **74**, 241—262 (1947).

Franssen, C. J. H., Ph. Levert u. J. A. van Duyvendijk: Over biologie en bestrijding van den randoekolfboorder (Mudaria variabilis RPKE). (On the biology and control of the kapok pod borer [M. variabiils RPKE].) Bergcultures **15**, 1728—1743 (1941).

Freeman, W. E.: „Samaru 26 C" a new strain of cotton bred in Northern Nigeria. Trop. Agriculture, Trin. **23**, 109—113 (1946).

Friend, W. H.: Here comes a giant sunn hemp resists nematode, drouth and root rot; good fiber and green manure crop. Sth. Seedsman **8**, 20, 43 (1945).

Fulton, H. J.: Some factors that influence the immediate effects of pollen on boll characters in cotton. J. Agric. Res. **63**, 469—480 (1941).

Gaddum, E. W.: Some observations on jassid at the Kenya coast. Emp. J. Exp. Agric. **10**, 133—145 (1942).

Gadkari, P. D., K. G. Deo u. N. T. Nadkerny: Studies in the genetic consistence of Gaorani 6. Indian Cott. Gr. Rev. **1**, 65—76 (1947).

Ganeshan, D.: Aceto-carmine smear technique for cotton cytology. Curr. Sci. **8**, 114—115 (1939).

— Hybrid vigour in cotton. I. The manifestation of hybrid vigour in the seed. Indian J. Genet. Pl. Breed. **2**, 134—150 (1942).

Ganguly, J. K.: A note on the secondary association of chromosomes and the basic number in Corchorus capsularis Willd. Sci. and Cult. **11**, 272 (1945).

Garrido, T. G.: Results of sisal and maguey hybridization Philipp. J. Agric. **10**, 233—248 (1939).

Gattenberger, P.: (Replace the cotton variety 1306 by better varieties.) Sovetskij Hlopok (Soviet Cotton) Nr. 11/12, 29—32 (1939). [Russisch.]

Gavrilov, G.: (Cotton breeding in the Kara-Kalpak ASSR.). Sovetskij Hlopok (Soviet Cotton) Nr. 3, 39 bis 40 (1940). [Russisch.]

Ghose, R. L. M.: The genetics of (Corchorus) jute. The inheritance of pod shape and its linkage relationship. Indian J. Genet. Pl. Breed. **2**, 128—133 (1942).

— The genetics of (Corchorus) jute. The inheritance of pod shape and its linkage relationship. Bull. Indian Cent. Jute Cttee. **6**, 366—369 (1943).

— u. B. Das Gupta: Floral biology, anthesis, and natural crossing in jute. Indian J. Genet. Plant. Breed. **4**, 80—84 (1945).

— u. J. Datta: A note on leaf mutations in jute. Sci. and Cult. **10**, 297—298 (1945).

— u. J. S. Patel: Jute breeding experimental technique. (Selection and handling of breeding material.) Agric. Res. Bull. Indian Cent. Jute Cttee. Nr. 2, 3 (1945).

Goldsmith, G. W. u. E. J. Moore: Field tests of the resistance of cotton to Phymatotrichum omnivorum. Phytopathology **31**, 452—463 (1941).

Gorčakov, A.: (The cotton research institutes in the third Five Year Plan.) Sovetskij Hlopok (Soviet Cotton) Nr. 11/12, 17—20 (1939). [Russisch.]

Govande, G. K.: Linkage relations of the white-pollen factor in Asiatic cottons. Indian J. Agric. Sci. **10**, 842 bis 843 (1940).

— Breeding for resistance to cotton root rot in Gujerat. Proc. 29th Indian Sci. Congr. Baroda, Part. III Sect. Agric. **43**, 217 (1942). (Abstr.)

— A new gene for lintlessness in Asiatic cottons. Curr. Sci. **13**, 15—16 (1944).

— Linkage relations of the lid gene for lintlessness in Asiatic cottons. Curr. Sci. **13**, 321—322 (1944).

— A new mutant in Asiatic cottons. Curr. Sci. **15**, 170 (1946).

Granhall, I.: Linodling och linberedning i Belgien och Holland. (Flax cultivation and processing in Belgium and Holland.) Sverig. Utsädesfören. Tidskr. **49**, 405 bis 422 (1939).
— Sveriges Utsädesförenings linlaboratorium, dess tillkomst och arbetsuppgifter. (The Flax Laboratory of the Swedish Seed Association-its future and functions.) Swerig. Utsädesfören. Tidskr. **50**, 365—368 (1940).
— Odling av spånadsväxter. (Growing textile plants). K. LandtbrAkad. Tidskr. **80**, 124—134 (1941).
— Svalöfs Atlaslin. Ny sort av oljelin för Skåne och angränsande områden samt Gotland. (Svalöf Atlas flax — A new variety of linseed for Scania and neighbouring districts and Gotland.) Sverig. Utsädesfören Tidskr. **53**, 403—404 (1943).
— Fem års försöksverksamhet med spånadsväxter. (Five years research on textile plants.) Sverig. Utsädesfören. Tidskr. **54**, 16—27, 66—83 (1944).
— Linrosten — en spanadslinets fiende. (Flax rust — an enemy of spinning flax.) Lantmannen **28**, 733, 735 bis 736 (1944).
— Det nya linlaboratoriet, dess utrustning och arbetsuppgifter. (The new flax laboratory, its equipment and tasks.) Sverig. Utsädesfören. Tidskr. **55**, 92—100 (1945).
— Spånadsväxternas odling och förädling. (Cultivation and breeding of textile plants.) Årsb. Jordbruksforskning, Stockholm 69—80 (1946).
— Lindag på Svalöf den 22 juli 1946. (Flax Day at Svalöf on 22 July 1946.) Sverig. Utsädesfören, Tidskr. **56**, 617—618 (1946).
Granick, E. B.: A karyosystematic study of the genus Agave. Amer. J. Bot. **31**, 383—398 (1944).
Grečuhin, E. I. u. N. A. Belovickaja: Monoecious hemp. C. R. (Doklady) Acad. Sci. URSS. **27**, 42—46 (1940).
Gregory, P. J. u. M. Ishaque: Cotton prospects of Bengal. Indian Cott. Gr. Rev. **1**, 26—27 (1947).
Grimes, M. A.: A comparison of five methods of measuring finesses of cotton fibres. Textile Res. 1941 **11**, 459—466 (1941).
Groof, G. de: L'Urena lobata jute Congolaise. (U. lobata, Congo jute.) Bull. Agric. Congo Belge **31**, 7—55 (1940).
Groszmann, A.: Cinco anos de melhoramento de algodão pelo Departamento de Genética, Estatística e Biometria da Escola Superior de Agricultura do Estado de Minas Gerais-Viçosa. (Five years of cotton breeding by the Department of Genetics, Statistics and Biometry of the Agricultural High School of the State of Minas Gerais-Viçosa.) Ceres, Brasil **5**, 94—114 (1943).
Gubin, A. F.: (The effect of pollination of flax by bees.) Jarovizacija Nr. 1 (34), 102—103 (1941). [Russisch.]
— (The pollination of fiber flax by honey bees.) Proc. Lenin Acad. Agric. Sci. USSR. Nr. 3/4, 30—32 (1942). [Russisch.]
— Cross pollination of fibre flax. Bee World **26**, Nr. 4, 30—31 (1945).
Guha, M. P.: White-flowered plant of Urena lobata Linn. — A new observation. Curr. Sci. **15**, 113 (1946).
Gulati, A. N.: Neppiness in Indian cotton yarns. Indian Cott. Gr. Rev. **1**, 60—63 (1947).
— u. N. Ahmad: The maturity of cotton fibre. Indian Fmg. **6**, 9—11 (1945).

H ....., A. G.: Manila hemp seedlings. A. Afr. Agric. J. **5**, 379 (1940).
H ....., J. B.: The classification and evolution of cotton. Trop. Agriculture, Trin. **16**, 82—83 (1939).
Haän, H. de: Ontstaan en veredeling van het vlas. (The origin and breeding of flax.) Cultivator Algem. Bond Oudleerl. Inricht Middelbaar Landbouwonderwijs, Wageningen Pp. 15 (1943).
Håkansson, A.: Die Chromosomenpaarung von zwei Salixbastarden. Hereditas, Lund **30**, 639—641 (1944).
Hammond, D.: The expression of genes for leaf shape in Gossypium hirsutum L. and Gossypium arboreum L. I. The expression of genes for leaf shape in Gossypium hirsutum L. II. The expression of genes for leaf shape in Gossypium arboreum L. Amer. J. Bot. **28**, 124—138, 138—150 (1941).
Hancock, H. A.: 23. Measures to check deterioration in Egyptian cotton varieties. Pt I-The Giza seed maintenance system. J. Text. Inst., Manchr. **36**, T267—T277 (1945).
— 23. Measures to check deterioration in Egyptian cotton varieties. Pt II-Development of new varieties. J. Text Inst., Manchr. **36**, T278—T292 (1945).
— 23. Measures to check deterioration in Egyptian cotton varieties. Pt III-Extent, nature, and causes of deterioration. J. Text. Inst., Manchr. **36**, T293—T310 (1945).
Hancok, N. I.: Factors in the breeding of cotton for increased oil and nitrogen content. Circ. Tenn. Agric. Exp. Sta. Nr. 79, 7 (1942).
— Length, fineness, and strenght of cotton lint as related to heredity and environment. J. Amer. Soc. Agron. **36**, 530—536 (1944).
— Variations in length, strength, and fineness of cotton fibers from bolls of known flowering dates, locks, and nodes. J. Amer. Soc. Agron. **39**, 122—134 (1947).
Hansford, C. G.: Vascular diseases of cotton in Uganda. E. Afr. Agric. J. **5**, 279—282 (1940).
Harland, S. C.: A história da evolução dos algodões cultivados do novo mundo. (The history of the evolution of the cultivated cottons of the New World.) An. Primeira Reunião Sul-Amer. Bot. **1**, 215—225 (1938) 1939.
— Some comments on Dr. Mason's article „A note on the technique of cotton breeding". Emp. Cott. Gr. Rev. **16**, 186—193 (1939).
— The genetics of cotton. Jonathan Cape, London (1939).
— Genetical studies in the genus Gossypium and their relationship to evolutionary and taxonimic problems. Proc. 7th Int. Genet. Congr. Edinburgh 23—30 August, 138—143, 1939 (1941).
— New polyploids in cotton by the use of colchicine. Trop. Agriculture, Trin. **17**, 53—54 (1940).
— Taxonomic relationship in the genus Gossypium. J. Wash. Acad. Sci. **30**, 426—432 (1940).
— Abstracts of papers 1915—1941. Soc. Nac. Agraria Inst. Cott. Genet., Peru, Nr. 1, 32 (1942).
— Breeding of a cotton immune from natural crossing. Nature, Lond. **151**, 307 (1943).
— The selection experiment with Peruvian Tangüis cotton. Bull. Inst. Cott. Genet. Lima, Peru, Nr. 1, 98 (1944).
— u. O. M. Atteck: The genetics of cotton. XVIII. Transference of genes from diploid North American wild cottons (Gossypium Thurberi, Tod., G. Armourianum Kearney, and G. aridum comb. nov. Skosted) to tetraploid new world cottons (G. barbadense and G. hirsutum L.). J. Genet. **42**, 1—19 (1941).
—,— The genetics of cotton. XIX. Normal alleles of the crinkled mutant of Gossypium barbadense L. differing in dominance potency, and an experimental verification of Fisher's theory of gominance. J. Genet. **42**, 21—47 (1941).
Harrison, G. J.: Breeding California cotton. Calif. Cultiv. **88**, 696 (1941).
Hermann, F. J.: The perennial species of Urtica in the United States east of the Rocky Mountains. Amer. Midl. Nat. **35**, 773—778 (1946).
Heyn, A. N. J.: Agel- en buntalvezel. (Fibre of the gebang palm and of the talipot palm.) Bergcultures **15**, 162—163 (1941).
Hoffmann, W.: Gleichzeitig reifender Hanf. Züchter **13**, 277—283 (1941).

Hoffmann, W.: Helle Stengel — eine wertvolle Mutation des Hanfes (Cannabis sativa. L.). Züchter **17/18**, 56—59 (1946).
— Die Vererbung der Geschlechtsformen des Hanfes (Cannabis sativa L.) I. Züchter **17/18**, 257—277 (1947).
— u. E. Knapp: Röntgenbestrahlungen beim Hanf. Züchter **12**, 1—9 (1940).
Houston, B. R.: Two important flax diseases in California in 1945. Plant Dis. Reporter **29**, 570—571 (1945).
— u. B. H. Stanford: Tests on flax varieties and hybrids for resistance to anthracnose. Plant Dis. Reporter **29**, 572—573 (1945).
Hu, C. L.: (A review of cotton research in China.) Chinese J. Sci. Agric. **1**, 147—158 (1943).
— Progress report of the cotton department of the National Agricultural Research Bureau in China for the years 1942 to 1943, inclusive. J. Amer. Soc. Agron. **37**, 610—615 (1945).
— Ten years results of experiments and extension work with Delfos cotton in China. J. Amer. Soc. Agron. **37**, 616—618 (1945).
Hultén, E.: Two new species of Salix from Alaska. Svensk Bot. Tidskr. **34**, 373—376 (1940).
Humphrey, L. M.: A preliminary report of the effects of inbreeding in cotton with special reference to staple length and lint percentage. Genetics **25**, 121—122 (1940). (Abstr.)
— Effects of inbreeding cotton with special reference to staple length and lint percentage. Bull. Ark. Agric. Exp. Sta., Nr. 387, 16 (1940).
— u. A. V. Tuller: Mejoras en la técnica de hibridazion del algodón. (Improvements in the technique of hybridization in cotton.) Rev. Fac. Nac. Agron., Colombia **5**, 37—41 (1943).
Husain, M. A. u. K. B. Lal: The bionomics of Empoasca devastans Distant on some varieties of cotton in the Punjab. Indian J. Ent. **2**, Pt. 2, 123—136 (1940).
Hutchinson, J. B.: Some problems in applied genetics. Chronica Botanica **5**, 403—404 (1939).
— The relationships of Gossypium Raimondii Ulb. Trop. Agriculture, Trin. **16**, 271—272 (1939).
— The quality of West Indian Sea Island cotton. A note on investigations made during the year 1938 bis 1939. Rep. 4th Ord. Gen. Mtg. W. Ind. Sea Island Cott. Ass. (Inc.) Antigua (1939), Pp. 25—26 (1940).
— A note on Gossypium brevilanatum Hoch. Trop. Agriculture, Trin. **20**, 4 (1943).
— The cottons of Jamaica. Trop. Agriculture, Trin. **20**, 56—58 (1943).
— Notes on the native cottons of Trinidad. Trop. Agriculture, Trin. **20**, 235—238 (1943).
— The effect of environment on hair characters and spinning value in Sea Island cotton. J. Text. Inst., Manchr. **34**, T61—T69 (1943).
— The cottons of Puerto Rico. J. Agric. Univ. P. R. **28**, 35—42 (1944).
— The crinkled dwarf allelomorph series in the New World cottons. J. Genet. **47**, 178—207 (1946).
— On the occurrence and significance of deleterious genes in cotton. J. Genet. **47**, 272—289 (1946).
— The inheritance of brown lint in New World cottons. J. Genet. **47**, 295—309 (1946).
— Letters to a critic. 1. Can the cotton breeder please the spinner? Emp. Cott. Gr. Rev. **24**, 96—100 (1947).
—, R. L. M. Ghose u. Bhola Nath: Further studies on the inheritance of leaf shape in Asiatic Gossypiums. Indian J. Agric. Sci. **9**, 765—786 (1939).
— u. H. L. Manning: The efficiency of progeny-row-breeding in cotton improvement. Emp. J. Exp. Agric. **11**, 140—154 (1943).
—, — The Sea Island cottons. Emp. J. Exp. Agric. **13**, 80—92 (1945).

Hutchinson, J. B. u. R. A. Silow: Gene symbols for use in cotton genetics. J. Hered. **30**, 461—464 (1939).
—, — u. S. G. Stephens: Evolution and domestication of cotton. Genetics **30**, 9—10 (1945). (Abstr.)
— u. S. G. Stephens: Note on the ,,French" or ,,Small-seeded" cotton grown in the West Indies in the 18th century. Trop. Agriculture, Trin. **21**, 123—125 (1944).
—, — u. K. F. Dodds: The seed hairs of Gossypium. Ann. Bot. Lond. **9**, 361—369 (1945).

Inozemcev, A.: (Long stapled varieties of Egyptian cotton in the Azerbaijan SSR.) Sovetskij Hlopok (Soviet Cotton), Nr. 9, 28—33 (1939). [Russisch.]
Iyengar, N. K.: Chromatin bridges in cotton. Indian J. Agric. Sci. **12**, 785—787 (1942).
— Chromosome conjugation in pentaploid cottons. Indian J. Genet. Pl. Breed. **3**, 99—107 (1943).
— Cytological investigations on auto- and allo-tetraploid Asiatic cottons. Indian J. Agric. Sci. **14**, 30—40 (1944).
— Cytological investigations on hexaploid cottons. Indian J. Agric. Sci. **14**, 142—151 (1944).
— Cytological investigations on some of the interspecific hybrids of (American x Asiatic) x American cottons and their progenies. Indian J. Genet. Pl. Breed. **5**, 32—45 (1945).
Iyengar, R. L. N.: Variation in the measurable characters of cotton fibres. II. Variation among seeds within a lock. Indian J. Agric. Sci. **11**, 703—735 (1941).
— Variation in the measurable characters of cotton fibres. III. Variation of maturity among the different regions of the seed surface. Indian, J. Agric. Sci. **11**, 866—875 (1941).
— Variation of fibre length in a bulk sample of cotton and in a single seed of the bulk. Technol. Leafl. Indian Cent. Cott. Comm. Tedhnol. Lab., Nr. 6, 3 (1944).
— A new method for finding the diameter of cotton fibres. Indian Cott. Gr. Rev. **1**, 32 (1947).

Jacob, K. T.: Certain abnormalities in the root tips of cotton. Curr. Sci. **10**, 174—175 (1941).
— Nuclear changes in the lint primordial cells of Gossypium arboreum var. typicum (Ki). Sci. and Cult. **7**, 512—513 (1942).
Jurion, F.: Quelques concidérations sur l'orientation de la sélection cotonnière au Congo Belge. Bull. Agric. Congo Belge **32**, 677—687 (1941).

Kadam, B. S.: New strains for old crops in Bombay. Part I. Cotton. Indian Fmg. **6**, 353—356 (1945).
Kaniskin, M. F.: (Hemp investigations at the Penza Experimental Station.) Bjulletenj Zernovogo Hozjajstva Jugo-Vostoka SSSR. (Bull. Inst. Grain Husb. S. E. USSR.) Saratov, Nr. 3, 21—28 (1945). [Russisch.]
Karam Rasul, Ch.: What's doing in All-India. The Punjab. Cotton improvement. Indian Fmg. **5**, 84 (1944).
Kara-Murza, L. H.: (On carbo-hydrate metabolism of cotton plant infested with virus.) Izvestija Akademii Nauk SSSR. (Bull. Acad. Sci. URSS. Sér. Biol.), Nr. 5, 491—496 (1946). [Russisch.]
Kasparjan, A. S.: A colchicine-induced amphidiploid — Upland x Egyptian cotton (Gossypium hirsutum L. x G. barbadense L.) C. R. (Doklady) Acad. Sci. URSS. **26**, 163—165 (1940).
— Overcoming non-crossability in cotton. C. R. (Doklady) Acad. Sci. URSS. **49**, 66—68 (1945).
Kearney, T. H.: Cotton breeding in relation to taxonomy. Proc. 8th. Amer. Sci. Congr., Wash. **3**, 251 bis 255 (1940).
— Egyptian-type cottons: their origin and characteristics Bur. Pl. Ind., Soils, Agric. Eng. Agric. Res. Admin., U. S. Dep. Agric., Pp. 23 (1943). [Mimeographed.]

Kearney, T. H., R. H. Peebles u. E. G. Smith: S x P cotton in comparison with Pima. Circ. U. S. Dep. Agric., Nr. 550, 16 (1940).
— u. I. E. Webber: Morphology of two American wild species of cotton and of their hybrid. J. Agric Res. **58**, 445—459 (1939).
Khadilkar, T. R.: A dwarf mutant in Neglectum verum cotton. Curr. Sci. **15**, 278—279 (1946).
— Breeding of high ginning Jarila cotton. Indian Cott. Gr. Rev. **1**, 64 (1947).
Khan, M. A.: What's doing in All-India. The Punjab. Indian Fmg. **3**, 286—288 (1942).
— What's doing in all-India. The Punjab. Indian Fmg. **4**, 99—101 (1943).
Khurshid, A. B. H.: Review of work done on cotton in Hyderabad. Indian Cott. G. Rev. **1**, 40—42 (1947).
Kihara, Y., H. Nakahara u. G. Kodera: (Studies of fiber from Cannabis sativa [I]. Relation of constituents of fiber to the growth periods and sex. On the value of the fiber from wooden parts of hemp as pulp.) Bull. Agric. Chem. Soc. Japan **16**, 731—738 (1940).
Kilany, M. A. El.: (Flax in Egypt. Part 1. Research and culture.) Bull. Minist. Agric. Egypt., Nr. 204, 36 (1939).
— (Jute and kindred fibres in Egypt. Research and culture.) Bull. Minist. Agric. Egypt., Nr. 215, 26 (1939).
Kime, P. H. u. R. H. Tilley: Hybrid vigor in Upland cotton. J. Amer. Soc. Agron. **39**, 308—317 (1947).
Knight, R. L.: The genetics of blackarm resistance. IV. Gossypium punctatum (Sch. and Thon.) crosses. J. Genet. **46**, 1—27 (1944).
— The theory and application of the back-cross technique in cotton-breeding. J. Genet. **47**, 76—86 (1945).
— Breeding cotton resistant to blackarm disease (Bact. malvacearum). Part 1. Introductory. Emp. J. Exp. Agric. **14**, 153—160 (1946).
— Breeding cotton resistant to blackarm disease (Bact. malvacearum) Part II: breeding methods. Emp. J. Exp. Agric. **14**, 161—174 (1946).
— The genetics of blackarm resistance V. Dwarf-bunched and its relationship to B_1. J. Genet. **48**, 43—50 (1947).
— u. T. W. Clouston: The genetics of blackarm resistance. I. Factors B_1 and B_2. J. Genet. **38**, 133—159 (1939).
—, — The genetics of blackarm resistance. 2. Classification, on their resistance, of cotton types and strains. 3. Inheritance in crosses within the Gossypium hirsutum group. J. Genet. **41**, 391—409 (1941).
Kommedahl, T. u. J. J. Christensen: Late wilt of flax. Phytopathology **37**, 13 (1947). (Abstr.)
Konstantinov, N. M.: On the influence of the stock upon the scion in cotton. C. R. (Doklady) Acad. Sci. URSS. **46**, 159—161 (1945).
Koshal, R. S., A. N. Gulati u. N. Ahmad: The inheritance of mean fibre-length, fibre-weight per unit length and fibre-maturity of cotton. Indian J. Agric. Sci. **10**, 975—989 (1940).
Krasovskij, I. R.: (Upbringing of élites of cotton and inheritance of yield of lint.) Jarovizacija, Nr. 2 (29), 85—89 (1940). [Russisch.]
— (Influences of enforced self-pollination in cotton on fruiting and yield.) Jarovizacija, Nr. 1 (34), 104 bis 106 (1941). [Russisch.]
— (Degeneration of industrial varieties of cotton.) Jarovizacija, Nr. 2 (35), 47—50 (1941). [Russisch.]
Kreibohm de la Vega, G. A.: Distribución de semilla de algodón de variedades mejoradas. (Distribution of cotton seed of improved varieties.) Rev. Industr. Agric. Tucumán **30**, 172—175 (1940).
Krüger, E.: Untersuchungen über zwei der bedeutendsten Leinparasiten-Colletotrichum lini Manns et Bolley und Septoria linicola (Speg.) Gar. (Sphaerella linorum Wr.). Arb. biol. Reichsanst. Land- u. Forstw. Berl. **23**, 163—188 (1941).
Kugler, W.: Obtención de dos cosechas por año de multiplicaciones de lino, en la Argentina. (Obtaining two harvests a year of flax undergoing multiplication in Argentina.) Rev. Argent. Agron. **8**, 261 (1941).
— u. C. Remussi: Algunas características morfológicas, fitopatológicas y de resistencia a las heladas en variedades agrícolas de lino cultivadas en la Estación Experimental de Pergamino, durante los años 1937 y 1938. (Certain characteristics of morphology, disease resistance and frost resistance in agricultural varieties of flax cultivated at the Pergamino Experimental Station in 1937 and „Granos" Semilla Selecta, B. Aires 1938 **3**, Nr. 3, 3—24, Nr. 4, 3—38 (1939).
Kuhk, R.: Vergleichende Untersuchungen an di- und tetraploidem Lein (Linum usitatissimum L.) Z. Pflanzenz. **25**, 92—111 (1943).
Kulebjaev, V.: (New coarse linted forms of cotton.) Sovetskij Hlopok (Soviet Cotton), Nr. 6, 46—48 (1939). [Russisch.]
— (New varieties of Soviet long linted cotton in Turkmenistan.) Sovetskij Hlopok (Soviet Cotton), Nr. 11 bis 12, 34—44 (1940). [Russisch.]

Lal, K. B. u. Husain M. Afzal: Hairiness of cotton leaves and antijassid resistance. Curr. Sci. **14**, 153 bis 154 (1945).
Lappo, A. I.: (A brief review of work of the Division of Plant-Breeding and Seed Culture for the years 1936—1937.) Ann. White Russian Agric. Inst., Gorki **8**, 50—53 (1939). [Russisch.]
Leding, A. R. u. L. R. Lytton: Cotton variety tests in the Rio Grande Valley of New Mexico 1940—1943. Bull. N. Mex. Agric. Exp. Sta., Nr. 319, 16; also Supplement to Bull. 319, 2 (1944).
Leggett, W. F.: The story of linen. Chemical Publishing Co., Inc. New York, Pp. XI + 103 (1945).
Levan, A.: The response of some flax strains to tetraploidy. Hereditas, Lund **28**, 246—248 (1942).
— Experimentally induced chlorophyll mutants in flax. Hereditas, Lund **30**, 225—230 (1944).
Lewis, D.: The science of plant breeding. Nature, Lond. **155**, 355—356 (1945).
Lopes, J. P.: Sôbre a cariologia da secção Coarctatae Berger do género Haworthia Duval. (The caryology of the section Coarcticae Berger of the genus Haworthia Duval.) Agron. Lusitana **6**, 129—212 (1944).
Lord, E.: A textile technologist in the cotton field — 1. Emp. Cott. Gr. Rev. **23**, 83—89 (1946).
— A textile technologist in the cotton field — II. Emp. Cott. Gr. Rev. **23**, 163—171 (1946).
— The production and characteristics of the world's cotton crops. J. Text. Inst., Manchr. **37**, T151—T179 (1946).
— The production and characteristics of the world's cotton crops. Part 2. Egypt. Shirley Inst. Mem. **20**, 21—65 (1946).
Lock, G. W.: Sisal Experimental Station report for the year 1938. Pamphl. Dep. Agric. Tanganyika, Nr. 25, 23 (1939).
— Report of the Sisal Experimental Station, Mlingano, for the year 1939. Pamphl. Dep. Agric. Tanganyika, Nr. 26, 21 (1941).
Luthra, J. C. u. R. S. Vasudeva: Studies on the root-rot disease of cottons in the Punjab. IX. Varietal susceptibility to the disease. Indian, J. Agric. Sci. **11**, 410—421 (1941).
Lutkov, A. N.: Mass production of tetraploid flax plants by colchicine treatment. C. R. (Doklady) Acad. Sci. **22**, 175—179 (1939).

McGregor, W. G.: The production of flaxseed in Canada. Fmrs' Bull. Dep. Agric. Canada, Nr. 23, (Publ. 545), Pp. 16 (1946).

Mackay, E. L.: Sex chromosomes of Cannabis sativa. Amer. J. Bot. **26**, 707—708 (1939).

McMichael, S. C.: Occurrence of the dwarf-red character in Upland cotton. J. Agric. Res. **64**, 477—481 (1942).

Mahabale, T. S. u. P. D. Bhate: Cytology of the common Bombay fibre aloe, Agave vivipara L. J. Univ. Bombay, Sect. B **10**, Pt. 3, 51—55 (1941).

Mahta, D. N.: Cotton growing in India. Indian Cott. Gr. Rev. **1**, 1—9 (1947).

Maksimenko, N. K.: (Cotton with naturally-coloured fibre.) Agrobiologija (Agrobiology), Nr. 1, 107—122 (1946). [Russisch.]

Malyh, P. V.: (New lines of flax for fibre.) Len i Konoplja, Nr. 4, 30—34 (1939). [Russisch.]

— (Breeding flax for resistance to loging.) Len i Konoplja (Flax and Hemp), Nr. 11/12, 40—42 (1940). [Russisch.]

Marchionatto, J. B.: Ensayos de laboratorio con el „Fusarium vasinfectum". (Laboratory trials with F. vasinfectum.) Minist. Agric. Nac. Direcc. Gen. Invest. Inst. Sanid. Veg., B. Aires **2**, Ser. A, Nr. 22, 7 (1946).

Marriott, S.: Breeding jassid-resistant cotton varieties. Qd. Agric. J. **57**, 204—206 (1943).

Martins, R. F. C.: Algodão. (Cotton.) Rev. Soc. Bras. Agron. **3**, 214—226 (1940).

Martins, R. G.: A seleção individual no melhoramento do algodoeiro. (Individual plant selection in cotton breeding.) Bol. Minist. Agric., Rio de J. **28**, 47—60 (1939).

Medina, J. C.: Nota preliminar sobre a reprodução sexual do sisal. (Preliminary note on sexual reproduction in sisal.) Rev. Agric. Piracicaba **17**, 41—50 (1942).

Medvedev, P. F.: (Results of the introduction of Americen forms of Apocynum. Vestnik Socialističeskogo Rastenievodstva (Soviet Plant Industry Record), Nr. 3, 107—116 (1940). [Russisch.]

Medvedeva, G. B.: (On the embryology of Hibiscus cannabinus.) J. Bot. URSS. **29**, 264—273 (1944). [Russisch.]

Mendes, A. J. T.: Polyploid cottons obtained through use of colchicine. I. Cytological observations in octoploid Gossypium hirsutum. Bot. Gaz. **102**, 287—294 (1940).

— Algodões poliplóides obtidos pela colchicina. Observações citológicas em Gossypium hirsutum, octoplóide. (Polyploid cottons obtained by colchicine. Cytological observations on octoploid G. hirsutum.) Bragantia, São Paulo **2**, 101—110 (1942).

Merrill, G. R., A. R. Macormac u. H. R. Mauersberger: American cotton handbook. N. Y., Pp. 1024, illus. tables. charts. (1941).

Metha, C. V.: Cotton growing. Trend of development. Times Engng. Suppl. **44**, XIII (1939).

Miège, E.: Les recherches cotonnières en Afrique du Nord. Un. Coton. Emp. Français, Pp. 15—43 (1942).

Miljan, A.: „Eliit" Jõgeva Sordikasvanduse kiulinasort. („Eliit" new variety of fibre flax of the Plant Breeding Station Jõgeva.) Jõgeva Sordikasvanduse toimetised, Tallinnas, Nr. 93, 8 (1939).

Miller, J. H., M. G. Burton u. T. Manning: A statistical study of the relations between flax fiber numbers and diameters and sizes of stems. J. Agric. Res. **70**, 269—281 (1945).

Miller, R. W. R.: Experimental work. Ann. Rep. Dep. Agric. Tanganyika, 31st December, p. 8, 1944 (1945).

Millikan, C. R.: Wilt disease of flax. J. Dep. Agric. Vict. **43**, 305—313, 354—361 (1945).

Millington, A. J.: Wada, a rust-resistant flax variety. J. Aust. Inst. Agric. Sci. **12**, 50—51 (1946).

Mohammad Afzal, C.: A new strain of desi cotton for south western tract of the Punjab. Seas. Notes Punjab Agric. Dep. **18**, 9—12 (1939).

— u. Sarup Singh: The genetics of a petaloid mutant in cotton. Indian J. Agric. Sci. **9**, 787—790 (1939).

Moore, J. H.: The distribution and relation of fiber population, length, breaking load, weight, diameter, and percentage of thinwalled fibres on the cottonseed in five varieties of American upland cotton. J. Agric. Res. **62**, 255—302 (1941).

— The influence of any internal genetic change in a standard variety of cotton upon fibre length. J. Amer. Soc. Agron. **33**, 679—683 (1941).

— Correlation of combed staple lenght on the cottonseed with commercial staple lenght in American upland cotton. J. Amer. Soc. Agron. **35**, 491—498 (1943).

— Cotton fiber: strains grown influence quality. Res. & Fmg. Raleigh, N. C. **3**, Progr. Rep., Nr. 3, 3—4 (1945).

Morgenroth, E.: Die Kultur der Baumwolle in Brasilien. Forschungsdienst **13**, 341—346 (1942).

— Die geschichtliche Entwicklung und Fortschritte der Kultur und Züchtung der Baumwolle in Rußland. Kühn-Archiv **60**, 315—336 (1943/1944).

Mudra, A.: Câteva observatiuni asupra cromozomilor la buretele vegetal (Luffa cylindrica Roem). (Some observations on the chromosomes of the vegetable sponge [L. cylindrica Roem.].) Bul. Fac. Agron. Cluj **7**, 169—171 (1939).

— Tetraploide Mitosen beim Hanf. Bul. Fac. Agron. Cluj. **8**, 269—272 (1939).

Murley, M. R.: A pocynum in Iowa. Iowa St. Coll. J. Sci. **21**, 229—235 (1947).

Muskett, A. E., J. Colhoun, E. L. Calvert u. C. W. R. McCreary: Investigations by the Plant Disease Division, Ministry of Agriculture. Diseases of flax. 19th Rep. Agric. Res. Inst. N. Ire, Pp. 25 (1945—1946).

Nakatomi, S.: (Induced polyploidy in Asiatic varieties of cotton plant by colchicine treatment.) Proc. Crop. Sci. Soc. Japan **12**, 16—20 (1940).

Nanda, D. N. u. M. Afzal: A statistical study of the boll formation in cotton. Indian J. Agric. Sci. **15**, 116—119 (1945).

—, — u. V. G. Panse: A statistical study of flower production in cotton. Indian J. Agric. Sci. **14**, 78—88 (1944).

Nascimento Filho, A. C.: Os cromosomios do genero Sida. (The chromosomes of the genus Sida.) Bol. Soc. Brasil. Agron. Rio de J. **4**, 67—71 (1941).

Nassonov, V. A.: (Anatomical characteristics of the geographical races of hemp.) Soviet Plant Industry Record, Nr. 4, 107—120 (1940). [Russisch.]

Nath, B.: Genetics of petal colour in Asiatic cottons. Indian J. Genet. Pl. Breed. **2**, 43—49 (1942).

— u. G. K. Govande: On the occurrence of the complementary gene for crumpled. Cpa in Rozi cotton. Indian J. Genet. Pl. Breed. **3**, 133—134 (1943).

Nayak, H. R.: Studies on the quality of Jaywant cotton grown from seeds obtained from different stages of propagation. Indian J. Agric. Sci. **12**, 865—872 (1942).

Neal, D. C. u. H. B. Brown: Wilt resistance of the new cottons. Better Crops with Plant Food **24**, Nr. 10; 16, 46 (1940).

Necati Turgay, S.: La question cotonnière en Turquie. Cot. et Cult. Coton **13**, 51—53 (1939).

Neely, J. W.: The effect of genetical factors, seasonal differences and soil variations upon certain characteristics of Upland cotton in the Yazoo-Mississippi Delta. Tech. Bull. Miss. Agric. Exp. Sta., Nr. 28 (1941).

— Inheritance of cluster habit and its linkage relation with anthocyanin pigmentation in Upland cotton. J. Agric. Res. **64**, 105—117 (1942).

— Relation of green lint to lint index in Upland cotton. J. Agric. Res. **66**, 293—306 (1943).

Neely, J. W. u. S. G. Brain: Studies with recently developed cotton strains in the Yazoo-Mississippi Delta, 1943. Circ. Miss. Agric. Exp. Sta., Nr. 121, 8 (1944).

—, — Cotton variety tests in the Yazoo-Mississippi Delta. Bul. Miss. Agric. Exp. Sta., Nr. 398, 8, Nr. 416, 11 (1944/1945).

Neuer, H., E. Prieger u. R. v. Sengbusch: Hanfzüchtung. I. Die Steigerung des Faserertrages von Hanf. Züchter 17/18, 33—39 (1946).

— u. R. v. Sengbusch: Die Geschlechtsvererbung bei Hanf und die Züchtung eines monöcischen Hanfes. Züchter 15, 49—62 (1943).

Newcombe, H. B.: A note on the relation of Gossypium Raimondii Ulbrich to other American species. J. Hered. 30, 530 (1939).

Nilsson, H.: Eine segregate Form von Salix caprea, die durch Stecklinge vermehrt werden kann. Hereditas, Lund 27, 309—312 (1941).

Nodder, C. R. u. A. S. Gillies: A technique for spinning yarn samples from small quantities of fibre. Tech. Res. Mem. Indian Cent. Jute Cttee., Nr. 1, 1—23 (1940).

Novikov, V. A.: Causes underlying bud and boll shedding in cotton and means to control it. C. R. (Doklady) Acad. Sci. URSS. 32, 148—151 (1941).

— (Investigations on the salt resistance of cotton.) Bull. Acad. Sci. URSS., Sér. Biol., Nr. 6, 307—331 (1943). [Russisch.]

— (The shedding of buds and capsules in the cotton plant, as influenced by the length of day. Bull. Acad. Sci. URSS., Sér. Biol., Nr. 1, 29—37 (1944). [Russisch.]

Nowell, W.: A review of the work of the Experiment Stations, seasons 1939—1940 to 1941—1942. Emp. Cott. Gr. Corp. Lond., p. 30 (1943).

— u. J. W. Munroe: A review of the work of the Experiment Stations, season 1938—39, with a note on the entomological work. Emp. Cott. Gr. Corp. Lond., p. 18 (1940).

Obaton, F.: Sur l'anthèse de deux Salix. Bull. Soc. Bot. Fr. 91, 77—78 (1944).

O'Kelly, J. F.: Degeneration within cotton varieties. J. Amer. Soc. Agron. 34, 782—796 (1942).

— Cotton varieties in the hill section of Mississippi 1943. Bull. Miss. Agric. Exp. Sta., Nr. 396, 7 (1944).

— Cotton varieties in the hill section of Mississippi 1944. Bull. Miss. Agric. Exp. Sta., Nr. 411, 8 (1944).

—, E. B. Ferris u. T. E. Ashley: Tests of cotton varieties in the hill section of Mississippi 1942. Bull. Miss. Agric. Exp. Sta., Nr. 386, 7 (1943).

Oljšanskij, M. A.: (Variations in the length of cotton fibre in the direction of selection.) Jarovizacija, Nr. 2 (35), 101—103 (1941). [Russisch.]

Opitz, K.: Ökologie und Züchtung der Faserpflanzen Hanf und Lein. Forschungsdienst, Sonderh. 16, 390 (1942).

— u. G. Maurmann: Untersuchungen über die Wirkung der Dürre auf Lein und Sommergerste. Landw. Jb. 91, 576—617 (1942).

Oudot, G.: Plantes à fibres. Bull. Écon. Indochine 43, 77—91 (1940).

Ovčinnikov u. Razin: (The 30th anniversary of the Leningrad Flax Research Station.) Len i Konoplja (Flax and Hemp), Nr. 10, 10—12 (1940). [Russisch.]

Paiva, R. M.: Uma nova técnica para determinação de ,,fineza" das fibras de algodão, para uso de selecionador de plantas. (A new technique for determining the fineness of cotton fibres for the use of plant breeders.) Rev. Agric., S. Paulo 15, 486—491 (1940).

Panov, A. V.: (Preliminary rapprochement of cotton species.) Jarovizacija, Nr. 4 (31), 106—110 (1940). [Russisch.]

Panse, V. G.: Application of genetics to plant breeding. II. The inheritance of quantitative characters and plant breeding. J. Genet. 40, 283—302 (1940).

— A statistical study of the relation between quality and return per acre in cotton. Indian J. Agric. Sci. 11, 546—565 (1941).

— Plot size in yield surveys on cotton. Curr. Sci. 15, 218—219 (1946).

— u. K. Ramiah: Competition in mixed cotton crops. Proc. 28th Indian Sci. Congr. Benares Part III: Sect. Agric.: abst. 19, 251 (1941).

— u. V. B. Sahasrabudhe: A rapid method of sampling for fibreweight determination in cotton. Indian J. Genet. Pl. Breed. 3, 28—44 (1943).

Pardo Pascual, M.: Cañamos Turcos y del país. (Turkish and local hemps.) Agricultura, Madrid 11, 319 bis 322 (1942).

— Comentarios a un ensayo de variedades de lino para fibra. (Comments on a trial of flax varieties.) Bol. Inst. Invest. Agron. Madr. Nr. 12, 349—369 (1945).

Pasković, F.: (Morphological and biological investigation of some of our varieties of hemp.) Archiv Minist. Poljopr. 6, Nr. 14, 62—82 (1939). [Serbisch.]

— Tehnološka ispitivanja naših važnijih sorata konoplje. (Technological investigation of our main hemp varieties.) Arhiv Minist. Poljopr. 7, 89—110 (1940).

Patel, J. S.: Annual Report of the Agricultural Research Scheme for the year 1939—1940. Indian Cent. Jute Comm. Calcutta p. 50 (1940).

Patel, G. B.: Cotton improvement in South Gujerat (Bombay Province). Indian Cott. Gr. Rev. 1, 19—21 (1947).

—, R. L. M. Ghose u. B. Das Gupta: The genetics of (Corchorus) jute. Part II. Inheritance of anthocyanin pigmentation. Agric. Res. Mem., Indian Cent. Jute Cttee Nr. 3, 42 (1944).

—, — u. A. T. Sanyal: The genetics of Corchorus (jute). III. The inheritance of corolla colour, branching habit, stipule character, and seed coat colour. Indian J. Genet. Pl. Breed. 4, 75—79 (1945).

Pavate, V.: Cultivation of Gadag 1 (upland) cotton. Indian Fmg. 6, 112—113 (1945).

— Cultivation of Jayavant cotton in the Bombay Province. Indian Fmg. 7, 392—394 (1946).

Pavluhin, A. J.: (Methods of evaluation of resistance against lodging of strains of flax for fibre production during the first stages of selection.) Proc. Lenin Acad. Sci. USSR. Nr. 3, 17—20 (1940). [Russisch.]

Pearson, N. L.: Neps in cotton yarns as related to variety, location, and season of growth. Tech. Bull. U. S. Dep. Agric. Nr. 878, 18 (1944).

Peebles, R. H.: Pure-seed production of Egyptian-type cotton. Circ. U. S. Dep. Agric. Nr. 646, 20 (1942).

— u. E. G. Smith: Inheritance of oil glands in Pima cotton. J. Agric. Res. 66, 447—452 (1943).

Persijaninova, A.: (Change in proportion of male and female plants in hemp sown from old seeds.) Soviet Plant Industry Record Nr. 4, 146—148 (1940). [Russisch.]

Peters, R. W.: Cotton culture. Cotton breeding. Qd. Agric. J. 57, 142—146 (1943).

Pilar Rodrigo, A. del.: Especies de ,,Sida" espontáneas en la Argentina que pueden utilizarse como textiles. (Wild species of Sida in Argentina that can be used for textiles.) Rev. Argent. Agron. B. Aires 10, 373—377 (1943).

Pittery, R. u. M. Engelbeen: Note sur l'expérimentation cotonnière. Rapport Annual pour l'Exercice 1938 (2me Partie) Publ. Inst. Agron. Congo Belge 1939, Hors Sér. Pp. 129—156, 1939 (1938).

Pope, O. A., D. M. Simpson u. E. N. Duncan: Effect of corn barriers on natural crossing in cotton. J. Agric. Res. 68, 347—361 (1944).

Pope, O. A. u. J. O. Ware: Effect of variety, location and season on oil, protein and fuzz of cottonseed and on fiber properties of lint. Tech. Bull. U. S. Dep. Agric. Nr. 903, 41 (1945).

Poperekov, M.: (Apply the theory and methods of work of Dr. T. D. Lysenko to work on flax and cotton.) Len i Konoplja (Flax and Hemp) Nr. 1, 6—12 (1939). [Russisch.]

— (Pedigree flax in the third 5-year plan.) Len i Konoplja Nr. 4, 10—13 (1939). [Russisch.]

— (Achievements of Soviet science in the field of flax breeding.) Len i Konoplja (Flax and Hemp) Nr. 8, 16—17 (1939). [Russisch.]

— (Newly bred varieties of linen flax.) Len i Konoplja (Flax and Hemp) Nr. 7, 10—11 (1940). [Russisch.]

Postma, W.: Opmerkingen over de cytologie van normale en van met colchicine behandelde Cannabis-planten. (Observations on the cytology of normal and of colchicine-treated Cannabis plants.) Erfelijkheid in Praktijk, Leiden 4, 171—173 (1939).

— Some remarks on the cytology of normal and colchicine-treated hemp-plants. (Cannabis sativa L.). Rec. Trav. Bot. Néerland 36, 672—676 (1939).

Prado, F. V.: 4a. Contribución al conocimiento to de la histología y citología del maguey (Agave). (4th Contribution to the knowledge of the histology and cytology of the agave.) An. Inst. Biol. Univ. Méx. 13, 43—46 (1942).

Prayag, S. H.: Karnatak cotton and its improvement. Indian Fmg. 3, 488—491 (1942).

Presley, J. T. u. C. J. King: A description of the fungus causing cotton rust, and a preliminary survey of its hosts. Phytopathology 33, 382—389 (1943).

Pressley, E. H.: Estudio sobre el efecto del pólen en el largo de la fibra del agodón. (A study of the effect of pollen upon the length of cotton fibres.) Rev. Fac. Nac. Agron., Colombia 6, Nr. 22, 65—117 (1946).

Prohanov, J. I.: (The conspectus of a new system of cotton.) Botaničeskij Žurnal SSSR. (Journal botanique de l'URSS.) 32, Nr. 2, 61—78 (1947). [Russisch.]

Protzman, C. M.: The new jute-production industry of Brazil. Foreign Agriculture 9, 55—64 (1945).

R....., Z.: Erfolgreiche deutsche Ölleinzüchtung. Wien. landw. Zeitg. 92, 15 (1942).

— Aus einer Pflanze werden zwei. Dtsch. landw. Pr. 69, 40 (1942).

Ramiah, K.: A short review of genetical and plant breeding work in cotton with suggestions for the future. Indian Cent. Cott. Comm., 2nd Conf. Sci. Res. Wkrs. Cott. India, Genet. and Pl. Breed. Paper Nr. 1, 7.

— Description of cotton varieties. Indian Cent. Cotton Cttee. Pp. 28. (ohne Jahr).

— Anthocyanin genetics of cotton and rice. Indian J. Genet. Pl. Breed. 5, 1—14 (1945).

— u. P. D. Gadkari: A note on deterioration and acclimatization of strains. Indian Cent. Cott. Comm., 2nd Conf. Sci. Res. Wkrs. Cott. India, Genet. and Pl. Breed. Paper Nr. 16, 4.

—, — Further observations on sterility in cotton. Indian J. Agric. Sci. 11, 31—36 (1941).

—, — Further studies on the Punjab hairy lintless gene in cotton. Proc. 28th Indian Sci. Congr. Benares, Part III: Sect. Agric.: abst. 42, p. 258 (1941).

— u. S. R. Kaiwar: Studies on the Punjab hairy lintless cotton mutant. Indian J. Genet. Pl. Breed. 2, 98—110 (1942).

— u. B. Nath: Note on a new gene affecting leaf shape in Asiatic cottons. Curr. Sci. 10, 490—491 (1941).

—, — A new gene affecting anthocyanin pigmentation in Asiatic cottons. Proc. 28th Indian Sci. Congr. Benares Part III: Sect. Agric.: abst. 41, p. 258 (1941).

—, — Genetics of single lobe leaf mutant in cotton. Indian J. Genet. Pl. Breed. 3, 89—98 (1943).

Ramiah, K. u. B. Nath: „Red leaf' in G. hirsutum cotton. Proc. 31st Indian Sci. Congr. Delhi Pt. III, p. 163 (1944). (Abstr.)

—, — Genetics of two new anthocyanin patterns in Asiatic cottons. Indian J. Genet. Pl. Breed. 4, 23—42 (1944).

— u. V. N. Paranjpe: The occurence and inheritance of a new type of hairiness in Asiatic cottons. Curr. Sci. 13, 158—160 (1944).

Ramo, I. u. M. V. de Morais: O comportamento de linhagens de algodão Delfos e Stoneville em Ribeirão Preto. (The behaviour of lines of Delfos and Stoneville cotton in Ribeirão Preto.) Bragantia, S. Paulo 3, 553 bis 595 (1943).

Ranganatha, Rao, V. N.: Hybridization between two hybrids. Proc. 7th Int. Genet. Congr. Edinburgh 23— bis 30 August, p. 244—245, 1939 (1941). (Abstr.)

— A study of the inheritance of locular composition in Mysore-American cotton fruit and its relation to yield. J. Mysore Agric. Exp. Un. 18, 1—11 (1940).

Ray, C. R.: Chromosome studies in the genus Linum. Amer. J. Bot. 27, Nr. 10 Suppl. (1940). (Abstr.)

— Cytological studies on the flax genus, Linum. Amer. J. Bot. 31, 241—248 (1944).

Ray, C. (jr.): Cytology and genetics in the flax genus, Linum. Proc. Va. Acad. Sci. 2, 182—183. 1940—1941 (1941).

— Anthracnose resistance in flax. Phytopathology 35, 688—694 (1945).

Ray, W. W.: Varietal reaction of cotton to bacterial blight. Phytopathology 36, 409 (1946). (Abstr.)

Regen, L.: The development of the embryo sac in Agave virginica. Bull. Torrey Bot. Cl. 68, 229—236 (1941).

Renard, K. G.: (Study of the flax stem as material for breeding flax for fibre and producing high yields of fibre.) Theses and scientific papers read at the 4th District Conference of Workers of Universities and Research Institutions, Omsk Nr. 1, Agron. Sect. 19—25 (1941).

Resende, F.: Suculentas Africanas III. Contribuição para o estudo da morfologia, da fisiologia da floração e da geno-sistemática das Aloinàe. (African succulent plants. III. Contribution to the study of the morphology, physiology, flowering and geno-systematics of the Aloinae.) Mem. Soc. Brot., Coimbra 2, 119 (1943).

Reyes, C. A.: Cultivo, industrialización y posibilidades futuras del abacá en Costa Rica. (Cultivation, industrial processing and future possibilities of abacá in Costa Rica.) Dep. Nac. Agric. San José, C. R. Pp. 36 (1945).

Richmond, T. R.: Competition in cotton variety tests. J. Amer. Soc. Agron. 35, 606—612 (1943).

— Inheritance of green and brown lint in Upland cotton. J. Amer. Soc. Agron. 35, 967—975 (1943).

Riley, H. P.: Chromosome studies in a hybrid between Gasteria and Aloe. Genetics 32, 102 (1947). (Abstr.)

Roberts, I.: Cotton .... stronger than steel. Sth. Seedsman 9, 15 (1946).

Roberts, W.: Estate farming in India. III. B. C. G. A. farm Khanewal. Indian Fmg. 3, 174—175 (1942).

Robinson, B. B.: Marihuana investigations. IV. A study of marihuana toxicity on goldfish applied to hemp breeding. J. Amer. Pharm. Ass. 30, 616—619 (1941).

Rogach, A. R.: (Interspecific hybrids in Linum.) Vestnik Gibridizacii (Hybridization) Nr. 2, 84—88 (1941).

Ross, J. G. u. J. W. Boyes: Tetraploidy in flax. Canad. J. Res. 24, Sect. C., 4—6 (1946).

Rothmaler, W.: Sôbre a sistemática e a sociologia dos linhos de Portugal. (The systematics and synecology of the flaxes of Portugal.) Agron. Lusitana 6, 253—279 (1944).

Rudakova, M. M.: (Diagnosis and prediction of earliness and sex in hemp on the basis of the theory of age cycles.) Pap. St. Inst. Agric. Gorkii, USSR. **4**, 93—121 (1943). [Russisch.]

Rumjancev, A. H.: (Fibre quality of simultaneous ripening hemp.) Len i Konoplja Nr. 6, 46—47 (1939). [Russisch.]

Rybin, W. A.: Erzeugung von tetraploiden Pflanzen beim Hanf durch Colchicinbehandlung. C. R. (Doklady) Acad. Sci. URSS. **24**, 586—591 (1939).

— (Chromosome doubling in hemp.) Vestnik Gibridizacii (Hybridization) Nr. 1, 99—110 (1941). [Russisch.]

S...., G.: A new rubber- and fibre-yielding plant. Int. Rev. Agric. **35**, 32 T. (1944).

Sabnis, T. S. u. T. R. Mehta: Some observations on the genetics of linseed. Indian J. Agric. Sci. **15**, 263 bis 265 (1945).

Sankaran, R.: Long staple cotton in Sind. Indian Fmg. **6**, 257—259 (1945).

Šaronov, W. A.: (Wild ramie in Kolkhida. Its history, investigation and introduction into breeding work.) Trudy Botaničeskogo Sada, Moskva Nr. 4, 61—118 (1941). [Russisch.]

Sawhney, K.: Cotton growing in Hyderabad State. Volume 1 (Being a report on a survey of the cotton crop 1931—1935.) Indian Cent. Cott. Comm., Bombay Pp. 90 (1939).

Schad, C., Hughes, R. Mayer u. G. Meneret: Recherches sur le lin en 1940. Ann. Phytogénét. Paris **7**, 109—141 (1941).

Schilling, E.: Die Faserleistung und Ölleistung verschiedener Leinformen. Angew. Bot. **24**, 194—220 (1942).

— Bemerkungen zur Kombinationszüchtung bei Lein. Z. Pflanzenz. **25**, 380—391 (1943).

Schlösser, L.-A.: Über das Fertilwerden autoploider Leinsippen. Züchter **16**, 3—8 (1944).

Sen, D. L.: Technological reports on trade varieties of Indian cottons, 1945. Technol. Bull. Indian Cott. Comm. Ser. A. Nr. 64, 96 (1945).

— Technological reports on trade varieties of Indian cottons, 1946. Technol. Bull. Indian Cent. Cott. Cttee., Nr. 65, Ser. A., 100 (1946).

Šendereckij, E.: (Extending new bast fibre crops into more northerly regions.) Socialističeskoe Seljskoe Hozjajstvo (Socialistic Agriculture) Moskva Nr. 7/8, 92 (1941) . [Russisch.]

Sengbusch, R. v.: Beitrag zum Geschlechtsproblem bei Cannabis sativa. (Vorläufige Mitteilung.) Z. Vererbungsl. **80**, 617—618 (1942).

Sen Gupta, J. C.. u. N. Kumar Sen: On photoperiodic effect of jute plants. Indian J. Agric. Sci. **14**, 196 bis 202 (1944).

— u. N. R. Sen: On photoperiodic effect on jute plants. Proc. 31st Indian Sci. Congr. Delhi Pt. III, p. 89—90 (1944). (Abstr.)

Serdjukov, V. K.: (The technical analysis of fibre plants by means of small samples.) Vestnik Socialističeskogo Rastenievodstva (Soviet Plant Industry Record) Nr. 2, 124—142 (1940). [Russisch.]

Sethi, B. L.: A note on the cultivation of improved varieties of cotton in the United Provinces. Bull. Dep. Agric. Unit. Prov. Nr. 84, 5 (1941).

— Review of work done on cotton in the U. P. Indian Cott. Gr. Rev. **1**, 34—38 (1947).

— u. M. A. A. Ansari: Improved types of cotton in the United Provinces. Indian Fmg. **4**, 461—462 (1943).

Severson, H.: A cotton empire geared to get results. Delta and pine land company of Mississippi, once plagued by floods and boll weevils, is now world's largest cottonseed dealer. Sth. Seedsman **5**, Nr. 11; 22, 26 (1942.)

Sikka, S. M., Khan, Ihsan-Ur-Rahman u. M. Afzal: Study of somatic chromosomes of some wild and cultivated species of Gossypium. Indian J. Genet. Pl. Breed. **4**, 55—68 (1944).

Silow, R. A.: The genetics of leaf shape in diploid cottons and the theory of gene interaction. J. Genet. **38**, 229—276 (1939).

— The genetics and taxonomic distribution of some specific lint quantity genes in Asiatic cottons. J. Genet. **38**, 277—298 (1939).

— The comparative genetics of Gossypium anomalum and the cultivated Asiatic cottons. J. Genet. **42**, 259 bis 358 (1941).

— The inheritance of lint colour in Asiatic cottons. J. Genet. **46**, 78—115 (1944).

— The genetics of species development in the Old World cottons. J. Genet. **46**, 62—77 (1944).

— Further data on the inheritance of lint color. J. Hered. **36**, 62—64 (1945).

— Evidence on chromosome homology and gene homology in the amphidiploid New World cottons. J. Genet. **47**, 213—221 (1946).

— The exchange of attributes between alleles. Pap. Genet. Soc., Lond. (4 January): Pp. 3 (1946). [Mimeographed.]

— u. S. G. Stephens: „Twinning" in cotton. J. Hered. **35**, 76—78 (1944).

— u. C. P. Yu: Anthocyanin pattern in Asiatic cottons. J. Genet. **43**, 249—284 (1942).

Simlote, K. M.: Improvement of cotton in Central India. Indian Fmg. **7**, 68—71 (1946).

Simpson, D. M. u. K. L. Hertel: Environmental modification of fiber properties as a source of error in cotton experiments. J. Agric. Res. **73**, 97—111 (1946).

— u. R. Weindling: Bacterial blight resistance in a strain of Stoneville cotton. J. Amer. Soc. Agron. **38**, 630—635 (1946).

Singh, A.: Cytology of Linum spp. Proc. 27th Indian Sci. Congr. Madras Pt III, Sect. Bot., Abstr. 37, p. 140 (1940).

Sinnott, E. W. u. R. Bloch: Fiber development in Luffa, the sponge gourd. Amer. J. Bot. Suppl. **29**, 17s. (1942). (Abstr.)

—, — Development of the fibrous net in the fruit of various races of Luffa cylindrica. Bot. Gaz. **105**, 90—99 (1943).

Sizov, I. A.: (Flax breeding news.) Len i Konoplja (Flax and Hemp.) Nr. 1, 54—55 (1939). [Russisch.]

— (Productive forms of flax obtained by hybridization.) Vestnik Socialističeskogo Rastenievodstva (Soviet Plant Industry Record) Nr. 2, 51—56 (1940). [Russisch.]

— (Initial material for flax breeding.) Soviet Plant Industry Record Nr. 4, 87—106 (1940). [Russisch.]

Skovsted, A.: Some hybridization experiments in the tribe Hibisceae. C. R. Lab. Carlsberg **24**, Sér. Physiol., 1—30. (1944).

Smith, A. L.: The reaction of cotton varieties to Fusarium wilt and root-knot nematode. Phytopathology **31**, 1099—1107 (1941).

Smith, E. C.: Sex expression in willows. Bot. Gaz. **101**, 851—861 (1940).

— The willows of Colorado. Amer. Midl. Nat. **27**, 217 bis 252 (1942).

Smith, E. G.: Inheritance of smooth and pitted bolls in Pima cotton. J. Agric. Res. **64**, 101—103 (1942).

Smith, H. P.: Strom-proof (Macha) cotton. Sth. Agric. **74**, Nr. 4, 24 (1944).

—, D. T. Killough, D. L. Jones u. M. H. Byrom: Mechanical harvesting of cotton as affected by varietal characteristics and other factors. Bull. Tex. Agric. Exp. Sta. Nr. 580, 49 (1939).

Smith, W. S., W. J. Martin u. N. L. Pearson: Relationship of certain characteristics of seed cottons to ginning. J. Agric. Res. **66**, 249—260 (1943).

Spangenberg, J.: Ensayos de linos de fibra en el Uruguay. Resultados obtenidos. (Trials of fibre flax in Uruguay. Results obtained.) Rev. Fac. Agron. Univ. Montevideo Nr. 37, 83—151 (1944).

Sparrow, F. K.: Types of pods of Asclepias syriaca found in Michigan. J. Agric. Res. **73**, 65—80 (1946).
Stephens, S. G.: Colchicine treatment as a means of inducing polyploidy in cotton. Trop. Agriculture, Trin. **17**, 23—25 (1940).
— Colchicine-produced polyploids in Gossypium. I. An autotetraploid Asiatic cotton and certain of its hybrids with wild diploid species. J. Genet. **44**, 272—295 (1942).
— Grafting experiments with cotton. Trop. Agriculture Trin. **20**, 33—39 (1943).
— Meiosis of a triple species hybrid in Gossypium. Nature. Lond. **153**, 82—83 (1944).
— Phenogenetic evidence for the amphidiploid origin of New World cottons. Nature, Lond. **153**, 53—54 (1944).
— The genetic organization of leaf-shape development in the genus Gossypium. J. Genet. **46**, 28—51 (1944).
— Colchicine-produced polyploids in Gossypium. II. Old World triploid hybrids. J. Genet. **46**, 303—312 (1945).
— A genetic survey of leaf shape in New World cottons — a problem in critical identification of alleles. J. Genet. **46**, 313—330 (1945).
— The modifier concept. A developmental analysis of leaf-shape, modification' in the New World cottons. J. Genet. **46**, 331—344 (1945).
— Canalization of gene action in the Gossypium leaf-shape system and its bearing on certain evolutionary mechanisms. J. Genet. **46**, 345—357 (1945).
— Some observations on leaf shape expression in the Malvaceae. Amer. Nat. **79**, 380—384 (1945).
— The genetics of ,,corky". I. The new world alleles and their possible role as an interspecific isolating mechanism. J. Genet. **47**, 150—161 (1946).
— Some recent trends in cotton research in the United States. Emp. Cott. Gr. Rev. **24**, 28—35 (1947).
— Cytogenetics of Gossypium and the problem of the origin of New World cottons. Advances Genet., N. Y. **1**, 431—442 (1947).
Steyaert, R. L.: La sélection du cotonnier pour la résistance aux stigmatomycoses. Publ. Inst. Agron. Congo Belge Sér. Sci. Nr. 16, 29 (1939).
Stoa, T. E.: Which flax varieties to grow in 1942. Bi-m. Bull. N. Dak. Agric. Exp. Sta. **4**, Nr. 4, 2—6 (1942).
— Varieties of flax that resist rust. Bi-m. Bull. N. Dak. Agric. Exp. Sta. **5**, Nr. 1, 29—31 (1942).
— Sheyenne — a new variety of flax. Bi-m. Bull. N. Dak. Agric. Exp. Sta. **7**, 17 (1945).
— Varieties of flax and disease resistance. Bi.-m. Bull. N. Dak. Agric. Exp. Sta. **7**, 18—23 (1945).
— Dakota — a new flax. Bi-m. Bull. N. Dak. Agric. Exp. Sta. **9**, Nr. 2, 53 (1946).
Straib, W.: Beiträge zur Epidemiologie und Bekämpfung des Flachsrostes. Angew. Bot. **24**, 16—30 (1942).
Strogonov, B. P.: (Über die Anpassung der Baumwolle an salzreichen Boden.) C. R. (Doklady) Acad. Sci. URSS. **54**, 457—460 (1946). [Russisch.]
Stroman, G. N.: A heritable female-sterile type in cotton. J. Hered. **32**, 167—168 (1941).
— Diameter of fibre in different strains of Acala cotton. J. Agric. Res. **64**, 243—255 (1942).
Sukačev, V. N.: (Work on willow breeding.) Lesnoe Hozjajstvo (Forestry) Nr. 3, 24—34 (1939). [Russisch.]

Tariman, M. C.: Türkiye ketenlerinin morfolojik ve teknolojik vasiflari ve bunlardan faydalanma imkânlari. (Morphological and technological characteristics of Turkish flaxes and possibility of their utilization.) T. C. Yüksek Ziraat Enstitüsü Basimevi, Ankara Nr. 145, 108 (1944).
Tavares, H.: O indice de oleo nas seleções do algodão. (The oil index in cotton selections.) Bol. Sec. Agric. Pernambuco **4**, 168—174 (1939).
Ter-Avanesjan, D. V.: (The biology of flowering of cotton.) Vestnik Socialističeskogo Rastenievodstva (Soviet Plant Industry Record) Nr. 5, 181—183 (1940). [Russisch.]
Ter-Avanesjan, D. V.: (The effectiveness of intravarietal crosses in cotton.) Vestnik Socialističeskogo Rastenievodstva (Soviet Plant Industry Record) Nr. 1, 35—40 (1941).]Russisch.]
— Genetic diversity of gametes in the flower of cottonplant. C. R. (Doklady) Acad. Sci. URSS. **44**, 345—347 (1944).
— (Grafting methods in cotton breeding.) Doklady Vsesojuz. Akad. Seljsk. Nauk. im. V. I. Lenina (Proc. Lenin Acad. Agric. Sci. USSR.) Nos. 4/5, 34—35 (1945). [Russisch.]
Thomas, I. u. A. J. Millington: Flax and linseed breeding in W. A. Wada, a new rust resistant flax variety. J. Dep. Agric. W. Aust. **23**, 39—42 (1946).
Thomas, R.: American cotton growing in Sind. Indian Fmg. **5**, 557—559 (1944).
Thoria, L. u. N. Ahmad: The analysis, grading and utilization of Indian linters. Technol. Bull. Indian. Cent. Cott. Cttee., Nr. 34, 24 (1943).
Tiflova, A. M.: (Breeding simultaneous ripening hemp of the ,,masculinized" type.) Len i Konoplja (Flax and Hemp), Nr. 10/11, 42—44 (1939). [Russisch.]
Tisdale, H. B. u. J. B. Dick: Cotton wilt in Alabama as affected by potash supplements and as related to varietal behaviour and other important agronomic problems. J. Amer. Soc. Agron. **34**, 405—426 (1942).
Tobler, F.: Flachsrost und Bastfaser. Faserforschung **15**, 132—135, also Zbl. Bact. 1943, **106**, 74 (1941).
Torres, J. P. u. T. G. Garrido: Progress report on the breeding of abacá (Musa textilis Née). Philipp. J. Agric. **10**, 211—232 (1939).
Tschanev, K. siehe Čanev.
Tseitlin, G. siehe Ceitlin.
Tsinda, K. siehe Cinda.
Turner, J. H. jr.: The effect of potash level on several characters in four strains of upland cotton which differ in foliage growth. J. Amer. Soc. Agron. **36**, 688—698 (1944).
Turner, T. W.: Seven-year experiment in cotton breeding at Hampton Institute. Proc. Va. Acad. Sci. **2**, 181, 1940—1941 (1941).

Vallega, J.: Especialización fisiológica de Melampsora lini, en Argentina. (Physiological specialization of M. lini in Argentina.) An. Inst. Fitotéc. Santa Catalina **4**, 59—74 (1942).
Vansell, G. H.: Cotton nectar in relation to bee activity and honey production. J. Econ. Ent. **37**, 528 bis 530 (1944).
— Some western nectars and their corresponding honeys. J. Econ. Ent. **37**, 530—533 (1944).
Vanterpool, T. C.: Selenophoma linicola sp. nov. on flax in Saskatchewan. Mycologia **39**, 341—348 (1947).
Varuncjan, I.: (Artificial polyploidy in cotton.) Agrobiologija (Agrobiology), Nr. 1, 148—157 (1946). [Russisch.]
Veli-Zade, I.: (New wilt resistant lines of Upland.) Sovetskij Hlopok (Soviet Cotton), Nr. 4, 39—40 (1940). [Russisch.]
Verma, P. M. u. M. Afzal: Studies on the cotton jassid (Empoasca devastans Distant) in the Punjab. 1. Varietal susceptibility and development of the pest on different varieties of cotton. Indian J. Agric. Sci. **10**, 911—926 (1940).
Visweswara Rao, K. u. R. Vasudeva Sarma: Occurrence of gossypol. Curr. Sci. **14**, 270—271 (1945).
Volkov, A.: (The structure of flax fibre in connexion with its place of origin.) Socialističeskoe Seljskoe Hozjajstvo (Socialistic Agriculture) Moskva, Nos. 1/2, 54—56 (1946). [Russisch.]

Waelkens, M.: La sélection pédigrée à Bambesa pendant la période 1934—1938. Bull. Com. Coton. Cong. **4**, 3—10 (1939).
— u. M. Lecomte: Le choix de la variété de coton dans les districts de l'Uelé et de l'Ubangui. Publ. Inst. Agron. Congo Belge: Ser. Tech., Nr. 29, 31 (1941).
Ware, J. O.: Relation of fuzz pattern to lint in an upland cotton cross. J. Hered. **31**, 489—496 (1940).
— Genetic relations of sparse lint, naked seeds and some other characters in upland cotton. Bull. Ark. Agric. Exp. Sta., Nr. 406, 32 (1941).
— Seed cover and plant color and their inter-relations with lint and seed in upland cotton. J. Amer. Soc. Agron. **33**, 420—436 (1941).
— u. D. C. Harrell: Inheritance of strength of lint in upland cotton. J. Amer. Soc. Agron. **36**, 976—987 (1944).
—, W. H. Jenkins u. D. C. Harrell: Inheritance of green fuzz, fiber length, and fiber length uniformity in upland cotton. J. Amer. Soc. Agron. **35**, 382—392 (1943).
—, —, — Seed characters and lint production. J. Hered. **35**, 153—160 (1944).
Warmke, H. E. u. A. F. Blakeslee: Effect of polyploidy upon the sex mechanism in dioecious plants. Genetics **24**, 88—89 (1939). (Abstr.)
Waterhouse, W. L. u. I. A. Watson: A note on determinations of physiological specialization in flax rust. J. Roy. Soc. N. S. W. **75**, 115—117 (1942).
—, — Further determinations of specialization in flax rust caused by Melampsora lini (Pers.) Lév. J. Roy. Soc. N. S. W. **77**, 138—144 (1944).
Webber, H. H.: A tropical winter plant breeding station enabling two field generations a year. J. Amer. Soc. Agron. **37**, 859—861 (1945).
Webber, J. M.: Relationships in the genus Gossypium as indicated by cytological data. J. Agric. Res. **58**, 237—261 (1939).
Welch, F. J.: The current cotton research program in production and related fields. Miss. Agric. Exp. Sta., Pp. 103 (1945). [Mimeographed.]
Wells, W. G.: A review of the 1943—44 cotton growing season. Qd Agric. J. **59**, 264—267 (1944).
Wettstein, F. v.: Untersuchungen zur plasmatischen Vererbung I. Linum. Biol. Zbl. **65**, 149 (1946).
Wilkinson, J.: The cytology of the cricket bat willow (Salix alba var. caerulea). Ann. Bot. London **5**, 149 bis 165 (1941).
— The cytology of Salix in relation to its taxonomy. Ann. Bot. Lond. **8**, 269—284 (1944).
Wilson, I. M.: Observations on wilt disease of flax. Trans. Brit. Mycol. Soc. **29**, 221—231 (1946).
Wouters, W.: Au sujet du spécimen-type de Gossypium obtusifolium Roxb. Bull. Jard. Bot. Brux. **17**, 245 bis 251 (1945).
Wulf, E. V. (Editor): (Flora of cultivated plants. V. Fiber plants. Part I.) Moskva-Leningrad, Pp. 308 (1940). [Russisch.]

Xavier, L. P.: O caroá, história, cultura e distribuiçao geográfica. (Caroá, its history, cultivation and geographical distribution.) Minist. Agric. Dep. Nac. Prod. Veg., Div. Fom. Prod. Veg., Serv. Inform. Agric. Rio de J., Pp. 270 (1942).

Yamada, N.: (Hybridization between cultivated Asiatic and cultivated American cotton species. A review.) Jap. J. Genet. **16**, 79—86 (1940).
Yamashita, K.: (Cotton plants treated with colchicine.) Jap. J. Genet. **16**, 267—270 (1940).
Yang, M. N.: On the genetical behaviour of some mutants of Upland cotton. Acta Brevia Sinensia, Nr. 9, 18 (1945). (Abstr.) [Mimeographed.]
Young, P. A.: Cottons resistant to wilt and root knot and the effect of potash fertilizer in east Texas. Bull. Tex. Agric. Exp. Sta., Nr. 627, 26 (1943).
Young, V. H. u. L. M. Humphrey: Varietal resistance to the Fusarium wilt disease of cotton. Bull. Univ. Ark. Agric. Exp. Sta., Nr. 437, 23 (1943).
Yu, C.-P.: The genetical behavior of three virescent mutants in Asiatic cotton. J. Amer. Soc. Agron. **33**, 756—758 (1941).
Žebrak, A. R. u. M. M. Ržaev: Mass production of amphidiploids by colchicine treatment in cotton. C. R. (Doklady) Acad. Sic. URSS. **26**, 159—162 (1940).
Zemit, V. E.: (The influence of wide spacing on the hereditary variation of varieties of fibre flax.) Len i Konoplja (Flax and Hemp), Nr. 1, 40—43 (1940). [Russisch.]
— (Achievements in breeding fibre flax in the ten years 1929—1939.) Len i Konoplja (Flax and Hemp), Nr. 5, 30—33 (1940). [Russisch.]
Zhebrak, A. R. siehe Žebrak.
Zhurbin, A. I. siehe Žurbin.
Žurbin, A. I.: Polyploids in cotton experimentally produced by colchicine treatment. C. R. (Doklady) Acad. Sci. URSS. **30**, 524—526 (1941).
— Influence of grafting upon the generative sphere in cotton-plant. C. R. (Doklady) Acad. Sci. URSS **46**, 375—378 (1945).

f) Zuckerpflanzen.

Abbott, E. V.: Cytospora rot of sugarcane in Louisiana. Proc. 6th Congr. Int. Soc. Sug. Cane Tech., La, Pp. 447 bis 457 (1938) 1939.
— A progress report on the study of chlorotic streak of sugarcane in Louisiana. Sug. Bull., N. O. **18**, Nr. 19, 3—6 (1940).
— Disease testing and initial seedling selection work at the Houma Station during 1940 and 1941. Sug. Bull., N. O. **20**, 137—141 (1942).
— The present situation with respect to chlorotic streak in Louisiana. Sug. Bull., N. O. **21**, 207—208 (1943).
— u. E. M. Summers: Disease testing and initial seedling selection work at the Houma station during 1943. Sug. Bull., N. O. **22**, 144—148 (1944).
—, — Disease testing and initial seedling selection work at the Houma Station during 1945. Sug. Bull., N. O. **24**, 137A—138A (1946).
Abegg, F. A.: The induction of polyploidy in Beta vulgaris L. by colchicine treatment. Proc. Amer. Soc. Sug. Beet Technol., Pp. 118—119 (1940). [Mimeographed.]
— Evaluation of polyploid strains derived from curly-top resistant and leafspotresistant sugar-beet varieties. Proc. 3rd Amer. Soc. Sug. Beet Technol. 1942, Pp. 309 bis 320 (1943).
Abraham, A.: Natural and artificial polyploids in tapioca (Manihot utilissima). Proc. 31st Indian Sci. Congr. Delhi, Pt III, p. 91 (1944). (Abstr.)
Adamenko, F. I. et al.: (The cultivation of sugar beet for seed in irrigated districts of the Kirgiz and Kazah S.S.R.'s.) Vsesojuznyj Naučno-issledovateljskij Institut Saharnoj Promyšlennosti (VNIS) (State Sugar Beet Research Institute) Frunze, Pp. 105 (1942). [Russisch.]
Agete, F.: Buscando nuevas cañas para Cuba. (Searching for new canes for Cuba.) Rev. Minist. Agric. Cuba **29**, 72—76 (1946).
Alvarez, A. S.: Purificación del jugo de las nuevas variedades de caña. (Purification of the juice of the new varieties of cane.) Bol. Estac. Agric. Tucumán, Nr. 49, 25 (1944).
— El jugo de caña despues de clarificado aumenta o disminuye en pureza? (Does cane juice become more

pure or less pure after refining?) Bol. Estac. Agríc. Tucumán, Nr. 57, 11 (1946).
Anderson, W. S.: Hodo, new sorghum for sirup, described. Miss. Fm. Res. 7, Nr. 3, 1 (1944).
Anonymus: Varietal trials, 1938 season. Cane Gr. Quart. Bull. 6, 123—136 (1939).
— Les croisements de sucrière et de fourragères. (Crosses of sugar and fodder beets.) Publ. Inst. Belge Amélior. Better. 7, 332—333 (1939).
— Release of sugarcane varieties C. P. 29/103 and C. P. 29/120. Sug. Bull., N. O. 18, Nr. 1, 4 (1939).
— Science marches on. Sug. Bull., N. O. 18, Nr. 4, 1 (1939).
— Experiment station notes. S. Afr. Sug. J. 24, 453 bis 455 (1940).
— Some practical results of sugarcane research in India. Misc. Bull. Imp. Coun. Agric. Res. New Delhi, Nr. 34, 41 (1940).
— Seed breeding and seed production. Sug. Beet Manual, London Chapter 13, Pp. 60—62 (1940).
— Cold resistant sugar cane. Aust. Sug. J. 31, 568 bis 570 (1940).
— Cane breeding. Aust. Sug. J. 32, 251—254 (1940).
— (Beet Growing. Vol. I. Biology, genetics and breeding of sugar beet.) Vsesojuznyj N—I Institut Saharnoj Promyšlennosti (Sci. Inst. Sug. Ind.) Kiev, Pp. 918 (1940). [Russisch.]
— Sugar beet investigational work United States and Canada. The American Society of Sugar Beet Technologists, Colorado (1941). [Mimeographed.]
— Sugar beet investigational work-United States and Canada. Pp. 20 + 7 (1941). [Mimeographed.]
— Zijn bloeigevallen te verwachten ook in nieuwe rietsoorten, welke tot dusver geen bloeiverschijnselen hebben vertoond? (Are instances of flowering to be expected even in new cane varieties which so far have shown no signs of flowering?) Arch. Suikerind. Nederland Ned.-Ind., Nr. 2, 422 (1941).
— Notes on new cane varieties in south Queensland. Aust. Sug. J. 34, 215 (1942).
— A seccessful field day. Promising new cane varieties. S. Afr. Sug. J. 26, 259 (1942).
— Sugar — food for man and gun. S. Afr. Sug. J. 26, 391—395 (1942).
— Summit (Panama, C. Z.): Experiment Gardens. Chronica Botanica 7, 177 (1942).
— Experimental varieties of sugar cane. A warning. Aust. Sug. J. 34, 337—338 (1942).
— What the scientists are doing: New Co canes. Indian Fmg 4, 364 (1943).
— „The Sugar Experiment Stations Acts 1900 to 1941". List of varieties of sugar cane approved for planting in 1943. Aust. Sug. J. 43, 370—371, 373 (1943).
— Dr. Brandes explains the sorgo situation. Sug. Bull. N. O. 21, 175—176 (1943).
— „Trojan" and „Eros" canes. Aust. Sug. J. 36, 139 (1944).
— The South African sugar year book 1943—1944. Published by South African Sugar Journal, Durban, Pp. 199, Tables. Illus. (1944).
— Industrial crops. 2. Sugar cane in the Soviet Union. Agriculture, Moscow, Nr. 10, 3—5 (1945). [Mimeographed.]
— Story of college sugar beet seed research is WJR radio feature. Sug. Beet J. Michigan 10, 134—135 (1945).
— What the scientists are doing. The 1945 batch of Co. canes. Indian Fmg 6, 176 (1945).
— Bureau of Sugar Experiment Stations. Distribution of new varieties. Aust. Aug. J. 37, 163 (1945).
— Higher yields foreseen from MSC-developed sugar beet hybrids. Sug. Beet J. Michigan 10, 107 (1945).
— New improvements due in resistant seed types. U. and I. Cultiv. Utah 5, Nr. 3, 28 (1945).

Anonymus: Strains of sugar beet. Fmrs' Leafl. Nat. Inst. Agric. Bot., Cambridge (Revised Ed.), Nr. 5, 4 (1945).
— Hybrid vigor possible for sugar beet seed? Sug. Beet J., Michigan 11, 256—257 (1946).
— What the scientists are doing. The 1946 batch of new Co. canes. Indian Fmg. 7, 252 1946.
Arceneaux, G.: Report of the committee on technique of field experiments. Proc. 6th Congr. Int. Soc. Sug. Cane Tech., La, Pp. 387—290 (1938) 1939.
— Some varietal stalk characters of importance in the mechanical harvesting of sugarcane. Sug. Bull. N. O. 21, 17—19 (1942).
— Varietal factors for calculating yield of sugar per ton of cane. Sug. Bull. N. O. 21, 76—77 (1943).
— The variety census for 1942. Sug. Bull. N. O. 21, 109 (1943).
— Varietal factors for calculating yield of sugar per ton of cane. Int. Sug. J. 96, 125—126 (1944).
— Results of sugarcane variety tests in Louisiana during 1944. Sug. Bull. N. O. 23, 143—149 (1945).
— Generalized factors for computing varietal yields of sugar from results of field tests with sugarcane. Sug. Bull., N. O. 23, 186—191 (1945).
— u. L. G. Davidson: Milling tests of promising unreleased varieties of sugar. Sug. Bull., N. O. 20, 180—181 (1942).
— u. L. P. Herbert: A statistical analysis of varietal yields of sugarcane obtained over a period of years. J. Amer. Soc. Agron. 35, 148—160 (1943).
—, — u. C. C. Krumbhaar: Results of sugarcane variety tests in Louisiana during 1941. Sug. Bull., N. O. 20, 149—155 (1942).
—, —, — Results of sugarcane variety tests in Louisiana during 1943. Sug. Bull., N. O. 22, 159—165 (1944).
—, — u. L. C. Mayeaux jr.: Results of sugarcane variety tests in Louisiana during 1945. Sug. Bull., N. O. 24, 139—142, 159—161 (1946).
—, —, — Results of sugarcane variety tests in Louisiana during 1946. Sug. Bull., N. O. 25, 141—146, 155 (1947).
—, —, — Some stalk characters of importance in sugarcane varietal comparison. Sug. Bull., N. O. 24, 176, 178 (1946).
—, I. E. Stokes, B. A. Belcher u. R. T. Gibbens jr.: Border competition in sugarcane variety tests under Louisiana and Georgia conditions. Proc. 6th Congr. Int. Soc. Sug. Cane Tech., La, Pp. 403—420 (1938) 1939.
Arhimovič, A.: (A new type of isolator for groups of beetroot plantings.) J. Inst. Bot. Acad. Sci. Ukraine, Nr. 23 (31), 167—175 (1940). [Russisch.]
Artschwager, E.: Indications of polyploidy in sugar beets induced by colchicine. Proc. Amer. Soc. Sug. Beet Technol., Pp. 120—121 (1940). [Mimeographed.]
— Morphology of the vegetative organs of sugarcane. J. Agric. Res. 60, 503—549 (1940).
— A comparative analysis of the vegetative characteristics of some variants of Saccharum spontaneum. Tech. Bull. U.S. Dep. Agric., Nr. 811, 55 (1942).
— Pollen degeneration in male-sterile sugar beets with special reference to the tapetal plasmodium. Amer. J. Bot. Suppl. 33, 817 (1946). (Abstr.)
Avanzi, E.: Cenni sui primi resultati della selezione del sorgo zuccherino. (Notes on the first results of selection in sweet sorghum.) Ann. Fac. Agrar. Univ. Pisa 5 (N. S.), 198—207 (1942).

B....., A. F.: Cold resistant sugar cane. Cane Gr. Quart. Bull. 7, 175—176 (1940).
— New seedlings for field trial at Mackay. Cane Gr. Quart. Bull. 9, 23 (1941).
Badami, V. K.: New sugarcane seedlings for Orissa. Proc. 28th Indian Sci. Congr. Benares Part. III. Sec. Agric. abst. 27, p. 253 (1941).

Bain, F. M.: Field experiments on sugar cane in Trinidad. Annual report for 1944. Sugar-cane Investigation Committee of Trinidad, Pp. 1—44 (1945).
— u. R. Ross: The sugar-cane variety situation in Trinidad in 1944. Sugar-cane Investigation Committee of Trinidad, Pp. 45—53 (1945).
Balch, R. T.: Report on aconitic acid studies at Houma, La., Station. Sug. Bull., N. O. **23**, 197—198 (1945).
— u. C. B. Broeg: Sugarcane wax studies, 1942—1943. Sug. Bull., N. O. **22**, 106—109, 117—119, 123—127 (1944).
Beauchamp, C. E.: La variedad Mayagüez-63, caña de grandes promesas para el colono Cubano. (The variety Mayagüez 63, a cane of great promise for the Cuban colonist.) Bol. Of. Asoc. Tech. Azucar, Cuba **1**, 81—85 (1942).
— Nourishment of sugar cane and its importance in the production of varieties. Proc. 16th Annu. Conf. Asoc. Tech. Azucareros, Cuba, Pp. 71—74 (1942).
— Effects of drought on the cane yield. Proc. 17th Annu. Mtg. Asoc. Técn. Azucareros Cuba, Pp. 13—28 (1943).
— The upper and lower Brix of sugar cane as an index of its maturity. Proc. 18th Annu. Conf. Asoc. Técn. Azucareros, Cuba, Pp. 15—34 (1944).
Bechard, R. M.: Juice retention value of varieties of cane. As handled at Amatikulu during seasons 1940 and 1941. Proc. 16th Congr. S. Afr. Sug. Tech. Ass., Pp. 40—42 (1942).
Bell, A. F.: The selection of second year seedlings. Proc. 6th Congr. Int. Soc. Sug. Cane Tech., La, Pp. 710 bis 714 (1938) 1939.
— Report of the committee on seedling propagation. 41st Rep. Bur. Sug. Exp. Stas Brisbane 1941, Pp. 16 bis 18 (1941).
Bell, G. D. H.: Induced bolting and anthesis in sugar beet and the effect of selection of physiological types. J. Agric. Sci. **36**, 167—183 (1946).
— u. A. B. Bauer: Experiments on growing sugar beet under continuous illumination. J. Agric. Sci. **32**, 112—141 (1942).
—, — Experiments on growing sugar beet under continuous illumination. III. The production of a seed crop in the field and the resolution of a heterogeneous population. J. Agric. Sci. **33**, 85—94 (1943).
Bennett, C. W., E. Carsner, G. G. Coons u. E. W. Brandes: The Argentine curly top of sugar beet. J. Agric. Res. **72**, 19—48 (1946).
Bentancur, M. O.: El sorgo azucarado (Andropogon Sorghum var. saccharatum Korn). (Sweet sorghum [A. Sorghum var. saccharatum Korn].) Rev. Fac. Agron. Univ. Montevideo, Nr. 41, 87—105 (1945).
Beregovaja, M. M.: (The question of artificial infection of sugar beet with Cercospora beticola Sacc. in connexion with the production of resistant varieties.) Naučnye Zapiski Saharnoj Promyšlennosti (Sci. Trans. Sug. Ind.), Nr. 1, 128—137 (1939). [Russisch.]
Besser, A. A.: (The maple sugar industry.) Priroda (Nature), Nr. 3, 46—57 (1943).
Bhagavanthi, Kutty, P. R. Amma u. T. Ehambaram: Sugarcane x bamboo hybrids. J. Indian Bot. Soc. **18**, 209—229 (1940).
Björling, K.: Pleospora betae n. sp., die Schlauchfruchtform von Phoma betae (Oud.) Fr. (P. betae, n. sp., the ascigerous form of Ph. Betae [Oud.] Fr.). Bot. Notiser, Pp. 215—222 (1944).
Bockstahler, H. W.: Resistance to Fusarium yellows in sugar beets. Proc. Amer. Soc. Sug. Beet Technol., Pp. 191—198 (1940). [Mimeographed.]
— u. R. F. Seamans: Threshing and cleaning equipment for sugar beet seed. J. Amer. Soc. Agron. **32**, 794—802 (1940).
Bonne, C.: Die Züchtung von Rüben zur Verwendung als Zucker- oder als Futterrüben. Zuckerrübenbau **23**, 25—30 (1941).

Booberg, G.: Verslag der cultuurafdeeling van het Proefstation voor de Java-Suikerindustrie te Pasoeroean over het jaar 1939. (Report of the crop section of the Experiment Station for the Java Sugar Industry at Pasoeroean 1939.) Jversl. Veer. Proefst. Java-Suikerind., Pp. 9—24 (1939).
— Anplantkwesties. (Planting problems.) Arch. Suikerind. Nederland Ned.-Ind. **2**, 101—109 (1941).
— Factoren, welke den rietbloei beïnvloeden. (Factors which affect flowering of cane.) Arch. Suikerind. Nederland Ned.-Ind. **2**, 418—421 (1941).
Boonstra, A. E. H. R.: Rasverschillen bij bieten. V. Het verband tusschen de hoeveelheid drogestof, suiker, strikstof, aschbestanddeelen en water bij ons bietensortiment. (Race differences in beets. V. The relation between the amount of dry matter, sugar, nitrogen, ash and water in our beet collection.) Meded. Inst. Suikerbiet., Bergen-o.-Z., Nr. 8, 301—324 (1940).
— De veredeling van de suikerbiet, gezien in het licht van nieuwe enderzockingen. (The improvement of the sugar beet, viewed in the light of new investigations.) Institute of Plantbreeding, Wageningen, Holland, Pp. 3 (1941).
— Grootere opbrengst bij bieten door verlenging van den groeiduur. (Larger yield from beets by lengthening the growing period.) Meded. Inst. Suikerbiet., Bergen-o.-Z., Nr. 3, 29—48 (1941).
Borden, R. J.: Replication: The safeguard for uncontrolled variation. Hawaii Plant Rec. **47**, 135—153 (1943).
— Variety differences in nitrogen utilization. Hawaii. Plant. Rec. **50**, 39—49 (1946).
Bordonos, M. G.: (Sugar beet forms producing one-seeded balls.) Proc. Lenin Acad. Agric. Sci. USSR., Nr. 11, 3—4 (1941). [Russisch.]
Bose, S. S., K. L. Khanna u. P. C. Mahalanobis: Statistical notes for agricultural workers. Note on the optimum shape and size of plots for sugarcane experiments in Bihar. Indian J. Agric. Sci. **9**, 807—816 (1939).
Bougy, E.: Fluctuations dans les hybrides des betteraves fourragères et sucrières. (Fluctuations in the hybrid between forage and sugar beets.) Bull. Ass. Chim. Sucr., Pp. 305—312 (1939).
— Dissociation d'hybrides de betteraves sucrières et fourragères. Publ. Inst. Belge Amélior. Better. **10**, 35—43 (1942).
— Dissociation en F_3 de l'hybride Vauriac x Vilmorin A. Publ. Inst. Belge Amélior. Better. **11**, 207—213 (1943).
— Dissociation en F_4 de l'hybride Vauriac x Vilmorin A. Publ. Inst. Belge Amélior. Better. **13**, 195—200 (1945).
Brandes, E. W.: Three generations of cold resistant sugarcane. Sug. Bull., N. O. **18**, Nr. 4, 3—5 (1939).
— Research on sugar plants and some practical adaptations. Proc. Amer. Soc. Sug. Beet Technol., Pp. 158 bis 165 (1940). [Mimeographed.]
— Research on sugar plants and some practical adaptations. S. Afr. Sug. J. **24**, 307—313 (1940).
— Survival of wild sugarcane buds exposed to below zero (F.) temperatures. Sug. Bull., N. O. **18**, Nr. 16, 3—4 (1940).
— A progress report on development of cold tolerance on sugarcane varieties. Sug. Bull., N. O. **19**, 46—47 (1941).
— u. G. H. Coons: Climatic relations of sugar cane. Aust. Sug. J. **34**, 89—95 (1942).
—, — Climatic relations of sugarcane and sugar beet. S. Afr. Sug. J. **26**, 179—183, 251—255 (1942).
— u. J. Matz: Problems and progress in breeding. Temperate Zone sugar cane. S. Afr. Sug. J. 1940, **24**, 69—71; also Sug. J., N. O. 1939 **2**, 3—6 (1940).
—, G. B. Sartoris u. C. O. Grassl: Assembling and evaluating wild forms of sugarcane and closely related

plants. Proc. 6th Congr. Int. Soc. Sug. Cane Tech. La, Pp. 128—154 (1938) 1939.
Brandes, E. W., W. G. Taggart u. G. L. Billeaud: Release of new variety seed cane C. P. 36/13. Sug. Bull., N. O. **24**, 167 (1946).
—, — u. W. F. Giles: Release of new sugar cane variety. Sug. Bull., N. O. **19**, 117—118 (1941).
—, — u. W. C. Kemper: Release of new sugar cane variety. Sug. Bull., N. O. **20**, 169 (1942).
—, — u. J. J. Shaffer jr.: Release of new variety seed cane C. P. 36—105. Sug. Bull., N. O. **23**, 169 (1945).
Brett, P. G. C.: Seed setting of sugar cane in South Africa. Nature, Lond. **157**, 657—658 (1946).
Brewbaker, H. E.: Performance of direct increases of pedigreed and commercial lots of sugar beets. Proc. Amer. Soc. Sug. Beet Technol., Pp. 147—148 (1940). [Mimeographed.]
— Principal features of the seed accession system in use by the Great Western Sugar Company. Proc. Amer. Soc. Sug. Beet Technol., Pp. 181—184 (1940). [Mimeographed.]
— u. H. L. Bush: Generation studies of sugar-beet varieties. Proc. 3rd Amer. Soc. Sug. Beet Technol. 1942, Pp. 342—348 (1943).
Bush, H. L.: The three dimensional quasi-factorial experiment with three groups of sets for testing sugar beet breeding strains. Proc. Amer. Soc. Sug. Beet Technol., Pp. 113—116 (1940). [Mimeographed.]
— Further studies in newer designs for large-scale variety tests. Proc. 3rd Amer. Soc. Sug. Beet Technol. 1942, Pp. 365—372 (1943).
Buzacott, J. H.: The relationship between hardness of sugar cane and varietal resistance to the beetle borer (Rhabdocnemis obscura Boisd.). Tech. Commun., Bur. Sug. Exp. Sta., Dep. Agric., Brisbane, Nr. 8, 127—152 (1940).
Buzanov, I. F.: (The state congress on sugar beet breeding 5th—9th July 1939.) Naučnye Zapiski Saharnoj Promyšlennosti (Sci. Trans. Sug. Ind.), Nr. 2/3, 35—53 (1939). [Russisch.]
— (Breeding sugar beet in new beet growing regions). Naučnye Zapiski Saharnoj Promyšlennosti (Sci. Trans. Sug. Ind.), Nr. 1/2, 3—12 (1940). [Russisch.]

Cabanos, J. B.: The Granja sugar cane experiment station. Philipp. J. Agric. **10**, 35—68 (1939).
Carsner, E. u. B. Tolman: Relationship of top growth to sugar beet crown temperatures and induction of flowering. Amer. J. Bot. **29**, 691—692 (1942). (Abstr.)
Chandraratna, M. F. u. K. D. S. S. Nanayakkara: Studies in cassava. I. A classification of races occurring in Ceylon. Trop. Agriculturist 1944 **100**, 219 bis 230, 1945 **101**, 3—12 (1944/1945).
—, — Studies in cassava. I. A classification of races occurring in Ceylon (continued). Trop. Agriculturist **101**, 214—222 (1945).
Charter, C. F.: A tentative grouping of sugar-cane soils on the basis of their moisture relationships. Rep. Sug. Cane Invest. Comm., Trinidad, Pp. 224—232 (1942).
Chatterji, S. N.: Sugarcane in Bengal. Indian Fmg. **4**, 132—133 (1943).
Chevalier, A.: L'Abbé Henri Colin (1880/1943). Rev. Bot. Appl. **23**, 70—73 (1943).
Chona, B. L.: Red-rot of sugarcane and its control. Indian Fmg. **4**, 27—32 (1943).
— Sugarcane smut and its control. Indian Fmg. **4**, 401—404 (1943).
— u. G. W. Padwick: More light on the red-rot epidemic. Indian Fmg. **3**, 70—73 (1942).
Clydesdale, C. S.: Saccharine sorghums. Qd. Agric. J. **57**, 197—201 (1943).

Coelho, M. u. C. Coelho: Comportamento cultural-econômico de canas ,,P.O.J." e ,,Co" na zona da mata de Pernambuco. (Agricultural and economic behaviour of P.O.J. and Co. canes in the Pernambuco forest zone.) Bol. Sec. Agric. Pernambuco **6**, 89—103 (1940).
Coke, J. E.: Disease and bolting resistance in varieties. Spreckels Sug. Beet Bull., Calif. **6**, 4 (1942).
Colin, H.: En attendant la betterave polyploïde. Publ. Inst. Belge Amélior. Better. **9**, 219—226 (1941).
— u. E. Bougy: Les croisements de sucrières et de fourragères. Dissociation, en F_2, de l'hybride Kuhn x Mangold. Publ. Inst. Belge Amélior. Better. **7**, 28 bis 47 (1939).
Coons, G. H.: U. S. sugar beet seed meets war crisis. Sugar 1943 (January-February), Pp. 1—6, 7—13 (1943).
— et al.: Report of 1939 tests of U. S. 200×215. Proc. Amer. Soc. Sug. Beet Technol., Pp. 165—168 (1940). [Mimeographed.]
—, — Report on 1941 tests of U. S. 200×215, U. S. 215×216, and other varieties arising in leafspot — resistance breeding investigations of the U. S. Department of Agriculture. Proc. 3rd Amer. Soc. Sug. Beet Technol. 1942, Pp. 356—364 (1943).
— u. D. Stewart: U. S. 200×215, a new sugar beet variety resistant to leaf spot. Sug. J. U. S. Dep. Agric., Washington, D. C., July, p. 4 (1940).
—, — u. J. O. Gaskill: A new leaf-spot resistant beet variety. Sugar, Chicago **36**, Nr. 7, 4 (1941).
Copp, L. G. L.: Sugar-beet variety trials. N. Z. J. Sci. Tech. **27**, Sect. A, 376—380 (1946).
Cornelison, A. H.: Vegetative differences influence the composition of sugar cane. Hawaii. Plant. Rec. **48**, 125—164 (1944).
Cottrell-Dormer, W.: Some notes on sugar cane in Tonga. Proc. Qd. Soc. Sug. Cane Technol., Pp. 67 bis 71 (1944).
Craig, N., G. C. Stevenson u. H. Evans: Annual report of the Sugar Cane Research Station for the year 1940. Rev. Agric. Maurice **20**, 358-363 (1941).
Croizat, L.: A study of Manihot in North America. J. Arnold Arbor. **23**, 216—225 (1942).
— Preliminari per uno studio del genere ,,Manihot" nell' America meridionale. (Preliminary notes for a study of the genus Manihot in South America.) Rev. Argent. Agron. **10**, 213—226 (1943).
— Manihot Tweediana Mueller is unacceptable. Rev. Argent. Agron. **11**, 173—174 (1944).
Cross, W. E.: Las cañas ,,Tucumanas" de semillero. Resultados obtenidos en los últimos años. (The ,,Tucumán" seedling canes. Results obtained in recent years.) Rev. Industr. Agric. Tucumán **31**, 119—168 (1941).
— La rapida multiplicación de caña de las nuevas variedades. (Rapid multiplication of cane of new varieties.) Circ. Estac. Exp. Agric. Tucumán, Nr. 106, 2 (1942).
— Variedadas de caña importadas: resultados obtenidos en los ultimos años. (Imported cane varieties: results obtained in recent years.) Rev. Industr. Agric. Tucumán **32**, 193—273 (1942).
— Variedades de caña convenientes para plantar. (Cane varieties suitable for planting.) Bol. Estac. Agric. Tucumán, Nr. 35, 17 (1942).
— Observaciones y ensayos culturales relacionados con el ,,carbón" de la caña de azucar. (Observations and cultural tests concerned with sugar cane smut.) Bol. Estac. Agric. Tucumán, Nr. 37, 12 (1942).
— Nuevas observaciones sobre el ,,carbón" en las distintas variedades de caña de azucar. (Further observations on smut in the different sugar cane varieties.) Bol. Estac. Exp. Agric. Tucumán, Nr. 39, 15 (1943).
— Datos adicionales sobre el ,,carbón" en las distintas variedades de caña de azúcar. (Additional data on smut in the different varieties of sugar cane.) Bol. Estac. Exp. Agric. Tucumán, Nr. 43, 13 (1943).

Cross, W. E.: Las cañas „tucumans" de semillero. Resultados obtenidos hasta la cosecha de 1942 inclusive. (The Tucumán seedling canes. Results obtained up to the 1942, harvest inclusive.) Rev. Industr. Agríc. Tucumán 33, 151—341 (1943).
— Memoria anual del año 1942. (Annual report for 1942.) Rev. Industr. Agric. Tucumán 33, 33—137 (1943).
— Variedades de caña resistentes al „carbón". (Cane varieties resistant to smut.) Bol. Estac. Exp. Agric. Tucumán, Nr. 45, 24 (1944).
— Nuevos datos sobre el „carbón" en las distintas variedades de caña de azúcar. (New data on smut in the different varieties of sugar cane.) Bol. Estac. Agric. Tucumán, Nr. 50, 35 (1944).
— Las cañas „tucumanas" de semillero. Resultados obtenidos hasta la cosecha de 1944 inclusive. (The Tucumán canes. Results obtained up to the 1944 harvest inclusive.) Rev. Industr. Agríc. Tucumán 35, 55—221 (1945).
— Variedades de caña importadas. (Imported sugar cane varieties.) Rev. Industr. Agríc. Tucumán 35, 241—297 (1945).
— El efecto del „carbón" en las cañas de distintas variedades durante el año agrícola 1944—1945. (The effect of smut on the canes of different varieties during the agricultural year 1944—45.) Bol. Estac. Exp. Agric. Tucumán, Nr. 55, 31 (1945).
— La rapida multiplicación de caña de las nuevas variedades. (The rapid multiplication of new varieties of sugar cane.) Circ. Estac. Exp. Agríc. Tucumán, Nr. 134, 2 (1946).
Culbertson, J. O.: Inheritance of factors influencing sucrose percentage in Beta vulgaris. J. Agric. Res. 64, 153—172 (1942).

D....., H. H.: Experiment station notes. Valuable fertilizer trials: seedling successes. S. Afr. Sug. J. 24, 505—507 (1940).
— Experiment station notes. Wide range of activity: gratifying achievements. S. Afr. Sug. J. 24, 559—561 (1940).
— Experiment station notes. Harvesting of experimental varieties: interesting variety trials. S. Afr. Sug. J. 24, 617—621 (1940).
— Sugarcane varieties in Mauritius. A commentary. S. Afr. Sug. J. 25, 541 (1941).
— Experiment Station notes. Important fertilizer experiments. S. Afr. Sug. J. 26, 365—369 (1942).
— Transport of sugarcane pollen by airplane. S. Afr. Sug. J. 26, 497 (1942).
— Experiment Station notes. Effects of red rot. S. Afr. Sug. J. 26, 569—573 (1942).
— Alternative cane varieties as a measure of protection against plant disease. S. Afr. Sug. J. 26, 597—599 (1942).
— Experiment Station notes. A mixed bag fertiliser experiments. S. Afr. Sug. J. 27, 99, 101 (1943).
— Experiment Station notes-Release of cane variety N : Co. 310. S. Afr. Sug. J. 29, 505 (1945).
— New sugarcane varieties. Methods of introduction. S. Afr. Sug. J. 30, 343, 345 (1946).
D....., N. L.: A review of the genus Saccharum. Indian Fmg. 6, 581 (1945).
D....., L. R.: The action and use of colchicine in the production of polyploid plants. A. Afr. Agric. J. 5, 369 (1940).
Dahlberg, H. W.: A study of sugar beet hybrids. Proc. Amer. Soc. Sug. Beet Technol., Pp. 143—144 (1940). [Mimeographed,]
— How your beets are being improved. Through the Leaves, Colorado 31, 20—26 (1943).
— Non-sugar relationships in breeding high-purity beeds. Proc. 3rd Amer. Soc. Sug. Beet Technol. 1942, Pp. 322—325 (1943).

Dahlberg, H. W., A. C. Maxon u. H. E. Brewbaker: Breeding for resistance to leaf spot and other characters. Proc. Amer. Soc. Sug. Beet Technol., Pp. 169—180 (1940). [Mimeographed.]
Das, C. M.: What's doing in All-India. United Provinces. Indian Fmg. 3, 342—344 (1942).
Decoux, L., G. Roland u. M. Simon: Etude de la variabilité des variétés de betteraves. Publ. Inst. Belge Amélior. Better. 9, 163—167 (1941).
—, —, — Résultats préliminaires en vue d'étudier l'action de la colchicine sur le développement de la betterave. Publ. Inst. Belge Amélior. Better. 10, 45—55 (1942).
—, —, — Etude de la variabilité des variétés de betterave. IIe Communication. Publ. Inst. Belge Amélior. Better. 10, 199—212 (1942).
— u. M. Simon: Contribution à l'étude de la constitution des échantillons destinés à la determination des principaux critères d'appréciation des betteraves. Publ. Inst. Belge. Amélior. Better. 9, 169—174 (1941).
—, — u. W. van Goideshoven: Etude de la variabilité des variétés de betterave. IIIe Communication. Publ. Inst. Belge Amélior. Better. 11, 127—147 (1943).
—, —, — Etude de la variabilité des variétés de betterave. IVe Communication. Publ. Inst. Belge Amélior. Better. 12, 451—463 (1944).
—, J. Vanderwaeren, G. Roland u. M. Simon: Les variétés de betterave sucrière en Belgique de 1935 à 1939. Publ. Inst. Belge Amélior. Better. 9, 139—161 (1941).
—, — u. M. Simon: Les variétés de betterave sucrière en Belgique de 1934 à 1938. Publ. Inst. Belge Amélior. Better. 7, 675—708 (1939).
—, —, — L'influence variétale sur la maturation de la betterave sucrière. II. Communication. Publ. Inst. Belge Amélior. Better. 9, 19—42 (1941).
—, —, — Les variétées de betterave sucrière en Belgique de 1937 à 1941. Publ. Inst. Belge Amélior. Better. 10, 159—189 (1942).
—, —,— La culture associée des variétés de betterave. Publ. Inst. Belge Amélior. Better. 10, 213—218 (1942).
—, —, — Variétés de betterave à grand et à petit bouquet foliaire. Publ. Inst. Belge Amélior. Better. 10, 219—228 (1942).
—, —, — Is de verbetering der beet mogelijk in oorlogstijd ? (It is possible to improve beet in war-time ?) Publ. Belg. Inst. Verbetering der Beet, Tienen-België 11, 32 (1943).
—, —, — Les variétés de betterave sucrière en Belgique de 1938 à 1942. Publ. Inst. Amélior Better. 11, 103—126 (1943).
—, —, — Les variétés de betterave sucrière en Belgique de 1938 à 1942. IIe. Communication. Publ. Inst. Belge Amélior. Better. 11, 551—568 (1943).
—, —, — u. W. van Goideshoven: Les variétés de betterave sucrière en Belgique de 1939 à 1943. Publ. Inst. Belge Amélior. Better. 12, 395—450 (1944).
—, —, —, — La susceptibilité variétale de la betterave à la montaison précoce et tardive. Note préliminaire. Publ. Inst. Belge Amélior. Better. 12, 465 bis 475 (1944).
—, —, — Variétés de betterave à grand et à petit bouquet foliaire. II. Communication. Publ. Inst. Belge Amélior. Better. 12, 477—483 (1944).
—, —, — Un essai d'irradiation de graine de betterave en ultra-violets et en hautes fréquences. Publ. Inst. Belge Amélior. Better. 12, 509—515 (1944).
—, —, — u. G. Roland: Recherches sur la betterave d'hiver effectuées à Tirlemont de 1932 à 1937. Publ. Inst. Belge Amélior. Better. 10, 229—253 (1942).
Dedek, J. u. D. Ivančenko: L'amélioration par la sélection de la valeur industrielle de la batterave: Publ. Inst. Belge Amélior. Better. 9, 227—250 (1941).

Dedek, J., D. Ivančenko, J. Novak u. J. Vasatko: L'influence du sol et de la variété sur le comportement de la betterave pendant la fabrication. Publ. Inst. Belge Amélior. Better. **7**, 599—619 (1939).

Deming, G. W.: Comparison of some advanced generations of a hybrid strain of sugar beet with the original third generation selection. Proc. Amer. Soc. Sug. Beet Technol. Pp. 149—154 (1940). [Mimeographed.]

— Use of red garden beet in sugar-beet top crosses. Proc. 3rd Amer. Soc. Sug. Beet Technol. 1942 Pp. 336 bis 341 (1943).

Dillewijn, C. van: Indrukken van een reis naar Turkije. (Impressions of a journey to Turkey.) Landbouwk. Tijdschr., Wageningen **52**, 309—319 (1940).

— Wild and noble Saccharum in Asia Minor. Int. Sug. J. **42**, 165 (1940).

— Fortschritte der Zuckerrohrzüchtung. Z. Pflanzenz. **24**, 569—591 (1942).

— Die Bedeutung des Feldversuches für die Zuckerindustrie, dargestellt an der Entwicklung in Java. Zuckerrübenbau **25**, 25—32 (1943).

Dodds, H. H.: Notes on the present sugarcane variety position in South Africa, 1941. Proc. 15th Annu. Congr. S. Afr. Sug. Tech. Ass., Pp. 78—94 (1941).

— Notes on the sugarcane variety position in South Africa, 1941. S. Afr. Sug. J. **25**, 353—361 (1941).

— Further developments of sugarcane varieties in South Africa. Proc. 18th Annu. Congr. S. Afr. Sug. Tech. Ass., Pp. 38—43 (1944).

Doxtator, C. W.: Harvest sampling studies with five varieties of sugar beets. Proc. Amer. Soc. Sug. Beet Technol., Pp. 103—106 (1940). [Mimeographed.]

— Breeding methods with sugar beets: greenhouse and field technique. Proc. Amer. Soc. Sug. Beet Technol. Pp. 141—143 (1940). [Mimeographed.]

— u. A. W. Skuderna: Some crossing experiments with sugar beets. Proc. 3rd Amer. Soc. Sug. Beet Technol. 1942, Pp. 325—335 (1943).

Drewes, H.: What's the news about garden beets? Quicker and more even in germination tests than regular seed, sheared seed also reducts costly labor of thinning seedlings. Sth. Seedsman **6**, Nr. 2; 10, 35, 38 (1943).

Dutt, N. L.: The present position of thick Coimbatore sugarcanes. Indian Fmg. **3**, 473—477 (1942).

— Control of flowering in sugarcane. Indian Fmg. **4**, 11—13 (1943).

— Varietal composition of the sugarcane crop in India in 1941—1942. Indian J. Agric. Sci. **13**, 471—477 (1943).

— u. M. K. Krishnaswami: A note on photoperiodism experiments on sugarcane at Coimbatore. Proc. 27th Indian Sci. Congr., Madras Pt III, Sect. Agric. Abst. **47**, Fp. 227—228.

—, — Protogeny in Uganda spontaneum. Curr. Sci. **12**, 24—26 (1943).

Dymond, G. C.: Varietal milling results in Natal, 1941. Proc. 16th Congr. S. Afr. Sug. Tech. Ass., Pp. 37—40 (1942).

Eggebrecht, H.: Die Echtheitsbestimmung von Zucker- und Futterrübensaatgut im Keimlingsstadium. Forschungsdienst **11**, 42—50 (1941).

Emmerez de Charmoy, D. d': La Station de Sélection des Cannes à Sucre de la Réunion. Agron. Trop., Nr. 11/12, 617—620 (1946).

Engelke: Der Zuckerrübenbau in Böhmen und Mähren. Zuckerrübenbau **24**, 61—70 (1943).

Ernould, L.: La cytologie de la betterave. Publ. Inst. Belge Amélior. Better. **12**, 39—54 (1944).

— Les espèces botaniques du genre Beta. Publ. Inst. Belge Amélior. Better. **13**, 219—253 (1945).

Evans, H.: Recherches sur la résistance aux Phytalus. (Research on the resistance to Phytalus.) Rev. Agric. Maurice **19**, 49—51 (1940).

Evans, H.: Part III — Research botany. 11th Rep. Sug. Cane Res. Sta. Mauritius 1940, Pp. 17—29 (1941).

—. u. G. C. Stevenson: A botanical and agricultural description of some sugarcane varieties raised by the sugarcane Research Station, Mauritius. Bull. Dep. Agric. Mauritius, Nr. 16, 10 (1939).

Fawcett, G. L.: La determinación de algunas nuevas variedades de caña plantadas en le pais. (The determination of certain new cane varieties planted in the country.) Rev. Industr. Agric. Tucumán **33**, 5—10 (1943).

Fennah, R. G.: A summary of experimental work on varietal resistance of sugarcane to Tomaspis saccharina 1936—1939. Trop. Agriculture, Trin. **16**, 233—240 (1939).

Fife, J. M. u. E. Carsner: Tip burn of sugar beet with special reference to some light and nitrogen relations. Phytopathology **35**, 910—920 (1945).

Finn, V. V.: (The teratology of the sugar beet flower.) Jarovizacija, Nr. 3 (36), 144—145 (1941). [Russisch].

Finney, D. J.: The relationship of plant number and yield in sugar-beet and mangolds. Emp. J. Exp. Agric. **9**, 57—64 (1941).

Forbes, I. L. u. P. J. Mills: Disappearance of virus from mosaicdiseased sugarcane plants. Phytopathology **33**, 713—718 (1943).

—, — u. P. H. Dunckelmann: The role of red rot in the windrowing for seed of present day sugarcane varieties in Louisiana. Sug. Bull., N. O. **22**, 148—149 (1944).

Gaskill, J. O.: Selection of sugar beets for size of root under wide and normal spacings. Proc. 3rd Amer. Soc. Sug. Beet Technol. 1942, Pp. 372—377 (1943).

— Sugar-beet leaf spot is being controlled through breeding of resistant varieties. Colo Fm. Bull. **5**, 10—13 (1943).

Gaviria, J. E.: Creación de nuevas variedades de caña de azúcar en Colombia. (Production of new varieties of sugar cane in Colombia.) Rev. Fac. Nac. Agron., Colombia 1940 **3**, 851—893 (1940).

Geerts, J. M.: De europeesche suikerrietcultuur in Nederlandsch-Indie. (European sugar cane cultivation in the Dutch East Indies.) Landbouwk. Tijdschr., Wageningen **53**, 223—239 (1941).

Gericke, S.: Wirkung verschiedener Wachstumsfaktoren auf den Ertrag der Zuckerrübe. Zuckerrübenbau **26**, 21—24 (1944).

Gibbens, R. T. jr., I. E. Stokes u. C. C. Krumhaar: Results of tests of C. P. 29—103 and C. P. 29—120. Sug. Bull., N. O. **18**, Nr. 3, 4—5 (1939).

Giddings, N. J.: Some factors influencing curly top virus concentration in sugar beets. Phytopathology **36**, 38—52 (1946).

— Mass action as a factor in curly-topvirus infection of sugar beet. Phytopathology **36**, 53—56 (1946).

Ginneken, P. J. H. van: De onderscheiding van suikerbietenrassen op grond van de minerale samenstelling van loof en wortel. (How to distinguish sugar beet strains by the mineral composition of leaf and root.) Meded. Inst. Suikerbiet., Bergen-o.-Z., Nr. 6, 179 bis 276 (1940).

— Instituut voor Suikerbietenteelt te Bergen op Zoom. (The Institute for Sugar Beet Cultivation at Bergen op Zoom.) Vakbl. Biol. **21**, 65—69 (1941).

Gondö, A.: (Pollen studies of sugar cane II. Pollen-preservation I.) Agric. and Hort., Japan **14**, 2673 bis 2684 (1939).

Gouaux, C. B.: Sugar cane test field report season for 1941. Sug. Bull., N. O. **20**, 142—146 (1942).

— Sugar cane test field report season of 1943. Sug. Bull., N. O. **22**, 166—170 (1944).

Gouaux, C. B.: Sugar cane test fields-season of 1944. Sug. Bull., N. O. **23**, 150—155 (1945).
— Milling tests of some important commercial canes. Sug. Bull., N. O. **24**, 131 (1946).
— Some comparative averages and results of sugar cane test fields. Sug. Bull., N. O. **24**, 200—203 (1946).
— u. W. G. Taggart: Sugar cane test field-Season of 1945. Sug. Bull., N. O. **24**, 147—154 (1946).
—, — Sugar cane test fields-Season of 1946. Sug. Bull., N. O. **25**, 147—157 (1947).
Graner, E. A.: Tratamento de mandioca pela colchicina. I. Nota preliminar sôbre poliploidia indicada pela diferença de tamanho dos estômatos. (Treatment of cassava with colchicine. I. Preliminary note on polyploidy indicated by the difference in size of stomata.) J. Agron., S. Paulo **3**, 83—98 (1940).
— Notas sobre florescimento e fruticação da mandioca. (Notes on flowering and friut formation in cassava.) Bragantia, São Paulo **2**, 1—12 (1942).
— Genética de manihot. I. Hereditariedade da forma da folha e da coloração da película externa das raizes em Manihot utilissima Pohl. (Genetics of cassava. I. Inheritance of leaf form and coloration of the root epidermis in M. utilissima Pohl.) Bragantia, São Paulo **2**, 13—22 (1942).
— Tratamento de mandioca pela colchicina. II. Formas poliplóides obtidas. (Treatment of cassava by colchicine. II. Polyploid forms obtained.) Bragantia, São Paulo **2**, 23—54 (1942).
— Uma forma tetraploide de mandioca Vassourinha de provável valor hortícola. (A tetraploid form of Vassourinha cassava of probable horticultural value.) Rev. Agric. Piracicaba **19**, 380—391 (1944).
Grassl, C. O.: Saccharum robustum and other wild relatives of „noble" sugar canes. J. Arnold Arbor. **27**, 234—252 (1946).
Greaves, C. u. G. Molinet: Proyecto de la caña de azucar. (The sugar cane project.) 3a Conf. Interamericana Agric., Caracas, Pp. 71 (1945).
Groot, G. J. de, jr.: Identificatie van rietvariëteiten en de methode Jeswiet. (Identification of varieties of cane and Jeswiet's method.) Arch. Suikerind. Nederland, Ned.-Ind. **2**, 438—444 (1941).

H....., V. M.: The keeping qualities of Co. 281 and Co. 301. An experiment. S. Afr. Sug. J. **27**, 39 (1943).
Haudricourt, A.: Les colocasiées alimentaires (taros e yautias.) (Edible Colocasiae [taros and yautias].) Rev. Bot. Appl. **21**, 40—65 (1941).
Herman, I. V.: (The extent of natural crossing in sugar beet.) Vestnik Gibridizacii (Hybridization), Nr. 2, 102 (1941). [Russisch.]
Hidegheti Bittera, N. von: Der Zuckerrübenbau in Ungarn. Zuckerrübenbau **24**, 130—136 (1942).
Holme, R. V.: Annual report of the Research Department of the Sugar Manufacturers' Association Ltd. 1942—1943. J. A. S. T. Quart. **7**, 1—30 (1943).
— Annual report of the Agronomy Section of the Research office of the Sugar Manufacturers' Association (of Jamaica) Ltd. J. A. S. T. Quart. **9**, 1—36 (1945).
Hughes, C. G.: The remarkable expansion of Q. 28. Cane Gr. Quart. Bull. **10**, 22—23 (1946).
Hull, R. u. M. A. Watson: Virus yellow of sugar beet. J. Minist. Agric. **52**, 66—70 (1945).

Ingram, J. W. et al: Recent findings in the control of sugarcane insects at the Houma Laboratory of the Bureau of Entomology and Plant Quarantine. Sug. Bull., N. O. **20**, 146—148 (1942).
—, E. K. Bynum, R. Mathes u. T. E. Holloway: Report on research on sugarcane-insect control by the Houma, La., Laboratory of the U. S. Bureau of Entomology and Plant Quarantine during 1942. Sug. Bull., N. O. **21**, 208—211 (1943).

Innes, R. F.: Simple methods of field experimentation for the investigation of sugar cane problems. Jamaican Ass. Sug. Tech. **3**, 30—46 (1939).
Innes, R. F.: A survey of the yields of sugar cane in Jamaica during the 1940—1941 crop. Bull. Dep. Sci. Agric., Jamaica, Nr. 31 (N. S.), 31 (1942).
— u. M. S. Goodman: Field experiments on sugar cane in Jamaica (1940). Jamaican Ass. Sug. Technol. Quart. **4**, 2—68 (1941).
Inniss, B. De L.: Sugar cane mosaic disease in Jamaica and Barbados. Bull. B. W. I. Cent. Sugar Cane Breed. Sta. Barbados, Nr. 26, 12—21 (1944).
Isaac, P. V.: How mid-rib hardness affords resistance to the sugarcane topborer Scirpophaga nivella F., in India. Curr. Sci. **8**, 211—212 (1939).

Jakuškin, I.: Selection and seed-raising. Agricultural Chron., Moscow, Nr. 4, 1 (1946). [Mimeographed.]
Janaki, E. K.: Triplopolyploidy and the production of fertile intergeneric hybrids in Saccharum. Proc. 7th Int. Genet. Congr. Edinburgh 23—30 August, p. 166 bis 167, 1939 (1941).
Janaki Ammal, E. K.: Triplo-polyploidy in Saccharum spontaneum L. Curr. Sci. **8**, 74—77 (1939).
— Chromosome numbers in sclerostachya fusca. Nature, Lond. **145**, 464 (1940).
— Intergeneric hybrids of Saccharum. J. Genet. **41**, 217—253 (1941).
— The breakdown of meiosis in a male-sterile Saccharum. Ann. Bot. London **5**, 83—87 (1941).
— Intergeneric hybrids of Saccharum. IV. Saccharum-Narenga. J. Genet. **44**, 23—32 (1942).
Jancke, A. u. E. Mikschik: Über den „schädlichen Stickstoff" der Zuckerrübe und dessen Verteilung in derselben nebst praktischen Folgerungen. Forschungsdienst **13**, 138—149 (1942).
Joachim, A. W. R. u. D. G. Pandittesekere: Investigations of the hydrocyanic acid content of manioc (Manihot utilissima). Trop. Agriculturist **100**, 150 bis 163 (1944).

K....., H. G.: Resistance of Eros to grub attack. Cane Gr. Quart. Bull. **10**, 117 (1947).
K....., H. W.: The variety Q. 20 in the Mackay area. Cane Gr. Quart. Bull. **6**, 190 (1939).
Keller, A. G.: Milling tests of C. P. 36—105. Sug. Bull., N. O. **24**, 97—98 (1946).
Kerr, H. W.: Field experimentation with sugar cane. Tech. Commun., Bur. Sug. Exp. Sta., Dep. Agric., Brisbane, Nr. 11, 177—232 (1939).
— Bureau of Sugar Experiment Stations. New variety trials and distribution. Aust. Sug. J. **34**, 153—155 (1942).
— u. A. F. Bell: The Queensland Cane Growers' Handbook. Bur. Sug. Exp. Sta., Dep. Agric. Brisbane (1939).
Khan, M. A. u. P. C. Raheja: Frosts and sugarcane culture. Indian Fmg. **4**, 465—467 (1943).
Khushi Mohammad: A new early ripening variety of sugarcane Co. 396. Seas. Notes Punjab Agric. Dep. **18**, 40—42 (1940).
— A new heavy-yielding variety of sugarcane. Indian Fmg. **2**, 140—141 (1941).
King, N. J.: Seedling raising at Bundaberg and some notes on Q. 25. Cane Gr. Quart. Bull. **8**, 29—32 (1940).
— C. P. 29/116 and Q. 49 in South Queensland. Cane Gr. Quart. Bull. **10**, 24—26 (1946).
Kiryu, T.: (On a method of varietal resistance trials of sugar cane to red rot.) Ann. Phyto. Soc., Japan **10**, 156—170 (1940).
Kohls, H. L.: A method of correcting tonnage of sugar beets for variation in per cent stand. Proc. Amer. Soc. Sug. Beet Technol., Pp. 128—132a (1940). [Mimeographed)

Krasočkin, V. T.: (On the influence of the region where the seeds are grown on the succeeding generation of beets.) Vestnik Socialističeskogo Rastenievodstva (Soviet Plant Industry Record), Nr. 1, 61—70 (1941). [Russisch.]

L....., H. M.: Sugar cane x bamboo hybrids. Int. Sug. J. **41**, 95; also Trop. Agriculture; Trin. **16**, 118 (1939).
— Varietal problems of the sugar cane. Int. Sug. J. **41**, 335—337 (1939).
— Collecting wild forms of sugar cane. Int. Sug. J. **41**, 377—379 (1939).

Ladell, W. R. S.: Cane varieties. Part I. Jamaican Ass. Sug. Tech. Quart. **4**, 4—23 (1940).

Lakshmikantham, M.: Pith in sugarcane. Curr. Sci. **15**, 284—285 (1946).

Lanjouw, J.: Two interesting species of Manihot L. from Suriname. Rec. Trav. Bot. Néerland **36**, 543 bis 549 (1939).

Lantz, E. M.: Effect of doubling the number of chromosomes of the sugar beet on the carotene and ascorbic acid contents of the leaves. Pr. Bull. N. Mex. Agric. Exp. Sta., Nr. 973, 2 (1943).

Laubscher, F. X.: A new sweet sorghum variety. Fmg. S. Afr. **18**, 841—842, 856 (1943).

Lauritzen, J. I.: Testing varieties of sugarcane for resistance to inversion of sucrose and for windrowing qualities in Louisiana. Sug. Bull., N. O. **20**, 210—215 (1942).
— Windrowing qualities of C. P. 34—120 and certain other commercial varieties of sugarcane in Louisiana. Sug. Bull., N. O. **21**, 200—202 (1943).
—, R. T. Balch and C. A. Fort: Resistance to inversion of sucrose in harvested sugarcane in Louisiana. Proc. 6th Congr. Int. Soc. Sug. Cane Tech., L, Pp. 809—819 (1938) 1939.
—, E. W. Brandes u. J. Matz: Influence of light and temperature on sugar-cane and Erianthus. J. Agric. Res. **72**, 1—18 (1946).

Leach, L. D.: Effect of downy mildew on productivity of sugar beets, and selection for resistance. Hilgardia **16**, 317—334 (1945).

Lepa, P.: (The summer sowing of sugar beet in the Uzbek S. S. R.) Proc. Lenin Acad. Agric. Sci. USSR., Nr. 1, 20—21 (1944). [Russisch.]

Levan, A.: The effect of chromosomal variation in sugar beets. Hereditas, Lund **28**, 345—399 (1942).
— Jämförande undersökning över utvecklingen av diploid och tetraploid sockerbeta och foderbeta. (Comparative investigation on the development of diploid and tetraploid sugar beet and fodder beet.) Sverig. Utsädesfören. Tidskr. **53**, 215—238 (1943).
— On the normal occurrence of chromosome doublings in second-year root tips of sugar beets. Hereditas, Lund **30**, 161—164 (1944).
— A haploid sugar beet after colchicine treatment. Hereditas, Lund **31**, 399—410 (1945).
— u. P. A. Olson: On the decreased tendency to bolting in tetraploids of mangels and sugar beets. Hereditas, Lund **30**, 253—254 (1944). (Abstr.)

Lima Romero, J. J.: The bacterial flora of sugar. Proc. 15th Annu. Mtg. Asoc. Tecn. Azucareros Cuba, Pp. 155—167 (1941).

Lintner, J. L. u. S. M. Maritz: A description of apparatus for weighing cane on experimental plots. Proc. 14th Annu. Congr. S. Agr. Sug. Tech. Ass. Durban 2nd—4th April, Pp. 48—51 (1940).

Lynes, F. F.: Polyploidy in sugar beets induced by storage of treated seed. J. Amer. Soc. Agron. **37**, 402—404 (1945).
— u. C. E. Cormany: Studies on some F_1 sugar beet hybrids. Proc. Amer. Soc. Sug. Beet Technol., Pp. 185 bis 190 (1940). [Mimeographed.]

Lynes, F. F. u. C. E. Cormany: A method of forming a permanent pedigree record for breeding strains of sugar beets. J. Amer. Soc. Agron. **33**, 368—370 (1941).
—, — Refinements in the technique of isolating by bags and cages. Proc. 3rd Amer. Soc. Sug. Beet Technol. 1942, Pp. 399—405 (1943).
— u. C. D. Harris: Polyploidy in sugar beets induced by the use of colchicine, ethyl mercury phosphate, and other chemicals. Proc. 3rd. Amer. Soc. Sug. Beet Technol. 1942, Pp. 304—309 (1943).

McDougall, W. A.: Notes on the use of varieties in lessening grub damage to cane. Cane Gr. Quaet. Bull. **8**, 48—49 (1940).

McIntosh, A. E. S.: Report on sugar cane breeding and seedling testing for the year 1938—1939. Agric. J. Barbados **8**, 87—112 (1939).
— Sixth Annual Report of the British West Indies Central Sugar Cane Breeding Station, Barbados, for the year ending September 30, 1939. Pp. 39 (1939).
— Report on a third visit to Jamaica. Bull. B. W. I. Cent. Sug. Cane Breed. Sta., Nr. 22, 17 (1940).
— British West Indies Central Sugar Cane Breeding Station. A progress report on the work of the station for Jamaica. J. A. S. T. Quart. **4**, Nr. 3, 42—62 (1941).
— Progress in contributing colonies of seedlings released by the British West Indies Central Sugar-Cane Breeding Station. Trop. Agriculture, Trin. **18**, 232—237 (1941).
— Recent developments in sugar-cane breeding in Barbados. Emp. J. Exp. Agric. **10**, 31—42 (1942).
— Nobilisation in cane breeding at the British West Indies Central Sugar Cane Breeding Station and its practical results to date. Proc. Mtg B.W.I. Sug. Technol. Barbados, Pp. 26—40 (1944).
— Report on a visit to British Guiana, July and August 1944. Bull. B.W.I. Cent. Sug. Cane Breed. Sta., Nr. 27, 8 (1944).

McM....., A.: A newly released cane. Some notes on N: Co. 310. S. Afr. Sug. J. **30**, 91 (1946).

McMartin, A.: The sugar cane varieties of Natal. Proc. 14th Annu. Congr. S. Afr. Sug. Tech. Ass. Durban, 2nd—4th April, Pp. 33—38 (1940).
— Bud sports of sugarcane in Natal. Proc. 15th Annu. Congr. S. Afr. Sug. Tech. Ass., Pp. 95—99 (1941).
— Pineapple disease of sugarcane cuttings and its control. Proc. 18th Annu. Congr. S. Afr. Sug. Tech. Ass., Pp. 44—46 (1944).
— A review of the present sugar-cane variety and disease position. Proc. 20th Congr. S. Afr. Sug. Tech. Ass., Pp. 73—76 (1946).
— Sugarcane variety and diseases. The present position. S. Afr. Sug. J **30**, 499—503 (1946).

Martens, P., L. Decoux u L. Ernould: Obtention, par la colchicine, de betteraves sucrières triploides et tetraploides. Publ. Inst. Belge Amélior. Better. **12**, 251—256 (1944).

Martin, J. P.: Varietal differences of sugar cane in growth, yields, and tolerance to nutrient deficiencies. Hawaii. Plant. Rec. **5**, 79—91 (1941).
— Stem galls of sugar-cane induced with insect extracts. Science **96**, 39 (1942).

Mathes, R. u. J. W. Ingram: Development and use of sugarcane varieties resistant to the sugar-cane borer. J. Econ. Ent. **35**, 638—642 (1942).
—, — Investigations of sugarcane borer control by the use of resistant varieties. Sug. Bull. N. O. **22**, 189 bis 192 (1944).
—, — u. W. E. Haley: Preliminary report on studies of progenies of sugarcane crosses for susceptibility to sugarcane borer injury in Louisiana Proc. 6th Congr. Int. Soc. Sug. Cane Tech., La, Pp. 581—589 (1938) 1939.

Mercado, T.: A comparative study of two sports of cassava and their parent varieties. Philipp. Agric. **28**, 308—320 (1939).

Miščenko, A. S. (Editor): (Symposium on the research work of the State Research Institute for the Sugar Industry (VNIS) on beet cultivation and beet seed production). Vsesojuznyj N-I Institut Saharnoj Promyšlennosti Harkov-Kiev, Pp. 318 (1939). [Russisch.]

— (Main conclusions from the scientific research work of the State Research Institute for the Sugar Industry [VNIS] for 1938.) Vsesojuznyj N-I Institut Saharnoj Promyšlennosti (Sci. Inst. Sug. Ind.) Moskva-Leningrad, Pp. 300, Orlovskij, N. I. (Breeding work with sugar beet at the Stations of the Sugar Trust in 1938) (Pp. 171—177) (1940). [Russisch.]

Moberly, G. S.: The cane testers' handbook. Sugar. Industry Central Board, Durban, Pp. VIII+85, Figs. 5 tables. (1943).

— The replacement of Uba by new variety canes from 1936 to 1944. Proc. 19th Ann. Congr. S. Afr. Sug. Technol. Ass., Natal, Pp. 29—34 (1945).

Molegoda, W.: The Kitul palm. Trop. Agriculturist **101**, 251—257 (1945).

Moriya, A.: Contributions to the cytology of genus Saccharum. 1. Observations on the F_1 progeny of sugar cane and sorghum hybrids. Cytologia, Tokyo **11**, 117—135 (1940).

— (Preliminary note on the chromosome numbers of sugarcane varieties F108 and some others.) Jap. J. Genet. **17**, 62—64 (1941).

Munerati, O.: Beobachtungen über die Teilung von Beta-Rüben. Züchter **13**, 290 (1941).

— Die Vererbung der Weißblättrigkeit bei Beta vulgaris L. Züchter **14**, 214—215 (1942).

— The duration of the beet cycle. Int. Rev. Agric. **33**, 177T—214T (1942).

Mungomery, R. W.: Bureau of Sugar Experiment Stations. Report of the Division of Entomology and Pathology — Nr. 2. Aust. Sug. J. **37**, 433—436 (1946).

— u. J. H. Buzacott: Varietal resistance to cane grubs. Cane Gr. Quart. Bull. **8**, 45—47, also Aust. Sug. J. **32**, 325—326 (1940).

Narain, R. u. A. Singh: Sampling of sugarcane for chemical analysis, II. Indian J. Agric. Sci. **12**, 822 bis 836 (1942).

Nichols, R. F. W.: Breeding cassava for virus resistance. E. Afr. Agric. J. **12**, 184—194 (1947).

Noeldechen: Standweite und Sortentyp. Kühn-Archiv **60**, 178—188 (1943/1944).

— Standweite und Sortentyp. Zuckerrübenbau **26**, 3—9 (1944).

Nolla, J. A. B.: Sugar cane investigations in progress at the Agricultural Experiment Station of the University of Puerto Rico. Proc. 15th Annu. Mtg. Asoc. Tecn. Azucareros Cuba, Pp. 31—37 (1941).

Normanha, E. S. u. O. J. Brook: Ensaios de variedades de mandioca na Estação Experimental de Ubatuba. (Tests of manihot varieties at the Ubatuba Experimental Station.) Bragantia, São Paulo **2**, 521—559 (1942).

—, — u. J. B. de Castro: Observações de campo como contribuição ao estudo do superbrotamento ou envassuramento da mandioca. (Field observations contributing to the study of excessive proliferation or witches' broom of manioc.) Rev. Agric. Piracicaba **21**, 271—302 (1946).

North, D. S.: Sugarcane improvement work in New South Wales. Proc. 6th Congr. Int. Soc. Sug. Cane Tech., La, Pp. 79—88 (1938) 1939.

Okanenko, A. V. u. V. F. Savickij: (Distribution of saccharose in roots of different sorts of sugar beet.) Proc. Lenin Acad. Agric. Sci. URSS., Nr. 12, 3—5 (1940). [Russisch.]

Opsomer, J.-E.: Technique et premiers résultats de l'amélioration du manioc à Yangambi (Congo Belge). Agric. Elev. Congo Belge **13**, 4—7 (1939).

Orjuela Navarrete, J. E.: La enfermedad „red stripe" de las hojas de la caña de azucar en Colombia. (The red stripe disease of the leaves of cane sugar in Colombia.) Agricultura Triop., Bogotá (Suppl.) **2**, Nr. 4, 23—37 (1946).

Orlovskij, N. I.: (The present position and future problems in breeding high yielding sugar beet varieties free from bolting.) Naučnye Zapiski Saharnoj Promyšlennosti (Sci. Trans. Sug. Ind.), Nr. 2/3, 76—91 (1939). [Russisch.]

Owen, F. V. et al.: Curly-top-resistant sugar-beet varieties in 1938. Circ. U. S. Dep. Agric., Nr. 513, 10 (1939).

— Inheritance of cross- and self- sterility and self-fertility in Beta vulgaris. J. Agric. Res. **64**, 679—698 (1942).

— Male sterility in sugar beets produced by complementary effects of cytoplasmic and Mendelian inheritance. Amer. J. Bot. **29**, 692 (1942). (Abstr.)

— Variability in the species Beta vulgaris L. in relation to breeding possibilities with sugar beets. J. Amer. Soc. Agron. **36**, 566—569 (1944).

— Cytoplasmically inherited male-sterility in sugar beets. J. Agric. Res. **71**, 423—440 (1945).

—, E. Carsner u. M. Stout: Photothermal induction of flowering in sugar beets. J. Agric. Res. **61**, 101—124 (1940).

— u. A. Murphy: Progress with curly-top-resistant varieties of sugar beets. Fm. Home Sci. Utah **4**, Nr. 1, 13—14 (1943).

— u. G. K. Ryser: Some mendelian characters in Beta vulgaris and linkages observed in the Y-R-B group. J. Agric. Res. **65**, 155—171 (1942).

Pal, H. N.: Coimbatore canes in Assam. Indian Fmg. **4**, 576—578 (1943).

Panasjuk, M. P. (Editor): (Main conclusions from the scientific research work of the State Research Institute for the Sugar Industry [VNIS] for 1937.) Vsesojuznyj N-I Institut Saharnoj Promyšlennosti Moskva-Leningrad, Pp. 484 (1939). [Russisch.]

Patlan, P. I., P. I. Semuškin, A. D. Rjabuško, F. K. Demčenko u. I. Ja. Iščuk: (Brief conclusions from the work of the experimental field of the Verkhnjačka breeding station for 1938.) Naučnye Zapiski Saharnoj Promyšlennosti (Sci. Trans. Sug. Ind.) **16**, Nr. 2/3, 131—138 (1939). [Russisch.]

Paul, L. B.: Studies on the performance of some Coimbatore canes regarding maturing-period, yield, habit, etc., under the improved methods of cultivation. Proc. 9th Annu. Convent Sugar. Technol. Ass. India Part I, Pp. 41—49 (1940).

Pázler, J.: Richtlinien und Aufgaben der heimischen Zuckerrübenzucht. Z. Zuckerindustr. Čsl. Repub. **65**, 51—59 (1943).

Pedersen, A.: Om Bederoernes Farver. (On the colours of beetroots.) K. VetHøjsk. Aarsskr., Pp. 60—111 (1944).

Pedrosa, R.: A Vd. debe interesárle que ... (It should interest you that ...). Bol. Ofic. Asoc. Tec. Azucar., Cuba **4**, 172—174 (1945).

Perak, J. T.: Los cromosomas de Manihot Tweedieana. (The chromosomes of Manihot Tweedieana.) Rev. Argent. Agron. **7**, 364—365 (1940).

Peschel, V.: Zuckerrübenzüchtung und Rübenernte. Mitt. Landw. **59**, 839—840 (1944).

Peto, F. H. u. J. W. Boyes: Comparison of diploid and triploid sugar beets. Canad. J. Res. **18**, Sect. C, 273—282 (1940).

Peto, F. H. u. K. W. Hill: Colchicine treatments of sugar beets and the yielding capacity of the resulting polyploids. Proc. 3rd Amer. Soc. Sug. Beet Technol. 1942, Pp. 304—309 (1943).

Pickles, A.: A discussion of researches on the sugarcane froghopper (Homop., Cercopidae). Trop. Agriculture Trin. 19, 116—123 (1942).

Posnette, A. F.: Root-rot of cocoyams (Xanthosoma sagittifolium Schott). Trop. Agriculture, Trin. 22, 164—170 (1945).

Puertas, R. P.: Que variedad de caña me conviene sembrar ? (What variety of sugar cane should I sow ?) Bol. Ofic. Asoc. Tec. Azucar., Cuba 3, 162—166 (1944).

— M-L-3-18 cane and its practical results. Proc. 19th Mtg. Asoc. Técn. Azucareros, Cuba, Pp. 81—91 (1945).

Rands, R. D., E. V. Abbott u. E. M. Summers: Disease resistance tests and seedling selections in 1938 and 1939. Sug. Bull., N. O. 18, Nr. 15, 5—9 (1940).

Ramos, R. M.: The role of the research in the Puerto Rican cane industry. Proc. 8th. Amer. Sci. Congr., Washington 5, 201—222 (1940).

Rasmusson, J.: Quantitative inheritance in root crops. Proc. 7th Int. Genet. Congr. Edinburgh 23—30 August, Pp. 245—246, 1939 (1940).

— The field trials in sugar-beet breeding. Proc. 7th Int. Genet. Congr. Edinburgh 23—30 August, p. 245, 1939 (1941). (Abstr.)

— u. A. Levan: Tetraploid sugar beets from colchicine treatments. Hereditas, Lund 25, 97—102 (1939).

Reinboth, G.: Die Cercosporabekämpfung bei den Rüben. Zuckerrübenbau 23, 10—12 (1941).

Rosenfeld, A. H.: The deterioration of harvested sugar cane. A preliminary study of varietal trends in Egypt. Trop. Agriculture, Trin. 19, 133—138 (1942).

Roubaix, J. de u. O. Lazar: Chlorophylle et betterave. Publ. Inst. Belge Amélior. Better 10, 191—198 (1942).

S....., H. H.: Cassava research. E. Afr. Agric. J. 5, 401—403 (1940).

S....., S. O.: Resistance of new seedlings to frenchi grub attack. Cane Gr. Quart. Bull. 8, 142—143 (1941).

Sampaio, S. C., A. Conagin u. L. T. Moraes: Contribuição para o estudo das variedades de cana de açúcar, cultivadas no estado de São Paulo. (Contribution towards the study of the sugar cane varieties cultivated in the state of São Paulo.) Rev. Agric., Piracicaba 20, 403—408 (1945).

Sanockaja, E. I.: (Breeding sugar beet in new sugar beet growing regions.) Naučnyj Otčet Vsesojuznogo Naučno-Issled. Inst. Sveklovičnogo Polevodstva za 1941—1942 gg. (Sci. Rep. All-Union Res. Inst. Sugar Beet Husbandry for 1941—1942), Pp. 86—101 (1945).

Sartoris, G. B.: The behavior of sugarcane in relation to length of day. Sug. News. 20, 421—424, also Proc. 6th Congr. Int. Soc. Sug. Cane Tech., La, Pp. 796 bis 802 (1938) 1939.

— Necrotic stripes in sugarcane. J. Hered. 31, 515 bis 520 (1940).

Schlehuber, A. M.: Atypic dwarfing in sorgo. J. Amer. Soc. Agron. 36, 361—364 (1944).

— Inheritance of stem characters in certain sorghum varieties and their hybrids. J. Hered. 36, 219—222 (1945).

Schlösser, L. A.: Physiologische Untersuchungen an polyploiden Pflanzen-Reihen. Forschungsdienst 10, 28—40 (1940).

Schwanitz, F.: Eine somatische Mutation an der Rübenwurzel. Züchter 13, 87—88 (1941).

Seale, C. C.: The results of variety experiments on sugar cane at Frome Central, Jamaica, 1943—44. Proc. Mtg. B.W.I. Sug. Technol. Barbados, Pp. 108—112 (1944).

Sedlmayr, C.: A cukorrépafajtakérdés megoldása. (The solution of the problem of sugar beet varieties.) Rep. Hung. Agric. Exp. Sta. 43, 23—42 (1940).

Senaratna, J. E.: Bisexual flowers in the manioc, Manihot esculenta Crantz (M. utilissima Pohl). Ceylon J. Sci. 12, Sect. A, 169 (1945).

Siegumfeldt, G. H.: Progress in genetics-new methods in plant breeding. Proc. Amer. Soc. Sug. Beet Technol., Pp. 155—157 (1940). [Mimeographed.]

Simon, E. C.: Report on new varieties. Sug. Bull., N. O. 24, 145—146 (1946).

Singh, S. B.: Viable sugarcane seed produced in the United Provinces. Curr. Sci. 15, 253 (1946).

Singh, Harbans, Mohammad Khushi u. Hardial Singh: Sugarcane and wheat in one year. Indian Fmg. 4, 358—361 (1943).

Skuderna, A. W.: A comparison of three methods of harvesting sugar beet plots. Proc. Amer. Soc. Sug, Beet Technol, Pp. 122—127 (1940). [Mimeographed.]

— Use of colchicine in nutrient solution with sugar beets. Proc. 3rd Amer. Soc. Sug. Beet Technol. 1942. p. 321 (1943).

— Sugar beets in the war and post-war periods from the standpoint of the beet sugar industry. J. Amer. Soc. Agron. 36, 576—583 (1944).

— u. C. W. Doxtator: Comparison of quasi-factorial and randomized block designs for testing sugar beet varieties. Proc. Amer. Soc. Sug. Beet Technol., Pp. 116—118 (1940). [Mimeographed.]

—, —, E. Swift, R. L. Bowman u. A. Deschamps: A study of varietal adaptation with sugar beets — 1937 to 1941, inclusive. Proc. 3rd Amer. Soc. Sug. Beet Technol. 1942, Pp. 349—356 (1943).

Sloan, W. J. S.: Varietal trials, 1945 and 1946 seasons. Cane Gr. Quart. Bull. 10, 133—142 (1947).

Sobrinho, V.: Considerações geraes sobre o genero Manihot. (General considerations on the genus Manihot.) Bol. Sec. Agric. Industr. Com. Pernambuco 4, 54—58 (1939).

Solano, J. A.: Las nuevas variedades de caña de azúcar en el ingenio. (The new sugar cane varieties in the factory.) Vida Agr. 19, 873—891 (1942).

Sorenson, H.: The behavior of mosaic on certain soils and mosaic in regard to cane breeding. Proc. 6th Congr. Int. Soc. Sug. Cane Tech., La, Pp. 357—360 (1938) 1939.

Sornay, A. de: Brief description of a crossing-lantern used by the sugarcane research station, Mauritius. Proc. 6th Congr. Int. Soc. Sug. Cane Tech., La, Pp. 170 bis 171 (1938) 1939.

— Seedling selection in Mauritius. Proc. 6th Congr. Int. Soc. Sug. Cane Tech., La, Pp. 718—726 (1938) 1939.

— Estimation of cane yields by means of random rows and stools. Rev. Agric. Maurice 21, 107—113 (1942).

— u. C. Mazery: Selection of sugarcane seedlings from first-ratoon populations. Rev. Agric. Maurice 22, 94—101 (1943).

Stevenson, G. C.: Breeding and testing sugarcane seedlings for mosaic disease resistance at the British West Indies Central Sugar Cane Breeding Station, Barbados. Proc. 6th Congr. Int. Soc. Sug. Cane Tech., La, Pp. 71—75 (1938) 1939.

— Breeding and testing sugarcane seedlings for gumming disease resistance at the British West Indies Central Sugar Cane Breeding Station, Barbados. Proc. 6th Congr. Int. Soc. Sug. Cane Tech., La, Pp. 75 bis 78 (1938) 1939.

— The inheritance of gumming disease resistance in sugarcane breeding. Bull. Sugarcane Res. Sta., Mauritius, Nr. 15, 9 (1939).

— An investigation into the origin of the sugarcane variety Uba Marot. Bull. Dep. Agric. Mauritius, Nr. 17, 3—10 (1940).

Stevenson, G. C.: Sugar-cane varieties in mauritius. An historical review, with particular reference to present breeding problems. Emp. J. Exp. Agric. **8**, 301 bis 310 (1940).
— L'adaptation des nouvelles variétés de cannes aux différentes localités de Maurice. Rev. Agric. Maurice **20**, 3—6 (1941).
— Obtention et sélection de nouvelles variétés de cannes. Rev. Agric. Maurice **20**, 135—139 (1941).
— Part I — Cane breeding. 10th Rep. Sug. Cane Res. Sta. Mauritius, Pp. 6—16 1939 (1940).
— Part I — Geneticist's Division. 11th Rep. Sug. Cane Res. Sta. Mauritius 1940, Pp. 6—11 (1941).
— Une nouvelle variété de canne: la M. 112/34. Rev. Agric. Maurice **20**, 325—328 (1941).
— The effect of different localities on the growth of sugarcane varieties. Rev. Agric. Maurice **21**, 167—173 (1942).
— The present and potential value of sugarcane breeding. Emp. J. Exp. Agric. **11**, 38—48 (1943).
— Note on a case of twin arrows in a sugarcane seedling. Rev. Agric. Maurice **23**, 162 (1944).
— Trois nouvelles variétés de canne à sucre. Rev. Agric. Maurice **24**, 253—258 (1945).
— Sugarcane varieties produced by the Sugarcane Research Station, and their value to the sugar industry of Mauritius. Bull. Mauritius Sugar Res. Sta., Nr. 18, 24 (1946).
Stevenson, M. G. C.: The effect of different localities on the growth of sugarcane varieties. S. Afr. Sug. J. **26** 433—437 (1942).
Stewart, D., C. A. Lavis u. G. H. Coons: Hybrid vigor in sugar beets. J. Agric. Res. **60**, 715—738 (1940).
Stocker, O., S. Rehm u. H. Schmidt: Der Wasser- und Assimilationshaushalt dürreresistenter und dürreempfindlicher Sorten landwirtschaftlicher Kulturpflanzen. II. Zuckerrüben. Jb. Bot. **91**, 278 (1943).
Stokes, I. E. u. T. E. Ashley: Sugarcane production in Mississippi. Bull. Miss. Agric. Exp. Sta., Nr. 395, 17 (1943).
Stout, M.: The relation of temperature to reproduction in sugar beets. Amer. J. Bot. **29**, 692 (1942). (Abstr.)
— u. F. V. Owen: Vernalization of sugar-beet seed. Proc. 3rd Amer. Soc. Sug. Beet Technol. 1942, Pp. 386 bis 395 (1943).
Summers, E. M.: Effect of mosaic caused by different strains of virus upon yields of susceptible sugarcane varieties. Sug. Bull, N. O. **21**, 181—183 (1943).
— u. E. V. Abbott: Disease testing and initial seedling selection work at the Houma Station during 1942. Sug. Bull., N. O. **21**, 156—158 (1943).
Suzuki, E.: (Cytological observations on some sugar cane varieties.) Jap. J. Genet. **16**, 276—278 (1940).
— Cytological studies of sugar cane. I. Observations on some P. O. J. varieties. Cytologia, Tokyo **11**, 507—514 (1941).

Taggart, W. G. u. G. Arceneaux: An evaluation of C. P. 29—103 and C. P. 29—120. Sug. Bull., N. O. **18**, Nr. 2, 2—3 (1939).
— u. E. C. Simon: The rapid multiplication of a promising new sugar cane variety. Sug. Bull., N. O. **19**, 73—75 (1941).
Thuljaram Rao, J.: Rind hardness as a possible factor in resistance of sugarcane varieties to the stem borer. Curr. Sci. **10**, 365—366 (1941).
— u. T. S. Venkatraman: Hard leaf mid-rib in sugarcanes and resistance to top-borer (Scirpophaga nivella F.). Curr. Sci. **10**, 171—172 (1941).
Troje, E.: Futter- und Zuckererträge bei verschiedenen Zuckerrübensorten unter verschiedenen Wachstumsbedingungen. Zuckerrübenbau **26**, 31—32 (1944).

Uichanco, L. B.: The College of Agriculture and the Philipine sugar industry. Sug. New. **20**, 385—386, 394 (1939).

Vasconcélos, D. de M.: Competição de variedades de mandioca. (Competition of cassava varieties.) Bol. Sec. Agric. Pernambuco **6**, 164—173 (1940).
Vasudevamurthy, M.: What's doing in All-India. Mysore. Indian Fmg. **3**, 219—221 (1942).
Vendrih, J. V.: (Cultural methods of plant nutrition and mass selection as a means of improving varieties and increasing yields in sugar beet.) Trudy Moskovskoj ordena Lenina Seljskohozjajstvennoj Akademii imeni K. A. Timirjazeva **5**, Nr. 2, 5—31 (1940). [Russisch.]
Venkatraman, T. S.: Cane hybridization work in Coimbatore. Proc. 6th Congr. Int. Soc. Sug. Cane Tech., La, Pp. 731—732 (1938) 1939.
— Sugar cane x bamboo hybrids. S. Afr. Sug. J. **23**, 447—449 (1939).
— Cultivated and wild plants. Indian Fmg. **2**, 358—359 (1941).
— The message of the sugarcane. Indian J. Genet. Pl. Breed. **2**, 3—10 (1942).
— u. N. Parthasarathy: Chromosome counts in the sugarcane and its hybrids. Curr. Sci. **11**, 194—195 (1942).
Venkoba Rao, B.: List of promising sugarcane varieties evolved under the Imperial Council of Agricultural Research cheme which are being issued for trial to the raiyats. Mysore Agric. Cal., Pp. 70—72 (1940).
— A note on the two most outstanding sugarcane varieties H. M. 661 and H. M. 647. Mysore Agric. Cal., Pp. 19—20 (1941).
Venner, A. K.: Science and sugar. Fmrs. Wkly, Lond. **23**, Nr. 19; 34, 37 (1945).
Vijayasaradhy, M.: „Thermo"' or „vacuum" flasks for preserving sugarcane pollen. Curr. Sci. **8**, 554—555 (1939).
Višnevskij, V. P.: (The quality of catalase in the beetroot, and the resistance of sugar beet to Botrytis cinerea during storage.) Biohimija **5**, 408—416 (1940). [Russisch.]

Walawalkar, D. G.: Improved sugarcane varieties in India. Indian Fmg. **3**, 425—427 (1942).
Wang, Chi Chu: (Introduction of sugar cane varieties and improvement of sugar cane production.) Fukien Agric. J. **5**, 40—57 (1944).
Weller, D. M.: Colchicine in relation to sugar cane breeding. Presented before the Agricultural Section of the Hawaiian Sugar Technologists, December 12, p. 11 (1939). [Mimeographed.]
Whitney, L. D., F. A. I. Bowers u. M. Takahashi: Taro varieties in Hawaii. Bull. Hawaii Agric. Exp. Sta., Nr. 84, 86 (1939).
Wiehe, M. P. O.: La morve rouge de la canne à sucre. (Red rot of sugar-cane.) Rev. Agric. Maurice **20**, 189—202 (1941).
Wiehe, P. O.: La sensibilité de quelques variétés de cannes aux principales maladies existant à Maurice. Rev. Agric. Maurice **21**, 225—226 (1942).
— Red rot and M 134/32. Rev. Agric. Maurice **23**, 242—243 (1944).
Wightman, G. M. u. W. Ryle-Davies: Annual report of the Agronomy Section of the Research Office of the Sugar Manufacturers' Association (of Jamaica) Ltd. J. A. S. T. Quart. 10, Nr. 2, 75 (1946).
Williams, C. H. B.: The variety and fertiliser position of the sugar industry, VI. Sug. Bull. Dep. Agric. Brit. Guiana, Nr. 9, 55—62 (1940).
— Report on the sugar experiment stations for the year 1944. Sug. Bull. Dep. Agric. Brit. Guiana, Nr. 13, 37—43 (1945).

Williams, C. H. B.: Report on the sugar experiment stations for the year 1945. Sug. Bull. Dep. Agric. Brit. Guiana, Nr. 14, 29—35 (1946).
— The variety and fertilizer position of the sugar industry, XI. Sug. Bull. Dep. Agric. Brit. Guiana, Nr. 14, 23—28 (1946).
— u. C. Cameron: Field experiments with sugar cane, VIII. Sug. Bull. Dep. Agric. Brit. Guiana, Nr. 8, 108 (1939).
—, — Field experiments with sugar cane, IX. Sug. Bull. Dep. Agric. Brit. Guiana, Nr. 9, 1—54 (1940).
—, — Field experiments with sugar cane, X. Sug. Bull. Dep. Agric. Brit. Guiana, Nr. 10, 1—22 (1941).
—, — Field experiments with sugar cane, XI. Sug. Bull. Dep. Agric. Brit. Guiana, Nr. 11, 1—37 (1942).
—, — Field experiments with sugar cane, XII. Sug. Bull. Dep. Agric. Brit. Guiana, Nr. 12, 1—46 (1944).
—, — Field experiments with sugar cane, XIII. Sug. Bull. Dep. Agric. Brit. Guiana, Nr. 13, 1—29 (1945).
—, — Field experiments with sugar cane, XIV. Sug. Bull. Dep. Agric. Brit. Guiana, Nr. 14, 1—22 (1946).
— u. R. R. Follett-Smith: Sugar agronomy in British Guiana. Proc. Mtg B.W.I. Sug. Technol. Barbados, Pp. 96—107 (1944).
Wilson, J. W.: Correlation of sugar yields with the per cent of joints bored by Diatraea saccharalis (F.). Sugarcane borer studies — I. Florida Ent. 25, Nr. 2, 19—24, 1942, 1943.
Wimmer, G.: Die Möglichkeit einer besseren Verwertung der Zuckerrüben im Zusammenhange mit der Entwicklung ihrer Blätter. Zuckerrübenbau 23, 91—98 (1941).

Yakushkin, I. siehe Jakuškin.
Yamasaki, M.: (Sugar cane varieties in Formosa.) Proc. Crop. Sci. Soc. Japan 12, 94—97 (1940).
— u. H. Arikado: The cell-sap concentration of sugarcane varieties in relation to their resistance to the attack of the white leaf louse. Proc. 6th Congr. Int. Soc. Sug. Cane Tech., La, Pp. 475—478 (1938) 1939.
Yen, C. H.: (Studies on the life history of the sugarcane borer, Argyris sp., with special reference to the damage caused by the borer to the cane.) New Agric. J. Fukien 3, 97—111 (1943).
Yusuf, N. D.: Inducing flowering in non-flowering sugarcanes. Curr. Sci. 15, 164—166 (1946).
— The effect of long periods of darkness on flowering in Saccharum spontaneum Soraparai, 270. Curr. Sci. 16, 62 (1947).
— u. N. L. Dutt: Photoperiod in relation to flowering in sugarcane. Curr. Sci. 14, 304—306 (1945).

Zosimovič, V. P.: (New hybrids between wild and sugar beet, that are resistant to Cercospora.) Selekcija i Semenovodstvo, Nr. 1, 12—16 (1939). [Russisch.]
— (Interspecific hybridization in beet [I. Beta vulgaris L. x Beta lomatogona E. et M. II. Beta vulgaris L. × B. macrorhiza Stev.].) Vestnik Gibridizacii (Hybridization), Nr. 1, 46—65 (1941). [Russisch.]
— u. N. I. Orlovskij: (Breeding and seed production of sugar beet.) Naučnye Zapiski Saharnoj Promyšlennosti (Sci. Trans. Sug. Ind., Nr. 2/3, 19—32 (1939). [Russisch.]

g) Stärkepflanzen.

Boiteau, P.: Nouvelles observations cytologiques sur le manioc cultivé. Chronica Botanica 6, 388 (1941).

Correia, F. A. u. C. G. Fraga Júnior: Tecnologia da mandioca. (Technology of manioc). Bragantia, São Paulo 5, 213—226 (1945).

Graner, E. A.: Polyploid cassava induced by colchicine treatment. J. Hered. 32, 281—288 (1941).

Springensguth, W.: Die Kultur des Manioks, seine Krankheiten und Schädlinge im Litoral des Staates Sta. Catharina (Brasilien). Tropenpflanzer 43, 286 bis 306 (1940).

h) Buchweizen.

Couch, J. F., J. Naghski u. C. F. Krewson: Buckwheat as a source of rutin. Science 103, 197—198 (1946)

Frolova, S. L., V. V. Saharov u. V. V. Mansurova: Cytological basis of high fertility in autotetraploid buckwheat. Nature, London 158, 520 (1946).
—, —, — Homostyly of the flowers of buckwheat as a morphological manifestation of sterility. Nature, London 158, 520 (1946).

Gluščenko, I. E.: (The retention of typical varietal characters of buckwheat after inter-varietal pollination). Jarovizacija, Nr. 2 (35), 68—75 (1941). [Russisch.]

Kobeljkov, V. V.: (The effect of free pollination on the yield of buckwheat.) Jarovizacija, Nr. 3 (36), 120 bis 121 (1941). [Russisch.]
Kopeljkievskij, G. V.: (Breeding and seed production in buckwheat.) Selekcija i Semenovodstvo, Nr. 5, 34—36 (1939). [Russisch.]
— u. K. N. Mihajlov: (Experiments on the vegetative hybridization of buckwheat.) Jarovizacija, Nr. 2 (35), 111—114 (1941). [Russisch.]
Krotov, A.: (Cross pollination of buckwheat varieties.) Proc. Lenin Acad. Agric. Sci. URSS., Nr. 5/6, 13—15 (1944). [Russisch.]

Meyer, K.: Normale und anormale Buchweizensamen. Angew. Bot. 25, 354—358 (1943).
Modilevskij, J. S.: (Die Bestäubung bei Buchweizen.) C. R. (Doklady) Acad. Sci. URSS. 53, 165 (1946). [Russisch.]

Nishina, Y., Y. Sinotô u. D. Satô: Effects of fast neutrons upon plants, III. Cytological observations on the abnormal forms of Fagopyrum and Cannabis. Cytologia, Tokyo 10, 458—466 (1940).

Omeljčenko, V. K.: (On the conditions determining reliable and high yields of buckwheat.) Vestnik Socialističeskogo Rastenievodstva (Soviet Plant Industry Record), Nr. 3, 59—66 (1940). [Russisch.]

Saharov, V. V., S. L. Frolova u. V. V. Mansurova: High fertility of buckwheat tetraploids obtained by means of colchicine treatment. Nature, Lond. 154, 613 (1944).
—, —, — Tetraploidy in cultivated buckwheat (Fagopyrum esculentum). C. R. (Doklady) Acad. Sci. URSS. 43, 213—216 (1944).
—, —, — Production of highly fertile tetraploid buckwheat (Fagopyrum esculentum). C. R. (Doklady) Acad. Sci. URSS. 44, 254—256 (1944).
—, —, — Autotetraploidy in different varieties of buckwheat (Fagopyrum esculentum). C. R. (Doklady) Acad. Sci. URSS. 46, 79—82 (1945).
Saltykovskij, A. I.: (Immunity of buckwheat to bacteria.) Selekcija i Semenovodstvo (Breeding and Seed Growing), Nr. 8, 25—27 (1939). [Russisch.]
— On the taxonomy of common buckwheat (Fagopyrum esculentum Moench) and Fagopyrum emarginatum Roth. C. R. (Doklady) Acad. Sci. URSS. 26, 180—182 (1940).
Sando, W. J.: A colchicin-induced tetraploid in buckwheat. J. Heredity 30, 271—272 (1939).

Sinotô, D. u. D. Satô: Poliploidi da colchicina in Fagopyrum. (Colchicine polyploids in Fagopyrum.) Scientia Genetica 1, 354—369 (1940).

Stoletova, E. A.: (Ecological and geographical classification of buckwheat.) Proc. Lenin Acad. Agric. Sci. USSR, Nr. 2, 7—10 (1940). [Russisch.]

i) Stimulantien.

A....., P. V.: Planters' day at Lyamungu. Mon. Bull. Coffee Bd. Kenya 11, 140 (1946).

Abeele, M. van den: La culture du théier. Bull. Agric. Congo Belge 33, 124—173 (1942).

Aguirre, E. F.: Magnifica labor que realiza la Estación Experimental del Tabaco. (The magnificent work that is being carried on by the Tobacco Experiment Station.) Agronomia, Habana 3, 148 (1943).

Akkoyunlu, Z. u. F. Ipekoglu: Tütün çeşidlerimizin standardlanmasina dogru. (Towards the standardization of our tobacco varieties.) Tütün Institüsü Raporlari 3, 99—112 (1944).

Alcaraz Mira, E.: La Estación de Estudios del Tabaco y la investigacion sobre el Tabaco en España. (The Station of Tobacco Studies and the investigations on tobacco in Spain.) Tabac, Rome 6, Nr. 1/4, 92—99 (1943).

— Breves notas acerca de las variedades de tabaco cultivadas en España. (Short notes on the tobacco varieties cultivated in Spain.) Tabac, Rome 6, Nr. 1/4, 114—119 (1943).

— Obtención de razas de tabaco resistentes al mosaico ordinario. (The production of races of tobacco resistant to common mosaic.) Bol. Inst. Nac. Invest. Agron., Madr., Nr. 11, 89—120 (1944).

— u. R. de la Borbolla y Alcalá: Contribución al estudio químico de los tabacos españoles. 1¹ part. Tabacos exóticos. (Contribution to the study of Spanish tobaccos. I. Exotic tobaccos.) Bol. Inst. Invest. Agron. Madr., Nr. 8, 61—80 (1943).

— u. A. Izquierdo Tamayo: Obtención de plantas tetraploides de N. rústica y N. tabacum mediante la colchicina. (The production of tetraploid plants of N. rustica and N. tabacum by means of colchicine.) Bol. Inst. Nac. Invest. Agrion. Madr., Nr. 11, 49—87 (1944).

— u. J. M. Sequeiros Bores: Poder de absorción de las principales variedades de tabaco cultivades en España. (The absorption capacity of the principal varieties of tobacco cultivated in Spain.) Bol. Inst. Invest. Agron. Madr., Nr. 12, 285—328 (1945).

Alibert, H.: Note préliminaire sur une nouvelle maladie du cacaoyer le „swollen shoot". Agron. Trop., Nr. 1/2, 34—43 (1946).

Allard, H. A.: Some aspects of the phyllotaxy of tobacco. J. Agric. Res. 64, 49—55 (1942).

— Length-of-day behavior of Nicotiana gossei. J. Agric. Res. 67, 459—464 (1943).

Allemann, O.: Einige Notizen über den schweizerischen Tabakanbau. Schweiz. landw. Mh., Nr. 6, 161/168 (1945).

Alvarado, J. A.: Es el arábigo el rey de los cafés finos y remunerativos para explotarse? Rev. Agríc. Guatemala 15, 143—176, 16, 14—19 (1939).

Anderson, E.: Recombination in species crosses. Genetics 24, 668—698 (1939).

Andrés, J. M. u. F. Saura: Los cromosomas de la yerba mate y otras especies del género „Ilex". (The chromosomes of maté and other species of the genus „Ilex".) Fac. Agron. Vet. Inst. Genet. Univ. B. Aires 2, 161—168 (1945).

Anonymus: Quarterly notes of the Coffee Research and Experiment Station, Lyamungu, Moshi. Dep. Agric. Tanganyika, June, Nr. 10, 15 (1939).

Anonymus: A seleção dos cafeeiros nas Indias Neerlandesas. (Coffee selection in the Netherlands Indies.) Rev. Dep. Nac. Café Rio de J. 7, 347—349 (1939).

— The control of plant diseases. The value of immune or resistant types. Gold Cst Fmr. 8, 81—82 (1939).

— Tabacco mosaic. Rhod. Agric. J. 36, 172—173 (1939).

— Tobacco research, Programme for present season. Rhod. Agric. J. 36, 825—826 (1939).

— New „nicotineless" burley leaf has been developed in Kentucky. Sth. Tob. J. 54, Nr. 9, 7 (1940).

— Nogmaals het versnellen van de kieming van theezaden. (Once again — on increasing the rapidity of germination of tea seeds.) Bergcultures 14, 685—687 (1940).

— Onderzoekingen betreffende Virginia-tabak. (Investigations on Virginia tobacco.) Landbouw 16, 621—694 (1940).

— University of Kentucky breeds two new disease-resistant types. Sth. Tob. J. 54, Nr. 9, 5 (1940).

— Descubrimiento de un tabaco arboreo. (Discovery of a tree tobacco.) Ciencia, Mexico, D. F. 2, 364 (1941).

— Swollen-shoot disease of cacao. Trop. Agriculture, Trin. 18, 86 (1941).

— Two new varieties of flue-cured leaf resistant to black root-rot. Sth. Tob. J. 56, 4 (1942).

— Better hops. Mon. Sci. News., Nr. 22, 2—3 (1943).

— Eelworm resistance. Rhod. Agric. J. 40, 213 (1943).

— New hope for healthy tobaccos. Lighter, Ottawa 13, 4—5 (1943). [Mimeographed.]

— Resistant tobacco seed now available. Ext. Fm. News 29, 4 (1943).

— Tobacco varieties resistant to ordinary mosaic. Int. Rev. Agric. 34, 17M—18M (1943).

— A supressão do Instituto de Café do E. de São Paulo e a sua significação. (The suppression of the Coffee Institute of the State of São Paulo and its significance.) Rev. Dep. Nac. Café, Rio de J. 22, 792—794 (1944).

— (Disease resistant varieties of tobacco.) Socialisticeskoe Zemledelie (Socialist Agriculture), Nr. 153, 2. [Russisch.]

— Observation on „low nicotine" tobacco. Lighter, Ottawa 14, 10—11 (1944).

— Thrips. Mon. Sci. News, Nr. 33, 3 (1944).

— O levantamento botânico das variedades cafeeiras. (The botanical improvement of coffee varieties.) Bol. Superint. Serv. Café, São Paulo, Nr. 217, 322—323 (1945).

— Tea selection. III. The vegetative propagation of selected bushes. Tea Quart. 18, 91—94 (1946).

Antill, R. N.: Notes on tobacco seed selection. Nyasald Agric. Quart. J. 4, Nr. 3, 18—22 (1944).

Arghirescu, V.: Les variétiés des tabacs Roumains. Acta Nicotiana 1, 155—168 (1939).

— Tutunul Ialomita. (The Ialomita tobacco.) Bul. Cultiv. Ferment. Tutun. 28, 3—11 (1939).

Bacchi, O.: Observações citológicas em Coffea. VII. A macrosporogênese na variedade „Monosperma". (Cytological observations in Coffea. VII. Macrosporogenesis in the variety Monosperma.) Bragantia, São Paulo 1, 483—490 (1941).

Badenhuizen, N. P.: Colchicine-induced tetraploids obtained from plants of economic value. Nature, London 147, 577 (1941).

— De toepassingsmogelijkheden van verdubbeling van het aantal chromosomen door middel van colchicine-behandeling bij tabak en Java-jute. (A study of the possibilities offered by doubling the number of the chromosomes in tobacco and Java-jute by means of colchicine.) Landbouw 17, 231—251 (1941).

Bahtadze, K. E.: (The technique of tea breeding.) Soviet Subtropics, Nr. 1 (65), 13—21. [Russisch.]

— (Hybridization as a method of tea breeding.) Vestnik Gibridizacii (Hybridization), Nr. 2, 67—83 (1941). [Russisch.]

Baker, R. E. D.: Witches' broom disease investigations. III. Notes on the occurence of witches' broom disease of cacao at River Estate, 1939—1942. Trop. Agriculture, Trin. **20**, 5—12 (1943).
— Witches' broom disease investigation. IV. Further notes on the susceptibility of I. G. selections at River Estate to witches' broom disease of cacao. Trop. Agriculture, Trin. **20**, 156—158 (1943).
— u. S. H. Crowdy: Studies in the witches' broom disease of cacao caused by Marasmius perniciosus Stahel. Part II. Field studies and control methods. Mem. Imp. Coll. Trop. Agric. Trin., Pp. 28 (1944).
— u. W. T. Dale: Notes on a virus disease of cacao. Ann. Appl. Biol. **34**, 60—65 (1947).
Bakhtadze, K. E. siehe Bahtadze.
Bald, C.: Indian tea. A textbook on the culture and manufacture of tea. W. Thacker and Co., London (1940).
Bartolucci, A.: Il fenomeno della fotoperiodicità ed il tabacco. La N. tomentosa Ruiz et Pavon pianta a giorno preve. (The phenomena of photoperiodism and tobacco. N. tomentosa Ruiz et Pavon a short-day plant.) Boll. Tec. Tab. **36**, 5—12 (1939).
— Il fenomeno della poliploidia ed il tabacco. 1ª nota-L'uso della colchicina e della centrifugazione dei semi per trasformare gl'ibridi sterili in ibridi fertili. (Polyploidy and tobacco-Sterile hybrids made fertile by colchicine and centrifuging treatments.) Boll. Tec. Tab. **36**, 141—148 (1939).
— Poliploidia e tabacco. L'uso della colchicina e della centrifugazione dei semi, per transformare gli ibridi sterili in ibridi fertili. (Polyploidy and tobacco. The use of colchicine and centrifuging the seeds to transform sterile hybrids into fertile hybrids.) Acta Nicotiana, Berlin **1**, 240—246 (1939).
— Il fenomeno della poliploida e il tabacco: 2ª nota-Nuove specie fertili di tabacco prive o quasi di nicotina e loro caratteristiche morfologiche. (The phenomenon of polyploidy and tobacco: 2nd note-New fertile species of tobacco free or almost free of nicotine and their morphological characteristics.) Bol. Tec. Tab. **37**, 157—174 (1940).
Batt, R. F. u. H. Martin: The home-production of nicotine. Rep. Agric. Hort. Res. Sta., Long. Ashton, Bristol, Pp. 140—144 (1944).
Battaglia, E.: Contributo alla cariologia della Papaveraceae (Subfam. Fumarioideae) con particolae riguardo ai generi Dicentra, Corydalis, Cysticapnos ed Adlumia. Scient. Genetica **2**, 1—25 (1940).
Beard, F. H.: Commercial varieties of hops: a preliminary comparison of their chief characteristics at East Malling. Ann. Rep. E. Malling Res. Sta. 1942, Pp. 75 bis 83 (1943).
— Hops: their varieties and cultivation. J. Inst. Brew. **49**, 118—125 (1943).
— Hop growing in Great Britain with special reference to research work. Wallerstein Laboratories Communications **8**, 83—98 (1945).
— Observations on the incidence of mould (Spaerotheca humulis) on the seedling hops at East Malling in 1945. 33rd Rep. E. Malling Res. Sta. 1945, Pp. 107—114 (1946).
— u. D. J. Wilson: Propagation trials with hops. II. Preliminary trials in propagation by soft-wood cuttings. 33rd Rep. E. Malling Res. Sta. 1945, Pp. 96—103 (1946).
Benincasa, M.: Le principali varietà di tabacco cultivate in Italia. (The principal varieties of tobacco cultivated in Italy.) Acta Nicotiana, Berlin **1**, 169 bis 206 (1939).
Bernardini, L.: Tabac sans nicotine. (Tobacco without nicotine.) Tabac, Rome, **2**, Nr. 1, 67—70 (1939).
Berthault, P.: L'amélioration du tabac et de sa culture en Kabylie. C. R. Acad. Agric. Fr. **32**, 775—779 (1946).

Billes, D. J.: Pollination of Theobroma cacao L. in Trinidad, B. W. I. Trop. Agriculture, Trin. **18**, 151 bis 156 (1941).
Bingefors, S.: Svensk havanna och hur den odlas. (Swedish havanna and how it is grown.) Lantmannen **28**, 483—485 (1944).
Bojarskij, J.: (Bud mutations of the genus Nicotiana.) J. Inst. Bot. Acad. Sci. Ukraïne, Nr. 21/22 (29/30), 217—219 (1939). [Russisch.]
— (On the cultivation of high-yield and high-quality varieties of yellow makhorkas.) J. Bot. Acad. Sci. Ukraine **1**, 267—274 (1940). [Russisch.]
Bolsunov, I. I.: (An experimentally obtained haploid in Nicotiana rustica L.) J. Inst. Bot. Acad. Sci. Ukraine, Nr. 21/22 (29/30), 197—199 (1939). [Russisch].
— (A valuable hybrid of Nicotiana rustica devoid of inflorescence and upper suckers.) Selekcija i Semenovodstvo, Nr. 2/3, 40—41 (1939). [Russisch.]
— Zur Untersuchung der Heterosis bei Nicotiana rustica L. V. Fortgesetzte Selbstbefruchtung und Ernteertrag. Z. Pflanzenz. **26**, 223—244 (1944).
Bond, J.: Seedlings and cuttings. Tea Quart. **17**, 20—21 (1944).
Bond, T. E. T.: The present position of tea selection in Java. (Review.) Tea Quart. **14**, 23—30 (1941).
— Clone testing in the field. (Review.) Tea Quart. **14**, 61—64 (1941).
— The ‚phloem necrosis' virus disease of tea in Ceylon. I. Introductory accounts, symptoms, and transmission by grafting. Ann. Appl. Biol. **31**, 40—46 (1944).
Botti, G.: I problemi del tabacco e l'opera dell' Istituto Scientifico Sperimentale per i Tabacchi. (The problems of tobacco and the work of the Scientific Research Institute for Tobacco.) Tabacco **51**, Nr. 570, 19—22 (1947).
Bradley, M. V. u. T. H. Goodspeed: Colchicine-induced allo-and autopolyploidy in Nicotiana. Proc. Nat. Acad. Sci., Wash. **29**, 295—301 (1943).
Bredemann, G.: Beiträge zur Züchtung des Mohnes (Papaver somniferum L.) auf hohen Alkaloidgehalt. C. R. VI. Congr. Int. Techn. Chim. Industr. Agric., Budapest **2**, 378—386 (1939).
Brieger, F. G. u. R. Forster: Tumores em certos híbridos do gênero Nicotiana. (Tumours in certain hybrids of the genus Nicotiana.) Bragantia, São Paulo **2**, 259—274 (1942).
—, — Modificação da dominância em N. Tabacum petiolaris. (Modification of dominance in N. Tabacum petiolaris.) Rev. Agric. Piracicaba **18**, 446—447 (1943).
—, A. R. Lima u. R. Forster: Comportamento de variedades e progênies de fumo na resistência ao ,,vira-cabeça". (Behaviour of varieties and progenies of tobacco in resistance to ,,vira-cabeça".) Bragantia, São Paulo **2**, 275—294 (1942).
—, —, —, A. S. Costa u. S. Ribeiro dos Santos: Ensaio de épocas de transplante para o fumo. (Experiments on the times of planting out in tobacco.) Bragantia, São Paulo **2**, 295—310 (1942).
Brink, R. u. D. C. Cooper: Incomplete seed failure as a result of somatoplastic sterility. Genetics **26**, 487 bis 505 (1941).
Bullock, J. F. u. E. G. Moss: Strains of flue-cured tobacco resistant to black shank. Circ. U. S. Dep. Agric., Nr. 682, 9 (1943).

Callan, E. McC.: Thrips resistance in cacao. Trop. Agriculture, Trin. **20**, 127—135 (1943).
Carletto, G. M.: A polonização controlada na flor do cacaueiro. (Controlled pollination in the flower of the cacao tree.) Bol. Téc. Inst. Cacau, Bahia, Nr. 6, 39 (1946).
Carvalho, A.: Genética de Coffea. IV. Instabilidade do par de alelos Na-na de Coffea arabica L. (Genetics of Coffea. IV. Instability of the pair of alleles Na-na

of C. arabica L.) Bragantia, São Paulo **1**, 453—466 (1941).
Carvalho, A.: Trabalhos de seleção de Coffea arabica realizados em Tanganyika. (Work on breeding C. arabica in Tanganyika.) Rev. Inst. Café S. Paulo **16**, 868 bis 872 (1941).
— Sementes selecionadas de café. (Selected seeds of coffee.) Rev. Inst. Café, S. Paulo **16**, 1418—1426 (1941).
— Distribuição geográfica e classificação botânica do gênero Coffea com referência especial á espécie arabica. (Geographical distribution and botanical classification of the genus Coffea with special reference to the species arabica.) Bol. Superintend. Serv. Café, São Paulo 1945, **20**, 1138—1146; 1946, **21**, 6—10, 69—73, 127 bis 130, 174—184; (1945/1946).
Cassagnol, P. L., F. Boncy u. A. Denis: Etude sur la qualité des cafés d'Haiti. Bull. Dep. Agric. Haiti, Nr. 25, 53 (1943).
Castellani, E.: A proposito della presunta resistenza di varietà brasiliane do Caffè alla Hemileia vastatrix. (Regarding the supposed resistance of Brazilian varieties of coffee to H. vastatrix.) Agricoltura Colon **36**, 100—101 (1942).
Castetter, E. F.: New evidence on the antiquity of tobacco cultivation in the South-west. Amer. J. Bot. **31**, 2s (1944). (Abstr.)
Černicin, N. G.: (Investigation of the stage of maturity of seed plants of Nicotiana rustica.) Vsesojuznyj Naučno-Issledovateljskij Institut Tabačnoj i Mahoročnoj Promyšlennosti imeni A. I. Mikojana (VITIM). (All-Union Mikojan Research Institute of the Tobacco and Makhorka Industry.) Krasnodar, Nr. 143, 241 bis 247 (1941). [Russisch.]
Chavarriaga Misas, E.: Cuál variedad de cacao debemos sembrar? (Which variety of cacao should we sow?) Rev. Fac. Nac. Agron. Colombia **4**, 1325—1329 (1941).
Cheesman, E. E.: The case for long range research in cacao production. Trcp. Agriculture, Trin. **17**, 203 bis 207 (1940).
— Notes on the nomenclature, classification and possible relationships of cacao populations. Trop. Agriculture, Trin. **21**, 144—159 (1944).
— Results of cacao experiments in 1944—1945. Trop. Agriculture, Trin. **23**, 63—65 (1946).
Cheney, R. H.: China tea substitutes in the New York area. Amer. J. Bot. Suppl. **29**, 5s (1942). (Abstr.)
— Tea substitutes in the United States. J. N. Y. Bot. Gdn **43**, 117—124 (1942).
Chetty, C. V. S.: Methods of production and marketing of cigarette and cigar tobacco in the United States of America, Canada, Java and Sumatra. Misc. Bull. Imp. Coun. Agric. Res. Delhi, Nr. 48, 91 (1942).
Chevalier, A.: A propos du Lembyrea de Madagascar. Rev. Bot. Appl. **19**, 250 (1939).
— Sur quelques caféiers et faux caféiers de l'Angola et du Mayombe portugais. Rev. Bot. Appl. **19**, 396—407 (1939).
— Les problèmes de la caféiculture dans les colonies françaises. Rev. Bot. Appl. **20**, 229—251 (1940).
— Note sur les caféiers sauvages de l'Afrique austro-orientale. Rev. Bot. Appl. **20**, 529—540 (1940).
— Nouveau groupement des espèces du genre Coffea et spécialement de celles de la section Eucoffea. (New classification of the species of the genus Coffea and especially of those of the section Eucoffea.) C. R. Acad. Sci. Paris **210**, 357—361 (1940).
— Le statut actuel du genre Coffea L. Rev. Bot. Appl. **22**, 129—150 (1942).
— Notes sur le houblon. Rev. Bot. Appl. **23**, 225—242 (1943).
— Révision du genre Theobroma d'après l'herbier du Muséum national d'Histoire naturelle de Paris. Rev. Bot. Appl. **26**, 265—285 (1946).

Chiarugi, A.: Nota sulla cariologia del Caffè d'Etiopia (Coffea arabica L.). N. Giorn. Botan. Ital., n. s. **47**, 515—519 (1940).
Christoff, M.: Sur l'amélioration de la production du tabac en Bulgarie. (On the improvement of tobacco production in Bulgaria.) Tabac, Rome, **2**, Nr. 1, 38—40 (1939).
Christoph, M.: Die wichtigsten Tabaksorten. Acta Nicotiana, Berlin **1**, 141—153 (1939).
Clausen, R. E.: Monosomic analysis in Nicotiana Tabacum. Genetics **26**, 145 (1941). (Abstr.)
— II. Symposium on theoretical and practical aspects of polyploidy in crop plants. Polyploidy in Nicotiana. Biol. Symp. **4**, 95—110 (1941).
— Symposium on „Theoretical and practical aspects of polyploidy in crop plants". Polyploidy in Nicotiana (with discussion by H. H. Smith). Amer. Nat. **75**, 291—309 (1941).
— u. D. R. Cameron: Inheritance in Nicotiana Tabacum. XVIII. Monosomic analysis. Genetics **29**, 447—477 (1944).
Clayton, E. E.: Resistance to root knot nematode in Nicotiana. Proc. 41st Annu. Conv. Ass. Sth Agric. Wkrs, Birmingham, Ala., February 7—9, p. 200 (1940). (Abstr.)
— Resistance of tobacco to blue mold (Peronospora tabacina). J. Agric. Res. **70**, 79—87 (1945).
— Transfer of wildfire resistance from Nicotiana longiflora to N. tabacum. Phytopathology **37**, 4—5 (1947). (Abstr.)
— A wildfire resistant tobacco. J. Hered. **38**, 35—40 (1947).
— u. T. W. Graham: Tobacco resistant to root knot and nematode root rot. Phytopathology **36**, 684 (1946). (Abstr.)
— u. H. H. Mc Kinney: Resistance to the common mosaic disease of tobacco. Phytopathology **31**, 1140 bis 1142 (1941).
— u. T. E. Smith: Resistance of tobacco to bacterial wilt (Bacterium solanacearum). J. Agric. Res. **65**, 547—554 (1942).
Comrie, A. A. D.: Colorimetric determination of the preservative value of hops: standard colour values of soma hybrid hops. J. Inst. Brew. **46**, 255—256 (1940).
Coolhaas, C.: Proeftuin Kedoengpani van het Proefstation Midden- en Oost-Java. (Experimental plantation Kedoengpani of the Central and East-Java Experiment Station.) Bergcultures **15**, 376 (1941).
Cooper, D. C. u. R. A. Brink: Somatoplastic sterility as a cause of seed failure following interspecific hybridization. Genetics **25**, 114 (1940). (Abstr.)
—, — Somatoplastic sterility as a cause of seed failure after interspecific hybridization. Genetics **25**, 593—617 (1940).
Cope, F. W.: Some factors controlling the yield of young cacao. II. 8th Annu. Rep. Cacao Res., Trin., Pp. 4—15 (1938) 1939.
— A note on the range of compatibility in cacao. 8th Annu. Rep. Cacao Res., Trin., Pp. 16—17 (1938) 1939.
— Compatibility and fruit setting in cacao. 8th Annu. Rep. Cacao Res., Trin., Pp. 17—20 (1938) 1939.
— Studies in the mechanism of self-imcompatibility in cacao. I. 8th Annu. Rep. Cacao Res., Trin., Pp. 20—21 (1938) 1939.
— Some factors controlling the yield of young cacao. III. 9th Rep. Cacao Res., Trinidad, Pp. 6—12 (1939) 1940.
— Studies in the mechanism of self-incompatibily in cacao. II. 9th Rep. Cacao Res., Trinidad, Pp. 19—23 (1939) 1940.
Costa, A. S. u. A. R. Lima: Sôbre variedades de fumo que localizam o virus do mozaico comum (Nicotiana virus 1). [Varieties of tobacco that localize the common mosaic virus (Nicotiana virus 1).] Rev. Agric., S. Paulo **15**, 209—213 (1940).

Croesen, V. R. Ij.: Tabakscultur in Nederland. (Tobacco cultivation in Holland). Landbouw. Tijdschr. Groningen 52, 145—164, 213—233 (1940).

Dale, W. T.: Witches' broom disease investigations. XII. Further studies on the infection of cacao pods by marasmius perniciosus Stahel. Trop. Agriculture, Trin. 23, 217—221 (1946).

Dawson, R. F.: Accumulation of anabasine in reciprocal grafts of Nicotiana glauca and tomato. Amer. J. Bot. 31, 351—355 (1944).

— An experimental analysis of alkaloid production in Nicotiana: the origin of nornicotine. Amer. J. Bot. 32, 416—423 (1945).

Determann, W.: Über Zusammenhänge zwischen Alkaloidgehalt und Zahl und Größe der Milchröhren in den Kapseln von Papaver somniferum L. Ein Beitrag zur Züchtung des Mohnes auf hohen Alkaloidgehalt. Z. Pflanzenz. 23, 371—410 (1940).

Diachun, S. u. W. D. Valleau: Reaction of 35 species of Nicotiana to tobacco-streak virus. Phytopathology 37, 7 (1947). (Abstr.)

Dobrunov, L. G.: (Ontogenetische und metamere Variabilität der Tabakblätter.) C. R. (Doklady) Acad. Sci. URSS. 54, 553—556 (1946). [Russisch.]

Dufrénoy, J.: Etudes génétiques sur le tabac en France. Chronica Botanica 7, 277—278 (1942).

Dusseau, A.: L'institut expérimental des tabacs de Bergerac. Rev. Hort., Paris 115, 348—351 (1943).

— Anomalies de germination du pollen chez deux hybrides amphidiploïdes de Nicotiana. Bull. Soc. Bot. Fr. 91, 59—60 (1944).

— u. A. Fardy: Comportement cytogénétique de l'hybride interspécifique Nicotiana rustica L. var. Zlag (n = 24) × N. paniculata L. (n = 12), hautement stérile, transformé en hybride amphidiploïde fertile après traitement à la colchicine. C. R. Soc. Biol., Paris 137, 235—236 (1943).

Eden, T.: Report of the plant physiologist for 1939. Bull. Tea Res. Inst. Ceylon, Nr. 21, 47—52 (1939).

— The selection of high-yielding tea bushes for vegetative propagation. Tea Quart 14, 98—102 (1941).

Egiz, S. A.: On the problem of the origin of Nicotiana rustica. C. R. (Doklady) Acad. Sci. URSS. 26, 952 bis 956 (1940).

— u. F. A. Fatalizade: Cyto-genetic experiments with Nicotiana trigonophylla Dun. C. R. (Doklady) Acad. Sci. URSS. 22, 124—126 (1939).

Emden, J. H.: Opzet en inrichting van de proeven tot locale toetsing van een aantal geselecteerde theecloonen. (Outlay of some experiments for the local testing of a number of thea clones.) Arch. Theecult. Ned-Ind. 14, 91—115 (1940).

Engledow, F.: Tea. Chem. Ind., Lond., Nr. 123, 220 bis 221 (1946).

Esteva, C., jr.: Third Annual Report of the Tobacco Institute of Puerto Rico. San Juan, p. 50 (1940).

— El Instituto del Tabaco de Puerto Rico, sus actividades del presente, sus orientaciones para el futuro. (The Puerto Rico Tobacco Institute, its present activities and plans for the future) Rev. Agric., P. Rico 33, 22—26 (1941).

— Annual Report of the Tobacco Institute of Puerto Rico for the fiscal years 1939—1940, 1940—1941, Pp. 98 (1942).

Evans, G.: Research and training in tropical agriculture. J. R. Soc. Arts. 87, 333—350 (1939).

Fagerlind, F.: Perisperm oder Endosperm bei Coffea? (Perisperm or endosperm in Coffea?) Svensk. Bot. Tidskr. 33, 303—309 (1939).

Fardy, A.: Etude cytologique et génétique du croisement interspécifique Nicotiana Tabacum L. (var. purpurea) × N. sylvestris Speg. et Comes et de sa descendance. E. Drouillard, Bordeaux, Pp. 108, 24 figs, 7 pls. tables (1941).

Fardy, A.: L'evolution des noyaux somatiques de Nicotiana tabacum. L. et N. sylvestris Speg. et Comes, et les rapports entre nucléole et chromosomes à satellites. C. R. Soc. Biol., Paris 135, 587—589 (1941).

Fatalizade, F. A.: Acenaphthene-induced polyploidy in Nicotiana. C. R. (Doklady) Acad. Sci. URSS. 22, 180—183 (1939).

— Cytogenetic studies on the genus Nicotiana and the origin of N. rustica. C. R. (Doklady) Acad. Sci. URSS. 25, 770—772 (1939).

Ferwerda, F. P.: Resultaten van eenige mengproeven met koffiecloonen op Bangelan. (Results of some trials of mixed plantings with coffee clones at Bangelan.) Bergcultures 15, 557—566 (1941).

Fisher, R. A.: The design of experiments. Oliver & Boyd Ltd., London and Edinburgh 3 rd ed. 12s. 6d., Pp. XI + 236. tables (1942).

— u. F. Yates: Statistical tables for biological, agricultural and medical research. Oliver & Boyd, Ltd., London and Edinburgh 2 nd. ed. 13s. 6d., Pp. VIII + 98. 34 tables (1943).

Fletcher, L.: Brewing trials with three new varieties of hops raised by Prof. E. S. Salmon, at Wye College, Kent; 1944 growths. J. Inst. Brew. 51, 232 (1945).

Fluiter, H. J. de: Proeven en waarnemingen in verband met de bestrijding van het bruinvlek, Alternaria longipes (Ell. et Ev.) Mason. [Experiments and observations in connexion with the combating of brown spot A. longipes (Ell. et Ev.) Mason.] Meded. Besoek. Proefst., Nr. 65, 1—40 (1939).

Ford, J. S., L. Fletcher u. T. Manson: Contributions to the Institute of Brewing Research Scheme. Brewing trials with new varieties of hops raised by Prof. E. S. Salmon, at Wye College, Kent, and grown at various places, 1941 growths. J. Inst. Brew. 48, 136—137 (1942).

Forster, H. H.: Resistance in the genus Nicotiana to Phytophthora parasitica Dastur var. nicotianae Tucker. Phytopathology 33, 403—404 (1943).

Frahm-Leliveld, J. A.: Nieuwe gezichtspunten voor het selectieonderzoek bij soortskruisingen van koffie naar aanleiding van cytologische gegevens. (New viewpoints for research on breeding in the case of species crosses of coffee on the basis of cytological data.) Bergcultures 14, 380—386 (1940).

— Onstaan en voorkomen van rondboon en voosboon bij koffie. (Origin and occurrence of round beans and spongy beans in coffee.) Bergcultures 14, 1358—1362 (1940).

Franco, C. M.: Relation between chromosome number and stomata in Coffea. Bot. Gaz. 100, 817—827 (1939).

François, E.: L'avenir de la culture des caféiers à Madagascar. Etat actuel de la question. (The future of the cultivation of coffee in Madagascar-Present position.) Rev. Bot. Appl. 20, 153—164 (1940).

Frimmel, F.: Vererbung der Blattform des Tabaks. Acta Nicotiana, Berlin 1, 247—250 (1939).

Gage, C. E.: The tobacco industry in Puerto Rico. Circ. U. S. Dep. Agric., Nr. 519, 54 (1939).

Gandrup, J.: Hedendaagsch pionierswerk. (Modern pioneering.) Bergcultures, Pp. 312—318 (1940).

García Fortuño, M.: Apuntes sobre la genética del tabaco. Descripción del método de cruzamientos utilizado por el Instituto del Tabaco. (Notes on tobacco genetics. Description of the method of crossing employed by the Tobacco Institute.) Rev. Agric. P. Rico 31, 39—41 (1939).

Garner, W. W.: The production of tobacco. Blakiston Co., Pa., Pp. XIII + 516. 81 figs. 11 tables (1946).

Gärtner, K.: Über den Nikotingehalt der ungarischen Tabake. Acta Nicotiana, Berlin 1, 500—501 (1939).

Gelin, O.: Vallmoförsök på Weibullsholm. (Poppy research at Weibullsholm.) Weibulls Ill. Årsb. 37, 31—36 (1942).

— u. N. Schwanbom: Erfarenheter fran årets vallmoodling. (Results from the year's poppy cultivation.) Lantmannen 25, 739—740 (1941).

—, — Studier över vallmons odling och förädling. (Studies on poppy cultivation and breeding.) Agri Hortique Genetica, Landskrona 1, 34—56 (1943).

Gerstel, D. U.: Inheritance in Nicotiana tabacum. XVII. Cytogenetical analysis of glutinosa-type resistance to mosaic disease. Genetics 28, 533—536 (1943).

— Inheritance in Nicotiana Tabacum. XIX. Identification of the Tabacum chromosome replaced by one from N. glutinosa in mosaicresistant Holmes Samsoun tobacco. Genetics 30, 448—454 (1945).

— Inheritance in Nicotiana Tabacum. XX. The addition of Nicotiana glutinosa chromosomes to tobacco. J. Hered. 36, 197—206 (1945).

— Inheritance in Nicotiana Tabacum. XXI. The mechanism of chromosome substitution. Genetics 31, 421 bis 427 (1946).

Gilbert, S. M.: Selection within Coffea arabica in Tanganyika Territory. E. Afr. Agric. J. 4, 249—253 (1939).

— The Coffee Research and Experiment Station, Tanganyika Territory: a brief survey of the first ten years' work. Emp. J. Exp. Agric. 13, 113—124 (1945).

Gillett, S.: Coffee team services. Progress report on the cultural work. Mon. Bull. Coffee Bd. Kenya 6, 56 bis 57, 59 (1940).

— Summary of the annual report on cultural work conducted by the Coffee Services of the Agricultural Department during 1943. Mon. Bull. Coffee Bd. Kenya 9, 108—110 (1944).

Gisquet, P., J. Dufrénoy u. A. Dusseau: Introduction à l'étude d'hybrides interspécifiques de Nicotiana. Rev. Cytol. et Cytophysiol. végét. 4, 86—91 (1939).

—, —, — Interspecific hybrids among Nicotiana; hybrids between N. Tabacum var. purpurea and N. petunioides var. sylvestris. Proc. 7th Int. Genet. Congr. Edinburgh 23—30 August, p.130, 1939 (1941). (Abstr.)

—, —, — Hybrides interspécifiques de Nicotiana. Scientia Genetica 2, 67—78 (1940).

—, A. Dusseau u. H. Hitier: Premier hybride stabilisé en une varieté nouvelle, issu du croisement Nicotiana Tabacum var. purpurea y N. sylvestris. C. R. Acad. Sci, Paris 209, 356—357 (1939).

Goenaga, A.: Notes on diseases of tobacco observed in Puerto Rico during the 1944—1945 season. Plant Dis. Reporter 29, 311—314 (1945). [Mimeographed.]

Golubinskij, I. N.: Numerical relations of sexes in various hybrid combinations in hop. C. R. (Doklady) Acad. Sci. URSS. 25, 414—418 (1939).

— Influence of temperature alternation on the germinable power of hop seeds (Humulus lupulus L.). C. R. (Doklady) Acad. Sci. URSS. 32, 85—86 (1941).

— Meiotic abnormalities in hops induced by atmospheric electricity. C. R. (Doklady) Acad. Sci. URSS. 46, 247 bis 249 (1945).

— u. N. I. Golubinskaja: Effect of heat upon the mutation rate in hop (Humulus Lupulus L.). C. R. (Doklady) Acad. Sci. URSS. 25, 773—776 (1939).

— u. M. S. Šlos: A contribution to the problem of parthenogenesis in Humulus Lupulus L. C. R. (Doklady) Acad. Sci. URSS. 23, 486—487 (1939).

Goodspeed, T. H.: The South American genetic groups of the genus Nicotiana and their distribution. Proc. 8th Amer. Sci. Congr., Wash. 3, 231—238 (1940).

— A fourth new species of Nicotiana from Peru. Univ. Calif. Publ. Bot. 18, 195—204 (1941).

Goodspeed, T. H.: El tabaco y ostras especies del género Nicotiana. (Tobacco and other species of Nicotiana.) Bol. Fac. Agron. Vet. B., Aires, Nr. 22, 21 (1942).

— Nicotiana Arentsii a new, naturally occurring amphidiploid, species. Proc. Calif. Acad. Sci. 25, 291—306 (1944).

— A fifth species of Nicotiana from Peru. Rev. Univ., Cuzco 33, 65—70 (1944).

— Cytotaxonomy of Nicotiana. Bot. Rev. 11, 533—592 (1945).

— Studies in Nicotiana. III. A taxonomic organization of the genus. Univ. Calif. Publ. Bot. 18, 335—344 (1945).

— Chromosome number and morphology in Nicotiana. VII. Karyotypes of fifty-five species in relation to a taxonomic revision of the genus. Univ. Calif. Publ. Bot. 18, 345—368 (1945).

— Meiotic prophase studies in Nicotiana. Genetics 31, 217 (1946). (Abstr.)

— Karyotypes, meiotic behavior, and systematics in Nicotiana. Genetics 31, 217 (1946). (Abstr.)

— Meiotic prophase phenomena in species and interspecific hybrids of Nicotiana. J. Arnold Arbor. 27, 453—469 (1946).

— u. P. Avery: Trisomic and other types in Nicotiana sylvestris. J. Genet. 38, 381—458 (1939).

—, — The twelfth primary trisomic type in Nicotiana sylvestris. Proc. Nat. Acad. Sci., Wash. 27, 13—14 (1941).

Gornik, R.: Die Tabaksorten Jugoslawiens. Acta Nicotiana 1, 207—219 (1939).

Goulden, C. H.: Methods of statistical analysis. John Wiley and Sons, Inc., New York 17s. 6d. Pp. VII + 277 (Chapman and Hall, London) (1939).

Greenleaf, W. H.: Induction of polyploidy and sterility in amphidiploids induced by heteroauxin treatment. Amer. J. Bot. 26, 673 (1939). (Abstr.)

— Genic sterility in tabacum-like amphidiploids of Nicotiana. Amer. J. Bot. 28, 726 (1941). (Abstr.)

— Sterile amphidiploids: their possible relation to the origin of Nicotiana Tabacum. Amer. Nat. 75, 394—399 (1941).

— Sterile and fertile amphidiploids: their possible relation to the origin of Nicotiana Tabacum. Genetics 26, 301 bis 324 (1941).

— The Probable explanation of low transmission ratios of certain monosomic types of Nicotiana Tabacum. Proc. Nat. Acad. Sci., Wash. 27, 427—430 (1941).

— Genic sterility in Tabacum-like amphidiploids of Nicotiana. J. Genet. 43, 69—96 (1942).

Greewood, M.: Report on the Central Cocoa Research Station, Tafo, 1938—1942. Dep. Agric., Gold Coast, Pp. 63 (1943).

Griffith, R. B., W. D. Valleau u. R. N. Jeffrey: Chlorophyll and carotene content of eighteen tobacco varieties. Plant Physiol. 19, 689—693 (1944).

Haan, I. de: De ontwikkeling en de vertakking van oculaties bij thee in verband met de eigenschappen van het oculatiehout en den onderstam. (The development and the branching of buddings in tea in relation to the properties of the bud wood and the stock.) Bergcultures 15, 1623—1628 (1941).

Håkansson, A.: Some observations on the seed development in Ecuadorian cacao. Hereditas, Lund 33, 526—538 (1947).

Halcrow, M.: Individual tree recording for yield and quality. Mon. Bull. Coffee Bd. Kenya 5, 210—211 (1939).

Hall, C. J. J. van: Coffee selection in the Netherlands Indies. J. Mysore Agric. Exp. Un. 18, 42—48 (1940).

Harlan, J. D.: N. Y. State's experiments in growing new varieties of hops. Amer. Brewer, Nr. 72 (8), 27—29 (1939).

Harlan, J. D.: What new varieties of hops did in N. Y. in 1939. Amer. Brewer **73** (6), 29; (7), 21—22 (1940).

Haslam, R. J.: A burley variety makes its debut. Lighter, Ottawa **12**, Nr. 2, 9 (1942). [Mimeographed.]

— Harrow Resistant Broadleaf Burley. Lighter, Ottawa **15**, 14 (1945). [Mimeographed.]

— Another new variety of burley released at Harrow. Lighter, Ottawa **17**, 12 (1947). [Mimeographed.]

Hatton, R. G. u. F. H. Beard: A report of hop research work at East Malling Research Station. J. Inst. Brew. **45**, 556—559 (1939).

Heeger, E. F.: Sortenkundliche Untersuchungen zur Frage der Opiumgewinnung in Deutschland. Forschungsdienst **8**, 513—514 (1939).

— u. K. H. Bauer: Untersuchungen über den Morphingehalt der zum Handel zugelassenen und einiger anderer Mohnsorten und die Möglichkeit der Opiumgewinnung im Deutschen Reich. Landw. Jb. **90**, 397—429 (1940).

Heggestad, H. E.: Varietal variation and inheritance studies on natural water-soaking in tobacco. Phytopathology **34**, 1002 (1944). (Abstr.)

— Varietal variation and inheritance studies on natural water-soaking in tobacco. Phytopathology **35**, 754 bis 770 (1945).

Heierle, E.: Der Tabakanbau in der Schweiz. Jahresbericht der Forschungsstelle der Einkaufsgenossenschaft für Inland-Tabak ,,Sota", Zürich. Tabac, Rome **2**, Nr. 3, 20—40 (1939).

Henderson, R. G.: Field and greenhouse studies on 16 tobacco hybrids and varieties. Phytopathology **36**, 400—401 (1946). (Abstr.)

Hendrickx, F. L. u. P. C. Lefevre: Observations préliminaires sur la résistance de lignées de Coffea arabica L. à quelques ennemis. Bull. Agric. Congo. Belge **37**, 783—800 (1946).

Heusser, C.: Das White Burley-Problem im schweizerischen Tabakanbau. Tabac, Rome **4—5**, Nos. 2/4, 140—143 (1942).

Hill, D. D. u. D. E. Bullis: Summary of hopgrade investigations for 1941—1942. Circ. Brew. Hop. Res. Inst., Nr. 6 (1943).

Hills, C. H. u. H. H. McKinney: The effect of mosaic virus infection on the protein content of susceptible and resistant strains of tobacco. Phytopathology **32**, 857—866 (1942).

Hills, K. L.: The suitability of a number of varieties of opium poppy for the production of morphine from the ripe capsule. J. Coun. Sci. Industr. Res. Aust. **19**, 177—186 (1946).

Hoed, F. u. P. Elsocht: Contribution à l'étude de l'amélioration du houblon en Belgique. Ann. Gembl. **45**, 65—119 (1939).

—, — Etude sur les variétés de houblon-résumant les recherches entreprises en 1938 par le Cercle d'Etudes et de Recherches pour l'Amélioration du Houblon. Petit J. Brass. Nr. 1044, 7 (1939).

Holubinski, I. N. siehe Golubinskij.

Honing, J. A.: Nicotiana Tabacum crosses. The Kloempany dwarf factor. Polymery as to single and double flowers. Interaction of factors. A necrotic dwarf. A dwarf without ovules. Genetica **21**, 109—152 (1939).

— Types of Nicotiana Tabacum, grown in the Netherlands East-Indies, in subsequent generations partly constant and partly variable as to the need of light for germination. Genetica **23**, 1—21 (1942).

Houk, W. G.: Nomenclatura dos cafeeiros. Lista de referências e bibliografia. (Coffee nomenclature. List of references and bibliography.) Bol. Téc. Inst. Agron. Campinas, Nr. 63, 49 (1939).

Howes, F. N.: The early introduction of cocoa to West Africa. Trop. Agriculture, Trin. **23**, 172 (1946).

Inoue, T. u. S. Arima: (Meiosis in Thea sinensis L. I. Causasus tea.) Proc. Crop. Sci. Soc. Japan **12**, 134 bis 140 (1940).

Ipekoglu, F.: Tütünün korelatif vasiflari üzerinde bir tedkik. (A study on correlative characters of tobacco.) Tütün Institüsü Rapolari **3**, 5—8 (1943).

Jakovuk, A. S.: (Intra-varietal crossing in Nicotiana Tabacum L.) Vsesojuznyj Naučno-Issledovateljskij Institut Tabačnoj i Mahoročnoj Promyšlennosti imeni A. I. Mikojana (VITIM). (All-Union Mikojan Research Institute of the Tobacco and Makhorka Industry), Krasnodar, Nr. 143, 29—37 (1941). [Russisch.]

— (Selective fertilization in tobacco plants.) Proc. Lenin. Acad. Agric. Sci. USSR., Nr. 4, 3—7 (1941). [Russisch.]

— (The occurrence of interchanged dominance in some tobacco hybrids.) Jarovizacija, Nr. 2 (35), 120 (1941).

— u. E. N. Psareva: (A handbook on the appraisal of cigarette and cigar varieties of tobacco.) Vsesojuznyj Naučno-Issledovateljskij Institut Tabačnoj i Mahoročnoj Promyšlennosti imeni A. I. Mikojana. (Mikojan Research Institute of the Tobacco and Mahorka Industry) Krasnodar, Pp. 136 (1941). [Russisch.]

Johnson, J. u. W. B. Ogden: Tobacco mosaic and its control. Bull. Wis. Agric. Exp. Sta., Nr. 445, 22 (1939).

— Tobacco varieties and strains in Wisconsin. Bull. Wis. Agric. Exp. Sta., Nr. 448, 30 (1939).

—, — u. O. J. Attoe: Experiments on the leaf-burn of tobacco. Circ. Wis. Agric. Exp. Sta. Res. Bull., Nr. 153, 75 (1944).

Journée, C. u. L. Vaitzman: Quelques aspects du problème du tabac en Belgique. Importance économique — Technique de production. Organisation de la vente et du travail d'amélioration. J. Duculot, Gembloux (1939).

Jouvenel-Marcillac, M.: La cytomixie chez les Papavéracées. C. r. Soc. Biol. Paris **136**, 468—469 (1942).

— Le noyau des papavéracées. C. r. Acad. Sci. Paris **215**, 67—69 (1942).

Kaden, O. F.: Untersuchungen des Gerbstoffgehaltes der Kakaobohnen. Versuch einer chemischen Zuchtwahl der Kakaobäume. Tropenpflanzer **42**, 409—418 (1939).

Kajitch, M.: Dévelopment et conditions oecologiques de la culture du tabac Yougoslave; mesures pour l'améliorer. Tabac, Rome **2**, Nr. 4, 43—53; also in German Pp. 54—62 (1939).

Kasparjan, A. S.: Tetraploidy in tea (Thea sinensis L.). C. R. (Doklady) Acad. Sci. URSS **27**, 1017—1019 (1940).

Kausche, G. A.: Über Transplantations- und Kreuzungsversuche zur Frage der natürlichen und erworbenen Infektreaktion bei virusinfizierten Tabakpflanzen. Naturwissenschaften **29**, 404—405 (1941).

Kaznowski, L.: Problèmes fondamentaux de la sélection du tabac. Tabac, Rome **2**, Nr. 2, 81—83 (1939).

Kendall, M. G.: The advanced theory of statistics. Charles Griffin and Co. Ltd., London, **1**, 42s, Pp. XII+457, 16 illus., 79 tables. (1943).

— The advanced theory of statistics. Charles Griffin and Co., Ltd., London, 2nd Ed., Pp. VII+521, 31 illus., 52 tables. (1946).

Kern, E. M. u. C. Alper: Multi-dimensional graphical representation for analysing variation in quantitative characters. Ann. Mo. Bot. Gdn. **32**, 279—281 (1945).

Keyworth, W. G.: Hop diseases in Great Britain. Wallerstein Laboratories Communications **8**, 99—109 (1945).

Kightlinger, C. V.: Black root rot resistant strains of Havana seed tobacco for the Connecticut Valley. Bull. Mass. Agric. Exp. Sta., Nr. 432, 20 (1946).

Kisfaludy, M. v.: Die in Ungarn angebauten wichtigsten Tabakgattungen. Acta Nicotiana, Berlin 1, 220—227 (1939).

Klinkowski, M.: Die Wanderungswege des Kaffeebaumes. Ein Beitrag zur Wanderungsgeschichte kolonialer Nutzpflanzen. Züchter 17/18, 247—255 (1947).

Knapp, A. W. u. J. F. Hearne: The presence of leucoanthocyanins in Criollo cacao. Analyst 64, 475—480 (1939).

Koenig, G.: Die Entwicklung der Reichsanstalt für Tabakforschung in Forchheim in zwölfjähriger Tätigkeit (1927—1938). Landw. Jb. 89, 651—668 (1939).

Koenig, P. u. L. Rave: Fortschrittsbericht über Anbau und Züchtung des Tabaks. Chronica Nicotiana, Bremen 3, Nr. 2, 21—27 (1942).

Kosmodemjanskij, V. N.: (Transgression in different generations of hybrids of Nicotiana Tabacum.) Vsesojuznyj Naučno-Issledovateljskij Institut Tabačnoj i Mahoročnoj Promyšlennosti imeni A. I. Mikojana (VITIM). (All-Union Mikojan Research Institute of the Tobacco and Makhorka Industry.) Krasnodar, Nr. 143, 3—38 (1941). [Russisch.]

Kostoff, D.: Abnormal mitosis in tobacco plants forming hereditary tumours. Nature, London 144, 599 (1939).

— Indicii citogenetice pentru aplicarea hibridizării interspecifice în scopul creiării de forme dorite de tutun. (Cytogenetic indices for applying interspecific hybridization in breeding desirable tobacco forms.) Bul. Cultiv. Ferment. Tutun 28, 165—176 (1939).

— Nicotine and citric acid content in the progeny of the allopolyploid hybrid Nicotiana rustica L. × N. glauca Grah. Curr. Sci. 8, 59—62; also C. R. (Doklady) Acad. Sci. URSS. 22, 121—123 (1939).

— The origin of the tetraploid Nicotiana from Bathurst. Curr. Sci. 8, 110—112 (1939).

— Lethality of gametes conditioned by exchange of segments between partially homologous chromosomes in a Nicotiana species hybrid. Curr. Sci. 8, 260 (1939).

— Cytogenetic determination of a new Australian tobacco species and its mode of origin. C. R. (Doklady) Acad. Sci. URSS. 22, 267—269 (1939).

— Lethality of gametes as dependent on the exchange of segments between partly homologous chromosomes in a Nicotiana species hybrid. C. R. (Doklady) Acad. Sci. URSS. 24, 372—373 (1939).

— Some cytogenetic indices for applying interspecific hybridization in breeding desirable tobacco forms. Tabac, Rome 2, Nr. 4, 31—42 (1939).

— Nicotine and citric acid content in the progeny of the allopolyploid hybrid N. rustica L. × N. glauca Grah. Tabac, Rome 2, Nr. 4, 73—78 (1939).

— Some cytogenetic indices for applying interspecific hybridization in breeding desirable tobacco forms. Acta Nicotiana, Berlin 1, 251—256 (1939).

— Relation degrees in phylesis of certain Nicotiana species determined by cytogenetic analysis. Genetica 22, 215—230 (1940).

— Some cytogenetic indices for applying interspecific hybridization in breeding desirable tobacco forms. Acta Nicotiana 1, 251—256 (1940).

— Hybridization in Nicotiana before rediscovery of Mendel's laws. Tabac, Rome 4—5, Nos. 2/4, 26—34 (1942).

— Tabakpflanzen mit gefüllten Blüten durch Artkreuzungen erhalten. Züchter 14, Nr. 1 (1942).

— Cytogenetics of the genus Nicotiana. Karyosystematics, genetics, cytology, cytogenetics and phylesis of tabaccos. States Printing House, Sofia 1941—1943, Pp. XXVIII + 1071, 345 figs. tables. (1943).

— (Resistance to tobacco mosaic virus. I. Resistant tobacco varieties mosaic virus experimentally produced.) Centralena Zemedelski Izsledovatelski i Kontrolena Instituta, Sofija. (Cent. Agric. Res. Control Inst.) Sofia, Pp. 3—12, 35—45 (1944). [Bulgarisch.]

Kostoff, D. u. R. Georgieva: (Resistance to tobacco mosaic virus II. Inheritance of necrotic reaction and plant breeding value of the form Nicotiana tabacum var. virii.) Centralen Zemedelski Izsledovatelski i Kontrolena Instituta, Sofija. (Cent. Agric. Res. Control Inst.) Sofia, Pp. 13—34, 46—54 (1944). [Bulgarisch.]

—, A. Gorbačeva u. P. Dimitroff: Die Vergrößerung der Zellen in auto- und allopolyploiden Tabakpflanzen. Z. Pflanzenz. 25, 112—116 (1943).

— u. M. Sarana: Heritable variations in Nicotiana Tabacum L. induced by abnormal temperatures, and their evolutionary significance. J. Genet. 37, 499—547 (1939).

Kotte, W.: Die durch Tylenchus dipsaci Kühn verursachte „Umfällerkrankheit" des Tabaks. Z. Pfl. Krankh. 53, 37—42 (1943).

Krug, C. A.: A importância da genética e da citologia para o melhoramento do cafeeiro. (The importance of genetics and cytology for breeding coffee.) Bol. Superintend. Serv. Café, São Paulo 19, 746—751 (1944).

— Melhoramento do cafeeiro. (Improvement of coffee.) Bol. Superintend. Serv. Café, São Paulo 20, 979—992 (1945).

— Melhoramento do cafeeiro. (Improvement of coffee.) Bol. Superintend. Serv. Café, São Paulo 20, 1038 bis 1046 (1945).

— Un clone tétraploïde du Coffea excelsa à haut rendement. Rev. Bot. Appl. 26, 578 (1946).

— u. A. Carvalho: Genetical proof of the existence of coffee endosperm. Nature, London 114, 515 (1939).

—, — Prova genética da existência de endosperma na semente de café. (Genetical proof of the existence of endosperm in the coffee seed.) J. Agron, S. Paulo 2, 381—384 (1939).

—, — Genética de Coffea. V - Hereditariedade da coloração bronzeada das folhas de Coffea arabica L. (Genetics of Coffea. V - Inheritance of the bronze coloration in the young leaves of Coffea arabica L.) Bragantia, São Paulo 2, 199—220 (1942).

—, — Genética de Coffea. VI - Independência dos fatores xc xc (xanthocarpa) e Br Br (Bronze) em Coffea arabica L. (Genetics of Coffea. VI — Independence of the factors xc xc (xanthocarpa) and Br Br (bronze) in Coffea arabica L.) Bragantia, São Paulo 2, 221—230 (1942).

—, — Genética de Coffea. VII - Hereditariedade dor caracteres de Coffea arabica L. var. maragogipe Hort ex Froehner. Genetics of Coffea. VII - Inheritance of the characters of Coffea arabica L. var. maragogipe Hort ex Froehner.) Bragantia, São Paulo 2, 231—247 (1942).

—, — Genética de Coffea. VIII - Hereditariedade dos caractereres de C. arabica L. var. anomala K. M. C. (Genetics of Coffea. VIII - The mode of inheritance of the characters of C. arabica L. var. anomala K. M. C.) Bragantia, São Paulo 5, 781—791 (1945).

— u. A. J. T. Mendes: Cytological observations in Coffea. IV. J. Genet. 39, 189—203 (1940).

—, — Observações citológicas em Coffea. IV. (Cytological observations in Coffea. IV.) Bragantia, São Paulo 1, 467—482 (1941).

—, — Conhecimentos gerais sôbre a genética e a citologia do gênero Coffea. (General state of knowledge on the genetics and cytology of the genus Coffea.) Rev. Agric. Piracicaba 18, 399—408 (1943).

—, J. E. T. Mendes u. A. Carvalho: Taxonomia de Coffea arabica L. Descrição das variedades e formas encontradas no Estado de São Paulo. (Taxonomy of C. arabica L. Description of the varieties and forms occurring in the state of São Paulo. Bol. Téc. Inst. Agron., Campinas, Nr. 62, 57 (1939).

Kunkel, L. O.: Genetics of viruses pathogenic to plants. Publ. Amer. Ass. Advanc. Sci., Nr. 12, 22—37 (1940).

Kuzmenko, A. A. u. V. D. Tihvinskaja: (Inheritance of nicotine and anabasine content by Nicotiana tabacum x N. glauca hybrids and interaction of stock and scion when these species are grafted.) Bull. Acad. Sci. URSS. Sér. Biol., Nr. 4, 564—576 (1940). [Russisch.]

Lamb, J.: A note on the manufacture of leaf from selected individual bushes. Tea Quart. 12, 183—185 (1939).

Lambers, M. Hille Ris: Indrukken van de koffiecultuur in Oosten Midden-Afrika. (Impressions of coffee cultivation in East and Central Africa.) Bergcultures 13, 1800—1810 (1939).

— Impressions of the coffee-growing in East and Central Africa. I, II. E. Afr. Agric. J. 6, 32—33, 74—76 (1940).

— Koffieselectie en koffie-plantmateriaal. (Coffee selection and coffee planting material.) Bergcultures 15, 1522—1532 (1941).

Lang, A.: Beiträge zur Genetik des Photoperiodismus. I. Faktorenanalyse des Kurztagcharakters von Nicotiana tabacum „Maryland-Mammut". Z. indukt. Abstamm.- u. VererbLehre 80, 210—219 (1942).

Larroque, P.: De l'utilisation de la statistique mathématique pour la sélection rapide des plantes (applications à la sélection des maïs, des ricins et des abrasins). Inst. Rech. Agron. For., Gouv.Gén. Indochine (1939).

Lebrun, J.: Recherches morphologiques et systématiques sur les caféiers du Congo. Publ. Inst. Nat. Agron. Congo Belge, Pp. 183 (1941).

Lefèvre, J. u. R. Ferrary: Observations sur les caractères de quelques nouvelles espèces amphidiploïdes dans le genre Nicotiana. C. R. Acad. Agric. Fr. 32, 109—112 (1946).

Leliveld, J. A.: Vruchtzetting bij koffie. (Fruit setting in coffee.) Arch. Koffiecult. Ned.-Ind. 1938, 12, 127 bis 161. (From. Landbouwk. Tijdschr., Wageningen 51, 495) (1938) 1939. (Abstr.)

Ligt, N. M. de: Conuga-koffie. (Conuga coffee.) Bergcultures 13, 726—732 (1939).

Lima, A. R., F. G. Brieger u. S. R. dos Santos: Seleção de fumo „Amarelinho" para estufa. (Selection of the flue — cured tobacco Amarelinho.) Bragantia, São Paulo 4, 523—540 (1944).

— u. A. S. Costa: Variedades de fumo resistentes a „Viracabeça". (Varieties of tabacco resistant to wilt.) Rev. Agric. S. Paulo 15, 133—140 (1940).

— u. R. Forster: Uma forma gigante em Nicotiana tabacum L. (A gigant form in N. Tabacum L.) Bragantia, São Paulo 2, 111—113 (1942).

Loeff, C.: Tabaksselectie in de Vorstenlanden. (Tobacco breeding in the Vorstenlanden.) Erfelijkheid in Praktijk, Leiden 4, 160 (1939).

McKinney, H. H.: Studies on genotypes of tobacco resistant to the common-mosaic virus. Phytopathology 33, 300—313 (1943).

— Reaction of resistant tobaccos to certain strains of Nicotiana virus 1 and other viruses. Phytopathology 33, 551—568 (1943).

— u. E. E. Clayton: Genotype and temperature in relation to symptoms caused in Nicotiana by the mosaic virus. J. Hered. 36, 323—331 (1945).

McLean, R. A.: Observations on Cerospora leaf spot of tobacco and the question of varietal resistance. Phytopathology 33, 354—362 (1943).

McMaster, P. G. W.: An impression of Lyamungo Coffee Research Station. Mon. Bull. Coffee Bd. Kenya 11, 57—58 (1946).

McMurtrey, J. E. jr, C. W. Bacon u. D. Ready: Growing tobacco as a source of nicotine. Tech. Bull. U. S. Dep. Agric., Nr. 820, 39 (1942).

MacRae, N. A.: Breeding for mosaic resistance. Lighter, Ottawa 11, Nr. 4, 9—11 (1941). [Mimeographed].

— Genetic analysis of Nicotiana triplex segregation products and their relationship to existing N. tabacum types. Tech. Bull. Dep. Agric. Can., Nr. 33, 36 (1941).

Magie, R. O.: The epidemiology and contral of downy mildew on hops. Tech. Bull., N. Y. St. Agric. Exp. Sta., Nr. 267, 48 (1942).

— Disease resistance in New York hops urgently needed. Fm. Res. 10, Nr. 3; 5, 6 (1944).

Mallah, G. S.: Inheritance in Nicotiana Tabacum. XVI. Structural differences among the chromosomes of a selected group of varieties. Genetics 28, 525—532 (1943).

Marjanovič, O.: (On the cytology and embryloogy of the hybrid F_t Nicotiana rustica [variety Unterwalden] x Nicotiana quadrivalvis.) J. Inst. Bot. Acad. Sci. Ukraine, Nr. 21/22 (29/30), 187—195 (1939). [Russisch.]

Markwood, L. N.: Nornicotine as the predominating alkaloid in certain tobaccos. Science 92, 204—205 (1940).

Marsh-Smith, E. C.: Vegetative reproduction. Tea Quart. 18, 107—111 (1947).

Matthews, E. M. u. R. G. Henderson: Yellow Special tobacco, a new flue-cured variety resistant to black root-rot. Bull. Va Polyt. Inst. Agric. Exp. Sta., Nr. 346, 7 (1943).

Mayne, W. W.: Annual report of the coffee scientific officer, 1938—39, 1939—40, 1940—41. Bull. Mysore Coffee Exp. Sta., Nr. 19, 16, Nr. 21, 21, Nr. 23, 17 (1941).

— Annual Report of the Coffee Scientific Officer, 1942 bis 1943. Bull. Mysore Coffee Exp. Sta., Nr. 25, 19 (1943).

— Report of the U.P.A.S.I. Coffee Scientific Officer, 1943—1944. Bull. Mysore Coffee Exp. Sta., Nr. 26, 14 (1944).

— Report of the U.P.A.S.I. Coffee Scientific Officer, 1944—1945. Bull. Mysore Coffee Exp. Sta., Nr. 27, 15 (1944).

Mendes, A. J. T.: Duplicação do número de cromosômios em café, algodão e fumo, pela ação da colchicina. (Duplication of the chromosome number in coffee, cotton and tobacco by the action of colchicine.) Bol. Téc. Sec. Agric. Indústr. Com. S. Paulo, Nr. 57, 21 (1939).

— Observações citológicas em Coffea. VI - Desenvolvimento do embrião e do endosperma em Coffea arabica L. (Cytological observations in Coffea. VI - Development of embryo and endosperm in C. arabica L.) Bragantia, São Paulo 2, 115—128 (1942).

— Observações citológicas em Coffea. VIII. Poliembrionia. (Cytological observations in Coffea. VIII. Polyembryony.) Bragantia, São Paulo 4, 693—708 (1944).

— Sementes de café poliembriônicas e desprovidas de embrião. (Polyembryonic seeds and seeds without embryo in coffee.) Bol. Sperintend. Serv. Café, São Paulo 19, 618—620 (1944).

— Citologia de Coffea. (Cytology of Coffea.) Rev. Agric., Piracicaba 20, 412—415 (1945).

— u. O. Bacchi: Observações citológicas em Coffea. V. Uma variedade haploide („dihaploide") de C. arabica L. (Cytological observations on Coffea. V. A. haploid variety („di-haploid") of C. arabica L.) J. Agron., S. Paulo 3, 183—206; also Bol. Téc. Inst. Agron. Estado, Campinas, Nr. 77, 26 (1940).

Mendes, J. E. T.: Ensaio de variedades de cafeeiros. (Coffee variety test.) Bol. Téc. Inst. Agron. Campinas, Nr. 65, 36 (1939).

Mendes, J. E. T., F. G. Brieger, C. A. Krug u. A. Carvalho: Melhoramento de Coffea arabica L. var. Bourbon. (Improvement of C. arabica L. var. Bourbon.) Bragantia, São Paulo 1, 1—76, also Rev. Inst. Café S. Paulo 1941, 28, 1167—1176 (1941).

—, C. A. Krug u. J. Bergamin: Relatório de uma viagem de estudos sôbre a lavoura cafeeira nos estados do Rio de Janeiro e Espírito Santo. Bol. Superintend. Serv. Café, São Paulo 20, 1025—1034, 1094—1104 (1945).

Middelburg, H. A.: Proefstation voor Vorstenlandsche Tabak. Jaarverslag 1938—1939. (Experiment Station for Vorstenland Tobacco. Annual report 1938 bis 1939.) Meded. Proefst. Vorstenl. Tab. Klaten (Java), Nr. 88, 43 (1940).

Miny, M. P.: La culture du cacaoyer au Congo belge. Situation actuelle. — Perspectives d'avenir. Bull. Agric. Congo Belge 33, 385—444 (1942).

Miranda, S.: Problemas da produção de cacau fino na Bahia. (Problems surrounding the production of high quality cacao in Bahia.) Bol. Minist. Agric., Rio d. J. 33, Nr. 1, 61—69 (1944).

— u. P. Silva: Mutações em Theobroma leiocarpa Bern. var. Comum. (Mutations in T. leiocarpa Bern. var. Comum.) Bahia Rur. Oct.-Nov.; also Trop. Agriculture, Trin. 17, 139 (1939) 1940.

Modilewskij, J.: (Cytogenetical investigation of the genus Nicotiana. VII. Crossing amphidiploid Nicotiana disualovii with some species of the genus Nicotiana.) J. Inst. Bot. Acad. Sci. Ukraine, Nr. 21/22 (29/30), 107—137 (1939). [Russisch.]

— (Cytogenetic investigation of the genus Nicotiana. IX. Cytology and embryology of the haploid N. rustica.) J. Inst. Bot. Acad. Sci. Ukraine, Nr. 21/22 (29 bis 30, 201—215 (1939). [Russisch.]

— (Cytogenetic investigations on the genus Nicotiana. X. Cytology and embryology of the amphidiploid Nicotiana Dubek 44/39 x N. silvestris.) J. Bot. Acad. Sci. Ukraine 1, 189—213 (1940). [Russisch.]

— u. L. Dzubenko: (Cytogenetical investigation of the genus Nicotiana. VIII. Embryology and cytology of the amphidiploid N. rustica var. pumila x N. paniculata.) J. Inst. Bot. Acad. Sci. Ukraine, Nr 23 (31), 3—12 (1940). [Russisch.]

Moffett, A. A.: The cytological basis of variation in varieties of Nicotiana Tabacum. Trans. Roy. Soc. S. Afr. 30, 235—243 (1945).

Monnier, P.: Une nouvelle maladie à virus du cacaoyer en Afrique occidentale: le swollen shoot. Rev. Bot. Appl. 26, 166—173 (1946).

Moss, E. G. u. J. F. Bullock: Two new varieties of flue-cured tobacco, 400 and 401. Bull. N. C. Agric. Exp. Sta., Nr. 337, 8 (1942).

Mothes, K. u. K. Hielke: Die Tabakwurzel als Bildungsstätte des Nikotins. Naturwissenschaften 31, 17—18 (1943).

Muir, R. M.: Growth hormones as related to the setting and development of fruit in Nicotiana tabacum. Amer. J. Bot. 29, 716—720 (1942).

Müntzing, A.: Some observations on pollination and fruit setting in Ecuadorian cacao. Hereditas, Lund 33, 397—404 (1947).

Mutovkina, T. D.: (Elimination and selection of tea seedlings.) Soviet Subtropics, Nr. 5 (57), 45—49 (1939). [Russisch.]

Myer, S.: New varieties of hops. Brew. Tr. Rev. 59, 57—58 (1945).

Nagel, L.: Morphogenetic differences between Nicotiana alata and N. langsdorffii as indicated by their response to indoleacetic acid. Ann. Mo. Bot. Gdn. 26, 349—372 (1939).

Narasimha Swamy, R. L.: Genetical studies in Coffea arabica L. A preliminary study with young leaf colour and ripe pericarp colour. Indian J. Agric. Sci. 10, 414—421 (1940).

Narasimha Swamy, R. L.: Brief note on somatic variation in „Kents" strain of Coffea arabica L. Curr. Sci. 15, 80—81 (1946).

Negodi, G.: Contributo alla cariologia delle Papaveracee, subfam. Fumarioideae, con particolare riguardo ai generi Dicentra, Corydalis, Cisticapnos ed Adlumia. Sci. Genetica 2, 1—25 (1940).

Nelson, N. T.: Tobacco research in Canada. Emp. J. Exp. Agric. 9, 265—276 (1941).

Nikollov, J.: (On the botanical composition of the poppy in the new regions [of Bulgaria].) Rev. Inst. Rech. Agron. Bulg. 9, Nr. 1, 63—73 (1939). [Bulgarisch.]

Noguti, Y., H. Oka u. T. Otuka: Studies on the polyploidy in Nicotiana induced by the treatment with colchicine. II. Growth rate and chemical analysis of diploid and its autotetraploid in Nicotiana rustica and N. Tabacum. Jap. J. Bot. 10, 343—364 (1940).

—, K. Okuma u. H. Oka: Studies on the polyploidy in Nicotiana induced by the treatment with colchicine. I. General observations on the autotetraploid of Nicotiana rustica and N. Tabacum. Jap. J. Bot. 10, 309—319 (1939).

Nolla, J. A. B.: Second Annual Report of the Tobacco Institute of Puerto Rico. San Juan, p. 36 (1939).

— Studies on disease resistance. II. Development of tobacco strains resistant to the ordinary tobacco mosaic. Acta Nicotiana, Berlin 1, 408—413 (1939).

Norris, R. V.: The work of the Tea Research Institute. Tea Quart. 12, 5—15 (1939).

Nott, J.: Report on new (Wye) varieties of hops-season 1944. Worcs. Agric. Quart. Chron. 12, 259—263 (1944).

— Trial of new varieties of hops-1944 season. Final report. Worcs. Agric. Quart. Chron. 13, 115, 117 (1945).

Novikov, V.: Essais d'amélioration de la combustibilité des tabacs tunisiens. Ann. Serv. Bot. Tunis 18, 211—254 (1941).

Ochoa, H. u. E. Chavarriaga: Apuntes sobre el cultivo del cacao. (Notes on cacao cultivation.) Rev. Fac. Nac. Agron, Colombia 2, 442—478 (1940).

Oka, T.: (Interspecific hybridization in Nicotiana. Breeding of nicotine-free tobacco by hybridization - N. Tabacum x N. glauca.) Jap. J. Genet. 16, 87—88 (1940).

Ono, T.: (Sex behaviour of triploid interesexes in Humulus japonicus). Bot. Mag. Tokyo 55, 94—102 (1941).

Ostendorf, F. W.: De bloeislaging van de cacao. (Fruit setting in cacao flowers.) Bergcultures 13, 1539—1544 (1939).

Pacheco Herrarte, M.: El mejoramiento del cafeto. (Improvement of coffee.) Rev. Agric., Guatemala 16, 83—84 (1939).

Pal, B. P.: A brief note on the seed supply of Virginia tobacco. Agric. Live-Stk. India 9, 42—43 (1939).

— u. B. V. Nath: The accumulation and movement of nicotine in reciprocal grafts between tobacco and tomato plants. Proc. Indian Acad. Sci. 20, Sect. B, 79—87 (1944).

—, S. Ramanujam u. Narayana Rao T.: An improved variety of flue-cured tobacco. Indian Fmg. 6, 154—155 (1945).

Patel, B. S.: What's doing in All-India-Bombay. Indian Fmg. 3, 601 (1942).

Perera, P. R.: A non-fermenting type of the tea plant, Camellia Thea, Link. Curr. Sci. 10, 485 (1941).

Perucci, E.: La production industrielle de la semence de tabac de première génération. Tabac, Rome 2, Nr. 3, 71—75 (1939).

Perucci, E.: Una varietà italiana di tabacco di grande reddito Kentucky „Perucci". (A high yielding Italian tobacco Kentucky Perucci.) Tabacco 51, Nr. 750, 9—18 (1947).

Petrenko, A. G.: (Physiologico-biochemical characteristics of some peculiarities in the process of starvation of different botanical varieties of tobacco.) Zbornik Rabot po Posleurožajnoj Obrabotke Tabaka. Edited by M. F. Maškovtsev. Vsesojuznyji N. I. Institut Tabačnoj i Mahoročnoj Promyšlennosti imeni A. I. Mikojana (VITIM) (All-Union Mikojan Research Institute of the Tobacco and Makhorka Industry). Krasnodar, Nr. 142, 3—82 (1940). [Russisch.]

Pieper, H.: Vergleichende Untersuchungen an Varietäten des Kulturmohns (Papaver somniferum L.). (Ergebnisse der Deutschen Hindukusch-Expedition III.) Landw. Jb. 89, 333 (1939).

Piescu, A. u. N. Neagu: Hibrizi Nicotiana Tabacum × N. sylvestris. (Hybrids of N. Tabacum × N. silvestris.) Bul. Cultiv. Ferment. Tutun. 29, 259—278 (1940).

Poddubnaja-Arnoljdi, V. A. u. M. M. Lodkina: (Embryogenesis in hybridization of distant species and polyploidy in the genus Nicotiana.) Botaničeskij Žurnal (J. Bot. URSS.) 30, 195—216 (1945). [Russisch.]

Poel, J. van der: Kort overzicht van het slijmziektevraagstuk bij de Deli-tabak. (Short review of the slime disease problem in Deli-tobacco.) Meded. Deli-Proefst. Ser. 3, Nr. 2, 5—16 (1939).

Portères, R.: Étude sur les caféiers spontanés de la section „des Eucoffeae". Leur répartition, leur habitat, leur mise en culture et leur sélection en Côte d'Ivoire. Ann-Agric. Afr. Franc. Etrangère, T. 1, Nr. 1, also Portuguese translation, Rev. Dep. Nac. Café, Rio de J. 12, Ano 6, 193—210, 299—306, 451—473, 625—635, 12, Ano 7, 8—26 (1938 bis 1939).

— Le choix des variétés et l'amélioration des caféiers en Côte d'Ivoire. Rev. Bot. Appl. 19, 18—29 (1939).

Posnette, A. F.: Self-incompatibility in cocoa (Theobroma spp.). Trop. Agriculture, Trin. 17, 67—71 (1940).

— Natural pollination of cocoa, Theobroma leiocarpa, on the Gold Coast. Trop. Agriculture, Trin. 19, 12—16 (1942).

— Cacao selection on the Gold Coast. Trop. Agriculture, Trin. 20, 149—155 (1943).

— Virus diseases of cacao in Trinidad. Trop. Agriculture, Trin. 21, 105—106 (1944).

— Pollination of cacao in Trinidad. Trop. Agriculture, Trin. 21, 115—118 (1944).

— Incompatibility in Amazon cacao. Trop. Agriculture, Trin. 22, 184—187 (1945).

— Inter-specific pollination in Theobroma. Trop. Agriculture, Trin. 22, 188—190 (1945).

— u. M. Palma: Observations on cacao on the Paria Peninsula, Venezuela. Trop. Agriculture, Trin. 21, 130—132 (1944).

Pound, F. J.: The present position of fertilisers and cocoa. Lecture to the Agricultural Society of Trinidad and Tobago. Proc. Agric. Soc. Trin. Tob. 39, 98—105 (1939).

— Search for resistance to witchbroom in cocoa. Proc. Agric. Soc. Trin. Tob. 40, 35—37 (1940).

— The quest for witches' broom resistant trees. Proc. Agric. Soc. Trin. Tob. 43, 55—63 (1943).

Prakken, R.: A spontaneous haploid of Nicotiana Tabacum. Genetica 23, 63—76 (1942).

Prasad, J.: A short note on the beedi leaf industry. Indian For. Leafl, Nr. 60, 12 (1943).

Pratt, A. M.: Coffee team services. Progress report on the cultural work at the Scott Agricultural Laboratories. Mon. Bull. Coffee Bd. Kenya 6, 73—76 (1940).

Psareva, E. N.: (System and methods for studying tobacco varieties [collections].) Vsesojuznyj Naučno-Issledovateljskij Institut Tabačnoj i Mahoročnoj Promyšlennosti imeni A. I. Mikojana (VITIM). (All-Union Mikojan Research Institute of the Tobacco and Makhorka Industry) Krasnodar, Nr. 143, 38—86 (1941). [Russisch.]

— (Instructions for describing, observing and assessing different morphological characters in Nicotiana Tabacum.) Vsesojuznyj Naučno-Issledovateljskij Institut Tabačnoj i Mahoročnoj Promyšlennosti imeni A. I. Mikojana (VITIM). (All-Union Mikojan Research Institute of the Tobacco and Makhorka Industry), Krasnodar, Nr. 143, 87—98 (1941). [Russisch.]

Psareva, M. M.: (Results of research on poppy breeding.) Selekcija i Semenovodstvo, Nr. 1, 16—19 (1939). [Russisch.]

Rabak, F.: The effect of seeds on the quality of hops. Wallerstein Laboratories Communications 6, 160—166 (1943).

Raczyński, F.: Kilka obserwacji nad nowymi odmianami: krzyżówka Kentucky × Szamoszaty oraz machorka Kómarno. (Some observations on the new varieties Kentucky × Szamoszaty hybrid and Komarno.) Przegl. Uprawy Tytoniu 6, 44—45 (1939).

Rădulescu, M.: Studiul morfologic și biometric al semintelor de „Nicotiana Tabacum" var. Virginia Bright. (Morphological and biometric study of the seeds of N. Tabacum var. Virginia Bright.) Bul. Cultiv. Ferment. Tutun. 30, 107—112 (1941).

Raghavan, T. S. u. A. R. Srinivasan: Cytogenetical studies in Nicotiana. Part I. Cytology of Nicotiana glutinosa and N. Tabacum var. macrophylla and the F_1 hybrid between them. J. Indian Bot. Soc. 20, 307—340 (1941).

—, — Cytogenetical studies in Nicotiana. Part II. Morphological features of Nicotiana glutinosa and the hybrid between Nicotiana glutinosa and N. tabacum. Proc. Indian Acad Sci. 14, Sect. B, 35—46 (1941).

Ramanujam, S. u. A. B. Joshi: Interspecific hybridization in Nicotiana. A cytogenetical study of the hybrid N. glauca Grah. × N. plumbaginifolia Viv. Indian J. Genet. Pl. Breed. 2, 80—97 (1942).

Rapin, J.: Plantes oléagineuses indigènes. Rev. Hort. Suisse 14, 40—43, 49—53 (1941).

— Le rôle de la culture du tabac dans la vallée de la Broye dans l'intensification de la production agricole. Ber. Schweiz. Bot. Ges. 53 A, 116—123 (1943).

Rave, L.: Die Deutschen Tabaksorten. Acta Nicotiana, Berlin 1, 233—238 (1939).

— Die wichtigsten Varietäten der Tabakpflanze. Acta Nicotiana, Berlin 1, 228—232 (1939).

— Compte rendu des progrès accomplis dans la culture et l'amélioration du tabac. Chronica Nicotiana, Bremen 2, Nr. 2, 10—19 (1941).

Renskij, M. D.: (Nicotine in the stem of Nicotiana rustica.) Vsesojuznyj Naučno-Issledovateljskij Institut Tabačnoj i Mahoročnoj Promyšlennosti imeni A. I. Mikojana (VITIM). (All-Union Mikojan Research Institute of the Tobacco and Makhorka Industry), Krasnodar, Nr. 143, 234—240 (1941). [Russisch.]

Roelofsen, P. A.: Beïnvloeding en behoud van de kwaliteit van robusta-marktkoffie. (Influencing and conserving the quality of Robusta market coffee.) Bergcultures 13, 812—817 (1939).

— Verslag van het Deli Proefstation over het jaar 1939. (Report of the Deli Experiment Station for the year 1939.) Meded. Deli-Proefst. 3e Serie, Nr. 7, 75 (1940).

— Recent research at the Deli Tobacco Experiment Station, Medan, Sumatra. Emp. J. Exp. Agric. 11, 15—22 (1943).

Rogozinski, A.: Nowa odmiana typu Kentucky. (A new variety of the Kentucky type.) Przegl. Uprawy Tytoniu **6**, 40—42 (1939).

Rossi, U.: Die Äthylenisierung des Tabaks. Auswirkungen von Äthylengas auf einige Sorten von Rauchtabak. Tabac, Rome **4/5**, Nr. 2/4, 57—85 (1942).

Ryan, C. E. V.: Some notes on the selection of highyielders on Doombagastalawa Estate, Kotmale. Tea Quart. **16**, 45—51 (1943).

S....., R. L.: Notes on a visit to some Java Experimental Stations. Mon. Bull. Coffee Bd. Kenya **5**, 44—45 (1939).

Sadik, H. G. u. Mohammad Noor: A promising type of Hukka tobacco (Jullundur Selection Nr. 5). Seas, Notes Punjab Agric. Dep. **18**, Nr. 2, 45—46 (1939).

Salmon, E. S.: Hops. J. S.-E. Agric. Coll. Wye, Nr. 43, 24—26 (1939).

— On a tricussate arrangement of the leaves and flowering branches in certain seedlings of the hop. J. S.-E. Agric. Coll. Wye, Nr. 44, 9—11 (1939).

— Twenty-third report on the trial of new varieties of hops, 1939. J. Inst. Brew. **46**, 367—375 (1940).

— Trial of new varieties of hops, 1940. 24th Rep. E. Malling Res. Sta. 1941, Pp. 21 (1941).

— Twenty-fourth report on the trial of new varieties of hops. J. Inst. Brew. **47**, 371—379 (1941).

— Twenty-fifth report on the trial of new varieties of hops, 1941. E. Mall. Res. Sta., Kent, Pp. 22 (1942).

— Trials of new varieties of hops. Worcs. Agric. Quart. Chron. **11**, 111—113 (1943).

— Twenty-fifth report on the trial of new varieties of hops, 1941. J. Inst. Brew. **49**, 9—19 (1943).

— Twenty-sixth report on the trial of new varieties of hops, 1942. J. Inst. Brew. **50**, 28—36 (1944).

— Four seedings of the Canterbury Golding. J. Inst. Brew. **50**, 244—250 (1944).

— Twenty-seventh report on the trial of new varieties of hops, 1943. J. Inst. Brew. **50**, 270—279 (1944).

— Twenty-eighth report on the trial of new varieties of hops, 1943. J. Inst. Brew. **51**, 224—231 (1945)

— Twenty-ninth report on the trial of new varieties of hops, 1945. J. Inst. Brew. **52**, 294—301 (1946).

—, F. H. Beard u. R. G. Hatton: The merits of the new varieties of hops. Ann. Rep. E. Malling Res. Sta. 1942, Pp. 108—112 (1943).

—, —, — The merits of the new varieties of hops. J. Inst. Brew. **49**, 29—33 (1943).

— u. A. H. Burgess: Reports received from brewers on recent brewing trials with certain new varieties of hops: II. J. Inst. Brew. **53**, 100—110 (1947).

Samarina, A. P. u. M. V. Kolelišvili: (The anatomical method in tea breeding.) Soviet Subtropics, Nr. 9 (72), 30—33 (1940). [Russisch.]

Sather, J. D. u. D. D. Hill: Some problems in measuring certain hop qualities. Wallerstein Laboratories Communications **7**, 87—100 (1944).

Savelli, R. u. C. Caruso: Stimulation mutuelle dans la germination des grains de pollen de Nicotiana. (Mutual stimulation in the germination of the pollen grains of Nicotiana.) C. R. Acad. Sci. Paris **210**, 184 bis 186 (1940).

Schultes, R. E.: Plantae Colombianae IV. Una planta estimulante del Putumayo. (Colombian plants IV. A stimulant from Putumayo.) Rev. Fac. Nac., Agron., Colombia **5**, 59—79 (1943).

Schweizer, J.: Besoekisch Proefstation (Proefstation voor rubber, koffie en tabak). Jaarverslag tabak over Juli 1938 t/m Juni 1939. (Besoeki Experiment Station (Experiment Station for rubber, coffee and tobacco.) Annual report on tobacco from July 1938 to June 1939.) Meded. Besoek. Proefst., Nr. 64, 64 (1939).

Schweizer, J.: Besoekisch Proefstation (Proefstation voor rubber, koffie en tabak). Jaarverslag tabak over Juli 1939 t/m Juni 1940. (Besoeki Experiment Station [Experiment Station for rubber, coffee and tobacco]. Annual report on tobacco from July 1939 to June 1940.) Meded. Besoek. Proefst., Nr. 66, 1—30 (1940).

— Bijdrage tot de chemische kwaliteitsbepaling van Virginia-tabak in de tropen. (Contribution on the chemical determination of quality of Virginia tobacco in the tropics.) Meded. Besoek. Proefst., Nr. 66, 43—53 (1940).

Shmuck, A. A. siehe Šmuk.

Siljvestrov, V. G.: (Obtaining seeds of Nicotiana rustica on experimental plots.) Proc. Lenin Acad. Agric. Sci. USSR., Nr. 11, 7—10 (1941). [Russisch.]

Silva, P.: Instituto de Cacau da Bahia. Relatorio 1939. (Bahia Cacao Institute. Report for 1939.) Bahia, Pp. 76 (1939).

Simonet, M. u. A. Fardy: Compartement cytogénétique d'un hybride amphidiploïde fertile. Nicotiana tabacum L. var. pupurea Anast. (n = 24) × N. sylvestris Speg. et Comes (n = 12) obtenu après traitement à la colchicine. C. R. Acad. Sci. Paris **215**, 378—380 (1942).

Simura, T.: (Studies on the resistance to brown blight in tea plants.) Jap. J. Genet. **14**, 243—247 (1939).

— (Further studies on the resistance to brown blight in tea plants.) Jap. J. Genet. **16**, 246—256 (1940).

— (On the frost resistance of tea plants.) Proc. Crop Sci. Soc. Japan **12**, 98—114 (1940).

Smith, H. H.: Induction of polyploidy in Nicotiana species and species hybrids by treatment with colchicine. Genetics **24**, 85—86 (1939). (Abstr.)

— The induction of polyploidy in Nicotiana species and species hybrids by treatment with colchicine. J. Hered. **30**, 291—306 (1939).

— Heteroploid types of Nicotiana resulting from colchicine treatment. Collecting Net **15**, 173 (1940). (Abstr.)

— II. Symposium on theoretical and practical aspects of polyploidy in crop plants. Discussion. Biol. Symp. **4**, 111—113 (1941).

— Effects of different proportions of specific Chromosomal complements on size in Nicotiana. Genetics **28**, 89—90 (1943). (Abstr.)

— Effects of genome balance, polyploidy, and single extra chromosomes on size in Nicotiana. Genetics **28**, 227—236 (1943).

— Induced heteroploids of Nicotiana. Amer. J. Bot. **30**, 121—129 (1943).

— u. C. W. Bacon: Increased size and nicotine production in selections from intraspecific hybrids of Nicotiana rustica. J. Agric. Res. **63**, 457—467 (1941).

— u. C. R. Smith: Alkaloids in certain species and interspecific hybrids of Nicotiana. J. Agric. Res. **65**, 347—359 (1942).

Smith, T. E.: Investigations of control measures for Granville wilt of tobacco. J. Elisha Mitchell Sci. Soc. **58**, 133 (1942). (Abstr.)

— Control of bacterial wilt (Bacterium solanacearum) of tobacco as influenced by crop rotation and chemical treatment of the soil. Circ. U. S. Dep. Agric., Nr. 692, 16 (1944).

— The problem in tobacco breeding work. Res. Fmg., N. C.: Progr. Rep., **5**, Nr. 1, 1—2 (1946).

—, E. E. Clayton u. E. G. Moss: Flue-cured tobacco resistant to bacterial (Granville) wilt. Circ. U. S. Dep. Agric., Nr. 727, 7 (1945).

Smits, C.: De verbouw van tabak in Nederland. (Tobacco growing in Holland.) Tijdschr. ned. Heidemaatsch. **53**, 144—158 (1941).

Šmuk, A. A. u. A. Borozdina: Alkaloids of various plant species within the genus Nicotiana. C. R. (Doklady) Acad. Sci. URSS. **32**, 62—65 (1941).

Snoep, W.: Productiegegevens van robusta-cloonen in het ressort van het Proefstation Midden-en Oost-Java. (Yield data of Robusta clones in the district of the Central and East Java Experiment Station.) Bergcultures 13, 652—661 (1939).
— Toepassing van takenten in de practijk. (Utilization of lateral scions in practice.) Bergcultures 14, 482 bis 492 (1940).
— Productiegegevens van koffie-cloonen in het ressort van het Proefstation Midden- en Oost-Java. (Yield data of coffee clones in the district of the Central and East Java Experiment Station.) Bergcultures 15, 399—409 (1941).
Solly, N.: A planter's impression of the work being conducted at the Scott Agricultural Laboratories by members of the coffee team. Mon. Bull. Coffee Bd. Kenya 6, 44—45 (1940).
Srinivasan, K. H. u. R. L. Narasimhaswamy: A review of coffee breeding work done at the Government Coffee Experiment Station, Balehonnur. Bull. Mysore Coffee Exp. Sta., Nr. 20, 16 (1940).
Stehlé, H.: La culture du café à la Martinique et son amélioration. Bull. Agric. Martinique 10, 98—139 (1941).
Stino, K. R.: Inheritance in Nicotiana Tabacum — XV. Carmine-white variegation. J. Hered. 31, 19—24 (1940).
Stoffels, E. H. J.: La sélection du caféier arabica à la station de Mulungu. Publ. Inst. Agron. Congo Belge: Sér. Sci., Nr. 25, 72 (1941).
Strydom, H. L.: Production of Turkish tobacco in the Western Cape Province. Bull. Dep. Agric. S. Afr., Nr. 244, 28 (1944).
Sugiura, T.: Chromosome studies on Papaveraceae with special reference to the phylogeny. Cytologia, Tokyo 10, 558—576 (1940).
Sysakjan, N. u. A. Kobjakova: (The trends of certain biochemical processes in plants during a period of 24 hours.) Biohimija 5, 301—308 (1940). [Russisch.]

T....., A. S.: Coffee selection. E. Afr. Agric. J. 4, 241—243 (1939).
Taunay, A. de E.: Confronto de variedades. (Comparison of varieties.) Rev. Inst. Café S. Paulo 28, 10—16 (1941).
Tedin, O., G. Pettersson u. J. E. Lindberg: Erfarenheter av och försök met odling av nikotinstark tobak i Skåne 1942. (Experiences and experiments with the cultivation of tobacco of high nicotine content in Scania in 1942.) Sverig. Utsädesfören. Tidskr. 53, 283—298 (1943).
Teixeira Mendes, J. E.: Seleção do cafeeiro. Literatura estrangeira. (Coffee selection. Foreign literature.) Rev. Inst. Café S. Paulo 16, 740—744 (1941).
Temme, J.: Natrot bij tabak (Nicotiana Tabacum). (Wet rot in tobacco [N. Tabacum].) Tijdschr. PlZiekt. 49, 113—116 (1943).
Ternovskij, M. F.: (Interspecific hybridization in tobacco and Makhorka [N. rustica breeding].) Vsesojuznyj Naučno-Issledovateljskij Institut Tabačnoj i Mahoročnoj Promyšlennosti imeni A. I. Mikojana (VITIM). (All-Union Mikojan Research Institute of the Tobacco and Makhorka Industry.) Krasnodar, Nr. 143, 99—125 (1941). [Russisch.]
— (Methods for breeding tobacco varieties resistant to tobacco mosaic and powdery mildew.) Vsesojuznyj Naučno-Issledovateljskij Institut Tabačnoj i Mahoročnoj Promyšlennosti imeni A. I. Mikojana (VITIM). (All-Union Mikojan Research Institute of the Tobacco and Makhorka Industry.) Krasnodar, Nr. 143, 126 bis 141 (1941). [Russisch.]
— (The role of wild growing Nicotiana species in tobacco breeding.) Vestnik Gibridizacii (Hybridization), Nr. 2, 61—66 (1941). [Russisch.]

Thomas, A. S.: The wild Arabica coffee on the Boma plateau, Anglo-Egyptian Sudan. Emp. J. Exp. Agric. 10, 207—212 (1942).
— The wild coffees of Uganda. Emp. J. Exp. Agric. 12, 1—12 (1944).
— The cultivation and selection of robusta coffee in Uganda. Emp. J. Exp. Agric. 15, 65—81 (1947).
Thorold, C. A.: Elgon dieback disease of coffee. Mon. Bull. Coffee Bd. Kenya 10, 85—86, 95—98 (1945).
— A study of yields, preparation out-turns, and quality in Arabica coffee. Part I. Yields. Emp. J. Exp. Agric. 15, 96—106 (1947).
Tihvinskaja, V. D.: Some data on inheritance of chemical composition by hybrids between N. tabacum and N. glauca. C. R. (Doklady) Acad. Sci. URSS. 24, 791—793 (1939).
Tollenaar, D.: De beteekenis van het oderstamvraagstuk voor de cacao-cultuur. (The importance of the stock problem for cacao cultivation.) Bergcultures 15, 553—556 (1941).
Tomur, K.: Tütün tohum cesidlerinin intas muyaenelerinde ziya faktorunun ehemmiyeti. (The importance of light in the germination of different kinds of tobacco seeds. Tütün Institüsü Rapolari 3, 9—12 (1943).
Treloar, A. E.: Random sampling distributions. Burgess Publishing Company, Minneapolis, Minn., Pp. 94 (1942). [Mimeographed.]
Trotter, A.: Una interessante mutazione teratologica nel tabacco „Aja Soluk" affine al „Kroepoek" riscontrata nel Leccese. (An interesting teratological mutation, like „Kroepoek", in Aja Soluk tobacco.) Boll. Tec. Tab. 36, 193—202 (1939).
Tschepourkovsky, E.: An experimental biometrical study on the inheritance of the number of rays in Papaver somniferum. Proc. 6th Pacific Sci. Congr. 4, 737—738 (1940).
Tubbs, F. R.: The improvement of planting material. Tea Quart. 12, 38—47 (1939).
— Address to the Ceylon Association in London. Tea Quart. 12, 160—166 (1939).
— Tea selection. I. The present position. Tea Quart. 18, 59—60 (1946).
— Tea Selection. II. Selection of mother bushes. Tea Quart. 18, 60—65 (1946).

Valleau, W. D.: Breeding tobacco for resistance to mosaic. J. Bact. 43, 272 (1942).
— Control of the common mosaic disease of tobacco by breeding. Phytopathology 32, 1022—1025 (1942).
— The relative positions of the N and N'factors on Nicotiana tabacum chromosomes. Phytopathology 33, 14 (1943). (Abstr.)
— Breeding tobacco varieties resistant to mosaic. Phytopathology 36, 412 (1946). (Abstr.)
— u. S. Diachun: Burley tobacco diseases in central Kentucky in 1940. Plant Dis. Reporter 25, 18—20 (1941). [Mimeographed.]
—, E. M. Johnson u. S. Diachun: Tobacco diseases. Bull. Ky Agric. Exp. Sta., Nr. 437, 60 (1942).
Veen, R. van der: Koffieproducties. (Coffee production.) Bergcultures 15, 278—285 (1941).
Voelcker, O. J.: Growth rate of cross and self-fertilized cacao. Trop. Agriculture, Trin. 16, 203—205 (1939).
— The degree of cross pollination in cacao in Nigeria. Trop. Agriculture, Trin. 17, 184—186 (1940).
— A review of cacao selection in the Cameroons. Trop. Agriculture, Trin. 17, 223—225 (1940).
Vollema, J. S.: Het versnellen van de kieming van theezaden. (Increasing the rapidity of germination of tea seeds.) Bergcultures 13, 1012 (1939).
— Overzicht van de voornaamste cultuurtechnische werkzaamheden van het Proefstation West-Java in de laatste jaren. (Survey of the main agronomic

activities of the West-Java Experiment Station in recent years.) Bergcultures 13, 1034—1042 (1939).
Volotov, E. N.: Polyploids in Papaver somniferum L. induced by treatment with colchicine. C. R. (Doklady) Acad. Sci. URSS. 31, 261—263 (1941).
— (Einfluß der Polyploidie auf das Tapetum bei Mohn.) C. R. (Doklady) Acad. Sci. URSS. 54, 565 (1946). [Russisch.]
Voroncov, V. E.: (The art of tea scenting.) State Tea and Sub-Tropical Crops Res. Inst. Tiflis, Pp. 125 (1939). [Russisch.]

Wahnon, J. S.: Subsídios para o estudo dos cacaus de S. Tomé e Principe. (Notes for the study of the cacaos of S. Tomé and Principe.) Sér. Invest. Minist. Econ., Serv. Edit. Repart. Estud., Inform. Prop., Lisboa, Nr. 9, 84 (1941).
Ward, G. M.: Nicotine-a product of tobacco. Tech. Bull. Dep. Agric. Canada, Nr. 38, 21 (1941).
Warmke, H. E. u. A. F. Blakeslee: Induction of simple and multiple polyploidy in Nicotiana by colchicine treatment. J. Hered. 30, 419—432 (1939).
—, — Induction of tetraploidy in Nicotiana sanderae and in the sterile hybrid N. tabacum × N. glutinosa by colchicine treatment. Genetics 24, 109—110 (1939). (Abstr.)
Weij, H. G. van der: Selection of Sumatra tobacco. Acta Nicotiana, Berlin 1, 257—258 (1939).
Wellensiek, S. J.: Biologisch en landbouwkundig werk in Nederlandsch Indië. 19. Theeselectie. (Biological and agricultural work in the Dutch. Indies. 19. Tea selection.) Vakbl. Biol. 20, 201—210 (1939).
— De stand der theeselectie. (The position of tea selection.) Bergcultures 14, 223—226 (1940).
— Genetical observations with the tea-plant. Genetica 22, 435—452 (1940).
— De veredeling van de theeplant. (The breeding of the tea plant.) Grondslagen Algemeene Plantenveredeling, Haarlem, Pp. 305—347 (1943).
Werkman, J. W.: Het keuren van theecloonen. (The judging of tea clones.) Bergcultures 13, 520—524 (1939).
Wheeler, H.-M.: A contribution to the cytology of the Australian-South Pacific species of Nicotiana. Proc. Nat. Acad. Sci. Wash. 31, 177—185 (1945).
Wight, W. u. P. K. Barua: The tea plant in industry: some general principles. Memor. Tocklai Exp. Sta., Indian Tea Ass., Nr. 7, 13 (1939).
Wildeman, E. de: Etudes sur le genre Coffea L. Classification, caractères morphologiques, biologiques et chimiques, sélection et normalisation. Bruxelles, Pp. VI + 495, 104 figs. 7 pls. tables. (1941).
Wolf, F. A.: Growth curves of oriental tobacco and their significance. Bull. Torrey Bot. 74, 199—214 (1947).
Woodmansee, C. W., K. E. Rappu. J. S. McHargue: Nicotine content of Nicotiana rustica grown in Kentucky and of 15 selections of dark tobacco. Bull. Ky Agric. Exp. Sta., Nr. 470, 11 (1944).

Yasui, K.: Cytogenetic studies in artificially raised interspecific hybrids of Papaver. IX. On the bivalents-association in the meiosis of the PMC of Papaver somniferum. Cytologia, Tokyo 10, 551—557 (1940).
— Cytogenetic studies in artificially raised interspecific hybrids of Papaver. VIII. F_1 plants of P. bracteata × P. lateritium. Cytologia, Tokyo 11, 452—464 (1941).

Zenkevič, E.: (Influence of X-rays upon the formation of leaves in Nicotiana rustica.) Bull. Acad. Sci. URSS. Sér. Biol., Nr. 4, 621—627 (1940). [Russisch.]
Žukov, N. I.: Inheritance of nicotine and anabasine in interspecific hybrids Nicotiana rustica L. × N. glauca Grah. C. R. (Doklady) Acad. Sci. URSS 22, 116—118 (1939).

Žukov, N. I.: (Alteration of the nature of plants by means of interspecific hybridization in the genus Nicotiana.) Vsesojuznyj Naučno-Issledovateljskij Institut Tabačnoj i Mahoročnoj Promyšlennosti imeni A. I. Mikojana (VITIM). (All-Union Mikojan Research Institute of the Tobacco and Makhorka Industry.) Krasnodar, Nr. 143, 142—233 (1941). [Russisch.]

k) Gewürzpflanzen.

Althaus, W. G.: 100th. anniversary varieties. Sth. Seedsman 9, Nr. 12; 12, 38 (1946).
Anonymus: El Linaloé y su aprovechamiento. (Linaloe and its utilization.) Méx. For. 18, 53—55 (1940).
— Mutation in chilli. Indian Fmg 1, 178 (1940).
— Mededeelingen voor en uit de practijk. (Information from and for the planter.) Landbouw 17, 56—58, 61 bis 62 (1941).
— Leptospermum citratum. New physiological forms. Perfum. Essent. Oil Rec. 34, 6—7 (1943).

Barnes, W. C.: No more hot pods in this pepper. Paprika pepper acreage expands in South Carolina as growers find it outpays cotton. Sth. Seedsman 5, Nr. 11; 11, 35 (1942).
— Varietal characters of importance in paprika breeding. Proc. Amer. Soc. Hort. Sci. 42, 575—578 (1943).
Baur, G.: Bestäubungs- und Befruchtungsverhältnisse bei weißem Senf (Sinapis alba L.). Züchter 12, 189 bis 193 (1940).
Beckenbach, J. R.: A pendant fruited pepper. Sth. Seedsman 7, 18 (1944).
Bouriquet, G.: Le Laboratoire de Phytopathologie et de Mycologie de Tananarive. Sa contribution à la vie économique de Madagascar. Agron. Trop. 2, 36—46 (1947).
Bregman, A.: De pepercultuur en-handel op Bangka. (Cultivation and trade of pepper [Piper nigrum] in Bangka Island.) Landbouw 16, 139—256 (1940).

Chevalier, A.: Moutardes et vignes à verjus. Rev. Bot. Appl. 21, 93—110 (1941).
— Les moutardes d'Orient. Possibilité de leur culture en France et dans les colonies. Rev. Bot. Appl. 22, 467—473 (1942).
— La systématique du genre Vanilla et l'amélioration des plants de vanille. Rev. Bot. Appl. 26, 107—111 (1946).
Chowdhury, S.: Cultivation of pan in Sylhet. Indian Fmg 5, 122—124 (1944).
Christensen, H. M. u. R. Bamford: Haploids in twin seedlings of pepper, Capsicum annuum L. J. Hered. 34, 99—104 (1943).
Cochran, H. L.: A chlorophyll deficient pimiento. J. Hered. 30, 81—83 (1939).
— Characters for the classification and identification of varieties of Capsicum. Bull. Torrey Bot. Cl. 67, 710 bis 717 (1940).
— Georgia pimientos. Market Gr. J. 71, 289, 297 (1942).
— Perfecting „Perfection". Sth. Seedsman 6, Nr. 6; 15, 26 (1943).
— The Truhart Perfection pimiento. Bull. Ga Exp. Sta., Nr. 224, 18 (1943).
Coulter, F. C.: Story of garden vegetables. XIII. Peppers native Americans now known the world over. Seed World 52, Nr. 10, 10—11 (1942).

Deshpande, R. B.: A case of variegation in Capsicum annuum L. Curr. Sci. 8, 313—314 (1939).
— Studies in Indian chillies. 5. Inheritance of anther colour and its relation to colour in petal and node in Capsicum annuum L. Indian J. Agric. Sci. 9, 185—192 (1939).
— Inheritance of bunchy habit in chilli. (Capsicum annuum L.) Indian J. Genet. Pl. Breed 4, 54 (1944).

Dorasami, L. S. u. D. M. Gopinath: Vine chilly-a comparative morphological and cytological study. Proc. Indian Acad. Sci. **20**, Sect. B., 40—42 (1944).

Dugand, A.: El genero Capparis en Colombia. (The genus Capparis in Colombia.) Caldasia **2**, 29—54 (1941).

Fernie, L. M.: Preliminary trials on the rooting of clove cuttings. E. Afr. Agric. J. **12**, 135—136 (1946).

Garcia, J. G.: Contribuição para o estudo cario-sistemático do género Lavandula L. (Contribution towards the karyo-systematic study of the genus Lavandula L.) Bol. Soc. Brot. **16**, 183—193 (1942).

Giral, F. u. J. Senoiain: Contenido en ácido ascórbico de algunas variedades de chiles mexicanos. (Ascorbic acid content of certain Mexican varieties of chilli.) Ciencia. Mexico, D. F. **1**, 258—259 (1940).

Guenther, E. S.: Botany, origin and application of essential oils. Amer. Perfum., Pp. 15 (1943).

Györffy, B.: (Tetraploid paprika.) Acta Univ. Szeged. **5**, 30—38 (1939).

— A paprika C-vitaminjáról. II. Különböző paprikafajták C-vitamin tartalma. (Vitamin C in paprika. II. Vitamin C contents of several varieties of paprika.) Arb. Ung. Biol. Forsch.-Inst. **14**, 297—313 (1942).

— (Vitamin C in rotem Pfeffer. 3. Polyploidie und Vitamin C in Paprika.) (Vorl. Mitt.) Mat. természett. Értes **61**, 329—335 (1942). [Ungarisch.]

Hagiwara, T. u. Y. Oomura: (Plastid inheritance of variegation in Capsicum annuum.) Jap. J. Genet. **15**, 328—330 (1939).

Hubbard, C. E.: Malayan grasses. Kew Bull., Nr. 1, 24—25 (1941).

Hurel-Py, G.: Étude des noyaux végétatifs de Vanilla planifolia. Rev. Cytol. Cytophysiol. Vég. **3**, 129—133 (1939).

Lantz, E. M.: The carotene and asorbic acid contents of peppers. Bull. N. Mex. Agric. Exp. Sta., Nr. 306, 14 (1943).

Leroy, J.-F.: Les piments. Rev. Bot. Appl. **23**, 196—218 (1943).

Mándy, G.: A paprika alaktana mint a fajtaleirás alapja. (The morphology of red pepper as a basis of varietal description.) Bull. Hung. Coll. Hort. **10**, 93 bis 133 (1944).

— Adatok a paprika (Capsicum annuum L.) fajtarendszertanához. (Data on the taxonomy of red pepper [C. annuum L.].) Mesogazdaságtudományi Közlemények **1**, 21—32 (1944).

Molegode, W.: The arecanut in Ceylon. Trop. Agriculturist **100**, 102—105 (1944).

Mostovoj, K.: (Constitutional sterility in paprika [C. annuum L.].) Ann. Tschech. Akad. Landw. **16**, 308 bis 316 (1942).

Narasimha Swamy, R. L.: Notes on breeding Elettaria cardamomum Maton. Mysore Agric. J. **21**, 66—69 (1942—1943).

Nishiyama, I.: (Studies on artificial polyploid plants. I. Production of tetraploids by treatment with colchicine.) Agric. Hort. **14**, 1411—1422 (1939).

Nissley, C. H.: New Jersey vegetables. Hort. News, N. J. **23**, 1420 (1942).

Odland, M. L.: Pennwonder. Sth. Seedsman **9**, 11 (1946).

— u. A. M. Porter: Inheritance of the immature fruit color of peppers. Proc. Amer. Soc. Hort. Sci. **36**, 647 bis 651, 1939 (1938).

—, — A study of natural crossing in peppers (Capsicum Frutescens). Proc. Amer. Soc. Hort. Sci. **38**, 585—588 (1941).

Oláh, L.: A kalocsai népies fűszerpaprika származása. (The origin of the capsaicin-free red pepper variety Kalocsa.) Mesogazdaságtudományi Közlemények **1**, 7—15 (1944).

Pal, B. P. u. S. Ramanujam: Induction of polyploidy in chilli (Capsicum annuum L.) by colchicine. Nature, Lond. **143**, 245—246 (1939).

—, — Asynapsis in chilli (Capsicum annuum L.). Curr. Sci. **9**, 126—128 (1940).

—, — u. A. B. Joshi: Colchicine-induced polyploidy in crop plants. II. Chilli (Capsicum annuum L.). Indian J. Genet. Pl. Breed. **1**, 28—40 (1941).

Parry, J. W.: The spice handbook. Spices, aromatic seeds and herbs. Brooklyn, N. Y., Pp. XVII + 254. photos. tables (1945).

Penfold, A. R., F. R. Morrison u. S. Smith-White: The occurrence of two physiological forms of Leptospermum citratum (Challinor, Cheel and Penfold) as determined by chemical analysis of the essential oils. J. Proc. Roy. Soc. N. S. W. **76**, 93—95 (1942).

Pennington, C.: Vanilla pollination is no mystery. Rev. Agric. Puerto Rico **35**, 225—233 (1944).

Popoff, P.: Untersuchungen über den Einfluß einiger genetischer und ökologischer Faktoren auf Ertrag und biologischen Wert von Paprika (Capsicum annuum L.) unter besonderer Berücksichtigung des Ascorbinsäuregehaltes. Gartenbauwiss. **17**, 446—492 (1943).

Raghavan, T. S. u. K. R. Venkatasubban: Studies in the South Indian chillies I. A description of the varieties, chromosome numbers and the cytology of some X-rayed derivatives in Capsicum annuum Linn. Proc. Indian Acad. Sci. **12**, Sect. B, 29—46 (1940).

—, — Cytological studies in the family Zingiberaceae with special reference to chromosome number and cyto-taxonomy. Proc Indian Acad. Sci. **17**, Abt. C, 118—132 (1943).

Roque, A. u. J. Adsuar: Studies on the mosaic of peppers (Capsicum frutescens) in Puerto Rico. J. Agric. Univ. P. R. **25**, 40—50 (1941).

Rudorf, W.: Deutscher Speise- und Gewürzpfeffer. Forschungsdienst **15**, 57 (1943).

Schermerhorn, L. G.: Rutgers World Beater Nr. 13 pepper. N. J. Hort. Soc. News **22**, 1348 (1941).

Schuphan, W.: Gemüsepaprika, eines unserer wertvollsten Gemüse. Forschungsdienst **12**, 615—617 (1941).

Schwanitz, F.: Über den Einfluß des Entfernens der Keimblätter auf die Entwicklung und den Ertrag von diploidem und autotetraploidem gelbem Senf (Sinapis alba). Züchter **14**, 86—93 (1942).

Singh, M. P. u. R. S. Roy: Inheritance of fruit position in a chillie cross. Curr. Sci. **14**, 133 (1945).

Swamirao, R., M. P. Narasimharao u. S. T. Ramaswami: Emasculation in chillies (Capsicum genus). Curr. Sci. **10**, 296 (1941).

Takahasi, N. u. T. Osumi: (On the inheritance of the colour of stem and fruit in Panax Ginseng, C. A Meyer.) Jap. J. Genet. **16**, 273—276 (1940).

Thomas, K. M. u. K. Krishna Menon: The present position of pollu disease of pepper in Malabar. Madras Agric. J. **27**, 348—356 (1939).

Tompos, A.: A fűszerpaprikahüvely héjvastagságának mérése. (On the determination of the thickness of the pericarp of the spice paprika.) Rep. Hung. Agric. Exp. Sta. **42**, 285—288 (1939).

Toole, M. G. u. R. Bamford: The formation of diploid plants from haploid peppers. J. Hered. **36**, 67—70 (1945).

Walker, R. I.: Chromosome number of Zanthoxylum americanum. Bot. Gaz. **103**, 625—626 (1942).

1) Ölpflanzen.

Abraham, A.: Cytological studies in Sesamum. Proc. 31st Indian Sci. Congr. Delhi Pt III, p. 91 (1944). (Abstr.)

Adriaens, L.: Les oléagineux du Congo Belge. Bull. Agric. Congo Belge 34, 397—536 (1943).

Alam, Z.: Nomenclature of oleiferous brassicas cultivated in the Punjab. Indian J. Agric. Sci. 15, 173—181 (1945).

Alibert, H.: Pourquoi et comment on fait la fécondation artificielle sur le palmier à huile. Fm and For., Nigeria 6, 27—30 (1945).

D'Amato, F.: Ricerche embriologiche e cariologiche sul genere Euphorbia. (Embryological and cytological researches on the genus Euphorbia.) Nuovo G. Bot. Ital. 46, 470—509 (1939).

Ananjeva, S. V.: (Vegetative hybridization of sunflower.) Socialistic Grain Farming, Saratov, Nr. 4, 168—170 (1940). [Russisch.]

— (The mentor method applied to the sunflower.) Socialistic Grain Farmin, Saratov, Nr. 2, 118—133 (1941). [Russisch.]

— (The vegetative hybridization of sunflowers.) Bjulletenj Instituta Zernovogo Hozjajstva Jugo-Vostoka SSSR. (Bull. Inst. Grain Husb. S. E., USSR.) Saratov, Nr. 2, 19—25 (1944). [Russisch.]

Andersson, G.: Odling av oljeväxter. (Growing oil plants.) Lantmannen 25, 175—177. (1941).

— Oljeväxternas odling och förädling. (Cultivation and breeding of oil plants.) Sverig. Utsädesfören. Tidskr. 51, 256—270 (1941).

— Oljeväxterna. Odlingens omfattning och odlingsmaterial i Sverige och grannländerna. (Oil crops. Extent of cultivation and types grown in Sweden and the neighbouring countries.) Årsb. Jordbruksforskning, Stockholm, Pp. 94—105 (1945).

— u. C. M. Björklund: Skördetidsförsök i höstraps och vitsenap sommaren 1944. (Experiments on the time of harvesting of winter rape and white mustard, summer 1944.) Sverig. Utsädesfören. Tidskr. 55, 20—25 (1945).

André, É. u. M. Kogane-Charles: Sur la possibilité d'accroître la production d'huile des cultures de colza. C. R. Acad. Sci., Paris 214, 636—638 (1942).

—, — Sur quelques caractères chimiques des graines de colza utilisables en vue de sélectionner les meilleures variétés. C. R. Acad. Sci., Paris 215, 587—588 (1942).

Angelo, E.: The breeding and improvement of tung. Proc. 41st Annu. Conv. Ass. Sth. Agric. Wkrs, Birmingham, Ala., February 7—9, p. 162 (1940). (Abstr.)

— Some outstanding seedling progenies of tung. Proc. Amer. Soc. Hort. Sci. 42, 315—317 (1943).

—, R. T. Brown u. H. J. Ammen: Pollination studies with tung trees. Proc. Amer. Soc. Hort. Sci. 41, 176 bis 180 (1942).

Anonymus: Aleurites montana (Lour) Wils. en de daaruit gewonnen hout-of tung-olie. (A. montana [Lour] Wils. and the woodoil or tung oil.) Landbouw 15, 3—94 (1939).

— Perilla-olie. (Mededeeling van het secretariaat van de Commissie voor Handelgewassen.) (Perilla oil — Contribution from the Secretariat of the Committee for Commercial Crops.) Bergcultures 14, 768—770 (1940).

— Flax variety trials on Dominion Experimental Farms and Stations 1942. Mimeogr. Rep. Cent. Exp. Fm, Cereal Div., Ottawa, Nr. 61, 19 (1942).

— Sesamos con mayor riqueza en aceites. (Sesame with higher oil content.) Ciencia, Mexico, D. F. 3, 173—174 (1942).

— Land Settlement in the Gold Coast.-Tung. Crown Colonist, Lond. 13, Nr. 135, 135 (1943).

— Mexico finds a substitute for tung oil. Agric. Amer. 3, 38 (1943).

Anonymus: Texas flax to extend northward to Waco. Sth. Seedsman 6, 20 (1943).

— More oil in new safflower strains. Sth. Seedsman 7, 31 (1944).

— Gute Erfolge der Raps- und Rübsenzüchtung. Forschungsdienst 17, 51 (1944).

— Neue Sonnenblumenzüchtung in Bulgarien. Mitt. Landw. 59, 77 (1944).

— Sunflowers as oilseed crop. Mon. Sci. News, Nr. 40, 3 (1944).

— What the scientists are doing. Improvement of toria in the Punjab. Indian Fmg 5, 523—524 (1944).

— New varieties of oil-bearing plants. Soviet News, Nr. 1624, 3 (1947).

— Three new strains of rape developed by H. G. Neufeld. Canad. Grain J. 2, Nr. 6, 16 (1947).

Arny, A. C.: Flax varieties registered. I. J. Amer. Soc. Agron. 35, 823—824 (1943).

Ashby, M.: The tung oil industry of the United States. Report of an inquiry carried out in Florida, Louisana and Mississippi. Bull. Imp. Inst., Lond. 38, 5—32 (1940).

Aubréville, A.: Les deux stations expérimentales du palmier à huile. Rev. Bot. Appl. 19, 1—14 (1939).

Banerji, I.: A contribution to the morphology and cytology of Carthamus tinctorius Linn. Proc. Nat. Inst. Sci. India 6, 73—86 (1940).

Baranskij, D. I.: (A new method of selecting the sunflower for oil content.) Selekcija i Semenovodstvo (Breeding and Seed Growing), Nr. 3, 44—45 (1940.)

Barbera, G.: Il ricino nel piano autarchico. (The castor oil plant in the plan for autarchy.) Ital. Agric. 78, 49—57 (1941).

Beirnaert, A.: Le problème de la stérilité chez le palmier à huile. Bull. Agric. Congo Belge 31, 95—110 (1940).

— u. R. Vanderweyen: Contribution à l'étude génétique et biométrique des variétés d'Elaeis guineensis Jacquin. Publ. Inst. Nat. Agron. Congo Belge 1941: Sér. Sci., Nr. 27, 101 (1941).

—, — Les graines sélectionnées livrées par la station de Yangambi. Publ. Inst. Agron. Congo Belge Ser. Techn. Nr. 28, 41 (1941).

—, — Influence de l'origine variétale sur les rendements. Publ. Inst. Agron. Congo Belge Ser. Tech., Nr. 30, 26 (1941).

Blackman, G. E.: Sunflowers as an oil seed crop. J. Minist. Agric. 50, 517—521 (1944).

Blackmon, G. H.: The tung-oil industry. Bot. Rev. 9, 1—40 (1943).

Blaringhem, L.: Sur un cas nouveau de xénie chez le tournesol. C. R. Acad. Sci., Paris 215, 337—339 (1942).

— Xénie et fascies florales du tournesol (Helianthus annuus L.). Ann. Sci. Nat. (11 Sér. Botanique) 4, 103 bis 118 (1943).

Bondar, G.: Peñao, Cnidoscolus Marcgravii Polh. Novo recurso oleifero da Bahia. (C. Marcgravii Polh., a new source of oil in Bahia.) Bol. Inst. Cent. Fom Econ. Bahia, Nr. 12, 16 (1942).

— A piassaveira e outras palmeiras Attaleaineas na Bahia. (Piassava and other Attalea palms in Bahia.) Bol. Inst. Cent. Fom. Econ. Bahia, Nr. 13, 76 (1942).

Brieger, F. G. u. J. T. A. Gurgel: Experiencias preliminares sobre a mamoneira (Ricinus communis, L.). (Preliminary researches on the castor oil plant [R. communis, L.].) Rev. Agric., S. Paulo 15, 229—248 (1940).

Brown, R. T. u. E. Fisher: Period of stigma receptivity in flowers of the tung tree. Proc. Amer. Soc. Hort. Sci. 39, 164—166 (1941).

Buchinger, A.: Erfahrungen im Safloranbau. Mitt. Landw. 59, 334—335 (1944).

Bustarret, J. u. P. Jonard: Observations sur la culture et la sélection de quelques plantes oléagineuses. Ann. Agron., Paris, Pp. 1—21 (1944).

C......, C.: Les soleils annuels. Rev. Hort. Suisse 15, 279—281 (1942).

Caidze, I.: (The tung tree [Aleurites fordii].) Socialisticeskoe Zemledelie (Socialist. Agriculture), Nr. 170, 4 (1945). [Russisch.]

Chevalier, A.: Les sapotacées à graines oléagineuses et leur avenier en culture. Rev. Bot. Appl. 23, 97—159 (1943).

— Taxonomie, biogéographie et sélection des palmiers du genre Elaeis. Rev. Bot. Appl. 23, 295—307 (1943).

Chopinet, R.: Hybrides intergénériques ,,Raphano-Brassica". Rev. Hort., Paris 116, 98—100 (1944).

Claassen, C. E. u. T. A. Kiesselbach: Experiments with safflower in Western Nebraska. Bull. Neb. Agric. Exp. Sta., Nr. 376, 28 (1945).

Constancio Lázaro, R.: Observáciones citogenéticas sobre girasol (Helianthus annuus). (Cytogenetic observations on sunflower [H. annuus].) Rev. Fac. Agron. Univ. Montevideo, Nr. 23, 59—67 (1941).

Cook, L. J. u. C. W. Hooper: Sunflower cultivation trials with local and imported varieties. J. Dep. Agric. S. Aust. 50, 133—135 (1946).

Cornell, D. S.: The African oil palm: its history, cultivation and importance. Lloydia, Cincinnati 7, 101 bis 120 (1944).

Crane, M. B. u. D. Lewis: Genetical studies in pears. II. A classification of cultivated varieties. J. Pomol. 18, 52—60 (1940).

Demidenko, T. T.: Varieties of sunflower a characterized according to their drought resistance. C. R. (Doklady) Acad. Sci., URSS. 47, 513—516 (1945).

Deshpande, R. B.: A sterile mutant in safflower (Carthamus tinctorius L.). Curr. Sci. 9, 370—371 (1940).

— A case of chlorophyll deficiency in safflower (Carthamus tinctorius L.). Curr. Sci. 12, 273—274 (1943).

Dickey, R. D.: The importance of tung seed selection. Proc. Amer. Soc. Hort. Sci. 41, 127—130 (1942).

Dillman, A. C.: Breeding flax resistant to rust. Oil Paint Drug Rep. 142, 3, 40 B (1942).

Dimitri, M. J.: La resistencia del girasol (Helianthus annuus L.) a las sequías estivales. (The resistance of the sunflower [H. annuus L.] to summer droughts.) ,,Granos" Semilla Selecta, B. Aires 8, Nr. 1/3, 29—31 (1944).

Dionigi, A.: Il miglioramento genetico del ricino. La creacione della varietà ,,M. 6". (The genetic improvement of the castor oil plant and the creation of the variety M. 6.) Genetica Agraria, Roma 1, 9—31 (1946).

Dobrev, K.: (Precautions at sunflower sowing-time.) Zemledelie, Sofia 45, 53—55 (1941). [Bulgarisch.]

Domingo, W. E.: Amount of natural out-crossing in the castor oil plant. J. Amer. Soc. Agron. 36, 360—361 (1944).

— Flowerless castor-bean plants. J. Hered. 36, 116—120 (1945).

— u. D. M. Crooks: Investigations with the castor bean plant: I. Adaptation and variety tests. J. Amer. Soc. Agron. 37, 750—762 (1945).

Doná dalle Rose, A.: Tre razze di lini da olio Marsic I-Capace-Bianco Nano. (Three races of linseed, Marsic I-Capace-Bianco Nano.) Ital. Agric. 76, 845—849 (1939).

— Trois races de lin à huile. (Marsic I, Capable, Blanc nain.) Rev. Bot. Appl. 20, 633—635 (1940).

Drobinskij, B. N.: (Segregation in the F_2 generation of grey mustard as indicated by colour of seed and Mendel's ,,Law".) Jarovizacija, Nr. 3 (36), 117—118 (1941). [Russisch.]

Džidžavadze, S. Š.: (Further data on the oil bearing plants of the Soviet humid subtropics.) Sovetskaja Botanika (Soviet Botany), Nr. 5/6, 43—48 (1941). [Russisch.]

El-Kilany, M. A.: Flax in Egypt. Part III. Flower and anther colour inheritance. Bull. Minist. Agric. Egypt, Nr. 205, 5—7 (1939).

Engelbeen, M.: Les Aleuritis. Bull. Agric. Congo Belge 37, 255—342 (1946).

Ermakov, A. I.: (A universal and quick method for refractometric determination of oil in seeds.) Proc. Lenin Acad. Agric. Sci. USSR, Nr. 18, 27—34 (1939). [Russisch.]

Etchecopar, J. A.: La biologia floral del girasol y su relación con la técnica del mejoramiento. (The floral biology of the sunflower and its relation to the technique of improvement. Rev. Argent. Agron. 11, 11—19 (1944).

— u. S. E. Foulcault: Influencia de épocas de siembra en las caracteristicas cualitativas y cuantitativas del fruto, en variedades y selecciones de girasol, en la Estación Experimental de Pergamino. (Influence of sowing dates on the qualitative and quantitative characteristics of the fruit in sunflower varieties and selections at the Pergamino Experimental Station.) ,,Granos" Semilla Selecta, B. Aires 5, 3—10 (1941).

— u. E. Sívori: Notas sobre el mejoramiento del girasol. (Notes on the improvement of the sunflower.) Rev. Argent. Agron. 8, 252—260 (1941).

—, — Algunos aspectos en el mejoramiento del girasol. (Some aspects of the improvement of the sunflower.) ,,Granos" Semilla Selecta, B. Aires 6, Nr. 10/12, 36—37 (1942).

Fernholz, D. L. u. G. F. Potter: Preliminary experiments on the resistance of the tung tree to low temperature. Proc. Amer. Soc. Hort. Sci. 39, 167—172 (1941).

Ferrand: Découvertes récentes dans la génétique du palmier à huile (Elaeis Guineensis) et leurs conséquences quant à la sélection de ce végétal. C. R. Acad. Agric. Fr. 32, 75—79 (1946).

Fickendey, E.: Die Züchtung der Ölpalme. (Elaeis guineensis Jacquin.) Z. Pflanzenz. 26, 136—162 (1944).

Fuelleman, R. F. u. W. L. Burlison: Castor beans-an industrial war crop. Seed World 53, 12—13 (1943).

Garnett, C. B.: Report of the Department of Agriculture, Nyasaland Protectorate, for the year 1943, Pp. 15 (1943).

Granhall, I.: Svalöfs Regina vårraps. Ny oljeväxt för Skånes och Gotlands kraftigare jordar. (Svalöfs Regina, spring rape. A new oil plant for the heavier soils of Sciania and Gotland.) Sverig. Utsädesfören. Tidskr. 51, 341—343 (1941).

Grieben, H.: Resultado de los ensayos comparativos de rendimiento ,,standard" entre variedades de lino para grano, realizados en el año agrícola 1942—1943. (Result of the comparative trials for ,,standard" yield of linseed varieties obtained in the agricultural year 1942 bis 1943.) ,,Granos" Semilla Selecta, B. Aires 8, Nr. 10 bis 12, 3—36 (1944).

— u. G. Cipolla: Resultados de los ensayos comparativos de rendimiento ,,standard" entre variedades de lino para grano realizados en el año agrícola 1941/42. (Results of the comparative ,,standard" yield trials of varieties of linseed effected in the agricultural year 1941/42.) ,,Granos" Semilla Selecta, B. Aires 6, Nr. 10 bis 12, 3—32 (1942).

Grinčak, A. L.: (Urgent problems in breeding and seed production in regard to oil bearing crops.) Selekcija i Semenovodstvo (Breeding and Seed Growing), Nr. 3, 31—32 (1940). [Russisch.]

Hackbarth, J.: Fragen des Anbaues und der Züchtung von Ölpflanzen in Deutschland. Phosphorsäure 10, 131—139 (1941).

Harada, M.: (Quantitative investigations of the tissue of Rhus fruits found in Japan with special reference to their wax content.) Bult. Sci. Fak. Terk. Kyušu Univ. 9, 327—336 (1941).

Harland, S. C.: An alteration in gene frequency in Ricinus communis L. due to climatic conditions. Heredity 1, 120—125 (1947).

Heggeness, O. A.: Viking flax. Bi-m. Bull. N. Dak. Agric. Exp. Sta. 3, Nr. 4, 7—8 (1941).

Henry, A. J. u. D. N. Grindley: Investigation of the oil of the seeds of Cephalocroton cordofanus. J. Soc. Chem. Ind., Lond. 62, 60 (1943).

Hill, A. G.: Oil plants in East Africa. (1) Groundnuts, (2) sesame and (3) sunflowers. E. Afr. Agric. J. 12, 140—152 (1947).

Honig, F.: Höhere Rapserträge durch Fremdbestäubung Mitt. Landw. 55, 81 (1940).

Hopper, T. H.: Seed flax production and flax research. Bi-m. Bull. N. Dak. Agric. Exp. Sta. 1, Nr. 3, 17 bis 19 (1939).

Hosszu, M. de: Linseed. N. F. U. Rec. 23, 157 (1945).

Hwang, Shui-Lwen u. Lee, You-Kai: (Studies of the quality of tung oil fruits of Kwangsi Province.) Kwangsi Agric. 2, 343—370 (1941).

I. S.: (Seminar for sunflower breeders.) Jarovizacija, Nr. 3 (36), 133 (1941). [Russisch.]

Igel, G.: Erfahrungen mit dem Ölfruchtbau in der Ukraine. Mitt. Landw. 59, 16—17 (1944).

Joffe, E. J.: Polymorphism of karyotype in Helianthus annuus L. C. R. (Doklady) Acad. Sci. URSS. 30, 76—78 (1941).

John, C. M. u. U. Narasinga Rao: Chromosome number of Sesamum radiatum Schum and Thonn. Beskr. Curr. Sci. 10, 364—365 (1941).

Kadam, B. S. u. V. K. Patankar: Natural cross-pollination in safflower. Indian J. Genet. Pl. Breed. 2, 69—70 (1942).

Kaden, O. F.: Die Ölpalmbestände Bahias und ihre Nutzungsmöglichkeiten. Tropenpflanzer 43, 177—183 (1940).

Kirjuhin, I.: (The influence of wounding the pistils of sunflowers on pollination.) Jarovizacija, Nr. 3 (30), 213—215 (1940). [Russisch.]

Klemm, M.: Der Kürbis und seine Bedeutung als Ölpflanze. Forschungsdienst 11, 676—698 (1941).

Klimočkina, L. V.: Chromosome morphology in Helianthus annuus L. C. R. (Doklady) Acad. Sci. URSS. 27, 584—586 (1940).

Knapp, O.: Sonnenblumenzüchtung in Ungarn. Züchter 12, 193—199 (1940).

— Versuche zur Züchtung einer giftfreien Ricinussorte. Züchter 15, 97—100 (1943).

Kostoff, D.: Autosyndesis and structural hybridity in F_1-hybrid Helianthus tuberosus L. × Helianthus annuus L. and their sequences. Genetica 21, 285—300 (1939).

Kostov, D. siehe Kostoff.

Kozo-Poljanskij, B. M.: (Der Blütenmechanismus bei Sesam.) C. R. (Doklady) Acad. Sci. URSS. 58, 919 (1947). [Russisch.]

Krijthe, J. M.: (On the influence of colchicine upon the anthers of Carthamus tinctorius L.) Proc. Acad. Sci. Amst. 45, 283—287 (1942).

Krug, C. A., P. T. Mendes u. O. F. de Sousa: Melhoramento da mamoneira (Ricinus communis L.). I. Plano geral dos trabalhos em execução nas secções de genética e plantas oleaginosas do Instituto Agronômico do Estado de São Paulo. II. Observações gerais sobre a variabilidade do gênero Ricinus. III. Primeira série de ensaios de variedades (1937/38 bis 1938/39). (Improvement of castor oil plant [Ricinus communis L.]. I. General plan of work in the sections of genetics and oil plants of the State Agronomic Institute of São Paulo. II. General observations on the variability of the genus Ricinus. III. First series of variety tests [1937/38 bis 1938/39].) Bragantia, São Paulo 2, 129 bis 154, 155—197; 3, 85—122 (1942/1943).

Kugler, W. F. u. C. Remussi: Descripción de variedades agricolas de lino por sus caracteres morfológicos. (Description of agricultural varieties of flax by their morphological characters.) „Granos" Semilla Selecta B. Aires 3, 131—149 (1939).

Kumar, L. S. S. u. A. Abraham: A cytological study of sterility in Sesamum orientale L. Indian J. Genet. Pl. Breed. 1, 41—60 (1941).

— u. N. M. Patel: Natural cross-fertilization in Linum usitatissimum L. (linseed) in the Bombay Deccan. J. Univ. Bombay 8, Pt. 5 (Biol. Sci., Nr. 8), 105 bis 110 (1940).

— u. D. S. Ranga Rao: Studies in blooming in three Punjab types of sesamum in the Bombay Deccan. J. Univ. Bombay 9, Nr. 5, 69—77 (1941).

—, — Studies in the development of seed characters in linseed (Linum usitatissimum). J. Univ. Bombay 11, Part. 5, Sect. B, 113—119 (1943).

—, — Inheritance of sterility in Sesamum orientale L. Indian J. Genet. Pl. Breed. 5, 58—59 (1945).

Kursell, C. von: Zuchtarbeiten an der neuen Ölpflanze Saflor. Pflanzenbau 15, 463—482 (1939).

L......, A.: Anbauversuche mit Saflor im Donauland. Eine neue Ölpflanze. Obst u. Gem. 6, 264 (1942).

Langham, D. G.: Fertile tetraploids of sesame, Sesamum indicum Loew, induced by colchicine. Science 95, 204 (1942).

— Un método nuevo para efectuar polinización controlada en el ajonjolí (Sesamum indicum Loew) y una estimación de la hibridización natural. (A new method of artificially pollinating S. indicum Loew. and an estimation of natural crossing.) Circ. Minist. Agric. Cría, Dep. Genét. Inst. Exp. Agric. Zootec., El Valle, D. F., Nr. 4, 7 (1943).

— Natural and controlled pollination in sesame. J. Hered. 35, 255—256 (1944).

— The inheritance of glands in sesame (Sesamum indicum L.). Amer. J. Bot. 31, 10s (1944). (Abstr.)

— Sesame breeding in Venezuela. Amer. J. Bot. 31, 10s—11s (1944). (Abstr.)

— Variación en mel número de hojas, càpsulas e hileras de semillas por nudo en el ajonjolí (Sesamum indicum Loew). (Variation in the number of leaves, capsules and seed rows, per node, in S. indicum, Loew.) Circ. Minist. Agric. Cría, Dep. Genét. Inst. Exp. Agric. Zootec., El Valle, D. F., Nr. 6, 4 (1944).

— El carácter glabro en el ajonjolí (Sesamum indicum L.). (The character „glabrous" in sesame [S. indicum L.].) Circ. Minist. Agric. Cría, Dep. Genét. Inst. Exp. Agric. Zootec., El Valle, D. F., Nr. 8, 5 (1945).

— Ajonjolí (Sesamum indicum L.) sin glándulas en el haz de la hoja. (Sesame [S. indicum L.] without glands on the surface of the leaf.) Circ. Minist. Agric. Cría, Dep. Genét. Inst. Exp. Agric. Zootec., El Valle, D. F., Nr. 9, 6 (1945).

— El modo de herencia del numéro de cápsulas por axila en el ajonjolí (Sesamum indicum L.). (The mode of inheritance of the number of capsules per axil in sesame [S. indicum L.].) Circ. Minist. Agric. Cría, Dep. Genét. Inst. Exp. Agric. Zootec., El Valle, D. F. Nr. 10, 7 (1945).

— Plantas de ajonjolí (Sesamum indicum L.) que se tornan amarillas antes de la madurez. (Plants of sesame [S. indicum L.] that turn orange before maturity.) Circ. Minist. Agric. Cría, Dep. Genét. Inst. Exp. Agric. Zootec., El Valle, D. F., Nr. 11, 5 (1945).

Langham, D. G.: Hoja fruncida, un carácter indeseable en el ajonjolí (Sesamum indicum L.). (Crinkle leaf, an undesirable character in sesame [S. indicum L.].) Circ. Minist. Agric. Cría, Dep. Genét. Inst. Exp. Agric. Zootec., E. Valle, D. F., Nr. 12, 6 (1945).
— Genetics of sesame. J. Hered. 36, 135—142 (1945).
— Genetics of sesame. II. Inheritance of seed pod number, aphid resistance, ,,yellow leaf", and wrinkled leaves. J. Hered. 36, 245—253 (1945).
— Genetics of sesame. III. ,,Open sesame" and mottled leaf. J. Hered. 37, 149—152 (1946).
— Genetical studies of the sesame flower. Science 103, 280 (1946).
— Variations of the sesame flower. Genetics 31, 221 bis 222 (1946). (Abstr.)
— Open sesame. Genetics 31, 222 (1946). (Abstr.)
— Initiation of a linkage map for sesame (Sesamum indicum L.). Genetics 32, 94 (1947). (Abstr.)
— Seedling characters in sesame (Sesamum indicum L.). Genetics 32, 94 (1947). (Abstr.)
— u. M. Rafael Cortes: Herencia del número de glándulas foliares en el ajonjolí (Sesamum indicum L.). (Inheritance of the number of foliar glands in sesame [S. indicum L.].) Circ. Minist. Agric. Cría, Dep. Genét. Inst. Exp. Agric. Zootec., El Valle, D. F., Nr. 7, 6 (1945).
— u. M. Rodriguez: Resistencia al áfido (Myzus persicae Sulz.) en el ajonjolí (Sesamum indicum L.). (Resistance to aphid [M. persicae Sulz.] in sesame [S. indicum L.].) Circ. Minist. Agric. Cría, Dep. Genét. Inst. Exp. Agric. Zootec., El Valle, D. F., Nr. 13, 7 (1945).

Lazaro, R.: Efectos de colchicina en Helianthus annuus (Effect of colchicine in H. annuus.) Rev. Fac. Agron. Univ. Montevideo, Nr. 32, 163—168 (1943).

Ljaščenko, I. F.: (Influence of external conditions on dominance in sunflower.) Jarovizacija, Nr. 4 (31), 111—112 (1940). [Russisch.]
— Cases of no segregation in sunflower hybrids. C. R. (Doklady) Acad. Sci. URSS. 27, 821—822 (1940).
— Cytoplasmic heredity in sunflower. C. R. (Doklady) Acad. Sci. URSS. 27, 823—824 (1940).
— Some results obtained by crossing cultivated and wild sunflower. C. R. (Doklady) Acad. Sci. URSS. 30, 242 bis 244 (1941).
— (The influence of external conditions on segregation in the F_2 generation of sunflower hybrids.) Jarovizacija Nr. 3 (36), 114—117 (1941). [Russisch.]

Lowig, E. u. O. Baumgartner: Untersuchungen über den Ölgehalt von Rapssamenherkünften. Forschungsdienst 9, 496—502 (1940).

Lubovskaja, N.: Plant husbandry I. Vegetative hybrids of sunflowers and Jerusalem artichokes. Agricultural Chron., Moscow, Nr. 10, 2—3 (1946). [Mimeographed.]

Lutikov, I.: (Broomrape on sunflowers, and how to combat it.) Kolhoznoe Proizvodstvo (Collective Farming), Nr. 12, 29 (1946). [Russisch.]

McCann, L. P.: Development of the pistillate flower and structure of the fruit of tung (Aleurites fordii). J. Agric. Res. 65, 361—378 (1942).
— Embryology of the tung tree. J. Agric. Res. 71, 215 bis 229 (1945).

McGregor, W. G.: Safflowers in Canada. Cereal Div., Cent. Exp. Fm, Ottawa, Nr. 62, 8 (1943). [Mimeographed.]

Majdrakov, P.: Die Züchtung der Ölpflanzen in Bulgarien. Mitt. Landw. 59, 122 (1944).

Marčenko, I. I.: (Ways of producing perennial and tuberous sunflowers.) Selekcija i Semenovodstvo (Breeding and Seed Growing), Nr. 7, 37—39 (1939). [Russisch.]

Mendes, P. T.: Nota preliminar sôbre a hibridação do tungue. (Preliminary note on the hybridization of tung.) Rev. Agric., Piracicaba 20, 274—276 (1945).
— u. O. F. de Sousa: Melhoramento da mamoneira (Ricinus communis L.). IV. — Segunda e terceira séries de ensaios de variedades anãs (1940/41 e 1941/42). (Improvement of the castor oil plant [R. communis L.]. IV. — Second and third series of trials of dwarf varieties [1940/41 and 1941/42].) Bragantia, São Paulo 5, 351—358 (1945).
—, — Melhoramento da mamoneira (Ricinus communis L.). V. — Primeira série de ensaios de linhagens e variedades (1938/39 e 1939/40). (Improvement of the castor oil plant [R. communis L.]. V. — First series of trials of races and varieties [1938/39 and 1939/40].) Bragantia, São Paulo 5, 359—380 (1945).
—, — Melhoramento da mamoneira (Ricinus communis L.). VI. — Segunda e terceira séries de ensaio de linhagens e variedades (1940/41 e 1941/42). (Improvement of the castor oil plant [R. communis L.]. VI. — Second and third series of trials of races and varieties [1940/41 and 1941/42].) Bragantia, São Paulo 5, 381—396 (1945).

Merrill, S. jun.: The budding of tung (Aleurites fordii Hemsl.). Proc. Amer. Soc. Hort. Sci. 44, 227—235 (1944).

Minkevič, I. A.: (The Soviet collection of linseed and its importance for breeding work.) Selekcija i Semenovodstvo (Breeding and Seed Growing), Nr. 1, 25 bis 26 (1940). [Russisch.]

Mohammad, A.: Breeding of improved varieties of oilseeds in the Punjab. Seas. Notes Punjab Agric. Dep. 18, Nr. 2, 21—24 (1939).
— u. N. D. Gupta: Inheritance of alternate and opposite arrangement of leaves in Sesamum indicum DC. Indian J. Agric. Sci. 11, 659—661 (1941).
— u. A. R. Khan: Some breeding investigations on linseed (Linum usitatissimum L.) in the Punjab. Indian J. Agric. Sci. 11, 432—445 (1941).
— u. S. M. Sikka: Pseudogamy in genus Brassica. Curr. Sci. 9, 280—282 (1940).
—. — Improvement of toria (Brassica napus L. var. dichotoma Prain) and taramira (Eruca sativa L.) by group-breeding. Indian J. Agric. Sci. 11, 589—596 (1941).
—, — u. M. A. Aziz: Inheritance of seed colour in some oleiferous Brassicae. Indian J. Genet. Pl. Breed. 2, 112—127 (1942).

Morozov, V. K.: (Breeding sunflower by the method of inbreeding.) Socialistic Grain Farming, Saratov, Nr. 2, 29—50 (1940). [Russisch.]
— (New varieties of sunflowers.) Bjulletenj Instituta Zernovogo Hozjajstva Jugo-Vostoka SSSR. (Bull. Inst. Grain Husb. S. E. USSR.) Saratov, Nr. 2, 6 bis 10 (1944). [Russisch.]
— (Methods of breeding sunflowers.) Bjulletenj Instituta Zernovogo Hozjajstva Jugo-Vostoka SSSR. (Bull. Inst. Grain Husb. S. E. USSR.) Saratov, Nr. 2, 11 bis 18 (1944). [Russisch.]
— (Seed production and breeding of the sunflower.) Naučnyj Otčet Inst. Zernovogo Hozjajstva Jugo-Vostoka SSSR. za 1941/42 gg. (Sci. Rep. Inst. Grain Husbandry South-Eastern USSR. for 1941/42), Pp. 190 bis 208 (1944). [Russisch.]
— u. A. M. Galaktionova: (Spatial isolation in breeding and seed production of sunflowers.) Jarovizacija, Nr. 2 (35), 43—46 (1941). [Russisch.]

Murray, D. B.: The Oil Palm Research Station of Nigeria. Trop. Agriculture, Trin. 22, 93—96 (1945).

Nabi, M. M.: Future of oil-seeds in the Rawalpindi Circle. Seas. Notes Punjab Agric. Dep. 18, 15—16 (1939).

Nesbitt, L. L. u. E. P. Painter: Does linseed oil contain conjugated double bonds? Bi-m. Bull. N. Dak. Agric. Exp. Sta. 6, 31—35 (1944).

Nicolaisen, W.: Probleme des Anbaus und der Züchtung von Raps und Rübsen. Forschungsdienst 11, 286 bis 299 (1941).

Oort, A. J. P.: Bepaalt het gehalte aan mosterd-olie de resistentie van Cruciferen tegen knolvoet? (Does the mustard oil content determine the resistance of Cruciferae to club root? Tijdschr. PlZiekt. 51, 117—119 (1945).

Pančenko, A. J.: (Die Rolle der Wurzel und des überirdischen Pflanzenteils der Sonnenblume in der Resistenz zum Sommerwurz.) C. R. (Doklady) Acad. Sci. URSS. 58, 911 (1947). [Russisch.]

Paul, W. R. C. u. P. M. Gaywala: The cultivation of gingelly in Ceylon. Trop. Agriculturist 97, 321—326 (1941).

Perry, B. A.: Chromosome number relationships in the genus Euphorbia. Amer. J. Bot. Suppl. 28, 4s—5s (1941). (Abstr.)

— Cytological relationships in the Euphorbiaceae. Va J. Sci. 3, 140—144 (1942).

Pietsch, A.: Beitrag zur photographischen Darstellung, Farbbestimmung und Bedeutung der ölhaltigen Samen von in Deutschland wachsenden Pflanzen. Landw. Jb. 91, 369—417 (1941).

Plotnikov, A. I.: (Selectivity of sunflower at fertilization.) Jarovizacija, Nr. 2 (29), 94—96 (1940). [Russisch.]

Popova, G. M.: (Breeding oil crops.) Selekcija i Semenovodstvo (Breeding and Seed Growing), Nr. 12, 5—7 (1939). [Russisch.]

Potter, G. F., E. Angelo, J. H. Painter u. R. T. Brown: A statistical study of variation in tung fruits. Proc. Amer. Soc. Hort. Sci. 37, 515—517 (1939) 1940.

Pustovojt, V.: (New varieties of sunflower.) Socialističeskoe Seljskoe Hozjajstvo (Socialistic Agriculture), Moscow, Nr. 2, 88—89 (1941). [Russisch.]

Putt, E. D.: Observations on morphological characters and flowering processes in the sunflower (Helianthus annuus L.). Sci. Agric. 21, 167—179 (1940).

— Investigations of breeding technique for the sunflower (Helianthus annuus L.). Sci. Agric. 21, 689—702 (1941).

— Association of seed yield and oil content with other characters in the sunflower. Sci. Agric. 23, 377—383 (1943).

— u. J. Unrau: The influence of various cultural practices on seed and plant characters in the sunflower. Sci. Agric. 23, 384—398 (1943).

Raghavan, T. S. u. K. V. Krishnamurthy: Chromosome number of Sesamum lacininiatum, Klein. Curr. Sci. 14, 152—153 (1945).

Ramanujam, S.: Autotriploidy in toria (Brassica campestris L.). Curr. Sci. 9, 325—326 (1940).

— A haploid plant in toria (Brassica campestris L.). Proc. Indian Acad. Sci. 14, Sect. B., 25—34 (1941).

— Chromosome number of Sesamum prostratum Retz. Curr. Sci. 10, 439—440 (1941).

— An interspecific hybrid in Sesamum - S. orientale L. × S. prostratum Retz. Curr. Sci. 11, 426—428 (1942).

— The cytogenetics of an amphidiploid Sesamum orientale × S. prostratum. Curr. Sci. 13, 40—41 (1944).

Ramella, R.: El lino P. 330 M. A. Su origen, historia y valor agrícola. (Linseed P. 330 M. A. Its origin, history and agricultural value.) "Granos" Semilla Selecta, B. Aires 3, Nr. 8, 3—15 (1939).

Raptopoulos, T.: Pollen germination tests in cherries. J. Pomol. 18, 61—67 (1940).

Rasul, C. K.: What's doing in All-India. Punjab. Indian Fmg 6, 180—181 (1945).

Regel, C.: In Mitteleuropa wildwachsende und angebaute Ölpflanzen. Angew. Bot. 22, 400—413 (1940).

— Beiträge zur Kenntnis von mitteleuropäischen Nutzpflanzen. Angew. Bot. 23, 137—151 (1941).

Reyntens, H.: Proeven op Witerkoolzaad. (Trials with colza.) Meded. LandbHoogesch. Gent 11, 287—303 (1946).

Richharia, R. H. u. W. J. Kalamkar: Chromosome number in Indian linseed, Linum usitatissimum L. Indian J. Agric. Sci. 9, 561—564 (1939).

— u. J. P. Kotval: Chromosome numbers in safflower (Carthamus tinctorius Linn.). Curr. Sci. 9, 73—74 (1940).

— u. D. P. Persai: Tetraploid til (Sesamum orientale L.) from colchicine treatment. Curr. Sci. 9, 542 (1940).

Ross, A. M.: Some morphological characters of Helianthus annus (sic) L., and their relationship to the yield of seed and oil. Sci. Agric. 19, 372—379 (1939).

Rybin, V. A.: Colchicine-induced tetraploidy in Helianthus annuus L. C. R. (Doklady) Acad. Sci. URSS. 24, 368—371 (1939).

Sabnis, T. S. u. T. R. Mehta: Improvement of linseed in the United Provinces. Indian Fmg 5, 224—226 (1944).

—, — A missing type of Brassica campestris recovered. Curr. Sci. 15, 171 (1946).

Sallans, H. R.: Canadian linseed. I. The effect of variety and environment on the composition of linseed. Canad. J. Res. 22, Sect. F., 119—131 (1944).

— u. G. D. Sinclair: Canadian linseed. II. Relations between iodine value and fatty acid composition of linseed. Canad. J. Res. 22, Sect. F., 132—145 (1944).

Scheibe, A.: Die Ölrauke (Eruca sativa Lam.), eine für Deutschland neue Ölpflanze. Landw. Jb. 91, 199—233 (1941).

Schilling: Die Ölleinzüchtung in Deutschland. Mitt. Landw. 56, 859—860 (1941).

Schwarze, P.: Zur Methodik der Auslese von senfölfreien Rapssorten. Züchter 17/18, 19—22 (1946).

Ščibrja, N. A.: (A cross between Jerusalem artichoke [Helianthus tuberosus] and sunflower [Helianthus annuus].) Vestnik Gibridizacii (Hybridization), Nr. 1, 66—84 (1941).

Scott, F. M. u. M. Reynolds: Traumatic acid and mitosis in Ricinus communis. Amer. J. Bot. Suppl. 28, 5s (1941). (Abstr.)

Sebto, A. G. u. M. I. Gilev: (An early variety of the castor oil plant for the littoral region.) Selekcija i Semenovodstvo (Breeding and Seed Growing), Nr. 9, 25 (1939). [Russisch.]

Sergejev, L. I.: (Frostresistenz der Ölbäume und Feichoa.) C. R. (Doklady) Acad. Sci. URSS. 58, 1203 (1947). [Russisch.]

— u. K. A. Sergejeva: (Anatomisch-physiologische Eigenschaften der Blätter des Ölbaumes [Olea europea] im Zusammenhang mit seiner Standhaftigkeit.) C. R. (Doklady) Acad. Sci. URSS. 57, 727—730 (1947). [Russisch.]

Sessous: Fortschritte in der Züchtung fettliefernder Pflanzen. Mitt. Landw. 54, 812—813 (1939).

Sheffield, A. F. W.: First annual report of the Oil Palm Research Station for 1939—1940. Nigeria: Part A, Pp. 8; Part B, Pp. 13 (1940).

Shehibrya, N. A. siehe Ščibrja.

Sidorov, B. N. u. N. N. Sokolov: Production of tetraploids in Ricinus communis treated with colchicine. C. R. (Doklady) Acad. Sci. URSS. 31, 264—265 (1941).

—, — (Eine weibliche Form von Ricinus communis.) C. R. (Doklady) Acad. Sci. URSS. 57, 497 (1947). [Russisch.]

Sikka, S. M.: Cytogenetics of Brassica hybrids and species. J. Genet. 40, 441—509 (1940).

Sikka, S. M.: Species formation and economic utility of amphidiploids in Brassica. Proc. 27th Indian Sci. Congr., Madras Pt III, Sect. Agric. **34**, 221 (1940). (Abstr.)

Smith-White, S.: Studies on the tung oil tree (Aleurites Fordii Hemel). J. Roy. Soc. N. S. W. **74**, 42—73 (1940).

Spangenberg, G. E.: A cultura da mamoneira. (The cultivation of the castor oil plant.) Bol. Agron. Brazil. **10**, 207—210 (1946).

Srinivasan, A. R.: Contribution to the morphology of Pedalium murex Linn. and Sesamum indicum D. C. Proc. Indian Acad. Sci. **16**, Abt. B, 155—164 (1942).

Stahmann, M. A., K. P. Link u. J. C. Walker: Mustard oils in crucifers and their relation to resistance to clubroot. J. Agric. Res. **67**, 49—63 (1943).

Stankov, S. S.: (Wild oil plants of the USSR. and their practical utilization.) Seljhozgiz, Pp. 78 (1944). [Russisch.]

Taran, E. N.: (Oil extracted from the first fruits of the hybrid Aleurites Fordii × Aleurites cordata.) Proc. Lenin Acad. Agric. Sci. USSR., Nr. 7, 34—36 (1941). [Russisch.]

Tavčar, A.: Somatische Mutationen an Blättern und im Perikarp von Sonnenblumen (Helianthus annuus L.). Poljod. Znanstvena Smotra, Zagreb, Nr. 5, 40 (1942).
— Boja perikarpa i sadržaj ulja u sjemenu sunčanice (Helianthus annuus L.). (Pericarp colour and oil content in seeds of the sunflower [H. annuus L.].) Poljodjelska Znanstvena Smotra, Zagreb, Nr. 8, 72—87 (1944).

Toovey, F. W. u. G. K. G. Campbell: Second annual report of the Oil Palm Research Station, Nigeria for 1940—1941. Part B. Selection, breeding and other botanical investigations, Pp. 8 (1941).
—, — Third annual report of the Oil Palm Research Station, Nigeria. Part B. Selection, breeding, and other botanical investigations. 1941—1942, Pp. 8 (1942).

Troll, H.-J.: Ölpflanzen. Anbau und Züchtungsprobleme bei erprobten und neuen Arten. Umschau, Pp. 61—65 (1943).

Turesson, G. u. H. Nordenskiöld: Chromosome doubling and cross combinations in some Cruciferous plants. Ann. Agric. Coll., Sweden **11**, 201—206 (1943).

Unrau, J. u. W. J. White: The yield and other characters of inbred lines and single crosses of sunflowers. Sci. Agric. **24**, 516—525 (1944).

Vachhani, M. V.: Phyllody of til in relation to date of sowing. Curr. Sci. **14**, 238 (1945).
— u. R. B. Deshpande: A case of partial dialysis of carpels in linseed. Indian J. Genet. Pl. Breed. **2**, 178 bis 180 (1942).

Vakulin, D. J.: A new narrow-leaved large-seeded form of Lallemantia iberica. C. R. (Doklady) Acad. Sci. URSS. **25**, 781—783 (1939).
— (On the possibility of determination of varieties of Lallemanzia according to absolute weight and size of seeds.) Proc. Lenin Acad. Agric, Sci. USSR., Nr. 10, 16—18 (1940). [Russisch.]

Vanderweyen, R.: L'élimination des pisifera ou stériles dans les palmeraies issues de croisements Tenera × Tenera. Bull. Agric. Congo Belge **33**, 114—122 (1942).

Vasiljev, V. F. (Herausgeber): Krambe [Crambe abyssinica Hochst.] — eine neue Ölpflanze.) Voronež, Pp. 86 (1941). [Russisch.]

Vorobjeva, N. F.: (A method of estimating the oil content of sunflower seeds during breeding.) Socialistic Grain Farming, Saratov, Nr. 1, 171—175 (1941). [Russisch.]

Vydrin, V. I.: (New promising varieties of sunflower for Siberia.) Selekcija i Semenovodstvo (Breeding and Seed Growing), Nr. 10/11, 27—29 (1939).

Walster, H. L.: Advancing the flax front. Bi-m. Bull. N. Dak. Agric. Exp. Sta. **5**, Nr. 5, 2—11 (1943).

Webster, C. C.: A note on a uniformity trial with oil palms. Trop. Agriculture, Trin. **16**, 15—19 (1939).
— Recent progress in the cultivation of tung oil trees (Aleurites fordii and A. montana.) Trop. Agriculture, Trin. **16**, 267—271 (1939).
— Notes on the cultivation of tung oil trees. Part 3: Possibilities for the production of improved planting material. Nyasald Tea Ass. Quart. J. **5**, Nr. 2, 6 bis 10 (1940).
— A note on the possible production of hybrid seed by natural cross pollination between Aleurites Fordii and A. montana. Nyasald Agric. Quart. **1**, Nr. 2, 14 bis 15 (1941).
— Notes on the progress of tung experimental work. The selection and yield recording of promising trees. Nyasald Agric. Quart. **1**, Nr. 2, 16—18 (1941).
— A note on the yield of tung trees in Nyasaland. E. Afr. Agric. J. **6**, 160—163 (1941).
— A note on pollination in budded plantations of tung trees. (Aleurites montana.) Nyasald Agric. Quart. J. **3**, Nr. 3, 17—19 (1943).
— Observations and experiments on flowering and pollination of the tung tree. E. Afr. Agric. J. **9**, 136—143 (1944).
— Improved planting material of the tung tree: a progress report. E. Afr. Agric. J. **11**, 165—169 (1946).
— The cultivation of the tung tree in Nyasaland. Emp. J. Exp. Agric. **14**, 18—24 (1946).
— u. M. F. H. Selby: A progress report on the selection and vegetative propagation of promising tung trees (Aleurites Montana). Nyasald Agric. Quart. J. **4**, 1—6 (1944).

Weibel, R. O. u. C. M. Woodworth: Use of the natural crossing plot in making castor bean hybrids. J. Amer. Soc. Agron. **38**, 563—565 (1946).

Yazicioğlu, T.: Türkiye'nin nebatî yağ zenginliği. (Turkey's riches in oil-bearing plants.) T. C. Yüksek Ziraat Enstitüsü Basimevi, Ankara, Nr. 150, 119 (1945).

Zagorodnyj, G. P. u. S. M. Haritonova: (Xanthium Strumarium as a new oil plant.) Proc. Daghestan Agric. Inst., Nr. 1, 270—276 (1939). [Russisch.]

Zaharov, B. S.: (The lenght of the daylight period, and its influence on the incidence of infection by Orobancle in the sunflower.) Doklady Akad. Nauk SSSR. **27**, 265—267 (1940). [Russisch.]
— Influence of daylength on susceptibility of sunflower to affection with broom-rape. C. R. (Doklady) Acad. Sci. URSS. **32**, 446—447 (1941).
— Resistance of sunflower to broomrape in relation to photoperiod. C. R. (Doklady) Acad. Sci. URSS. **34**, 262—264 (1942).

m) Kampfer-Pflanzen.

Golubinskij, I. N.: (New experimental tetraploid specimen of Ocimum canum Sims.) J. Bot. URSS. **24**, 104—107 (1939). [Russisch.]

Kravčenko, J.: (Variety trials of Ocimum gratissimum.) Soviet Subtropics, Nr. 2/3, 82—84 (1939). [Russisch.]

Lapin, V. K.: Production of an amphidiploid basil Ocimum canum Sims. × Ocimum gratissimum L. by colchicine treatment. C. R. (Doklady) Acad. Sci. URSS. **23**, 84—87 (1939).

Nilov, V. I. (Editor): (The biochemistry and physiology of southern trees and shrubs.) Vsesojuznaja Akademija Seljskohozjajstvennyh Nauk im. V. I. Lenina. Gosudarstvennyj Nikitskij Botaničeskij Sad im. Molotova (Lenin Academy of Agricultural Sciences, Molotov's Nikita Botanic Garden), Jalta 21, Nr. 2, 176 (1939). [Russisch.]

n) Gerbstoffpflanzen.

Andersson, G.: Ny förädlingsgren. (A new sphere in plant breeding.) Sverig. Utsädesfören. Tidskr. 57, 70 (1947).

Kokina, S. I. u. A. J. Kokin: (On the contents of tannins in Calligonum species.) Botaničeskij Žurnal SSSR. (Journal botanique de l'URSS) 32, Nr. 1, 23—31 (1947). [Russisch.]

Philp, J. u. S. P. Sherry: The degree of natural crossing in green wattle, Acacia decurrens Willd. and its bearing on wattle breeding. J. S. Afr. For. Ass., Nr. 14, 1—28 (1946).

Wind, R.: De looistoffenindustrie in Nederlandsch-Indië. (The tanning substances industry in the Netherlands East Indies.) Bergcultures 15, 569—574; also Econ. Weekblad 10, Nr. 16 (1941).

o) Farbpflanzen.

Ganguly, J. K.: The somatic and meiotic chromosomes of Commelina benghalensis Linn. Curr. Sci. 15, 112 (1946).

Leggett, W. F.: Ancient and medieval dyes. Brooklyn, N. Y., Pp. VI+95 (1944).

Rao, L. N.: Studies in Santalaceae. Ann. Bot., Lond. 6, 151—175 (1942).

p) Heilpflanzen.

Anonymus: Kinaproeftuin „Cinchona". (Cinchona Experimental Plantation „Cinchona".) Bergcultures 14, 322 (1940).
— Note sur la culture du quinquina aux Indes Néerlandaises. Agric. Élev. Congo Belge 14, 10—11 (1940).
— Economy in the use of drugs in war-time. Med. Res. Coun. Ther. Req. Comm. London War Memor, Nr. 3, 18 (1941).
— Cinchona breeding. J. N. Y. Bot. Gdn. 44, 192 (1943).
— Quinine content of newly discovered cinchona stands. Science Suppl. 97, Nr. 2518, 12 (1943).
— Medicinal plants in the USSR. Agriculture, Moscow, Nr. 10, 9—11 (1945). [Mimeographed.]
Asana, J. J. u. R. N. Sutaria: On the number of chromosomes of some Indian Araceae. J. Univ. Bomb. 7, 58—62 (1939).

Ball, R. S.: Pyrethrum cultivation in Kenya. Tea Quart. 17, 28—36, also Nyasald Agric. Quart. J. 4, 7—18 (1944).
Barnard, C. u. H. Finnemore: Drug plant investigations. 1. Progress report. J. Coun. Sci. Industr. Res. Aust. 18, 277—285 (1945).
Boeiman, H. A. C.: De mogelijkheden van de cultuur van eenige minder bekende geneeskruiden en hare toepassing. (The possibilities of cultivating some of the less wellknown medicinal plants and their application.) Bergcultures 14, 1160—1166 (1940).
Bonisteel, W. J.: Polyploidy in relation to chemical analysis. J. Amer. Pharm. Ass. 29, 404—408 (1940).
Brewer, W. R. u. A. Laurie: Culture studies of the drug plant Atropa belladonna. Proc. Amer. Soc. Hort. Sci. 44, 511—517 (1944).

Camp, W. H.: Biochemical clines, polymorphic populations, and the problems of specific delimitation on the Cinchona population of Ecuador. Amer. J. Bot. (Suppl.) 33, 234 (1946). (Abstr.)
Carvalho, A. u. C. A. Krug: A quineira (Cinchona sp.) — origem, classificação, exploração econômica no mundo e tentativas de sua aclimatação no Brasil. (Cinchona, its origin, classification, economic utilization in the world, and attempts at acclimatizing it in Brasil.) Instituto Agron. Campinas, S. Paulo, Pp. 141 (1944).
Cheney, R. H.: Medicinal herbaceous species in the north-eastern United States. Bull. Torrey Bot. Cl. 73, 61—72 (1946).
Chevalier, A.: Quelques Strychnos africains inoffensifs ou peu toxiques. Rev. Bot. Appl. 27, 353—377 (1947).
Croizat, L.: New species of Croton from Guatemala. Field Museum of Natural History, Chicago Bot. Ser. 22, Nr. 8, Publ. 516, 445—453 (1942).

Davidson, J.: The Cascara tree in British Columbia. Bull. Prov. B. C., Nr. 108, 32 (1942).
Dawson, R. F.: Translocation of hyoscyamine in Datura Stramonium L. Amer. J. Bot. 31, Nr. 8, 9s. (1944). (Abstr.)

Efimenko, O. M.: (A micro-chemical study of the cinchona tree.) Vestnik Socialističeskogo Rastenievodstva (Soviet Plant Industry Record, Nr. 2, 151—158 (1940).
Esdorn, I.: Beiträge zur Heilpflanzenforschung. Ber. dtsch. bot. Ges. 57, 166—175 (1939).

Fernie, L. M.: The vegetative propagation of cinchona by cuttings. E. Afr. Agric. 12, 228—236 (1947).
Fischer, A. F.: Growing cinchona under American control. Torreya 44, 1—5 (1944).
Flück, H.: Probleme der Arzneipflanzenproduktion in der Schweiz. Vjschr. Naturf. Ges. Zürich 86, Nos 3/4, XXIII—XXIV (1941).
Fourment u. Roques: Répertoire des plantes médicinales et aromatiques d'Algérie. Bull. Agric., Algérie. Nr. 61, 159 (1942).

Georgieff, C.: (Cultivation and collection of medicinal nda aromatic plants in Bulgaria.) Minist. Agric. St. Lands. St. Exp. Field Sta.-Kazanlak, Pp. 271 (1940).
Glotov, V.: Amphidiploid fertile form of Mentha piperita L. produced by colchicine treatment. C. R. Acad. Sci. URSS, n. s. 28, 450—453 (1940).
Greenway, P. J.: Empire production of drugs. IV. Strophanthinum. E. Afr. Agric. J. 11, 184—185 (1946).
Gregory, W. C.: Cytology and phylogeny in the Ranunculaceae. Proc. Va Acad. Sci. 1, 215—216 (1939—1940).

Heeger, E. F.: Sortenkundliche Untersuchungen zur Kenntnis der deutschen Baldriansorten. Heil- u. Gewürzpfl. 21, 1—35 (1942).
Hills, K. L. u. G. P. Kelenyi: A preliminary report upon the cultivation of Duboisia spp. J. Coun. Sci. Industr. Res. Aust. 19, 359—375 (1946).
Hodge, W. H.: Alkaloid distribution in the bark of some Peruvian cinchonas. Caribbean Forester, Puerto Rice 7, 79—86 (also in Spanish 86—97) (1946). [Mimeographed.]

Jacob, K. T.: Chromosome numbers and the relationship between satellites and nucleoli in Cassia and certain other Leguminosae. Ann. Bot., London 4, 201—226 (1940).
Jayaweera, D. M. A.: Drug plants (indigenous and exotic) that can be grown in Ceylon-Part 1. Trop. Agriculturist 101, 130—135 (1945).
Jensen, H. W.: Heterochromosome formation in Benzoin aestivale. J. Elisha Mitchell Sci. Soc. 57, 202—203 (1941).

Krug, C. A., C. S. N. Antunes, J. B. C. Nery Sobrinho u. A. Carvalho: Pesquisas de aclimatação de quineiras (Cinchona sp.) no estado de São Paulo. (Investigations on acclimatizing Cinchona sp. in the state of São Paulo.) Inst. Agron. Estad. A. Paulo Fundos Univ. Pesquisas, Pp. 97 (1945).

Litardière, R. de: Recherches taxonomiques et caryologiques sur le Melissa officinalis L. Rev. Bot. Appl. **25**, 16—18 (1945).

Little, R. D.: Histology of barks of Cinchona and some related genera occurring in Colombia. For. Econ. Admin., Gen. Commodities Div. Cinchona Sect., Washington, D. C., p. 73 (1945).

Longacre, D. J.: Somatic chromosomes of Aconitum noveboracense and A. uncinatum. Bull. Torrey Bot. Cl. **69**, 235—239 (1942).

Madueño Box, M.: Contribuciones al estuduio de plantas medicinales productoras de alcaloides. (Contributions towards the study of medicinal plants producing alkaloids.) Bol. Inst. Inwest. Agron. Madr., Nr. 10, 137—176 (1944).

Manciot, A.: Les plantes médicinales. Paris, Pp. 160. illus. (ohne Jahr).

Marañon, J. u. H. H. Bartlett: Cinchona cultivation and the production of totaquina in the Philippines. Nat. Appl. Sci. Bull. Univ. Philipp. **8**, 111—187 (1941).

Martin, W. E. u. J. A. Gandara: Alkaloid content of Ecuadoran and other American cinchona barks. Bot. Gaz. **107**, 184—199 (1945).

Marzell, H.: Die deutschen Namen der heimischen Heilpflanzen. Dtsch. Heilpflanzen **10**, Nr. 3, 17—20 (1944).

Mehra, P. N.: Chromosome number and morphology in some species of the genus Ephedra. Proc. 31st. Indian Sci. Congr. Delhi, Pt III, p. 92 (1944).(Abstr.)

— Colchicine and sulphanilamide effect on the mitotic division of body nucleus in the pollen grains in the genus Ephedra. Proc. 31st Indian Sci. Congr., Delhi, Pt III, p. 93 (1944). (Abstr.)

Melville, R.: The botanical source of Indian Belladonna. J. Bot., Lond. **80**, 54—55 (1942).

Mendes, A. J. T.: Estudo citológico da quineira (Cinchona spp.). (Cytological study of Cinchona spp.) J. Agron., S. Paulo **2**, 43—48: also Bol. Téc. Inst. Agron. Campinas, Nr. 58, 8 (1939).

Mirimanoff, A.: Procédés modernes de culture de plantes officinales. Rev. Bot. Appl. **27**, 417—418 (1947).

Moreau, R. E.: An annotated bibliography of cinchona-growing from 1883—1943. East. Afr Agric. Res. Inst., Amani, Pp. 41 (1945).

Mukerji, B. u. S. K. Ghosh: Lobelia nicotianaefolia Heyne as substitute for Lobelia inflata Linn., B. P. Curr. Sci. **14**, 198—199 (1945).

Murthi, S. N.: Studies in the Labiatae. IV. Contribution to the morphology of Orthosiphon stamineus Benth. J. Indian. Bot. Soc. **26**, 87—94 (1947).

Oudot, G.: Notes sur le quinquina (extrait du compte rendu de mission aux Indes Néerlandaises de M. G. Oudot, ingénieur des travaux d'agriculture de l'Indochine). Bull. Écon. Indochine **42**, 777—788 (1939).

Pantulu, J. V.: A note on the chromosome numbers of Cassia. Curr. Sci. **9**, 416—417 (1940).

— Chromosome number in Cassia sophera Linn. Curr. Sci. **15**, 77 (1946).

— Chromosome number of Cassia fistula. Curr. Sci. **15**, 255 (1946).

Peacock, S. M., jr., D. B. Leyerle u. R. F. Dawson: Alkaloid accumulation in reciprocal grafts of Datura Stramonium with tobacco and tomato. Amer. J. Bot. **31**, 463—466 (1944).

Pereira, J. R.: Contribuicão para o estudo das plantas alucinatórias particularmente da maconha (Cannabis sativa L.). (Contribution towards the study of plants producing hallucinations, particulaly hashish [C. sativa L.].) Rev. Flora Medicinal, S. Paulo **12**, 127 (1945).

Pfeiffer, N. E.: Prolonging the life of Cinchona pollen by storage under controlled conditions of temperature and humidity. Contr. Boyce Thompson Inst. **13**, 281 bis 293 (1944).

Popenoe, W.: Quinine from the „fever-tree". Agriculture in the Americas **2**, 43—47 (1942).

— Cultivo de la quina (cinchona) en Guatemala. (The cultivation of quinine (cinchona) in Guatemala.) Rev. Fac. Nac. Agron Colombia **4/5**, Nr. 18, 314—332 (1942).

Porterfield, W. M. jr.: China's contribution in medicinal herbs. J. N. Y. Bot. Gdn **43**, 223—230 (1942).

Rowson, J. M.: Increased alkaloidal contens of induced polyploids of Datura. Nature, Lond. **154**, 81—82 (1944).

Shimoya, C.: Observacões citológicas em chaulmoogra. (Cytological observations on chaulmoogra.) Ceres, Brasil **6**, 76—81 (1944).

Sievers, A. F. u. E. C. Higbee: Medicinal plants of tropical and subtropical regions. Foreign Agric. Rep. U. S. Dep. Agric., Nr. 6, 47 (1942).

Sirks, M. J.: Genotypical predetermination in Datura. Genetica **22**, 197—214 (1940).

Steere, W. C.: The discovery and distribution of Cinchona pitayensis in Ecuador. Bull. Torrey Bot. Cl. **72**, 464—471 (1945).

— The botanical work of the cinchona missions in South America. Science **101**, 177—178 (1945).

Stoffels, E. H. J.: Le quinquina. (Cinchona.) Publ. Inst. Agron. Congo Belge, Sér. Tech., Nr. 24, 51 (1939).

— Le quinquina. (Cinchona.) Publ. Inst. Agron. Congo Belge, 2me Ed. Sér. Tech., Nr. 24, 57 (1945).

Stuhr, E. T.: The distribution, abundance and uses of wild drug plants in Oregon and Southern California. Econ. Bot. **1**, 57—68 (1947).

Swen, M. S. D.: Importance of yellow chang-shan and its plan for mass production. Curr. Sci. **14**, 334 (1945).

Swirlowsky, Ed.: Hybridologische Studien in der Gattung Digitalis. J. Genet. **38**, 533 (1939).

Thomas, A. S.: Cinchona in Uganda. Emp. J. Exp. Agric. **14**, 75—84 (1946).

Vasiljčenko, I. T.: (The phylogenetic significance of seedling morphology in the Umbelliferae.) Sovetskaja Botanika (Soviet Botany), Nr. 3, 30—40 (1941). [Russisch.]

Wit, F.: De biologie van de vruchtvorming bij Strophantus gratus. (The biology of fruitsetting of S. gratus.) Landbouw **17**, 98—105 (1941).

Wulff, H. D.: Über die Ursache der Sterilität des Kalmus (Acorus calamus L.). Planta **31**, 478—491 (1940).

Žukovskij, P. M.: (New kinds of plant material of special interest.) Moskovskaja ordena Lenina Seljskohozjajstvennaja Akademija imeni K. A. Timirjazeva. Naučnaja Konferencija 3—10 ijunja 1944 g. Doklady. [Proc. Sci. Conf. Timirjazev Agric. Acad. (3—10 June, 1944).] Nr. 1, 7—13 (1945). [Russisch.]

q) Kautschukpflanzen.

Abbe, L. B.: A rapid histological technic for staining latex in roots of Taraxacum kok-saghyz. Stain Tech. **21**, 19—22 (1946).

Adriaens, L.: Contribution à l'étude chimique de quelques gommes du Congo Belge. M. Hayez, Bruxelles [Mém. Inst. Roy. Colon. Belge (Sect. Sci. Nat. Méd.) Tome VIII].

A-n, G.: Utsädesföreningens extra möte under Lantbruksveckan 1945. (Special meeting of the Seed Association during the Agricultural Week 1945.) Sverig. Utsädesfören. Tidskr. **55**, 144—146 (1945).

Andersson, G.: Möjligheterna för en svensk produktion av naturgummi. (The possibilities of natural rubber production in Sweden.) K. LantbrAkad. Tidskr. **85**, 269—282 (1946).

— u. G. Bengtsson: Redogörelse för första årets arbeten med gummimaskros vid Sveriges Utsädesförening. (Report on the first year's work with koksaghyz by the Swedish Seed Association.) Sverig. Utsädesfören. Tidskr. **55**, 174—200 (1945).

Anonymus: Identificatiekenmerken van de voornaamste in de praktijk aangeplante Hevea-cloonen. (Distinctive characteristics of the main Hevea clones grown by planters.) Archipel Drukk, Buitenzorg (1939).

— Buddings and clonal seedlings. 1st Quart. Circ.Ceylôn Rubb. Scheme **16**, 32—40 (1939).

— Aanbevolen Hevea-plantmateriaal 1939/40 (recommended Hevea planting material 1939—1940). Bergcultures **13**, 1641—1646 (1939).

— Selectie bij jonge hevea, op grond van den latexvloei ten gevolge van insnijdingen in den bast door middel van een hoeking wielmes. (Selection of young hevea on the basis of latex flow resulting from incisions in the bark by means of an angular wheel knife.) Bergcultures **14**, 582—583 (1940).

— Travaux de la Division des Hevea de l'Inéac. Agric. Lev. Congo Belge **14**, 8—10 (1940).

— Eenige aanvullende gegevens van het Proefstation Midden- en Oost-Java betreffende het hiervoor behandelde onderwerp. (Some supplementary data from the Central and East Java Experiment Station regarding the preceding topic treated.) Bergcultures **14**, 86—87 (1940).

— Aanbevolen Hevea-plantmateriaal 1940—1941. (Recommended Hevea planting material 1940—1941.) Bergcultures **14**, 1578—1584 (1940).

— Gewijzigde internationale standaardiseering van aanduidingen voor tapsystemen bij rubber (1940). (The international notation for tapping systems [Revised version: 1940].) Arch. Rubbercult. Ned.-Ind. **24**, 315 bis 332 (1940).

— Wickham did not „smuggle" Hevea seeds from Brazil. Rubb. Res. Inst. Plant. Bull., Nr. 12, 12—13 (1940).

— Hevea seeds from Brazil. Rubb. Res. Inst. Plant. Bull., Nr. 13, 8—9 (1940).

— Ressort Lampongsche Districten van het Suid- en West-Sumatra Syndicaat. (The Lampong districts area of the South and West Sumatra Syndicate.) Bergcultures **15**, 1200—1205 (1941).

— Rubbercultuur in tropisch Amerika. (Rubber cultivation in tropical America.) Bergcultures **15**, 1373 bis 1382 (1941).

— Uniformiteit in de aanduiding van Hevea-plantmateriaal. (Uniformity in the description of Hevea planting material.) Bergcultures **15**, 1560—1563 (1941).

— Een aanstaande rubberboom in tropisch Amerika? (A coming rubber tree in tropical America.) Bergcultures **15**, 1638 (1941).

— Aanbevolen Hevea-plantmateriaal 1941—1942. (Recommended Hevea planting material 1941—1942.) Bergcultures **15**, 1692—1709 (1941).

— (Kok saghyz cultivation.) Seljhozgiz, Moskva, p. 98 (1941). [Russisch.]

— Résistance au vent des clones d'hévéa dans la région de Yangambi. Bull. Agric. Congo Belge **32**, 69—82 (1941).

— U.S.A.: Russian dandelion seed arrives for rubber test. Chronica Botanica **7**, 180—181 (1942).

— Rubber from guayule. Trop. Agriculture, Trin. **20**, 161—163 (1943).

Anonymus: Schutz und Meliorierung der Wälder. Intersylva **3**, 553—554 (1943).

— Cultivation and breeding of Russian rubber-bearing plants. Imperial Bureau of Plant Breeding and Genetics, Cambridge, Pp. 57 (1944) [Mimeographed.]

— Recommended planting material (1944). Adv. Circ. Res. Scheme Ceylon. Nr. 20, 4 (1944).

— Hevea hybrids for the western hemisphere. Agric. Amer. **4**, 37 (1944).

— The control of bark rot and canker. Adv. Circ. Rubb. Scheme, Ceylon, Nr. 21, 4 (1944).

— Improved kok-saghyz variety. Soviet News, Nr. 1568, 3 (1946).

Arias, A. C.: Estudios climatológicos. Areas geográficas de dispersión. Parthenium argentatum, Hevea brasiliensis, Castilloa elastica. (Climatological studies. Geographical areas of distribution. P. argenteum, H. brasiliensis, C. elastica.) Secretaría de Agricultura y Fomento, Dirección de Geografía, Meteorología e Hidrología, México, Pp. 112 (1942).

Artschwager, E.: Contribution to the morphology and anatomy of guayule (Parthenium argentatum). Tech. Bull. U. S. Dep. Agric., Nr. 842, 34 (1943).

— u. R. C. McGuire: Contribution to the morphology and anatomy of the Russian dandelion (Taraxacum kok-saghyz). Tech. Bull. U. S. Dep. Agric., Nr. 843, 24 (1943).

Balandin, D. A. u. B. P. Kolesnikov: (The content of guttapercha in the species of Euonymus in the Maritime Province.) Sovetskaja Botanika (Soviet Botany), Nr. 4, 42—54 (1943).

Baldwin, J. T. jr.: A first interpretation. Amer. J. Bot. (Suppl.) **33**, 215—216 (1946). (Abstr.)

— Hevea rigidifolia. Amer. J. Bot. **34**, 261—266 (1947).

— Hevea: a first interpretation. J. Hered **38**, 54—64 (1947).

Bangham, W. N. u. A. d'Angremond: Tapresultaten van eenige nieuwe A. V. R. O. S.-Hevea-cloonen, welke door kunstmatige kruis-bestuiving ontstaan zijn. (Tapping results on some new A. V. R. O. S. Hevea clones which originated in cross-pollinations.) Arch. Rubber. cult. Ned.-Ind. **23**, 191—231 (1939).

Bannan, M. W.: Tetraploid Taraxacum kok-saghyz. I. Characters of the leaves and inflorescences in the parental colchicine-induced generation. Canad. J. Res. **23**, Sect. C., 131—143 (1945).

— Tetraploid Taraxacum kok-saghyz. II. Characters of F_1 plants grown in pots. Canad. J. Res. **24**, 81—97 (1946).

— Tetraploid Taraxacum kok-saghyz. III. Achene weight, flowering and plant development. Canad. J. Res. **25**, Sect. C., 59—72 (1947).

Baranovskij, A. L. u. T. N. Makarevič: (Spindle tree Euonymus europea a valuable gutta-percha bearing plant.) Proc. Lenin Acad. Agric. Sci. USSR., Nos. 3/4, 23—25 (1942). [Russisch.]

Belikov, P. S.: (Der Einfluß des Heteroauxins auf die „Beschleunigung" und Frostresistenz der Kok-Saghyzstecklinge.) C. R. (Doklady) Acad. Sci. URSS. **58**, 143 bis 145 (1947). [Russisch.]

—, B. L. Lipman u. I. I. Olejnikova: (Physiologischbiochemische Charakteristik der Zuchtsorten von KokSaghyz.) C. R. (Doklady) Acad. Sci. URSS. **54**, 545 bis 584 (1946). [Russisch.]

— u. V. I. Olejnikov: (Zur Physiologie der Kautschukspeicherung in den Wurzeln der Zuchtsorten der KokSaghyz.) C. R. (Doklady) Acad. Sci. URSS. **58**, 1191 (1947). [Russisch.]

Bergner, A. D.: Guayule plants with low chromosome numbers. Science **99**, 224—225 (1944).

— Polyploidy and aneuploidy in guayule. Tech. Bull. U. S. Dep. Agric., Nr. 918, 36 (1946)

Bie, G. J. van der: Een vergelijkend onderzoek van de eigenschappen van de rubber, verkregen van oculaties op onderstam van Hevea Spruceana hybride en van Hsvea brasiliensis. (A comparison of the properties of rubber, obtained from buddings on stock of Hevea Spruceana hybrids and of Hevea brasiliensis.) Arch. Rubbercult. Ned.-Inp. **25**, 271—293 (1941).
— Onderzoek van latex, verkregen van oculaties op Hevea spruceana hybride onderstam en op Hevea brasiliensis onderstam. (Investigation of the properties of latex, obtained from buddings on stock of H. Spruceana hybrid and of H. brasiliensis.) Arch. Rubbercult. Ned.-Ind. **25**, 503—509 (1941).
Blanchard, A. J., Revised by A. Avakian u. R. W. Moats: Guayule. A list of references. Libr. List U. S. Dep. Agric. Libr., Washington, D. C., Nr. 10, 61 (1944). [Mimeographed.]
Blandin, J. J.: Why rubber is coming home. Agric. Amer. **1**, Nr. 4, 1—7, 10 (1941).
Böhme, R. W.: Anbau und Züchtung von Kautschuk- und Guttaperchapflanzen in der gemäßigten Zone. (Sammelreferat.) Z. Pflanzenz. **23**, 411—453 (1940).
Brandes, E. W.: Rubber on the rebound-East to West. Agric. Amer. **1**, Nr. 3; 1—4, 6—7, 10—11 (1941).
— Rubber from the Russian dandelion. Agric. Amer. **2**, 127—131 (1942).
— A borracha de torna-viagem à América. (Rubber on the rebound to America.) Bol. Minist. Agric., Rio de J. **33**, Nr. 2, 101—116 (1944).
Budhiraja, K. L. u. R. Beri: Common latex bearing woody plants of India. Indian For. Leafl., For. Res. Inst. Dehra Dun, Nr. 70, 18 (1944).
Camargo, F.: Uma nova planta Brasileira produtora de borracha. (A new Brazillian plant producing rubber) Bol. Minist. Agric. Rio de J. **32**, 45—55 (1943).
Chevalier, A.: Études botaniques sur le genre Hevea. Rev. Bot. Appl. **22**, 1—12 (1942).
Colenbrander, G. H.: Het splitsen van Hevea-kiemplanten. (The splitting of Hevea seedlings.) Bergcultures **14**, 84—85 (1940).
Coster, C.: Het keuren op cloonechtheid van heveazaden. (Examination of rubber seeds for clonal authenticity.) Bergcultures **15**, 192 (1941).
Cramer, P. J. S.: Een proef met Testatex-selectie en over vroeg uitdunen. (An experiment on selection by the Testatex method and early thinning out.) Arch. Rubbercult. Ned.-Ind. **24**, 333—352 (1940).
Demmon, E. L.: Rubber production opportunities in the American tropics. J. For. **40**, 207—210 (1942).
Dijkman, M. J.: Verdere resultaten der Heveaselectie in West-Java. (Further results of Hevea selection in West Java.) Bergcultures **13**, 492—503 (1939).
— Selectieve uitdunning in jonge rubberzaailingen-aanplanten. (Selective thinning in young rubber seedling plantations.) Bergcultures **15**, 11—15 (1941).
— u. F. W. Ostendorf: Zaailingentoetstuin Pangkalan 1929. (Seedling test-garden Pangkalan 1929.) Arch. Rubbercult. Ned.-Ind. **25**, 435—465 (1941).
— u. J. S. Vollema: Resultaten van een hoog gelegen heveacloonentoetstuin. (Results from a high lying plantation for resting rubber clones.) Arch. Rubbercult. Ned.-Ind. **24**, 557—570 (1940).
Doten, S. B.: Rubber from rabbit brush (Chrysothamnus nauseosus). Bull. Nev. Agric. Exp. Sta., Nr. 157, 22 (1942).
Ducke, A.: Novas contribuições para o conhecimento des seringueiras („Hevea") da Amazônia brasileira. (Further contributions towards the knowledge concerning the Hevea species of Brazilian Amazonia.) Arq. Serv. Florestal, Rio de Janeiro **2**, 25—43 (1943).
Dunin-Barkovskij, V. N.: (Wild kendyrj [Apocynum, sibiricum P.], its utilization and preservation.) Moskva Pp. 176 (1941).

Eikema, J. S.: Kiem- en bewaarproeven van hevea zaden. (Germination and storage tests of Hevea seeds.) Bergcultures **15**, 1049—1060 (1941).
Esau, K.: Apomixis in guayule. Proc. Nat. Acad. Sci., Wash. **30**, 352—355 (1944).
Federer, W. T.: Studies on sample size and number of replicates for guayule investigations. J. Amer. Soc. Agron. **37**, 469—478 (1945).
Ferrand, M.: Note sur la sélection de l'Hevea en pépinière. Rapport Annuel pour l'Exercice 1938 (2me Partie) Publ. Inst. Agron. Congo Belge Hors Sér., Pp. 99—113 (1939).
Ferrand: Phytotechnie de l'Hevea brasiliensis. Paris, Pp. 435. 67 figs. tables (1944).
Ferwerda, F. P.: Gegevens tot medio 1939 betreffende de toetstuinen voor cloonen en zaaisels aangelegd in de periode 1926—1932. (Data up to the middle of 1939 on the test plantations for clones and seedlings laid down in the period 1926—1932.) Arch. Rubbercult. Ned.-Ind. **24**, 353—394 (1940).
Filippov, D. I.: (Some problems of breeding kok-saghyz.) Jarovizacija, Nr. 3 (36), 21—28 (1941). [Russisch.]
— (Root cuttings for sowing kok-saghyz.) Socialističeskoe Zemledelie (Socialist Agriculture), Nr. 114, 2 (1944). [Russisch.]
Ford, C. E.: Ceylon clones VII. 1st Quart. Circ. Ceylon Rubb. Res. Scheme **16**, 1—11 (1939).
— Ceylon clones VIII. Quart. Circ. Ceylon Rubb. Res. Scheme **17**, 1—13 (1939) 1940.
— Planting material. 2nd Quart. Circ. Ceylon Rubb. Res. Scheme **17**, 142—158 (1940).
— Ceylon clones — IX (1940). Quart. Circ. Ceylon Rubb. Res. Scheme **18**, 13—23 (1941).
— The performance of imported clones in Ceylon-V. Quart. Circ. Rubb. Res. Scheme (Ceylon) **18**, 70—86 (1941).
— Ceylon clones — XI (1944—1945). Quart. Circ. Rubb. Res. Scheme (Ceylon) **22**, 22—25 (1945).

Gardner, E. J.: Sexual plants with high chromosome number from an individual plant selection in a natural population of guayule and mariola. Genetics **31**, 117 bis 124 (1946).
— Wind pollination in guayule, Parthenium argentatum Gray. J. Amer. Soc. Agron. **38**, 264—272 (1946).
— Studies on the inheritance of apomixis and sterility in the progeny of two hybrid plants in the genus Parthenium. Genetics **32**, 262—276 (1947).
— Insect pollination in guayule, Parthenium argentatum Gray. J. Amer. Soc. Agron. **39**, 224—233 (1947).
Giljarov, M. S. u. F. N. Pravdin: (The ecology of pollination in kok-saghyz.) Bull. Acad. Sci. URSS., Sér. Biol., Nr. 6, 243—260 (1943). [Russisch.]
Giraldo, H. C.: Adelanto en la recolección de semillas resistentes de caucho. (Progress in collecting resistant seed of rubber.) Agricultura Trop., Bogotá **2**, Nr. 4, 21 (1946).
Gorham, P. G. u. M. L. Landes: Investigations on rubber-bearing plants. I. Propagation of Taraxacum kok-saghyz by means of leaf cuttings. Bot. Gaz. **107**, 260—267 (1945).
— Investigations on rubber-bearing plants. II. Carbohydrates in the roots of Taraxacum koksaghyz Rod. Canad. J. Res. **24**, Sect. C, 47—53 (1946).
Gorjainov, M. N.: (A method for determining the content of gum in live roots of caoutchouc bearing plants.) Proc. Lenin Acad. Agric. Sci. USSR., Nr. 11, 5—6 (1941). [Russisch.]
Greenway, P. J.: Wild rubber in East Africa. E. Afr. Agric. J. **7**, 224—227 (1942).

Haasis, F. W.: Some staining technics for guayule study. Stain Technol. **20**, 37—44 (1945).

Haines, W. B.: A method for foliage comparisons in field experiments with Hevea. Emp. J. Exp. Agric. **10**, 117—124 (1942).

Hall, H. M. u. F. L. Long: Rubber-content of North American plants. Carnegie Institution of Washington, Washington, D. C., Pp. 65, 3 pls. 14 tables (1921).

Howes, F. N.: Russian rubber plants. J. R. Hort. Soc. **68**, 305—306 (1943).

Jong, W. H. de: Productiecijfers van den Hevea-cloon Gondang Tapen I. (Production figures of the Hevea clone Gondang Tapen I.) Bergcultures **13**, 1087—1089 (1939).

— Productiecijfers van hevea-cloon Gondang Tapen I. (Yield figures of the rubber clone Gondang Tapen I.) Bergcultures **14**, 1223—1226 (1940).

Kalinkevič, A. F.: (Zur Frage über die Rolle des Kalziums, Kaliums und Natriums bei der Bildung des Kautschuks in den Wurzeln von Kok-Saghyz.) C. R. (Doklady) Acad. Sci. URSS. **58**, 89—91 (1947). [Russisch.]

Kazakevič, L. I.: (The cultivation of rubber-bearing plants in the South East.) Naučnyj Otčet Inst. Zernovogo Hozjajstva Jugo-Vostoka SSSR. za 1941—42 gg. (Sci. Rep. Inst. Grain Husbandry South-Eastern USSR. for 1941—42), Pp. 66—85 (1944). [Russisch.]

Klippert, W. E.: The cultivation of Hevea rubber in tropical America. Chronica Botanica **6**, 199—200 (1941).

Koroleva, V. A.: Interspecific hybridization in the genus Taraxacum. C. R. (Doklady) Acad. Sci. URSS. **24**, 174—176 (1939).

— (Breeding Taraxacum kok-saghyz.) Soviet Plant Industry Record, Nr. 1, 104—105 (1940). [Russisch.]

— (Biological features of kok-saghyz and of the non-rubber-bearing dandelions infesting plantations.) Vestnik Socialističeskogo Rastenievodstva (Soviet Plant Industry Record), Nr. 2, 12—31 (1940). [Russisch.]

— (On the role of selection in kok-saghyz.) Proc. Lenin. Acad. Agric. Sci. USSR., Nr. 13, 3—6 (1940). [Russisch.]

Kostoff, D. u. E. Tiber: A tetraploid rubber plant Taraxacum kok-saghyz obtained by colchicine treatment. C. R. (Doklady) Acad. Sci. URSS. **22**, 119—120 (1939).

Kramer, H. H.: The evaluation of individual plant selections from a natural population of guayule, Parthenium argentatum Gray. J. Amer. Soc. Agron. **38**, 22—31 (1946).

Krotkov, G.: A review of literature on Taraxacum kok-saghyz Rod. Bot. Rev. **11**, 417—461 (1945).

Kuljtjasov, M. V.: (The questions of the origin of S. tau-saghyz in the light of new facts.) Sovetskaja Botanika (Soviet Botany), Nr. 3, 21—29 (1941). [Russisch.]

Kupcov, A. I.: Über die relative Größe der Wurzel bei im ersten Lebensjahre blühenden Exemplaren von Kok-saghyz. C. R. (Doklady) Acad. Sci. URSS. **30**, 649—651 (1941).

— (Division of the leaf lamina in T. kok-saghyz as a significant selective character.) C. R. (Doklady) Acad. Sci. URSS. **34**, 26—30 (1942). [Russisch.]

— Division of leaf blade in kok-saghyz as a character of breeding value. C. R. (Doklady) Acad. Sci. URSS. **39**, 22—26 (1942).

— (Preliminary results of breeding tau-saghyz.) Proc. Lenin Acad. Agric. Sci. USSR., Nos. 11/12, 18—23 (1942). [Russisch.]

— (The utilization of natural selection in plant-breeding.) J. Gen. Biol. USSR. **4**, 87—115 (1943). [Russisch.]

Lammers, R. P.: Eenige verdere gegevens over het splitsen van Hevea-kiemplanten. Some further data on the splitting of Hevea seedlings. Bergcultures **15**, 397—398 (1941).

Langford, M. H.: Science's fight for healthy Hevea. Agric. Amer. **4**, 151—153, 158 (1944).

— Regional differences in resistance of Hevea selections to South American leaf blight. Phytopathology **36**, 686 (1946). (Abstr.)

Lloyd, F. E.: Guayule (Parthenium argentum Gray): a rubber-plant of the Chinuahuan Desert. Carnegie Institution of Washington, Washington, D.C., Pp. VIII + 213. 46 pls. (1911).

Loomis, H. F.: Methods of splitting Hevea seedlings. J. Agric. Res. **65**, 97—124 (1942).

Lysenko, T. D.: (Good germination of kok-saghyz ensures a high yield.) Jarovizacija, Nr. 3 (36), 3—11 (1941). [Russisch.]

— Industrial crops. 1. A new method of growing kok-saghyz. Agriculture, Moscow, Nr. 10, 2—3 (1945). [Mimeographed.]

Mann, C. E. T.: Annual Report. Botanical Division. Rep. Rubb. Res. Inst. Malaya (1938), Pp. 59—114 (1939).

— Recommendations on the choice of planting material, clones and clonal seedlings. J. Rubb. Res. Inst. Malaya **9**, 79—88; also 14th Annu. Conf. Inc. Soc. Plant Penang 23rd to 25th September 1938 (1939).

— Improvement of yield in Hevea brasiliensis. Proc. 7th Int. Genet. Congr. Edinburgh 23—30 August, 1939 (1941), p. 209 (1939). (Abstr.)

— Improvement in the quality of rubber planting material. J. Rubb. Res. Inst. Malaya **10**, 108—125 (1940).

— The work of the botanical division of the Rubber Research Institute. Rubb. Res. Inst. Plant. Bull., Nr. 17, 1—4 (1941).

Manskaja, S. M. u. G. I. Popov: (Increasing the rubber content of kok-saghyz.) Bull. Acad. Sci. URSS., Sér. Biol., Nr. 4, 187—192 (1944). [Russisch.]

Markova, L. G.: (An embryological study of guayule and related species.) Botaničeskij Žurnal (J. Bot. URSS.) **31**, 19—26 (1946). [Russisch.]

Martin, G.: Competitive rubber plants. J. R. Soc. Arts. **92**, 146—155 (1944).

Martínez, M.: Plantas huliferas. (Rubber plants.) Ediciones Botas, Mexico, Pp. 110. 12 illus. (1943).

Maštakov, S. M.: (Über die Methodik zur Prüfung des qualitativen Zustandes des Pflanzenkautschuks.) C. R. (Doklady) Acad. Sci. URSS. **58**, 489 (1947). [Russisch.]

Medvedev, P. F.: (The first varieties of Asclepias.) Soviet Plant Industry Record, Nr. 1, 106 (1940). [Russisch.]

— (The rubber content of North American Apocynum species and some factors in latex accumulation.) Vestnik Socialističeskogo Rastenievodstva (Soviet Plant Industry Record), Nr. 5, 101—108 (1940). [Russisch.]

Mehta, C. R. u. T. P. Mehta: Chemical examination of Ocimum canum Sims. Curr. Sci. **12**, 300—301 (1943).

Mitchell, J. H., M. A. Rice u. D. B. Roderick: Rubber analysis of plants in South Carolina. Science **95**, 624—625 (1942).

Molotkovskij, G. H.: On the vegetative propagation of the rubber plants kok-saghyz, tau-saghyz and krym-saghyz. C. R. (Doklady) Acad. Sci. URSS. **40**, 291 bis 293 (1943).

— (Einfluß der Dekapitation auf Wachstum und Entwicklung von Kok- und Krym-Saghyz.) C. R. (Doklady) Acad. Sci. URSS. **56**, 969—971 (1947). [Russisch.]

— (Dekapitation und Veränderungsfaktoren der Natur der Kok- und Krym-Saghyz.) C. R. (Doklady) Acad. Sci. URSS. **58**, 1527 (1947). [Russisch.]

Moore, R. J.: Investigations on rubber-bearing plants. III. Development of normal and aborting seeds in Asclepias syriaca L. Canad. J. Res. **24**, Sect. C, 55—65 (1946).
— Investigations on rubber-bearing plants. IV. Cytogenetic studies in Asclepias (Tourn.) L. Canad. J. Res. **24**, Sect. C, 66—73 (1946).
Murray, R. K. S.: Report of the botanist and mycologist for 1939. Report of the Work of the Rubber Research Board in 1939, Ceylon 1940, Pp. 48—71 (1940).
— Report of the botanist and mycologist for 1940. 10th Rep. Rubb. Res. Scheme Ceylon, Pp. 46—63 (1941).
Mynbaev, K.: (New forms of Taraxacum kok-saghyz.) Soviet Plant Industry Record, Nr. 1, 105 (1940). [Russisch.]
— (Intra-clonal variation of morphological features and rubber content in kok-saghyz in relation to conditions of growth.) Vestnik Socialističeskogo Rastenievodstva (Soviet Plant Industry Record), Nr. 2, Pp. 32—39 (1940). [Russisch.]
— (Conditions of rearing and how to increase the productivity of koksaghyz.) Vestnik Socialłstičeskogo Rastenievodstva (Soviet Plant Industry Record), Nr. 5, 87—100 (1940). [Russisch.]
— (The effect of time of harvesting of seed on the capacity for rubber accumulation in the progeny.) Jarovizacija, Nr. 3 (36), 29—31 (1941). [Russisch.]
— (New methods of breeding of kok-saghyz.) Proc. Lenin Acad. Sci. USSR., Nr. 1/2, 24—32 (1942). [Russisch.]

Navašin, M. S. u. H. Gerasimova: Production of tetraploid rubber-yielding plant, Taraxacum koksaghyz Rodin, and its practical bearing. C. R. (Doklady) Acad. Sci. URSS. **31**, 43—46 (1941).
—, H. N. Gerasimova u. A. F. Čeredničenko: Tetraploid kok-saghyz as a variety of improved productivity. C. R. (Doklady) Acad. Sci. URSS. **47**, 432—435 (1945).
Ničiporovič, A. A., L. A. Ostapenko u. N. G. Vasiljeva: (Anatomical peculiarities of diploid and polyploid forms of rubber yielding plant Taraxacum koksaghyz Rodin.) Bull. Acad. Sci. URSS., Sér. Biol., Nr. 2, 309—331 (1941). [Russisch.]
Nylov, V. I. u. L. A. Mihelson: (New chemical compounds obtained through distant hybridization. Vestnik Gibridizacii (Hybridization), Nr. 2, 95—98 (1941). [Russisch.]

O....., F. W.: Moet onze rubberboom herdoopt worden? (Must our rubber tree be renamed?) Bergcultures **15**, 1438—1440 (1941).
O'B....., T. E. H.: Planting notes. Promising local clones. 1st Quart. Circ. Ceylon Rubb. Res. Scheme **16**, 19—20 (1939).
O'Brien, T. E. H.: Rubber investigations in Ceylon. Emp. J. Exp. Agric. **10**, 61—76 (1942).
Ostendorf, F. W.: Rubber-plantmateriaal, 1936—1941. (Rubber planting material, 1936—1941.) Bergcultures **15**, 852—859 (1941).
— Resultaten van toetstuin 1934 (proeftuin Tjiomas) tot en met het tweede tapjaar. (Results from the test plantation Tjiomas [experimental plantation] up to and including the second tapping year.) Bergcultures **15**, 960—964 (1941).
— Resultaten van toetstuin 1930 (Proeftuin Tjiomas) tot en met het zevende tapjaar. (Results from the trial plantation [Tjiomas experimental plantation] up to and including the 7th tapping year.) Bergcultures **15**, 1630—1637 (1941).

Paddock, E. F.: On the number of chromosomes in Hevea. Chronica Botanica **7**, 412—413 (1943).
Perry, B. A.: Chromosome number relationships in the genus Euphorbia. Chronica Botanica **7**, 413—414 (1943).
Perry, B. A.: Chromosome number and phylogenetic relationships in the Euphorbiaceae. Amer. J. Bot. **30**, 527—543 (1943).
Perry, E. L.: Growing rubber in California. Rep. Smithson. Instn 1945. Publ. 3817, Pp. 351—362 (1946).
Pilaar, C. J.: Het splitsen van Hevea-kiemplanten op de R. O. Pasir Koppo. (The splitting of Hevea seedlings on the Rubber Estate Pasir Koppo.) Bergcultures **14**, 503—505 (1940).
Poddubnaja-Arnoldi, V.: Development of pollen and embryosac in interspecific hybrids of Taraxacum. C. R. (Doklady) Acad. Sci. URSS. **24**, 374—377 (1939).
— Embryogenesis in remote hybridization in the genus Taraxacum. C. R. (Doklady) Acad. Sci. URSS. **24**, 382—385 (1939).
— (Interspecific hybridization in the genus Taraxacum.) Bull. Soc. Nat., Moscow **48**, 87—98 (1939). [Russisch.]
— (Ein Experiment zur Gewinnung von tetraploiden Kok-Saghyz.) C. R. (Doklady) Acad. Sci. URSS. **56**, 873—876 (1947). [Russisch.]
Popov, G. I.: (The influence of cultivation and management on the anatomy of the kok-saghyz root.) Proc. Lenin Acad. Agric. Sci., Nr. 11/12, 14—17 (1942). [Russisch.]
Powers, L.: Fertilization without reduction in guayule (Parthenium argentatum Gray) and a hypothesis as to the evolution of apomixis and polyploidy. Genetics **30**, 323—346 (1945).
— u. E. J. Gardner: Frequency of aborted pollen grains and microcytes in guayule, Parthenium argentatum Gray. J. Amer. Soc. Agron. **37**, 184—193 (1945).
— u. R. C. Rollins: Reproduction and pollination studies on guayule, Parthenium argentatum Gray and P. incanum H. B. K. J. Amer. Soc. Agron. **37**, 96—112 (1945).
Presley, J. T.: Observations on Phymatotrichum root rot of guayule. Plant Dis. Reporter **28**, 998—1000 (1944). (Mimeographed.)
Procenko, D. F.: (Über die Besonderheiten der Regeneration bei Kok-Saghyz während der Stecklingbereitung.) C. R. (Doklady) Acad. Sci. URSS. **53**, 277 (1946). [Russisch.]
— u. M. D. Artemenko: (Die Entwicklung der Nitrabakterien in der Rizosphäre des Kok-Saghyz.) C. R. (Doklady) Acad. Sci. URSS. **53**, 359 (1946). [Russisch.]
Prokofjeva, A. A.: (De l'origine plastidique du caoutchouc.) Botaničeskij Žurnal (Journ. Botanique de l'URSS.) **31**, Nr. 2, 5 (1946). [Russisch.]

Rands, R. D.: El cultivo del caucho (Hevea brasiliensis) en la América tropical. (The cultivation of rubber [H. brasiliensis] in tropical America.) Publ. Agríc. Ofic. Cooper. Agríc. Unión Panamericana, Washington, D. C., Nr. 147/148, 44 (1944).
Regel, C.: Beiträge zur Kenntnis von mitteleuropäischen Nutzpflanzen. Angew. Bot. **23**, 117—123 (1941).
Rollins, R. C.: Analagous plant types from natural populations of guayule (Parthenium argentatum) and mariola (P. incanum) and certain derivatives of crosses between these species. Amer. J. Bot. **31**, 13s—14s (1944). (Abstr.)
— Evidence for natural hybridity between guayule (Parthenium argentatum) and mariola (Parthenium incanum). Amer. J. Bot. **31**, 93—99 (1944).
— Some known and probable levels of reciprocal introgression between guayule (Parthenium argentatum) and mariola (P. incanum). Genetics **30**, 18—19 (1945). (Abstr.)
— Evidence for genetic variation among apomictically produced plants of several F_1 progenies of guayule (Parthenium argentatum) and mariola (P. incanum). Amer. J. Bot. **32**, 554—560 (1945).

Rollins, R. C.: Interspecific hybridization in Parthenium. I. Crosses between guayule (P. argentatum) and mariola (P. incanum). Amer. J. Bot. **32**, 395—404 (1945).
— Interspecific hybridization in Parthenium. II. Crosses involving P. argentatum, P. incanum, P. stramonium, P. tomentosum and P. hysterophorus. Amer. J. Bot. **33**, 21—30 (1946).
— The occurrence of sublethal dwarfed hybrids in Parthenium and their chimeric mutation to normalcy. Amer. J. Bot. Suppl. **33**, 825 (1946). (Abstr.)
— Can populations of plants showing the combined effects of polyploidy, interspecific hybridization, sexuality, and facultative apomixis be usefully organized into systematic entities? Amer. J. Bot. Suppl. **33**, 843 (1946). (Abstr.)

Šarova, N. L., N. V. Kojalovič u. P. A. Jakimov: (Sur les deux formes du caoutchouc dans les feuilles de Kok-Saghyz.) Sovetskaja Botanika **15**, Nr. 4, 226 bis 227 (1947). [Russisch.]
Ščepotjev, F. L.: (The biology of flowering of Euonymus nana M. B.) Sovetskaja Botanika (Soviet Botany), Nr. 3, 130—135 (1941). [Russisch.]
Schachameyer, C.: Contribution à l'étude de la variabilité de deux caractères principaux de l'hévéa cultivé. Bull. Agric. Congo Belge, Nr. 3/4, 24 (1942).
Schmöle, J. F.: Verslag tot ult. 1938 van de toetstuinen van cloonen en zaailingfamilies in Polonia. (Report to end of 1938 on the test plantations of clones and seedling families in Polonia.) Arch. Rubbercult. Ned.-Ind. **23**, 141—161 (1939).
— De invloed van den onderstam op de productie van oculaties. II. (The influence of the stock on the production of budgrafts. II.) Arch. Rubbercult. Ned.-Ind. **24**, 305—314 (1940).
— Kort verslag tot ult. 1939 van de toetstuinen van cloonen en zaailingfamilies in Polonia. (Short report up to the end of 1939 on the test plantations of clones and seedling families in the Polonia plantation.) Arch. Rubbercult. Ned.-Ind. **24**, 455—469 (1940).
— Zeer dicht planten en intensief uitdunnen van zaailingen-aanplanten. (Very dense planting and intensive thinning of seedling plantations.) Bergcultures **15**, 620 bis 621 (1941).
— Hevea brasiliensis en Hevea Spruceana-hybride als onderstam voor oculaties. II. (Hevea brasiliensis and Hevea Spruceanahybrids as stock for budgrafts. II.) Arch. Rubbercult. Ned.-Ind. **25**, 149—165 (1941).
Schultes, R. E.: Estudio preliminar del género Hevea en Colombia. (Preliminary study of the genus Hevea in Colombia.) Rev. Acad. Colombiana Cienc. Exactas, **6**, 331—338; also Rev. Fac. Nac. Agron., Colombia 1946, **6**, Nr. 22, 18—45 (1945).
— The genus Hevea in Colombia. Bot. Mus. Leafl. Harv. **12**, 1—19 (1945).
Schultz, E. F.: La vara de oro ,,Solidago" spp. (Golden rod, Solidago spp.) Rev. Industr. Agric. Tucumán **32**, 340 (1942).
Schweizer, J.: Hevea-plantmateriaal in Besoeki. (Hevea planting material in Besoeki.) Bergcultures **15**, 598—609 (1941).
— Bijdrage tot de physiologie van den z. g. ,,Hoogtap". (Contribution to the physiology of the so-called ,,hightapping".) Bergcultures **15**, 1214—1228 (1941).
Sharp, C. C. T.: Progress of breeding investigations with Hevea brasiliensis. The Pilmoor crosses 1928—1931 series. J. Rubb. Res. Inst. Malaya **10**, 34—66 (1940).
Soliva, M.: L'évolution de la culture du caoutchouc au cours des dix dernières années. Rev. Bot. Appl. **22**, 47—69 (1942).
Sorensen, H. G.: Crown budding for healthy Hevea. Agric. Amer. **2**, 191—193 (1942).

Sorensen, T. u. G. Gudjónsson: Spontaneous chromosome-aberrants in apomictic Taraxaca. Morphological and cyto-genetical investigations. Det Kongelige Danske Videnskabernes Selskab, Biol. Skrift., København **4**, 48 (1946).
Stebbins, G. L. jr. u. M. Kodani: Chromosomal variation in guayule and mariola. J. Hered. **35**, 163—172 (1944).
Stevens, O. A.: Asclepias syriaca and A. speciosa distribution and mass collections in North Dakota. Amer. Midl. Nat. **34**, 368—374 (1945).
Symontowne, R.: Notes on the Cryptostegia plant. Trop. Agriculture Trin. **20**, 195—197 (1943).

Tobler, F.: Ein weiterer Beitrag zur Kenntnis des sog. Gummihanfes. Ber. Dtsch. Bot. Ges. **59**, 46—51 (1941).
Tollenaar, D.: De belangrijkheid van het onderstamvraagstuk voor de rubbercultuur. (The importance of the question of the stock for rubber cultivation.) Bergcultures **15**, 1014—1019 (1941).

Veljtiščev, P. A.: (The pollinators of Parthenium argentatum in USSR.) Priroda (Nature), Nr. 4, 52—53 (1939). [Russisch.]
Vollema, J. S. u. M. J. Dijkman: Resultaten der toetsing van Heveacloonen in den proeftuin Tjiomas. II. (Results of the testing of Hevea clones in experimental garden Tjiomas. II.) Arch. Rubbercult. Ned.-Ind. **23**, 47—129 (1939).
Vonk, H.: De voorlichting der zg. small holders in Malaka op het gebied van de rubbercultuur. (The instruction of the so-called small holders in Malaya as regards rubber cultivation.) Landbouw. **17**, 259—290 (1941).

Warmke, H.: The cytology and breeding behavior of the Russian dandelion, Taraxacum kok-saghyz. Amer. J. Bot. Suppl. **29**, 19s (1942). (Abstr.)
— Macrosporogenesis, fertilization, and early embryology of Taraxacum kok-saghyz. Bull. Torrey Bot. Cl. **70**, 164—173 (1943).
— The effect of tetraploidy on root weight and rubber content in the Russian dandelion. Amer. J. Bot. **31**, Nr. 8, 6s—7s (1944). (Abstr.)
— Self-fertilization in the Russian dandelion, Taraxacum kok-saghyz. Amer. Nat. **78**, 285—288 (1944).
— Experimental polyploidy and rubber content in Taraxacum kok-saghyz. Bot. Gaz. **106**, 316—324 (1945).
Whaley, W. G.: Rubber, heritage of the American tropics. Sci. Mon. N. Y. **21**—**31** (1946).
Whittenberger, R. T.: Oil blue NA as a stain for rubber in sectioned or ground plant tissues. Stain Technol. **19**, 93—98 (1944).
Woodward, C. H. u. D. T. MacDougal: What is guayule? J. N. Y. Bot. Gdn. **43**, 168—170 (1942).

Zotov, V. D. u. E. P. White: Rubber production in New Zealand. Trials with kok-saghyz. N. Z. J. Agric. **67**, 75—78 (1943).
Zweede, J. C.: Het splitsen van Hevea-kiemplanten. (The splitting of Hevea seedlings.) Bergcultures **14**, 1271—1275 (1940).

r) Obst.
1. Obst, Allgemeines.

Adema, J.: Groeistoffen en late val. (Growth substances and late drop.) Fruitteelt **31**, 177—182 (1941).
Alderman, W. H. u. F. E. Haralson: Minnesota fruit breeding farm. Report for 1941. Minn. Hort. **70**, 106 bis 107, 111—112 (1942).
Alvarez-Ladiada, M.: Temas de genética frutal. La teoría de la incompatibilidad de las variedades frutales. (Topics of fruit genetics. The theory of incompati-

bility of fruit tree varieties.) Rev. Agric. P. Rico **32**, 355—362 (1940).
Anderson, J. P.: Two notable plant hybrids from Alaska. Proc. Iowa Acad. Sci. **50**, 155—157 (1943).
Anonymus: Nederlandsche Fruitsoorten. (Dutch varieties of fruits.) S. Gouda Quint-D. Brouwer en Zoon, Uitgevers, Arnhem. (Undated).
— Die amerikanischen Pflanzenpatente Nr. 89—94. Züchter **11**, 47—48 (1939).
— Die amerikanischen Pflanzenpatente Nr. 95—144. Züchter **11**, 192—200 (1939).
— Research on fodder crops and pastures. Imperial College of Agricultural Research, St. Augustine, Trinidad. Chronica Botanica **5**, 490 (1939).
— Fruit research in Madras. Indian Fmg **1**, 82—83 (1940).
— Fruit breeding an aid to truth. N. J. Hort. Soc. News **21**, 1182 (1940).
— Enquête betreffende de toepassing van bespuitingen met α-naphthaleenazijnzuur met het doel den laten val te doen verminderen. (Enquiry on the use of spraying with α-naphthalene acetic acid to reduce the amount of late drop.) Fruitteelt **31**, 143—148 (1941).
— Amerikanische Pflanzenpatente Nr. 336—353. Züchter **13**, 68—69 (1941).
— Annales agricoles Vaudoises. 17me et 18me années. Lausanne, Pp. 103 (1941).
— New fruits. Amer. Fruit Gr. **62**, Nr. 4, 8 (1942).
— Comptes rendus de expérimentation fruitière en Algérie. Bull. Agric., Algérie, Nr. 92, 47 (1942).
— Annales agricoles Vaudoises. 19ne année. Lausanne, Pp. 64 (1942).
— A new type of fruit. Soviet War News, Nr. 575, 4 (1943).
— Annales agricoles Vaudoises. 21me année. Lausanne, Pp. 66 (1943).
— Vitamin variation in fruits, plants, du chiefly to variety and sunlight. Calif. Citogr. **29**, 128—129 (1944).
— Horticulture in Byelorussia. Agriculture, Moscow, Nr. 3, 3—4 (1945).
— I. V. Michurin, the great reorganizer of nature. Agriculture, Moscow, Nr. 6, 1—3 (1945). [Mimeographed.]
— Fruit growing. Agriculture, Moscow, Nr. 8, 4—5 (1945). [Mimeographed.]
— Horticulture. Agriculture, Moscow, Nr. 10, 5—7 (1945). [Mimeographed.]
— Les cultures fruitières en URSS. Rev. Bot. Appl. **25**, 113—116 (1945).
— State news. Texas. Amer. Fruit Gr. **67**, Nr. 7, 227 (1947).
Astrego, J. J.: Stuifmeelonderzoek bij enkele fruitsoorten onder glas. (Pollen studies of some species of fruit under glass.) Landbouw. Tijdschr. **55**, 181—190 (1943).

Barrett, C.: Report of the Variety Committee — 1939. Yearb. Calif. Avocado Ass, Pp. 24—29 (1939).
Black, M. W.: Uneconomical deciduous fruit varieties. Fmg. S. Afr. **18**, 382—387 (1943).
Blake, M. A.: A Russian nectarine proves promising for breeding. Hort. News, N. J. **26**, 1737 (1945).
Blattny, C. u. B. Starý: Atlas škodlivých činitelů našich ovocných plodin. (Atlas of the injurious agencies affecting our fruit trees and shrubs.) Praha, Pp. 375. 130 pls. (1944).
Borisoglebskij, A. D.: (In memory of I. A. Efremov.) Plodovo-Jagodnye Kuljtury (Fruit Crops), Nr. 4, 82 bis 84 (1940). [Russisch.]
Brooks, R. M. u. H. P. Olmo: Register of new fruits and nut varieties-list, Nr. 2. Proc. Amer. Soc. Hort. Sci. **47**, 544—569 (1946).
Budagovskij, V. I.: (I. V. Mičurin on root stocks for fruit trees.) Agrobiologija (Agrobiology), Nr. 2, 111 bis 119 (1946). [Russisch.[

Bullock, R. M. u. J. C. Snyder: Some methods of tree fruit pollination. Proc. 42nd Mtg. Wash. St. Hort. Ass., Pp. 215—226 (1946).

Celikov, P. N.: (V. A. Mokrušin, a disciple of Michurin in the Urals.) Sadovodstvo (Horticulture), Nr. 3, 21 bis 22 (1940). [Russisch.]
— (Michurinites of the Sverdlovsk region participators in the Soviet Agricultural Exhibition.) Sadovodstvo (Horticulture), Nr. 6, 27—30 (1940). [Russisch.]
Crane, M. B.: The production of fruit. J. Roy. Hort. Soc. **66**, 350—357 (1941).
— u. A. G. Brown: The causal sequence of fruit development. J. Genet. **44**, 160—168 (1942).

Day, L. H. u. W. P. Tufts: Nematode-resistant rootstocks for deciduous fruit trees. Circ. Calif. Agric. Exp. Sta., Nr. 359, 16 (1944).
Dickinson, F.: CThe introduction and propagation of fruit trees. W. China Un. Univ. Agric. Sch. Agric. Res. Inst. May 22, p. 1 (1943). [Mimeographed].
Dobrunov, L. G. u. O. M. Gladyševa: (Die qualitativen Unterschiede der Früchte aus verschiedenen Teilen der Krone.) C. R. (Doklady) Acad. Sci. URSS. **55**, 655—658 (1947). [Russisch.]
Doortjes, J. A.: Vruchtvorming zonder bevruchting (parthenocarpie). (Fruit formation without fertilization [parthenocarpy].) Fruitteelt **32**, 43—45 (1942).
Drain, B. D.: Breeding plants for adaptation, freedom from insect injury, and disease resistance. Proc. Amer. Soc. Hort. Sci. **45**, 225—228 (1944).
Dugand, A.: Noticias bótanicas colombianas. II. Especies nuevas y críticas. (Colombian botanical notes. II. New and critical species.) Caldasia, Bogota, Nr. 8, 285—299 (1943).

Elazari-Volcani, I.: Activities of the Agricultural Research Station, Rehovoth. Proc. Conf. Middle East Agric. Develop. Cairo February 7th—10th 1944, Agric. Rep., Nr. 6, 124—131 (1944).
Evreinoff, V.-A.: Les cognassiers à gros fruit. Rev. Hort., Paris **28**, 114 (1942).

Faes, H.: Station fédérale d'essais viticoles et arboricoles à Lausanne et Domaine de Pully. Landw. Jb., Schweiz **55**, 703—738 (1941).
— Rapports annuels 1943 et 1944. Station fédérale d'essais viticoles et arboricoles à Lausanne et Domaine de Pully. Landw. Jb., Schweiz **46**, 671—707 (1945).
Filatov, F.: (Let us profit by the legacy of I. V. Michurin in work on crop plants.) Socialistic Grain Farming, Saratov, Nr. 3, 6—12 (1940). [Russisch.]
Flory, W. S. jr.: Breeding new fruits for Virginia growers. Proc. 50th Annu. Mtg Va St. Hort. Soc. **34**, 50—56 (1946).
Fridström, A.: Sortvalet beroende av tillfälligheter? (Variety selection based on chance?) Fruktodlaren, Nr. 4, 106—108 (1942).

Gardner, V. R., F. C. Bradford u. H. D. Hooker jr.: The fundamentals of fruit production. McGraw-Hill Publishing Company, Ltd., London (1939).
Gerritsen, J. B.: Bestuiving en vruchtzetting van de Schoone van Boskoop. (Pollination and fruit setting of Belle de Boskoop.) Fruitteelt. **36**, 38—39 (1946).
Gill, N. T.: The origins of our garden crops. Gdnrs' Chron. **105**, 57—58, 268 (1939).
— The origins of our garden crops. Gdnrs' Chron. **105**, 356, 390, 406 (1939).
Gorškov, I. S.: (From the work of the Central Breeding and Genetics Laboratory of I. V. Michurin. Jarovizacija, Nr. 3 (30), 152—156 (1940). [Russisch.]
— (The work of the I. V. Michurin Central Fruit Genetics Laboratory.) Agrobiologija (Agrobiology), Nr. 2, 75 bis 86 (1946). [Russisch.]

Gourley, J. H. u. F. S. Howlett: Modern fruit production. Macmillan and Co., New York, Pp. 579. 50 tables. 87 figs. (1941).

Grigorjev, I. J.: (A follower of Michurin.) Sady i Ogorody (Fruit and Vegetable Gardens), Nr. 4, 4 bis 5 (1941).

Grosjean, J.: Het vraagstuk van de loodglansziekte bij vruchtboomen. (The problem of silver leaf in fruit trees.) Tijdschr. PlZiekt., 49, 172—178 (1943).

Gross, E.: Untersuchungen über den Einfluß von Kältebehandlung auf einige Obstarten. Gartenbauwiss. 17, 295—303 (1943).

Haas, P. G. de: Über die Beziehungen zwischen Herkunft und Standort im Obstbau. Kühn-Archiv 60, 189—203 (1943/1944).

Hahn, G. G.: Field tests with a staminate clone of alpine currant immune from blister rust under greenhouse conditions. Plant Dis. Reporter 25, 476—478 (1941). [Mimeographed.]

Hansen, N. E.: Northern plant novelties for 1941. 38th Rep. S. Dak. St. Hort. Soc. 1941, Pp. 30—34 (1941).

— Northern plant novelties for 1945. 42nd Rep. S. Dak. State Hort. Soc., Pp. 57—60 (1945).

— Breeding hardy fruits for northwest prairie. Amer. Fruit Gr. 67, Nr. 7; 12, 22, 31 (1947).

Hartman, F. O. u. F. S. Howlett: An analysis of the fruit characteristics of seedlings of Rome Beauty, Gallia Beauty, and Golden Delicious parentage. Proc. Amer. Soc. Hort. Sci. 40, 241—244 (1942).

Hašba, L. H.: (Collection and study of somatic mutations.) Sadovodstvo (Horticulture), Nr. 7, 28—30 (1939). [Russisch.]

Hildebrand, E. M. u. P. V. Weber: Varietal susceptibility of currants to the cane blight organism, and to currant mosaic virus. Plant Dis. Reporter 28, 1031 bis 1035 (1944). [Mimeographed.]

Hilkenbäumer, F.: Resistenzzüchtung bei Obstunterlagen. Kühn-Archiv 60, 455—461 (1943/1944).

Hills, J. L.: Fifty-fourth Annual Report 1940—1941. Bull. Vt. Agric. Exp. Sta., Nr. 475, 40 (1941).

Hoblyn, T. N.: Testing new varieties of fruit plants. Proc. 7th Inst.Genet.Congr., Edinburgh 23—30 August, p. 147, 1939 (1941). (Abstr.)

Hsü, Shao Hua: (A list of varieties of vegetables and fruits of the Shao Wu horticultural station of Fukien University.) Fukien Agric. J. 3, 370—379 (1941).

Hudson, J. P.: Winter fruits and vegetables in New Zealand. Gdnrs' Chron. 120, 297, 307—308 (1947).

Hülsmann, B.: Morphologische Beobachtungen an Unterlagenquitten aus Wageningen. Gartenbauwiss. 17, 201—210 (1942).

— Selektion von Obstunterlagen. Züchter 17/18, 224 bis 232 (1947).

Iljinskij, A. A.: (The wild fruit trees of the Southern Daghestan.) Plodovo-Jagodnye Kuljtury (Fruit Crops) Nr. 3, 40—48 (1940). [Russisch.]

Ingram, C.: Apparent manifestation of a paternal habit. Gdnrs' Chron. 115, 178 (1944).

Isaev, S. I.: (The role of the maternal plant in the formation of the hereditary constitution in hybrids.) Agrobiologija (Agrobiology), Nr. 2, 87—96 (1946). [Russisch.]

Jakovlev, P. N.: (New Michurin varieties.) Jarovizacija, Nr. 3 (30), 45—52 (1940). [Russisch.]

— (Michurin varieties in the tests of the 1939—1940 winter.) Jarovizacija, Nr. 6 (33), 51—63 (1940). [Russisch.]

— (To the memory of I. V. Michurin [on the fifth anniversary of his death].) Plodovo-Jagodnye Kuljtury (Fruit Crops), Nr. 4, 3—8 (1940). [Russisch.]

Jakovlev, P. N.: (Development of I. V. Michurin's doctrine by the research institutes.) Sady i Ogorody (Fruit and Vegetable Gardens), Nr. 6, 2—8 (1941). [Russisch.]

Johansson, E.: Nyare undersökningar på fruktodlingens område. (Recent investigations in the sphere of fruit cultivation.) Sverig. Pomol Fören. Årsskr. 41, 72—84 (1940).

— Askorbinsyrehalt hos frukt och köksväxter i färskt och torkat tillstand. (The ascorbic acid content of fruits and vegetables in fresh and dried condition.) Årsskr. Alnarps. Landbr. Mej.-och TrädInst., Pp. 1 bis 26 (1940).

— Frostskador i Svenska fruktträdgårdar vintern 1939 bis 1940. Sammanställing av rapporter från trädgårdskonsulenter och fruktodlare. (Frost damage in Swedish orchards in the winter 1939/1940 - Combined reports from horticultural advisors and from fruit growers.) Årsskr. Alnarps Landbr. Mej.-och TrädInst., Pp. 1—23 (1941).

— Om uppdragning av nya fruktsorter. (On the production of new varieties of fruit.) Sverig. Pomol. Fören. Årsskr. 43, 121—130 (1942).

— Nyare undersökningar på fruktodlingens område. (Recent investigations on fruit growing.) Sverig. Pomol. Fören. Årsskr. 45, 174—180 (1944).

Kalmykov, S. S.: (A large-fruited species of hawthorn.) Bjull. Kazah. Naučno-issled. Inst. Zemled. im. Akad. V. R. Viljjamsa (Bull. V. R. Williams Kazah. Inst. Agric.), Nr. 7/8, 19—22 (1940). [Russisch.]

Kemmer, E.: Die Bedeutung der Standortsfragen für die Obstzüchtung. Forschungsdienst 9, 511—517 (1940).

— Über die Regenerationsfähigkeit der Obstgehölzwurzeln. Gartenbauwiss. 18, 101—117 (1944).

— u. F. Schulz: Die Bedeutung des Sämlings als Unterlage. Landw. Jb. 89, 114—139 (1939).

—, — Die Bedeutung des Sämlings als Unterlage. (Baumschulstadium.) Gartenbauwiss. 18, 59—97 (1944).

Kessler, H.: Die Bedeutung der Sortenwahl bei der Erstellung von Gefrierkonserven aus Obst und Gemüse. Schweiz. Z. Obst- u. Weinb. 52, 548—564 (1943).

Khašba, L. Kh. siehe Hašba, L. H.

Kobel, F.: Jahresbericht 1944 der Eidgen. Versuchsanstalt für Obst-, Wein- und Gartenbau in Wädenswil. Landw. Jb., Schweiz 46, 159—175 (1945).

Konovalov, I. N.: (The great Russian scientist, I. V. Mičurin.) Sovetskaja Botanika (Soviet Botany) 13, Nr. 5, 3—6 (1945). [Russisch.]

Kovalev, N. V.: Immunity of fruit trees to fungus -diseases. C. R. (Doklady) Acad. Sci. URSS. 27, 176 bis 179 (1940).

— (Principal regularities in ecological differentiation of fruit trees.) Proc. Lenin Acad. Agric. Sci. USSR., Nr. 1, 5—9 (1940). [Russisch.]

— (Wild fruit plants of the Caucasus and their role in the people's economy.) Priroda (Nature), Nr. 5, 44 bis 55 (1941). [Russisch.]

Krjukov, F. A.: (Let us make use of the effect of the severe winter of 1939—1940 for the selection of winterhardy varieties.) Sadovodstvo (Horticulture), Nr. 4, 21 (1940). [Russisch.]

— (A great citizen of the USSR. — Ivan Vladimirovič Michurin.) Vestnik Socialističeskogo Rastenievodstva (Soviet Plant Industry Record), Nr. 3, 15—16 (1940). [Russisch.]

Kuenen, D. J.: Spint op vruchtboomen. (Red spider on the fruit trees.) Tijdschr. PlZiekt. 49, 130—131 (1943).

Lantz, H. L.: New seedling fruits productions of 1940. Trans. Iowa St. Hort. Soc. **75**, 124—128 (1940).
— Fruit breeding contributes to better living. N. S. Dak. Hort. **17**, 142—144, 155, 160 (1944).
— Fruit breeding contributes to better living. Virginia Fruit **33**, 18—24 (1945).
Lemasson, J.: La recherche de variétés fruitières nouvelles. Applications aux arbres fruitiers des principes de la génétique. Rev. Hort., Paris **115**, 378—379 (1943).
Lesjuk, E. A.: (Michurin's varieties of fruits.) Jarovizacija, Nr. 3 (30), 53—80 (1940). [Russisch.]
— (Results of the industrial and biological study of I. V. Mičurin's varieties.) Agrobiologija (Agrobiology), Nr. 2, 120—122 (1946). [Russisch.]
Lewis, D.: Chemical control of fruit formation. J. of Pomology and Horticult. Science **22**, 175—182 (1946).
Lijftoft, G.: Over den invloed van het stuifmeel op de kwaliteit van het fruit. (On the influence of the pollen on the quality of the fruit.) Fruitteelt **29**, 107 bis 113 (1939).
Lind, E.: Intryck fron en studieresa i Tyskland, Frankrike och England. (Impressions from a study tour in Germany, France and England.) Sverig. Pomol. Fören Årsskr. **40**, 108—122 (1939).
Long, E. M.: Developmental anatomy of the fruit of the Deglet Noor date. Bot. Gaz. **104**, 426—436 (1943).
Lundin, Y.: Frukträdens härdighet vintern 1941—1942. (Hardiness of fruit trees in the winter 1941/42.) Fruktodlaren, Nr. 6, 168—171 (1942).
— S. P. F.: s nya fruktsortlistor. (The Swedish Pomological Association's new lists of fruit varieties.) Sverig. Pomol. Fören. Årsskr. **44**, 176—181 (1943).
— Hemträdgårdens frukter och bär. (Horticultural fruits and berries.) Stockholm, Pp. 131. illus (1944).

McCrory, S. A.: Winter injury and hardiness of fruit trees. 38th Rep. S. Dak. St. Hort. Soc. 1941, Pp. 56 bis 60 (1941).
Mack, W. B.: American horticultural science today. Proc. Amer. Soc. Hort. Sci. **47**, 533—543 (1946).
Masefield, G. B.: Some recent observations on the plantain crop in Buganda. E. Afr. Agric. J. **10**, 12—17 (1944).
Meier, K.: Bericht der Eidg. Versuchsanstalt für Obst-, Wein- und Gartenbau in Wädenswil für das Jahr 1943. Landw. Jb., Schweiz **45**, 891—953 (1944).
Mičurin, I. V.: (To fruit-growers of the Urals and Siberia.) Plodovo-Jagodnye Kuljtury (Fruit Crops), Nr. 4, 9—11 (1940). [Russisch.]
— (How to rear fruit trees in the Urals.) Plodovo-Jagodnye Kuljtury (Fruit Crops), Nr. 4, 12—14 (1940). [Russisch.]
— (Results of sixty years of plant breeding work and further prospects.) Vestnik Gibridizacii (Hybridization), Nr. 1, 5—9 (1941).
Miéville, R., E. Poilane u. A. Chevalier: Les possibilités de l'Indochine du Nord en cultures fruitières. Rev. Bot. Appl. **22**, 363—391 (1942).
Mihajlov, A. J.: (Fruit-growing in the Karelian-Finnish SSR.) Plodovo-Jagodnye Kuljtury (Fruit Crops), Nr. 5, 102 (1940). [Russisch.]
— Siberian fruit — at 50 degrees below zero. Soviet News, Nr. 1356, 2 (1946).
Miller, C. D. u. K. Bazore: Fruits of Hawaii. Description, nutritive value, and use. Bull. Univ. Hawaii Agric. Exp. Sta., Nr. 96, 129. 21 figs (1945).

Nebel, B. R.: Longevity of pollen in apple, pear, plum, peach, apricot, and sour cherry. Proc. Amer. Soc. Hort. Sci. **37**, 130—132 (1939) 1940.
— Inducing changes in plants with colchicine shows progress. Results obtained with flowers indicate what may be expected with fruits in time-chief difficulty is in perfecting technic. Fm Res., N. Y. St. Sta. **6**, 10, 15 (1940).

Nilsson, F.: Odling och förädling av fruktträd och bärbuskar i USA. Berättelse över en studieresa 1937 bis 1938. (Cultivation and improvement of fruit tress and bush fruits in USA. Report on a study tour 1937 bis 1938.) Sverig. Pomol. Fören. Årsskr. **40**, 7—80 (1939).
— Nagra erfarenheter och intryck av trädgardsodling i Norrland. (Some discoveries and impressions of horticulture in Norrland.) Fruktodlaren, Nr. 1, 16—19 (1940).
— Försöksverksamhetens betydelse för fruktodlingen. (The importance of research for fruit culture.) Sverig. Pomol. Fören Årsskr. **43**, 155—161 (1942).
— Specialkursen för trädgårdskonsulenter m. fl. på Alnarp. — Den lokala försöksverksamheten. (Special course for horticultural advisors etc. at Alnarp-Local experimental research.) Sverig. Pomol. Fören Årsskr. **43**, 224—231 (1942).
— Förutsättningar och arbetsuppgifter för växtförädlingen av fruktträd. (The essential basis and tasks of fruit tree breeding.) Sverig. Pomol. Fören. Årsskr. **44**, 5—16 (1943).
— Fruktodlarna och försöksverksamheten. (Fruit growers and research.) Sverig. Pomol. Fören. Årsskr. **44**, 230—236 (1943).
— Balsgårds Fruktträdsförädlingsanstalt. (The Balsgård Institute for Fruit Tree Improvement.) Sverig. Pomol. Fören. Årsskr. **45**, 58—59 (1944).
— Växtförädlingen och fruktodlarna. (Plant breeding and the fruit growers.) Fruktodlaren, Nr. 1, 11—15 (1946).

Östlind, N.: Kromosomer. Diploida och triploida Fruktsorter. (Chromosomes. Diploid and triploid varieties of fruits.) Fruktodlaren, Nr. 2, 47—50 (1939).
Ozbek, S.: Çiçek tomurcuğu teşekkülü esas tutularak Kastamonu dolaylarindaki en önemli meyve türlerinin verimliliğine tesir eden biyolojik faktörler üzerinde araştirmalar. (Flower bud formation and investigations on the biological factors influencing productiveness in the chief fruit varieties in Kastamonu.) T. C. Ziraat Vekâleti Ankara Yüksek Ziraat Enstitüsü, Nr. 143, 88 (1943).

Parrott, P. J.: The Research Program of the Geneva Experiment Station. Proc. 86th Annu. Meet. N. Y. St. Hort. Soc., Pp. 248—260 (1941).
Paškevič, V. V.: (An urgent problem in fruit culture.) Sadovodstvo (Horticulture), Nr. 1, 36—37 (1939). [Russisch.]
Passecker, F.: Untersuchungen über die Befruchtungsverhältnisse von Kern- und Steinobstsorten. Gartenbauwiss. **15**, 532—558 (1941).
— Jugend- und Altersformen bei den Obstgehölzen. Gartenbauwiss. **18**, 219—230 (1944).
Pedersen, A.: Danmarks Frugtsorter. I. Aebler. (Danish fruit varieties. I. Apples.) Gartnerforenings Bogforlag, København, Nr. 3, 129—192.
— Danmarks Frugtsorter. II. Paerer, blommer, kirsebaer. (Danish varieties of fruits. II. Pears, plums and cherries.) Udgivet af Faellesudvalget for Frugtavlsøkonomi, Paa Alm. Dansk Gartnerforenings Bogforlag, Nr. 1/2, 128. photos. (Undated.)
Pellett, K.: Pioneers in Iowa horticulture. Trans Iowa St. Hort. Soc. **75**, 5—71 (1940).
Persson-Ferlenius, G. R.: Praktiska synpunkter pa växtförädlingen av fruktträd. (Practical aspects of the breeding of fruit trees.) Sverig. Pomol. Fören. Årsskr. **46**, 92—97 (1945).
Petrov, M. P.: (On wild fruit-trees of Turkmenian highlands.) Priroda, Nr. 5, 55—68 (1939). [Russisch.]
Puškarskij, S. D.: (Collection of pollen for hybridization.) Plodovo-Jagodnye Kuljtury (Fruit Crops), Nr. 3, 95—96 (1940). [Russisch.]

Raheja, P. C.: What's doing in All-India. North-west Frontier Province. Indian Fmg, **5**, 529—531 (1944).

Rakitin, J. V.: (Neues über die Beschleunigung der Fruchtreife.) Priroda **7**, 70—73 (1940). [Russisch.]

— — (Eine praktische Anleitung für die Beschleunigung der Fruchtreife.) Moskva-Leningrad (Verlag Akad. d. Wiss. UdSSR.), Pp. 92 (1942). [Russisch.]

Rehder, A.: New species, varieties and combinations from the collections of the Arnold Arboretum. J. Arnold Arbor. **20**, 85—101, 409—431 (1939).

— Notes on some cultivated trees and shrubs. II. J. Arnold Arbor **26**, 472—481 (1945).

Rietsema, I. (Editor): Vierde beschrijvende rassenlijst voor fruit 1938—1939. (Fourth descriptive list for varieties of fruit 1938—1939.) Published by the Nederlandsch Algemeen Keuringsdienst (N. A. K.), Pp. 86 (1938—1939).

Rogers, W. S. u. R. J. Garner: Report on fruit plant nurseries and fruit research stations in western Germany and the German official fruit plant certification scheme. British Intelligence Objectives Sub-Committee London, Final Rep., Nr. 1072, Item, Nr. 22, Trip. Nr. 1515, 17 (1945). [Mimeographed.]

Rubcov, G. A.: (Wild fruit trees as a basis of seed production for nurseries.) Sadovodstvo (Horticulture), Nr. 1, 13—16 (1940). [Russisch.]

Rudnickij, N. V.: (Problems of northern fruit culture.) Sadovodstvo (Horticulture), Nr. 1, 32—35 (1939). [Russisch.]

Rudorf, W.: Die Züchtung frostresistenter Obstsorten mit besonderer Berücksichtigung der Resistenz gegen Spätfrostschäden. Forschungsdienst **9**, 266—276 (1940).

— Züchtungsforschung an den Obstarten (Edelsorten). Forschungsdienst, Sonderh. **16**, 472—481 (1942).

— Die Pflanzenzüchtung im Dienste der Konservierung von Obst und Gemüse. Forschungsdienst **17**, 583—590 (1944).

Rygg, G. L.: Compositional changes in the date fruit during growth and ripening. Tech. Bull. U. S. Dep. Agric., Nr. 910, 51 (1946).

Schmidt, M.: Die Frage der frostfesten Sämlingsunterlagen als züchterisches Problem. Dtsch. Obstbau, A **57**, 153—155 (1942).

— Beiträge zur Züchtung frostwiderstandsfähiger Obstsorten. Züchter **14**, 1—19 (1942).

Sengbusch, R. von: Probleme und Zielsetzung der Gemüse- und Obstzüchtung in Verbindung mit der Konservierung. Züchter **15**, 78—90 (1943).

Siddappa, G. S. u. A. M. Mustafa: Preparation and preservation of fruit and vegetable products. Misc. Bull. Imp. Coun. Agric. Res., Delhi, Nr. 63, 24 (1946).

Širjaeva, K. A.: (The application of the mentor in bringing-up hybrid seedlings.) Sadovodstvo (Horticulture), Nr. 7, 24—25 (1940). [Russisch.]

Skard, O. u. G. Vallevik: Skaden på frukttrea vinteren 1939—1940. (Damage done to fruit trees during the winter 1939—1940.) Meld. Norg. LandbrHøgsk. **22**, 1—56 (1942).

Solovjeva, M. A.: (Determination of the frost resistance of fruit trees.) Sovetskaja Botanika (Soviet Botany), Nr. 1/2, 133—144 (1941). [Russisch.]

Sonesson, N.: Balsgårds Fruktträdsförädlingsanstalt. (Balsgard Institute for the Breeding of Fruit Trees.) Sverig. Pomol. Fören. Årsskr. **43**, 5—24 (1942).

Stark, A. L.: Fruit pollination- a problem in Uhta. Fm Home Sci. Utah **5**, 5—6 (1944).

Stepanov, P. A.: (The efficacy of Michurin's principles.) Sadovodstvo (Horticulture), Nr. 12, 22—23 (1939). (Russisch.]

— (A half century of work by a disciple of Michurin, A. I. Oloničenko.) Plodovo-Jagodnye Kuljtury (Fruit Crops), Nr. 5, 115—117 (1940). [Russisch.].

Strunck, R. u. A. Waschneck: Untersuchungen über den Einfluß des Standortes und der Pflege auf die Frostschäden an Obstbäumen. Gartenbauwiss. **17**, 273 bis 294 (1943).

Stuivenberg, J. H. M. van: Parthenocarpie (Parthenocarpy.) Vakbl. Biol. **24**, 25—31 (1943).

Tatarincev, A. S.: (Further notes on change of fruit form under the influence of foreign pollen.) Jarovizacija, Nr. 5 (32), 130—131 (1940). [Russisch.]

— (Research work at the plant breeding faculty of the Michurin Horticultural Institute.) Plodovo-Jagodnye Kuljtury (Fruit Crops), Nr. 3, 78—81 (1940). [Russisch.]

Teterev, F. K.: (The scientific legacy of V. V. Paskevich). Soviet Plant Industry Record, Nr. 1, 9—19 (1940). [Russisch.]

— (The great reformer of nature — Ivan Vladimirowič Michurin.) Vestnik Socialističeskogo Rastenievodstva (Soviet Plant Industry Record), Nr. 3, 3—14 (1940). [Russisch.]

Timofeev, N. N.: (I. V. Mičurin and his work.) Moskovskaja ordena Lenina Seljskohozjajstvennaja Akademija imeni K. A. Timirjazeva. Doklady Vypusk III. Naučnaja Konferencija 4—11 Ijunja 1945 g. (Timirjazev, Agric. Acad. Moscow Proc., Nr. III., Sci. Conf. 4—11 June 1945.) Pp. 59—64 (1946). [Russisch.]

Tolmačev, A. I.: (Scientific notes — A study of vitaminbearing plants in the northern regions.) Sovetskaja Botanika (Soviet Botany **6**, 48—50 (1942). [Russisch.]

Tolmačev, I. A.: (An experiment on overcoming incompatibility.) Jarovizacija, Nr. 5 (32), 125—126 (1940). [Russisch.]

Trenkle, E.: Primitivsorten und deren Bedeutung für die Leistungssteigerung im Obstbau. Dtsch. Obstbau, Nr. 6, 101—106 (1942).

Trenkle, R.: Beitrag zur Unterlagen- und Stammbildnerfrage. Klarstellungen. Dtsch. Obstbau **57**, 3 bis 9 (1942).

Turrell, F. M.: Estimating fruit surfaces. Amer. J. Bot. (Suppl.) **33**, 224 (1946). (Abstr.)

Tydeman, H. M.: A progress report on breeding work with the tree fruits. 33rd Rep. E. Malling Res. Sta. 1945, Pp. 63—66 (1946).

Uljaniščev, M.: (On the path of a great horticulturist.) Sady i Ogorody (Fruit and Vegetable Gardens), Nr. 6, 13—15 (1941). [Russisch.]

— (Zur Frage der Wiederherstellung der Fruchtbarkeit bei Art- und Gattungsbastarden von Obstbäumen.) C. R. (Doklady) Acad. Sci. URSS. **58**, 1175 (1947). [Russisch.]

Usatov, S. P.: (Fruit growing in the north.) Plodovo-Jagodnye Kuljtury (Fruit Crops), Nr. 4, 80—81 (1940). [Russisch.]

Venjaminov, A. N.: (Improvement of old varieties of fruit trees by selection of vegetative mutations.) Plodovo-Jagodnye Kuljtury (Fruit Crops), Nr. 5, 15 bis 20 (1940). [Russisch.]

Vikla, J.: Fruit and tree improvement. 38th Rep. S. Dak. St. Hort. Soc., Pp. 61—62 (1941).

Villforth, F.: Die Aromastoffe im Obst und Gemüse. Gartenbauwiss. **17**, 382—396 (1943).

Vyvyan, M. C.: Fruit fall and it control by synthetic growth substance. Imperial Bureau of Horticulture and Plantation Crops, East Malling 3s. 6d. Tech. Comm., Nr. 18, 72 (1946).

W....., H.: Von der Sortenwahl unter besonderer Berücksichtigung Zürichscher Verhältnisse. Schweiz. Z. Obst- u. Weinbau **53**, 152—156 (1944).

W-m....., H.: En kuriositet. (A curiosity.) Fructodlaren, Nr. 1, 17—18 (1944).
Wellington, R.: Report on Geneva. New York Experiment Station has introduced 137 varieties. Amer. Fruit Gr. **67**, Nr. 7, 13, 24, 30—31 (1947).
Wilcox, A. N.: Breeding small fruits. Amer. Fruit Gr. **66**, 16, 39—40 (1946).
Wormald, H.: Diseases of fruits and hops. Crosby Lockwood and Son, Ltd., London (1939).

2. Kernobst.

Adametz, L.: Vierzigjährige Erfahrungen über frost- und schorfresistente Apfelsorten im Altvatergebiet. Gartenbauwiss. **15**, 487—508 (1941).
Alderman, D. C. u. H. L. Lantz: Apple breeding: inheritance and statistical studies on the fruits of crossbred seedlings with Antonovka parentage. Proc. Amer. Soc. Hort. Sci. **36**, 279—283 (1938) 1939.
Anonymus: Test many new apple varieties. News Lett., Idaho **22**, Nr. 2, 1 (1939).
— Zwei Apfelneuzüchtungen für die Obstpraxis. Forschungsdienst **11**, 122 (1941). (Abstr.)
— Tetraploide Apfelbäume in Schweden. Schweiz. Z. Obst- u. Weinb. **52**, 480 (1943).
— The Wrixparent apple. News Lett. Ill. St. Hort. Soc., Nr. 1, 4 (1943).
— Aufstellung einer Reichssortenliste für Äpfel. Forschungsdienst **17**, 53 (1944).
— Try a pear-apple! Soviet News, Nr. 1438, 3 (1946).
— Peter Gideon and the Wealthy apple. Amer. Fruit Gr. **67**, Nr. 7 20, 40 (1947).
— Nationwide fruits-apples. Amer. Fruit Gr. **67**, Nr. 9, 33 (1947).
Aschan, K.: Fortsatta undersökningar rörande avkomman av triploida äpplesorter. (Further investigations on the offspring of triploid apple varieties.) Sverig. Pomol. Fören. Årsskr. **44**, 61—70 (1943).

Ball, E.: Cider orchard restoration in Herefordshire. J. Minist. Agric. **54**, 107—111 (1947).
Barker, B. T. P.: The production of cider fruit on bush trees. Vintage quality trials. Progress report, Nr. 1, 1942 crop. Rep. Agric. Res. Sta. Long Ashton, Pp. 124—135 (1943).
— u. L. F. Burroughs: The production of cider fruit on bush trees. Vintage quality trials. Progress report, Nr. 2, 1943 crops. Rep. Agric. Hort. Res. Sta., Long Ashton, Bristol, Pp. 170—178 (1945).
—, — The production of cider fruit on bush trees. Vintage quality trials. Progress report, Nr. 3, 1944 crops. Rep. Agric. Hort. Res. Sta., Long Ashton, Bristol, Pp. 178—184 (1945).
Birjukov, A. P.: (New apple varieties in the forest steppes of the Trans-urals.) Plodovo-Jagodnye Kuljtury (Fruit Crops), Nr. 4, 45—49 (1940). [Russisch.]
Blake, M. A.: Dwarf apple trees appear in progeny of apple crosses. N. J. St. Hort. Soc. News **22**, 1321 (1941).
—, L. J. Edgerton u. O. W. Davidson: Standards for judging the growth status of apples in New Jersey. Bull. N. J. Agric. Exp. Sta., Nr. 715, 36 (1945).
Blaser, H. W. u. J. Einset: Leaf development in a periclinal chimera of „Spy" apple. Amer. J. Bot. Suppl. **33**, 818 (1946). (Abstr.)
Börner, C. u. F. Gollmick: Blutlausimmune Naumburger Edelapfelzüchtungen. Angew. Botanik **25**, 144—149 (1943).
Botez, I.: Felul de a se comporta în procesul de fecundare a câtorva varietăți de păr. (The behaviour of certain pear varieties in the process of fertilization.) Anal. Inst. Cerc. Agron. Român. **10**, 395—415, 1938 (1939).

Bowman, F. T.: Inheritance and use of vigour in pear seedlings. J. Aust. Inst. Agric. Sci. **9**, 24—29 (1943).
— Root types among apple clones raised from root cuttings. J. Aust. Inst. Agric. Sci. **9**, 127—129 (1943).
Bradford, F. C.: Second-year changes in apparent vigor of apple varieties of prospective value as trunkformers. Proc. Amer. Soc. Hort. Sci. **44**, 215—220 (1944).
Bremer, H. u. H. Işmen: Elma ve armutların en tehlikeli hastalığı olan Fusicladium üzerinde etütler (Studies on Fusicladium, the most dangerous disease of apples and pears.) Ziraat Dergisi **5**, 10—18 (1944).
Brown, A. G.: The order and period of blossoming in pear varieties. J. Pomol. Hort. Sci. **20**, 107—110 (1943).
Bryner, W.: Zur Wahl des Saatgutes für die Anzucht von Apfelunterlagen. Schweiz. Z. Obst- u. Weinb. **52**, 546 (1943).

Caldas, J. P.: Estudos citológicos em variedades culturais de macieiras. (Cytological studies of cultivated varieties of apple.) Lisboa, Pp. 48 (1945).
Černenko, S. F.: (Apple breeding in connexion with the choice of initial forms.) Agrobiologija (Agrobiology), Nr. 2, 97—110 (1946). [Russisch.]
Černjaev, I. P.: (Take notice of sports.) Sadovodstvo (Horticulture), Nr. 1, 37—41 (1939). [Russisch.]
Chatot, J.: Remarques sur la résistance comparée de quelques variétés de poires à l'occasion des gelées printanières. Rev. Hort., Paris **114**, 191 (1942).
Collins, J. L. u. K. R. Kerns: Inheritance of three leaf types in the pineapple. J. Hered. **37**, 123—128 (1946).
Crane, H. H.: Some good American apples. J. R. Hort. Soc. **71**, 172—173 (1946).
Crane, M. B.: The mystery of Lord Lambourne. Grower, Lond. **22**, 10—11, 12 (1944).
— Origin of viruses. Nature, Lond. **155**, 115—116 (1945).
— u. D. Lewis: Genetical studies in Pears, II. A. Classification of Cultivated Varieties. Journal of Pomology **18**, 52—60 (1940).
—, — Genetical studies in pears. III. Incompatibility and sterility. J. Genet. **43**, 31—43 (1942).
— u. P. T. Thomas: Genetical studies in pears. I. The origin and behaviour of a new giant form. J. Genet. **37**, 287—299 (1939).
Dibrova, P. A.: (New apple varieties from the northern Urals.) Plodovo-Jagodnye Kuljtury (Fruit Crops), Nr. 4, 30—41 (1940). [Russisch.]
— (New varieties of apple of the North Urals.) Michurin's Sverdlovsk Fruit Exp. Sta., Pp. 14, Also Plodovo-Jagodnye Kuljtury (Fruit Crops), Nr. 4 (1940). [Russisch.]
Dorsey, M. J.: Inbreeding experiments with the apple. Proc. Amer. Soc. Hort. Sci. **36**, 292 (1938) 1939. (Abstr.)
— A progress report on the apple breeding project. Trans. Ill. Hort. Soc. **78**, 98—106 (1944).
Drain, B. D.: Southern pear breeding. Proc. Amer. Soc. Hort. Sci. **42**, 301—304 (1943).
Dubovik, N. V.: (A weeping form of the Chinese apple [Malus prunifolia].) Plodovo-Jagodnye Kuljtury (Fruit Crops), Nr. 5, 109 (1940). [Russisch.]
Duka, S. H.: (The influence of nutrition and the production of apple seedlings with the characteristics of cultivated varieties.) Jarovizacija, Nr. 2 (29), 90—92 (1940). [Russisch.]
Dustman, R. B.: Sugar distribution in the blossoms and fruit of three varieties of apple. Proc. W. Va Acad. Sci. **17**, 42—46 (1945).

Einset, J.: The occurrence of a tetraploid and two triploid apple seedings in progenies of diploid parents. Science **99**, 345 (1944).

Einset, J.: The spontaneous origin of polyploid apples. Proc. Amer. Soc. Hort. Sci. **46**, 91—93 (1945).

—, H. W. Blaser u. B. Imhofe: A chromosomal chimera of the Northern Spy apple. J. Hered. **37**, 265 bis 266 (1946).

Ellenwood, C. W.: Red varieties of early apples. Bi-m. Bull. Ohio Agric. Exp. Sta. **27**, 124 (1942).

Evreinoff, V. A.: L'importance et la diversité du genre Prunus (au sens large) dans l'arboriculture fruitière. Rev. Hort., Paris **26**, 540—542 (1939).

— Notes sur l'origine botanique et génétique de nos arbres fruitiers. I. Les pommiers. Rev. Hort., Paris **116**, 11—13 (1944).

— Notes sur l'origine botanique et génétique de nos arbres fruitiers. II. Les poiriers. Rev. Hort., Paris **116**, 18—19 (1944).

Ewert: Eigene Erfahrungen über die Fruchtbarkeit triploider Apfelsorten. Dtsch. Obstbau **57**, 81—83 (1942).

Field, C. P.: Low temperature injury to fruit blossom. II. A comparison of the relative susceptibility and effect of environmental factors on three commercial apple varieties. 29th Rep. E. Malling Res. Sta. Pp. 29—35, 1941 (1942).

Fish, V. B., R. B. Dustman u. R. S. Marsh: The ascorbic acid content of several varieties of apples grown in West Virginia. Proc. Amer. Soc. Hort. Sci. **44**, 196—200 (1944).

Fitzpatrick, R. E., F. C. Mellor u. M. F. Welsh: Crown rot of apple trees in British Columbia- rootstock and scion resistance trials. Sci. Agric. **24**, 533—541 (1944).

Fleckinger, J.: Caractères morphologiques et végétatifs en relation avec la triploïdie chez le pommier et le poirier. Proc. 7th Int. Genet. Congr. Edinburgh 23—30 August, 121, 1939 (1941). (Abstr.)

Gardner, V. R.: A study of the Sweet-and Sour apple chimera and its clonal significance. J. Agric. Res. **68**, 383—394 (1944).

Geiger-Vifian, A.: Der Vitamin C-Gehalt in schweizerischen Apfelsorten. Schweiz. landw. Mh., Nr. 11, 280—286 (1945).

Gorškova, T. P.: (Hybrids between apple and pear.) Agrobiologija (Agrobiology), Nr. 1, 130—133 (1946) [Russisch.]

Grant, E. P.: Apples as a source of vitamin C. Sci. Agric. **27**, 162—164 (1947).

Griggs, W. H. u. A. L. Schrader: Comparison of certain varieties as pollenizers for the Delicious apple. Proc. Amer. Soc. Hort. Sci. **40**, 87—90 (1942).

Grüner, M. N.: (On the development of the apple seedling.) Proc. Lenin Acad. Sci. USSR., Nr. 23/24, 6—12 (1939). [Russisch.]

Hall, D.: Apple Lane's Prince Albert. J. R. Hort. Soc. **66**, 96—97 (1941).

Haller, M. H. u. J. R. Magness: Picking maturity of apples. Circ. U. S. Dep. Agric., Nr. 711, 23 (1944).

Hansen, E.: Quantitative study of ethylene production in apple varieties. Plant Physiol. **20**, 631—635 (1945).

Heilborn, O.: Pollen tube growth in apple styles after inter-varietal cross-pollination. Lantbruks-Högskolans Annaler, Uppsala **7**, 171—183 (1939).

— On some effects of primary and secondary polyploidy in apples and pears. Ann. Agric. Coll. Sweden **9**, 116—126 (1941).

Henning, W.: Morphologisch-systematische und genetische Untersuchungen an Arten und Artbastarden der Gattung Malus. Züchter **17/18**, 290—349 (1947).

Hilton, R. J.: Fire blight in Alberta. A serious scourge of apple trees. Pr. Bull. Univ. Alberta **32**, 2—3 (1947).

Hoblyn, T. N.: The dessert apple plantation of tomorrow. What to plant and how to plan it in the light of present day knowledge. 33rd Rep. E. Malling Res. Sta. 1945, Pp. 115—120 (1946).

Hockey, J. E. u. C. C. Eidt: Resistance of apple seedlings to scab. Sci. Agric. **24**, 542—550 (1944).

Hough, L. F.: The new pear breeding project. Trans. Ill. Hort. Soc. **78**, 106—113 (1944).

— A survey of the scab resistance of the foliage on seedings in selected apple progenies. Proc. Amer. Soc. Hort. Sci. **44**, 260—272 (1944).

Howlett, F. E.: Pollination and fruit setting of the apple tree in the North Central States. Trans. Ill. Hort. Soc. **78**, 325—336 (1944).

— The Melrose apple. News Lett. Ill. St. Hort. Soc., Nr. 2, 4 (1947).

Hülphers, A.: Husmodersäpple kontra Holländare. (Husmoder apple versus Holländare.) Fruktodlaren, Nr. 1, 22—23 (1943).

Hülsmann, B.: Veredlungsversuche auf verschieden stark bedornten Abrissen der Apfelunterlage Ketziner Ideal. Gartenbauwiss. **17**, 171—175 (1942).

Husz, B.: Az alma metaxeniájaról. (A case of metaxenia in apples.) Bull. Hung. Coll. Hort. **8**, 128—135 (1942).

Isely, D.: Early maturing varieties in codling moth control. J. Econ. Ent. **36**, 757—759 (1943).

Jakovlev, L. I.: (The Baškirskii Krasavec apple.) Plodovo-Jagodnye Kuljtury (Fruit Crops), Nr. 5, 75 (1940). [Russisch.]

Johansson, E.: Ett remonterande päronträd. (An ever bearing pear tree.) Fruktodlaren, Nr. 5, 139—141 (1942).

— Undersökningar rörande högkromosomiga äpple- och päronträd. (Investigations on apple and pear trees with high chromosome numbers.) Sverig. Pomol. Fören. Årsskr. **44**, 55—60 (1943).

— Fortsatta undersökningar beträffande en en 68-kromosomig äpplesort. (Further investigations regarding a 68-chromosome variety of apple.) Sverig. Pomol. Fören. Årsskr. **45**, 135—140 (1944).

Kalašnikov, V. M.: (Cultivated varieties of pear of the Far East.) Plodovo-Jagodnye Kuljtury (Fruit Crops), Nr. 4, 50—56 (1940).

Karmacsi, B.: Egy almakiméra ismertetése. (A report of an apple chimaera.) Bull. Hung. Coll. Hort. **8**, 169 (1942).

Kemmer, E. u. F. Schulz: Versuche mit Pirus baccata-Unterlagen. Gartenbauwiss. **15**, 526—531 (1941).

—, — Kärnstammens betydelse som underlag. (The importance of the seedling as a stock.) Sverig. Pomol. Fören. Årsskr. **45**, 141—145 (1944).

Kidson, E. B.: The vitamin C content of Nelson apples. N. Z. J. Sci. Tech. **25**, Sect. B, 134—136 (1943).

Kieser, M. E. u. A. Pollard: Vitamin C in English apples. Nature, Lond. **159**, 65 (1947).

Kiper, N. Ö.: Orta Anadolu armutculugu ve en münim armut çeşitleri. (Central Anatolian pear culture and the most important pear varieties.) T. C. Yüksek Ziraat Enstitüsü Çalişmalarindan, Ankara, Nr. 123, 98 (1941).

Klang, C. A.: Ingrid Marie, en värdefull äpplenyhet. (Ingrid Marie, a useful new apple.) Fruktodlaren, Nr. 2, 41—43 (1943).

Kobel, F.: Der Stand und die Aufgabe der Apfelsortenzüchtung. Schweiz. Ztschr. f. Obst- und Weinbau, Nr. 1, 6—10 (1945).

Konovalov, I. N.: (Sur la présence des cellules pierreuses dans les fruits de quelques espèces du genre Malus Mill.) Sovetskaja Botanika **14**, Nr. 4, 262—269 (1946). [Russisch.]

Kruft, F.: Die Apfelsorten-Neuzüchtungen „Erwin Junge" und „Franz Späth". Dtsch. landw. Pr. **68**, 104 (1941).

Krumbholz, G.: Beiträge zur Morphologie der Apfelblüte. II. Über die Eignung der Blütenmerkmale zur Sortenbeschreibung. Gartenbauwiss. **13**, 1—65 (1939).

Kudagovskij, V. A.: (Breeding vegetatively reproduced rootstocks.) Sadovodstvo (Horticulture), Nr. 5 14—17 (1939). [Russisch.]

Kuznecov, P. V.: (The role of Pirus salicifolia Pall. in the development of fruit growing in arid regions.) Sovetskaja Botanika (Soviet Botany), Nr. 1/2, 103 bis 107 (1941). [Russisch.]

Lantz, H. L. u. B. S. Pickett: Apple breeding: variation within and between progenies of Delicious with respect to freezing injury due to the November freeze of 1940. Proc. Amer. Soc. Hort. Sci. **40**, 237—240 (1942).

Levošin, V. K.: (Form development within a clone in apples.) Socialistic Grain Farming, Saratov, Nr. 1, 142—158 (1940). [Russisch.]

— (The origin of forms within a clone in the apple.) Plodovo-Jagodnye Kuljtury (Fruit Crops), Nr. 3, 3—13 (1940). [Russisch.]

Lewis, D.: Parthenocarpy induced by frost in pears. J. Pomol. Hort. Sci. **20**, 40—41 (1942).

— u. I. Modlibowska: Genetical studies in pears. IV. Pollen-tube growth and incompatibility. J. Genet. **43**, 211—222 (1942).

Lincoln, F. B.: Chromosome counts on Hibernal and Northwestern Greening apples. Proc. Amer. Soc. Hort. Sci. **37**, 217 (1939) 1940.

Lukašev, A. M.: (On the path of Michurin.) Jarovizacija, Nr. 3 (30), 179—183 (1940). [Russisch.]

McMunn, R. L.: The Red Bird apple self-unfruitful. News Lett. Ill. St. Hort. Soc., Nr. 3, 3—4 (1944).

McVaugh, R.: The status of certain anomalous native crap-apples in eastern United States. Bull. Torrey Bot. Cl. **70**, 418—429 (1943).

Maier, J.: Developing new varieties of apples. Proc. 85th Annu. Mtg. N. Y. St. Hort. Soc., Pp. 148—151 (1940).

Makarov, D. D.: (A pear tree producing three crops of fruit a year.) Plodovo-Jagodnye Kuljtury (Fruit Crops), Nr. 5, 113—114 (1940). [Russisch.]

Mansvetov, V. I.: (In the path of Michurin.) Sadovodstvo (Horticulture), Nr. 12, 24—25 (1939). [Russisch.]

Martin-Lecointe, G.: La sélection des sujets porte-greffes. Rev. Hort., Paris **115**, 377—378 (1943).

Maurer, E. u. W. Redecker: Der Einfluß einiger vegetativ vermehrter Unterlagen auf das Wachstum von fünf Apfelsorten in der Baumschule. Forschungsdienst **12**, 324—337 (1941).

Meader, E. M. u. C. H. Blassberg: Blossom hardiness of forty-five apple varieties. Proc. Amer. Soc. Hort. Sci. **47**, 58—60 (1946).

Möckel, W., J. Wolf u. U. Degen: Weitere Untersuchungen über den Ascorbinsäure- und Vitamin C-Gehalt deutscher Apfelsorten. Gartenbauwiss. **17**, 176—185 (1942).

Modlibowska, I.: Pollen tube growth and embryo-sac development in apples and pears. J. Pomol. **21**, 57—89 (1945).

Moore, R. C.: Developing new varieties of apples at the Virginia Experiment Station. Virginia Fruit **27**, 14—15 (1939).

— A study of the inheritance of susceptibility and resistance to apple cedar rust. Proc. Amer. Soc. Hort. Sci. **37**, 242—244 (1939) 1940.

— Apple breeding program-cedar rust inheritance. Virginia Fruit **28**, 163—166 (1940).

Moore, R. C.: Some results of apple breeding at V. P. I. Virginia Fruit **33**, 63—67 (1945).

— Inheritance of fire blight resistance in progenies of crosses between several apple varieties. Proc. Amer. Soc. Hort. Sci. **47**, 49—57 (1946).

Murneek, A. E.: Factors affecting size and color of fruit (with reference to apples and peaches.) Bull. Mo. Agric. Exp. Sta., Nr. 428, 19 (1941).

Murray, H. R., R. J. M. Reid, P. O. Roy u. D. S. Blair: Some new apple varieties. Rep. Pomol. Soc. Quebec, Pp. 12—15 (1944).

Natividade, J. V.: Mutações somáticas em variedades Portuguesas de pomóideas. (Somatic mutations in Portuguese varieties of pome fruits.) Agron. Lusitana **1**, 7—21 (1939).

Near, R.: Our apple varieties. Proc. 85th Annu. Mtg. N. Y. St. Hort. Soc., Pp. 273—275 (1940).

Newcomer, E. H.: Studies in the nature of the clonal variety. IV. Cytological studies of bud sports of McIntosh, Stark and Baldwin apples. Tech. Bull. Mich. Agric. Exp. Sta., Nr. 187, 23 (1943).

Nilsson, F.: Tetraploidi hos päronplantor framkallad med hjälp av colchicin. (Tetraploidy in young pear trees produced with the aid of colchicine.) Sverig. Pomol. Fören. Årsskr. **41**, 103—107 (1940).

— u. G. Larsson: Nya tetraploider av äpple vid Balsgård. (New tetraploids of the apple at Balsgård.) Sverig. Pomol. Fören. Årsskr. **45**, 123—129 (1944).

Nilsson-Ehle, H.: Fortsatta arbeten på framställande av tetraploida äpplen. (Further work on the production of the tetraploid apples.) Sverig. Pomol. Fören. Årsskr. **43**, 25—28 (1942).

— Några nya rön rörande tetraploida äpplesorter och deras användning och roll vid växtförädlingen nos frukträd. (Some new information about tetraploid apple varieties and their use and role in the breeding of fruit treed.) Sverig. Pomol. Fören. Årsskr. **45**, 229—237 (1944).

Nyquist, J.: Späsperudsäpplet. (The Späsperud apple.) Fruktodlaren, Nr. 1, 16—17 (1944).

Oldén, E. J.: Några nya högkromosomiga äppletyper. (Some new high-chromosome types of apples.) Sverig. Pomol. Fören. Årsskr. **46**, 105—115 (1945).

Oloničenko, A. I.: (A new apple variety for Siberia.) Plodovo-Jagodnye Kuljtury (Fruit Crops), Nr. 4, 28—29 (1940). [Russisch.]

Osmanov, V. O. u. S. V. Sosnin: (New élites of apple and pear from the Crimean Fruit Station.) Sadovodstvo (Horticulture), Nr. 12, 27—30 (1939). [Russisch.]

Östlind, N.: Om grundstamar för äpple och päron. (On stocks for apple and pear.) Sverig. Pomol. Fören. Årsskr. **45**, 5—15 (1944).

Påhlman, A.: Anteckningar om äpplenamnet pipping. (Notes on the name Pippin in apples.) Sverig. Pomol. Fören. Årsskr. **42**, 78—87 (1941).

Perlberger, J.: The occurience of apple and pear scab in Palestine in relation to weather conditions. Palest. J. Bot. R. Ser. **4**, 157—161 (1944).

Pickett, B. S. u. H. L. Lantz: Apple varieties: behavior of two hundred varieties following the freeze of November, 1940. Proc. Amer. Soc. Hort. Sci. **40**, 212 bis 214 (1942).

Pirovano, A.: Induzione di precocità ed anomalie in meli innestati su soggetti adulti a maturazione precoce. (Induction of earliness and anomalies in apples grafted on adult early maturing rootstocks.) Ital. Agric. **78**, 797—805 (1941).

Pollard, A., M. E. Kieser u. J. D. Bryan: The apple as a source of vitamin C. Rep. Agric. Hort. Res. Sta., Long Ashton, Bristol, Pp. 200—202 (1945).

Roberts, R. H.: Factors affecting apple setting. Trans. Ill. Hort. Soc. **78**, 118—124 (1944).
— Blossom structure and setting of Delicious and other apple varieties. Proc. Amer. Soc. Hort. Sci. **46**, 87—90 (1945).
— Cold injury of apple blossom, 1945. Proc. Amer. Soc. Hort. Sci. **47**, 61—63 (1946).
Rodionov, A. u. M. Zelenskij: (Winter-hardiness of apple and plum hybrids in relation to soil conditions. Jarovizacija, Nr. 1 (34), 108—109 (1941). [Russisch.]
Rodrigues, A.: Algumas relações entre o número de sementes, a forma e as dimensões dos frutos, em variedades culturaids de pereiras. (Some relations between the number of seeds, and the shape and dimensions of the fruits, in cultivated varieties of pears.) Agron. Lusitana **7**, 121—157 (1945).
Rubcow, G. A.: (Origin and evolution of the pear genus Pirus Tourn.) Proc. Lenin Acad. Agric. Sci. USSR., Nr. 13, 14—17 (1939). [Russisch.]
— Polymorphismus and centers of formation of Pyrus species in the USSR. C. R. (Doklady) Acad. Sci. USSR. **24**, 81—84 (1939).
— Origin and evolution of the cultivated pear. C. R. (Doklady) Acad. Sci. URSS. **28**, 350—353 (1940).
— Ontogeny, age modifications and anomalies in the development of the pear. C. R. (Doklady) Acad. Sci. URSS. **30**, 79—81 (1941).
— Geographical distribution of the genus Pyrus and trends and factors in its evolution. Amer. Nat. **78**, 358—366 (1944).
Rubtsov, G. A. siehe Rubcov.
Rudloff, C. F. u. G. Wundrig: Zur Physiologie des Fruchtens bei den Obstgehölzen. I. Die Aufblühfolge bei einigen Birnensorten. Gartenbauwiss. **12**, 530 bis 554 (1939).

Šablovskij, B. I.: (Root grafts of hybrids on mentors.) Jarovizacija, Nr. 4 (31), 87—94 (1940). [Russisch.]
Sager, R.: Relative fire-blight injury of 117 varieties of apple. Hort. News. **25**, 1619 (1944).
Schmidt, M.: Ein Fall gehäufter Chimärenbildung beim Apfel. Züchter **14**, 112—117 (1942).
— Beiträge zur Züchtungsforschung beim Apfel. I. Phaenologische, morphologische und genetische Studien an Nachkommenschaften von Kultursorten. Züchter **17/18**, 161—224 (1947).
Shaw, J. K.: Descriptions of apple varieties. Bull. Mass. Agric. Exp. Sta., Nr. 403, 187 (1943).
— Bud sports of McIntosh. News Lett. Ill. St. Hort. Soc., Nr. 8, 4 (1944).
— u. L. Southwick: Somatic mutations in the apple. Science **97**, 202 (1943).
—, — Certain stock-scion incompatibilities and uncongenialities in the apple. Proc. Amer. Soc. Hort. Sci. **44**, 239—246 (1944).
Simmonds, A.: The origin of apple „Bramley's Seedling". J. R. Hort. Soc. **70**, 99—103 (1945).
— The origin of apple „Ellison's Orange". J. R. Hort. Soc. **70**, 150—151 (1945).
Simonov, I. N.: (Biochemical and biological differences between apple varieties.) Trudy Moskovskoj ordena Lenina Seljskohozjajstvennoj Akademii imeni K. A. Timirjazeva **4**, Pt. 1, 71—101 (1940). [Russisch.]
Sinha, A. C. u. M. C. Vyvyan: Studies on the vegetative propagation of fruit tree rootstocks. II. By hardwood cuttings. J. Pomol. Hort. Sci. **20**, 127—135 (1943).
Smith, M. A.: Blister spot, a bacterial disease of apple. J. Agric. Res. **68**, 269—298 (1944).
Soder, A. R.: The Soder apple. Trans. Iowa Hort. Soc. **74**, 35—36 (1939).
Souilijaert, G.: Sélection de poires photographiées au verger Orléanais. Direction Régionale des Services Agricoles, Pp. 27, photos. (1945).

Southwick, L., A. P. French u. O. C. Roberts: The identification of pear varieties from non — bearing trees. Bull. Mass. Agric. Exp. Sta., Nr. 421, 51 (1944).
Štejnbok, V. D. u. N. V. Čukanova: (Michurin's variety „Northern Bužbon".) Sadovodstvo (Horticulture), Nr. 7, 26—27 (1940). [Russisch.]
Struckmeyer, B. E.: Hand pollination of Delicious in the Wenatchee, Washington orchards. Amer. Fruit Gr. **66**, 14, 32—33 (1946).
Stuart, N. W.: Cold hardiness of seedlings from certain apple varieties as determined by freezing tests. Proc. Amer. Soc. Hort. Sci. **38**, 315 (1941). (Abstr.)
Sudds, R. H.: The effect of the rootstocks on ten years' growth and yield of the Gallia Beauty apple. Proc. Amer. Soc. Hort. Sci. **44**, 236—238 (1944).
Sugawara, T.: (The vitamin C content of apples.) J. Hort. Ass. Jap. **12**, 109—112 (1941).
Swanson, C. P. u. V. R. Gardner: A pomological and cytological study of a russeted sport of the Stark apple. J. Agric. Res. **68**, 307—315 (1944).

Tarasenko, G. G.: (Malus prunifolia, Borkh. und ihre Abstammung.) Sov. Plant. Ind. Record., Nr. 3, 31—38 (1940). [Russisch.]
Thomas, L. A.: Stock and scion investigations. III. The root-systems of some ownrooted apple trees. J. Coun. Sci. Industr. Res. Aust. **17**, 167—178 (1944).
Thomas, P. T.: A useful abnormality of the pollen in a pear. Nature, Lond. **149**, 168—169 (1942).
Toenjes, W.: The new Close apple. Quart. Bull. Mich. Agric. Exp. Sta. **24**, 321—322. News Lett. Ill. St. Hort. Soc., Nr. 1, 2—3 (1943).
Trebušenko, P. D.: Inheritance of anthocyanine coloration in apple-tree. (A contribution to the problem of breeding red-fruited and redfleshed varieties of apples.) C. R. (Doklady) Acad. Sci. URSS. **23**, 939—943 (1939).
Tukey, H. B. u. K. D. Brase: Differences in congeniality of two sources of McIntosh apple budwood propagated on rootstock USDA 227. Proc. Amer. Soc. Hort. Sci. **45**, 190—194 (1944).
Tydeman, H. M.: The inheritance of susceptibility to sulphur damage in families of seedling apples. J. Pomol. **19**, 137—145 (1941).
— Further studies on new varieties of apple rootstocks. J. Pomol. Hort. Sci. **20**, 116—126 (1943).
— A preliminary account of experiments in breeding early and midseason dessert apples. Ann. Rep. E. Malling Res. Sta. 1943, **51**, 34—42 (1944).
— Two new apple varieties bred at East Malling. 33rd Rep. E. Malling Res. Sta. 1945, Pp. 121—122 (1946).

Varencov, I. J.: (Chemical and technological characteristics of the Reinette apples of the Krasnojarsk zone.) Plodovo-Jagodnye Kuljtury (Fruit Crops), Nr. 4, 69—76 (1940). [Russisch.]
Veh, R. von: Über Entwicklungsbereitschaft und Wüchsigkeit der Embryonen von Apfel, Pfirsich u. a. Züchter **11**, 249—255 (1939).
Verner, L.: The Idared apple. Circ. Idaho Agric. Exp. Sta., Nr. 84, 3 (1942).
— The Payette and Idagold apples. Circ. Univ. Idaho Agric. Exp. Sta., Nr. 89, 3 (1944).

Wallace, T., T. Swarbrick u. L. Ogilivie: Some new troubles in apples with special reference to the variety Lord Lambourne. Grower, Lond. **22**, 12—13, also Fruit Grower **98**, 427 (1944).
Wanscher, J. H.: Contributions to the cytology and life history of apple and pear. K. VetHøjsk. Aarsskr., Pp. 21—70 (1939).
Weeks, W. u. L. P. Latimer: Incompatibility of Early McIntosh and Cortland apples. Prog. Amer. Soc. Hort. Sci., Pp. 284—286 (1938) 1939.

Wellington, R.: Notes on new varieties of apples. Proc. 90th Annu. Mtg. N. Y. St. Hort. Soc., Pp. 232 bis 236 (1945).
— u. G. H. Howe: The performance of seedlings derived from selfing and crossing the McIntosh apple. Proc. Amer. Soc. Hort. Sci. **44**, 273—279 (1944).
Welsh, M. F.: Studies of crown rot of apple trees. Canad. J. Res. **20**, Sect. C, 457—490 (1942).
Winter, J. D.: Cold storage studies with Minnesota-grown apples. Proc. Amer. Soc. Hort. Sci. **46**, 143 bis 144 (1945).
Wolf, J.: Über den Vitamin C-Gehalt deutscher Äpfel. Gartenbauwiss. **16**, 292—313 (1941).
Wyman, D.: A few hybrid crabapples. Nat. Hort. Mag. **20**, 52—54 (1941).
Z....., L.: Varietá precocissime di melo a Tripoli. (Excessively early varieties of apple in Tripoli.) Note Fruttic. **18**, 73—75 (1940).
Žavoronkov, P. A.: (The pear in the Čeljabinsk province.) Plodovo-Jagodnye Kuljtury (Fruit Crops), Nr. 4, 57—59 (1940). [Russisch.]
— (Forms of the Sibirian crab apple.) Plodovo-Jagodnye Kuljtury (Fruit Crops), Nr. 5, 53—59 (1940). [Russisch]
Zorin, F. M.: (A pear forming vegetative fruits.) Agrobiologija (Agrobiology), Nr. 2, 135—139 (1946). [Russisch.]
Zwintzscher, M.: Experimentelle Untersuchungen zur Züchtung von Obstgehölzen mit frostwiderstandsfähigen Fruchtknospen und Blüten. I. Malusformen. Z. Pflanzenz. **26**, 245—352 (1944).

3. Steinobst.

Anonymus: Two peaches are named. N. J. St. Hort. Soc. News **20**, 1148 (1939).
— New twentieth century peaches bred and tested in New Jersey. N. J. Peach Coun. New Brunswick. Pp. 12 (1939—1940).
— New twentieth century peaches bred and tested in New Jersey. Propagated and Distributed by The New Jersey Peach Council, Inc., New Brunswick, N. J., Pp. 9 (1940—1941).
— Partners produce new York Imperial Cherry. Amer. Fruit Gr. **61**, Nr. 3, 33 (1941).
— Finger printing peach trees. Development of early bearing peach tree. Amer. Fruit. Gr. **61**, Nr. 7, 20 (1941).
— The „Ron's" seedling cherry. Agric. Gaz. N. S. W. **55**, 297—298 (1944).
— A quick test for fertility. Mon. Sci. News., Nr. 2, 3—4 (1945).
— A polinização nas ameixeiras. (Pollination in plums.) Bol. Agron. Brasil **9**, Nr. 101—102, 28 (1945).
— Southland, new peach developed by USDA. Canner **103**, Nr. 12, 24 (1946).
— Nationwide fruits. Peaches. Amer. Fruit Gr. **67**, Nr. 5, 18—19 (1946).
— Nationwide fruits. Peaches. Amer. Fruit Gr. **67**, Nr. 6, 25 (1946).
— Cherry day at East Malling. Gdnrs' Chron. **120**, 19—20 (1946).
Bailey, J. S.: The beach plum in Massachusetts. Bull. Mass. Agric. Exp. Sta., Nr. 422, 16 (1944).
— u. A. P. French: The inheritance of blossom type and blossom size in the peach. Proc. Amer. Soc. Hort. Sci. **40**, 248—250 (1942).
Becker, J.: Adatok fontosabb meggyfajtáink kémiai összetételéhez. (Data on the chemical composition of some Hungarian sour cherries.) Bull. Hung. Coll. Hort. **7**, 3—8 (1941).
— Adatok fontosabb barackjaink kémiai összetételéhez. (Data on the chemical composition of some Hungarian apricots.) Bull. Hung. Coll. Hort. **7**, 9—13 (1941).

Beketovskij, D. N. u. M. I. Šeljutko: (The biological characterization of the first generation [F_1] of the heterogeneous form of the steppe-cherry [Prunus chamaecerasus var. pendula Dipp, foliis variegatis].) Botaničeskij Žurnal (J. Bot. URSS.) **30**, 77—94 (1945). [Russisch.]
Blake, M. A.: A quarter century of peach breeding. Amer. Fruit Gr. **59**, 22—23, 26—27 (1939).
— Some results of crosses of early ripening varieties of peaches. Proc. Amer. Soc. Hort. Sci. **37**, 232—241 (1939) 1940.
— An acquaintance with peach varietal types is essential in peach breeding to secure improved varieties. Proc. Amer. Soc. Hort. Sci. **38**, 144—147 (1941).
— Rare variation of peach. N. J. St. Hort. Soc. News. **22**, 1319 (1941).
— Some present day problems and essentials in peach production and marketing. Proc. 74th Annu. Meet. Ohio State Hort. Soc., Pp. 87—92 (1941).
— Better peaches from scientific breeding. N. J. Agric. **25**, Nr. 4, 7—8 (1943).
— Brief plain facts about some peach varieties. Hort. News, N. J. **24**, 1532—1533 (1943).
— Classification of fruit bud development on peaches and nectarines and its significance in cultural practice. Bull. N. J. Agric. Exp. Sta., Nr. 706, 24 (1943).
— Some methods used in breeding peaches in New Jersey. Proc. Amer. Soc. Hort. Sci. **45**, 220—224 (1944).
— A little known Chinese species of peach is being used in breeding work. Hort. News, N. J. **25**, 1576, 1586 (1944).
— Four new varieties of peaches show promise. Hort. News, N. J. **27**, 1788, 1791 (1946).
— Four new early varieties of peaches tested hardy. Hort. News, N. J. **27**, 1824 (1946).
— Three new late varieties of peaches for New Jersey. Hort. News, N. J. **27**, 1894 (1946).
— Some present and future needs in peach breeding. Amer. Fruit Gr. **67**, Nr. 2, 11, 17 (1946).
— Four new early varieties of peaches tested hardy. Virginia Fruit **34**, 16—17 (1946).
— Five New Jersey Station peach varieties were named during January, 1947. Hort. News, N. J. **28**, 1917 (1947).
— u. C. H. Steelman jr.: Preliminary investigations of the cold resistance of peach fruit buds at the pink bud stages of development. Proc. Amer. Soc. Hort. Sci. **45**, 37—41 (1944).
—, — What is the cold resistance of fruit buds of peaches. Hort. News, N. J. **25**, 1577—1578, 1588. (1944).
Bogoljubova, O. P.: (The Ussururian White Plum.) Vestnik Socialističeskogo Rastenievodstva (Soviet Plant Industry Record), Nr. 1, 192 (1941). [Russisch.]
Börner, C.: Die Frage der züchterischen Bekämpfung der schwarzen Blattläuse der Kirschen. Z. Pfl. Krankh. **53**, 129—141 (1943).
Burkholder, C. L.: Plum variety trials at the Purdue Farm Bedford, Indiana. Bull. Ind. Agric. Exp. Sta Nr. 458, 20 (1941).

Caldwell, J. S. u. C. W. Culpepper: Further studies in utilization of eastern freestone peaches. I. Studies of canning quality. Fruit Prod. J. **23**, 170—173, 186—187, 191 (1944).
Carlson, R. F. u. H. B. Tukey: Differences in afterripening requirements of several sources and varieties of peach seed. Proc. Amer. Soc. Hort. Sci. **46**, 199 bis 202 (1945).

Christov, A.: (Crown-gall on fruit trees in Bulgaria.) Rev. Inst. Rech. Agron. Bulg. **10**, 3—27 (1940). [Bulgarisch.]
— (Contribution to the study of leaf spot on the plum.) Yearb. Agric. Exp. Stas., Bulgaria **1**, 63—69 (1943). [Bulgarisch.]
Coe, F. M.: Comparative hardiness of peach varieties and selections. Fm. Home Sci. Utah **5**, 12 (1944).
Constantinescu-Ismail, N.: Contribuțiuni la studiul autofertilitatei si interfertilitatei la diferitele varietați de pruni. (Contributions to the study of self and interfertility in different plum varieties.) Hort. Romaneascâ **17**, Nr. 9/10, 2—4 (1939).
Cosmo, I.: Tre casi di probabile mutazione gemmaria osservati sù piante di pesco. (Three cases of probable bud mutation observed in peach plants.) Riv. Frutticult. **4**, 125—134 (1940).
Crane, M. B. u. A. G. Brown: Incompatibility and sterility in the gage and dessert plums. J. Pomol. **17**, 51—66 (1939).
Crist, J. W.: Variation and correlation in bud mutants of the Montmorency cherry. J. Agric. Res. **59**, 393 bis 395 (1939).
— Photosynthetic studies of mutational barrenness in the Montmorency cherry. J. Agric. Res. **59**, 547—553 (1939).
— Twig growth in mutants of Montmorency cherry. Proc. Amer. Soc. Hort. Sci. **37**, 245—249, 1940 (1939).
Culpepper, C. W. u. J. S. Caldwell: Further studies in utilization of eastern freestone peaches. II. Studies of quality for preserve making. Fruit Prod. J. **23**, 200—205 (1944).
—, — Studies in preservation of eastern freestone peaches. Proc. Amer. Soc. Hort. Sci. **46**, 241—245 (1945).
Cuni, L.: Como se ha iniciado el estudio de las variedades de melocotonero en el Venéto. (How the study of the peach varieties in Veneto has been started.) Agriculture, Madrid **12**, 291—292 (1943).
Currey, E. A.: New peach varieties found good in Delta. Miss. Fm. Res. **5**, Nr. 8, 8 (1942).

Dahl, C. G.: Plommonstenarnas betydelse för bestämming av plommonsorter. (The importance of the stones of the plum for determining plum varieties.) Sverig. Pomol. Fören. Årsskr. **41**, 96—102 (1940).
Dermen, H.: Periclinal and total polyploidy in peaches induced by colchicine. Genetics **26**, 147 (1941). (Abstr.)
— Simple and complex periclinal tetraploidy in peaches induced by colchicine. Proc. Amer. Soc. Hort. Sci. **38**, 141 (1941).
— Inducing polyploidy in peach varieties. J. Hered. **38**, 77—82 (1947).
— u. D. H. Scott: A note on natural and colchicine-induced polyploidy in peaches. Proc. Amer. Soc. Hort. Sci. **36**, 299 (1938) 1939.
Detjen, L. R.: Peach varieties by comparison. Circ. Del. Agric. Exp. Sta., Nr. 17, 10 (1945).
— Fruitfulness in peaches and its relationship to morphology and physiology of pollen grains. Bull. Univ. Del. Agric. Exp. Sta., Nr. 257 (Tech.-Nr. 34), 24 (1945).
Duhan, K.: Untersuchungen über die Blüh- und Befruchtungsverhältnisse bei Marillen. Gartenbauwiss. **18**, 253—265 (1944).
Duka, S. H.: (Hybrid between plum and apricot.) Jarovizacija, Nr. 3 (30), 215 (1940). [Russisch.]
Dumont, H. u. G. Valdeyron: Le verger d'essais de Sbeïtla. Ann. Serv. Bot. Tunis. **18**, 3—38 (1941).

Enikeev, H. K.: (An example of the influence of the stock on the formation of characters in inter-species cherry hybrids.) Jarovizacija, Nr. 2 (35), 103—104 (1941). [Russisch.]
Enikeev, H. K.: (Formation of characters in interspecific cherry hybrids when grafted.) Agrobiologija (Agrobiology), Nr. 1, 95—106 (1946). [Russisch.]
Erikson, D.: Certain aspects of resistance of plum trees to bacterial canker. Part II. On the nature of the bacterial invasion of Prunus sp. by Pseudomonas mors-prunorum Wormald. Ann. Appl. Biol. **32**, 112—117 (1945).
Evreinoff, V.-A.: Le prunier Japonais (Prunus salicina Lindl.) Rev. Hort., Paris **115**, 296—298 (1943).
— Le prunier japonais (Prunus salicina Lindl.), son origine, ses variétés, sa culture. Rev. Hort., Paris **115**, 307—310, 323—325 (1943).
— Notes sur l'origine botanique et génétique de nos arbres fruitiers. III. Le pêcher. Rev. Hort., Paris **116**, 55—56 (1944).
— Notes sur l'origine botanique et génétique de nos arbres fruitiers. IV. Le prunier. Rev. Hort., Paris **116**, 69—70 (1944).

Fikry, A.: Nematode disease of stone-fruits. Bull. Minist. Agric. Egypt., Nr. 217, 9 (1939).
Filosofova, T. P. u. H. K. Enikeev: (Michurin's blackthorn x plum hybrids and the prospects of utilizing them in breeding.) Sovetskaja Botanika, Nr. 2, 29—38 (1939). [Russisch.]
Flory, W. S. jr.: Varietal rating of plums with reference to canker resistance. Progr. Rep. Tex. Agric. Exp. Sta., Nr. 753, 4 (1941). [Mimeographed.]
— Pollen studies with plums representing certain species and interspecific hybrids. Proc. 42nd Annu. Conv. Ass. Sth. Agric. Wkrs., Atlanta, Ga., February 5—7, 171—172 (1941).
— Meeting the peach variety problem. Virginia Fruit **33**, 142—149 (1945).
— The present situation with reference to new peach varieties. Va Fruit **35**, 31—42 (1947).
— u. M. L. Tomes: Studies of plum pollen, its appearance and germination. J. Agric. Res. **67**, 337—358 (1943).
Forsberg, N.-G.: Vilka plommon odlas i vårt land? (Which plums are grown in our country.?) Fruktodlaren, Nr. 2, 53—55 (1941).
— Körsbärens härstamning och indelning. (The origin and classification of the cherry.) Fruktodlaren, Nr. 3, 77—78 (1941).
French, A. P.: Plant characters of cherry varieties. Bull. Mass. Agric. Exp. Sta., Nr. 401, 23 (1943).

Gardner, V. R.: Studies in the nature of the clonal variety. III. Permanence of strain and other differences in the Montmorency cherry. Tech. Bull. Mich. Agric. Exp. Sta., Nr. 186, 20 (1943).
Goluško, I. L.: (The origin of the black apricot.) Vestnik Socialističeskogo Rastenievodstva (Soviet Plant Industry Record), Nr. 1, 175—179 (1941). [Russisch.]
Graves, G.: The beach plum, its written record. Nat. Hort. Mag. **23**, 73—97 (1944).
Grünberg, I. P.: Variedades de durazneros y ciruelos que se cultivan en el país. (Varieties of peaches and plums which are cultivated in the country.) Buenos Aires **6**, 453, 78 figs. (1944).
Gurjev, P. G.: (New kinds of apple and plum for the Leningrad Territory.) Plodovo-Jagodnye Kuljtury (Fruit Crops), Nr. 3, 14—18 (1940). [Russisch.]
Gustafson, F. G.: Concentration of growth hormone and fruitfulness in the Montmorency cherry. Proc. Nat. Acad. Sci. Wash. **38**, 131—133 (1942).

Harper, R. S.: Some new canning peach varieties. J. Dep. Agric. Vict. **42**, 181—186 (1944).
Hesse, C. O.: Variation in resistance to brown rot in apricot varieties and seedling progenies. Proc. Amer. Soc. Hort. Sci. **36**, 266—268 (1938) 1939.

Hildebrand, E. M.: Prune dwarf. Phytopathology **32**, 741—751 (1942).
— The cherry virus complex in New York. Phytopathology **34**, 1003 (1944). (Abstr.)
Hofmann, F. W.: Some foliar characters for peach breeding. Proc. Va Acad. Sci. **1**, 208—209 (1939 bis 1940).
Hrubý, K.: The cytology of the Duke cherries and their derivatives. J. Genet. **38**, 125—131 (1939).
Husz, B.: Elomunkálatok a meggy terméshozamának fokozásához. (Investigations to increase the productivity of the Hungarian sour cherry.) Bull. Coll. Hort. **9**, 60—86 (1943).

Ingram, C.: Cherry hybrids. Gdnrs' Chron. **112**, 163 (1942).
— The. Yingtao or Chinese fruiting cherry. J. R. Hort. Soc. **68**, 307—309 (1943).
— A revised classification of the deciduous cherries. Gdnrs' Chron. **119**, 196—197 (1946).
— The rock cherries. Amygdalocerasus. Gdnrs' Chron. **120**, 138—139, 150—151, 162, 174, 186 (1946).

Jakovlev, P. N.: (Influence of the mentor on the transference of pigments.) Jarovizacija, Nr. 1 (28), 11—13 (1940). [Russisch.]
— (Intergeneric hybridization in stonefruit trees.) Vestnik Gibridizacii (Hybridization), Nr. 1, 34—38 (1941). [Russisch.]
Jenkins, E. W.: Interrelation of pollination, position on cluster and set of pears. Bull. Vt. Agric. Exp. Sta., Nr. 493, 27 (1942).
Jensen, H.: Växtförädling av plommon. (Plum breeding). Sverig. Pomol. Fören. Årsskr. **46**, 98—104 (1945).
Johnston, S.: The Redhaven peach. Quart. Bull. Mich. Agric. Exp. Sta. **23**, 93—95 (1940).
— Peach culture in Michigan. Circ. Bull. Mich. Agric. Exp. Sta., Nr. 177, 88 (1941).
— The Fairhaven peach. Quart. Bull. Mich. Agric. Exp. Sta. **29**, 86—87 (1946).

Kelli, A. Č.: (The sour cherry Ural'skaja Krasavitsa.) Plodovo-Jagodnye Kuljtury (Fruit Crops), Nr. 4, 60—61 (1940).
King, J. R.: Cytological studies on some varieties frequently considered as hybrids between the plum and the apricot. Proc. Amer. Soc. Hort. Sci. **37**, 215—217 (1939) 1940.
Kobel, F.: Die Befruchtungsverhältnisse der Kirschen. Schweiz. Ztschr. f. Obst- und Weinbau **48**, 64—69, 87—94 (1939),
— Study on the conditions of fecundation of pomaceous and stone fruit species. Int. Rev. Agric. **34**, 398T bis 407T (1943).
Kohanovskaja, L. N.: Increasing crossability between Prunus ssp. by temperature treatment. C. R. (Doklady) Acad. Sci. URSS. **27**, 155—159 (1940).
Kostina, K.: (The origin and evolution of cultivated apricots). Vestnik Socialističeskogo Rastenievodstva (Soviet Plant Industry Record), Nr. 1, 165—174 (1941). [Russisch.]
Kovalev, N. V.: (Ecological differentiation of the cherry plum Prunus cerasifera Ehrh. [s. l. Kov.].) C. R. (Doklady) Acad. Sci. URSS. **23**, 280—283 (1939).
— New species of plums. C. R. (Doklady) Acad. Sci. URSS. **23**, 284—286 (1939).
— (Myrobalan [Prunus cerasifera Ehrh.] its varieties and the cultivated types.) Plodovo-Jagodnye Kuljtury (Fruit Crops), Nr. 3, 32—39 (1940). [Russisch.]
Krjukov, F. A.: (Influence of the 1939—40 winter on the collection of plum species and varieties at the Krasnyi Pakhar'· Experiment Station.) Soviet Plant Industry Record, Nr. 4, 75—78 (1940). [Russisch.]

Kruft, F.: Die Selektion bei der Deutschen Hauszwetsche. Dtsch. Obstbau, Nr. 9, 169—171 (1940).

Lammerts, W. E.: A Royal apricot sport of short chilling requirement: origin and transmission of characteristics to seedlings. Proc. Amer. Soc. Hort. Sci. **38**, 175—178 (1941).
— An evaluation of peach and nectarine varieties in terms of winter chilling requirements and breeding possibilities. Proc. Amer. Soc. Hort. Sci. **39**, 205—211 (1941).
— Embryo culture an effective technique for shortening the breeding cycle of deciduous trees and increasing germination of hybrid seed. Amer. J. Bot. **29**, 166 bis 171 (1942).
— The breeding of ornamental edible peaches for mild climates. I. Inheritance of tree and flower characters. Amer. J. Bot. **32**, 53—61 (1945).
Lantz, H. L. u. T. J. Maney: Peach breeding: a study of inheritance in some cross-bred seedlings. Proc. Amer. Soc. Hort. Sci. **38**, 184—186 (1941).
Lesley, J. W.: Five new peach varieties especially adapted to mild winters. Bull. Calif. Agric. Exp. Sta., Nr. 632, 19 (1939).
— A genetic study of saucer fruit shape and other characters in the peach. Proc. Amer. Soc. Hort. Sci. **37**, 218—222 (1939) 1940.
— New peach varieties for a subtropical climate. Calif. Citrog. **29**, 138 (1944).
— Peach breeding in relation to winter chilling requirements. Proc. Amer. Soc. Hort. Sci. **45**, 243—250 (1944).
Leslie, W. R.: Manitoba news letter. N. S. Dak. Hort. **16**, 16 (1943).
— Manitoba news letter. N. and S. Dak. Hort. **19**, 37 (1946).
— Manitoba news letter. N. and S. Dak. Hort. **19**, 53, 59 (1946).
Lewis, D.: Breakdown of self-incompatibility by α-naphthalene acetamide. Nature, Lond. **149**, 610 (1942).

MacDanields, L. H.: Notes on the pollination of the Italian prune. Proc. Amer. Soc. Hort. Sci. **40**, 84—86 (1942).
McWhorter, O. T.: The eastern Oregon wild plum. 35th Rep. 58th Ann. Mtg. Ore. St. Hort. Soc. and 29th Ann. Mtg. West. Nut Growers Ass., Pp. 45—46 (1943).
Maliga, P.: Adatok a meggyfajták megtermékenyülési viszonyaihoz különös tekintettel a Pándy-meggyre. (Data on the fruiting of different sour cherry varieties with special regard to the variety Pándy.) Bull. Hung. Coll. Hort. **10**, 287—319 (1944).
Marani, M. u. C. Gerbaldi: Sei varietà di pesco raccomandabili. (Six recommended varieties of peaches.) Riv. Frutticult. **4**, 113—123 (1940).
Markov, N. V.: Wild apricots in the Alma Ata region. Plodovo-Jagodnye Kuljtury (Fruit Crops), Nr. 3, 49—51 (1940).
Meader, E. M. u. M. A. Blake: Some plant characteristics of the progeny of Prunus persica and P. kansuensis crosses. Proc. Amer. Soc. Hort. Sci. **36**, 287 bis 291 (1939).
—, — Some plant characteristics of the second generation progeny of Prunus persica and Prunus kansuensis crosses. Proc. Amer. Soc. Hort. Sci. **37**, 223—231 (1939) 1940.
—, — Further studies on identification of peach varieties by leaf characteristics. Proc. Amer. Soc. Hort Sci. **39**, 177—182 (1941).
—, O. W. Davidson u. M. A. Blake: A method for determining the relative cold hardiness of dormant peach fruit buds. J. Agric. Res. **70**, 283—302 (1945).

Mittmann-Maier, G.: Untersuchungen über die Moniliaresistenz von Sauerkirschen. Z. PflKrankh. 50, 84—94 (1940).

Moore, M. H.: Bacterial canker and leaf spot of plum and cherry. A summary of present knowledge on control measures in Britain. 33rd Rep. E. Malling Res. Sta. 1945, 134—137 (1946).

Morettini, A.: Nuove varietà di peschi d'incrocio presentate alla settima mostra delle pesche di Verona. (New varieties of hybrid peaches presented at the seventh peach show at Verona.) Ital. Agric. 77, 709—718 (1940).

— Due nuove varietà di susine precoci. Gli incroci Morettini. 1. °Florenzia×Beauty n °355. 2. °Shiro° S. Rosa n. °243. (Two new varieties of early plums. Morettini's hybrids Florenzia×Beauty No. 355 and Shiro×S. Rosa No. 243.) Riv. Soc. Tosc. Ortic. 31, 73—87 (1946).

Muraro, I.. A.: Ameixeiras-Especialidades de suas variedades. (Plums-Characteristics of the varieties.) Bol. Agron. Brasil 10, 147—148 (1946).

Palmiter, D. H. u. E. M. Hildebrand: The yellow-red virosis of peach: its identification and control. Bull. N. Y. St. Agric. Exp. Sta. Nr. 704, 17 (1943).

Passecker, F.: Lebensdauer und Veredlungsunterlage bei der Marille (Aprikose). Gartenbauwiss. 18, 231 bis 252 (1944).

— Vermehrungs- und Züchtungsfragen bei der Aprikose. Züchter 17/18, 277—284 (1947).

Pirovano, A.: Ibridi di pruni con peschi. (Hybrids between plums and peaches.) Ital. Agric. 78, 681 bis 685 (1941).

Pojarkova, A. J.: (Critical review of the species of cherry belonging to the cycle of Cerasi prostratae (Labill.) Ser. of Central and Anterior Asia.) J. Bot. URSS. 24, 225—246 (1939). [Russisch.]

Pratasenja, G. D.: Production of polyploid plants. Haploid and triploids in Prunus persica. C. R. (Doklady) Acad. Sci. URSS. 22, 348—351 (1939).

Ramírez Cantú, D.: Algunas plantas notables de Tepoztlán, Mor. (Some notable plants of Tepoztlán Mor.) An. Inst. Biol. Univ. Méx. 16, 353—357 (1945).

Raptopoulos, T.: Pollen-tube growth studies in cherries. J. Genet. 42, 73—89 (1941).

— Chromosomes and fertility of cherries and their hybrids. J. Genet. 42, 91—114 (1941).

Riggoti, R.: Un nuovo portainnesto del pesco? Prunus persica×Prunus davidiana. (A new peach stock? P. persica×P. davidiana.) Ital. Agric. 21, 664—667 (1942).

Rjabov, I. N.: (Classification of peaches.) Vsesojuznaja Akademija Seljsohozjajstvennyh Nauk. im. V. I. Lenina. Gosudarstvennyj Nikitskij Botaničeskij Sad im. Molotova (Lenin Academy of Agricultural Science, Molotov's Nikita Botanic Garden) Jalta, Pp. 32 (1939). [Russisch.]

Rodrigues, A.: Filometria e carpometria nas Pomóideas e Prunóideas. Generalização à pomologia sistemática de um método de determinaçao da forma. (Phyllometry and carpometry in the Pomoideae and Prunoideae. Generalization to systematic pomology of a method to determine the shape of leaves and fruits.) Agron. Lusitana 5, 251—277 (1943).

Roy, B.: Studies on pollen tube growth in Prunus. J. Pomol. 16, 320—328 (1939).

Rozanova, M. A.: (Contribution à l'étude de la corrélation entre la teneur en vitamine C des fruits et la forme des sepals chez les espèces de l'englantier.) Sovetskaja Botanika 14, Nr. 4, 287—288 (1946). [Russisch.]

Rožkov, M. I.: (An excellent plum variety.) Sadovodstvo (Horticulture), Nr. 12, 37—38 (1939). [Russisch.]

Salamatov, M. P.: (The plum in the Čeljabinsk province.) Plodovo-Jagodnye Kuljtury (Fruit Crops), Nr. 4, 62—63 (1940). [Russisch.]

Samsonov, I. M.: (The valuable qualities of Michurin's Ideal cherry.) Sady i Ogorody (Fruit and Vegetable Gardens), Nr. 6, 15—16 (1941). [Russisch.]

Savage, E. F. u. F. F. Cowart: Factors affecting peach tree longevity in Georgia. Bull. Ga. Exp. Sta., Nr. 219, 15 (1942).

Schelhorn, M. v.: Über eine triploide Vogelkirsche. Züchter 17/18, 232—235 (1947).

Schmidt, M.: Untersuchungen über den züchterischen Wert von Sämlingen der Kirschpflaume, Prunus cerasifera Ehrh. Gartenbauwiss. 15, 247—311 (1940).

— Forschungsaufgaben der Züchtung bei Kirschen. Dtsch. Obstbau 57, 41—46 (1942).

Schrader, A. L. u. I. C. Haut: Peach varieties: what's in the future? Trans. Peninsula Hort. Soc. 33, 32—37 (1943).

Scott, D. H. u. F. P. Cullinan: The inheritance of wavy-leaf character in the peach. J. Hered. 33, 293 bis 295 (1942).

—, — Some factors affecting the survival of artificially frozen buds of peach. J. Agric. Res. 73, 207—236 (1946).

— u. J. H. Weinberger: Inheritance of pollen sterility in some peach varieties. Proc. Amer Soc. Hort. Sci. 45, 229—232 (1944).

Skard, O. u. E. Weydahl: Plommetorking. (Prune drying.) Meld. Norg. LandbrHøgsk. 22, 87—104 (1942).

Smith, C. O. u. L. C. Cochran: A noninfectious heritable leaf-spot and shot-hole disease of the Beaty plum. Phytopathology 33, 1101—1103 (1943).

Smith, W. H.: Some observations on the ripening of plums by ethylene. J. Pomol. 21, 53—56 (1945).

Southwick, L. u. A. P. French: The identification of plum varieties from non-bearing trees. Bull. Mass. Agric. Exp. Sta., Nr. 413, 51 (1944).

Teterev, F. K.: (Michurin's varieties of sour and sweet cherries in the Leningrad province.) Vestnik Socialističeskogo Rastenievodstva (Soviet Plant Industry Record), Nr. 3, 17—20 (1940). [Russisch.]

Tikhonow, N. N.: (The Ussuri plum [Prunus triflora].) Plodovo-Jagodnye Kuljtury (Fruit Crops), Nr. 4, 15—27 (1940). [Russisch.]

Tukey, H. B.: The Imperial Epineuse. Amer. Fruit. Gr. 66, 16 (1946).

— u. R. F. Carlson: Morphological changes in peach seedlings following after-ripening treatments of the seeds. Bot. Gaz. 106, 431—440 (1945).

Veldhuis, M. K. u. A. M. Neubert: Freestone peach varieties for canning in Washington. Fruit Prod. J. 23, 229—233 (1944).

Venjaminov, A. N.: (Studium der vegetativen Mutationen bei der Kirsche „Ljubskaja".) Sadovodstvo 11/12, 24—33 (1940). [Russisch.]

Verner, L.: The Lamida, Ebony and Spalding sweet cherries. Circ. Idaho Agric. Exp. Sta., Nr. 109, 4 (1946).

Wanscher, J. H.: Partial pollen sterility as a somatic character of the peach. K. VetHøjsk. Aarsskr., Pp. 91 bis 105 (1941).

Weeks, W. D.: Chromosome number of the beach plum (Prunus maritima). Proc. Amer. Soc. Hort. Sci. 38, 141 (1941).

Weimarck, H.: Bidrag till Skånes Flora. 15. En spontan hybrid mellan slån och terson. (Contribution to the Flora of Scania. 15. A spontaneous hybrid between sloe and bullace.) Bot. Notiser, Pp. 218—226. Medd. Lunds Bot. Mus., Nr. 57 (1942).

Weimarck, H.: Om pollenkorn och klyvöppningar hos Prunus Insititia, P. spinosa och hybriden dem emellan. (On pollen grain and stomata of P. insititia, P. spinosa and the hybrid between them.) Bot. Notiser, Pp. 389 bis 398. Medd. Lunds Bot. Mus., Nr. 70 (1943).
— Fältstudier 1943 inom släktet Prunus. (1943 field studies within the genus Prunus.) Sverig. Pomol. Fören. Årsskr. **44**, 45—54 (1943).
Weinberger, J. H.: Characteristics of the progeny of certain peach varieties. Proc. Amer. Soc. Hort. Sci. **45**, 233—238 (1944).
— Newer peach varieties. Eastern Fruit Gr. **7**, 10, 16, 19 (1944).
—, P. C. Marth u. D. H. Scott: Inheritance study of root knot nematode resistance in certain peach varieties. Proc. Amer. Soc. Hort. Sci. **42**, 321—325 (1943).
Wellington, B.: Trends in varieties and culture of cherries, pears and peaches. Proc. 90th Annu. Mtg. N. Y. St. Hort. Soc., Pp. 249—257 (1945).
Wellington, R.: Promising plum varieties. Amer. Fruit Gr. **66**, 10—11, 26—29 (1946).
Wenjaminoff, A. N. siehe Venjaminov.
Wenzl, H.: Untersuchungen über die Kräusel-(Sternflecken-)Krankheit von Prunus armeniaca und anderen Prunaceen. Phytopath. Z. **13**, 588—623 (1941).
Wet, A. F. de: A preliminary study of peach varieties, on peach and plum roots. Sci. Bull. Dep. Agric. S. Afr., Nr. 226, 19 (1941).
— The new dessert peach „Boland". Fmg. S. Afr. **19**, 381—382, 390 (1944).
Whitehouse, W. E.: Peach rootstocks resistant to root knot nematode. Agric. Tec. Chile **4**, 145—150 (1944).
Wight, W. F.: Seven new peaches and a new plum for the Western States. Circ. U. S. Dep. Agric., Nr. 552, 23 (1940).
Wood, M. N.: Two new varieties of almond: the Jordanolo and the Harpareil. Circ. U. S. Dep. Agr., Nr. 542, 13 (1939).

Zobrist, L., R. Conrad, H. Zogg u. R. Maag: Untersuchungen über die Gloeosporium-Fruchtfäule an Kirschen. Schweiz. Z. Obst- u. Weinbau **53**, 161—169 (1944).

4. Schalenobst.

Anonymus: A blight resistant chestnut. Sci. Amer. Ba. **161**, 93, also Madras Agric. J. 1939, **27**, 375—376 (1939).
— Blight resistant chestnuts. Amer. Fruit Gr. **66**, 35 (1946).
Babu, C. N.: Chromatin bridges in the root tip of groundnut. Curr. Sci. **10**, 173—174 (1941).
— Certain abnormalities in the root tip of groundnut. Curr. Sci. **10**, 291—292 (1941).
Batchelor, L. D., O. L. Braucher u. E. F. Serr: Walnut production in California. Circ. Calif. Agric. Exp. Sta., Nr. 364, 34 (1945).
Batten, E. T.: Peanuts go pleasingly plump. Plumper pods are sought in breeding work with Valencia, Spanish, Virginia types. Sth. Seedsman **6**, Nr. 3, 11, 34 (1943).
— Two new strains of Virginia type peanuts. Bull. Va. Polytechn. Inst., Agric. Exp. Sta., Nr. 370, 4 (1945).
— High-yielding Hollands enter the peanut picture. Sth. Seedsman **9**, Nr. 1, 13 (1946).
Beattie, R. K.: The search for blight-resistant chestnuts in the Orient. Rep. Proc. 32nd Annu. Mt. Nth. Nut. Gr. Ass. Hershey, Pa., Pp. 18—23 (1941).
Becker, J.: A nemes gesztenye táplálkozási értékéröl. (The nutritive value of the chestnut.) Bull. Hung. Coll. Hort. **8**, 122—127 (1942).
Burkart, A.: Los frutos de las especies silvestres de Arachis. (The fruits of the wild species of Arachis.) Proc. 8th Amer. Sci. Congr., Wash. **3**, 175—178 (1940).
Burkett, J. H.: Pecan breeding technique. Proc. 22nd Annu. Mtg. Tex. Pecan Grs' Ass. Pp. 66—68 (1943).
Bush, C. D.: Nut grower's handbook. A practical guide to the successful propagation planting, cultivation, harvesting and marketing of nuts. Orange Judd Co., New York, p. 189 (1941).
Čapljaev, S. K.: (From work on breeding nuts.) Jarovizacija, Nr. 3 (30), 157—159 (1940). [Russisch.]

Chevalier, A.: Variabilité et hybridité chez les noyers. Notes sur des Juglans peu connus, sur l'Annamocarya et un Carya d'Indochine. Rev. Bot. Appl. **21**, 477 bis 509 (1941).
Clapper, R. B.: New chestnuts for our forests? Amer. Forest **49**, 331—333, 365 (1943).
Clos, E. C.: Los tipos de mani („Arachis hypogaea") cultivados en la Argentina y su distribución geográfica. (The types of groundnut [A. hypogaea] cultivated in Argentina and their geographical distribution.) Physis. B. Aires **18**, 317—329 (1939).
Colby, A. S.: The Crath Carpathian walnut in Illinois. Ill. Hort **32**, 4—5 (1943).
— The Crath Carpathian walnut in Illinois. 34th Rep. Nth. Nut. Gr. Ass., Pp. 107—119 (1943).
Colwell, W. E. u. N. C. Brady: The effect of calcium on certain characteristics of peanut fruit. J. Amer. Soc. Agron **37**, 696—708 (1945).
Cooper, J. F.: Once in a blue moon — a new goober that clicks. Sth. Seedsman **8**, 11, 51 (1945).
Costa, A. S. u. O. F. de Souza: Nota sobre a verrugose do amendoinzeiro. (A note on groundnut scab.) O Biologico, S. Paulo **7**, 347—349 (1941).
Cox, L. G.: Preliminary studies on catkin forcing and pollen storage of Corylus and Juglans. 34th Rep. Nth. Nut. Gr. Ass., Pp. 58—60 (1943).
Cranmer, C. B.: Asiatic chestnut. Flor. Exch. **101**, 9 (1943).
Crath, P. C.: Cross-pollination the cause of hardiness, earliness and sweetness in Carpathian Persian walnuts. Rep. Proc. 30th Annu. Mtg. Nth. Nut Gr. Ass., Pp. 85 bis 89 (1939).

Dahlgren, K. V. O.: En ny varietet av hassel med ascidieblad samt om Corylus avellana L. var. Zimmermanni Hahne. (A new variety of hazel with an „ascidium" leaf-and on C. avellana L. var. Zimmermanni Hahne.) Svensk Bot. Tidskr. **35**, 353—360 (1941).
Danielsson, B.: Polyploida hasseltyper. (Polyploid Types of hazel.) Sverig. Pomol. Fören Årsskr. **46**, 116—122 (1945).
Dodge, F. N.: Pecan varieties. Proc. 23rd Annu. Mtg. Tex. Pecan Grs' Ass., Pp. 7—16 (1944).

Farrior, J. W. u. G. K. Middleton: The improvement of Forginia type peanuts by mass selection. Proc. 42nd Annu. Conv. Ass. Sth. Agric. Wkrs, Atlanta, Ga., February 5—7, Pp. 102 (1941).
Fedorako, B. I.: (On the form manifoldness of hazel in the forest-steppe zone of the western Pre-Urals.) Priroda (Nature), Nr. 10, 71 (1940). [Russisch.]
Fernandes, C. T.: A castanha e suas caracteristicas da ponta de vista comercial. (The chestnut and its characteristics from the commercial point of view.) Junta Nac. Frutas, Lisboa **6**, 558—563 (1946).
— O castanheira na distrito de Vila Real. (Concelhos das zonas Centro e Leste.) (The chestnut in the district of Vila Real [Communes of the central and eastern zones]). Junta Nac. Frutas, Lisboa **6**, 816—845 (1946).

Gellatly, J. U.: Varieties and plantings of nut trees in British Columbia. Rep. Proc. 31st Annu. Mtg. Nth. Nut. Gr. Ass., Pp. 74—80 (1940).

Geraldes, C. de M.: Subsidios para o estudo das características dos amendoins de Angola. (Notes for the study of the characteristics of the Angola groundnuts.) An. Inst. Sup. Agron. Lisboa **10**, 157—165 (1939).

Giessen, C. van der u. A. Govers: De aardnootvariëteitSchwarz 21en haar resistentie tegen slijmziekte. (The groundnut Schwarz 21 and its resistance to slime disease.) Landbouw **15**, 183—188 (1939).

Glenn, E. M.: Variation in non-clonal Franquette and Mayette walnuts. 33rd Rep. E. Malling Res. Sta. 1945, Pp. 67—69 (1946).

— u. A. W. Witt: Progress report on the walnut variety collection at East Malling. 33rd Rep. E. Malling Res. Sta. 1945, Pp. 70—74 (1946).

Gravatt, G. F.: Blight-resistant chestnuts. Proc. Ann. Meet. Nut. Gr. Ass., Pp. 17—19, January, Pp. 3 (1940).

— u. B. S. Crandall: The Phytophthora root disease of chestnut and chinkapin. 35th Rep. Nth. Nut Gr. Ass., Pp. 83—87 (1944).

Graves, A. H.: Reports on research for 1940. Chestnut breeding work in 1940. Brooklyn, Bot. Gdn Rec. **30**, 87—93 (1941).

— Report on chestnut breeding. For. Leaves **32**, 15—16 (1942).

— Report on research for 1941: Chestnut breeding work in 1941. Brooklyn, Bot. Gdn. Rec. **31**, 94—99 (1942).

— Chestnut breeding work in 1942. Brooklyn, Bot. Gdn. Rec. **32**, 78—80 (1943).

— Chestnut breeding work in 1943. Brooklyn, Bot. Gdn. Rec. **33**, 11—13 (1944).

— The Brooklyn Botanic Garden Chestnut Breeding Project. 35th Rep. Nth. Nut Gr. Ass., Pp. 23—35 (1944).

Greeves-Carpenter, C. F.: A blight-resistant chestnut. Sci. Amer. **161**, 93 (1939).

Guerreiro, M. G.: Sôbre a caracterização das formas de castanheira „Longal" et „Judia" por meio da análise biométrica dos frutos e das fôlhas. (The characteristization of the Longal and Judia chestnuts by means of biometrical analysis of the fruits and leaves. Bol. Junta Nac. Frutas, Portugal **5**, 3—11 (1945).

Guest, P.: The relationship between chlorosis of macadamia seedlings and certain chemical constituents of macadamia seeds. Proc. Amer. Soc. Hort. Sci. **41**, 61—64 (1942).

Hacquart, A.: L'amélioration des semences d'arachide au Sénégal. Bull. Agric. Congo Belge **30**, 106—125 (1939).

Hansen, E.: Asorbic acid content of walnut hulls. Proc. Amer. Soc. Hort. Sci. **42**, 265—266 (1943).

Harvey, P. H. u. E. F. Schultz, jun.: Multiplying peanut hybrids by vegetative propagation. J. Amer. Soc. Agron. **35**, 637—639 (1943).

Higgins, B. B.: Inheritance of seed-coat color in peanuts. Amer. J. Bot. **26**, 4s (1939). (Abstr.)

— Inheritance of seed-coat color in peanuts. J. Agric. Res. **61**, 745—752 (1940).

—, K. T. Holley, T. A. Pickett u. C. D. Wheeler: Peanut breeding and characteristics of some new strains. Bull. Ga. Exp. Sta. Nr. 213, 3—11 (1941).

Humphrey, N.: A groundnut wilt disease on the coast of Kenya. E. Afr. Agric. J. **5**, 110—113 (1939).

— A note on groundnut selection work. E. Afr. Agric. J. **7**, 220—221 (1942).

Iliev, P.: (What variety of pea-nut should we sow in our country?) Zemledelie, Sofia **45**, 79—82 (1941). [Bulgarisch.]

— Genetische Untersuchungen an der Erdnuß (Arachis hypogaea L.). Bohnenfarbe- und Fruchtgrößenvererbung. Züchter **14**, 141—145 (1942).

Jablokov, A. S.: (New facts in the sphere of controlling dominance.) Jarovizacija, Nr. 1 (28), 81—83 (1940). [Russisch.]

Kalašnikov, V. M.: (Nut-bearing species of the Far East.) Plodovo-Jagodnye Kuljtury (Fruit Crops), Nr. 5, 60—68 (1940).

Kalmykov, S. S.: (The pistachio tree in Kazakstan.) Soviet Subtropics, Nr. 11/12 (75/76), 53—55 (1940). [Russisch.]

Kline, L. V.: A method of evaluating the nuts of black walnut varieties. Proc. Amer. Soc. Hort. Sci. **41**, 136 bis 144 (1942).

— u. Chase, S. B.: Compilation of data on nut weight and kernel percentage of black walnut selections. Proc. Amer. Soc. Hort. Sci. **38**, 166—174 (1941).

Kumar, L. S. S. u. W. V. Joshi: Inheritance of flower colour in Arachis hypogea L. (groundnut). Indian J. Genet. Pl. Breed. **3**, 59—60 (1943).

Lebedeva, T. A.: (The lenght of the daylight period, and its influence on the growth of Arachis.) Doklady Akad. Nauk SSSR **27**, 262—264 (1940). [Russisch.]

León, H.: Comentarios sobre el articulo de J. P. Carabia: „Notas sobre la nomenclatura de algunas palmas cubanas." (Comments on J. P. Carabia's article on „Notes on the nomenclature of certain Cuban palms"). Caribbean Forester, P. R. **6**, 165—170 (1945). [Mimeographed.]

Little, E. L. (jr.): Notes on the nomenclature of Carya Nut. Amer. Midl. Nat. **29**, 493—508 (1943).

Lochrie, J. V.: The groundnut (Arachis hypogaea). Swaziland Agric. Rev. Notes, Nr. 6, 9—14 (1939). [Mimeographed.]

Lounsberry, C. C.: Black walnuts of Iowa. Trans. Iowa St. Hort. Soc. **75**, 210—212 (1940).

MacDaniels, L. H. u. J. E. Wilde: Further tests with black walnut varieties. 34th Rep. Nth. Nut Gr. Ass., Pp. 64—82 (1943).

Machado, O.: Estudos novos sobre uma planta velha o cajueiro (Anacardium occidentale L.). (New studies on an old plant, the cashew nut [A. occidentale L.]). Rodriguésia, Rio de J., **8**, 19—48 (1944).

McKay, J. W.: Male sterility in Castanea. Proc. Amer. Soc. Hort. Sci., **37**, 509—510 (1939) 1940.

— Infertility in hybrid walnuts. Rep. Proc. 32nd Annu. Mt. Nth. Nut. Gr. Ass. Hershey, Pa., Pp. 84—86 (1941).

— u. H. L. Crahe: The immediate effect of pollen on the fruit of the chestnut. Proc. Amer. Soc. Hort. Sci., **36**, 293—298 (1938) 1939.

—, — Xenia in the chestnut. Science, **89**, 348—349 (1939).

— u. H. H. McKay: Microsporogenesis in Juglans intermedia Carr. Amer. J. Bot. Suppl., **28**, 4s (1941). (Abstr.)

Manning, W. E.: A leaf character for shagbark hickory. Amer. J. Bot. Suppl., **29**, 13s (1942). (Abstr.)

Marloth, R. H.: The pecan in South Africa. Fmg S. Afr., **21**, 665—676 (1946).

Mehta, T. R.: Extension of United Provinces groundnut cultivation. Indian Fmg, **3**, 85—86 (1942).

Middleton, G. K., W. E. Colwell, N. C. Brady u. E. F. Schultz jr.: The behavior of four varieties of peanuts as affected by calcium and potassium variables. J. Amer. Soc. Agron. **37**, 443—457 (1945).

—, P. H. Harvey, H. H. F. Robinson u. J. W. Farrior: Peanut breeding and variety studies. (A progress report.) Agron. Inform. Circ. N. C. Agric. Exp. Sta., Nr. 125, 9 (1940) [Mimeographed].

Miller, G. A.: Tree breeding. Rep. Proc. 31st Annu. Mtg. Nth. Nut Gr. Ass., Pp. 99—103 (1940).

Muenscher, W. C. u. B. I. Brown: A key to some seedlings of walnuts. 34th Rep. Nth. Nut. Gr. Ass., Pp. 62—63 (1943).

Mustafa, A. M. u. N. A. Janjua: Almond growing in Baluchistan. Indian Fmg., **3**, 539—542 (1942).

Pascual, A.: Almond growing throughout the world: (3) The United States. Int. Rev. Agric., **33**, 165 T bis 170 T (1942).

Reed, C. A.: The 1946 status of Chinese chestnut growing in the eastern United States. Nat. Hort. Mag. **26**, 83—93 (1947).

Rehder, A.: Carya alba proposed as nomen ambiguum. J. Arnold Arbor., **25**, 482—483 (1944).

Romberg, L. D. u. C. L. Smith: Effects of cross-pollination, self-pollination, and sib-pollination on the dropping, the volume, and the kernel development of pecan nuts and on the vigor of the seedlings. Proc. Amer. Soc. Hort. Sci., **47**, 130—138 (1946).

Santos, E.: O dendezeiro. (The oil palm.) Bol. Minist. Agric., Rio de J., **34**, 97—102 (1945).

Schreiber, W. R.: The Amazon Basin brazil nut industry. Foreign Agric. Rep. U. S., Dep. Agric., Nr. 4, 36 (1942).

Schuster, C. E.: Notes on the history of nut production in the Pacific Northwest. 36th Rep. and 59th Mtg. Ore. St. Hort. Soc. (30th Mtg. West. Nut Gr. Ass.), **99**, 104 (1944).

Sellschop, J.: Sound advice. S. Afr. Sug. J., **26**, 507 (1942).

Slate, G. L.: Filberts. Circ. N. Y. St. Agric. Exp. Sta., Nr. 192, 14 (1941).

Smith, B. W.: Macrosporogenesis and embryogeny in Arachis hypogaea L. as related to seed failure. Amer. J. Bot. Suppl., **33**, 826 (1946). (Abstr.)

Smith, C. L. u. L. D. Romberg: Stigma receptivity and pollen shedding in some pecan varieties. J. Agric. Res. **60**, 551—564 (1940).

—,— Relative behavior of four varieties of pecan during the development of an orchard at the U. S. Pecan Field Station, Brownwood, Texas. Proc. 24th Annu. Mtg, Tex. Pecan Grs' Ass., Pp. 10—15 (1945).

Snyder, D. C.: The chestnut in Iowa. Rep. Proc. 31st Annu. Mtg. Nth. Nut Gr. Ass., p. 129 (1940).

Steinbauer, C. E., J. H. Beattie u. E. T. Batten: Influence of mass selection within certain large-seeded Virginia-type peanut varieties. Proc. Amer. Soc. Hort. Sci., **37**, 685—688 (1939) 1940.

—, J. McD. McClown, E. T. Batten u. E. E. Hall: Performance of some large-seeded and small-seeded peanut varieties and selections in Virginia and South Carolina. Proc. Amer. Soc. Hort. Sci., **41**, 240—244 (1942).

Stoke, H. F.: Chestnuts in Eastern United States. Rep. Proc. 31st Annu. Mtg. Nth. Nut Gr. Ass., Pp. 119—123 (1940).

Stone, C. L., L. E. Jones u. W. E. Whitehouse: Longevity of pistache pollen under various conditions of storage. Proc. Amer. Soc. Hort. Sci., **42**, 305—314 (1943).

Talbert, T. J.: Nut tree culture in Missouri. Bull. Mo. Agric. Exp. Sta., Nr. 454, 32 (1942).

Trosjko, I.: (How to distinguish female pistachio plants from male [in the winter period].) Soviet Subtropics, Nr. 3 (67), 58 (1940). [Russisch.]

Urquijo, P.: Aspectos de la obtención de híbridos resistentes a la enfermedad del castaño. (Aspects of the procurement of hybrids resistant to the chestnut disease.) Bol. Pat. Veg. Ent. Agríc., Madr., **13**, 447 bis 462 (1944).

Vilkomerson, H.: Flowering habits of the chestnut. Rep. Proc. 31st Annu. Mtg. Nth. Nut Gr. Ass., Pp. 114—116 (1940).

Whitehouse, W. E. u. C. L. Stone: Some aspects of dichogamy and pollination in pistache. Proc. Amer. Soc. Hort. Sci., **39**, 95—100 (1941).

Williams, H. A.: Edible varieties of almonds. J. R. Hort. Soc., **68**, 62—65 (1943).

Wills, J. M.: The Queensland nut. Qd. Agric. J., **60**, 342—351; **61**, 8—16 (1945).

5. Beerenobst.

Atanasjev, M. M. u. H. E. Morris: Yellowing in everbearing Progressive strawberries. Phytopathology, **31**, 1 (1941). (Abstr.)

Allander, H.: Om den Svenska s. k. Rubus nemorosus. (On the Swedish so-called R. nemorosus.) Svensk Bot. Tidskr. **35**, 287—295 (1941).

Anderson, H. W.: Red stele root rot of the strawberry. Trans. Ill. Hort. Soc. **74**, 383—393 (1940).

Anderson, O. C.: A cytological study of resistance of Viking currant to infection by Cronartium ribicola. Phytopathology, **29**, 26—40 (1939).

Anonymus: American Fruit Grower presents annual nationwide variety survey: strawberries. Amer. Fruit. Gr., **60**, Nr. 10, 8 (1940).

— Blueberries-three new varieties. Amer. Fruit Gr., **60**, Nr. 3, 10, 24 (1940).

— Plant recording. Strawberry „Victory", raised by Arthur G. Sainsbury, Mangere, via Otahuhu. J. N. Z. Inst. Hort. **11**, 13 (1941).

— Small fruits. Amer. Fruit. Gr., **62**, Nr. 5, 8 (1942).

— Strawberry „Melody", raised by Arthur G. Sainsbury, Mangere, via Otahuhu. J. N. Z. Inst. Hort., **12**, 74 (1943).

— Thornless gooseberries. Gdnrs' Chron. **118**, 125 (1945).

— Breeding strawberries resistant to „red core" root rot. Mon. Sci. News, Nr. 5, 3—4 (1945).

— Nationwide fruits. Jamberries. Amer. Fruit Gr., **67**, Nr. 2, 13 (1946).

Araratjan, A. G.: Mixoploidie bei Hyppophaë rhamnoides L. C. R. (Doklady) Acad. Sci. URSS., **27**, 857 bis 861 (1941).

Bailey, L. H.: Species Batorum. The genus Rubus in North America (north of Mexico). Gentes Herbarum, N. Y., **5**, 1—64 (1941).

— Species Batorum. The genus Rubus in North America (north of Mexico). Gentes Herbarum, N. Y., **5**, 201 bis 228 (1941).

— Species Batorum. The genus Rubus in North America (north of Mexico). Gentes Herbarum, N. Y., **5**, 231 bis 422 (1943).

— Species Batorum. The genus Rubus in North America VI. Cuneifolii. Gentes Herbarum, N. Y., **5**, 425—461 (1943).

— Species Batorum Boreali-Americana. VII. Canadenses. Gentes Herbarum, Ithaca, N. Y., **5**, 465—503 (1944).

— Species Batorum Boreali-Americana. VIII. Alleghenienses. Gentes Herbarum, Ithaca, N. Y., **5**, 507—588 (1944).

— Species Batorum Boreali-Americana. IX. Arguti, Rubi Europae peregrini. Gentes Herbarum, Ithaca, N. Y., **5**, 591—856 (1945).

— The genus Rubus in North America. X. Subg. V. Idaeobatus. Subg. VI. Anoplobatus. Gentes Herbarum, Ithaca, N. Y., **5**, 859—918 (1945).

Bailey, J. S. u. A. P. French: Identification of blue berry varieties by plant characters. Bull. Mass. Agric. Exp. Sta., Nr. 431, 20 (1946).

Bain, H. F.: Cranberry tetraploids. Proc. 74th. Annu. Conv. Amer. Cranberry G. Ass., Pemberton, N. J., **74**, 12—13, 16 (1943).

— u. J. B. Demarée: Red stele root disease of the strawberry caused by Phytophthora fragariae. J. Agric. Res. **70**, 11—30 (1945).

Banga, O.: Het belang van de bloemkenmerken voor de indentificeering van roode bessen rassen. (The significance of the flower characters for the identification of varieties of red currants.) Fruitteelt 32, 187—194 (1942).

Beakbane, A. B.: Trials of loganberries, blackberries and hybrid berries at East Malling. Sci. Hort. 7, 64 bis 70 (1939).

— Studies of cultivated varieties of Rubus and their hybrids. II. Description and selection of clonal races of some cultivated blackberries and hybrid berries, including loganberries. J. Pomol. 18, 368—378 (1941).

Benedict, H. G.: The Boysenberry without thorns, an early ripener. Market Gr. J., 64, 154 (1939).

Berkeley, G. H. u. G. C. Chamberlain: Diseases of the raspberry. Fmrs' Bull. Nr. 123 Dep. Agric. Can. Publ. Nr. 760, 11 (1944).

Blair, D. S. u. M. B. Davis: Bush fruits. Fmrs' Bull. Dep. Agric. Dom. Canad. Nr. 131 (Publ. 775) 22 (1945).

Bologovskaja, R. P.: (Breeding small fruits in Siberia and the Far East.) Plodovo-Jagodnye Kuljtury (Fruit Crops) Nr. 4, 64—68 (1940). [Russisch.]

— (The raspberry „Kuz'min's Novelty".) Plodovo-Jagodnye Kuljtury (Fruit Crops) Nr. 5, 73—74. [Russisch.]

Brierley, W. G. u. R. H. Landon: Winter behavior of strawberry plants. Bull. Minn. Agric. Exp. Sta. Nr. 375, 24 (1944).

—, — A study of cold resistance of the roots of the Latham red raspberry. Proc. Amer. Soc. Hort. Sci. 47, 215—218 (1946).

—, — Some relationships between rest period, rate of hardening, loss of cold resistance and Winter injury in the Latham raspberry. Proc. Amer. Soc. Hort. Sci., 47, 224—234 (1946).

Brown, C. A.: History of the wild strawberry in Louisiana. Home Gdng. New Orleans, 2, 168, 174, 180 (1942).

Bunyard, E. A.: The strawberry and its history. Gdnrs' Chron., 105, 154—155 (1939).

Burkhart, L.: Firmness of strawberries as measured by a penetrometer. Plant Physiol., 18, 693—698 (1943).

C.: Goed opletten! (Jets over een krachtig groeiende lamtorosoort.) (Keep a good look out! [Something about a vigorous species of wild tamarind].) Bergcultures, 14, 526—527 (1940).

Čajlahjan, M. C.: Content of vitamin C in wild roses of Armenia. C. R. (Doklady) Acad. Sci. URSS, 40, 369—371 (1943).

Camp, W. H.: On the structure of populations in the genus Vaccinium. Brittonia, 4, 189—204 (1942).

— A survey of the American species of Vaccinium subgenus Euvaccinium. Brittonia, 4, 205—247 (1942).

— Studies in the Ericales: a new name in blueberries. Bull. Torrey Bot. Cl., 63, 240 (1942).

— Studies in the Ericales: a review of the North American Gaylussacieae; with remarks on the origin and migration of the group. Bull. Torrey Bot. Cl., 68, 531—551 (1942).

— Another new name in Vaccinium. Bull. Torrey Bot. Cl., 71, 179—180 (1944).

— The North American blueberries with notes on other groups of Vacciniaceae. Brittonia N. Y., 5, 203—275 (1945).

Chandler, F. B.: Blueberry storage. Science, 95, 603 bis 604 (1942).

Cicin, N. V. u. E. M. Darevskaja: (Ein Bastard von Rubus flagellaris × Rubus saxatilis L.) C. R. (Doklady) Acad. Sci. URSS, 57, 829 (1947). [Russisch.]

Clark, J. H.: Prevalence of certain diseases affecting the foliage in some strawberry progenies. Proc. Amer. Soc. Hort. Sci., 36, 455—460 (1938) 1939.

Clark, J. H.: Leaf characters as a basis for the classification of blueberry varieties. Proc. Amer. Soc. Hort. Sci., 38, 441—446 (1941).

— Another New Jersey strawberry. N. J. St. Hort. Soc. News 23, 1385 (1942).

— Four New Jersey strawberry varieties receive names. N. J. St. Hort. Soc. News 24, 1465, 1470 (1943).

— The place of the Lupton variety in the New Jersey strawberry breeding program. Proc. Amer. Soc. Hort. Sci. 45, 263—266 (1944).

— u. S. G. Gilbert: Selection of criterion leaves for the identification of blueberry varieties. Proc. Amer. Soc. Hort. Sci., 40, 347—351 (1942).

Coe, F. M.: Promising new everbearing strawberry originated in Utah. Amer. Fruit Gr. 60, Nr. 5, 18 (1940).

— Red raspberry varieties for freezing, local market, and home use. Fm. Home Sci. Utah, 5, Nr. 2, 3, 11 (1944).

— Strawberry varieties old and new. Fm. Home Sci., Utah, 6, 7, 10 (1945).

Colby, A. S.: The Brainerd blackberry. Amer. Fruit Gr., 60, Nr. 3, 11, 35 (1940).

— Some recent developments in strawbery growing. Trans. Ill. Hort. Soc., 74, 393—405 (1940).

— Plan now to eradicate raspberry anthracnose. News Lett. Ill. St. Hort. Soc., Nr. 2, 1—3 (1945).

Crane, M. B.: Reproductive versatility in Rubus. I. Morphology and inheritance. J. Genet., 40, 109 bis 118 (1940).

— The origin of new forms in Rubus. II. Theloganberry, R. loganobaccus Bailey. J. Genet., 40, 129 bis 140 (1940).

— u. P. T. Thomas: Segregation in asexual (apomictic) offspring in Rubus. Nature, Lond., 143, 684 (1939).

Crété, P.: Polyembryonie chez l'Actinidia chinensis Planch. Bull. Soc. Bot. Fr., 91, 89—92 (1944).

Čuev, N. V. u. V. P. Matveev: (Hybridization of strawberries under artificial light under the conditions of the Murman province.) Plodovo-Jagodnye Kuljtury (Fruit Crops), Nr. 4, 77—79 (1940). [Russisch.]

Culpepper, C. W. u. J. S. Caldwell: Varietal adaptability of strawberries to preservation in sulphur dioxidecalcium solution. Fruit Prod. J., 23, 5—9, 25, 27, 29, 46—51 (1943).

Curl, A. L. u. E. K. Nelson: The occurrence of citric and isocitric acid in blackberries and in dewberry hybrids. J. Agric. Res., 67, 301—303 (1943).

Darrow, G. M.: Seed size in strawberries. Proc. Amer. Soc. Hort. Sci., 37, 564—566 (1939) 1940.

— The maytime, starbright and redstar strawberries. Circ. U. S. Dep. Agric., Nr. 597, 5 (1940).

— Strawberry varieties old and new. Amer. Fruit. Gr. 61, Nr. 3, 11, 22 (1941).

— Blackberry growing. Fmr's Bull. U.S. Agric., Nr.1399, 18 (1942).

— The Midland and Fairpeake strawberries. Circ. U.S. Dep. Agric., Nr. 694, 4 (1944).

— u. W. H. Camp: Vaccinium hybrids and the development of new horticultural material. Bull. Torrey Bot. Cl., 72, 1—19 (1945).

—, —, H. E. Fischer u. H. Dermen: Studies on the cytology of Vaccinium species. Proc. Amer. Soc. Hort. Sci., 41, 187—188 (1942). (Abstr.)

—, —, —, — Chromosome numbers in Vaccinium and related groups. Bull. Torrey Bot. Cl., 71, 498—506 (1944).

— u. J. H. Clark: The Sunrise red raspberry. Circ. N. J. Agric. Exp. Sta. Nr. 397, 4 (1939).

—,— The Atlantic, Pemberton, and Burlington blueberries. Circ. U.S. Dep. Agric., Nr. 589, 8 (1940).

—, — u. E. B. Morrow: The inheritance of certain characters in the cultivated blueberry. Proc. Amer. Soc. Hort. Sci., 37, 611—616 (1939) 1940.

Darrow, G. M. u. E. B. Morrow: Breeding new strawberry varieties. Bull. N. C. Agric. Exp. St., Nr. 320, 12 (1939).
—, — The Massey strawberry. Bull. N. Carolina Exp. Sta., Nr. 327, 3 (1940).
—, R. B. Wilcox u. C. S. Beckwith: Blueberry growing. Fmr's Bull. U. S. Dep. Agric., Nr. 1951, 38 (1944).
—, O. Woodward u. E. B. Morrow: Improvement of the rabbiteye blueberry. Proc. Amer. Soc. Hort. Sci., 45, 275—279 (1944).
Demarée, J. B. u. M. S. Wilcox: Blueberry cane canker. Phytopathology, 32, 1068—1075 (1942).
Dermen, H.: The mechanism of colchicine-induced cytohistological changes in cranberry. Amer. J. Bot., 32, 387—394 (1945).
— Periclinal cytochimeras and histogenesis in cranberry. Amer. J. Bot., 34, 32—43 (1947).
— u. H. F. Bain: Periclinal and total polyploidy in cranberries induced by colchicine. Genetics, 26, 147 bis 148 (1941). (Abstr.)
—,— Periclinal and total polyploidy in cranberries induced by colchicine. Proc. Amer. Soc. Hort. Sci. 38, 400 (1941).
—, — A general cytohistological study of colchicine polyploidy in cranberry. Amer. J. Bot., 31, 451—463 (1944).
— u. G. M. Darrow: Colchicine-induced tetraploid and 16-ploidstrawberries. Proc. Amer. Soc. Hort. Sci. 36, 300—301 (1938) 1939.
Dogadkina, N. A.: A contribution to the question of genome relations in some species of Fragaria. C. R. (Doklady) Acad. Sci. URSS., 30, 166—168 (1941).
Drain, B. D.: Red raspberry breeding for southern adaptation. Proc. Amer. Soc. Hort. Sci. 36, 302—304 (1938) 1939.
— Developing of new horticultural varieties for Upper South. Raspberries and strawberries given attention. Sth. Flor., 49, Nr. 21, 7, 15 (1940).
— Tennessee Autumn red raspberry. Circ. Tenn. Agric. Exp. Sta., Nr. 70, 4 (1940).
— How new varieties are developed. New Tennessee fruits. Years and much outlay required by station. Sth. Flor. 52, Nr. 23, 5—6 (1942).
— Breeding for adaption of the strawberry and raspberry. Trans. Ill. St. Hort. Soc., 77, 216—218 (1943).
— Tennessee Luscious red raspberry. Circ. Tenn. Agric. Exp. Sta. Nr. 92, 4.
Duis, W. H.: Selection of the low-bush blueberry in West-Virginia. Proc. Amer. Soc. Hort. Sci., 38, 434 bis 437 (1941).
Duka, S. H.: (An new form of berry. Blackcurrant × gooseberry.) Jarovizacija, Nr. 3 (30), 119—122 (1940). [Russisch.]
— (New forms of hybrids between Fragaria grandiflora and F. elatior.) Jarovizacija, Nr. 5 (32), 131—132 (1940). [Russisch.]

Eaton, E. L., C. C. Eidt, A. D. Pickett u. J. F. Hockey: The blueberry. Fmrs' Bull. Canad. Dep. Agric., Nr. 120, 30 (1943).
Evreinoff, V.: Origines et ancêtres des groseilliers à grappes cultivés. Bull. Soc. Nat. Hort. Fr., 6, Sér. 6, 135—142 (1939).
— Le groseillier bleu: Ribes dikuscha Fisch. Rev. Hort. Paris, 28, 46—47 (1942).

Fagerlind, F.: Kompatibilität und Inkompatibilität in der Gattung Rosa. Acta Hort. Berg. 13, 247—302 (1944).
— Pollenkonkurrenz und Bastardierungsschwierigkeiten in der Gattung Rosa. Svensk Bot. Tidskr., 40, 284 bis 292 (1946).
Fedorova, N. J.: Cytology of polyploid hybrids Fragaria grandiflora × F. elatior and their fertility. C. R. (Doklady) Acad. Sci. URSS., 52, Nr. 8, 711 bis 712 (1946).
Filosofova, T. P.: Bridging species method in hybridization of the garden strawberry, F. grandiflora, with other Fragaria species. C. R. (Doklady) Acad. Sci. URSS, 31, 924—926 (1941).
Fischer, H. E., G. M. Darrow u. F. Perlmutter: Raspberry and blackberry breeding: production of tetraploid raspberries. Proc. Amer. Soc. Hort. Sci., 42, 447—456 (1943).
—,— u. G. F. Waldo: Further chromosome studies of some varieties of blackberries. Proc. Amer. Soc. Hort. Sci., 38, 401—404 (1941).
Fister, L. A.: Tennessee Shipper strawberry. Circ. Tenn. Agric. Exp. Sta., Nr. 76, 4 (1941).
— u.B. D. Drain: Tennessee Supreme strawberry. Cir. Tenn. Agric. Exp. Sta., Nr. 68, 4 (1940).
—, — Tennessee Beauty strawberry. Circ. Univ. Tenn. Agric. Exp. Sta. Nr. 81, 4 (1942).
Floor, J. u. W. S. Rogers: Key for the identification of the commonly cultivated commercial varieties of strawberries. J. Pomol., 21, 34—40 (1945).

Gerritsen, J. D.: Vragen rondom de bladvalziekte van de roode bes. (Problems about the leaf-spot of red currants.) Tijdschr. PlZiekt., 52, 119—120 (1946).
Grubb, N. H.: Five new raspberry seedlings. Ann. Rep. E. Malling Res. Sta. 1942, Pp. 31—33 (1943).
— Malling promise raspberry. 33rd Rep. E. Malling Res. Sta. 1945, Pp. 133 (1946).
— u. R. V. Harris: The planting and maintenance of raspberry cane nurseries. Gdnrs' Chron., 117, 27, 33 bis 34 (1945).
Guillaumin, A.: Fructification de nouvelles espèces d'Actinidia. Rev. Hort. Paris, 114, 204—205 (1942).
— u. C. Guinet: Actinidia chinensis Planchon, Liane fruitière d'extrême-orient intéressante pour nos cultures et pour l'hygiène alimentaire. Rev. Hort. Paris, 27, 315—319 (1941).
Gustafsson, Å.: Differential polyploidy within the blackberries. Hereditas, Lund, 25, 33—47 (1939).
— The origin and properties of the European blackberry flora. Hereditas, Lund, 28, 249—277 (1942).
— The genesis of the European blackberry flora. Acta Univ. Lund, 54, 199 (1943).
— The constitution of the Rosa canina complex. Hereditas, Lund, 30, 405—418 (1944).
— u. Håkansson A.: Meiosis in some Rosa-hybrids. Bot. Notiser 331—343 (1942).
— u. J. Schröderheim: Ascorbic acid and hip fertility in Rosa species. Nature, Lond., 153, 196—197 (1944).
—, — Ascorbic acid in Rosa hybrids. Hereditas, Lund. 31, 489—497 (1945).

Hahn, G. G.: Blister rust relations of cultivated species of red currants. Phytopathology, 33, 341—353 (1943).
Harrison, J. W. H. u. G. A. D. Jackson: Ascorbic acid and hip fertility in Rosa species. Nature, Lond., 153, 403—404 (1944).
Havis, A. L.: A developmental analysis of the strawberry fruit. Amer. J. Bot., 30, 311—314 (1943).
Herold, G.: Sortenprüfung bei Erdbeeren unter besonderer Berücksichtigung der Befruchtungsverhältnisse. Gartenbauwiss. 16, 216—262 (1941).
— Sortenprüfung bei Erdbeeren. Schweiz. Z. Obst- u. Weinb. 52, 383—384 (1943). (Abstr.)
Hickman, C. J.: The red core root disease of the strawberry caused by Phytophthora Fragariae n. sp. J. Pomol. 18, 89—118 (1940).
Hildreth, A. C. u. L. Powers: The Rocky Mountain strawberry as a source of hardiness. Proc. Amer. Soc. Hort. Sci. 38, 410—412 (1941).

Hunter, A. W. S.: Rus resistent black currants. Canad. Hort. and Home Mag. (1943).
— u. M. B. Davis: Breeding rust resistant black currants. Proc. Amer. Soc. Hort. Sci. **42**, 467—468 (1943).

Iliev, P.: (A varietal experiment with strawberries.) Zemledelie, Sofia **43**, 122—125 (1939). [Bulgarisch.]

Jacoby, F. C. u. F. Wokes: Carotene and lycopene in rose hips and other fruits. Biochem. J. **38**, 279—282 (1944).

Jeffers, W. F.: Further progress in breeding strawberries for resistance to red stele. Trans. Peninsula Hort. Soc. **32**, Nr. 5, 70—71 (1942).
— u. G. M. Darrow: Promising strawberry crosses resistant to the red stele disease. Trans. Peninsula Hort. Soc. **31**, Nr. 4, 20—23 (1941).
—, — The red-stele-resistent Temple strawberry. Trans. Peninsula Hort. Soc. **33**, 43—44 (1943).
—, — u. C. E. Temple: Progress in breeding for resistance to the red stele root disease of the strawberry in Maryland. Trans-Peninsula Hort. Soc. **30**, Nr. 4, 49—52 (1940).

Johansson, E.: Nyare erfarenheter om bärodling. (Recent results relating to the growing of bush fruits.) Årsb. Jordbruksforskning, Stockholm, Pp. 183—194 (1945).

Johnston, S.: Observations on the inheritance of horticulturally important characteristics in the highbush blueberry. Proc. Amer. Soc. Hort. Sci. **40**, 352 bis 356 (1942).
— Observations on hybridizing lowbush and highbush blueberries. Proc. Amer. Soc. Hort. Sci. **47**, 199—200 (1946).

Kanér, R.: Rubusstudier i nordvästra Skåne. (Rubus studies in north-western Scania.) Bot. Notiser, Pp. 367—374 (1941).

Katinskaja, J. K.: (The best varieties of strawberry for the northern and central zones of the Union.) Vestnik Socialističeskogo Rastenievodstva (Soviet Plant Industry Record) Nr. 5, 47—53 (1940). [Russisch.]

Kemmer, E. u. G. Herold: Zur Sortenfrage bei Erdbeeren. Forschungsdienst **7**, 353—360 (1939).

Kičunov, N. I.: (A valuable rose for the production of rose oil.) Vestnik Socialističeskogo Rastenievodstva (Soviet Plant Industry Record), Nr. 3, 145—146 (1940). [Russisch.]

Knowles, D., O. Grottodden u. T. E. Long: Freezing preservation of North Dakotagrown raspberries. Bi-m. Bull. N. Dak. Agric. Exp. Sta. **5**, Nr. 5, 17—19 (1943).

Kobel, F. u. F. Schütz: Erdbeer-Neuzüchtung der Eidg. Versuchsanstalt Wädenswil. Schweiz. Z. Obst- u. Weinb. **52**, 483—485 (1943).

Kobozev, V. V.: (A cranberry with large berries from the Turuhan region.) Plodovo-Jagodnye Kuljtury (Fruit Crops), Nr. 5, 76 (1940). [Russisch.]

Kondratjeva, M. N.: (New varieties of gooseberry.) Sadovodstvo (Horticulture), Nr. 12, 42 (1939). [Russisch.]

Kronenberg, H. G.: Virusziekten in aardbeien. (Virus diseases in strawberries.) Tijdschr. PlZiekt. **49**, 74—76 (1943).

Kuzmin, A. J.: (The role of the intermediary plant in obtaining the first and the second generation between black-currant and gooseberry.) Proc. Lenin Acad. Sci. USSR, Nr. 5, 3—12 (1940). [Russisch.]
— (The role of the intermediary in hybridizing currants with gooseberries. Sovetskaja Botanika (Soviet Botany), Nr. 1/2, 14—23 (1941). [Russisch.]
— (Hybrid of raspberry and blackberry.) Vestnik Gibridizacii (Hybridization), Nr. 1, 116—118 (1941). [Russisch.]

Latimer, L. P.: Performance of strawberry varieties in New Hampshire. Bull. N. H. Agric. Exp. Sta., Nr. 368, 12 (1946).

Ledeboer, M. u. I. Rietsema: Unfruitfulness in black currants. J. Pomol **18**, 177—181 (1940).

Leslie, W. R.: Manitoba news letter. N. S. Dak. Hort. **15**, 136, 139 (1942).
— Growing small fruits in the prairie provinces. Bull. Line Elevators Fm Serv., Winnipeg, Nr. 6, 52 (1945).
— Manitoba news letter. N. and S. Dak. Hort. **19**, 133, 143 (1946).

Lewis, D.: Genetical studies in cultivated raspberries. I. Inheritance and linkage. J. Genet. **38**, 367—379 (1939).
— Differential fertilization in Rubus Idaeus L. Proc. Genet. Soc. (Supplement to J. Genet. **37**, 3) 3 (1939). (Abstr.)
— The relationship between polyploidy and fruiting habit in the cultivated raspberry. Proc. 7th Int. Genet. Congr. Edinburgh 23—30 August, Pp. 190 1939 (1941). (Abstr.)
— Genetical studies in cultivated raspberries. II. Selective fertilization. Genetics **25**, 278—286 (1940).

Lineberry, R. A. u. L. Burkhart: The vitamin C content of small fruits. Proc. Amer. Soc. Hort. Sci. **41**, 198—200 (1942).
—, — Nutrient deficiencies in the strawberry leaf and fruit. Plant Physiol. **18**, 324—333 (1943).

Loomis, N. H. u. G. M. Darrow: Suwannee — a new home — garden strawberry. Circ. Miss. Agric. Exp. Sta, Nr. 123, 3 (1945).

Matveeva, E.: (New varieties of raspberry for the northern zone.) Sadovodstvo (Horticulture), Nr. 12, 39—40 (1939). [Russisch.]

Meader, E. M. u. G. M. Darrow: Pollination of the rabbit-eye blueberry and related species. Proc. Amer. Soc. Hort. Sci. **45**, 267—274 (1944).

Melville, R.: Ascorbic acid and hip fertility in Rosa species. Nature, Lond. **153**, 404—405 (1944).

Merrill, T. A. u. S. Johnston: Further observations on the pollination of the highbush blueberry. Proc. Amer. Soc. Hort. Sci. **37**, 617—619 (1939) 1940.

Meurman, O.: Edeltäviä tietoja karviaismarjapensasko keista. (Preliminary information on gooseberry bush trials.) Valt. Maatalousk. Tiedon, Nr. 166, 14 (1939).

Miller, J. C.: Strawberry breeding in Louisiana. Amer. Fruit Gr. **59**, Nr. 3, 11, 24, 28 (1939).
— Strawberry breeding problems. Proc. Amer. Soc. Hort. Sci. **45**, 259—262 (1944).

Mindlina, S.: (Die Heterogamie der Arten der Sectio Caninae der Gattung Rosa.) C. R. (Doklady) Acad. Sci. URSS **57**, 289 (1947). [Russisch.]

Morris, H. E. u. M. M. Afanasjev: Yellows, a noninfections disease of the progressive everbearing strawberry in Montana. Tech. Bull. Mo. Agric. Exp. Sta., Nr. 424, 11 (1944).
—, — Montana Progressive strawberry. A yellows-resistant, everbearing variety developed during research on yellows. Circ. Mont. Agric. Exp. Sta., Nr. 181, 2 (1945).

Morrow, E. B.: Some effects of cross-pollination versus self-pollination in the cultivated blueberry. Proc. Amer. Soc. Hort. Sci. **42**, 469—472 (1943).
— u. G. M. Darrow: Inheritance of some characteristics in strawberry varieties. Proc. Amer. Soc. Hort. Sci. **39**, 262—268 (1941).

Murrill, W. A.: More about white blackberries. Science **99**, 513—514 (1944).
— A white blackberry. J. Hered. **36**, 217—218 (1945).

Newcomer, E. H.: Chromosome numbers of some species and varieties of Vaccinium and related genera. Proc. Amer. Soc. Hort. Sci. **38**, 468—470 (1941).

Newton, W.: Combining beauty with utility in rose breeding. Sci. Agric. **24**, 304—306 (1944).

Nilsson, F.: Tetraploida typer inom släktet Ribes. (Tetraploid types in the genus Ribes.) Sverig. Pomol. Fören. Årsskr. **45**, 130—134 (1944).

— u. E. Johansson: Nya typer och hybrider inom släktet Fragaria. (New types and hybrids within the genus Fragaria.) Sverig. Pomol. Fören. Årsskr. **45**, 146—151 (1944).

Offord, H. R., C. R. Quick u. V. D. Moss: Self-incompatibility in several species of Ribes in the western states. J. Agric. Res. **68**, 65—17 (1944).

Oldham, C. H.: Strawberries. Bull. Minist. Agric. Lond., Nr. 95, 67 (1944).

— The cultivation of berried fruits in Great Britain. Crosby Lockwood and Son, Ltd., London 21s. Pp. 374. 41 figs. 28 tables (1946).

Overcash, J. F., L. A. Fister u. B. D. Drain: Strawberry breeding and the inheritance of certain characteristics. Proc. Amer. Soc. Hort. Sci. **42**, 435—440 (1943).

Palmer, E. J.: The species concept in Crataegus. Chronica Botanica **7**, 373—375 (1943).

Pavlova, N. M.: (Initial material for the breeding and cultivation of bush fruits.) Vestnik Socialističeskogo Rastenievodstva (Soviet Plant Industry Record) Nr. 5, 33—46 (1940). [Russisch.]

Petrov, A. V.: (New varieties of strawberry.) Sady i Ogorody (Fruit and Vegetable Gardens), Nr. 2, 53—54 (1941). [Russisch.]

Petrov, D. F.: On the ocurrence of facultative pseudogamy in a triploid variety of rapsberries, Immertragende (R. Idaeus). C. R. (Doklady) Acad. Sci. URSS, **22**, 352—353 (1939).

Philosophova, T. P. siehe Filosofova.

Pilin, G. M.: (An experiment on changing the nature of the gooseberry.) Jarovizacija, Nr. 5 (32), 129—130 (1940). [Russisch.]

Porpázy, A.: Málna-szeder namesitési kisérletek. (Crosses between raspberry and dewberry.) Bull. Hung. Coll. Hort. **10**, 143—145 (1944).

Powers, L.: Meiotic studies of crosses between Fragaria ovalis and × F. ananassa. J. Agric. Res. **69**, 435—448 (1944).

— Strawberry breeding studies involving crosses between the cultivated varieties (× Fragaria ananassa) and the native Rocky Mountain strawberry (F. ovalis). J. Agric. Res. **70**, 95—122 (1945).

Randoin, L. u. J. Boisselot: La valeur anti-scorbutique de fruit comestible d'Actinidia chinensis (Planchon) source exceptionnelement riche en vitamine C. Rev. Hort. Paris **27**, 319 (1941).

Ratsek, J. C., W. S. Flory jr. u. S. H. Yarnell: Crossing relations of some diploid and polyploid species of roses. Proc. Amer. Soc. hortic. **38**, 637—654 (1941).

Rietsema, I.: Oplossing van het mozaïekvraagstuk bij de frambozen. (A solution of the mosaic problem in raspberries.) Landbouwk. Tijdschr. Wageningen **51**, 14—25 (1939).

Roberts, O. C. u. A. S. Colby: Identification of certain red and purple raspberry varieties by means of primocanes. Proc. Amer. Soc. Hort. Sci. **42**, 457—462 (1943).

Rosanova, M. A. siehe Rozanova.

Rozanova, M. A.: Rôle of autopolyploidy in the origin of Siberian raspberry. C. R. (Doklady) Acad. Sci. URSS **24**, 58—60 (1939).

— Evolution of cultivated raspberry. C. R. (Doklady) Acad. Sci. URSS **24**, 179—181 (1939).

— On genotypic differences between races of Rubus caesius L. C. R. (Doklady) Acad. Sci. URSS **27**, 590—593 (1940).

Rozanova, M. A.: Autosyndesis in the genus Rubus. C. R. (Doklady) Acad. Sci. URSS **29**, 143—145 (1940).

— (Distant hybridization within the genus Rubus.) Vestnik Gibridizacii (Hybridization) Nr. 2, 21—50 (1941). [Russisch.]

— (On some species, subspecies and varieties within the conspecies Rubus Idaeus L.) J. Bot. URSS **30**, 44—48 (1945). [Russisch.]

S......i: Neue Stachelbeersorten. Land und Frau **24**, 33 (1940).

Šaškin, I. N.: (The regulation of the time of flowering and hybridization in Actinidia.) Jarovizacija, Nr. 3 (30), 218—219 (1940). [Russisch.]

Šaškina, L.: (Strawberry grafts.) Jarovizacija, Nr. 6 (33), 108—109 (1940). [Russisch.]

— u. Šaškin, I. N.: (Experiments and results of breeding strawberries and wild strawberries.) Jarovizacija, Nr. 3 (30), 123—138 (1940). [Russisch.]

Schiemann, E.: Artkreuzungen bei Fragaria III. Flora, Jena **37**, 166—192 (1943).

Slate, G. L.: Breeding autumn-fruiting raspberries. Proc. Amer. Soc. Hort. Sci. **37**, 574—578 (1939) 1940.

— Raspberries. Proc. 86th Annu. Meet. N. Y. St. Hort. Soc., Pp. 274—278 (1941).

— „Milton"-a new red raspberry. Fm. Res. N. Y. St. Sta. 1942 **8**, 7,9 (1942).

— Methods and problems in raspberry breeding. Proc. Amer. Soc. Hort. Sci. **45**, 255—258 (1944).

— Newer small fruits. Amer. Fruit Gr. **64**, Nr. 6, 8, 21 (1944).

— Newer varieties of small fruits. Amer. Fruit Gr. **66**, 10—11, 32—33 (1946).

— u. W. B. Robinson: Ascorbic acid content of strawberry varieties and selections at Geneva, New York in 1945. Proc. Amer. Soc. Hort. Sci. **47**, 219—223 (1946).

— u. R. F. Suit: A second report on the breeding of autumn-fruiting red raspberries. Proc. Amer. Soc. Hort. Sci. **44**, 283—288 (1944).

Srinivasachar, D.: Embryological studies of some members of Rhamnaceae. Proc. Indian Acad. Sci. **11**, Sect. B, 107—116 (1940).

Stahl, J. L.: Chimeras of Bowen blackberry. An unstable patented plant. J. Hered. **37**, 51—55 (1946).

Šurakov, F.: Berries on the Kola Peninsula. Agricultural Chron., Moscow, Nr. 11, 5—6 (1946). [Mimeographed.]

Temple, C. E.: Red stele root rot of strawberry. Trans. Peninsula Hort. Soc. **29**, Nr. 5, 141—149 (1939).

Thomas, A. S.: Food crops as indicator plants in Uganda. E. Afr. Agric. J. **8**, 136—140 (1943).

Thomas, H. E. u. E. V. Goldsmith: The Shasta, Sierra, Lassen, Tahoe, and Donner strawberries. Bull. Calif. Agric. Exp. Sta. Nr. 690, 12 (1945).

Thomas, P. T.: Reproductive versatility in Rubus. II. The chromosomes and development. J. Genet. **40**, 119—128 (1940).

— The origin of new forms in Rubus, III. The chromosome constitution of R. loganobaccus Bailey, its parents and derivatives. J. Genet. **40**, 141—156 (1940).

Tolmačev, I. A.: (Hybrids of Crandall-currant with black-currant.) Vestnik Gibridizacii (Hybridization) Nr. 1, 115 (1941). [Russisch.]

Vaarama, A.: Cytological studies on some Finnish species and hybrids of the genus Rubus L. Maataloust. Aikakausk. **11**, 72—85 (1939).

Vincent, C. L.: Vegetable and small fruit growing in toxic ex-orchard soils of central Washington. Bull. Wash. Agric. Exp. Sta. Nr. 437, 31 (1944).

Waldo, G. F.: The Brightmore strawberry. Sta. Circ. Ore. Agric. Exp. Sta. Nr. 263, 3 (1942). [Mimeographed.]

Waldo, G. F. u. G. M. Darrow: Breeding autumn-fruiting raspberries under Oregon conditions. Proc. Amer. Soc. Hort. Sci. **39**, 274—278 (1941).
—,—, W. F. Jeffers, J. B. Demarée u. E. M. Meader: Breeding strawberries for resistance to red-stele root disease. Trans. Peninsula Hort. Soc. **36**, Nr. 5, 22 bis 33 (1946).
— u. E. H. Wiegand: Two new varieties of blackberry the Pacific and the Cascade. Sta. Circ. Ore. Agric. Exp. Sta. Nr. 269, 4 (1942). [Mimeographed.]
—, — u. H. Hartman: New berries from Oregon's plant breeding research. Sta. Bull. Ore. Agric. Exp. Sta. Nr. 416, 11 1943.
Wasscher, J.: De in ons land inheemsche en gekweekte Vacciniumsoorten. (The indigenous and the cultivated species of Vaccinium in our country.) Tijdschr. Ned. Heidemaatsch. **55**, 148—155 (1943).
White, E. u. J. H. Clark: Some results of self-pollination of the Highbush Blueberry at Whitesbog, New Jersey. Proc. Amer. Soc. Hort. Sci. **36**, 305—309 (1938) (1939).
Wilcox, A. N.: Recent progress in strawberry breeding. Minn. Hort. **69**, 83 (1941).
Williams, C. F.: Breeding raspberries for North Carolina. Res. and Fmg, Raleigh, N. C. **3**, Progr. Rep. Nr. 2; 9, 12 (1945).
— u. G. M. Darrow: The trailing raspberry-Rubus parvifolius L.: characteristics and breeding. Tech. Bull. N. C. Agric. Exp. Sta. Nr. 65, 13 (1940).
Wokes, F., E. H. Johnson, J. G. Organ u. F. C. Jacoby: Vitamins in rose hips. Nature, Lond. **151**, 279 (1943).

6. Agrumen (Citrus).

Bacchi, O.: Observações citológicas em citrus. I. Número de cromosômios de algumas espécies e variedades. (Cytological observations in citrus. I. The chromosome numbers of certain species and varieties.) J. Agron. S. Paulo **3**, 249—258; also Bol Téc. Inst. Agron. Estado, Campinas, Nr. 85 (1940).
— Cytological observations in citrus. III. Megasporogenesis, fertilization, and polyembryony. Bot. Gaz. **105**, 221—225 (1943).
— Indentificação colorimétrica em Citrus. (Colorimetric identification of Citrus.) Bragantia, São Paulo **3**, 179 bis 187 (1943).
— Observações citológicas em citrus. III. Megasporogênese, fertilização e poliembronia. (Cytological observations in citrus. III. Megasporogenesis, fertilization, and polyembryony.) Bragantia, São Paulo **4**, 405—412 (1944).
— Observações citológicas em Citrus. VI. Resultados preliminares do efeitoda colchicina sôbre semetes em germinação. (Cytological observations in Citrus. VI. Preliminary results on the effect of colchicine on germinating seeds.) Bragantia, São Paulo **4**, 679—691 (1944).
Batchelor, L. D. u. M. B. Rounds: Effect of rootstocks on lemon decline and yield in two experimental orchards. Calif. Citrogr. **29**, 242—243, 265—269 (1944).
Bedford, E. C. G.: The biology and economic importance of the South African citrus thrips, Scirtothrips aurantii Faure. Publ. Univ. Pretoria, Ser. II. Nat. Sci., Nr. 7, 68 (1943).
Benton, R. J.: Citrus improvement. Bud selection and propagation of varieties. Ten years' progress. Agric. Gaz. N. S. W. **50**, 143—145, 174, 214—218 (1939).
Bou Bono, B.: Alguna observaciones sobre el número de semillas de la mandarina „Clementina". (Some observations on the number of seeds in the Clementine mandarin.) Bol. Inst. Invest. Agron. Madr. **2**, Nr. 3, 195—207 (1939).
Brieger, F. G. u. J. T. A. Gurgel: Influência do cavalo sobre a fertilidade do polen no cavaleiro, em Citrus. (Influence of the rootstock on the pollen fertility of the scion in Citrus.) Bragantia, São Paulo **1**, 713—757 (1941).
Brieger, F. G., S. Moreira u. Z. Leme: II. Estude sobre o melhoramento da laranja Baía. (II. Study on the improvement of the Bahia orange.) J. Agron., S. Paulo **2**, 161—182 (1939).
—, —, — Estudo sobre o melhoramento da laranja „Baía" III (Conclusão). (Study on the improvement of the Bahia orange. III [Conclusion].) Bragantia, São Paulo **1**, 567—610 (1941).

Caryl, R. E. u. J. C. Johnston (reviser): Citrus culture in California. Circ. Calif. Agric. Ext. Serv. (revised 1946), Nr. 114, 46 (1940 u. 1946).
Chen, T.-Y.: (Cytological observations on twenty-two citrus fruits.) J. Agric. Ass. China, Nr. 177, 1—8 (I) (1944).
Chen, Wen Hsien: (The Citrus Experiment Station of Fukien University.) Fukien Agric. J. **2**, 231—238 (1940).
Chevalier, A.: L'origine botanique d'un agrume hybride: le Clémentinier. Rev. Bot. Appl. **19**, 428—430 (1939).
— Subdivision et composition actuelle du genre Citrus. Rev. Bot. Appl. **23**, 11—15 (1943).
— L'origine géographique des Aurantiacées (agrumes) cultivées et les étapes de leur amélioration spécialement en Indochine. Rev. Bot. Appl. **23**, 15—25 (1943).

Esinovskaja, V. N.: (An early mutant of the Unshiu mandarin.) Soviet Subtropics, Nr. 4 (56), 89 (1939). [Russisch.]
Etchandy, A. M.: Estudio sobre el limón (Citrus Limonum, Risso). Su conservación frigorífica y su industrialización. (Study on the lemon [C. Limonum Risso]. Its cold storage and industrial utilization.) Rev. Fac. Agron. Univ. Montevideo, Nr. 34, 205—274 (1943).

Fauvel, J. H.: Quelle est l'origine exacte du Clémentinier? Rev. Hort. Agric, Afrique Nord **43**, 62—65 (1939).
— Les pays d'origine des espèces du genre Citrus et les contrées d'expansion en Extrême-Orient d'après le Professeur Tyôsaburo Tanaka. Rev. Hort. Agric. Afrique Nord **46**, 33—39 (1942).
— Les meilleurs Citrus d'Extrême-Orient. Rev. Hort. Agric. Afrique Nord **46**, 62—68 (1942).
Friend, W. H. u. S. H. Yarnell: Clonal selection of grape fruit with respect to yield. Proc. Amer. Soc. Hort. Sci. **38**, 358—362 (1941).
Frost, H. B.: The Pearl tangelo-a new citrus variety. Calif. Citrogr. **25**, 346 (1940).
— u. C. A. Krug: Diploid-tetraploid periclinal chimeras as bud variants in Citrus. Genetics **27**, 619—634 (1942).
—, — Quimeras periclinais diplóides-tetraplóides surgidas em forma de variações somáticas em Citrus. (Diploid-tetraploid periclinal chimaeras arising as somatic variations in Citrus.) Bragantia, São Paulo **4**, 449—474 (1944).

Gondell, M. A.: La susceptibilidad de diferentes especies y variedades citricas a la Phytophthora citrophthora (Sm. y Sm.) Leon., P. parasitica Dastur y P. megasperma Leon. (The susceptibility of different citrus species and varieties to P. citrophthora [Sm. et Sm.] Leon., P. parasitica Dastur and P. megasperma Leon.) Minist. Agric. Nac. Direcc. Gen. Invest. Inst. Sanid. Veg., B. Aires **2**, Ser. A, Nr. 19, 24 (1946).

Hall, W. J.: A decade of citrus research in Southern Rhodesia. Proc. Trans. Rhodesia Sci. Ass. **38**, 74—87 (1941).

Herrero Egaña, M. u. A. Acerete Lavilla: Variedades del naranjo. (Orange varieties.) Bol. Inst. Invest. Agron. Madr., Nr. 6, 125—131 (1942).

Hu Chang-Chih u. Wu Chien-Chi: Selection studies of Hwang-kuo (sweet orange) at Kin-tang and Kiang-Tsing in Szechuan, Paper II. Nanking, J. 11, 51—72 (1942).

Jarry-Desloges, R.: L'oranger ? King of Siam. Rev. Hort. Paris 28, 133 (1942).

Krug, C. A.: Chromosome numbers in the subfamily Aurantioideae with special reference to the genus citrus. Bot. Gaz. 104, 602—611 (1943).

— Observações citológicas em Citrus. IV. Números de cromosômios na subfamília Aurantioideae com referência especial ao gênero Citrus. (Cytological observations in Citrus. IV. Chromosome number in the subfamily Aurantioideae with special reference to the genus Citrus.) Bragantia, São Paulo 4, 413—428 (1944).

— u. O. Bacchi: Triploid varieties of citrus. J. Hered. 34, 277—283 (1943).

—, — Observacões citológicas em Citrus. II. Variedades triplóides. (Cytological observations in Citrus. II. Triploid varieties.) Bragantia, São Paulo 4, 393—403 (1944).

—, — Observações citológicas em Citrus. V. Poliploidia em relação a densidade e ao tamanho dos estomas em Citrus e outros gêneros das Aurantioideae. (Cytological observations in Citrus V. Polyploidy in relation to the density and size of the stomata in Citrus and other genera of the Aurantioideae.) Bragantia, São Paulo 4, 429—447 (1944).

Lapin, V. K. u. V. G. Teluh: Size and number of stomata in diploid and polyploid forms in Citrus, Poncirus and Fortunella. C. R. (Doklady) Acad. Sci. URSS. 27, 365—368 (1940).

Larsen, G.-H.: Improductivité du Clémentinier. Bull. Doc. Renseign. Agric., Serv. Arb. Algér., Nr. 92, 15 bis 18 (1942).

Lees, P. M., G. Bergeret u. A. M. Etchandy: Observaciones sobre frutas cítricas de la región de San Antonio Dep. del Salto de la cosecha de 1940 y 1942 conservadas en cámara frigorífica. (Observations on citrus from the region of San Antonio, Department of Salto, from the harvests of 1940 and 1942 after cold storage.) Rev. Fac. Agron. Univ. Montevideo, Nr. 34, 159—193 (1943).

Levitt, E. C.: The need for care in selection of Late Valencia buds. Avoid strains which tend to sport. Agric. Gaz. N. S. W. 55, 382—383 (1944).

Levitt, J. u. R. C. Nelson: The relative resistance of morphologically different orange-peel cells to various inury factors. Biodynamica 4, 57—64 (1942).

Lykov, T. G. u. P. I. Lykova: (Our citrus breeding plot.) Soviet Subtropics, Nr. 11 (63), 43—44 (1939). [Russisch.]

Moreira, S.: Cavalos para citros em São Paulo. (Rootstocks for citrus fruits in São Paulo.) Rev. Agric. Piracicaba. 21, 206—226 (1946).

— u. J. T. A. Gurgel: A fertilidade do polen e sua correlação com o número de sementes, em espécies e formas do gênero Citrus. (Pollen fertility and its correlation with number of seeds in species and forms of the genus Citrus.) Bragantia, São Paulo 1, 669—711 (1941).

Mortensen, E. u. C. R. Riecker: Seed production and seedling yields of some citrus varieties of possible value for rootstock purposes. Proc. Amer. Soc. Hort. Sci. 41, 145—148 (1942).

Nandi, H. K., S. C. Bhattacharya u. S. Dutt: Nursery behaviour of five indigenous citrus rootstock varieties with Khasi orange as scion in Assam. Indian J. Agric. Sci. 13, 489—493 (1943).

Oppenheimer, H. R.: (Unfruitfulness of clementines.) Hameshek Hahaklai 1943—1944, Nr. 1, 8—9; Nr. 2/3, 12—13; Nr. 4, 23—33; Nr. 6, 12 (1944).

— (Some more investigations on unfruitfulness of clementines.) Hameshek Hahaklai 6, 13—15 (1945).

Rebour, H.: Anomalies de la floraison du Clémentinier. Bull. Doc. Renseign. Agric., Serv. Arb. Algér., Nr. 118, 4—6 (1945).

Reig Feliú, A.: Análisis cuantitativo de las esencias del fruto de los agrios más cultivados en España. (Quantitative analysis of the essences of the citrus fruits most cultivated in Spain.) Bol. Inst. Invest., Agron., Madrid, Nr. 9, 161—191 (1943).

Ruggieri, G.: Le mutazioni vegetative nella patologia degli agrumi. (Vegetative mutations in the pathology of the citrus fruits.) Bol. Staz. Pat. Veg. Roma 22, 119—130 (1942).

Sigurd Arendsen, S.: Estudio de la susceptibilidad presentada por diversas especies y variedades de citrus al ataque de Phythophthora citrophthora (Sm. and Sm.) Leon. (Study of the susceptibility presented by several species and varieties of citrus to the attack of Phythophthora citrophthora (Sm. and Sm.) Leon. Bol. Sanid. Veg., Santiago 2, 54—57 (1942).

Simonneau, P.: Action de la pollinisation du Clémentinier par les autres espèces d'agrumes. Bull. Doc. Renseign. Agric., Serv. Arb. Algér., Nr. 118, 10—11 (1945).

Sokoljskaja, V. P.: (A new method of producing parthenocarpic citrus fruits.) Soviet Subtropics, Nr. 10 (74), 36—39 (1940). [Russisch.]

— (Parthenocarpy as a result of pistil removal in flowers of citrus species.) Proc. Lenin Acad. Agric. Sci. URSS., Pp. 23—24, 12—16 (1940). [Russisch.]

Swingle, W. T.: Three new varieties and two new combinations in Citrus and related genera of the orange subfamily. J. Wash. Acad. Sci. 32, 24—26 (1942).

Torres, J. P.: Progress report on citrus hybridization propagation. Philipp. J. Agric. 10, 95—119 (1939).

Trías, A.: El problema citrícola en el país. (The problem of growing citrus fruits in our country.) Rev. Fac. Agron. Univ. Montevideo, Nr. 24, 67—85 (1941).

Webber, H. J. u. L. D. Batchelor (Editors): The citrus industry. Vol. 1. History, botany, and breeding. Berkeley and Los Angeles, Pp. XX + 1028. 233 figs. 52 tables. (1943).

Wei, C. T.: (Storage diseases of sweet oranges in Szechuan.) Nanking J. 9, 239—268 (1940).

Wong, Cheong-Yin: The influence of pollination on seed development in certain varieties of citrus. Proc. Amer. Soc. Hort. Sci. 37, 161—164 (1939) 1940.

Yarnell, S. H.: Leaf segregation in Citrus-Poncirus hybrids. Proc. Amer. Soc. Hort. Sci. 40, 259—263 (1942).

Zorin, F. M.: (The seed progeny of a vegetative hybrid of Shiva-Mikan with the Citrangequat.) Jarovizacija, Nr. 5 (32), 128—129 (1940). [Russisch.]

— (First hybrid citrus-fruits.) Vestnik Gibridizacii (Hybridization), Nr. 1, 118 (1941). [Russisch.]

7. Palmen.

Anonymus: Yield recording and the issue of planting material by the Coconut Research Scheme. Leafl. Coconut Res. Scheme, Ceylon, Nr. 11, 8 (1941).

Bailey, L. H.: Palms of the Seychelles. Gentes Herbarum, Ithaca, N. Y. 6 (1), 1—48 (1942).
— Palms of the Mascarenes. Gentes Herbarum, Ithaca, N. Y. 6 (2), 51—104 (1942).
— Studies in palms. 4. Brahea, and one Erythea. 5. New palms in Panama, and other. Gentes Herbarum, N. Y. 6, 177—264 (1943).
— Revisio Palmettorum. Revision of the American palmettoes. Gentes Herbarum, Ithaca, N. Y. 6, 367—459 (1944).
Bertoni, G. T.: El mbokayá o coco Paraguay (Acrocomia totai Mart.). (Mbokayá or the Paraguay coconut [A. totai Mart.].) Rev. Minist. Agric. Comercio Industr., Paraguay 1, 36—50 (1941).
Bondar, G.: Palmeiras da Bahia. (The palms of Bahia.) Bon. Inst. Cent. Fom. Econ. Bahia, Nr. 6, 22 (1939).
— O coqueiro (Cocos nucifera L.) no Brasil. (The coconut [C. nucifera L.] in Brazil.) Bol. Inst. Cent. Fom. Econ. Bahia, Nr. 7, 100 (1939).
— Palmeiras do genero Cocos e descrição de duas especies novas. (Palms of the genus Cocos and a description of two new species.) Bol. Inst. Cent. Fom. Econ. Bahia, Nr. 9, 53 (1941).
— As cêras no Brasil e o licuri Cocos coronata Mart. na Bahia. (Waxes in Brazil and C. coronata Mart. in Bahia.) Bol. Inst. Cent. Fom. Econ. Bahia, Nr. 11, 86 (1942).
Briton-Jones, H. R.: The diseases of the coconut palm. Baillière, Tindall and Cox, London (1940).
Burret, M.: Um caso de hibridação entre Arecastrum Romanzoffianum e Butia capitata. (A case of hybridisazion between A. Romanzoffianum und B. capitata.) Rodriguesia, Rio de J. 4, 277 (1940).
— Die Palmen Arabiens. Bot. Jb. 73, 175—190 (1943).

Carabia, J. P.: Notas sobre la nomenclature de algunas palmas cubanas. (Notes on the nomenclature of certain Cuban palms.) Caribbean Forester, P. R. 6, 159—164 (1945). [Mimeographed.]
Child, R.: Coconut research scheme. Programme of experiments for 1942. Trop. Agriculturist 97, 232—235 (1941).
Cook, O. F.: A new commercial oil palm in Ecuador. Nat. Hort. Mag. 21, 70—85 (1942).
Covas, G. u. A. Ragonese: Las palmeras argentinas del género ,,Acrocomia''. (The Argentine palms of the genus Acrocomia.) Rev. Argent. Agron. 8, 1—7 (1941).

Dugand, A.: Una palma nueva del' género Desmoncus. (A new palm of the genus Desmoncus.) Caldasia, Bogota, Nr. 6, 75—76 (1943).
— Palmas nuevas o críticas colombianas. (New or critical Colombian palms.) Caldasia, Bogota, Nr. 9, 387 bis 395 (1944).
— Palmas nuevas o críticas colombianas. II. (New or critical Colombian palms. II.) Caldasia, Bogota, Nr. 10, 443—458 (1944).

Haas, A. R. C.: Chlorine accumulation in date palm varieties. Bot. Gaz. 106, 179—184 (1944).
Hodge, W. H.: A synopsis of the palms of Dominica. Caribbean Forester, Puerto Rico 3, 103—109 (1942). [Mimeographed.]
Huyskamp, A. H.: De invloed van de droogte op de vruchtdracht van den aanplant dar K. O. Talisse. (The influence of drought on the bearing of the plantation of the Talisse Coconut Estate.) Bergcultures 15, 1110—1111 (1941).

Jacob, K. C.: A new variety of coconut palm (Cocos nucifera L. var. spicata K. C. Jacob). J. Bombay Nat. Hist. Soc. 41, 906—907 (1940).
Joshi, A. C.: A palm suitable for cultivation in India Bactris utilis Benth. et Hook. F. Indian Fmg 7, 237 bis 239 (1946).

Menon, S. R. K.: Notes on the fall of immature coconuts in Ceylon. J. Coconut Industr., Ceylon 5, 87—91 (1941).

Tammes, P. M. L.: Over de ontwikkeling van de vrucht van den klapper en de factoren, welke van invloed zijn op de hoeveelheid copra per noot. (On the development of the fruit of the coconut and the factors that influence the quantity of copra per nut.) Bergcultures 14, 1101—1107; also Landbouw 1940, 16, 385—395 (1940).

8. Sonstiges tropisches und subtropisches Obst.

Addison, G.: Sexo em mamão (Carica papaya L.). (Sex in papaya [C. papaya L.].) Rev. Agric. Piracicaba 18, 448—449 (1943).
Adsuar, J.: Transmission of papaya bunchy top by a leaf hopper of the genus Empoasca. Science 103, 316 (1946).
Agete, F.: La piña. (The pineapple.) Rev. Minist. Agric. Cuba 28, Nr. 2, 24—32, 65—67 (1945).
Agnew, G. W. J.: Notes on the papaw and its improvement in Queensland. Qd Agric. J. 56, 358—373 (1941).
Alderman, W. H.: Pioneer goals passed in fruit breeding. Minn. Fm & Home Sci. 1, Nr. 3, 2—3 (1944).
Aliev, D.: (The black pomegranate.) Soviet Subtropics, Nr. 11/12 (75/76), 56 (1940). [Russisch.]
Almeida, F. J. de: Safra e contra-safra na Oliveira. (Cropping and crop failure in the olive.) Min. Agric. Lisboa Invest. Sér., Nr. 7, 154 (1940).
— Formas leucocarpas de oliveire Olea europaea L. (White fruited forms of olive, O. europaea L.) Bol. Soc. Broteriana 19, 617—634 (1945).
Anonymus: L'olivier dans le monde. Superficie-Production-Commerce de ses produits. Inst. Int. Agric., Rome (1939).
— An interesting new Fuerte sport. Yearb. Calif. Avocado Ass., p. 115 (1939).
— Pawpaw sex-linked inheritance. Fmr's Wkly, Bloemfontein 57, 1395 (1939).
— Banana mutation. — A sectorial chimera caused by mutation (in this case loss) in a colour factor. New Guinea Agric. Gaz. 6, 50—52 (1940).
— Papaya production in the Hawaiian Islands. Bull. Hawaii Agric. Exp. Sta., Nr. 87, 64 (1941).
— Avocado research. Calif. Cultiv. 89, 195 (1942).
— Developing the disease-resistant banana. Crown Colonist 12, 250 (1942).
— Brazil developing new pineapple varieties. Agric. Amer. 4, 137 (1944).
— Yearbook of the California Avocado Society for the year 1943, Pp. 103 (1944).
— Yearbook of the California Avocado Society for the year 1944, Pp. 89 (1945).
— Yearbook of the California Avocado Society for the year 1945, Pp. 123 (1946).
— Culture of the feijoa. N. Z. J. Agric. 73, 465—470 (1946).

Bailey, L. H.: Quidam Rubi tropicales: Certain Rubi of the occidental tropics. Gentes Herbarum, Ithaca, N. Y. 6, 325—364 (1944).
Baker, K. F. u. J. L. Collins: Notes on the distribution and ecology of Ananas and Pseudananas in Soutn America. Amer. J. Bot. 26, 697—702 (1939).
Baldwin, J. T. jr. u. R. Culp: Polyploidy in Diospyros virginiana L. Amer. J. Bot. 28, 942—944 (1941).

Barrett, M. F.: Ficus tsjahela. Bull. Torrey Bot. Cl. 73, 86—90 (1946).
— Ficus in Florida. 1. Australian species. Amer. Midl. Nat. 36, 412—430 (1946).
Boeuf, F.: Role de la génétique dans l'amélioration des cultures fruitières coloniales. Conf. Inst. Fruits Agrumes Coloniaux, Paris, Nr. 7, 8 (1944).
Bosman, F. H.: Spineless cactus. Fmg. S. Afr. 17, 665 bis 667 (1942).
Brieger, F. G. u. J. T. A. Gurgel: Poliembrionia em mangueira Mangifera indica, L. (Polyembryony in mango M. indica, L.) Bragantia, São Paulo 2, 481—487 (1942).
Budagovskij, V. I.: (Marga hndzôr, a dwarf apple.) Plodovo-Jagodnye Kuljtury (Fruit Crops), Nr. 3, 19 bis 23 (1940). [Russisch.]

Camp, A. F. u. H. Mowry: The cultivated persimmon in Florida. Bull. Fla Agric. Ext. Serv., Nr. 124, 36 (1945).
Carter, W.: The geographical distribution of mealybug wilt with notes on some other insect pests of pineapple. J. Econ. Ent. 35, 10—15 (1942).
Chasset, L.: La sélection des greffons. Rev. Hort, Paris 115, 378 (1943).
Chaudhuri, K.: A note on the morphology and chromosome number of Litchi chinensis Sonner. Curr. Sci. 9, 416 (1940).
Cheesman, E. E. u. K. S. Dodds: Genetical and cytological studies of Musa. IV. Certain triploid clones. J. Genet. 43, 337—357 (1942).
Chen, Wen Hsiun: (Cultivation of litchi in Putien and suggestions for improvement.) Fukien Agric. J. 3, 255 bis 267 (1941).
Chevalier, A.: Trois arbres précieux de France à améliorer: olivier, noyer, chataignier. Utilité d'en étendre la culture et de la modivserniser. Rev. Bot. Appl. 21, 206—221 (1941).
— u. J. Leroy: Les fruits coloniaux. Paris, Pp. 126 (1946).
Coit, J. E.: Germplasms of the Mexican avocado. Yearb. Calif. Avocado Ass., p. 47 (1939).
Condit, I. J.: Parthenocarpy in the fig. Proc. Amer. Soc. Hort. Sci. 36, 401—404 (1938) 1939.
— A bibliography on the avocado (Persea americana Miller). Citrus Exp. Sta., Calif., p. 293 (1939). [Mimeographed.]
— Fig characteristics useful in the identification of varieties. Hilgardia 14, 1—68 (1941).
— The Brunswick (Magnolia) fig. Proc. Amer. Soc. Hort. Sci. 39, 143—146 (1941).
— El olivo. Publ. Agric. Unión Panamericana Washington, D. C., Nos. 142/143, 46 (1942).
— The fig-variety character, flattened neck. Proc. Amer. Soc. Hort. Sci. 42, 255—258 (1943).
— San Piero, the Brown Turkey fig of California. Proc. Amer. Soc. Hort. Sci. 44, 211—214 (1944).
Cooper, W. C.: Effect of growth substances on flowering of the pineapple under Florida conditions. Proc. Amer. Soc. Hort. Sci. 41, 93—98 (1942).
Crandall, B. S.: A new species of Cephalosporium causing persimmon wilt. Mycologia 37, 495—498 (1945).
Cronquist, A.: Studies in the Sapotaceae. IV. The North American species of Manilkara. Bull. Torrey Bot. Cl. 72, 550—562 (1945).
— Studies in the Sapotaceae. VI. Miscellaneous notes. Bull. Torrey Bot. Cl. 73, 465—471 (1946).
Cruess, W. V., L. A. Hohl, M. A. Jiménez, S. Nichols-Roy u. R. Torres y Zorrilla: Estudios de laboratorio sobre guayaba. (Laboratory studies on the guava.) Alm. Minist. Agric. Argentina 2, 81—86 (1946).

Daji, J. A.: What's doing in All-India-Bombay. Indian Fmg. 5, 324—325 (1944).

Díaz y Muñoz, J. u. P. Burgos Peña: Estudio de algunas variedades de aceitunas en la cosecha 1934 bis 1935. (Study of certain olive varieties of the 1934 bis 1935 crop.) Bol. Inst. Invest. Agron. Madr. 2, Nr. 3, 241—252 (1939).
Dodds, K. S.: Genetical and cytological studies of Musa. V. Certain edible diploids. J. Genet. 45, 113—138 (1943).
— The genetic system of banana varieties in relation to banana breeding. Emp. J. Exp. Agric. 11, 89—98 (1943).
— Genetical and cytological studies of Musa. VI. The development of female cells of certain edible diploids. J. Genet. 46, 161—179 (1945).
— Musa Fehi, the indigenous banana of Fiji. Nature, Lond. 157, 729—730 (1946).
—, u. C. S. Pittendrigh: Genetical and cytological studies of Musa. VII. Certain aspects of polyploidy. J. Genet. 47, 162—177 (1946).
— u. N. W. Simmonds: Genetical and cytological studies of Musa. VIII. The formation of polyploid spores. J. Genet. 47, 223—241 (1946).
Donno, G.: L'olivicoltura in provincia di Benevento con particolare riguardo alle principali razze di olivo coltivate. (Olive cultivation in the Benevento province with particular reference to the main varieties of olive cultivated. Ann. Fac. Agrar Portici, Univ. Napoli 1942 bis 1943, 14, 308—356 (1943).
Dugand, A.: Nuevas especies de Ficus de Colombia y del Ecuador. (New species of Ficus from Colombia and Ecuador.) Caldasia, Bogota, Nr. 6, 77—80 (1943).
— Nuevas nociones sobre el género Ficus en Colombia (New ideas on the genus Ficus in Colombia.) Caldasia, Bogota, Nr. 8, 265—283 (1943).
— Nuevas nociones sobre el género Ficus en Colombia. II. (New ideas on the genus Ficus in Colombia. II.) Caldasia, Bogota, Nr. 9, 375—386 (1944).
— Nuevas nociones sobre el género Ficus en Colombia. III. (New ideas on the genus Ficus in Colombia. III.) Caldasia, Bogota, Nr. 10, 439—442 (1944).
— Nuevas nociones sobre el género Ficus en Colombia. V. (New ideas on the genus Ficus in Colombia. V.) Caldasia, Bogota 4, 112—120 (1946).
— Nuevas nociones sobre el género Ficus en Colombia. VI. (New ideas on the genus Ficus Colombia. VI.) Caldasia, Bogota 4, 229—230 (1946).
Dyal, S.: Tropical fruits. Chemical Publishing Company, Inc. Brooklyn, N. Y., Pp. 257 + X. 11 figs. tables. (1942).

Ellis, W. J. u. F. G. Lennox: Some observations on Australian-grown Papain. J. Coun. Sci. Industr. Res. Aust. 16, 166—172 (1943).
Evreinoff, V. A.: Le plaqueminier du Japon ou kaki, botanique, variétés, culture, utilisation. Rev. Hort. Paris 27, 343—346, 367—369, 399—402, 435—436, 463—466, 495—497 (1941).
— Le goumi du Japon, Eleagnus multiflora Thunbg. Rev. Hort. Paris 28, 131—132 (1942).

Fletcher, W. F.: The native persimmon. Firms' Bull. U. S. Dep. Agric., Nr. 685, 22 (1942).
Francolini, F.: Ancora sulla riproduzione agamica dell, olivo. (Once again the question of agamic reproduction in the olive.) Ital. Agric. 21, 538—542 (1942).
Fries, R. E.: Einige Gesichtspunkte zur systematischen Gruppierung der amerikanischen Annonaceen-Gattungen. Ark. Bot. 30A, Nr. 8, 1—31 (1943).

Gaddini, L. u. R. Ciferri: Il banano nell'Oasi di Derna. (The banana of the Derna Oasis.) Relaz. Monogr. Agr.-Golon., Firenze, Nr. 59, 34 (1940).
Godson, J. u. M. Channin: La guayaba. Una nueva fuente de vitamina. (The guava. A new source of vitamin C.) Rev. Minist. Agric. Cuba 29, 93—95 (1946).

Gorter, G. J. M. A.: A leaf-spot disease of the olive. Fmg S. Afr. **18**, 795—798, 801 (1943).

Graner, E. A.: Irregularidades na meiose de uma forma diploide esteril de bananeira. (Meiotic irregularities of a sterile diploid form of banana.) J. Agron., S. Paulo **2**, 1—8 (1939).

— Observações sôbre a distribuição do sexo no mamão. (Obersvations on the distribution of sex in papaya.) Rev. Agric. S. Paulo **16**, 341—357 (1941).

Großmann, H. M.: Pineapple plant selection, with special reference to the elimination of inferior types. Qd Agric. J. **52**, 27—42 (1939).

— Pineapple plant selection, with special reference to the elimination of inferior types. Qd Agric. J. **61**, 203 bis 215 (1945).

— Fruit culture. Avocado varieties. Qd Agric. J. **64**, 8—12 (1947).

Guadagnin, L.: A banancira. (The banana.) Ceres, Brasil **6**, 316—326 (1945).

Guha, M. P.: On the occurrence of a mixed inflorescence in Artocarpus integrifolia, Linn. Sci. and Cult. **11**, 99 bis 100 (1945).

Gunaratnam, S. C.: The cultivation of the mango in the dry zone of Ceylon. Trop. Agriculturist **102**, 23—30 (1946).

Haigh, J. C.: The improvement of papaw by selection-II. Trop. Agriculturist **102**, 17—22 (1946).

Hodgson, R. W.: Floral situation, sex condition and parthenocarpy in the oriental persimmon. Proc. Amer. Soc. Hort. Sci. **37**, 250—252 (1939) 1940.

— Avocado research at the University of California, Los Angeles - progress and plans. Yearb. Calif. Avocado Ass. Pp. 33—35 (1940).

Hofmeyr, J. D. J.: Sex-linked inheritance in Carica papaya L. S. Afri. J. Sci. **36**, 283—285 (1939).

— The use of colchicine in horticulture, with special reference to Carica papaya L. Fmg S.Afr. **16**, 311—312, 332 (1941).

— Sex reversal in Carica papaya L. S. Afr. J. Sci. **36**, 286—287 (1939).

— The genetics of Carica papaya. Chronica Botanica **6**, 245—247 (1941).

— Further studies of tetraploidy in Carica papaya L. S. Afr. J. Sci. **41**, 225—230 (1945).

— u. H. van Elden: Tetraploidy in Carica papaya L. induced by colchicine. S. Afr. J. Sci. **38**, 181—185 (1942).

Horn, C. L.: Existence of only one variety of cultivated mangosteen explained by asexually formed „seed". Science **92**, 237—238 (1940).

— The frequency of polyembryony in twenty varieties of mango. Proc. Amer. Soc. Hort. Sci. **42**, 318—320 (1943).

Kadam, B. S.: The newer knowledge of sex inheritance in the papaya. Indian J. Genet. Pl. Breed. **2**, 66—68 (1942).

Kalmykov, S. S.: (Extend the use of Michurin's methods in breeding subtropical crops.) Soviet Subtropics Nr. 4 (68), 26—27 (1940). [Russisch.]

— (The myrobalan in Kazahstan.) Bjull. Kazah. Naučno-issled. Inst. Zemled. im. Akad. V. R. Viljjamsa (Bull. V. R. Williams Kazah. Inst. Agric.), Nr. 1/2, 15 bis 17 (1940). [Russisch.]

Kienle, I.: Kaki, ein neues vitaminreiches Obst. Obst. u. Gem. **6**, 269 (1942).

Klein, J.: More and better cherimoyas. Calif. Cultiv. **90**, 274—275 (1943).

Kovalev, N. V.: (New varieties of myrobalan.) Sadovodstvo (Horticulture), Nr. 12, 44—45 (1939). [Russisch.]

Kumar, L. S. S.: Reserve mutation in Opuntia decumana. Curr. Sci. **11**, 338—339 (1942).

Kumar, L. S. S. u. A. Abraham: Chromosome number in Carica. Curr. Sci. **11**, 58 (1942).

—, — The papaya, its botany, culture and uses. J. Bombay Nat. Hist. Soc. 1943, Pp. 252—256 (1943).

—, — u. V. K. Srinivasan: The cytology of Carica papaya Linn. Indian J. Agric. Sci. **15**, 242—253 (1945).

— u. K. Ranadive: A cytological study of the genus Anona. J. Univ. Bombay, Sect. B. **10**, Pt. 3, 1—8 (1941).

— u. V. K. Srinivasan: Chromosome number of Carica dodecaphylla Vell Fl. Flum. Curr. Sci. **13**, 15 (1944).

Lammerts, W. E.: Progress report on avocado breeding. Yearb. Calif. Avocado Soc., Pp. 36—41 (1942).

Leroy, J. F.: Fruits tropicaux et subtropicaux d'importance secondaire. Rev. Bot. Appl. **24**, 34—50 (1944).

Leslie, W. R.: Variety notes on some tree fruits grown in prairie orchards. Fmrs' Bull. Canad. Dep. Agric., Nr. 135 (Publ., Nr. 780), 27 (1946).

Lewcock, H. K.: Pineapple culture in Queensland. Qd. Agric. J. 1939, **52**, 614—632; 1940, **53**, 6—44, 266—277 (1939—1940).

Lynch, S. J.: The Dade white-sapote. Pr. Bull. Fla Agric. Exp. Sta., Nr. 581, 4; also Calif. Cultiv. 1943, **90**, 227 (1943).

McVaugh, R.: To make tough meat tender. Agric. Amer. **3**, 134—136 (1943).

Marloth, R. H.: The mango in South Africa: I. Soil and climatic requirements, and varieties. Fmg. S. Afr. **22**, 457—463 (1947).

Matsubara, S., M. Ikeda u. K. Tezima: (On the sterility of mume.) J. Hort. Ass. Jap. **12**, 113—122 (1941).

Mauri, N.: Les figuiers cultivés en Algérie. Bull. Agric. Algérie, Nr. 93, 56 (1942).

— Essais de mise, au point d'une méthode pratique de détermination des variétés de figuier par mensuration du feuillage. Rev. Hort. Agric. Afrique Nord **47**, 8—11 (1943).

— Les figuiers cultivés en Algérie. Bull. Doc. Renseign. Agric., Serv. Arb. Algér., Nr. 93, 103 (1944).

Mazzolani, G.: La patria dell'olivo. Alcuni aspetti dello studio botanico dell'olivo (O. europaea L.). (The home of the olive. Some aspects of the botanical study of the olive [O. europaea L.].) Ital. Agric. **79**, 367—372 (1942).

Mendiola, N. B.: Introduction of tsampedak and suspected case of natural hybridization in Artocarpus. Philipp. Agric. **28**, 789—796 (1940).

— u. T. Mercado: Introduction and trial culture of Ambon bananas in the College of Agriculture. Philipp. Agric. **29**, 415—430 (1940).

Merrill, E. D.: Records of Indo-Chinese plants, III. J. Arnold Arbor. **23**, 156—197 (1942).

Miyabayashi, T.: (Varietal differences in the tannin cells of kaki fruits.) J. Hort. Ass. Jap. **12**, 143—154 (1941).

Morettini, A.: L'incremento produttivo negli olivi „Moraiolo" e „Frantoio", con l'impiego di adatte varietà impollinatrici. (Increased yield in the Moraiolo and Frantoio olives by the use of suitable pollinating varieties.) Ital. Agric. **78**, 631—639 (1941).

Mouat, H. M.: A progress report on a collection of Japanese persimmons (Diospyros kaki L. f.) at Mount Albert, Auckland. N. Z. J. Sci. Tech. **28**, Sect. A, 94—96 (1946).

Mowry, H. u. L. R. Toy (Bearbeitet von H. S. Wolfe): Miscellaneous tropical and sub-tropical Florida fruits. Bull. Agric. Ext. Serv. Fla, Nr. 109 (bearbeitet im Bull. 85), 94 (1941).

Murri, N. M.: (The best varieties of the oriental persimmon.) Soviet Subtropics, Nr. 11/12 (75/76), 33—36 (1940). [Russisch.]

Mustard, M. J.: The ascorbic acid content of some Malpighia fruits and jellies. Science 104, 230—231 (1946).
— u. S. J. Lynch: Effect of various factors upon the ascorbic acid content of some Florida-grown mangos. Bull. Fla Agric. Exp. Sta., Nr. 406, 12 (1945).
—, — Flower-bud formation and development in Mangifera indica. Bot. Gaz. 108, 136—140 (1946).

Nixon, R. W.: Date culture in the United States. Circ. U. S. Dep. Agric., Nr. 728, 44 (1945).

Ortega Nieto, J. M.: Estudio preliminar sobre la variabilidad de las características industriales de la aceituna, según su posición en el árbol y de un árbol a otro (una aplicación del ánalisis de la ,,varianza" y de la correlación). Consecuencias prácticas. (Preliminary study of the variation in the industrial characteristics of the olive according to position on the tree and from tree to tree. [An application of the analysis of variance and of correlation.] Practical consequences.) Bol. Inst. Invest. Agron, Madr., Nr. 8, 81—99 (1943).

P....., A.: Congress on olive-growing studies in Italy. Int. Rev. Agric. 33, 304T—307T (1942).
Poerck, R. de: Note contributive à l'amélioration des agrumes au Congo belge. Publ. Inst. Nat. Agron. Congo Belge, Sér. Tech., Nr. 33, 80 (1944).
Popenoe, W.: The avocado — a horticultural problem. Trop. Agriculture, Trin. 18, 3—7 (1941).
— The Avocado — A horticultural problem. Yearb. Calif. Avocado Soc., Pp. 79—85 (1941).
— Origin of the name ,,Fuerte". Calavo News 15, 5 (1941).
— Aguacates de China. Yearb. Calif. Avocado Soc., Pp. 27—32 (1942).
— El problema de la variedad en la horticultura tropical. (The variety problem in tropical horticulture.) Rev. Fac. Nac. Agron. Colombia 4/5, Nr. 16/17, 157—165 (1942).

Riche, F. G. H. le: Guava varieties in South Africa. Fmg S. Afr. 21, 9—17 (1946).
Robinson, R. A.: Diospyros Kaki. Gdnrs' Chron. 121, 4 (1947).
Rounds, M. B.: Observations on the Avocado variety situation. Yearb. Calif. Avocado Soc., Pp. 30—34 (1941).
Ruehle, G. D.: The cause and control of avocado scab. Pr. Bull. Fla Agric. Exp. Sta., Nr. 580, 4 (1943).
— The Kent and Zill mangos. Pr. Bull. Fla Agric. Exp. Sta., Nr. 614, 4 (1945).
Rževkin, A. A.: (The olive.) Vsesojuznaja Akademija Seljskohozjajstvennyh Nauk im. V. I. Lenina. Gosudarstvennyj Nikitskij Botaničeskij Sad im. Molotova Jalta 14, Nr. 3, 40 (1939). [Russisch.]

Saljkova, A. K.: (The biochemical study of the olives of Eastern Transcaucasia.) Soviet Subtropics, Nr. 8 (72) 28—30 (1940). [Russisch.]
Schrader, O. L.: Contribuicão ao estudo da tamareira no Brasil (Phoenix dactylifera L.). (Contribution to the study of the date palm in Brazil [Ph. dactylifera L.].) Bol. Minist. Agric., Rio. de J. 34, 1—101 (1945).
Schroeder, C. A.: Pollen germination in the avocado. Proc. Amer. Soc. Hort. Sci. 41, 181—182 (1942).
— Multiple embryos in the avocado. J. Hered. 35, 209 bis 210 (1944).
Schwarzenberg, F. C.: Polinización artificial del chirimoyo. (Artificial pollination of the cherimoyer.) Agric. Tec. Chile 6, 156—172 (1946).
Sen, P. K.: Black-tip disease of the mango. Indian J. Agric. Sci. 13, 300—333 (1943).

Simonet, M., R. Chopinet u. J. Baccialone: Contribution à l'étude de quelques variétés de figuiers des Alpes-Maritimes et du Var. Rev. Bot. Appl. 25, 44—72 (1945).
Snell: ,,Kaki", eine neue Obstart in Italien. Angew. Bot. 23, 124—125 (1941).
Srikantia, C. u. N. L. Kantiengar: Analysis of Raspuri and Badami varieties of mango (Mangifera indica) grown in Mysore. Proc. Indian Acad. Sci. 15, Sect. B, 280—284 (1942).
Stambaugh, S. U.: Disease resistant papaya will aid industry. Florida Gr. 49, 5 (1941).
Stevens, H. E. u. R. B. Piper: Enfermedades del aguacate en La Florida. (Diseases of the avocado in Florida.) Rev. Fac. Nac. Agron. Colombia 4/5, Nr. 16 bis 17, 11—51 (1942).
Storey, W. B.: A genetical interpretation of sex determination in Carica Papaya L. Cornell Univ. Abstr. Thes., Pp. 358—360, 1940 (1941).

Traub, H. P.: Polyembryony in Myrciaria cauliflora. Bot. Gaz. 101, 233 (1939).
—, C. S. Pomeroy, T. R. Robinson u. W. W. Aldrich: Avocado production in the United States. Circ. U. S. Dep. Agric., Nr. 620, 28 (1941).
—, T. R. Robinson u. H. E. Stevens: Papaya production in the United States. Circ. U. S. Dep. Agric., Nr. 633, 36 (1942).

Vélez, I.: Wild pineapples in Venezuela. Science 104, 427—428 (1946).
Venkataramani, K. S.: Studies of Indian bananas. I. A descriptive study of twenty-four varieties. Proc. Indian Acad. Sci. 23, Sect. B, 113—128 (1946).
— ,,Kaio" an imported banana variety. Curr. Sci. 15, 110 (1946).
Vidal, V. Canhoto u. I. Costa Netto: Azeites elementares. (Individual olive oils.) Agron. Lusitana 7, 109 bis 120 (1945).
Vinson, C. G. u. F. B. Cross: Vitamin C content of persimmon leaves and fruits. Science 96, 430—431 (1942).

Wardlaw, C. W.: Cercospora leaf spot disease of bananas Nature, Lond. 144, 11—14 (1939).
— The banana in Central America. 1. Cultivation. Nature, Lond., 313—316 (1941).
— Banana research at the Imperial College of Tropical Agriculture, Trinidad, B. W. I. J. R. Soc. Arts. 90, 644—655 (1942).
Webber, H. J.: The vitamin C content of guavas. Proc. Amer. Soc. Hort. Sci. 45, 87—94 (1944).
Wilson, G. B.: Cytological studies in the Musae. I. Meiosis in some triploid clones. Genetics 31, 241—258 (1946).
— Cytological studies in the Musae. II. Meiosis in some diploid clones. Genetics 31, 475—482 (1946).
— Cytological studies in the Musae. III. Meiosis in some seedling clones. Genetics 31, 483—493 (1946).
Wolfe, H. S. u. S. J. Lynch: Papaya culture in Florida. Bull. Agric. Ext. Serv. Fla, Nr. 113, 35 (1942).
—, L. R. Toy u. A. L. Stahl: Avocado production in Florida. Bull. Fla. Agric. Ext. Serv., Nr. 129, 107 (1946).

Young, T. W.: Investigations on the unfruitfulness of the Haden mango. (Mangifera indica, Linn.) in Florida. Abstr. Thes. Cornell Univ. 1942, Pp. 483—487 (1943).

Zimmerman, G. A.: Further report on the papaw. Rep. Proc. 31st Annu. Mtg. Nth Nut Gr. Ass. 1940, Pp. 133 bis 136 (1940).
— Hybrids of the American papaw. J. Hered. 32, 83—91 (1941).

Zirkle, C.: The origin of the banana. Morris Arbor. Bull. Univ. Pa. **3**, 69—71 (1941).
Zorin, F. M.: (Michurinites in the Soviet Subtropics.) Jarovizacija, Nr. 3 (30), 164—170 (1940). [Russisch.]

9. Weinbau.

Akman, A.: Ortaanadolu ve bilhassa Ankara mintakasi saraplari üzerinde araştirmalar. (Researches on viticulture in Central Anatolia, with special reference to the Ankara region.) T. C. Yüksek Ziraat Enstitüsü Çalismalarindan, Ankara, Nr. 116, 63 (1941).

Alstyne, L. M. van: Promising new grapes: some of the best of the new red kinds. Fm. Res., N. Y. St. Sta. **6**, 12 (1940).

Amerine, M. A. u. A. J. Winkler: Grape varieties for wine production. Circ. Calif. Agric. Exp. Sta., Nr. 356, 15 (1943).

—, — Composition and quality of musts and wines of California grapes. Hilgardia **15**, 493—673 (1944).

Anliker, J. u. F. Kobel: Wuchsstoffversuche mit Rebveredlungen. Landw. Jb., Schweiz **46**, 203—248 (1945).

Anonymus: Amerikanische Pflanzenpatente Nr. 195, 145—155. Züchter **11**, 255—257 (1939).

— Grape cultivation in India. Curr. Sci. **11**, 78 (1942).

— Aussichten der Züchtung frostresistenter Rebenformen. Forschungsdienst **15**, 288 (1943).

— New varieties of fruits. Gdnrs' Chron. **113**, Nr. 2927, 42 (1943).

— White Niagara grape produces a „sport". Agric. Amer. **4**, 97 (1944).

— Grape breeding. J. N. Y. Bot. Gdn. **46**, 252 (1945).

— Hybrid grapes. J. N. Y. Bot. Gdn. **47**, 280 (1946).

— Nationwide fruits. Grapes. Amer. Fruit Gr. **67**, Nr. 6, 24 (1946).

Batista, A. de J.: Santiago-Uma bôa uva para córte. (Santiago-a good grape for mixing.) Rev. Agron., S. Paulo **7**, 554 (1943).

Bergman, H. F. u. C. A. Magoon: The tartrate content of Maryland-grown American grape varieties. Proc. Amer. Soc. Hort. Sci. **46**, 253—255 (1945).

Billeau, A.: Experiente si cercetări asupra înrădăcinării port-altoilor Americani si Europeo-American. (Experiments and researches on root formation in American and European-American rootstocks.) Anal. Inst. Cerc. Agron. Român **10**, 464—472, 1938 (1939).

Börner, C.: Parasitäre Spezialisation und pflanzliche Immunität nach Untersuchungen über die Reblaus. Verh .VII Int. Kongr. Ent. **4**, 2279—2290 (1939).

— Anfälligkeit, Resistenz und Immunität der Reben gegen Reblaus. Allgemeine Gesichtspunkte zur Frage der Spezialisierung von Parasiten; die harmonische Beschränkung des Lebensraumes. Z. hyg. Zool. **31**, 274—285, 301—308, 325—334 (1939).

— Die ersten reblausimmunen Rebenkreuzungen. Angew. Bot. **25**, 126—143 (1943).

Breider, H.: Untersuchungen zur Vererbung der Widerstandsfähigkeit von Weinreben gegen die Reblaus, Phylloxera vastatrix Planch. I. Das Verhalten von F_3-Generationen, die aus Selbstungen von widerstandsfähigen und anfälligen F_2-Bastarden gewonnen wurden. Z. Pflanzenz. **23**, 145—168 (1939).

— Morphologisch-anatomische Merkmale der Rebenblätter als Resistenzeigenschaften gegen die Reblaus, Phylloxera vastatrix Planch. Züchter **11**, 229—244 (1939).

— Über Pollenfertilität der Rebenarten und ihrer F_1-Bastarde. Züchter **12**, 209—212 (1940).

Constantinescu-Ismail, G.: Selectiunea în viticultură cu privire specială la varietățile Românești. (Selection in viticulture with particular respect to the Rumanian varieties.) Anal. Inst. Cerc. Agron. Român **10**, 416—434, 1938 (1939).

Cosmo, I.: Rilievi ampelografici comparativi su varietà di „Vitis vinifera". I vitigni bordolesi. (Comparative ampelographic notes on varieties of V. vinifera. The Bordeaux vines.) Ital. Agric. **77**, 473—482 (1940).

— Varietà di uve da tavola da consigliare nelle Venezie. (Dessert grape varieties recommended for the Venetian provinces.) Ital. Agric. **78**, 589—601 (1941).

Covas, G. u. J. R. Christensen: Determinación del tamaño de parcelas para ensayos comparativos de rendimientos en la vid. (Determination of the size of plots for comparative yield tests in vine.) Rev. Argent. Agron. **12**, 26—29 (1945).

Cowart, F. F. u. E. F. Savage: An evaluation of certain grape varieties for use as rootstocks. Proc. Amer. Soc. Hort. Sci. **44**, 315—318 (1944).

Dalmasso, G.: Mutabilità delle diverse varietà di viti. (Variability of the different vine varieties.) Annu. Staz. Sper. Vitic. Enologia Conegliano 1938—1939, **9**, 135—152 (1939).

— La genetica nella citicoltura moderna. (Genetics in modern viticulture.) Genetica Agraria, Roma, **1**, 187—203 (1947).

— u. I. Cosmo: II. Contributo allo studio della biologia fiorale della vite. (Second contribution towards the study of the floral biology of the vine.) Annu. Staz. Sper. Vitic. Conegliano 1940—41, **10**, 311—379 (1941).

Depardon, L. u. P. Buron: Les vins d'hybrides producteurs directs. C. R. Acad. Agric. Fr. **32**, 182—183 (1946).

Dubois, J.: Le vigneron vaudois et ses vins. Lausanne, Pp. 261, 4 tables. plates. (1944).

Dunne, T. C.: Pollen-containing sprays for the cross-pollination of Ohanez grapes. J. Dep. Agric. W. Aust. **19**, 210—213 (1942).

Faes, H.: Lexique viti-vinicole international Francais-Italien, Espagnol-Allemand. Office International du Vin, Paris, Pp. 278 (1940).

— Les producteurs directs à raisins rouges dans notre vignoble. Publ. Sta. Fédérale Essais Viticoles Arboricoles, Montagibert, Lausanne, Nr. 338, 12 (1944).

Fennell, J. L.: Future „ideal" grapes. J. Hered. **32**, 193—197 (1941).

— Timing and production of grape pollen by grafting. J. Hered. **36**, 183—185 (1945).

— The tropical grape. Sci. Mon. N. Y., Pp. 465—468 (1945).

Ferraz do Amaral, J.: A videira „Golden Queen". (The vine „Golden Queen".) O Biológico, Sao Paulo **8**, 47—50 (1942).

— Videira híbrida com qualidades de „Vitis vinifera". (A hybrid vine with the qualities of V. vinifera). O Biológico, Sao Paulo **8**, 80—82 (1942).

— „Vitis vinifera"- para vinhos brancos de tipo e gosto europeu. (V. vinifera for white wines of European type and flavour). O Biológico, Sao Paulo **8**, 110—112 (1942).

Flerov, A. F. u. E. I. Kovalenko: (Der Einfluß von Wuchsstoffen und Alkaloiden auf die Entwicklung der Stecklinge und die Keimung des Weines.) C. R. (Doklady) Acad. Sci. URSS. **58**, 677 (1947). [Russisch.]

Gollmick, F.: Über die Lebensdauer des Rebenpollens. Angew. Bot. **24**, 221—232 (1942).

— u. F. Hilpert: Untersuchungen über die Frosthärte der Reben und Obstgewächse. Wiss. Jber. Biol. Reichsanst. Land- u. Forstwirtsch. 1940. Mitt. Biol. Reichsanst. 1941, Heft 65, 61—62 (1941).

Gračanin, M.: Da li je propadanje loze na podlozi Aramon × Rupestris Ganzin 1 edafski uvjetovano. (Is the degeneration of the vines grafted on Aramon × Rupestris Ganzin 1 edaphically conditioned.) Arhiv Minist. Poljopr. **6**, 3—26 (1939).

Gray, G. F.: Resistance of grape varieties to the freeze of November 1940. Proc. Amer. Soc. Hort. Sci. **40**, 329—331 (1942).

Husfeld, B.: Genetik und Rebenzüchtung. Agron., Lusitana **1**, 200—235 (1939).
— Forschungsdienst und Ergebnisse im Weinbau unter besonderer Berücksichtigung üer Rebenzüchtung. Forschungsdienst, Sonderh. 16, 519 (1942).
— Die züchterischen Möglichkeiten in Menge und Güte des Ertrages bei interspezifischen Vitis-Kreuzungen. Wein u. Rebe **25**, 4—28 (1943).

Jacob, H. E.: Grape growing in California. Circ. Calif. Agric. Exp. Sta., Nr. 116, 80 (1940).

Kaczmarek, A. u. R. Weise: Physikalisch-chemische Untersuchungsmethoden für die Rebenzüchtung. 1. Über das Absorptionsspektrum als Hilfsmittel bei der Rotweinzüchtung. Gartenbauwiss. **16**, 314—357 (1941).

Kawe, A. u. O. v. Veh: Ist die Gesamtsäure ein zuverlässiger Anhaltspunkt bei der Auslese in der Rebenzüchtung? Weinland **11**, 72 (1939).

Kertesz, Z. I.: The chemical composition of maturing New York State grapes. Tech. Bull. N. Y. St. Agric. Exp. Sta., Nr. 274, 13 (1944).

Kobel, F.: Ein Beitrag zur Sortenzüchtung bei der Europäerrebe. Landw. Jb. Schweiz **54**, 807—815 (1940).

Krimbas, B. D.: (Ampelography of Greece.) Ministère de l'Agriculture, Athènes, **1**: 266. 150 pls.; 1944: **2**: 189, 135 pls.

Kuzmin, A. J.: (Training vine seedlings.) Plodovo-Jagodnye Kuljtury (Fruit Crops), Nr. 5, 46—52 (1940). [Russisch.]
— "Training" of grape seedlings. Bull. Acad. Sci. URSS. Sér. Biol., Nr. 5, 802—809 (1940).

Lattin, G. de: Über den Ursprung und die Verbreitung der Reben. Züchter **11**, 217—225 (1939).
— Spontane und induzierte Polyploidie bei Reben. Züchter **12**, 225—231 (1940).

Le Roux, M. S.: Grape varieties for the local market. Fmg. S. Afr. **18**, 397—400 (1943).

Leyvraz, H.: Les vignes hybrides dites producteurs directs (PD). Publ. Sta. Fédérale Essais Viticoles Arboricoles Lausanne-Pully, Nr. 322, 16 (1943).
— Expériences dans la culture des raisins de table. Publ. Sta. Fédérale Essais Viticoles Arboricoles Chimic Agric., Montagibert, Lausanne, Nr. 350, 12 (1946).

Loomis, N. H.: Drying of American-type grapes. Proc. Amer. Soc. Hort. Sci. **47**, 195—198 (1946).

Luttrell, E. S.: Black rot of muscadine grapes. Phytopathology **36**, 905—924 (1946).

Magoon, C. A. u. E. Snyder: Grapes for different regions. Fmrs' Bull. U. S. Dep. Agric., Nr. 1936, 38 (1943).
—, — Grape regions of the United States. Proc. Amer. Soc. Hort. Sci. **42**, 425—431 (1943).

Makarevskaja, E. A. u. K. M. Iluridze-Molčan: Catalase in vine schoots. C. R. (Doklady) Acad. Sci. URSS. **31**, 614—616 (1941).

Massey, A. B.: Native grapes of Virginia. Proc. Va. Acad. Sci. **2**, 179, 1940—1941 (1941).

Moisejev, K. A.: (Regarding forms of Vitis amurensis Rupr. with hermaphrodite flowers.) Proc. Lenin Acad. Sci. USSR., Nr. 3, 16—18 (1941). [Russisch.]

Moog, H.: Warum ist eine amtliche Sortenbezeichnung im Weinbau notwendig? Weinland **11**, 89—93 (1939).

Moore, R. C.: The grape breeding program at V. P. I. Virginia Fruit **34**, 20—22 (1946).
— The present situation with reference to new grape varieties. Va Fruit **35**, 27—31 (1947).

Negrulj, A. M.: (The evolution of cultivated grapes.) Priroda (Nature), Nr. 4, 37—46 (1940). [Russisch.]
— u. J. N. Kondo: (Heredity of cold resistance in buds of hybrid grapes.) Proc. Lenin Acad. Sci. USSR., Nr. 23/24, 13—17 (1939). [Russisch.]

Nikolaišvili, D.: The Viticulture and Wine-making Research Institute (Georgian Academy of Sciences). Agricultural Chron. Moscow, Nr. 12, 8—11 (1946). [Mimeographed.]

Nicolas, G.: Résistance des vignes américaines au phylloxéra et sensibilité de leurs cellules aux phytohormones. C. R. Acad. Agric. Fr. **32**, 444—446 (1946).

Oberle, G. D.: An evaluation of the use of French hybrid wine grapes in breeding hardy grapes for the eastern United States. Proc. Amer. Soc. Hort. Sci. **42**, 413 bis 417 (1943).
— New grapes bred to order. Rural New-Yorker **103**, 436, 476 (1944).

Olmo, H. P.: Breeding new grape varieties. Wine Rev. **7**, 8—10, 32 (1939).
— La caryologie des Vitis et ses applications à la création de nouvelles variétés. Rev. Hort. **26**, 557—558 (1939).
— Somatic mutation in the vinifera grape. III. The Seedless Emperor. J. Hered. **31**, 211—213 (1940).
— The use of seed characters in the identification of grape varieties. Proc. Amer. Soc. Hort. Sci. **40**, 305—309 (1942).
— Choice of parent as influencing seed germination in fruits. Proc. Amer. Soc. Hort. Sci. **41**, 171—175 (1942).
— Storage of grape pollen. Proc. Amer. Soc. Hort. Sci. **41**, 219—224 (1942).
— Breeding new tetraploid grape varieties. Proc. Amer. Soc. Hort. Sci. **41**, 225—227 (1942).
— Pollination of the Almeria grape. Proc. Amer. Soc. Hort. Sci. **42**, 401—406 (1943).
— u. A. D. Rizzi: Selection for fruit color in the Emperor grape. Proc. Amer. Soc. Hort. Sci. **42**, 395—400 (1943).

P....., E.: Tokayer oder Ruländer? Schweiz, Z. Obst- u. Weinb. **51**, 25—26 (1942).

Pirovano, A.: Mitteilungen über europäische Reben, die sich als resistent gegenüber Phylloxera und teilweise resistent gegenüber Peronospora erwiesen haben. Wein u. Rebe **21**, 144—153 (1939).
— Nuove uve da tavola redditizie senza semi. (Profitable new seedless dessert grapes.) L'Ortofruttic. Italiana **8**, 173—176 (1939).
— Origini della vite e possibilità di raggruppamenti continentali. (Abstammung [Ursprung] der Rebe und Möglichkeit der kontinentalen Gruppierung.) Italia agricola **80**, 61—65 (1943).

Popenoe, W.: Grapes for tropical America. Trop. Agriculture Trin. **19**, 23—28 (1942).

Potapov, A.: (Growing grapes in the province of Kharkov.) Socialističeskoe Zemledelie (Socialist Agriculture), Nr. 180, 2 (1945). [Russisch.]

Pozzi-Escot, E.: Un cas de xénie chez la vigne. C. R. Acad. Sci., Paris **208**, 1046 (1939).

Prosperi, V.: Contributo alla produzione e studio di nuove varietà di uve da tavola. (Contribution towards the production and study of new varieties of dessert grapes.) Ital. Agric. **77**, 755—766 (1940).
— Nuovi vitigni portainnesti di produzione italiana. (New vine root stocks produced in Italy.) Ital. Agric. **78**, 165—170 (1941).

Rodrian u. Binstadt: Bericht über Frostschäden an Reben im Winter 1939/40 in den deutschen Weinbaugebieten. Wein und Rebe **23**, 231—277 (1941).

Rodrigues, A.: A contribução da histologia para a resolução dos problemas taxonómicos no género „Vitis". (The contribution of histology to the solution of taxonomic problems in the genus Vitis.) Rev. Agron., Lisboa **27**, 200—213 (1939).

Rodrigues, A.: Sôbre a caracterização das espécies e híbridos do género Vitis. Um novo método ampelométrico. (Describing the characteristics of species and hybrids of the genus Vitis. A new ampelometric method.) Agron. Lusitana 1, 315—326 (1939).
— Variações do recorte da fôlha da videira. (Variations in the outline of the vine leaf.) Agron. Lusitana 3, 189—193 (1941).
— Acérca do valor taxonómico do número de dentes da fôlha na separaçao de dois hibridos do género Vitis L. (On the taxonomic value of the number of teeth on the leaf in separating two hybrids of the genus Vitis L.) Agron., Lusitana 3, 325—340 (1941).
— Sôbre o recorte e assimetria da fôlha da videira. (The outline and asymmetry of the vine leaf.) Agron., Lusitana 4, 137—153 (1942).
— O número e a disposição dos feixes foliares nas suas relações com a forma fôlha da videira. (The number and disposition of the foliar bundles in their relations to the form of the leaf of the vine.) Bol. Soc. Broteriana 19, 635—654 (1945).
Roux, M. S. le: Sunscald in table grapes. Fmg. S. Afr. 21, 506—510 (1946).

Sartorius, O.: Vererbungsstudien an der Weinrebe mit besonderer Berücksichtigung des Blattes. Gartenbauwiss. 16, 12—23 (1941).
Schellenberg, H.: Der Riesling × Sylvaner. Schweiz, Z. Obst- u. Weinbau 53, 24—32 (1944).
Scherz, W.: Sind selbstfertile hermaphrodite Weinreben obligat autogam? Züchter 11, 244—249 (1939).
— Die Mutationen der Rebe, ihre Bedeutung und Auswertung für die Züchtung. Wein u. Rebe 22 (4), 73 bis 86 (1940).
— Über somatische Genommutanten der Vitis vinifera-Varietät „Moselriesling". Züchter 12, 212—225 (1940).
— Die Aussichten züchterischer Bekämpfung von Winterfrostschäden der Weinrebe. Wein u. Rebe 25, 43—60 (1943).
— u. J. Seemann: Schäden an Reben durch Spätfröste und die Aussichten ihrer züchterischen Bekämpfung. Züchter 16, 25—35 (1944).
Scheu, G.: Die Frostschäden des Jahres 1939/40. Ein Beitrag zur Frage der Frostfestigkeit unserer Rebsorten vom Standpunkt des Rebenzüchters. Wein und Rebe 24, 47—64, 79—87 (1942).
Scheu, H.: Beobachtungen an F_2-Populationen interspezifischer Rebenkreuzungen. Züchter 11, 225—229 (1939).
Ségal, L.: Sélection de vignes résistantes aux parasites en Allemagne. Rev. Vitic., Paris 90, 473—475 (1939).
Smith, M. B. u. H. P. Olmo: The pantothenic acid and riboflavin in the fresh juice of diploid and tetraploid grapes. Amer. J. Bot. 31, 240—241 (1944).
Snyder, E. u. F. N. Harmon: Grape progenies of self-pollinated vinifera varieties. Proc. Amer. Soc. Hort. Sci. 37, 625—626 (1939) 1940.
—, — „Synthetic" Zante currant grapes. Breeding investigations indicate possible origin, and point way toward production of new varieties. J. Hered. 31, 315—318 (1940).
—, — Drying different vinifera grape varieties for raisins. Proc. Amer. Soc. Hort. Sci. 44, 201—204 (1944).
—, — Temperature and maturity in relation to raisin production. Proc. Amer. Soc. Hort. Sci. 46, 249—252 (1945).
Sousa, J. S. I. de: É possivel a cultura de uvas finas para mesa em São Paulo? (Is it possible to grow high quality dessert grapes in Sao Paulo?) Rev. Agric. Piracicaba 21, 249—263 (1946).
Sousa, L. de G. M. da C. e: Casos de sui produtividade nalgumas casta de uvas de mesa. (Cases of self-fertility in certain varieties of dessert grapes.) Rev. Agron., Lisboa 30, 446 (1942).

Steuk, W. K.: Variations of the Catawba grape. Bi-m. Bull. Ohio Agric. Exp. Sta. 30, 31—33 (1945).
Stout, A. B.: Progress in breeding for seedless grapes. Proc. Amer. Soc. Hort. Sci. 37, 627—629 (1939) 1940.
— The Interlaken Seedless grape. J. N. Y. Bot. Gdn. 48, 92—94 (1947).
Stummer, A.: Züchtungsergebnisse mit der Vinifera-Rebensorte Früher Malingre. Gartenbauwiss. 16, 358—370 (1941).

Teodorescu, I. C.: Clasificarea și recunoașterea soiurilor de vițe roditoare din podgoriile Românesti. (Classification and identification of varieties of cines of Rumanian vineyards.) Anal. Inst. Cerc. Agron. Român. 11, 222—237, 1939 (1940).
Tuneu, R. u. A. Guerra: Primera contribución al estudio ampelográfico de las variedades de vid cultivadas en el Uruguay. (First contribution to the ampelographic study of the vine varieties cultivated in Uruguay.) Rev. Fac. Agron. Univ. Montevideo Nr. 22, 67—106 (1940).

Vasconcellos, J. de Carvalho e, L. Santa Barbara u. A. Baptista: Castas de videira. Seu estudo botanico. (Varieties of vine. Their botanical study.) Rev. Agron., Lisboa 29, 177—227, 353—403, 486 bis 523, 1942: 20, 91—141 (1941/1942).
— Castas de videira: seu estudo botanico. (Varieties of vine: their botanical study.) Rev. Agron., Lisboa 30, 214—275 (1942).
Vitolović, V.: Ispitivanje sorata vinove loze u kalničkokriževačkom vinogorju. (Testing the vine varieties in the Kalnik vineyard regions of Western Croatia.) Arhiv Minist. Poljopr. 6, 69—79 (1939).

Wellington, R.: The Ontario grape and its seedlings as parents. Proc. Amer. Soc. Hort. Sci. 37, 630—634 (1939) 1940.
Wilcox, A. N., W. H. Alderman u. F. E. Haralson: New grape varieties named. Minn. Hort. 72, 4—6 (1944).

Zillig, H.: Die Frostwiderstandsfähigkeit der Rieslingrebe und anderer wirtschaftlich wichtiger Vitis-Varietäten. Wein u. Rebe 23, 99—114 (1941).
Zimmermann: Die Aufgabe der Rebenzüchtung in Baden. Forschungsdienst 11, 631—635 (1941).

s) Gemüse.
1. Gemüse, Allgemeines.
Andersen, E. M. et al.: Commercial vegetable varieties for Florida. Bull. Fla. Agric. Exp. Sta., Nr. 405, 30 (1944).
Anonymus: The way to better seeds. J. E. Ohlsens Enke, Copenhagen, Pp. 43.
— Objectives in vegetable breeding. Trop. Agriculturist 95, 344—349 (1940).
— (The history of a vegetative hybrid.) Jarovizacija, Nr. 6 (33), 102—103 (1940). [Russisch.]
— Vegetable and flower variety and breeding field days August 14 and 15, 1941. Market Gr. J. 69, 390 (1941).
— Research work and workers. Seed World 52, Nr. 10, 36 (1942).
— Restrict vegetable varieties. Seedsmen accept voluntary program to cut down on number of varieties as a war measure. USDA asks help of whole trade. Seed World 53, Nr. 1, 12—14 (1943).
— The production of seed of root crops and vegetables. Imperial Agricultural Bureaux Joint Publication, Great Britain, Nr. 5, 93 (1943).
— Research work and workers. Seed World 55, 24 (1944).

Anonymus: Breeding better vegetables for the South at the U.S. Regional Vegetable Breeding Laboratory. Misc. Publ. U. S. Dep. Agric., Nr. 578, 34 (1945).
— Gribovo selectionists. Agricultural Chron., Moscow, Nr. 4, 2—3 (1946). [Mimeographed.]
— A Bulgarian Burbank. Free Bulgaria 2, Nr. 1, 14 (1947).
Arróniz Sala, C.: Manera de realizarse la fecundación de las flores de algunas plantas hortícolas autógamas. (Mode of fertilization in the flowers of certain self-pollinated horticultural plants.) Bol. Inst. Invest. Agron., Madr., Nr. 9, 129—147 (1943).
Auchter, E. C.: A background for victory. Research in nutritional values of vegetables provides a basis for avoiding mistakes of „war garden" campaign of 1917—1918. Seed World 51, 34—35 (1942).

Barrons, K. C.: New varieties for special crops. Rep. Proc. 24th Annu. Conv. Mich. Muck Fmrs' Assoc., P. 31—33 (1942).
— What's cookin in hybrid vegetables. Sth. Seedsman 6, 14, 42, 54 (1943).
— Vegetable varieties for commercial production in Michigan. Circ. Mich. Agric. Exp. Sta., Nr. 191, 35 (1944).
Bindloss, E. A.: Cell lengths in the terminal meristematic region of the stem as related to tallness and dwarfness. Amer. J. Bot. 27, Nr. 10, Suppl. 2s (1940). (Abstr.)
Binkley, A. M.: Plant breeding problem of station. Background of breeding work in horticulture section is outlined. Colo. Fm. Bull. 2, 8—10 (1940).
Bremer, A. H.: Daglengd og grønsakdyrking. (Length of day and vegetable growing.) Meld. Stat. Forsøksgard i Grønsakdyrking Kvithamar i Stjørdal, Oslo 1943, 24, G5—G48 (1944).
Bugakov, A.: (The creators of new varieties.) Socialističeskoe Zemledelje (Socialist Agriculture), Nr. 119, 2 (1944). [Russisch.]

Connolly, F., M. C. Hiltz u. A. D. Robinson: Thiamin in Manitoba vegetables. Canad. J. Res. 25, 43—53 (1947).
Cooper, J. F.: Florida crops „styled" to modern design. New disease-resistant strains, developed to meet state's requirements, gain favor with growers and consumers. Sth., Seedsman 6, Nr. 1; 13, 16—17 (1943).

Elden, H. van: The production of vegetable seed. Fmg. S. Afr. 17, 425—430 (1942).
— Production of vegetable seed. Fmg. S. Afr. 17, 807—808 (1942).
Ermakov, A. I.: (Biochemical changes in grafted plants.) Vestnik Socialističeskogo Rastenievodstva (Soviet Plant Industry Record), Nr. 2, 57—67 (1940). [Russisch.]

Fischer, A.: Über die Herkunft der Gemüsearten. Naturwissenschaften 27, 205—209 (1939).
Fridström, A.: Odlingsvärda köksväxtsorter. (Vegetable varieties worth cultivating.) Fruktodlaren, Nr. 1, 10—13 (1942).
Frimmel, F. v.: Die Heterosis-Frage im Gemüsebau. Obst- u. Gemüsebau 88, 24—25 (1942).
Frisak, A.: Frøavl av grønnsaker og rotvekster. (The raising of seed of green vegetables and root crops.) Grøndahl and Søns Forlag Oslo, Pp. 252, 145 Figs., 4 plates. 1943 (1945).
Frolov, I.: (Experiments at the State farm „Lesnoï".) Sovhoznoe Proizvodstvo (State Farming), Nr. 8/9, 29—30 (1944). [Russisch.]

Garcia, J. A.: Plant breeding a very important factor in the growing of garden seed. Seed World 47, Nr. 6, 24—25 (1940).

Gimesi, N.: A növénysejt osztódásának mechanikája. (Mechanics of plant-cell division.) Bull. Hung. Coll. Hort. 8, 142—155 (1942).
Godunova, P. M.: (Breeding, seed production and cultivation of legumes.) Naučnyj Otčet Inst. Zernovogo Hozjajstva Jugo-Vostoka SSSR. za 1941—42 gg. (Sci. Rep. Inst. Grain Husbandry South-Eastern USSR. for 1941—42), Pp. 215—225 (1944). [Russisch.]

Haskell, R. J.: Vegetables made to order. Sth. Seedsman 7, 13, 38—39 (1944).
— u. V. R. Boswell: Disease-resistant varieties of vegetables for the home garden. Leafl. U. S. Dep. Agric. (Revised 1943), Nr. 203, 8, 1940 (1943).
Hastings, D.: Seedsmen visit vegetable laboratory. Seed World 50, Nr. 1, 36—37 (1941).
Hastings, W. R.: The All-Americas... they are important. 1943 winners should be featured by dealers on counters, in cataloga, in advertising. Sth. Seedsman 5, Nr. 11, 20, 24 (1942).
Hawkins, F.: Pollen control in vegetable crops. E. Afr. Agric. J. 10, 90—91 (1944).
Hawthorn, L. R.: Tests of vegetable varieties for the winter garden region, 1937—1941. Bull. Tex. Agric. Exp. Sta., Nr. 626, 50 (1943).
Herklots, G. A. C.: Vegetable cultivation in Hong Kong. Hong Kong, Pp. 208, 86 figs. tables. (1947).
Hoare, A. H.: Vegetable crops for market. London, Pp. 188, 34 figs. tables. (1945).
Hoffman, I. C.: Progress in greenhouse vegetable breeding. Proc. 25th Annu. Mtg. Ohio Veg. Potato Gr. Ass., Pp. 103—106 (1940).

Jansen, J. A. u. I. Rietsema: Spruitkoolvariëteiten op zandgrond. Verslag eener vergelijkende proef in 1938 in den proeftuin der R. K. Landen Tuinbouw School te Breda. (Brussels sprout varieties on sandy soil. Report of a comparative trials in 1938 in the experimental plots of the R. K. Agricultural and Horticultural College at Breda.) Meded. Tuinbouw VoorlichtDienst, 's-Gravenhage, Nr. 12, 28 (1939).
Jenkins, J. M. jr.: Vegetable improvement problems in the South. Market Gr. J. 66, 115—116 (1940).
Jones, H. A.: Progress in the development of vegetable varieties resistant to disease. Chronica Botanica 5, 381—383 (1939).

Kämpfer, M.: Entwicklung und gegenwärtiger Stand des Qualitätsbegriffs bei Gemüse. Landw. Jb. 93, 523—662 (1944).
Kappert, H.: Die Vererbungswissenschaft in der gärtnerischen Pflanzenzüchtung unter besonderer Berücksichtigung der Blumenzüchtung. Forschungsdienst 10, 533—545 (1940).
Koopman, C.: De bestrijding der vetvlekkenziekte. (The control of grease spot disease.) Tijdschr. PlZiekt. 50, 62—68 (1944).
Kopetz, L. M.: Die praktischen Auswirkungen bisheriger photoperiodischer Untersuchungen bei Gemüse. Gartenbauwiss. 16, 178—187 (1941).
Krevčenko, L. E.: (A case of heterogeneity in the first hybrid generation.) Jarovizacija, Nr. 3 (36), 111—112 (1941). [Russisch.]
Krickl, M.: Spätaustreiben — relativ geringer Gewichtsverlust — hoher osmotischer Wert. Gartenbauwiss. 17, 51—90 (1942).
Kristensen, R.: Indskraenkning af Antallet af Sorter og Stammer af Køkkenurter. (Limiting the number of varieties and strains of vegetables.) Nord. JordbrForskn., Nos. 5/6, 170—176 (1943).

Lamm, R.: Årsskrift från Alnarps Lantbruks-, Mejerioch Trädgårdsinstitut 1938. Redogörelse för stamförsök och statskontroll av köksväxtstammar vid statens trädgårdsförsök ar 1937. (Annual Report of the Alnarp Agricultural, Dairying and Horticultural Institute 1938 — Report on trials of strains and state control of vegetable strains in the state horticultural trials in the year 1937.) Malmö, p. 166 (1939).

Lamm, R.: Sammanfattning över de jämförande sort och stamförsöken med köksväxter vid Alnarp ar 1938. (Summary of the comparative trials with varieties and strains of vegetables at Alnarp in the year 1938. Arsskrift från Alnarps Lantbruks-, Mejeri- och Trädgårdsinstitut, Malmö (1939) 1940, Pp. 1—24 (1940).

— Redogörelse för stamförsök och statskontroll av köksväxtstammar vid statens trädgårdsförsök år 1938. (Report on trials of strains and state control of vegetable strains in the state horticultural trials in the year 1938.) Årsskrift från Alnarps Lantbruks-, Mejeri- och Trädgårdsinstitut, Malmö (1939) 1940, Pp. 25 bis 134 (1940).

— Nyare undersökningar på köksväxtodlingens område. (Recent investigations on vegetable culture.) Sverig. Pomol. Fören. Årsskr. **43**, 87—106 (1942).

— Nyare undersögningar på köksväxtodlingens område. (Recent investigations in the sphere of vegetable growing.) Sverig. Pomol. Fören. Årsskr. **44**, 88—98 (1943).

— Nyare danska sort- och stamförsök med köksväxter. (Recent Danish trials of varieties and strains of vegetables.) Årsb. Jordbruksforskning, Stockholm, Pp. 136 bis 146 (1946).

— u. S. E. Lenander: Redogörelse för stamförsök och statskontroll av köksväxtstammar vid Statens trädgårdsförsök år 1939. (Report on trials of strains and state control of vegetable strains in the State horticultural experiments, 1939.) Årsskr. Alnarps Landbr. Mej.-och TrädInst., Pp. 1—92 (1940).

—, — u. B. Hylmö: Sort- och stamförsök med köksväxter vid Alnarp 1937—1940. (Variety and strain trials with vegetables at Alnarp 1937—40.) Årsskr. Alnarps Landbr. Mej.-och TradInst., Pp. 3—80 (1941).

— u. A. Nyhlén: Sammanfattning av sort- och stamförsök med köksväxter ur Statens trädgårdsförsöks meddelanden av år 1940. (Summary of variety and strain trials with vegetables from the report of the State horticultural experiments of the year 1940.) Årsskr. Alnarps Landbr. Mej.-och TrädInst., Pp. 93 bis 114 (1940).

Leslie, W. R. u. W. Godfrey: Vegetables for prairie farms. Fmrs' Bull. Dep. Agric. Ottawa, Bull. **83**, 78 (1940).

Luthra, J. C., A. Sattar u. K. S. Bedi: Further studies on the control of gram blight. Indian Fmg. **4**, 413 bis 416 (1943).

McRae, L.: Garden newcomer's you'll like. Bett. Homes & Gdns. **22**, Nr. 8, 12 (1944).

Maher, F. A.: Vegetable planting guide for Victoria. J. Dep. Agric. Vict. **40**, 424—430 (1942).

Meier, K.: Bericht der Eidgenössischen Versuchsanstalt für Obst-, Wein- und Gartenbau in Wädenswil für die Jahre 1935/1937. Landw. Jb. Schweiz **54**, 389—464 (1940).

— Bericht der Eidgenössischen Versuchsanstalt für Obst-, Wein- und Gartenbau in Wädensnil für das Jahr 1940. Landw. Jb. Schweiz **57**, 419—461 (1943).

Milsum, J. N. u. D. H. Grist: Vegetable gardening in Malaya. Dep. Agric. S. S. and F. M. S., Kuala Lumpur. (1941).

Morrison, G.: The fundamentals of seed breeding. Nat. Seedsman February, Pp. 3 (1940).

— Behind the scenes and beneath the surface in vegetable seed breeding. Market Gr. J. **68**, 300—303, 331, 340—342 (1941).

Munger, H. M.: The place for hybrid types of vegetables. Trans. Peninsula Hort. Soc. **36**, Nr. 5, 49—57 (1946).

Myers, C. E.: Story of a variety. A biographical sketch of Penn State Ballhead, one of the few varieties of which the complete history can be traced. Seed World **52**, Nr. 9, 10—11, 36 (1942).

Naundorf, G. u. E. Haase: Der Wuchsstoffhaushalt polyploider Pflanzenformen nach Colchicin-Behandlung. Naturwissenschaften **31**, 570 (1943).

Nilsson, E.: Inkorsningsfaran vid köksväxtfröodling. (The danger of cross-pollination in the production of vegetable seed.) Svensk Frötidn. **9**, 48—50 (1940).

Nilsson, F.: Statens Trädgårdsförsök (State Horticultural Research). Fruktodlaren, Nr. 5, 152—156 (1944).

— Växtförädling av köksväxter. (Breeding of vegetables). Årsb. Jordbruksforskning, Stockholm, Pp. 171—182 (1945).

Norwood, J. W.: Back stage work toward better vegetables. Market Gr. J. **65**, 439—441 (1939).

Nyhlén, Å.: Odlingsresultat från lokala försök med köksväxter under åren 1936—1937. (Results of local cultivation trials with vegetables during 1936—37.) Arsskr. Alnarps Landbr. Mej.-och TrädInst. Pp. 1 bis 47 (1940).

Orton, C. R.: Epistle to the farm. Bull. W. Va. Agric. Exp. Sta., Nr. 307, 56 (1943).

Pepkowitz, L. P., R. E. Larson, J. Gardner u. G. Owens: The carotene and ascorbic acid concentration of vegetable varieties. Plant Physiol. **19**, 615—626 (1944).

Piettre, L.: Action de la colchicine sur les végétaux. C. R. Soc. Biol. **131**, 1095—1197 (1939).

Poole, C. F.: The amazing science of plant breeding. Sth. Seedsman **6**, Nr. 6; 9, 54—55 (1943).

Porte, W. S.: Pan America gets the flag. Seed World **53**, Nr. 7, 14 (1943).

Preobraženskij, G. N.: (15 years' work of the Birjučii Kut Vegetable Breeding Station.) Ovoščevodstvo (Vegetable Growing), Nr. 10, 20—24 (1940). [Russisch.]

Quarrel, C. P.: Intensive salad production including some vegetables. London, Pp. 250, 36 figs. (1945).

Ramsey, G. B. u. J. S. Wiant: Market diseases of fruits and vegetables. Asparagus, onions, beans, peas, carrots, celery, and related vegetables. Misc. Publ. U. S. Dep. Agric., Nr. 440, 70 (1941).

Reinhold, J.: Zuchtziele im Gemüsebau. Forschungsdienst **8**, 287—298 (1939).

— Die Haltbarkeit von Gemüse ein wichtiger Qualitätsfaktor. Z. Pflkrankh. **53**, 175—199 (1943).

Ritchie, T. F.: Horticultural dividends-from vegetable plant breeding. Canad. Hort. Home Mag., Gr. Ed. **66**, 203—204 (1943).

Russek, L. M.: Experimentos sobre partenocarpia. I. Solanáceas y Cucurbitáceas. (Experiments on parthenocarpy. I. Solanaceae and Cucurbitaceae.) Ciencia, Mexico, D. F., **5**, 34—36 (1944).

Salvo, C.: Regional vegetable breeding for the Southeast. Agrarian, Clemson, S. C. **1**, Nr. 2, 19, 36 (1939).

Schmidt, M.: Aufgaben der Züchtung im Gemüsebau Ertragssteigerung und Sicherung der Erträge stehen im Fordergrund. Obst- und Gemüsebau **87**, 117—118 (1941).

Shifriss, O.: On developmental reversal of „dominance". Genetics **32**, 103 (1947). (Abstr.)

Sinnott, E. W. u. W. G. Whaley: The developmental basis of inherited size differences in plant organs. Genetics **25**, 136 (1940). (Abstr.)

Thompson, H. C.: Vegetable crops. McGraw-Hil Publishing Company, Ltd., London (1939).

Timofeev, N. N.: (The inheritance of characters in vegetables and its relationship to the length of the vegetative period.) Moskovskaja ordena Lenina Seljskohozjajstvennaja Akademija imeni K. A. Timirjazeva, Naučnaja Konferencija 6—13 Dekabrja 1944 g. II. (Timirjazev Agric. Acad. Moscow Rep. Sci. Conf. 6—13 December) 1944, Nr. 2, 90—93 (1945). [Russisch.]

Timofeev, N. N.: (The quality of the head and the length of the vegetative period in an assortment of head forming plants [cabbage, lettuce, etc.].) Moskovskaja ordena Lenina Seljskohozjajstvennaja Akademija imeni K. A. Timirjazeva. Doklady Vypusk III. Naučnaja Konferencija 4—11 Ijunja 1945 g. (Timirjazev Agric. Acad. Moscow Proc. No. III. Sci. Conf. 4—11 June 1945), Pp. 69—73 (1946). [Russisch.]

Tucker, C. M.: Controlling plant diseases in the home garden. Circ. Mo. Agric. Exp. Sta., Nr. 238, 8 (1942).

W....., P.: Ivan Claude Jagger. Market Gr. J. 64, 180 (1939).

Walker, J. C.: Diseases of vegetable crops. Edwards Brothers, Inc., Ann Arbor, Michigan (1939).
— Disease resistance in crucifers. Proc. 7th Int. Genet. Congr. Edinburgh 23—30 August, Pp. 311—312, 1939 (1941). (Abstr.)
— Disease resistance in vegetable crops. Bot. Rev. 7, 458—506 (1941).

Watts, R. L. u. G. S. Watts: The vegetable growing business. Orange Judd Publishing Company, Inc., New York (1939).

Weidhaas, H.: Experimentelle Studien an Gemüse über die Entnahme von Durchschnittsproben zur chemischen Qualitätsbestimmung unter Anwendung statistischer Methoden. Bodenkunde u. Pflanzenernährung, 1941, 30, 1 (1942).

Wellensiek, S. J.: Selectieschema's voor groentegewassen. (Selection schemes for vegetables.) Meded. Inspecteur Tuinbouw Tuinbouwonderwijs, Wageningen, Pp. 605—610 (1943).

Winter, F. L.: Prospect and retrospect in vegetable development. Seed World 47, Nr. 6, 26—27 (1940).
— ,,Indispensables" of the seed industry. Breeders and research workers shoulder responsibility of keeping highli-bred varieties true to type and acceptable to customers. Sth. Seedsman 6, Nr. 1; 12, 40—41 (1943).

Work, P.: Vegetable variety field days held in New York. Seed World 48, Nr. 7, 14—15 (1940).
— Vegetable variety field days. USDA and University of Maryland play hosts to seedsmen and experiment station workers. Seed World 50, Nr. 5, 8—9 (1941).

2. Leguminosen.

Abeele, M. van den: La sélection du soja. Bull. Agric. Congo Belge 32, 361 (1941).

Adler, E.: New soybean and cowpea varieties, Promising results of Natal tests. Fmr's Wkly 56, 1371 (1939).
— New varieties of soybeans and cowpeas. S. Afr. Sug. J. 23, 205 (1939).

Aganjan, V. N.: (Some data on experiments in raising the protein and oil content of soya bean.) Selekcija i Semenovodstvo (Breeding and Seed Growing) Nr. 6, 10—12 (1940). [Russisch.]

Aitken, Y. u. J. M. Haughton: The species Pisum sativum in relation to Australian agriculture. J. Aust. Inst. Agric. Sci. 11, 35—40 (1945).

Allard, H. A. u. H. F. Allard: The wild bean Phaseolus polystachus (L.) B. S. P.: Its chromosome number. J. Wash. Acad. Sci. 30, 335—337 (1940).
— u. W. J. Zaumeyer: Responses of beans (Phaseolus) and other legumes to length of day. Tech. Bull. U. S. Dep. Agric. Nr. 867, 24 (1944).

Allen, D. I.: Differential growth response of certain varieties of soybeans to varied mineral nutrient conditions. Res. Bull. Mo. Agric. Exp. Sta. Nr. 361, 43 (1943).

Allington, W. B.: Soybean disease investigations at the U. S. Regional Soybean Laboratory. Soybean Digest, 4, Nr. 11; 60, 65 (1944).

Amargos, J. L.: Un nuevo tipo de frijol de carita. (A new type of Carita bean.) Rev. Minist. Agric. Cuba 27, Nr. 28, 50—54 (1944).

Anderson, M. E.: Sensation Refugees, two new mosaic-resistant varieties. Canner 92 (7), 14—15 (1941).
— Bean improvement. West. Cann. Pack. 33 (12), 19 bis 20 (1942).
— Breeding for post-war years. A look into the future at the type of peas and beans we will grow. Seed World 53, 45, 48 (1943).
— Two new wilt-resistant pea varieties for processors. Canner 102, Nr. 6, 22 (1946).

Andersson, G.: Redogörelse för arbetena med soja vid Sveriges Utsädesförening åren 1938—1940. Report on the work with soya beans by the Swedish Seed Association for the years 1938—1940. Sverig. Utsädesfören. Tidskr. 51, 94—122 (1941).
— Redogörelse för arbetena med soja vid Sveriges Utsädesförening åren 1941—1943. (Report on the work with soya bean by the Swedish Seed Association 1941—1943.) Sverig. Utsädesfören. Tidskr. 54, 279 bis 293 (1944).

Andrus, C. F.: The factorial interpretation of anthracnose resistance in beans. Phytopathology 29, 1 (1939). (Abstr.)
— The factorial interpretation of anthracnose resistance in beans. Tech. Bull. U. S. Dep. Agric. Nr. 810, 29 (1942).

Anonymus: Agricultural research in Idaho. Nr. 1 of a series. Beans that resist mosaic and curly top. News Lett. Idaho 22, Nr. 13, 1 (1939).
— The Clemson soy bean. Agrarian, Clemson, S. C. 2, 28 (1939).
— Canning peas in Wisconsin. Bull. Wis. Agric. Exp. Sta. Nr. 444, 24 (1939).
— Breeders succeed in producing new lima bean varieties for freezing. Canner 90 (25), 12 (1940).
— New Plentiful bean variety. N. J. Hort. Soc. News 21, 1176 (1940).
— Pigeon-peas. Indian Fmg 1, 178—179 (1940).
— Foreign countries conducting many soybean research problems. Soybean Digest 1, Nr. 2, 12 (1940).
— ,,Hybrid" soybean varieties. Seed World 49, Nr. 6, 32 (1941).
— New varieties released at three stations. Soybean Digest 1, Nr. 5, 3 (1941).
— Dakotan believes soybean can be pushed northward... Soybean Digest 1, Nr. 7, 4—5 (1941).
— Soybeans for farm gardens. Soybean Digest 2, Nr. 4, 11—12 (1942).
— Genetic studies. Soybean Digest 2, Nr. 7, 4 (1942).
— New Indiana varieties. Soybean Digest 2, Nr. 10; 2, 11 (1942).
— Frozen beans retain vitamin C. Seed World 52, Nr. 12, 35 (1942).
— One plant's poison is another's perfect food. Sth. Seedsman 6, 22 (1943).
— This new snap bean for the South thrives when going is rough. Sth. Seedsman 6, Nr. 4, 24 (1943).
— A feijão guando. (The cowpea.) Bol. Minist. Agric., Rio de J. 32, Nr. 1, 95—96 (1943).
— Cayuga soybeans to Russia. Soybean Digest 3, Nr. 5, 13 (1943).
— Wilt-resistant canning pea named for Prof. E. J. Delwiche. Canner 101, 11 (1945).
— What is the Delwiche Commando pea like? Canner 102, Nr. 2, 13—14 (1945).
— The green-pea tree. Soviet News Nr. 1445, 3 (1946).
— Adapted varieties. Soybean Digest 6, 22 (1946).
— New variety by Russians. Soybean Digest 6, Nr. 12, 27 (1946).

Anonymus: Westan: nematode-resistant lima bean from California. Sth. Seedsman 10, Nr. 1, 44 (1947).
— Two more All-America's. Sth. Seedsman 10, Nr. 1, 15 (1947).
— New varieties of soy bean. Soviet News, Nr. 1636, 4 (1947).
Arnold, H. C.: Soya beans. Notes on cultivation. Rhod. Agric. J. 37, 588—606 (1940).
— Soya beans (continued). Rhod. Agric. J 39, 418—432 (1942).
Atabekova, A. I.: (Sporogenesis and gametogenesis in cultivated pea.)Moskovskaja odena Lenina Seljskohozjajstvennaja Akademija imeni K. A. Timirjazeva. Naučnaja Konferencija 6—13 Dekabrja 1944 g. II. (Timirjazeva Agric. Acad. Moscow Rep. Sci. Conf. 6—13. December 1944, Nr. 2, 72—73 (1945). [Russisch.]

B......, A. C.: Cambridge Multipod pea. Gdnr's Chron. 116, 53 (1944).
Babb, M. F., J. E. Kraus, B. L. Wade u. W. J. Zaumeyer: Drought tolerance in snap beans. J. Agric. Res. 62, 543—553 (1941).
Bacher, T.: Sektionen for Foraedlings- og Forsøgsvirksomhed med Haveplanter. Referat fra N. J. F.'s Havebrugssektions Møde i København, Fredag den 26. Februar 1942. (The Section for Breeding and Research Work with Horticultural Plants. Report from the Scandinavian Agricultural Association's Meeting in Copenhagen. Friday 26 February 1942.) Nord. JordbrForskn. 24, 7—10 (1942).
Barrons, K. C.: Natural crossing in beans at different degrees of isolation. Proc. Amer. Soc. Hort. Sci. 36, 637—640 (1938) 1939.
— Breeding better beans for south. Alabama tests a variety which outstrips all current market favorites. Sth. Seedsman 2, 4 (1939).
— Root-knot resistance in beans. J. Hered. 31, 35—38 (1940).
Bhaduri, P. N.: A study of the effects of different forms of colchicine on the roots of Vicia faba L. J. R. Micros, Soc. 59, 245—276 (1939).
— A study of the relation of chromosomes to nucleoli in species of Scilla, Vicia and Oenothera. Proc. 7th Int. Genet. Congr. Edinburgh 23—30 August, Pp. 64 bis 65, 1939 (1941). (Abstr.)
Bjälfve, G.: Ärternas kokbarhet och fosfathalt. (Cooking quality of peas and phosphate content.) Lantmannen 28, 44—46, 68—69 (1944).
Bolhuis, G. G.: Natuurlijke kruisbestuiving bij kedelee. (Natural crossing in soybeans and the results of experiments on this subject made in Java.) Landbouw 16, 119—128 (1940).
Bonney, V. B. u. H. Fischbach: Comparative chemical studies on pea seed and canned soaked dry peas. J. Ass. Off. Agric. Chem., Wash. 28, 409—417 (1945).
Bose, R. D.: Studies in Indian pulses. IX. Contributions to the genetics of mung (Phaseolus radiatus Linn. syn. Ph. aureus Roxb.). Indian J. Agric. Sci. 9, 575—594 (1939).
Boswell, V. R.: Growing field beans in humid areas. Leafl. U. S. Dep. Agric. Nr. 223, 8 (1942).
Boyenval, J.: Essais préliminaires sur la normalisation des variétés de haricots. I. La proportion des téguments dans le poids des grains secs. C. R. Acad. Agric. Fr. 33, 90—93 (1947).
Boyes, J. u. G. Bond: The effectiveness of certain strains of the soya-bean nodule organism when associated with different varieties of the host plant. Ann. Appl. Biol. 29, 103—108 (1942).
Bremer, A. H.: Sukkererter med store skolmer. (Sugar peas with large pods.) 18. og 19. Arbeidsåret Meld. St. Forsøkstasjon Grønsakdyrk. Kvithamar i Stjørdal, Oslo G 38—42 (1937—1938) 1939.

Brovcyn, A. N.: (Variation in pea varieties and their admixtures.) Vestnik Ovoščevodstvo i Kartofelj (Vegetable and Potato Journal) Nr. 2, 27—38 (1941). [Russisch.]
Brumfield, R. T.: Cell-lineage studies in root meristems by means of X-ray induced chromosome rearrangements. Genetics 27, 135 (1942). (Abstr.)
Buchinger, A.: Ein Beitrag zur Züchtung der Ackererbse (Pisum arvense). Forschungsdienst 12, 298—309 (1941).
Burkholder, P. R.: Vitamins in edible soybeans. Science 98, 188—190 (1943).
— u. I. McVeigh: Vitamin content of some mature and germinated legume seeds. Plant Physiol. 20, 301—306 (1945).
Burkholder, W. H. u. E. T. Bullard: Varietal susceptibility of beans to Xanthomonas phaseoli var. fuscans. Plant Dis. Reporter 30, 446—448 (1946).
Burlingham, E. V.: N. W. Wonder pea for Dixie. Sth. Seedsman 7, 11, 46—47 (1944).

Calder, R. A.: Field peas. The development and performance of a new blue and a new white field pea. N. Z. J. Agric. 65, 347—349 (1942).
— A new garden pea: Greencrop (Greenfeast × Greatcrop). N. Z. J. Sci. Tech. 25, Sect. A, 165—169 (1943).
— Field and garden peas. A survey of the selection and breeding work undertaken by the Agronomy Divison, 1930—1943. N. Z. J. Sci. Tech. 25, Sect. A., 242—255 (1944).
— Garden pea varieties. J. Roy. N. Z. Inst. Hort. 15, 19—23 (1946).
Caldwell, J. S. u. C. V. Culpepper: Snap-bean varieties suited to dehydration. Part 1 and 2. Canning Age 24, 309—311, 313, 363—364, 366, 368 (1943).
—, G. W. Culpepper, B. D. Ezell, M. S. Wilcox u. M. C. Hutchins: The dehydration of peas. Canner 103, Nr. 13; 13—16, 32, 34; Nr. 14; 30, 34, 36—38, 40; Nr. 15; 20, 22—25 (1947).
—, —, M. C. Hutchins, B. D. Ezell u. M. S. Wilcox: Studies of varietal suitability for dehydration in snap beans. Canner 99, Nr. 9, 13—15, 22; Nr. 10, 12—15, 30; Nr. 11, 16—18, 30; Nr. 12, 17—18, 38, 40 (1944).
Camara, A. de Sousa da: Die Wirkung von Röntgenstrahlen auf die meiotischen Chromosomen der Vicia faba L. Bol. Soc. Broteriana 13, 187—209 (1939).
Carles, J.: La lentille du Puy et son cru. Bull. Soc. Bot. Fr. 92, 169—172 (1945).
Carr, R. B.: Soybean varieties in the Yazoo-Mississippi Delta. Soybean Digest, 6, Nr. 9, 12—13 (1946).
Carson, C. M.: Preview of '48... All-Americas. Sth. Seedsman, 10, Nr. 3, 14—15, 54 (1947).
Carter, J. jr.: Pinto bean tests in northeastern New Mexico, 1940—1942. Pr. Bull. N. Mex. Coll. Agric. Mech. Arts Agric. Exp. Sta. Nr. 962, 2 (1943). [Mimeographed.]
Cartter, J. L.: What is the U. S. Regional Soybean Laboratory doing? Soybean Digest 4, Nr. 11; 22, 62 (1944).
— u. T. H. Hopper: Influence of variety, environment, and fertility level on the chemical composition of soybean seed. Tech. Bull. U. S. Dep. Agric. Nr. 787, 66 (1942).
Castan, R.: Diversité des réactions d'inhibition entre bourgeons parmi les variétés de pois. Bull. Soc. Bot. Fr. 91, 10—11 (1944).
Chamberlain, E. E.: Varieties of garden and field peas immune to pea-mosaic. N. Z. J. Sci. Tech. 21, 178—183A (1939).
Chang, S. C. u. M. K. Tsu: (Nitrogen fixation among different strains of Rhizobium leguminosarum.) Kwangsi Agric. 3, 231—240 (1942).

Chao, R. Y. u. S. L. Yu: (Anthesis of soya bean.) New Agric. J. Fukien **3**, 34—49 (1943).

Chevalier, A.: Le dolique de Chine en Afrique. Son histoire. Ses affinités. Les formes sauvages et cultivées. Son rôle dans l'alimentation indigène et en agriculture tropicale et subtropicale. Rev. Bot. Appl. **24**, 128—152 (1944).

Clausen, R. T.: A botanical study of the yam beans (Pachyrrhizus). Mem. Cornell Agric. Exp. Sta., Nr. 264, 38 (1944).

Coon, B. F.: Resistance of soybean varieties to Japanese beetle attack. J. Econ. Ent. **39**, 510—513 (1946).

Costa, A. S. u. R. Forster: Duas molestias de virus do feijoeiro (Phaseolus vulgaris L.). (Two virus diseases of the bean P. vulgaris L.) O Biologico, S. Paulo **7**, 177—182 (1941).

Cottier, K.: Investigations concerning varieties of dwarf beans in New Zealand. N. Z. J. Sci. Tech. **23**, 12A bis 23A. (1942).

Coulter, F. C.: The story of garden vegetables. XII. Peas: cultivated from the stone age onward. Seed World **52**, Nr. 8, 14—15, 30 (1942).

Crnjaković, D.: Prinos poznavanju kvalitativnih svojstava našeg graha (Phaseolus vulgaris L.). Postotak i debljina ljuske te sposobnost razkuhavanja. (Contribution to the study of the qualities of our beans [Ph. vulgaris L.]. Percentage and thickness of skin and cooking quality.) Poljodjelska Znanstvena Smotra, Zagreb, Nr. 8, 57—71 (1944).

Cutler, G. H. u. A. H. Probst: Earlyana an early soybean for Northern Indiana. Circ. Ind. Agric. Exp. Sta., Nr. 286, 8 (1943).

—, — Earlyana-an early soybean for northern Indiana. Soybean Digest. **3**, 4, 14 (1943).

Cyrus, W. F.: Dixie Wonder pea. Sth., Seedsman **8**, Nr. 7; 11, 40 (1945).

— Dixie Wonder. Sth. Seedsman **9**, 18, 44, 48 (1946).

Dahl, E.: Brytmärgärter. (Marrow fat peas with thick pods.) Fructodlaren, Nr. 2, 60—61 (1946).

Dana, B. F.: Resistance and susceptibility to curly top in varieties of common bean, Phaseolus vulgaris. Phytopathology **30**, 786 (1940). (Abstr.)

— The Pioneer bean-resists curly top. Seed World **55**, 46—47 (1944).

Daniel, L.: L'hérédité chez le haricot xenić. C. R. Acad. Sci., Paris **209**, 389—392 (1939).

— Sur les variations de la couleur des gousses et des graines des haricots. C. R. Acad. Sci., Paris **209**, 499 bis 501 (1939).

Dean, L. L. u. C. W. Hungerford: A new bean mosaic in Idaho. Phytopathology **36**, 324—326 (1946).

Delwiche, E. J.: Wisconsin canning pea trials 1937 bis 1941. Res. Bull. Wis. Agric. Exp. Sta., Nr. 144, 36 (1942).

Delwiche, E. T.: Extending the soybean belt northward. Proc. Amer. Soybean Ass. 19th Annu. Mtg. Madison, Wisc. Sept., Pp. 11—12, 22—26 (1939).

Dermen, H.: Intranuclear polyploidy in bean induced by naphthalene-acetic acid. J. Hered. **32**, 133—138 (1941).

Dimmock, F.: Canada includes many excellent soybean acres. Soybean Digest **1**, Nr. 7, 5 (1941).

Dodonova, E. W.: Über den Unterschied der Erbsensamen nach dem Gehalt an Dehydrases. Enzymologia, Hague **9**, 373—379 (1941).

Domingo, W. E.: Inheritance of number of seeds per pod and leaflet shape in the soybean. J. Agric. Res. **70**, 251—268 (1945).

Dorchester, C. S.: Seed and seedling characters in certain varieties of soybeans. J. Amer. Soc. Agron. **37**, 223—232 (1945).

Dundas, B.: Inheritance of resistance to powdery mildew in runner beans (Phaseolus coccineus), tepary beans (P. acutifolius), yard long beans (Vigna sesquipedalis) and cowpeas (Vigna sinensis). Phytopathology **29**, 824 (1939). (Abstr.)

Dundas, B.: A new factor for resistance to powdery mildew (Erysiphe polygoni) in beans (Phaseolus vulgaris). Phytophatology **30**, 786 (1940). (Abstr.)

— A preliminary report on the inheritance of resistance to rust (Uromyces appendidulatus) in beans (Phaseolus vulgaris). Phytopathology **30**, 786 (1940). (Abstr.)

— Further studies on the inheritance of resistance to powdery mildew of beans. Hildgardia **13**, 551—565 (1941).

Dusseau, A. u. C. Magnant: Étude caryologique et dénombrement chromosomique chez une phaséolée, Voandzeia subterranea Thouars. C. R. Acad. Sci. Paris **212**, 455—456 (1941).

Earley, E. B.: Minor element studies with soybeans: I. Varietal reaction to concentrations of zinc in excess of the nutritional requirement. J. Amer. Soc. Agron. **35**, 1012—1023 (1943).

Ekbote, R. B.: Genetics of two mutations in Cicer. Indian J. Genet. Pl. Breed. **2**, 50—65 (1942).

Ekinci, A. S.: Türkiye fasulya soy ve çeşitlerinin sistematik ve morfolojik tetkiki ve standardizasyona başlamak için ilk mesai. (The beans of Turkey and a detailed systematic and morphological examination of the several varieties and a preliminary standardization for initial work.) T. C. Yüksek Ziraat Enstitüsü Çalişmalarindan, Ankara Nr. 69, 207 (1939).

— Systematische und morphologische Untersuchung der Bohnenrassen und -sorten der Türkei. Gartenbauwiss. **14**, 358—432 (1940).

Elgueta, M. u. L. Baillon: Ensayo de fecundación ajena en frejoles. (Test of cross fertilization in Phaseolus.) Agric. Tec. Chile **4**, 38—40 (1944).

Eliasson, S. u. G. Jacobson: Sortförsök med ärter och baljväxtblandsäd. Sammanställningar av resultaten av de av Jordbruksförsöksanstalten ledda försöken med ärter och baljväxtblandsäd under åren 1940 (1912)—1944. (Variety trials with peas and mixed legumes. A summary of results of experiments during 1940[1912]—1944 with peas and with mixed legumes, conducted by the Agricultural Research Institute.) Medd. Lantbrukshögskolan Jordbruksförsöksanstalten, Stockholm, Nr. 17, 108 (1946).

Ellisor, L. O.: Notes on the biology and control of the velvetbean caterpillar, Anticarsia gemmatilis Hbn. Entomological Progress, Nr. 3, Bull. La Agric. Exp. Sta., Nr. 350, 17—23 (1942).

Enin, T. K.: (Hybridization of peas in breeding new varieties.) Selekcija i Semenovodstvo, Nr. 2/3, 32 bis 35 (1939). [Russisch.]

Enken, V. B.: (Infection of beans with bacterial diseases.) Selekcija i Semenovodstvo (Breeding and Seed Growing), Nr. 9, 17—20 (1939).

Enzie, W. D.: New pea varieties compared for yield and plant characters. Fm Res. N. Y. **9**, Nr. 1 (Repr., Nr. 34), 3 (1943).

Evans, R. J., et al: Some factors influencing the protein, cystine, and methionine content of dry peas. Cereal Chem. **24**, 150—156 (1947).

Faller, A.: Mitosestörungen und Wachstumshemmung durch Methyltestosteron im Wurzelmeristem der Feuerbohne. Z. Zellforsch. A, Pp. 32, 173—193 (1942).

Fernando, M.: The relative resistance of some cowpea varieties to Agromyza phaseoli Coq. Trop. Agriculturist **96**, 221—224 (1941).

Ferrée, C. J.: The soya bean and the new soya flour. London, Pp. XI + 79. 14 figs. tables (1939).

Fomin, P.: (Vegetative hybridization in legumes.) Socialistic Grain Farming, Saratov, Nr. 4, 170—171 (1940). [Russisch.]

Fuchs, W. H.: Untersuchungen über die Qualität der Gemüseerbsen. Züchter **14**, 94—96 (1942).
— u. R. von Sengbusch: Kleine Maschine zum Entpalen von grünen Erbsen für züchterische Zwecke. Züchter **14**, 285—289 (1942).

Gagarin, G. D.: (The technique of intravarietal crossing in peas.) Selekcija i Semenovodstvo (Breeding and Seed Growing), Nr. 5, 18 (1940). [Russisch.]
Garbuzova, D. A.: (Züchtung der Askohitosis-resistenten Erbsen.) Selekcija i Semenovodstvo, Nr. 9/10, 30—31 (1946). [Russisch.]
Garrigues, R.: Action de quelques substances huileuses sur les racines de Pisum sativum. C. R. Soc. Biol., Paris **135**, 1065—1068 (1941).
Gelin, O.: Weibulls gula kokärter, Original Munk, Ambrosia II och Kloster (720—729). (Weibull's yellow culinary peas-Original Munk, Ambrosia II and Kloster [720—729].) Weibulls Ill. Årsb. **39**, 24—27 (1944).
— Weibulls Klosterpea. Ny förädling. (Weibulls Klostererbse, eine Neuzüchtung.) Agri Hort. Gen. **3**, 33—37 (1945).
— u. N. Schwanbom: Om kokegenskapen hos ärter och dess natur. (On cooking quality in peas and its nature.) Agri Hortique Genetica, Landskrona **1**, 75—96 (1943).
—, — Ärternas kokbarhet. (Cooking quality of peas.) Lantmannen **28**, 209—210 (1944).
Generalov, G. F.: (New high yielding varieties of peas.) Selekcija i Semenovodstvo (Breeding and Seed Growing), Nr. 9, 12—14 (1939). [Russisch.]
Genter, C. F. u. H. M. Brown: X-ray studies on the field bean. J. Hered. **32**, 39—44 (1941).
Gibson, R. M., R. L. Lovvorn u. B. W. Smith: Response of soybeans to experimental defoliation. J. Amer. Soc. Agron. **35**, 768—778 (1943).
Giles, W. F.: Interesting types of beans. J. R. Hort. Soc. **68**, 73—82 (1943).
Gobs-Sonnenschein, C.: Die experimentelle Erzeugung polyploider Sojabohnen mit Alkaloidgemischen in Verbindung mit Kreuzungen polyploider Rassen. Züchter **15**, 62—68 (1943).
Graham, G. H.: The Hopi bean. Rep. Neb. Bd. Agric., Pp. 426—428 (1942).
Granhall, I.: Växtförädlingsstudier beträffande sojaböna, lin m. m. i Österjöländerna och Mellaneuropa. (Plant breeding studies concerning soya beans, flax, etc. in the Baltic countries and Central Europe.) Sverig. Utsädesfören. Tidskr. **49**, 161—179, 336—350 (1939).
Gréen, S.: Förändringar a artsortimentet. (Changes in the available assortment of peas.) Fruktodlaren, Nr. 3, 99—101 (1944).
— En titt på trädgårdsärterna. (A glance at the garden peas.) Fruktodlaren, Nr. 2, 68—70 (1945).
Gromik, I. U.: (New botanical varieties of pea.) Naučnye Zapiski Saharnoj Promyšlennosti (Sci. Trans. Sug. Ind.), Nr. 1/2, 139—140 (1940). [Russisch.]
— u. T. A. Stegajlo: (The investigation of natural hybridization and selective fertilization in peas.) Naučnye Zapiski Saharnoj Promyšlennosti (Sci. Trans. Sug. Ind.), Nr. 2/3, 101—103 (1939). [Russisch.]
Guillochon, L.: Les Légumineuses alimentaires de Tunisie. Rev. Bot. Appl. **20**, 389—402 (1940).

Hähne, H.: Beiträge zur Frage der Bekämpfung der durch Pseudomonas medicaginis var. phaseolicola Burkh. verursachten Fettfleckenkrankheit der Bohne. Angew. Bot. **24**, 31—61 (1942).
Håkansson, A. u. A. Levan: Nucleolar conditions in Pisum. Hereditas, Lund **28**, 436—440 (1942).
Hardenburg, E. V.: Experiments with field beans. Bull. Cornell Univ. Agric. Exp. Sta., Nr. 776, 28 (1942).

Hare, W. W. u. J. C. Walker: Ascochyta diseases of canning pea. Res. Bull. Wis. Agric. Exp. Sta., Nr. 150, 31 (1944).
Harrington, C. D.: Influence of aphid resistance in peas upon aphid development, reproduction, and longevity. J. Agric. Res. **62**, 461—466 (1941).
— Biological races of the pea aphid. J. Econ. Ent. **38**, 12—22 (1945).
—, E. M. Searles, R. A. Brink u. C. Eisenhart: Measurement of the resistance of peas to aphids. J. Agric. Res. **67**, 369—387 (1943).
Harter, L. L. u. W. J. Zaumeyer: Differentiation of physiologic races of Uromyces phaseoli typica on bean. J. Agric. Res. **62**, 717—731 (1941).
—, — A monographic study of bean diseases and methods for their control. Tech. Bull. U. S. Agric., Nr. 868, 160 (1944).
—, — u. B. L. Wade: Pea diseases and their control. Fmrs' Bull. U. S. Dep. Agric. (Revised), Nr. 1735, 28 (1945).
Haskell, R. J.: All-Southern snap bean. Sth. Seedsman **9**, 16, 64, 67 (1946).
Hastings, W. R.: All-America awards. Sth. Seedsman **7**, 16—17, 40 (1944).
Hawthorne, P. L.: The breeding and improvement of edible cowpeas. Proc. Amer. Soc. Hort. Sci. **42**, 562 bis 564 (1943).
Hedges, F.: Association of Xanthomonas phaseoli and the common bean-mosaic virus, Marmor phaseoli. I. Effect of pathogenicity of the seed-borne infective agents. Phytopathology **34**, 662—693 (1944).
Heim, R.: Faculté germinative et conservation des graines de soja. C. R. Acad. Agric. Fr. **32**, 412—415 (1946).
Henson, P. R.: Southern soybean program of the U. S. Regional Soybean Laboratory. Soybean Digest **5**, Nr. 11; 47, 60 (1945).
— The southern regional soybean variety program. Soybean Digest **6**, Nr. 11, 37—39 (1946).
— u. R. S. Carr: Soybean varieties and dates of planting in the Yazoo-Mississippi Delta. Bull. Miss. Agric. Exp. Sta., Nr. 428, 12 (1946).
Hiorth, G.: Eiweißreiche Wurzelknollen bei niedrigen Feuerbohnen. Züchter **14**, 43—47 (1942).
Holmberg, S. A.: Sojaodlingen i Förenta Staterna. (Soya bean cultivation in the United States.) K. Lantbr Akad. Tidskr., Stockh. **78**, 261—272 (1939).
— Från sojaväxtförädlingen vid Fiskeby. (On soya bean breeding at Fiskeby.) K. LantbrAkad. Tidskr. **85**, 373—384 (1946).
Hopkins, T. T. u. J. L. Sawin: Characteristics desirable in peas, beans and corn for canning and freezing. West. Cann. Pack. **31**, Nr. 12, 13—16, 22 (1939).
Howard, F. L. u. E. M. Andersen: Susceptibility of Logan and Florida Belle beans to Fusarium yellows. Phytopathology **35**, 655 (1945). (Abstr.)
Hua, H. N.: Genetical study of Vicia Faba. A preliminary report. Acta Brevia Sinensia, Nr. 4, 18—20 (1943). [Mimeographed.]
— (Natural crossing in Vicia faba.) Chinese J. Agric. Sci. **1**, 63—65 (1943).
Huelsen, W. A.: Three new varieties of bush lima beans: Baby Potato, Early Baby Potato, and Illinois Large Podded. Bull. Ill. Agric. Exp. Sta., Nr. 461, 107 bis 120 (1939).
Humphrey, L. M.: New high yielding soybean varieties for the south. Soybean Digest **7**, Nr. 2, 11—12 (1946).
— Steel jacket soybeans. Sth. Seedsman **10**, Nr. 2, 12 (1947).
Husås, Ø.: Ertegallmyggen. (The pea midge.) Tidskr. Norske Landbr. **47**, 123—137 (1940).

Isbell, C. L.: Alabama's new bean: Mild, White and Giant. Sth. Seedsman **9**, 11, 38 (1946).

Iyengar, N. K.: Cytological investigations on the genus Cicer. Ann. Bot., London **3**, 271—305 (1939).

Izarlišvili, S. J.: (Establishing the degree of resistance of different dwarf bean varieties to anthracnose [Colletotrichum Lindemuthianum Br. et Cav.] under the conditions of the Georgian S. S. R.) Izv. Gruzinsk. Opytn. Stanc. Zašč. Rast. (Bull. Georgian Pl. Protection Res. Sta.) **2**, 227—252 (1940). [Russisch.]

Jaretzky, R. u. G. Schenk: Versuche mit Acenaphthen und Colchicin an Gramineen- und Leguminosenkeimlingen. Jb. wiss. Bot. **89**, 13—19 (1940).

Jensen, J. H.: The Scottsbluff Pinto bean. Circ. Neb. Agric. Exp. Sta., Nr. 78, 6 (1944).

— u. R. W. Goss: Physiological resistance to halo blight in beans. Phytopathology **32**, 246—253 (1942).

— u. J. E. Livingstone: Variation in symptoms produced by isolates of Phytomonas medicaginis var. phaseolicola. Phytopathology **34**, 471—480 (1944).

Joubert, T. G. la G.: Hard-skin in peas. Fmg. S. Afr. **17**, 767—768, 791—792 (1942).

Kadam, B. S., S. M. Patel u. V. K. Patankar: Two new genes for corolla colour in Cicer Arietinum L. Curr. Sci. **10**, 78—79 (1941).

Kendrick, J. B. u. W. C. Snyder: Fusarium yellows in beans. Phytopathology **32**, 1010—1014 (1942).

Khan, A. R. u. M. P. Bhatnagar: Cowpea varieties and culture. Indian Fmg **6**, 212—213 (1945).

King, B. M. u. D. I. Allen: Soybean production in Missouri. Boll. Mo. Agric. Exp. Sta., Nr. 445, 31 (1942).

Kišpatić, J.: Einleitende Versuche über Rassenbildung bei Uromyces fabae (Pers.) de By. Phytopath. Z. **14**, 475—483 (1943).

Knapp, O.: Beitrag zur Frage der Qualitäts- und Immunitätszüchtung bei Buschbohnen. Züchter **13**, 19—21 (1941).

Knowles, D., O. Grottodden u. T. E. Long: Variety tests of vegetables for freezing preservation. The comparative suitability of varieties of green beans, lima beans, wax beans, sweet corn and peas for freezing, preservation. Bull. N. Dak. Agric. Exp. Sta., Nr. 322, 22 (1943).

Kobel, F.: Die Stangenbohnensorten der Schweiz. Landw. Jb., Schweiz **55**, 133—175 (1941).

Koch, L.: Proeven met sojaboonen. (Tests with soya beans.) Erfelijkheid in Praktijk, Leiden H. 4, 148 bis 150 (1939).

— Breeding soybeans for Holland. Soybean Digest **7**, Nr. 2; 13, 24 (1946).

Kopetz, L. M.: Über den Einfluß der Temperatur auf Wachstum und Entwicklung einiger Pflückerbsensorten. Gartenbauwiss. **17**, 255—262 (1943).

Koseljkova, N. N.: (New varieties of lentils.) Selekcija i Semenovodstvo (Breeding and Seed Growing), Nr. 9, 14—15 (1939). [Russisch.]

Kovarskij, A. E.: (The process of form development in interspecific hybrids of cowpeas under the influence of different conditions of environment.) Selekcija i Semenovodstvo (Breeding and Seed Growing), Nr. 10 bis 11, 9—13 (1939). [Russisch.]

Kramer, A.: Relation of maturity to yield and quality of raw and canned peas, corn and lima beans. Proc. Amer. Soc. Hort. Sci. **47**, 361—367 (1946).

Kumar, L. S. S.: A comparative study of autotetraploid and diploid types in mung (Phaseolus radiatus Linn.). Proc. Indian Acad. Sci. **21**, Sect. B, 266—268 (1945).

— u. A. Abraham: A study of colchicine polyploidy in Phaseolus radiatus L. J. Univ. Bombay **11**, (N. S.), Pt. 3, Nr. 12, 30—37 (1942).

—, — Induction of polyploidy in crop plants. Curr. Sci. **11**, 112—113 (1942).

Kurgatnikov, M. M.: (The properties of the starch of different varieties of pea.) Vestnik Socialističeskogo Rastenievodstva (Soviet Plant Industry Record), Nr. 5, 142—148 (1940). [Russisch.]

— u. A. I. Lebedeva: (Breeding peas for chemical composition.) Proc. Lenin Acad. Agric. Sci. USSR, Nr. 9, 17—20 (1939). [Russisch.]

Kutz, A. L.: (Broad beans [Vicia Faba L.] of the Daghestan mountains.) Proc. Daghestan Agric. Inst., Nr. 1, 131—179 (1939).

Lackey, C. F.: Relative concentrations of two strains of curly-top virus in tissues of susceptible and resistant beans. Phytopathology **32**, 910—912 (1942).

Lamm, R.: Studies on linkage relations of the Cy factors in Pisum. Hereditas, Lund **33**, 405—419 (1947).

Lamprecht, H.: Zur Genetik von Phaseolus vulgaris. XIV. Über die Wirkung der Gene P, C, J, Ins, Can, G, B, V, Vir, Och und Flav. Hereditas, Lund **25**, 255 bis 288 (1939).

— Translokation, Genspaltung und Mutation bei Pisum. Hereditas, Lund **25**, 431—458 (1939).

— Genstudien an Pisum sativum. IV. Über Vererbung von Wachslosigkeit und ein neues Gen für lokale Ausbildung von Wachs, Wsp. Hereditas, Lund. **25**, 459 bis 471 (1939).

— Über Blüten- und Komplex-Mutationen bei Pisum. Z. Vererbungsl. **77**, 177—185 (1939).

— The limit between Phaseolus vulgaris and Ph. multiflorus from the genetical point of view. Proc. 7th Int. Genet. Congr. Edinburgh 23—30 August, Pp. 179—180, 1939 (1941). (Abstr.)

— Zur Genetik von Phaseolus vulgaris. XV. Über die Vererbung der Mehrfarbigkeit der Testa. Hereditas, Lund **26**, 65—99 (1940).

— Zur Genetik von Phaseolus vulgaris. XVI. Weitere Beiträge zur Vererbung der Teilfarbigkeit. Hereditas, Lund **26**, 277—291 (1940).

— Zur Genetik von Phaseolus vulgaris. XVII. Zwei neue Gene für Abzeichen auf der Testa, Punc und Mip, sowie über die Wirkung von V und Inh. Hereditas, Lund **26**, 292—302 (1940).

— Zur Genetik von Phaseolus vulgaris. XVIII. Über matte Samenschale und ihre Vererbung. Hereditas, Lund **26**, 302—304 (1940).

— Die Artgrenze zwischen Phaseolus vulgaris L. und multiflorus Lam. Hereditas, Lund **27**, 51—175 (1941).

— Aktuella uppgifter vid förädlingen av bönor och ärter. (Current tasks in breeding beans ans peas.) Weibulls, Ill. Årsb. **36**, 33—38 (1941).

— Über Genlabilität bei Pisum. Züchter **13**, 97—105 (1941).

— Die Koppelungsgruppe Gp-Cp-Fs-Ast von Pisum. Hereditas, Lund **28**, 143—156 (1942).

— Genstudien an Pisum sativum. V. Multiple Allele für Punktierung der Testa: Fs_{ex}-Fs-fs. Hereditas, Lund **28**, 157—164 (1942).

— Genstudien an Pisum sativum. VI. Weitere Ergebnisse zur Vererbung der Wachslosigkeit. Hereditas, Lund **30**, 613—620 (1944).

— Genstudien an Pisum sativum. VII. Tragantflecken zwischen Keimblättern und Testa sowie ihre Vererbung. Hereditas, **30**, 621—627 (1944).

— Genstudien an Pisum sativum. VIII. Das Testmerkmal griseostriata und seine Vererbung. Hereditas, Lund **30**, 627—630 (1944).

— Om sambandet mellan fröproduktion och baljfärg hos trädgårdsbönor, särskilt med hänsyn till effekten av klorofyll och xantofyll. (On the relation between seed production and pod colour in garden beans, especially with reference to the effect of chlorphyll and xantophyll.) Agri Hortique Genetica, Landskrona **2**, 1—19 (1944).

Lambrecht, H.: En ny prydnadsväxt ur kor ningen mellan trädgardsoch blomsterböna, jämte ett klarläggande av artbegreppet. (A new ornamental plant from a cross between the garden bean and the scarlet runner, with an elucidation of the species concept.) Agri Hortique Genetica, Landskrona **3**, 14—32 (1945).
— The linkage group N-Z-Fa-Td of Pisum. Hereditas, Lund **31**, 347—382 (1945).
— Die Koppelungsgruppe Oh-Ar-S-Wb-K von Pisum. Hereditas, Lund **32**, 41—59 (1946).
— Die Koppelungsgruppe Uni-M-Mp-F-St-B-Gl von Pisum. Agri Hort. Gen. **4**, 15—42 (1946).
— En-, två- och treblommighetens praktiska betydelse vid förädlingsarbete med ärter. (Die praktische Bedeutung der Ein-, Zwei- und Dreiblütigkeit bei der Züchtungsarbeit mit Erbsen.) Agri Hort. Gen. **4**, 79 bis 98 (1947).
— Studien über die Zeitigkeit bei Pisum. I. Die Begriffe Zeitigkeit und Lebensdauer. Agr. Hort. Gen. **4**, 105 bis 118 (1947).
— u. E. Åkerberg: Über neue Gene für die Ausbildung von Testafarbe bei Pisum. Hereditas, Lund **25**, 323 bis 348 (1939).
— u. V. Svensson: Weibulls Original Balder. Ny ärtsort, som kommer i marknaden våren 1944. (Weibull's Original Balder. A new pea which will come on the market in spring 1944.) Weibulls Ill. Årsb. **39**, 27—30 (1944).
Leandri, J.: Sur la mitose dans une plantule tricotyle de haricot. Bull. Soc. Bot. Fr. **88**, 421—424 (1941).
Lebedeff, G. A.: Heredity and environment in the production of hard seeds in common beans (Phaseolus vulgaris). Res. Bull. P. R. Agric. Exp. Sta., Nr. 4, 27 (1943).
— Seed viability in field beans. Rec. Genet. Soc. America, Nr. 12, 50—51 (1943). (Abstr.)
— Inheritance of hard-seed production in common beans (Phaseolus vulgaris). Genetics **28**, 80 (1943). (Abstr.)
— Inheritance of hard-shell in beans. Genetics **30**, 12—13 (1945). (Abstr.)
— Studies on the inheritance of hard seeds in beans. J. Agric. Res. **74**, 205—215 (1947).
— u. J. Adsuar: Mejoramiento de las habichuelas en Puerto Rico. (Improvement of beans in Puerto Rico.) Agric. Exp. P. R. **1**, 3 (1941).
Lefebvre, C. L. u. H. S. Sherwin: Observations on the bacterial canker of cowpea. Phytopathology **35**, 487 (1945). (Abstr.)
Ligon, L. L.: Mungbeans. A legume for seed and forage production. Bull. Okla. Agric. Exp. Sta., Nr. 284, 12 (1945).
Lippisch, A.: Sojaanbau in den Alpen- und Donaugauen. Obst- und Gemüsebau **60**, 142—143 (1942).

McFarlane, J. S. u. G. H. Rieman: Leafhopper resistance among the bean varieties. J. Econ. Ent. **36**, 639 (1943).
Mackie, W. W.: Breeding for resistance in Blackeye cowpeas to cowpea wilt, charcoal rot, and root-knot nematode. Phytopathology **29**, 826 (1939). (Abstr.)
— Origin, dispersal, and variability of the lima bean, Phaseolus lunatus. Hilgardia **15**, 1—29 (1943).
— Mung beans ... bonanza crop if you know their culture. Sth. Seedsman **8**, 22, 28, 44 (1945).
— Blackeye beans in California. Bull. Calif. Agric. Exp. Sta., Nr. 696, 56 (1946).
McIlroy, G. G.: Where to in cornwelt soybean production. Soybean Digest **4**, Nr. 11, 29—30 (1944).
McNew, G. L.: Which varieties of peas need treatment? Canner **98**, 14, 16, 26, 28, 30 (1944).
Magruder, R.: Preliminary observations on some of the newer varieties of bush lima beans. Canner **92** (12), 58 (1941).
Magruder, R.: Recent advances in the breeding and culture of lima beans. Proc. Ohio Veg. Potato Gr's Ass. **27**, 80—86 (1942).
— u. R. E. Wester: Natural crossing in lima beans in Maryland. Proc. Amer. Soc. Hort. Sci. **37**, 731—736 (1939) 1940.
—, — Green cotyledon, a new character in the mature lima bean (Phaseolus lunatus L.). Proc. Amer. Soc. Hort. Sci. **38**, 581—584 (1941).
—, — Inheritance of green cotyledon in the lima bean. Proc. Amer. Soc. Hort. Sci. **40**, 410 (1942). (Abstr.)
—, — Prevention of field hybridization in the lima bean. Proc. Amer. Soc. Hort. Sci. **40**, 413—414 (1942).
—, — Natural crossing of lima beans in Maryland during. 1941. Proc. Amer. Soc. Hort. Sci. **42**, 557—561 (1943).
—, — Two new large-podded bush limas. Seed World **53**, Nr. 5; 14, 45 (1943).
—, — A preview of two new bush lima beans. Canner **101**, Nr. 2, 20 (1945).
—, — Birth of a bean. Sth. Seedsman **9**, 16, 52, 57 (1946).
Marshak, A.: Chromosome abnormalities produced in interphase nuclei with X-rays and neutrons. Genetics **26**, 161 (1941). (Abstr.)
Mastenbroek, C.: De vatbaarheid van boonenrassen voor de vetvlekkenziekte. (Varietal susceptibility of beans to halo blight.) Tijdschr. PlZiekt. **49**, 135—162 (1943).
Matagrin, A.: Le soja et les industries du soja. Produits alimentaires. Huile de soja. Lécithine végétale. Caséine végétale. Gauthier-Villars, Paris (1939).
Mayo, J. K.: Soya beans in Nigeria. Trop. Agriculture, Trin. **22**, 226—229 (1945).
Menezes, O. B. de: Estudos para a genética do guando. (Studies on the genetics of pigeon pea.) Bol. Minist. Agric. Rio de J. **32**, 69—83 (1943).
— Estudos para o melhoramento do guandoespaçamento e competição de variedades. (Studies on breeding pigeon peas, spacing and competition of varieties.) Rev. Agric. Piracicaba **19**, 399—412 (1944).
Milner, R. T.: News from the Regional Soybean Laboratory. Proc. Amer. Soybean Ass., August, 18, 19, 20, Dearborn, Mich, Pp. 36—38 (1940).
Mogilev, L. M. u. N. A. Rjahovskij: (Selection of pure lines resistant to Ascochyta from the pea variety Heine's Viktoria.) Naučnye Zapiski Saharnoj Promyšlennosti (Sci. Trans. Sug. Ind.), Nr. 1, 124—127 (1939). [Russisch.]
—, — (Pure lines of pea variety „Victoria Heine" resistant to Ascochyta pisi Lib.) Proc. Lenin Acad. Agric. Sci. USSR., Nr. 3, 11—12 (1941). [Russisch.]
Mogileva, A.: (Intravarietal crossing in peas.) Jarovizacija, Nr. 1 (34), 104 (1941). [Russisch.]
Morrison, G.: The story of beans. Some of the effort that goes into new varieties-descriptive list. Nat. Seedsman **8**, 20—21, 45 (1941).
Morse, W. J.: Soybean variety registered. I. J. Amer. Soc. Agron. **35**, 834—835 (1943).
— Registration of varieties of soybeans. II. J. Amer. Soc. Agron. **36**, 458—460 (1944).
Morton, C. V.: Taxonomic studies of tropical American plants. Notes on Phaseolus. Contr. U. S. Nat. Herb. **29**, 84—85 (1944).
Moyer, L. S. u. N. M. Fishman: The chlorophyll-protein complex. II. Species relationships in certain legumes as shown by electric mobility curves. Bot. Gaz. **104**, 449—454 (1943).
Munn, M. T.: Producing soy sprouts. Soybean Digest **6**, 27 (1946).
Murphy, D. M.: A Great Northern bean resistant to curlytop and common beanmosaic viruses. Phytopathology **30**, 779—784 (1940).

Murphy, D. M.: Bean improvement and bean diseases in Idaho. Bull. Idaho Agric. Exp. Sta., Nr. 238, 22 (1940).
— State bean improvement progam produces new resistant variety. News Lett. Coll. Agric. Univ. Idaho 23, Nr. 4, 1 (1940).

Nilsson, E.: Erblichkeitsversuche mit Pisum. X. Die Koppelungsgruppen Pa-R-Tl-Btb und Wlo-P-Pl. (Inheritance experiments with Pisum. X. The linkage groups Pa-R-Tl-Btb and Wlo-P-Pl.) Hereditas, Lund, 25, 48—64 (1939).
Noll, W.: Über weitere Befallsymptome und Maßnahmen zur Verhütung von Schäden durch Ascochyta pinodella Jones, A. pisi Lib. und Mycosphaerella pinodes (Berk. u. Blox.) Stone bei Erbsen. Z. PflKrankh. 50, 49—71 (1940).
Novikov, V. V.: (Pea-bean hybrids.) Jarovizacija, Nr. 3 (30), 106—118 (1940). [Russisch.]
— (Hybrid peas obtained by pollinating, situated on different parts of the stem.) Sovetskaja Botanika (Soviet Botany) 13, Nr. 5, 24—27 (1945). [Russisch.]

Obermayer, E.: A szójabab termesztésének jelentösége hazánkban. (The importance of growing soya beans in Hungary.) Rep. Hung. Agric. Exp. Sta. 42, 153 bis 156 (1939).
Ochoa, L.: Mosaico amarillo. Nota sobre las semillas de los frijoles en relación con esta enfermedad. (Yellow mosaic. Note on bean seeds in relation to this disease.) Rev. Minist. Agric. Cuba 27, Nr. 28, 17—19 (1944).
Overpeck, J. C.: Soybean production in Mexico. Pr. Bull. N. Mex. Coll. Agric. Mech. Arts Agric. Exp. Sta. Nr. 963, 2 (1943). [Mimeographed].

P...., P.: Procès-verbal. Séance du 7. Décembre 1941. Rev. Hort. Agric. Afrique Nord 46, 27—32 (1942).
Pal, B. P. u. T. Narayana Rao: Ovule mortality in gram (Cicer arietinum L.). Proc. Indian Acad. Sci. 12, Sect. B, 50—61 (1940).
Pavlova, A. M. u. M. P. Kozak: (Breeding gram.) Selekcija i Semenovodstvo, Nr. 1, 19—21 (1939). (Russisch.]
Pellew, C.: Genetical studies on the first reciprocal translocation found in Pisum sativum. J. Genet. 39, 363 bis 390 (1940).
Pesola, V. A.: Sinikka, uusi vihreä talousherne. (Sinikka, a new green cooking pea.) Valt. Maatalousk. Tiedon, Nr. 158, 15 (1939).
Phadnis, B. A.: Xenia in cotyledon colour of gram (Cicer arietinum). Curr. Sci. 15, 256 (1946).
Pinckard, J. A.: Root-knot. Circ. Miss. Agric. Exp. Sta., Nr. 104, 4 (1942).
Pollard, L. H., E. B. Wilcox u. H. B. Peterson: Maturity studies with canning peas. Bull. Utah Agric. Exp. Sta., Nr. 328, 16 (1947).
Popesco, C.: Obtention par greffe d'une race tardive et vivace de Haricot de Soissons. C. R. Acad. Sci., Paris 210, 446—447 (1940).
Portères, R.: Observations sur les possibilités de culture du soja en Guinée forestière. Bull. Agron. Minist. France d'Outre Mer., Nr. 1, 82 (1946).
Portheim, L.: Further studies on the action of heterauxin on Phaseolus vulgaris. Ann. of Bot. N. s. 5, Nr. 17, 35—46 (1941).
Prakken, R.: Inheritance of colours in Phaseolus vulgaris L. I. Genetica 22, 331—408 (1940).
Probst, A. H.: Border effect in soybean nursery plots. J. Amer. Soc. Agron. 35, 662—666 (1943).
— Influence of spacing on yield and other characters in soybeans. J. Amer. Soc. Agron. 37, 549—554 (1945).
— u. J. L. Cartter: A portable soybean nursery thresher and its operation. J. Amer. Soc. Agron. 33, 673—675 (1941).

Pryor, D. E. u. R. E. Webster: Relative resistance and susceptibility of U. S. 243 and U. S. 343. lima beans to lima-bean mosaic. Phytopathology 36, 170—172 (1946).
Psarev, G. M. u. H. A. Veselovskaja: Growth response of soya to daylength in relation to development. C. R. (Doklady) Acad. Sci. URSS. 30, 844—847 (1941).
Puhaljskaja, E. P.: (Inheritance by the Irlandets pea of characters acquired as a result of growing in situations geographically different.) Jarovizacija, Nr. 2 (35), 104—107 (1941). [Russisch.]

Ramanujam, S. u. A. B. Joshi: Colchicine-induced polyploidy in crop plants. I. Gram (Cicer arietinum L.). Indian J. Agric. Sci. 11, 835—849 (1941).
— u. H. Singh: Narrow leaf-another leaf mutation in gram (Cicer arietinum L.). Indian J. Genet. Pl. Breed. 5, 46—50 (1945).
Rangaswami Ayyangar, G. N. u. K. Kunhi Krishnan Nambiar: Lablab — the garden bean. Indian Fmg. 2, 469—472 (1941).
—, — Studies in Dolichos lablab (Roxb.) and (L.), the Indian field and garden bean. IV. Proc. Indian Acad. Sci. 14, Sect. B, 95—113 (1941).
Ranninger, R.: Die vegetative Vermehrung der Buschbohne. Gartenbauwiss. 17, 250—254 (1943).
Raphael, T. D.: Canning peas trials, 1944. Tasm. J. Agric. 16, 24—29 (1945).
— u. W. F. Walker: French beans. Summary of trials. Tasm. J. Agric. 17, 270—278 (1946).
— u. N. H. White: Varietal resistance to halo blight in beans. J. Aust. Inst. Agric. Sci. 10, 76—77 (1944).
Reid, W. D.: Resistance of beans against bacterial-wilt anthracnose, and bean-mosaic. N. Z. J. Agric. 67, 411—412 (1943).
— The resistance of beans against bean-wilt and anthracnose and notes on occurence of bean-mosaic. N. Z. J. Sci. Tech. 25, Sect. A, 125—128 (1943).
— Resistance of beans to halo-blight and anthracnose and the occurence of beanmosaic and bean-weevil. N. Z. J. Sci. Tech. 1945, 27, Sect. A, 331—335 (1946).
— u. A. Hastings: Bean varieties. Descriptions of bean varieties used in trials of resistance to bean diseases. N. Z. J. Sci. Tech. 1945, 27, Sect. A, 320—330 (1946).
Richards, B. L. u. W. H. Burkholder: A new mosaic disease of beans. Phytopathology 33, 1215—1216 (1943).
Rietsema, C.: De boonenteelt. (Bean growing.) Meded. Tuinbouw-Voorl Dienst, 's Grav., Nr. 13, 32 (1939).
Riollano, A.: El mejoramiento de nuestras habichuelas (Datos preliminares). (The improvement of our beans [Preliminary data].) Estac. Exp. Agríc. Dep. Agrón. Puerto Rico, Nr. 25, 4 (1944). [Mimeographed.]
Rosen, G. von: Linkage studies in Pisum sativum × P. abyssinicum. Hereditas, Lund 28, 136—142 (1942).
— Eine physiologische Wirkung der Blütengröße bei Pisum. Hereditas, Lund 28, 240—241 (1942). (Abstr.)
— Röntgeninduzierte Mutationen bei Pisum sativum. Hereditas, Lund 28, 313—338 (1942).
— Artkreuzungen in der Gattung Pisum. Insbesondere zwischen P. sativum L. und P. abyssinicum Braun. Hereditas, Lund 30, 261—400 (1944).
— Artkreuzungen in der Gattung Pisum, insbesondere zwischen P. sativum L. und P. abyssinicum Braun. Berlingska Boktryckeriet, Lund, Pp. 261—400 (1944).
Rozynski, H. von: Erfahrungen mit Soja-Anbau. Mitt. Landw. 58, 889—890 (1943).
Rubin, B. A. u. O. T. Lutikova: (Enzyme action in peas in relation to the development and productivity of the plants.) Doklady Akad. Nauk. SSSR 27, 34—37 (1940). [Russisch.]
Rybin, V. A.: Tetraploid plants of Vicia Faba produced by colchicine treatment. C. R. (Doklady) Acad. Sci. URSS. 24, 483—485 (1939).

S....., E. H. G.: The development of the soya bean crop in the United States. Bull. Imp. Inst., Lond. **43**, 88—93 (1945).

Šakurov, V. Z.: (A new resistant variety of pea.) Selekcija i Semenovodstvo (Breeding and Seed Growing), Nr. 2, 23—24 (1940). [Russisch.]

— (Intravarietal crossing in peas.) Jarovizacija, Nr. 4 (31), 112—114 (1940). [Russisch.]

Sansome, E. R.: Abnormal meiosis in Pisum sativum. Proc. 7th Int. Genet. Congr. Edinburgh 23—30 August, p. 255 1939 (1941). (Abstr.)

— u. F. W. Sansome: Genetical sterility in Pisum sativum. Nature, Lond. **145**, 226 (1940).

Schreiber, F.: The genetics of partial coloration in beans (Phaseolus vulgaris). Proc. 7th Int. Genet. Congr. Edinburgh 23—30 August, p. 256, 1939 (1941). (Abstr.)

— Die Genetik der Teilfärbung der Bohnensamen (Phaseolus vulgaris). Z. indukt. Abstamm.- u. Vererbungsl. **78**, 59—114 (1940).

Schroeder, W. T. u. J. C. Walker: Influence of controlles environment and nutrition on the resistance of garden pea to fusarium wilt. J. Agric. Res. **65**, 221 bis 248 (1942).

Schultz, H. K. u. L. L. Dean: Inheritance of curly top disease reaction in the bean, Phaseolus vulgaris. J. Amer. Soc. Agron. **39**, 47—51 (1947).

Scully, N. J., M. W. Parker u. H. A. Borthwick: Relationship of photoperiod and nitrogen nutrition to initiation of flower primordia in soybean varieties. Bot. Gaz. **107**, 218—231 (1945).

Seemann, J.: Über die Bedeutung der Unterkühlung für die Selektion frostresistenter Bohnenpflanzen. Züchter **14**, 258—264 (1942).

Šenjavskij, A. L.: (Pollination with a mixture of pollen in a self-pollinated plant [pea].) Selekcija i Semenovodstvo (Breeding and Seed Growing), Nr. 5, 19 bis 22 (1940).

Senn, H. A.: The relation of anatomy and cytology to the classification of the Leguminosae. Chronica Botanica **7**, 306—308 (1943).

Sessous, G.: Stand und Ziel von Anbau und Züchtung der Soja. Forschungsdienst, Sonderh. 16, 400 (1943).

— u. K. Schiller: Grundsätzliches zur chemischen Auslese bei der Sojazüchtung. Züchter **11**, 1—14 (1939).

Severson, H.: Viva Guatemala. It gave us the lima bean. Sth. Seedsman **6**, Nr. 5; 11, 38—39, 36 (1943).

Shorrock, R. W.: Dried peas. J. Minist. Agric. **52**, 530 bis 534 (1946).

Singletary, B. H.: The Singletary pea ... a versatile legume. Sth. Seedsman **5**, Nr. 9, 9, 29, 36 (1942).

Slugin, P. T.: (The chemical composition of the new varieties of soya bean of the Ussuri Breeding Station.) Selekcija i Semenovodstvo (Breeding and Seed Growing), Nr. 9, 20—21 (1939). [Russisch.]

Smith, F. L.: A genetic analysis of red seed-coat color in Phaseolus vulgaris. Hilgardia **12**, 553—621 (1939).

— Inheritance of red seed coat colour in common beans. Amer. J. Bot. **26**, 675 (1939). (Abstr.)

Sneep, J.: De Ascochyta-vlekkenziekte van de boon (Phaseolus). (The Ascochyta spot diseases of Phaseolus.) Tijdschr. PlZiekt. **51**, 1—16 (1945).

Sonnenschein, Cl.: Neuere Forschungen über die Erzeugung polyploider Formen von Sojabohnen. (Vorl. Mitt.) Forsch.dienst **12**, 532—537 (1941).

Stark, F. C. jr. u. C. H. Mahoney: A study of the time of development of the fibrous sheath in the sidewall of edible snap bean pods with respect to quality Proc. Amer. Soc. Hort. Sci. **41**, 353—359 (1942).

Staten, H. W.: Mung beans for Oklahoma. Circ. Okla Agric. Exp. Sta., Nr. C—104, 7 (1942).

Straib, W.: Untersuchungen zur Biologie und Bekämpfung des Bohnenrostes Uromyces phaseoli (Pers.) Wint. Gartenbauwiss. **17**, 397—445 (1943).

Strand, A. B.: Species crosses in the genus Phaseolus. Proc. Amer. Soc. Hort. Sci. **42**, 569—573 (1943).

Stroman, G. N.: Results of 1941 bean strain tests at Deming and Estancia, New Mexico. Pr. Bull. N. Mex. Agric. Exp. Sta., Nr. 934, 2 (1941). [Mimeographed.]

— Albinos in pinto beans. J. Hered. **37**, 59—60 (1946).

—, J. Carter jr. u. J. C. Overpeck: Pinto bean improvement. Bull. Agric Exp. Sta. New Mexico Coll. Agric. Nr. 270, 3—17 (1940).

Sutton, E.: Trisomics in Pisum sativum derived from an interchange heterozygote. Genetics **24**, 88 (1939). (Abstr.)

Sze, L. C.: Structural changes of the meiotic chromosomes of ,,Vicia faba". Acta Brevia Sinensia, Nr. 8, 14 (1944). [Mimeographed.]

Tervet, I. W.: Soybean diseases in Minnesota. Plant Dis. Rep. **27**, 135—138 (1943).

Thadani, K. I. u. R. T. Mirchandani: Studies on soy bean in Sind. Proc. 28th Indian Sci. Congr. Benares, Part III: Sect. Agric. abst. 20, p. 251 (1941).

Thomas, P. T. u. S. H. Revell: Secondary association and heterochromatic attraction. I. Cicer arietinum. Ann. Bot. **10**, 159—164 (1946).

Ting, C. L.: Genetic studies on the wild and cultivated soybeans. J. Amer. Soc. Agron. **38**, 381—393 (1946).

Toit, J. J. du: The cultivation of soybeans. Fmg. S. Afr. **17**, 9—16, 53 (1942).

Torssell, R.: Svalöfs Malmärt (Vrm 01020). Ny foderärtsort för Värmland, Dalarne och angränsande områden av Svealand och Norrland. (Svalöfs Malm pea [Vrm 01020]. A new forage pea variety for Värmland, Dalarne and neighbouring districts of Svealand and Norrland.) Sverig. Utsädesfören. Tidskr. **51**, 71—82 (1941).

— Kvaliteten hos kokärter v 1940 och 1941 års skörd samt hos bruna bönor av 1941 års skörd. Sverig. Utsädesfören. Tidskr. **52**, 419—432 (1942).

Townsend, G. R. u. B. L. Wade: Close-up of something new in snap beans. Two varieties developed for Florida resist rust, mildew, common mosaic and are heat and drouth tolerant. Sth. Seedsman **6**, Nr. 3; 9, 40 (1943).

Trail, F.: Selection of pea strains to meet western needs. West. Cann. Pack. **31**, Nr. 12; 9, 23 (1939).

Tschermak-Seysenegg, E. von: Die Bildung von Traganth, eine Parallelvariation in den Samenschalen der Erbse und der wilden Kicher. Züchter **12**, 161—164 (1940).

— Über einige selbst beobachtete Parallelvariationen der Samenschalenfarbe und Samenform bei Hülsenfrüchten. Züchter **13**, 73—77 (1941).

— Über Linsenzüchtung. Züchter **14**, 81—83 (1942).

— Über Bastarde zwischen Fisole (Phaseolus vulgaris L.) und Feuerbohne (Phaseolus multiflorus Lam.) und ihre eventuelle praktische Verwertbarkeit. Züchter **14**, 153 bis 164 (1942).

— Untersuchungen zur Erklärung des Ertragsunterschiedes zwischen groß- und kleinsamigen Linsen und über Möglichkeiten, den Ertrag der großsamigen Heller-Linsen zu steigern. Züchter **16**, 1—3 (1944).

Vachhani, M. V.: Further inheritance studies of two mutations in Cicer. Indian J. Genet. Pl. Breed. **2**, 173—177 (1942).

Vik, K.: Forsøk med erter og blandinger av erter og korn. (Experiments with peas and mixtures of peas and oats.) Meld. Norg. LandbrHøgsk. **26**, 1—62 (1946).

Wade, B. L.: Frost tolerance of strains of market garden peas. Proc. Amer. Soc. Hort. Sci. **38**, 530—534 (1941).

— Genetic studies of variegation in snap beans. J. Agric. Res. **63**, 661—669 (1941).

— Snap beans for marketing, canning and freezing. Fmr's Bull. U. S. Dep. Agric., Nr. 1915, 14 (1942).

Wade, B. L.: New wax beans have what is takes. Sth. Seedsman 5, Nr. 10; 9, 18, 26 (1942).
— Wando pea has that „someting extra". High quality plus cold tolerance makes pea valuable as home garden type and as parent in breeding projects. Sth. Seedsman 5, Nr. 11; 7, 31, 34 (1942).
— A key to pea varieties. Circ. U. S. Dep. Agric., Nr. 676, 12 (1943).
— Logan, a new, hardy snap bean. Seed World 53, Nr. 5, 12—13, 40—41 (1943).
— u. C. F. Andrus: A genetic study of common bean mosaic under conditions of natural field transmission. J. Agric. Res. 63, 389—393 (1941).
—, P. H. Heinze, M. S. Kanapaux u. C. F. Gaetjens: Inheritance of ascorbic acid content in snap beans. J. Agric. Res. 70, 170—174 (1945).
— u. M. S. Kanapaux: Ascorbic acid content of strains of snap beans. J. Agric. Res. 66, 313—324 (1943).
— u. W. J. Zaumeyer: Genetic studies of resistance to alfalfa mosaic virus and of stringiness in Phaseolus vulgaris. J. Amer. Soc. Agron. 32, 127—134 (1940).
Walker, J. C.: Disease resistant pea varieties. Canner 88 Nr. 12, 2, Pp. 89 (1939).
—, E. J. Delwiche u. W. W. Hare: A major gene for resistance to near-wilt in pea. Phytopathology 34, 1013 (1944). (Abstr.)
— u. J. P. Jolivette: Productivity of mosaic-resistant Refugee beans. Phytopathology 33, 778—788 (1943).
Walls, E. P.: Edible soybeans. Food Packer 25, Nr. 11, 47—49 (1944).
Weber, C. R.: More, better soybeans. Fm Sci. Reporter, Iowa 5, 3—6 (1944).
Weimer, J. L.: Methods of value in breeding Austrian winter field peas for disease resistance in the South. Phytopathology 30, 155—160 (1940).
Weiss, M. G.: Field seed cleander for soybeans. J. Amer. Soc. Agron. 33, 849—850 (1941).
— Inheritance physiology of efficiency in iron utilization in soybeans. Genetics 28, 253—268 (1943).
Wellensiek, S. J.: De invloed der bladeren aan het steriele stengeldeel van Pisum op de ontwikkeling van de plant. (The influence of the leaves on the sterile portion of the stem of Pisum upon the development of the plant.) Hand. XXIIIe Ned. Nat.-en Geneesk. Congr., Pp. 161—164.
— Oogstanalyse. I. Oriëntatie bij erwten. (Yield analysis. I. Orientation work with peas.) Meded. Landb-Hoogesch. Wageningen 45, 29 (1941).
— Pisum-crosses VI: seed surface. Genetica 23, 77—92 (1942).
— Het rassenvraagstuk in het algemeen, geïllustreerd aan erwten. (The variety problem in general, illustrated by peas.) Tech. Ber. Peulvruchten Studiecombinatie, Nr. 26, 4 (1946?).
Went, F. W.: Transplanation experiments in peas. III. Bot. Gaz. 104, 460—474 (1943).
Wester, R. E. u. R. Magruder: Yield results of Lima bean trials in 1945. Canner 103, Nr. 9; 44, 46 (1946).
White, C. T.: The varieties of Tonga or Tongan bean cultivated in Queensland. Qd. Agric. J. 57, 215—216 (1943).
White, N. H. u. T. D. Raphael: The reaction of green pea varieties to downy mildew and two viruses. Tasm. J. Agric. 15, 92—93, 97, 104 (1944).
Williams, L. F.: The breeding work of the U. S. Regional Soybean Laboratory. Soybean Digest 4, Nr. 11; 34, 64 (1944).
— Off-colored seeds in the Lincoln soybean. Soybean Digest 5, Nr. 11; 50, 61 (1945).
Wingard, S. A.: New rust-resistant pole beans of superior quality. Bull. Va. Agric. Exp. Sta., Nr. 350, 31 (1943).
— Victory beans ... ten timely new varieties. Sth. Seedsman 6, Nr 4, 9, 36—37 (1943).
Wingard, S. A.; High tide of rust resistance. Sth. Seedsman 6, Nr. 5; 18, 42 (1943).
Woodbury, G. W.: Selection of beans and peas for processing. Canner 100, Nr. 8, 48—50 (1945).
Woodworth, C. M.: Inhibiting factors in soy beans. Proc. 7th Int. Genet. Congr. Edinburgh 23—30 August, Pp. 318, 1939 (1941). (Abstr.)
York, H. A.: Growing soybeans. Miss. Fm Res. 2, Nr. 6, 3—5; Nr. 7, 3—5 (1939).
— Growing soybeans in the Yasoo-Mississippi Delta. Bull. Miss. Exp. Sta., Nr. 331, 32 (1939).
Youngman, W. H.: America — home of the bean. Agric. Amer. 3, 228—232 (1943).
Zahnley, J. W.: Soybean production in Kansas. Bull. Kans. Agric. Exp. Sta., Nr. 306, 31 (1942).
Zaumeyer, W. J.: The inheritance of a leaf variegation in beans. Phytopathology 31, 26 (1941). (Abstr.)
— Inheritance of a leaf variegation in beans. J. Agric. Res. 64, 119—127 (1942).
— u. L. L. Harter: Inheritance of resistance to six physiologic races of bean rust. J. Agric. Res. 63, 599 bis 622 (1941).
—, — Genetic studies of resistance to six physiologic races of bean rust. Phytopathology 31, 26 (1941). (Abstr.)
—, — Genetic studies of symptom expression of bean mosaic virus 4. Phytopathology 33, 16 (1943). (Abstr.)
—, — Inheritance of symptom expression of bean mosaic virus 4. J. Agric. Res. 67, 295—300 (1943).
—, — Two new virus disease of beans. J. Agric. Res. 67, 305—328 (1943).
—, — Pintos 5 and 14. Sth. Seedsman 9, Nr. 8; 14, 50, 54 (1946).
Žogolev, A. M.: (A new variety of pea — „Novo-Urenskii 01253".) Selekcija i Semenovodstvo (Breeding and Seed Growing), Nr. 10/11, 30 (1939). [Russisch.]

3. Tomaten.

Aberdeen, J. E. C.: Diseases of the tomato and their control. Qd. Agric. J. 60, 277—299 (1945).
— Experiments in the control of bacterial wilt of tomatoes in south-eastern Queensland. Qd. J. Agric. Sci. 3, 87—91 (1946).
Alexander, L. J.: A new strain of the tomato leaf-mould fungus (Cladosporium fulvum). Phytopathology 32, 901—904 (1942).
— Development of a Fusarium-wilt-resistant glasshouse tomato variety. Phytopathology 37, (1947). (Abstr.)
—, R. E. Lincoln u. V. Wright: A survey of the genus Lycopersicon for resistance to the important tomato diseases occurring in Ohio and Indiana. Plant Dis. Rep. Suppl., Nr. 136, 51—85 (1942).
— u. C. M. Tucker: Physiologic specialization in the tomato wilt fungus Fusarium oxysporum f. lycopersici. J. Agric. Res. 70, 303—313 (1945).
Almeida, A. O. de, M. D. Goulart, M. Ielpo u. A. V. Pinto: Estudo da ação inhibidora do suco de „Solanum lycopersicum" sobre a germinação de sementes e crescimento de plantas. (Study of the inhibiting action of the juice of S. lycopersicum on seed germination and plant growth.) Rev. Brasil. Biol. 1, 345—354 (1941).
Alpatjev, A. V.: (Intra-varietal crossing in tomatoes.) Vestnik Ovoščevodstvo i Kartofelj (Vegetable and Potato Journal), Nr. 1, 62—69 (1941). [Russisch.]
— (Production of cold resistant tomatoes.) Vestnik Ovoščevodstvo i Kartofelj (Vegetable and Potato Journal), Nr. 1, 88—92 (1941). [Russisch.]
— (Breeding frost resistant tomatoes for outdoor culture in Central Russia.) Proc. Lenin Acad. Agric. Sci. USSR., Nr. 8/9, 3—11 (1944). [Russisch.]

Androsova, M. P.: (Cold resistance of tomato plants.) Vestnik Ovoščevodstvo i Kartofelj (Vegetable and Potato Journal), Nr. 2, 116—118 (1940). [Russisch.]
— (Cold resistance of the tomato crop.) Ovoščevodstvo (Vegetable Growing), Nr. 2, 21—22 (1940). [Russisch.]
Andrus, C. F.: Resistance of tomato varieties to late blight in South Carolina. Plant Dis. Reporter 30, 269—270 (1946). [Mimeographed.]
— u. G. B. Reynard: Resistance to Septoria leaf spot and its inheritance in tomatoes. Phytopathology 35, 16—24 (1945).
—, —, H. Jorgensen u. J. Eades: Collar rot resistance in tomatoes. J. Agric. Res. 65, 339—346 (1942).
—, — u. B. L. Wade: Relative resistance of tomato varieties, selections, and crosses to defoliation by Alternaria and Stemphylium. Circ. U. S. Dep. Agric., Nr. 652, 23 (1942).
Anonymus: Disease resistant tomatoes. Agric. Gaz. N. S. W. 50, 368—369 (1939).
— Tomato leaf mould. Adv. Leafl. Minist. Agric. Fish., Lond., Nr. 263, 4 (1941).
— Bounty tomato. Bi-m. Bull. N. Dak. Agric. Exp. Sta. 4, Nr. 4, 12 (1942).
— Orange-colored .... no stem-yet it's tomato: „Penn-orange". Sth. Seedsman 6, 18 (1943).
— Seedless tomatoes. Science Suppl. 97, Nr. 2509, 6 (1943).
— Science helps tomatoes. N. J. Agric. 26, 2—3 (1944).
— Wild tomato of Peru has improved world food supply. Agric. Amer. 4, 38 (1944).
— Turrialba, new Latin American tomato, is a „quickie". Canner 101, Nr. 26, 16 (1945).
— New tomato variety. Canad. Grain J. 2, Nr. 8, 14 (1947).
Ashby, E.: Correlation between seed weight and „adult" weight in tomatoes. Nature, London 144, 712 (1939).
Bailey, D. M.: The Essary tomato. Circ. Univ. Tenn. Agric. Exp. Sta., Nr. 71, 4 (1940).
— The seedling test method for root-knotnematode resistance. Proc. Amer. Soc. Hort. Sci. 38, 573—575 (1941).
Banga, O.: Een vergelijking van het voor meeldauw onvatbare tomatenras „Vetomold" met enkele nederlandsche rassen van kastomaten. (A comparison of the mildew immune tomato variety Vetomold with some Dutch varieties of greenhouse tomatoes.) Meded. Tuinbouw. Voorlicht Dienst, 'S-Gravenhage, Nr. 24, 40 (1941).
Barratt, R. W. u. M. C. Richards: Investigations in tne relationship between Alternaria blight and „physiological" maturity in the tomato plant. Phytopathology 33, 1 (1943). (Abstr.)
—, — Alternaria blight versus the genus Lycopersicon. Tech. Bull. N. H. Agric. Exp. Sta., Nr. 82, 24 (1944).
Barrons, K. C.: Spartan Hybrid-a first generation hybrid tomato for greenhouse production. Proc. Amer. Soc. Hort. Sci. 42, 524—528 (1943).
— u. H. E. Lucas: The production of first-generation hybrid tomato seed for commercial planting. Proc. Amer. Soc. Hort. Sci. 40, 395—404 (1942).
Beckenbach, J. R.: Florida AES to release new high yield tomatoes; limited seed samples available by next spring. Sth. Seedsman 8, Nr. 12; 18, 48 (1945).
Bewley, W. F.: New tomato „Vetomold". Leaf mould immune. Fruitgrower 92, 430 (1941).
Bindloss, E. A.: A developmental analysis of cell length as related to stem length. Amer. J. Bot. 29, 179—188 (1942).
Blaydes, G. W.: Chimeras of Lycopersicon and Sanseveria. Amer. J. Bot. Nr. 8 13s (1944).
Blood, H. L.: Breeding disease-resistant tomato varieties for the Intermountain States and the Pacific Coast. Nat. Cann. Ass., Washington, D. C. January 25, Pp. 2 (1939). [Mimeographed.]
Blood, H. L.: Breeding disease resistant tomato varieties. 2. For the Intermountain States and Pacific Coast. Canner 88, Nr. 12, Pt. 2, 87—88 (1939).
— Breeding technique for disease-resistant tomatoes. Study of suitable plant varieties for Pacific Coast and Intermountain States. West. Cann. Pack. 31, Nr. 2 50 (1939).
— Curly top, the most serious menace to tomato production in Utah. Fm. Home Sci. Utah 3, Nr. 1, 8—9, 11 (1942).
— Breeding of wilt resistance in the tomato. Fm. Home Sci., Utah 7, Nr. 3; 3, 14—16 (1946).
— Scientists seek tomato varieties resistant to Verticillium wilt. Reduction of tomato diseases important in program of increased production. Fm. & Home Sci. 2, 5, 8 (1941).
Boddy, F. A.: Varietal resistance to tomato disease. Gdnrs' Chron. 112, 219 (1942).
Bohn, G. W. u. D. H. Scott: A second gene for uniform unripe fruit color in the tomato. J. Hered. 36, 169 bis 172 (1945).
— u. C. M. Tucker: Immunity to Fusarium wilt in the tomato. Science 89, 603—604 (1939).
—, — Studies on Fusarium wilt of the tomato. 1. Immunity in Lycopersicon pimpinellifolium Mill. and its inheritance in hybrids. Res. Bull. Mo. Agric. Exp. Sta., Nr. 311, 82 (1940).
Boldyrev, A. N.: (The question of changing the nature of plants by grafting.) Selekcija i Semenovodstvo (Breeding and Seed Growing), Nr. 6, 5—7 (1940). [Russisch.]
Boswell, V. R.: Disease-resistant and hardy varieties of vegetables. Nat. Hort. Mag. 24, 268—273 (1945).
Bowser, P. H.: The Early Chatham tomato. Quart. Bull. Mich. Agric. Exp. Sta. 25, 245—248 (1943).
Brasher, E. P.: More tomatoes for the war effort. Trans. Peninsula Hort. Soc. 32, Nr. 5, 71—78 (1942).
Bremer: Das Blattrollen der Tomaten. Phytopath. Ztschr. 13, 945 (1941).
Brežnev, D. D.: (Intravarietal crossing as a method of increasing yield in tomatoes.) Ovoščevodstvo, Nr. 6, 38—41 (1939). [Russisch.]
— Selective fertilization and urgent tasks of breeding and seed-growing.) Vestnik Socialističeskogo Rastenievodstva (Soviet Plant Industry Record), Nr. 1, 20—23 (1941). [Russisch.]
— (The utilization of the world collection of tomatoes for breeding work.) Vestnik Socialističeskogo Rastenievodstva (Soviet Plant Industry Record), Nr. 1, 77—88 (1941). [Russisch.]
Brown, F. C.: Dwarf and bush tomatoes. J. R. Hort. Soc. 70, 81—83 (1945).
Brown, G. B. u. G. W. Bohns: Ascorbic acid in fruits of tomato varieties and F_1 hybrids forced in the greenhouse. Proc. Amer. Soc. Hort. Sci. 47, 255 bis 261 (1946).
Brown, H. D.: Miscellaneous field tests 1940. Fertilizers for vegetables. Proc. 26tn Ann. Mtg. Ohio Veg. Potato Grow. Ass., Pp. 5—15 (1941).
Burgess, I. M.: Hybrid vigor in some tomato crosses. Proc. Amer. Soc. Hort. Sci. 38, 570—572 (1941).
Butler, H. F.: Tomato importations for breeding. Proc. Amer. Soc. Hort. Sci. 36, 674—676 (1938) 1939.
Butler, L.: The inheritance of fruit size in the tomato. Canad. J. Res. 19, Sect. C, 216—224 (1941).
Buzulin, G. S.: (Production of early varieties of tomato.) Agrobiologija (Agrobiology), Nr. 2, 140—143 (1946). [Russisch.]

Caldwell, R. M.: Breeding disease resistant tomato varieties. 1. For the Middle West and East. Canner 88, Nr. 12, Pt. 1, 87 (1939).

Campbell, J. A.: Truck Crops Branch Station seeks answer to industry problems. Spezialized research unit, though only two years in operation, has determined and published information vital to growers. Miss. Fm. Res. **3**, Nr. 11, 1 (1940).

Carson, C. M.: This tomato grew up. Sth. Seedsman **10**, Nr. 2, 11 (1947).

Chang, Shien Ta: (Experiments on the grafting of tomatoes.) Fukien Agric. J. 1941, **3**, 37—40, 1942: **4**, 80—81 (1941/1942).

Clarke, E. J. u. G. O. Sherrard: A leaf-spot on tomatoes and its relationship to the variety Vetomold. Gdnrs' Chron. **117**, 71 (1945).

Cooper, D. C. u. R. A. Brink: Seed collapse following matings between diploid and tetraploid races of Lycopersicon pimpinellifolium. Genetics **30**, 376—401 (1945).

Corbett, W.: Experiments on the production of tomatoes in the open. J. Pomol. Hort. Sci. **22**, 1—10 (1946).

Corns, J. B.: Hybrids in vegetable breeding with special reference to sweet corn and tomatoes. Market Gr. J. **66**, 117, 120—124 (1940).

Cottrell-Dormer, W.: An electric pollinator for tomatoes. Qd. J. Agric. Sci. **2**, 157—169 (1945).

Coulter, F. C.: Tomato-the universal favorite was long under suspicion. Seed World. **54**, 10—11, 38 (1943).

Crane, M. B.: „Rogues" and segregation in tomatoes. Gdnrs' Chron. **105**, 92—93, 110—111 (1939).

— Inheritance of resistance to leaf-mould in tomatoes. Gdnrs' Chron. **117**, 123 (1945).

Cunningham, G. H.: Disease-free seed for tomato growers. Scientific experiment promises freatly reduced costs and freedom from pests. Orchard. N. Z. **14**, Nr. 9, 23—24 (1941).

Currence, T. M.: The Mingold tomato and the Duluth snap bean. Market Gr. J. **64**, 216 (1939).

— A combination of semisterility and two simply inherited characters that can be used to reduce the cost of hybrid tomato seed. Rec. Genet. Soc. America, Nr. 12, 47 (1943). (Abstr.)

— A combination of semi- sterility with two simply inherited characters that can be used to reduce the cost of hybrid tomato seed. Proc. Amer. Soc. Hort. Sci. **44**, 403—406 (1944).

— Commercial advantages of hybrid tomatoes. Seed World **55**, Nr. 9, 8—9 (1944).

— u. J. M. Jenkins jr.: Natural crossing in tomatoes as related to distance and direction. Proc. Amer. Soc. Hort. Sci. **41**, 273—276 (1942).

—, R. E. Larson u. A. A. Virta: A comparison of six tomato varieties as parents of F_1 lines resulting from the fifteen possible crosses. Proc. Amer. Soc. Hort. Sci. **45**, 349—352 (1944).

Daskaloff, Chr.: Ergebnisse aus Kreuzungen: Sol. racemigerum × Sarja und Plondiwer. Züchter **14**, 105—111 (1942).

— Neue Ergebnisse aus Kreuzungen: Sol. racemigerum × Sarya und Plowdiwer. Züchter **15**, 92—97 (1943).

Deslandes, J. A.: Fatos sôbre doenças do tomateiro. (Data on the diseases of the tomato plant.) Bol. Minist. Agric., Rio de J. **33**, Nr. 2, 1—70 (1944).

Doolittle, S. P.: Tomato diseases. Fmr's Bull. U. S. Dep. Agric., Nr. 1934, 83 (1943).

—, W. S. Porte u. F. S. Beecher: High resistance to common tobacco mosaic in certain lines of Lycopersicon hirsutum. Phytopathology **36**, 685 (1946). (Abstr.)

Douglass, J.: Selection of tomato seedlings. Discard „buck" plants. Agric. Gaz. N. S. W. **50**, 305—306 (1939).

Eastman, M. G.: Agricultural research in New Hampshire. Rep. Dir. N. H. Agric. Exp. Sta. Bull. **330**, 42, 1940 (1941).

Ekstrand, H.: Ärftliga missbildningar av tomatfrukter. (Heritable malformations of tomato fruits.) Växtskyddsnotiser, Nr. 4/5, 55—57 (1939).

Ellis, D. E.: Root knot resistance in Lycopersicon peruvianum. Plant Dis. Reporter **27**, 402—404 (1943).

Enin, T. K.: Results of an analysis of segregation of tomato hybrids conducted for each family separately. C. R. (Doklady) Acad. Sci. URSS. **24**, 177—178 (1939).

Epps, W. M.: Tomato late blight in South Carolina. Plant Dis. Reporter, Suppl. **165**, 310—312 (1946). [Mimeographed.]

Ermolaeva, N. I.: (An experiment in vegetative hybridization in order to obtain economically valuable types of tomato.) Jarovizacija, Nr. 2 (35), 31—34 (1941). [Russisch.]

Faulkner, R. P.: Tomato Vetomold. Gdnrs' Chron. **112**, 184 (1942).

Fennell, J. L.: A new tomato for the tropics. Agric. Amer. **5**, 233—234 (1945).

— — El „tomate Turrialba". Una nueva variedad para tierras cálidas. (The Turrialba tomato. A new variety for the tropics.) Rev. Inst. Defensa, Café Costa Rica **16**, 443—445 (1946).

Foster, H. H.: Preliminary report on certain tomato lines, selections and varieties, under observation at the Mississippi Truck Crops Branch Station. Plant Dis. Reporter **29**, 396—400 (1945). [Mimeographed.]

— A report on certain tomato lines, selections and varieties under observation at the Mississippi Truck Crops Branch Station during 1945 and 1946. Plant Dis. Reporter **30**, 410—416 (1946).

— u. J. A. Campbell: Observations on Mississippi tomato diseases during 1946. Plant. Dis. Reporter **30**, 339—342 (1946). [Mimeographed.]

Foster, R. E. u. J. C. Walker: Predisposition of tomato to Fusarium wilt. J. Agric. Res. **74**, 165—185 (1947).

Frazier, W. A., K. Kikuta, J. S. McFarlane u. J. W. Hendrix: Tomato improvement in Hawaii. Proc. Amer. Soc. Hort. Sci. **47**, 277—284 (1946).

Frimmel, F. v.: Welcher Artbildungsvorgang hat zur Domestikationsform Lycopersicum esculentum Mill. geführt? Z. Pflanz. **25**, 437—442 (1943).

— u. K. Lauche: Neue Wege der Züchtung auf Frühreife der Tomaten. Z. Pflanzenzüchtung **24**, 374—382 (1941).

Gorostiaga, A.: Acido asorbico (Vitamin C) en tomates, pimientos morrones y ajies. (Ascorbic acid [vitamin C] in tomatoes, sweet peppers and chillies.) Rev. Fac. Agron. Univ. Montevideo, Nr. 35, 99—112 (1944).

Guba, E. F.: A red forcing tomato resistant to Cladosporium leaf mold. Phytopathology **29**, 9 (1939). (Abstr.)

— Bay State, a red forcing tomato bred for resistance to leaf mold. Bull. Mass. Agric. Exp. Sta., Nr. 393, 8 (1942).

Gustafson, F. G.: B-naphthoxyacetic acid as an inductor of parthenocarpy in tomatoes. Proc. Amer. Soc. Hort. Sci. **40**, 387—389 (1942).

Györffy, B.: A paradicsom C-vitamin tartalmáról. (Ascorbic acid content of tomatoes.) Arb. Ung. Biol. Forsch.-Inst. **15**, 441—449 (1943).

Haber, E. S.: Outstanding sweet corn and tomato varieties at Ames in 1939. Trans. Iowa Hort. Soc. **74**, 93—98 (1939).

Hackbarth, J.: Dreijährige Ertragsprüfungen mit Tomaten-Zuchtstämmen aus Kreuzungen mit Sol. racemigerum. Gartenbauwiss. **15**, 36—47 (1940).

Hackbarth, J.: Vererbungsversuche mit Tomaten. Biologe 12, 130—134 (1943).

Hallsworth, E. G. u. V. M. Lewis: Variation of asorbic acid in tomatoes. Nature, Lond. 154, 431—432 (1944).

—, — Some factors affecting the ascorbic-acid content of tomatoes in New South Wales. Emp. J. Exp. Agric. 15, 132—147 (1947).

Hammer, K. C. u. L. A. Maynard: Factors influencing the nutritive value of the tomato. A review of the literature. Misc. Publ. U. S. Dep. Agric., Nr. 502, 23 (1942).

Hardin, M.: Breeding drought-resistant tomatoes. Market Gr. J. 70, 129 (1942).

Hargrave, P. D.: Tomato varietal yield tests. Sci. Agric. 23, 322—326 (1943).

Harrison, A. L.: Recent research on the control of tomato diseases. Proc. 42nd Annu. Cony. Ass. Sth. Agric. Wkrs, Atlanta, Ga., February 5—7, Pp. 164 bis 165 (1941).

Haškova, O.: (Grafting as a means of plant transformation.) Proc. Lenin Acad. Agric. Sci. USSR., Nr. 8/9, 12—18 (1944). [Russisch.]

Hatcher, E. S. J.: Hybrid vigour in the tomato. Nature, Lond. 143, 523 (1939).

— Studies in the inheritance of physiological characters. V. Hybrid vigour in the tomato. Part III. A critical examination of the relation of embryo development to the manifestation of hybrid vigour. Ann. Bot., London 4, 735—764 (1940).

— Studies in the inheritance of physiological characters. VI. Hybrid vigour in the tomato. Part IV. The effect of flower removal on the manifestation of hybrid vigour. Ann. Bot., London 5, 501—508 (1941).

Hawthorn, L. R.: Breeding summer tomatoes for increased size. Proc. Amer. Soc. Hort. Sci. 40, 390—394 (1942).

Heinze, P. H. u. C. F. Andrus: Apparent localization of Fusarium wilt resistance in the Pan American tomato. Amer. J. Bot. 32, 62—66 (1945).

Hendrix, J. W., K. Kikuta u. W. A. Frazier: Breeding tomatoes for resistance to gray leaf spot in Hawaii. Proc. Amer. Soc. Hort. Sci. 47, 294—300 (1946).

Hofmeyr, J. D.: Good news for tomato growers. A new tomato resistant to fusarium wilt disease. Fmg. S. Afr. 17, 356 (1942).

Holmes, F. O.: The Chilean tomato, Lycopersicon chilense, as a possible source of disease resistance. Phytopathology 29, 215—216 (1939). (Abstr.)

— Derivatives of tomato that tend to escape tobacco-mosaic disease. Phytopathology 33, 19 (1943). (Abstr.)

— A tendency to escape tobacco-mosaic disease in derivatives from a hybrid tomato. Phytopathology 33, 691—697 (1943).

Hontschik, A.: Beitrag zur Morphologie der Tomate. Z. Pflanzenz. 26, 127—135 (1944).

Houghtaling, H. B.: Stem morphogenesis in Lycopersicum: a quantitative study of cell size and number in the tomato. Bull. Torrey Bot. Cl. 67, 33—55 (1940).

Howlett, F. S.: Fruit set and development from pollinated tomato flowers treated with indolebutyric acid. Proc. Amer. Soc. Hort. Sci. 41, 277—281 (1942).

Huelsen, W. A.: Wilt-resistant tomato varieties released by the Illinois Station. Circ. Ill. Agric. Exp. Sta., Nr. 490, 22 (1939).

Hughes, H. M.: Outdoor tomato trials, 1944. 32nd Rep. E. Malling Res. Sta., Pp. 38—43, 1944 (1945).

Hutton, E. M.: Present and future trends of tomato varieties in Australia. J. Aust. Inst. Agric. Sci. 11, 128—134 (1945).

Ipatjev, A. N. u. A. V. Gaenko: (Analysis of earliness in the tomato.) Trudy Omsk. Seljskohozjajstvennogo Inst. imeni S. M. Kirova (Trans. Kirov Inst. Agric. Omsk, USSR.) 4 (17), 127—131 (1939). [Russisch.]

Irving, G. W. jr., T. D. Fontaine u. S. P. Doolittle: Lycopersicin, a fungistatic agent from the tomato plant. Science 102, 9—11 (1945).

Isbell, C. L.: Pass the word to gardeners: graft tomatoes, onto weeds. Sth. Seedsman 7, 14, 42 (1944).

Jacynina, K. N.: Breeding for tomato variety resistant to bacterial canker Aplanobacter michiganense E. F. Smith. C. R. (Doklady) Acad. Sci. URSS. 32, 372 bis 373 (1941).

Janes, B. E.: Some chemical differences between artificially produced parthenocarpic fruits and normal seeded fruits of tomato. Amer. J. Bot. 28, 639—646 (1941).

Kajewski, S. F.: Cross breeding experiments in the Bowen district. Qd. Agric. J. 56, 473—476 (1941).

Kangas, J. T.: Produce vegetable varieties adapted to New England. Market Gr. J. 67, 105 (1941).

Kerkis, J. J. u. N. N. Pigulevskaja: Interaction between Lycopersicum esculentum and Datura stramonium in the case of grafting. C. R. (Doklady) Acad. Sci. URSS. 32, 505—508 (1941).

Kidson, E. B.: Vitamin C content of different tomato varieties grown in the Nelson district. N. Z. J. Sci. Tech. 25, Sect. B, 129—134 (1943).

Kikuta, K. u. W. A. Frazier: Breeding tomatoes for resistance to spotted wilt in Hawaii. Proc. Amer. Soc. Hort. Sci. 47, 271—276 (1946).

—, J. W. Hendrix u. W. A. Frazier: Pearl Harbor. A tomato variety resistant to spotted wilt in Hawaii. Circ. Hawaii Agric. Exp. Sta., Nr. 24, 4 (1945).

Koleff, N.: Neues Zuchtziel bei der Tomatenzüchtung. Züchter 14, 264—266 (1942).

Kreutzer, W. A. u. L. R. Bryant: Varietal resistance and mulching show promise in control of tomato fruit rot. Colo. Fm. Bull. 5, 12—15 (1943).

Kuljčenko, N. I.: (Intravarietal crossing of tomatoes in the greenhouse.) Ovoščevodstvo, Nr. 6, 41—42 (1939). [Russisch.]

Lamm, R.: A case of abnormal meiosis in Lycopersicum esculentum. Hereditas, Lund 30, 253 (1944). (Abstr.)

Lande, O.: Tomatoes on firing line. The way to give „staying power" to our fighting yanks. Market Gr. J. 70, 162—163, 168—169 (1942).

Larson, R. E.: The F_1 combining ability of certain tomato varieties. Proc. Amer. Soc. Hort. Sci. 39, 313—314 (1941).

— u. T. M. Currence: The extent of hybrid vigor in F_1 and F_2 generations of tomato crosses. Tech. Bull. Minn. Agric. Exp. Sta., Nr. 164, 32 (1944).

— u. W. L. Marchant: The response of three F_1 lines and ten strains of tomatoes to two distinct soil types. Proc. Amer. Soc. Hort. Sci. 45, 341—347 (1944).

— The ring rot bacterium in relation to tomato and eggplant. J. Agric. Res. 69, 309—325 (1944).

Lee, F. A. u. C. B. Sayre: Factors affecting the acid and totals solids content of tomatoes. Tech. Bull. N. Y. St. Agric. Exp. Sta., Nr. 278, 28 (1946).

Lesley, J. W. u. M. Lesley: Unfruitfulness in the tomato caused by male sterility. J. Agric. Res. 58, 621—630 (1939).

—, — An hereditary variegation in tomatoes associated with sterility. Amer. J. Bot. 28, 727 (1941). (Abstr.)

—, — An hereditary variegation in tomatoes. Genetics 27, 550—560 (1942).

—, — Parthenocarpy in a deficient tomato plant and in its aneuploid progeny. Genetics 26, 159—160 (1941). (Abstr.)

—, — Parthenocarpy in a tomato deficient for a part of a chromosome and in its aneuploid progeny. Genetics 26, 374—386 (1941).

Lesley, J. W. u. M. Lesley: Hybrids of the Chilean tomato. Sterile and fertile plants of Lycopersicon peruvianum Dun. (L. chilense Dun.) and diploid and tetraploid hybrids with cultivated tomatoes. J. Hered. 34, 199—205 (1943).
—, — A genetic study of flesh color of fruit in hybrids of Lycopersicon esculentum Mill. and L. hirsutum Humb. and Bonpl. Genetics 31, 222—223 (1946). (Abstr.)
Leslie, W. R.: Manitoba news letter. N. S. Dak. Hort. 18, 52, 60 (1945).
— Manitoba news letter. N. S. Dak. Hort. 20, 21, 29 (1947).
Lincoln, R. E., F. P. Zscheile, J. W. Porter, G. W. Kohler u. R. M. Caldwell: Provitamin A and vitamin C in the genus Lycopersicon. Bot. Gaz. 105, 113—115 (1943).
Lindstrom, E. W.: Genetic stability of tomato diploids and tetraploids derived from haploid. Proc. Iowa Acad. Sci. 47, 75 (1940).
— Genetic stability of haploid, diploid, and tetraploid genotypes in the tomato. Genetics 26, 387—397 (1941).
Locke, S. B.: Resistance in South American Lycopersicon species to early blight and Septoria blight. Phytopathology 32, 12 (1942). (Abstr.)
Luckwill, L. C.: Heterosis in Lycopersicum crosses in relation to seed weight. Nature 144, 908 (1939).
— Observations on heterosis in Lycopersicum. J. Genet. 37, 421—440 (1939).
— The evolution of the cultivated tomato. J. R. Hort. Soc. 68, 19—25 (1943).
— The genus Lycopersicon. An historical, biological, and taxonomic survey of the wild and cultivated tomatoes. Aberd. Univ. Stud., Nr. 120, 44 (1943).
Lyon, C. B.: Inheritance of stages of earliness in an interspecific cross between Lycopersicon esculentum and L. pimpinellifolium. J. Agric. Res. 63, 175—182 (1941).
— Responses of two species of tomatoes and the F_1 generation to sodium sulphate in the nutrient medium. Bot. Gaz. 103, 107—122 (1941).
—, K. C. Beeson u. M. Barrentine: Macro-element nutrition of the tomato plant as correlated with fruitfulness and occurence of blossom-end rot. Bot. Gaz. 103, 651—667 (1942).

MacArthur, J. W.: Size inheritance in tomato fruits. J. Hered. 32, 291—295 (1941).
— u. L. P. Chiasson: Cytogenetic notes on tomato species and hybrids. Genetics 32, 165—177 (1947).
McFarlane, J. S., E. Hartzler u. W. A. Frazier: Breeding tomatoes for nematode resistance and for high vitamin C content in Hawaii. Proc. Amer. Soc. Hort. Sci. 47, 262—270 (1946).
Maher, F. A.: „Tatura Dwarf Globe". A new tomato suitable for market and canning. J. Dep. Agric. Vict. 40, 241—244 (1942).
Mattson, H.: „Bounty" tomato. Preliminary report and description. Bi-m. Bull. N. Dak. Agric. Exp. Sta. 3, Nr. 3, 11—14 (1941).
— Bounty tomato in standard yield trials in 1940 and 1941. Bull. N. Dak. Agr. Exp. Sta., Nr. 310, 7 (1942).
Meyer, A. u. N. D. Peacock: Heterosis in the tomato as determined by yield. Proc. Amer. Soc. Hort. Sci. 38, 576—580 (1941).
Miller, J. C.: New breeding theory is boon to southland. Length-of-day experiments result in prolific „Dixie" tomato and explain shortcomings of many northern-bred varieties. Sth. Seedsman 5, Nr. 12; 7, 35 (1942).
— u. J. J. Mikell: Linkage studies of certain characters in tomatoes. Proc. Amer. Soc. Hort. Sci. 37, 884—885 (1939) 1940.

Mills, M. u. E. M. Hutton: Fusarium wilt of tomato in Australia. 1. The relationship between different isolates of the pathogen and resistance in varieties of Lycopersicon esculentum Mill. and other Lycopersicon species. J. Coun. Sci. Industr. Res. Aust. 19, 376—386 (1946).
Mühlendyck, E.: Beiträge zur Morphologie der Frucht der Tomate (Solanum Lycopersicum). Z. Pflanzenz. 25, 117—163 (1943).
Muller, C. H.: A revision of the genus Lycopersicon. Misc. Publ. U. S. Dep. Agric., Nr. 382, 29 (1940).
— The taxonomy and distribution of the genus Lycopersicon. Nat. Hort. Mag. 19, 157—160 (1940).
Myers, C. E.: Something new in tomatoes. Seed World 55, 20, 36 (1944).

Newcomer, E. H.: A colchicine-induced homozygous tomato obtained through doubling clonal haploids. Proc. Amer. Soc. Hort. Sci. 38, 610—612 (1941).
Nicolaisen, N.: Fremdbefruchtung bei den Kulturtomaten und Stangenbohnen. Forschungsdienst 13, 124—129 (1942).

Oba, G. I., M. E. Riner u. D. H. Scott: Experimental production of hybrid tomato seed. Proc. Amer. Soc. Hort. Sci. 46, 269—276 (1945).

Pal, B. P. u. H. B. Singh: A note on the economic possibilities of the cross, Lycopersicon esculentum × L. pimpinellifolium. Indian J. Genet. Pl. Breed. 3, 115—120 (1943).
Paščenko, T. E.: (Biology of tomato-flowering.) Proc. Lenin. Acad. Agric. Sci. USSR., Nr. 12, 15—19 (1940). [Russisch.]
Poljankova, T.: (Changes in the chromosome number and morphology in tomatoes under the influence of grafting.) Agrobiologija (Agrobiology), Nr. 2, 128—130 (1946). [Russisch.]
Pollard, A., M. Kieser u. J. D. Bryan: Factors influencing the vitamin C content of tomatoes. Rep. Agric. Hort. Res. Sta., Long Ashton, Bristol, Pp. 171 bis 179 (1944).
—, —, — Factors influencing the composition of the tomato. A comparison of varieties and of indoor and outdoor culture. Rep. Agric. Hort. Res. Sta., Long Ashton, Bristol, Pp. 203—208 (1945).
Porte, W. S., S. P. Doolittle u. F. L. Wellman: Hybridization of a mosaic-tolerant, wiltresistant Lycopersicon hirsutum with Lycopersicon esculentum. Phytopathology 29, 757—759 (1939). (Abstr.)
— u. H. B. Walker: The Pan America tomato, a new red variety highly resistant to fusarium wilt. Circ. U. S. Dep. Agric., Nr. 611, 6 (1941).
—, — A cross between Lycopersicon esculentum and disease-resistant L. peruvianum. Phytopathology 35, 931—933 (1945).
— u. F. L. Wellman: Development of interspecific tomato hybrids of horticultural value and highly resistant to Fusarium wilt. Circ. U. S. Dep. Agric., Nr. 584, 19 (1941).
Porter, J. W. u. F. P. Zscheile: Carotenes of Lycopersicon species and strains. Arch. Biochem. 10, 537—545 (1946).
Potašnikova, B. G.: (Promoting early ripening and increasing the yield of the seed generation of the tomato grafted on Black Nightshade.) Jarovizacija, Nr. 2 (35), 114—116 (1941). [Russisch.]
Powers, L.: Studies on the nature of the interactions of the genes differentiating quantitative characters in a cross between Lycopersicon esculentum and L. pimpinellifolium. J. Genet. 39, 139—170 (1939).
— Inheritance of quantitative characters in crosses involving two species of Lycopersicon. J. Agric. Res. 63, 149—174 (1941).

Powers, L.: The nature of the series of environmental variances and the estimation of the genetic variances and the geometric means in crosses involving species of Lycopersicon. Genetics **27**, 561—575 (1942).
— Relative yields of inbred lines and F_1 hybrids of tomato. Bot. Gaz. **106**, 247—268 (1945).
— u. C. B. Lyon: Inheritance studies on duration of developmental stages in crosses within the genus Lycopersicon. J. Agric. Res. **93**, 129—148 (1941).

Reynard, G. B.: „Dunk" tomatoes. Seed World **52**, Nr. 10, 35 (1942).
— „Dunking" tomatoes for their health. Sth. Seedsman **5**, Nr. 1; 10, 30 (1942).
— Polycotyledony in the genus Lycopersicon. Abstr. Diss. Univ. Md. **41**, Nr. 6, 17—18 (1944). (Abstr.)
— u. C. F. Andrus: Inheritance of resistance to the collar-rot phase of Alternaria solani on tomato. Phytopathology **35**, 25—36 (1945).
— u. M. S. Kanapaux: Ascorbic acid (vitamin C) content of some tomato varieties and species. Proc. Amer. Soc. Hort. Sci. **41**, 298—300 (1942).
Richards, M. C. u. R. W. Barratt: A partial survey of the genus Lycopersicon for resistance to Phytophthora infestans. Plant Dis. Reporter **30**, 16—20 (1946). [Mimeographed.]
Riche, F. J. H. le: Asorbic acid content of tomato varieties. Fmg. S. Afr. **20**, 105—110 (1945).
Rick, C. M.: A new male-sterile mutant in the tomato. Science **99**, 543 (1944).
— A survey of cytogenetic causes of unfruitfulness in the tomato. Genetics **30**, 347—362 (1945).
— Field identification of genetically male-sterile tomato plants for use in producing F_1 hybrid seed. Proc. Amer. Soc. Hort. Sci. **46**, 277—283 (1945).
— The development of sterile ovules in Lycopersicum esculentum Mill. Amer. J. Bot. **33**, 250—256 (1946).
— A hair-suppressing gene that indirectly affects fruitfulness and the proportion of cross-pollination in the tomato. Genetics **32**, 101—102 (1947). (Abstr.)
Robbins, W. J.: Growth of excised roots and heterosis in tomato. Amer. J. Bot. **28**, 216—225; also Science 1940, **92**, 416 (1941). (Abstr.).
— u. V. Kavanagh: Growth, of excised roots of polyploid tomatoes. Amer. J. Bot. **30**, 602—605 (1943).
Romshe, F. A.: Experiments with greenhouse tomatoes: varieties, cultural methods, and relationship between yield and vegetative vigor. Bull. Okla. Agric. Exp. Sta., Nr. B-260, 30 (1942).
— Nematode resistance test of tomatoes. Proc. Amer. Soc. Hort. Sci. **40**, 423 (1942).
Ross, A. A.: Studies of growth correlations in the tomato. Qd. J. Agric. Sci. **3**, 121—156 (1946).
Ryžkov, V. L. u. P. V. Mihajlova: (On ethyology of the leaf-roll in tomatoes.) Izvestija Akademii Nauk SSSR. (Bull. Acad. Sci. URSS. Sér. Biol.), Nr. 5, 487—489 (1946). [Russisch.]

Safir, S. A.: (Production of frost-resistant tomatoes.) Ovoščevodstvo (Vegetable Growing), Nr. 2, 22—23 (1940). [Russisch.]
Šapovalov, M. u. J. W. Lesley: Wilt resistance of the Riverside variety of tomato to both Fusarium and Verticillium wilts. Phytopathology **30**, 760—768 (1940).
— u. B. A. Rudolph: Essar — a new Verticillium wilt resistant canning tomato Seed World **46**, Nr. 13, 12—13 (1939).
Schaper, P.: Arbeiten und Probleme zur züchterischen Bekämpfung des Kartoffelkäfers. IV. Untersuchungen über das Verhalten von Tomaten gegen den Befall und Fraß des Kartoffelkäfers. Z. Pflanzenz. **23**, 454 bis 475 (1940).

Scott, G W : Tomato: King of processed vegetables — And the West held the Coronation. West. Cann. Pack. **35**, 17—19 (1943).
Selivanova, A. N. u. A. V. Alpatjev: (Early varieties of tomato for the conserving industry.) Ovoščevodstvo (Vegetable Growing), Nr. 12, 34—36 (1939). [Russisch].
Sengbusch, R. v.: Tomatenzüchtung. Frostwiderstandsfähigkeit, Lagerfähigkeit, Hochglanz der Fruchtschale und Zwergformen. Pflanzenbau **17**, 143—152 (1940).
Shapovalov, M. siehe Šapovalov.
Shifriss, O.: Hybrid tomato. Sth. Seedsman **8**, 15, 26, 30 (1945).
Shilova, S. N. siehe Šilova.
Šilova, S. N.: (Intravarietal hybridization in tomatoes.) Vestnik Socialističeskogo Rastenievodstva (Soviet Plant Industry Record.) Nr. 1, 24—30 (1941). [Russisch.]
Sivrina, A. N.: (Inheritance of chemical characters in tomatoes.) Vestnik Socialističeskogo Rastenievodstva (Soviet Plant Industry Record), Nr 5, 133—141 (1940). [Russisch.]
Smith, K: The Rutgers tomato. N. J. Hort. Soc. News **20**, 1127 (1939).
Smith, P. G.: Embryo culture of a tomato species hybrid. Proc. Amer. Soc. Hort. Sci. **44**, 413—416 (1944).
— Reaction of Lycopersicon spp. to spotted wilt. Phytopathology **34**, 504—505 (1944).
Snyder, W. C., K. F. Baker u. H. N. Hansen: Interpretation of resistance to Fusarium wilt in tomato. Science **103**, 707—708 (1946).
Sokolova, A. M.: (Tetraploid tomatoes.) Vestnik Ovoščevodstvo i Kartofelj (Vegetable and Potato Journal), Nr. 2, 39—47 (1941). [Russisch.]
Solovjeva, N. A.: (Changes in the chemical composition of tomatoes as a result of grafting.) Jarovizacija. Nr. 2 (35), 109—111 (1941). [Russisch.]
Somos, A.: Adatok a „Turul" paradicsomfajta ismeretéhez. (A new tomato variety „Turul".) Bull. Hung. Coll. Hort. **9**, 87—94 (1943).
Stair, E. C. u. R. K. Showalter: Tetraploidy in tomatoes induced by the use of colchicine. Proc. Amer. Soc. Hort. Sci. **40**, 383—386 (1942).
Stearn, W. T.: Where did the tomato originate? Gdnrs' Chron, **118**, 6, 18—19, 30—31 (1945).
Strachan, C. C. u. F. E. Atkinson: Ascorbic acid content of tomato varieties and its retention in processed products. Sci. Agric. **26**, 83—94 (1946).

Tucker, C. M.: The development of wilt resistant tomatoes. Proc. Mo. Acad. Sci. **7**, 90—91 (1942).

Vincent, C. L.: Washington State: a new forcing tomato. Bull. Wash. Agric. Exp. Sta., Nr. 436, 12 (1944).
Virgin, W. J.: The Chilean tomato, Lycopersicon chilense, found resistant to curly top. Phytopathology **30**, 280 (1940). (Abstr.)
— u. J. C. Maloit: The use of the seedling inoculation technique for testing tomatoes for resistance to Verticillium wilt. Phytopathology **37**, 22—23 (1947). (Abstr.)

Walker, W. F.: Tomato production trials and recommendations. Tasm. J. Agric. **16**, 138—143 (1945).
— u. K. W. Pierce: Tomato production investigations, 1943—1944. Tasm. J. Agric. **15**, 71—82 (1944).
Wallace, J. M.: Acquired immunity from curly top in tobacco and tomato. J. Agric. Res. **69**, 187—214 (1943).
Weaver, J. G.: Seeking a tomato resistant to bacterial wilt. Res. and Fmg., N. C. **3**, Progr. Rep., Nr. 1, 11 (1944).

Wenholz, H.: Spotted wilt of tomatoes. Breeding for resistance. Hawkesbury Agric. Coll. J. **36**, 103 (1939).
Werner, H. O.: Two new tomato varieties. Rep. Neb. Bd. Agric., Pp. 350—358 (1944).
Whaley, W. G.: The relation of organ size to meristem size in the tomato. Proc. Amer. Soc. Hort. Sci. **37**, 910—912 (1939) 1940.
Wilson, K. S. u. C. L. Withner jr.: Stock-scion relationships in tomatoes. Amer. J. Bot. **33**, 796—801 (1946).
Wokes, F. u. J. G. Organ: Oxidizing enzymes and vitamin C in tomatoes. Biochem. J. **37**, 259—265 (1943).

Yamada, T.: A new tomato. Hawaii Fm. and Home **7**, Nr. 4, 7 (1944).
Yeager, A. F.: The Victor tomato. Quart. Bull. Mich. Agric. Exp. Sta. **23**, 3—6 (1940).
— Do you know tomatoes? Nat. Hort. Mag. **24**, 126 bis 127 (1945).
Young, P. A.: Resistance of tomato varieties to Fusarium lycopersici. Phytopathology **29**, 25 (1939). (Abstr.)
— White-flower character from x-ray treatment of tomato seed. J. Hered. **31**, 78—79 (1940).
— New genetic characters of the tomato. Amer. Nat. **75**, 280—282 (1941).
— Wilt-resistant tomatoes with new genetic characters. Phytopathology **32**, 24 (1942). (Abstr.)
— Varietal resistance to blossom-end rot in tomatoes. Phytopathology **32**, 214—220 (1942).
— Two genetic characters of tomato fruits that might be mistaken for symptoms of disease. Phytopathology **32**, 436—438 (1942).
— The Rainbow tomato. Sth. Seedsman **8**, Nr. 10; 14, 52 (1945).

Zielinski, Q.: Fasciation in horticultural plants with special reference to the tomato. Proc. Amer. Soc. Hort. Sci. **46**, 263—268 (1945).

4. Kohl.

Anonymus: Sortenbereinigung bei Kopfkohl. Mitt. landw. **56**, 880 (1941).
— Brussels sprouts. Adv. Leafl. Minist. Agric. Fish., Lancs., Nr. 209, 4 (1943).
— Cabbage for dehydration. J. Minist. Agric. **50**, 656 bis 667 (1944).
— (Cauliflower varieties and the new variety Early Beth Alpha.) Hassadeh **25**, 317 (1945).

Bailey, L. H.: Certain noteworthy Brassicas. Gentes Herbarum, Ithaca, N. Y. **4**, Fasc. 9, Art 26, Pp. 319 bis 330 (1940).
Barr, C. G. u. E. H. Newcomer: Physiological aspects of tetraploidy in cabbage. J. Agric. Res. **67**, 329—336 (1943).
Bishop, C. J.: The genetical basis of sterility in tetraploid broccoli. Genetics **32**, 78 (1947). (Abstr.)
Boswell, V. R.: Disease-resistant and hardy varieties of vegetables. Nat. Hort. Mag. **23**, 138—143 (1944).
— Commercial cabbage culture. Circ. U. S. Dep. Agric. 1933 (Revised 1945), Nr. 252, 60 (1945).
Brieger, F. G. u. J. T. A. Gurgel: Seleção e produção de sementes em hortaliças (com referência especial ao gênero Brassica). (Selection and seed production in vegetables [with special reference to the genus Brassica].) Bragantia, São Paulo **2**, 449—480 (1942).

Calder, R. A.: Marrow-stem kale (B. oleracea L.) Investigations leading up to selection and seed production. N. Z. J. Sci. Tech. **21**, 223—229 A (1939).
Caldwell, J. S., C. W. Culpepper, B. D. Ezell, M. C. Hutchins u. M. S. Wilcox: The dehydration of kale. Canner **101**, Nr. 25, 13—14, 16, 28, 30, 32 (1945).
Claridge, J. H. u. R. A. Calder: Chou moellier and thousand-headed kale. N. Z. J. Agric. **68**, 347—349 (1944).
Coulter, F. C.: The story of garden vegetables. V. Cabbage: planted the world over for peasant and for king. Seed World **50**, Nr. 6, 18—19 (1941).
Crane, M. B.: The origin and relationship of the Brassica crops. J. R. Hort. Soc. **68**, 172—174 (1943).

Davis, J. F.: A comparison of methods for harvesting experimental plots of cabbage. Proc. Amer. Soc. Hort. Sci. **47**, 327—330 (1946).
Detjen, L. R.: Fixation of cabbage varieties. Proc. Amer. Soc. Hort. Sci. **45**, 362—366 (1944).
— u. W. H. Phillips: Relative effects of superior vs. inferior seed-branch positions in cabbage on time of seedstalk initiation in the immediate progenies of inbred plants. Bull. Del. Agric. Exp. Sta., Nr. 245 (Tech. Nr. 30), 21 (1943).
Drewes, H.: Cabbage breeding. Market Gr. J. **65**, 486—488 (1939).

Foster, R. E. u. J. C. Walker: Improvement of ascorbic acid content in yellows-resistant cabbage. Phytopathology **36**, 398 (1946). (Abstr.)

Giles, W. F.: Cauliflower and broccoli. What they are and where they come from. J. Roy. Hort. Soc. **66**, 265—277 (1941).
Gurgel, J. T. A.: Experimentos sôbre hortaliças. (Experiments with vegetables.) Rev. Agric. Piracicaba **18**, 450—454 (1943).

Himič, R. E.: (Cauliflower breeding.) Vestnik Ovoščevodstvo i Kartofelj (Vegetable and Potato Journal), Nr. 2, 48—50 (1941). [Russisch.]
Howard, H. W.: The cytology of autotetraploid kale, Brassica oleracea. Cytologia, Tokyo **10**, 77—87 (1939).
— The size of seeds in diploid and autotetraploid Brassica oleracea L. J. Genet. **38**, 325—340 (1939).
— Experimental polyploidy in the genera Raphanus and Brassica. Abstr. Diss. Univ. Camb. (1938—1939), Pp. 23—25 (1940).
— The effect of polyploidy and hybridity on seed size in crosses between Brassica chinensis, B. carinata, amphidiploid B. chinensis-carinata and autotetraploid B. chinensis. J. Genet. **43**, 105—119 (1942).
— Heteroauxin and the production of tetraploid shoots by the callus method in Brassica oleracea. J. Genet. **44**, 1—9 (1942).

Inoue, Y.: (Colchicine-induced tetraploid in Chinese cabbage [Brassica pekinensis Rupr.].) Jap. J. Genet. **15**, 318—319 (1939).
Isbell, C. L.: Propagating cabbage by root cuttings. Proc. Amer. Soc. Hort. Sci. **46**, 341—344 (1945).
— Further observations on and the application of propagating cabbage by leaf cuttings. Proc. Amer. Soc. Hort. Sci. **47**, 335—339 (1946).

Janes, B. E.: The relative effect of variety and environment in determining the variations of per cent dry weight, ascorbic acid, and carotene content of cabbage and beans. Proc. Amer. Soc. Hort. Sci. **45**, 387—390 (1944).
Jannaccone, A.: Selezione del cavolfiore Gigante Tardivo di Napoli. (Selection of the cauliflower Gigante Tardivo di Napoli.) Servizio delle Sementi, Pp. 71—88 (1939).

Koloberdina, Z. I.: (The use of heterosis for increasing the yield of cabbage.) Vestnik Ovoščevodstvo i Kartofelj (Vegetable and Potato Journal), Nr. 1, 79—84 (1941). [Russisch.]

Kraus, J. E.: Chinese cabbage varieties, their classification, description, and culture in the Central Great Plains. Circ. U. S. Dep. Agric., Nr. 571, 20 (1940).
Krickl, M.: Möglichkeiten im Adventgemüsebau. Züchtung der Sommersorten von Kopfkohl auf Wintersorten. Züchter 13, 197—207 (1941).
— Züchtungsversuche zur Beeinflussung der Kopfbildung bei Kopfkohlarten. Züchter 14, 185—196 (1942).

Lang, J. M. S.: A trial of broccoli varieties in Scotland. Scot. J. Agric. 24, 113—117 (1943).
Levy, F.: (Cauliflower and the development of new varieties in Palestine.) Hassadeh 25, 222—223 (1945).
Lorenz, O. A.: Response of Chinese cabbage to temperature and photoperiod. Proc. Amer. Soc. Hort. Sci. 47, 309—319 (1946).

Menezes, O. B. de: Número de cromossômios em Brassica e Lactuca. (Chromosome number in Brassica and Lactuca.) Rev. Agric., Piracicaba 18, 277—278 (1943).
Miller, J. C.: Improved cabbage technique benefits trade. Sth. Seedsman 7, 13, 38 (1944).
Myers, C. E.: The Penn State Ballhead cabbage. Some problems encountered in its development. Bull. Agric. Exp. Sta., Pa., Nr. 430, 52 (1942).
Myers, C. H. u. W. 1. Fisher: Experimental methods in cabbage breeding and seed production. Mem. Cornell Agric. Exp. Sta., Nr. 259, 29 (1944).

Newcomer, E. H.: A colchicine-induced tetraploid cabbage. Amer. Nat. 75, 620 (1941).
— An F_2 colchicine-induced tetraploid cabbage and some comparisons with its diploid progenitor. J. Elisha Mitchell Sci. Soc. 59, 69—72 (1943).

Poole, C. F.: Vitamin C in cabbage. Sth. Seedsman 6, 10, 47, 50 (1943).
— u. P. C. Grimball: Ringing in the vitamins. Sth. Seedsman 7, 11, 46 (1944).
—, — u. M. S. Kanapaux: Factors affecting the ascorbic acid content of cabbage lines. J. Agric. Res. 68, 325—329 (1944).
—, P. H. Heinze, J. E. Welch u. P. C. Grimball: Differences in stability of thiamin, riboflavin, and ascorbic acid in cabbage varieties. Proc. Amer. Soc. Hort. Sci. 45, 396—404 (1944).
Pound, G. S.: Cabbage varietal reactions to mosaic viruses. Phytopathology 36, 408—409 (1946). (Abstr.)
Priobus, K.: Die Keimlingsfarbe bei Herbstrüben- u. Kohlsorten als Mittel zur Sortenunterscheidung. Gartenbauwiss. 18, 27—31 (1944).

Ramis, J. de Ros: Variedades comerciales de la col. (Commercial cabbage varieties.) Agricultura, Madrid 12, 62—65 (1943).
Raphael, T. D.: Cabbage trials with notes on seed production. Tasm. J. Agric. 16, 46—52 (1945).
Reinking, O. A. u. W. O. Gloyer: Work progressing on new cabbage strains. Fm. Res. N. Y. St. Sta. 6, Nr. 2, 11 (1940).

Schmidt, M.: Die Kopfkohlsorten der Sortenliste. Obst- u. Gemüsebau 1941, 87, 132—134; 1942, 88, 4 (1941 bis 1942).
Sikka, S. M.: Cytogenetics of Brassica hybrids and species. J. Genet. 40, 441—509 (1940).

Thomas, P. T. u. M. B. Crane: Genetic classification of Brassica crops. Nature, Lond. 150, 430 (1942).

Vasiljev, V. L.: (Choosing winter varieties of white cabbage and cauliflower for the regions of the far North.) Vestnik Socialističeskogo Rastenievodstva (Soviet Plant Industry Record), Nr. 3, 67—76 (1940). [Russisch.]

Walker, J. C.: Progress in combination of yellows and mosaic resistance with high ascorbic acid content in cabbage. Phytopathology 34, 1012—1013 (1944). (Abstr.)
— u. R. E. Foster: The inheritance of ascorbic acid content in cabbage. Amer. J. Bot. 33, 758—761 (1946).
— u. G. S. Pound: Improvement of cabbage for disease resistance. Phytopathology 37, 23 (1947). (Abstr.)
Wark, D. C.: Some observations on the magnitude of agronomic variation within cabbage varieties and a description of varieties. J. Coun. Sci. Industr. Res. Aust. 19, 347—358 (1946).
— A method of selection within a variety of cabbage. J. Aust. Inst. Agric. Sci. 12, 150—152 (1946).
Whitcomb, W. D.: The cabbage maggot. Bull. Mass. Agric. Exp. Sta., Nr. 412, 28 (1944).

5. Zwiebeln.

Anonymus: Breeding for resistance to downy mildew of onions. Agric. Gaz. N. S. W. 51, 562 (1940).
— The cytology of the species hybrid Allium Cepa × fistulosum and its polyploid derivatives. Hereditas, Lund 27, 253—272 (1941).

Becker, Th.: Blütenbiologische Studien an Zwiebeln, Möhren, Sellerie und Petersilie. Kühn-Archiv 60, 466—492 (1943/1944).
Beekom, C. W. C. van: Vatbaarheidsverschillen voor kopiot (Botrytis spp.) in het Nederlandsche uiensortiment. (Differences in the susceptibility to neck rot [Botrytis spp.] of the onion varieties grown in the Netherlands.) Tijdschr. PlZiekt. 46, 208—211 (1940).
— Verslag der proefvelden en proefnemingen met uien en sjalotten over 1940. (Report on the experimental fields and trials with onions and shallots during 1940.) Meded. Landbvoorlicht Dienst. Wageningen, Holland, Nr. 15, 92 (1941).
— De resultaten van drie jaren onderzoek van ui en sjalot. (The results of three years' study on onions and shallots.) Meded. Landbvoorlicht Dienst, Wageningen, Nr. 27, 76 (1942).
— u. G. Veenstra: Uienproefvelden en proefnemingen met uien 1939. (Onion experimental plots and trials with onions 1939.) Ned. Uien-Federatie, Wageningen, Holland, Pp. 51 (1939).
Berger, C. A.: Experimental studies on the cytology of allium. Biol. Bull. Wood's Hole 87, 163 (1944). (Abstr.)
— Naturally occurring polyploidy in the development of Allium cepa L. Biol. Bull. Wood's Holle 91, 217 (1946). (Abstr.)
— u. E. R. Witkus: Polyploid mitosis as a normally occurring factor in the development of Allium cepa L. Amer. J. Bot. 33, 785—787 (1946).
Binkley, A. M. u. H. A. Jones: A comparison of Sweet Spanish hybrids with commercial Sweet Spanish onion strains. Proc. Amer. Soc. Hort. Sci. 44, 485—487 (1944).
— u. W. C. Sparks: Station breeding onions for resistance to insekts, purple blotch, and pink root. Colo. Fm. Bull. 6, 8—11 (1944).
Brierley, P. u. F. F. Smith: Reaction of onion varieties to yellowdwarf virus and to three similar viruses isolated from shallot, garlic and narcissus. Phytopathology 36, 292—296 (1946).
— u. N. W. Stuart: Influence of nitrogen nutrition on susceptibility of onions to yellowdwarf virus. Phytopathology 36, 297—301 (1946).

Castan, R.: Variation dans la morphologie de la racine d'Allium cepa sous l'action de l'acide β indol-acétique. C. R. Soc. Biol., Paris 135, 765—766 (1941).

Clarke, A. E., H. A. Jones u. T. M. Little: Inheritance of bulb color in the onion. Genetics 29, 569—575 (1944).
— u. H. H. McKay: A cytological study of some triploid onion plants. J. Hered. 37, 131—136 (1946).
Coulter, F. C.: The story of garden vegetables. XI: Onion, valued from time immemorial. Seed World 52, Nr. 4, 30—31, 35 (1942).

Davis, G. N.: „It's a natural" for the onion grower. Mild flavor and higher yields of Early Grano make it increasingly popular with trade and consumers. Sth. Seedsman 6, Nr. 2; 11, 31 (1943).
— A newcomer named „Red" is first hybrid onion. Sth. Seedsman 7, 13, 52 (1944).
— u. H. A. Jones: „San Joaquin". Sth. Seedsman 9, 17 (1946).

Ehrenburg, P. M.: (A brief account of work in improving the Dungan onion.) Bjull. Kazah. Naučnoissled. Inst. Zemled. im. Acad. V. R. Viljjamsa (Bull. V. R. Williams Kazah. Inst. Agric.), Nr. 7/8, 11—19 (1940). [Russisch.]
Ellenhorn, J. E.: (Difference in the chromomeres of allelomorphic satellited chromosomes in Allium Cepa.) Doklady Acad. Nauk SSSR. 27, 357—360 (1940). [Russisch.]
Emsweller, S. L. u. H. A. Jones: Further studies on the chiasmata of the Allium cepa × A. fistulosum hybrid and its derivatives. Amer. J. Bot. 32, 370 bis 379 (1945).

Fedorov, G. V.: (Frost resistant onion hybrids.) Trudy Omsk. Seljskohozjajstvennogo Inst. imeni S. M. Kirova (Trans. Kirov Inst. Agric. Omsk, USSR.) 4 (17), 133—136 (1939). [Russisch.]
Feinbrun, N.: Allium sectio Porrum of Palestine and the neighbouring countries. Palest. J. Bot. J. Ser. 3, 1—21 (1943).

Geitler, L.: Natürliches diploides Allium carinatum. Ber. dtsch. bot. Ges. 61, 210—211 (1944).

Haber, E. S.: Onion breeding. Trans. Iowa Hort. Soc. 74, 246—249 (1939).
Hawthorn, L. R.: Behavior of certain characters in breeding Yellow Bermuda onions. Proc. Amer. Soc. Hort. Sci. 36, 668—673 (1938) 1939.
— Texas Grano- new strain is earlier and yields mor U. S. No. 1 onions than Early Grano. Sth. Seedsman 7, 12, 52 (1944).
Holdsworth, M.: A comparative study of onion varieties in relation to bolting and yield when grown from sets. Ann. Appl. Biol. 32, 22—34 (1945).

Ivanoff, S. S.: Expression of certain hereditary factors in Yellow Bermuda onions induced by unseasonable planting in the greenhouse. Bot. Gaz. 106, 411—420 (1945).

Jones, H. A.: Onion breeding. Market Gr. J. 64, 53, 56 (1939).
— u. A. E. Clarke: A natural amphidiploid from an onion species hybrid-Allium cepa L. × Allium fistulosum L. J. Hered. 33, 25—32 (1942).
—, — u. F. J. Stevenson: Studies in the genetics of the onion Allium cepa L. Proc. Amer. Soc. Hort. Sci. 44, 479—484 (1944).
— u. G. N. Davis: Inbreeding and heterosis and their relation to the development of new varieties of onions. Tech. Bull. U. S. Dep. Agric., Nr. 874, 28 (1944).
—, B. A. Perry u. G. N. Davis: The new Excel. Sth. Seedsman 10, Nr. 3; 13, 57, 60 (1947).

Jones, H. A., D. R. Porter u. L. D. Leach: Breeding for resistance to onion downy mildew caused by Peronospora destructor. Hilgardia 12, 531—550 (1939).
—, J. C. Walker, T. M. Little u. R. H. Larson: Relation of color-inhibiting factor to smudge resistance in onion. J. Agric. Res. 72, 259—264 (1946).

Krickl, M.: Neue Zuchtziele bei Küchenzwiebel im Hinblick auf die Marktversorgung. Züchter 11, 321 bis 324 (1939).
— Spätaustreiben — relativ geringer Gewichtsverlust — hoher osmotischer Wert. Ein Beitrag zur Züchtung besonders lagerfester Speisezwiebeln. Gartenbauwiss. 17, 51—90 (1942).
— Neue Möglichkeiten der Züchtung von Speisezwiebeln. Forschungsdienst 16, 227—239 (1943).
Krivenko, A. A.: (Interspecific hybridization in onions.) Vestnik Ovoščevodstvo i Kartofelj (Vegetable and Potato Journal) 1941, Nr. 1, 70—78 (1941). [Russisch.]

Leggieri, L.: Selezione di alcune varietà di cipolle da serbo. (Selection of certain varieties of onions for keeping.) Servizio delle Sementi, Pp. 53—70 (1939).
Levan, A.: Amphibivalent formation in Allium cernuum and its consequences in the pollen. Bot. Notiser, Pp. 256—260 (1939).
— The effect of acenaphthene and colchicine on mitosis of Allium and Colchicum. Hereditas, Lund 26, 262 bis 276 (1940).
— The cytology of Allium amplectens and the occurrence in nature of its asynapsis. Hereditas, Lund 26, 353 bis 394 (1940).
— Meiosis of Allium Porrum, a tetraploid species with chiasma localisation. Hereditas, Lund 26, 454—462 (1940).
— Notes on a progeny plant of asynaptic Allium amplectens. Hereditas, Lund 30, 468 (1944). (Abstr.)
Liu, T. T.: Cytological studies of „Allium odorum". Acta Brevia Sinensia, Nr. 8, 15 (1944). [Mimeographed.]

Magruder, R. et al.: Descriptions of types of principal American varieties of onions. Misc. Publ. U. S. Dep. Agric., Nr. 435, 87 (1941).
Mann, L. K. u. B. J. Hoyle: Use of the refractometer for selecting onion bulbs high in dry matter for breeding. Proc. Amer. Soc. Hort. Sci. 46, 285—292 (1945).
Mensinkai, S. W.: Cytogenetic studies in the genus Allium. J. Genet. 39, 1—45 (1939).
Miani, G.: Trabanti e nucleoli nel gen. Allium. (Trabants and nucleoli in the genus Allium.) Ann. Bot. Roman 22, Nr. 2, 11—27 (1941).
Mol, W. E. de: Über den Einfluß hochkomprimierter Gase auf Zellteilung (Meiosis) und Zellstreckung bei Zwiebelgewächsen. Gartenbauwiss. 16, 207—215 (1941).
Murphy, J. B.: Megasporogenesis and development of the embryo sac of Allium cernuum. Bot. Gaz. 108, 129—136 (1946).

Nichols, C.: Spontaneous chromosome aberrations in root tips of Allium. Collecting Net 15, 171—172 (1940). (Abstr.)
— Spontaneous chromosome aberrations in Allium. Genetics 26, 89—100 (1941).
Nichols, C. jr.: The effects of age and irradiation on chromosomal aberrations in Allium seed. Amer. J. Bot. 29, 755—759 (1942). (Abstr.)
Nybom, N.: Accessory chromosomes in Allium. Hereditas, Lund 33, 571—572 (1947). (Abstr.)

Poljakova, T. F.: Effect of high and low temperature upon chiasma formation in Allium cepa L. C. R. (Doklady) Acad. Sci. URSS. 27, 594—597 (1940).

Sax, K.: The behaviour of X-ray induced chromosomal aberrations in Allium root tip cells. Genetics **26**, 418—425 (1941).

Scott, G. W.: Meet the persevering onion breeders. Sth. Seedsman **6**, Nr. 4; 12, 33, 37, 40 (1943).

Sievers, A. F., M. S. Lowman u. M. L. Ruttle: Investigations of the yield and quality of the oils from some hybrid and tetraploid mints. J. Amer. Pharm. Ass. Sci. Ed. **34**, 225—231 (1945).

Sparks, W. C. u. A. M. Binkley: Natural crossing in Sweet Spanish onions as related to distance and direction. Proc. Amer. Soc. Hort. Sci. **47**, 320—322 (1946).

Stearn, W. T.: The Welsh onion and the ever-ready onion. Gdnrs' Chron. **114**, 86—88 (1943).

Sze, L. C.: On the nucleolus in the pollen grain divisions of „Allium fistulosum". Acta Brevia Sinensia, Nr. 8, 15—16 (1944). [Mimeographed.]

Trofimec, N. H.: (The biology of flowering and fertilization in Allium.) Vestnik Socialističeskogo Rastenievodstva (Soviet Plant Industry Record), Nr. 5, 76—86 (1940). [Russisch.]

— (Selecta fertilization in Allium fistulosum.) Vestnik Socialističeskogo Rastenievodstva (Soviet Plant Industry Record), Nr. 1, 31—34 (1941). [Russisch.]

Ustinova, E. I.: (A comparative-embryological study of normal and viviparous onion species [Allium].) J. Bot. URSS. **29**, 232—239 (1944). [Russisch.]

Walker, J. C., H. A. Jones u. A. E. Clarke: Smut resistance in an Allium species hybrid. J. Agric. Res. **69**, 1—8 (1944).

Yarwood, C. E.: Onion downy mildew. Hilgardia **14**, 595—691 (1943).

6. Kürbisgewächse.

Abraham, G.: Bantam size watermelons find market. New variety „Little Midget" fits icebox and is shipped in crates. Market Gr. J. **47**, 473, 475—476 (1941).

Afify, A.: Cytological studies in the Cucurbitaceae and their evolutionary significance. J. Genet. **46**, 116—123 (1944).

Alekseeva, M. V. u. I. Prezent: (Maintaining the varietal type in melons.) Jarovizacija, Nr. 5 (32), 139 (1940). [Russisch.]

Anderson, W. S.: Growing cucumbers for pickling in Mississippi. Bull. Miss. Agric. Exp. Sta., Nr. 355, 17 (1941).

Anonymus: (Cultivation of cucurbits in the USSR. Breeding, seed production, cultivation and mechanization of cucurbits.) Lenin Academy of Agricultural Sciences, Moscow, Pp. 99 (1939). [Russisch.]

— Hawkesbury wilt-resistant water melon. Hawkesbury Agric. Coll. J. **36**, 23 (1939).

— Introduces new squash. Seed World **47**, Nr. 1, 7 (1940).

— Breeding mildew resistant cantaloupe. Seed World **48**, Nr. 13, 17 (1940).

— New hybrid watermelon developed in Iowa. Seed World **49**, Nr. 1, 15 (1941).

— New muskmelon wilt-resistant. Seed World **49**, Nr. 4, 24 (1941).

— Cantaloup lovers. Seed World **54**, 16 (1943).

— „Naked" squash seeds better than peanuts. Sth. Seedsman **6**, Nr. 2, 37 (1943).

— (Letter from a reader.) Sovhoznoe Proizvodstvo (State Farming), Nr. 3, 49 (1945). [Russisch.]

Araratian, A. G.: A contribution to karyological knowledge of melons. C. R. (Doklady) Acad. Sci. URSS. **25**, 777—780 (1939).

Arenkova, D. N.: Acenaphthene-induced tetraploidy in muskmelons. C. R. (Doklady) Acad. Sci. URSS. **27**, 1028—1029 (1940).

Bailey, R. M.: Progress in breeding cucumbers resistant to scab (Cladosporium cucumerinum). Proc. Amer. Soc. Hort. Sci. **36**, 645—646 (1938) 1939.

Bailey, L. H.: Species Cucurbitae. Gentes Herbarum, Ithaca, N. Y. **6**, 267—322 (1943).

Barnes, W. C., C. N. Clayton u. J. M. Jenkins jr.: The development of downy mildew-resistant cucumbers. Proc. Amer. Soc. Hort. Sci. **47**, 357—360 (1946).

Barrows, F. L.: Inheritance in Cucurbita pollen. Genetics **26**, 137 (1941). (Abstr.)

Beattie, W. R.: Cucumber growing. Fmr's Bull. U. S. Dep. Agric., Nr. 1563, 25 (1942).

Behr, L.: Die Wirkung übernormaler Dosen des Beizmittels „Ceresan" auf den Keimvorgang von Cucumis sativus L. Züchter **17/18**, 44—50 (1946).

Berkner, F.: Der schalenlose Kürbis, ein Fett- und Eiweißlieferant. Züchter **12**, 123—126 (1940).

Boswell, V. R.: Disease-resistant and hardy varieties of vegetables. Nat. Hort. Mag. **23**, 203—208 (1944).

Braum, A. E.: Resistance of watermelon to the wilt disease. Amer. J. Bot. **29**, 683—684 (1942).

Buchinger, A.: Die wichtigsten europäischen Kürbisarten. Gartenbauwiss. **18**, 311—332 (1944).

Burr, H. S.: Potential differences and fruit form in cucurbits. Amer. J. Bot. Suppl. **29**, 4s (1942). (Abstr.)

Cárdenas, M.: Las cucurbitas cultivadas de Bolivia. (The cultivated cucurbits of Bolivia.) Rev. Agric. Bolivia **2**, Nr. 2, 3—12 (1944).

— Notas sobre taxonomía de plantas económicas de Bolivia. Una Cucurbita nueva. (Notes on the taxonomy of economic plants of Bolivia. A new Cucurbita. Rev. Agric., Bolivia **2**, Nr. 3, 76—77 (1945).

Carson, C. M.: Icebox Number Two. Sth. Seedsman **9**, Nr. 12, 11 (1946).

Černyšev, D.: (The fodder watermelon.) Sovhoznoe Proizvodstvo (State Farming), Nr. 3, 24—25 (1944). [Russisch.]

Chakravarty, H. L.: Studies on Indian Cucurbitaceae. Indian J. Agric. Sci. **16**, 1—90 (1946).

Čižov, S. T.: (Yields of hybrid cucumbers in glasshouses.) Moskovskaja ordena Lenina Seljskohozjajstvennaja Akademija imeni K. A. Timirjazeva. Naučnaja Konferencija 3—10 ijunja 1944 g. Doklady. [Proc. Sci. Conf. Timirjazev Agric. Acad. (3—10 June, 1944)], Nr. 1, 42—43 (1945). [Russisch.]

Contardi, H. G.: Estudios genéticos en „Cucurbita y consideraciones agronómicas. (Genetical studies in Cucurbita and agronomic considerations.) Physis, B. Aires **18**, 331—347 (1939).

— Inventario de las variaciones observadas en los frutos de una población de zapallos (Cucurbita pepo L.). (Inventory of the variations observed in the fruits of a population of pumpkins [C. Pepo L.].) An. Inst. Fitotéc. Santa Catalina **1**, 187—199 (1939) 1940.

Cook, H. T. u. T. J. Nugent: The Hawkesbury watermelon, a promising wilt-resistant variety. Phytopathology **29**, 5 (1939). (Abstr.)

—, — Developing wilt resistant watermelons for Virginia. Proc. Va. Acad. Sci. **2**, 182, 1940—1941 (1941).

Cooper, J. F.: Exit fusarium wilt, villain; enter Blacklee, the watermelon hero. Sth. Seedsman **6**, Nr. 2, 18—19, 30 (1943).

Coulter, F. C.: The story of garden vegetables. X. Muskmelon: the ancient records are obscure. Seed World. **52**, Nr. 2, 12—13 (1942).

— Watermelon-the summer favorite is of African origin Seed World **55**, 12—13, 35 (1944).

Culpepper, C. W. u. H. H. Moon: Differences in the composition of the fruits of Cucurbita varieties at

different ages in relation to culinary use. J. Agric. Res. **71**, 111—136 (1945).
Currence, T. M., C. J. Eide u. J. G. Leach: The Golden Gopher muskmelon. Market Gr. J. **68**, 14—16 (1941).
—, R. E. Lawson u. R. M. Brown: A rapid method for finding the volume and density of muskmelon fruits. J. Agric. Res. **68**, 427—440 (1944).
Curtis, L. C.: Heterosis in summer squash (Cucurbita pepo) and the possibilities of producing F_1 hybrid seed for commercial planting. Proc. Amer. Soc. Hort. Sci. **37**, 827—828; also Market Gr. J. **67**, 312 (1939) 1940.
— Yankee hybrid. A new first generation hybrid of summer squash. Market Gr. J. **67**, 472 (1940).
— Development of the All-America squash winner. Seed World **48**, Nr. 9, 13 (1940).
— Comparative earliness and productiveness of first and second generation summer squash (Cucurbita Pepo) and the possibilities of using the second generation seed for commercial planting. Proc. Amer. Soc. Hort. Sci. **38**, 596—598 (1941).
— Yankee hybrid summer squash. An early, productive first generation cross. Circ. Conn. Agric. Exp. Sta., Nr. 152, 61—65 (1942).

Davis, G. W.: A new melon, Baby Persian, Makes bow. California produces a uniform-sized fruit for commercial growers; also a potential for southern gardens. Sth. Seedsman **4**, Nr. 2; **7**, 31 (1941).
Doolittle, S. P., F. S. Beecher u. W. S. Porte: A hybrid cucumber resistant to bacterial wilt. Phytopathology **29**, 996—998 (1939).

Enzie, W. D.: The Geneva Delicata squash. Fm Res.; N. Y. St. Sta. **6**, Nr. 1, 1, 13 (1940).
Ervin, C. D.: Polysomaty in Cucumis Melo. Proc. Nat. Acad. Sci., Wash. **25**, 335—338 (1939).
— A study of polysomaty in Cucumis Melo. Amer. J. Bot. **28**, 113—124 (1941).

Fernando, M. u. S. B. Udurawana: The relative resistance of some strains of bitter-gourd to the cucurbit fruitfly. Trop. Agriculturist **96**, 347—352 (1941).
Filov, A. I.: (Reaction of hermaphrodite varieties of cucurbits to self-pollination.) Proc. Lenin Acad. Agric. Sci. USSR., Nr. 12, 6—10 (1939). [Russisch.]
— (The new water-melon variety „Im. XVII Parts'ezd".). Ovoščevodstvo (Vegetable Growing), Nr. 1, 43—44 (1939). [Russisch.]
— A tentative classification of cucumbers based on their ecological evolution. C. R. (Doklady) Acad. Sci. URSS. **26**, 811—814 (1940).
— (The connexion between the morphological characters of cucumber fruits, and their economically valuable qualities.) Vestnik Socialističeskogo Rastenievodstva (Soviet Plant Industry Record), Nr. 1, 126—132 (1941). [Russisch.]
French, M. H.: Feeding values of different varieties of pumpkins for livestock. E. Afr. Agric. J. **9**, 221—224 (1944).

Gabaev, S. G.: Experiments on colchicine and acenaphthene treatment of the cucumber for the production of polyploids. C. R. (Doklady) Acad. Sci. URSS. **28**, 164—166 (1940).
Goldhausen, M.: (Breeding drought-resistant [desert] water-melons.) Proc. Lenin Acad. Agric. Sci. URSS., Nr. 13, 18—21 (1939). [Russisch.]
Griffiths, A. E.: With cantaloupes, „13" is a lucky number. Sth. Seedsman **6**, 13, 37, 44 (1943).

Haltern, F. V.: No more bad years for watermelons. New wilt-resistant variety developed at Georgia A.E.S. Sth. Seedsman **6**, Nr. 1, 10 (1943).

Hartmair, V.: Über asiatische Formen von Cucumis sativus L. Züchter **13**, 125—144 (1941).
— Cytologische Untersuchungen an asiatischen Formen von Cucumis sativus L. Züchter **14**, 132—136 (1942).
— Colchicininduzierte Polyploidie bei Gurken. Züchter **15**, 13—16 (1943).
Hartman, J. D. u. F. C. Gaylord: The Purdue 44 muskmelon. Circ. Ind. Agric. Exp. Sta., Nr. 295, 8 (1944).
Hawthorne, P. L.: A polyploid watermelon. Proc. Amer. Soc. Hort. Sci. **45**, 348 (1944).
Heinze: Prüfung von Gurkensorten auf ihre Resistenz gegen das Gurkenmosaikvirus. Nr. 1. Wiss. Jber. Biol. Reichsanst. Land- u. Forstwirtsch. 1940. Mitt. Biol. Reichsanst. 1941, Heft 65, 23 (1941).
Hoffman, J. C. u. H. D. Brown: Preliminary studies on a new muskmelon hybrid. Proc. Amer. Soc. Hort. Sci. **38**, 535—536 (1941).
Hutchins, A. E.: Some examples of heterosis in the cucumber, Cucumis sativus L. Proc. Amer. Soc. Hort. Sci. **36**, 660—664 (1938) 1939.
— Inheritance in the cucumber. J. Agric. Res. **60**, 117 bis 128 (1940).
— A male and female sterile variant in squash, Cucurbita maxima, Duchesne. Proc. Amer. Soc. Hort. Sci. **44**, 494—496 (1944).
— u. F. E. Croston: Productivity of F_1 hybrids in the squash, Cucurbita maxima. Proc. Amer. Soc. Hort. Sci. **39**, 332—336 (1941).
— u. L. Sando: Gourds-their culture, uses identification, and relation to other cultivated Cucurbitaceae. Bull. Minn. Agric. Exp. Sta., Nr. 356, 35 (1941).
Hwang, Tsung-Chen: Photoperiodic responses of the growth and the blooming of pumpkin. Acta Brevia Sinensia, Nr. 8, 22 (1944). [Mimeographed.]

Isbell, C. L.: Taking the „Punk out of southern pumpkins. Sth. Seedsman **7**, 11, 51 (1944).
Ivanoff, S. S.: Breeding cantaloupes for resistance to downy mildew and other diseases and pests. Phytopathology **32**, 10 (1942). (Abstr.)
— Resistance of cantaloupes to downy mildew and the melon aphid. J. Hered. **35**, 35—39 (1944).
— Texas cantaloupe resists aphids, downy mildew. Sth. Seedsman **8**, 11, 28 (1945).

Jannaccone, A.: Selezione del popone invernale di Capua (varietà „Palermitano"). (Selection of the winter Capua melon [variety „Palermitano"].) Servizio delle Sementi, Pp. 89—99 (1939).
— Risultati di un sessennio di lavoro per la selezione del popone invernale di Capua (var. Palermitano). (Results of some six years selection work on the Capua winter melon [var. Palermitano].) Ann. Fac. Agrar. Portici, Univ. Napoli 1942—1943, **14**, 176—183 (1943).
Jenkins, J. M. jr.: Natural self-pollination in cucumbers. Proc. Amer. Soc. Hort. Sci. **40**, 411—412 (1942).
— Downy mildew resistance in cucumbers. J. Hered. **33**, 35—38 (1942).
— Studies on the inheritance of downy mildew resistance and of other characters in cucumbers. J. Hered. **37**, 267—271 (1946).
— Mildew resistance needs maintenance! Sth. Seedsman **10**, Nr. 6, 64—65 (1947).

Kazakevič, L. I.: (Methods of growing the fodder pumpkin and water-melon.) Naučnyj Otčet Inst. Zernovogo Hozjajstva Jugo-Vostoka SSSR. za 1941—42 gg. (Sci. Rep. Inst. Grain Husbandry South-Eastern USSR. for 1941—42), Pp. 86—95 (1944). [Russisch.]
Kobjakova, J. A.: (Alteration of varietal characters in cucumbers when open pollinated and when pollinated with a limited range of pollen parents.) Vestnik Ovoščevodstvo i Kartofelj (Vegetable and Potato Journal), Nr. 1, 85—87 (1941).

Koot, Y. van: Enkele onderzoekingen betreffende de Fusarium-ziekte bij de komkommer. (Some investigations on the Fusarium wilt of cucumbers.) Tijdschr. PlZiekt. **49**, 52—73 (1943).

Krebčenko, L. E.: (Breeding cucurbits.) Ovoščevodstvo (Vegetable Growing), Nr. 12, 30—34 (1939). [Russisch.]

Kulikov, P. I.: (Revise the rules for spatial isolation of cucurbits.) Ovoščevodstvo (Vegetable Growing), Nr. 9, 22—23 (1940). [Russisch.]

Lutohin, S. N.: (Mit Phytohormon und Glyzerin induzierte Parthenocarpie bei Cucurbita maxima L.) C. R. (Doklady) Acad. Sci. URSS. **58**, 1525 (1947). [Russisch.]

Mahoney, C. H.: Superb Golden. A new hybrid muskmelon. Quart. Bull. Mich. Agric. Exp. Sta. **21**, 225 bis 227 (1939).

Mann, L. K.: Fruit shape of watermelon as affected by placement of pollen on stigma. Bot. Gaz. **105**, 257 bis 262 (1943).

Millán, R.: Los zapallos Bugango y Angola. (The Bugango and Angola squashes.) Rev. Argent. Agron. **10**, 192—196 (1943).

— Variaciones del zapallito amargo „Cucurbita andreana" y el origen de Cucurbita maxima. (Variations of the bitter gourd C. andreana and the origin of C. maxima.) Rev. Argent. Agron. **12**, 86—93 (1945).

Miller, J. C.: Longfellow pumpkin. Bred for south, it combines desirable qualities of its parents, cushaw and African squash. Sth. Seedsman **8**, 13 (1945).

Müller, K. O.: Zur Züchtung krätze- und blattbrandwiderstandsfähiger Gurken. Kranke Pflanze **18**, Nr. 5/6 (1942).

Munger, H. M.: Breeding muskmelons, with special reference to the possible utilization of first generation hybrids. Cornell Univ. Abstr. Thes., Pp. 350—353, 1941 (1942).

— The possible utilization of first generation muskmelon hybrids and an improved method of hybridization. Proc. Amer. Soc. Hort. Sci. **40**, 405—410 (1942).

— Iroquois muskmelon is resistant to Fusarium wilt. Fm Res. N. Y. St. Sta. **10**, Nr. 2, 20 (1944).

Pangalo, K. I.: (Concerning hybrid synthesis of new exceptionally drought-resisting varieties of table watermelon.) Priroda, Nr. 6, 61—64 (1939). [Russisch.]

— (Sex and flowering in cultivated Cucurbitaceae.) J. Bot. URSS. **28**, 10—23 (1943). [Russisch.]

— (A new genus of the Cucurbitaceae, Praecitrullus, m.- an ancestor of the contemporary watermelon [Citrullus Forsk.].) J. Bot. URSS. **29**, 200—204 (1944). [Russisch.]

—, u. M. K. Goldhausen: Interspecific hybridization in the genus Cucurbita. C. R. (Doklady) Acad. Sci. URSS. **24**, 61—64 (1939).

Petrova, N.: (White fruited forms found among the Nerosimyi cucumber.) Vestnik Ovoščevodstvo i Kartofelj (Vegetable and Potato Journal), Nr. 3, 101—102 (1940). [Russisch.]

Poole, C. F.: Genetics of cultivated cucurbits. Proc. Amer. Soc. Hort. Sci. **40**, 386 (1942). (Abstr.)

— Genetics of cultivated cucurbits. J. Hered. **35**, 122 bis 128 (1944).

—, u. P. C. Grimball: Inheritance of new sex forms in Cucumis melo L. J. Hered. **30**, 21—25 (1939).

—, — Interaction of sex, shape, and weight genes in watermelon. J. Agric. Res. **71**, 533—552 (1945).

—, — u. D. R. Porter: Inheritance of seed characters in watermelon. J. Agric. Res. **63**, 433—456 (1941).

Powers, L.: Early Cheyenne Pie pumpkin. Circ. U. S. Dep. Agric., Nr. 537, 4 (1939).

Pryor, D. E.: The influence of vitamin B_1 on the development of cantaloupe powdery mildew. Phytopathology **32**, 885—895 (1942).

Pryor, D. E. u. T. W. Whitaker: The reaction of cantaloupe strains to powdery mildew. Phytopathology **32**, 995—1004 (1942).

—, — u. G. N. Davis: The development of powdery mildew resistant cantaloupes. Proc. Amer. Soc. Hort. Sci. **47**, 347—356 (1946).

Rietberg, H.: De fusariose van komkommers en meloenen. (The fusariosis affections of cucumbers and melons.) Meded. Tuinbouw-Voorlichtingsdienst, 's Gravenhage, Nr. 20, 48 (1940).

Riollano, A.: Nueva variedad de melones para Puerto Rico. (A new variety of melon for Puerto Rico.) Agric. Exp. P. R. **2**, 7—8 (1942).

Roque, A.: El pepinillo „Puerto Rico, Nr. 39". Nueva variedad de alta producción y resistente al mildeu. (The cucumber Puerto Rico, Nr. 39. A new variety of high yield and resistant to mildew.) Agric. Exp. P. R., Nr. 6, 5—6 (1941).

Schnack, B. u. C. E. Cavia: Un tipo sexual anormal en Cucurbita Pepo. (An abnormal sex type in C. Pepo.) An. Inst. Fitotec. Santa Catalina 1941 **3**, 21—28 (1943).

Scott, D. H. u. M. E. Riner: A mottled-leaf character in winter squash inherited as a dominant Mendelian character. J. Hered. **37**, 27—28 (1946).

—, — Inheritance of male sterility in winter squash. Proc. Amer. Soc. Hort. Sci. **47**, 375—377 (1946).

Scott, F. M.: Nuclear size, in Echinocystis and in Cucurbita. Amer. J. Bot. Suppl. **28**, 6s (1941). (Abstr.)

Seaton, H. L. u. J. C. Kremer: The relation of certain floral abnormalities to the pollination of Cucurbita. Proc. Amer. Soc. Hort. Sci. **36**, 626 (1938) 1939. (Abstr.)

—, — The influence of climatological factors on anthesis and anther dehiscence in the cultivated cucurbits. A preliminary report. Proc. Amer. Soc. Hort. Sci. **36**, 627—631 (1938) 1939.

Shifriss, O.: Artificially induced polyploids in the genus Cucumis. Cornell Univ. Abstr. Thes., Pp. 363—365, 1941 (1942).

— Polyploids in the genus Cucumis. Preliminary account J. Hered. **33**, 144—152 (1942).

— Male sterilities and albino seedlings in cucurbits. J. Hered. **36**, 47—52 (1945).

— Prolific ‚Cuke' joins hybrid hits. Sth. Seedsman **8**, 15, 26 (1945).

—, C. H. Myers u. C. Chupp: Resistance to mosaic virus in the cucumber. Phytopathology **32**, 773—784 (1942).

Sinnot, E. W.: Developmental factors affecting inherited differences in fruit size in cucurbits. Genetics **24**, 85 (1939). (Abstr.)

— The relation between growth rate and fruit size in cucurbits. Amer. J. Bot. **31**, 5s (1944). (Abstr.)

— The relation of growth to size in cucurbit fruits. Amer. J. Bot. **32**, 439—446 (1945).

—, A. F. Blakeslee u. A. Franklin: A comparative study of fruit development in diploid and tetraploid cucurbits. Genetics **26**, 168—169 (1941). (Abstr.)

—, — u. H. E. Warmke: The effect of colchicine-induced polyploidy on fruit shape in cucurbits. Genetics **24**, 84—85 (1939). (Abstr.)

— u. A. H. Franklin: A developmental analysis of the fruit in tetraploid as compared with diploid races of cucurbits. Amer. J. Bot. **30**, 87—94 (1943).

Somers, L. A.: To save an industry. Keep your eyes open for a new type watermelon plant. Market Gr. J. **70**, 84—85 (1942).

Tkačenko, N. N.: (Conditions for the best pollination and fruit development of cucumbers.) Proc. Lenin. Acad. Agric. Sci. USSR., Nr. 10, 37—43 (1940). [Russisch.]

Tuljženkova, F. F.: (Identification of water-melon varieties in the laboratory.) Vestnik Ovoščevodstvo i Kartofelj (Vegetable and Potato Journal), Nr. 5, 56—61 (1940). [Russisch.]

Walker, M. N.: The breeding of wilt resistant water-melons in the south. Proc. 41st Annu. Conv. Ass. Sth. Agric. Wkrs, Birmingham, Ala., February 7—9, Pp. 34 bis 35 (1940). (Abstr.)
— Fusarium wilt of watermelons. I. Effect of soil temperature on the wilt disease and the growth of water-melon seedlings. Bull. Fla Agric. Exp. Sta., Nr. 363, 29 (1941).
— A useful pollination method for watermelons. J. Hered. 34, 11—13 (1943).
— The Blacklee watermelon. A new Fusarium wilt-resistant variety for Florida. Pr. Bull. Fla Agric. Exp. Sta., Nr. 605, 4 (1944).

Walker, W. F.: Cucurbit investigations. Tasm. J. Agric. 17, 335—341 (1946).

Ware, G. W.: The world's largest watermelons. Market Gr. J. 66, 223—224 (1940).

Welch, A. u. I. E. Melhus: Wilt resistance in F. hybrid watermelons. Phytopathology 32, 181—182 (1942).

Whitaker, T. W. u. G. F. Carter: Critical notes on the origin and domestication of the cultivated species of Cucurbita. Amer. J. Bot. 33, 10—15 (1946).
— u. D. E. Pryor: Genes for resistance to powdery mildew in Cucumis Melo. Proc. Amer. Soc. Hort. Sci. 41, 270—272 (1942).
—, — The reaction of 21 species in the Cucurbitaceae to artificial infection with cantaloupe powdery mildew (Erysiphe cichoracearum D. C.). Phytopathology 35, 533—534 (1945).

White, O. E.: Genetic studies on wild and cultivated watermelons (Citrullus). Proc. Va. Acad. Sci. 2, 183, 1940—1941 (1941).

Wilson, J. D. u. J. J. Wilson: A mosaic-tolerant, pickling-type cucumber. Bi.-m. Bull. Ohio Agric. Exp. Sta. 29, 110—113 (1944).

Wolf, E. A. u. J. D. Hartman: Plant- and fruit-pruning as a means of increasing fruit set in muskmelon breeding. Proc. Amer. Soc. Hort. Sci. 40, 415—420 (1942).

Wong, C. Y.: Induced parthenocarpy of water-melon, cucumber and pepper. Science 89, 417—418 (1939).
— Chemically induced parthenocarpy in certain horticultural plants, with special reference to the watermelon. Bot. Gaz. 103, 64—86 (1941).

Work, P.: More new varieties for 1941. Cos Endive; Honey Gold Melon; 3 Pascal celery strains; Essary canning tomato. Market Gr. J. 67, 503—504 (1940).

Young, R. E.: The Butternut squash. Flower Gr. 32, 192 (1945).

7. Mohrrüben.

Anonymus: A new disease of carrots. Agric. Gaz. N. S. W. 55, 493—494 (1944).
— (Carrots of the Nantes variety from seeds grown in Palestine.) Hassadeh 25, 265—267 (1945).

Coulter, F. C.: The story of garden vegetables. VI. Carrot: one of France's great contributions to the garden. Seed World 50, Nr. 7, 20—21 (1941).

Harper, R. H. u. F. P. Zscheile: Carotenoid content of carrot varieties and strains. Food Res. Illinois 10, 84—97 (1945).

Klawitter, G. u. R. v. Sengbusch: Zielsetzung und Probleme in der Speisemöhrenzüchtung. Züchter 15, 16—22 (1943).

Klawitter, G. u. R. v. Sengbusch: Züchterische Untersuchungen des Aufbaues, der Färbung, des Refraktometerwertes und des Geschmacks von Speisemöhren. Züchter 15, 44—46 (1943).
—, — Methodisches zur Möhrenzüchtung: Querschneidemethode als Voraussetzung für die Verarbeitung einer großen Zahl von Einzelrüben zwecks Auslese auf Qualitätseigenschaften. Züchter 15, 91 (1943).

Kullander, S.: Undersökningar över karotinhalten hos morötter och brännässlor jämte några observationer över karotinets hållbarhet. (Investigations on the carotin content in carrots and nettles with some observations on the keeping property of the carotin.) Årsskr. Alnarps Landbr. Mej.-och Trädlnst., Pp. 1—11 (1941).

Lachman, W. H.: A quantitative study of form and size in five varieties of carrots. Proc. Amer. Soc. Hort. Sci. 36, 623—625 (1938) 1939.

Michelly, G. u. R. v. Sengbusch: Züchterisch brauchbare chemische Auslesemethode auf hohen Zuckergehalt bei Möhren. Züchter 17/18, 78—79 (1946).

Morris, H. J., C. A. Weast u. H. Lineweaver: Seasonal variation in the enzyme content of eleven varieties of carrots. Bot. Gaz. 107, 362—372 (1946).

Schuphan, W.: Biochemische Sortenprüfung an Gartenmöhren als neuzeitliche Grundlage für planvolle Züchtungsarbeit. Züchter 14, 25—43 (1942).
— Biochemische Sortenprüfung an Gartenmöhren als neuzeitliche Grundlage für planvolle Züchtungsarbeit. Züchter 15, 90 (1943).
— u. E. Euen: Über die Beziehungen zwischen Färbung, Carotingehalt und Geschmack bei Gartenmöhren. Züchter 16, 11—25 (1944).

Stubbs, L. L. u. B. J. Grieve: A new virus disease of carrots. J. Dep. Agric. Vict. 42, 411—412, 415 (1944).

Svensson, V.: Morotsodlingen. Betydelsen av förstklassiga stammar och av rationell drift. (Carrot cultivation. The importance of first class strains and sound management.) Weibulls Ill. Årsb. 37, 36—41 (1942).
— Morotsodling. En värdefull kultur för de lättare jordana. (Carrot growing. A valuable crop for lighter soils.) Weibulls Ill. Årsb. 39, 31—35 (1944).
— „Krussjuka" å morötter. („Curl disease" in carrots.) Weibulls Ill. Årsb. 40, 40—41 (1945).

Wilson, J. D.: Relative susceptibility of carrot varieties to nematode demage, yellows and defoliation by blights. Bi-m. Bull. Ohio Agric. Exp. Sta. 31, 35—39 (1946).

Woodbury, G. W. u. H. K. Schultz: Crown division of roots as an adjunct to carrot breeding and seed production studies. Proc. Amer. Soc. Hort. Sci. 44, 488—490 (1944).

Zagorodskih, P.: New data on the origin and taxonomy of cultivated carrot. C. R. (Doklady) Acad. Sci. URSS. 25, 520—523 (1939).

Zinn, F.: Zur Methodik der Bestimmung der Wurzelform, des Herz-Rinden-Verhältnisses und des Zuckergehaltes bei Möhren. Z. Pflanzenz. 26, 119—126 (1944).

8. Salat.

Andersen, S.: Om Slaegten Valerianella (Vaarsalat) i Danmark. (The genus Valerianella [spring salad] in Denmark.) Bot. Tidsskr. 46, 43—44 (1942).

Anonymus: Celtuce-a new vegetable for 42. Seed World 50, 22 (1941).
— New lettuce enjoys heat. Sth. Seedsman 7, 51 (1944).
— Slobolt, new summer leaf lettuce, lasts longer. Sth. Seedsman 8, 51 (1945).
— Great Lakes lettuce. A useful imported variety. Agric. Gaz. N. S. W. 58, 61 (1947).

Boswell, V. R.: Disease-resistant and hardy varieties of vegetables. Nat. Hort. Mag. **25**, 158—164 (1946).

Chevalier, A.: Laitues, chicorées et pissenlits. L'origine des formes cultivées. Rev. Bot. Appl. **23**, 273—281 (1943).

Childers, N. F. u. P. S. Robles: Slobolt lettuce for the tropics. Agric. Amer. **7**, 93—95 (1947).

Coulter, F. C.: Story of garden vegetables. IX. Lettuce, the queen of salad plants. Seed World **51**, 16—19 (1942).

Einset, J.: Cytological basis for sterility in induced autotetraploid lettuce (Lactuca sativa L.). Amer. J. Bot. **31**, 336—342 (1944).

— Aneuploidy in relation to partial sterility in autotetraploid lettuce (Lactuca sativa L.). Amer. J. Bot **34**, 99—105 (1947).

Garman, H. R. u. L. V. Barton: The response of lettuce seeds to thiourea treatments as affected by variety and age. Amer. J. Bot. (Suppl.) **33**, 229 (1946). (Abstr.)

Hoffman, I. C.: Selecting leaf lettuce. Proc. 29th Annu. Mtg. Ohio Veg. and Potato Gr. Ass., Pp. 152—156 (1944).

Howard, H. W.: Wild and cultivated watercress types. J. Minist. Agric. **53**, 453—456 (1947).

— u. I. Manton: Allopolyploid nature of the wild tetraploid watercress. Nature, Lond. **146**, 303—304 (1940).

Jagger, I. C.: Brown blight of lettuce. Phytopathology **30**, 53—64 (1940).

— u. T. W. Whitaker: The inheritance of immunity to mildew (Bremia lactucae) in lettuce. Proc. 7th Int. Genet. Congr. Edinburgh 23—30 August, Pp. 166 1939 (1941). (Abstr.)

—, — The inheritance of immunity from mildew (Bremia lactucae) in lettuce. Phytopathology **30**, 427—433 (1940).

—, —, J. J. Uselman u. W. M. Owen: The Imperial strains of lettuce. Circ. U. S. Dep. Agric., Nr. 596, 16 (1941).

Knott, J. E. u. A. A. Tavernetti: Production of head lettuce in California. Circ. Calif. Agric. Ext. Serv., Nr. 128, 51 (1944).

Koleff, N.: Untersuchungen über die Keimruhe bei Salatsamen (L. sativa var. capitata, L.). Gartenbauwiss. **17**, 263—272 (1943).

Kopetz, L. M.: Strunkuntersuchungen an Kopfsalaten. Züchter **11**, 277—278 (1939).

Krickl, M.: Züchtungsversuche beim Salat. Züchter **12**, 243—249 (1940).

Ling, L. u. M. C. Tai: On the specialization of Bremia lactucae on Compositae. Trans. Brit. Mycol. Soc. **28**, 16—25 (1945).

Macpherson, N. J.: Glasshouse lettuce trials in Lancashire. J. Minist. Agric. **52**, 117—120 (1945).

Martin, J. P. u. C. E. Pemperton: Disease symptoms in lettuce and celtuce, caused by the bean leaf hopper Empoasca Solana Del. Hawaii Plant. Rec. **46**, 111 bis 118 (1942).

Morrison, G.: The story of lettuce. Some of the interesting developments behind the improved varieties of today. Nat. Seedsman **8** (3), 22—25 (1941).

Ogilvie, L.: Downy mildew of lettuce: further investigations on strains of Bremia Lactucae occurring in England. Rep. Agric. Hort. Res. Sta., Long Ashton, Bristol, Pp. 147—150 (1945).

Pryor, D. E.: A unique case of powdery mildew on lettuce in the field. Plant Dis. Reporter **25**, 74 (1941). [Mimeographed.]

Raleigh, G. J.: Lettuce varieties for New York undergoing improvement. Fm Res. **10**, Nr. 2, 9, 15 (1944).

Reimers, F. E.: Phasic development in various biological groups of Lactuca sativa var. capitata. C. R. (Doklady) Acad. Sci. URSS. **25**, 790—793 (1939).

Schreiber, F.: Erfahrungen über Salatzüchtung. Kühn-Archiv **60**, 462—465 (1943/1944).

Stebbins, G. L. jr.: Notes on some Indian species of Lactuca. Indian For. Rec. **1** (N. S.), Bot., 237—245 (1939).

Thompson, R. C.: An amphidiploid Lactuca produced through non-reduction in F_1 hybrids. J. Hered. **33**, 253—264 (1942).

— Further studies on interspecific genetic relationships in Lactuca. J. Agric. Res. **66**, 41—48 (1943).

— Inheritance of seed color in Lactuca sativa. J. Agric. Res. **66**, 441—446 (1943).

— Reaction of Lactuca species to the aster yellows virus under field conditions. J. Agric. Res. **69**, 119—125 (1944).

— Slobolt for home gardens. Sth. Seedsman **8**, Nr. 8, 11 (1945).

— u. W. F. Kosar: Polyploidy in lettuce induced by colchicine. Proc. Amer. Soc. Hort. Sci. **36**, 641—644 (1938) 1939.

—, T. W. Whitaker u. W. F. Kosar: Interspecific genetic relationships in Lactuca. J. Agric. Res. **63**, 91—107 (1941).

Wehlmann: Gruppeneinteilung und Sortenmerkmale beim Kopfsalat. Gartenbauwiss. **15**, 585—589 (1941).

Wheeler, W.: Why not have the best in lettuce? Horticulture **21**, 457 (1943).

Whitaker, T. W.: The inheritance of chlorophyll deficiencies in cultivated lettuce. J. Hered. **35**, 317—320 (1944).

— u. I. C. Jagger: Cytogenetic observations in Lactuca. J. Agric. Res. **58**, 297—306 (1939).

— u. D. E. Pryor: The inheritance of resistance to powdery mildew (Erysiphe cichoracearum in lettuce. Phytopathology **31**, 534—540 (1941).

—, — Demonstrating downy mildew (Bremia Lactucae) in lettuce. Stain Technol. **18**, 121—123 (1943).

— u. R. C. Thompson: Cytological studies in Lactuca. Bull. Torrey Bot. Cl. **68**, 388—394 (1942).

9. Sonstige Gemüse.

Anonymus: Announce new variety of asparagus. Seed World **47**, Nr. 2, 32 (1940).

— Eine neue Kartoffelfrucht in Böhmen. Obst- u. Gem. **6**, 223 (1942).

Araratjan, A. G.: Heterochromosome in the wild spinach. C. R. (Doklady) Acad, Sci. URSS. **24**, 56—57 (1939).

Averjanova, O. P.: (Intra-varietal crossing in the eggplant.) Jarovizacija, Nr. 1 (34), 106—108 (1941). [Russisch.]

Belval, H.: L'amélioration du topinambour. Bull. Soc. Bot. Fr. **91**, 108—111 (1944).

Berger, C. A.: Reinvestigation of polysomaty in Spinacia. Bot. Gaz. **102**, 759—769 (1941).

— Some criteria for judging the degree of polyploidy of cells in the resting stage. Amer. Nat. **75**, 93—95 (1941).

— A new criterion of the degree of polyploidy of „resting" nuclei. Genetics **26**, 137—138 (1941). (Abstr.)

Berger, C. A. u. E. R. Witkus: A cytological study of c-mitosis in the polysomatic plant Spinacia oleracea, with comparative observations on Allium cepa. Bull. Torrey Bot. Cl. **70**, 457—466 (1943).

Binkley, A. M. u. W. A. Kreutzer: Strain of Giant Pascal celery resistant to ,,yellows" in being developed at station. Colo. Fm Bull. **7**, 3—4 (1945).

Caldwell, J. S., C. W. Culpepper, M. C. Hutchins, B. D. Ezell u. M. S. Wilcox: The dehydration of okra: variety and stage of maturity as factors in determining quality. Canner **101**, Nr. 17, 14—16, 22—24, 26 (1945).

—, B. D. Ezell, C. W. Culpepper, M. S. Wilcox u. M. C. Hutchins: Variety, its effect on dehydrated spinach. Canner **101**, Nr. 9, 12—14, 22; Nr. 10, 22—24; 26, 32; Nr. 11, 20, 22 (1945).

Carpenter, C. D. u. E. W. Friedlander: Occurrence of vitamins in fungi. Science **95**, 625 (1942).

Cheema, G. S., B. Nazareth u. S. R. Dhareshwar: Improvement of brinjals (Solanum melongena, L.) by selektion in the Bombay Province. Proc. Indian Acad. Sci. **16**, Sect. B, 25—48 (1942).

Chevalier, A.: Les rhubarbes cultivées en Europe et leurs origines. Rev. Bot. Appl. **22**, 474—485 (1942).

Chin, T. C. u. H. W. Youngken: The cytotaxonomy of Rheum. Amer. J. Bot. Suppl. **33**, 840 (1946). (Abstr.)

Christiansen, E.: (Acht Jahre Züchtungsversuche mit Spargelsorten, 1933—1940.) Tidsskr. Planteavl **46**, 704—718 (1942). [Dänisch.]

Coulter, F. C.: Story of garden vegetables. IX. Egg plant-and its travels from ancient India. Seed World **51**, Nr. 1, 36—37 (1942).

— The story of garden vegetables. XVII. Radish, still going strong after thirty centuries of development. Seed World. **53**, 10—11 (1943).

— The story of garden vegetables. XVIII. Rhubarb-from a costly oriental drug to a popular pie. Seed World **53**, 10—11 (1943).

— The story of garden vegetables. XIX. Spinach-not one of the most ancient, but one of the most popular. Seed World **54**, 28, 57 (1943).

Cowley, F. J.: El cultivo del ajo. (The cultivation of garlic.) Rev. Minist. Agric. Cuba **27**, Nr. 28, 25—27 (1944).

Crane, M. B. u. K. Mather: The natural cross-pollination of crop plants with particular reference to the radish. Ann. Appl. Biol. **30**, 301—308 (1943).

Currence, T. M.: Progeny tests of asparagus plants. J. Agric. Res. **74**, 65—76 (1947).

Daskaloff, Ch.: Beitrag zum Studium der Heterosis bei der Eierfrucht (Sol. Melongena L.) und die Möglichkeit einer praktischen Ausnutzung. Forschungsdienst **12**, 617—618 (1941).

Drewes, H.: Spinach growing in the U. S. Market Gr. J. **66**, 51—53 (1940).

— America takes over spinach seed production. Sth. Seedsman **6**, 9, 33, 41 (1943).

Filov, A. I.: An agro-ecological classification of eggplants and a study of their characters. C. R. (Doklady) Acad. Sci. URSS. **26**, 815—818 (1940).

Gentcheff, G. u. Å. Gustafsson: The double chromosome reproduction in Spinacia and its causes. I. Normal behaviour. II. An X-ray experiment. Hereditas, Lund. **25**, 349—358, 371—386 (1939).

—, — The double chromosome reproduction in Spinacia and its causes 2. An x-ray experiment. Hereditas **25**, 372—386 (1940).

Glotov, V.: Amphidiploid fertile form of Mentha piperita. L. produced by colchicine treatment. C. R. (Doklady) Acad. Sci. URSS. **28**, 450—453 (1940).

Grêne, G.: Une culture d'ail sélectionné. Rev. Hort. Suisse **14**, 323—324 (1941).

Hagiwara, T. u. H. Kusamitu: (Distinction of males and females in Asparagus by using a solution of potassium chlorate.) Agric. Hort. **11**, 990—992 (1939).

Hanna, G. C.: Yield studies as related to asparagus breeding. Proc. Amer. Soc. Hort. Sci. **36**, 677—679 (1938) 1939.

— Correlation studies of asparagus comparing yields of various shorter periods with ten-year yields. Proc. Amer. Soc. Hort. Sci. **41**, 321—323 (1942).

Härdh, H.: Tutkimuksia kromosomimorfologiasta ja polysomatia-ilmiöstä Spinacia oleracea L.. (Studies of the chromosome morphology and polysomaty in Spinacia oleracea L.) Maataloust. Aikakausk. **11**, 317 bis 332 (1939).

Hartmair, V.: Cytologische Untersuchungen an Rettichen. Züchter **12**, 120—122 (1940).

Heidt, K.: Meerkohlarten (Crambe hispanica L., Cr. abyssinica Hochst., Cr. maritima L., Cr. tatarica Jacq.) als ertragreiche Öl-, Gemüse- und Futterpflanzen. Pflanzenbau **20**, 170—176 (1945).

Hopper, W. E. R.: Seed formation, germination, and postgermination development in certain Cichorieae. Trans. Ill. Acad. Sci. **34**, 70—72 (1941). (Abstr.)

Howard, F. L. u. R. Desrosiers: Studies on the resistance of eggplant varieties to Phomopsis blight. Proc. Amer. Soc. Hort. Sci. **39**, 337—340 (1941).

Howard, H. W.: Seed size in crosses between diploid and autotetraploid Nasturtium officinale and allotetraploid N. uniseriatum. J. Genet. **48**, 111—118 (1947).

Hylmö, B.: Einwirkung durch Hormonbehandlung auf das Geschlecht der Spinacia. Bot. Notiser, Nr. 4, 389 bis 394 (1940).

Kadam, B. S., R. M. Kulkarni u. S. M. Patel: Natural crossing in Cajanus cajan (L.) Millsp. in the Bombay-Deccan. Indian J. Genet. Pl. Breed. **5**, 60—62 (1945).

—, V. K. Patnakar, S. M. Patel u. B. B. Chaudhari: Chafa-a new variety of gram from Bombay. Indian Fmg **6**, 444—446 (1945).

Kasparova, S.: (Resistance of chicory.) Bull. Acad. Sci. URSS., Sér. Biol., Nr. 3, 353—370 (1941).

Keppler, E.: Inzuchtleistungen und Bastardierungseffekt beim Radies (Raphanus sativus). Z. Pflanzenz. **23**, 661—684 (1941).

Kligman, A. M.: Some cultural and genetic problems in the cultivation of the mushroom, Agaricus campestris Fr. Amer. J. Bot. **30**, 745—763 (1943).

Klušnikova, E. S.: (The experimental and cytological study of the bispored form of Psalliota campestris Fr.) Učenye Zapiski Moskovskogo Gosudarstvennogo Universiteta. Trudy Instituta Botaniki (Scientific Proceedings of the Moscow State University. Transactions of the Institute of Botany) **36**, 136—171 (1940). [Russisch.]

Kopetz, L. M.: Zeitstufenversuche mit Kohlrabisorte ,,Roggli". Ein Beitrag zur Frage der Auslösung von Schossern. Vorl. Mitt. Züchter **14**, 136—137 (1942).

Kremer, J. C.: Influence of honey bee habits on radish seed yield. Quart. Bull. Mich. Agric. Exp. Sta. **27**, 413—420 (1945).

Krickl, M.: Zur Züchtung von frostwiderstandsfähigen und nicht holzig werdenden frühen Kohlrabi. Gartenbauwiss. **18**, 185—203 (1944).

Kumar, L. S. S., A. Abraham u. V. K. Srinivasan: Preliminary note on autotetraploidy in Cajanus indicus Spreng. Proc. Indian Acad. Sci. **21**, Sect. B, 301—306 (1945).

Kvasnikov, B. B.: (Selecting chicory not subject to bolting.) Proc. Lenin Acad. Agric. Sci. USSR., Nr. 20, 3—9 (1940). [Russisch.]

Levine, M.: The effect of colchicine and acenaphthene in combination with X-rays on plant tissue. II. Bull. Torrey Bot. Cl. **73**, 34—59 (1946).

Lipšic, S. J.: (Contributary items towards a monograph of the genus Scorzonera. Part 2.) The Moscow Society of Naturalists, Moscow, Pp. 168 (1939). [Russisch.]

Marčenko, I. I.: (Distant hybridization in Jerusalem artichokes.) Vestnik Gibridizacii (Hybridization), Nr. 1, 114—115 (1941). [Russisch.]

Montelaro, J.: Easy-pickin's: new spineless okra. Sth. Seedsman **9**, Nr. 1, 15 (1946).

Moskalenko, S. S.: (Capparis as a conqueror of the desert.) Ovoščevodstvo (Vegetable Growing), Nr. 1, 25—28 (1940). [Russisch.]

Mukerji, B.: Indian rhubarb as substitute for „official" rhubarb. Curr. Sci. **12**, 275 (1943).

Mündler, M. u. F. Schwanitz: Über einen Ertrags- und Düngungsversuch mit diploidem und autotetraploidem Münchener Bierrettich. (Raphanus sativus var. major/L.A Voss.) Züchter **14**, 137—140 (1942).

Nelson, R.: Progress in the control of peppermint wilt by resistance. Rep. Proc. 24th Annu. Conv. Mich. Muck. Fmrs' Ass., Pp. 38—42 (1942).

— Present status of wilt-resistant hybrid mints and plans for their further development. Proc. Mich. Muck. Fmrs' Ass. **25**, 38, 40 (1943).

— Production of mint species hybrids resistant to Verticillium wilt. Phytopathology **37**, 16—17 (1947). (Abstr.)

Nicolaisen, N. u. R. Hanow: Bestimmung der Geschlechtsverhältnisse bei Spinat. Z. Pflanzenz. **23**, 476 bis 486 (1940).

Nugent, T. J. u. H. T. Cook: Developing a wilt resistant spinach variety for Virginia. Proc. Va. Acad. Sci. **2**, 182, 1940—1941 (1941).

Overcash, J. P.: Propagation and culture of garden sage in Tennessee. Proc. Amer. Soc. Hort. Sci. **46**, 345—349 (1945).

Pal, B. P. u. H. B. Singh: Floral characters and fruit formation in the eggplant. Indian J. Genet. Pl. Breed. **3**, 45—58 (1943).

Park, M. u. M. Fernando: A variety of brinjal (Solanum Melongena Linn.) resistant to bacterial wilt. Trop. Agricultirist **94**, 19—21 (1940).

Peeling, B. A.: Clemson spineless okra. Agrarian, Clemson, S. C. **1**, Nr. 2, 11 (1939).

Person, L. H.: The occurence of a variant in Rhizoctonia solani. Phytopathology **34**, 715—717 (1944).

Pillai, S. M.: Preliminary studies in coriander (Coriandrum sativum L.). Madras Agric. J. **27**, 79—84 (1939).

Pollard, L. H. u. F. B. Wann: Variety studies give information on yields and other characteristics of celery strains. Fm Home Sci., Utah **6**, 6—10 (1945).

Poole, C. F. u. P. C. Grimball: A heyday for corn-on-the-cobbers. Sth. Seedsman **6**, 11, 50 (1943).

Popov, P.: (Contribution to the study of okra [Hibiscus esculentus L.] found in Bulgaria.) Rev. Inst. Rech. Agron. Bulg. **9**, Nr. 1, 3—15 (1939).

Raghavendra, Rao, M. R. u. M. Sreenivasaya: Asparagine from Indian pulses. Curr. Sci. **15**, 25—26 (1946).

Randall, T. E. u. C. M. Rick: Preliminary cyto-genetic studies on polyembryony in Asparagus officinalis L. Amer. J. Bot. Suppl. **28**, 5s (1941). (Abstr.)

—, — A cytogenetic study of polyembryony in Asparagus officinalis L. Amer. J. Bot. **32**, 560—569 (1945).

Rangaswami Ayyangar, G. N. u. Kunhi Krishnan Nambiar, K.: Albinism in lablab. Curr. Sci. **10**, 255 (1941).

Rangaswami Ayyangar, G, N. u. Kunhi Krishnan Nambiar, K.: Tricotyledony in lablab. Curr. Sci. **10**, 255—256 (1941).

Reinhold, J.: Der Einfluß der Tageslänge und der Lichtintensität auf das Wachstum der Reichsspinatsorten. Gartenbauwiss. **18**, 266—287 (1944).

Richardson, A. L. u. T. M. Currence: The relation of yield of staminate and pistillate asparagus plants to the rate of growth of progenies in the young stage. Proc. Amer. Soc. Hort. Sci. **38**, 613—617 (1941).

Roux: Ein Beitrag zur Erhöhung des Anbauerfolges der Treibzichorie. Gartenbauwiss. **15**, 559—564 (1941).

Sagar Roy, R.: A new variety of brinjal (Solanum melongenum L.). Curr. Sci. **13**, 287—288 (1944).

Sain, S.: (The possibility of producing new varieties of Jerusalem artichoke by vegetative means.) Doklady Vsesojuz. Akad. Seljsk. Nauk im V. I. Lenina (Proc. Lenin Acad. Agric. Sci. USSR.), Nr. 11/12, 39—41 (1944). [Russisch.]

Salaman, R. N.: Why „Jerusalem" artichoke? J. R. Hort. Soc. **65**, 338—348, 376—383 (1940).

Singh, D. N., R. K. Bansal u. S. P. Mital: Cajanus obcordifolia Singh. A new species of Cajanus. Indian J. Agric. Sci. **12**, 779—784 (1942).

Singh, H. B.: A naturally-occurring tetraploid brinjal. Indian J. Genet. Pl. Breed. **2**, 71—72 (1942).

Solis, M. A.: Estudio botánico farmacognósico de la achicoria de Quito: Achyrophorus quitensis Schultze Bip., var. de flores blancas. (Pharmacological study of the Quito chicory: A. quitensis Schultze Bip., variety with white flowers.) Flora, Ecuador **2**, Nr. 5/6, 79—97 (1942).

Steiner, E.: Cytogenetic studies on Talinum and Portulaca. Bot. Gaz. **105**, 374—379 (1944).

Stolej, V. Ja.: (The Chinese radish.) Ovoščevodstvo Nr. 4, 31 (1939). [Russisch.]

Stoletova, E. A.: (The geographical groups of coriander as initial material for breeding.) Selekcija i Semenovodstvo, Nr. 5, 12—13 (1939). [Russisch.]

Sundar, Rao, Y.: Chromosomes of Erythrina indica Lamk. J. Indian Bot. **24**, 42—44 (1945).

Tatebe, T.: (On inheritance of color in Solanum melongena Linn.) Jap. J. Genet. **15**, 261—271 (1939).

Taylor, H. V. u. E. E. Skillman: Rhubarb. Bull. Minist. Agric. Lond., Nr. 113, 24 (1944).

Thompson, E. G.: Brussels sprouts trials, 1940—1942. J. Nat. Inst. Agric. Bot. **5**, 45—59 (1944).

Thompson, H. C.: Asparagus production. New York, Pp. 124, 12 figs. 4 tables. (1946).

Thomson, C. L. u. O. J. Robb: Asparagus selections and certain cultural practices compared for yield, earliness and sex ratios. Sci. Agric. **26**, 289—299 (1946).

Todhunter, E. N.: The ascorbic acid (vitamin C) content of rhubarb. Proc. Amer. Soc. Hort. Sci. **40**, 437 bis 440 (1942).

Townsend, G. R., R. A. Emerson u. A. G. Newhall: Resistance to Cercospora apii Fres. in celery (Apium graveolens var. dulce). Phytopathology **36**, 980—982 (1946). (Abstr.)

Venkataramani, K. S.: Breeding brinjals (Solanum melongena) in Madras. I. Hybrid vigour in brinjals. Proc. Indian Acad. Sci. **23**, Sect. B, 266—273 (1946).

Virabhadra Rao, J.: Chemical examination of Erythrina indica (white variety). Curr. Sci. **14**, 198 (1945).

Wager, V. A.: Bacterial wilt of the egg-plant. Fmg. S. Afr. **19**, 661—664 (1944).

— Egg-plants resistant to bacterial wilt. Fmg S. Afr. **21**, 410—412 (1946).

Whitaker, T. W.: The occurrence of a spontaneous triploid celery. Proc. Amer. Soc. Hort. Sci. **39**, 346—348 (1941).

Wohlers, C.: La sélection de la chicorée Witloof pour la graine. Rev. Hort. Suisse **15**, 65—66 (1942).

Wolf, J.: Untersuchungen an Spargel. II. Mitteilung: Vitamin C. Gartenbauwiss. **15**, 590—598 (1941).

Woolford, B. C.: The five big winners ... of the 1944 awards. Sth Seedsman **6**, 12—13, 45, 49 (1943).

Yamaguti, Y.: (On the inheritance of the anomalous „tomoe" shape of flower in Portulaca grandiflora.) Jap. J. Genet. **15**, 357—358 (1939).

— (A few abnormal characters in Portulaca grandiflora.) Jap. J. Genet. **16**, 307—308 (1940).

Yeager, A. F. u. D. H. Scott: Studies of mature asparagus plantings with special reference to sex survival and rooting habits. Proc. Amer. Soc. Hort. Sci. **36**, 513—514 (1938) 1939.

—, — Asparagus. 38th Rep. S. Dak. St. Hort. Soc., Pp. 52—55 (1941).

t) Forstwirtschaft.

Adamson, R. S., E. Esterhuysen ü. E. P. Phillips: Some changes in nomenclature. IV. J. Sth. Afr. Bot. **9**, 137—140 (1943).

Åhlman, S.: Hur tidigt kan anlag för smal- eller bredkronighet konstateras? (How early can the tendency to narrow or broad crown be determined?) Skogen **30**, 231 (1943).

Aljbenskij, A. V.: (Changing the aspen from a monoecious into a dioecious plant.) Proc. Lenin. Acad. Agric. Sci. USSR., Nr. 19, 27—29 (1940). [Russisch.]

— Specific characters in F_1 of the inter-specific Larix hybrids. Proc. Lenin Acad. Agric. Sci. USSR., Nos. 23 bis 24, 20—23 (1940).

— (Poplar hybridization in the USSR.) J. Bot. URSS. **29**, 86—90 (1944). [Russisch.]

Allen, G. S.: Parthenocarpy, parthenogenesis, and self-sterility of Douglas fir. J. For. **40**, 642—644 (1942).

Andersson, E.: Brundbergsfilialen i verksamhet-dess förädlingsuppgifter. (The Brunsberg Branch Station starts work-its tasks in breeding.) Svensk PappTidn. **46**, 343—351, 400—402 (1943).

— Växtförädlingen och skogsbruket. Ett diskussionsinlägg. (Plant breeding and forestry-A contribution to the discussion.) Skogen **30**, 178—180 (1943).

— Verksamheten vid värmlands-och norrlands-filialerna. (Work at the Värmland and Norrland branch stations.) Svensk Papp Tidn. **47**, Nr. 15, 376—378, Nr. 17, 427 bis 433, Nr. 18, 449—451 (1944).

— A case of asyndesis in Picea Abies. Hereditas, Lund **33**, 301—347 (1947).

Anonymus: A handbook of home-grown timbers. (Second edition.) His Majesty's Stationery Office, London (1939).

— Nomenclature of Australian timbers. Pt III. Trade Circ. Coun. Sci. Industr. Res. Div. For. Products, Aust., Nr. 47, 45—82 (1940).

— Plantforoedlingi Skovbruget. (Plant breeding in forestry.) Naturhistorisk Tidende, København **4**, 115—120 (1940).

— Viaje de estudios silvícolas en Europa del Ing. Agr. Lucas A. Tortorelli. (Study tour on forestry in Europe by Lucas A. Tortorelli.) Rev. Argent. Agron. **7**, 141 bis 143 (1940).

— Bör kamgranen gynnas vid bestandsvårdshuggningarna? (Should the comb spruce be spared in sylvicultural felling?) Skogsägaren **17**, 145 (1941).

— Improvement of forest trees in Sweden. Anglo-Swedish Rev., Pp. 93—94 (1941).

— Skogsträdens förädling. (Forest tree breeding.) Skogen **28**, 57—60, 81—83 (1941).

— The Commission on forest seeds and tree race problems of the International Union of Forest Research Organization. Chronica Botanica **6**, 207 (1941).

Anonymus: Canada: tree breeding and propagation. News Bull. Emp. For. Departments, Pp. 39 (1942).

— Inventering av elitgranbestånd. (Survey of élite spruce stands.) Skogsägaren **18**, 184—187 (1942).

— Tillvaratagande och förädling av mindervärdigt virke samt avfalls- och biprodukter i skoksindustrien. (Exploitation and improvement of inferior wood and waste and by-products in the timber industry.) Industriens Utredningsinstitut Norrlandsutredningen, Stockholm, Pp. 328 (1942).

— Tropical forest research. J. For. **40**, 169—172 (1942).

— Växtförädling och rationell skogsdrift. Ur ett föredrag av professor Nils Sylvén. (Plant breeding and rational forest management-from a lecture by Professor Nils Sylvén.) Skogen **29**, 347—349 (1942).

— Forest Research in India and Burma 1942—43. Part I. The Forest Research Institute. For. Res. Inst. Dehra Dun, Pp. 144 (1942—43).

— Frågan om Skogsförsöksanstaltens omorganisation. (The question of the reorganization of the Forestry Research Institute.) Svensk PappTidn. **46**, 472—473 (1943).

— Inventering av elitstammar. (Survey of élite trees.) Skogen **30**, 74 (1943).

— Arter, hybrider och skogar av sydbokssläktet (Nothofagus) i Nya Zeeland och Australien. (Species, hybrids and woods of southern beech Nothofagus in New Zealand and Australia.) Svensk Bot. Tidskr. **38**, 123—126 (1944).

— Plan utarbetad för frö- och plantförsörjning. (Plan worked out for supplying seed and plants.) Svensk PappTidn. **48**, 149—150 (1945).

— Skogsodling och fröproveniens. (Forest planting and seed provenance.) Skogen **32**, 93—94 (1945).

Arcybašev, D. D.: (Work on introducing valuable species into northern zones.) Lesnoe Hozjajstvo (Forestry), Nr. 9, 16—22 (1939). [Russisch.]

Arnborg, T.: Busktallen. En inventering. (The bush pine: A survey.) Skogen **28**, 174 (1941).

Atchison, E.: Chromosome numbers in the Myrtaceae. Amer. J. Bot. **34**, 159—164 (1947).

Austin, L.: Forest genetics. Rep. Smithson. Instn. Publ. 3491, Pp. 433—440 (1938) 1939.

B......, R.: Skogsfröförsörjningen i Gävleborgs län. (Supplying forest tree seed in the Gavleborg district.) Skogen **30**, 307—308 (1943).

Babcock, E. B.: The probable center of origin of the genus Pinus. Amer. J. Bot. (Suppl.) **33**, 233 (1946). (Abstr.)

Baldwin, H. I.: Forest tree seed of the north temperate regions with special reference to North America. Chronica Botanica Company, Waltham, Mass. (1942).

Beard, J. S.: The importance of race in teak, Telectona grandis L. Carrbean Forester **4**, 135—139 (1943).

Bedwell, J. L. u. T. W. Childs: Susceptibility of white bark pine to blister rust in the Pacific Northwest. J. For. **41**, 904—912 (1943).

Beketovskij, D. N.: (Some characteristics of the atypical form Robinia pseudoacacia L. var. monophylla Kirschn.) J. Bot. URSS. **19**, 29—35 (1944).

Berg, A.: A rust-resistant red cedar. Phytopathology **30**, 876—878 (1940).

Bergstrom, I.: On the progeny of diploid x triploid Populus tremula with special reference to the occurrence of tetraploidy. Hereditas, Lund **26**, 191—201 (1940).

Beversluis, J. R.: De micrografische identificatie van conifere houtsoorten. (Micrographical identification of types of conifer wood.) Meded. LandbHoogesch., Wageningen **43**, Nr. 2, 39 (1943).

Black, R. A.: Description of new eucalypt hybrid × Eucalyptus radiodives („Butter-cup Peppermint")

(radiata × dives) R. A. Black, hybrid now. Victorian Nat. **60**, 175 (1944).
Blin, H.: L'hybridation dans le genre Eucalyptus. Rev. Hort., Paris **26**, 555—556 (1939).
Bray, M. W. u. B. H. Paul: Pulping studies on selected hybrid poplars. Paper Tr. J. **115** (16), 33—38 (1942).
Buchholz, J. T.: Multi seeded acorns. Trans. Ill. Acad. Sci. **34**, 99—101 (1941).
— The cause of sterility in cross-pollinations between certain species of pines. Amer. J. Bot. **31**, 2s (1944). (Abstr.)
— Embryological aspects of hybrid vigor in pines. Science **102**, 135—142 (1945).
Burger, H.: Holz, Blattmenge und Zuwachs. V. Mitteilung. Fichten und Föhren verschiedener Herkunft auf verschiedenen Kulturorten. Mitt. Schweiz. Ant. Forstl. Versuchsw. **22**, 10—62 (1941).

Castro, D. de: Nota sôbre o número de cromosomas da ,,Beluta celtiberica" Rothm. et Vasc. (Note on the chromosome number of B. celtiberica Rothm. et Vasc.) Broteria **13**, 73—74 (1944).
Champion, H. G.: Seed selection in forestry. Nature, Lond. **152**, 354 (1943).
— Genetics in forestry. Emp. For. J. **24**, 12—13 (1945).
— Genetics and forestry. Quart. J. For. **39**, 74—81 (1945).
Chevalier, A.: Les ormes de France. Rev. Bot. Appl. **22**, 429—459 (1942).
— Notes sur les conifères de l'Indochine. Rev. Bot. Appl. **24**, 7—34 (1944).
— Un Eucalyptus résistant au froid venu probablement par mutation. Rev. Bot. Appl. **26**, 232—234 (1946).
Cook, D. B.: Characteristics of Dunkeld larch and its parent species. J. For. **40**, 884—885 (1942).
— An abnormal balsam fir. Torreya **45**, 13 (1945).
Cook, W. H.: Division of applied biology. 25th Ann. Rep. Nat. Res. Counc. Canada 1941—1942, Pp. 16—18 (1942).
Cox, H. A. (Editor): A handbook of Empire timbers. His Majesty's Stationery Office, London (1939).
Curry, J. R.: Selection, propagation, and breeding of high-yielding Southern Pines for naval stores production. J. For. **41**, 686—687 (1943).

Dadswell, H. E.: The card sorting method applied to the identification of the commercial timbers of the genus Eucalyptus. J. Coun. Sci. Industr. Res. Aust. **14**, 266—280 (1941).
Dangeard, P.: Sur l'existence d'un système de fibrilles préfusoriales dans la mitose somatique du pin maritime. C. R. Acad. Sci., Paris **211**, 657—659 (1940).
— Sur les différences de taille entre chromosomes appartenant à différents tissus dans la plantule de pin maritime. C. R. Soc. Biol., Paris **135**, 581—583 (1941).
Davis, S. H. jr.: Poplar canker. A note on the susceptibility of varius poplar species. Morris Arbor. Bull. **4**, 28 (1942).
Dayton, W. A.: Kelsey Locust, Robinia kelseyi Hort. ex Hutchins. Amer. Midl. Nat. **30**, 504—509 (1943).
Degelius, G.: Tall med försenad klorofyllbildning. (Pine with retarded chlorophyll formation.) Skogen **30**, 79 (1943).
Dengler, A.: Über die Entwicklung künstlicher Kiefernkreuzungen. Z. Forst- u. Jagdw. **71**, 457—485 (1940).
— Über die Befruchtungsfähigkeit der weiblichen Kiefernblüte. Ztschr. f. Forst- u. Jagdw. **72**, 48—55 (1940).
— Bericht über Kreuzungsversuche zwischen Trauben- und Stieleiche (Quercus sessiliflora Smith und Quercus pedunculata Ehrh. bzw. Robur L.) und zwischen europäischer und japanischer Lärche (Larix europaea D. C. bzw. decidua Miller und Larix leptolepis Murray bzw. Kämpferi Sargent). Mitt. H. Göring-Akad. dtsch. Forstwiss. **1**, 87—109 (1941).
Deuber, C. G.: The vegetative propagation of eastern white pine and other five-needled pines. J. Arnold Arbor. **23**, 198—213 (1942).
Devoto, F. E.: Las hibridaciones entre especies forestales y sus frecuentes hibridaciones naturales en nuestro pais. (Un hibrido artificial entre ,,Chorisia insignis" y ,,Chorisia speciosa".) (Resumen.) (Hybridization between forest species and their frequent natural hybridization in our country. [An artificial hybrid between C. insignis and C. speciosa].) Physis, B. Aires **18**, 369—374 (1939). [Abstr.]
Dillewijn, C. van: Cytologie en veredeling van Populus. (Cytology and breeding of Populus.) Ned. Boschb.-Tijdschr. **12**, 470—481 (1939).
— Zytologische Studien in der Gattung Populus L. Genetica **22**, 131—182 (1940).
Dorman, K. W., C. S. Schopmeyer u. A. G. Snow: Top bracing and guying in the breeding of southern pines. J. For. **42**, 140—141 (1944).
Doyle, J.: Naming of the redwoods. Nature, Lond. **155**, 254—257 (1945).
— u. A. Kane: Pollination in Tsuga Pattoniana and in species of Abies and Picea. Sci. Proc. R. Dublin Soc. **23**, 57—70 (1943).
Duffield, J. W.: Chromosome couts in Quercus. Amer. J. Bot. **27**, 787—788 (1940).
— The cytological basis of forest tree improvement. J. For. **40**, 859—864 (1942).
— Polyploidy in Acer rubrum L. Chronica Botanica **7**, 390—391 (1943).
— u. A. G. Snow jr.: Pollen longevity of Pinus strobus and Pinus resinosa as controlled by humidity and temperature. Amer. J. Bot. **28**, 175—177 (1941).
—, — Effect of storage conditions on pollen longevity of Pinus strobus and Pinus resinosa. J. For. **39**, 410—411 (1941).

Ekdahl, I.: Die Entwicklung von Embryosack und Embryo bei Ulmus glabra Huds. Svensk. Bot. Tidskr. **35**, 143—156 (1941).
Eklundh, C.: Artkorsningar inom sl. Abies, Pseudotsuga, Picea, Larix, Pinus och Chamaecyparis, tillhörande fam. Pinaceae. (Species crosses in the genera Abies, Pseudotsuga, Picea, Larix, Pinus and Chamaecyparis belonging to the family Pinaceae.) Svensk PappTidn. **46**, 55—61 (1943).
— Artkorsningar inom sl. Abies, Pseudotsuga, Picea, Larix, Pinus och Chamaecyparis, tillhörande fam. Pinaceae. (Species crosses within the genera Abies, Pseudotsuga, Picea, Larix, Pinus and Chamaecyparis, belonging to the family Pinaceae.) Svensk. PappTidn. **46**, 101—105, 130—133 (1943).
Erdtman, H.: Die phenolischen Inhaltsstoffe des Kiefernholzes. IV. Svensk PappTidn. **46**, 226—228 (1943).
— Tallkärnvedens fenoliska substanser. IV. (Phenol substances of heart wood of pine. IV.) Svensk PappTidn. **46**, 532 (1943).
— Die Konstitution der Harzphenole und ihre biogenetischen Zusammenhänge. VIII. Zur Kenntnis des Conidendrins (Sulfitlaugenlactons) und dessen Verbreitung unter verschiedenen Coniferen. Svensk PappTidn. **47**, 155—159 (1944).
Esson, J. G.: Fastigiate oak reproduced from seed. J. N. Y. Bot. Gdn. **47**, 275—278 (1946).

Fassett, N. C.: Juniperus virginiana, J. horizontalis and J. scopulorum. II. Hybrid swarms of J. virginiana and J. scopulorum. Bull. Torrey Bot. Cl. **71**, 475—483 (1944).
— Juniperus virginiana, J. horizontalis, and J. scopulorum. III. Possible hybridization of J. horizontalis

and J. scopulorum. Bull. Torrey Bot. Cl. **72**, 42—46 (1945).

Fassett, N. C.: Juniperus virginiana, J. horizontalis and J. scopulorum. IV. Hybrid swarms of J. virginiana and J. horizontalis. Bull. Torrey Bot. Cl. **72**, 379—384 (1945).

— Juniperus virginiana, J. horizontalis, and J. scopulorum. V. Taxonomic treatment. Bull. Torrey Bot. **72**, 480—482 (1945).

Fedorova-Sarkisova, O. V.: (Über die Chromosomenzahl einiger Arten der Weiden und Pappeln.) C. R. (Doklady) Acad. Sci. URSS. **54**, 357—360 (1946). [Russisch.]

Fischer, F.: Nachzucht und Erziehung der Eiche im bernischen Bucheggberg. Mitt. Schweiz. Anst. Forst. Versuchs. **23**, 375—470 (1944).

Fleischmann, R.: Weitere Ergebnisse auf dem Gebiete der Robinienzüchtung. Züchter **11**, 90—94 (1939).

Florin, B.: Stolp-huggning. Skona elitbestånd och elitträd. (Post production. Spare élite stands and élite trees,) Skogsägaren, Nr. 10, 231—233 (1944).

Flory, W. S. jr. u. F. R. Brison: Propagation of the Ness hybrid oaks. Proc. Amer. Soc. Hort. Sci. **40**, 590 (1942).

—, — Propagation of a rapid growing semievergreen hybrid oak. Bull. Tex. Agric. Exp. Sta., Nr. 612, 32 (1942).

Forsaith, C. C.: Statistics for foresters. N. Y. **16**, 69 (1943).

Fouarge, J.: Note sur la caryocinèse chez les chênes pédonculé et rouvre. Bull. Inst. Agron. Gembloux **8**, 111—113 (1939).

Freeman, O. M.: A red maple, silver maple hybrid. J. Hered. **32**, 11—14 (1941).

Gardner, V. R.: Winter hardiness in juvenile and adult forms of certain conifers. Bot. Gaz. **105**, 408—410 (1944).

Gavris, V. P.: (Selection of immune forms of the common pine.) Lesnoe Hozjajstvo (Forestry), Nr. 8, 5—8 (1939). [Russisch.]

Geete, E.: Ormgranar och deras förekomst. (Worm spruces and their incidence.) Skogen **31**, 263—265 (1944).

Gertz, O.: Fagus silvatica L. f. osbyensis. Bot. Notiser, Pp. 75—83 (1942).

Goidànich, G. u. F. Azzaroli: Relazione sulle esperienze di selezione di olmi resistenti alla grafiosi e di inoculazioni artificiali di ,,Graphium ulmi" eseguite nel 1938. (Report on experiments on selecting elms resistant to elm disease and on artificial inoculations with G. ulmi in 1938.) Boll. Staz. Pat. Veg. Roma **19**, 222—240 (1939).

—, — Relazione sulle esperienze di selezione di olmi resistenti alla grafiosi e di inoculazioni artificiali di ,,Graphium ulmi" eseguite nel 1939—1940. (Report on experiments of selecting elms resistant to dutch elm disease and of artificial inoculations of ,,G. ulmi" carried out in 1939—1940.) Bol. Staz. Pat. Veg., Roma **21**, 287—306 (1941).

Gopal-Ayengar, A. R.: Structure and behaviour of meiotic chromosomes in gymnosperms. Genetics **27**, 143 (1942). (Abstr.)

Goršenin, N. M. (Editor): (Summary of research work for the year 1938.) Vsesojuznyj N. I. Agrolesomeliorativnyj Inst. (VNIALMI), Moscow, p. 172 (1940). [Russisch.]

Gram, K., C. Muhle Larsen, C. Syrach Larsen u. M. Westergaard: Contributions to the cytogenetics of forest trees. II. Alnus studies. K. VetHøjsk. Aarsskr., Pp. 44—58 (1941).

Graves, A. H.: Breeding work towards the development of a timber type of blightresistant chestnut. Report for 1939. Bull. Torrey Bot. Cl. **67**, 772—777 (1940).

Graves, A. H.: Breeding work toward the development of a timber type of blightresistant chestnut: Report for 1940. Bull. Torrey Bot. Cl. **68**, 667—674 (1941).

— Breeding work toward the development of a timber type of blightresistant chestnut: report for 1941. Amer. J. Bot. **29**, 622—626 (1942).

— Breeding for development of a timber type of diseaseresistant chestnut. Phytopathology **33**, 18 (1943). (Abstr.)

Grossheim, A. A.: (A new pyramidal poplar Populus Schischkini.) J. Bot. URSS. **29**, 124 (1944).

Gruenhagen, R. H.: Hypoxylon pruinatum and its pathogenesis on poplar. Phytopathology **35**, 72—80 (1945).

Guenther, E.: Australian Eucalyptus oils. Drug and Cosmetic Ind. **51**, 160—167, 203, 281—285, 404—409 (1942).

Guerreiro, M. G. u. C. T. Fernandes: O castanheiro no distrito de Bragança. (The chestnut in the district of Braganza.) Bol. Junta Nac. Frutas, Portugal **5**, 5—32 (1945).

Guinier, P.: Arbres et forêts. Notes botanico-forestières. II. Sapins sans branches. Rev. Eaux For., Pp. 316 bis 318 (1944).

Hall, R. C.: Control of the locust borer. Circ. U. S. Dep. Agric., Nr. 626, 20 (1942).

Haque, A.: Chromosome numbers in Sesbania spp. Curr. Sci. **15**, 78 (1946).

— Haploid-haploid polyembryony in Sesbania aculeata Pers. Curr. Sci. **15**, 287 (1946).

Haupt, A. W.: Oogenesis and fertilization in Pinus lambertiana and P. monophylla. Bot. Gaz. **102**, 482—498 (1941).

Heiberg, H. H. H.: En oversikt over proveniensproblemet hos våre viktigste skogstraer, furu, gran og bjørk. (A survey of the provenance problem of our most important forest trees, pine, spruce and birch.) Medd. Norske Skogforsøksvesen **6**, Nr. 20/23, 51—109 (1937—1939).

Heimburger, C.: Report on Poplar hybridization. II. 1937 and 1938. For. Chron. **16**, 149—160 (1940).

— Report on strain tests of Scots pine and Norway spruce, Petawawa projects 52 and 53. Scots pine (Pinus silvestris). Rep. Inter. Un. For. Res. Organizations, Stockholm, Pp. 17—20 (1945).

Helms, A. D.: A visit to the Danish arboretum and forest botanical gardens, June 1939. Aust. For. **5**, 16—20 (1940).

Hesselman, H.: Den naturvetenskapliga avdelningens versamhet under åren 1902—1938 och avdelningens framtida uppgifter. (The activities of the natural science division during the years 1902—1938 and the future tasks of the Division.) Medd. Skogsförsöksanst. Stockh. **31**, 163—170, 1938—39 (1939).

Holdridge, L. R.: Trees of Puerto Rico. Occ. Pap. For. Serv. U. S. Dep. Agric. Vol. I, Nr. 1, 105, Vol. II, Nr. 2, 105 (1942).

Hopp, H.: Multiple measures for distinguishing closely related plant forms. Chronica Botanica **7**, 402—403 (1943).

Houtzagers, G.: Onze bosschen na vyf jaren oorlog Nr. VIII. Groote beteekenis van het gebruik van beter plantmateriaal. (Our woods after five years of war. Nr. VIII. Great importance of the use of better planting material.) Tidschr. Ned. Heidemaatsch **58**, 107—110 (1947).

Jablokov, A. S.: (Differences in the F_1 interspecific hybrids of woody forest trees.) Jarovizacija Nr. 4 (31), 33—40 (1940). [Russisch.]

— (Pro Darwinism in breeding woody forest trees.) Lesnoe Hozjajstvo (Forestry), Nr. 4, 75—77 (1940). [Russisch.]

Jablokov, A. S.: (Interspecific hybridization in sylviculture.) Vestnik Gibridizacii (Hybridization) Nr. 1, 39—45 (1941).
— (A giant form of Populus tremula in woodlands of the USSR.) Trudy Vsesojuz. Naučno-Issled. Inst. Lesnogo Hozjajstva (Trans. All-Union For. Res. Inst.), Pp. 52 (1941). [Russisch.]

Jackson, A. B.: The identification of conifers. London, Pp. VII + 152, 48 figs. (1946).

Jensen, H.: Flaskympningsmetoden och dess användbarhet inom skogsträdsförändlingen. (Bottle grafting technique and its application in the improvement of forest trees.) Svensk PappTidn. 45, 33—36 (1942).
— Nya vägar för produktion av högklassigt skogsträdsfrö. (New ways for producing high grade seed of forest trees.) Skogen 29, 73—77 (1942).
— Plantagemäßig produktion av högvärdig skogsfrö. (Plantation production of high grade forest seed.) Skogen 30, 53—56 (1943).
— Modern teknik vid växtförädling med träd och buskar. (Modern technique in plant breeding applied to trees and shrubs.) Kungl. Lantbruks. Akad. Tidskr. 82, 330—340 (1943).
— Om elitfröplantager. (On élite seed plantations.) Skogen 32, 74—77 (1945).
— Inavelsrisken vid skogsfröplantager. (The risk of inbreeding in forest tree seed plantations.) Svenska SkogsvFören. Tidskr. 43, 178—180 (1945).
— Um estudo citológico comparativo de Thalictrum e Ilex. (A comparative cytological study of Thalictrum and Ilex.) Bragantia, São Paulo 3, 199—222 (1943).
— Heterochromosome formation in the genus Ilex. Amer. Nat. 78, 375—379 (1944).
— u. A. Levan: Colchicine-induced tetraploidy in Sequoia gigantea. Hereditas, Lund, Pp. 222—224 (1941).

Johanson, Å.: Eken. Trollväsens och Vaskungs skötebarn. The oak. The trolls' and Vasaking's favourite.) Skogsägaren 20, 169—172 (1944).

Johnsson, H.: Cytological studies of diploid and triploid Populus tremula and of crosses between them. Hereditas, Lund 26, 321—352 (1940).
— Växtförädling av björk-mål och medel. (Improvement of birch-objects and means.) Svensk PappTidn. 43, 450—456, 44, 4—7, 20—22 (1940).
— Generativ och vegetativ förökning av Populus tremula. (Generative and vegetative multiplication of P. tremula.) Svensk Bot. Tidskr. 36, 177—199 (1942).
— Die Chromosomenzahl von Carpinus betulus L. Hereditas, Lund 28, 228—230 (1942). (Abstr.)
— Cytological studies of triploid progenies of Populus tremula. Hereditas, Lund 28, 306—312 (1942).
— Poppelkulturer-Intryck från en studieresa till Belgien och synpunkter på poppel i svenskt skogsbruk. (Poplar plantations: impressions from a study tour to Belgium and views on the poplar in Swedish forestry.) Svensk. PappTidn., Nr. 9, 208—214 (1943).
— Triploidy in Betula alba L. Bot. Notiser, Pp. 85 bis 96 (1944).
— Interspecific hybridization within the genus Betula. Hereditas, Lund 31, 163—176 (1945).
— The triploid progeny of the cross diploid × tetraploid Populus tremula. Hereditas, Lund 31, 411—440 (1945).
— Chromosome numbers of the progeny from the cross triploid × tetraploid Populus tremula. Hereditas, Lund 31, 500—501 (1945). (Abstr.)
— Chromosome numbers of twin plants of Quercus robur and Fagus silvatica. Hereditas, Lund 32, 469—472 (1946).
— Progeny of triploid Betula verrucosa Erh. Bot. Notiser, Nr. 2, 285—290 (1946).
— u. C. Eklundh: Colchicinbehandling som metod vid växtförädling av lövträd. (Colchicine treatment as a technique in the improvement of broad leaved trees.) Svensk PappTidn. 43, 373—377 (1940).

Johnson, L. P. V.: The breeding of forest trees. For. Chron. 15, 139—151 (1939).
— A descriptive list of natural and artificial interspecific hybrids in North American forest-tree genera. Canad. J. Res. 17, Sect. C, 411—444 (1939).
— The breeding of forest trees. For. Chron. 15, 139 bis 151 (1939).
— Studies on the relation of growth rate to wood quality in Populus hybrids. Canad. J. Res. 20, Sect. C, 28 bis 40 (1942).
— The storage and artificial germination of forest tree pollens. Canad. J. Res. 21, Sect. C, 332—342 (1943).
— Reduced vigour, chlorophyll deficiency, and other effects of selffertilization in Pinus. Canad. J. Res. 23, Sect. C, 145—149 (1945).
— Forest tree breeding. New Trail 3, 143—150 (1945).
— Development of sexual and vegetative organs on detached forest tree branches cultured in the greenhouse. For. Chron. 21, 130—136 (N. R. C., Nr. 1281) (1945).
— Fertilization in Ulmus wiht special reference to hybridization procedure. Canad. J. Res. 24, Sect. C, 1—3 (1946).
— A note on inheritance in F_1 and F_2 hybrids of Populus alba L. × P. grandidentata Michx. Canad. J. Res. 24, Sect. C, 313—317 (1946).
— u. E. C. Bradley: Hybridization technique for forest trees. Canad. J. Res. 24, Sect. C, 305—307 (1946).
— u. C. Heimburger: Preliminary report on interspecific hybridization in forest trees. Canad. J. Res. 24, Sect. C, 308—312 (1946).
— u. G. A. Young: Genetics and cytology. Rev. Activ. Nat. Res. Coun. Canad., Nr. 1021, 22—23 (1941).

Koehler, A.: Heredity versus environment in improving wood in forest trees. J. For. 37, 683—687 (1939).

Komisarov, D. A.: (Die Besonderheiten einer, mit Colchicin polyploidisierten, Kiefer (Pinus silvestris L.). C. R. (Doklady) Acad. Sci. URSS. 58, 2077—2080 (1947). [Russisch.]

Komisarova, M. V.: (Cytologische Untersuchungen an polyploiden Kiefern [Pinus silvestris L.].) C. R. (Doklady) Acad. Sci. URSS. 58, 887 (1947). [Russisch.]

Koning, H. C.: Verslag over het onderzoek naar den populierenkanker over 1939. (Report on the investigation on canker of poplar during 1939.) Tijdschr. ned. Heidemaatsch. 52, 326—333 (1940).
— u. A. J. Ter Pelkwijk: Verslag van het onderzoek naar den populierenkanker in 1942—1943. VI. (Report on the investigation on canker of poplar in 1942 bis 1943. VI.) Tijdschr. ned. Heidemaatsch. 56, 50—54 (1944).

Kramer, P. J.: Amount and duration of growth of various species of tree seedlings. Plant Physiol. 18, 239—251 (1943).

Krijthe, N. u. J. C. Went: Inoculaties van iepensbastaarden verricht in 1938. (Inoculations of elm hybrids, carried out in 1938.) Tijdschr. PlZiekt. 45, 71—74 (1939).

Krupenikov, I. A.: Growth of Pinus silvestris L. in solonchak soils. C. R. (Doklady) Acad. Sci. URSS. 41, 255—258 (1943).

Laing, E. V.: Studies on the genus Larix with particular reference to the hybrid larch. (Larix eurolepis A. Henry). Scot For. J. 58, 6—32 (1944).

Lammers, R. P.: Een en ander over nieuwe lamtorovormen. (A thing or two about new lamtoro forms.) Bergcultures 14, 1168—1171 (1940).

Langlet, O.: Om utvecklingen av granar ur frö efter självbefruktning och efter fri vindpollinering. Hittills framkomna resultat av ett försök, anlagt år 1909 av fil. Dr. Nils Sylvén. (On the development of spruce from seed after self-pollination and after wind pollination-Results to date of an experiment, instituted in 1909 by Dr. N. Sylvén.) Medd. Skogsförsöksanst., Stockh. **32**, 1—22 (1940).

Langlet, O.: Om utvecklingen av granar ur frö efter självbefruktning och efter fri vindpollinering. (On the development of spruce raised from seed after self-pollination and after open-pollination by the wind.) Medd. Skogsförsöksanst., Stockh., H. 32, Nr. 1, 22 (1940).

— Kulturförsök med tysk gran av första och andra generationen. (Cultural experiments with German spruce of the first and second generation.) Medd. Skogsförsöksanst., Stockh. 1940—1941, Nr. 32, 361 bis 380 (1941).

— Kvalitetsbeteckning å skogsfrö. (Marking the quality of forest tree seed.) Skogen **29**, 315—318 (1942).

— Photoperiodismus und Provenienz bei der gemeinen Kiefer (Pinus silvestris L.) Medd. Skogsförsöksanst., Stockh. 1942—1943, **33**, 295—327 (1944).

— Om möjligheterna att skogsodla med gran-och tallfrö av ortsfrämmande proveniens. (On the possibilities of forest cultivation with spruce and pine seed of distant provenance.) Svenska SkogsvFören. Tidskr. **43**, 68—78 (1945).

Langner, W.: Züchtung auf Wüchsigkeit. (Theorie einer Auslese als Beitrag zur Forstpflanzenzüchtung.) Forstwiss. Zbl. **61**, 313—318 (1939).

— Le problème des races forestières dans la forêt Européenne. Intersylva, Munich **2**, 479—488 (1942).

Lantelmé, W.: Phototropismus und Provenance. Ein Nachwort und eine Mahnung zur „phototropistischen Methode" des Herrn Professor Schmidt. Allg. Forst- u. Jagdztg. **115**, 69—85 (1939).

Larsen, C. S.: De enkelte Arters Anvandelse, Proveniens og Foraedling. (The use, provenance and breeding of the various tree species.) Svenska Skogs-Fören. Tidskr. **41**, 182—199 (1943).

Laurie, M. V. u. A. L. Griffith: The problem of the pure teak plantation. Indian For. Rec. (N. S.) Silviculture **5**, 13—121 (1941).

Lindquist, B.: Tallens roll i svensk skogsträdsförädling. (The role of the pine in Swedish forest tree-breeding.) Tidsskr. Skogbr. **48**, 10—17, 40—46 (1940).

— Die Rolle der Kiefer in der schwedischen Waldbaumveredelung. Z. wiss. tech. Fortschr. Forstw. **18**, 293—302 (1942).

— Skoglig rasförädling-några arbetsuppgifter. (Race improvement of forest trees-some tasks.) Svensk Snickeritidskr., Nr. 1, 4 (1945).

— Betula callosa Notö, a neglected species in the Scandinavian subalpine forests. Svensk Bot. Tidskr. **39**, 161—186 (1945).

— On the variation in Scandinavian Betula verrucosa Ehrh. with some notes on the Betula series Verrucosae Sukacz. Svenska Bot. Tidskr. **41**, 45—80 (1947).

— u. E. Malmberg: Bättre frö för framtidens tallskogar. (Better seed for future pine woods.) Stockholm, Pp. 11, 9 figs. (1943).

— u. E. Runquist: Krontypsvariationen hos den smalkroniga tallen i Norrland. (The variation in crown type in the narrow crowned pine in Norrland.) Skogen **30**, 69—72 (1943).

Little, E. L. jr.: New names in Quercus and Osmanthus. J. Wash. Acad. Sci. **33**, 8—11 (1943).

— Ochroma lagopus Swartz, the name of the balsa of Ecuador. Caribbean Forester **5**, 108—114 (1944). [Mimeographed.]

Lobanov, N. V.: (Die Methodik der Forschung über das Wachstum der Holzgewächse bei verschiedenen Feuchtigkeitsgraden des Bodens.) C. R. (Doklady) Acad. Sci. URSS. **55**, 549 (1947). [Russisch.]

Lofting, E. C. L.: VIII. Laerkearternes Udvikling i Hedeplantagerne og Japansk Laerks Anvendelighed som Hjaelpettrae ved Opbygning af Hedeskov. (VIII. The development of the larch species in heath plantation and the use of the Japanese larch as an adjunct in forming a heathland wood.) Det forstlige Forsogsvaesen i Danmark **16**, 323—364. (Beretning Nr. 147.) (1945).

Looby, W. J. u. J. Doyle: Formation of gynospore, female gametophyte, and archegonia in Sequoia. Sci. Proc. R. Dublin Soc. (N. S.) **23**, 35—54 (1942).

Lubjako, M. N.: (Selection of quick-growing forms among local forest species.) Lesnoe Hozjajstvo (Forestry), Nr. 5, 28—32 (1941). [Russisch.]

Lybye, S.: Skovtraeforaedling. (Forest tree breeding.) Hedeselsk. Tidsskr. Aarhus **68**, 49—52 (1947).

McNair, J. B.: Some chemical properties of Eucalyptus in relation to their evolutionary status. Madroño **6**, 181—190 (1942).

Manciot, A.: Les arbres de nos forêts. Paris, Pp. 159, illus (ohne Jahr).

Martinez, M.: Una nueva pinácea mexicana: Picea chihuahuana sp. nov. (A new Mexican member of the Pinaceae: P. chihuahuana sp. nov.). An. Inst. Biol. Méx. **13**, 31—34 (1942).

— Tres espécies nuevas mexicanas del género Abies. (Three new Mexican species of the genus Abies.) An. Inst. Biol. Univ. Méx. **13**, 621—634 (1942).

— Una nueva espécie del género Pinus: Pinus michoacana. (A new species of the genus Pinus: Pinus michoacana.) An. Inst. Biol. Univ. Méx. **15**, 1—6 (1944).

— El Pinus macrophylla Engelm. y su variedad Blancoi. (P. macrophylla Engelm. and its variety Blancoi.) An. Inst. Biol. Univ. Méx. **15**, 341—348 (1944).

Mason, H. L. u. W. P. Stockwell: A new pine from Mount Rose, Nevada. Madroño **8**, 61—63 (1945).

Melville, R.: The British elm flora. Nature, Lond. **153**, 198—199 (1944).

— Typification and variation in the smoothleaved elm, Ulmus carpinifolia Gleditsch. J. Linn. Soc. (Bot.) **53**, 83—90 (1946).

Minckler, L. S.: Genetics in forestry. J. For. **37**, 559 bis 564 (1939).

— One-parent heredity tests with loblolly pine. J. For. **40**, 505—506 (1942).

Mirov, N. T.: Forest genetics: present status and outlook for the future. Proc. 6th Pacific Sci. Congr. **4**, 727—730 (1940).

— Possibility of simple biochemical tests for differentiation between species of genus Pinus. J. For. **40**, 953—954 (1942).

— Effect of the crown on the composition of oleoresin in pines. J. For. **43**, 345—348 (1945).

— u. P. Stockwell: (Colchicinbehandlung von Kiefersamen.) J. Hered. **30**, 389 (1939).

Mitchell, H. L.: The development of a high yielding strain of naval stores pine-to increase the output of oleoresin per tree. Naval Stores Rev. **52** (7), 10, 12 (1943).

—, C. S. Schopmeyer u. K. W. Dormann: Pedigreed pine for naval stores production. Science **96**, 559—560 (1942).

Muller, C. H.: Hybridism, ecotypes, and perpipheral race variants in Quercus. Amer. J. Bot. Suppl. **28**, 17s (1941). (Abstr.)

Navarro de Andrade, E.: The eucalyptus in Brazil. J. Hered. **32**, 215—220, 240 (1941).

Newcomer, E. H.: Induced parthenocarpy in Ginkgo. Amer. Nat. **79**, 186—187 (1945).

Nilsson, T.: Skogsträdsförädling enligt kemisktkvalitativa riktlinjer. (Forest tree breeding on the basis of chemical estimations of quality.) Svensk PappTidn. **46**, 507—513 (1943).

Ording, A.: Om vekstforedling av skogstraer i Danmark. Dr. Syrach Larsens arbeid i Hørsholm. (Forest tree breeding in Denmark: Dr. S. Larsen's work at Hørsholm.) Tidsskr. Skogbr. **47**, 209—221 (1939).
— Vektsforedling av skogstraer. (Tree bredding.) Tidsskr. Skogbr. **51** (1943).
Osborn, A.: An interesting hybrid conifer: Cupressocyparis Leylandii. J. R. Hort. Soc. **66**, 54—55 (1941).

Palmer, E. J.: The red oak complex in the United States. Amer. Midl. Nat. **27**, 732—740 (1942).
— Quercus Prinus Linnaeus. Amer. Midl. Nat. **29**, 783 bis 784 (1943).
Parker, R. N.: Two natural hybrids. Indian Forester **65**, 585—586 (1939).
Parsons, T. H.: Balsa as a commercial crop. Trop. Agriculturist **101**, 120—126 (1945).
Pauley, S. S.: Early selection for heterosis in poplar hybrids. Genetics **32**, 100 (1947). (Abstr.)
Pearson, G. A.: Applied genetics in forestry. Sci. Mon. N. Y. **58**, 444—453 (1944).
Penfold, A. R. u. F. R. Morrison: Commercial Eucalyptus oils. Bull. Tech. Educ. Branch, Technol. Mus. Sydney, Nr. 2, 36 (1944).
—, — Eucalyptus. The essence of Australia. Sidney Technological Museum, 30th April 1945, Pp. 8 (1945).
Pescott, E. E.: Germination of the seed of Mallee Eucalypts. Vict. Nat. **58**, 8 (1941).
Peter-Contesse, J.: Hérédité et sélection. J. For. Suisse **92**, 121—126 (1941).
Petre, A.: Björken. (The birch.) Stockholm, Pp. 56. photos. (1942).
Petrini, S.: De internationella tallproviensförsöken av år 1907. (The international provenance experiments with pine of the year 1907.) Medd. Skogsförsöksanst. Stockh. **33**, 247—266 (1942).
Petterson, H.: Växtförädlingen och skogsbruket. (Plant breeding and forestry.) Skogen **30**, 31—37 (1943).
— Skogsskötsel och växtförädling. (Forest management and plant breeding. Skogen **30**, 195—196 (1943).
—, C. Malmström u. I. Trägårdh: Redogörelse för verksamheten vid Statens Skogsförsöksanstalt under år 1939. (Report on the work of the State Institute of Experimental Forestry during 1939.) Medd. Skogsförsöksanst. Stockh., Nr. 32, 381—389 (1940—1941).
—, —, — Redogörelse för verksamheten vid Statens Skogsförsöksanstalt under år 1940. (Report on the work of the state Institute of Experimental Forestry during 1940.) Medd. Skogsförsöksanst. Stockh., Nr. 390 bis 395 (1940—1941).
Pjatnickij, S. S.: (Hybridization in oaks.) Lesnoe Hozjajstvo (Forestry), Nr. 7, 38—43 (1939). [Russisch.]
— Neue experimentell erhaltene Hybridformen der Eichen × Quercus Miczurinii und × Quercus Wyssotzkyi. C. R. (Doklady) Acad. Sci. URSS. **30**, 851 bis 853 (1941).
Pommerleau, R.: Les maladies de l'érable à sucre et leur prévention. Forêt Québécoise **9**, 311—348 (1945).

Redman, K.: Nomenclature confusion of Populus candicans Aiton. J. Amer. Pharm. Ass. **31**, 140—141 (1942).
Régnier, R.: Le chancre suinant et les différents types de peupliers. C. R. Acad. Agric. Fr. **29**, 335—340 (1943).
Reinecke, O. S. H. u. C. L. Wicht: A semi-evergreen form of Lombardy poplar. J. S. Afr. For. Ass., Nr. 9, 19—22 (1942).
Richens, R. H.: Forest tree breeding and genetics. Imperial Bureau of Plant Breeding and Genetics, Cambridge; Imperial Forestry Bureau, Oxford, 5s. Pp. 79 (1945).
Richter, F. I.: Pinus: the relationship of seed size and seedling size to inherent vigor. J. For. **43**, 131—137 (1945).

Richter, F. I.: New perspectives in forest tree breeding. Science **104**, 1—3 (1946).
Riker, A. J. u. T. F. Kouba: White pine selected in blister-rust areas. Phytopathology **30**, 20 (1940). (Abstr.)
—, —, W. H. Brener u. L. Byam: White pine selections tested for resistance to blister rust. Phytopathology **33**, 11 (1943). (Abstr.)
—, —, —, — White pine selections tested for resistance to blister rust. J. For. **41**, 753—760 (1943).
Robertson, W. M.: Canada: silvicultural research. News Bull. Emp. For. Departments, Pp. 45—46 (1944).
Rockwell, F. I.: What kind of trees shall I plant? N. S. Dak. Hort. **18**, 40 (1945).
Rohmeder, E.: Ergebnisse der forstlichen Saatgutforschung als Mittel zur Ertragssteigerung im Walde. Z. wiss. techn. Fortschr. Forstw. **18**, 165—176 (1942).
Romanova, E. A.: (Einige Angaben über die experimentelle Forschung des intraspecifischen Kampfes ums Dasein bei Holzgewächsen.) C. R. (Doklady) Acad. Sci. URSS. **78**, 1505 (1947). [Russisch.]
Roper, G. D.: Seed selection. Quart. J. For. **38**, 114 bis 115 (1944).
Rouleau, E.: Populus balsamifera of Linnaeus not a nomen ambiguum. Rhodora **48**, 103—110 (1946).
Rubner, K.: Vorläufige Mitteilung über einen neuen Fichten-Provenienzversuch. Z. ges. Forstwesen **76**, 28—32 (1944).

S...., J. R.: Forest genetics research at the Arboretum. Morris Arbor. Bull. **4**, 49—50 (1946).
Sampath, S.: Chromosome numbers in Sesbania spp. Curr. Sci. **16**, 30—31 (1947).
Schädelin, W.: Ergebnisse der Lärchenforschungen von Prof. Dr. Ernst Münch und waldbauliche Folgerungen. Schweiz. Z. Forstw. **92**, 41—49, 67—79 (1941).
Scheffer, T. C., H. G. Lachmund u. H. Hopp: Relation between hot-water extractives and decay resistance of black locust wood. J. Agric. Res. **68**, 415 bis 426 (1944).
Schenck, C. A.: Forstliche Genetik in den Vereinigten Staaten. Allg. Forst- u. Jagdztg. **118**, 126—128 (1942).
Schmidt, W.: Erbforschungsfragen. Allg. Forst- u. Jagdztg. **117**, Nr. 1, 1—15 (1941).
— Das Ostwestgefälle der Kiefernrassen, neue Einblicke und Methodenvorschläge für internationale Versuche. Intersylva **3**, 473—494 (1943).
Schramm, J. R.: An interesting variant of white spruce. Morris Arbor. Bull. Univ. Pa. **3**, 67—69 (1941).
— An experiment in greenhouse benches. Morris Arbor. Bull. **4**, 25—26 (1942).
Schreiner, E. J.: Some ecological aspects of forest genetics. J. For. **37**, 462—465 (1939).
— u. J. W. Duffield: Metaxenia in an oak species cross. J. Hered. **33**, 97—98 (1942).
— u. M. A. Huberman: Induced flowering- a tool for mass selection, progeny tests and forest management. J. For. **38**, 491—492 (1940).
Schumacher, F. X. u. R. A. Chapman: Sampling methods in forestry and range management. Bull. Duke Univ. Sch. For. Durham, N. C., Nr. 7 (1942).
Schweizer, J.: Over lamtoro-soorten als houtleveranciers in een koffieaanplant. (On lamtoro varieties as sources of wood in a coffee plantation.) Bergcultures **14**, 1069—1077 (1940).
— u. C. Coolhaas: Levering van oculatiehout van de nieuwe lamtoro-soorten. (Supply of bud wood of the new lamtoro species.) Bergcultures **14**, 1557 (1940).
Sherry, S. P.: A note on the vegetative propagation of Pinus insignis. J. S. Afr. For. Ass., Nr. 9, 23—25 (1942).
Sibilia, C.: L'Ulmus pumila e la grafiosi. (U. pumila and the Dutch elm disease.) Bol. Staz. Pat. Veg. Roma **20**, 147—149 (1940).

Silva, A. R. P. da: Uma forma Dolichocarpa d azinheira. (A dolichocarpa form in the holm oak.) Rev. Agron. Lisboa **30**, 462 (1942).

Skinner, F. L.: Cedar variation. N. S. Dak. Hort. **17**, 112 (1944).

Skinner, H. T.: The origin of horticultural forms in cultivated conifers. Exhibit of the Morris Arboretum at the Philadelphia Spring Flower Show. Arboretum Bull. Ass. Morris Arboretum Univ. **3**, 51—52 (1941).

Smith, E. C.: A study of cytology and speciation in the genus Populus L. J. Arnold Arbor. **24**, 275—305 (1943).

— u. C. Nichols jr.: Species hybrids in forest trees. J. Arnold Arbor. **22**, 443—454 (1941).

Smith-White, S.: Cytological studies in the Myrtaceae. I. Microsporogenesis in several genera of the trible Leptospermoideae. Proc. Linn. Soc. N. S. W. **67**, 335 bis 342 (1942).

Snow, A. G. jr.: Sex and vegetative propagation. J. For. **40**, 807—808 (1942).

— u. J. W. Duffield: Genetics in forestry. J. For. **38**, 404—408 (1940).

Sprague, T. A.: A triploid aspen-Populus tremula-from Sweden. Proc. Linn. Soc. London **152**, 111—113 (1940). (Abstr.)

Spurr, S. H.: Effect of seed weight and seed origin on the early development of eastern white pine. J. Arnold Arbor. **25**, 467—480 (1944).

Steinbauer, G. P.: Frequency of polyembryony in Fraxinus seeds. Bot. Gaz. **105**, 285 (1943).

Stockwell, P. u. F. I. Richter: Pinus: the fertile species hybrid between knobcone and Montery pines. Madroño **8**, 157—160 (1946).

Stockwell, W. P.: Pre-embryonic selection in the pines. J. For. **37**, 541—543 (1939).

Stout, A. B.: The Redding hemlock tree. J. N. Y. Bot. Gdn **40**, 233—235 (1939).

Suessenguth, K.: Einige neue Gattungen und Arten der Cyperaceae aus Südamerika. Bot. Jb. Systematik usw. **73**, 113—125 (1943).

Sundar, Rao, Y.: Chromosome numbers in Sesbania. Curr. Sci. **15**, 78 (1946).

Sutô, T.: Meiotic Chromosomes in Populus nigra and Toisusu (Salix) cardiophylla. Japan, Journ. Genet. **16**, 304—306 (1940).

Svoboda, P.: Křiženci lesních dřevin a cesty k jejich využití. (Hybrids of forest trees and ways of utilizing them.) Lesnicka Práce (Forestry Practice) **19**, 373—408 (1940).

Swingle, R. U.: Phloem necrosis-a virus disease of the American elm. Circ. U. S. Dep. Agric., Nr. 640, 8 (1942).

—, B. S. Meyer u. C. May: Phloem necrosis research during 1944. Phytopathology **35**, 489 (1945). (Abstr.)

Sylvén, N.: Den svenska „Föreningen för växtförädling av skogsträd". (The Swedish Forest Tree-Breeding Association.) Tidsskr. Skogbr. **47**, 149—160 (1939).

— Föreningen för växtförädling av skogsträd. Styrelseberättelse för år 1939. (Association for Forest Tree Breeding-Board's report for 1939.) Svensk Papp Tidn. **43**, 106—107 (1940).

— Insänd uppgifter om det senaste vinterhalvårets frostskador å avenskodlade träd och buskar! (Information received about frost injuries to trees and bushes grown in Sweden during the winter-half of last year.) Svensk Papp Tidn. **43**, 281—285 (1940).

— Lång- och kortdagstyper av de svenska skogsträden. (Long- and shortday types of Swedish forest trees.) Svensk Papp Tidn. **43**, 317—324, 332—342 (1940).

— Föreningen för växtförädling av skogsträd. Styrelseberättelse för år 1941. (Association for Breeding of Forest Trees-Board's report for 1941.) Svensk Papp Tidn. **45**, 136—138 (1942).

— Årsberättelse över föreningens för växtförädling av skogsträd verksamhet under år 1941. (Annual report on the work of the Association for Forest Tree Improvement during the year 1941.) Svensk Papp Tidn. **45**, 404—409, 519—522, 542—543 (1942).

Sylvén, N.: La sélection des arbres forestiers en Suède. Intersylva, Munich **2**, 471—478 (1942).

— Skogsträdsförädlingen under 1941. Från Föreningens för växtförädling av skogsträd verksamhet. (Forest tree breeding during 1941 — From the work of the Association for Forest Tree Breeding.) Skogen **29**, 173 bis 175 (1942).

— Svensk Skogsträdsförädling. Riktlinjer och önskemål. (Swedish forest tree breeding. Lines of work and objectives.) Stockholm **53**, 43 figs. (1943).

— Riktlinjer och önskemål vid Svensk skogsträdsförädling. (Lines of work and objektives in Swedish forest tree breeding.) Svenska SkogsvFören. Tidskr. **41**, 40 bis 78 (1943).

— Föreningen för växtvörädling av skogsträd. Styrelseberättelse för år 1942. (The Association for Forest Tree Breeding: Report of the Board of Management for the year 1942.) Svensk Papp Tidn. **46**, 187—189 (1943).

— Skogsskötsel och växtförädling. (Forest management and plant breeding.) Skogen **30**, 148—149 (1943).

— Skogsskötsel och växtförädling. (Forest Danagement and plant breeding.) Skogen **30**, 148—149 (1943).

— Årsberättelse över Föreningens för växtförädling av skogsträd verksamhet under år 1943. (Annual report on the work of the Association for Forest Tree Breeding during the year 1943.) Svensk Papp Tidn. **47**, 38 (1944).

— Föreningen för växtförädling av skogsträd. Styrelseberättelse för år 1943. (The Association for Forest Tree Breeding: Report of the Board for 1943.) Svensk Papp Tidn. **47**, 104—106 (1944).

— Om ekens lövspricknings-och lövfällningsdata. Ett bidrag till kännedomen om ekens mångformighet. (On the dates of leaf emergence and leaf fall in the oak. A contribution to the knowledge of the diversity of the oak.) Svensk Papp Tidn. **47**, 167—174 (1944).

— Eken i det svenska skogsbruket. (The oak in Swedish forestry.) Svensk Papp Tidn. **47**, 522 (1944).

— Årsberättelse över Föreningens för växtförädling av skogsträd verksamhet under år 1944. (Annual report on the work of the Association for Forest Tree Breeding during the year 1944.) Medd. Fören. Växtföräd. Skogstr., Nr. 39, 50 (1945).

— Föreningen för växtförädling av skogsträd. Styrelseberättelse för år 1944. (The Association for Forest Tree Breeding. Report of the Board, 1944.) Svensk Papp Tidn. **48**, 113—116 (1945).

Syrach Larsen, C.: Skovtrae foraedling i Gävleborg Län. (Forest tree breeding in Gävleborg district.) Dansk Skovforen. Tidsskr. **9**, 394—397 (1945).

— Estimation of the genotype in forest trees. K. Vethøjsk. Aarsskr., Pp. 87—128 (1947).

Tartakowsky, H. S. G. u. S. S. T. Arentsen: La roya del álamo en Chile. (Poplar rust in Chile.) Bol. Sanid. Veg., Chile **1**, 21—32 (1941).

Thaarup, P.: Bastarden Sitkagran × Hvidgran. (The hybrid Sitka spruce × white spruce.) Dansk Skovforen Tidsskr., Nr. 9, 381—384 (1945).

Thimann, K. V. u. A. L. Delisle: Notes on the rooting of some conifers from cuttings. J. Arnold Arbor. **23**, 103—109 (1942).

Tirén, L.: Skogsträdens Fruktsättning år 1943. (Set of fruit of forest trees in the year 1943.) Flygblad, Nr. 57, 12 (1943).

— Tallfröets grobarhet i Norrland. Preliminärt meddelande. (Germination capacity of pine seed in Norrland-Preliminary communication.) Skogen **31**, 16 (1944).

— Tallfröets grobarhet in Norrland. (Germination capacity of pine seed in Norrland.) Skogen **31**, 365 (1944).

Tucker, C. M.: Phloem necrosis, a destructive disease of the American elm. Circ. Mo. Agric. Exp. Sta., Nr. 305, 15 (1945).
Tucker, J. M. u. C. H. Muller: Additions to the oak flora of El Salvador. Madroño 8, 111—117 (1945).

Ullén, G.: Skogsvårdsföreningens exkursion till Skåne. (Excursion of the Sylvicultural Society to Scania.) Skogen 30, 255—257 (1943).

Vasiljev, V. N.: (Systematics and geography of the far eastern birches.) J. Bot. URSS. 27, 3—19 (1942). [Russisch.]
Vogelsang, P.: Hybrid poplars. Paper Tr. J. 113, Nr. 20, 38—40 (1941).

W—N, J. M.: Skogsdagar i Värmland. (Forestry meetings in Värmland.) Skogen 31, 279—281 (1944).
Wagener, W. W.: The canker of Cupressus indiced by Coryneum cardinale N. Sp. J. Agric. Res. 58, 1—46 (1939).
Wakeley, P. C.: Geographic cource of loblolly pine seed. J. For. 42, 23—32 (1944).
Walter, J. M., C. May u. C. W. Collins: Dutch elm disease and its control. Circ. U. S. Dep. Agric., Nr. 677, 12 (1943).
Waterman, A. M.: Canker of hybrid poplar clones in the United States, caused by Septoria misiva. Phytopathology 36, 148—156 (1946).
Watrous, R. C. u. H. V. Barnes: Bibliography on cork oak. Bibliogr. Bull. U. S. Dep. Agric., Nr. 7, 66 (1946).
Weidman, R. H.: Evidences of racial influence in a 25-yean test of ponderosa pine. J. Agric. Res. 59, 855—887 (1939).
Welch, D. S. u. D. L. Collins: Dutch elm disease and its control. Cornell Ext. Bull., Nr. 437, 19 (1940).
Went, J. C.: Verslag van de onderzoekingen over de iepenziekte, verricht op het Phytopathologisch Laboratorium „Willie Commelin Scholten" te Baarn, gedurende 1938. (Report on the investigations on elm disease, carried out at the Willie Commelin Scholten Phytopathological Laboratory at Baarn, during 1938.) Tijdschr. PlZiekt. 45, 52—62 (1939).
Wettstein, W. von: Die Vermehrung und Kultur der Pappeln. Forstarchiv 15, 164—168 (1939).
— Züchtung von Birke und Pappel. Umschau 1940, Pp. 741—745 (1940).
— Wuchssteigerung durch Kombinationszüchtung und Chromosomenvermehrung. Forstarchiv 17, 80—83 (1941).
— Züchtungsmöglichkeiten auf höheren Zellulosegehalt. Holz als Roh- u. Werkstoff 5, 373—375 (1942).
— Über den gegenwärtigen Stand der forstlichen Pflanzenzüchtung. Ein Rückblick und Ausblick. Allg. Forst- u. Jagdztg. 118, 128—131 (1942).
— Unterschiede bei Nachkommen von Alpenlärchen. Allg. Forst- u. Jagdztg. 118, 157—161 (1942).
— Möglichkeiten der Züchtung neuer Ökotypen nach Kreuzung. Züchter 14, 282—285 (1942).
— Saatgut und Züchtung. Forstwiss. Zbl. 64, 135—143 (1942).
— Züchtung von Cellulosepflanzen. Der Papier-Fabrikant Nr. 2, 54—57 (1943).
— u. Ch. Daubinet: Luxurierende Kreuzungen bei Pinus silvestris und die Grundlagen für ihre Durchführung. Züchter 13, 207—208 (1941).
— u. H. Propach: Sichtungsarbeit zur Birkenzüchtung. Züchter 11, 279—280 (1939).
Wilkinson, J.: The cytology of the cricket bat willow (Salix alba var. caerulea). Ann. of. Bot. N. s. 5, Nr. 17, 149—165 (1941).
Wolf, C. B.: The Gander oak, a new hybrid oak from San Diego County, California. Proc. Calif. Acad. Sci. 25 177—188 (1944).

Wright, J. W.: Genotypic variation in white ash. J. For. 42, 489—495 (1944).
— Ecotypic differentiation in red ash. J. For. 42, 591 bis 597 (1944).

Young, H. E.: Fused needle disease and its relation to the nutrition of Pinus. Qd. Agric. J. 53, 45—54, 156 bis 177, 278—315 (1940).

u) Berichte der Versuchsstationen, Kongresse, Konferenzen usw.
Genetik.

Russian contributions to the 1939 Genetics Congress. A. Plant Genetics. American Documentation Institute, Document 1563.
The Seventh International Congress of Genetics. Nature, Lond. 144, 813 (1939).
Report of the President of the Carnegie Institution of Washington for the year ending October 31, 1940. Yearb. Carneg. Instn., Nr. 39, 326, 1939—1940 (1940).
Genetics in the Soviet Union. Three speeches from the 1939 Conference on Genetics and Selection. Sci. and Soc. 4, 183—233 (1940).
Report of the President of the Carnegie Institution of Washington for the year ending October 31, 1941. Yarb. Carneg. Instn., Nr. 40, 346, 1940—1941 (1941).
Lunds Universitets Årsberättelse 1940—1941. (Lund University Annual Report 1940—1941.) Lund, Pp. 79 (1941).
Lunds Universitets Årsberättelse 1941—1942. (Lund University Annual Report 1941—1942.) Lund, Pp. 80 (1942).
Lunds Universitets Årsberättelse 1942—1943. (Annual Report of Lunds University 1942—1943.) Håkan Ohlssons Boktryckeri, Lund, Pp. 79 (1943).
Lunds Universitets Årsberättelse 1943—1944. (Annual Report of Lunds University 1943—1944.) Håkan Ohlsson Boktryckeri, Lund, Pp. 82 (1944).
Lunds Universitets Årsberättelse 1944—1945. (Annual Report of Lunds University 1944—1945.) Håkan Ohlssons Boktryckeri, Lund, Pp. 84 (1945).
Inauguración del Laboratorio de Citogenética del Departamento de Genética y Fitotecnia del Ministerio de Agricultura. (Inauguration of the Laboratory of Cytogenetics of the Department of Genetics and Plant Breeding of the Ministry of Agriculture.) Agric. Tec. Chile 6, 181 (1946).
International genetics conference. Nature, Lond. 157, 35—38 (1946).
Genetic research in Britain, 1939—1945, a summary by various workers. Heredity 1, 1—17 (1947).

Allgemeine Pflanzenzüchtung.

(Bulgarian Agricultural Association, Sofia-Report on work of the Association during 1938.) Zemledelie, Sofia 43, Nr. 4, 1—25 (1939). [Bulgarisch.]
Fiftieth annual report of the Agricultural Experiment Station of the Alabama Polytechnic Institute, Auburn, January 1 to December 31, 1939, Pp. 43 (1939).
Mededeelingen van het Algemeen Proefstation der A.V.R.O.S. Algemeene serie Nr. 59. Verslag van den Directeur van het Algemeen Proefstation der A.V.R.O.S. over het tijdvak 1 Januari 1937—31 December 1939. (Contributions from the General Experiment Station of A.V.R.O.S. General series Nr. 59, Report of the Director of the General Experiment Station of A.V.R.O.S. for the period 1 January 1937—31 December 1939.) Nr. 59, 76 (1939).
Memoria de la Estación Experimental Agrícola de la Sociedad Nacional Agraria Lima, Peru. (Report of the

Agricultural Experiment Station of the National Society of Agriculture, Lima, Peru.) Nr. 11a, 363 (1938) 1939.

Rapport annuel exercices 1936—1937 et 1937—1938. Bull. Serv. Nat. Prod. Agric. Enseign. Rur., Haiti, Nr. 15, 152 (1939).

(The All-Union Congress on Breeding and Seed Production.) Selekcija i Semenovodstvo, Nr. 4, 26—37 (1939). [Russisch.]

Verslag van de Commissie van Advies inzake de Bevordering van de Cultuur van Handelsgewassen over het jaar 1939. (Report of the Advisory Committee on the Promotion of the Cultivation of Industrial Plants for the year 1939.) Secretariaat, Bericht Nr. 83, Departement van Economische Zaken 1939, Pp. 41 (1939).

Verslag over het jaar 1939 van Het Algemeen Landbouw Syndicaat-Het Zuid- en West-Sumatra Syndicaat-De Centrale Vereeniging tot beheer van Proefstations voor de Overjarige Cultures in Nederlandsch-Indië en van de onder deze Organisaties ressorteerende Vereenigingen Instellingen. (Report for the year 1939 of the General-Agriculture Syndicate-The South and West Sumatra Syndicate-The Central Association for the Management of Experiment Stations for Perennial Crops in the Netherlands Indies and of the Association and Institutes under the jurisdiction of these Organizations.) Batavia, Pp. 249 (1939).

Verslag over het jaar 1938 van Het Algemeen Landbouw Syndicaat — Het Zuid-en West-Sumatra Syndicaat — De Centrale Vereeniging tot beheer van Proefstations voor de Overjarige Cultures in Nederlandsch-Indië en van de onder deze Organisaties ressorteerende Vereenigingen en Instellingen. (Report for the year 1938 of the General Agriculture Syndicate — The South and West Sumatra Syndicate — The Central Association for the Management of Experiment Stations for Perennial Crops in the Netherlands Indies and of the Associations and Institutes under the jurisdiction of these Organizations.) Batavia, Pp. 248 (1939).

Publications de l'Institut National pour l'Étude Agronomique du Congo Belge (I. N. E. A. C.). Rapport Annuel pour l'exercice 1938 (1re Partie). Congo Belge Hors Sér., p. 269 (1939).

Reports of subordinate officers of the Department of Agriculture, Madras, for 1939—1940 (1940), p. 167.

Report on the operations of the Department of Agriculture Madras Presidency for the year 1939—1940 (1941), p. 81.

Rapport Annuel. Bull. Dep. Agric. Haiti, Nr. 27, 185 (1939—1940).

Commissie van advies inzake de bevordering van de cultuur van handelsgewassen. (Advisory Committee on the promotion of the cultivation of industrial plants.) Bergcultures 14, 1427—1442 (1940).

Estract from an address delivered by H. R. Stewart, Esquire, C. I. E., I. A. S., Director of Agriculture, Punjab, at the Rotary Club, on ,,Thirty years of Punjab Agriculture". Seas. Notes Punjab Agric. Dep. 18, 1—9 (1940).

La Estación Experimental de Guatraché (La Pampa): breve reseña sobre su orígen y desenvolvimiento. (The Guatraché Experiment Station: brief outline of its origin and development.) ,,Granos" Semilla Selecta, B. Aires 4, Nr. 7/8, 14—15 (1940).

Research aids farm progress. 53rd Rep. Purdue Univ. Agric. Exp. Sta. Lavayette, Indiana, Pp. 112 (1940).

Science for the farmer. 53rd Rep. Pa. Agric. Exp. Sta. 1940, Bull., Nr. 399, 76 (1940).

Annual Report of the East African Agricultural Research Station, Amani. 1939, Pp. 26 (1940).

50th Annual Report of the Washington Agricultural Experiment Station. Bull. Wash. Agric. Exp. Sta., Nr. 394, 124 (1940).

Fifty-first annual report of the Agricultural Experiment Station of the Alabama Polytechnic Institute, Auburn, January 1 to December 31, 1940, Pp. 45 (1940).

Informe anual correspondiente al año agricola 1939—1940. (Annual Report for the agricultural year 1939—1940.) Bol. Chacra Exp. ,,La Previsión" 3, 3—33 (1940).

Informes del Director del Departamento de Agricultura, doctor ·Eduardo Mejia Vélez. (Report of Dr. E. M. Vélez, Director of the Department of Agriculture.) Mem. Minist. Econ. Nac. Bogotá. 2, 112 (1940).

Informe anual de la Estación Agricola Experimental de Palmira. Junio 10. de 1939, a Junio 10. de 1940. (Annual report of the Palmira Agricultural Experimental Station. 1st June, 1939 to 1st June 1940.) Mem. Minist. Econ. Nac. Bogota 3, 171 (1940).

Memoria anual de 1940 de la Estación Experimental. Agrícola de la Molina. (Annual report of the Estación Experimental Agricola de la Molina for 1940.) Lima, Peru, p. 30 (1940).

Memoria de la Estación Experimental Agrícola de La Molina, correspondiente al año 1939. (Report of the Agricultural Experiment Station of La Molina. Report for the year 1939.) Lima, Peru, Nr. 12a, 371 (1940).

Report of the Agricultural Department, St. Kitts-Nevis, for the year ended 31st December, 1939, Pp. 38 (1940).

Report of the Agricultural Department, St. Kitts-Nevis, for the year ended 31st December, 1940, Pp. 8 (1940).

To Colorado legislators. A report from the Colorado Experiment Station, Fort Collins. Pp. 26 (1940).

Annual report of the Agricultural Experiment Station, Puerto Rico. Pp. 70 (1940—1941).

Rapport Annuel 1940—1941, 1941—1942. Bull. Dep. Agric. Haiti, Nr. 31, 377 (1940—1942).

Annual Report of the Department of Agriculture, Uganda for the period 1st January, 1939 to 30th June, 1940. Pp. 8 (1941).

Annual Report of the Department of Agriculture, Bengal for the year 1940—1941, Pt. I, Pp. 33 (1941).

Annual report of the Department of Agriculture in the Province of Bombay, 1939—1940. Pp. 276 (1941).

Fifty-Third Annual Report Rhode Island State College Agricultural Experiment Station. Kingston, R. I., Contr. 586, Pp. 71 (1941).

Tätigkeitsbericht der Kaiser Wilhelm-Gesellschaft zur Förderung der Wissenschaften für das Geschäftsjahr 1940—1941. Naturwiss. 29, 425—433 (1941).

Informes sobre las labores desarrolladas en los diferentes departamentos tecnicos de la Estacion Agricola Experimental de Palmira. (Reports on the work carried out in the different technical departments of the Palmira Agricultural Experiment Station.) Agricultura Ganad. Bogotá 13, 989—1095 (1941).

Verslag van de Commissie van Advies inzake de Bevordering van de Cultuur van Handelsgewassen over het jaar 1940. (Report of the Advisory Committee on the Promotion of the Cultivation of Industrial Plants during the year 1940.) Secretariaat Bericht Nr. 90, Departement van Economische Zaken, Pp. 30, also Bergcultures 15, 1091—1104 (1941).

Research and farming 1941. 64th. Ann. Rep. N. C. Agric. Exp. Sta., Pp. 83 (1941).

Agriculture and animal husbandry in India 1938—1939. Imp. Counc. Agric. Res., Delhi, Pp. 422 (1941).

Primera Reunión Argentina de Agronomía, Abril 1941. Resoluciones y resúmenes de los trabajos presentados. (First Argentine Agronomy Meeting, April 1941. Resolutions and summaries of the works presented.) Soc. Argent. Agron., B., Aires, Pp. 150 (1941).

The All Union Agricultural Fair. Chronica Botanica 6, 417—418 (1941).

O que é o que tem feito a Estação Experimental Fitotécnica de Bagé. (What the Experiment Station for Applied Botany at Bagé is and what it has done.) Rev. Agron., Rio Grande do Sul 5, Nr. 49, 65—67 (1941).

Fifty-first annual report for the fiscal year ended June 30, 1941. Bull. Wash. Agric. Exp. Sta., Nr. 410, 112 (1941).
Rapport annuel pour l'exercice 1939. Publ. Inst. Agron. Congo Belge, Hors. Sér., Pp. 305 (1941).
Rapport sommaire sur les travaux poursuivis en 1940 par les stations d'amélioration des plantes. Ann. Phytogénét., Paris 7, 143—156 (1941).
Fifty-second annual report of the Agricultural Experiment Station of the Alabama Polytechnic Institute, Auburn, January 1 to December 31, 1941, Pp. 32 (1941).
Fifty-fourth annual report of the Agricultural Experiment Station of the University of Tennessee, for 1941. Knoxville, Pp. 100 (1941).
Fifty-fourth annual report of the South Carolina Experiment Station of Clemson Agricultural College for the year ended June 30, 1941. Pp. 182 (1941).
Reports of subordinate officers of the Department of Agriculture, Madras for 1940—1941. Pp. 178 (1941).
Verslag over het jaar 1940 van Het Algemeen Landbouw Syndicaat-Het Zuid-en West-Sumatra Syndicat — De Centrale Vereeniging tot beheer van Proefstations voor de Overjarige Cultures in Nederlandsch-Indië en van de onder deze Organisaties ressorteerende Vereenigingen en Instellingen. (Report for the year 1940 of the General Agriculture Syndicate-The South and West-Sumatra Syndicate — The Central Association for the Management of Experiment Stations for Perennial Crops in the Netherlands Indies and of the Associations and Institutes under the jurisdiction of these Organizations.) Batavia, Pp. 282 (1941).
Report of the Secretary of Agriculture 1941. U. S. Dep. Agric., Washington D. C., Pp. 245 (1941).
East African Agricultural Research Station, Amani. Thirteenth Annual Report 1940. H. M. Stationery Offic. London, price 6d. (Colonial Nr. 181), Pp. 22 (1941).
Annual Report of the Department of Agriculture, Bengal, for the year 1939—1940 (1940), Teil I, Pp. 22; (1941), Teil II, Pp. 390 (1940) 1941.
(Conference on breeding and seed production in the South-East.) Socialistic Grain Farming, Saratov, Nr. 1, 184—211 (1941). [Russisch.]
Fifty-second annual report of the University of Wyoming Agricultural Experiment Station 1941—1942, Pp. 60 (1941—1942).
Annual Report of the Department of Agriculture, Bengal for the year 1941—1942, Pt. II, Pp. 26 (1942).
Annual Report of the Department of Agriculture, Bengal for the year 1940—1941, Pt. II, Pp. 548 (1942).
Fifty-fifth Annual Report of the Agricultural Experiment Station of the University of Tennessee, for 1942, Pp. 111 (1942).
Annual report of the Massachusetts Agricultural Experiment Station for the fiscal year ending November 30, 1941. Bull. Mass. Agric. Exp. Sta., Nr. 388, 108 (1942).
Fifty-third annual report of the Agricultural Experiment Station of the Alabama Polytechnic Institute, Auburn, January 1 to December 31, 1942, Pp. 30 (1942).
The Scottish Plant Breeding Station, Craigs House, Corstorphine, Edinburgh. Trans. Highl. Agric. Soc. Scot. 54, 141—143 (1942).
A year's progress in solving farm problems of Illionis 1937—1938. Rep. Ill. Agric. Exp. Sta. 1938 (1942), Pp. 350 (1942).
Agricultural research in Utah. Bull. Utah Agric. Exp. Sta., Nr. 306, 110 (1942).
High lights in agricultural research in Idaho. 49th. Rep. Univ. Idaho Agric. Exp. Sta. 1941: Bull. Nr. 244, Pp. 63 (1942).
(For the increase of food and raw material resources of the country. [Programme of research work of Breeding and Genetics Institute of USSR. for 1942].) Proc. Lenin Acad. Agric. Sci. USSR., Nr. 3/4, 15—22 (1942). [Russisch.]

Report on the operations of the Department of Agriculture, Madras Presidency for the year 1940—1941. Pp. 74 (1942).
Twenty-second annual report 1941—1942. Bull. Ga Cst. Pl. Exp. Sta., Nr. 35, 154 (1942).
Fifty-fourth annual report Georgia Experiment Station, Experiment, Georgia, of the University system of Georgia for the year 1941—1942. Pp. 111 (1942).
Sixteenth annual report of the Council for Scientific and Industrial Research, for the year ended 30th June, Canberra 1942, Pp. 74 (1942).
Eleventh Biennial Report of the Director 1940—1942 of the Agricultural Experiment Station, Kansas 1942. Pp. 79 (1942).
Report on agricultural research for the year ending June 30, 1942. 7th Rep. Ia. Corn. Res. Inst. Agric. Exp. Sta., Part I, 293 (1942).
Fifty-Fourth Annual Report Rhode Island State College Agricultural Experiment Station. Kingston, R. I., Contr. 614, 62 (1942).
Scientific reports of the Imperial Agricultural Research Institute, New Delhi, for the year ending 30th June, 1941, Pp. 73 (1942).
Annual Report of the Department of Agriculture, Jamaica, for the year ended 31st March, 1942. Pp. 18 (1942).
Report of the Minister of Agriculture for the Dominion of Canada for the year ended March 31, 1942, Ottawa, Pp. 162 (1942).
Annual report of the president. University of Saskatchewan. Academic Year 1941—1942. Saskatoon, Pp. 105 (1942).
Administration report of the Agriculture Departement, Government of Travancore. 1116 M E. Trivandrum, Pp. 52 (1942).
Informe anual correspondiente al año agricola 1941—1942. (Annual report for the agricultural year 1941—1942.) Bol. Chacra Exp. „La Previsión" 3, 115—181 (1942).
Scottish Society for Research in Plant-Breeding. Report by Director of Research. I. Research Programme. Rep. Scot. Soc. Res. Pl. Breed, Pp. 33 (1942).
Progress reports from experiment stations Season 1940 bis 1941. Programmes of experiments. Season 1941 bis 1942. Emp. Cott. Gr. Corp. Lond., Pp. 216 (1942).
Field seed developments. Bureau of Plant Industry reports on research work going on to improve varieties and create new ones. Seed World 51, 8—9 (1942).
Agricultural research in South Dakota. 56th Rep. S. Dak. Agric. Exp. Sta., Pp. 58 (1942—1943).
Research and farming 1943. 66th Ann. Rep. N. C. Agric. Exp. Sta., Pp. 122 (1943).
(Collated plan of research for 1943 at the Institute of Grain Husbandry of the South-Eastern USSR. and its subsidiary stations.) Bjulletenj Instituta Zernovogo Hozjajstva Jugo-Vostoka SSSR. (Bull. Inst. Grain Husb. SEUSSR.) Saratov, Nr. 1, 5—20 (1943). [Russisch.]
Progress through agricultural research, Louisiana 1942 bis 1943. Rep. La Agric. Exp. Sta., Pp. 142 (1943).
What's new in farm science. 59th Rep. Dir. Wisc. Agric. Exp. Sta., Pt 2, Bull., Nr. 460, 87, 1942 (1943).
Agricultural research in New Hampshire. Rep. Dir. N. H. Agric. Exp. Sta. Bull., Nr. 351, 66 (1943).
Science solves farm problems and aids agricultural production. 56th Rep. Dir. Purdue Univ. Ind. Agric. Exp. Sta., Pp. 121 (1943).
Greater farm production for Nebraska. 56th Rep. Neb. Agric. Exp. Sta., Pp. 95 (1943).
Utsädesföreningens extra möte under Lantbruksveckan 1943. (Seed Associations extra meeting during the Agricultural Week 1943.) Sverig. Utsädesfören. Tidskr. 53, 112—115 (1943).
Försöksledarmötet. (The research directors' meeting.) Lantmannen 27, 1028—1030 (1943).

Annual Report of the Imperial Council of Agricultural Research, Delhi 1941—1942, Pp. 130 (1943).
Thirty-third Annual Report of the John Innes Horticultural Institution for the year 1942 (1943), Pp. 18 (1943).
Twenty-third annual report of the Georgia Coastal Plain Experiment Station, Tifton for 1942—1943. Bull. Ga. Cst. Pl. Exp. Sta., Nr. 36, 114 (1943).
Report of the acting president, University of Saskatchewan, for the academic year 1942—1943. Saskatoon. Pp. 79 (1943).
Report of the Minister of Agriculture for the Dominion of Canada for the year ended March 31, 1943. Pp. 155 (1943).
Annual administration report of the Department of Agriculture, United Provinces, for the year ending 30th June, 1942. Allahabad, Pp. 52 (1943).
Årsberättelse avgiven den 28. januari 1943 av Akademiens Sekreterare. (Annual report issued 28th January 1943 by the Secretary of the Academy.) K. Lantbr. Akad. Tidskr. 82, 11—34 (1943).
Seventeenth annual report of the Department of Scientific and Industrial Research, New Zealand 1943. Pp. 44 (1943).
General rewiev of research work with lists of papers published during the year. Ann. Rep. E. Malling Res. Sta. 1942, Pp. 14—30 (1943).
Report (abridged) by the Directors and report by the Director of Research to the annual general meeting 15th July, 1943. Scott. Soc. Res. Pl. Breed, Pp. 33 (1943).
Report on agricultural research for year ending June 30, 1943. Rep. Ia Agric. Exp. Sta., Pt I, Pp. 298 (1943).
Rapport pour les exercices 1940 et 1941. Inst. Agron. Congo Belge, Pp. 152 (1943).
Annual report of the Imperial Council of Agricultural Research for 1942—1943. New Delhi, Pp. 38 (1943).
Report of the Waite Agricultural Research Institute, South Australia 1941—1942. Pp. 84 (1943).
Årets försöksledarmöte. (Annual meeting of directors of research.) Lantmannen 27, 1007—1011 (1943).
Vorbildliche Leistungen deutscher Saatgutzüchter. Forschungsdienst 15, 101 (1943).
Sixty-eighth Annual Report of the Ontario Agricultural College and Experimental Farm 1942. Pp. 71 (1943).
Report on the operations of the Department of Agriculture, Burma, for the years 1941—1942 and 1942 bis 1943. Pp. 26 (1943).
Ninth Progress Report of the University of Alaska Agricultural Experiment Stations 1942—1943. Pp. 59 (1943).
Annual Report of the University of Florida Agricultural Experiment Station, for the fiscal year ending June 30, 1943. Pp. 224 (1943).
Annual Report for the year ending October 31, 1942. 66th Rep. Conn. Agric. Exp. Sta. 1942, Bull., Nr. 468, 53—95 (1943).
Fifty-third Annual Report of the Washington Agricultural Experiment Station, for the fiscal year ended June 30, 1943: Bull., Nr. 435, 163 (1943).
Fifty-fifth Annual Report of the Georgia Experiment Station for the year 1942—1943. Pp. 75 (1943).
Fifty-sixth Annual Report of the Agricultural Experiment Station of the University of Tennessee for 1943. Pp. 115 (1943).
Annual report of the Director for the fiscal year ending June 30, 1943. Bull. Del. Agric. Exp. Sta., Nr. 244, 44 (1943).
Annual report of the Massachusetts Agricultural Experiment Station for the fiscal year ending November 30, 1942. Bull. Mass. Agric. Exp. Sta., Nr. 398, 64 (1943).
Fifty-third annual report for the year ending June 30, 1942, of the Agricultural Experiment Station, University of Arizona, Tucson. Pp. 93 (1943).
Farm science in war and peace. 65th Rep. N. J. Agric. Exp. Sta., Pp. 64 (1943—1944).
Agricultural research in South Dakota. 57th Rep. S. Dak. Agric. Exp. Sta., Pp. 48 (1943—1944).
Rapport Annuel 1943—1944. Bull. Dep. Agric. Haiti, Nr. 36, 140 (1943—1944).
Fifty-Fifth Annual Report of the Agricultural Experiment Station of the New Mexico College of Agriculture and Mechanic Arts., Pp. 76 (1943—1944).
Director's Annual Report for the fiftyseventh fiscal year 1943—1944. Ann. Rep. Colo. Agric. Exp. Sta., Pp. 43 (1943—1944).
Fifty-fourth Annual Report of the University of Wyoming Agricultural Experiment Station 1943—1944. Pp. 47 (1944).
Fifty-seventh Annual Report of the Tennessee Agricultural Experiment Station, 1944. Pp. 107 (1944).
Fifty-fourth annual report for the fiscal year ended June 30, 1944. Bull. Wash. Agric. Exp. Sta., Nr. 455, 168 (1944).
(A meeting of the active members and directors of institutes of the Lenin Academy of Agricultural Sciences.) Proc. Lenin Acad. Agric. Sci. USSR., Nr. 1, 42—47 (1944). [Russisch.]
Thirteenth annual report of the Minister for Agriculture, Eire 1943—1944. Dublin, Pp. 159, 77 (1944).
Annual Report on the Agricultural Department of Nigeria for the year 1943: (1944), S. P., Nr. 12, 34 (1944).
Seventeenth annual report of the Council for Scientific and Industrial Research, for the year ended 30th June, 1943. Canberra, Pp. 76 (1944).
Annual report of the Botanical Division, Department of Agriculture, Ashanti for 1943—1944. Pp. 4 (1944).
Eighteenth Annual Report of the Department of Scientific and Industrial Research, New Zealand, Pp. 57 (1944).
Annual Report of the Department of Agriculture for the year ended 31st March, 1943. Jamaica, Pp. 20 (1944).
Scottish Society for Research in Plant Breeding. Report by the Directors and Report by the Director of Research to the Annual General Meeting. 20. July, Pp. 34 (1944).
Thirty-fourth Annual Report of the John Innes Horticultural Institution for the year 1943. Pp. 26 (1944).
(Short report for 1944 on scientific research at the Institute for Grain Husbandry [awarded the Workers, Red Banner] of the South East Region of the USSR.) Bjulletenj Instituta Zernovogo Hozjajstva Jugo-Vostoka SSSR. (Bull. Inst. Grain Husb. S. E. USSR.) Saratov, Nr. 4, 3—28 (1944). [Russisch.]
Fifty-seventh Annual Report of the Purdue University. Agricultural Experiment Station, Lafayette, Indiana for the year ending June 30, 1944. Pp. 102 (1944).
Progress report of the Institute of Plant Industry, Indore, Central India, for the year ending 31st May 1944. Pp. 32 (1944).
Annual Administration Report of the Agricultural Department, Bangalore, for the year 1942—1943. Pp. 19 (1944).
Annual Report of the University of Florida Agricultural Experiment Station, for the fiscal year ending June 30, 1944. Pp. 214 (1944).
Report of the Director for the year ending October 31, 1943. 67th Rep. Conn. Agric. Exp. Sta. 1943. Bull., Nr. 477, 39—82 (1944).
Report of progress of the Maine Agricultural Experiment Station, Orono, for year ending June 30, 1944. Bull., Nr. 426, 409 (1944).
Report of the Secretary of Agriculture 1944. Washington, D. C., Pp. 196 (1944).
Sixty-ninth Annual Report of the Ontario Agricultural College and Experimental Farm, Department of Agriculture 1943. Pp. 86 (1944).

Research in agriculture, Louisiana 1943—1944. Rep. La Agric. Exp. Sta., Pp. 179 (1944).
What's new in farm science. 60th Rep. Dir. Wisc. Agric. Exp. Sta., Pt 2, Bull., Nr. 463, 80, 1943 (1944).
Agricultural research in New Hampshire. Rep. Dir. N. H. Agric. Exp. Sta. Bull., Nr. 354, 68 (1944).
Wartime agricultural research. 51st Rep. Idaho Agric. Exp. Sta. Bull., Nr. 255, 63 (1944).
Science points the way. 56th Rep. Ark. Agric. Exp. Sta. Bull., Nr. 453, 35 (1944).
Nebraska agriculture 1943. 57th Rep. Neb. Agric. Exp. Sta., Pp. 117 (1944).
Highlights of the work of the Mississippi Experiment Station. 57th Rep. Dir. Miss. Agric. Exp. Sta., Pp. 52 (1944).
I växtförädlingens tjänst. Fran Utsädesföreningens årsmöte i Svalöv. (In the service of plant breeding. From the Seed Associations Annual Meeting in Svalöf.) Lantmannen **28**, 713—714 (1944).
Improvement of native food crops. A précis of the more important work done in East Africa during 1943. Rep. E. Afr. Agric. Res. Inst., Tanganyika Territory 1944, Nr. DF/5/2, 11 (1944).
Fifty-fourth Annual Report of the Washington Agricultural Experiment Station, for the fiscal year ended June 30, 1944: Bull., Nr. 455, 168 (1944).
Fifty-sixth Annual Report of the South Carolina Experiment Station of Clemson Agricultural College, for the year ended June 30, 1943. Pp. 173 (1944).
Sixty-Third Annual Report of the New York State Agricultural Experiment Station Geneva, N. Y. for the year ended June 30, 1944. Pp. 62 (1944).
Report of the Administrator of Agricultural Research, 1944. U. S. Dep. Agric., Agric. Res. Admin. Wash., Pp. 234 (1944).
Annual Report of the Director for the fiscal year ending June 30, 1944. Bull. Del. Agric. Exp. Sta., Nr. 251, 46 (1944).
Fifty-seventh Annual Report of Cornell University Agricultural Experiment. Station 1944. Pp. 188 (1944).
Rapport pour les exercices 1942 et 1943. Publ. Inst. Agron. Congo Belge: Hors Sér., Pp. 154 (1944).
A summarized report of the experimental and extension work by the Szechuan Provincial Agricultural Improvement Instistute. 1936-Present. Szechuan Prov. Agric. Improv. Inst., Pp. 12 (1944).
(Plant breeding and seed production.) Naučnyj Otčet Narymskoj Gosudarstvennoj Selekcionnoj Stancii za 1941—1942 gg. (Sci. Rep. of the Narym State Plant Breeding Station for 1941—1942.) Moskva, Pp. 9—49 (1944). [Russisch.]
(A description of the products, soils and climate of the station.) Naučnyj Otčet Narymskoj Gosudarstvennoj Selekcionnoj Stancii za 1941—1942 gg. (Sci. Rep. of the Narym State Plant Breeding Station for 1941 bis 1942.) Moskva, Pp. 5—8 (1944). [Russisch.]
Annual Report for the year. Veterinary and Agricultural Department, Swaziland, Pp. 26 (1944). [Mimeographed.]
Fifty-fourth annual report for the year ending June 30, 1943 of the Agricultural Experiment Station, University of Arizona, Tucson, Pp. 95 (1944).
Annual report of the Department of Science and Agriculture, Barbados, for the year ending 31st March, 1944. Pp. 14 (1944).
Annual report of the Director of the Kentucky Agricultural Experiment Station for the year 1944. 57th Ann. Rep. Ky Agric. Exp. Sta., Pp. 68 (1944).
Report on agricultural research for year ending June 30, 1944. Rep. Ia Agric. Exp. Sta. Pt I. Pp. 298 (1944).
Twelfth biennial report of the director, Kansas Agricultural Experiment Station, for the biennium July 1, 1942, to June 30, 1944, Pp. 99 (1944).

Eighteenth Annual Report of the Council for Scientific and Industrial Research, Australia, for the year ended 30th June, 1944, Pp. 78 (1944).
Shaping the future of Hawaii's agriculture. Rep. Univ. Hawaii Agric. Exp. Sta, Pp. 115, 1944 (1945).
Science for the farmer. 58th Rep. Pa Agric. Exp. Sta. Bull., Nr. 475, 48 (1945).
Wartime agricultural research. 52nd Rep. Idaho Agric. Exp. Sta. Bull., Nr. 264, 51 (1945).
Farm science and practice. 63rd Rep. Ohio Agric. Exp. Sta. 1944. Bull., Nr. 659, 197 (1945).
Progress through research. 58th Rep. Md Agric. Exp. Sta. 1944—1945, Pp. 36 (1945).
Agricultural research in South Dakota. 58th Rep. S. Dak. Agric. Exp. Sta. 1944—1945, Pp. 47 (1945).
Research in Agriculture. Rep. La Agric. Exp. Sta. 1944 bis 1945, Pp. 144 (1945).
Annual Report of the Massachusetts Agricultural Experiment Station for the fiscal year ending June 30, 1945. Bull., Nr. 428, 71 (1945).
Report of the Administrator of Agricultural Research, 1945. U. S. Dep. Agric., Agric. Res. Admin. Wash., Pp. 238 (1945).
Annales agricoles Vaudoises. 22me année. Lausanne, Pp. 155 (1945).
25 aar i planteforaedlingens tjeneste. (25 years in the service of plant breeding.) Pajbjergfondens Foraedlingsvirksomhed, Børkop, Pp. 46, illus. (1945).
Annual Report of the Department of Science and Agriculture, Barbados for the year 1944—1945, Pp. 17 (1945).
Seventieth Annual Report of the Ontario Agricultural College and Experimental Farm, Department of Agriculture 1944, Pp. 97 (1945).
Annual Report of the Department of Agriculture, Mauritius 1944, Pp. 30 (1945).
Report on the Department of Agriculture, Gold Coast, for the year 1944—1945. Pp. 8 (1945).
Annual Report of the Department of Agriculture, Sierra Leone for the year 1944. Pp. 15 (1945).
Annual Report of the Imperial Council of Agricultural Research, New Delhi 1944—1945. Pp. 45 (1945).
Fifty-fifth annual report for the year ending June 30, 1944 of the Agricultural Experiment Station, University of Arizona 1945, Pp. 100 (1945).
Report of the Federal Experiment Station in Puerto Rico 1944. Agric. Res. Admin., U. S. Dep. Agric., Pp. 44 (1945).
Fifty-fifth Annual Report of the Washington Agricultural Experiment Station for the fiscal year ended June 30, 1945. Bull., Nr. 470, 167 (1945).
Research and farming. 68th Rep. N. C. Agric. Exp. Sta. **4**, Progr. Rep., Nr. 4, 108 (1945).
Report of progress of the Maine Agricultural Experiment Station, Orono, for year ending June 30, 1945. Bull., Nr. 438, 698 (1945).
Agricultural research in New Hampshire. Rep. Dir. N. H. Agric. Exp. Sta., Bull., Nr. 363, 61 (1945).
Panorama of agricultural research. 58th Rep. Tex. Agric. Exp. Sta., Pp. 76 (1945).
Fifth Annual Report of the Institute of Agriculture, Anand, for the period ended 31 March, 1945. Bull., Nr. 5, 45 (1945).
Fifty-sixth Annual Report of the University of Arizona Agricultural Experiment Station, for the year ending June 30, 1945, Pp. 81 (1945).
Fifty-eighth Annual Report of the Purdue University Agricultural Experiment Station, Lafayette, Indiana, for the year ending June 30, 1945, Pp. 102 (1945).
Report of the College of Agriculture, University of Vermont, July 1, 1944 bis June 30, 1945, Nr. 1, 64 (1945).
Silver Anniversary Annual Report of the Georgia Coastal Plain Experiment Station, Tifton, Georgia 1944—1945. Bull., Nr. 42, 156 (1945).

Administration Report of the Director of Agriculture, Trinidad and Tobago, for the year 1944. Coun. Pap., Nr. 47, 16 (1945).
Annual Report of the Livestock and Agricultural Department, Swaziland, 1945, Pp. 38 (1945). [Mimeographed.]
Progress report of the Institute of Plant Industry, Indore, Central India, for the year ending 31st May 1945. Pp. 30 (1945).
Annual Report of the University of Florida Agricultural Experiment Station, for the fiscal year ending June 30, 1945. Pp. 229 (1945).
58th Annual Report of the Agricultural Experiment Station of the University of Kentucky, 1945. Pp. 68 (1945).
Report on agricultural research for the year ending June 30, 1945. Rep. Ia Agric. Exp. Sta., Pt. I, Pp. 355 (1945).
Nineteenth Annual Report of the Council for Scientific and Industrial Research, Australia, for the year ended 30th June 1945. Nr. 32, 164 (1945).
Fifty-sixth Annual Report of the New Mexico Agricultural Experiment Station 1944—1945. Pp. 58 (1945).
58th Annual Report of the Colorado Agricultural Experiment Station, 1944. Pp. 52 (1945).
Fifty-seventh Annual Report of Georgia Experiment Station of the University System of Georgia, July 1, 1944 to June 30, 1945, Pp. 91 (1945).
+ Årsberättelse över Sveriges utsädesförenings verksamhet under år 1944. (Annual Report on the work of the Swedish Seed Association 1944.) Sverig. Utsädesfören, Tidskr. **55**, 247—328 (1945).
Annual Report of the Department of Agriculture and Stock, Queensland, for the year 1944—1945. Pp. 40 (1945).
Fifty-Seventh Annual Report of the South Carolina Experiment Station of Clemson Agricultural College for the year ended June 30, 1944. Pp. 154 (1945).
Sixty-fourth Annual Report of the New York State Agricultural Experiment Station, for the year ended June 30, 1945. Pp. 74 (1945).
Annual Report of the Department of Agriculture, Jamaica, for the year ended 31st March, 1944. Kingston, Pp. 16 (1945).
The fourth annual report of the governing body of the Institute of Agriculture, Animal Husbandry and Dairying, Anand, for the period ended 31st March 1944. Gen. Ser. Bull. Inst. Agric. Anand, Nr. 4, 43 (1945).
35th Annual Report of the John Innes Horticultural Institution 1944. Pp. 24 (1945).
Report of the National Agricultural Research Bureau, Ministry of Agriculture and Forestry 1932—1944. Sino-British Co-operation Office, 19th April, Pp. 36 (1945).
Annual Report on the Department of Agriculture, Uganda, for the period 1st July, 1943 bis 30th June, 1944. Pp. 10 (1945).
Fifty-seventh Annual Report Rhode Island State College Agricultural Experiment Station 1945. Contr. 674, Pp. 39 (1945).
Årsberättelse för 1944 avgiven den 28 januari 1945. (Annual report for 1944 issued 28 January 1945.) K. LantbrAkad. Tidskr. **84**, 11—34 (1945).
Nineteenth annual report of the Department of Scientific and Industrial Research, New Zealand, Pp. 65 (1945).
Report of the Department of Agriculture, Nyasaland Protectorate, for the year 1944. Pt II. Experimental work., Pp. 12 (1945).
Scottish Society for Research in Plant-Breeding. Report by the Directors and Report by the Director of Research to the Annual General Meeting 19th July, 1945. Pp. 34 (1945).
Report of the East African Agricultural Research Institute, Amani for the years 1942—1945, Pp. 43 (1946).

Fourth annual report of the Director of the Institute of Tropical Agriculture, University of Puerto Rico. Fiscal year 1945—1946, Pp. 38 (1946).
Annual Report of the Department of Agriculture and Stock Queensland, for the year 1945—1946. Pp. 88 (1946).
Administration report of the Director of Agriculture, Trinidad and Tobago, for the year 1945. Pp. 20 (1946).
Annual Report of the Massachusetts Agricultural Experiment Station for the fiscal year ending June 30, 1946. Bull., Nr. 436, 70 (1946).
59th Annual Report of the New York State College of Agriculture at Cornell University, Ithaca, New York, and of the Cornell University Agricultural Experiment Station 1946. Pp. 183 (1946).
Sixty-fifth annual report of the New York State Agricultural Experiment Station, Geneva, New York, 1946. Pp. 73 (1946).
The experiment station reports. 57th Rep. Ark. Agric. Exp. Sta., Bull., Nr. 464, 23 (1946).
Agricultural research in Scotland in 1945. Scottish Plant-Breeding Station. Trans. Highl. Agric. Soc. Scot. **58**, 123—125 (1946).
In Scientific Institutions. Agricultural Chron., Moscow, Nr. 11 9—12 (1946). [Mimeographed].
Annual report of the Livestock and Agricultural Department, Swaziland. Pp. 45 (1946). [Mimeographed].
Annual report of the Department of Agriculture, Basutoland, for the year ended 30 September, 1946. Pp. 28 (1946).
Annual session of the Indian Academy of Sciences held at Udaipur. Curr. Sci. **15**, 7—8 (1946).
Thirteenth Biennial Report of the Director of the Kansas Agricultural Experiment Station, for the biennium July 1, 1944, to June 30, 1946. Pp. 96 (1946).
Report of the Minister of Agriculture for the Dominion of Canada for the year ended March 31, 1946. Pp. 235. (1946).
Annual Report of the President of the University of Saskatchewan, Saskatoon, academic year 1945—1946. Pp. 171 (1946).
Annual Report of the Department of Agriculture, Northern Rhodesia 1945. Pp. 28 (1946).
Annual Report of the Department of Agriculture, Colony and Protectorate of Kenya 1945. Pp. 125 (1946).
Annual Report of the National Northwestern College of Agriculture, Wukung, Shensi, China, 1946. Pp. 74 (1946).
Twenty years of plant cultivation in Central Asia. Soviet News, Nr. 1356, 4 (1946).
Annual Report of the Department of Agriculture, Mauritius 1945. Pp. 34 (1946).
Annual Report of the Department of Agriculture, Uganda for the period 1st July, 1944 bis 30th June, 1945. Pt. II, Pp. 60 (1946).
Report on the administration of the Department of Agriculture, Travancore 1946. Pp. 38 (1946).
Abridged Scientific Reports of the Imperial Agricultural Research Institute, New Delhi for the triennium ended 30th June, 1944. Pp. 85 (1946).
General review of research work with lists of papers sublished during the year. 33rd Rep. E. Malling Res. Sta. 1945. Pp. 17—44 (1946).
Report of the Federal Experiment Station in Puerto Rico, 1945. U. S. Dep. Agric., Washington, Pp. 62 (1946).
Annual Report on the Agricultural Department, Nigeria, for the year 1944. Sess. Pap., Nr. 13, 47 (1946).
20th Annual Report of the Department of Scientific and Industrial Research, New Zealand. Pp. 110 (1946).
Administration Report of the Acting Director of Agriculture for 1944. Part IV. — Education, science, and art (D). Ceylon, Pp. D. 23 (1946).
Scottish Society for Research in Plant-Breeding. Report by the Directors and Report by the Director of Rese-

arch to the Annual General Meeting 18th July, 1946. Pp. 41 (1946).

Scientific Reports of the Imperial Agricultural Research Institute, New Delhi, for the year ended 30th June, 1945. Pp. 93 (1946).

59th Annual Report of the Nebraska Agricultural Experiment Station 1946. Pp. 119 (1946).

Report of the Secretary of Agriculture 1945. Washington, D. C., Pp. 167 (1946).

Akademiens sekreterare: Årsberättelse för 1945. (Academy of Agriculture, Secretary's report for 1945.) K. LandbrAkad. Tidskr. 85, 13—32 (1946).

Annual Report of the Department of Agriculture, Jamaica, for the year ended 31st March, 1945. Pp. 14 (1946).

Proceedings of the 33rd Indian Science Congress, Bangalore 1946. Pt. II., Sect. VI, Botany. Pp. 16 (1946).

Thirty-sixth Annual Report of the John Innes Horticultural Institution for the year 1945. Pp. 28 (1946).

Science for the farmer. 59th Rep. Pa. Agric. Exp. Sta., Bull., Nr. 480, 66 (1946).

Research and the farmer. IX. Plant breeding. Mon. Rep. Minist. Agric. N. Ire 20, 258—260 (1946).

Agricultural research in Idaho. 53rd Rep. Idaho Agric. Exp. Sta., Bull., Nr. 268, 60 (1946).

Thirty-seventh Annual Report of the John Innes Horticultural Institution. Pp. 32, 1946 (1947).

Annual Administration Report of the Agriculture Department United Provinces for the year ending 30 June, 1945. Pp. 86 (1947).

Annual report of the Department of Agriculture, Colony of Sierra Leone for the year 1945. Pp. 24 (1947).

Office of the Agricultural Attaché. British Embassy, Washington 8, D. C. February, 1947 — April, Tech. Digest, Nr. 4, 26 (1947). [Mimeographed.]

Report of the FAO Mission for Greece. Food and Agricultural Organization of the United Nations, Washington D. C. March 1947. Pp. 188 (1947).

Annual Report of the Department of Agriculture for the year ended 31 August, 1946. Fmg S. Afr. 22, 77—360 (1947).

Akademiens sekreterare: Arsberättelse för 1946. (Academy of Agriculture, Secretary's report for 1946.) K. LentbrAkad. Tidskr. 86, 13—35 (1947).

Botanik, Angewandte Botanik, Entwicklungsphysiologie.

Bericht über die dreiundfünfzigste Generalversammlung der Deutschen Botanischen Gesellschaft in Graz, August 1939. Ber. dtsch. bot. Ges. 57, (1)—(37) (1939).

Recent research on Empire products. A record of work conducted by Government. Technical Departments overseas. Bull. Imp. Inst. Lond. 37, 368—387 (1939).

De Commissie voor de Handelsgewassen. (The Committee for Industrial Plants.) Bergcultures 14, 1291 (1940).

Weibullsholm fyller 70 år och inviger samtidigt landets modernaste renserier och lagerlokaler. (Weibullsholm completes 70 years and inaugurates at the same time the country's most modern winnowing and storage plant.) Lantmannen 25, 585—586 (1941).

De opening van de nieuwe laboratoria van het Proefstation West-Java. (The opening of the new laboratories of the West Java Experiment Station.) Bergcultures 15, 1342—1352 (1941).

Twenty-fourth report and accounts of the National Institute of Agricultural. Botany, Cambridge, 1942 bis 1943, Pp. 16 (1943).

Twenty-fifth report and accounts of the National Institute of Agricultural Botany, Cambridge, 1943 bis 1944, Pp. 14 (1944).

Annual Report of the Canadian Seed Growers' Association, Ottawa 1944—1945, Pp. 78 (1945).

(Letter from L. Zonštein.) Sovhoznoe Proizvodstvo (State Farming), Nr. 3, 49 (1945). [Russisch.]

Proceedings of the Association of Applied Biologists. Ann. Appl. Biol. 32, 277—282 (1945).

Berättelse över Jordbrukets forskningsråds verksamhet under budgetåret 1945/46. (Report on the work of the Agricultural Research Council during the financial year 1945/46). K. LantbrAkad. Tidskr. 85, 477—484 (1946).

Annual Report of the Canadian Seed Growers' Association, Ottawa 1945—1946, Pp. 50 (1946).

Botanists in conference. Soviet News, Nr. 1746, 4 (1947).

Pflanzenkrankheiten.

(Plant virus diseases and their control. Transaction of the conference on plant virus diseases. Moscow, 4—7 February, 1940.) Microbiol. Inst. Acad. Sci. USSR., Pp. 340 (1941). [Russisch.]

Weizen.

Fifth Annual Report for the year 1938. Wheat Research Institute, Christchurch, New Zealand. C. Wheatbreeding. Bull. Dep. Sci. Industr. Res. N. Z., Nr. 68, 15—21 (1939).

Report of the First Conference of the Hard Red Winter Wheat Quality Advisory Committee, March 30—31, 1939. Div. Cer. Crops Dis., Bur. Pl. Ind., U. S. Dep. Agric. May 24, Pp. 5 (1939). [Mimeographed.]

Sixth Annual Report for the year 1938—39. Wheat Research Institute, Christchurch, New Zealand. C. Wheat-breeding. Bull. Dep. Sci. Industr. Res. N. Z., Nr. 81, 20—25 (1939).

State Research Farm. Annual Field Day. J. Dep. Agric. Vict. 37, 459—462, 484—500 (1939).

Guide Book. The State Research Farm, Werribee. Dep. Agric. Vict., Pp. 49 (1939).

Das Lehr- und Demonstrationsgut Roßberg-Kempttal. Schweiz. landw. Mh., Nr. 5, 44 (1939).

BPI reports on results of Michels grass tests. U. S. Dep. Agric. Pr. Release, Pp. 2 (1940). [Mimeographed.]

Fourteenth Annual Report of the Council for Scientific and Industrial Research, for year 1939—1940. Gov. Commonwealth Aust., Pp. 102 (1940).

Seventh Annual Report for the year 1939—1940. Wheat Research Institute, Christchurch, New Zealand. Bull. Dep. Sci. Industr. Res. N. Z., Nr. 84, 26 (1940).

Wheat Research Institute. Eighth Annual Report, May, 1941. Bull. Dep. Sci. Industr. Res. N. Z., Nr. 87, 24. 1940—1941 (1942).

Wheat breeding, Department of Botany. Report of the Seventh Eastern Wheat Conference, Lafayette, Indiana, June 19—20, Pp. 7—9 (1941).

Soft winter wheat breeding programs and problems. Report of the Seventh Eastern Wheat Conference, Lafayette, Indiana, June 19—20, Pp. 10—11 (1941).

Meddelelser fra Statens Forsøgsvirksamhed i Plantekultur. Forsøg med Hvedesorter. 1936—1940. (Reports on state research in plant cultivation. Experiments with varieties of wheat. 1936—1940.) Tidsskr. Planteavl. 46, 165—169 (1941).

Annual Report of the Department of Agriculture, Kenya 1944, Pp. 8 (1945).

Roggen.

Foreløbig Meddelelse om Forsøg Rugsorter 1939—1941. (Preliminary reports on experiments with varieties of rye 1939—1941.) Tidsskr. Planteavl. 47, 187—188. (327 Meddelelse. A.) (1942).

Hafer.

Report on oat co-operative rod row tests and supplementary rod row tests. Rep. Dominion Rust Res. Lab., Winnipeg, Nr. 39, 11 (1940).
The Cressy Research Farm. 1. General description and Algerian oat improvement work. Tasm. J. Agric. 11, 117—121 (1940).

Gerste.

Visit of the Institute of Brewing to the Plant-Breeding Institute, Cambridge University, on July 14th. J. Inst. Brew. 45, 467—468 (1939).
Twelfth annual report of the Minister for Agriculture, Department of Agriculture, Eire 1942—1943. Dublin, Pp. 77 (1943).
Société d'Encouragement de la Culture des Orges de Brasserie et des Houblons en France SECOBRAH. Rapport sur la campagne 1942. Paris, Pp. 60 (1944).
Société d'Encouragement de la Culture des Orges de Brasserie et des Houblons en France SECOBRAH. Rapports sur la campagne 1943. Paris, Pp. 75 (1944).
Annual report of the Council of the Institute of Brewing for the year ended 31st December, 1944. Advisory Sub-Committee on barley. J. Inst. Brew. 51, 72—73 (1945).
Annual Report of the Council of the Institute of Brewing for the year ended 31st December, 1945. Pp. 23 (1945).
Report of the seed propagation division, 1945. J. Dep. Agric., Eire 43, 74—87 (1946).

Mais.

Report on agricultural research for the year ending June 30, 1939. Part II. Iowa Corn Research Institute fourth annual report. Ia Agric. Exp. Sta., Pp. 88 (1939).
Report of the Second Corn Improvement Conference (First Field Meeting) held at the University of Wisconsin, Madison, Wisconsin. September 16—17, 1938. Washington, D.C., Pp. 53 (1939). [Mimeographed.]
Report of the First Southern Corn Improvement Conference (Organization Meeting) held at New Orleans, Lousiana. November 24, 1939. Washington, D. C. February 25, Pp. 37 (1940). [Mimeographed.]
Report on the Third Corn Improvement Conference held at the University of Missouri, Columbia, Missouri, November 27, and 28, 1939. Washington, D. C., Pp. 21 (1940). [Mimeographed.]
Report on Agricultural research for the year ending June 30, 1942. 7th Rep. Ia Corn. Res. Inst. Agric. Exp. Sta., Teil II, Pp. 85 (1942):
Plowshares and swords. 64th Rep. N. J. St. Agric. Exp. Sta.; 56th Rep. N. J. Agric. Coll. Exp. Sta. 1942—1943. Pp. 64 (1943).
Report on agricultural research for the year ending June 30, 1943. Part II. Iowa Corn Research Institute Eight Annual Report. Iowa, Pp. 79 (1943).
Report on agricultural research for year ending June 30, 1944. Rep. Ia Agric. Exp. Sta., Pt. II, Pp. 92 (1944).
Report on agricultural research for the year ending June 30, 1945. Rep. Ia Agric. Exp. Sta. 1945, Pt II, Pp. 91 (1945).
Report on agricultural research for the year ending June 30, 1946. Rep. Ia Agric. Exp. Sta., Pt. II, Pp. 70 (1946).

Reis.

State of North Borneo. III. Production. 1. Agriculture. H. E. Governor's Administr. Rep. St. N. Borneo, Pp. 5 (1939).
Rice Research Station. Hawkesbury Agric. Coll. J. 37, 126—127 (1940).
Report of the Department of Agriculture, British Guiana, for the year 1942. Pp. 10 (1942).

O arroz no Brasil. (Rice in Brazil.) Bol. Minist. Agric., Rio d J. 32, Nr. 11, 138—139 (1943).
Administration report of the Agricultural Department, Travancore. 1943. Pp. 39 (1943).
Administration report of the Department of Agriculture. 1118 M. E. Trivandrum, Pp. 39 (1944).
Administration report of the Agricultural Department, Government of Travancore. 1119 M. E. Trivandrum, Pp. 36 (1945).

Futtergräser.

Fourth Annual Report of the U. S. Regional Pasture Research Laboratory. State College, Pa., Pp. 106 (1940).
5th Annual Report of the U. S. Regional Pasture Research Laboratory, State College, Pa., Pp. 120 (1941).
Annual Report of the Department of Agriculture, New Zealand for 1941—1942. Pp. 12 (1942).
Sixth Annual Report of the U. S. Reginal Pasture Research Laboratory. State College, Pa., Pp. 113. 2 figs. (1942). [Mimeographed.]
Seventh Annual Report of the U. S. Regional Pasture Research Laboratory. State College, Pa., Pp. 121 (1943). [Mimeographed.]
Eighth Annual Report of the U. S. Regional Pasture Research Laboratory State College, Pa. 1944, Pp. 115 (1944). [Mimeographed.]
Ninth Annual Report of the U. S. Regional Pasture Research Laboratory State College, Pa. 1945. Pp. 116 (1945). [Mimeographed.]
Tenth annual report of Pasture Research in the Northeastern United States State College, Pennsylvania 1946. Pp. 75 (1946). [Mimeographed.]

Futterleguminosen.

1943 report of the Uniform Alfalfa Nurseries. Agric. Res. Admin. Bur. Pl. Ind. Soils, Agric. Engng, Div. For. Crops, Dis., U. S. Dep. Agric., Pp. 20 (1944). [Mimeographed.]

Wurzel- und Knollengewächse.

The Scottish plant breeding station. Craigs House, Corstorphine, Edinburgh. Trans. Highl. Agric. Soc. Scot. 51, Ser. 5, 181—183.
Station develops new Irish potato. Tests show heavy production qualities and resistance to insects and diseases. Ext. Farm-News 24, Nr. 12, 4 (1939).
Australia's potato growing industry. Agric. Gaz., N. S. W. 51, 264—268, 295—296 (1940).
The Scottish Plant Breeding Station, Craigs House, Corstorphine, Edinburgh. Trans. Highl. Agric. Soc. Scot. 52, Ser. 5, 121—123 (1940).
Report on the marketing of potatoes in India and Burma. Agric. Market, India Market, Ser., Nr. 22, 332 (1941).
The Scottish Plant Breeding Station, Craigs House, Corstorphine, Edinburgh. Trans. Highl. Agric. Soc. Scot. 55, 100—102 (1943).
Agricultural research in Scotland in 1944. The Scottish Plant-Breeding Station. Trans. Highl. Agric. Soc. Scot. 57, 75—77 (1945).
Redogörelse för resultaten av Skånes Fröodlingsförenings stamförsök med foderrotfrukter 1929—1941. (Report on the results of the Scanian Seed Production Association's tests of strains of fodder roots 1929—1941.) K. LantbrAkad. Tidskr. 85, 113—217 (1946).

Faserpflanzen.

Annual report of the Indian Central Cotton Committee, for the year ending 31st August, 1939, Bombay. Pp. 191 (1939).
(Breeding report of the cotton plant „Ryo-yo No. 2".) Res. Bull. Agric. Exp. Sta. Kung-Chu-Ling, Manchoukuo, Nr. 28, 11 (1939).

Cotton productivity in the State of São Paulo. Bull. Chamb. Comm., S. Paulo, Nr. 24 (1939).

Report of the Administrative Council of the Corporation submitted to the Eighteenth Annual General Meeting on May 19th, 1939. Rep. Emp. Cott. Gr. Corp., Pp. 55 (1937/1938) 1939.

Progress reports from experiment stations, season 1937 bis 1938. Emp. Cott. Gr. Corp. London, Pp. 150 (1939).

Summary proceedings of the thirty-ninth meeting of the Indian Central Cotton Committee, Bombay, held on the 31st March, 1939, Pp. 28 (1939).

Summary proceedings of the fortieth meeting of the Indian Central Cotton Committee, Bombay, held on the 3rd and 4th August, 1939. Pp. 62 (1939).

Report of the Administrative Council of the Corporation submitted to the Twentieth Annual General Meeting on May 27th, 1941. Annu. Rep. Emp. Cott. Gr. Corp. 1939—1940 (1941), Pp. 34 (1939—1940).

Annual Report of the Indian Central Cotton Committee for the year ending 31st August, 1940. Bombay, Pp. 183 (1940).

Progress reports from experiments stations. Season 1938 bis 1939 Programmes of experiments, Season 1939 bis 1940. Emp. Cott. Gr. Corp. Lond., Pp. 198 (1940).

Summary proceedings of the forty-first meeting of the Indian Central Cotton Committee, Bombay, held on the 19th and 20th January, 1940 Pp. 59 (1940).

20th Annual report of the Indian Central Cotton Committee, for the year ended 31st August, 1941. Pp. 174 (1941).

Indian Central Cotton Committee. Curr. Sci. 10, 93 (1941).

Progress reports from experiment stations. Season 1939 bis 1940. Programmes of experiments. Season 1940 bis 1941. Emp. Cott. Gr. Corp. Lond., Pp. 176 (1941).

Quarterly notes of the Sisal Experimental Station, Mlingano, Tanga. Dep. Agric. Sisal Board, Tanganyika Nr. 21/22, 28 (1941).

(Results of work of the Plant Protection Station of the Cotton Research Institute on pests and diseases of cotton and lucerne.) Sojuzniki, Taškent, p. 72 (1941). [Russisch.]

Second conference on cotton growing problems in India. January, 1941. Rep. Indian Cent. Cott. Comm., Bombay, p. 177 (1941).

Indian Central Cotton Committee. Annual report for the season 1941—1942. Pp. 103 (1942).

Progress report of the Cotton Genetics Research Scheme, Indore, for 1941—1942. Pp. 31 (1942).

Annual report of the Director, Technological Laboratory, Indian Central Cotton Committee for the year ending 31st May, 1943. Bombay, Pp. 36 (1943).

Progress report of the Cotton Genetics Research Scheme, Indore, 1942—1943. Pp. 20 (1943).

Progress reports from experiment stations. Season 1941 bis 1942. Programmes of experiments. Season 1942 bis 1943. Emp. Cott. Gr. Corp., Lond., Pp. 183 (1943).

Sixteenth annual report of the Agricultural Research Institute of Northern Ireland, Hillsborough, Co. Down. 1942—1943. Pp. 30 (1943).

Twenty-second Annual Report of the Indian Central Cotton Committee for the year ended 31st August 1943. Pp. 132 (1943).

54th and 55th Annual Reports of the Agricultural Experiment Station, Alabama Polytechnic Institute January 1, 1943 bis December 31, 1944. Pp. 39 (1944).

Annual report of the Department of Agriculture for the period 1st July 1942 to 30th June 1943. Uganda, Pp. 12 (1944).

Annual report on the Agricultural Department, St Vincent 1943. Pp. 12 (1944).

Progress report of the Cotton Genetics Research Scheme, Indore, 1943—1944. Pp. 27 (1944). [Mimeographed.]

Progress reports from experiment stations. Season 1942 bis 1943. Programmes of experiments. Season 1943 bis 1944. Emp. Cott. Gr. Corp. Lond., Pp. 181 (1944).

Report of the Administrative Council of the Corporation submitted to the twenty-third Annual General Meeting on June 6th, 1944. Emp. Cott. Gr. Corp., Pp. 16 (1944).

Report on the staple length of the Indian cotton crop of 1943—1944 season. Statist. Leafl. Indian Cent. Cott. Comm., Nr. 1, 9 (1944).

Twenty-third annual report of the Indian Central Cotton Committee for the year ended 31st August, 1944. Bombay, Pp. 128 (1944).

Seventeenth annual report of the Agricultural Research Institute of Northern Ireland, Hillsborough, Co. Down. 1943—1944. Pp. 35 (1944).

Investigations by the Plant Disease Division. 18th Ann. Rep. Agric. Res. Inst. N. Ire., p. 15 (1944—1945).

Annual Report on the Agricultural Department, St. Vincent 1944. Pp. 19 (1945).

Progress Report of the Cotton Genetics Research Scheme, Indore, Central India, for the year 1944—1945. Pp. 37 (1945). [Mimeographed.]

Report of the Administrative Council of the Corporation submitted to the Twenty-Fourth Annual General Meeting on June 5th, 1945. Rep. Emp. Cott. Gr. Corp. 1943—1944, Pp. 18 (1945).

Progress reports from experiment stations. Season 1943 bis 1944. Programmes of experiments. Season 1944 bis 1945. Emp. Cott. Gr. Corp. Lond., Pp. 176 (1945).

Twenty-fourth Annual Report of the Indian Central Cotton Committee 1945. Pp. 139 (1945).

Progress reports from experiment stations. Season 1944 bis 1945. Programmes of experiments. Season 1945 bis 1946. Emp. Cott. Gr. Corp. Lond., Pp. 142 (1946).

Report of the Administrative Council of the Corporation submitted to the twenty-fifth Annual General Meeting on June 25, 1946. Rep. Emp. Cott. Gr. Corp. Lond. 1944—1945, Pp. 20 (1946).

Indian Central Cotton Committee. Annual Report of the Director, Technological Laboratory, for the year ending 31 May, 1946. Pp. 35 (1946).

Zuckerpflanzen.

Thirty-ninth annual report of the Bureau of Sugar Experiment Stations. Report of the Director to the Hon. the Secretary for Agriculture and Stock, Brisbane, p. 66 (1939).

Annual report of the Director for the year ended 30th June 1939. Cane Gr. Quart. Bull. 7, 84—106 (1939).

The yield of sugar cane in Barbados in 1938. Bull. B. W. I. Cent. Sug. Cane Breed. Sta., Nr. 20, 10 (1939).

Report of Committee in Charge of the Experiment Station. Hawaiian Sugar Planters' Association for the year ending September 30, p. 121 (1939).

Ninth Annual Report of the Sugarcane Research Station, for the year 1938. Dep. Agric. Mauritius, Pp. 65 (1939).

Fortieth Annual Report of the Bureau of Sugar Experiment Stations, Queensland. Brisbane, Pp. 22 (1940).

(Resolutions of the meeting of breeders technologists, phytopathologists and biochemists at the Research Institute for the Sugar Industry [VNIS] on 3—5 th July, 1940 at Kiev.) „Principles of sugar beet variety production in connexion with securing the initial stages of seed production." Naučnye Zapiski Saharnoj Promyšlennosti (Sci. Trans. Sug. Ind.), Nr. 1/2, 141—144 (1940). [Russisch.]

Proceedings of the fourteenth annual conference. Asociacion de técnicos azucareros de Cuba. Pp. 362 (1940).

Report of Committee in charge of the Experiment Station, Hawaiian Sugar Planters' Association for the year ending September 30, 1940. Pp. 130 (1940).

Seventh Annual Report of the British West Indies Central Sugar Cane Breeding Station, Barbados for the year ending September 30, 1940. p. 30 (1940).

Sixth Progress Report on experiments at Umfolozi. Proc. 14th Annu. Congr. S. Afr. Sug. Tech. Ass. Durban 2nd—4th April, p. 52—55 (1940).

Verslag der cultuurafdeeling van het proefstation voor de Java-Suiker-Industrie te Pasoeroean over het jaar 1940. (Report of the crop section of the Java Sugar Industry experiment Station at Pasoeroean for 1940.) Soerabaja, Pp. 28 (1940).

Editorial. Jamaican Ass. Sug. Technol. Quart. **4**, 1—2 (1941).

Eighth Annual Report of the British West Indies Central Sugar Cane Breeding Station, ending September 30th, 1941. Pp. 37 (1941).

The yield of sugar cane in Barbados in 1940. Bull. B. W. I. Cent. Sug. Cane Breed. Sta., Nr. 23, 12 (1941).

A beterraba sacarina. Resultados dos ensaios culturais, económicos e analíticos realizados em 1941. (Sugar beet. Results of cultural, economic and analytical tests carried out in 1941.) Sér. Estud. Inform. Téc., Minist. Econ, Serv. Edit. Repart. Estud. Inform. Prop., Lisboa, Nr. 17, 98 (1942).

Bureau of Sugar Experiment Stations. Annual Report. Aust. Sug. J. **33**, 376—381 (1942).

Distribution of Q. 28 in the Mackay district. Aust. Sug. J. **34**, 47 (1942).

Ninth annual report of the British West Indies Central Sugar Cane Breeding Station, ending September 30, 1942. Barbados, Pp. 42 (1942).

Review of sugarcane research. Curr. Sci. **11**, 451 (1942).

The yield of sugar cane in Barbados in 1941. Bull. B. W. I. Cent. Sug. Cane Breed. Sta., Nr. 24, 10 (1942).

Annual report on the Department of Agriculture for the year ended 31st December 1943. Zanzibar, Pp. 6 (1943).

Forty-third Annual Report of the Bureau of Sugar Experiment Stations, Queensland 1943. Pp. 22 (1943).

Resumé of the twelfth annual report of the sugarcane research station, 1941. Rev. Agric. Maurice **22**, 22—26 (1943).

Reports on research work 1943. B. W. I. Sug. Ass., Barbados, Pp. 49 (1943).

Tenth annual report of the British West Indies Central Sugar Cane Breeding Station, ending September 30, 1943. Barbados, Pp. 41 (1943).

Eleventh Annual Report ending September 30, of the British West Indies Central Sugar Cane Breeding Station, Barbados 1944, Pp. 41 (1944).

Planning production of the Louisiana sugarcane crop. Sug. Bull., N. O. **23**, 44—46 (1944).

Reports on research work 1944. B. W. I. Sug. Ass., Barbados, Pp. 54 (1944).

Report on the Sugar Experiment Stations for the year 1943. Sug. Bull. Dep. Agric. Brit. Guiana, Nr. 12, 65—73 (1944).

What the scientists are doing. The 1944 batch of Co. canes. Indian Fmg **5**, 420 (1944).

Bureau of Sugar Experiment Stations. Gumming disease in Mossman. Aust. Sugg. J. **37**, 251—253 (1945).

Dix années de recherches à l'Institut Belge pour l'Amélioration de la Betterave à Tirlemont de 1932 à 1941. Bruxelles, Pp. 436 (1945).

Fifteenth Annual Report of the Sugarcane Research Station, Department of Agriculture, Mauritius 1944. Pp. 23 (1945).

Forty-fifth Annual Report of the Bureau of Sugar Experiment Stations, Queensland 1945. Pp. 27 (1945).

Résumé du rapport de la Station de Recherches sur la canne à sucre pour l'année 1944. Rev. Agric. Maurice **24**, 203—208 (1945).

Forty-sixth Annual Report of the Director of the Bureau of Sugar Experiment Stations. Queensland 1946. Pp. 46 (1946).

Sixteenth Annual Report of the Sugarcane Research Station, Department of Agriculture. Mauritius 1945. Pp. 17 (1946).

Stimulantien.

Annual report of the board for the year ended 30th June, 1939. Coffee, Bd. Kenya, Pp. 57 (1939).

Annual Report of the Tocklai Experimental Station, 1938. Sci. Dep. Indian Tea Ass., Calcutta, Pp. 40 (1939).

Progress report of work done on the Coffee Experiment Station, Balehonnur, for the period 1932 to 1936. Bull. Mysore Coffee Exp. Sta., Nr. 18, 41 (1939).

Report on the marketing of tobacco in India and Burma. Agric. Market. India. Market. Ser., Nr. 10, 503 (1939).

(Symposium on breeding, genetics and seed production of tobacco and Nicotiana rustica.) Vsesojuznyj Naučno-Issledovateljskij Institut Tabačnoj i Mahoročnoj Promyslennosti imeni A. I. Mikojana (VITIM) (All-Union Mikojan Research Institute of the Tobacco and Makhorka Industry) Krasnodar, Nr. 139, 275 (1939). [Russisch.]

15th Annual Report of the Board of the Tea Research Institute of Ceylon for 1940. Bull. Tea Res. Inst., Ceylon, Nr. 2, 86 (1940).

Fifth Annual Report of the Coffee Research and Experiment Station, Lyamungu, Moshi 1938. Dep. Agric. Tanganyika, p. 39 (1940).

Report of the Agricultural Officer, Mlanje Experimental Station 1939. Nyasald Tea Ass. Quart. J. **5**, Nr. 2, 1—5 (1940).

Tocklai Experimental Station Annual Report 1939. Indian Tea Ass. Sci. Dep., p. 33 (1940).

(Symposium on diseases of tobacco and Nicotiana rustica.) Vsesojuznyj Naučno-issledovateljskij Institut Tabačnoj i Mahoročnoj Promyšlennosti imeni A. I. Mikojana (VITIM) (All-Union Mikojan Research Institute of the Tobacco and Makhorka Industry) Krasnodar, Nr. 141, 196 (1940). [Russisch.]

Report of the Coffee Board of Kenya for the year ended 30th June, 1941, p. 31 (1941).

Sixteenth annual report of the Board of the Tea Research Institute of Ceylon for 1941. Bull. Tea Res. Inst. Ceylon, Nr. 23, 81 (1941).

Annual Report for the year of the Tea Research Institute of Ceylon 1942. Bull., Nr. 24, 55 (1942).

Eighteenth Annual Report of the Board of the Tea Research Institute of Ceylon for 1943. Bull., Nr. 25, 63 (1943).

Miscelánea agrícola 19. (Agricultural miscellany. 19.) Bol. Agric. Territorios Españoles del Golfo de Guinea, Nr. 7, 83—84 (1943).

Ninth annual report of the Coffee Research and Experiment Station, Lyamungu, Moshi 1942. Dep. Agric., Tanganyika, Pp. 9 (1943).

Trelawney Tobacco Research Station. Annual Report for 1943. Tob. Res. Bd. S. Rhodesia. Publ., Nr. 7, 39 (1943).

Annual report of the Department of Agriculture, Colony of Mauritius, 1943. Port Louis, Pp. 30 (1944).

Eleventh report on cacao research 1941—1943. Imp. Coll. Trop. Agriculture, Trin., Pp. 38 (1944).

Nineteenth Annual Report of the Board of the Tea Research Institute of Ceylon for 1944. Bull. Tea Res. Inst. Ceylon, Nr. 26, 66 (1944).

Indian Tea Association, Scientific Department, Annual Report by the Director, Tocklai Experimental Station 1944. Pp. 11 (1944).

Trelawney Tobacco Research Station. Annual report for 1944. Publ. Tob. Res. Bd. S. Rhod., Nr. 8, 59 (1944).

West African Cacao Research Institute, Tafo. Quarterly report. October to December, 1944. Pp. 15 (1944). [Mimeographed.]

West African Cacao Research Institute, Tafo. Quarterly Report. July to September, 1944. Pp. 22 (1944). (Mimeographed.)

Annual Report of the Tobacco Institute of Puerto Rico for the fiscal years 1941—1942, 1942—1943, Pp. 65 (1945).

Annual Report of the West African Cacao Research Institute, Tafo, April, 1944 to March, 1945. Pp. 33 (1945). [Mimeographed.]
Annual report of the Council of the Institute of Brewing for the year ended 31st December, 1944. Advisory Sub-Committee on hops. J. Inst. Brew. 51, 74—77 (1945).
Annual Report of the Indian Tea Association, Tocklai Experiment Station, 1945. Pp. 18 (1945).
Report and Proceedings of the Cocoa Research Conference held at the Colonial office, London, May-June 1945, Nr. 192, 168 (1945).
Twentieth Annual Report of the Board of the Tea Research Institute of Ceylon for 1945. Bull., Nr. 27, 52 (1945).
Trelawney Tobacco Research Station. Annual report for 1945. Publ. Tob. Res. Bd. S. Rhod., Nr. 9, 87 (1945).
Twelfth Annual Report of the Coffee Research and Experiment Station, Lyamungu Moshi 1945. Pamphl., Nr. 44, 14 (1946).
Annual Report of the West African Cacao Research Institute, Tafo, April 1945, to March, 1946. Pp. 58 (1946). [Mimeographed.]
Istituto Scientifico Sperimentale per i Tabacchi. (The Tobacco Research Institute.) Tabacco 50, Nr. 567, 2 (1946).
Tenth Quarterly Report of the West African Cacao Research Institute, Tafo, October-December, 1946. Pp. 23 (1946). [Mimeographed.]
Progress report of the Dominion Experimental Substation Delhi, Ontario 1937 to 1945. Dep. Agric. Dom. Canada, Pp. 16 (1947).
Istituto Scientifico Sperimentale per i Tabacchi. (The Scientific Research Institute for Tobacco.) Tabacco 51, 2—5 (1947).

Ölpflanzen.
Report of the Department of Agriculture, Nyasaland Protectorate, for the year 1942. Pp. 16 (1942).
Annual Report of Tung Experiment Station, Nyasaland. Pp. 17 (1943).
Fifth annual report of the Oil Palm Research Station, Nigeria, 1943—1944. Pp. 31 (1944). [Mimeographed.]

Heilpflanzen.
The world's cinchona bark industry-I, II. Bull. Imp. Inst., Lond. 37, 18—31, 183—196; also Trop. Agriculturist 93, 288—308 (1939).
Report of the Puerto Rico Experiment Station 1943 (1944). Pp. 38 (1944).

Kautschukpflanzen.
Report of the work of the Rubber Research Board in 1938. (Established unter Ceylon ordinance, Nr. 10 of 1930.) Rubb. Res. Scheme (Ceylon) March, p. 100 (1939).
West Java Experimental Station, Buitenzorg. Bull. Rubb. Gr. Ass. 21, 240—248 (1939).
Annual Report. Botanical division. III. Selection and breeding. Rubb. Res. Inst. Malaya, Pp. 130—141 (1939) 1940.
De overbrenging van Heveazaden uit Brazilië naar Engeland in het jaar 1876 foor Henri A. Wickham. (The conveyance of Hevea seed from Brazil to England in the year 1876 by Henry A. Wickham.) Bergcultures 14, 609 (1940).
Abridged report on the work in 1940 of the Rubber Research Institute of Malaya. Pp. 19 (1941).
Report of the work of the Rubber Research Board in 1941. Rubb. Res. Scheme, Ceylon, Pp. 44 (1942).
Report of the work of the Rubber Research Board in 1944. Ceylon, March 1945, Pp. 31 (1945).
Report of the work of the Rubber Research Board in 1945. Rubber Res. Scheme, Ceylon, November 1946, Pp. 44 (1946).

Obst.
Report and accounts of the Coconut Research Scheme for 1938. Sess. Paper XI Coconut Res. Scheme, Ceylon, Pp. 22 (1939).
Report and accounts of the Coconut Research Scheme for 1940. Sess. Paper Coconut Res. Sch. Ceylon, Nr. XVI, 18 (1941).
Report of the Variety Committee 1942. Yearb. Calif. Avocado Soc. 1942, Pp. 13—17 (1942).
Annual Report of the Coconut Research Scheme for 1941. Ceylon, Sessional Paper XI, Pp. 19 (1942).
Research work and workers. Seed World 1942 52, Nr. 12, 44 (1942).
Thirty-ninth Annual Report of the South Dakota State Horticultural Society for the year ending June 30, 1942 to the Governor of South Dakota. Pp. 100. Describes the activities of the State Horticultural Society and records among other things new varieties of apples, pears and other fruit trees developed in the state and the diseases recently reported.
Miscelánea agrícola. 10. (Agricultural miscellany. 10.) Bol. Agric. Territorios Españoles del Golfo de Guinea Nr. 7, 81—82 (1943).
Miscelánea agrícola. 28. (Agricultural miscellany. 28.) Bol. Agric. Territorios Españoles del Golfo de Guinea, Nr. 7, 87—88 (1943).
Royal Horticultural Society Report. Fruit and Vegetable Committee. Gdnrs' Chron. 116, 196 (1944).
Annual Report of the Coconut Research Scheme for 1943. Ceylon, May 1945. Sess. Pap., Nr. 4, 22 (1945).
General review of research work with lists of papers published during the year. 32nd Rep. E. Malling Res. Sta. 1944, Pp. 17—33 (1945).
Report of the Ontario Horticultural Experiment Station for 1943 and 1944. Pp. 61 (1945).
Horticulture, 1. A prominent horticulturist. Agricultural Chron., Moscow, Nr. 8, 7—8 (1946). [Mimeographed.]
Relatório dos trabalhos realizados pelo Departamento de Pomologia da Estação Agronómica Nacional, durante 1945, e subsidiados pela Junta Nacional das Frutas. (Report of the work done by the Pomological Department of the National Agronomic Station during 1945, and subsidized by the National Fruit Board.) Bol. Junta Nac. Frutas, Lisboa 6, 117—126 (1946).
150th birthday of an old timer. Amer. Fruit Gr. 67, Nr. 7; 21, 40 (1947).

Gemüse.
Burlingham Horticultural Station. Tomato variety trial. Horticulture, Norfolk C. C. 2, 30—31 (1939).
Twenty-eighth annual report of the Experimental and Research Station, Turner's Hill, Cheshunt, Herts. 1942. Pp. 78 (1943).
Wartime agricultural research. 56th Rep. R. I. Agric. Exp. Sta. Contr. 659, Pp. 46 (1944).

Forstwirtschaft.
Redogörelse för verksamheten vid Statens Skogsförsöksanstalt under tiden 1932—31/10 1937 jämte förslag till arbetsuppgifter under den kommande femårsperioden. (Report on the work at the State Forestry Institute during 1932—31/10 1937 with a programme for work during the coming 5 year period.) Medd. Skogsförsöksanst. Stockb. 31, 109—162. 1938—1939 (1939).
Föreningen för växtförädling av skogsträd. (The Swedish Forest Tree Breeding Association.) Svensk Papp Tidn. 43, 90—92 (1940).
Årsberättelse över Föreningens för växtförädling av skogsträd verksamhet under år 1939. (Annual Report on the work of the Association for Forest Tree Breeding during the year 1939.) Svensk Papp Tidn 43, 130—135, 153—158 (1940).

Årsberättelse över Föreningens för växtförädling av skogsträd verksamhet under år 1941. (Annual report on the work of the Association for Breeding of Forest Trees during 1941.) Svensk Papp Tidn. 45, 300—307, 324—327 (1942).

Föreningen för växtförädling av skogsträd. (The Association for Forest Tree Breeding.) Svensk Papp Tidn. 46, 137 (1943).

Årsberättelse över Föreningens för växtförädling av skogsträd verksamhet under år 1942. (Annual report on the work of the Association for Forest Tree Breeding during the year 1942.) Svensk Papp Tidn. 46, 451—458, 477—483, 501—506, 535—541, 572—577 (1943).

Report of further discussion on post-war forest policy at a general meeting of the Society held in London on 23rd September, 1943. Roy. Engl. For. Soc., Lond., Pp. 43 (1943).

Rationalisering av frö-och plantanskaffningen vid skogsodling. Föredrag av jägmästare Harald Sjöström vid Svenska skogsvårdsforeningens årsmöte. (Rationalizing the supply of seed and plants in sylviculture. Lecture by Forest Officer Harald Sjöström at the Annual Meeting of the Swedish Sylvicultural Association.) Skogen 30, 207—209 (1943).

Föreningen för växtförädling av skogsträd. (Association for the Breeding of Forest Trees.) Skogen 31, 112 (1944).

Ordnarie årsmöte med Föreningen för växtförädling av skogsträd. (Ordinary meeting of Association for Forest Tree Breeding.) Svensk Papp Tidn. 47, 107 (1944).

The Forest Research Institute. Forest Research in India and Burma 1944—1945. Pt. 1, Pp. 149.

Redogörelser för verksamheten vid Statens Skogsförsöksanstalt under åren 1941—1944. (Report on the work of the State Institute of Experimental forestry during the years 1941—1944.) Medd. Skogsförsöksanst., Stockh. 34, 417—450 (1945).

Soeben erschien:

Lehrbuch der Pflanzenphysiologie

Zweiter und Dritter Band

Entwicklungs- und Bewegungsphysiologie der Pflanze

Von

Dr. Erwin Bünning

o. Professor an der Universität Tübingen

Mit 404 Abbildungen.   XII, 464 Seiten.   1948   Halbleinen DM 29.70

SPRINGER-VERLAG / BERLIN · GÖTTINGEN · HEIDELBERG

PLANTA
ARCHIV FÜR WISSENSCHAFTLICHE BOTANIK

Unter Mitwirkung von

E. BÜNNING-TÜBINGEN, **A. ERNST**-ZÜRICH, **H. v. GUTTENBERG**-ROSTOCK
R. HARDER-GÖTTINGEN, **W. SCHUMACHER**-BONN, **G. TISCHLER**-KIEL

Herausgegeben von

WILHELM RUHLAND UND **OTTO RENNER**

36. Band, 1./2. Heft   mit 81 Textabbildungen   1948   DM 22.—

INHALTSVERZEICHNIS

Wilhelm Ruhland zum 70. Geburtstag. Von W. SCHUMACHER. Beiträge zur Entwicklungsgeschichte und zur Physiologie panaschierter Blätter. Von CH. THIELKE. *Spartina Townsendii* an der Westküste von Schleswig-Holstein. Von D. KÖNIG. Beiträge zur photometrischen Messung von Chlorophyll-Lösungen. Von K. WENDEL. Adventivwurzelbildung an der Infloreszenz. Von A. BEYER. Über die Fachverhältnisse der Früchte von *Cornus* L. und verwandter Gattungen. Von F. KIRCHHEIMER. Über den Einfluß der Askorbinsäure auf die Auxinaktivierung. Von E. RAADTS. Ein Beitrag zur Konstitutionsspezifität der Heteroauxinwirkung. Von M. STEINER. Beiträge zur Entwicklungsgeschichte unifazialer Blätter. Von CH. THIELKE. Weitere Untersuchungen über die formative Wirkung des Lichtes und mechanischer Reize auf Pflanzen. Von E. BÜNNING, L. HAAG und G. TIMMERMANN. Zur wissenschaftlichen und praktischen Kenntnis der Kautschukpflanze *Taraxacum Kok Saghyz Rod.* Von CH. MORGENSTERN und F. TOBLER. Über die geotropische Erregung. Von E. BÜNNING und D. GLATZLE.

SPRINGER-VERLAG / BERLIN · GÖTTINGEN · HEIDELBERG

If you have any concerns about our products,
you can contact us on
ProductSafety@springernature.com

In case Publisher is established outside the EU,
the EU authorized representative is:
**Springer Nature Customer Service Center GmbH
Europaplatz 3, 69115 Heidelberg, Germany**

Printed by Libri Plureos GmbH
in Hamburg, Germany